Atomic Masses of the Elements and Their Symbols

Element	Symbol	Atomic number	Atomic mass (amu)
actinium	Ac	89	[227]
aluminum	Al	13	26.98
americium	Am	95	[243]
antimony	Sb	51	121.76
argon	Ar	18	39.95
arsenic	As	33	74.92
astatine	At	85	[210]
barium	Ba	56	137.33
berkelium	Bk	97	[247]
beryllium	Be	4	9.012
bismuth	Bi	83	208.98
bohrium	Bh	107	[264]
boron	B	5	10.81
bromine	Br	35	79.91
cadmium	Cd	48	112.41
calcium	Ca	20	40.08
californium	Cf	98	[251]
carbon	C	6	12.01
cerium	Ce	58	140.12
cesium	Cs	55	132.91
chlorine	Cl	17	35.45
chromium	Cr	24	52.00
cobalt	Co	27	58.93
copper	Cu	29	63.55
curium	Cm	96	[247]
darmstadtium	Ds	110	[281]
dubnium	Db	105	[262]
dysprosium	Dy	66	162.50
einsteinium	Es	99	[252]
erbium	Er	68	167.26
europium	Eu	63	151.96
fermium	Fm	100	[257]
fluorine	F	9	19.00
francium	Fr	87	[223]
gadolinium	Gd	64	157.25
gallium	Ga	31	69.72
germanium	Ge	32	72.64
gold	Au	79	196.97
hafnium	Hf	72	178.49
hassium	Hs	108	[277]
helium	He	2	4.002
holmium	Ho	67	164.93
hydrogen	H	1	1.0079
indium	In	49	114.82
iodine	I	53	126.90
iridium	Ir	77	192.22
iron	Fe	26	55.85
krypton	Kr	36	83.80
lanthanum	La	57	138.91
lawrencium	Lr	103	[262]
lead	Pb	82	207.21
lithium	Li	3	6.941
lutetium	Lu	71	174.97
magnesium	Mg	12	24.31
manganese	Mn	25	54.94
meitnerium	Mt	109	[268]
mendelevium	Md	101	[258]
mercury	Hg	80	200.59
molybdenum	Mo	42	95.94
neodymium	Nd	60	144.24
neon	Ne	10	20.18
neptunium	Np	93	[237]
nickel	Ni	28	58.69
niobium	Nb	41	92.91
nitrogen	N	7	14.01
nobelium	No	102	[259]
osmium	Os	76	190.23
oxygen	O	8	16.00
palladium	Pd	46	106.42
phosphorus	P	15	30.97
platinum	Pt	78	195.08
plutonium	Pu	94	[244]
polonium	Po	84	[209]
potassium	K	19	39.10
praseodymium	Pr	59	140.91
promethium	Pm	61	[145]
protactinium	Pa	91	231.04
radium	Ra	88	[226]
radon	Rn	86	[222]
rhenium	Re	75	186.21
rhodium	Rh	45	102.91
roentgenium	Rg	111	[272]
rubidium	Rb	37	85.47
ruthenium	Ru	44	101.07
rutherfordium	Rf	104	[261]
samarium	Sm	62	150.36
scandium	Sc	21	44.96
seaborgium	Sg	106	[266]
selenium	Se	34	78.96
silicon	Si	14	28.09
silver	Ag	47	107.87
sodium	Na	11	22.99
strontium	Sr	38	87.62
sulfur	S	16	32.07
tantalum	Ta	73	180.95
technetium	Tc	43	[98]
tellurium	Te	52	127.60
terbium	Tb	65	158.93
thallium	Tl	81	204.38
thorium	Th	90	232.04
thulium	Tm	69	168.93
tin	[illegible]	[illegible]	118.71
titani[illegible]	[illegible]	[illegible]	47.87
tungs[illegible]	[illegible]	[illegible]	183.84
urani[illegible]	[illegible]	[illegible]	238.03
vanad[illegible]	[illegible]	[illegible]	50.94
xenon	[illegible]	[illegible]	131.29
ytterbi[illegible]	[illegible]	[illegible]	173.04
yttrium	Y	39	88.91
zinc	Zn	30	65.41
zirconium	Zr	40	91.22

D1119541

Note: The names of elements 112–118 are provisional; brackets [ ] denote the most stable isotope of a radioactive element.
Online at: http://www.iupac.org/publications/pac/2003/pdf/7508x1107.pdf

About the Authors

Ira Blei was born and raised in Brooklyn, New York, where he attended public schools and graduated from Brooklyn College with B.S. and M.A. degrees in chemistry. After receiving a Ph.D. degree in physical biochemistry from Rutgers University, he worked for Lever Brothers Company in New Jersey, studying the effects of surface-active agents on skin. His next position was at Melpar Incorporated, in Virginia, where he founded a biophysics group that researched methods for the detection of terrestrial and extraterrestrial microorganisms. In 1967, Ira joined the faculty of the College of Staten Island, City University of New York, and taught chemistry and biology there for three decades. His research has appeared in the *Journal of Colloid Science*, the *Journal of Physical Chemistry*, and the *Archives of Biophysical and Biochemical Science*. He has two sons, one an engineer working in Berkeley, California, and the other a musician who lives and works in San Francisco. Ira is outdoors whenever possible, overturning dead branches to see what lurks beneath or scanning the trees with binoculars in search of new bird life, and has recently served as president of Staten Island's local Natural History Club.

George Odian is a tried and true New Yorker, born in Manhattan and educated in its public schools, including Stuyvesant High School. He graduated from The City College with a B.S. in chemistry. After a brief work interlude, George entered Columbia University for graduate studies in organic chemistry, earning M.S. and Ph.D. degrees. He then worked as a research chemist for 5 years, first at the Thiokol Chemical Company in New Jersey, where he synthesized solid rocket propellants, and subsequently at Radiation Applications Incorporated in Long Island City, where he studied the use of radiation to modify the properties of plastics for use as components of space satellites and in water-desalination processes. George returned to Columbia University in 1964 to teach and conduct research in polymer and radiation chemistry. In 1968, he joined the chemistry faculty at the College of Staten Island, City University of New York, and has been engaged in undergraduate and graduate education there for three decades. He is the author of more than 60 research papers in the area of polymer chemistry and of a textbook titled *Principles of Polymerization*, now in its fourth edition, with translations in Chinese, French, Korean, and Russian. George has a son, Michael, who is an equine veterinarian practicing in Maryland. Along with chemistry and photography, one of George's greatest passions is baseball. He has been an avid New York Yankees fan for more than five decades.

Ira Blei and George Odian arrived within a year of each other at the College of Staten Island, where circumstances eventually conspired to launch their collaboration on a textbook. Both had been teaching the one-year chemistry course for nursing and other health science majors for many years, and during that time they became close friends and colleagues. It was their habit to have intense, ongoing discussions about how to teach different aspects of the chemistry course, each continually pressing the other to enhance the clarity of his presentation. Out of those conversations developed their ideas for this textbook.

Organic and Biochemistry

Connecting Chemistry to Your Life

SECOND EDITION

Ira Blei

George Odian

College of Staten Island

City University of New York

OCT 24 2007

W. H. Freeman and Company · New York

DOUGLAS COLLEGE LIBRARY

Senior Acquisitions Editor: Clancy Marshall
Senior Marketing Manager: Krista Bettino
Developmental Editor: Donald Gecewicz
Publisher: Craig Bleyer
Media Editor: Victoria Anderson
Associate Editor: Amy Thorne
Photo Editor: Patricia Marx
Photo Researcher: Elyse Rieder
Design Manager: Diana Blume
Project Editor: Jane O'Neill
Illustrations: Fine Line Illustrations
and Imagineering Media Services, Inc.
Illustration Coordinator: Bill Page
Production Coordinator: Julia DeRosa
Composition: Schawk, Inc.
Printing and Binding: RR Donnelley

Library of Congress Control Number: 2005935007

ISBN 0-7167-7072-5
EAN 9780716770725

©2006 by W. H. Freeman and Company
All rights reserved

Printed in the United States of America

First printing

W. H. Freeman and Company
41 Madison Avenue
New York, NY 10010
Houndmills, Basingstoke RG21 6XS, England
www.whfreeman.com

Contents in Brief

Contents

Preface

Organic and Biochemistry: Connecting Chemistry to Your Life is designed to be used in a one-semester course presenting organic and biochemistry to students who intend to pursue careers as nurses, dieticians, physician's assistants, physical therapists, or environmental scientists.

Goals of This Book

Our chief objective in writing both editions of this book is to promote a better understanding of chemical *principles*—the comprehensive laws that explain how matter behaves—through the use of real-world examples. Students who merely memorize today's scientific information without understanding the basic underlying principles will not be prepared for the demands of the future. On the other hand, students who have a clear understanding of basic physical and chemical phenomena will have the tools that they need to understand new facts and ideas and will be able to incorporate new knowledge into their professional practices in appropriate and meaningful ways.

The other central goal of our book is to introduce students to how the human body works at the level of molecules and ions—that is, to the chemistry of physiological function. Throughout the book, we take every opportunity to illustrate chemical principles with specific examples of biomolecules and with real-world applications having physiological or medical contexts. Part 1 provides a brief review of the basic chemical principles underlying the properties of the organic compounds explored in Part 2, "Organic Chemistry." These properties are then put to work in Part 3, "Biochemistry."

In the process of exploring and using chemical principles, we emphasize two major themes throughout: (1) the ways in which intermolecular and intramolecular forces influence the properties of substances, and (2) the relations between molecular structures within the body and their physiological function.

New to This Edition

- Chapter 3, "Saturated Hydrocarbons," has been revised to help students in mastering the different families of organic compounds more readily. The treatment of enzymes and nutrition in Chapters 14 and 18, respectively, has been expanded because of the importance of these topics.
- Because visuals are so important to chemistry as a discipline and to chemistry textbooks, we have taken particular care with the illustrations in this new edition.
- In line with the second major goal of this textbook—showing students how the human body works at the level of molecules and ions—we changed the Pictures of Health that appear in most chapters. Each Picture of Health combines a photograph of an actual person with a drawing of the body and its processes in action, thus showing students how "macroscopic" everyday

activities relate to the molecular and ionic activity that goes on within the body. We think that the Picture of Health feature will engage students and that each Picture of Health helps to visually reinforce the concepts described in words in the main text. At the same time, the range of activities shown—from eating cotton candy to farming to playing tennis—highlights chemistry's central role in life.

- We know that students rely on a textbook for review and for test preparation. For that reason, we changed the format of the Summary at the end of each chapter. The new format—a list of short bulleted paragraphs—will make it easier for a student to identify the most important concepts in each chapter. The reviews of key reactions serve the same purpose, and they follow the chapter summaries in some of the chapters.
- We enhanced the more conceptual questions in each chapter. The Expand Your Knowledge category within the Exercise sets will show the students how to synthesize the concepts in the chapter—getting the students to think more like chemists.
- There are three kinds of boxes in this textbook: Chemistry in Depth, Chemistry Within Us, and Chemistry Around Us. Each of these kinds of boxes is designed to give the student more information and an awareness of the myriad applications of chemistry. To enhance the role of these boxes in the classroom, and to reinforce their purpose, we added "box exercises" to the Expand Your Knowledge category in the Exercises at the ends of chapters. The box exercises relate to the boxes and the applications in them, and these exercises will draw student's attention to this interesting feature. Look for the flask icons in the Exercise sections. Further, we added new applications or updated information to many of these boxes—demonstrating the dynamism of chemistry and its constant effects on our lives.
- Finally, the design of the new edition brightens the Concept Checklists, making them easier for students to find. The various lists of rules (such as the rules for naming certain compounds) are now that much easier to find, too, inasmuch as they follow a similar checklist format. We wanted our readers to be able to navigate our book easily, and its clean and logical design will help them to do so.

Pedagogical Features

The features of this book are **applications, problem-solving strategies, visualization,** and **learning tools,** in a real-world context to connect chemistry to students' lives.

Making Connections with Applications

Students are motivated to learn a subject if they are convinced of its fundamental importance and personal relevance. Examples of the relevance of chemical concepts are woven into the text and emphasized through several key features.

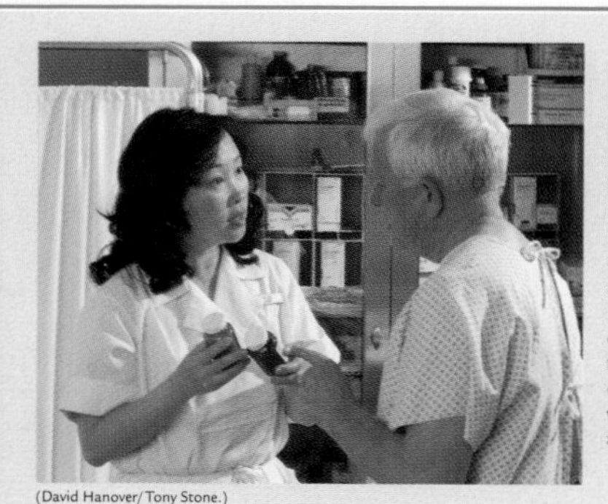

(David Hanover/Tony Stone.)

Chemistry in Your Future

On his way to see the doctor for an examination, a patient pauses to discuss his high-blood-pressure drug with you. He says that he's been meaning to ask why his drug suddenly underwent an increase in price a few months ago. You explain that the FDA recently approved a "single enantiomer" version of the drug. The new version is purer and more effective and causes fewer side effects, and so the clinic thinks that it is worth the cost of the more expensive manufacturing process. "Purer in what way?" he asks. This chapter helps you to answer his question.

Chemistry in Your Future A scenario at the beginning of each chapter describes a typical workplace situation that illustrates a practical, and usually professional, application of the contents of that chapter. A link to the book's Web site leads the student to further practical information.

A Picture of Health This completely revised series of drawings and photographs shows how chapter topics apply to human physiology and health.

A PICTURE OF HEALTH
Examples of Stereoselective Molecules

Proteins such as collagen (in tendons) and hemoglobin (in blood) are composed of L-α- amino acids.
The saccharide in DNA is D-deoxyribose.
D-Glucose is a source of energy.
L-Dopa is used in treating Parkinson disease, which results from a degeneration of certain brain cells.
A conformational change in *cis*-11-retinal is the basis of vision.
L-Enantiomers of lipids are used to construct cell membranes.
Cholesterol is one of 256 stereoisomers.

(Photodisc/Photosearch.)

Three Categories of Boxes A total of 50 boxed essays, divided into three categories, broaden and deepen the reader's understanding of basic ideas. Icons in the exercise sets reinforce the use of these practical essays.

Chemistry Within Us These boxes describe applications of chemistry to human health and well-being.

Chemistry Around Us These boxes describe applications of chemistry to our everyday life (including commercial products) and to biological processes in organisms other than humans.

Chemistry in Depth These boxes provide a more detailed description of selected topics, ranging from spectroscopic analysis to the mechanisms of key organic reactions.

Making Connections Through Problem Solving

Learning to work with chemical concepts and developing problem-solving skills are integral to understanding chemistry. We help students develop these skills.

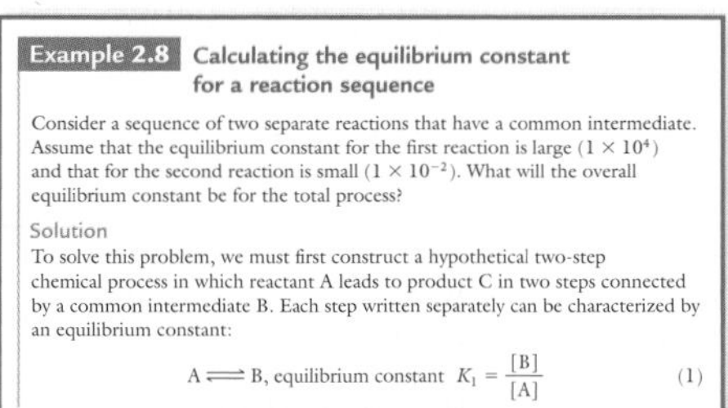

Example 2.8 Calculating the equilibrium constant for a reaction sequence

Consider a sequence of two separate reactions that have a common intermediate. Assume that the equilibrium constant for the first reaction is large (1×10^4) and that for the second reaction is small (1×10^{-2}). What will the overall equilibrium constant be for the total process?

Solution

To solve this problem, we must first construct a hypothetical two-step chemical process in which reactant A leads to product C in two steps connected by a common intermediate B. Each step written separately can be characterized by an equilibrium constant:

$$A \rightleftharpoons B, \text{ equilibrium constant } K_1 = \frac{[B]}{[A]} \quad (1)$$

In-Chapter Examples Nearly 150 in-chapter examples with step-by-step solutions, each followed by a similar in-chapter problem, allow students to verify and practice their skills.

End-of-Chapter Exercises The nearly 1400 end-of-chapter exercises are divided into three categories:

- **Paired Exercises** are arranged according to chapter sections; each odd-numbered paired exercise is followed by an even-numbered exercise of the same type.
- **Unclassified Exercises** do not reference specific chapter sections but test the student's overview of chapter concepts.
- **Expand Your Knowledge Exercises** challenge students to expand their problem-solving skills by applying them to more-complex questions or to questions that require the integration of material from different chapters.

Answers to Odd-Numbered Exercises are supplied at the end of the book. Step-by-step solutions to the odd-numbered exercises are supplied in the *Student Solutions Manual.* Step-by-step solutions to even-numbered exercises are supplied in the *Instructor's Resource Manual.* Step-by-step solutions to in-chapter problems are supplied in the *Study Guide.*

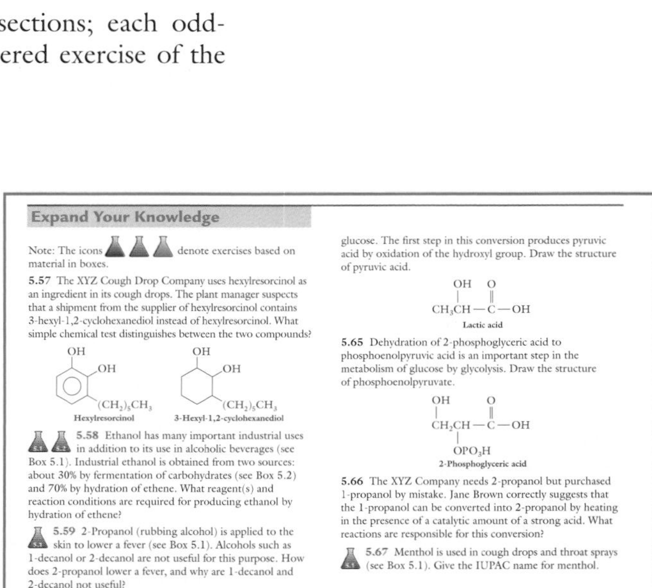

Expand Your Knowledge

Note: The icons denote exercises based on material in boxes.

5.57 The XYZ Cough Drop Company uses hexylresorcinol as an ingredient in its cough drops. The plant manager suspects that a shipment from the supplier of hexylresorcinol contains 3-hexyl-1,2-cyclohexanediol instead of hexylresorcinol. What simple chemical test distinguishes between the two compounds?

Hexylresorcinol | 3-Hexyl-1,2-cyclohexanediol

5.58 Ethanol has many important industrial uses in addition to its use in alcoholic beverages (see Box 5.1). Industrial ethanol is obtained from two sources: about 30% by fermentation of carbohydrates (see Box 5.2) and 70% by hydration of ethene. What reagent(s) and reaction conditions are required for producing ethanol by hydration of ethene?

5.59 2-Propanol (rubbing alcohol) is applied to the skin to lower a fever (see Box 5.1). Alcohols such as 1-decanol or 2-decanol are not useful for this purpose. How does 2-propanol lower a fever, and why are 1-decanol and 2-decanol not useful?

glucose. The first step in this conversion produces pyruvic acid by oxidation of the hydroxyl group. Draw the structure of pyruvic acid.

Lactic acid

5.65 Dehydration of 2-phosphoglyceric acid to phosphoenolpyruvic acid is an important step in the metabolism of glucose by glycolysis. Draw the structure of phosphoenolpyruvate.

2-Phosphoglyceric acid

5.66 The XYZ Company needs 2-propanol but purchased 1-propanol by mistake. Jane Brown correctly suggests that the 1-propanol can be converted into 2-propanol by heating in the presence of a catalytic amount of a strong acid. What reactions are responsible for this conversion?

5.67 Menthol is used in cough drops and throat sprays (see Box 5.1). Give the IUPAC name for menthol.

Making Connections Through Visualization

Illustrations Illustrations and tables have been carefully chosen or designed to support the text and are carefully labeled for clarity. Special titles on certain illustrations—Insight into Properties, Insight into Function, and Looking Ahead—emphasize the use of secondary attractive forces and molecular structure as unifying themes throughout the book and remind readers that the concepts learned in Parts 1 and 2 will be applied to the biochemistry in Part 3.

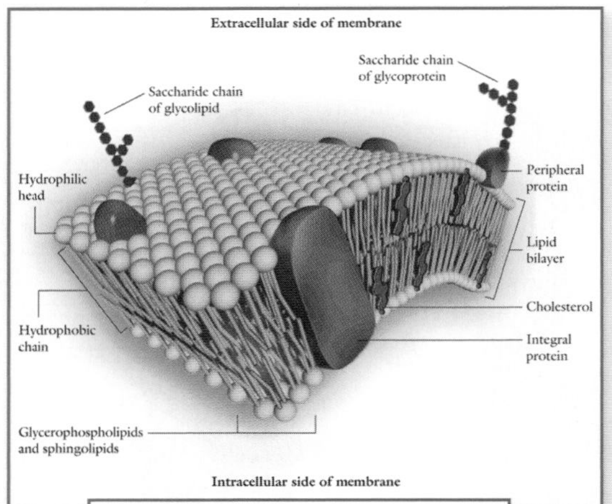

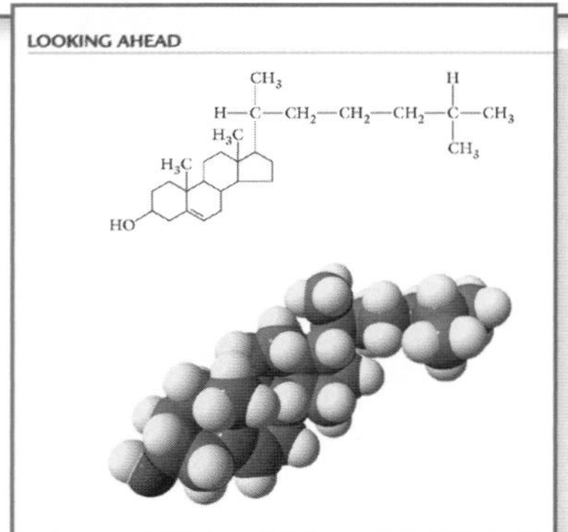

Ball-and-Stick and Space-Filling Molecular Models Molecular structures of compounds, especially organic compounds, offer considerable interpretive challenge to students. Throughout the book, two-dimensional molecular structures are supported by generous use of ball-and-stick and space-filling molecular models to aid in the visualization of three-dimensional structures of molecules.

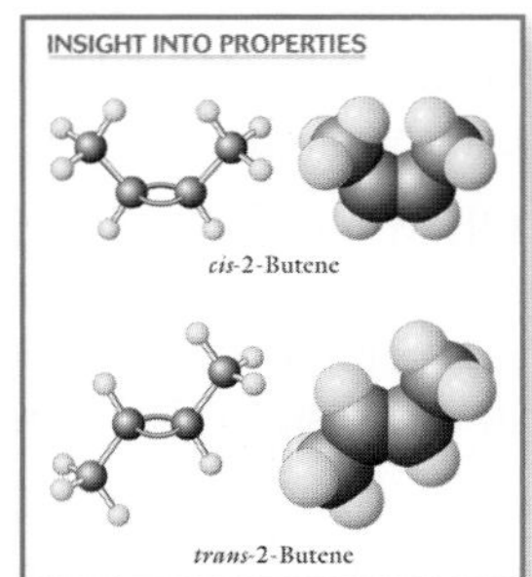

Functional Use of Color Color is used functionally and systematically in schematic illustrations and equations to draw attention to key changes or components and to differentiate one key component from another. For example, in molecular models, the carbon, hydrogen, oxygen, and nitrogen atoms are consistently illustrated in black, white, red, and blue, respectively. In structural representations of chemical reactions, color is used to highlight the parts of the molecule undergoing change. The strategic use of color makes diagrams of complex biochemical pathways less daunting and easier to understand.

$$CH_3CH_2{-}\overset{\displaystyle O}{\overset{\|}{C}}{-}H \underset{H^+\,(-\,CH_3OH)}{\overset{CH_3OH,\ H^+}{\rightleftharpoons}} \underset{\text{Hemiacetal}}{CH_3CH_2{-}\overset{OH}{\underset{OCH_3}{C}}{-}H} \underset{H^+,\ H_2O\,(-\,CH_3OH)}{\overset{CH_3OH,\ H^+\,(-\,H_2O)}{\rightleftharpoons}} \underset{\text{Acetal}}{CH_3CH_2{-}\overset{OCH_3}{\underset{OCH_3}{C}}{-}H}$$

Making Connections by Using Learning Tools

Learning Objectives Each chapter begins with a list of learning objectives that preview the skills and concepts that students will master by studying the chapter. Students can use the list to gauge their progress in preparing for exams.

Concept Checklists The narrative is punctuated with short lists serving to highlight or summarize important concepts. They provide a periodic test of comprehension in a first reading of the chapter, as well as an efficient means of reviewing the chapter's key points.

Concept checklist

- ✓ When the attractive forces between different molecules are similar, solutions will form.
- ✓ When the attractive forces between different molecules are different, solutions will not form.

Rules Rules for nomenclature, balancing reaction equations, and other important features are highlighted so that students can find them easily when studying or doing homework.

Cross-References Cross-referencing in the text and margins alerts students to upcoming topics, suggests topics to review, and draws connections between material in different parts of the book.

» **Water-soluble vitamins are considered in Chapter 18.**

Chapter Summaries Serving as a brief study guide, the Summary at the end of each chapter points out the major concepts presented in each section of the chapter.

Summaries of Key Reactions At the end of most organic chemistry chapters, this feature summarizes the important reactions of a given functional group.

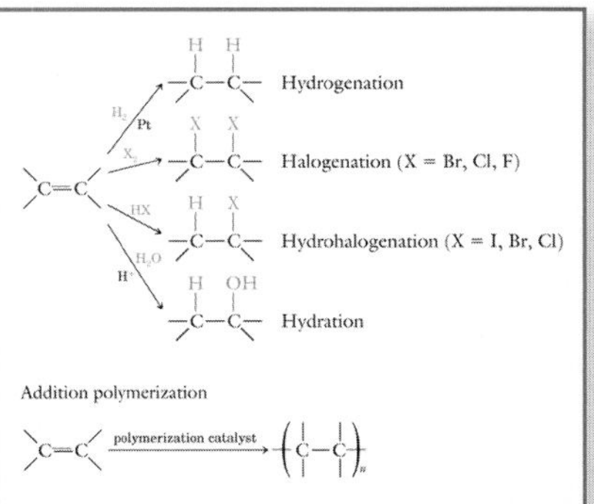

Key Words Important terms are listed at the end of each chapter and keyed to the pages on which their definitions appear.

Organization

Part 1: A Review of General Chemistry (Chapters 1 and 2)

To understand the molecular basis of physiological functioning, students must have a thorough grounding in the fundamental concepts of general chemistry and organic chemistry. Part 1 reviews the structure and properties of atoms, ions, and molecules. Chapter 1 describes the structure of atoms and how they bond together to form molecules, as well as the physical properties of molecules and the nature of interactions between them. Chapter 2 examines the major types of chemical reactions, chemical kinetics and equilibria, and acids and bases. Readers of this book are assumed to have had at least an introductory course in high school chemistry. Therefore, the first two chapters will not necessarily be assigned reading for the course in which the students are enrolled but may serve as reference material supporting the main topics of the book.

Part 2: Organic Chemistry (Chapters 3 Through 9)

Part 2 uses the fundamental principles reviewed in Part 1 as a starting point for exploring the physical and chemical properties of carbon-containing compounds. Chapter 3 provides a general introduction to organic chemistry and then examines saturated hydrocarbons. Unsaturated hydrocarbons are the subject of Chapter 4. Chapter 5 begins the study of oxygen-containing organic compounds by examining alcohols, phenols, ethers, and related compounds; together with Chapter 6, on aldehydes and ketones, it lays the foundation for the subsequent study of carbohydrates. Chapter 7 examines carboxylic acids and esters, preparing students for the subsequent study of lipids and nucleic acids. Amines and amides are considered in Chapter 8, a prelude to the subsequent examination of amino acids, polypeptides, proteins, and nucleic acids. Chapter 9 describes the concepts of stereochemistry and their importance in biological systems.

Part 3: Biochemistry (Chapters 10 Through 18)

Biochemistry is the study of the biomolecules and the chemical processes that govern life functions. Chapters 10 through 13 present the principal biomolecules: carbohydrates, proteins, lipids, and nucleic acids. The structural features of these biomolecules are described in regard to the relations between their chemical structures and their physiological functions. Chapters 14 through 18 focus on those functions—specifically, on metabolism, the extraction of energy from the environment, and the use of energy to synthesize biomolecules. Chapter 14 provides a general survey of cell structures, metabolic systems, and enzymes, whereas Chapters 15 through 17 describe the key features of carbohydrate, lipid, and amino acid metabolism, respectively. Chapter 18 demonstrates how these principal metabolic pathways are integrated into the overall functions of the body. It does so by examining digestive processes and nutrition and then comparing the responses of the body under moderate and severe physiological stress.

Flexibility for Chemistry Courses

We recognize that all introductory courses are not alike. For that reason, we offer this text in three versions, so you can choose the option that is right for you:

- *General, Organic, and Biochemistry* (ISBN 0-7167-4375-2)—the comprehensive 26-chapter text
- *An Introduction to General Chemistry* (ISBN 0-7167-7073-3)—10 chapters that cover the core concepts in general chemistry
- *Organic and Biochemistry* (ISBN 0-7167-7072-5)—16 chapters that cover organic and biochemistry plus two introductory chapters that review general chemistry

For further information on the content in each of these versions, please visit our Web site: http://www.whfreeman.com/bleiodian2e.

Supplements

A mouse icon in the margins of the textbook indicates that a resource on the book's companion Web site (www.whfreeman.com/bleiodian2e) accompanies that section of the book. Animations, simulations, videos, and more resources found on the book's companion site help to bring the book to life. Its practice tools such as interactive quizzes help students review for exams.

For Students

Student Solutions Manual, by Mark D. Dadmun of the University of Tennessee–Knoxville, contains complete solutions to the odd-numbered end-of-chapter exercises.

Study Guide, by Marcia L. Gillette of Indiana University, Kokomo, provides reader friendly reinforcement of topics covered in the text. Includes chapter outlines, hints, practice exercises with answers, and more.

General, Organic, and Biochemistry Laboratory Manual, Second Edition, by Sara Selfe of Edmonds Community College.

Web Site, www.whfreeman.com/bleiodian2e, offers a number of features for students and instructors including online study aids such as quizzes, molecular visualizations, chapter objectives, chapter summaries, Web review exercises, flashcards, Web-linked exercises, molecules in the news, and a periodic table.

For Instructors

Instructor's Resource Manual, by Mark D. Dadmun of the University of Tennessee–Knoxville, contains complete solutions to the even-numbered end-of-chapter exercises, chapter outlines, and chapter overviews.

New! Enhanced Instructor's Resource CD-ROM To help instructors create lecture presentations, Web sites, and other resources, this CD-ROM allows instructors to search and export the following resources by key term or chapter: all text images; animations, videos, PowerPoint, and more found on the Web site; and the printable electronic Instructor's Manual (available in Microsoft Word format), which can be fully edited and includes answers to even-numbered end-of-chapter questions.

Test Bank, by Margaret G. Kimble of Indiana University–Purdue University, contains more than 2500 multiple-choice, fill-in-the-blank, and short-answer questions, available in both print and electronic formats.

More than 200 **Overhead Transparencies.**

Instructor's Web Site, which is password-protected, contains student resources, laboratory information, and PowerPoint files.

Course Management Systems (WebCT, Blackboard) As a service to adopters, electronic content will be provided for this textbook, including the instructor and student resources in either WebCT or Blackboard formats.

Acknowledgments

We are especially grateful to the many educators who reviewed the manuscript and offered helpful suggestions for improvement. For the first edition, we thank the following persons:

Brad P. Bammel, Boise State University; George C. Bandik, University of Pittsburgh; Bruce Banks, University of North Carolina, Greensboro; Lorraine C. Brewer, University of Arkansas; Martin L. Brock, Eastern Kentucky University; Steven W. Carper, University of Nevada, Las Vegas; John E. Davidson, Eastern

Kentucky University; Geoffrey Davies, Northeastern University; Marie E. Dunstan, York College of Pennsylvania; James I. Durham, Blinn College; Wes Fritz, College of DuPage; Patrick M. Garvey, Des Moines Area Community College; Wendy Gloffke, Cedar Crest Community College; T. Daniel Griffiths, Northern Illinois University; William T. Haley, Jr., San Antonio College; Edwin F. Hilinski, Florida State University; Vincent Hoagland, Sonoma State University; Sylvia T. Horowitz, California State University, Los Angeles; Larry L. Jackson, Montana State University; Mary A. James, Florida Community College, Jacksonville; James Johnson, Sinclair Community College; Morris A. Johnson, Fox Valley Technical College; Lidija Kampa, Kean College; Paul Kline, Middle Tennessee State University; Robert Loeschen, California State University, Long Beach; Margaret R. R. Manatt, California State University, Los Angeles; John Meisenheimer, Eastern Kentucky University; Frank R. Milio, Towson University; Michael J. Millam, Phoenix College; Renee Muro, Oakland Community College; Deborah M. Nycz, Broward Community College; R. D. O'Brien, University of Massachusetts; Roger Penn, Sinclair Community College; Charles B. Rose, University of Nevada, Reno; William Schloman, University of Akron; Richard Schwenz, University of Northern Colorado; Michael Serra, Youngstown State College; David W. Seybert, Duquesne University; Jerry P. Suits, McNeese State University; Tamar Y. Susskind, Oakland Community College; Arrel D. Toews, University of North Carolina, Chapel Hill; Steven P. Wathen, Ohio University; Garth L. Welch, Weber State University; Philip J. Wenzel, Monterey Peninsula College; Thomas J. Wiese, Fort Hays State University; Donald H. Williams, Hope College; Kathryn R. Williams, University of Florida; William F. Wood, Humboldt State University; Les Wynston, California State University, Long Beach.

We also wish to thank the students of George C. Bandik, University of Pittsburgh; Sharmaine Cady, East Stroudsburg University; Wes Fritz, College of DuPage; Wendy Gloffke, Cedar Crest Community College; Paul Kline, Middle Tennessee State University; Sara Selfe, Edmonds Community College; Jerry P. Suits, McNeese State University; and Arrel D. Toews, University of North Carolina, Chapel Hill, whose comments on the text and exercises provided invaluable guidance in the book's development.

For the second edition, we thank the following persons:

Kathleen Antol, Saint Mary's College; Clarence (Gene) Bender, Minot State University–Bottineau; Verne L. Biddle, Bob Jones University; John J. Blaha, Columbus State Community College; Salah M. Blaih, Kent State University, Trumbull; Laura Brand, Cossatot Community College; R. Todd Bronson, College of Southern Idaho; Charmita Burch, Clayton State University; Sharmaine Cady, East Stroudsburg University; K. Nolan Carter, University of Central Arkansas; Jeannie T. B. Collins, University of Southern Indiana; Thomas G. Conally, Alamance Community College; Loretta T. Dorn, Fort Hays State University; Daniel Freeman, University of South Carolina; Laura DeLong Frost, Georgia Southern University; Edwin J. Geels, Dordt College; Marcia L. Gillette, Indiana University, Kokomo; James K. Hardy, University of Akron; Harvey Hopps, Amarillo College; Shell L. Joe, Santa Ana College; James T. Johnson, Sinclair Community College; Margaret G. Kimble, Indiana University–Purdue University, Fort Wayne; Richard Kimura, California State University, Stanislaus; Robert R. Klepper, Iowa Lakes Community College; Edward A. Kremer, Kansas City, Kansas Community College; Jeanne L. Kuhler, Southern Illinois University; Darrell W. Kuykendall, California State University, Bakersfield; Jennifer Whiles Lillig, Sonoma State University; Robert D. Long, Eastern New Mexico University; David H. Magers, Mississippi College; Janet L. Marshall, Raymond Walters College–University of Cincinnati; Douglas F. Martin, Penn Valley Community College; Craig P. McClure, University of

Alabama at Birmingham; Ann H. McDonald, Concordia University Wisconsin; Robert P. Metzger, San Diego State University; K. Troy Milliken, Waynesburg College; Qui-Chee A. Mir, Pierce College; Cynthia Molitor, Lourdes College; John A. Myers, North Carolina Central University; E. M. Nicholson, Eastern Michigan University; Naresh Pandya, Kapiolani Community College; John W. Peters, Montana State University; David Reinhold, Western Michigan University; Elizabeth S. Roberts-Kirchhoff, University of Detroit, Mercy; Sara Selfe, Edmonds Community College; David W. Smith, North Central State College; Sharon Sowa, Indiana University of Pennsylvania; Koni Stone, California State University, Stanislaus; Erach R. Talaty, Wichita State University; E. Shane Talbott, Somerset Community College; Ana M. Q. Vande Linde, University of Wisconsin–Stout; Thomas J. Wiese, Fort Hays State University; John Woolcock, Indiana University of Pennsylvania.

Special thanks are due to Irene Kung, University of Washington; Stan Manatt, California Institute of Technology; and Mark Wathen, University of Northern Colorado, who checked calculations for accuracy for the first edition; and Mark D. Dadmun and Marcia L. Gillette, who checked calculations for accuracy for the second edition.

Finally, we thank the people of W. H. Freeman and Company for their constant encouragement, suggestions, and conscientious efforts in bringing this second edition of our book to fruition. Although most of these people are listed on the copyright page, we would like to add some who are not and single out some who are listed but deserve special mention. We want to express our deepest thanks to Clancy Marshall for providing the opportunity, resources, and enthusiastic support for producing this second edition; to Jane O'Neill and Patricia Zimmerman for their painstaking professionalism in producing a final manuscript and published book in which all can feel pride; and to Moira Lerner (first edition) and Donald Gecewicz (second edition) whose creativity, cheerful encouragement, and tireless energy were key factors in the manuscript's evolution and preparation.

The authors welcome comments and suggestions from readers at: irablei@bellsouth.net; odian@mail.csi.cuny.edu.

PART 1

A REVIEW OF GENERAL CHEMISTRY

This book is intended for a one-semester course in organic chemistry and biochemistry. Organic chemistry, the focus of Part 2, concerns the chemical and physical properties of carbon-containing compounds and is preparatory to the study of biochemistry—the chemistry of life processes and the focus of Part 3. Part 1 is a brief, two-chapter review of the chemical principles underlying these two areas of chemistry: Chapter 1 surveys atomic structure, chemical reactivity, bonding, and molecular structures and properties; Chapter 2 reviews key characteristics of chemical reactions. The readers of this book are assumed to have had at least an introductory course in high-school chemistry. Therefore, the first two chapters will not necessarily be assigned reading for the course in which you are currently enrolled but may serve as reference material supporting the main topics of the book.

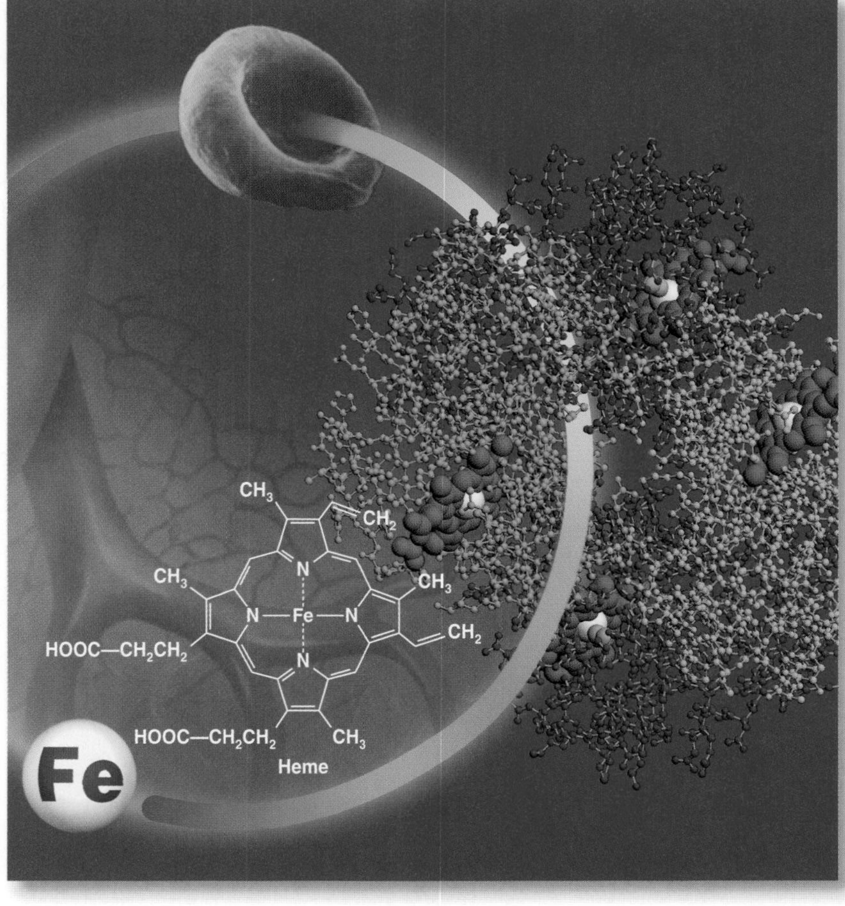

CHAPTER 1 THE PROPERTIES OF ATOMS AND MOLECULES

Learning Objectives

- Describe the structure of the atom in regard to its principal subatomic particles.
- Identify main-group and transition elements, metals, nonmetals, and metalloids.
- Correlate the arrangement of the periodic table with the electron configurations of the valence shells of the elements.
- Describe the octet rule and its effect on the formation of chemical bonds.
- Draw Lewis structures of compounds.
- Use VSEPR theory to predict molecular shape.
- Describe the difference between polar and nonpolar covalent bonds.
- Describe secondary forces, and correlate chemical structure with types of secondary forces.

Chemistry is the study of matter and its transformations, and no aspect of human activity is untouched by it. The discoveries of chemistry have transformed the foods that we eat, the homes that we live in, and the manufactured objects that we use in our daily lives. In addition to explaining and transforming the chemical world outside our bodies, chemists have established a detailed understanding of the chemical processes taking place inside our bodies.

Chemists deal with two types of substances: elements and compounds. An **element** is a substance that cannot be transformed into simpler substances nor can it be created by combining simpler substances. Elements combine to form **compounds** in fixed proportions. For example, glucose, also called dextrose, is a chemical combination of the elements carbon, oxygen, and hydrogen. One hundred grams (100 g) of glucose will always contain 40.00 g of carbon, 53.33 g of oxygen, and 6.67 g of hydrogen, whether it is extracted from rose hips or synthesized in the laboratory.

Chapter 1 reviews the structure of atoms and the relation between atomic structure and the periodic table, the nature of the chemical bond, quantitative relations in chemical reactions, the structures of molecules and their representations, and, finally, the weak interactions between molecules that underlie solution formation and other phenomena.

1.1 ATOMIC STRUCTURE

The science of chemistry rests on the principle that all matter is ultimately composed of atoms. What do we know about atoms? We know that, although they are very small, **atoms** have mass and are composed of three types of particles—**protons, neutrons,** and **electrons**—whose properties are listed in Table 1.1. The protons and neutrons are located in a **nucleus,** which occupies a very small part of the atom's volume. About 10 million average-sized nuclei would equal the volume of one average atom. The volume of an atom is therefore defined by a dense cloud of negative electricity (the electrons) surrounding a very small positively charged nucleus. Electrons are very low in mass: 1840 of them are equal in mass to one proton. Therefore, the mass of an atom is almost equal to the sum of protons and neutrons in the nucleus. The **atomic mass** of each element can be found in the table inside the back cover of this book.

The atoms of elements are characterized by the number of protons in the nucleus—each element having a different number of protons called the **atomic number.** All the atoms of any given element have the same number of protons, and the number of protons is balanced by an equal number of electrons. Some elements can lose electrons, and such elements form positively charged atoms called **cations.** Other elements can gain electrons, and such elements form negatively charged atoms called **anions.**

Although there must be a fixed number of protons in the atoms of an element, the number of neutrons can vary. So, in a sample of any naturally

TABLE 1.1 Properties of Subatomic Particles

Particle	Electrical charge	Symbol	Mass (amu)	Location
proton	1+	p	1.00728	nucleus
neutron	0	n	1.00867	nucleus
electron	1−	e	0.0005486	outside nucleus

occurring element, there can be a family of atoms all having the same number of protons in the nucleus but with varying numbers of neutrons. Such atoms are called **isotopes.** A convenient symbolic notation for an isotope is to write its elemental symbol followed by its atomic mass, as, for example, C-12, the carbon isotope of mass 12. The atomic mass of a sample of a naturally occurring element (see the table inside the back cover of this book) is a mass average of the natural abundances of its family of isotopes; that is, averaged according to the percentage of each isotope present in a natural sample. Some isotopes are unstable and emit high-energy radiation. These isotopes are called **radioisotopes,** and some are used in medical diagnosis and therapy.

1.2 THE PERIODIC LAW AND THE PERIODIC TABLE

An element's identity depends on the number of protons in its nucleus—its atomic number. When the elements are arranged in order of increasing atomic number, their physical and chemical properties do not change smoothly and continuously but instead repeat in a periodic manner. This observation is known as the **periodic law.**

The periodic law is embodied in the **periodic table,** displayed inside the front cover of this book. In all versions of the periodic table, the elements are arranged so that those with similar chemical properties are aligned in the same vertical column, called a **group** or a **family.** Each horizontal row is called a **period.** There are a number of different ways that the columns can be numbered, and they are shown in the version inside the front cover. Each box in this version of the periodic table shows, from top to bottom, the atomic number, the element's symbol, and the atomic mass to two decimal places. The symbols of the elements and their corresponding atomic numbers are a feature of every version of the periodic table.

Figure 1.1 presents an abbreviated version of the periodic table for your reference as you read about the relations between an element's position in the table and its reactivity. This simplified version of the table focuses on the elements most likely to be encountered in a biological context.

The elements can be grouped into three major categories: metals, nonmetals, and transition elements. The elements belonging to the first two and last five groups in the table, denoted by Roman numerals in Figure 1.1, are called the **main-group elements.** Note that, in Figure 1.1, the columns of transition elements, beginning in period 4, are not numbered.

The elements in Group I—lithium, sodium, potassium, rubidium, and cesium—are examples of metals. **Metals** are shiny, malleable substances that can be melted and cast into desired shapes and are excellent conductors of heat and electricity. When metals react chemically, they tend to lose electrons to form cations. The Group I metals react vigorously with water, producing hydrogen and chemical products called bases (for example, the metal potassium reacts with water to form potassium hydroxide). Because bases added to water form caustic, or alkaline, solutions, these elements are called **alkali metals.** The group to the right of the alkali metals—beryllium, magnesium, calcium, strontium, and barium—are called the **alkaline earth metals.**

The elements in the group that directly precedes the last column on the right are called the **halogens:** fluorine, chlorine, bromine, iodine, and astatine. They are examples of **nonmetals,** elements that cannot be cast into shapes and do not conduct electricity. When nonmetals react, they tend to gain electrons to form anions.

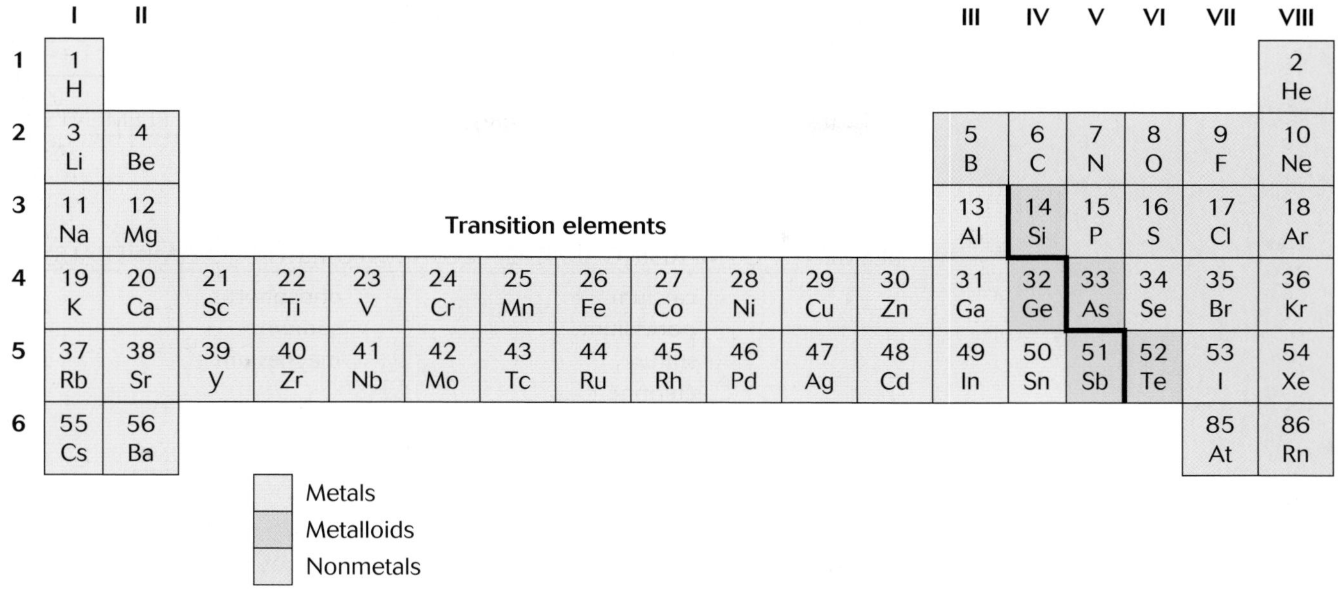

	I	II											III	IV	V	VI	VII	VIII
1	1 H																	2 He
2	3 Li	4 Be											5 B	6 C	7 N	8 O	9 F	10 Ne
3	11 Na	12 Mg					Transition elements						13 Al	14 Si	15 P	16 S	17 Cl	18 Ar
4	19 K	20 Ca	21 Sc	22 Ti	23 V	24 Cr	25 Mn	26 Fe	27 Co	28 Ni	29 Cu	30 Zn	31 Ga	32 Ge	33 As	34 Se	35 Br	36 Kr
5	37 Rb	38 Sr	39 y	40 Zr	41 Nb	42 Mo	43 Tc	44 Ru	45 Rh	46 Pd	47 Ag	48 Cd	49 In	50 Sn	51 Sb	52 Te	53 I	54 Xe
6	55 Cs	56 Ba															85 At	86 Rn

Metals
Metalloids
Nonmetals

Figure 1.1 A short form of the periodic table. Elements are identified by symbol and atomic number. The metals are separated from the nonmetals by a zigzag line. The elements bordering that line have properties of both groups and are called metalloids. The main-group elements are denoted by Roman numerals, and periods 1 through 6 are identified at the left.

Figure 1.1 identifies an intermediate category of elements, called **metalloids** or **semimetals,** which have properties between those of metals and nonmetals and lie between the two larger classes.

The elements in the last column on the right-hand side of the table, Group VIII, are called the **noble gases:** helium, neon, argon, krypton, xenon, and radon. These elements are called noble because they are virtually inert (nonreactive) and do not easily form compounds, as most of the other elements do. No compounds of helium, neon, or argon are known. Before 1962, the noble gases were called inert gases, but, at that time, xenon was discovered to form compounds with fluorine and oxygen; later, krypton was found to form compounds with fluorine.

The first period, or row, of the periodic table has only two members: hydrogen and helium. Hydrogen is not an alkali metal, but it does have some properties in common with that family and is often placed in the first position in Group I. Helium is inert; so it is placed in the first position in Group VIII.

The second and third periods each contain eight elements, and, if there were no other kinds of elements, the periodic table would consist of eight groups. In period 4, however, a new set of ten elements appears. The properties of these elements are intermediate between those of Groups II and III of the preceding period 3. These intervening elements are called **transition elements** or **transition metals.** An important property of transition elements is that they can form cations of more than a single electrical charge. For example, iron can form cations of charge 2+ and charge 3+, and copper can form cations of 1+ and 2+.

The transition metals are found in biological systems in which electrons are transferred in the course of metabolic processes. Iron is a transition element of great biological importance—the reason for the deep red color of

TABLE 1.2 Essential Elements in the Human Body (Listed by Their Relative Abundance)

ELEMENTS COMPRISING 99.3% OF TOTAL ATOMS (MAJOR ELEMENTS)	
hydrogen	carbon
oxygen	nitrogen
ELEMENTS COMPRISING 0.7% OF TOTAL ATOMS (MAJOR MINERALS)	
calcium	phosphorus
potassium	sodium
sulfur	magnesium
chlorine	
ELEMENTS COMPRISING LESS THAN 0.01% OF TOTAL ATOMS (TRACE ELEMENTS)	
iron	iodine
copper	zinc
manganese	cobalt
chromium	selenium
molybdenum	fluorine
tin	silicon
vanadium	

oxygenated blood. Molybdenum is a key factor in the biological process of atmospheric nitrogen fixation. Cobalt is the central feature of the structure of vitamin B_{12}. Table 1.2 is a list of elements essential in human nutrition. The physiological consequences of nutritional deficiencies of those elements are listed in Table 1.3.

TABLE 1.3 Elements Required for Human Nutrition and the Consequences of Their Deficiencies

Element	Result of nutritional deficiency
calcium	bone weakness, osteoporosis, muscle cramps
magnesium	calcium loss, bowel disorders
potassium	muscle weakness
sodium	muscle cramps
phosphorus	muscle and bone weakness
iron	anemia
copper	anemia
iodine	goiter
fluorine	tooth decay
zinc	poor growth rate
chromium	hyperglycemia
selenium	pernicious anemia
molybdenum	poor growth rate
tin	poor growth rate
nickel	poor growth rate
vanadium	poor growth rate

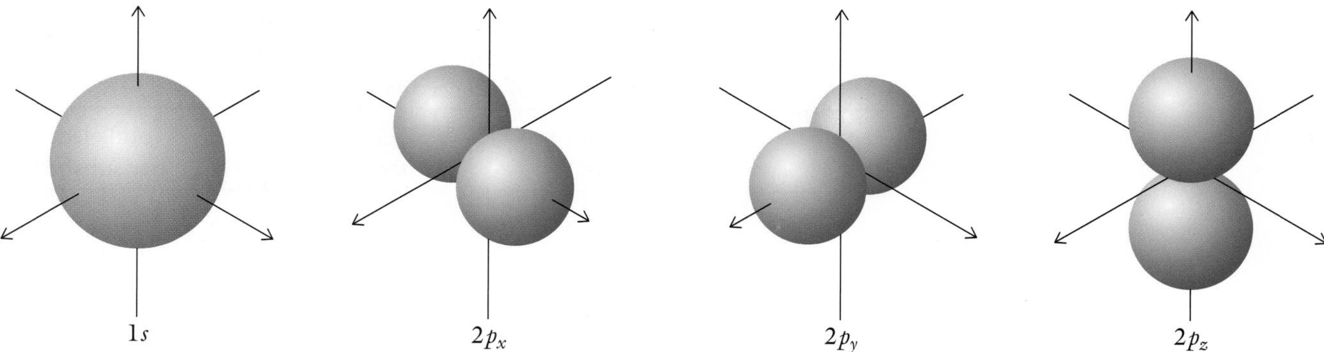

Figure 1.2 Atomic *s* and *p* orbitals. Their properties allow the representation of regions of space in which there is a greater than 90% probability of finding an electron.

1.3 ATOMIC STRUCTURE AND THE PERIODIC TABLE

The electrons in atoms are not randomly arranged. Instead, they are organized into a series of **shells** found at successively greater distances from the nucleus. Each shell is identified by a **principal quantum number,** 1, 2, 3, . . . , *n*, which corresponds to the shell's energy level, or energy state. The higher the number, the greater the energy and the farther the electrons are from the nucleus. The electrons reside in **orbitals,** regions of space in which the probability of finding an electron is highest. Orbitals are best visualized as clouds of negative electricity surrounding the nucleus. They have a variety of shapes, depending on the principal quantum number (overall energy of the shell), but each one can accommodate no more than two electrons. The types of orbitals, called **subshells,** are identified by lowercase letters (*s, p, d,* and so forth). To summarize, the electrons of an atom are organized into shells, subshells within shells, and orbitals within subshells. The *s* and *p* types of atomic orbitals are illustrated in Figure 1.2.

Later, we shall see that the formation of a covalent chemical bond requires the sharing, between atoms, of electrons contained in the orbitals. Because the orbitals are restricted to certain positions in space around the nucleus, the connections that they create between atoms give rise to the unique shapes, and therefore the characteristic properties, of molecules of different substances.

The quantitative relations between principal quantum number, number of subshells, and subshell types are tabulated in Table 1.4. You can see that, as the principal quantum number increases, the number of subshells increases. The number of orbitals and the types of orbitals within each subshell also

TABLE 1.4 Relation Between Principal Quantum Number and Number and Type of Orbital Within a Subshell

Principal quantum number	Number of subshells	Subshell type
1	1	*s*
2	2	*s, p*
3	3	*s, p, d*

TABLE 1.5 Relation Between Subshell Type, Number of Orbitals, Number of Electrons per Subshell, and Total Number of Electrons per Subshell

Subshell	Number of orbitals	Electrons per orbital	Total electrons per subshell
s	1	2	2
p	3	2	6
d	5	2	10

increase. The first shell contains an *s* orbital; the second shell contains *s* and *p* orbitals; the third shell contains *s, p,* and *d* orbitals; and so forth. The numbers of orbitals within each subshell are listed in Table 1.5.

The final feature of electron organization is that electrons have been found to have a property called **spin,** which is envisioned as being analogous to the rotation of a planet. As stated earlier, one orbital can contain no more than two electrons. A single electron within an orbital is called an **unpaired electron** or a **lone electron.** For an orbital to accommodate two electrons, their spins must be opposite. The electrons are then said to be **spin-paired.**

1.4 ATOMIC STRUCTURE AND PERIODICITY

We can account for the arrangement of elements in the periodic table by the following exercise. The elements will be "created," starting with hydrogen, by adding electrons one at a time around an atomic nucleus. To keep the resulting atom neutral, the atomic number of the nucleus will increase simultaneously every time that we add a new electron. This imaginary exercise has been given the German name ***Aufbau,*** which means "buildup." The result of the aufbau procedure is the complete description of the electron organization of the atom, also called the atom's **electron configuration.**

Here is a list of rules for constructing an atom. We will begin to put them into practice in Example 1.1.

Aufbau rules for constructing an atom

1. The larger the principal quantum number, the greater the number of subshells, as listed in Table 1.4.
2. Each subshell has a unique number of orbitals (see Table 1.5).
3. The order in which electrons enter the available shells and subshells depends on the orbital energy levels. The lowest-energy orbitals (closest to the nucleus), the 1*s* and 2*s* orbitals, fill before the 2*p* orbitals and those more distant.
4. There can be no more than two electrons per orbital, and the two electrons can occupy the same orbital only if they are spin-paired (their spins are opposite).
5. In a subshell with more than one orbital, electrons entering that subshell will not spin-pair until every orbital in the subshell contains one electron.

The method used for writing electron configuration identifies electrons, first, by the principal quantum number—1, 2, and so forth; next, by the orbital type—*s, p,* and so forth; and, finally, by the number of electrons in each orbital (by using a superscript)—$1s^2$, $2p^3$, for example.

Example 1.1 Diagramming electron configurations

Diagram the electron configuration of elements 1, 2, 3, 4, 5, and 6. (Remember, the atomic number and the number of electrons are identical.)

Solution

Use Table 1.5 to determine the principal quantum number (shell), available subshells, and number of electrons per subshell, and construct a table as shown here for each of these elements. An *s* subshell can contain a maximum of two electrons. Because we fill the orbitals one electron at a time, the first two elements will exhaust the first *s* subshell, 1*s*. In the next two elements, numbers 3 and 4, electrons will fill the second *s* subshell, 2*s*, and electrons in the fifth and sixth elements must go into the next available (empty) subshell, 2*p*, which can accommodate a maximum of six electrons in its three orbitals.

Element	Shell	Subshell	Number of electrons	Electron configuration
hydrogen	1	*s*	1	$1s^1$
helium	1	*s*	2	$1s^2$
lithium	1	*s*	2	
	2	*s*	1	$1s^22s^1$
beryllium	1	*s*	2	
	2	*s*	2	$1s^22s^2$
boron	1	*s*	2	
	2	*s*	2	
	2	*p*	1	$1s^22s^22p^1$
carbon	1	*s*	2	
	2	*s*	2	
	2	*p*	2	$1s^22s^22p^2$

Hydrogen's electron configuration is $1s^1$; that is, there is one electron (indicated by the superscript) in the *s* orbital of the first shell (indicated by the number 1 preceding *s*). Helium's configuration is $1s^2$: helium's two electrons fill and complete the *s* orbital of the first shell. Lithium, atomic number 3, with three electrons, follows helium. Its third electron must enter the second shell, where the first available empty subshell is the 2*s* orbital. Thus, the configuration of lithium is $1s^22s^1$. With the fourth element, beryllium, the 2*s* subshell is complete. Beryllium's electron configuration is $1s^22s^2$. Boron, atomic number 5, follows beryllium. Its fifth electron must enter the next available empty orbital. The electron configuration of boron is $1s^22s^22p^1$. Carbon's electron configuration is $1s^22s^22p^2$.

Problem 1.1 Write the electron configuration for fluorine.

In Example 1.1, the electron configuration for carbon must be looked at more closely. The three *p* orbitals are of equal energy and, as shown in Figure 1.2, lie perpendicularly to one another and are identified according to the Cartesian axes, *x*, *y*, and *z*, on which they lie. This designation is added as a subscript. Thus, the three *p* orbitals are known as p_x, p_y, and p_z. According to rule 5, these orbitals are occupied singly until each contains one electron, after which an electron of opposite spin may enter each one. Although it is not always necessary for the electron configuration to reveal the orbital spin states, to do so will be important in your study of organic chemistry. Therefore, taking the spin states into account, the precise electron configuration of carbon is $1s^22s^22p_x^{\,1}2p_y^{\,1}$.

TABLE 1.6 Electron Configurations of Some of the Alkali Metals, the Halogens, and the Noble Gases

Alkali metals (Group I)		Halogens (Group VII)		Noble gases (Group VIII)	
Li	[He] $2s^1$	F	[He] $2s^2 2p^5$	Ne	[He] $2s^2 2p^6$
Na	[Ne] $3s^1$	Cl	[Ne] $3s^2 3p^5$	Ar	[Ne] $3s^2 3p^6$
K	[Ar] $4s^1$	Br	[Ar] $3d^{10} 4s^2 4p^5$	Kr	[Ar] $3d^{10} 4s^2 4p^6$
Rb	[Kr] $5s^1$	I	[Kr] $4d^{10} 5s^2 5p^5$	Xe	[Kr] $4d^{10} 5s^2 5p^6$

Table 1.6 presents the electron configurations of the elements in Groups I, VII, and VIII for periods 2 through 5. Because every period in Table 1.6 ends with a noble gas, whose outer shell contains a maximum number of electrons for that shell (or row), the bracketed name of a noble-gas element is used as a kind of shorthand in writing the electron configurations for elements 3 and higher.

For example, the full electron configuration of element 19, potassium, is $[1s^2 2s^2 2p^6 3s^2 3p^6]4s^1$. Brackets have been placed around the part of the electron configuration that matches the configuration of argon (symbol Ar), the preceding noble gas. This configuration is called the argon **core,** and the electron configuration of potassium can be abbreviated as $[\text{Ar}]4s^1$.

The electrons of the outermost shell are printed in red in Table 1.6. Notice that (1) all the noble gases have eight electrons in the outermost shell, (2) all the alkali metals have a single electron in the outermost shell, and (3) all the halogens have seven electrons in the outermost shell. The aufbau exercise therefore demonstrates that all the main-group elements in a vertical column have the same number of outer-shell electrons and that the number of outer-shell electrons is the same as the main-group elements' group number in Figure 1.1. Therefore, in Figure 1.1, Roman numeral I means one electron in the outer shell, II means two electrons in the outer shell, and so forth.

In a symbolic representation called **Lewis symbols** (after American chemist Gilbert N. Lewis, who first proposed them), the nucleus and inner-shell electrons are represented by the element's symbol, and dots around it represent the outer-shell electrons. The Lewis symbols for elements in periods (rows) 1, 2, and 3 of the periodic table are shown in Table 1.7. With the exception of helium, the number of outer-shell electrons is equal to the group number of the main-group element.

1.5 CHEMICAL BONDS

In general, chemical reactions of the elements result in the formation of a noble-gas electron configuration in the atoms' electron shells. Because the

TABLE 1.7 Lewis Symbols of the Elements in the First Three Periods of the Periodic Table

I	II	III	IV	V	VI	VII	VIII
H·							:He
Li·	·Be·	:B·	·C·	·N·	·O·	:F·	:Ne:
Na·	·Mg·	:Al·	·Si·	·P·	·S·	:Cl·	:Ar:

noble gases are inert and chemically stable, chemists have concluded that a filled outermost electron shell is the most stable configuration that an atom can have. Because a noble gas has eight electrons in its outer shell, this conclusion is summarized as the **octet rule.** The exception to this rule is hydrogen, which requires only two electrons to achieve the noble-gas configuration of helium. The outermost electron shell (in a given period) is called the **valence shell.** The electrons in the outer shell are called the **valence electrons.**

Concept check

✓ The octet rule states that all elements, with the exception of the noble gases, react chemically to achieve the same outer electron configuration as that of the noble gases.

The result of chemical reactions is the formation of compounds in which the atoms are connected to each other by chemical bonds.

There are two major types of chemical bonds: ionic and covalent. An **ionic bond** is a strong electrical attraction between a positive ion and a negative ion. It results when electrons are transferred from the valence shell of one atom into the valence shell of another. A **covalent bond** is a bond formed when electrons are shared between atoms. In general, ionic bonds are formed between metals and nonmetals, and covalent bonds are formed between nonmetals.

The kind of bond—ionic or covalent—that will form as a result of a chemical reaction depends on the relative abilities of the reacting atoms to attract electrons. When one atom has a much greater ability to attract electrons than the other atom does, electron transfer will take place, and the resulting bond will be ionic. Such is the case when metals react with nonmetals. When two atoms are similar in their ability to attract electrons, electron transfer is no longer possible. In those cases, octet formation is accomplished through electron sharing, and the resulting bond is covalent. This kind of bonding generally takes place in reactions between the nonmetals.

The elements differ in their ability to attract valence electrons or to retain their own valence electrons. The degree of attraction depends on the size of the positively charged nucleus. Within a period, the nuclei increase in positive charge, and the distance between the valence-shell electrons and the nucleus remains nearly the same. As a result, the intensity of electrical attraction between nuclei and outer electrons increases with atomic number across each period, and it becomes more and more difficult to remove electrons from the valence shell.

On the other hand, the attraction for valence electrons exerted by an atom's nucleus decreases from the top to the bottom of any group of the periodic table. This decrease is because the number of electron shells increases down each group, and the outer electrons become increasingly insulated from the influence of the positively charged nucleus. The net result is that the elements in the upper right-hand side of the table—N, O, and F—have the greatest ability to attract electrons. The ability of an atom within a molecule to attract electrons toward itself is called its **electronegativity.** The effects of electronegativities of atoms within molecules will be considered in Section 1.9.

Concept checklist

✓ An ionic bond will form if one of the bonding partners has a much greater electron-attracting power than the other.

✓ A covalent bond will form if the electron-attracting powers of the bonding partners are similar in strength.

TABLE 1.8 Relation Between Group Number and Charge of the Ions of the Main-Group Elements*

group number	I	II	III	V	VI	VII
ion charge	1+	2+	3+	3−	2−	1−

*The ionic charge of a metal is equal to its group number, and the ionic charge of a nonmetal is equal to its group number minus 8.

Ionic Bonds

The processes in the formation of an ionic bond can be visualized by noting the changes in the electron configurations of reacting elements. For example, the chemical reaction between sodium and chlorine atoms that leads to the production of sodium chloride can be shown by using Lewis symbols:

$$\mathrm{Na}\cdot + \cdot\ddot{\underset{\cdot\cdot}{\mathrm{Cl}}}\colon \longrightarrow \mathrm{Na}^+ + \colon\ddot{\underset{\cdot\cdot}{\mathrm{Cl}}}\colon^-$$

In this reaction, the single electron in the outer shell of sodium is transferred to the outer shell of the chlorine atom. The sodium atom becomes positively charged and is now a sodium ion (a cation). The chlorine atom becomes negatively charged and is now a chloride ion (an anion). An ionic bond is formed as a result of the strong electrical attraction between the ions. The electron configuration of the sodium ion is identical with that of the noble gas preceding elemental sodium in the periodic table—neon. The electron configuration of the chloride ion is that of the noble gas immediately to the right of elemental chlorine in the table—argon. By reacting with one another, both atoms have achieved a noble-gas electron configuration.

The elements in Groups I, II, and III are metals. A metal loses electrons from its valence shell to form a cation whose positive charge is equal to the metal's group number. In contrast, elements in Groups V, VI, and VII gain electrons to form anions whose charge is equal to the group number minus 8. Table 1.8 illustrates the relation between a main-group element's group number and its ionic charge.

Example 1.2 Predicting the electrical charges of ions

What are the electrical charges of ions formed from the elements sodium, calcium, aluminum, nitrogen, sulfur, and chlorine?

Solution

Remember that the elements in Groups I, II, and III are metals. A metal loses electrons from its valence shell to form a cation whose positive charge is equal to the metal's group number. In contrast, Groups V, VI, and VII gain electrons to form anions whose charge is equal to the group number minus 8. According to Table 1.8, sodium forms Na^+, calcium forms Ca^{2+}, aluminum forms Al^{3+}, nitrogen forms N^{3-}, sulfur forms S^{2-}, and chlorine forms Cl^-.

Problem 1.2 Predict the charges of ions formed from the elements (a) potassium; (b) magnesium; (c) indium; (d) phosphorus; (e) oxygen; (f) bromine.

A chemical compound is identified by its **formula.** In a formula, the symbols of a compound's elements have subscripts that tell us the relative numbers of atoms of each element. The formula of a covalent compound consists of the actual number of atoms of each kind in the fundamental unit called a molecule. A **molecule** is the smallest unit that retains the chemical and physical properties of a substance. Ionic solids consist of three-dimensional arrays of positive ions surrounded by negative ions and negative ions surrounded by positive ions. This arrangement makes it impossible to identify a unique structural unit (such as a molecule) that can be represented by the formula of the solid. Therefore, formulas of ionic compounds show only the combining ratios of the constituent elements.

The combining ratios in the formula of an ionic compound depend on the requirement that the net charge of the component ions must be zero. Thus, when lithium reacts with fluorine, the 1+ charge on the resulting lithium cation and the 1− charge on the fluoride anion lead to a 1:1 ratio of atoms in the resulting product. The formula of the newly formed compound is therefore LiF. (When the formula contains only one cation or anion, its subscript is omitted, and the 1 is understood.) The 1:2 ratio of ions in the product of the magnesium-fluorine reaction is described in the formula MgF_2. Similarly, the 1:3 ratio of reacting aluminum and fluorine atoms leads to a product of formula AlF_3. Two-element compounds such as these are called **binary compounds.** Rules for naming ionic compounds are presented in Appendix 1.

Ionic solids are crystalline and melt at very high temperatures. They are also insulators—they cannot conduct electricity. However, when ionic compounds are melted or dissolved in water, both the resulting liquid state of the compound and the solution conduct electricity. The reason is that, in the melted solid and in the solution, the ions are free to move about independently of one another. The individual ions can migrate to positive and negative electrodes and thus carry electric current.

Covalent Bonds

The formulas of compounds such as water (H_2O), carbon dioxide (CO_2), and oxygen (O_2), whose chemical bonds are covalent, represent not only the combining ratios of component elements, as in ionic compounds, but also the actual ultimate structural units that we call molecules. The formulas of covalent compounds are therefore called **molecular formulas.**

One of the simplest covalent bonds unites two hydrogen atoms to form the hydrogen molecule, H_2. Hydrogen atoms possess only one electron, and they can achieve the noble-gas configuration of helium (two electrons) by each gaining one more electron. A single covalent bond uniting two hydrogen atoms—or any two atoms—consists of two electrons shared by both atoms. As a result of the sharing, each atom achieves a noble-gas configuration—in this case, helium.

For elements capable of forming covalent bonds, the number of such bonds formed, called the **combining power,** depends on their position in the periodic table. The combining power of an element is the number of covalent bonds that it forms to complete its octet. For example, carbon, in Group IV, requires four additional electrons to fill its octet and is called tetravalent. In methane, CH_4, the carbon atom fills its octet by forming four covalent bonds with four hydrogen atoms. In carbon tetrachloride, CCl_4, it forms four covalent bonds with four atoms of chlorine. Nitrogen and other elements in Group V are trivalent. They require three additional electrons and form three covalent bonds. The Group VI elements, such as oxygen and sulfur, are divalent. They require two additional electrons and form two bonds to complete their octets. Group VII elements are monovalent and form one bond. Rules for naming covalent compounds are presented in Appendix 1.

Example 1.3 Predicting formulas of simple molecular compounds

Predict the formulas for the products of the reactions between the following pairs of elements: (a) carbon and bromine; (b) nitrogen and iodine; (c) carbon and sulfur.

Solution

(a) Carbon is tetravalent and bromine is monovalent; therefore, carbon will react with four bromine atoms, and the compound formed is CBr_4.

(b) Nitrogen is trivalent and iodine is monovalent; therefore, nitrogen will react with three iodine atoms, and the compound formed is NI_3.

(c) Carbon is tetravalent and sulfur is divalent; therefore, carbon will react with two sulfur atoms, and the compound formed is CS_2.

Problem 1.3 Predict the formulas for the products of the reactions between the following pairs of elements: (a) carbon and hydrogen; (b) phosphorus and chlorine; (c) carbon and oxygen.

Concept checklist

✓ The formula of an ionic compound represents only the combining ratios of the component ions; it does not represent a unique structural unit.

✓ The formulas of compounds whose chemical bonds are covalent represent not only the combining ratios of component elements, as in ionic compounds, but also the ultimate structural units that we call molecules.

Formula Mass and Molar Mass

A balanced chemical equation, such as

$$H_2SO_4 + 2\ KOH \longrightarrow K_2SO_4 + 2\ H_2O$$

in which the numbers of atoms of each kind on the left-hand side are equal to the numbers of atoms of each kind on the right-hand side, allows us to calculate the actual masses of substances taking part in the reaction. This calculation is accomplished by relating the balancing coefficients to the masses of constituents.

The formula of a compound is a statement of how many atoms of each element there are in a fundamental unit of the compound. A general name for the fundamental unit of any kind of compound is the **formula unit.** The formula unit for ionic compounds is the formula; for covalent compounds, it is the molecular formula. Examples of formula units of ionic compounds are NaCl, $CaBr_2$, and Al_2O_3 and, of covalent compounds, are H_2O, CO_2, and CBr_4.

The atomic masses in the periodic table are relative masses (relative to one another) and are given in **atomic mass units,** abbreviated **amu.** We assign a relative mass to a compound by adding the relative masses of all the atoms in the formula unit. This sum is called the **formula mass.** The term **molecular mass** is often used for covalent compounds.

As pointed out earlier, there is a distinction between the structures of ionic compounds, such as NaCl, and those of compounds that exist as molecules, such as H_2O. The definition of formula mass using the idea of a formula unit permits us to ignore those distinctions when we are concerned primarily with mass relations. All we need to know from a formula unit is what elements are present in the compound and in what proportions.

To determine the formula mass of water, H_2O, we add the relative masses (taken from the inside back cover of this book) of its two hydrogen atoms and one oxygen atom: 1.008 amu + 1.008 amu + 16.00 amu = 18.02 amu.

A compound's formula mass can be calculated only if its formula is known, as illustrated in Example 1.4. A discussion of the use of significant figures can be found in Appendix 2.

Example 1.4 Calculating the formula mass of a compound

Calculate to four significant figures the formula masses of the following compounds: NaCl, $CaCl_2$, Al_2O_3, $C_6H_{12}O_6$ (glucose, a simple sugar).

Solution

Formula mass is defined as the sum of the atomic masses of a compound's constituent atoms. Therefore,

$$\begin{aligned}
\text{Formula mass of NaCl} &= \text{atomic mass of Na} + \text{atomic mass of Cl} \\
&= 22.99 \text{ amu} + 35.45 \text{ amu} = 58.44 \text{ amu} \\
\text{Formula mass of CaCl}_2 &= \text{atomic mass of Ca} + 2 \times \text{atomic mass of Cl} \\
&= 40.08 \text{ amu} + (2 \times 35.45 \text{ amu}) = 111.0 \text{ amu} \\
\text{Formula mass of Al}_2\text{O}_3 &= 2 \times \text{atomic mass of Al} + 3 \times \text{atomic mass of O} \\
&= (2 \times 26.98 \text{ amu}) + (3 \times 16.00 \text{ amu}) = 102.0 \text{ amu} \\
\text{Formula mass of C}_6\text{H}_{12}\text{O}_6 &= 6 \times \text{atomic mass of C} + 12 \times \text{atomic mass of H} \\
&\quad + 6 \times \text{atomic mass of O} \\
&= (6 \times 12.01 \text{ amu}) + (12 \times 1.008 \text{ amu}) \\
&\quad + (6 \times 16.00 \text{ amu}) = 180.2 \text{ amu}
\end{aligned}$$

Problem 1.4 Calculate to four significant figures the formula masses of the following compounds: (a) $Zn(HPO_4)_2$, (b) Fe_3O_4, (c) H_2O, (d) H_2SO_4, (e) $C_3H_6O_2N$ (the amino acid alanine, a component of proteins).

Samples of different elements contain equal numbers of atoms when the mass of each sample, in grams, is identical with the element's relative mass in atomic mass units. Thus, there are equal numbers of atoms in 55.85 g of iron, 22.99 g of sodium, 16.00 g of oxygen, and so forth. The definition of relative mass of an element allows us to make the same statement about the formula mass of a compound. That is, the mass of any compound in grams that is equal to its formula mass in atomic mass units contains the same number of formula units as the mass of any other compound in grams that is equal to its formula mass in atomic mass units. Therefore, the number of formula units in 58.44 g of NaCl is equal to the number of formula units in 180.2 g of glucose.

Example 1.5 Converting formula units into mass

Calculate to four significant figures the masses, in grams, of NaCl, $CaCl_2$, Al_2O_3, and $C_6H_{12}O_6$ that will contain equal numbers of formula units of each compound.

Solution

The formula masses of these compounds were calculated in Example 1.4, and those masses in grams will all contain the same numbers of formula units. The masses are 58.44 g of NaCl, 111.0 g of $CaCl_2$, 102.0 g of Al_2O_3, and 180.2 g of $C_6H_{12}O_6$.

Problem 1.5 Calculate to four significant figures the masses, in grams, of (a) Na_2SO_4, (b) KBr, (c) MgO, and (d) C_6H_{14} that will contain equal numbers of formula units of each compound.

A **mole** (abbreviated **mol**) is the number of atoms contained in the atomic mass (in grams) of an element. It is also the number of formula units contained in the formula mass (in grams) of a compound. The mole also specifies

TABLE 1.9 Relations Between Molar Quantities

Substance	Formula	Formula mass (amu)	Molar mass (g/mol)	Particles per mole	Moles
atomic hydrogen	H	1.008	1.008	6.022×10^{23} hydrogen atoms	1 mol H atoms
molecular hydrogen	H_2	2.016	2.016	6.022×10^{23} hydrogen molecules 12.044×10^{23} hydrogen atoms	1 mol H_2 molecules 2 mol H atoms
water	H_2O	18.02	18.02	6.022×10^{23} water molecules 6.022×10^{23} oxygen atoms 12.044×10^{23} hydrogen atoms	1 mol H_2O molecules 1 mol O atoms 2 mol H atoms
calcium chloride	$CaCl_2$	111.0	111.0	6.022×10^{23} $CaCl_2$ formula units 6.022×10^{23} Ca^{2+} ions 12.044×10^{23} Cl^- ions	1 mol $CaCl_2$ formula units 1 mol Ca^{2+} ions 2 mol Cl^- ions

the number of formula units in a sample of a given mass. The number of fundamental particles in 1 mol is called the **Avogadro number,** 6.022×10^{23}.

A mole of any element or compound contains the same number of atoms or formula units as a mole of any other element or compound. The mass in grams of 1 mol of a compound is called its **molar mass.** The molar mass of sodium chloride weighs 58.44 g. It has the same number of formula units as 1 mol of glucose, which weighs 180.2 g.

The balancing coefficients in a balanced chemical equation can be read as formula units, moles, or molecules. Table 1.9 illustrates the relations between formula mass (atomic mass units), molar mass (grams per mole), particles per mole, and number of moles for hydrogen atoms, hydrogen molecules, water, and calcium chloride.

1.6 LEWIS STRUCTURES AND STRUCTURAL FORMULAS

The Lewis electron-dot notation that we used previously to represent the structure of the valence shells of the elements can be used to represent covalent bonds as well. The resultant diagrams are called **Lewis structures.** The key concept in building Lewis structures is to satisfy the octet rule—the rule stating that—with the exception of hydrogen—an element achieves stability by filling its outer shell with eight electrons.

It is important to recognize that in fact hydrogen is one of several exceptions to the octet rule. Hydrogen requires only two electrons, or a **duet,** to achieve the nearest noble-gas configuration, helium. For this reason, hydrogen will always form a covalent bond to only one other atom and therefore will never be found between two other atoms in Lewis structures.

Other exceptions to the octet rule are encountered with boron and beryllium, which form compounds such as BeH_2 and BF_3 that contain four and six electrons, respectively, in the central atoms' completed valence shells. However, the octet rule applies to all the other elements in the first and second periods of the periodic table, the principal focus of our interest.

Let's use the Lewis method of representing molecular structure to examine how two fluorine atoms react to form a diatomic molecule:

$$:\ddot{\underset{..}{F}}\cdot + \cdot\ddot{\underset{..}{F}}: \longrightarrow :\ddot{\underset{..}{F}}:\ddot{\underset{..}{F}}:$$

On the left side of the arrow, the reacting atoms are represented with numbers of valence electrons equal to their group numbers. On the right side of the arrow, we see the fluorine molecule, with each atom surrounded by a completed octet. (Overlapping circles have been drawn around the bonded fluorine atoms to indicate that the shared pair of electrons is part of each atom's octet.) Each fluorine atom shares one electron with the other, and so, in effect, they both have eight electrons. In this way, each fluorine atom can achieve the stable electron configuration of neon. Note that the result of their bonding is the sharing of a pair of electrons between them.

The Lewis structure (the molecule drawn with Lewis symbols) is simplified by representing the bonding pair of electrons as a line between the atoms and the other pairs as dots surrounding the atoms:

$$:\ddot{\underset{..}{F}}-\ddot{\underset{..}{F}}:$$

The pairs of electrons not shared in the covalent bond are called **nonbonded electrons** or **lone pairs.** Lewis structures can be further simplified by leaving out all nonbonded electrons, in which case they are called **structural formulas,** F—F.

Drawing Lewis Structures

The derivation of a Lewis structure requires (1) knowledge of a compound's molecular formula and (2) a set of simple rules (illustrated in Example 1.6). Although the molecular formula (for example, C_3H_8) tells us the number of atoms of each element in a molecule of the compound, it does not tell us, as the Lewis structure does, how they are connected to one another.

Example 1.6 Drawing Lewis Structures: I

Draw the Lewis structure of methane, CH_4.

Solution

Step 1. Place the symbols for the bonded atoms into an arrangement that will allow you to begin distributing electrons. In methane (as well as in many other compounds), there is only one atom of one kind (here, C) and several of another (here, four H atoms); so, as a first approximation, choose C as the central atom to which the rest are bonded. It is most likely that the carbon atom is central, with the hydrogens terminal:

```
   H
H  C  H
   H
```

Step 2. Calculate the total number of valence electrons. To do so, add the valence electrons contributed by each one of the atoms in the molecule.

Element	Valence electrons	Atoms per molecule	Number of electrons
carbon	4	1	4
hydrogen	1	4	4
		total valence electrons available =	8

Step 3. Represent shared pairs of electrons by drawing a line between the bonded atoms. The eight electrons are accounted for by drawing four single bonds between all atoms of the molecule.

Step 4. Because there are no lone pairs, the final structure is

```
     H
     |
  H—C—H
     |
     H
```

The octet rule is satisfied for carbon, and each hydrogen possesses its duet.

Problem 1.6 Draw the Lewis structure for silicon tetrachloride, $SiCl_4$.

Example 1.7 Drawing Lewis structures: II

Construct the Lewis structure for ammonia, NH_3.

Solution

Step 1. Place the symbols for the bonded atoms into an arrangement that will allow you to begin distributing electrons. In ammonia, there is only one atom of one type and several of another; so, as a first approximation, choose N as the central atom to which the rest are bonded.

```
    H

 H  N  H
```

Step 2. Determine the total number of valence electrons in the molecule. To do so, add the valence electrons contributed by each one of the atoms in the molecule.

Element	Valence electrons	Atoms per molecule	Number of electrons
nitrogen	5	1	5
hydrogen	1	3	3
		total valence electrons available =	8

Step 3. Represent shared pairs of electrons by drawing a line between the bonded atoms. For ammonia,

```
     H
     |
  H—N—H
```

Each line (bond) represents two electrons, accounting for six of the eight available valence electrons.

Step 4. Position the remaining valence electrons (two in this case) as a lone pair to satisfy the octet rule for each atom in the compound (remember that hydrogen's "octet" consists of only two electrons):

```
     H
     |
  H—N—H
     ··
```

Problem 1.7 Construct the Lewis structure for phosphorus trichloride, PCl_3.

Example 1.8 Drawing Lewis structures: III

Construct the Lewis structure for water, H_2O.

Solution

Step 1. Position the bonded atoms in a likely arrangement. In water, there is only one atom of oxygen and two of hydrogen, so a good guess is that the oxygen is the central atom to which the others are bonded:

```
 H  O  H
```

Step 2. Determine the total number of valence electrons in the molecule by adding up the valence electrons contributed by each of the atoms in the molecule.

Element	Valence electrons	Atoms per molecule	Number of electrons
oxygen	6	1	6
hydrogen	1	2	2
		total valence electrons available	= 8

Step 3. Represent electron-pair bonds by drawing a line between bonded atoms. For water:

H—O—H

This structure accounts for four of the eight available valence electrons.

Step 4. Position the remaining four valence electrons as lone pairs around the atoms to satisfy the octet rule for each:

$$\mathrm{H{-}\ddot{\underset{\cdot\cdot}{O}}{-}H}$$

Problem 1.8 Construct the Lewis structure for sulfur dichloride, SCl_2.

The next step in learning to use Lewis structures is to diagram a compound in which there is no unique central atom.

Example 1.9 Writing Lewis structures for more-complex compounds: I

Write the Lewis structure for ethane, C_2H_6. Ethane is one of a large family of organic compounds called alkanes. Remember that hydrogen cannot be inserted between the two carbon atoms, because it is not capable of participating in more than one bond at a time.

Solution

Step 1. Place the carbon atoms so that they are joined together, with the hydrogens in terminal positions:

```
    H  H
 H  C  C  H
    H  H
```

Step 2. Calculate the total number of valence electrons.

Element	Valence electrons	Atoms per molecule	Number of electrons
carbon	4	2	8
hydrogen	1	6	6
		total valence electrons available	= 14

Step 3. All these electrons are accommodated by drawing the seven bonds required to connect all the atoms of the molecule.

Step 4. Because there are no lone pairs, the final structure is

```
   H  H
   |  |
H—C—C—H
   |  |
   H  H
```

Note that the octet rule is satisfied for all C and H atoms. Because no lone pairs are shown, this Lewis structure is also a structural formula.

Problem 1.9 Write the Lewis structure for C_3H_8.

Lewis Structures Containing Multiple Bonds

Some interesting cases arise when there appear to be too few electrons to satisfy the octet rule. This apparent deficiency is found when we try to write the Lewis structures for compounds such as ethylene (C_2H_4), acetylene (C_2H_2), nitrogen (N_2), carbon dioxide (CO_2), and hydrogen cyanide (HCN).

Example 1.10 Writing Lewis structures for compounds with multiple bonds: I

Write the Lewis structure for ethylene, C_2H_4, a member of the alkene family of organic compounds.

Solution

Step 1. As we did for ethane, we can assume that the carbon atoms are joined and the hydrogen atoms occupy terminal positions:

```
H       H
   C  C
H       H
```

Step 2. Calculate the total number of valence electrons

Element	Valence electrons	Atoms per molecule	Number of electrons
carbon	4	2	8
hydrogen	1	4	4
		total valence electrons available =	12

Step 3. We can insert 10 of the 12 electrons by drawing five bonds to connect all the atoms, with the result that two electrons are left over:

```
H       H
 \     /
  C—C
 /     \
H       H
```

Step 4. The hydrogen-atom duets are satisfied. However, if we are to satisfy the octet rule, an additional pair of electrons is needed for each carbon atom. That would seem to mean that four more electrons are required, but only two electrons are available. This apparent dilemma can be solved by adding the two remaining electrons as an additional bonding pair between the carbon atoms, to form what is called a **double bond.** The shared double bond provides an octet around each carbon atom:

```
H       H
 \     /
  C=C
 /     \
H       H
```

Problem 1.10 Write the Lewis structure for carbon dioxide, CO_2.

Example 1.11 Writing Lewis structures for compounds with multiple bonds: II

Write the Lewis structure for nitrogen, N_2.

Solution

Step 1. Begin the construction of the gaseous diatomic element nitrogen by joining the two nitrogen atoms with a covalent bond:

$$N—N$$

Step 2. Calculate the total number of valence electrons.

Element	Valence electrons	Atoms per molecule	Number of electrons
nitrogen	5	2	10
		total valence electrons available =	10

Step 3. When we attempt to place one bond between the atoms and add the remaining electrons as lone pairs on each nitrogen atom,

$$:\ddot{N}—\ddot{N}:$$

we find that we are short two pairs of electrons for the necessary octets.

Step 4. However, we can move two of the lone pairs to a position between the nitrogen atoms to form two more bonds: then we add the remaining four electrons as lone pairs on each of the atoms to obtain

$$:N{\equiv}N:$$

Three pairs of electrons are shared to complete the octets, and the bond between the nitrogen atoms is called a **triple bond.**

Problem 1.11 Write the Lewis structure for the rocket fuel hydrazine, N_2H_4.

When you begin your study of organic chemistry (Chapter 3), you will find that many important organic molecules contain single, double, or triple bonds. Eyesight depends on the properties of a double bond, and its story is told in Box 4.2.

Lewis Structures of Polyatomic Ions

The Lewis structures for polyatomic ions (ions made up of two or more atoms) are developed in the same way as the Lewis structures in the preceding worked examples. However, in calculating the total number of valence electrons, we must take the charge on the ion into account. If it is a negative ion, electrons equal to the negative charge must be added to the sum of the valence electrons. If the polyatomic ion has a positive charge, a number of electrons equal to its positive charge must be subtracted. For example, the ammonium ion possesses a charge of 1+, and the carbonate ion has a charge of 2−, and so the total number of valence electrons is calculated as follows:

Ion	Total number of valence electrons
NH_4^+	$N(1 \times 5) + H(4 \times 1) - 1 = 8$
CO_3^{2-}	$C(1 \times 4) + O(3 \times 6) + 2 = 24$

Using the steps outlined in the preceding examples, we find the Lewis structures of the ammonium ion and the carbonate ion to be

[H—N(—H)(—H)—H]⁺ and [C(—O)(—O)(=O)]²⁻

Ammonium ion Carbonate ion

1.7 THREE-DIMENSIONAL MOLECULAR STRUCTURES

Lewis structures provide us with two kinds of information about molecular structure:

- How atoms within the molecule are connected to one another
- Whether the connecting bonds are single, double, or triple

However, Lewis structures cannot describe the actual spatial arrangements of atoms within molecules. The important point to remember is that, although Lewis structures allow us to decide which atoms are connected and what the nature of the bonds holding them together is (single, double, or triple), Lewis structures cannot reveal the three-dimensional arrangements of the atoms. The three-dimensional, or spatial, arrangement of atoms in a molecule gives rise to a molecule's unique characteristics. Those spatial arrangements are of central importance in all physiological functions, such as enzymatic activity, membrane transport, nerve transmission, and antigen–antibody interactions. See Box 9.2 on smell and taste, Box 9.3 on chiral drugs, Section 9.5 on chiral recognition, Box 4.2 on vision, Box 4.3 on pheromones, and Section 14.5 on enzymes for examples.

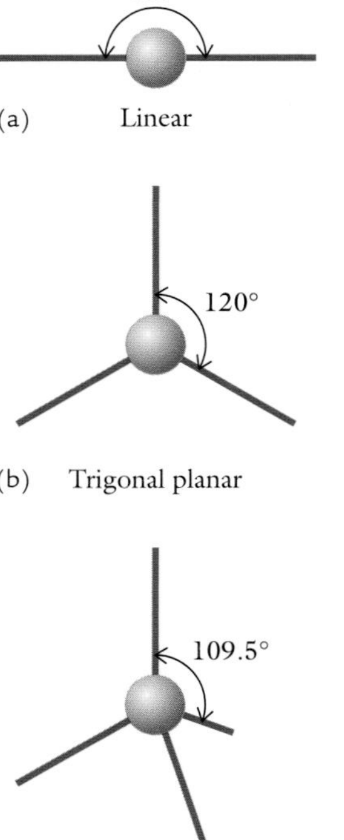

Figure 1.3 The positions assumed by pairs of electrons around a central atom are represented by straight lines emanating from the central atom. (a) Two pairs form a linear molecule; (b) three pairs, an equilateral triangle; and (c) four pairs, a regular tetrahedron.

VSEPR Theory

A simple and very useful theory allows us to predict the shapes of both simple and complex molecules. The theory states that the shape of a molecule is affected by each of the valence-electron pairs surrounding a central atom, not just the pairs participating in bonds.

The central idea of the theory is that, because electrons tend to repel one another, the mutual repulsion of all the electron pairs forces the atoms within the molecule to take positions in space that minimize those mutual repulsions. In other words, the electrons strive to get as far away from one another as possible. The name of the theory is the **valence-shell electron-pair repulsion theory,** or the **VSEPR theory.**

Because the most stable arrangement for the pairs of electrons in a molecule is to be as far away from one another as possible, the electron pairs must occupy points in space that define symmetrical three-dimensional geometrical figures. They are illustrated in Figure 1.3. Beryllium chloride, $BeCl_2$ takes the shape of a linear molecule so that the two pairs of electrons can be as far apart as possible. Boron trifluoride, BF_3, takes the shape of an equilateral triangle for the same reason. The tetrahedron is the result of the symmetrical arrangement of four pairs of electrons around a central atom. Each of the four pairs is as far from the others as possible (while still remaining attached to the central atom).

To illustrate how considerations of the repulsion of the electron pairs of covalent bonds affect molecular shapes, let's apply VSEPR theory to the three-dimensional structure of methane, CH_4.

Example 1.12 Using VSEPR theory to predict molecular structure: I

Use VSEPR theory to predict the three-dimensional structure of methane, CH_4.

Solution

The Lewis structure of methane was determined in Example 1.6 and is

```
     H
     |
  H—C—H
     |
     H
```

The Lewis structure shows four pairs of electrons surrounding the central carbon atom. The three-dimensional figure that allows maximum separation of four pairs of electrons is the tetrahedron. Therefore, VSEPR theory predicts that methane will have a tetrahedral structure, as illustrated in Figure 1.4.

Problem 1.12 Use VSEPR theory to predict the three-dimensional structure of carbon tetrachloride, CCl_4.

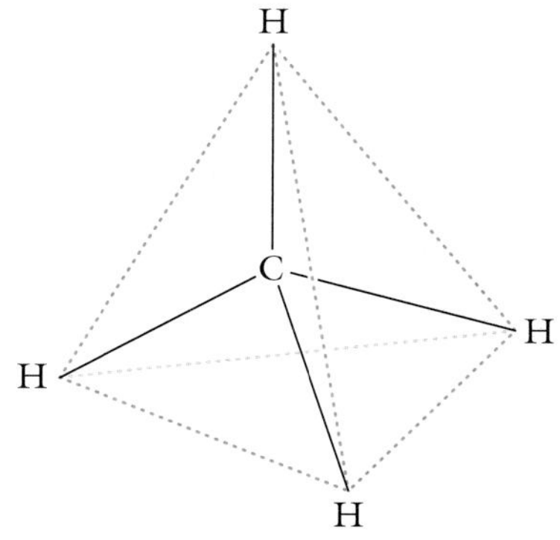

Figure 1.4 Methane has a tetrahedral molecular shape. The dotted lines indicate the outline of a regular tetrahedron, and the solid lines represent the C—H bonds.

A **bond angle** is the angle between two bonds that share a common atom. Figure 1.4 shows the carbon atom of CH_4 in its central location, with single bonds, separated by bond angles of 109.5°, connecting it to each of four hydrogen atoms. The dotted lines indicate the outline of the hypothetical solid figure enclosing the carbon atom, and the solid lines represent the carbon–hydrogen single bonds.

Nonbonding Electrons

In contrast with the structure of methane, there are situations in which the arrangement of four electron pairs is not symmetrical. Specifically, when one or more of an atom's electron pairs do not participate in a bond, those nonbonding pairs take up considerably more space, causing the bonding electron pairs to be squeezed more closely together.

VSEPR theory explains the symmetry-altering effect of nonbonding electron pairs by pointing out that electron pairs in bonds are localized because they are fixed between two positively charged nuclei. Nonbonded pairs are not subjected to a similar localizing influence and, as a result, they occupy a greater volume of space than a bonded pair does. In doing so, they push the bonded pairs more closely to one another than the tetrahedral geometry would predict. As an example, let's see what VSEPR theory predicts for the structure of ammonia.

Example 1.13 Using VSEPR theory to predict molecular structure: II

Use VSEPR theory to predict the structure of the ammonia molecule.

Solution

The Lewis structure for ammonia, NH_3, was worked out in Example 1.7.

```
     H
     |
  H—N—H
     ··
```

As in methane, there are four pairs of electrons surrounding the central nitrogen atom. VSEPR theory therefore predicts that the structure will be tetrahedral; but, in this case, one of the pairs is nonbonded, which alters the perfect symmetry exhibited by methane. Experiments have shown that the bond angles between the hydrogen atoms and the central nitrogen atom are 107°, rather than the 109.5° of a perfect tetrahedron.

Although all the bonding and nonbonding electron pairs influence a molecule's shape, only the atoms in the molecule are considered in naming that shape. Therefore, the ammonia molecule takes the shape of a shallow pyramid: the three hydrogen atoms form the base, with the nitrogen centered above them. This shape is called trigonal pyramidal because the pyramid has a triangle for its base, as shown in the margin.

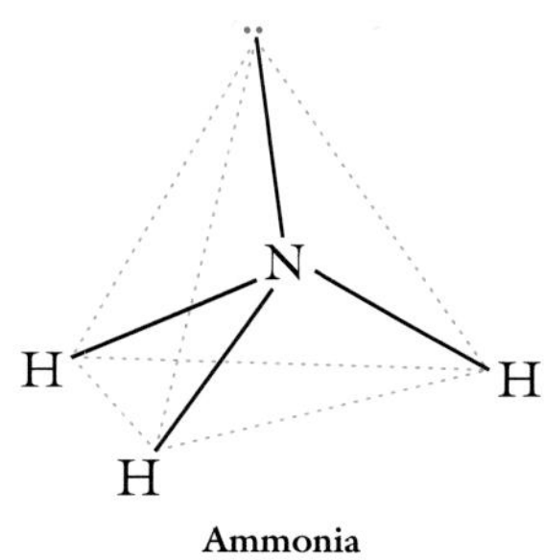

Ammonia

Problem 1.13 Use VSEPR theory to predict the structure of nitrogen triiodide, NI_3.

Example 1.14 Using VSEPR theory to predict molecular structure: III

Use VSEPR theory to predict the three-dimensional structure of water, H_2O.

Solution

The Lewis structure for water with its two lone pairs on the oxygen atom was worked out in Example 1.8.

$$H-\underset{\cdot\cdot}{\ddot{O}}-H$$

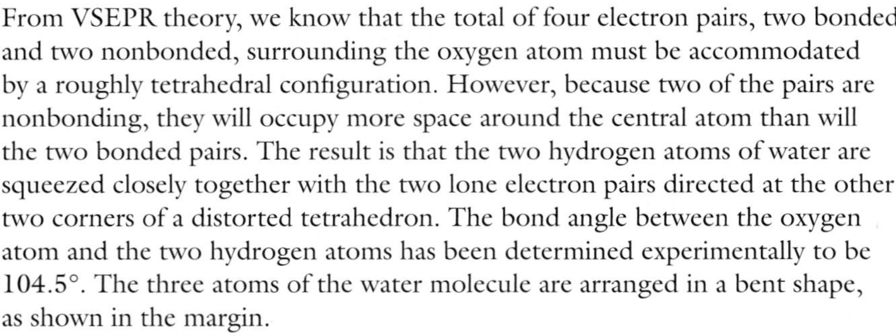

From VSEPR theory, we know that the total of four electron pairs, two bonded and two nonbonded, surrounding the oxygen atom must be accommodated by a roughly tetrahedral configuration. However, because two of the pairs are nonbonding, they will occupy more space around the central atom than will the two bonded pairs. The result is that the two hydrogen atoms of water are squeezed closely together with the two lone electron pairs directed at the other two corners of a distorted tetrahedron. The bond angle between the oxygen atom and the two hydrogen atoms has been determined experimentally to be 104.5°. The three atoms of the water molecule are arranged in a bent shape, as shown in the margin.

Water

Problem 1.14 Use VSEPR theory to predict the three-dimensional structure of oxygen difluoride, OF_2.

VSEPR theory treats double and triple bonds as though they were single bonds. The Lewis structure for acetylene, C_2H_2, is $H-C{\equiv}C-H$, and the molecule possesses a triple bond. Because VSEPR theory treats a triple bond as though it were a single bond, we proceed as though there were only two bonding pairs of electrons to be accommodated around each carbon atom. The geometrical arrangement that allows maximal separation of two bonding pairs is a straight line separating the bonding pairs by 180°. The VSEPR prediction of acetylene's geometry has been verified by experiment.

The steps for determining the shapes of molecules by VSEPR theory are:

Rules for predicting molecular shape by VSEPR theory

- Determine the Lewis structure.
- Determine the number of electron pairs (bonded and nonbonded) surrounding a central atom.
- Use VSEPR theory to decide what symmetrical three-dimensional figure will accommodate all bonded and nonbonded electron pairs.
- Classify the shape of the molecule by considering the arrangement of the joined atoms only, ignoring the nonbonded electron pairs. Thus, although the CH_4 and H_2O molecules each have four pairs of electrons around the central atom, the shape of methane is tetrahedral and that of water is bent.

Three-Dimensional Representations of Molecules

There are two fundamental ways of illustrating three-dimensional molecular structure on a two-dimensional page. One is called the ball-and-stick model, and the other is the space-filling model. Both are illustrated in Figure 1.5.

In the ball-and-stick model of methanol, the atoms are represented by balls and are differentiated by color; the bonds are represented by sticks and are about equal in length. The bond angles are correct, but the atom size relative to the bond length does not represent the actual ratio of atom size to bond length. Actual bond lengths are much shorter than represented by the ball-and-stick models. In the space-filling models, the atoms and bond lengths are drawn to precise relative sizes and are disposed in space at the correct angles with respect to one another. The surfaces of the space-filling "atoms" in the models represent, in relative terms, the closest that the atoms can approach one another. However, using space-filling models for instruction presents difficulties, because the bonds cannot be seen (the atoms hide them).

When it seems important to illustrate a particular molecular structure in three dimensions, one of these two models will be used. For example, the space-filling model of cholesterol shown in Figure 1.6 helps us to see the molecular shape of an important biochemical. However, most of the molecular structures that you will encounter in this book will be two-dimensional Lewis structures minus the nonbonded electrons. We have chosen this style because the Lewis structures are easy to draw and well within the capability of those of us who are not artists.

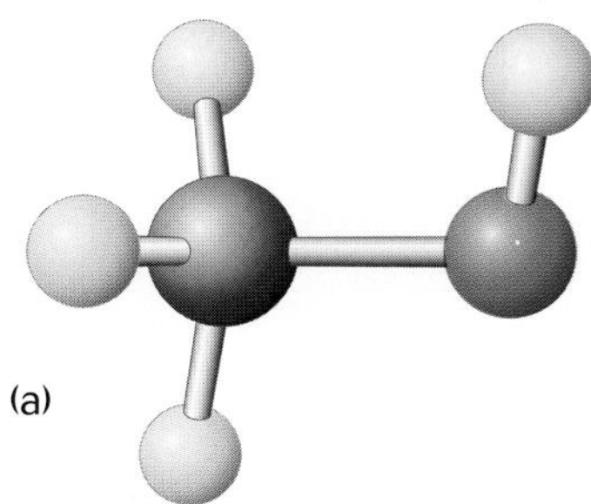

(a)

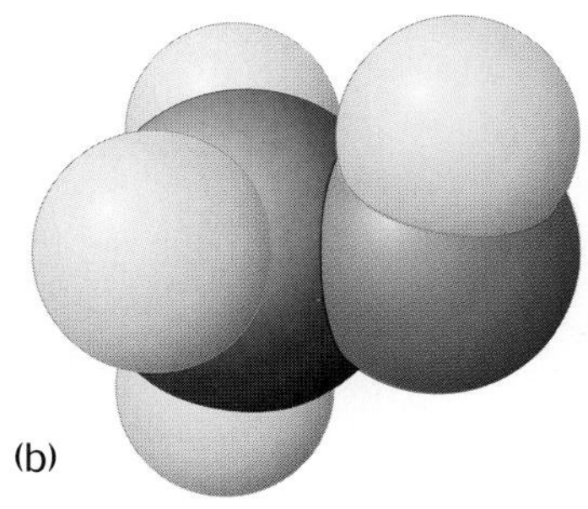

(b)

Figure 1.5 (a) A ball-and-stick and (b) a space-filling model of methanol, CH_3OH.

1.8 MOLECULAR INTERACTIONS

So far, we have concentrated on the forces that hold atoms together to form chemical compounds—covalent and ionic bonds. Now, we shall study the forces that operate not within molecules but between them. Walking on water—when it is frozen, of course—is a good reminder that water can exist as a solid (ice) or a liquid or a gas (steam). Water is not unique in this regard. All substances can exist in any of the three states of matter, depending on the conditions of temperatures and pressure.

Substances exist as solids and liquids because of attractive forces between the fundamental particles. This is easy to explain when the substance is an ionic compound, such as sodium chloride or potassium nitrate, because the component particles are positive and negative ions that are strongly attracted to one another. But the molecules of water, carbon dioxide, ammonia, and many other compounds are electrically neutral, yet can exist as liquids or solids. What forces cause water and other neutral molecules to attract one another?

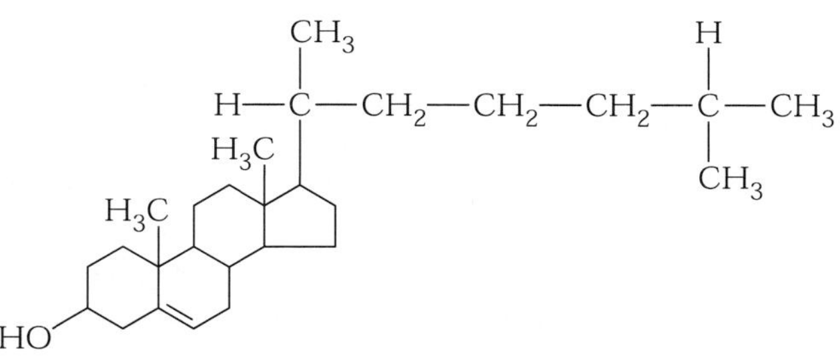

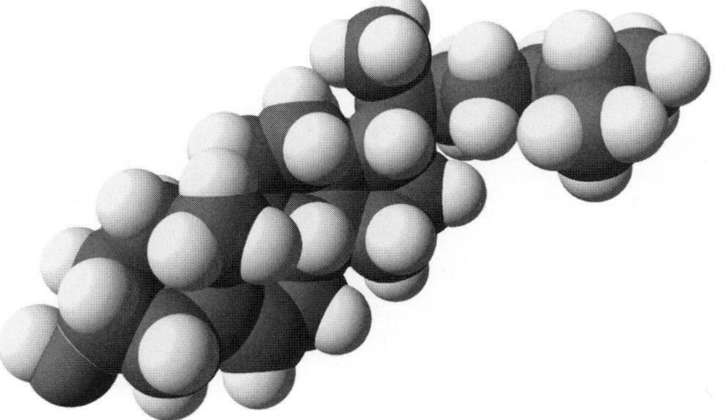

Figure 1.6 A space-filling molecular model of cholesterol, along with its planar molecular representation.

The existence of liquid and solid states in substances consisting of neutral molecules such as water or atoms such as helium is explained by a number of different types of attractive forces that have their origin in electrical asymmetry within molecules. Although these forces are much weaker than ionic or covalent chemical bonds, they underlie not only the formation of solids and liquids, but many physiological phenomena as well, including the structure and properties of biomolecules such as proteins, nucleic acids, carbohydrates, and lipids. The attractive forces between molecules are often called **intermolecular forces** or **van der Waals forces.** In this book, we will use the term **secondary forces** because it includes both intermolecular and intramolecular forces, and intramolecular forces are very important for biologically important molecules such as proteins, polysaccharides, and DNA.

At this point, it is useful to review the general properties of solids, liquids, and gases. A solid has a fixed volume and shape, both of which are independent of its container. A liquid also has a fixed volume but does not have a fixed shape. It can be poured from one container to another, and its shape depends on the shape of its container. The volume of a gas depends on the volume of its container. A gas always occupies all of its container. The volumes of both liquids and solids change very little when pressure is applied. They are said to be incompressible. On the other hand, a gas is easily compressed into a smaller volume. Figure 1.7 is a sketch of the molecular features of the three states of matter. There is little increase in volume when a solid melts, but when a liquid is vaporized, there is a large change in volume. The volume of 1 mol of liquid water at 100°C is 18.8 mL, and the volume of 1 mol of gaseous water (steam) at the same temperature and pressure is about 3.1×10^4 mL. (The use of scientific notation is considered in Appendix 3.)

In comparison with a gas, a solid cannot be compressed, because the constituent molecules are in contact, as close to one another as they can be. A solid holds its shape because strong attractive forces hold the molecules in fixed positions from which they cannot escape.

The compressibility of liquids is similar to that of solids, and so the molecules (or ions) are assumed to be as close to one another as they can be. The fact that a liquid flows and cannot hold its shape is explained by assuming that the attractive forces between molecules in the liquid state are weaker than those existing in solids. The internal structure of liquids is less orderly than that of a typical solid.

In a gas, the particles move at high speeds and change directions only on collision with the walls of their container or with other particles. Because the molecules in gases are so far apart, they have virtually no opportunity to interact; so secondary attractive forces play a small role in gases. Because the molecules of gases are not subject to such forces, the physical properties of gases are independent of the nature of the individual molecules. For example, in the gaseous state, there is virtually no difference in the physical behavior of

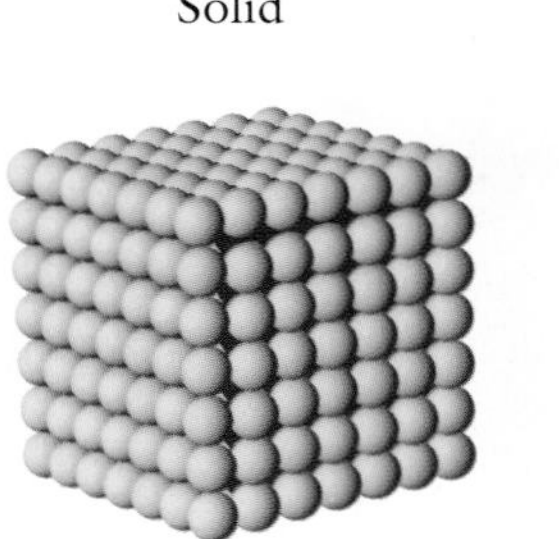

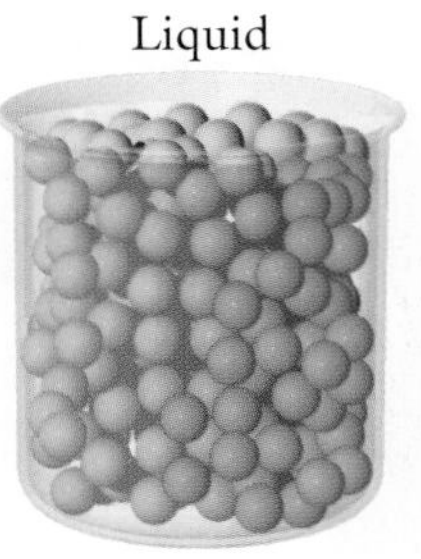

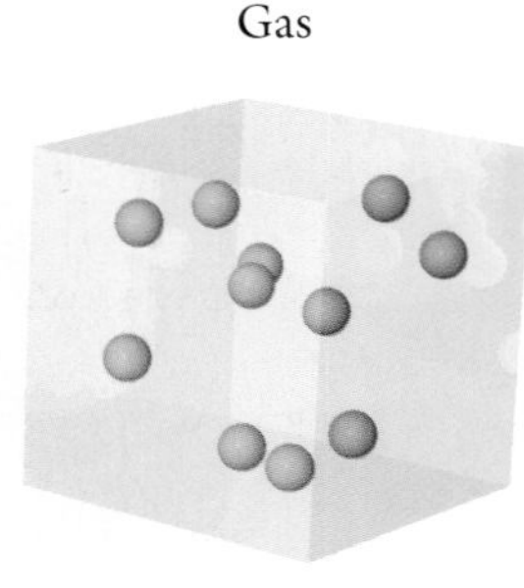

Figure 1.7 A molecular view of the three states of matter. The molecules in both the solid and the liquid states are in contact, but those in the liquid state are somewhat disordered. Molecules in the gaseous state are very far apart and totally disordered.

hydrogen, methane, or any other gaseous compound. As a consequence, all gases can be quantitatively described by only a few simple mathematical laws.

The energy of a moving body, or "energy of motion," is called **kinetic energy** (KE). It is equivalent to the work necessary to bring that body to rest or to impart motion to a resting body. Kinetic energy at the molecular level depends only on temperature. As the temperature increases, so does kinetic energy and, as a result, molecules move faster. In hot systems, molecules move about much faster than they do in cold systems. Kinetic energy tends to oppose the attractive forces between molecules.

Most solids melt and are transformed into liquids on addition of sufficient heat; they can revert to the solid state if the same amount of heat is removed. The transition from solid to liquid is called **melting,** and the temperature at which the transition takes place is called the **melting point.** The reverse process, transforming liquid into solid, is called **freezing,** and the temperature at which that transition takes place is called the **freezing point.** The melting point and the freezing point of a pure solid are identical. The transition from liquid to gas also requires heat and is called **vaporization.** The reverse of this process is called **condensation.** Continuous vaporization from the surface of a liquid into the atmosphere is called evaporation. The changes from one state to another as a result of the addition or removal of heat are called **phase transitions.** Phase transitions can be in either direction; that is, they are **reversible.**

Heat added to a system in transition from solid to liquid or from liquid to gas does not cause the temperature to increase. Instead, this heat is used to overcome the attractive forces between molecules. Only when the phase transition is complete—that is, when all the molecules of a solid have been converted into the liquid state—will the added heat cause the temperature to increase. The temperature of a sample of liquid water remains at 100°C until all the water is converted into steam at 100°C. Any heat then added to the steam will raise its temperature. The temperature at which the gaseous form of the liquid (bubbles) form throughout the liquid in a container open to the atmosphere is called the **normal boiling point.**

1.9 SECONDARY FORCES

The nature of molecular interactions depends on the secondary forces operating between molecules. The origin and characteristics of secondary forces will be explored in this section.

Polar Covalent Bonds

When an electron pair is shared by two atoms of the same element, as in molecules such as H_2 and F_2, the electrons are equally attracted to both atomic nuclei and will be found midway between them. Other examples are O_2, N_2, and Cl_2. This type of covalent bond is called a **nonpolar covalent bond.** When molecules consist of different elements, however, such as the molecules HBr and HCl, one atom participating in the bond is likely to exert a greater attraction for the electrons than the other atom. In these cases, the electrons tend to be closer to the more electronegative atom—the atom having the greater ability to attract the electrons. A bond of this type is called a **polar covalent bond.**

Polar bonds are distributed along a continuum, as shown in Figure 1.8 on the following page. At one extreme of this continuum is the ionic bond, in which both electrons reside on one of the ions of a cation–anion pair. At the other extreme is the pure covalent bond, in which the electron pair spends most of its time halfway between the bonded atoms. Most covalent chemical bonds are polar, however, and lie somewhere between these two extremes.

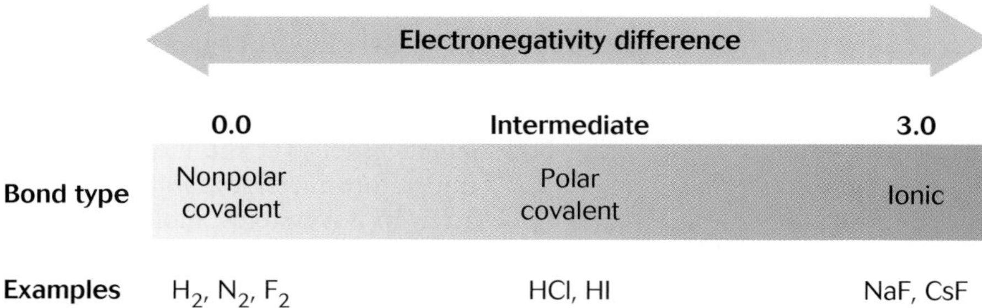

Figure 1.8 A continuum showing the effect of electronegativity difference on the type of bond formed between elements. At one extreme is the nonpolar covalent bond, which can form only between identical atoms (electronegativities are identical). At the other extreme is the ionic bond, which forms when the difference between the elements' electronegativities exceeds about 2.0.

As described in Section 1.5, the ability of an atom within a molecule to pull electrons toward itself is called its electronegativity. Figure 1.9 provides an abbreviated list of the electronegativities of various elements; the larger the number, the greater the electronegativity. Figure 1.10 is a three-dimensional plot of the main-group elements' electronegativity values in Figure 1.9. Note the increasing values from the minimum of cesium to the maximum of fluorine.

Dipole–Dipole Interactions

When the two bonded atoms of a binary molecule (a two-atom molecule) have different electronegativities, the electrical asymmetry that is produced causes one end of the molecule to possess a negative charge and the other end a positive charge. The electrical charges are partial charges rather than full charges and are indicated by lowercase Greek symbols δ^+ or δ^-. The H—F molecule provides a good example of a polar bond. When we want to show the existence of partial charges on its atoms, we write the formula $^{\delta^+}$H—F$^{\delta^-}$. Because it possesses only two atoms, the polar covalent bond of HF makes it a **polar molecule** as well. Although the HF molecule is electrically asymmetrical, it is electrically neutral overall because the collective positive charges of the combined nuclei are exactly balanced by the collective negative charges of the combined electrons.

H 2.2							He
Li 0.98	Be 1.6	B 2.0	C 2.6	N 3.0	O 3.4	F 4.0	Ne
Na 0.93	Mg 1.2	Al 1.6	Si 1.9	P 2.2	S 2.6	Cl 3.2	Ar
K 0.88	Ca 1.0	Ga 1.8	Ge 2.0	As 2.2	Se 2.6	Br 2.8	Kr
Rb 0.82	Sr 0.95	In 1.8	Sn 2.0	Sb 1.9	Te 2.1	I 2.7	Xe
Cs 0.79	Ba 0.89	Tl 1.8	Pb 1.9	Bi 1.9	Po 2.0	At 2.2	Rn

Figure 1.9 An abbreviated table of electronegativities.

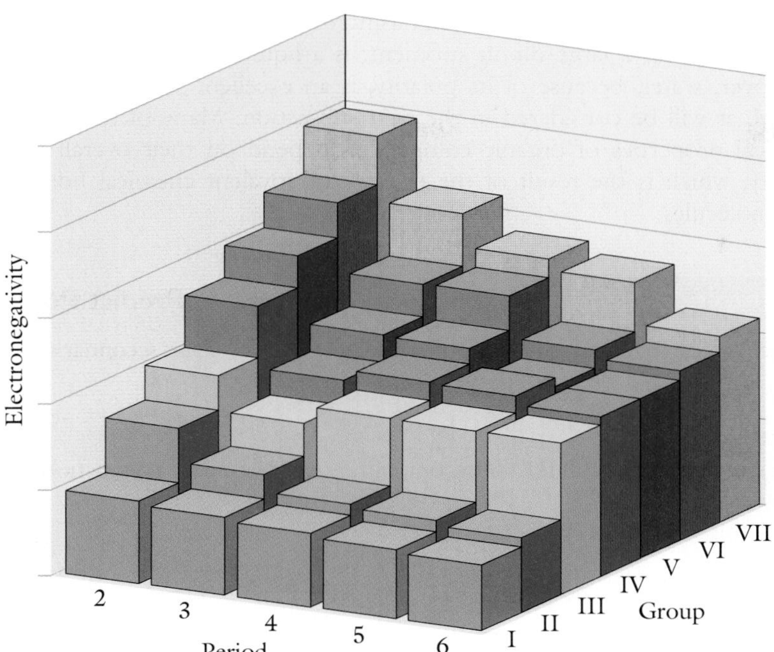

Figure 1.10 A three-dimensional plot of the main-group elements' electronegativity values in Figure 1.9.

A polar molecule is attracted to a neighboring polar molecule, positive end to negative end. This attractive force operates between molecules, in contrast with the chemical bonds acting within molecules, and therefore it is one of the secondary forces. The electrical force causing the attraction between polar molecules is called a **dipole–dipole force** and is measured by a quantity called the **dipole moment.** The greater the difference in electronegativities, the larger the dipole moment. The dipole moment is customarily represented by an arrow pointing along the bond from the positive end toward the negative end, $\mapsto$; that is, from the less electronegative element to the more electronegative element.

A molecule may contain two or more polar bonds and not be polar. Its overall shape may be such that the dipole moments cancel each other. As an example, imagine a generalized molecule consisting of three atoms of formula XY_2. In this molecule, X is the central atom with covalent bonds to each Y atom. Because the joined atoms X and Y are different, they must have different electronegativities. Therefore each individual bond in the molecule must be polar and must contribute to the overall polarity of the molecule. There are two possibilities for the molecular shape, linear and bent (see margin).

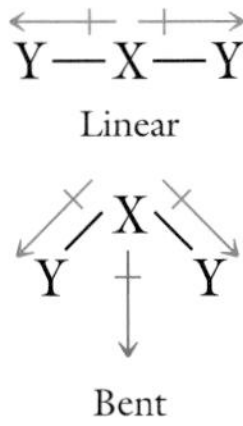

In the linear molecule, the two internal dipoles cancel each other's effect because the bonds are identical and therefore equal in magnitude and acting in directly opposite directions. Neither end of the molecule is more positive or negative than the other. Therefore the linear molecule will possess no overall dipole moment—it is not polar. An example is carbon dioxide, O=C=O. If the polar bonds in a linear molecule are different, as for example in Cl—Be—F, the molecule will be polar because the two internal dipoles are different and therefore do not completely cancel each other. In the bent molecule, the two internal dipoles do not cancel, and their interaction produces a new net overall dipole moment represented by the arrow shown in red.

The consequences of molecular polarity can be seen primarily in physical properties. Molecules with dipole moments tend to interact with one another but not with molecules possessing no dipole moment—nonpolar molecules. As a result, substances composed of polar molecules have higher melting and boiling points than do those composed of nonpolar molecules. At room temperature, methane, CH_4, which has no dipole moment, is a gas, but water, H_2O, which has a large dipole moment, is a liquid at the same temperature. Moreover, water, because of its polarity, is an excellent solvent for polar substances, as will be considered in the next subsection. Many of the physical and chemical properties of organic compounds depend on their overall molecular polarity, which is the result of the polarity of covalent chemical bonds within their molecules.

Example 1.15 Using Structural Formulas to Predict Polarity

Which of the following substances are characterized by polar secondary forces?

F_2 CH_4 HF CH_3Cl

Solution

The structural formulas for these compounds were developed in Section 1.6. They are

F—F H—C(H)(H)—H H—F H—C(H)(H)—Cl

Both F_2 and CH_4 have electrically symmetrical structures and therefore possess no dipole moments. They cannot interact through polar secondary forces. HF and CH_3Cl have significantly polar bonds and have electrically asymmetrical structures. Therefore, they possess permanent dipole moments and interact through polar secondary forces.

Problem 1.15 Which of the following compounds interact through polar secondary forces?

(a) H—H (b) Cl—C(Cl)(Cl)—Cl (c) H—C≡N (d) I—I

The Hydrogen Bond

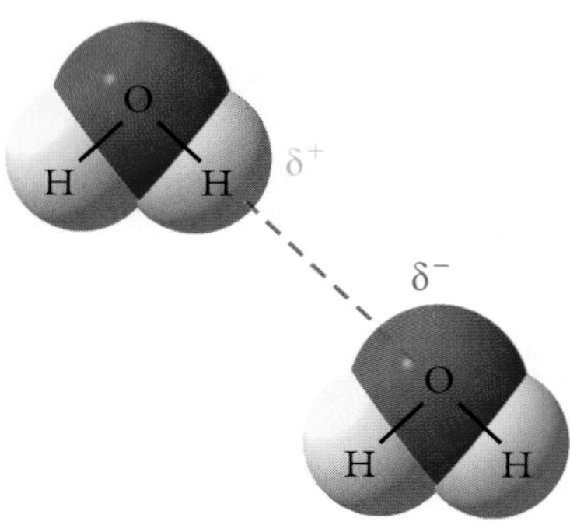

Figure 1.11 Space-filling model of hydrogen bonding between two water molecules. The hydrogen bond is represented by a dashed line to differentiate it from the covalent bonds (solid lines) of the water molecules.

A covalent bond between hydrogen and any one of the three elements oxygen, nitrogen, and fluorine possesses a special character. The differences in electronegativity between hydrogen and these elements and therefore the polarity within such a bond are very large. These differences lead to a unique and very strong attractive force, called the **hydrogen bond,** in molecules such as H_2O, NH_3, and HF. The hydrogen bond between two water molecules is shown in Figure 1.11. We shall show all secondary forces, such as the hydrogen bond in Figure 1.11, as dashed lines to differentiate them from the covalent bonds within molecules. The hydrogen bond is the result of the attraction between the partial positive charge of the hydrogen atom and the partial negative charge of the oxygen atom. Because there are two lone electron pairs per oxygen atom, two hydrogen bonds can form per water molecule. Therefore, in ice, for example, every oxygen atom, and, in liquid water, nearly every oxygen atom, is located at the center of a tetrahedron formed by four other oxygen atoms, with hydrogen bonds linking oxygen atoms together in a three-dimensional array.

A hydrogen bond, sometimes abbreviated H-bond, is a special case of a dipole–dipole attraction because it is stronger than any other dipolar secondary force. It not only forms between similar molecules, as in liquid water or ammonia, but also between different molecules, allowing mixtures—for example, of water and ammonia.

The hydrogen bond allows the hydrogen atom to act as "glue" between oxygen, nitrogen, or fluorine atoms of two different molecules. Remember, though, that not all molecules containing hydrogen atoms undergo hydrogen bonding.

Concept check

✓ Look for hydrogen bonding when a molecule contains a hydrogen atom covalently bonded to one of the three elements oxygen, nitrogen, and fluorine.

Example 1.16 Recognizing Molecules That Can Form Hydrogen Bonds

Which of the following molecules can form hydrogen bonds?

H_2 CH_4 NH_3 HF CH_3OH

Solution

A hydrogen bond can form if a compound possesses a hydrogen atom covalently bonded to either oxygen, nitrogen, or fluorine. Three of the preceding compounds fit that criterion: NH_3, ammonia; HF, hydrogen fluoride; and CH_3OH, methyl alcohol. Neither H_2 (hydrogen) nor CH_4 (methane) can form hydrogen bonds.

Problem 1.16 Which of the following molecules can form hydrogen bonds? (a) SiH_4; (b) CH_3CH_3; (c) CH_3NH_2; (d) H_2Te; (e) C_2H_5OH.

The importance of water in biological systems is due to its hydrogen-bonding properties. Hydrogen bonding also plays a key role in the regulation of body temperature. The characteristics of many important biological substances—proteins, nucleic acids, and carbohydrates—strongly depend on their ability to form hydrogen bonds within their own molecules and external hydrogen bonds with other molecules, particularly water. Chapters 12 and 13 illustrate specifically how hydrogen bonds are responsible in large part for the properties of biomolecules such as hemoglobin and the double-helical DNA of chromosomes.

London Forces of Interaction

Even in molecules such as methane that have no permanent polarity, temporary dipoles exist for brief moments. They exist because the erratic motions of electrons result in uneven distributions of electrical charge within such a molecule at any given time. The attractive force resulting from this temporary dipole is called a **London force.** London forces exist between all molecules, but their contribution to physical properties is particularly significant in nonpolar molecules.

London forces are the weakest type of secondary force. The next in strength are dipole–dipole forces, and the strongest are hydrogen bonds. This range of strength is made evident by comparing the boiling points listed in Table 1.10 on the following page for ethane, formaldehyde, and methanol—compounds of about equal molecular mass that interact by London forces, dipole–dipole forces, and hydrogen bonds, respectively.

London forces increase with an increase in the number of electrons within molecules and therefore increase with molecular mass. Because nonpolar molecules of large molecular mass possess many electrons, they exert greater attractive force than do nonpolar or even polar molecules of small molecular mass. Compare the boiling points of methane, ethane, butane, hexane, and octane (Table 1.10). To see how the molecular mass effect of the London force can override the strong secondary force of dipole–dipole interaction or hydrogen bonding, compare the boiling point of water with that of octane.

TABLE 1.10 Boiling Points Characteristic of the Three Types of Secondary Forces

Secondary force	Substance	Boiling point (°C)
London force	helium (He)	−269
	hydrogen (H_2)	−253
	methane (CH_4)	−164
	ethane (C_2H_6)	−89
	butane (C_4H_{10})	−1
	hexane (C_6H_{14})	69
	octane (C_8H_{18})	126
dipole–dipole force	methyl fluoride (CH_3F)	−78
	fluoroform (CHF_3)	−82
	formaldehyde (HCHO)	−21
hydrogen bond	methanol (CH_3OH)	65
	water (H_2O or HOH)	100
	ethylene glycol (HOC_2H_4OH)	198

Molecular Mixtures

The fact that oil and water do not mix is well known. Other substances—for example, ethyl alcohol and water or table sugar and water—mix so readily that their individual components cannot be distinguished. A uniform mixture of two or more substances is called a **solution.** Solution concentration based on chemical mass units (moles) is called **molarity** and is discussed in Appendix 4.

The key to the formation of this molecular mixture is that the secondary forces between the molecules in pure samples of each substance are about the same. Solutions will not form between substances whose secondary forces are very different. An old rule of thumb that expresses that idea is "like dissolves like."

To explore this idea, let's represent the secondary forces between the molecules of liquid A by the symbol A---A and those of liquid B by the symbol B---B. (The dashed lines represent the secondary forces operating between molecules.) For liquids A and B to mix thoroughly (form a solution), the molecules must be able to change partners to form A---B and A---B. If the forces represented by A---A and B---B are about the same, then molecules of A will be just as likely to interact with molecules of B as with each other, and a solution will probably form. Expressing this idea in the form of an equation, we write

$$\text{A---A} + \text{B---B} \longrightarrow \text{A---B} + \text{A---B}$$

Figure 1.12 shows the process in greater detail. In this sketch, a beaker of A molecules (red) is added to a beaker of B molecules (blue), and the two substances mix to form a solution (the forces between A molecules being about the same as those between B molecules) whose volume is equal to the sum of the volumes of the two individual components.

An understanding of the nature of secondary forces will enable you to predict whether two substances will form a molecular mixture or solution. You need only to examine the structure of the molecules in question and determine whether they are polar or nonpolar or whether H-bonding is possible between them.

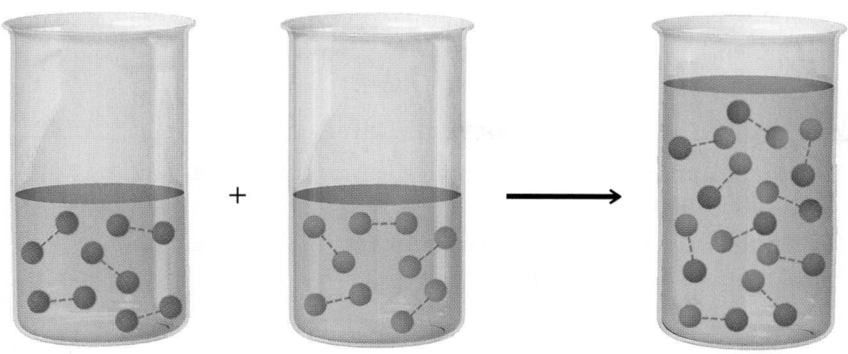

Figure 1.12 Schematic representation of the formation of a solution. A volume of liquid A is added to an equal volume of liquid B. In this case, the secondary forces indicated by the dashed lines are similar, so A---B attractions form, as does a solution.

Example 1.17 Using Molecular Structures to Predict Whether Solutions Will Form: I

Can water and ethanol form a molecular mixture? The structural formulas for ethanol and water are

```
    H  H
    |  |
H—C—C—O—H      H—O—H
    |  |         Water
    H  H
   Ethanol
```

Solution

Both molecules are polar, and the presence of the hydroxyl (OH) group in each molecule indicates that both can form H-bonds not only between their own molecules, but also between each other's molecules. These observations suggest that the secondary forces within each individual substance and those that would form within a mixture of the substances are about the same. Therefore a solution should form.

Problem 1.17 Can water and carbon tetrachloride form a solution? Explain your answer.

```
     Cl
     |
Cl—C—Cl
     |
     Cl
  Carbon
tetrachloride
```

Example 1.18 Using Molecular Structures to Predict Whether Solutions Will Form: II

Will the hydrocarbon hexane, C_6H_{14}, and water, H_2O, form a solution? The structures for hexane and water are

```
   H  H  H  H  H  H
   |  |  |  |  |  |
H—C—C—C—C—C—C—H      H—O—H
   |  |  |  |  |  |     Water
   H  H  H  H  H  H
       Hexane
```

Solution

The structural formulas must be examined to see how the secondary forces within each substance compare with those of the other. Hexane is a hydrocarbon with no permanent dipole; so its molecules interact through London forces. Water molecules interact through hydrogen bonds. Therefore the secondary forces between hexane and water molecules would be significantly different from those that exist between water molecules and those that exist between hexane molecules. The result is that water will be attracted to water and hexane to hexane. Hexane will not mix with water nor dissolve in it.

Problem 1.18 Explain why the following pairs of substances form or do not form solutions. (a) CCl_4 and H_2O; (b) C_5H_{12} and CCl_4.

Concept checklist

✓ When the attractive forces between different molecules are similar, a solution will form.

✓ When the attractive forces between different molecules are different, a solution will not form.

Summary

Atomic Structure

• Atoms contain three kinds of particles: positively charged protons, electrically uncharged (neutral) neutrons, and negatively charged electrons.

• The protons and neutrons are located together in the nucleus at the center of the atom. The electrons lie outside the nucleus and occupy most of the volume of the atom.

• Naturally occurring elements consist of isotopes—atoms with the same numbers of protons but different numbers of neutrons.

The Periodic Law and the Periodic Table

• Elements of similar chemical properties are aligned in the periodic table in vertical columns called groups.

• The horizontal rows, called periods, end on the right-hand side with a noble gas.

• Metals occupy most of the left-hand side of the table, and nonmetals occupy the extreme right-hand side.

• Elements are designated as either main-group or transition elements.

Atomic Structure and the Periodic Table

• Electrons in atoms are confined to different shells around the nucleus.

• Lewis symbols emphasize the number of electrons in an atom's valence shell.

• The tendency of atoms to attain noble-gas outer-shell electron configurations is expressed as the octet rule.

Chemical Bonds

• Compounds are the result of the stable association between atoms called a chemical bond.

• The two kinds of chemical bonds are ionic bonds, in which electrostatic forces bind ions together, and covalent bonds, in which atoms share one or more pairs of electrons.

Lewis Structures

• Lewis structures are used to predict how atoms in a molecule are connected to one another.

• The key concept in building Lewis structures is satisfaction of the octet rule.

• Bonding pairs of electrons are represented as lines between the atoms, and the lone pairs are represented as dots surrounding the atoms.

Three-Dimensional Molecular Structures

• The shapes of molecules are predicted by the VSEPR theory.

• This theory predicts that molecular shape depends on the mutual repulsion of all the electron pairs around a central atom.

• Repulsion causes bonded atoms to form symmetrical three-dimensional geometrical configurations.

Molecular Interactions and Secondary Forces

• Attractive forces between molecules are called secondary forces.

• The three principal secondary forces, in order of increasing strength, are London forces, dipole–dipole forces, and hydrogen bonds.

• These forces are responsible for the existence of different phases of matter and for the interactions between molecules.

• Solutions between different substances will form if the secondary forces between them are of the same type.

Key Words

atom, p. 3
atomic mass, p. 3
atomic number, p. 3
bond angle, p. 23
covalent bond, p. 11
dipole–dipole force, p. 29
electronegativity, p. 11
hydrogen bond, p. 30
ionic bond, p. 11
Lewis structures, p. 16
Lewis symbols, p. 10
London force, p. 31
octet rule, p. 11
periodic law, p. 4
periodic table, p. 4
polar covalent bond, p. 27
secondary forces, p. 26
structural formula, p. 17
valence electron, p. 11
VSEPR theory, p. 22

Exercises

Atomic Structure

1.1 Describe the masses and electrical charges of the three principal subatomic particles making up the atom.

1.2 If an electrically neutral atom can be shown to possess 16 electrons, what must its atomic number be? Identify the element.

1.3 Explain the fact that the mass of a newly discovered element is about twice its atomic number.

1.4 Complete the following table by filling in the blanks.

Protons	Neutrons	Electrons	Element
19	20	____	____
____	45	34	____
____	6	6	____
11	____	11	____

1.5 What would be an important feature of an atom consisting of 23 protons, 28 neutrons, and 20 electrons?

1.6 What would be an important feature of an atom consisting of 9 protons, 10 neutrons, and 10 electrons?

The Periodic Law and the Periodic Table

1.7 Identify the period and group of the following elements: Li, Na, K, Rb, Cs.

1.8 Identify the period and group of the following elements: Si, Ge, As, Sb.

1.9 Can the elements in Exercise 1.7 be described as metals or nonmetals?

1.10 Can the elements in Exercise 1.8 be described as metals or nonmetals? Explain your answer.

1.11 Which group of the periodic table is known as the alkali metals?

1.12 Which group of the periodic table is known as the halogens?

Atomic Structure and the Periodic Table

1.13 What aspect of electron configuration do the elements in the group of the periodic table known as the alkali metals have in common?

1.14 What aspect of electron configuration do the elements in the group of the periodic table known as the halogens have in common?

1.15 What aspect of electron configuration do the elements in the group of the periodic table known as the alkaline earths have in common?

1.16 What aspect of electron configuration do the elements in the group of the periodic table known as the noble gases have in common?

1.17 True or false: The number of protons in an atom is called the atomic number.

Chemical Bonds

1.18 What are the ways in which elements can satisfy the octet rule?

1.19 What are the results of the two ways of satisfying the octet rule?

1.20 What are the combining ratios of elements in Group I with elements in Group VII of the periodic table?

1.21 What are the combining ratios of elements in Group I with elements of Group VI of the periodic table?

1.22 What are the combining ratios of elements in Group II with elements in Group VII of the periodic table?

1.23 What are the combining ratios of elements in Group III with elements in Group VI of the periodic table?

1.24 Write the formulas of the compounds formed when calcium reacts with each of the first three elements of Group VI.

1.25 Write the formulas of the compounds formed when aluminum reacts with each of the first three elements of Group VI.

1.26 Write the formulas of the compounds formed when aluminum reacts with each of the first four elements of Group VII.

Lewis Structures

1.27 Write the Lewis structure of carbon tetrachloride, CCl_4.

1.28 Write the Lewis structure of chloroform, $CHCl_3$.

1.29 Write the Lewis structure of H_2S.

1.30 Write the Lewis structure of NCl_3.

1.31 Write the Lewis structure of carbon dioxide, CO_2.

1.32 Hydrocarbons are compounds that contain only carbon and hydrogen. In many hydrocarbons, the carbon atoms are connected to one another to form a chain. Write the Lewis structure for C_4H_{10}.

Three-Dimensional Molecular Structures

1.33 Use VSEPR theory to predict the angular orientation of fluorine atoms around the central boron atom in the polyatomic ion BF_4^-.

1.34 Use VSEPR theory to predict the three-dimensional structure of H_2S.

1.35 For each of the following molecules or ions (1) describe the molecular shape and (2) predict whether it has a dipole moment: (a) CCl_4; (b) BF_4^-; (c) CO_2.

1.36 For each of the following molecules (1) describe the molecular shape and (2) predict whether the molecule has a dipole moment: (a) $CHCl_3$; (b) NCl_3; (c) H_2O.

Molecular Interactions and Secondary Forces

1.37 Why doesn't carbon tetrachloride, CCl_4, a molecule with four polar covalent bonds, have a dipole moment?

1.38 Why doesn't carbon dioxide, CO_2, a molecule with two polar covalent bonds, have a dipole moment?

1.39 Fill in the following table with a yes or no answer to the question, Does this secondary force operate between the molecules of the given compound?

	London force	Dipole–dipole	Hydrogen bond
CH_4	______	______	______
$CHCl_3$	______	______	______
NH_3	______	______	______

1.40 Fill in the following table with a yes or no answer to the question, Does this secondary force operate between the molecules of the given compound?

	London force	Dipole–dipole	Hydrogen bond
CCl_4	______	______	______
CHF_3	______	______	______
H_2O	______	______	______

1.41 Will acetic acid form a molecular mixture with (dissolve in) water? Explain your answer. Its structure is

$$CH_3-C(=O)-OH$$

1.42 Will pentane form a molecular mixture with (dissolve in) water? Explain your answer. Its structure is

$$H-CH_2-CH_2-CH_2-CH_2-CH_2-H$$

Unclassified Exercises

1.43 What is the principal characteristic of elements in the same group of the periodic table?

1.44 Is sulfur a metal or a nonmetal?

1.45 Is calcium a metal or a nonmetal?

1.46 What is the chief property of the elements at the right-hand end of every period of the periodic table?

1.47 What is the octet rule?

Expand Your Knowledge

1.48 If necessary, correct the following statement: Most of the elements listed in the periodic table have equal numbers of protons, electrons, and neutrons.

1.49 What are the similarities and differences between the two principal isotopes of carbon, atomic number 6, whose atomic masses are 12 and 13, respectively?

1.50 Element X consists of two isotopes, masses 25 amu and 26 amu, and its mass is reported as 25.6 in the table of atomic masses. What are the percent abundances of its two isotopes?

1.51 An organic compound containing an OH group is called an alcohol. Write the Lewis structure for the alcohol CH_3OH (methanol). (Hint: The OH group is connected to the carbon atom.)

1.52 What is the relation between molecular mass and the strength of the London forces exerted between molecules?

1.53 What is the most important relation between elements in the same group in the periodic table?

1.54 Give an example of: (a) an atom with a half-filled subshell; (b) an atom with a completed outer shell; (c) an atom with its outer electrons occupying a half-filled subshell; (d) an atom with a filled subshell.

1.55 Predict the order of increasing electronegativity in each of the following groups of elements: (a) C, N, O; (b) Mg, Ca, Sr; (c) C, Ge, Pb; (d) Se, Te, Po.

1.56 Write Lewis structures for each of the following molecules and ions. In each case, the first atom listed is the central atom. (a) PCl_3; (b) SO_4^{2-}; (c) PO_4^{3-}; (d) ClO_4^-.

CHAPTER 2 CHEMICAL CHANGE

Learning Objectives

- Predict whether a chemical reaction will take place in a given solution.
- Describe how the rates of chemical reactions are affected by the concentrations of reactants, the temperature, and the presence of catalysts.
- Describe how chemical reactions are the result of collisions that lead to the formation and decay of an activated complex.
- Explain how a chemical system reaches a state of dynamic equilibrium.
- Describe Le Chatelier's principle.
- Explain the differences between strong and weak acids and between strong and weak bases.
- Define pH and describe its use as a measure of acidity.
- Use the Brønsted-Lowry theory to explain the properties of salts of weak acids and bases and of buffers.

Figure 2.1 A solution of sodium chloride (NaCl) added to a solution of silver nitrate ($AgNO_3$) produces the white precipitate silver chloride (AgCl). (Chip Clark.)

Our first look at the principles underlying chemical change was the introduction of the octet rule. The octet rule states that elements become stable by gaining, losing, or sharing electrons to achieve an outer shell of eight electrons. In this chapter, we examine chemical reactions that transform molecules into new structures, either through electron transfer or molecular rearrangements.

2.1 CHEMICAL REACTIONS IN SOLUTION

Many of the important chemical reactions of interest to us will be either precipitation reactions, neutralization reactions, or oxidation–reduction reactions. All these types of reactions take place in solution.

Precipitate Formation

When two ionic compounds are added to water and form new compounds, one reason for the transformation may be that one of the products is insoluble and forms a precipitate. Reactions of this type are shown in Figures 2.1 and 2.2. Other examples are the reactions between $BaCl_2$ and K_2SO_4 or between $CaCl_2$ and K_3PO_4:

$$BaCl_2 + K_2SO_4 \longrightarrow BaSO_4 + 2\ KCl$$

$$3\ CaCl_2 + 2\ K_3PO_4 \longrightarrow Ca_3(PO_4)_2 + 6\ KCl$$

These equations are stoichiometrically correct (balanced) but do not represent the actual process taking place in solution. To understand why, we must first examine the solubility of ionic compounds.

In aqueous solution, all dissolved ionic compounds are present in the form of individual ions. For example, in solution, $BaCl_2$ separates into Ba^{2+} cations and Cl^- anions. Chemists have developed a set of empirical rules for making qualitative predictions of whether an ionic substance will be soluble in water. They are summarized in Table 2.1. The general rule of solubility or insolubility for a given class of compounds is stated in the left-hand column, and important exceptions to that general rule are presented in the right-hand column.

Figure 2.2 A yellow precipitate of lead(II) iodide (PbI_2) forms immediately when colorless aqueous solutions of lead(II) nitrate [$Pb(NO_3)_2$], and potassium iodide (KI), are mixed. (Chip Clark.)

Let's examine the details of one reaction in which a precipitate is formed—the result of the addition of $BaCl_2$ to a solution of K_2SO_4. Table 2.1 shows that (1) both $BaCl_2$, and K_2SO_4 are ionic solids soluble in water and (2) one of the products, $BaSO_4$, is an insoluble solid. We can represent the details of the process with the following equation, which is called a **complete ionic equation.**

$$Ba^{2+}(aq) + 2\ Cl^-(aq) + 2\ K^+(aq) + SO_4^{2-}(aq) \longrightarrow$$
$$BaSO_4(s) + 2\ K^+(aq) + 2\ Cl^-(aq)$$

The notations (aq), (s), and (l) are used to denote that the components of a reaction are in the form of ions in aqueous solution, a precipitate, or a pure liquid, respectively.

Notice that the $K^+(aq)$ and $Cl^-(aq)$ ions appear unaltered on both sides of the equation. We can therefore exclude them from the equation and consider only those ions that participate in the reaction:

$$Ba^{2+}(aq) + SO_4^{2-}(aq) \longrightarrow BaSO_4(s)$$

This equation is called a **net ionic equation.** The $K^+(aq)$ and $Cl^-(aq)$ ions omitted from the net ionic equation are called **spectator ions** because they do not take part in the reaction and would appear on both sides of the equation if included.

TABLE 2.1 Qualitative Solubility Rules for Ionic Solids in Water

Soluble	Important exceptions
Na, K, NH_4 salts	
nitrates and acetates	
sulfates	Ca, Sr, and Ba sulfates
chlorides and bromides	Ag, Pb, and Hg(I) chlorides and bromides
Insoluble	**Important exceptions**
hydroxides	Li, Na, K, Rb, Ca, Sr, and Ba hydroxides
phosphates, carbonates, and sulfides	Li, Na, K, Rb, and NH_4^+ phosphates, carbonates, and sulfides

Example 2.1 Finding the net ionic equation

Write the net ionic equation that describes the result of adding silver nitrate, $AgNO_3$, to a solution of NaCl, a reaction in which a precipitate is formed. Refer to Table 2.1 for solubility data.

Solution

Table 2.1 shows that silver chloride is an insoluble salt. The reaction equation is

$$Ag^+(aq) + NO_3^-(aq) + Na^+(aq) + Cl^-(aq) \longrightarrow AgCl(s) + Na^+(aq) + NO_3^-(aq)$$

The net ionic equation is

$$Ag^+(aq) + Cl^-(aq) \longrightarrow AgCl(s)$$

Problem 2.1 Write (a) the complete ionic equation and (b) the net ionic equation for the reaction that takes place when calcium chloride, $CaCl_2$, is added to sodium phosphate, Na_3PO_4. Refer to Table 2.1 for solubility data.

Precipitates are occasionally formed in the body and can have significant consequences. For example, most nitrogen-containing compounds are metabolized to form urea, which becomes the principal dissolved component of urine. However, a certain group of nitrogen-containing compounds, the nucleotides (from which the genetic molecules RNA and DNA are synthesized, Chapter 13), undergo metabolic degradation to form uric acid. Under normal conditions, uric acid and its sodium salt are excreted in the urine. A small number of people suffer from a disorder in nucleotide metabolism that results in an overproduction of uric acid and leads to the clinical condition known as gout. Uric acid and its sodium salt are thousands of times less soluble than urea, and their overproduction rapidly saturates body fluids. In consequence, needlelike uric acid crystals are deposited in the cartilaginous structures of the joints, causing inflammation and great pain.

Neutralization Reaction: The Formation of Water or a Gas

Another process through which reactions between ions can take place is an acid–base neutralization—a reaction between an acid and a base in which either water or a gas or both are formed. The net ionic equation for neutralization reactions depends on whether the product is water or a gas or both. For example, the complete ionic equation for the reaction between

Figure 2.3 Carbon dioxide is one of the products of the reaction of metal carbonates with acids. In this figure, an eggshell, largely composed of calcium carbonate, reacts with hydrochloric acid. The bubbles on the eggshell surface are carbon dioxide. (Chip Clark.)

hydrochloric acid (HCl) and sodium hydroxide (NaOH) in water (a neutralization reaction) is

$$H^+(aq) + Cl^-(aq) + Na^+(aq) + OH^-(aq) \longrightarrow H_2O(l) + Na^+(aq) + Cl^-(aq)$$

[Again, (aq) means aqueous or dissolved in water and (l) stands for pure liquid.] The net ionic equation is

$$H^+(aq) + OH^-(aq) \longrightarrow H_2O(l)$$

and emphasizes the formation of water. Nitric acid (HNO_3) added to a solution of calcium hydroxide, $Ca(OH)_2$, is described by the same net ionic equation.

A neutralization reaction can result in the formation of a gas, as occurs when sodium bicarbonate is taken orally to relieve the distress of excess stomach acid (HCl). The neutralization in this case results in the formation of carbonic acid (H_2CO_3), which is unstable and rapidly decomposes to form CO_2 and H_2O. The reaction between HCl and $CaCO_3$ is shown in Figure 2.3. The equation for the formation of carbonic acid is

$$NaHCO_3(aq) + HCl(aq) \longrightarrow H_2CO_3(aq) + NaCl(aq)$$

The net ionic equation is

$$H^+(aq) + HCO_3^-(aq) \longrightarrow H_2CO_3(aq)$$

The gas-forming step consists of the decomposition of the unstable carbonic acid:

$$H_2CO_3(aq) \longrightarrow H_2O(l) + CO_2(g)$$

Note that $CO_2(g)$ means that carbon dioxide is in the gaseous state. The net ionic equation showing the overall result is

$$H^+(aq) + HCO_3^-(aq) \longrightarrow CO_2(g) + H_2O(l)$$

Oxidation–Reduction Reactions

Chemical reactions in which changes in the valence shell of the reaction participants take place are called oxidation–reduction or redox reactions. The earliest oxidation reactions were considered to be those in which there is an increase in the oxygen content of the new compound. These compounds were called oxides. It was found that oxygen could be displaced from oxides by heating with hydrogen. An increase in a compound's oxygen content or a decrease in its hydrogen content or both is still considered to be an oxidation. Similarly, a decrease in a compound's oxygen content or an increase in its hydrogen content or both is considered to be a reduction. However, we now know that these reactions are a subset of reactions in which oxidation describes the loss of electrons and reduction describes the gain of electrons. The best generality is that oxidation–reduction reactions describe changes in the valence shells of atoms as a result of a chemical reaction.

The loss and gain of electrons. To balance a redox reaction, the number of atoms of each type on the reactant side must equal that on the product side, and the number of electrons lost by the oxidized reactant (the reductant) must equal that gained by the reduced reactant (the oxidant). To decide when a redox reaction has taken place, we assign a number to an atom that defines its **oxidation state,** called its **oxidation number.** If that number undergoes a change as a result of a chemical reaction, the reaction is a redox reaction. Furthermore, the changes in oxidation number of reductant and oxidant are used to balance redox reaction equations.

Oxidation states. The following set of rules will allow you to assign an oxidation number to atoms whether in ionic or covalent compounds. They are listed in priority of application: rule 1 takes priority over rule 2, which takes priority over rule 3, and so forth.

1. Free elements are assigned an oxidation number of zero. Because oxidation number defines oxidation state, free elements have an oxidation state of zero.
2. The sum of oxidation states of all the atoms in a species (compound or polyatomic ion) must be equal to the net charge of the species.
3. In compounds, the Group I metals—Na, K, and so forth—are assigned an oxidation state of +1. Note that the charge of an ion is written, for example, 1+ or 2+. The oxidation states are written as +1, +2, and so forth, to differentiate them from ionic charge.
4. In its compounds, fluorine is always assigned an oxidation state of −1, because it is the most electronegative element.
5. The Group II metals—Ca, Mg, and so forth—are always assigned an oxidation state of +2, and Group III ions—Al, Ga, and so forth—are assigned an oxidation state of +3. These states are identical with the ionic charges of these elements in compounds.
6. Hydrogen in compounds is assigned an oxidation state of +1.
7. Oxygen in compounds is assigned an oxidation state of −2.

These rules were developed by considering numbers of electrons and relative electronegativities, and so, for ions such as Na^+ or Ca^{2+}, the ionic charges and the oxidation states are identical. However, for covalent compounds such as NO_2 or CH_4, the combined atoms have no real charge; they are assigned oxidation states, which implies electrical charge.

Example 2.2 Determining oxidation states

Determine the oxidation states of each atom in the following compounds. (a) $CaCl_2$; (b) NaH; (c) SO_2; (d) H_2O_2.

Solution

(a) Calcium (Ca) is assigned an oxidation state of +2 (rule 5). Because $CaCl_2$ is a neutral species, the oxidation state of Cl must be −1. This oxidation state is established by using rule 2 and setting up an algebraic equation with one unknown quantity to solve for the oxidation state of Cl in $CaCl_2$.

$$(+2) + (2\ \text{Cl}) = 0$$
$$\text{Cl} = -1$$

The oxidation state of Cl in $CaCl_2$ is −1.

(b) Sodium (Na) is assigned an oxidation state of +1 (rule 3), and rule 2 informs us that, in NaH, H must have an oxidation state of −1. Rule 6 is not violated, because it is of lower priority than rule 3. Hydrogen forms compounds with metals of Groups I and II called hydrides.

(c) Oxygen (O) is assigned an oxidation state of −2 (rule 7). The oxidation state of sulfur (rule 2) is therefore

$$(\text{S}) + (2 \times -2) = 0$$
$$\text{S} = +4$$

The oxidation state of sulfur in SO_2 is +4.

(d) Rule 6 requires H to have an oxidation state of +1, therefore, by rule 2,

$$(2 \times +1) + 2\ \text{O} = 0$$
$$\text{O} = -1$$

The oxidation state of oxygen in H_2O_2 is −1, typical of peroxides. There is no conflict, because rule 6 has priority over rule 7.

Problem 2.2 Determine the oxidation number of carbon in the following compounds: (a) CH_4; (b) CO_2.

The reaction of sodium with chlorine can be written as two separate processes called **half-reactions** in which electrons lost by one of the components are gained by the other. The loss of electrons is an oxidation, and the gain in electrons is a reduction. The practice of separating the complete reaction equation into equations showing only reduction or oxidation allows balancing with respect to electrons as well as atoms.

$$2\ Na \longrightarrow Na^+ + 2\ e^- \qquad (1)$$

$$Cl_2 + 2\ e^- \longrightarrow 2\ Cl^- \qquad (2)$$

Reaction 1 is an oxidation, and reaction 2 is a reduction. The rules tell us that the oxidation state of sodium changed from 0 to +1 and that of chlorine changed from 0 to −1. We can generalize these observations to say that any change in oxidation state to a more positive state is an oxidation and that any change in oxidation state to a less positive (more negative) state is a reduction.

Reaction with oxygen Metals react directly with oxygen to form oxides; for example,

$$2\ Mg + O_2 \longrightarrow 2\ MgO$$

$$4\ Fe + 3\ O_2 \longrightarrow 2\ Fe_2O_3$$

$$4\ Al + 3\ O_2 \longrightarrow 2\ Al_2O_3$$

Each of these reactions is a redox reaction in which the metals are oxidized and oxygen is reduced.

An example of a reaction in which a compound loses oxygen is the reaction of an ore (in this case, iron ore) with a reducing agent (in this case, carbon):

$$2\ Fe_2O_3(s) + 3\ C(s) \longrightarrow 4\ Fe(s) + 3\ CO_2(g)$$

Note that, as Fe_2O_3 loses oxygen, carbon gains it to form CO_2. Most metals are obtained in elemental form by the reduction of ores, which, in many cases, are oxides of metals.

When the direct reaction of an element or compound with oxygen is accompanied by the familiar sight of burning, the reaction is called a **combustion reaction.** Hydrogen, sulfur, and nitrogen, for example, each burn in excess oxygen to produce H_2O, SO_2, and NO_2, respectively.

Compounds containing carbon react with oxygen to produce CO_2. For example, if an alkane (a type of molecule containing only carbon and hydrogen described in Chapter 3) such as C_3H_8 is burned, the products will be CO_2 and H_2O. In Chapters 14 through 18, the energy-yielding reactions of metabolism will be shown to be equivalent to slow, controlled combustion reactions.

If less than sufficient O_2 is available, the combustion product of an alkane will be CO (carbon monoxide). There are a significant number of deaths every year from the carbon monoxide generated when fossil fuels are burned in heating systems that have insufficient oxygen intake.

For combustion reactions of carbon compounds, you may assume that combustion takes place with sufficient oxygen to yield CO_2 instead of CO (carbon monoxide).

Example 2.3 Balancing a combustion equation: I

Write the balanced equation for the combustion of the alkane C_5H_{12}.

Solution

The products of the alkane's combustion are water and carbon dioxide:

$$C_5H_{12} + O_2 \longrightarrow CO_2 + H_2O \quad \text{(unbalanced)}$$

All the carbon is contained in one product, and all the hydrogen in the other; so the coefficients of the products can be quickly determined:

$$C_5H_{12} + O_2 \longrightarrow 5\ CO_2 + 6\ H_2O \quad \text{(unbalanced)}$$

The sum of O atoms on the right is 16; therefore, the coefficient of O_2 must be 8.

$$C_5H_{12} + 8\ O_2 \longrightarrow 5\ CO_2 + 6\ H_2O$$

Problem 2.3 Write the balanced equation for the combustion of the alkane C_4H_{10}. (Hint: To avoid balancing coefficients that are fractions, multiply all coefficients by an appropriate whole number—for example, 2 or 3—to change the fractions to whole numbers.)

Example 2.4 Balancing a combustion equation: II

Write the balanced equation for the combustion of the carbohydrate glucose ($C_6H_{12}O_6$).

Solution

Glucose (and any other carbohydrate) is composed of the elements carbon, hydrogen, and oxygen. Therefore, the products of the combustion of glucose are water and carbon dioxide, as was the case with alkanes:

$$C_6H_{12}O_6 + O_2 \longrightarrow CO_2 + H_2O \quad \text{(unbalanced)}$$

All the carbon is in one product, and all the hydrogen is in the other; so the coefficients of the products are quickly determined:

$$C_6H_{12}O_6 + O_2 \longrightarrow 6\ CO_2 + 6\ H_2O \quad \text{(unbalanced)}$$

The sum of O atoms on the right is 18, but, because glucose is a carbohydrate, both reactants contribute O atoms. Six O atoms are contributed by glucose; therefore, oxygen need contribute only 12, and its coefficient is 6.

$$C_6H_{12}O_6 + 6\ O_2 \longrightarrow 6\ CO_2 + 6\ H_2O$$

Although the metabolic oxidation of glucose in living cells is far more complex, its overall result is represented by this stoichiometric equation.

Problem 2.4 Write the balanced equation for the combustion of lactic acid ($C_3H_6O_3$), a product of carbohydrate metabolism.

Dehydrogenation One of the reactions through which the cell extracts energy from food is the oxidation of malic acid to oxaloacetic acid. This reaction is an example of oxidation by hydrogen loss, also called a **dehydrogenation** reaction. The hydrogen is transferred as a hydride ion ($H{:}^-$) to an intermediary compound called nicotinamide adenine dinucleotide, abbreviated NAD^+, in an enzyme-catalyzed reaction (Chapters 14 and 15). The oxidation can be written as follows:

$$C_4H_6O_5 + NAD^+ \longrightarrow C_4H_4O_5 + NADH + H^+$$

In this reaction, one hydrogen atom in the form of a hydride ion is transferred to NAD^+ along with two electrons, and the other hydrogen atom enters solution as a proton.

The reverse of this reaction, in which the oxaloacetic acid gains hydrogen to form malic acid, is a reduction:

$$C_4H_4O_5 + NADH + H^+ \longrightarrow C_4H_6O_5 + NAD^+$$

Concept checklist

- ✓ Oxidation takes place when an atom's oxidation state becomes more positive.
- ✓ Reduction takes place when an atom's oxidation state becomes more negative.
- ✓ Oxidation takes place when a compound's oxygen content increases or its hydrogen content decreases or both.
- ✓ Reduction takes place when a compound's oxygen content decreases or its hydrogen content increases or both.

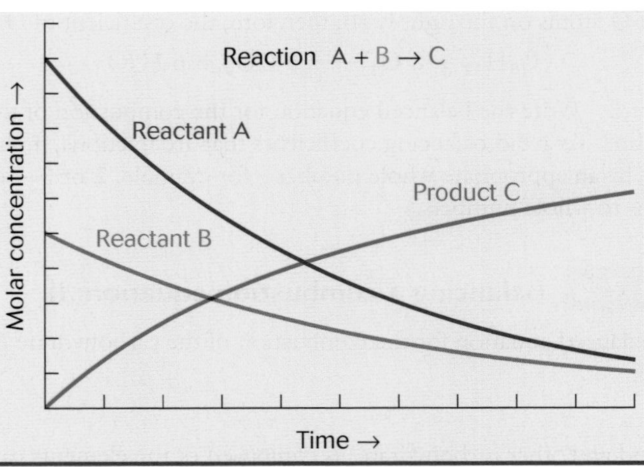

Figure 2.4 In a chemical reaction, the concentration of product increases and the concentration of reactant decreases with the passage of time.

2.2 CHEMICAL KINETICS

Consider reactions between substances such as nitrogen and hydrogen to form ammonia or between nitric oxide and bromine to form nitrosyl bromide. The balanced reaction equations for those processes are

$$3\ H_2 + N_2 \longrightarrow 2\ NH_3 \quad (1)$$

$$2\ NO + Br_2 \longrightarrow 2\ NOBr \quad (2)$$

The equations state that the concentrations of hydrogen and nitrogen are reduced and the concentration of ammonia increases in reaction 1 and that the concentrations of nitric oxide and bromine decrease and the concentration of nitrosyl bromide increases in reaction 2. Data representing these kinds of reactions are charted in Figure 2.4. As the reaction proceeds, the concentrations of reactants decrease and the concentrations of product simultaneously increase.

Chemical reactions take place by collisions between molecules or ions. The rate of reaction depends on three variables: (1) the concentration of reactants (the larger the concentrations, the greater the rate of collisions and, consequently, the greater the rate of reaction); (2) the spatial orientation of the molecules when they collide; and (3) the temperature at which the reaction takes place. The rates of some reactions are significantly increased by the presence of catalysts. The study of these aspects of chemical reactions is called **chemical kinetics.**

Molecular Collisions and Chemical Reactions

For a chemical reaction to take place between molecules, a collision must result in the breaking or the rearrangement of bonds. Bonds are broken or rearranged only if (1) the molecules collide with sufficient energy and (2) they collide in a favorable spatial orientation. A collision having both characteristics is called a **reactive collision.** In any assembly of molecules, only a small fraction of the total are moving about with velocities and, consequently, energies great enough to produce reactive collisions. The small fraction of fast-moving molecules increases rapidly with an increase in temperature. The minimum energy needed for a reactive collision is called the **activation energy, E_a.** The activation energy controls the rate of reaction: the greater the activation energy, the slower the rate.

When there is sufficient energy and the colliding molecules are in the proper spatial orientation with respect to each other, the collision forms a new reactive combination called an **activated complex.** The new structure is also known as the transition state. It is unstable and rapidly decays either into products or back into the original reactants.

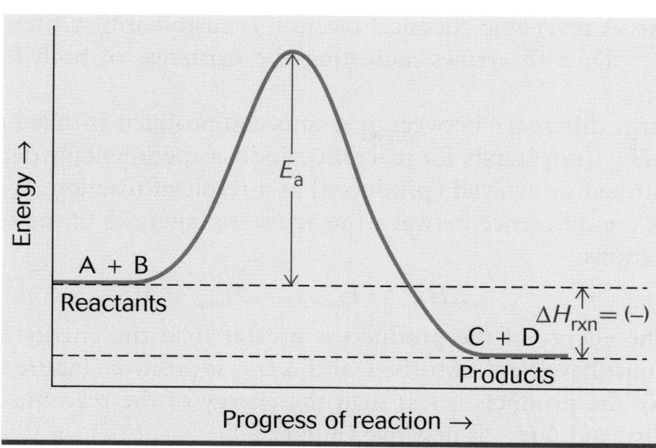

Figure 2.5 An activation-energy diagram plots the changes in the energy of an activated complex as an exothermic chemical reaction progresses from reactants to products.

Figure 2.5, called an activation-energy diagram, is a plot of the changes in energy (the *y*-axis) of a pair of colliding molecules as they form an activated complex that then proceeds to form products:

$$A + B \longrightarrow \text{activated complex} \longrightarrow C + D$$

The *x*-axis in Figure 2.5 represents the progress of the reaction as molecule A approaches molecule B, forms the activated complex, and proceeds to form products C and D. The difference between the energy of the activated complex and that of the reactants is the activation energy. In Figure 2.5, the difference in energy between the (A + B) level and the (C + D) level signifies that energy is lost when the reactants are chemically transformed into products. This difference in energy levels represents a loss of heat, and the reaction is characterized as **exothermic.** Exothermic reactions give off heat. The reaction as written, with reactants on the left and products on the right, is called the **forward reaction.**

There is no reason to suppose that the products of our hypothetical reaction, (C + D), cannot react to form (A + B); that is, the reaction is reversible. In a reversible reaction the forward and back reactions take place simultaneously. The reverse reaction,

$$C + D \longrightarrow \text{activated complex} \longrightarrow A + B$$

called the **back reaction,** is illustrated in Figure 2.6. It is the mirror image of the exothermic forward reaction and is therefore accompanied by an absorption of heat in an amount indicating the increased energy of the products (A + B). A reaction characterized by the absorption of heat is called

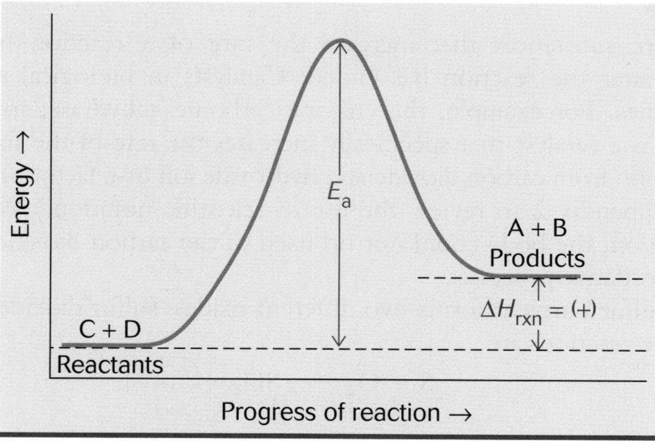

Figure 2.6 An activation-energy diagram plots the changes in the energy of an activated complex as an endothermic chemical reaction progresses from reactants to products.

endothermic. A reversible chemical reaction is customarily written as follows: $A + B \rightleftharpoons C + D$, with arrows indicating the existence of both forward and back reactions.

The energy difference between reactants and products is called the **heat of reaction, ΔH_{rxn}** (rxn stands for reaction), and is experimentally determined as the heat absorbed or evolved (produced) as a result of reaction. It can also be calculated as the difference between the activation energies of the forward and the back reactions:

$$\Delta H_{rxn} = E_{forward} - E_{back}$$

When the energy of the products is greater than the energy of the reactants, heat must have been absorbed, and ΔH_{rxn} is positive (Figure 2.6). When the energy of the products is less than the energy of the reactants, heat must have been lost, and ΔH_{rxn} is negative (Figure 2.5).

Concept checklist

- ✓ Increasing the reaction temperature increases the fraction of molecules that possess the required activation energy, and the result is an increased rate of reaction.
- ✓ In an exothermic reaction, heat is evolved (lost) and the heat of reaction is negative [$\Delta H_{rxn} = (-)$].
- ✓ In an endothermic reaction, heat is absorbed (gained) and the heat of reaction is positive [$\Delta H_{rxn} = (+)$].

Example 2.5 Using activation energy to calculate the heat of reaction

The activation energies characteristic of the reversible chemical reaction $A + B \rightleftharpoons C$ are $E_{forward} = 53$ kJ/mol and $E_{back} = 82$ kJ/mol. Calculate ΔH_{rxn}, the heat of the reaction. Is the reaction exothermic or endothermic? (1.0 cal = 4.184 J)

Solution

$$\begin{aligned}\Delta H_{rxn} &= E_{forward} - E_{back}\\ &= 53\text{ kJ/mol} - 82\text{ kJ/mol}\\ &= -29\text{ kJ/mol}\end{aligned}$$

The heat of reaction is negative, which means that heat has been lost and the reaction is exothermic.

Problem 2.5 The activation energies characteristic of the reversible chemical reaction $A + B \rightleftharpoons C$ are $E_{forward} = 74$ kJ/mol and $E_{back} = 68$ kJ/mol. Calculate ΔH_{rxn}, the heat of the reaction. Is the reaction exothermic or endothermic?

Catalysts

Catalysts are substances that increase the rate of a reaction but emerge unchanged after the reaction has ended. Catalysts in biological systems are called enzymes. For example, the enzyme carbonic anhydrase, found in red blood cells, is a catalyst that specifically increases the rate of the formation of bicarbonate ion from carbon dioxide and hydroxide ion by a factor of 3.5×10^6. (Refer to Appendix 2 to review the use of scientific notation.) Without this rapid a reaction, the body could not rid itself of the carbon dioxide generated by cellular oxidative processes.

When sulfur burns, it forms two different oxides: sulfur dioxide and sulfur trioxide. The reactions are

$$S + O_2 \longrightarrow SO_2 \qquad (1)$$

and

$$2\ SO_2 + O_2 \longrightarrow 2\ SO_3 \qquad (2)$$

Reaction 1 is much faster than reaction 2; so, when a sample of coal, which contains sulfur compounds, is burned, we might expect the bulk of the oxide produced to be SO_2. However, measurements have shown that, in coal combustion, large quantities of SO_3 are formed, which contributes significantly to the acid rain descending on our cities and forests. (Acid rain forms when SO_3 dissolves in water to form sulfuric acid: $SO_3 + H_2O \rightarrow H_2SO_4$.) The reason that SO_3 is formed in unexpected excess is that coal contains nitrogen compounds as well as sulfur, and so its combustion also produces nitric oxide (NO). Nitric oxide, in turn, reacts very rapidly with SO_2 in a series of steps to form SO_3:

$$2\ NO + O_2 \longrightarrow 2\ NO_2$$

$$2\ NO_2 + 2\ SO_2 \longrightarrow 2\ SO_3 + 2\ NO$$

$$\text{Sum: } 2\ SO_2 + O_2 \longrightarrow 2\ SO_3$$

As you examine this reaction sequence, notice that NO enters the first reaction as a reactant but emerges as a product at the end of the second reaction. In other words, nitric oxide speeds up the production of SO_3 and emerges unchanged; it is a catalyst. Notice, too, the role of NO_2, which is formed in the first step and used up in the second step (as emphasized by the connecting line). The NO_2 is called a **reaction intermediate.** The two steps are connected in a sequence because the product of the first step (NO_2) becomes a reactant in the second step. The final stoichiometry of the production of SO_3 is derived from an algebraic summation of the two steps in the reaction sequence.

All catalysts provide a new reaction pathway that has a lower activation energy, as illustrated in Figure 2.7. Although the activation energy is lowered by catalysts, the heats (endothermic or exothermic) of reaction and the products of a reaction are unchanged.

The protein molecules called **enzymes** are catalysts whose structures allow them to interact selectively only with particular molecules or families of molecules. The selectivity can be great enough that an enzyme can select one of thousands of different molecules in a cell for conversion into a product. Recall that carbonic anhydrase selects only CO_2 for its conversion into HCO_3^-.

The site, or location, where catalysis takes place on the enzyme—called an **active site**—is structurally related to the substance being converted—the **substrate**—analogous to the relation between a lock and a key. We will refer to this spatial relation as **complementarity.** The substrate is bound to the active site by secondary forces. The specifics of enzyme-catalyzed reactions will be considered in Chapters 14 through 18, but the important idea is that their catalytic properties depend on the formation of an activated complex. In all cases,

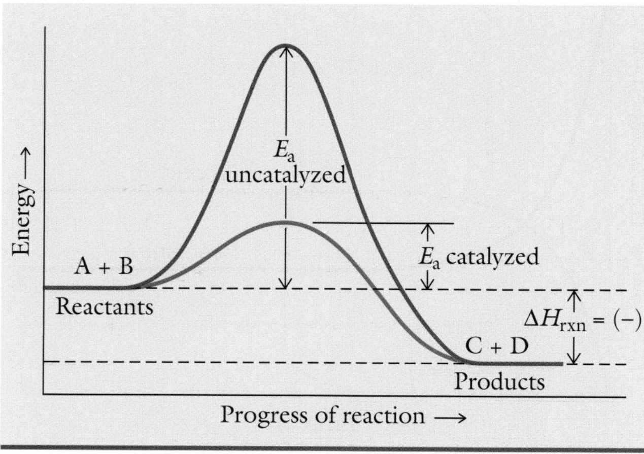

Figure 2.7 Reaction progress for a catalyzed reaction (compare Figure 2.6). The activation energy of the catalyzed reaction is significantly lowered, but the heat of reaction is unchanged.

the formation of a complex of enzyme and substrate provides a reaction pathway of lower overall activation energy and, consequently, a rapid reaction rate.

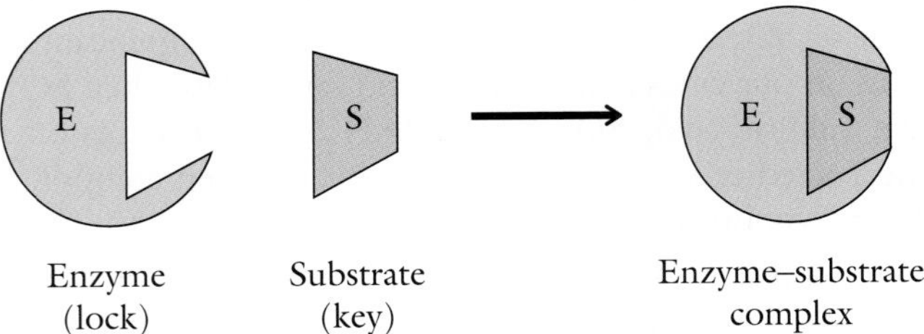

The activation energy controls the rate of the chemical reaction. The greater the activation energy, the smaller the percentage of colliding species that have the required activation energy and the slower the rate of reaction. The relation between activation energy, reaction rate, and temperature is an important feature of physiological function. The greater the activation energy, the greater is the increase in reaction rate with an increase in reaction temperature. In other words, in reactions with a large activation energy, the rate of reaction changes very rapidly with small increments in temperature increase. Conversely, in reactions with a small activation energy, the rates of reaction are relatively insensitive to changes in temperature.

Concept checklist

- ✓ A catalyst lowers the activation energy of a reaction by providing a different pathway leading to products.
- ✓ It does so by becoming an active participant in the chemical process, but it emerges unchanged.
- ✓ It does not alter the results of a reaction; rather, it alters only the speed at which the reaction takes place.

2.3 CHEMICAL EQUILIBRIUM

The time course of a reaction, such as that of hydrogen with nitrogen to produce ammonia, is plotted in Figure 2.4. Figure 2.8 plots a similar reaction over a long period of time. After sufficient time, the concentrations of product and reactants no longer change. At that point, the system has reached a state of **chemical equilibrium.** The amounts of reactants left at the end of a reaction can vary from large quantities to none, depending on the specific reaction.

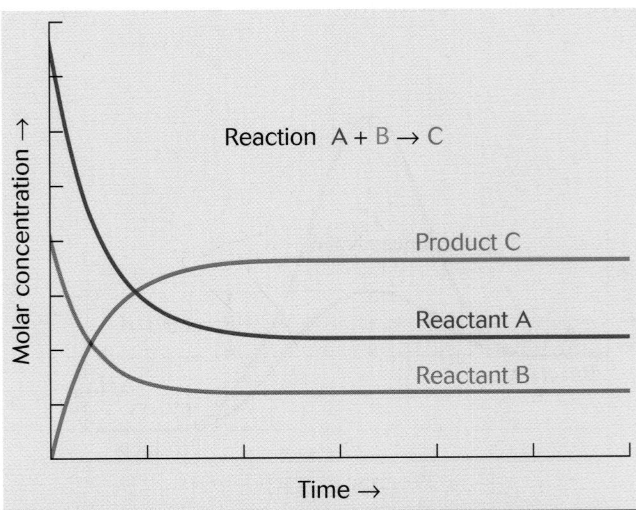

Figure 2.8 When the concentrations of products and reactants no longer change over time, a chemical reaction has reached equilibrium.

Chemical equilibrium is dynamic, a state or condition in which two opposing processes take place at the same rate. In Section 2.2, reversible chemical reactions were shown to consist of simultaneous opposing processes—the forward and back reactions. The two arrows used to identify a reversible reaction are also used to indicate a reaction in a state of equilibrium:

$$\mathrm{A} + \mathrm{B} \rightleftharpoons \mathrm{C} + \mathrm{D}$$

Therefore, whenever you see such a representation, you may assume that

- the chemical system is either at or is capable of reaching an equilibrium state; and
- the equilibrium state may be reached from either direction—that is, by adding either A to B or C to D.

In other words, equilibrium reactions are **reversible.** In theory, all reactions are reversible. However, in some reactions, the amounts of reactants left at the point of equilibrium are much too small to be of practical significance. In other reactions, such as the combustion of an alkane, reaction of the products (under most conditions) will not produce the original reactants. Such reactions are called **irreversible,** and, to emphasize that condition, balanced equations representing them will be written with a single arrow pointing to products. Such equations appear in different sections of this chapter and in subsequent chapters.

Equilibrium Constants

The quantitative study of chemical systems at equilibrium has shown that all chemical reactions can be described by an **equilibrium constant, K_{eq},** a mathematical function of the molar concentrations of reactants and products. If we propose that a reaction equation takes the form

$$a\,\mathrm{A} + b\,\mathrm{B} \rightleftharpoons c\,\mathrm{C} + d\,\mathrm{D}$$

where lowercase letters represent balancing coefficients and uppercase letters represent substances, the equilibrium constant is written, with brackets representing molar concentrations (see Appendix 4) at equilibrium, as follows:

$$K_{eq} = \frac{[\mathrm{C}]^c[\mathrm{D}]^d}{[\mathrm{A}]^a[\mathrm{B}]^b}$$

Pure solids or liquids do not appear in the equilibrium-constant expression for a reaction in which they take part. Therefore the equilibrium-constant expression must take the physical states of the reaction components into account. To do so, notations described in Section 2.1 are placed next to the components of the reaction equation: (*g*) for the gaseous state, (*l*) for the liquid state, (*s*) for the solid state, and (*aq*) for substances dissolved in water.

Example 2.6 Writing an equilibrium-constant expression

Write the equilibrium constant for the following reaction:

$$2\,\mathrm{N_2}(g) + 3\,\mathrm{Br_2}(g) \rightleftharpoons 2\,\mathrm{NBr_3}(g)$$

Solution

The reaction equation specifies the balancing coefficients of all components. Inserting them into the equilibrium-constant expression gives

$$K_{eq} = \frac{[\mathrm{NBr_3}]^2}{[\mathrm{N_2}]^2[\mathrm{Br_2}]^3}$$

Problem 2.6 Write the equilibrium constant for the balanced equation $\mathrm{H_2}(g) + \mathrm{I_2}(g) \rightleftharpoons 2\,\mathrm{HI}(g)$.

The most important qualitative information that we get from the equilibrium constant is how far the reaction of the components proceeds toward products. If the equilibrium constant is much larger than 1.0—for example, 1×10^3—then the concentrations of product (components in the numerator) must be much larger than the concentrations of the reactants (components in the denominator). Such a reaction therefore yields significant amounts of product, and the equilibrium is called **favorable.** Conversely, if the equilibrium constant is much smaller than 1.0—say, 1×10^{-3}—small amounts of products are the result, and the reaction is called **unfavorable.**

A favorable equilibrium constant indicates a favorable reaction only if there is a pathway that allows the reaction to proceed at a reasonable rate. For example, the equilibrium constant for the reaction between hydrogen and oxygen to form water is very large, but the reaction at room temperature takes many years to go to completion. However, with a little help from a spark (activation energy), the equilibrium state of large amounts of water can be rapidly attained.

Equilibrium concentrations obtained in the laboratory are used to calculate the value of K_{eq}.

Example 2.7 Calculating an equilibrium constant

For the following reaction at 25°C,

$$CO(g) + H_2O(g) \rightleftharpoons H_2(g) + CO_2(g)$$

the equilibrium concentrations are

$$[CO] = [H_2O] = 0.09 \text{ mol/L}$$
$$[H_2] = [CO_2] = 0.91 \text{ mol/L}$$

Calculate the value of the equilibrium constant.

Solution

The equilibrium-constant expression is

$$K_{eq} = \frac{[CO_2][H_2]}{[CO][H_2O]}$$

Substitute the equilibrium concentrations:

$$K_{eq} = \frac{[0.91][0.91]}{[0.09][0.09]} = 1 \times 10^2$$

Problem 2.7 In the reaction $2\ HI(g) \rightleftharpoons H_2(g) + I_2(g)$, the equilibrium concentrations at 25°C are: $[H_2] = [I_2] = 0.0011\ M$, and $[HI] = 0.0033\ M$. Calculate the value of the equilibrium constant.

Concept checklist

- ✓ When $K_{eq} > 1$, the reaction is favorable and, at equilibrium, the product-to-reactant ratio is greater than 1.
- ✓ When $K_{eq} < 1$, the reaction is unfavorable and the product-to-reactant ratio is less than 1.
- ✓ If we know the equilibrium concentrations of reactants and products of a reaction, we can calculate the value of the equilibrium constant.

The equilibrium constant can also be used to calculate the concentrations of the components of a reaction at equilibrium if the starting concentrations of the reactants are known.

Biochemical Reactions Are Connected in Sequences

Cellular metabolism consists of sequences of reactions. Many of the individual reactions have unfavorable equilibrium constants. However, each sequence as a whole produces adequate quantities of the necessary end product by incorporating highly favorable reactions as intermediate steps. This idea is best illustrated by an example.

Example 2.8 Calculating the equilibrium constant for a reaction sequence

Consider a sequence of two separate reactions that have a common intermediate. Assume that the equilibrium constant for the first reaction is large (1×10^4) and that for the second reaction is small (1×10^{-2}). What will the overall equilibrium constant be for the total process?

Solution

To solve this problem, we must first construct a hypothetical two-step chemical process in which reactant A leads to product C in two steps connected by a common intermediate B. Each step written separately can be characterized by an equilibrium constant:

$$A \rightleftharpoons B,\ \text{equilibrium constant}\ K_1 = \frac{[B]}{[A]} \qquad (1)$$

$$B \rightleftharpoons C,\ \text{equilibrium constant}\ K_2 = \frac{[C]}{[B]} \qquad (2)$$

The overall reaction, $A \rightarrow C$, is simply the result of adding reactions 1 and 2. This overall reaction is characterized by an equilibrium constant that also represents the overall process:

$$K_3 = \frac{[C]}{[A]}$$

The equilibrium constant for the overall reaction, K_3, is the product of the equilibrium constants for reactions 1 and 2:

$$K_3 = \frac{[B]}{[A]} \times \frac{[C]}{[B]} = \frac{[C]}{[A]} = K_1 \times K_2$$

Now substitute the K_1 value (1×10^{-2}) and the K_2 value (1×10^4). The overall equilibrium constant

$$K_3 = K_1 \times K_2 = 1 \times 10^2$$

The overall equilibrium constant is quite large, showing that an "unfavorable" reaction can still produce significant amounts of product when it is one of a sequence of more-favorable reactions. In Chapter 14, we shall see that sequences of this kind are characteristic of metabolic processes.

Problem 2.8 Consider a set of three hypothetical equilibrium reactions connected by common intermediates:

$$A \rightleftharpoons B \qquad K_1 = 1 \times 10^{-3}$$
$$B \rightleftharpoons C \qquad K_2 = 1 \times 10^{2}$$
$$C \rightleftharpoons D \qquad K_3 = 1 \times 10^{3}$$

Calculate the overall equilibrium constant for the reaction $A \rightleftharpoons D$.

Concept checklist

- ✓ In a reaction sequence, the products of one reaction become reactants in a second reaction; the products of the second reaction may become reactants in a third, and so forth.
- ✓ In such cases, the equilibrium constants can be combined into one that characterizes the overall process.

Le Chatelier's Principle

When a system in chemical equilibrium is temporarily displaced from equilibrium, it returns to its state of equilibrium through changes in the relative amounts of reactants and products. These amounts readjust so that the value of the equilibrium constant remains unchanged. In terms of the balanced chemical equation, there is a shift in mass from left to right or from right to left, depending on the nature of the imposed change, or stress.

Qualitative predictions of how a chemical system will respond to a disturbance in its equilibrium conditions are described by **Le Chatelier's principle:**

- If a system in an equilibrium state is disturbed, the system will adjust so as to neutralize the disturbance and restore the system to equilibrium.

For example, consider the following reaction to be at equilibrium:

$$N_2(g) + 3\ H_2(g) \rightleftharpoons 2\ NH_3(g)$$

How would changes in concentration of any of the components of the system affect the distribution of all the chemical components? Because the rate of a chemical reaction increases with an increase in the concentration of the reactants, adding nitrogen or hydrogen or both to the reaction mixture at equilibrium will increase the rate of the forward reaction, resulting in an increase in the concentration of ammonia and a new equilibrium mixture of N_2, H_2, and NH_3.

In other words, adding mass to the left-hand side of the reaction equation causes the equilibrium to shift to the right. Adding ammonia to the equilibrium mixture will increase the rate of the back reaction, resulting in an increase in the concentration of hydrogen and nitrogen and a new equilibrium mixture of H_2, N_2, and NH_3. Therefore, adding mass to the right-hand side of the reaction equation causes the equilibrium to shift to the left. The operation of Le Chatelier's principle is important in many biochemical processes (see, for example, Section 15.4).

Example 2.9 Using Le Chatelier's principle

Suppose that some $N_2O_4(g)$ were added to the following equilibrium system: $N_2O_4(g) \rightleftharpoons 2\ NO_2(g)$. What would the effect on the equilibrium mixture be?

Solution

The stress on the disturbed equilibrium mixture would be relieved and equilibrium would be reestablished by a reduction in the N_2O_4 concentration, which would lead to an increase in the NO_2 concentration. The reaction would shift to the right. However, note that both N_2O_4 and NO_2 concentrations would now be greater than in the preceding equilibrium state.

Problem 2.9 Suppose that some $H_2(g)$ were added to the following equilibrium system: $2\ HI(g) \rightleftharpoons H_2(g) + I_2(g)$. What would the effect on the equilibrium mixture be?

Many reactions yield not only new substances, but heat as well. In these cases, the equilibrium state of the chemical system can also be displaced by the addition or removal of heat. If heat is applied to an endothermic reaction, the equilibrium will shift to the product side. If heat is applied to an exothermic reaction, the equilibrium will shift to the reactant side.

In the next section, we apply the general principles of chemical equilibrium to systems in which acids and bases react with water to form new ionic species. The normal physiological function of cells and cellular systems depends strongly on these ionic equilibria.

2.4 ACIDS AND BASES

The terms acid and base have long been used to describe substances that have certain qualities in common. Some of them are listed in Table 2.2. Acids impart the typical tang to carbonated drinks and the tartness to citrus fruits, and bases are responsible for the bitterness of, for example, soapy water. When we eat the "wrong" foods and complain of heartburn (in reality, excess acid produced in the stomach), relief is available in the form of antacids (compounds that react with the excess acid and neutralize its effects). The health of cells critically depends on a balance between acids produced in metabolism and the body's ability to neutralize them.

Acids and Bases in Water

When an acid is added to water, it loses a hydrogen ion that reacts with the water to produce a **hydronium ion, H_3O^+.** For example, the addition of hydrogen chloride, HCl, to water produces the following reaction:

$$HCl(aq) + H_2O(l) \longrightarrow H_3O^+(aq) + Cl^-(aq)$$

All acids, whether molecules or polyatomic ions, produce hydronium ions in aqueous solution.

When a base is added to water, it produces a **hydroxide ion (OH^-),** either by adding ions directly to water, as NaOH does, or by reacting with water to remove a proton and leaving behind excess hydroxide ion, as in the addition of ammonia to water:

$$NH_3(aq) + H_2O(l) \rightleftharpoons NH_4^+(aq) + OH^-(aq)$$

(Remember, double arrows indicate an equilibrium reaction.) Because the ammonia causes an increase in the hydroxide ion concentration, it is classified as a base.

Some acids and bases are completely dissociated in aqueous solution. When hydrochloric acid is added to water, every molecule added to water leads to the formation of hydronium ions and hydrated chloride ions. Similarly, when the base sodium hydroxide (NaOH) is added to water, only hydrated $Na^+(aq)$ and $OH^-(aq)$ ions are found in the solution. Acids and bases that are completely ionized in water are called **strong acids** and **strong bases,** respectively. Table 2.3 on the following page lists the common strong acids and bases. Sulfuric acid, one of the strong acids listed in Table 2.3, is a polyprotic acid—an acid that has more than one dissociable proton. Other important polyprotic acids are phosphoric acid (H_3PO_4) and carbonic acid (H_2CO_3).

The strong acids and bases are called "strong" to differentiate them from the many acids and bases that do not fully dissociate in water. Acids and bases

TABLE 2.2 Properties of Acids and Bases

Acids	Bases
When dissolved in water, produce hydrogen ions.	When dissolved in water, produce hydroxide ions.
Neutralize bases to produce water and salts.	Neutralize acids to produce water and salts.
Solutions turn blue litmus paper red.	Solutions turn red litmus paper blue.

TABLE 2.3 Names and Formulas of All the Strong Acids and Bases

Strong acids		Strong bases	
Formula	Name	Formula	Name
$HClO_4$	perchloric acid	LiOH	lithium hydroxide
HNO_3	nitric acid	NaOH	sodium hydroxide
H_2SO_4	sulfuric acid*	KOH	potassium hydroxide
HCl	hydrochloric acid	RbOH	rubidium hydroxide
HBr	hydrobromic acid	CsOH	cesium hydroxide
HI	hydroiodic acid	TlOH	thallium(I) hydroxide
		$Ca(OH)_2$	calcium hydroxide
		$Sr(OH)_2$	strontium hydroxide
		$Ba(OH)_2$	barium hydroxide

*Only the first proton in sulfuric acid is completely dissociable. The first product of dissociation, HSO_4^-, is a weak acid.

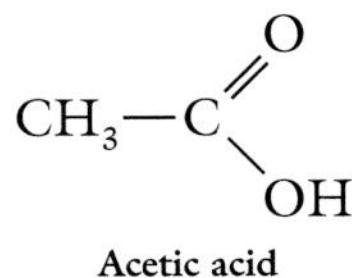

Acetic acid

that do not fully dissociate in water are called **weak acids** and **weak bases.** In this chapter, we principally use acetic acid as an example of a weak acid. Vinegar is an aqueous 5% acetic acid solution. The structural formula for acetic acid is presented in the margin. The hydrogen of the OH group is the dissociable hydrogen, or proton.

Acetic acid undergoes dissociation in water as follows:

$$CH_3COOH\,(aq) + H_2O(l) \rightleftharpoons CH_3COO^-(aq) + H_3O^+(aq)$$

Acetic acid is a weak acid because its aqueous solution contains mostly acetic acid molecules (CH_3COOH) and very few acetate (CH_3COO^-) and hydronium (H_3O^+) ions. If an acid is not listed in Table 2.3, it is a weak acid.

Ammonia is an example of a weak base. It reacts with water to produce hydroxide and ammonium ions:

$$NH_3(aq) + H_2O(l) \rightleftharpoons NH_4^+(aq) + OH^-(aq)$$

Ammonia is a weak base because its aqueous solution contains mostly hydrated ammonia molecules and very few ammonium and hydroxide ions. If a base is not listed in Table 2.3, it is a weak base.

Concept check

✓ The terms strong and weak do not denote the concentrations of acids and bases but tell us only that the substance does or does not fully dissociate in solution.

Because strong acids and strong bases are fully dissociated in water, we can easily calculate the hydronium or hydroxide ion concentrations in solutions of strong acids and bases as long as we know the solution concentrations.

Example 2.10 Calculating ion concentrations in a solution of a strong acid

What is the concentration of hydronium ion in a 0.010 *M* aqueous solution of HCl? What is the total concentration of all ions?

Solution

HCl, listed in Table 2.3 as a strong acid, is completely dissociated in aqueous solution; that is, 1 mol of HCl gives rise to 1 mol of H_3O^+ ion and 1 mol of Cl^- ion. In a 0.010 *M* solution, the concentrations of H_3O^+ and Cl^- are each 0.010 *M*. By adding the concentrations of both types of ions, we find that the total concentration of ions is 0.020 *M*.

Problem 2.10 What is the concentration of hydronium ion in a 0.050 *M* aqueous solution of HCl?

Example 2.11 Calculating ion concentrations in a solution of a strong base

What is the concentration of hydroxide ion in a 0.025 *M* aqueous solution of NaOH? What is the total concentration of all ions?

Solution

NaOH is listed in Table 2.3 as a strong base and therefore consists completely of ions in aqueous solution: 1 mol of NaOH gives rise to 1 mol of Na^+ ion and 1 mol of OH^- ion. In a 0.025 *M* solution, the concentrations of OH^- and Na^+ are each 0.025 *M*. By adding the concentrations of both types of ions, we find that the total concentration of ions is 0.050 *M*.

Problem 2.11 What is the concentration of hydroxide ion in a 0.050 *M* aqueous solution of NaOH?

A Measure of Acidity: pH

The acidity of a solution is measured in terms of the molar concentration of hydronium ions, a concentration that typically lies in the range of 1.0 *M* to 1.0×10^{-14} *M*. Because the use of scientific notation for these often very small concentrations can become quite cumbersome, a physiologist, S. P. L. Sørenson, devised the **pH** scale defined by the relation

$$[H_3O^+] = 10^{-pH}, \text{ or } pH = -\log [H_3O^+]$$

In pure water, the reaction between water molecules produces hydronium and hydroxide ions as follows:

$$H_2O(l) + H_2O(l) \rightleftharpoons H_3O^+(aq) + OH^-(aq)$$

and the concentration of hydronium ion is equal to the concentration of hydroxide ion. At 25°C, $[H_3O^+] = [OH^-] = 1 \times 10^{-7}$ *M*. When an acid is added to water, the hydronium ion concentration is raised to above 1×10^{-7} *M*; when a base is added, the hydronium ion concentration is reduced to less than 1×10^{-7} *M*. The equilibrium is described by a constant called the **ion-product of water, K_w**, and, at 25°C,

$$K_w = [H_3O][OH^-] = (1 \times 10^{-7})(1 \times 10^{-7}) = 1 \times 10^{-14}$$

- In a neutral solution, pH = 7, because $[H_3O^+] = [OH^-] = 1.00 \times 10^{-7}$ *M*.
- In an acidic solution, pH is less than 7, because $[H_3O^+]$ is greater than $[OH^-]$.
- In a basic solution, pH is greater than 7, because $[H_3O^+]$ is less than $[OH^-]$.

Example 2.12 Calculating pH from the hydronium ion concentration

Calculate the pH of an HCl solution whose concentration is 0.0010 *M*.

Solution

The solution requires the following definition:

$$pH = -\log [H_3O^+]$$

The H_3O^+ ion concentration substituted into the definition of pH gives

$$pH = -\log [0.0010]$$

The determination of pH is now a matter of obtaining the log of 0.001. Logarithms can be obtained directly with a hand calculator fitted with a logarithm function key.

Step 1 Enter the molar concentration of hydronium ion, enter 1, press EE, enter 3, press the change sign key (the key with the +/− sign), and then press the log key. The result is −3.

Step 2 Press the change sign key. The answer is now +3, so the pH = 3.00.

Problem 2.12 Calculate the pH of an aqueous 0.0030 *M* HCl solution.

Quantitative Aspects of Acid–Base Equilibria

The equation for the reaction of acetic acid with water is

$$CH_3COOH(aq) + H_2O(l) \rightleftharpoons H_3O^+(aq) + CH_3COO^-(aq)$$

and the equilibrium constant expression for its dissociation is

$$K_a = \frac{[H_3O^+][CH_3COO^-]}{[CH_3COOH]}$$

This equilibrium constant is called an **acid dissociation constant, K_a** ("a" for acid), because the reaction that it describes results in the formation of a hydronium ion. (Remember that pure solids or pure liquids do not appear in the equilibrium-constant expression).

Similarly, the dissociation of a weak base that results in the production of a hydroxide ion is characterized by a **base dissociation constant, K_b** ("b" for base). Consider the reaction of ammonia with water:

$$NH_3(aq) + H_2O(l) \rightleftharpoons NH_4^+(aq) + OH^-(aq)$$

The dissociation constant is

$$K_b = \frac{[NH_4^+][OH^-]}{[NH_3]}$$

The values of K_a and K_b are indicative of the relative amounts of products and reactants at equilibrium: the smaller the values of the dissociation constants, the smaller the amount of product relative to reactant—that is, the smaller the degree of dissociation. In other words, the value of this constant provides us with a quantitative measure of just how "weak" a weak acid or weak base is.

Concept check

✓ The smaller the dissociation constant, the weaker the acid or base.

Acid and base dissociation constants can be expressed in the negative logarithmic notation used for the pH. For weak acids,

$$K_a = 10^{-pK_a} \text{ and } -\log K_a = pK_a$$

TABLE 2.4 Values of K_a and pK_a for Some Weak Acids and Values of K_b and pK_b for Some Weak Bases

Acid	K_a	pK_a	Base*	K_b	pK_b
nitrous, HNO_2	4.47×10^{-4}	3.35	dimethylamine, $(CH_3)_2NH$	5.81×10^{-4}	3.24
cyanic, HCNO	2.19×10^{-4}	3.66	methylamine, CH_3NH_2	4.59×10^{-4}	3.34
formic, $HCHO_2$	1.78×10^{-4}	3.75	ammonia, NH_3	1.75×10^{-5}	4.76
hydrazoic, HN_3	1.91×10^{-5}	4.72	cocaine, $C_{17}H_{21}O_4N$	3.89×10^{-9}	8.41
acetic, $HC_2H_3O_2$	1.74×10^{-5}	4.76	aniline, $C_6H_5NH_2$	4.17×10^{-10}	9.38
carbonic, H_2CO_3	4.45×10^{-7}	6.35	morphine, $C_{17}H_{19}O_3N$	1.41×10^{-10}	9.85

*The nitrogens in all the bases have the same base properties as the nitrogen in ammonia.

For weak bases:

$$K_b = 10^{-pK_b} \text{ and } -\log K_b = pK_b$$

Table 2.4 lists various weak acids and bases, along with their dissociation constants and pK values. Note that, reading down each list, as the dissociation constants become smaller, the corresponding pK values become larger.

The Brønsted–Lowry Theory of Acids and Bases

In the early theory of acids and bases, the focus was on an acid's ability to add protons to water and a base's ability to add hydroxide ions to water. In the more broadly applicable **Brønsted–Lowry theory** of acids and bases,

- acids are described as **proton donors;** and
- bases are described as **proton acceptors.**

For example, hydroxide ion is a base because it is a proton acceptor. This focus on proton transfer permits a more inclusive classification of bases.

In our previous consideration of the reaction of HCl with H_2O,

$$HCl(g) + H_2O(l) \longrightarrow H_3O^+(aq) + Cl^-(aq)$$

we were interested only in the fact that the reaction of acid with water produces hydronium ions. However, the Brønsted–Lowry theory points out that the hydrochloric acid (HCl) is a proton donor, or an acid, and the water (H_2O) is a proton acceptor, or a base.

When acetic acid (a weak acid) reacts with water, the acid donates its proton to the water molecule, which acts as a base.

Conjugate acid–base pair

$$CH_3COOH(aq) + H_2O(l) \rightleftharpoons H_3O^+(aq) + CH_3COO^-(aq)$$

Conjugate acid–base pair

In this case, one molecule of protonated water (a hydronium ion) and one molecule of acetate ion are produced. As a result of acetic acid donating its proton, it becomes a base—acetate ion. When that base accepts a proton, it becomes acetic acid. According to the Brønsted–Lowry theory, the acetate ion is a base, just as is the hydroxide ion.

The Brønsted–Lowry theory links the two pairs, CH_3COOH/CH_3COO^- and H_3O^+/H_2O, and describes them as **conjugate acid–base pairs.**

Concept checklist

✓ Proton transfer requires a pair of substances: a donor and an acceptor.

✓ A substance other than water, such as acetate ion, can serve as a proton acceptor (base).

✓ When an acid donates a proton it becomes a base, and when a base accepts a proton it becomes an acid.

When ammonia (a weak base) reacts with water, the water donates its proton to the ammonia molecule, which acts as a base. The result is one molecule of deprotonated water (OH^-) and one molecule of ammonium ion (NH_4^+). As before, the Brønsted–Lowry theory links the two pairs, NH_4^+/NH_3 and H_2O/OH^-, as conjugate acid–base pairs:

Conjugate acid–base pair

$$NH_3(aq) + H_2O(l) \rightleftharpoons OH^-(aq) + NH_4^+(aq)$$

Conjugate acid–base pair

and the ammonium ion is an acid, just as is the hydronium ion.

In general, you will find that a conjugate base (a Brønsted–Lowry base) is the ion produced when its conjugate acid (a Brønsted–Lowry acid) donates a proton to water.

2.5 BUFFERS AND BUFFERED SOLUTIONS

The normal metabolic activity of cells results in the production of acids. Muscle tissue, for example, produces lactic acid. Such acids are a potential danger to proteins, whose structures and physiological functions depend on the maintenance of a proper pH. Even modest changes in pH can result in significant changes in the catalytic activity of enzymes, thereby causing cells to malfunction. The pH of blood must be maintained within very narrow limits that, if exceeded, can result in death. Fortunately, the body's buffer systems can effectively counter changes in pH.

A **buffered solution** is a solution that resists changes in pH when hydronium ion or hydroxide ion is added. Such a solution contains a **buffer system.** A buffer system always consists of a conjugate acid–base pair. The $H_2PO_4^-/HPO_4^{2-}$ and H_2CO_3/HCO_3^- systems are two examples of conjugate acid–base pairs that are used in the body as buffer systems. Many of the commercially available antacids create a buffer system in the stomach by providing a Brønsted–Lowry base such as citrate. When added to the low-pH environment of the stomach, citrate reacts with hydronium ion to increase the pH toward the pK value of the citric acid/citrate system.

If a small amount of hydroxide ion is introduced into a buffer solution, the conjugate acid will react with it, and, if a small amount of hydronium ion is added, the conjugate base will react with it. The conjugate acid behaves as a hydroxide ion sponge, and the conjugate base as a hydronium ion sponge, both constantly ready to absorb the appropriate ion and thus maintain a constant pH.

Each conjugate acid–base pair has a specific pH that it is able to maintain through its buffering action. This pH can be calculated by using the **Henderson–Hasselbalch equation:**

$$\text{pH} = \text{p}K_a + \log\left(\frac{[\text{proton acceptor}]}{[\text{proton donor}]}\right)$$

The equation states that the pH of a buffered solution depends on the pK_a of the proton donor of the acid–base pair (not the pK_b of the proton acceptor) and on the logarithm of the ratio of proton acceptor to proton donor. For maximum buffering effect, the value of the acceptor-to-donor ratio in a buffer solution must be kept as close to 1.0 as possible. By knowing the pK_a of the proton donor and the ratio of acceptor to donor, we can calculate the pH of a buffer solution.

Example 2.13 Calculating the pH of a buffer solution

Calculate the pH of an aqueous solution consisting of 0.00800 *M* acetic acid and 0.00600 *M* sodium acetate.

Solution

For this type of calculation, we use the Henderson–Hasselbalch equation, substituting the pK_a of acetic acid, 4.76 (Table 2.4), and the concentrations of acceptor and donor. (Note that the pK_a of acetic acid, not the pK_b of acetate, is used.) The acceptor (numerator) is acetate, and the donor (denominator) is acetic acid.

$$\text{pH} = 4.76 + \log\frac{[0.00600]}{[0.00800]}$$

$$= 4.76 - 0.125$$

$$\text{pH} = 4.64$$

Problem 2.13 Calculate the pH of an aqueous solution consisting of 0.00600 *M* acetic acid and 0.00800 *M* sodium acetate.

The information provided by the Henderson–Hasselbalch equation also tells us how to select a buffer system that will buffer at a specific pH. For example, a particular experiment on a biological reaction may require a buffer that maintains a solution at pH 7.00 ± 0.30.

The key to selecting a buffer system is to consider what the Henderson–Hasselbalch equation reveals when the concentrations of conjugate base and acid are equal; that is, when [acceptor] = [donor]. When that is the case,

$$\text{pH} = \text{p}K_a + \log 1.0$$

But log 1.0 = 0; therefore, when [acceptor] = [donor],

$$\text{pH} = \text{p}K_a$$

So, to prepare a buffer of specific pH, select a conjugate acid–base pair with a pK_a as close as possible to the desired pH, with ratios of acceptor to donor concentrations as close to 1.0 as possible.

Summary

Chemical Reactions in Solution

• Reactions in solution include those that are the result of oxidation–reduction or that form a precipitate, water, or a gas.

Chemical Kinetics

• The rate of a chemical reaction increases with increases in the concentrations of the reactants and in temperature.

• Catalysts are agents that also increase the rate of a reaction but emerge unchanged after the reaction has ended.

• Catalysts can't change the result of a reaction, only its rate.

Molecular Collisions and Chemical Reactions

• A chemical reaction takes place as the result of a reactive collision.

• In a reactive collision, two molecules combine to form an activated complex that then decomposes to form the reaction products.

• The energy required to form the activated complex is called the activation energy.

• The greater the activation energy, the slower the reaction.

• A catalyst lowers the activation energy of a reaction by providing a different route to products.

• Chemical processes within living cells are catalyzed by enzymes.

Chemical Equilibrium

• A chemical reaction is at equilibrium when the concentrations of reacting substances no longer change.

• The concentrations of reactants and products at that point are quantitatively described by the equilibrium constant.

Le Chatelier's Principle

• A system in chemical equilibrium can be displaced from equilibrium, but equilibrium can be reestablished through compensating changes in the relative amounts of reactants and products.

Acids and Bases

• Strong acids and bases become completely dissociated in water.

• Acids and bases that do not fully dissociate in water are called weak acids and bases.

A Measure of Acidity: pH

• Acidity is expressed by an acidity scale called pH.

• The three possible conditions in aqueous solutions with respect to acidity are pH equal to 7 in a neutral solution, pH less than 7 in an acidic solution, and pH greater than 7 in a basic solution.

Quantitative Aspects of Acid–Base Equilibria

• The dissociations of weak acids and weak bases are quantitatively described by equilibrium constants called dissociation constants.

• The smaller the dissociation constant, the weaker the acid or base.

The Brønsted–Lowry Theory of Acids and Bases

• The Brønsted–Lowry theory of acids and bases describes acids as proton donors and bases as proton acceptors.

• According to the theory, when an acid donates a proton, the acid becomes a base and, when a base accepts a proton, the base becomes an acid.

Buffers and Buffered Solutions

• A buffered solution resists changes in pH on the addition of hydronium ion or hydroxide ion.

• A buffered solution contains a buffer system consisting of a conjugate acid–base pair.

• The pH of a buffered solution depends on the pK_a of the proton donor of the acid–base pair and the ratio of the concentration of proton acceptor (base) to the concentration of proton donor (acid).

Key Words

activated complex, p. 44
activation energy, p. 44
Brønsted–Lowry theory, p. 57
buffer system, p. 58
catalyst, p. 46
chemical equilibrium, p. 48
dissociation constant, p. 56
equilibrium constant, p. 49
heat of reaction, p. 46
Henderson–Hasselbalch equation, p. 59
hydronium ion, p. 53
hydroxide ion, p. 53
ion-product of water, p. 55
Le Chatelier's principle, p. 52
net ionic equation, p. 38
pH, p. 55
strong acid, p. 53
weak acid, p. 54

Exercises

Chemical Reactions in Solution

2.1 Solutions of the following compounds are added together. Consult Table 2.1 to predict whether a reaction will take place. If a reaction will take place, write the net ionic equation that describes it. (a) $NaCl + AgNO_3$; (b) $KOH + MgCl_2$.

2.2 Solutions of the following compounds are added together. Consult Table 2.1 to predict whether a reaction will take place. If a reaction will take place, write the net ionic equation that describes it. (a) $Pb(NO_3)_2 + KCl$; (b) $K_3PO_4 + CaCl_2$.

2.3 Solutions of the following compounds are added together. Consult Table 2.1 to predict whether a reaction will take place. If a reaction will take place, write the net ionic equation that describes it. (a) $K_2SO_4 + BaCl_2$; (b) $AlCl_3 + KOH$.

2.4 Solutions of the following compounds are added together. Consult Table 2.1 to predict whether a reaction will take place. If a reaction will take place, write the net ionic equation that describes it. (a) $NaBr + AgNO_3$; (b) $KCl + Na_2CO_3$.

Chemical Kinetics

2.5 The activation energies for each of two reactions were found to be (a) 24 kJ and (b) 53 kJ (1.0 cal = 4.184 J). If the temperatures of both reactions are identical, which has the greater rate of reaction?

2.6 Hydrogen peroxide, H_2O_2, is unstable and slowly decomposes over time to form water and gaseous oxygen. However, in the presence of the enzyme catalase, its rate of decomposition was increased by a factor of more than 1×10^6, with no change in the H_2O_2 concentration. Explain.

2.7 The heat of a reaction, ΔH_{rxn}, is -21 kJ, and the activation energy of the forward reaction, $E_{forward}$, is $+37$ kJ. Calculate the activation energy of the reverse reaction, E_{back}.

2.8 The heat of a reaction, ΔH_{rxn}, is $+12$ kJ, and the activation energy of the forward reaction, $E_{forward}$, is -46 kJ. Calculate the activation energy of the reverse reaction, E_{back}.

2.9 Calcium carbonate is heated in a crucible open to the atmosphere. The reaction that takes place is a decomposition:

$$CaCO_3(s) \rightleftharpoons CaO(s) + CO_2(g)$$

Is the reaction reversible as performed?

2.10 The following decomposition reaction

$$2\ HCl(aq) + CaCO_3(s) \rightleftharpoons H_2O(l) + CO_2(g) + CaCl_2(aq)$$

is used as a field test for the detection of limestone. When it is conducted in the field in the open air, is the reaction reversible? Explain your answer.

2.11 For the reaction $CaCO_3(s) \rightleftharpoons CaO(s) + CO_2(g)$, which is the forward and which is the back reaction?

2.12 For the reaction $CaO(s) + CO_2(g) \rightleftharpoons CaCO_3(s)$, which is the forward reaction and which is the back reaction?

Chemical Equilibrium

2.13 In the reaction $H_2(g) + I_2(g) \rightleftharpoons 2\ HI(g)$, the equilibrium concentrations were found to be $[H_2] = [I_2] = 0.86\ M$ and $[HI] = 0.27\ M$. Calculate the value of the equilibrium constant.

2.14 In the reaction $NH_4Cl(s) \rightleftharpoons NH_3(g) + HCl(g)$ the equilibrium concentrations were found to be $[NH_3] = [HCl] = 3.71 \times 10^{-3}\ M$. Calculate the value of the equilibrium constant.

2.15 For the reaction $N_2(g) + 3\ H_2(g) \rightleftharpoons 2\ NH_3(g)$, in what direction will the equilibrium shift if some NH_3 is removed from the equilibrium mixture?

2.16 For the reaction $N_2(g) + 3\ H_2(g) \rightleftharpoons 2\ NH_3(g) +$ 92 kJ, what will be the effect of increasing the temperature on the extent of the reaction?

2.17 In the following aqueous equilibrium, a visual change can be observed when the equilibrium shifts. Predict the visual change when the indicated stress is applied. Add Cl^- to the following reaction:

$$\text{Heat} + Co^{2+}(aq, \text{pink}) + 4\ Cl^-(aq, \text{colorless}) \rightleftharpoons CoCl_4^{2-}(aq, \text{blue})$$

2.18 In the following aqueous equilibrium, a visual change can be observed when the equilibrium shifts. Predict the visual change when the indicated stress is applied. Add Fe^{3+} to the following reaction:

$$Fe^{3+}(aq, \text{brown}) + 6\ SCN^-(aq, \text{colorless}) \rightleftharpoons Fe(SCN)_6^{3-}(aq, \text{red})$$

2.19 Write the balanced equation corresponding to the following equilibrium constant:

$$K_{eq} = \frac{[CO][Cl_2]}{[COCl_2]}$$

2.20 Write the balanced equation corresponding to the following equilibrium constant:

$$K_{eq} = \frac{[CO][H_2]^3}{[CH_4][H_2O]}$$

2.21 Write the balanced equation corresponding to the following equilibrium constant:

$$K_{eq} \frac{[PCl_5]}{[Cl_2][PCl_3]}$$

2.22 Write the balanced equation corresponding to the following equilibrium constant:

$$K_{eq} = \frac{[O_3]^2}{[O_2]^3}$$

2.23 Write the equilibrium-constant expression for the following reaction:

$$N_2O_5(g) \rightleftharpoons NO_2(g) + NO_3(g)$$

2.24 Write the equilibrium-constant expression for the following reaction:

$$2\ NO(g) + 2\ H_2(g) \rightleftharpoons N_2(g) + 2\ H_2O(g)$$

Acids and Bases

2.25 The concentrations of hydronium ion and of hydroxide ion in pure water at 25°C are 1.00×10^{-7} *M*. What is the value of the ion-product for the ionization of water at that temperature?

2.26 The ion-product of pure water, K_w, is $1.00 \times 10^{-13.60}$ at body temperature, 37°C. What are the molar H_3O^+ and OH^- concentrations in pure water at body temperature?

2.27 What is the definition of a strong base? Give two examples.

2.28 What is the definition of a strong acid? Give two examples.

2.29 Calculate the H_3O^+ and Cl^- concentrations in a 0.30 *M* aqueous solution of HCl.

2.30 Calculate the H_3O^+ and ClO_3^- concentrations of a 0.28 *M* solution of perchloric acid, $HClO_3$.

2.31 What is the pH of pure water at 25°C?

2.32 What is the pH of pure water at 37°C? K_w is $1.00 \times 10^{-13.60}$ at 37°C.

2.33 What is the pH of a solution with $[H_3O^+] = 0.010\ M$?

2.34 What is the pH of a solution with $[H_3O^+] = 0.0026\ M$?

2.35 What is the definition of a weak acid? Give two examples.

2.36 What is the definition of a weak base? Give two examples.

2.37 Identify the conjugate acid–base pairs in each of the following equations:

(a) $HNO_2(aq) + H_2O(l) \rightleftharpoons NO_2^-(aq) + H_3O^+(aq)$

(b) $H_2PO_4^-(aq) + H_2O(l) \rightleftharpoons HPO_4^{2-}(aq) + H_3O^+(aq)$

2.38 Identify the conjugate acid–base pairs in each of the following equations:

(a) $HCOOH(aq) + NH_3(l) \rightleftharpoons HCOO^-(aq) + NH_4^+(aq)$

(b) $H_2O(l) + H_2O(l) \rightleftharpoons OH^-(aq) + H_3O^+(aq)$

Buffers and Buffered Solutions

2.39 Why does a solution of a conjugate acid–base pair behave as a buffered solution?

2.40 What is the pH of an aqueous solution consisting of 0.00600 *M* acetic acid and 0.00800 *M* sodium acetate?

2.41 What is the pH of a solution consisting of 0.0700 *M* formic acid (HCOOH) and 0.0700 *M* sodium formate (HCOONa)?

2.42 Calculate the pH of a solution consisting of 0.0750 *M* K_2HPO_4 and 0.0500 *M* KH_2PO_4. ($pK_a = 7.20$)

2.43 Calculate the pH of a solution consisting of 0.0500 *M* K_2HPO_4 and 0.0750 *M* KH_2PO_4. ($pK_a = 7.20$)

Unclassified Exercises

2.44 Calculate the pH of each of the following solutions: (a) 0.0031 *M* HNO_3; (b) 1.0 *M* HCl.

2.45 Calculate the pH of each of the following solutions: (a) 0.0069 *M* HI; (b) 0.019 *M* HBr.

2.46 Calculate the pH of a solution that is 0.0500 *M* in H_2CO_3 and 0.0750 *M* in $KHCO_3$.

2.47 What is the molar ratio of $HPO_4^{2-}/H_2PO_4^-$ in a buffered solution of pH 8.20?

2.48 The pH of an HCl solution is 3.20. Calculate the concentration of the HCl solution.

2.49 Calculate the pH and the hydroxide ion concentrations of each of the following solutions: (a) 0.0034 *M* HCl; (b) 0.025 *M* HNO_3.

2.50 Give three examples of buffer systems.

2.51 One of the ways in which ozone, O_3, is destroyed in the upper atmosphere is described by the following sequence of reactions:

$$O_3(g) + NO(g) \longrightarrow O_2(g) + NO_2(g)$$
$$NO_2(g) + O(g) \longrightarrow NO(g) + O_2(g)$$
$$\text{Sum:}\quad O_3(g) + O(g) \longrightarrow 2\ O_2(g)$$

Is one of the chemical components a catalyst? Explain your answer.

2.52 Magnesium hydroxide, $Mg(OH)_2$, is slightly soluble in water. The solubility can be described by the following equilibrium:

$$Mg(OH)_2(s) \rightleftharpoons Mg^{2+}(aq) + 2\ OH^-(aq)$$

What will happen to the equilibrium if hydrogen ion, H^+, is added to the solution?

2.53 State Le Chatelier's principle.

2.54 The ion-product of water increases as temperature increases. Is the ionization of water an exothermic or an endothermic reaction?

Expand Your Knowledge

2.55 75 mL of ethanol, a pure liquid, is dissolved in 60 mL of water. Which component is the solute and which is the solvent?

2.56 A chemist has 70.0 mL of 0.300 *M* HCl available and requires 135 mL of 0.180 *M* HCl. Will she be able to obtain the required quantity? Explain your answer.

2.57 What is the molarity of 225 mL of an aqueous solution containing 13.9 g of $MgCl_2$?

2.58 What is the meaning of the square brackets around the chemical components in equilibrium-constant expressions?

2.59 What is a conjugate acid–base pair?

2.60 Why is hydrogen sulfide, H_2S, a gas at $-10°C$, whereas water, H_2O, is a solid at the same temperature?

2.61 In what way is a liquid similar to a solid?

2.62 Hydrogen is produced commercially by the "water reaction":

$$CO(g) + H_2O(g) \rightleftharpoons H_2(g) + CO_2(g)$$

Decide how the equilibrium will shift in each of the following cases: (a) gaseous carbon dioxide is removed; (b) water vapor is added; (c) the pressure is increased by adding helium gas.

2.63 Under what circumstances can you calculate an overall equilibrium constant for a sequence of reactions whose individual equilibrium constants are known?

2.64 The Henderson–Hasselbalch equation is often used to calculate the pH of a buffer solution, but it also describes the pH when a strong base is used to partially or completely neutralize a weak acid such as acetic acid. Describe the situation when 50% of the acid has been neutralized. What species are present in the solution, and what is the relation of the pH to the pK_a of acetic acid?

2.65 Would the pH be 7.00 at the point of 50% neutralization as described in exercise 2.64? Explain your answer.

2.66 Your laboratory instructor provides you with four solutions. She tells you that they are equimolar solutions of monoprotic weak acids. You are to measure the pH of each solution, and correlate the pH values with their dissociation constants, smallest to largest. The list of pH values is as follows: (1) 6.65, (2) 3.41, (3) 4.82, (4) 2.85.

2.67 The activation energy, E_a, of a catalyzed reaction is 4 kcal/mol, and $\Delta H_{rxn} = -18$ kcal/mol. Is the ΔH_{rxn} of the uncatalyzed reaction larger, smaller, or the same as that of the catalyzed reaction? Explain.

2.68 The activation energy, E_a, of a catalyzed reaction is 4 kcal/mol, and $\Delta H_{rxn} = 18$ kcal/mol. Is the ΔH_{rxn} of the uncatalyzed reaction larger, smaller, or the same as that of the catalyzed reaction? Explain.

ORGANIC CHEMISTRY

Having completed our study of the basic structure and properties of atoms and molecules, we proceed to a consideration of organic chemistry, the study of compounds that contain the element carbon, usually together with hydrogen and often one or more other elements (oxygen, halogen, nitrogen, sulfur, or phosphorus). Both naturally occurring and synthetic organic compounds are important to our lives. An example of a naturally occurring organic molecule is heme—a component of the protein hemoglobin. Heme contains iron (Fe^{2+}) at its center. That is the site to which oxygen attaches when hemoglobin takes on oxygen at the lungs and stores it for subsequent delivery to cells and tissues throughout the body.

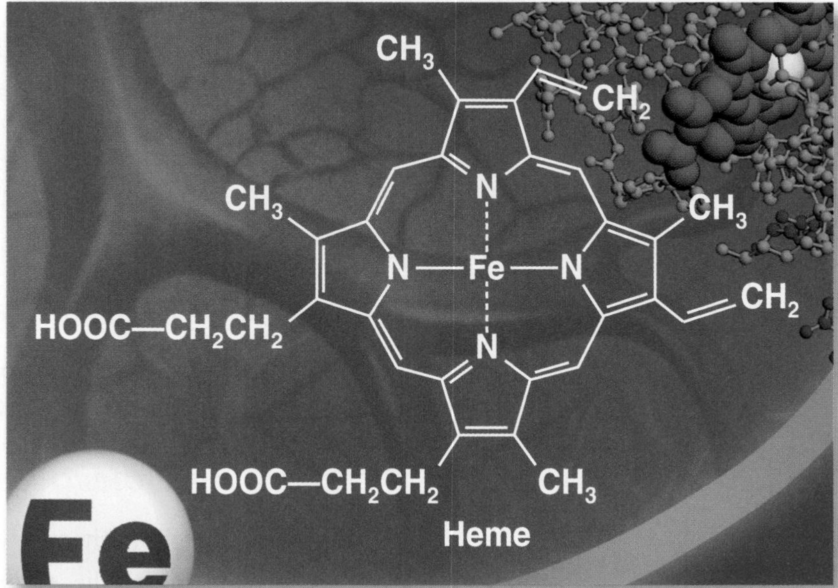

CHAPTER 3 SATURATED HYDROCARBONS

(David R. White, Richmond, Virginia.)

Chemistry in Your Future

No sooner do you clock in at the poison-control center than you receive a call from a frightened grandmother who thinks her tiny grandson may have drunk some lemon-scented furniture polish. The child appears well, but she believes she detects a smell of polish on his breath. You tell her to keep him warm and rush him to a hospital emergency room. You warn her above all not to induce vomiting, even though that is the recommended procedure when many other poisons are swallowed. The furniture polish contains petroleum distillates, which can do serious lung damage if aspirated during vomiting. This chapter describes the organic compounds called saturated hydrocarbons that are present in petroleum. Although they are used to make many important materials, including fuels, plastics, medicines, and furniture polish, they are poisons if significant amounts are swallowed.

For more information on this topic and others in this chapter, go to www.whfreeman.com/bleiodian2e

Learning Objectives

- Define organic chemistry.
- Identify the families of organic compounds.
- Describe the bonding in alkanes.
- Draw condensed and expanded structural formulas of alkanes.
- Draw constitutional isomers of alkanes.
- Name alkanes by the IUPAC nomenclature system.
- Draw and name cycloalkanes.
- Draw geometric (cis-trans) stereoisomers of cycloalkanes.
- Describe the physical properties of alkanes and cycloalkanes.
- Write equations for the halogenation of alkanes and cycloalkanes.
- Write equations for the combustion of alkanes and cycloalkanes.

The remainder of this book deals with organic chemistry and its extension into biochemistry. **Organic chemistry** is the study of **organic compounds,** compounds containing the element carbon. All other compounds are **inorganic** and are the subject matter of **inorganic chemistry.** However, chemists have traditionally assigned a small number of simple carbon compounds to the category of inorganic compounds. Carbon monoxide (CO), carbon dioxide (CO_2), and various carbonates (for example, H_2CO_3 and Na_2CO_3), bicarbonates (for example, $NaHCO_3$), and cyanides (for example, HCN and NaCN) are usually classified and studied with the inorganic compounds.

With more than 100 other elements in the periodic table, why does the study of carbon and its compounds merit so much time and effort? The answer lies in the number, complexity, and importance of organic compounds. There are close to 10 million organic compounds, more than 25 times the number of inorganic compounds. Carbon atoms bond to one another (only a few elements in the periodic table have this ability) as well as to other elements and are therefore uniquely capable of creating molecules that range from very small to very large. There are organic compounds containing tens, hundreds, thousands, and more carbon atoms bonded one to another.

Many organic compounds contain only carbon and hydrogen. A larger number also contain one or more other elements, principally oxygen, a halogen (fluorine, chlorine, bromine, iodine), nitrogen, sulfur, and phosphorus. The number of organic compounds is greatly enlarged because of the variety of structural arrangements in which carbon atoms bond to one another and to other elements.

The term **organic** originally conveyed the idea of compounds produced by a so-called vital force that chemists thought was present in living organisms, both plant and animal. Examples of such compounds were sugar, starch, plant oils, animal fats, and proteins. Early chemists assumed that organic compounds could not be synthesized in the laboratory. That assumption proved incorrect, however, and, by the early 1800s, the definition of organic chemistry had to be revised. Since then, an enormous number of organic compounds have been created or recreated in the laboratory. We study organic compounds, both naturally occurring and synthetic, because of their great importance to our lives:

- Our bodies are composed of organic compounds—carbohydrates, lipids, nucleic acids, proteins, and other organic molecules. Carbohydrates provide energy. Lipids are used to build the membranes that surround all living cells. They are also used as a storage form of energy and for the synthesis of steroid hormones and various fat-soluble vitamins. Nucleic acids—deoxyribonucleic acids (DNA) and ribonucleic acids (RNA)—direct and control the reproduction of an organism. Proteins, the key components of muscle and bone, also transport oxygen to tissues, catalyze chemical reactions throughout the body, protect against viruses and bacteria, and regulate bodily functions in general.
- Most of the energy to run automobiles, trains, and aircraft, to heat homes, offices, and factories, and to operate electrical equipment is obtained by burning organic compounds: gasoline, diesel oil, and other fuel oils, heating and cooking gases, and other petroleum products.
- Organic compounds in naturally occurring and synthetic medicines and medicaments—aspirin, penicillin, anesthetics, rubbing alcohol, and so forth—relieve pain and illness.
- Synthetic plastics, textiles, and rubbers are organic compounds. In beverage and detergent containers, antistick cookware, toys, polyester and nylon clothing, synthetic grass, and automobile tires, they have

A PICTURE OF HEALTH
Examples of Functional Groups

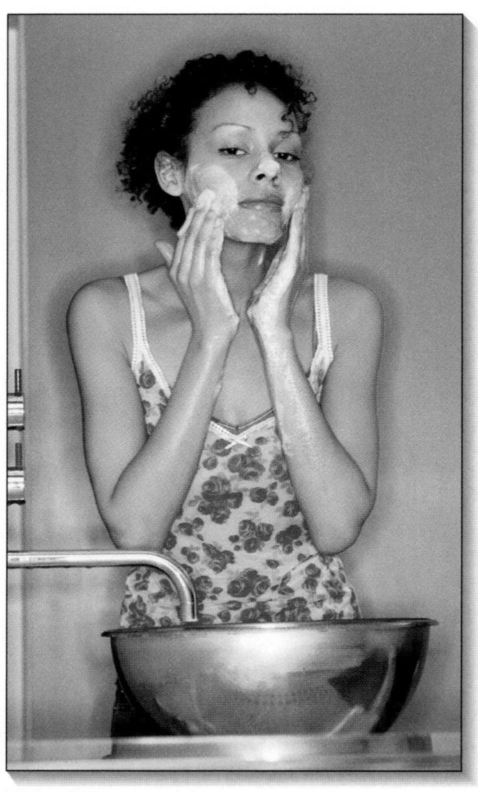
(IT Stock/Fotosearch.)

Hair is constructed from the protein α-keratin, containing the **amide** functional group.

The most commonly used drug, aspirin, often taken to reduce pain, contains the **carboxylic acid** and **ester** functional groups.

The soaps used to clean our skin surfaces are produced from animal fats and plant oils that contain the **ester** functional group.

The perception of vision depends on the compound *cis*-11-retinal, which contains the **aldehyde** and **alkene** functional groups.

Adrenalin, secreted by the adrenal glands in response to stress or threat, contains the **alcohol**, **amine**, **aromatic**, and **phenol** functional groups.

transformed modern life and contribute heavily to our high standard of living.

- The active ingredients in soaps, detergents, polishes, cosmetics, deodorants, and shampoos are organic compounds.
- The materials for manufacturing artificial body parts, such as hip and knee prostheses, heart valves, and dentures, are organic compounds.

We begin the study of organic chemistry with an overview of the formulas and families of organic compounds, followed by a detailed look at the group of compounds called saturated hydrocarbons.

3.1 MOLECULAR AND STRUCTURAL FORMULAS

As stated in Section 1.5, the **molecular formula** of a compound indicates the number of atoms of each element present in a molecule of the compound, but it does not indicate how the various atoms are connected to one another. Thus, although the molecular formula for methane, CH_4, shows that a molecule of methane contains one carbon atom and four hydrogen atoms, it does not describe how those atoms are connected or arranged. The **structural formula,** however, does show how the various atoms in a molecule are bonded together, as shown in the margin for methane.

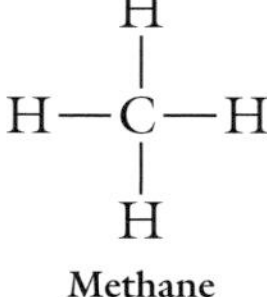
Methane

We can deduce the structural formula from the molecular formula by considering the **combining power**—the bonding requirements—of each element in the compound. The combining power of an element in a covalent compound is the number of covalent bonds that it typically forms to complete its octet of valence electrons (Section 1.5). The combining powers of the elements most commonly present in organic compounds are shown in Table 3.1. Carbon, with its four valence electrons, needs four more electrons to complete its octet, and so it is described as tetravalent; that is, it forms four covalent bonds. All carbon atoms in all compounds must have four bonds to other atoms, no more and no less. Nitrogen and phosphorus have five valence electrons, and so they need three electrons to complete their octets. They are trivalent. Oxygen and sulfur have six valence electrons, need two electrons to complete their octets, and are divalent. Hydrogen has one valence electron, needs one electron, and is monovalent. Each of the halogens (fluorine, chlorine, bromine, and iodine) has seven valence electrons, needs one electron, and is monovalent.

Sulfur and phosphorus are unique among the elements found in organic compounds. They exhibit variable combining powers because their octets can expand. Thus, phosphorus and sulfur are also pentavalent and hexavalent, respectively, in some compounds. Examples are the phosphorus atoms in phosphoric acids and esters (Sections 7.5 and 13.2) and the sulfur atoms in sulfuric acid (Table 2.3) and benzenesulfonic acid (Section 4.4).

Concept checklist

- ✓ A structural formula cannot be correct—cannot represent a real compound—unless the combining power of each atom is represented correctly. A structural formula showing an incorrect number of bonds for any atom is not a correct structural formula.
- ✓ A molecular formula is correct only if it can be translated into a correct structural formula.

We should note that, for compounds containing only carbon and hydrogen, a structural formula is the same as a Lewis structure (Section 1.6). (For compounds that also contain oxygen or nitrogen, a structural formula is the Lewis structure without the nonbonded electrons shown on oxygen or nitrogen.)

TABLE 3.1 Combining Powers of Elements Present in Organic Compounds

Element	Number of bonds	Bonding representation
C	4	—C— (one bond above, one below)
N	3	—N— (one bond above)
O	2	—O—
H, F, Cl, Br, I	1	H— F— Cl— Br— I—
P	3	—P— (one bond above)
	5	—P— (double bond above, single bond below)
S	2	—S—
	6	—S— (double bond above, double bond below)

Example 3.1 Determining correct molecular formulas

Which of the following molecular formulas are correct and which are incorrect? (a) CH_2F_2; (b) CH_3F_2; (c) CH_2F.

Solution

(a) CH_2F_2 is correct because we can draw a structural formula that shows the correct number of bonds for each of the atoms of the molecular formula. Carbon has four bonds, and hydrogen and fluorine each have one:

```
    H
    |
F—C—F
    |
    H
```

(b, c) We cannot draw a structure with the correct number of bonds for all atoms for either CH_3F_2 or CH_2F. We can draw incorrect structural formulas such as

```
     H                 F
     |            H\  /
F—C—H—F          C—F
     |            H/  \
     H                 H
     1                 2
```

for CH_3F_2, but they violate the bonding requirements of one or another atom. Structure 1 shows one H atom with two bonds, but H is monovalent. Structure 2 shows carbon with five bonds, but carbon is tetravalent. Structures 1 and 2 and other variations do not represent real compounds.

A structure such as 3 drawn to represent CH_2F also does not represent a real compound, because carbon must have four bonds not three.

```
    H
    |
H—C—F
    3
```

Problem 3.1 Which of the following molecular formulas are correct and which are incorrect? (a) CH_5N; (b) CH_5O; (c) C_2H_5Cl.

3.2 FAMILIES OF ORGANIC COMPOUNDS

Studying the chemistry of millions of organic compounds would be an impossible task were it not for the fact that very large groups of organic compounds have certain kinds of chemical behaviors in common. Thus organic compounds are organized into **families** (sometimes called **classes**), each consisting of a very large number of different compounds having a common characteristic pattern of chemical behavior. For example, all alkenes react with bromine (Br_2), all carboxylic acids are acidic, and all amines are basic. We study organic chemistry by studying the characteristic chemical pattern that identifies the members of each family.

The common physical and chemical properties of all compounds in a family result from the presence in their molecular structures of a common **functional group**—a specific atom or bond or a specific group of atoms in a specific bonding arrangement. The functional group dictates the behavior of a compound.

Table 3.2 shows the major families of organic compounds, each with its functional group. Note, as you examine the table, that:

- The only bonds found in **alkanes** are carbon–carbon and carbon–hydrogen single bonds.

TABLE 3.2 Families of Organic Compounds

Family	Functional group	Example
alkane	C—C and C—H single bonds	$CH_3{-}CH_3$ **Ethane**
alkene	$>C{=}C<$	$CH_2{=}CH_2$ **Ethylene**
alkyne	$-C{\equiv}C-$	$CH{\equiv}CH$ **Acetylene**
aromatic	C_6 ring: $-C{=}C-C{=}C-C{=}C-$	C_6H_6 ring (each C bonded to H) **Benzene**
alcohol	$-\overset{\vert}{\underset{\vert}{C}}-O-H$	$CH_3CH_2{-}O{-}H$ **Ethyl alcohol**
ether	$-\overset{\vert}{\underset{\vert}{C}}-O-\overset{\vert}{\underset{\vert}{C}}-$	$CH_3{-}O{-}CH_3$ **Dimethyl ether**
aldehyde	$-\overset{O}{\overset{\Vert}{C}}-H$	$CH_3{-}\overset{O}{\overset{\Vert}{C}}{-}H$ **Acetaldehyde**
ketone	$-\overset{\vert}{\underset{\vert}{C}}-\overset{O}{\overset{\Vert}{C}}-\overset{\vert}{\underset{\vert}{C}}-$	$CH_3{-}\overset{O}{\overset{\Vert}{C}}{-}CH_3$ **Acetone**
carboxylic acid	$-\overset{O}{\overset{\Vert}{C}}-OH$	$CH_3{-}\overset{O}{\overset{\Vert}{C}}{-}OH$ **Acetic acid**
ester	$-\overset{O}{\overset{\Vert}{C}}-O-\overset{\vert}{\underset{\vert}{C}}-$	$CH_3{-}\overset{O}{\overset{\Vert}{C}}{-}O{-}CH_3$ **Methyl acetate**
amine	$-\overset{\vert}{\underset{\vert}{C}}-\overset{H}{\overset{\vert}{N}}-H$	$CH_3{-}\overset{H}{\overset{\vert}{N}}{-}H$ **Methyl amine**
amide	$-\overset{O}{\overset{\Vert}{C}}-\overset{H}{\overset{\vert}{N}}-H$	$CH_3{-}\overset{O}{\overset{\Vert}{C}}{-}\overset{H}{\overset{\vert}{N}}{-}H$ **Acetamide**

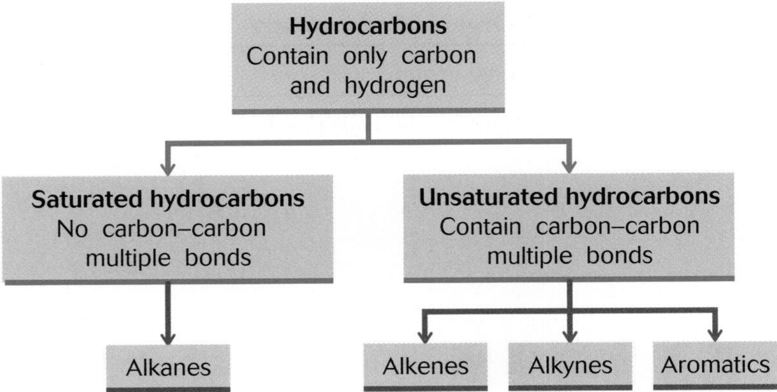

Figure 3.1 There are four families of hydrocarbons: alkanes, alkenes, alkynes, and aromatics.

- Alkenes, alkynes, and aromatics contain multiple bonds, two adjacent carbon atoms that share more than one bond between them. An **alkene** has a carbon–carbon double bond, two bonds between a pair of adjacent carbon atoms. An **alkyne** has a carbon–carbon triple bond, three bonds between a pair of adjacent carbon atoms. An **aromatic** has six carbon atoms in a cyclic arrangement with alternating single and double bonds.
- An **alcohol** contains a **hydroxyl group,** an —OH group attached to a carbon atom.
- An **ether** contains an oxygen attached directly to two different carbons.
- An **amine** contains an **amino group,** an $—NH_2$ group attached to a carbon atom. (In some amines, an NH or an N is attached to two or three carbon atoms, respectively.)
- Aldehydes, ketones, carboxylic acids, esters, and amides possess the **carbonyl group,** a carbon–oxygen double bond, but differ in the atom or group of atoms connected to the carbon of the carbonyl group. The carbonyl carbon of an **aldehyde** is directly connected to at least one hydrogen. The carbonyl carbon of a **ketone** is directly connected to carbon atoms, not hydrogen. The carbonyl carbon of carboxylic acids and esters is connected by a single bond to an oxygen, but these families differ in what is connected to that oxygen. In **carboxylic acids,** a hydrogen is connected to the oxygen; in **esters,** the oxygen is connected to a second carbon atom. An **amide** has a nitrogen connected to the carbonyl carbon.

Alkanes, alkenes, alkynes, and aromatics are also known as **hydrocarbons,** because they contain only carbon and hydrogen (Figure 3.1). Alkanes are **saturated hydrocarbons,** because they contain no carbon–carbon multiple bonds, only carbon–carbon single bonds. Alkenes, alkynes, and aromatics are **unsaturated hydrocarbons,** because they contain carbon–carbon multiple bonds.

Saturated and unsaturated hydrocarbons differ greatly in their ability to participate in chemical reactions. Saturated hydrocarbons undergo very few chemical reactions. Unsaturated hydrocarbons undergo many different chemical reactions.

Example 3.2 Identifying the family of a compound

Identify the family of each of the following compounds by referring to Table 3.2. Indicate how you arrived at the identification.

(a) $CH_3CH_2—O—CH_2CH_3$ (b) $CH_3CH_2CH_2—OH$

(c) $CH_3CH_2—\overset{\overset{\displaystyle O}{\|}}{C}—H$ (d) $CH_3CH_2—\overset{\overset{\displaystyle O}{\|}}{C}—OH$

Solution

(a) ether; (b) alcohol. Both an alcohol and an ether contain an oxygen atom, but its two single bonds are attached to different atoms. In an ether, oxygen is attached to two different carbon atoms. In an alcohol, one of oxygen's attachments is to a hydrogen.

(c) aldehyde; (d) carboxylic acid. Each compound has the carbonyl group, C=O, but the compounds differ in the atoms attached to the carbon of that group. In aldehydes, the carbonyl carbon is directly attached to an H; in carboxylic acids, the carbonyl carbon is attached to the O of an —OH group.

Problem 3.2 Identify the family of each of the following compounds by referring to Table 3.2:

(a) $CH_3CH_2-\overset{\overset{\displaystyle O}{\|}}{C}-CH_2CH_3$ (b) $CH_3CH_2-\overset{\overset{\displaystyle O}{\|}}{C}-O-CH_2CH_2CH_3$

(c) $CH_3-\overset{\overset{\displaystyle O}{\|}}{C}-NH_2$ (d) $CH_3CH_2-NH_2$

3.3 ALKANES

Alkanes, together with other hydrocarbons, are obtained from petroleum and natural gas. These materials are critical to our standard of living for two reasons: (1) they are used for cooking and heating and to generate power (Section 3.10); and (2) they are converted into other organic chemicals that are of commercial importance, such as plastics, textiles, rubbers, drugs, and detergents (see Box 3.1 on page 76).

Alkanes consist entirely of carbon–carbon and carbon–hydrogen single bonds. They have the general molecular formula C_nH_{2n+2}, where n, the number of carbon atoms, is an integer greater than 0. For example, for $n = 1$, $2n + 2 = 4$, and the molecular formula is CH_4; for $n = 2$, $2n + 2 = 6$, and the molecular formula is C_2H_6, and so forth. Each family of organic compounds can be described by a general molecular formula. In other words, all members of a given family contain the same ratios of the different kinds of atoms.

In all members of the alkane family, the C:H ratio is n:($2n + 2$). This ratio can be ascertained in Table 3.3, which lists the first ten unbranched alkanes, their molecular and structural formulas, names, and melting and boiling points.

TABLE 3.3 Formulas and Properties of Normal Alkanes

n	Molecular formula	Condensed structural formula	Name	Melting point (°C)	Boiling point (°C)
1	CH_4	CH_4	methane	−182	−162
2	C_2H_6	CH_3CH_3	ethane	−183	−89
3	C_3H_8	$CH_3CH_2CH_3$	propane	−190	−42
4	C_4H_{10}	$CH_3CH_2CH_2CH_3$	butane	−138	−1
5	C_5H_{12}	$CH_3CH_2CH_2CH_2CH_3$	pentane	−130	36
6	C_6H_{14}	$CH_3CH_2CH_2CH_2CH_2CH_3$	hexane	−95	69
7	C_7H_{16}	$CH_3CH_2CH_2CH_2CH_2CH_2CH_3$	heptane	−91	98
8	C_8H_{18}	$CH_3CH_2CH_2CH_2CH_2CH_2CH_2CH_3$	octane	−57	126
9	C_9H_{20}	$CH_3CH_2CH_2CH_2CH_2CH_2CH_2CH_2CH_3$	nonane	−51	151
10	$C_{10}H_{22}$	$CH_3CH_2CH_2CH_2CH_2CH_2CH_2CH_2CH_2CH_3$	decane	−30	174

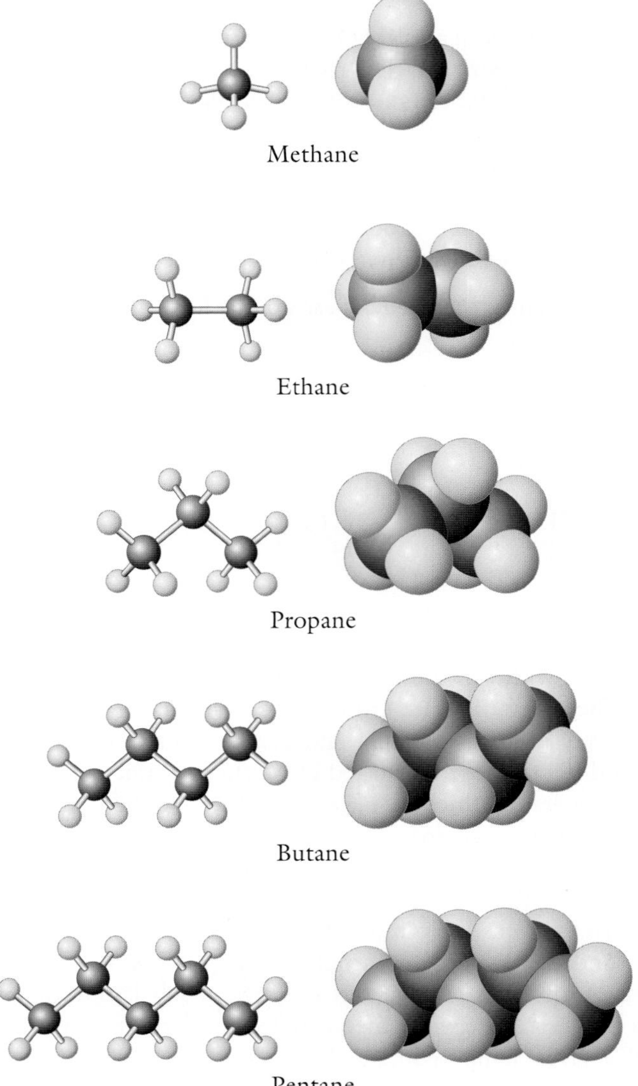

Figure 3.2 Ball-and-stick (*left*) and space-filling (*right*) models of alkanes.

The first four alkanes are gases, and the others are liquid at normal ambient temperatures (near 20°C). (The term unbranched will be defined in Section 3.5.)

Both ball-and-stick and space-filling models of the first five alkanes are shown in Figure 3.2 to illustrate the three-dimensional structures of alkanes. Space-filling models more correctly represent the dimensions of atoms relative to bonds than ball-and-stick models do—bonds are extremely short relative to atomic radii (Section 1.7). However, ball-and-stick models are more useful for the beginning student because they more clearly show the bonds between all the atoms in a molecule. We will generally use the ball-and-stick models.

Carbon Bonding in Alkanes

Carbon atoms in alkanes are distinguished by the following characteristics:

- Carbon is tetravalent: each carbon atom has four bonds.
- The four bonds of carbon possess **tetrahedral geometry:** each carbon atom is at the center of a tetrahedron with its bonds directed to the four corners of the tetrahedron (Figure 3.3).
- The four bonds are equivalent and have similar properties.

The angle between any two bonds of a tetrahedral (sp^3) carbon in any compound is 109.5°—the **tetrahedral bond angle** (see Figure 3.3). (A bond angle is the angle between two bonds that share an atom.) Valence-shell electron-pair repulsion (VSEPR) theory (Section 1.7) identifies the tetrahedral bond angle as the most stable bond angle for an atom forming four equivalent orbitals. In this arrangement, the electrons of the four orbitals are as far away from one another as possible and thus experience the least amount of repulsion.

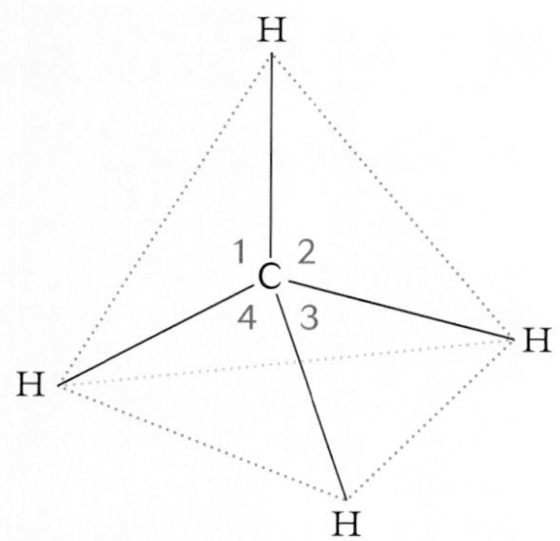

Figure 3.3 Tetrahedral sp^3-hybridized carbon. All angles (1 through 4) are 109.5°.

The equivalence of the four bonds of carbon is not at all what we would expect to find on the basis of the electron configuration of isolated (ground-state) carbon atoms. The ground-state carbon atoms possess four valence electrons distributed as follows: two paired electrons in the $2s$ orbital and one unpaired electron in each of two $2p$ orbitals. The third $2p$ orbital remains empty (left side of Figure 3.4). The ground-state electron configuration of carbon predicts that carbon is tetravalent but does not predict that the four bonds are equivalent.

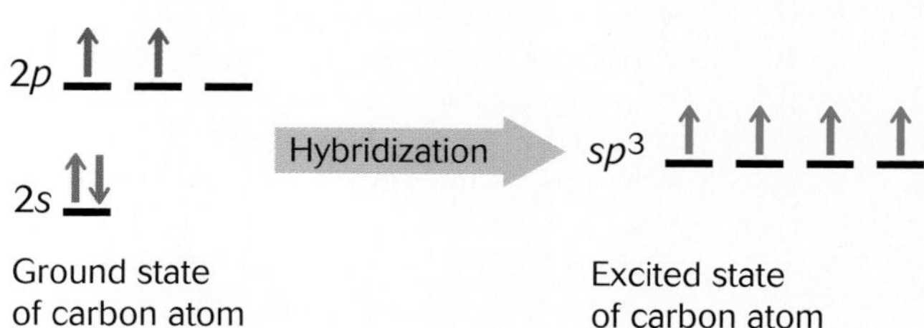

Figure 3.4 Formation of sp^3-hybridized carbon.

A modified theory of bonding, called **orbital hybridization theory,** explains discrepancies such as the properties of carbon atoms in alkanes. The orbital hybridization theory proposes that an excitation of ground-state carbon, during bond formation, creates a very different electron configuration for the element. The $2s$ and three $2p$ orbitals of carbon, together with its four electrons, mix ("hybridize") to produce four equivalent tetrahedral orbitals called ***sp*³ orbitals,** each containing one electron (right side of Figure 3.4). The sp^3 orbitals are teardrop shaped (Figure 3.5).

Carbon uses its four sp^3 orbitals when forming bonds to other atoms. Each sp^3 orbital forms a covalent bond by overlapping with an orbital from another atom. Figure 3.5 shows the formation of CH_4 from one carbon and four hydrogen atoms. Each sp^3 orbital of the carbon atom (containing one electron) overlaps with the $1s$ orbital of a hydrogen (also containing one electron). In alkanes containing more than one carbon, the carbon atoms bond to one another by the overlapping of their sp^3 orbitals. The single bonds formed with the use of sp^3 orbitals, like other single bonds, are called **sigma bonds** (σ bonds).

Conformations and Single-Bond Free Rotation

A single bond between any two atoms possesses **free rotation;** that is, the atoms in the bond are able to rotate freely relative to one another through 360°, and they do so continuously. There is little or no energy barrier to

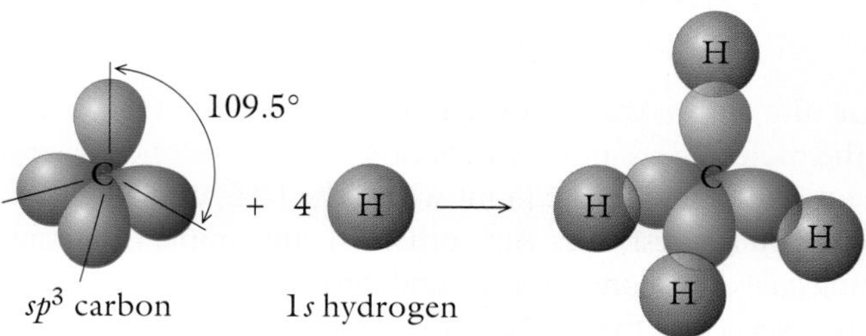

Figure 3.5 Formation of CH_4 from $1s$ orbitals of four hydrogens and sp^3 orbitals of one carbon.

3.1 CHEMISTRY AROUND US

Natural Gas and Petroleum

Petroleum (crude oil) and natural gas, the decay products of plant and animal remains, are dispersed together in porous rock formations below ground level throughout Earth.

Natural gas consists of alkanes of fewer than five carbons, mostly methane (typically 85%), with small amounts of ethane, propane, and butane. It is generally stored as a liquid in high-pressure tanks and separated into its various components for different uses. Methane, containing small amounts of ethane, is used to heat our homes (those that have a gas heating system) and for cooking. In highly industrial nations such as the United States, extensive underground gas pipelines distribute the gas over long distances (as far as from Texas to Maine). Propane, available in low-pressure cylinders, is the main source of heating and cooking gas in many sparsely populated agricultural and rural areas, where gas pipeline distribution networks are too expensive to install. Propane is often used as an economical substitute for gasoline and diesel fuel in tractor and other internal-combustion engines.

Petroleum—a thick, black liquid—contains hundreds of different hydrocarbons of five-carbon molecules and larger (mostly alkanes, branched and unbranched); but it also contains some cyclic compounds, alkenes, and aromatics. Petroleum is refined by fractional distillation, which separates the mixture into different products on the basis of differences in boiling points. The approximate compositions and boiling points of the different products are:

- Gasoline fuel — C_5–C_{10} — 30–200°C
- Kerosene and jet fuel — C_{10}–C_{18} — 180–275°C
- Diesel fuel and heating oil — C_{15}–C_{18} — 200–300°C
- Lubricating and mineral oils — C_{17}–C_{25} — 300–400°C
- Paraffin wax, asphalt — C_{20}–C_{40} — >400°C

The importance of the different products varies throughout the year. During the summer, the demand for gasoline is high, whereas more heating oil is consumed in the winter. Modern refineries can alter the relative amounts of the different fractions to meet demand by advanced chemical processes called **cracking** and **alkylation.** Cracking increases the amount of the lower-boiling-point fractions by thermally breaking larger molecules into smaller ones. Alkylation increases the amount of higher-boiling-point fractions by the reaction of smaller molecules to form larger ones. A full discussion of these reactions is beyond the scope of this book.

Petroleum is critical to life in modern, technological societies. Like natural gas, it is used to heat homes, schools, and factories, to power automobiles, trains, and aircraft, and to generate electricity. Various components of petroleum are used as lubricating oils and greases and as asphalt for roadways.

Just as important as the natural gas and petroleum used for energy are the approximately 5% and 20%, respectively, of the yearly yields of petroleum and natural gas that are converted into other organic chemicals. Although the percentage of the total amount of natural gas or petroleum being used is small, the volume of organic chemicals produced is enormous (trillions of pounds per year). Hydrocarbon compounds from natural gas and petroleum

rotation about single bonds. (Exceptions to this generalization will be discussed as the need arises.) This free rotation makes it possible for a molecule to have different **conformations**—that is, different orientations of the atoms of a molecule that result only from rotations about its single bonds.

Consider rotation about the C–C bond between the middle two carbons in butane (C_4H_{10}). This rotation occurs continuously, but we can visualize a butane molecule being frozen into different conformations at different points in time. Two conformations of butane are

CH_3 \ CH_2—CH_2 / CH_3 CH_3 \ CH_2—CH_2 \ CH_3

The first structure shows the conformation in which the CH_3 carbon atoms at the ends of the molecule are nearest each other. In the second structure, the middle C–C bond has rotated to the point where the CH_3 carbon atoms at the ends of the molecule are farthest from each other. In any sample of butane, the bonds of all the molecules are continuously undergoing rotation and changing their conformations. Never will they all have the same conformation at the same time.

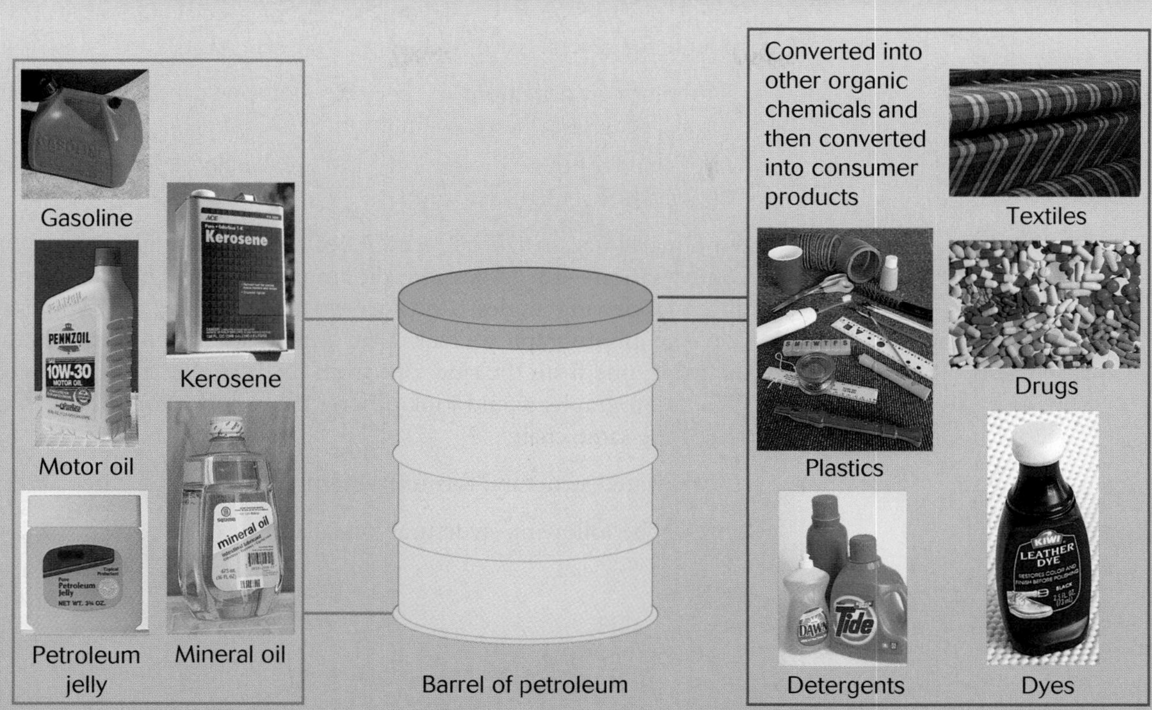

Utilization of a barrel of crude oil. (Gasoline, motor oil, kerosene, mineral oil, plastics, detergents, dyes, Tony Freeman/PhotoEdit; petroleum jelly, Young-Wolff/PhotoEdit; textiles, Frank Siteman/Picture Cube; drugs, J. C. Carton/Bruce Coleman.)

are converted by chemical reactions into a variety of organic chemicals—members of other organic families such as alcohols, aldehydes and ketones, carboxylic acids and esters, amines, and amides. These organic chemicals are in turn used to produce more than 90% of all plastics, textiles, rubbers, drugs, detergents, dyes, and a host of other products critical to the high standard of living that we enjoy today.

The situation for butane is even more complex. In a molecule such as butane, all the C–C and C–H single bonds are rotating simultaneously. The number of possible conformations for a molecule increases with the number of atoms. Ball-and-stick models of some of the conformations of butane are shown in Figure 3.6.

Conformation is not important for many of the small-sized molecules that we shall consider in Chapters 3 through 9. However, conformation is usually critical for large-sized biochemical molecules such as proteins, carbohydrates, and nucleic acids (Chapters 10, 12, and 13). One conformation is favored over others for a large-sized molecule, and this specific conformation determines the molecule's overall shape, which in turn is critical to its physiological function.

A physiological function is a function of a living organism or of an individual cell, tissue, or organ of which the organism is composed.

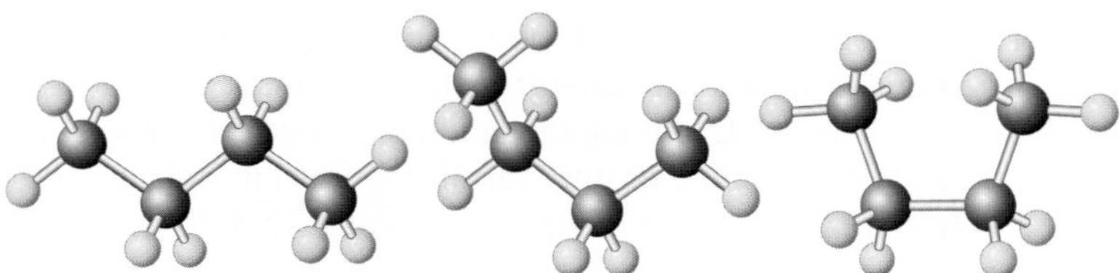

Figure 3.6 Ball-and-stick models of some conformations of butane.

3.4 TYPES OF STRUCTURAL FORMULAS

Writing and reading structural formulas present special difficulties for beginning students for several reasons:

- Different conformations of the same molecule can be easily mistaken for molecules of different compounds.
- Molecules are three dimensional but are usually represented by two-dimensional structural formulas.

Two-dimensional structural formulas of the same compound—even in the same conformation—can look quite different if drawn by different people or from different viewing angles. Consider what might happen if you asked several people to photograph a chair. One person might photograph the chair from the front, one from the side, one from the rear, one from the top, and so forth. The photographs would look different, but each would be a true representation of the same chair.

- Abbreviated structural formulas are often used to save time and space.

Consider the following structural formulas:

$$\begin{array}{ll} & \quad\;\; CH_3 \\ & \;\;/ \\ & CH_2 \\ & \;| \\ & CH_2 \\ & / \\ CH_3 & \end{array} \qquad\qquad \begin{array}{l} CH_3 \\ \quad \backslash \\ \quad CH_2{-}CH_2 \\ \qquad\qquad \backslash \\ \qquad\qquad CH_3 \end{array} \qquad\qquad \begin{array}{l} \qquad\;\; CH_2{-}CH_3 \\ \qquad\;\; | \\ CH_3{-}CH_2 \end{array} \qquad\qquad \begin{array}{l} CH_3 \\ \quad \backslash \\ \qquad CH_2 \\ \qquad | \\ \qquad CH_2 \\ \quad / \\ CH_3 \end{array}$$

$$\text{I} \qquad\qquad\qquad\qquad \text{II} \qquad\qquad\qquad\qquad \text{III} \qquad\qquad\qquad\qquad \text{IV}$$

$$CH_3{-}CH_2{-}CH_2{-}CH_3 \qquad \begin{array}{ccccccccc} & & H & & H & & H & & H & \\ & & | & & | & & | & & | & \\ H & - & C & - & C & - & C & - & C & - & H \\ & & | & & | & & | & & | & \\ & & H & & H & & H & & H & \end{array} \qquad CH_3CH_2CH_2CH_3$$

$$\text{V} \qquad\qquad\qquad\qquad\qquad \text{VI} \qquad\qquad\qquad\qquad\qquad \text{VII}$$

Each of these structural formulas represents the same compound, butane. Formulas I, II, and III represent the same conformation as well—the conformation in which the end carbons are farthest away from each other. I and II are exactly the same drawing except that II is rotated 90° counterclockwise relative to I. III represents the same conformation as I and II except that it does not depict the C–C–C tetrahedral (109.5°) bond angles as accurately. IV represents a conformation different from that of I, II, and III; it represents the conformation in which the end carbons are nearest each other. V represents butane without specifying anything about the conformation. The first four representations all translate into V when observed at an appropriate viewing angle.

The only bonds seen in formulas I through V are the carbon–carbon bonds; no bonds at all are seen in formula VII. These structures are varieties of abbreviated, or shorthand, structural formulas, differing in the extent to which bonds are understood to be present without being explicitly drawn. Formula VII is a **condensed structural formula.** Formula VI, in contrast, is a fully **expanded structural formula,** showing every carbon–hydrogen as well as every carbon–carbon bond in the molecule. Formulas I through V are partly expanded structural formulas, containing more detail than VII but less detail than VI. Formulas V and VII are shorthand versions of VI. When you have learned the combining powers of the various atoms in organic molecules (Table 3.1), you will not find it difficult to recognize that V, VI, and VII all represent the same molecule, butane.

Skeleton and **line structural formulas** also are used. Only the carbon atoms in a molecule are shown in a skeleton structural formula. In a line structural formula, carbon atoms are understood to be present at every intersection of two or more lines and wherever a line begins or ends.

C—C—C—C
Skeleton

Line

Example 3.3 Drawing structural formulas

Draw different structural formulas representing pentane, C_5H_{12}. Include at least two different conformations, one expanded structural formula, one condensed structural formula, one skeleton structural formula, and one line structural formula.

Solution

CH_3—CH_2—CH_2—CH_2—CH_3 $\quad$ CH_3—CH_2—CH_2—CH_2—CH_3

Different conformations

Expanded: H—C(H)(H)—C(H)(H)—C(H)(H)—C(H)(H)—C(H)(H)—H

Condensed: $CH_3CH_2CH_2CH_2CH_3$

Skeleton: C—C—C—C—C

Line

Problem 3.3 Draw different structural formulas representing hexane, C_6H_{14}. Include at least two different conformations, one expanded structural formula, one condensed structural formula, one skeleton structural formula, and one line structural formula.

3.5 CONSTITUTIONAL ISOMERS OF ALKANES

A compound called isobutane has the same molecular formula, C_4H_{10}, as that of butane. The two compounds have different structures as well as different properties.

CH_3—CH_2—CH_2—CH_3
Butane

CH_3—CH(CH_3)—CH_3
Isobutane

Butane and isobutane are examples of **isomers,** different compounds that have the same molecular formula. There are several different types of isomers. Butane and isobutane illustrate **constitutional (structural) isomers**—compounds that differ from each other in **connectivity**—that is, in the order in which the atoms are attached to one another. Other kinds of isomers will be introduced in Section 3.8 and in Chapters 4 and 9. Ball-and-stick models of butane and isobutane are shown in Figures 3.6 and 3.7, respectively.

›› Besides constitutional isomers, there are isomers called stereoisomers. Both types play key roles in the selectivity of biological processes, as described in Section 9.5.

Constitutional isomers, like all types of isomers, are different compounds with different chemical and physical properties. For example, butane has a boiling point of $-1°C$ and a melting point of $-138°C$, whereas the corresponding values for isobutane are $-12°C$ and $-159°C$. In later chapters when we consider the molecules of nature, such as carbohydrates, lipids, proteins, and nucleic acids, we will see that different isomers have different chemical and physiological behaviors. We often find that, although there are many possible isomers with the same molecular formula, only one isomer has physiological function. There is high selectivity in the molecules used in nature.

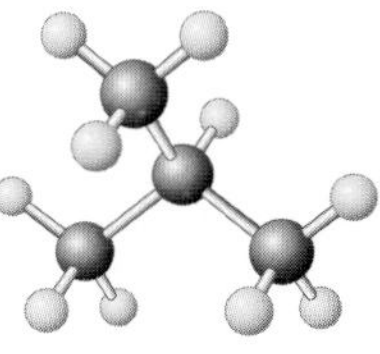

Figure 3.7 Ball-and-stick model of isobutane, the branched isomer of butane (see Figure 3.6).

When we look closely at the structures of constitutional isomers, we see that their basic difference lies in the number of carbon atoms connected in a successive manner (one after another). This number is called the **longest continuous carbon chain** (or, simply, the **longest chain**). Whereas four carbons are bonded in a successive manner in butane, the longest continuous chain in isobutane is three, not four, carbons long. In other words, three of isobutane's carbons are bonded one after another, and the fourth is a branch coming off the middle carbon of the longest continuous chain.

Butane is called an **unbranched** (or **linear** or **straight-chain) alkane,** and isobutane is called a **branched alkane.** In an unbranched alkane, all the carbon atoms are contained in the longest continuous chain. Such is not the case in a branched alkane, where the carbons not contained within the longest continuous chain are found instead on branches attached to the longest continuous chain.

The greater the total number of carbon atoms in an organic molecular formula, the higher the number of constitutional isomers represented by that formula. Thus, C_5H_{12} has more constitutional isomers than does C_4H_{10}, C_6H_{14} has more constitutional isomers than does C_5H_{12}, and so forth.

Example 3.4 Drawing constitutional isomers

Draw the different constitutional isomers of C_5H_{12}, taking care to draw only one structural formula for each isomer. It is incorrect to draw more than one structural formula for each isomer.

Solution

There are three constitutional isomers of C_5H_{12}:

$$CH_3{-}CH_2{-}CH_2{-}CH_2{-}CH_3$$

1

$$CH_3{-}\overset{\displaystyle CH_3}{\overset{|}{C}H}{-}CH_2{-}CH_3$$

2

$$CH_3{-}\underset{\underset{\displaystyle CH_3}{|}}{\overset{\overset{\displaystyle CH_3}{|}}{C}}{-}CH_3$$

3

The various isomers can be represented by first drawing the one that has all five carbons in the longest continuous chain (compound 1). Next, draw the branched isomers that have one carbon less in the longest continuous chain. Compound 2 has four carbons in the longest continuous chain, with the fifth carbon forming a branch on the next-to-last carbon from either end of the longest continuous chain. Compound 3 has three carbons in the longest continuous chain, with the fourth and fifth carbons forming separate branches on the middle carbon of the longest continuous chain. Figure 3.8 shows ball-and-stick models of the three isomers of C_5H_{12} (the conventions for naming these compounds will be considered in Section 3.6).

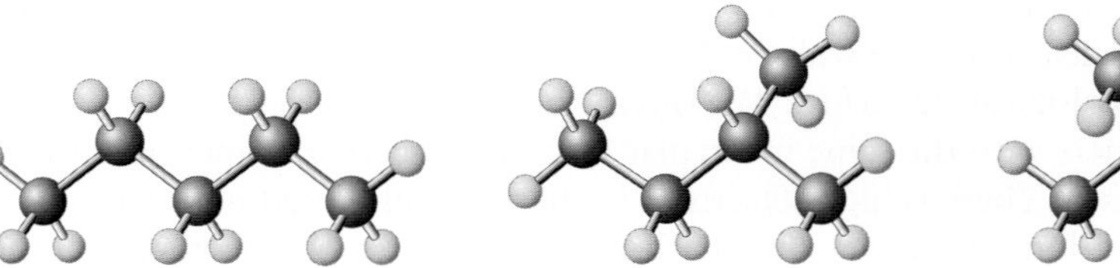

Pentane 2-Methylbutane 2,2-Dimethylpropane

Figure 3.8 Ball-and-stick models of isomers of C_5H_{12}.

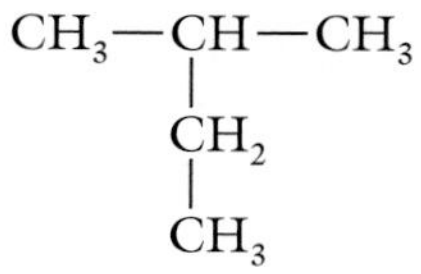

$$\begin{array}{c} CH_3-CH-CH_3 \\ | \\ CH_2 \\ | \\ CH_3 \end{array}$$
4

It is important in working out the answer to this problem not to incorrectly show two different structural formulas of the same isomer. For example, including structure 4 along with structures 1, 2, and 3 is a mistake because structures 2 and 4 represent the same compound. Either structure is a correct representation of the isomer having four carbons in a continuous chain, with the fifth carbon forming a branch on the next-to-last carbon of the longest continuous chain. If you include both structures, however, you convey the wrong idea; you imply that there are four constitutional isomers of C_5H_{12}. In fact, there are only three isomers, and your structural drawings must correctly express that fact.

As you solve problems such as that in this example, you must compare each new structure with the previously drawn structures to eliminate duplicates, looking for the longest continuous chain and analyzing the placement of branches along it. Use molecular models, if necessary.

Problem 3.4 Draw the different constitutional isomers of C_6H_{14}.

3.6 NAMING ALKANES

Compounds must be named by a generally accepted convention so that it is always clear what compound is being discussed.

Primary, Secondary, Tertiary, and Quaternary Classification

Chemists categorize a carbon atom in a compound on the basis of the number of other carbon atoms directly connected to it. This categorization is useful because it helps us to use the IUPAC nomenclature system and to understand the reactivity of a series of compounds. A carbon atom in a compound is categorized as a **primary** (1°), **secondary** (2°), **tertiary** (3°), or **quaternary** (4°) carbon, depending on whether it is directly bonded to a total of one, two, three, or four other carbon atoms, respectively.

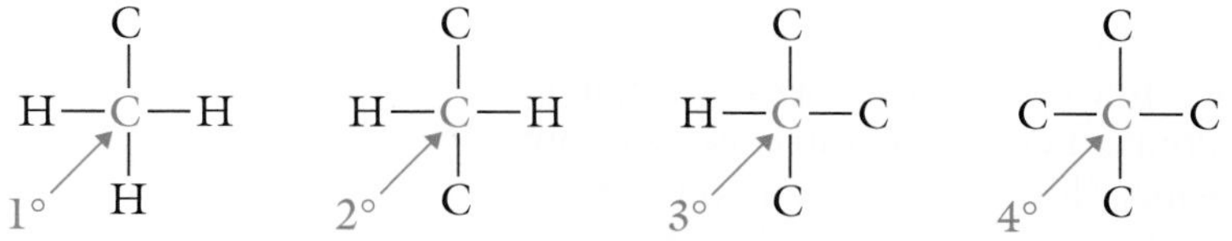

$$\begin{array}{c} C \\ | \\ H-C-H \\ | \\ H \end{array} \qquad \begin{array}{c} C \\ | \\ H-C-H \\ | \\ C \end{array} \qquad \begin{array}{c} C \\ | \\ H-C-C \\ | \\ C \end{array} \qquad \begin{array}{c} C \\ | \\ C-C-C \\ | \\ C \end{array}$$

1° 2° 3° 4°

Some functional groups, such as —OH, attached to a primary, secondary, or tertiary carbon also are categorized as primary, secondary, or tertiary, receiving the same categorization as the carbon to which it is attached. Reactivity of most compounds changes regularly, increasing or decreasing in the order primary, secondary, tertiary.

Note that there is no such thing as a quaternary functional group, because carbon is tetravalent. There is no way in which a functional group and four carbons can all bond at the same time to a single carbon atom.

Example 3.5 Identifying 1°, 2°, 3°, and 4° carbons

Indicate the 1°, 2°, 3°, and 4° carbons in the following compound:

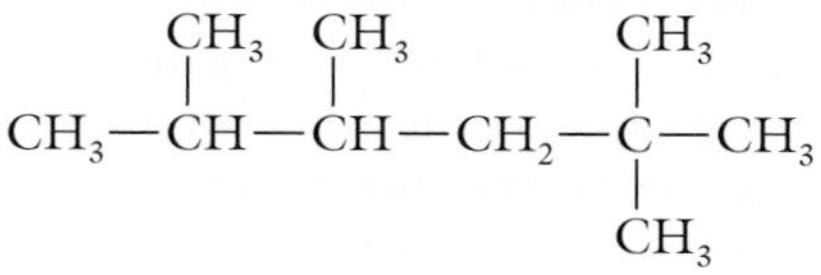

$$\begin{array}{c} \quad CH_3 \;\; CH_3 \qquad\quad CH_3 \\ \quad | \qquad\; | \qquad\qquad\; | \\ CH_3-CH-CH-CH_2-C-CH_3 \\ \qquad\qquad\qquad\qquad\; | \\ \qquad\qquad\qquad\qquad CH_3 \end{array}$$

Solution

Count the number of carbons to which each carbon atom is directly bonded.

```
           1°      1°              1°
          CH3     CH3             CH3
  1°       |       |       2°      | 4°  1°
 CH3 ─── CH ──── CH ──── CH2 ──── C ──── CH3
          3°      3°               |
                                1° CH3
```

Problem 3.5 Indicate the 1°, 2°, 3°, and 4° carbons in the following compound:

```
                        C(CH3)3
                          |
CH3 ─ CH2 ─ CH ─ CH ─ CH2 ─ CH2 ─ CH2 ─ CH3
            |
           CH3
```

Common Nomenclature System

Two nomenclature systems—the common nomenclature system and the IUPAC nomenclature system—are used to name alkanes. Both nomenclature systems name straight-chain (unbranched) alkanes by attaching the suffix **-ane** to a prefix denoting the total number of carbons in the compound (see Table 3.3). The alkanes from C_1 to C_{10} are named methane, ethane, propane, butane, pentane, hexane, heptane, octane, nonane, and decane. The suffix of each name is the ending of the family name—in this case, -ane for alkane.

Alkanes with more than three carbons have constitutional isomers. In the common nomenclature system, prefixes are added to the names of straight-chain alkanes to form the names of their branched isomers. For example, we noted in Section 3.5 that the branched C_4 alkane is distinguished from the straight-chain isomer by the prefix **iso-**. The straight-chain C_4 alkane is called butane, with no prefix, whereas the branched C_4 alkane is called isobutane. Some authors use the prefix *n*- or the word **normal** for straight-chain alkanes, as in *n*-butane or normal butane. We will not use these terms, because they are redundant. In this textbook, an alkane name with no prefix always means the straight-chain isomer.

The three C_5H_{12} isomers of Example 3.4 can be named by using an additional prefix, **neo-**. The three isomers are called pentane, isopentane, and neopentane. However, the common nomenclature system becomes useless beyond three isomers. The number of constitutional isomers increases rapidly with increasing numbers of carbons. There are 5 constitutional isomers of C_6H_{14} and 75 constitutional isomers of $C_{10}H_{22}$. To use the common nomenclature system for $C_{10}H_{22}$, we would need to devise and use 75 prefixes and memorize which prefix designates which isomer.

Organic chemists developed an alternate nomenclature system—the **IUPAC nomenclature system**—nearly a century ago. IUPAC stands for the International Union of Pure and Applied Chemistry, an international organization of chemists and other scientists that devised the new nomenclature system. Nomenclature, like other scientific activities, is ever evolving as new chemical structures are synthesized by chemists.

Alkyl Groups

The IUPAC nomenclature system requires the use of the names of **alkyl groups** that are commonly encountered in organic molecules. These groups are found attached to other groups or to some atom in a molecule. We need to learn the alkyl groups and their names before we can learn the IUPAC nomenclature system. We can visualize the derivation of various alkyl groups from alkanes. We will consider a limited number of alkyl groups in this book.

One-carbon and two-carbon alkyl groups, called **methyl** and **ethyl,** respectively, are derived from methane and ethane by loss of a hydrogen:

$$\underset{\text{Methane}}{CH_4} \xrightarrow{-H} \underset{\text{Methyl}}{CH_3—}$$

$$\underset{\text{Ethane}}{CH_3—CH_3} \xrightarrow{-H} \underset{\text{Ethyl}}{CH_3—CH_2—}$$

One bond from the alkyl group is incomplete, and that is the bond through which the alkyl group becomes attached to some other group or atom in a molecule. Alkyl groups do not have an independent existence but are a hypothetical means by which we understand the structure of compounds and name them.

Whereas only one alkyl group each is derived from methane and ethane, two different alkyl groups, **propyl** and **isopropyl,** are derived from propane. Removal of a hydrogen from one of the end carbons of propane yields the propyl group. The isopropyl group is obtained by removal of a hydrogen from propane's middle carbon:

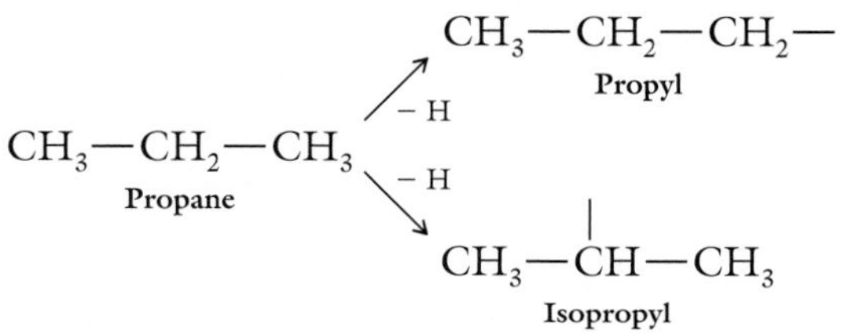

$$\underset{\text{Propane}}{CH_3—CH_2—CH_3} \xrightarrow{-H} \underset{\text{Propyl}}{CH_3—CH_2—CH_2—}$$

$$\underset{\text{Propane}}{CH_3—CH_2—CH_3} \xrightarrow{-H} \underset{\text{Isopropyl}}{CH_3—\overset{|}{C}H—CH_3}$$

The **butyl** and ***s*-butyl groups** are derived from butane. Removal of a hydrogen from one of the end carbons of butane yields the butyl group; removal of a hydrogen from one of the inner carbons yields the *s*-butyl group:

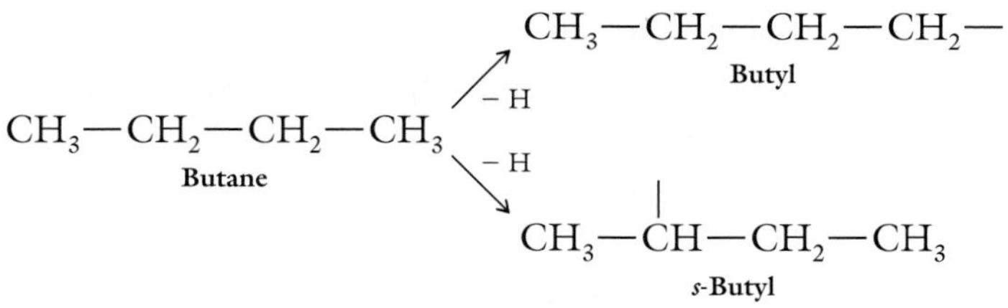

$$\underset{\text{Butane}}{CH_3—CH_2—CH_2—CH_3} \xrightarrow{-H} \underset{\text{Butyl}}{CH_3—CH_2—CH_2—CH_2—}$$

$$\underset{\text{Butane}}{CH_3—CH_2—CH_2—CH_3} \xrightarrow{-H} \underset{s\text{-Butyl}}{CH_3—\overset{|}{C}H—CH_2—CH_3}$$

The **isobutyl** and ***t*-butyl groups** are derived from isobutane. Removal of a hydrogen from one of the CH_3 carbons yields the isobutyl group; removal of a hydrogen from the CH carbon yields the *t*-butyl group:

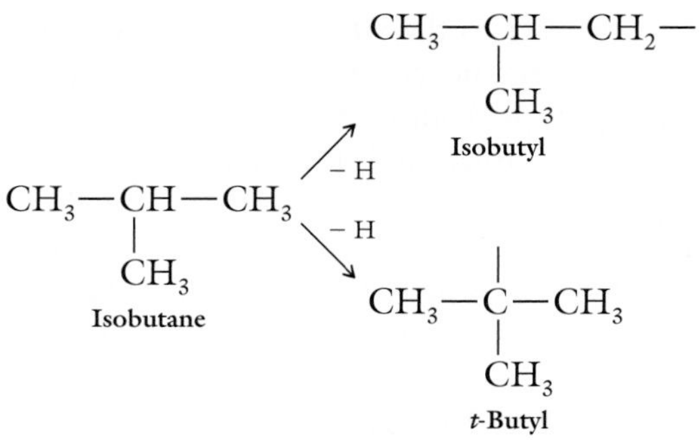

$$\underset{\text{Isobutane}}{CH_3—\underset{CH_3}{\underset{|}{C}H}—CH_3} \xrightarrow{-H} \underset{\text{Isobutyl}}{CH_3—\underset{CH_3}{\underset{|}{C}H}—CH_2—}$$

$$\underset{\text{Isobutane}}{CH_3—\underset{CH_3}{\underset{|}{C}H}—CH_3} \xrightarrow{-H} \underset{t\text{-Butyl}}{CH_3—\underset{CH_3}{\underset{|}{\overset{|}{C}}}—CH_3}$$

The prefixes *s-* or *sec-* and *t-* or *tert-* are abbreviations for secondary and tertiary, respectively. These prefixes are italicized when used. The prefix iso- is not italicized and is not abbreviated to *i-*. Recall that no prefix is used for the unbranched groups.

TABLE 3.4 Alkyl Groups

Name	Expanded structural formula*	Condensed structural formula
methyl	$CH_3—$	$CH_3—$
ethyl	$CH_3—CH_2—$	$CH_3CH_2—$ or $C_2H_5—$
propyl	$CH_3—CH_2—CH_2—$	$CH_3CH_2CH_2—$
isopropyl	$CH_3—\overset{\vert}{C}H—CH_3$	$(CH_3)_2CH—$
butyl	$CH_3—CH_2—CH_2—CH_2—$	$CH_3CH_2CH_2CH_2—$
s-butyl	$CH_3—\overset{\vert}{C}H—CH_2—CH_3$	$CH_3\overset{\vert}{C}HCH_2CH_3$
isobutyl	$CH_3—\underset{CH_3}{\underset{\vert}{C}H}—CH_2—$	$(CH_3)_2CHCH_2—$
t-butyl	$CH_3—\underset{CH_3}{\underset{\vert}{\overset{\vert}{C}}}—CH_3$	$(CH_3)_3C—$

*C–H bonds are not expanded.

You need to memorize these alkyl groups, including their condensed formulas, until you can recognize them easily (Table 3.4). Here are some hints to help you remember the different alkyl groups:

Guide to recognizing the alkyl groups

- An alkyl group with no prefix is unbranched and has its incomplete bond at the end of the chain of carbons.
- The *s*- and *t*-butyl groups have the incomplete bond at a secondary and tertiary carbon, respectively.
- The isopropyl and isobutyl groups each possess two methyl groups and fit the general formula

$$CH_3—\underset{CH_3}{\underset{\vert}{C}H}(CH_2)_n—$$

with $n = 0$ and 1 for isopropyl and isobutyl, respectively.

We will occasionally refer to the **simple alkyl groups** when describing the nomenclature rules for organic compounds. The simple alkyl groups are the unbranched alkyl groups from methyl, ethyl, propyl, and butyl through decyl and the branched three- and four-carbon alkyl groups (isopropyl, isobutyl, *s*-butyl, *t*-butyl).

In organic chemistry, molecules are often written in a very generalized form that uses the letter R to represent any alkyl group. For example, R–OH represents an —OH functional group attached to any alkyl group.

IUPAC Nomenclature System

There are only a few rules for naming alkanes by the IUPAC nomenclature system. The utility of the IUPAC rules is that, if they are followed correctly, they will always result in the same name for the same compound. Here are those rules and the order in which they must be used:

Rules for naming alkanes

1. The name of the longest continuous chain becomes the **base,** or **parent, name** of the compound. The suffix (ending) of the family name is added to the end of this base name. Note that, even if the subsequent rules are correctly applied, a failure to correctly identify the longest continuous chain will result in an incorrect name for the compound.

2. The base name accounts only for the carbons in the longest continuous chain. The carbons that are not part of the longest continuous chain—those attached as branches to the longest continuous chain and called **substituents** or **groups**—also must be included in the name. Substituents are included as follows:

- **a.** The name(s) of any alkyl group(s) in the compound is placed in front of the base name.
- **b.** Use the prefixes di-, tri-, tetra-, penta-, and hexa- before the name of the alkyl group when there are two, three, four, five, or six, respectively, of the same group.
- **c.** Alphabetize the names of alkyl groups when there are two or more different types of groups. Ignore all prefixes (both the branching prefixes *s*- and *t*- and the multiplying prefixes such as di-, tri-, and tetra-) in alphabetizing, with one exception: iso- is not ignored in alphabetizing.

3. Number the carbons in the longest continuous chain, starting from whichever end will result in the lowest number (or set of lowest numbers) for the alkyl group(s). An alternate rule is useful for most compounds: Number from the end nearest a branch.

4. In front of the name of each alkyl group, place the number of the carbon to which the group is attached.

5. Use hyphens to separate numbers from words; use commas to separate numbers.

In addition to naming organic compounds, the IUPAC nomenclature system helps us recognize whether two structural formulas represent the same compound or different compounds. As noted earlier, two different-looking structural formulas of the same compound will yield the same name if the nomenclature rules are correctly applied.

Example 3.6 Using the IUPAC nomenclature system to name alkanes: I

Give the IUPAC name for the following compound:

```
                          CH3
                          |
                          CH—CH3
                          |
CH3—CH2—CH2—CH—CH—CH3
                               |
                               CH2
                               |
                               CH3
```

Solution

As often occurs in the depiction of organic compounds, the longest continuous chain in this structural formula was not drawn in a horizontal line. Careful examination shows that the longest continuous chain in the compound contains seven carbons. Table 3.3 tells us that the base name for a seven-carbon chain is heptane. The longest continuous chain does not include the alkyl groups isopropyl (shown in blue in the structure on the following page) and methyl (shown in red). The names of these alkyl groups must be placed in front of the base name in alphabetical order, and each must be assigned a number to indicate its placement

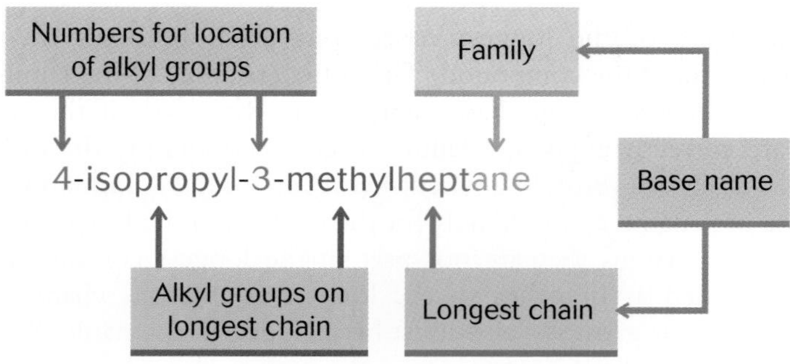

Figure 3.9 The IUPAC nomenclature system.

on the longest continuous chain. The numbering of the longest continuous chain begins with the terminal carbon nearest the methyl group.

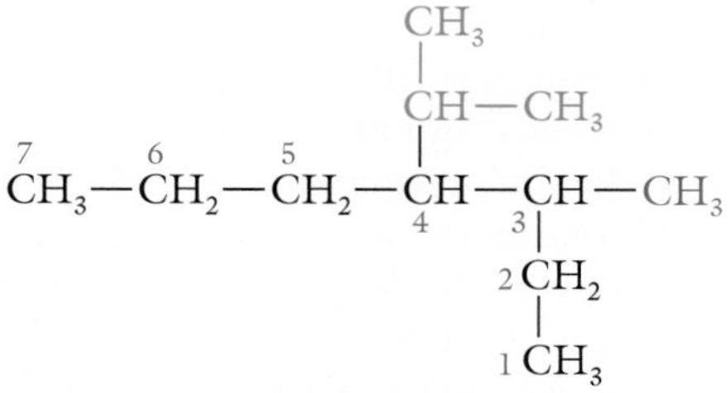

```
                               CH3
                               |
                               CH—CH3
 7        6        5           |
CH3—CH2—CH2—CH—CH—CH3
                               4      3|
                                     2CH2
                                       |
                                     1CH3
```

Numbering from the opposite end is incorrect, because that end is farther from an alkyl group (isopropyl). The IUPAC name of the compound is 4-isopropyl-3-methylheptane. Figure 3.9 shows the different segments of the IUPAC name.

Problem 3.6 Give the IUPAC name for each of the following compounds:

```
                                          CH3
                                          |
          CH3          CH2—CH2—CH—CH3
          |            |
(a) CH3—C—CH2—CH—CH2—CH3
          |
          CH—CH3
          |
          CH3
```

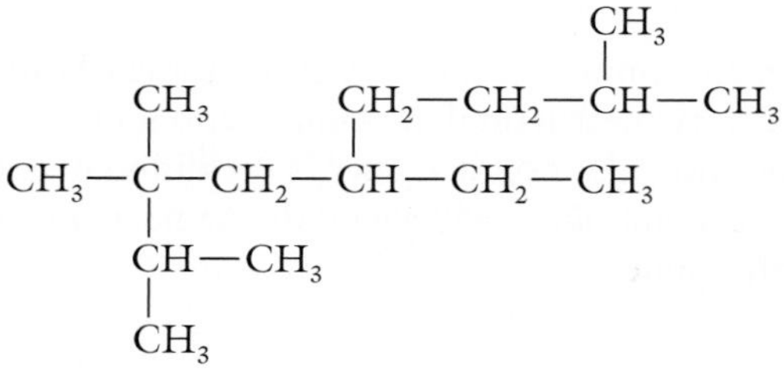

(b)

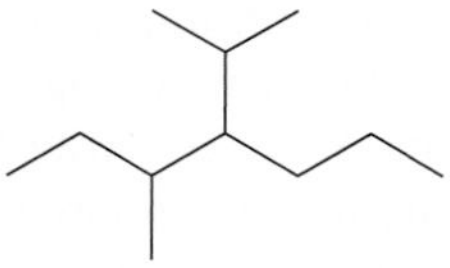

Example 3.7 Using the IUPAC nomenclature system to name alkanes: II

Give the IUPAC name for the following compound:

```
      CH3  CH3       CH3
      |    |         |
CH3—CH—CH—CH2—CH—CH3
```

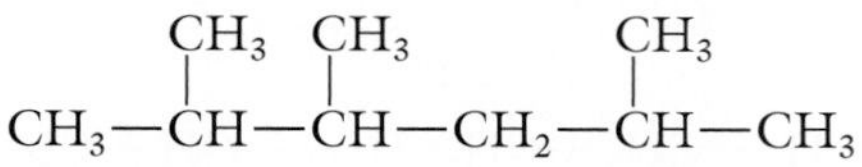

Solution

There are four longest continuous chains of the same length, six carbons, in this compound. These are the chains that begin with one of the carbon atoms marked with an asterisk in the following structure and continue toward one of the asterisk-marked carbons at the other end of the molecule:

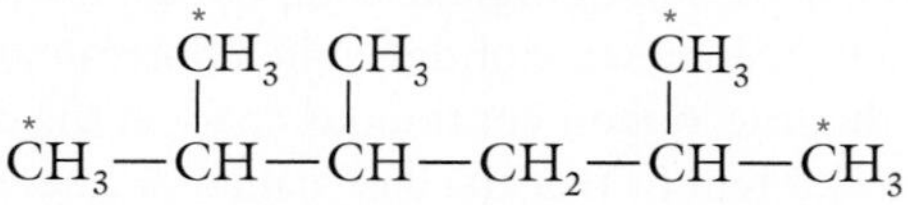

```
       *              *
      CH3  CH3       CH3
 *    |    |         |     *
CH3—CH—CH—CH2—CH—CH3
```

The correct name is the one that results in the set of lowest numbers for the alkyl groups. For many compounds, we can shorten this process because some or all parts of the molecule are equivalent. In this example, for instance, all four six-carbon longest continuous chains are equivalent: the two asterisk-marked carbons on the left side are equivalent because they are each part of a methyl group. Likewise, the two asterisk-marked carbons on the right side are equivalent because they, too, are each part of a methyl group. The compound's name can therefore be based on any one of those four longest continuous chains.

The correct name is the one with the set of lowest numbers. Numbering from left to right, the name is 2,3,5-trimethylhexane. Numbering from right to left, the name is 2,4,5-trimethylhexane. The numbers in the set 2,3,5 are lower than the numbers in the set 2,4,5. The IUPAC name, therefore, is 2,3,5-trimethylhexane.

Problem 3.7 Draw the structural formula of each of the following compounds: (a) 4-ethyl-3-methylhexane; (b) 4-*t*-butyl-3-methyloctane.

3.7 CYCLOALKANES

We have seen that carbon atoms can join together to form short or long chains. They can also form rings. In **cyclic,** or **ring,** molecules, there is a bond between the first and last carbons of a chain. Cyclic structures are often found in nature. Glucose, starch, cholesterol, and the sex hormones testosterone and progesterone are examples of biologically important ring compounds.

Alkanes are **acyclic** or **open-chain** molecules, with no bond connecting the first and last carbon atoms of the longest continuous chain. **Cycloalkanes,** the cyclic counterparts of the alkane family, possess the same characteristic functional groups as alkanes (C–H and C–C single bonds) but differ from alkanes in having a cyclic chain of carbon atoms. Starting at one of the carbon atoms in the cyclic chain and proceeding to successive carbons will return us to the starting carbon atom. Together, the alkanes and cycloalkanes constitute the saturated hydrocarbons, compounds that possess only C–C and C–H single bonds.

A cycloalkane is often represented by a condensed structural formula in the form of a geometrical shape—triangle, square, pentagon, hexagon, heptagon (Table 3.5 on the following page). The geometrical shapes are the cyclic equivalent of the line structural formulas introduced in Section 3.4. Recall that the intersection of two lines defines a carbon atom. Like other line structural formulas, the geometrical shapes representing cycloalkanes do not show any of the hydrogens attached to the carbons of the ring. One always assumes, when reading these structures, that each carbon is bonded to whatever number of hydrogens is necessary to complete the carbon atom's tetravalent requirement.

www.whfreeman.com/bleiodian2e

The general molecular formula describing cycloalkanes, C_nH_{2n}, has two fewer hydrogens than the general formula for alkanes, C_nH_{2n+2}. Any cyclic compound has two fewer hydrogen atoms than does the corresponding acyclic compound (the acyclic compound with the same number of carbon atoms). This correspondence is easy to understand if we imagine an alkane turning into a cycloalkane by losing a hydrogen from each of its two terminal carbons, which then form a bond with one another to close the ring.

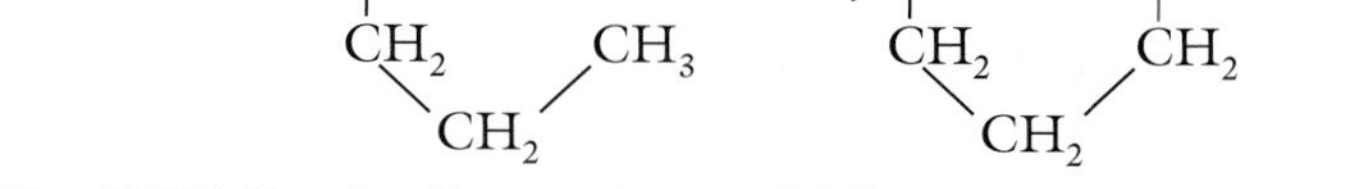

The IUPAC rules for naming cycloalkanes are straightforward variations on the rules for naming alkanes:

Rules for naming cycloalkanes

1. The base name of the ring structure consists of the prefix **cyclo-** followed, without hyphen or space, by the name of the straight-chain alkane that has the same number of carbons.

TABLE 3.5 Cycloalkanes

Name	Molecular formula	Structural formulas: Expanded*	Structural formulas: Condensed	Boiling point (°C)
cyclopropane	C_3H_6	CH_2 ring: CH_2 / CH_2-CH_2	△	−33
cyclobutane	C_4H_8	CH_2-CH_2 / CH_2-CH_2	□	12
cyclopentane	C_5H_{10}	ring of five CH_2	pentagon	49
cyclohexane	C_6H_{12}	ring of six CH_2	hexagon	81
cycloheptane	C_7H_{14}	ring of seven CH_2	heptagon	119

*C–H bonds are not expanded.

2. The names of substituents are placed in front of the base name. Numbers indicate the placement of substituents on the ring. The numbering of the ring carbons starts with a carbon holding a substituent and moves around the ring from there. When different sets of numbers can be obtained by starting at different carbons or counting in different directions, the correct name is the one with the set of lowest numbers.

Example 3.8 Using the IUPAC nomenclature system to name cycloalkanes

Give the IUPAC name for the following compound:

$(CH_3)_2CH$ — cyclopentane ring — CH_3; $CH_3CH_2CH_2CH_2$

Solution

The ring structure is cyclopentane. Adding the names of substituents in alphabetical order gives butylisopropylmethylcyclopentane. Numbering from the carbon holding the methyl group yields 1,3,4 as the set of numbers designating the positions of the substituents on the ring, irrespective of whether we proceed clockwise or counterclockwise. Starting at the carbon holding the isopropyl group yields 1,3,5 when we proceed clockwise and 1,2,4 counterclockwise. Starting at the butyl group yields 1,2,4 or 1,3,5, depending on the direction. The set of lowest numbers among these possibilities is 1,2,4, but there are two such sets, one starting at the carbon with the isopropyl group as C1 and the other starting at the carbon with the butyl group as C1. Which is the correct set? Alphabetical order wins; the butyl group has precedence over the isopropyl group, so the carbon with the butyl group is C1. 1-Butyl-2-isopropyl-4-methylcyclopentane is the correct IUPAC name.

Problem 3.8 Give the IUPAC name for each of the following compounds:

(a) $CH(CH_3)_2$, C_2H_5, $C(CH_3)_3$

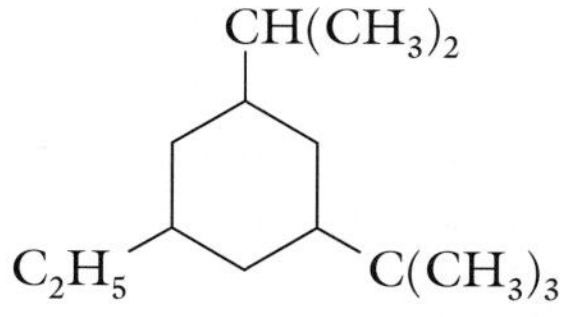

(b) CH_3, CH_3, $CH_3CHCH_2CH_3$

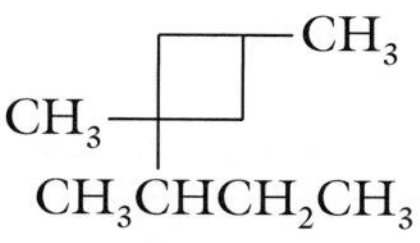

Cycloalkanes possess constitutional isomerism when the same atoms join in such a way as to form compounds with different-sized rings or when the placement of substituents around the ring is different or both. For example, there are five constitutional isomers of C_5H_{10} that include a ring:

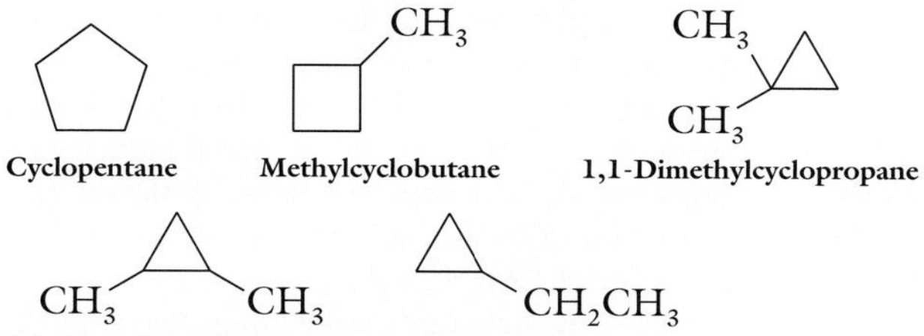

Cyclopentane — Methylcyclobutane (CH_3) — 1,1-Dimethylcyclopropane (CH_3, CH_3)

1,2-Dimethylcyclopropane (CH_3, CH_3) — Ethylcyclopropane (CH_2CH_3)

Different isomers of five-carbon cycloalkanes can have either three, four, or five of the carbons in the ring. The two isomers containing the cyclopropane ring can have their two additional carbons attached either at a single location or at two different locations on the ring.

Note that there is no need to use 1- in the names methylcyclobutane and ethylcyclopropane. The methyl and ethyl substituents are understood to be at C1. The designation 1- is needed only when there are two or more substituents on a ring structure.

Example 3.9 Drawing constitutional isomers of cycloalkanes

How many constitutional isomers are possible for C_4H_8? Give one structural formula for each isomer. Give the IUPAC name of each isomer.

Solution

Cyclobutane — Methylcyclopropane (CH_3)

Problem 3.9 How many constitutional isomers are possible for C_6H_{12}? Draw one structural formula for each isomer. Give the IUPAC name of each isomer.

Special aspects of the shape and stability of different-sized ring compounds are presented in Box 3.2 on the following page.

3.2 CHEMISTRY IN DEPTH

Stability and Shape of Cycloalkanes

Five- and six-membered ring structures are frequently found in nature in a variety of substances. Examples include starch, glucose, sex hormones, cholesterol, vitamins A and D, DNA, and some of the amino acids. Three- and four-membered ring structures, in contrast, are found very infrequently, because carbon rings of those sizes are much less stable than the five- and six-membered rings. The difference in stability results from the difference in bond angles for the different-sized rings.

Ring structures are drawn as if the rings were flat—that is, as if all the carbon atoms of the ring lay in a single plane. Although this portrayal is approximately correct for the three-, four-, and five-membered rings, it is incorrect for the six-membered ring. Cyclohexane's "ring" is a puckered, nonplanar shape called the **chair conformation:**

In the creation of three- and four-membered rings, the sp^3 orbitals of the participating carbon atoms must be forced out of their normal tetrahedral bond angles (109.5°) into angles of approximately 60° and 90°, respectively, to overlap one another and bond. This distortion of the normal bond angle is the reason for the instability of such compounds. Five-membered rings are stable because their formation occurs with only a very small distortion from the tetrahedral bond angle. Six-membered rings would be unstable if they were flat, because flat geometry would result in 120° bond angles, a large distortion from the tetrahedral bond angle. However, the chair conformation allows the six atoms of the ring to maintain their tetrahedral bond angles. As a result, six-membered rings are highly stable.

There are two types of bonds in the chair conformation of the cyclohexane ring—**axial** and **equatorial.** Axial bonds are those placed in the vertical direction, and equatorial bonds are those placed at a slight angle to the horizontal direction. For any substituted cyclohexane such as chlorocyclohexane, the substituent is larger than a hydrogen and is preferentially located in an equatorial position. Substituents are less stable in axial positions because their physical size puts them closer to hydrogens (or other substituents) in the other axial positions, and the result is destabilizing interferences. The term **steric hindrance** is sometimes used to describe this effect.

The physiological functions of molecules are strongly affected by molecular shape (Sections 10.6 and 12.5). One of the structural features that determines molecular shape is the existence of six-membered rings not as flat rings but as chair conformations. However, the exact conformation of six-membered rings is not critical at the level of this book. For the sake of simplicity, we will continue to use flat-ring structures rather than the chair conformation in our drawings. Any exceptions to this generalization will be noted as appropriate.

3.8 CIS-TRANS STEREOISOMERISM IN CYCLOALKANES

The carbon–carbon bonds of a ring, unlike those in acyclic structures, have **restricted rotation.** In other words, the carbon atoms of a ring are not free to rotate relative to each other. Certain cycloalkanes possess **geometrical,** or **cis-trans, stereoisomerism,** a type of isomerism different from constitutional isomerism, because of this restricted rotation. For example, there are actually six cycloalkanes of molecular formula C_5H_{10}, one more than the five constitutional isomers described in Section 3.7. More specifically, there are two different 1,2-dimethylcyclopropane compounds, not one:

cis-1,2-Dimethylcyclopropane *trans*-1,2-Dimethylcyclopropane

The two compounds are **geometric** (or **cis-trans**) **stereoisomers** and are distinguished one from the other by writing one of the italicized prefixes *cis-* and *trans-* before the IUPAC name.

Whereas constitutional isomerism results from differences in connectivity, **stereoisomerism** results from differences in configuration. Stereoisomerism is more subtle than constitutional isomerism. Cis and trans stereoisomers have the same substituents attached to the same carbon atoms of the ring—the same connectivity—but the compounds differ in **configuration**—that is, in the spatial orientation in which the substituents extend from the ring carbons into the regions lying on opposite sides of the plane of the ring. The cis isomer is the isomer in which two similar substituents (either hydrogens or methyls in 1,2-dimethylcyclopropane) on two different ring carbons are on the same side of the ring. The trans isomer has the two similar substituents on opposite sides of the ring. The difference between the cis and trans isomers can also be described by the proximity between substituents. Cis substituents are closer to each other than are trans substituents. (Cis-trans stereoisomers are one type of stereoisomers, often called **diastereomers** to distinguish them from a second type, called **enantiomers,** which will be considered in Chapter 9.)

Our convention for drawing and reading the two 1,2-dimethylcyclopropanes (and similar compounds) is that the ring is perpendicular to the plane of the paper with the H and CH_3 groups parallel to the plane of the paper. The front bonds of a ring are often drawn as thick lines to aid in visualizing this convention, especially when compounds are geometrical stereoisomers.

Cis and trans isomers are different compounds with different chemical and physical properties. For example, the boiling points of the cis and trans isomers of 1,2-dimethylcyclopropane are 37°C and 29°C, respectively. The physiological behaviors of cis and trans isomers are often different, another instance of the high selectivity of the molecules used in nature.

A cis isomer cannot easily convert into a trans isomer, and vice versa, because the carbon–carbon bonds of a cyclic structure, unlike those of an acyclic structure, are not free to rotate. You can prove this statement by using molecular models to build a cyclic compound and then trying to rotate the carbon–carbon bonds of the ring.

Cis and trans isomers occur only when each of two ring carbons has two different substituents. Such ring carbons are stereocenters. A **stereocenter** is an atom in a compound bearing substituents of such identity that a hypothetical exchange of the positions of any two substituents would convert one stereoisomer into another stereoisomer. In the structural drawings of the two 1,2-dimethylcyclopropanes, the difference between the cis and trans isomers is reversal of the configurations (the positions of H and CH_3) at the stereocenter on the right side. The ring carbon stereocenters are sometimes more precisely called **tetrahedral stereocenters** because the carbons have tetrahedral bond angles.

None of the following types of cycloalkanes possess geometrical isomerism:

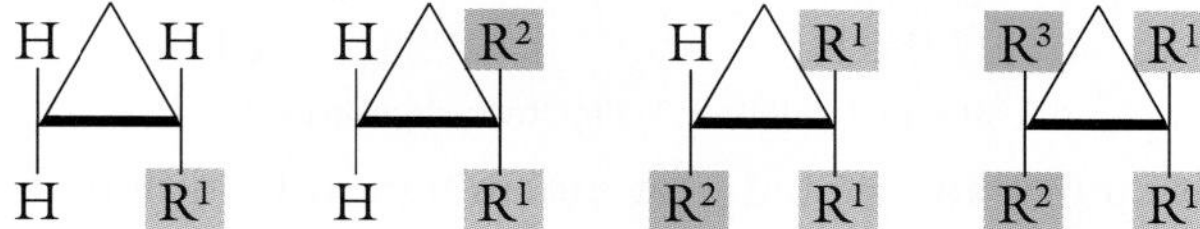

✓ Geometrical isomers are possible only when two of the ring atoms are stereocenters—that is, two ring atoms each bear two different substitutents.

Concept check

Example 3.10 Identifying and drawing cis and trans isomers

Which of the following compounds exist as cis and trans isomers? If cis and trans isomers are possible, show one structural drawing for each isomer. (a) Methylcyclopentane; (b) 1,3-dimethylcyclobutane; (c) 1,1,2,2-tetramethylcyclohexane.

Solution

(a, c) Two stereocenters are required for cis-trans isomerism. Neither methylcyclopentane nor 1,1,2,2-tetramethylcyclohexane exists as a pair of cis-trans isomers because neither has two stereocenters.

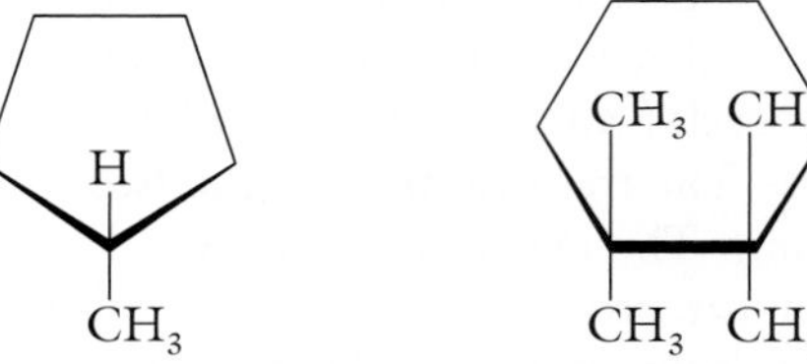

Methylcyclopentane **1,1,2,2,-Tetramethylcyclohexane**

Methylcyclopentane has only one ring carbon, C1, which has two different substituents (H and CH_3) attached to it. Each of the other ring carbon atoms has two of the same substituents (H). 1,1,2,2-Tetramethylcyclohexane has no stereocenters. Each of two of the ring carbons has two CH_3 groups, whereas each of the other four ring carbons has two H atoms.

(b) 1,2-Dimethylcyclobutane has two stereocenters and thus exists as two different compounds. Ring carbons C1 and C3 each have two different substituents (one H and one CH_3).

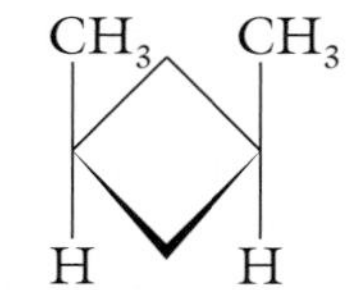

***cis*-1,3-Dimethylcyclobutane**

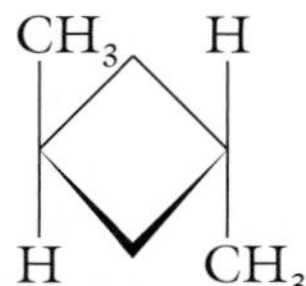

***trans*-1,3-Dimethylcyclobutane**

Problem 3.10 Which of the following compounds exist as cis and trans isomers? If cis and trans isomers are possible, show one structural drawing for each isomer. (a) 1,3-Dimethylcyclopentane; (b) 1-ethyl-1,2,2-trimethylcyclopropane; (c) 1,4-dimethylcyclohexane.

The use of the cis-trans nomenclature system is restricted to cycloalkanes in which each of two carbons of the ring has one hydrogen and one substituent other than hydrogen. Designation as a cis or trans isomer is ambiguous for cycloalkanes in which either one or two carbons of the ring have two different substituents but no hydrogen, such as 1-bromo-2-chloro-1,2-dimethylcyclopropane:

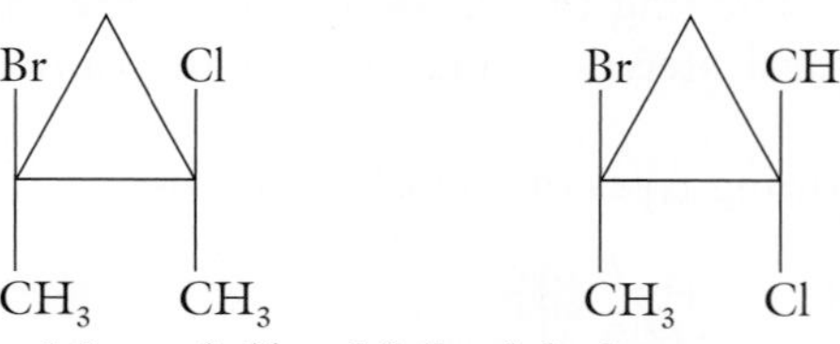

1-Bromo-2-chloro-1,2-dimethylcyclopropane

The terms cis and trans would not unambiguously differentiate the two stereoisomers. There is no convention in the IUPAC nomenclature system that specifies whether the terms cis and trans refer to the relation of the two methyl groups or the methyl and Br or the methyl and Cl. A different nomenclature system, the *E,Z* nomenclature system, is used to differentiate these stereoisomers but is beyond the scope of this book.

3.9 PHYSICAL PROPERTIES OF ALKANES AND CYCLOALKANES

The physical properties of organic compounds—their melting and boiling points, physical strength, and solubility—are just as critical to their use, both in biological systems and in our everyday life, as are their chemical properties. A lawn chair must be constructed from a solid substance, not a liquid, and the chair must be of sufficient strength to bear the weight of anyone who might sit in the chair. In the body, the melting points of molecules used in constructing bones must be above body temperature, and bones must be of sufficient strength to withstand a wide range of physical stresses.

Solubility in water is another important physical property because our bodies are aqueous systems. The physiological functioning of bones, skin, blood vessels, the heart, the liver, the lungs, and other tissues and organs requires them to be water insoluble. Conversely, the functioning of digestive enzymes, hormones, hemoglobin, and neurotransmitters depends on their solubility in aqueous systems so that they can be transported throughout the body in the blood and other bodily fluids.

In this chapter and subsequent ones, we will correlate the physical properties of organic compounds with the strength of their secondary forces. We shall see that the level of the secondary forces varies considerably from one family to another, and this variation results in significant differences in physical properties. Substances become useful as materials of construction, both in the body and outside, when their molecular masses and secondary forces are sufficiently high that they develop the required physical strength.

Melting and Boiling Points

Melting is the transition from the solid to the liquid state. Boiling causes the transition from the liquid to the gas state. Whether a substance has a high or low melting or boiling point depends on the strength of the secondary forces holding its constituent particles together. The greater the secondary forces, the higher the temperature required to overcome them and bring about the melting or boiling transition.

« Secondary forces are the attractive forces between molecules, introduced in Section 1.9.

In organic compounds, the strength of secondary forces is determined by three molecular features—family (and hence the type of secondary force), molecular mass, and molecular shape.

Family The type of secondary forces present in an organic compound is related to the compound's family because the family determines what chemical bonds are present and thus whether the bonds are polar or nonpolar and whether there is hydrogen bonding. The order of secondary forces, highest to lowest, is

$$\text{Hydrogen bonding} > \text{dipole–dipole} > \text{London}$$

Oil fires are difficult to fight with water because oil has a lower density than water. Water sinks below the oil and has limited effectiveness at shutting off the flames' access to oxygen. (Najlah Feanny/Stock Boston.)

All molecules possess London forces. This secondary force is the only one present in nonpolar molecules. Polar molecules have dipole–dipole forces as well as London forces; so they possess greater secondary forces than nonpolar molecules do. Hydrogen-bonding forces, the strongest of all secondary forces, are present in organic molecules only when O–H or N–H bonds are present.

« Electronegativity values are listed in Figure 1.9 and discussed in Section 1.9.

Alkanes and cycloalkanes contain only nonpolar bonds (C–C and C–H). Thus, they are nonpolar compounds with only London forces acting between the molecules. (There is a very slight electronegativity difference between carbon and hydrogen, but it is negligible from a practical viewpoint.) Therefore, alkanes and cycloalkanes have the lowest secondary forces of all organic families and the lowest melting and boiling points.

Molecular mass For any series of compounds within the same organic family, secondary forces increase with increasing molecular mass. London forces arise from uneven distributions of electrons within molecules. Larger molecules exert larger London forces because they possess larger numbers of electrons (Section 1.9). In accordance with this generalization, boiling and melting points increase with increasing molecular mass for all families of organic compounds. Tables 3.3 and 3.5 show this trend for straight-chain alkanes and cycloalkanes, respectively.

$$CH_3-\underset{\underset{CH_3}{|}}{CH}-CH_3$$

2-Methylpropane

−12°C

$$CH_3-CH_2-CH_2-CH_3$$

Butane

−1°C

□

Cyclobutane

12°C

Molecular shape For compounds within the same organic family and having the same or close to the same molecular mass, the order of boiling points, highest to lowest, is

Cycloalkane > straight-chain alkane > branched alkane

This trend is seen in the boiling points of the C_4 compounds (see margin).

Molecular shape affects the strength of secondary forces because molecular shape affects the surface area over which molecules attract each other, and this surface area determines the effectiveness of attractions between molecules. Some simple geometrical analogies to straight-chain, branched, and cyclic molecules are useful for understanding the boiling-point differences between the differently shaped molecules (Figure 3.10). Straight-chain molecules can be visualized as flexible (jointed) cylinders, something like a string of connected sausage links. Parts of a cylinder can move with respect to one another because of the free rotation about the C–C single bonds. Branched molecules are compact in shape and can be thought of as spheres. Pairs of adjacent cylindrical molecules contact one another over a much larger surface area than pairs of adjacent spherical molecules do (a line compared with a point). Contact over a larger surface area results in greater secondary forces and higher boiling points for the straight-chain alkane than for the branched alkane.

The higher boiling point of the cycloalkane relative to the straight-chain alkane is due to the cycloalkane's rigid structure. This structure can be imagined as a flat, square doughnut. The rigidity of cyclic structures allows the molecules to pack closely together, one on top of the other, achieving contact over a comparatively large surface area. The flexibility of the straight-chain compounds' jointed cylinders prevents those molecules from packing together in a similarly efficient way.

The effect of molecular shape on melting points differs somewhat from its effect on boiling points. Cycloalkanes do have higher melting points than the corresponding straight-chain alkanes, as one might expect from their shape. However, not all branched alkanes have lower melting points than the corresponding straight-chain alkanes. The reasons for this behavior are complex and beyond the scope of this book.

When compounds differ in more than one structural feature and the differences have opposing effects on secondary forces, we cannot reliably predict

INSIGHT INTO PROPERTIES

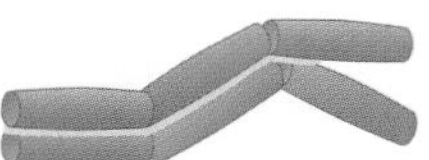

Straight-chain molecules

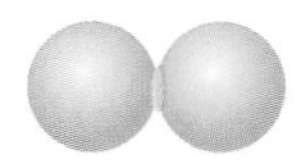

Branched molecules

Cyclic molecules

Figure 3.10 Straight-chain, branched, and cyclic molecules differ in their ability to pack closely together.

the order of increasing secondary forces. Consider compound A, a polar compound of molecular mass 60, and compound B, a nonpolar compound of molecular mass 128. Which has the higher boiling point? The molecular-mass factor favors compound B, but the family factor favors compound A. We are unable to make a reliable prediction.

Concept checklist

✓ Family, molecular mass, and molecular shape determine the strength of the secondary forces.

✓ The order of increasing secondary forces is difficult to predict for a set of compounds that differ in more than one structural feature when those differences have opposing effects on the secondary forces.

Example 3.11 **Predicting relative boiling points of various compounds**

Which compound has the higher boiling point in each of the following pairs? Explain each of your answers in regard to the difference in secondary forces in the two compounds. (a) Heptane and hexane; (b) nonane and 3-methyloctane; (c) cyclopentane and pentane.

Solution

(a) Heptane has the higher boiling point because of its larger molecular mass. Larger molecular mass results in greater secondary forces because of the greater number of electrons.

(b) Nonane has the higher boiling point. Branched molecules attract one another over a smaller surface area and have lower secondary forces relative to unbranched molecules of the same molecular mass.

(c) Cyclopentane has the higher boiling point. Its rigid structure results in tighter packing and greater secondary forces among molecules.

Problem 3.11 Which compound has the higher boiling point in each of the following pairs? Explain each of your answers in regard to the difference in secondary forces in the two compounds. (a) Cyclopentane and 2-methylbutane; (b) nonane and 3-methylnonane; (c) cyclobutane and cycloheptane.

Density

Greater secondary forces result in greater densities. Density increases with an increase in the number of molecules packed into a unit volume. In compounds with large secondary forces, the molecules attract one another strongly and pack closely together, and thus the density is quite high. Alkanes and cycloalkanes have lower densities than water because water possesses the stronger secondary forces—hydrogen bonding versus London forces. The density of water is about 1.0 g/mL. Liquid alkanes and cycloalkanes have densities less than 0.7 to 0.8 g/mL, depending on molecular mass. Some solid alkanes and cycloalkanes have higher densities, but still less than the density of water.

Solubility

A solute is soluble in a solvent only if the secondary attractive forces between molecules of solute are similar to the secondary attractive forces between molecules of solvent. When this requirement is met, the solute dissolves in solvent to form a solution because the solute molecules form sufficiently strong secondary attractive forces with solvent molecules (Section 1.9). As mentioned in Section 1.9, the requirement for solubility is often summarized by the phrase "like dissolves like."

Concept checklist

- ✓ Nonpolar compounds are soluble in nonpolar compounds.
- ✓ Polar or hydrogen-bonding compounds are soluble in polar or hydrogen-bonding compounds.
- ✓ Nonpolar compounds do not dissolve in polar or hydrogen-bonding compounds.

Example 3.12 Predicting the solubility of compounds

Explain why pentane is not soluble in water.

Solution

Pentane is a nonpolar compound and possesses only weak London forces. Water, in contrast, is highly polar, with very strong hydrogen-bonding forces. The energy needed to separate water molecules from one another is large. None of this energy can be retrieved by mixing, because water and pentane are so different in their types of secondary forces. The attractive secondary forces possible between the hydrogen-bonding water molecule and the nonpolar alkane are negligible.

Problem 3.12 Does sodium chloride (NaCl) dissolve in hexane? Why?

The physical properties considered so far are important both in the uses and in the health hazards associated with alkanes (Box 3.3).

3.3 CHEMISTRY AROUND US

Health Hazards and Medicinal Uses of Alkanes

Gaseous alkanes (C_1 to C_4) and the vapors from the volatile liquid alkanes (C_5 to C_7) can be harmful because alkanes are highly flammable and have an anesthetic effect. They can also kill by asphyxiation because their presence in the air reduces the concentration of oxygen.

Liquid alkanes also can be harmful if they are swallowed and then coughed or vomited up into the lungs. The alkanes, being nonpolar, dissolve the lipid molecules (also nonpolar) that make up the cell membranes of the air sacks in the lungs, and a condition called **chemical pneumonia** ensues because the lungs become less efficient at expelling gases and fluids. The term chemical pneumonia refers to a buildup of fluids in the lungs that is similar to that of bacterial or viral pneumonia. Hence, we are warned never to induce vomiting in someone who has swallowed gasoline or other liquid alkanes.

Another prudent warning in the handling of alkanes is to wear gloves when using liquid alkanes such as gasoline, paint thinner, and turpentine. These low-molecular-mass compounds dissolve the natural body oils that protect our skin.

There are cases in which the similarities of alkanes to human body oils allow them to be put to beneficial use. The high-molecular-mass alkanes (C_{18} to C_{25}) that make up mineral oil, for example, are quite similar to our natural body oils, enabling mineral oil to be used in hospitals as a laxative. It softens the stool because it passes through the digestive tract without being absorbed into the tissues.

Spray painting an automobile. The worker uses a respirator to prevent inhalation of hydrocarbon solvents and other chemicals present in the paint. (Michael Rosenfeld/Tony Stone.)

Prolonged use is not recommended, however, because the presence of mineral oil in the bowel prevents the absorption of the fat-soluble vitamins A, D, E, and K from foods during the digestive process. Hence, prolonged use of mineral oil depletes the body of these vitamins.

Petroleum jelly (petrolatum), a jellylike semisolid composed of C_{22} to C_{30} alkanes, is useful as both a lubricant and a solvent. It serves as the solvent in a variety of baby lotions, hand lotions, medicated salves, and cosmetics. Mineral oil is used for this purpose as well.

3.10 CHEMICAL PROPERTIES OF ALKANES AND CYCLOALKANES

As we have seen, alkanes and cycloalkanes together constitute the large organic category of saturated hydrocarbons, compounds possessing only C–C and C–H single bonds. Because of the stability of those bonds, the saturated compounds show low chemical reactivity relative to other families of organic compounds. They are inert to a wide range of reagents, including strong acids (for example, HCl or H_2SO_4), strong bases (for example, NaOH or KOH), reducing agents (such as H_2), and most oxidizing agents (for example, $KMnO_4$ or $K_2Cr_2O_7$). Alkanes and cycloalkanes undergo very few reactions. Each of the other families, in contrast, contains a functional group—specifically, some bond or set of bonds that are relatively weak—that gives rise to higher and more varied reactivity. For example, unsaturated hydrocarbons (alkenes, alkynes, and aromatics) are highly reactive because of their carbon–carbon multiple bonds, and they undergo reactions with many of the reagents to which the alkanes and cycloalkanes are unreactive.

In spite of the limited scope of their reactions, alkanes and cycloalkanes are extremely important chemicals: they are the ultimate reason for our high standard of living. Petroleum and natural gas are the sources of these hydrocarbons and others (see Box 3.1). Hydrocarbons from petroleum and natural gas are important not only for heating and energy generation by combustion, but also for the manufacture of other chemicals that are in turn converted into the myriad synthetic materials—medicines, plastics, rubbers, and textiles—that enhance our lives.

Combustion

All hydrocarbons, indeed, all organic compounds, undergo **combustion** (burning in air; Section 2.1). This reaction distinguishes organic compounds from most inorganic compounds, which do not undergo combustion.

Combustion requires oxygen and results in the breakage of all C–H and C–C bonds in the hydrocarbon. The reaction is exothermic (heat given off) and converts the carbon and hydrogen atoms of the hydrocarbon into carbon dioxide and water, respectively. The balanced equation for the combustion of propane is

$$\underset{\text{Propane}}{CH_3{-}CH_2{-}CH_3} + 5\ O_2 \longrightarrow 3\ CO_2 + 4\ H_2O + \text{heat}$$

Combustion requires ignition of the mixture of hydrocarbon and oxygen with a spark or flame, but it is self-sustaining once reaction starts because it is a very highly exothermic reaction.

Combustion is an example of an **oxidation reaction.** The oxidation of an organic compound is generally defined as an increase in the oxygen content or a decrease in the hydrogen content of a compound or both. More specifically, we describe the oxidation of an organic compound by an increase in the oxidation state of carbon. The greater the **oxidation state** of a carbon atom in a compound, the more bonds there are from that carbon to oxygen atoms or the fewer bonds there are from that carbon to hydrogen atoms or both (Section 2.1). Before combustion, each carbon in propane is in its lowest oxidation state—there are no oxygens bonded to the carbons. Only hydrogens and carbons are bonded to the carbons of propane. After combustion, no hydrogens or carbons are bonded to the carbons. Each carbon ends up in CO_2, where all four of its bonds are to oxygen—the highest oxidation state for a carbon atom.

The combustion of hydrocarbons is of enormous importance to human life. We depend on combustion for energy in the form of heat and for mechanical and electrical power. Heating our homes and cooking by combusting hydrocarbons is efficient because the reaction is highly exothermic, producing about 15,000 kJ per pound of fuel (1 cal = 4.184 J). In engines, the same large heat output results in a pressure buildup that is then converted into mechanical energy. At power plants, the heat of combustion is used to produce steam, which turns the blades of a turbine engine. The mechanical energy of the rotating turbine is converted into electrical energy because the turbine turns a coil positioned within a magnetic field. Some of the environmental consequences of the combustion reaction are considered in Box 3.4.

The combustion reaction that we have described is **complete combustion.** When there is insufficient oxygen or when combustion conditions (temperature, mixing of hydrocarbon and oxygen) are not optimized, **incomplete combustion** occurs and various amounts of carbon monoxide (CO) are formed instead of carbon dioxide. Thus, some of the carbon atoms in the fuel are not oxidized to their highest oxidation state.

Incomplete combustion is detrimental in two ways. First, less heat is generated per pound of fuel; so heat, power, and electricity become more costly. Second, the resulting carbon monoxide is highly toxic and especially dangerous because it has no odor. Carbon monoxide is selectively and strongly absorbed by hemoglobin. Even at low concentrations, it interferes with hemoglobin's ability to absorb oxygen in the lungs and deliver it to the tissues. Hence, we are warned never to run an automobile engine in a closed garage.

Throughout the remainder of this book, unless otherwise stated, the term combustion refers to complete combustion.

A reaction parallel to the combustion of saturated hydrocarbons is used by the body for energy generation. The body does not burn saturated hydrocarbons but uses instead the organic foodstuffs that we call carbohydrates, lipids, and proteins. These foodstuffs are oxidized in an indirect way, not by direct reaction with atmospheric oxygen, to produce energy needed by the body. The overall results are similar: the body takes in organic compounds and oxygen and produces energy, carbon dioxide, and water. There is one significant difference: in the body, the oxidation takes place by a complex multistep process that allows the energy to be captured and preserved in the form of new molecules that can perform useful work such as muscle contraction. This process is considered in Chapters 14 and 15.

Petroleum is the source of energy for airplanes and other modes of transportation as well as for lighting our buildings and highways. (Poulides/Thatcher/Tony Stone.)

3.4 CHEMISTRY AROUND US

The Greenhouse Effect and Global Warming

Earth has an average surface temperature comfortably between the boiling point and the freezing point of water and is thus suitable for our sort of life. This temperature range is based on one of the best understood processes in the atmospheric sciences, which is the ability of certain trace gases to be relatively transparent to incoming visible light from the sun, yet opaque to the energy radiated from Earth. This phenomenon, the **greenhouse effect,** is what makes Earth habitable for life.

The energy from the sun that is transformed into heat at Earth's surface would reradiate outward into space without our atmosphere. Such radiation doesn't happen, because Earth's atmosphere contains molecules called **greenhouse gases** that absorb the heat and reduce the heat radiated outward into space. Carbon dioxide is one of the greenhouse gases. It absorbs infrared radiation and emits the radiation again, to be absorbed by yet another greenhouse gas molecule. This absorption-emission-absorption cycle serves to keep the heat near Earth's surface, effectively insulating the surface from the coldness of space. Water vapor, methane, and nitrous oxide also are greenhouse gases.

Atmospheric scientists first used the term "greenhouse effect" in the early 1800s. At that time, it was used to describe the naturally occurring functions of trace gases in the atmosphere and did not have any negative connotations. Not until the mid-1950s was the term greenhouse effect coupled to concern about climate change. Since the 1800s, we've been burning vast quantities of fossil fuels to power our developing technological and global civilization. As a result, we've been releasing significant quantities of carbon dioxide back into the atmosphere. Carbon dioxide concentrations worldwide since that time have increased from approximately 280 ppm (or 0.0280%) to about 365 ppm (0.0365%). The increase might seem trivial, but it means that about 3 gigatons (3 billion metric tons) of carbon dioxide is being added to the atmosphere every year. Because carbon dioxide is a powerful greenhouse gas, a reasonable conclusion is that Earth's temperature should go up as concentrations increase. In fact, climatologists have detected a steady but small increase in global average temperatures in the past few decades, on the basis of weather data collected throughout the world. Six of the past ten years were the hottest on record.

Regardless of the cause of the warming, we understand enough about global climate to predict that, as the temperature goes up, the entire global climate system powered by heat energy also should change, although the magnitude and direction of the changes are uncertain. Are we seeing the end of the long period of comfortable climate since the last Ice Age? Will the climate change for the worse because of our actions? In fact, no one knows for sure. Most atmospheric scientists believe that the global climate is warming at least partly because of a buildup from fossil-fuel use, but what that warming means to humans and natural ecosystems is largely unknown. The climate is complex and influenced by many factors other than greenhouse-gas concentrations. This complexity makes it difficult to link any climatic events or characteristics to a single cause. As a result, controversy exists concerning the magnitude and danger of global warming induced by greenhouse gases. Many scientists take the issue very seriously and support efforts to slow or reverse the buildup of atmospheric gas concentrations with the expectation that global warming will slow as a result. Others, however, contend that these gases may not be affecting the climate and that the changes are part of natural, long-term climatic cycles. They suggest that efforts to reduce emissions are unnecessary and dangerous to economic growth and development. While the controversy rages, researchers throughout the world continue to gather atmospheric data, develop and refine predictive computer models, and try to reduce the uncertainty in our understanding of Earth's climate.

Example 3.13 Writing equations for the combustion of hydrocarbons

Write balanced equations for the combustion of (a) hexane; (b) 3-methylpentane; (c) 2,3-dimethylbutane.

Solution

Structural formulas are not needed to write combustion equations, because combustion results in complete dissection of the molecule. The molecular formula is sufficient. In this case, the same equation describes the combustion of all three compounds in the problem, because they are all constitutional isomers of C_6H_{14}. The unbalanced equation is

$$C_6H_{14} + O_2 \longrightarrow CO_2 + H_2O$$

Balance the elements in the order C followed by H followed by O. The 6 C and 14 H on the left side are balanced by 6 CO_2 and 7 H_2O on the right side. This change leads to 19 O on the right side that require $9\frac{1}{2}$ O_2 on the left side. The balanced equation is

$$C_6H_{14} + 9\tfrac{1}{2}\,O_2 \longrightarrow 6\,CO_2 + 7\,H_2O$$

All coefficients are multiplied by 2 to avoid fractional coefficients:

$$2\,C_6H_{14} + 19\,O_2 \longrightarrow 12\,CO_2 + 14\,H_2O$$

Problem 3.13 Write balanced equations for the combustion of (a) cyclopentane; (b) 2,2,3-trimethylhexane.

Halogenation (Halogen Substitution)

Molecular halogens (F_2, Cl_2, Br_2) react with alkanes and cycloalkanes, but only in the presence of heat or ultraviolet (UV) light. The reaction, called **halogenation,** substitutes a halogen atom (—F, —Cl, —Br) for a hydrogen atom in the alkane or cycloalkane to form an **alkyl halide.** The term alkyl halide encompasses any organic compound containing one or more halogen atoms. A number of halogen-containing products are important materials of commerce and everyday life (Box 3.5).

Halogenation can be illustrated by the chlorination of methane:

$$CH_4 + Cl_2 \xrightarrow{\text{UV or heat}} CH_3Cl + HCl$$

Organic chemists often write chemical equations in an abbreviated form such as

$$CH_4 \xrightarrow[\text{UV or heat}]{Cl_2} CH_3Cl$$

to emphasize the transformation that takes place in the organic reactant. The reagent (Cl_2) and reaction conditions (UV or heat) required for reaction to take place are written above and below the arrow, respectively. The nonorganic product (HCl) is omitted because it is not important in this context. In studying the properties of organic reactants, we will almost always be interested in the organic product(s) only, not the inorganic product(s). (The combustion reaction is the rare exception.)

Equations can also be written in a generalized form, for the purpose of emphasizing family behavior. In such cases, R is used to represent any alkyl group (methyl, ethyl, propyl, and so forth):

$$R{-}H \xrightarrow[\text{UV or heat}]{Cl_2} R{-}Cl$$

Halogenation usually yields a mixture of products because each C–H bond in an alkane or cycloalkane is susceptible to reaction. Halogenation of methane can proceed sequentially, as follows:

$$CH_4 \longrightarrow CH_3Cl \longrightarrow CH_2Cl_2 \longrightarrow CHCl_3 \longrightarrow CCl_4$$

Replacement of one hydrogen by halogen is called **monohalogenation** and replacement of two or more hydrogens by halogens is called **polyhalogenation.** Monohalogenation predominates if a large amount of methane is present relative to the halogen; polyhalogenation predominates if the halogen is present in excess.

Heat is shown as part of the reaction conditions in the equations for chlorination because a relatively high temperature (about 200°C) is required for a reaction to proceed. Some chemical reactions take place at ambient conditions, many require a moderate temperature (below about 100°C), whereas other reactions require higher temperatures, above 100°C and sometimes as high as 200°C or above. In general, heat will not be indicated as part of the reaction conditions in this book unless the higher temperatures are required for a reaction to take place.

DOUGLAS COLLEGE LIBRARY

3.5 CHEMISTRY AROUND US

Applications of Alkyl Halides and Some of the Problems That They Create

Many kinds of organic compounds possess anesthetic properties, but all too often they have other properties that interfere with their potential use in the control of pain. Cyclopropane (C_3H_6) and chloroform ($CHCl_3$) are two examples. Although they were used as anesthetics in the past, neither is used today. Cyclopropane is quite explosive, and chloroform causes liver damage. One of the most commonly used anesthetics at present is the alkyl halide 2-bromo-2-chloro-1,1,1-trifluoroethane (Halothane). It is nonflammable and nonexplosive and presents minimal health hazards.

$$CF_3-\underset{\displaystyle Cl}{\overset{\displaystyle H}{\underset{|}{\overset{|}{C}}}}-Br$$

Halothane

Alkyl halides have many other medical applications. Chloroethane (ethyl chloride) serves as a topical cooling agent. Athletes use it during competitions, to temporarily numb the pain from a bruise (see Box 6.3). Physicians apply it before performing minor surgery.

The alkyl halides have everyday uses as well. Tetrachloromethane (carbon tetrachloride) was once used widely as the solvent in dry-cleaning processes, but it caused health problems and has since been replaced by 1,1,2,2-tetrachloroethene (a chlorinated alkene). 1,2-Dichloroethane is used as a spot cleaner (clothes, piano keys). 1,2,3,4,5,6-Hexachlorocyclohexane (Lindane) is used as an insecticide to prevent insect infestation in stored seeds.

Cl, Cl, Cl, Cl, Cl, Cl (cyclohexane ring structure)

Lindane

The alkyl halide dichlorodifluoromethane (Freon) is one of several compounds that are known as chlorofluorohydrocarbons (CFCs) and have been used in refrigeration and air-conditioning equipment. CFCs have also been used extensively as the gaseous propellant in aerosol hair sprays and other similar products. However, CFCs cause damage to the ozone layer of Earth's atmosphere.

Ozone (O_3) in the stratosphere, a middle layer of the atmosphere, protects plant and animal life by absorbing much of the cancer-causing ultraviolet radiation from space

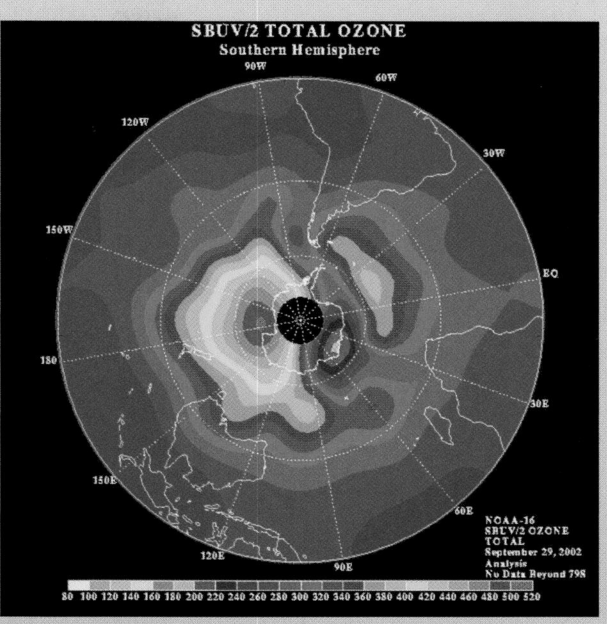

A view of the hole in the ozone layer over Antarctica in 2002. Ozone concentrations are expressed in Dobson units (higher concentrations are indicated by larger values). (NOAA/AP Photo.)

and preventing it from reaching Earth. Scientists have noted significant depletion of the ozone concentration in some parts of the atmosphere, especially over Antarctica, and have identified CFCs as the agent responsible. CFCs escape from Earth and concentrate in the ozone layer, where they participate in a complex set of reactions that convert ozone into oxygen. Oxygen is much less effective than ozone in protecting Earth from ultraviolet radiation. In 1995, Paul Crutzen, Mario Molina, and Sherwood Rowland were awarded a Nobel Prize in chemistry for their work on the effect of CFCs on the ozone layer.

Most industrialized nations have adopted the Montreal Protocol, an international treaty to protect the Earth's ozone layer. The production and use of CFCs are now severely curtailed. Certain alkanes, alcohols, and other compounds have replaced CFCs as the propellant in aerosol cans. Hydrofluorocarbons (HFCs) and hydrochlorofluorocarbons (HCFCs)—for example, 1,2,2,2-tetrafluoroethane and 1,1-dichloro-2,2,2-trifluoroethane—have been developed for use in refrigeration and air conditioning, as well as in aerosol cans and other applications. HFCs are more chemically reactive than CFCs, undergoing decomposition at lower atmospheric altitudes, and so much less survives to reach the upper atmosphere and damage the ozone layer. HCFCs are safer than CFCs, but not nearly as safe as HFCs, and are scheduled for phaseout.

Example 3.14 Writing equations for halogenation of hydrocarbons: I

Write an equation to show the bromination of cyclopentane—that is, the halogenation of cyclopentane with the use of bromine. (In this type of problem, the word halogenation refers to monohalogenation, not polyhalogenation, unless otherwise stated.)

Solution

$$\text{cyclopentane} \xrightarrow[\text{UV or heat}]{Br_2} \text{cyclopentyl}-Br$$

Problem 3.14 Write an equation for the chlorination of ethane.

Monohalogenation of some alkanes and cycloalkanes yields more than one monosubstituted product because substitutions at different C–H bonds yield constitutional isomers. For example, monochlorination of propane yields a mixture of 1-chloropropane and 2-chloropropane:

$$\underset{\textbf{Propane}}{CH_3—CH_2—CH_3} \xrightarrow[\text{UV or heat}]{Cl_2} \underset{\textbf{1-Chloropropane}}{CH_3—CH_2—CH_2—Cl} + \underset{\textbf{2-Chloropropane}}{CH_3—\overset{\overset{Cl}{|}}{C}H—CH_3}$$

1-Chloropropane is produced when reaction takes place at any one of the six hydrogens at the terminal carbons in propane. Substitution at either of the two hydrogens of the middle carbon yields 2-chloropropane.

Note that the preceding equation for the monochlorination of propane is unbalanced. The left side of the equation has half as many atoms of carbon, hydrogen, and chlorine as the right side. Unbalanced equations are often used to describe organic reactions for two reasons. First, the major emphasis is on describing the organic reactant(s) used, the reaction conditions needed, and the organic product(s) formed. Second, most organic reactions give less than a 100% yield. For example, in a typical monochlorination of propane, the yields of 1-chloropropane and 2-chloropropane are 39% and 44%, respectively, with the remainder of the product being a mixture of more highly chlorinated compounds. Only the major products, the monochloro products, are usually shown in the equation. The minor (polychloro) products are omitted.

Example 3.15 Writing equations for halogenation of hydrocarbons: II

Write an equation to show the monochlorination of butane.

Solution

$$CH_3—CH_2—CH_2—CH_3 \xrightarrow[\text{UV or heat}]{Cl_2}$$

$$CH_3—CH_2—CH_2—CH_2—Cl + CH_3—\overset{\overset{Cl}{|}}{C}H—CH_2—CH_3$$

Problem 3.15 Write an equation to show the monobromination of 2-methylpropane.

Naming Alkyl Halides

The IUPAC rules for naming alkyl halides are the same as those for alkanes and cycloalkanes except that they also provide for the identification of the halogens. The presence of F, Cl, Br, and I is signaled by the prefixes **fluoro-,**

chloro-, **bromo-**, and **iodo-**, respectively. Thus, the two products of the monochlorination of propane are 1-chloropropane and 2-chloropropane.

In the common nomenclature system, alkyl halides are named by attaching the halogen prefix to the name of the alkyl group followed by the suffix **-ide** (for **halide,** Appendix 1). The common names of 1-bromopropane and 2-bromopropane are propyl bromide and isopropyl bromide, respectively. This naming procedure is applicable only to alkyl halides with simple alkyl groups.

Example 3.16 Naming alkyl halides

Give the IUPAC and common names for the following alkyl halides:

(a) $CH_3{-}CH_2{-}CH_2{-}CH_2{-}Cl$

(b) $CH_3{-}CH(CH_3){-}CH(Cl){-}CH_2{-}CH_2{-}CH_3$

Solution

(a) IUPAC: 1-chlorobutane; common: butyl chloride.

(b) IUPAC: 3-chloro-2-methylhexane. Note that the halogen is alphabetized along with the methyl group. There is no common name, because the alkyl group attached to Cl is not a simple alkyl group.

Problem 3.16 Give the IUPAC and common names for the following compounds:

(a) $CH_3{-}CH(CH_3){-}CH_2{-}Br$ (b) cyclopentane ring with H, H on top and CH_3, Br below on adjacent carbons

Summary

Molecular and Structural Formulas

- The molecular formula of a compound indicates the elements present in the compound and the number of atoms of each element.
- The structural formula shows how various atoms are bonded together in a molecule and must be consistent with the combining powers of each of the atoms.

Families of Organic Compounds

- Organic chemistry is the chemistry of organic compounds, compounds of carbon, and its study is organized along family lines.
- A family contains compounds with a common set of physical and chemical properties, the result of possessing the same functional group.
- A functional group is a particular atom or type of bond or group of atoms in a particular bonding arrangement.

Alkanes

- Alkanes have the general formula C_nH_{2n+2} and contain only carbon–carbon and carbon–hydrogen single bonds.
- The carbon atoms in alkanes bond to hydrogens and other carbons through sp^3-hybrid orbitals, which are characterized by tetrahedral (109.5°) bond angles.
- The large number of organic compounds in each family is a consequence of two factors: (1) that carbon atoms bond to one another and (2) that there can be many more than one constitutional isomer for most organic molecular formulas.
- Constitutional isomers are compounds with the same molecular formula but different connectivity—that is, the order in which the atoms are attached to each other.
- Carbon atoms can bond to one another to form branched and unbranched acyclic molecules as well as cyclic molecules.
- Cyclic alkanes, called cycloalkanes, follow the general formula C_nH_{2n} and behave very much like the alkanes.

Naming Alkanes and Cycloalkanes

- By the IUPAC system, alkanes are named after the longest continuous carbon chain. Prefixes and numbers indicate substituents attached to the longest continuous chain.
- The base name of the ring in a cycloalkane consists of the prefix cyclo- followed by the name of the straight-chain alkane that has the same number of carbons. Prefixes and numbers indicate substituents attached to the ring.

Cis-Trans Stereoisomers in Cycloalkanes

- Cis and trans stereoisomers differ in configuration—the spatial orientation in which substituents extend from the ring carbons.
- Cis and trans isomers exist only when each of two ring carbons has two different substituents. Such ring carbons are called tetrahedral stereocenters.
- The cis isomer is the isomer in which two similar substituents on two ring carbons are on the same side of the ring.
- The trans isomer has the two similar substituents on opposite sides of the ring.

Physical Properties of Alkanes and Cycloalkanes

- The order of secondary forces is hydrogen bonding > dipole–dipole > London.
- Alkanes and cycloalkanes are nonpolar compounds with the weak London secondary forces.
- Their melting and boiling points are low relative to those of other organic families, and they are insoluble in water.
- Melting and boiling points increase with increasing molecular mass.
- Melting and boiling points vary with molecular shape in the order cycloalkane > straight-chain alkane > branched alkane.

Chemical Properties of Alkanes and Cycloalkanes

- Alkanes and cycloalkanes possess low chemical reactivity because their carbon–carbon and carbon–hydrogen single bonds are relatively strong compared with the bonds in other organic families.
- Alkanes and cycloalkanes undergo halogenation and combustion.
- Halogenation is a substitution reaction in which a hydrogen in the alkane or cycloalkane is replaced by a halogen.
- Halogenation takes place with a molecular halogen such as Cl_2 or Br_2 and requires ultraviolet radiation or heat.
- In combustion (burning), an oxidation reaction, alkanes and cycloalkanes react with oxygen to yield carbon dioxide and water.

Summary of Reactions

COMBUSTION

$$\text{Alkane or cycloalkane} + O_2 \longrightarrow CO_2 + H_2O + \text{heat}$$

HALOGENATION (X = F, Cl, Br)

$$C{-}H + X_2 \xrightarrow[\text{heat}]{\text{UV or}} C{-}X + HX$$

Key Words

alkane, p. 70
alkyl group, p. 82
branched alkane, p. 80
cis-trans isomers, p. 90
combining power, p. 68
combustion, p. 97
configuration, p. 91
conformation, p. 76
constitutional isomer, p. 79
cycloalkane, p. 87
family, p. 70
functional group, p. 70
geometrical stereoisomers, p. 90
halogenation, p. 99
hydrocarbon, p. 72
IUPAC nomenclature system, p. 82
line formula, p. 79
longest continuous carbon chain, p. 79
molecular formula, p. 68
orbital hybridization, p. 75
organic chemistry, p. 67
organic compound, p. 67
skeleton formula, p. 79
stereoisomer, p. 90
structural formula, pp. 68, 78
tetrahedral, p. 74
tetrahedral stereocenter, p. 91

Exercises

Molecular and Structural Formulas

3.1 Define combining power.

3.2 How many bonds do each of the following atoms form? (a) C; (b) H; (c) N; (d) O; (e) S; (f) Cl; (g) Br; (h) F.

3.3 Which of the following molecular formulas are correct? (a) CH_4; (b) CH_5; (c) CH_3.

3.4 Which of the following molecular formulas are correct? (a) C_2H_5; (b) C_2H_6; (c) C_2H_7.

3.5 Which of the following molecular formulas are correct? (a) C_2H_4Cl; (b) CH_4O; (c) CH_6N.

3.6 Which of the following molecular formulas are correct? (a) $C_4H_{10}Cl$; (b) C_4H_8Cl; (c) C_4H_9Cl.

Organization of Organic Chemistry into Families

3.7 Draw the functional group that characterizes each of the following families (refer to Table 3.2): (a) alkane; (b) carboxylic acid; (c) aldehyde.

3.8 Draw the functional group that characterizes each of the following families (refer to Table 3.2): (a) ester; (b) amide; (c) alkene.

3.9 To what family does each of the following compounds belong? (Refer to Table 3.2.)

(a) $CH_3-CH(OH)-CH_2-CH_3$

(b) $CH_2{=}CH-CH_3$

(c) $CH_3-C(=O)-CH_2CH_3$

(d) [benzene ring bearing CH_3]

(e) $CH_3-C(=O)-CH_3$

(f) $CH_3-O-CH(CH_3)-CH_3$

(g) $(CH_3)_2CH-C(=O)-OH$

(h) $CH_3CH_2-C(=O)-NH_2$

3.10 Some important compounds, especially those with physiological activity, belong simultaneously to more than one family. By referring to Table 3.2, identify the family for each of functional groups a–h in the following compound:

[Structure: HO—CH₂ (a); C=C (b); C=O (c); ring O (d); CH₂CH₂—NH₂ (e); aromatic ring (f); C=CH₂ (g); C(=O)—OH (h)]

Alkanes

3.11 Why are sp^3 orbitals proposed for the carbon atoms in alkanes?

3.12 What are the values of bond angles A, B, and C in the following compound?

$CH_3-CH_2-CH_3$ (expanded: H—C—C—C—H with H atoms above and below each carbon; angles A, B, C marked)

Types of Structural Formulas

3.13 Draw an expanded structural formula for each of the following compounds: (a) $(CH_3)_2CH_2$; (b) $CH_3CH_2CH_2CH_2CH_3$; (c) $CH_3CH_2CH_2CH(CH_3)_2$.

3.14 Draw a skeleton structural formula for each of the compounds in Exercise 3.13.

3.15 Draw a line structural formula for each of the following compounds:

(a) $CH_3-C(CH_3)_2-CH_2-CH(CH_3)-CH_2-CH_3$

(b) $CH_3-CH_2-CH(CH_3)-CH(C(CH_3)_3)-CH_2-CH_3$

(c) $CH_3-CH_2-CH_2-CH(CH(CH_3)-CH_3)-CH(CH_2CH_3)-CH_3$

(d) $CH_3-C(CH_3)(CH(CH_3)-CH_3)-CH_2-CH(CH_2-CH_2-CH(CH_3)-CH_3)-CH_2-CH_3$

3.16 Draw a condensed structural formula for each of the following compounds:

(a) [line structure] (b) [line structure]

Constitutional Isomers

3.17 For each of the following pairs of structural formulas, indicate whether the pair represents (1) the same compound or (2) different compounds that are constitutional isomers or (3) different compounds that are not isomers.

(Note: Two structural formulas represent the same compound when they portray different conformations of a single compound, when they differ in the extent to which they are condensed or expanded, or when they differ in the extent to which they show the correct bond angles.)

(a) $CH_3CH_2CH(CH_3)_2$ and $CH_3CH_2-CH(CH_3)-CH_3$

(b) $(CH_3)_3CCH_2-CH_3$ and $CH_3-CH_2-CH(CH_3)-CH_2-CH_3$

(c) $CH_3-CH_2-CH(CH_3)-CH_2-CH_3$ and [line structure]

3.18 For each of the following pairs of structural formulas, indicate whether the pair represents (1) the same compound or (2) different compounds that are constitutional isomers or (3) different compounds that are not isomers.

(Note: Two structural formulas represent the same compound when they portray different conformations of a single compound, when they differ in the extent to which they are condensed or expanded, or when they differ in the extent to which they show the correct bond angles.)

```
(a) CH3CH2—CH—CH3        and
           |
           CH3
                    (CH3)2CH—CH2—CH3

(b) CH3—CH2   CH3
        |     |
        CH2—CH—CH3        and
                    (CH3)2CHCH2CH2CH3

(c)  [line structure]     and     [line structure]
```

3.19 How many constitutional isomers are there for C_7H_{16}? Show one structural formula for each isomer.

3.20 Draw the structural formula for each of the following compounds: (a) an alkane, C_5H_{12}, that contains three methyl groups; (b) an alkane, C_5H_{12}, that contains four methyl groups; (c) an alkane, C_5H_{12}, that contains two methyl groups.

Naming Alkanes

3.21 Give the IUPAC names for each of the following compounds:

```
(a) CH3CH2CH(CH3)2        (b) [line structure]

         CH3       CH3
         |         |
(c) CH3—C—CH2—CH—CH2—CH3
         |
         CH3

                    CH(CH3)2
                    |
(d) CH3—CH2—CH—CH—CH2—CH3
               |
               CH3
```

3.22 Give the IUPAC names for each of the following compounds:

```
                 CH3
                 |
(a) CH3—CH2—C—CH2—CH3
                 |
                 CH3

                    CH3
                    |
(b) CH3—CH2—CH—CH—CH3
               |
               CH2CH3

(c) [line structure]

         CH3      CH2—CH2—CH—CH3
         |        |        |
         |        |        CH3 (on CH)
(d) CH3—C—CH2—CH—CH2—CH3
         |
         CH—CH3
         |
         CH3

(e)                 C(CH3)3
                    |
CH3—CH2—CH—CH—CH2—CH2—CH2—CH3
           |
           CH3
```

3.23 Draw the structural formula for each of the following compounds: (a) 2-methylbutane; (b) 2,2,4-trimethylhexane; (c) 3-isopropyl-2,3-dimethylhexane; (d) 4-ethyl-4-isopropyl-2,2-dimethyloctane; (e) 4-*s*-butyl-2-methylheptane.

3.24 Identify 1°, 2°, 3°, and 4° carbons in the compounds of Exercise 3.23.

Cycloalkanes

3.25 Draw the structural formula for each of the following compounds: (a) cyclooctane; (b) 4-isopropyl-1,2-dimethylcyclohexane; (c) 1-*t*-butyl-3-ethylcyclohexane.

3.26 Give the IUPAC names for each of the following compounds:

(a) (b) CH_3 CH_2CH_3 (c) $CH(CH_3)_2$ CH_3

3.27 Identify 1°, 2°, 3°, and 4° carbons in the compounds of Exercise 3.26.

3.28 Draw a line structural formula for each of the following compounds: (a) *t*-butylcyclopropane; (b) 1,2-diethylcyclopentane.

3.29 Draw the structural formula and give the IUPAC name for each of the following compounds: (a) a cycloalkane, C_6H_{12}, that contains one methyl group; (b) a cycloalkane, C_6H_{12}, that contains one ethyl group.

3.30 Draw the structural formula and give the IUPAC names for all cycloalkanes, C_6H_{12}, that contain two methyl groups.

Cis-Trans Isomerism in Cycloalkanes

3.31 Which of the following compounds exist as cis and trans isomers? If cis and trans isomers are possible, show one structural formula for each isomer. (a) 1-Chloro-1-methylcyclobutane; (b) 1-chloro-2-methylcyclohexane; (c) 1-chloro-2,2,3-trimethylcyclohexane.

3.32 For each of the following pairs of structural formulas, indicate whether the pair represents (1) the same compound or (2) different compounds that are constitutional isomers or (3) different compounds that are not isomers or (4) different compounds that are cis-trans isomers.

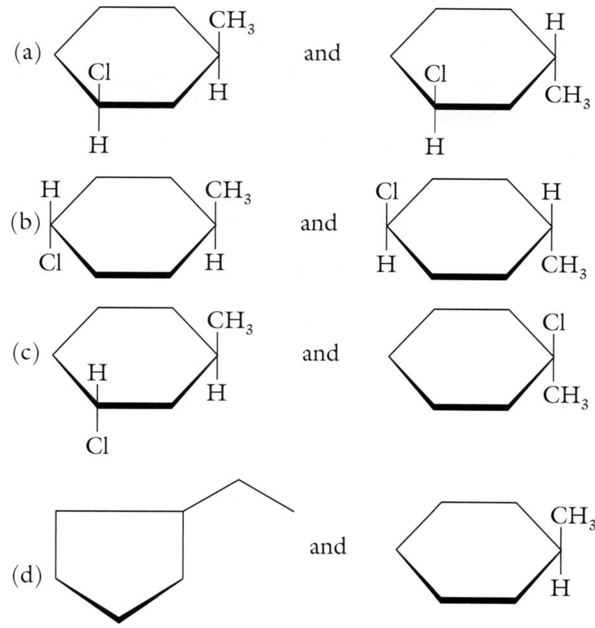

Physical Properties of Alkanes and Cycloalkanes

3.33 For each of the following pairs of compounds, indicate which compound in the pair has the higher boiling point. Explain each of your answers in regard to the difference in secondary forces in the two compounds. (a) Hexane and pentane; (b) cyclohexane and cyclopentane; (c) cyclohexane and methylcyclohexane; (d) hexane and 2-methylpentane; (e) cyclopentane and pentane.

3.34 For each of the following pairs of compounds, indicate whether a solution forms when the two compounds are mixed. Explain each of your answers in regard to the difference in secondary forces in the two compounds. (a) Cyclohexane and hexane; (b) hexane and water; (c) hexane and sodium chloride; (d) sodium chloride and water.

Chemical Properties of Alkanes and Cycloalkanes

3.35 Give the organic product(s) formed in each of the following reactions. If no reaction takes place, write "no reaction." If more than one product is formed, indicate only the major product.

(a) $CH_3{-}CH_2{-}CH_2{-}CH_3 \xrightarrow{H_2SO_4} ?$

(b) cyclopentane $\xrightarrow{NaOH} ?$

(c) cyclopentane $\xrightarrow{H_2} ?$

(d) $CH_3{-}CH_2{-}CH_2{-}CH_3 \xrightarrow{HCl} ?$

(e) $CH_3{-}CH_2{-}CH_2{-}CH_3 \xrightarrow[\text{UV or heat}]{Cl_2} ?$

3.36 Indicate whether reaction takes place when cyclohexane is treated with the following reagents. If reaction takes place, give the organic product(s) formed. If more than one product is formed, indicate which is the major product. If no reaction takes place, write "no reaction." (a) NaOH; (b) H_2SO_4; (c) H_2; (d) HCl; (e) Cl_2 (no heat, no UV); (f) Cl_2 (heat or UV).

3.37 Write the balanced equation for the complete combustion of each of the following compounds: (a) pentane; (b) 2-methylbutane; (c) cyclohexane; (d) methylcyclopentane.

3.38 Give the monochlorination product(s) of each of the following compounds: (a) pentane; (b) cyclopentane; (c) methylcyclopentane; (d) 2,2-dimethylpropane.

Unclassified Exercises

3.39 What property of carbon is responsible for the much larger number of organic compounds relative to the number of inorganic compounds?

3.40 Which of the following molecular formulas are correct? (a) C_5H_{10}; (b) C_5H_{14}; (c) C_5H_{12}

3.41 Draw structural formulas for the constitutional isomers of C_4H_9Cl, taking care to draw only one structural formula for each isomer.

3.42 Draw a condensed structural formula for each of the following compounds:

(a) pentane (b) 2-methylpentane

(c) (d)

3.43 For each of the following pairs of structural formulas, indicate whether the pair represents (1) the same compound or (2) different compounds that are constitutional isomers or (3) different compounds that are cis-trans isomers or (4) different compounds that are not isomers.

(a) $ClCH_2CH(CH_3)_2$ and $ClCH_2{-}CH{-}CH_2Cl$ (with CH_3 on the central CH)

(b) and

(c) CH_3CH_2, CH_3, CH_3 (substituents on cyclopentane) and CH_2, CH_2, $CH{-}CH(CH_3)_2$, CH_2, CH_2, CH_2 (ring)

(d) CH_2CH_3, CH_3 (substituents on cyclohexane) and CH_2, CH_2, $CH{-}CH(CH_3)_2$, CH_2, CH_2, CH_2 (ring)

(Exercise 3.43 continues on next page)

(e) 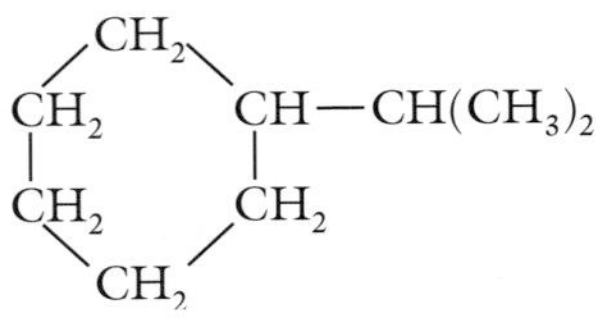 cyclohexane ring CH_2–CH_2–CH_2–CH_2–CH_2–CH–$CH(CH_3)_2$ and CH_3–CH_2–CH_2–CH_2–CH_2–CH(–CH_3)–$CH(CH_3)_2$

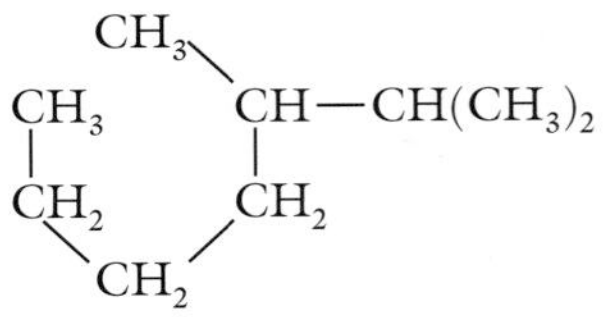

(f) $ClCH_2—CH(CH_3)—CH_3$ and $CH_3—CHCl—CH_2—CH_3$

(g) $(CH_3)_3CCH_2—Cl$ and $CH_3—CH_2—CH(Cl)—CH_2—CH_3$

(h) cyclohexane with Cl/H on two carbons and cyclohexane with Cl/H 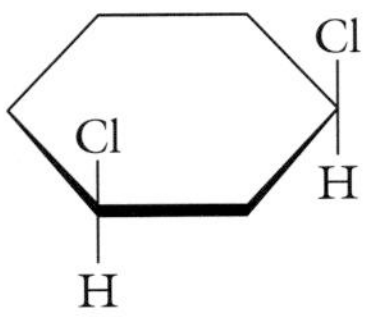 and H/Cl (chair drawings, H, Cl labels)

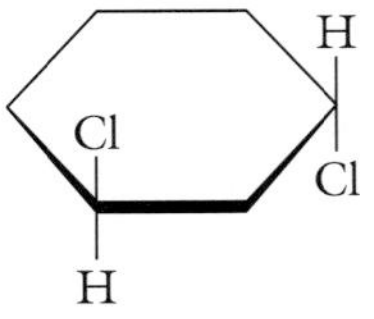

(i) 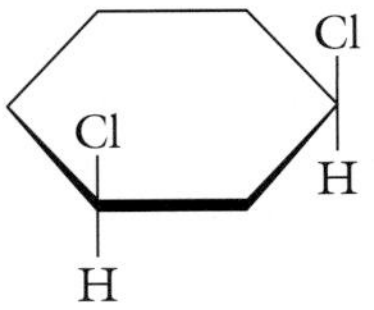 cyclohexane with Cl, H and Cl, H and cyclohexane with H, Cl and Cl, H (chair drawings)

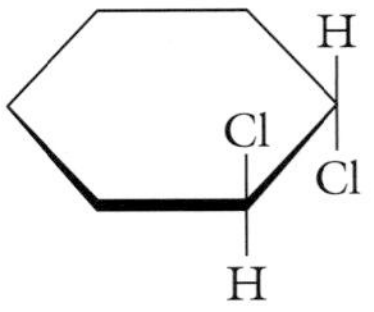

(j) $ClCH_2CH(CH_3)CH_2Cl$ and $ClCH_2—CH(CH_3)—CH_2Cl$

3.44 Each of the following names is an incorrect IUPAC name. State why the name is incorrect. Indicate, if possible, the correct name of the compound. (a) 4-Methylhexane; (b) 2-ethylpentane; (c) 1,3,4-trimethylcyclohexane; (d) 1-propyl-5-methylcyclopentane; (e) *cis*-1,1,2-trimethylcyclopentane.

3.45 An unknown compound has the formula C_5H_{10}. Is this a cyclic or acyclic compound?

3.46 An unknown compound has the formula $C_7H_{15}Br$. Is this a cyclic or acyclic compound?

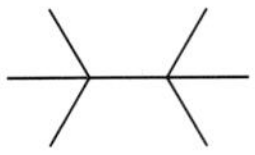

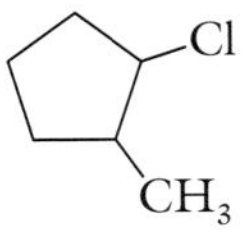

3.47 What is the molecular formula of each of the following compounds?

(a) $CH_3—CH_2—CH(CH_3)—CH_2—CH_3$

(b) [skeletal structure]

(c) cyclopentane with Cl and CH_3 (d) 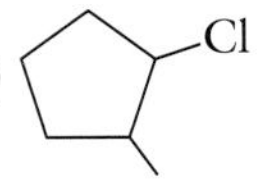 cyclopentane with Cl and methyl

(e) 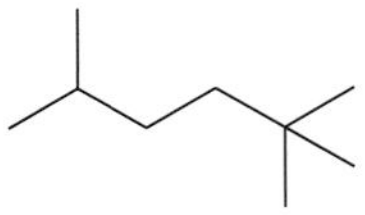 [skeletal structure] (f) 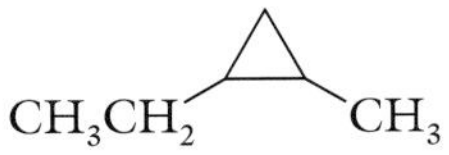 CH_3CH_2–cyclopropane–CH_3

3.48 Indicate whether reaction takes place when ethane is treated with each of the following reagents. If reaction takes place, give the organic product(s) formed. If more than one product is formed, indicate which is the major product. If no reaction takes place, write "no reaction." (a) NaOH; (b) H_2SO_4; (c) H_2; (d) HCl; (e) Cl_2 (no heat, no UV); (f) Cl_2 (heat or UV).

3.49 Write an equation to show the product(s) of each of the following reactions with 2,3-dimethylbutane: (a) combustion; (b) monobromination.

3.50 Give the perchlorination product of each of the following compounds. (In perchlorination, chlorine substitutes for all hydrogens in an alkane or cycloalkane.) (a) Pentane; (b) methylcyclopentane.

3.51 Chlorination of butane yields several dichlorobutane isomers (in addition to monochloro-, trichloro-, and other products). Draw structural formulas of the different dichlorobutane isomers.

3.52 Chlorination of methylcyclopropane yields several dichloromethylcyclopropane isomers (in addition to monochloro-, trichloro-, and other products). Draw structural formulas of the different dichloromethylcyclopropane isomers.

Expand Your Knowledge

Note: The icons denote exercises based on material in boxes.

3.53 Oil fires are difficult to fight with water because oil, having a lower density than that of water, floats on water. Why do the compounds in oil (mostly alkanes) have lower densities than water does?

3.5 **3.54** 2-Bromo-2-chloro-1,1,1-trifluoroethane (C_2HF_3BrCl) is the commonly used anesthetic Halothane (see Box 3.5). How many constitutional isomers are possible for C_2HF_3BrCl?

3.55 Approximately one-half of all synthetic medicines have petroleum as their starting material. How is this possible given that petroleum contains only hydrocarbons, whereas medicines contain a variety of oxygen- and nitrogen-containing functional groups such as —OH, $—NH_2$, and —COOH?

3.56 You are driving on an isolated back road when your car runs out of gas. Hours go by without help. It is now time to take your heart medication, for which you need some water. You have a gallon bottle of water in the trunk, but you also have a gallon bottle of gasoline in the trunk and neither bottle is labeled. How can you determine which bottle contains the water and which the gasoline? Gasoline has a distinctive smell but you have a cold and your sense of smell is completely absent this day. All that you have available to help you distinguish between the contents of the two bottles are some empty cups and some salt (NaCl). Describe two different tests that you can perform to determine which bottle contains the water.

3.3 **3.57** The major components of lipsticks are wax and other materials that are mostly alkanes. Why is lipstick easily removed by petroleum jelly (see Box 3.3) but not by water?

3.5 **3.58** The commonly used topical cooling agent ethyl chloride (see Box 3.5) is produced from ethane. What other reagent(s) and reaction conditions are required to produce ethyl chloride from ethane?

3.1 **3.59** Methane is the main component of the gas used to heat our homes and for cooking (see Box 3.1). The equation for the complete combustion of methane is

$$CH_4 + 2\ O_2 \longrightarrow CO_2 + 2\ H_2O + 803\ kJ$$

Calculate the heat generated when 160.43 g of methane is burned in a gas heating system or in a gas cooking oven.

3.1 **3.60** More gasoline (C_5–C_{10}) and less home heating oil (C_{15}–C_{18}) are used in the summer (see Box 3.1). Petroleum refineries increase the yield of gasoline and decrease the yield of home heating oil by a process called **cracking.** Cracking entails the heating of petroleum at high temperatures in the presence of hydrogen (H_2). Larger-sized alkanes are converted into smaller-sized alkanes. Write an equation to show the cracking of $CH_3(CH_2)_{13}CH_3$ to yield octane and heptane.

3.61 Incomplete combustion of an alkane—combustion without an adequate supply of oxygen—is dangerous because it produces the toxic gas carbon monoxide (CO) instead of the nontoxic carbon dioxide (CO_2). CO is toxic because it preferentially reacts with hemoglobin and prevents hemoglobin from picking up oxygen at the lungs and delivering it to cells and tissues (Section 12.7). Write a balanced equation to show the incomplete combustion of hexane.

3.4 **3.62** What is the probable cause of global warming, and why is it of concern to inhabitants of planet Earth (see Box 3.4)?

3.5 **3.63** Where is the ozone layer located? What is the cause of the ozone layer's depletion, and what are the resulting consequences (see Box 3.5)?

3.1 **3.64** What is the difference between petroleum and natural gas (see Box 3.1)?

3.65 Many fruits and vegetables such as apples and peppers have exterior coatings composed primarily of alkanes. Explain how such coatings prevent the loss of moisture from the inside of the fruits and vegetables.

3.66 The hydrocarbons in 1 gallon of gasoline contain about 700 moles of carbon atoms. How many moles of CO_2 are produced when 1 gallon of gasoline undergoes complete combustion? How many moles of water are produced?

3.67 Compare the complete and incomplete combustions of an alkane in the identities and relative amounts of reactants consumed and products produced. Is there a difference in the amount of heat produced? Is there any effect on the miles per gallon obtained in an automobile?

3.68 Bill Smith tells his friend Mary Jones that he makes sure to take a vitamin C pill everyday. Mary recommends that it is better to eat an orange to get his vitamin C. Bill says that it does not make any difference what source of vitamin C is used, because vitamin C, like any compound, has exactly the same structure irrespective of source, whether found in nature or synthesized in the laboratory. Mary counters, "You are correct that vitamin C is the same compound whatever its source, but it is still better to eat the orange." Who is correct and why?

3.69 Sodium chloride (NaCl), both in the molten state and dissolved in water, conducts electricity, whereas most organic compounds do not conduct electricity. Explain.

3.70 Engine knocking occurs when gasoline does not combust smoothly in the engine cylinders. The antiknocking quality of a gasoline is measured by its octane number. The octane scale is obtained by assigning octane numbers of 0 and 100 to heptane and 2,2,4-trimethylpentane (common name: isooctane), respectively. Draw the structure of isooctane. What structural feature results in a higher octane number for isooctane compared to heptane?

3.71 Chlorination of propane yields a mixture of 1-chloropropane and 2-chloropropane. The reaction proceeds in a **nonselective** manner; that is, the two products are not formed in equal amounts. The unequal products result because the hydrogens bonded to secondary carbons are four times as reactive as hydrogens bonded to primary carbons. The relative amounts of the two products depend not only on the difference in reactivities of primary and secondary hydrogens, but also on the relative numbers of primary and secondary hydrogens. Estimate the relative amounts of the two products from the product of the ratio of reactivities multiplied by the ratio of the numbers of different hydrogens.

3.2 **3.72** There are two chair conformations of *trans*-1,4-dimethylcyclohexane, one with both methyl groups axial and the other with both methyl groups equatorial (see Box 3.2). Draw the latter conformation, which is the more stable conformation. Why is it the more stable conformation?

3.73 Methane and propane are the fuels used for cooking in homes and outdoor barbecues, respectively. The heats of combustion of methane and propane are 213 and 531 kcal/mol, respectively. Compare the difference in the amounts of heat produced by the fuels in kilocalories per gram.

3.74 The boiling points of the cis and trans isomers of 1,2-chlorocyclohexane are 94°C and 75°C, respectively. Explain why the cis isomer has the higher boiling point.

3.75 Indicate the order of increasing oxidation state for the carbon atoms in methyl acetate (see Table 3.2) by placing a number near each carbon. Use number 1 to indicate the highest oxidation state (Section 2.1).

3.76 Place the following compounds in order of increasing oxidation state (Section 2.1): ethane, ethylene, ethyl alcohol, acetic acid, CO_2. Refer to Table 3.2 for the structures of the compounds.

CHAPTER 4 UNSATURATED HYDROCARBONS

(Lance Nelson/The Stock Market.)

Chemistry in Your Future

As a farmer and a consumer, you are interested in environmentally safe alternatives to pesticide use. You have studied a number of integrated pest-management programs whose purpose is to reduce pesticide use by combining pesticides with synthetic insect sex pheromones, or attractants. A major challenge to producing pheromones at reasonable costs is that the compounds usually have several isomeric configurations, and only one will work. Many pheromones belong to a family of unsaturated hydrocarbons called alkenes. Chapter 4 considers the chemistry of alkenes and their "configuration" problem.

For more information on this topic and others in this chapter, go to www.whfreeman.com/bleiodian2e

Learning Objectives

- Describe the bonding in alkenes.
- Draw and name constitutional isomers of alkenes.
- Draw cis-trans (geometrical) stereoisomers of alkenes.
- Describe and write equations for addition, polymerization, and oxidation reactions of alkenes.
- Draw and name alkynes.
- Describe the structure and stability of aromatics.
- Draw and name constitutional isomers of aromatics.
- Describe and write equations for substitution and other reactions of aromatics.

We began our exploration of organic chemistry in Chapter 3 with a look at the saturated hydrocarbons, compounds consisting of hydrogen and carbon in which all bonds are single bonds. Now we will examine the unsaturated hydrocarbons—the alkene, alkyne, and aromatic families—compounds consisting of hydrogen and carbon and containing carbon–carbon multiple bonds (double bonds or triple bonds).

The presence of a multiple bond greatly increases the chemical reactivity of an organic compound. As noted in Chapter 3, the reactivity of the saturated hydrocarbons is low. Alkenes and alkynes, on the other hand, are highly reactive, and the aromatic compounds, although generally more stable than the alkenes and alkynes, are also more reactive under certain conditions.

Many naturally occurring molecules contain carbon–carbon multiple bonds. Examples are animal fats and vegetable oils, natural rubber, and some proteins, nucleic acids (DNA, RNA), and vitamins. Molecules containing carbon–carbon double bonds also take part in sight and in the oxidation of fatty acids. The medical community's recommendation that we lower our dietary intake of saturated fats relative to unsaturated fats means that we should increase the proportion of fats containing carbon–carbon double bonds, which have less tendency to increase the cholesterol levels in blood. Unsaturated hydrocarbons are also the raw materials used to synthesize a wide variety of industrial and consumer products, including drugs, plastics, synthetic rubbers, dyes, soaps, and cosmetics.

A PICTURE OF HEALTH?

Examples of Unsaturated Hydrocarbons

Vitamin A, which affects skin, bones, mucosa, and teeth, as well as vision, contains the double bond. It is an essential nutrient in the human diet.

The alkene *cis*-11-retinal has a role in sight.

Many flavors and spices contain aromatic structures.

Cholesterol, regulated in the liver, contains the double bond of alkenes.

The female sex hormone estradiol, produced in the ovaries, contains an aromatic ring.

Because residues of the pesticide DDT, a chlorinated aromatic compound, can harm the body, its use has been banned in the United States.

Bent fused-ring aromatic compounds from tobacco smoke, automobile exhaust, and burnt barbecued meats are carcinogens.

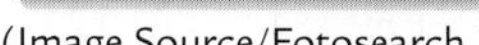

(Image Source/Fotosearch.)

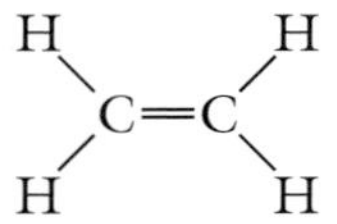

H H
C=C
H H
Ethene

4.1 ALKENES

Alkenes are unsaturated hydrocarbons containing a carbon–carbon double bond—that is, two adjacent carbon atoms joined together with two bonds. The general formula for an alkene is C_nH_{2n}, where n is an integer greater than 1. The first (smallest) member of the alkene family is C_2H_4 (IUPAC name, ethene; common name, ethylene). Table 4.1 shows the first nine members of the alkene family and their melting and boiling points. The first three members are gases at ambient temperatures; the others are liquids.

The alkenes are nonpolar compounds with physical properties similar to those of the alkanes. They are insoluble in water and have boiling points close to those of the alkanes. There is more variation in melting points among the alkenes than among the alkanes. The reasons are complex and will be considered in Section 11.2, which describes the differences between saturated and unsaturated fatty acids.

» Fatty acids are components of fats and oils as described in Section 11.3.

The general formula for an alkene, C_nH_{2n}, is the same as that for a cycloalkene. Both alkenes and cycloalkanes contain two fewer hydrogens than do alkanes, whose general formula is C_nH_{2n+2}. We can visualize the formation of an alkene as the loss by an alkane of one hydrogen from each of two adjacent carbons, which then complete their octets by forming a carbon–carbon double bond. The formation of a cycloalkane is visualized in a similar manner, except that the hydrogens are lost from (and the new bond formed between) two carbons that were nonadjacent.

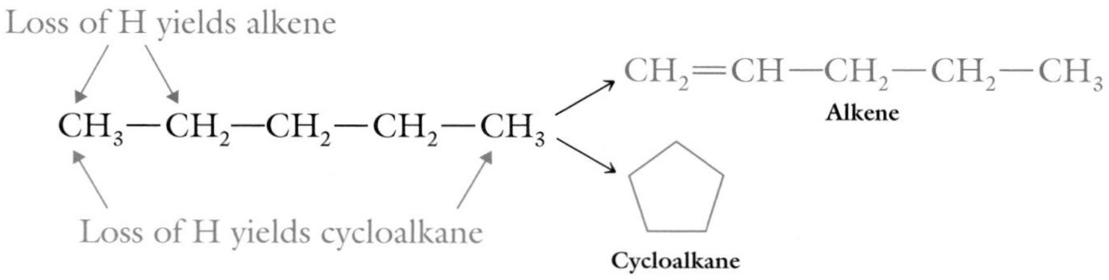

Loss of H yields alkene

$CH_3{-}CH_2{-}CH_2{-}CH_2{-}CH_3$ → $CH_2{=}CH{-}CH_2{-}CH_2{-}CH_3$ Alkene

Loss of H yields cycloalkane

Cycloalkane

Example 4.1 Distinguishing among alkanes, cycloalkanes, and alkenes

An unknown compound has the molecular formula C_7H_{14}. Is the compound an alkane, cycloalkane, or alkene?

Solution

This molecular formula does not fit the general formula for an alkane, C_nH_{2n+2}, but it does fit the general formula for both alkenes and cycloalkanes, C_nH_{2n},

TABLE 4.1 Formulas and Properties of 1-Alkenes

n	Molecular formula	Condensed structural formula	Name	Melting point (°C)	Boiling point (°C)
2	C_2H_4	$CH_2{=}CH_2$	ethene	−169	−104
3	C_3H_6	$CH_2{=}CHCH_3$	propene	−185	−47
4	C_4H_8	$CH_2{=}CHCH_2CH_3$	1-butene	−185	−6
5	C_5H_{10}	$CH_2{=}CHCH_2CH_2CH_3$	1-pentene	−138	30
6	C_6H_{12}	$CH_2{=}CHCH_2CH_2CH_2CH_3$	1-hexene	−140	63
7	C_7H_{14}	$CH_2{=}CHCH_2CH_2CH_2CH_2CH_3$	1-heptene	−119	94
8	C_8H_{16}	$CH_2{=}CHCH_2CH_2CH_2CH_2CH_2CH_3$	1-octene	−102	121
9	C_9H_{18}	$CH_2{=}CHCH_2CH_2CH_2CH_2CH_2CH_2CH_3$	1-nonene	−81	146
10	$C_{10}H_{20}$	$CH_2{=}CHCH_2CH_2CH_2CH_2CH_2CH_2CH_2CH_3$	1-decene	−66	171

4.1 CHEMISTRY AROUND US

Alkenes in Nature

Ethene, the simplest alkene, is produced as a hormone by plants such as bananas and tomatoes. Its function is to trigger the ripening process of the fruits. In the commercial fruit industry, fruit is picked before it ripens because fruit that is unripe and hard is much less damaged during shipment. On arrival at its destination, the fruit is artificially ripened by exposure to ethene.

Limonene

α-Pinene

Myrcene

Geraniol

Other alkenes are responsible for the distinguishing odors of some plants. Limonene, myracene, and α-pinene, for example, are found in the oils of citrus fruits, bay leaves, and pine trees, respectively. Geraniol is extracted from oil of roses and used to produce perfumes.

Vitamin A, another alkene-containing compound, is required for good vision. Humans either obtain it directly from animal sources or produce it within the body from plant sources of β-carotene such as carrots, spinach, and sweet potatoes. Lycopene, responsible for the red color and antioxidant properties of tomatoes, contains thirteen double bonds.

Geraniol is extracted from roses and used in perfume formulations. (Joy Spurr/Bruce Coleman.)

Vitamin A

Note that some of these molecules belong to more than one family: they contain two or more different functional groups. This multifunctionality is typical of many physiologically important molecules. Geraniol and vitamin A contain not only the carbon–carbon double bond of the alkene family, but also the —OH group of the alcohols.

with $n = 7$ in this case. The unknown compound is therefore either an alkene or a cycloalkane.

Problem 4.1 An unknown compound has the molecular formula C_7H_{16}. Is the compound an alkane, a cycloalkane, or an alkene?

Because alkenes and cycloalkanes follow the same general formula, the molecular formula alone does not allow us to distinguish between them. We need to know more about a compound, such as its chemical properties, to distinguish between the two possibilities.

Box 4.1 describes some of the alkenes found in nature.

4.2 BONDING IN ALKENES

Typical structural formulas of alkenes do not differentiate between the two bonds of a carbon–carbon double bond. The two lines of the double bond are drawn with equal weight and length. Note that this representation should not be taken to mean that the two bonds are equal in strength. In fact, the two bonds of the double bond in alkenes are not equivalent and have considerably different bond strengths. One bond is strong; about 355 kJ/mol (85 kcal/mol) are required to break it. The other bond is much weaker; its

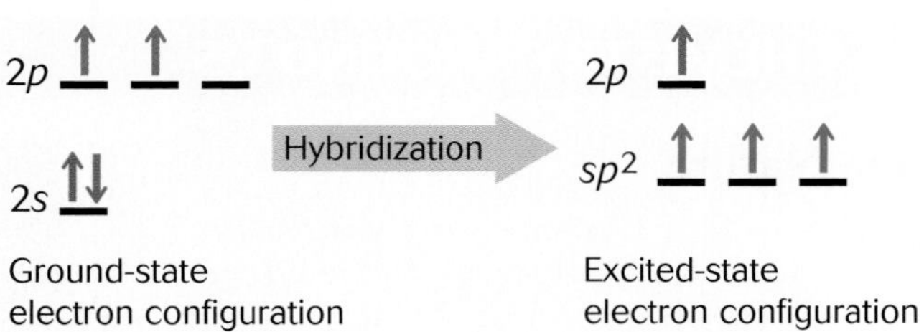

Figure 4.1 Formation of sp^2 orbitals for carbon–carbon double bonds.

energy is about 250 kJ/mol (60 kcal/mol). The strong and weak bonds are called **sigma (σ)** and **pi (π) bonds,** respectively. The weak π bond is responsible for the characteristic behavior of alkenes—their ability to undergo chemical reactions with a range of chemical reagents that have no effect on alkanes. The strong σ bond is similar in strength to the bonds in alkanes, which also are called σ bonds (Section 3.3).

The atomic orbitals used by carbon atoms to form the carbon–carbon double bond are not the sp^3 orbitals used in the single bonds of alkanes. True, the carbon atoms of a double bond are tetravalent, as are the carbon atoms in alkanes. Yet the four orbitals of a double-bonded carbon are not all equivalent, and so the bonds that they form have different strengths.

In a carbon–carbon double bond, only three of the four orbitals of each carbon (and the σ bonds that they form) are equivalent. One of those three equivalent orbitals is used to join one carbon to another in the formation of the strong σ bond of the double bond. The fourth orbital is different from the other three and is used to form the weak π bond of the carbon–carbon double bond.

In Chapter 3, we saw how carbon uses hybridized sp^3 orbitals in the formation of alkanes. The orbitals used in the formation of carbon–carbon double bonds, in contrast, are known as **sp^2 orbitals.** The sp^2 orbitals, like the sp^3 orbitals, form through an excitation process in which the atom's ground-state orbitals become hybridized. To form a carbon–carbon double bond, carbon's 2*s* orbital and two of its 2*p* orbitals hybridize (mix) to form three equivalent sp^2 orbitals, each containing one electron (Figure 4.1). One 2*p* orbital, containing one electron, is left unhybridized.

The spatial arrangement of the hybridized orbitals around the nucleus of the carbon atom is depicted on the far left in Figure 4.2. The three sp^2 orbitals lie in a flat plane at close to 120° bond angles, called the **trigonal bond angles.** The unhybridized 2*p* orbital is perpendicular to the plane of the sp^2 orbitals. This spatial arrangement of one 2*p* and three sp^2 orbitals fulfills the requirements of VSEPR theory (Section 1.7) for minimizing repulsions between electrons in those orbitals.

Figure 4.2 shows the bonding in ethene, C_2H_4, formed from two sp^2-hybridized carbon atoms and four hydrogen atoms. The strong σ bond of the double bond is formed by overlap of one sp^2 orbital from each of the two carbons. Each of the remaining two sp^2 orbitals of each carbon forms a σ bond by overlap with the 1*s* orbital of a hydrogen atom.

Lateral (sideward) overlap between the 2*p* orbital from each of the carbons results in the π bond of the double bond. The π bond is weak relative to the σ bond because the location of the 2*p* orbitals on each atom does not allow a high degree of overlap between them. It is even difficult to show the lateral

Figure 4.2 Double-bond formation in ethene. sp^2 and *p* orbitals are shown in green and blue, respectively.

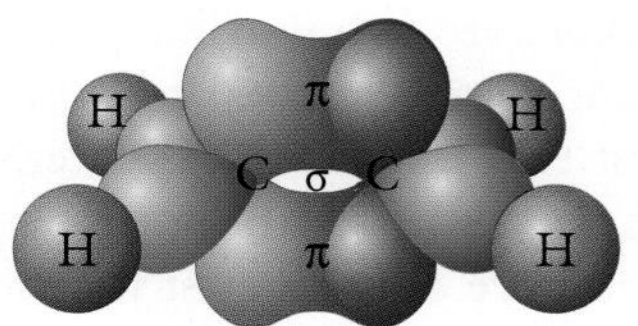

Figure 4.3 Pi-bond formation in ethene.

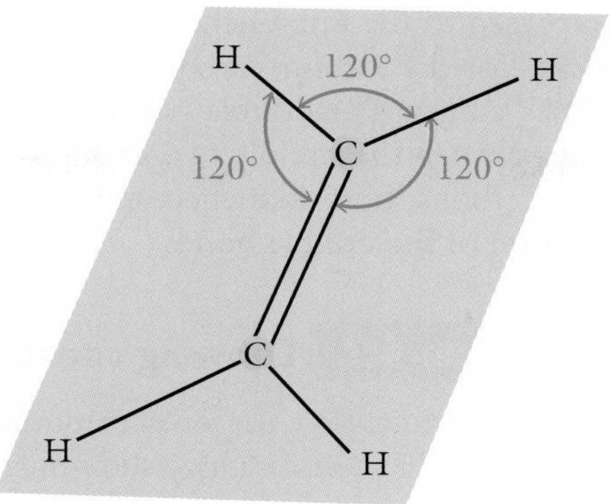

Figure 4.4 Trigonal bond angles of sp^2 carbons in the alkene double bond.

overlap of $2p$ orbitals pictorially. Dotted lines are used to indicate the π bond in Figure 4.2, and an alternate representation—one that deemphasizes the σ bond to show the π bond—is shown in Figure 4.3.

The two carbons of the double bond and the atoms to which they are attached all lie in the same plane (Figure 4.4). Note the 120° bond angles about the carbons of the double bond. The planar geometry of the atoms constituting and attached to the double bond is also evident in ball-and-stick and space-filling models (Figure 4.5).

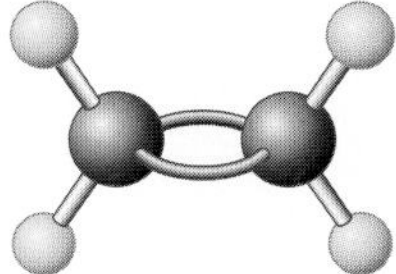

Ball-and-stick model

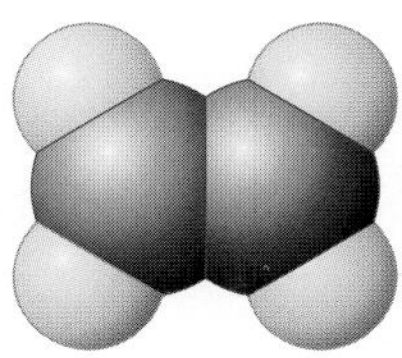

Space-filling model

Figure 4.5 Ball-and-stick and space-filling models of ethene.

4.3 CONSTITUTIONAL ISOMERS OF ALKENES

More constitutional isomers are possible for alkenes and other families than for alkanes because alkanes possess only one type of connectivity, whereas other families possess two:

- Different carbon skeletons are possible (the one option also open to alkanes).
- Different placements of the functional group (double bond for alkenes) along the carbon skeleton are possible (for all families except the alkane family).

Although there are only two C_4H_{10} alkanes, there are three C_4H_8 alkenes (Figure 4.6). In both the C_4 alkanes and the C_4 alkenes, the same two carbon skeletons are possible: one unbranched and one branched four-carbon skeleton.

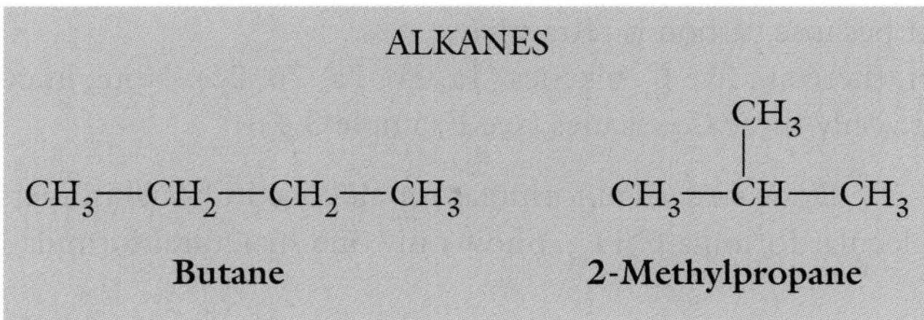

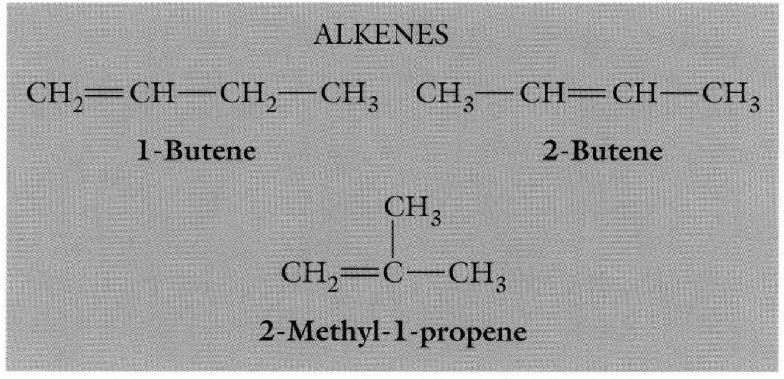

Figure 4.6 Structural isomers of C_4H_{10} and C_4H_8.

In regard to the unbranched C_4 alkene, however, there are two possible placements for the carbon–carbon double bond and, hence, an additional constitutional isomer. The double bond is between the first two carbons for 1-butene and between the middle two carbons for 2-butene. In alkenes with more than four carbons, both branched and unbranched, even more options exist for the location of the double bond.

Example 4.2 Drawing constitutional isomers

Draw structural formulas for all constitutional isomers of alkenes with the molecular formula C_5H_{10}. Be careful to show only one structural formula for each isomer.

Solution

There are two steps in solving this problem. First, draw the different carbon skeletons that are possible for branched and unbranched acyclic C_5 compounds:

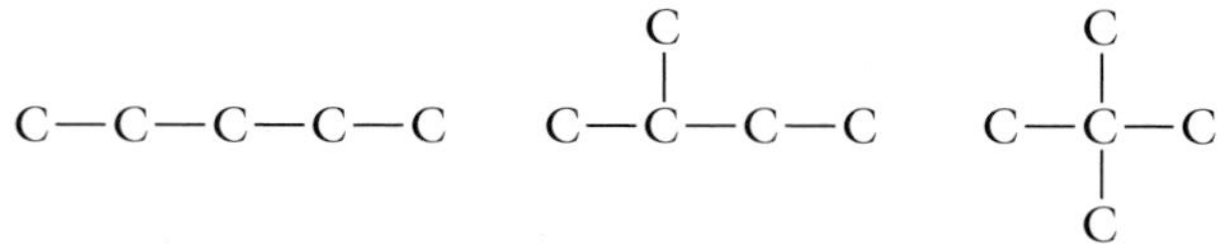

$$\underset{1}{C-C-C-C-C} \qquad \underset{2}{C-\overset{\overset{\displaystyle C}{|}}{C}-C-C} \qquad \underset{3}{C-\underset{\underset{\displaystyle C}{|}}{\overset{\overset{\displaystyle C}{|}}{C}}-C}$$

Second, consider each skeleton separately, looking for different possible locations for the double bond within that skeleton.

There are two C_5H_{10} alkenes with skeleton 1. The double bond can be located between the first two carbons of the chain (structure 1a) or between the second and third carbons (structure 1b):

$$\underset{1a}{CH_2{=}CH-CH_2-CH_2-CH_3} \qquad \underset{1b}{CH_3-CH{=}CH-CH_2-CH_3}$$

Do not draw duplicate structures. For example, placing the double bond between the first two carbons counting from the right side gives the same isomer as 1a. Placing the double bond between C2 and C3, counting from the right side, gives the same isomer as 1b.

Three constitutional isomers of C_5H_{10} have skeleton 2:

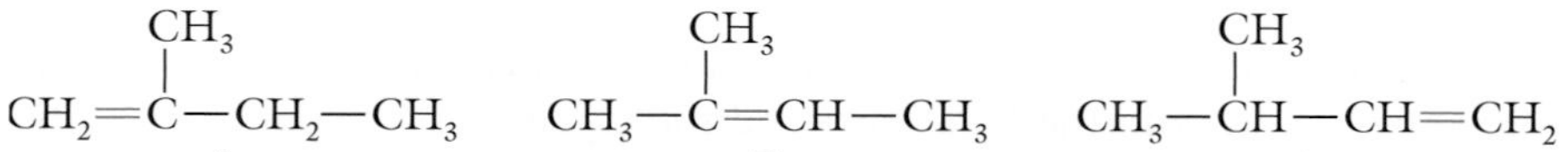

$$\underset{2a}{CH_2{=}\overset{\overset{\displaystyle CH_3}{|}}{C}-CH_2-CH_3} \qquad \underset{2b}{CH_3-\overset{\overset{\displaystyle CH_3}{|}}{C}{=}CH-CH_3} \qquad \underset{2c}{CH_3-\overset{\overset{\displaystyle CH_3}{|}}{CH}-CH{=}CH_2}$$

No alkenes are possible for skeleton 3. Placing a double bond in skeleton 3 would result in a total of five bonds to the central carbon—an impossible arrangement because carbon is tetravalent.

In short, there are five C_5 alkenes (1a, 1b, 2a, 2b, 2c). Note, in comparison, that there are only three C_5 alkanes (see Example 3.4).

Problem 4.2 Draw structural formulas for all constitutional isomers of alkenes with the molecular formula C_6H_{12}. Show only one structural formula for each isomer.

4.4 NAMING ALKENES

In the IUPAC nomenclature system, alkenes are named according to the same rules used for naming alkanes, but with a few adjustments:

Rules for naming alkenes

- For IUPAC nomenclature purposes, the longest continuous chain is redefined as the longest continuous chain containing both carbons of the double bond. For some alkenes, this chain may not be the longest continuous chain of carbons in the molecule.

- The family ending of the name is changed from **-ane** to **-ene.**
- The longest continuous chain is numbered from the end nearest the double bond. The position of the double bond is indicated by placing the lower of the two numbers for the double-bond carbons in front of the base name of the compound. (For ethene and propene, the number 1- is not required to indicate the position of the double bond, because there is only one possible placement of the double bond.)
- The names of substituents are used as prefixes, preceded by numbers that indicate their positions on the longest chain.
- A compound with two double bonds, called a **diene,** is given the ending **-adiene** instead of **-ene.** Two numbers are then provided, specifying the position of each double bond.
- A cyclic compound with a double bond in the ring is named as a **cycloalkene.** Numbering of the ring starts at one of the carbons of the double bond and proceeds through the other carbon of the double bond to substituents on the ring in the direction that yields the lower numbers for the substituents.

Example 4.3 Using the IUPAC nomenclature system to name alkenes

What is the IUPAC name for each of the following compounds?

(a) $CH_3—CH_2—\underset{\underset{CH_2}{\|}}{C}—CH_2—CH_2—CH_3$

(b) 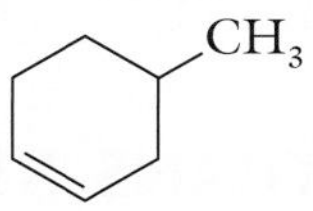 CH_3

(c) $CH_2{=}CH—CH{=}CH—CH_3$

Solution

(a) The longest continuous chain of all possible chains of carbon atoms in this molecule contains six carbons:

$$\overset{1}{CH_3}—\overset{2}{CH_2}—\overset{3}{\underset{\underset{CH_2}{\|}}{C}}—\overset{4}{CH_2}—\overset{5}{CH_2}—\overset{6}{CH_3}$$

However, this is not the longest chain for nomenclature purposes, because it contains only one of the carbons of the double bond. The longest chain for nomenclature purposes is the five-carbon chain that contains both carbons of the double bond. This chain yields pentene as the base name of the compound.

There are two possible directions for numbering this chain:

$$CH_3—CH_2—\overset{2}{\underset{\underset{\underset{1}{CH_2}}{\|}}{C}}—\overset{3}{CH_2}—\overset{4}{CH_2}—\overset{5}{CH_3} \qquad CH_3—CH_2—\overset{4}{\underset{\underset{\underset{5}{CH_2}}{\|}}{C}}—\overset{3}{CH_2}—\overset{2}{CH_2}—\overset{1}{CH_3}$$

The correct direction is the one shown on the left-hand side because it yields the lowest numbers (1,2 versus 4,5) for the carbons of the double bond. The lower of the two numbers is used in front of the base name: 1-pentene.

The complete name must also indicate the position of the ethyl group attached to the 1-pentene chain. Thus, the name of the compound is 2-ethyl-1-pentene.

(b) The numbering system for a cycloalkene always places the double bond between C1 and C2 of the ring. Thus, cycloalkene is automatically a 1-cycloalkene; we do not use the number "1." To find the position of the methyl group, we need to count through the two carbons of the double bond and toward the methyl group in the direction that assigns the lowest number to the methyl group. If we number in the counterclockwise direction, the methyl group is at C4.

Numbering in the clockwise direction places the methyl group at C5. The lower alternative is the correct one, and the compound is named 4-methylcyclohexene.
(c) With two double bonds, this compound is a diene. Numbering from left to right yields the set of lower numbers for the positions of the double bonds (1,3- versus 2,4-). The compound is named 1,3-pentadiene.

Problem 4.3 What is the IUPAC name for each of the following compounds?

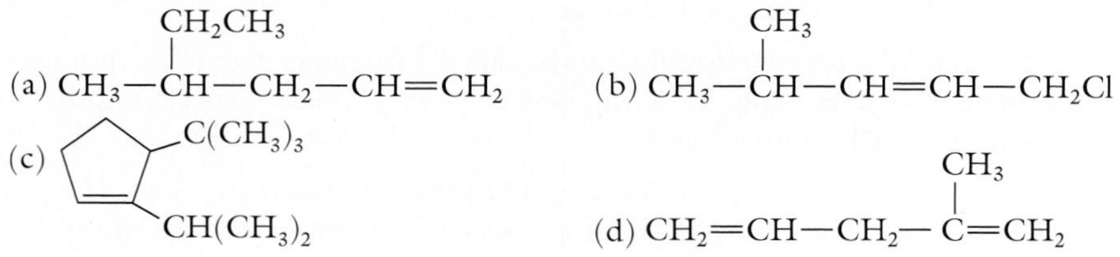

(a) $CH_3-CH(CH_2CH_3)-CH_2-CH=CH_2$ (b) $CH_3-CH(CH_3)-CH=CH-CH_2Cl$

(c) cyclopentene ring bearing $C(CH_3)_3$ and $CH(CH_3)_2$

(d) $CH_2=CH-CH_2-C(CH_3)=CH_2$

Alkenes containing no more than four carbons are often known by common names instead of IUPAC names. Common names, in use before the development of the IUPAC system, employ the ending **-ylene** instead of **-ene,** as in ethylene and propylene instead of ethene and propene, respectively. Ethylene and propylene are important industrial chemicals. More than 40 billion pounds of ethylene and 20 billion pounds of propylene are used each year in the United States, mostly to produce the plastics polyethylene and polypropylene (see Section 4.7).

4.5 CIS-TRANS STEREOISOMERISM IN ALKENES

« Single bonds in acyclic compounds have free rotation, as described in Section 3.3.

The carbon–carbon double bond, unlike the single bond in acyclic compounds, has **restricted rotation** (Section 3.8). In other words, the carbon atoms of the double bond are not free to rotate relative to each other. Certain alkenes possess geometrical, or cis-trans, stereoisomerism because of this restricted rotation. For example, there are two 2-butene compounds, not one as implied in Section 4.3. They are **cis-trans stereoisomers** that are distinguished one from the other by the italicized prefixes *cis-* and *trans-* before the IUPAC name.

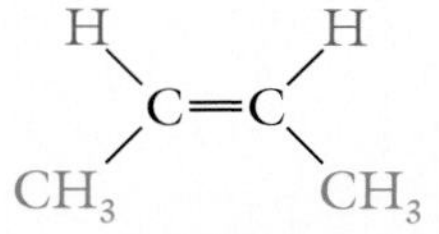

Cis-trans stereoisomerism in alkenes is analogous to cis-trans stereoisomerism in cycloalkanes (Section 3.8). In cis-trans steroisomers of cycloalkanes, the substituents on different carbons of a ring can be located either on the same side or on opposite sides of the plane of the ring. In cis-trans stereoisomers of alkenes, substituents on the carbons of a double bond are located either on the same side or on opposite sides of the double bond. Cis refers to the isomer that has the two similar substituents (hydrogens or methyls in 2-butene for example) on the same side of the double bond. The trans isomer has the two similar substituents on opposite sides of the double bond. The difference between the cis and trans isomers can also be described by the proximity between substituents. Cis substituents are closer to each other than are trans substituents. The structural differences are seen in Figure 4.7.

H, H / C=C / CH_3, CH_3

cis-2-Butene

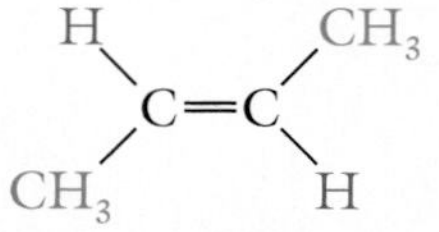

H, CH_3 / C=C / CH_3, H

trans-2-Butene

Cis and trans isomers are different compounds with differing chemical and physical properties. For example, the melting points of the cis and trans isomers of 2-butene are $-139°C$ and $-106°C$, respectively. Cis and trans isomers are different compounds because the π bond of the double bond presents a significant barrier to rotation of the two double-bonded carbons relative to each other. If rotation were free, as it is for acyclic carbon–carbon single bonds, cis and trans isomers would be easily converted one into the other, and separate compounds would not exist.

Cis and trans isomers have the same connectivity (the same order of attachment of the component atoms). Their difference resides in their **configurations,**

INSIGHT INTO PROPERTIES

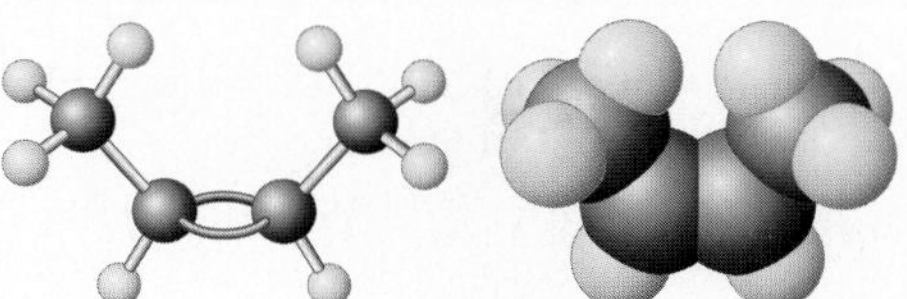

cis-2-Butene

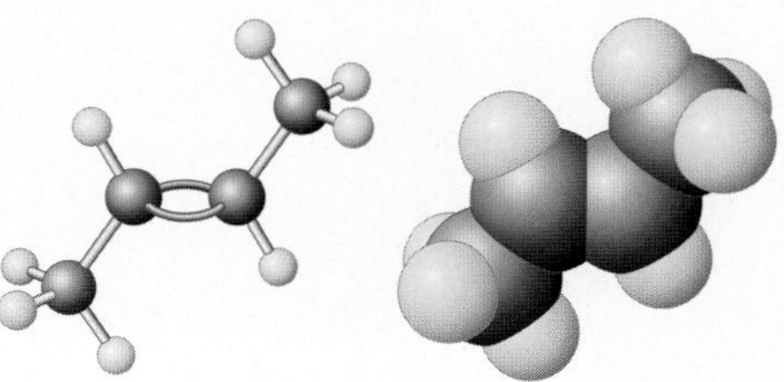

trans-2-Butene

Figure 4.7 Ball-and-stick and space-filling models of 2-butene. The cis and trans isomers are different compounds with different chemical and physical properties.

the arrangements of their atoms in space, at the carbon atoms of the double bond. Two different configurations are possible when each of the carbon atoms of the double bond is a stereocenter. (Recall from Section 3.8 that a stereocenter is an atom in a compound bearing substituents of such identity that a hypothetical exchange of the positions of any two substituents would convert one stereoisomer into another stereoisomer.) A carbon of a double bond is a stereocenter when it bears two different substituents. Such stereocenters are sometimes more precisely called **trigonal stereocenters** because the carbons have trigonal bond angles.

To better understand the relation between the cis and the trans configurations, imagine the hypothetical exchange of the groups at one or the other stereocenter. The result is the conversion of one stereoisomer into the other by reversal of the configuration at a stereocenter. Thus, an exchange of the H and CH_3 groups at either carbon of the double bond converts *cis*-2-butene into *trans*-2-butene and vice versa.

None of the following types of alkenes possesses geometrical isomerism:

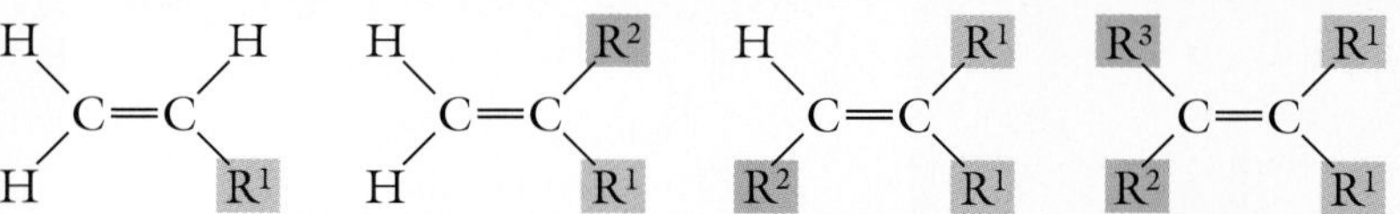

Concept check

✓ Geometrical isomers are possible only when each carbon of the double bond is a stereocenter—that is, each bears two different groups.

The chemical and physiological behaviors of cis and trans isomers often differ—another instance of the high selectivity of the molecules used in nature. Typically, the two geometrical isomers have very different physiological behaviors (Box 4.2 on the following page and Box 4.3 on page 121).

Example 4.4 Identifying and drawing cis and trans isomers

Which of the following compounds exist as a pair of cis and trans isomers? Show structural formulas of the cis and trans isomers. (a) 1-Chloropropene; (b) 1-chloro-2-methylpropene; (c) 2-pentene.

Solution

Examine each carbon of the compound's double bond. Cis-trans isomerism is possible only when each carbon of the double bond has two different groups.

4.2 CHEMISTRY WITHIN US

Vision and Cis-Trans Isomerism

Human eyesight begins with a chemical event, and cis-trans isomerism—the mechanism that allows us to perceive light—is its basis. The chemical event in question takes place in the rod and cone cells—distinctively shaped cells in the retina of the eye.

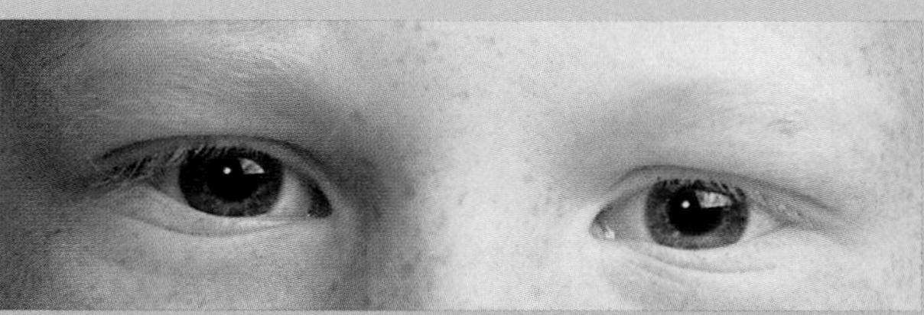

Cis-trans isomerism of 11-retinal is responsible for vision. (RubberBall Productions/PNI.)

Rod and cone cells contain rhodopsin, which is a complex of two molecules: the protein opsin and *cis*-11-retinal. The shape of *cis*-11-retinal allows it to fit snugly into a cavity (a so-called receptor site) in opsin. In this associated form, the pair is ready to receive light. *cis*-11-Retinal is produced in the body from vitamin A (see Box 4.1).

When light strikes a molecule of *cis*-11-retinal, its energy is sufficient to momentarily break the π bond of the double bond between C11 and C12. The σ bond between C11 and C12 is then, for a brief period, free to rotate. By the time the π bond becomes reestablished, the molecule has converted (isomerized) into *trans*-11-retinal, the more stable of the two isomers. This change has an important consequence: *trans*-11-retinal dissociates from the cavity of opsin because its elongated shape, relative to that of *cis*-11-retinal, no longer fits the cavity. The loss of *trans*-11-retinal results in the transmission of an electrical signal to the optic nerve, and this signal is interpreted as sight by the brain. Subsequently, the enzyme retinal isomerase forces an extremely rapid reversal of the isomerization reaction, changing *trans*-11-retinal back into *cis*-11-retinal. Rhodopsin is re-formed and ready to receive the next input of light energy.

The fit of *cis*-11-retinal and the nonfit of *trans*-11-retinal into opsin's cavity is an example of the **complementarity principle** commonly encountered in physiological processes (Sections 2.2 and 14.5). A remarkable aspect of this mechanism is the subtleness with which it operates: it takes a change in the configuration of only one of the five double bonds in *cis*-11-retinal to cause dissociation from opsin and the subsequent transmission of a light signal to the brain.

cis-11-Retinal

light breaks π bond between C11 and C12

rotation about σ bond between C11 and C12, followed by re-formation of π bond

***trans*-11-Retinal**

(a) 1-Chloropropene exists as a pair of cis-trans isomers. One carbon of the double bond has H and Cl; the other carbon has H and CH_3:

***cis*-1-Chloropropene** ***trans*-1-Chloropropene**

(b) 1-Chloro-2-methylpropene cannot exist as a pair of cis-trans isomers. One of the carbons of the double bond bears two groups (CH_3) that are the same:

1-Chloro-2-methylpropene

4.3 CHEMISTRY AROUND US

Cis-Trans Isomers and Pheromones

Pheromones are chemicals produced by insects and other animals for purposes of communication, such as attracting the opposite sex, sending a danger signal, or marking the location of food. Pheromones are specific for each species. Many pheromones contain carbon–carbon double bonds, and geometrical isomerism is critical to their physiological activities.

The specificity of its geometry makes a pheromone effective in an exceedingly small amount. The sex attractant for the common housefly is muscalure (IUPAC: *cis*-9-tricosene), and it effectively lures male houseflies when secreted by female houseflies in an amount less than 10^{-10} g.

Pheromone action depends on the pheromone's ability to bind at protein-receptor sites much like the complementarity principle for vision (see Box 4.2). Here, again, we see that there is a huge difference in physiological response to

$CH_3(CH_2)_6CH_2$, H on one carbon; $CH_2(CH_2)_{11}CH_3$, H on the other (C=C)

Muscalure

$CH_3CH_2CH_2$, H (C=C); H, $CH_2(CH_2)_7CH_2OH$ (C=C)

Bombykol

cis and trans structures. The trans isomer of muscalure has been synthesized in the laboratory and found to be totally inactive as a sex attractant for houseflies.

Bombykol (IUPAC: *trans*-10-*cis*-12-hexadecadien-1-ol) is the sex attractant for the silkworm moth. Pheromone action in this case is, again, critically dependent on geometrical isomerism. The physiologically active bombykol has a trans double bond between C10 and C11 and a cis double bond between C12 and C13 (carbons are numbered from the end of the chain bearing the —OH group). Other geometric isomers of bombykol (for example, the *cis*-10-*trans*-12 isomer instead of *trans*-10-*cis*-12) are incapable of eliciting a sexual response from male silkworm moths.

Pheromone action is being studied as an alternative to pesticide use in the control of insect populations. The critical step comprises identification and industrial synthesis of the sex pheromone for a particular species. The pheromone is placed in traps to lure male insects and stop the reproductive cycle. An alternate strategy is to spray the pheromone over a large area, such as an apple orchard, to confuse male insects and make it difficult for them to locate mates.

Bombykol's pheromone action as a sex attractant for the silkworm moth depends on geometrical isomerism. (Corbis.)

(c) 2-Pentene exists as a pair of cis-trans isomers. One carbon of the double bond has H and CH_3, the other carbon has H and CH_3CH_2:

H, CH_3 (C=C) H, CH_2CH_3 — *cis*-2-Pentene

CH_3, H (C=C) H, CH_2CH_3 — *trans*-2-Pentene

Problem 4.4 Which of the following compounds exist as a pair of cis and trans isomers? Show structural drawings of the cis and trans isomers. (a) 2-Methyl-2-pentene; (b) 3-methyl-2-pentene; (c) 4-methyl-2-pentene.

The use of the cis-trans nomenclature system is restricted to 1,2-disubstituted alkenes—that is, alkenes in which each carbon of the double bond has one hydrogen and one substituent other than hydrogen. This restriction is analogous to that of the cis-trans nomenclature system for cycloalkanes in which each of two carbons of the ring has one hydrogen and one substituent other than hydrogen (Section 3.8). Designation as a cis or trans isomer is ambiguous for alkenes in which either one or both carbons of the double bond have two different substituents but no hydrogen, such as 2-bromo-3-chloro-2-butene:

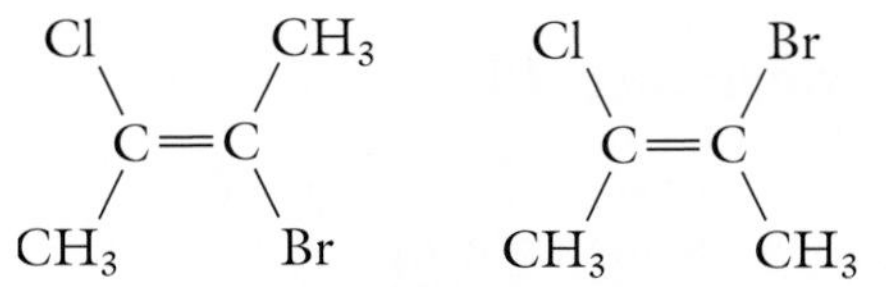

$$\underset{CH_3}{\overset{Cl}{\diagdown}}C{=}C\overset{CH_3}{\underset{Br}{\diagup}} \qquad \underset{CH_3}{\overset{Cl}{\diagdown}}C{=}C\overset{Br}{\underset{CH_3}{\diagup}}$$

2-Bromo-3-chloro-2-butene

The terms cis and trans would not unambiguously differentiate the two stereoisomers. No convention in the IUPAC nomenclature system specifies whether the terms cis and trans refer to the relation of the two methyl groups or of the methyl and Br or of the methyl and Cl. A different nomenclature system, the *E,Z*-nomenclature system, is used to differentiate these stereoisomers but is beyond the scope of this book.

4.6 ADDITION REACTIONS OF ALKENES

Although alkenes and alkanes have similar physical properties because they are both nonpolar compounds, the two families are very different in their chemical properties. Alkenes are highly reactive, whereas alkanes are not. The presence of the weak π bond in alkenes is responsible for their high reactivity. Alkanes possess only the stronger σ bond. Alkenes undergo reaction with a wide range of chemical reagents to which alkanes are entirely unresponsive.

The π bond of an alkene is sufficiently weak to undergo reaction with several chemical reagents:

- Hydrogen (H_2)
- Halogen (F_2, Cl_2, Br_2)
- Hydrogen halide (HCl, HBr, HI)
- Water

With each of these reagents, alkenes undergo the same general type of reaction, an **addition reaction.** All the elements present in the two reactant molecules end up "added together" in the one product molecule, as described by the general equation

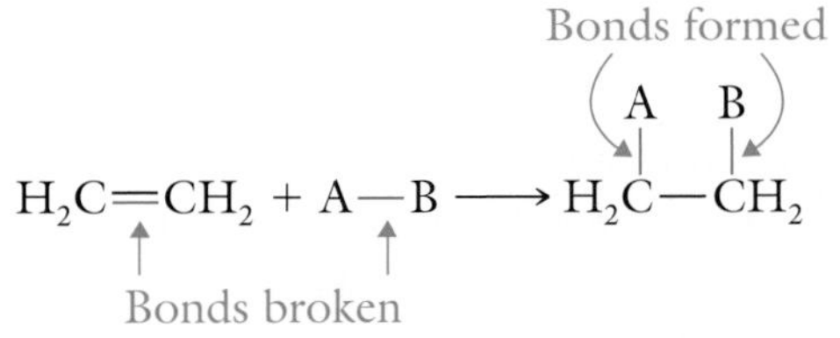

Bonds formed

$$H_2C{=}CH_2 + A{-}B \longrightarrow H_2\overset{A}{\overset{|}{C}}{-}\overset{B}{\overset{|}{C}}H_2$$

Bonds broken

where A—B symbolizes the chemical reagent. The reagent A—B breaks into two reactive fragments A and B, whose presence forces the π bond to break. There is an interim loss of completed octets for the carbons of the double bond and for the A and B fragments. Formation of the product, however, through formation of a σ bond between each carbon and an A or B fragment, once again completes all the atoms' octets.

Addition of Symmetrical Reagents (A = B)

Hydrogen and halogen are **symmetrical reagents**; that is, the two fragments of the A—B reagent molecule are the same. The addition of hydrogen (H_2) adds one hydrogen to each carbon of the double bond and converts the alkene into an alkane:

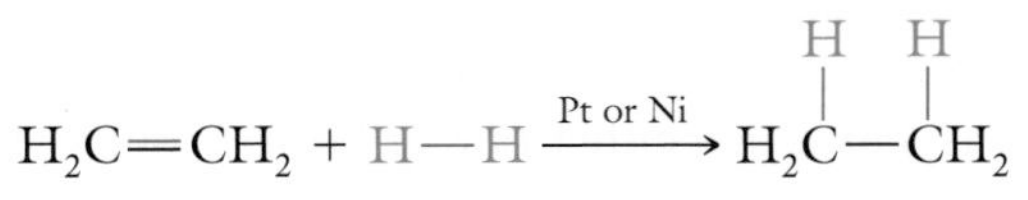

$$H_2C{=}CH_2 + H{-}H \xrightarrow{\text{Pt or Ni}} H_2\overset{H}{\overset{|}{C}}{-}\overset{H}{\overset{|}{C}}H_2$$

The addition of hydrogen proceeds readily at ambient temperatures, but only in the presence of a metal catalyst such as platinum (Pt) or nickel (Ni). For that reason, the reaction is called **catalytic hydrogenation.** Note that the need for a Pt or Ni catalyst is indicated on the arrow in the preceding equation. Writing the equation without specifying the presence of Pt or Ni is incorrect, because it conveys incorrect information, wrongly suggesting that the reaction will take place as long as both reactants are present. The addition of hydrogen to the double bond does not take place in the absence of Pt or Ni. Recall that a catalyst accelerates the rate of a chemical reaction by providing an alternate reaction pathway (with a lower activation energy) for the reaction.

« A catalyst increases the rate of a reaction but emerges unchanged after the reaction has ended, as described in Section 2.2.

Hydrogenation is a **reduction reaction** (Section 2.1). The reduction of an organic compound, the opposite of oxidation, corresponds to a decrease in the number of oxygen atoms or an increase in the number of hydrogen atoms (or both) bonded to the carbon atoms of the compound (Section 3.10). Each carbon of ethene's double bond is bonded to two hydrogens, whereas, after hydrogenation, each carbon is bonded to three hydrogens. Hydrogenation is used in the manufacture of margarine from vegetable oils, as we will see in Section 11.4.

The addition of halogens to the alkene double bond entails the addition of a halogen atom to each of the carbons of the double bond:

$$H_2C{=}CH_2 + Br{-}Br \longrightarrow \underset{\displaystyle H_2C{-}CH_2}{\overset{Br\quad Br}{|\quad\;\; |}}$$

The reaction, called **halogenation,** proceeds readily at ambient temperatures.

The rapid reaction of alkenes with hydrogen and halogens at ambient temperatures emphasizes the large difference in reactivity between alkenes and alkanes. Alkanes do not react at all with hydrogen. Although alkanes react with halogens, the reaction is of the substitution type (Section 3.10), not the addition type as in alkenes, and takes place only in the presence of ultraviolet radiation or at high temperatures (200°C).

« In the halogenation of alkenes, halogen adds to the double bond; in the halogenation of alkanes (Section 3.10), halogen substitutes for hydrogen.

Example 4.5 Writing equations for addition reactions of alkenes: I

Complete each of the following reactions by writing the structure of the organic product. If no reaction takes place, write "no reaction."

(a) $CH_3{-}CH{=}CH{-}CH_3 \xrightarrow{Cl_2} ?$

(b) $CH_3{-}CH{=}CH_2 \xrightarrow{H_2} ?$

(c) $CH_3{-}CH{=}CH_2 \xrightarrow[Ni]{H_2} ?$

Solution

The first step in solving this problem is to determine whether the reagent is one that adds to the double bond. If it is, proceed to break the reagent into its component fragments and add them to the carbons of the double bond.

(a) $CH_3{-}CH{=}CH{-}CH_3 \xrightarrow{Cl_2} CH_3{-}\overset{\overset{\displaystyle Cl}{|}}{C}H{-}\overset{\overset{\displaystyle Cl}{|}}{C}H{-}CH_3$

(b) No reaction, because a catalyst (Pt or Ni) is not present.

(c) The required catalyst is present, and addition takes place:

$$CH_3{-}CH{=}CH_2 \xrightarrow[Ni]{H_2} CH_3{-}CH_2{-}CH_3$$

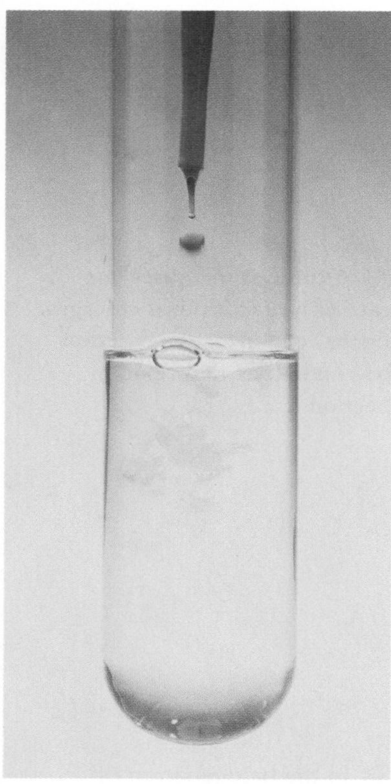

Figure 4.8 1-Hexene gives a positive test with Br_2. The deep red color of Br_2 is decolorized as it reacts with the double bond of 1-hexene. (Richard Megna/Fundamental Photographs.)

Problem 4.5 Complete each of the following reactions by writing the structure of the organic product. If no reaction takes place, write "no reaction."

(a) $CH_3—CH=CH—CH_3 \xrightarrow[Ni]{H_2} ?$

(b) $CH_3—CH=CH_2 \xrightarrow{Br_2} ?$

The ease with which bromine undergoes addition to alkenes provides chemists with a **simple chemical diagnostic test** for alkenes. A simple chemical diagnostic test is a reaction that takes place rapidly and gives a positive result that can be seen by the human eye. Most chemical reactions do not fit that description. Many chemical reactions proceed rapidly but give no visible indication of reaction. In most of them, colorless chemicals are mixed with other colorless chemicals to form colorless products. Bromine addition to alkenes is an exception. Bromine has a deep red color, but the product of bromine addition to an alkene is colorless. This color change affords us a ready means of differentiating between alkenes and alkanes. We add some bromine to an unknown compound and wait a few minutes (Figure 4.8):

- If the red color disappears, addition has taken place, and the unknown is an alkene.
- If the red color does not disappear, addition has not taken place, and the unknown is an alkane.

Example 4.6 **Identifying alkenes and alkanes**

An unknown compound is either hexane or 1-hexene. When bromine is added to the test tube, the red color of the bromine persists. Identify the unknown.

Solution

The compound is hexane because hexane does not react with bromine and decolorize it. If the unknown were 1-hexene, it would have reacted with bromine and decolorized it.

Problem 4.6 An unknown compound is pentane, cyclopentane, or 2-pentene. The red color of bromine is decolorized when bromine is added to the unknown. Identify the unknown.

Addition of Asymmetrical Reagents (A ≠ B)

Hydrogen halides (HCl, HBr, HI) and water (H_2O) are **asymmetrical reagents**—that is, A—B reagents whose two fragments are not the same. In all of these reagents, H is one of the fragments, and the other fragment is either a halogen or —OH.

The addition of a hydrogen halide, called **hydrohalogenation,** to an alkene adds hydrogen to one carbon of the double bond and a halogen to the other carbon:

$$H_2C=CH_2 + H—Cl \longrightarrow \underset{\text{H}}{\overset{}{}}H_2C(H)—CH_2(Cl)$$

The addition of water to an alkene adds hydrogen to one carbon of the double bond and OH to the other carbon:

$$H_2C=CH_2 + H—OH \xrightarrow{H^+} H_2C(H)—CH_2(OH)$$

Note that the addition of water, called **hydration,** requires the presence of a strong acid catalyst, such as sulfuric acid (H_2SO_4), as denoted by H^+ above the reaction arrow. The product of hydration is an alcohol, R—OH. Hydration is used industrially to produce various alcohols and is an important reaction in the metabolism of carbohydrates, fats, and proteins.

These reactions, like those with hydrogen and halogen, show the high reactivity of alkenes relative to alkanes. Alkanes do not undergo reaction with hydrogen halide or water.

Symmetrical alkenes are alkenes that have the same groups on each carbon of the double bond. Asymmetrical alkenes carry different groups on the two carbons of the double bond.

- Symmetrical alkenes: $CH_2{=}CH_2$, $RCH{=}CHR$, $R_2C{=}CR_2$
- Asymmetrical alkenes: $CH_2{=}CHR$, $CH_2{=}CR_2$, $RCH{=}CR_2$

The addition of an asymmetrical reagent (a hydrogen halide or water) to an asymmetrical alkene is more complex than the addition of either a symmetrical or an asymmetrical reagent to a symmetrical alkene. Only one product is possible for the last two cases. Two products are possible for the first case. Table 4.2 shows the different products formed from various combinations of symmetrical and asymmetrical reagents and alkenes.

As shown in Table 4.2, two products are possible for the addition of an asymmetrical reagent to an asymmetrical alkene because there are two ways for the addition to take place. Consider the addition of HCl to propene. One way is for HCl to add so that H adds to C1 and Cl adds to C2:

$$\overset{\text{H—Cl}}{CH_2{=}CH{-}CH_3} \longrightarrow \underset{\text{H}}{\overset{|}{}} \; CH_2{-}\overset{\text{Cl}}{\overset{|}{C}}H{-}CH_3$$

$$CH_2{=}CH{-}CH_3 \longrightarrow \overset{\text{H}}{\overset{|}{C}}H_2{-}\overset{\text{Cl}}{\overset{|}{C}}H{-}CH_3$$

2-Chloropropane
Major product

The other way is the reverse; HCl adds so that Cl adds to C1 and H adds to C2:

$$CH_2{=}CH{-}CH_3 \longrightarrow \overset{\text{Cl}}{\overset{|}{C}}H_2{-}\overset{\text{H}}{\overset{|}{C}}H{-}CH_3$$

1-Chloropropane
Minor product

The two possible products are 2-chloropropane and 1-chloropropane, respectively. This reaction is an example of one that has two competing pathways.

TABLE 4.2 Addition Reactions of Alkenes

	Symmetrical alkene ○═○	Asymmetrical alkene ○═□
	Products	
Symmetrical reagent ●—●	● ● on ○—○	● ● on ○—□
Asymmetrical reagent ●—■	● ■ on ○—○	● ■ on ○—□; ■ ● on ○—□

When two products are possible, some reactions proceed in a **nonselective** manner and give more or less equal amounts of both. Other reactions are **selective,** with one product formed in greater amounts than the other. The two products are called **major** and **minor** products, respectively. Additions of asymmetrical reagents to asymmetrical alkenes are selective reactions. One of the two possible products is formed in much greater abundance. 2-Chloropropane is the major product ($>90\%$) in the addition of HCl to propene.

Additions of asymmetrical reagents to asymmetrical alkenes follow the **Markovnikov rule** proposed in 1869 by Russian chemist Vladimir Markovnikov:

- The major product is the one formed when the hydrogen from the asymmetrical reagent adds to the carbon of the double bond that already had more hydrogens before the reaction.

The rule is often paraphrased as "the rich get richer": Of the two carbons in the alkene double bond, the one already relatively "rich" in H gets even "richer" in H. For the reaction of propene with HCl, C1 of propene has more hydrogens than does C2; and as a result, the H of HCl is more likely to become attached to C1.

Only one product is possible when either the reagent or the alkene is symmetrical, because there is only one way for the addition to take place.

The reason that reactions follow the Markovnikov rule has to do with the reactions' mechanism, described in Boxes 4.4 and 4.5.

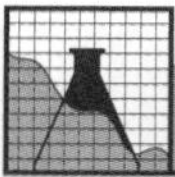

4.4 CHEMISTRY IN DEPTH

Mechanism of Alkene Addition Reactions

Our considerations of chemical reactions have used equations to describe what chemicals (reactants) undergo change and what new chemicals (products) are formed. We have not described the molecular-level details of how atoms are exchanged, how bonds are broken and reformed—in short, how the reactants change into the products. The "molecular-level how" of a reaction is called its **mechanism.** Chemists study reaction mechanisms so that they can better predict the outcome of chemical reactions and find ways of enhancing the efficiency with which important chemical products may be produced.

Most reaction mechanisms consist of more than one step. For example, addition reactions of alkenes, with the exception of hydrogenation, proceed by a common two-step mechanism in which ions play key roles. (The addition of hydrogen also proceeds by a multistep process, but neutral hydrogen atoms instead of ions have the key roles.) Consider the addition of HCl to ethene.

- **Step 1.** HCl interacts with the electron pair of the weak π bond of ethene. The H–Cl bond breaks to form H^+ and Cl^-, and the H^+ adds to ethene:

$$CH_2{=}CH_2 + H{-}Cl \longrightarrow CH_3{-}CH_2^+ + Cl^-$$

Carbocation

Ions rather than neutral atoms are formed when the H–Cl bond breaks, a form of bond breakage described as **heterolytic** (uneven). Because chlorine is more electronegative than hydrogen, chlorine carries away both electrons when the HCl bond breaks, thus forming Cl^-. Hydrogen becomes H^+ because it carries away no electrons from the bond undergoing cleavage.

The process takes place by a "flow" of electron pairs, as shown by the curved arrows. The electron pair of the π bond of ethene moves away from one of the carbons of the double bond and forms a bond to H^+, and, simultaneously, the electron pair of the H–Cl bond moves away from H and onto Cl to form Cl^-. This flow results in neutralization of the positive charge on H^+ and the creation of a new carbon–hydrogen bond. The carbon from which the electrons flowed is now electron deficient and carries a positive charge. An organic species carrying a positive charge on a carbon atom is called a **carbocation.**

- **Step 2.** Bond formation between the positive carbocation and the negative chloride ion yields the final product:

$$CH_3{-}CH_2^+ + Cl^- \longrightarrow CH_3{-}CH_2{-}Cl$$

Carbocation

4.5 CHEMISTRY IN DEPTH

Carbocation Stability and the Markovnikov Rule

We have seen that many organic reactions have more than one possible product. As in the addition of HCl (an asymmetrical reagent) to propene (an asymmetrical alkene), the two products are usually not formed in equal amounts. Each product is the result of a different mechanistic pathway, one of which will be favored over the other. Often the reason has to do with the presence of carbocation intermediates (see Box 4.4).

In the addition of HCl to propene, the two reaction pathways are

$$CH_2{=}CH{-}CH_3$$

PATHWAY 1: $CH_2{=}CH{-}CH_3 \xrightarrow{H^+} CH_3{-}\overset{+}{C}H{-}CH_3$ (2° carbocation) $\xrightarrow{Cl^-} CH_3{-}CH(Cl){-}CH_3$

PATHWAY 2: $CH_2{=}CH{-}CH_3 \xrightarrow{H^+} \overset{+}{C}H_2{-}CH_2{-}CH_3$ (1° carbocation) $\xrightarrow{Cl^-} CH_2(Cl){-}CH_2{-}CH_3$

Each pathway begins with the formation of a different carbocation, and there are differences in stability between them. These differences determine the winner of the competition between the two possible pathways for this reaction. The pathway proceeding through the more stable carbocation is the winner.

The stability of carbocations follows the order

$$\underset{3°}{R{-}\overset{+}{C}(R){-}R} > \underset{2°}{R{-}\overset{+}{C}(H){-}R} > \underset{1°}{R{-}\overset{+}{C}(H){-}H}$$

Tertiary, secondary, and **primary carbocations** are carbocations having three, two, and one nonhydrogen substituent, respectively, attached to the carbon carrying the positive (+) charge. The more stable a carbocation, the more easily and quickly it is formed. Reactions that proceed through more-stable carbocations take place at faster rates.

In the addition of HCl to propene, pathway 1 contains a secondary carbocation, whereas pathway 2 contains a primary carbocation. Because secondary carbocations are more stable and form faster than primary carbocations, more molecules of propene travel down pathway 1 than pathway 2. The overall result is that more 2-chloropropane than 1-chloropropane is formed. This result is predicted by the Markovnikov rule.

Example 4.7 Writing equations for addition reactions of alkenes: II

Complete each of the following reactions by writing the structure of the organic product. If no reaction takes place, write "no reaction." If there is more than one product, indicate which is the major and which is the minor product.

(a) $CH_2{=}CH{-}CH_3 \xrightarrow{Br_2} ?$

(b) $CH_3{-}CH{=}CH{-}CH_3 \xrightarrow{HCl} ?$

(c) $CH_2{=}CH{-}CH_2{-}CH_3 \xrightarrow[H^+]{H_2O} ?$

Solution

(a) Only one product is possible, because the reagent is symmetrical.

$$CH_2{=}CH{-}CH_3 \xrightarrow{Br_2} CH_2(Br){-}CH(Br){-}CH_3$$

(b) Only one product is possible, because the alkene is symmetrical.

$$CH_3{-}CH{=}CH{-}CH_3 \xrightarrow{HCl} CH_3{-}CH(H){-}CH(Cl){-}CH_3$$

(c) Two products are possible, because both the reagent and the alkene are asymmetrical. The Markovnikov rule tells you which product is the major product.

$$CH_2{=}CH{-}CH_2{-}CH_3 \xrightarrow[H^+]{HOH} \underset{\textbf{Major}}{CH_2(H){-}CH(OH){-}CH_2{-}CH_3} + \underset{\textbf{Minor}}{CH_2(OH){-}CH(H){-}CH_2{-}CH_3}$$

4.6 CHEMISTRY AROUND US

Synthetic Addition Polymers

Synthetic polymers are a key reason for the high standard of living in developed countries. The great utility of these materials is explained by a number of factors:

- Low cost
- Excellent combination of properties
- Ease of fabrication into products

The key property that makes polymers useful is their physical strength, the ability of the polymer in the form of a bottle, packaging film, a heart valve, or other product to carry a load without becoming distorted in shape or breaking. The physical strength of organic polymers is a consequence of their high molecular masses. In Section 3.9, we noted that secondary forces increase with increasing molecular mass, thereby causing corresponding increases in boiling and melting points. When molecular masses reach such magnitudes that a polymer's physical strength can be measured, we find that physical strength also increases with increasing molecular mass. Commercial polymers, ranging in molecular mass from 20,000 to 1,000,000, are chosen according to the strength needed for each particular application.

Synthetic polymers are inexpensive because the raw materials needed for making them, chiefly petroleum, are abundant and cheap. Aside from being a common source of fuel for generating heat and power, petroleum is converted by chemical reactions into the variety of monomers needed to produce useful polymers. Chemists have learned how to manage the reactions for producing and polymerizing monomers so that they run with maximum efficiency and economy to produce a polymer of the required molecular mass.

Another useful quality of synthetic polymers is their general inertness to broad ranges of conditions. Polymers used for food packaging are not affected by any moisture content in the food and therefore do not dissolve on contact with the food and contaminate it. Polymers used in medical devices (syringes, artificial joints) are inert to the physiological environment and therefore are less likely to trigger an allergic or other immune reaction. Furthermore, they can be sterilized.

Low cost and excellent properties would not guarantee the usefulness of synthetic materials were it not for the ease with which the materials are fabricated. The products made from polymers fall into three categories—plastics, rubbers, and fibers—all easily and economically fabricated into a variety of shapes and products. End products include sheets, films, bottles and other containers, hoses, gaskets, appliance housings, foam cushioning, fast-food containers, textile fabrics for clothes and carpeting, vehicle dashboards and tires, and rubber bands (see Figure 4.9).

Large parts of buildings and their furnishings and appliances, automobiles, trucks, aircraft, and industrial equipment are made from synthetic polymers. More than 100 billion pounds of synthetic polymers are produced and used each year in the United States. The worldwide annual production is more than 300 billion pounds.

Not represented among the addition polymers listed in Table 4.3 is an important class of addition polymers

Problem 4.7 Complete each of the following reactions by writing the structure of the organic product. If no reaction takes place, write "no reaction." If there is more than one product, indicate which is the major and which is the minor product.

(a) $CH_2{=}CH{-}CH_2{-}CH_3 \xrightarrow{Cl_2} ?$

(b) $CH_2{=}CH{-}CH_3 \xrightarrow[H^+]{H_2O} ?$

(c) $CH_2{=}CH{-}CH_2{-}CH_3 \xrightarrow{HCl} ?$

4.7 ADDITION POLYMERIZATION

The age we live in is often called the Age of Synthetic Polymers. This distinction results from the ability of alkenes as well as some other families of organic molecules to add to one another like links in a chain, thereby forming very large molecules (called **polymers**) consisting of hundreds or thousands of smaller units. Box 4.6 describes the many practical applications of synthetic polymers as plastics, fibers, and rubbers. In addition to the synthetic polymers produced commercially (Figure 4.9), there are the natural polymers (carbohydrates, proteins, and nucleic acids) produced by living organisms.

that are produced from dienes—that is, from monomers containing two double bonds. The dienes used for this purpose are 1,3-butadiene, isoprene (2-methyl-1,3-butadiene), and chloroprene (2-chloro-1,3-butadiene). Addition polymers formed from dienes are used as synthetic rubbers. Polybutadiene and polyisoprene are used for tires, footwear, gloves, and adhesives. Polychloroprene is used for adhesives, industrial belts and hoses, seals for building and highway joints, and roof coatings. The polymer structures obtained from diene monomers are more complicated than those from monomers containing only one double bond.

$$CH_2{=}CH{-}CH{=}CH_2$$
1,3-Butadiene

$$CH_2{=}C(CH_3){-}CH{=}CH_2$$
Isoprene

$$CH_2{=}C(Cl){-}CH{=}CH_2$$
Chloroprene

Synthetic polymers contribute greatly to our lives, but there are problems attendant on their use. Polymers are usually not biodegradable: the microorganisms in the ground do not attack and break them down. Instead, polymers accumulate in our landfills in ever-increasing amounts. Two strategies have been used to resolve this problem. Many communities have banned or restricted certain uses of polymers (for example, in fast-food packaging). Other communities have instituted recycling programs.

Unfortunately, the recycling of polymers is neither simple nor economical, because of the many different kinds of polymers being used. When all polymers are mixed together into a composite product, the result is a low-grade material of little practical use. Sorting the different polymers during recycling produces useful materials, but the process is expensive. Efforts to improve the economics of recycling continue.

The specific plastic used in a product is identified by a number and an abbreviated name to allow sorting in recycling: PETE or PET (1) for poly(ethylene terephthalate); HDPE (2) for high-density polyethylene; V or PVC (3) for poly(vinyl chloride); LDPE (4) for low-density polyethylene; PP (5) for polypropylene; PS (6) for polystyrene; OTHER (7) for other plastics. The number is usually located inside a triangle and the symbol outside the triangle.

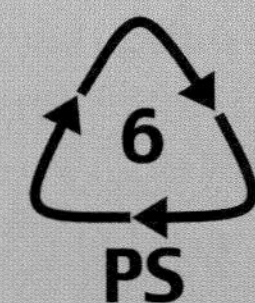

Recycling symbol for polystyrene.

Another problem with polymers is the toxic gases produced by many of them when they burn; for example, hydrogen chloride and hydrogen cyanide, respectively, are released when poly(vinyl chloride) and polyacrylonitrile burn. The toxic gases produced during a fire make it harder for occupants to escape and for firefighters to fight the fire, and they can cause health problems in survivors and firefighters.

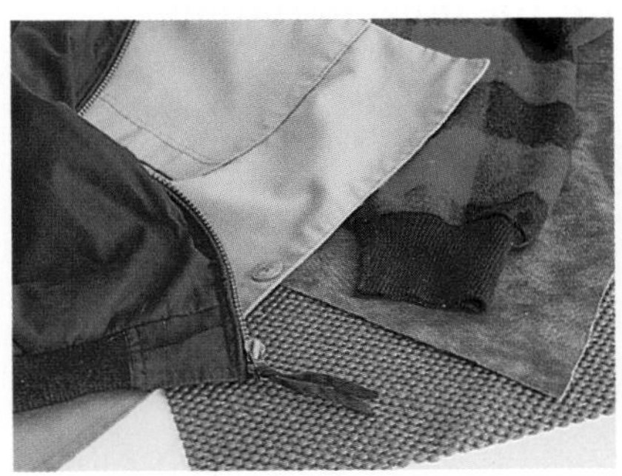

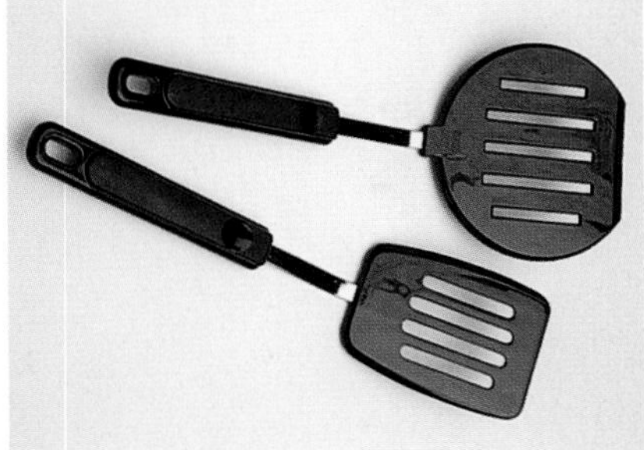

Figure 4.9 Commercial products created through addition polymerization. (Top row, left and center, and bottom row: Tony Freeman/PhotoEdit. Top row, right: Felicia Martinez/PhotoEdit.)

Addition polymerization is similar to the addition reactions described in Section 4.6 except that there is no reactant other than the alkene. Addition polymerization is a self-addition reaction in which many thousands of alkene molecules, individually called **monomers,** add to one another to form the polymer molecules. A representative part of such a process is shown here, illustrated by the polymerization of ethylene (IUPAC: ethene)

$$\underset{\text{Monomer}}{CH_2{=}CH_2 + CH_2{=}CH_2 + CH_2{=}CH_2} + \text{etc.} \longrightarrow$$

$$\underset{\text{Polymer}}{\sim CH_2{-}CH_2{-}CH_2{-}CH_2{-}CH_2{-}CH_2 \sim}$$

where ~ is used to indicate that the polymer structure goes on and on, with large numbers of monomers linked together. The linkage process relies on the breaking of π bonds in the monomer molecules, which are then linked together by new σ bonds. The resulting linked CH_2CH_2 units are called **repeat units.** Addition polymerization is one of two types of polymerization reactions; the other type is called condensation polymerization.

›› Condensation polymerization produces textile materials called polyesters and polyamides, as described in Sections 7.10 and 8.12.

The addition polymerization reaction is abbreviated as

$$\underset{\text{Ethylene}}{n\,CH_2{=}CH_2} \xrightarrow{\text{polymerization catalyst}} \underset{\text{Polyethylene}}{+\!\!\left(CH_2{-}CH_2\right)\!\!+_n}$$

where n is a large number, usually in the thousands. This notation emphasizes the structures of the monomer and repeat unit. Many alkenes, containing a range of substituents, undergo polymerization. Table 4.3 lists a few of the most important commercial addition polymers.

The polymerization reaction requires the presence of a catalyst. Moreover, polymerization will not take place in the presence of addition reagents (hydrogen, halogens, hydrogen halides, water, sulfuric acid), because they prevent polymerization by preferentially adding to the double bond. The type of catalyst required for polymerization is beyond the scope of this book and will be referred to simply as "polymerization catalyst."

Example 4.8 Writing equations for the polymerization of alkenes

Write the equation for the polymerization of 1-chloroethene (common name: vinyl chloride) to form poly(vinyl chloride).

Solution

$$n\,CH_2{=}\underset{\displaystyle Cl}{\underset{|}{CH}} \xrightarrow{\text{polymerization catalyst}} +\!\!\left(CH_2{-}\underset{\displaystyle Cl}{\underset{|}{CH}}\right)\!\!+_n$$

Problem 4.8 Write the equation for the polymerization of propene (common name: propylene) to form polypropylene.

4.8 THE OXIDATION OF ALKENES

Like other organic compounds, alkenes will combust, or burn in air (Section 3.10). Where sufficient oxygen is available, each carbon and hydrogen atom in the compound becomes oxidized to its highest oxidation state (CO_2 and H_2O, respectively). The balanced combustion equation for ethene is therefore

$$CH_2{=}CH_2 + 3\,O_2 \longrightarrow 2\,CO_2 + 2\,H_2O$$

Alkenes also undergo selective oxidations in which only the carbons of the double bond are oxidized. This reaction takes place with a variety of laboratory oxidizing agents, such as permanganate (MnO_4^-) and dichromate ($Cr_2O_7^{2-}$)

TABLE 4.3 Structure and Uses of Addition Polymers

Monomer name and structure	Polymer: Name and structure	Polymer: Uses
ethylene $CH_2{=}CH_2$	polyethylene $\text{+}CH_2{-}CH_2\text{+}_n$	food and detergent bottles; toys and housewares; electrical wire and cable insulation; plastic sheeting for agricultural use
propylene $CH_2{=}CH{-}CH_3$	polypropylene $\text{+}CH_2{-}CH(CH_3)\text{+}_n$	outdoor carpeting for home, sports stadiums; food packaging; housings for appliances
styrene $CH_2{=}CH{-}C_6H_5$	polystyrene $\text{+}CH_2{-}CH(C_6H_5)\text{+}_n$	fast-food containers, hot-beverage cups; food-packaging trays; hairbrush handles; toys
tetrafluoroethylene $CF_2{=}CF_2$	polytetrafluoroethylene $\text{+}CF_2{-}CF_2\text{+}_n$	nonstick cookware; high-performance mechanical parts and electrical insulation; chemical-resistant gaskets
vinyl chloride $CH_2{=}CH{-}Cl$	poly(vinyl chloride) $\text{+}CH_2{-}CH(Cl)\text{+}_n$	home vinyl siding, rain gutters; flooring (sheet, tile); garden hose; surgical gloves; wire and cable insulation
acrylonitrile $CH_2{=}CH{-}CN$	polyacrylonitrile $\text{+}CH_2{-}CH(CN)\text{+}_n$	acrylic textile fibers
methyl methacrylate $CH_2{=}C(COOCH_3){-}CH_3$	poly(methyl methacrylate) $\text{+}CH_2{-}C(COOCH_3)(CH_3)\text{+}_n$	factory and aircraft windows; bathtubs; contact lenses; dentures and dental fillings

ions, as well as with oxygen and ozone (O_3) from the atmosphere. The products of selective oxidation are alcohols, aldehydes, and ketones or carboxylic acids, depending on the oxidizing agent and the reaction conditions (such as temperature and pH). The selective oxidation of lipids left in the open air is responsible for the disagreeable odor and taste of rancid butter (Section 11.4). Ozone in the air that we breathe presents a health hazard because of the oxidation of double bonds in the membranes of the lungs, throat, and other tissues (Section 11.4).

The selective oxidations by permanganate (MnO_4^-) and dichromate ($Cr_2O_7^{2-}$) are useful as simple chemical tests for alkenes. A positive test is accompanied by a visible change:

- Permanganate oxidation: MnO_4^- (purple solution) is converted into MnO_2 (brown precipitate).
- Dichromate oxidation: $Cr_2O_7^{2-}$ (orange solution) is converted into Cr^{3+} (green solution).

Alkanes and cycloalkanes do not undergo permanganate and dichromate oxidations.

Example 4.9 Writing equations for the combustion of alkenes

Give the balanced equation for the combustion of 1-pentene.

Solution

Proceed by the method described in Example 11.13.

$$2\ C_5H_{10} + 15\ O_2 \longrightarrow 10\ CO_2 + 10\ H_2O$$

Problem 4.9 Give the balanced equation for the combustion of 2-octene.

4.9 ALKYNES

$H—C \equiv C—H$

Ethyne

Alkynes are unsaturated hydrocarbons containing the carbon–carbon triple bond, two adjacent carbon atoms bonded together with three bonds. Because of this triple bond, alkynes are more unsaturated than alkenes. Alkynes have two fewer hydrogens than the corresponding alkene or cycloalkane with the same number of carbon atoms. The general formula for alkynes is C_nH_{2n-2}, where n is an integer greater than 1. The first member of the alkyne family is C_2H_2 (IUPAC name: ethyne; common name: acetylene).

We will spend little time on alkynes, because few biological molecules contain the triple bond. Furthermore, the chemistry of alkynes closely resembles that of alkenes. For our purposes, it will suffice to note the following points:

- The nomenclature rules for alkynes are the same as those for alkenes, except that the ending is **-yne** instead of **ene.**
- Geometrical stereoisomerism is not possible for alkynes, because the bond angles at the carbons of the triple bond are 180° and only one group is attached to each carbon of the triple bond.
- The triple bond (formed from *sp*-hybridized carbon atoms) is composed of one strong σ bond and two weak π bonds. The π bonds are analogous to those in alkenes.

The triple bond reacts with the same reagents as the double bond does, but it does so with twice as much reagent because each π bond is reactive. Only the reactions with hydrogen, halogens, and hydrogen halides concern us here. Reactions with the other reagents give complex products.

Acetylene is an important industrial chemical, largely because of the oxyacetylene torch used in welding. The torch is supplied with acetylene and oxygen from separate high-pressure tanks. The combustion of acetylene produces a high-temperature flame capable of melting and vaporizing iron and steel. The high temperature achieved by the combustion of acetylene is a consequence of the relative instability of the two π bonds. This instability results in a much more exothermic combustion reaction than that of an alkane or alkene. Acetylene is also a precursor to some useful organic chemicals, such as acrylonitrile, which is subsequently polymerized to polyacrylonitrile (Box 4.6).

Example 4.10 Writing equations for addition reactions of alkynes

Show the sequential reaction of 1-butyne with excess HCl; that is, show the reaction of one molecule of 1-butyne with one HCl molecule and then with a second HCl molecule. Show only the major product for each addition.

Solution

Each addition of the asymmetrical reagent HCl follows the Markovnikov rule.

$$\underset{\text{1-Butyne}}{HC{\equiv}C{-}CH_2{-}CH_3} \xrightarrow{HCl} \underset{\text{Major product}}{\overset{H\quad Cl}{HC{=}C}{-}CH_2{-}CH_3} \xrightarrow{HCl} \underset{\text{Major product}}{\overset{H\quad Cl}{\underset{H\quad Cl}{HC{-}C}}{-}CH_2{-}CH_3}$$

Problem 4.10 Show the sequential reaction of 2-butyne with excess HCl. Show only the major product for each addition.

4.10 AROMATIC COMPOUNDS

Alkanes, alkenes, and alkynes are called **aliphatic hydrocarbons** to distinguish them from the family of hydrocarbons called **aromatic hydrocarbons.** The chemical properties of aromatic hydrocarbons are very different from those of the aliphatic hydrocarbons, although the physical properties are very similar for the two groups. Aromatic compounds are unsaturated hydrocarbons that do not behave like other unsaturated hydrocarbons. The term aromatic, originally used because many aromatic compounds have pleasant aromas, is now used to emphasize the property that distinguishes this family from other families—aromatic compounds have exceptional stability, far beyond what one would expect of unsaturated compounds.

The simplest aromatic hydrocarbon is benzene, C_6H_6, whose six carbons form a ring:

H H H H H H or

All carbons and hydrogens of benzene lie in one plane (Figure 4.10). The unsaturated ring system of benzene, called the **benzene ring** or **benzene system,** exists not only in benzene, but also in a wide variety of other compounds. Organic compounds whose structures contain a benzene ring are called **aromatic compounds.**

The benzene ring has exceptional stability, which it retains wherever it exists—whether as a constituent of a hydrocarbon or in some other family of organic compounds. The benzene ring, as depicted in the preceding structural formula, looks like a triene with alternating single and double bonds. We might expect it to react with three times the number of reagent molecules that would react with an alkene double bond. But such reactions do not take place.

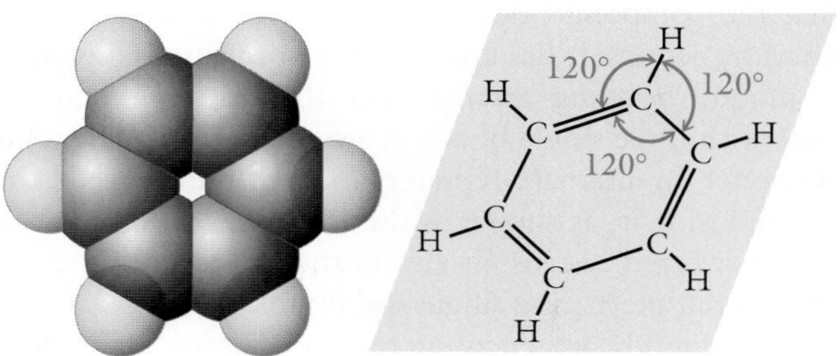

Figure 4.10 A space-filling model of benzene and a diagram showing the trigonal bond angles of its sp^2 carbons.

4.7 CHEMISTRY IN DEPTH

Bonding in Benzene

Each carbon atom of the benzene ring is sp^2 hybridized, as it is in a normal carbon–carbon double bond. That is, each carbon atom has three sp^2 orbitals and one $2p$ orbital. Each carbon atom uses its three sp^2 orbitals to form three σ bonds, one to each of the two adjacent carbons and one to a hydrogen atom (see the illustration at right). The key feature that distinguishes benzene from an alkene with three double bonds is the π bonding that results from sideward overlap of the six $2p$ orbitals. In an alkene double bond, $2p$ orbitals on two adjacent carbons overlap, thereby causing localization of the two electrons in the resulting π bond. In benzene, the $2p$ orbital on each carbon overlaps with two $2p$ orbitals, those from the adjacent carbon atoms on either side. This arrangement results in a continuous, circular overlap of six $2p$ orbitals (see the illustration on the left). The overlapping $2p$ orbitals of the benzene ring prove to have a remarkable stability, a feature manifested in benzene's characteristic chemical behavior.

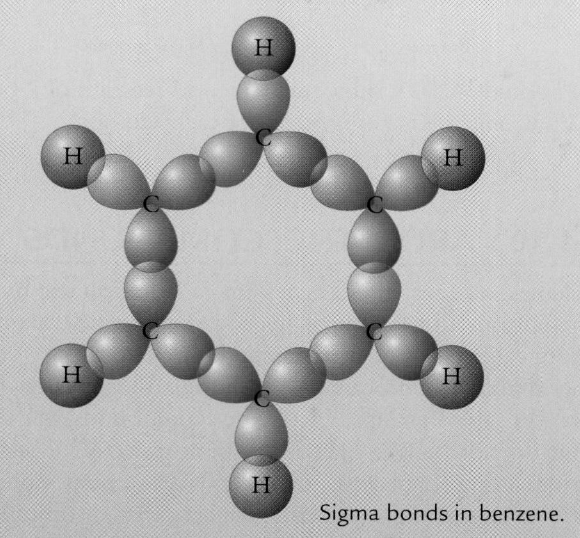

Sigma bonds in benzene.

INSIGHT INTO PROPERTIES

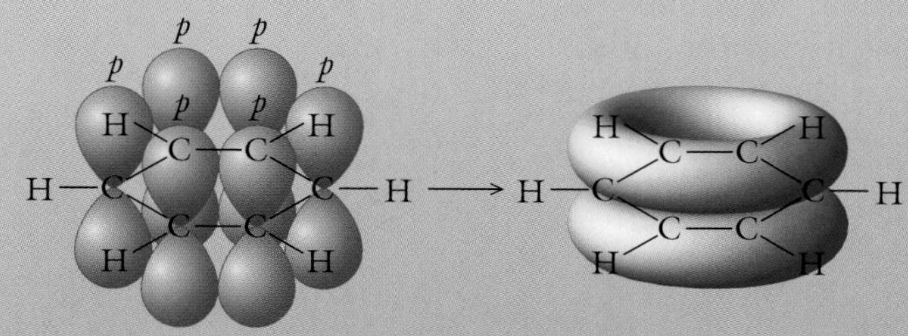

Overlap of $2p$ orbitals to form π bonding in benzene. The π bonding in benzene explains the benzene ring's great stability.

Benzene and other aromatic compounds resist reactions that would break into the benzene ring.

- Halogens, hydrogen halides, water, and sulfuric acid do not add to the double bonds of benzene under any conditions. These reagents readily add to the double bonds of alkenes.
- The combustion of aromatic compounds is less exothermic than the combustion of aliphatic compounds. (More-stable compounds give off less heat in undergoing reaction than do less-stable compounds.)

This resistance of the benzene structure to change is called **aromaticity** or **aromatic behavior.**

When you see the written structure of a compound containing a six-membered ring with three double bonds alternating with three single bonds, recognize that the compound does not behave as it looks; it is not like an alkene in behavior, because it has aromaticity's special stability. Aromaticity resides in the unique bonding arrangement that results when three double bonds alternate with three single bonds in a six-membered ring (Box 4.7).

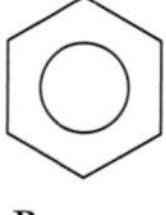

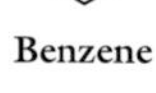
Benzene

Chemists prefer an alternate representation of the benzene ring—a hexagon with a circle inside it, as shown in the margin. This representation of the benzene ring is preferred because it clearly shows the ring as a unique structure, not simply three alternating single and double bonds.

Aromatic compounds are encountered throughout the biological world and in everyday life (Box 4.8). The benzene ring is not the only aromatic

4.8 CHEMISTRY AROUND US

Aromatic Compounds in Everyday Life

Benzene is a **carcinogen,** a substance that causes cancer, but many derivatives of benzene are not carcinogens. Many compounds encountered in everyday life contain benzene rings. Prominent examples are the many aromatic compounds found in flavorings and spices, such as the benzaldehyde in almonds, the vanillin in vanilla beans, and the anethole in anise seeds.

Benzaldehyde (Almond) Vanillin (Vanilla) Anethole (Anise)

Two other compounds containing the benzene ring are the female sex hormone estradiol and the synthetic pain killer aspirin.

Estradiol

Aspirin

A number of chlorinated aromatics have been used as insecticides (DDT) and as heat-exchanging and electrical-insulating fluids (PCB). DDT and PCB are no longer used in the United States, however, or in many other highly industrialized countries, because they are harmful to humans and nonbiodegradable. Nevertheless, the use of DDT continues in Third World countries where people are more concerned about short-term starvation than about long-term health effects.

Dichlorodiphenyltrichloroethane (DDT)

Polychlorinated biphenyl (PCB)

structure in chemistry. Several kinds of heterocyclic aromatic structures containing nitrogen in the ring are of major importance in biological systems. For example, nucleic acids such as DNA and RNA, the molecules responsible for heredity, contain the pyrimidine and purine aromatic structures (Section 13.1). Heterocyclic aromatic structures are also found in hemoglobin (Section 12.7) and chlorophyll.

Pyrimidine

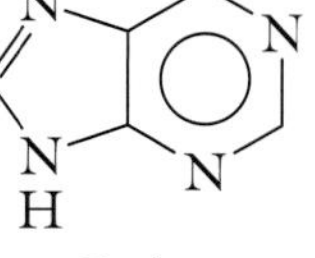

Purine

4.11 ISOMERS AND NAMES OF AROMATIC COMPOUNDS

In the IUPAC system, most monosubstituted benzenes are named by placing the name of the substituent in front of the parent name **-benzene:**

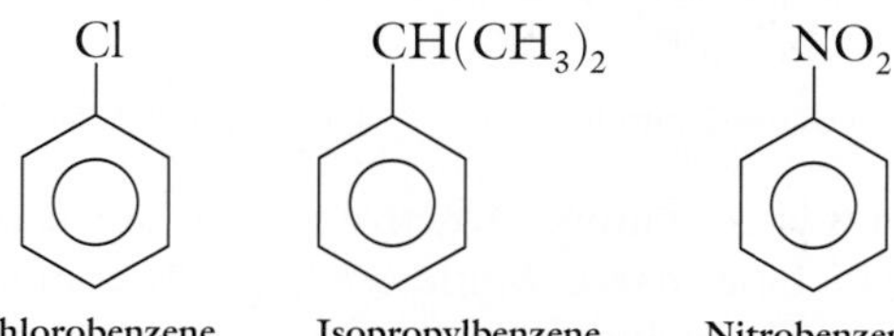

Chlorobenzene Isopropylbenzene Nitrobenzene

Some monosubstituted benzenes have common names that have been adopted by IUPAC as the preferred names. Examples include toluene for methylbenzene, phenol for hydroxybenzene, benzoic acid for carboxybenzene, and aniline for aminobenzene:

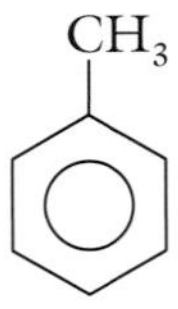

CH_3 OH O C—OH NH_2

Toluene (Methylbenzene) **Phenol** (Hydroxybenzene) **Benzoic acid** (Carboxybenzene) **Aniline** (Aminobenzene)

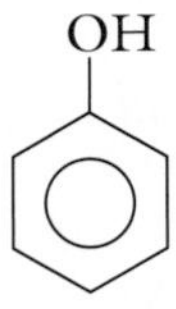

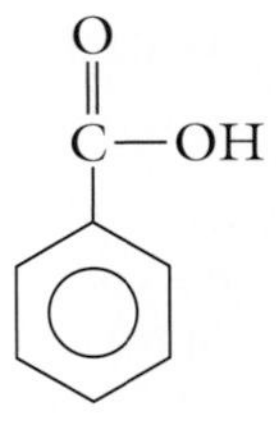

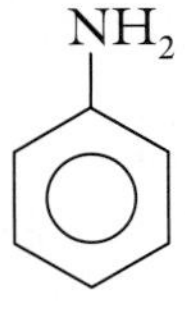

Disubstituted benzenes have three constitutional isomers, whose substituents take different positions around the benzene ring. There are two systems for indicating those positions: the numbering system (1,2-, 1,3-, 1,4-) or the *ortho-*, *meta-*, *para-* system. For the latter system, the prefixes *ortho-*, *meta-*, and *para-* (or their abbreviations *o-*, *m-*, and *p-*) are used for 1,2-, 1,3-, and 1,4-, respectively. Some disubstituted benzenes—for example, the dichlorobenzenes—are simply named as derivatives of benzene:

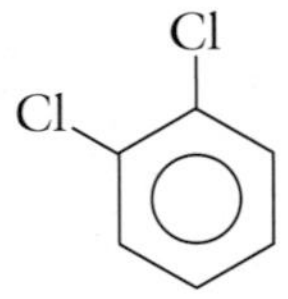

1,2-Dichlorobenzene (*o*-Dichlorobenzene) **1,3-Dichlorobenzene** (*m*-Dichlorobenzene) **1,4-Dichlorobenzene** (*p*-Dichlorobenzene)

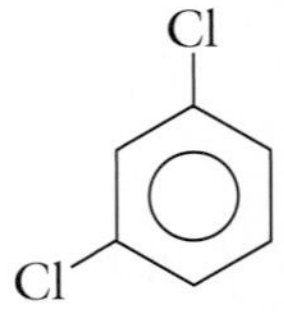

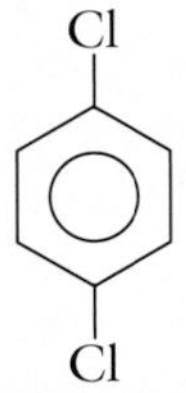

Other disubstituted benzenes are named as derivatives of the monosubstituted benzene when the latter has a preferred common name:

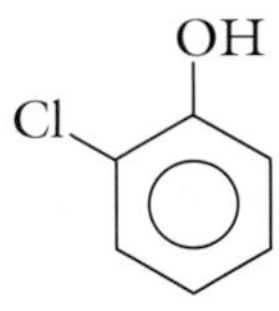

o-**Clorophenol** (2-Chlorophenol) *m*-**Chlorobenzoic acid** (3-Chlorobenzoic acid)

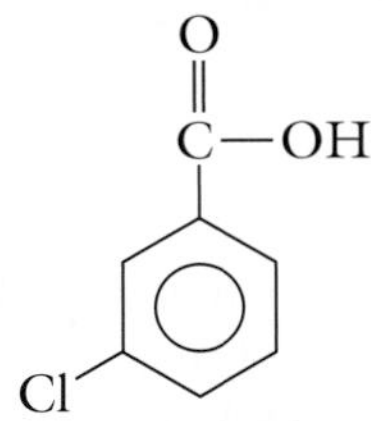

Trisubstituted benzenes are named similarly, except that numbers must be used (because the *ortho-*, *meta-*, and *para-* system is not applicable):

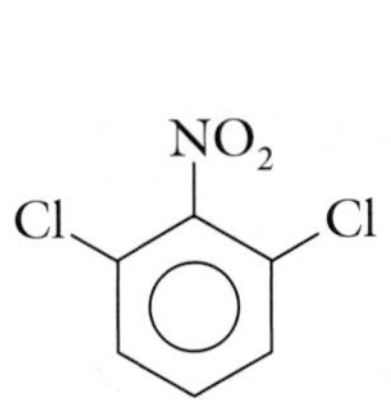

1,3-Dichloro-2-nitrobenzene 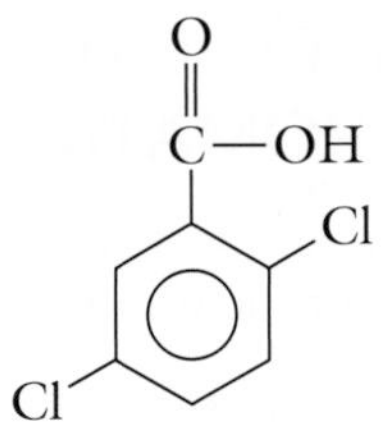 **2,5-Dichlorobenzoic acid**

Not all compounds containing a benzene ring are named by using benzene as the parent, or base, name. Some compounds contain an aliphatic part that is more complex than the aromatic component, and these compounds are usually given names derived from that of the aliphatic parent or base compound. In such cases, the prefix **phenyl** is added to indicate the presence of a benzene ring as a substituent:

CH_3 | CH_3—CH—CH—CH_3 (with benzene ring on C-3) or CH_3 | CH_3—CH—CH—CH_3 | C_6H_5

2-Methyl-3-phenylbutane

The phenyl group is often abbreviated by C_6H_5- or Ph- or the Greek letter ϕ (phi). We will use C_6H_5- in this book. A phenyl or substituted phenyl group is called an **aryl** group (abbreviation, Ar) to distinguish it from an alkyl group.

C_6H_5-CH_2- is known as a **benzyl** group:

C_6H_5—CH_2—(cyclohexane)

Benzylcyclohexane

Do not confuse phenyl and benzyl groups; they are different.

Unlike the nonaromatic cyclic compounds, substituted benzenes do not exist as cis and trans isomers, because the bonds extending from the carbons of the benzene ring are in the same plane as the ring (and only one bond extends from each of the carbons of the ring).

The basic approach to naming substituted benzenes is the same as that for substituted cycloalkanes (Section 3.7). The numbering of the ring carbons starts with a carbon holding a substituent and proceeds around the ring from there. When different sets of numbers can be obtained by starting at different carbons or by counting in different directions, the correct name is the one with the lowest set of numbers.

Example 4.11 Using the IUPAC nomenclature system to name aromatic compounds

Name each of the following compounds according to the IUPAC system:

(a) CH_2CH_3, CH_3CH_2 (b) CH_3, Cl (c) NH_2, Cl, $CH_2CH_2CH_3$

(d) NO_2, Cl, CH_2CH_3 (e) $C(CH_3)_3$, Cl (f) H, C_6H_5, $CH_2C_6H_5$, H

Solution

The numbering of the ring carbons starts with a carbon holding a substituent and proceeds around the ring from there. When different sets of numbers can be obtained by starting at different carbons or by counting in different directions, the correct name is the one with the set of lowest numbers.

(a) 1,3-Diethylbenzene, or *m*-diethylbenzene.

(b) 2-Chlorotoluene or *o*-chlorotoluene.

(c) 2-Chloro-4-propylaniline.

(d) The set of numbers is 1,3,5, with different possibilities that depend on which carbon is chosen as C1 and whether numbering is clockwise or counterclockwise. The correct choices are those that give the lower numbers to substituents whose names start lower in the alphabet. The name is 1-chloro-3-ethyl-5-nitrobenzene.
(e) 1-*t*-Butyl-2-chlorobenzene or *o*-*t*-butylchlorobenzene.
(f) *trans*-1-Benzyl-3-phenylcyclohexane.

Problem 4.11 Draw structural formulas for the following compounds: (a) *p*-ethylbenzoic acid, or 4-ethylbenzoic acid; (b) 2,4-dibromotoluene; (c) 4-phenyl-1-butene; (d) *m*-isopropylphenol, or 3-isopropylphenol; (e) 1-chloro-3-ethyl-4-propylbenzene; (f) benzyl chloride.

4.12 REACTIONS OF AROMATIC COMPOUNDS

By now you might have the impression that benzene is chemically inert, but it is not. Although it is true that the carbon–carbon double bonds making up the ring are highly resistant to addition, benzene undergoes substitution reactions in which some group substitutes for one of the hydrogen atoms extending from the ring. **Halogenation, sulfonation, nitration,** and **alkylation** are reactions in which substitution is by halogen (—F, —Cl, —Br), sulfonic acid (—SO_3H), nitro (—NO_2), and alkyl (—R) groups, respectively:

Halogenation: C_6H_5—H + Br_2 $\xrightarrow{\text{metal halide}}$ C_6H_5—Br + HBr

Bromobenzene

Nitration: C_6H_5—H + HNO_3 $\xrightarrow{H_2SO_4}$ C_6H_5—NO_2 + HOH

Nitrobenzene

Sulfonation: C_6H_5—H + SO_3 $\xrightarrow{H_2SO_4}$ C_6H_5—SO_3H

Benzenesulfonic acid

Alkylation: C_6H_5—H + RCl $\xrightarrow{\text{metal halide}}$ C_6H_5—R + HCl

Alkylbenzene

Halogenation and alkylation of an aromatic compound require a metal halide such as $FeCl_3$ or $AlCl_3$ as a catalyst. Nitration and sulfonation require sulfuric acid as a catalyst. The reactions do not take place unless the appropriate catalyst is present.

The substitution reactions of benzene do not contradict our earlier consideration of the high stability of the benzene ring; instead they serve to emphasize it. These reactions leave the benzene structure intact. Note, in contrast, that, although the same reagents induce halogenation, nitration, sulfonation, and alkylation of alkenes, they do so by a process of addition to the carbon–carbon double bonds. (Nitration, sulfonation, and alkylation of alkenes were not covered in Chapter 3, but they do take place, as addition reactions.)

Aromatic compounds, like all organic compounds, undergo combustion to form carbon dioxide and water. Selective oxidations of alkylbenzenes under conditions less extreme than combustion show the high stability of aromatics relative to aliphatics. Alkylbenzenes undergo selective oxidation under moderately strong oxidizing conditions—for example, in the presence of hot acidic $KMnO_4$ or $K_2Cr_2O_7$. The benzene ring remains intact because of its stability, whereas the aliphatic part is oxidized:

Bond not broken — Bonds broken

$C_6H_5-CH_2-CH_2-CH_2-CH_3 \xrightarrow{\text{hot acidic } MnO_4^- \text{ or } Cr_2O_7^{2-}} C_6H_5-C(=O)OH + CO_2 + H_2O$

Whatever the size of the alkyl group, one by one all of its carbon–carbon bonds break, and the carbons and hydrogens are oxidized to CO_2 and H_2O, respectively. However, the carbon–carbon bond between the benzene ring and the alkyl group is not broken, because that would destroy the aromatic structure by subsequent oxidation of ring carbons. The carbon of the alkyl group attached to the benzene ring is oxidized to the carboxyl group (COOH), the highest oxidation state (Sections 2.1 and 3.10) possible for carbon without breaking the bond to the aromatic ring.

It should be emphasized that the permanganate and dichromate oxidations of alkyl side chains of alkylaromatics occur only with hot reagents. Oxidation does not occur at ambient temperatures, which are the conditions described in Section 4.8 for the oxidation of alkenes.

Example 4.12 Writing equations for reactions of aromatic compounds

Write equations to show the product(s) formed in each of the following reactions. If no reaction takes place, write "no reaction." If more than one product is formed, indicate the major and minor products. (a) Benzene + Cl_2; (b) benzene + Cl_2 + $FeCl_3$; (c) benzene + SO_3 + H_2SO_4; (d) toluene + hot acidic $KMnO_4$.

Solution

(a) No reaction takes place, because halogenation requires a metal halide catalyst.

(b) $C_6H_6 \xrightarrow[FeCl_3]{Cl_2} C_6H_5Cl$

(c) $C_6H_6 \xrightarrow[H_2SO_4]{SO_3} C_6H_5SO_3H$

(d) $C_6H_5CH_3 \xrightarrow[H^+]{\text{hot } KMnO_4} C_6H_5-C(=O)OH$

Problem 4.12 Write equations to show the product(s) formed in each of the following reactions. If no reaction takes place, write "no reaction." If more than one product is formed, indicate the major and minor products. (a) Benzene + CH_3CH_2Cl; (b) benzene + CH_3CH_2Cl + $AlCl_3$; (c) isopropylbenzene + hot acidic $K_2Cr_2O_7$.

Substitution and side-chain oxidation reactions are used industrially to produce a range of raw materials that are converted into consumer products including detergents, dyes and pigments, and plastics.

To sum up what we have learned about reactions between organic molecules and halogens in this chapter and in Chapter 3, three different scenarios are possible:

Concept checklist

- ✓ Halogen substitution takes place at C–H bonds in alkanes in the presence of ultraviolet radiation or at high temperatures (Section 3.10).
- ✓ Halogen substitution takes place at C–H bonds of aromatic compounds in the presence of a metal halide catalyst (Section 4.4).
- ✓ Halogen addition takes place at C–C double bonds in alkenes with or without the presence of high temperatures, UV radiation, or a metal catalyst (Section 4.6).

Example 4.13 Writing equations for the halogenation of various compounds

Bromine is added to a mixture of cyclohexane, cyclohexene, and benzene in the presence of $FeBr_3$. Which compound(s) undergo reaction? Write equations to show the product(s) formed in any reaction. (Note: Reactions are carried out at ambient temperature in the absence of ultraviolet radiation unless otherwise stated.)

Solution

Cyclohexane does not undergo reaction, because the temperature is not high and there is no ultraviolet radiation. Cyclohexene and benzene undergo addition and substitution, respectively:

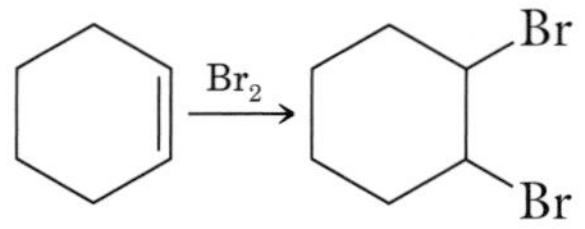

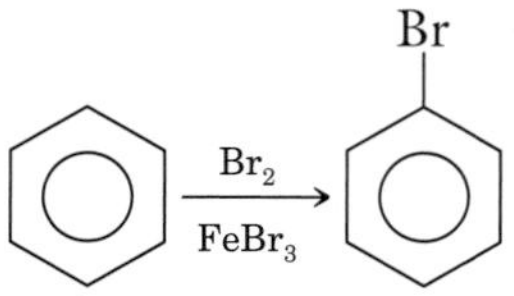

$$\text{cyclohexene} \xrightarrow{Br_2} \text{1,2-dibromocyclohexane (Br, Br)}$$

$$\text{benzene} \xrightarrow[FeBr_3]{Br_2} \text{bromobenzene (Br)}$$

Problem 4.13 Bromine is added to a mixture of cyclohexane, cyclohexene, and benzene in the presence of ultraviolet radiation but in the absence of $FeCl_3$. Which compound(s) undergo reaction? Write equations to show the product(s) formed in any reaction.

Fused-ring aromatic compounds are described in Box 4.9. Some of these compounds are used industrially. Others are implicated as carcinogens.

4.9 CHEMISTRY AROUND US

Fused-Ring Aromatics

A **fused-ring compound** contains rings that share two or more ring atoms. The compound estradiol (Box 4.8), for example, contains four fused rings, one of which is aromatic. A number of **fused-ring aromatic compounds,** compounds in which all rings are aromatic, are commercially important. Naphthalene is the active ingredient in moth balls. Anthracene is used to manufacture a variety of dyes.

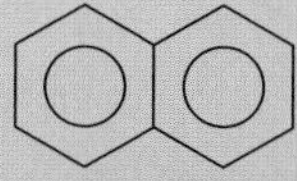

Naphthalene

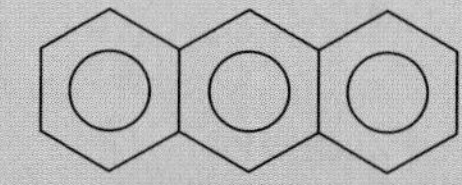

Anthracene

"Bent" fused-ring aromatics such as phenanthrene, benzanthracene, and benzopyrene are carcinogens, substances that cause cancer. These compounds are found in automobile exhaust, tobacco smoke, and burnt (especially grilled) meats. The carcinogenic behavior of bent fused-ring aromatics is discussed in Section 13.8.

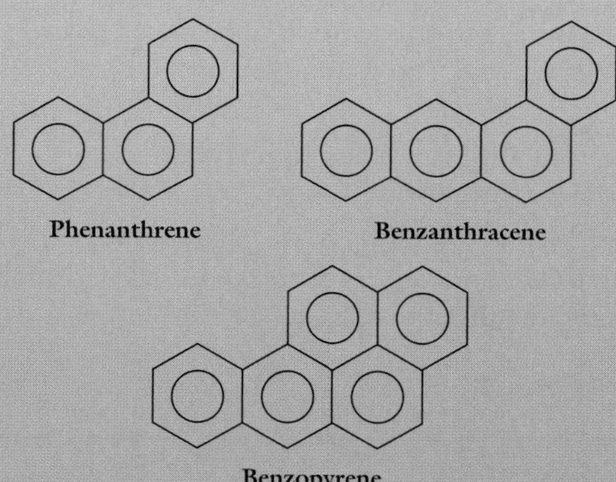

Phenanthrene

Benzanthracene

Benzopyrene

Tobacco smoke contains cancer-causing bent fused-ring aromatic compounds. (Network Production/The Image Works.)

Summary

- Unsaturated hydrocarbons, consisting of the alkene, alkyne, and aromatic families, contain carbon–carbon multiple bonds—that is, double or triple bonds between adjacent carbon atoms.

Alkenes

- Alkenes contain a double bond—that is, two adjacent carbons joined together with two bonds.
- The general formula for an alkene is C_nH_{2n}, which contains two fewer hydrogens than the general formula for an alkane.
- The carbon atoms of the double bond possess three sp^2-hybrid orbitals with trigonal (120°) bond angles, and one unhybridized $2p$ orbital.
- The orbitals overlap to form one strong bond (a σ bond) and one weak bond (a π bond).

Constitutional Isomers of Alkenes

- Constitutional isomerism exists in alkenes not only because a given formula allows different carbon skeletons, as in the alkane family, but also because it allows different placements of the double bond within a carbon skeleton.

Naming Alkenes

- In the IUPAC system, alkenes are named on the basis of the longest continuous chain containing the double bond.
- The base name is preceded by a number indicating the position of the double bond.
- The family ending of alkene names is -ene.
- Prefixes with numbers identify any substituents attached to the longest chain.

Cis-Trans Stereoisomerism in Alkenes

- Cis-trans (geometrical) stereoisomers exist in alkenes when each carbon atom of the double bond has two different substituents.
- The cis isomer has substituents on the same side of the double bond. The substituents are on opposite sides of the double bond in the trans isomer.

Addition Reactions of Alkenes

- The presence of the weak π bond in the double bond greatly increases the reactivity of alkenes relative to the ow reactivity of saturated hydrocarbons.
- Various addition reactions are possible: a reagent breaks into two fragments, each of which bonds (adds) to one of the carbons of the double bond.
- Hydrogen, halogens, hydrogen halides, and water are all able to add to the alkene double bond.
- The addition of asymmetrical reagents (hydrogen halides, water) to asymmetrical alkenes proceeds in a selective manner.
- The Markovnikov rule is followed, meaning that the hydrogen fragment from the reagent adds to the carbon of the double bond that already holds more hydrogens before reaction.

Addition Polymerization

- Addition polymerization is a self-addition reaction in which large numbers of alkene molecules add to one another to form high-molecular-mass molecules called polymers.

Oxidation of Alkenes

- Alkenes, like other organic compounds, can undergo combustion.
- Selective oxidation is also possible.

Alkynes

- Alkynes contain a triple bond, three bonds between two adjacent carbon atoms.
- Two of the three bonds are weak π bonds similar to those in the alkene double bond.
- The triple bond undergoes the same reactions as the double bond does but with twice as much of the reagent.

Aromatic Compounds

- Aromatic compounds contain the benzene ring, a six-membered ring with alternating double and single bonds.
- Substituted benzenes are named by IUPAC as derivatives of benzene, with prefixes indicating the substituents attached to the ring.
- Either a numbering system or the *ortho-*, *meta-*, *para-* system is used to indicate the positions of substituents for disubstituted benzenes.
- Only the numbering system is used for trisubstituted benzenes.
- Some monosubstituted benzenes have common names that are accepted by IUPAC and used as the basis for naming their di- and trisubstituted relatives.

Reactions of Aromatic Compounds

- The benzene ring has exceptional stability.
- Benzene is inert to the addition of halogens, hydrogen halides, sulfuric acid, and water—reagents that add to alkene double bonds.
- Benzene and other aromatic compounds undergo substitution reactions with halogens, nitric acid, sulfur trioxide, and alkyl halides.
- Alkylbenzenes undergo oxidation of the alkyl group, but the benzene ring itself remains intact.

Summary of Reactions

ALKENES

Addition

$$\mathrm{C{=}C} \xrightarrow{H_2,\ Pt} \mathrm{-\overset{H}{\underset{}{C}}-\overset{H}{\underset{}{C}}-} \quad \text{Hydrogenation}$$

$$\mathrm{C{=}C} \xrightarrow{X_2} \mathrm{-\overset{X}{C}-\overset{X}{C}-} \quad \text{Halogenation } (X = Br, Cl, F)$$

$$\mathrm{C{=}C} \xrightarrow{HX} \mathrm{-\overset{H}{C}-\overset{X}{C}-} \quad \text{Hydrohalogenation } (X = I, Br, Cl)$$

$$\mathrm{C{=}C} \xrightarrow[H^+]{H_2O} \mathrm{-\overset{H}{C}-\overset{OH}{C}-} \quad \text{Hydration}$$

Addition polymerization

$$\mathrm{C{=}C} \xrightarrow{\text{polymerization catalyst}} \mathrm{\left(-C-C- \right)_n}$$

AROMATICS

Substitution

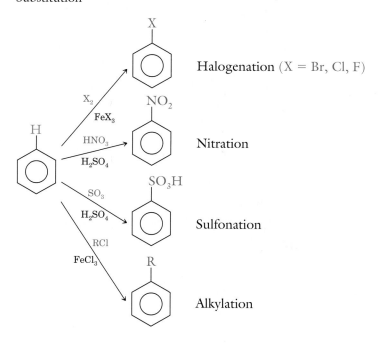

Benzene (H) + X_2, FeX_3 → X-benzene: Halogenation (X = Br, Cl, F)

Benzene (H) + HNO_3, H_2SO_4 → NO_2-benzene: Nitration

Benzene (H) + SO_3, H_2SO_4 → SO_3H-benzene: Sulfonation

Benzene (H) + RCl, $FeCl_3$ → R-benzene: Alkylation

Selective oxidation

R-benzene —(hot acidic MnO_4^- or $Cr_2O_7^{2-}$)→ COOH-benzene

Key Words

addition polymerization, p. 130
addition reaction, p. 122
aliphatic, p. 133
alkene, p. 112
alkylation, p. 138
alkyne, p. 132
aromatic, p. 133
asymmetrical reagent, p. 124
carbocation, p. 126
catalytic hydrogenation, p. 123
cis-trans stereoisomers, p. 118
halogenation, pp. 123, 138
hydration, p. 125
hydrohalogenation, p. 124
Markovnikov rule, p. 126
mechanism, p. 126
monomer, p. 130
nitration, p. 138
pi (π) bond, p. 114
polymer, p. 128
repeat unit, p. 130
restricted rotation, p. 118
sigma (σ) bond, p. 114
sp^2, p. 114
sulfonation, p. 138
symmetrical reagent, p. 122
trigonal bond angle, p. 114
trigonal stereocenter, p. 119

Exercises

Alkenes

4.1 An unknown compound has the molecular formula C_5H_{12}. Is the compound an alkane, cycloalkane, or alkene?

4.2 An unknown compound has the molecular formula C_5H_{10}. Is the compound an alkane, cycloalkane, or alkene?

4.3 Which of the following are correct molecular formulas? (a) C_6H_{13}; (b) C_6H_{12}; (c) $C_6H_{13}Br$.

4.4 Which of the following are correct molecular formulas? (a) C_5H_{12}; (b) C_5H_{11}; (c) C_5H_{10}.

4.5 Give the molecular formula for each of the following compounds:

(a) $CH_3{-}CH{=}CH{-}CH_2CH_3$

(b) 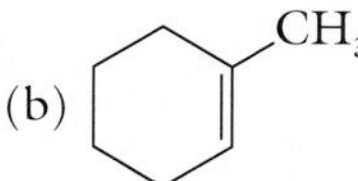

4.6 Give the molecular formula for each of the following compounds:

(a) $CH_2{=}C(CH_3){-}CH{=}C(CH_3)_2$

(b)

Bonding in Alkenes

4.7 Which (if any) of carbons 1 through 12 in the following compound are sp^2 hybridized?

CH_3(1) on a cyclohexene ring (carbons 2–7) bearing at C-5 the chain $\overset{8}{CH}{=}\overset{9}{C}(\overset{10}{CH_3}){-}\overset{11}{CH_2}{-}\overset{12}{CH_3}$

4.8 What are the values of bond angles A through E in the following compound?

CH_3 on a cyclohexene ring bearing $CH{=}C(CH_3){-}CH_2{-}CH_3$ (angles A, B, C, D, E marked)

Constitutional Isomers of Alkenes

4.9 Draw structural formulas of all constitutional isomers of C_4H_8. Show only one structural formula for each isomer. Note that the wording of this exercise requires you to include cycloalkanes as well as alkenes of molecular formula C_4H_8.

4.10 Draw structural formulas of all constitutional isomers of C_5H_{10}. Show only one structural formula for each isomer. Note that the wording of this exercise requires you to include cycloalkanes as well as alkenes of molecular formula C_5H_{10}.

4.11 Draw structural formulas for each of the following compounds: (a) an alkene, C_5H_{10}, containing all five carbons in a continuous chain; (b) an alkene, C_5H_{10}, containing three methyl groups; (c) a cyclic compound, C_5H_{10}, containing one methyl group and three CH_2 groups; (d) an alkene, C_5H_{10}, containing an isopropyl group.

4.12 Draw structural formulas for each of the following compounds: (a) a cyclic compound, C_5H_{10}, containing one ethyl group; (b) an alkene, C_6H_{12}, containing two methyl groups and one ethyl group; (c) an alkene, C_6H_{12}, containing two ethyl groups; (d) an alkene, C_6H_{12}, containing a *t*-butyl group.

4.13 For each of the following pairs of structural formulas, indicate whether the pair represents (1) the same compound or (2) different compounds that are constitutional isomers or (3) different compounds that are not isomers.

(a) $(CH_3)(Cl)C{=}C(CH_3)(CH_3)$ and $(CH_3)(Cl)C{=}C(Cl)(CH_3)$

(b) $(CH_3)(H)C{=}C(CH_3)(H)$ and $(CH_3)(H)C{=}C(H)(CH_2CH_3)$

(c) $(CH_3)(H)C{=}C(CH_3)(H)$ and $(CH_3)(CH_3)C{=}C(H)(H)$

4.14 For each of the following pairs of structural formulas, indicate whether the pair represents (1) the same compound or (2) different compounds that are constitutional isomers or (3) different compounds that are not isomers.

(a) $(CH_3)(H)C{=}C(CH_3)(H)$ and $(H)(CH_3)C{=}C(H)(CH_3)$

(b) $(CH_3)(H)C{=}C(CH_3)(H)$ and $(H)(H)C{=}C(H)(CH_2CH_3)$

(c) $(CH_3)(H)C{=}C(CH_3)(H)$ and $(CH_3)(H)C{=}C(Cl)(H)$

Naming Alkenes

4.15 Draw the structural formula for each of the following compounds: (a) 2-methyl-1-butene; (b) 3-ethyl-2-pentene; (c) 4-isopropyl-2,6-dimethyl-2-heptene; (d) 1,3-dimethylcyclohexene; (e) 3-*t*-butyl-2,4-dimethyl-1-pentene; (f) 5-methyl-1,4-hexadiene.

4.16 Name the following compounds by the IUPAC system:

(a) $CH_3{-}CH_2{-}CH_2{-}C({=}C(CH_3)_2){-}CH_2{-}CH_3$

(b) $CH_2{=}C(CH_3){-}CH{=}C(CH_3)_2$

(c) Cyclohexene ring bearing Cl, CH_3, and C_2H_5

(d) $CH_3{-}CH(CH_2CH_3){-}CH(CH_3){-}CH{=}CH{-}CH(CH_3)_2$

Cis-Trans Isomerism in Alkenes

4.17 Which of the following compounds exist as separate cis and trans isomers? Draw structural formulas of the cis and trans isomers where applicable. (a) 2-Methyl-2-hexene; (b) 3-methyl-2-hexene; (c) 4-methyl-2-hexene; (d) 2-methyl-1-hexene; (e) 1,2-dimethylcyclopentene; (f) 1,2-dimethylcyclopentane.

4.18 Draw structural formulas of the following compounds: (a) *cis*-1-chloro-1-butene; (b) *trans*-3-hexene; (c) *trans*-4,4-dimethyl-2-pentene; (d) *cis*-1,3-dichlorocyclohexane.

4.19 For each of the following pairs of structural formulas, indicate whether the pair represents (1) the same compound or (2) different compounds that are constitutional isomers or (3) different compounds that are cis-trans isomers or (4) different compounds that are not isomers.

(a) $(CH_3)(Cl)C{=}C(CH_3)(Cl)$ and $(CH_3)(Cl)C{=}C(Cl)(CH_3)$

(b) $\underset{Cl}{\overset{CH_3}{}}C{=}C\underset{Cl}{\overset{CH_3}{}}$ and $\underset{CH_3}{\overset{Cl}{}}C{=}C\underset{CH_3}{\overset{Cl}{}}$

(c) $\underset{H}{\overset{CH_3}{}}C{=}C\underset{H}{\overset{CH_3}{}}$ and $\underset{CH_3}{\overset{CH_3}{}}C{=}C\underset{H}{\overset{H}{}}$

(d) $\underset{H}{\overset{CH_3}{}}C{=}C\underset{H}{\overset{CH_3}{}}$ and $\underset{H}{\overset{CH_3}{}}C{=}C\underset{CH_2CH_3}{\overset{H}{}}$

4.20 For each of the following pairs of structural formulas, indicate whether the pair represents (1) the same compound or (2) different compounds that are constitutional isomers or (3) different compounds that are cis-trans isomers or (4) different compounds that are not isomers.

(a) $\underset{H}{\overset{CH_3}{}}C{=}C\underset{H}{\overset{CH_3}{}}$ and $\underset{CH_3}{\overset{H}{}}C{=}C\underset{CH_3}{\overset{H}{}}$

(b) $\underset{Cl}{\overset{CH_3}{}}C{=}C\underset{CH_3}{\overset{CH_3}{}}$ and $\underset{Cl}{\overset{CH_3}{}}C{=}C\underset{CH_3}{\overset{Cl}{}}$

(c) $\underset{H}{\overset{CH_3}{}}C{=}C\underset{H}{\overset{CH_3}{}}$ and $\underset{CH_3}{\overset{H}{}}C{=}C\underset{H}{\overset{CH_3}{}}$

(d) $\underset{H}{\overset{CH_3}{}}C{=}C\underset{H}{\overset{CH_3}{}}$ and $\underset{CH_3}{\overset{CH_3}{}}C{=}C\underset{H}{\overset{H}{}}$

Addition Reactions of Alkenes

4.21 Give the organic product(s) formed in each of the following reactions. If no reaction takes place, write "no reaction." If there is more than one product, indicate only the major product(s).

(a) CH_3 $\xrightarrow{Br_2}$?

(b) CH_3 $\xrightarrow[H^+]{H_2O}$?

(c) $CH_2{=}C(CH_3)_2 \xrightarrow[Ni]{H_2}$?

(d) $CH_2{=}C(CH_3)_2 \xrightarrow{HCl}$?

(e) $\xrightarrow{Cl_2}$?

(f) $CH_3{-}CH{=}CH{-}CH_3 \xrightarrow{HCl}$?

(g) $CH_3{-}CH{=}CH{-}CH_2CH_3 \xrightarrow{HBr}$?

(h) $CH_3{-}CH{=}CH{-}CH_2CH_3 \xrightarrow{Br_2}$?

(i) $CH_3{-}CH{=}CH{-}CH_3 \xrightarrow{Cl_2}$?

(j) $CH_3{-}CH{=}CH{-}CH_2CH_3 \xrightarrow[H^+]{H_2O}$?

(k) $CH_3{-}CH{=}CH{-}CH_2CH_3 \xrightarrow[Ni]{H_2}$?

4.22 Indicate whether a reaction takes place when 1-hexene is treated with each of the following reagents: (a) NaOH; (b) H_2O, H^+; (c) H_2/Ni; (d) HCl; (e) Cl_2.

If a reaction takes place, give the organic product(s) formed. If more than one product is formed, indicate only the major product(s). If no reaction takes place, write "no reaction."

4.23 An unknown compound is hexane, methylcyclopentane, or 1-hexene. The red color of bromine is not decolorized when bromine is added to the unknown. Identify the unknown.

4.24 An unknown compound is hexane, methylcyclopentane, or 1-hexene. The red color of bromine is decolorized when bromine is added to the unknown. Identify the unknown.

Addition Polymerization

4.25 Give the equation for the polymerization of acrylonitrile, $CH_2{=}CH{-}CN$.

4.26 Give the equation for polymerization of styrene, $CH_2{=}CH{-}C_6H_5$.

Oxidation of Alkenes

4.27 Give the balanced equation for the combustion of 3-hexene.

4.28 Give the balanced equation for the combustion of 2-methyl-1-hexene.

4.29 An unknown compound is pentane, cyclopentane, or 2-pentene. The orange color of dichromate is not decolorized when dichromate is added to the unknown. Identify the unknown.

4.30 An unknown compound is pentane, cyclopentane, or 2-pentene. The purple color of permanganate is decolorized and a brown precipitate forms when permanganate is added to the unknown. Identify the unknown.

Alkynes

4.31 Draw the structural formula of 3,4-dimethyl-1-pentyne.

4.32 For each of the following systems, show the sequential reaction of the alkyne with the reagent. Show only the major product for each step in the reaction. (a) 1-Butyne + excess Cl_2; (b) propyne + excess HCl; (c) 2-butyne + excess Cl_2.

Aromatics

4.33 What structural features are possessed by aromatic compounds? What chemical property distinguishes aromatic compounds from aliphatic compounds?

4.34 Which of the following compounds are aromatic?

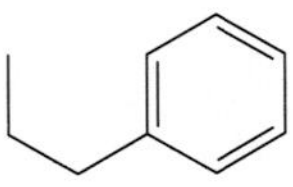

1 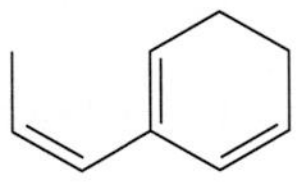 2 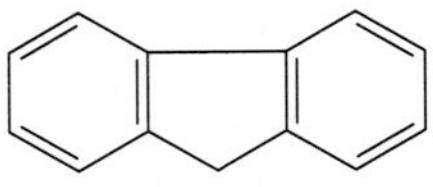 3 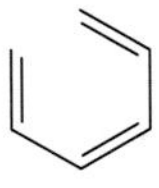 4 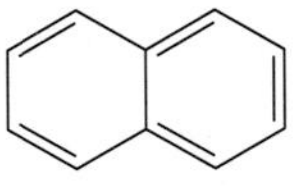 5 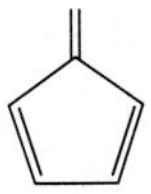 6  7 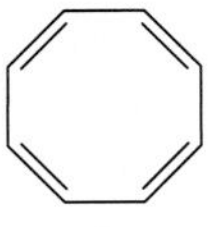 8

Isomers and Names of Aromatic Compounds

4.35 For each of the following pairs of structural formulas, indicate whether the pair represents (1) the same compound or (2) different compounds that are constitutional isomers or (3) different compounds that are cis-trans isomers or (4) different compounds that are not isomers.

(a) 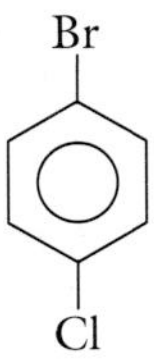 Br, Cl and  Cl, Br
(b) 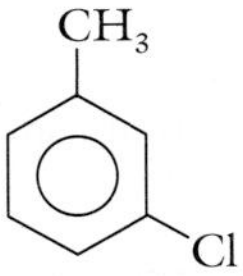 CH_3, Cl and CH_2Cl
(c) 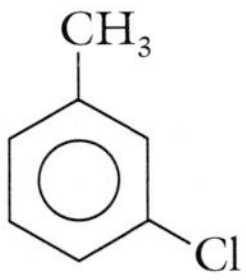 CH_3, Cl and Cl, CH_3

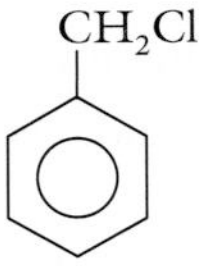

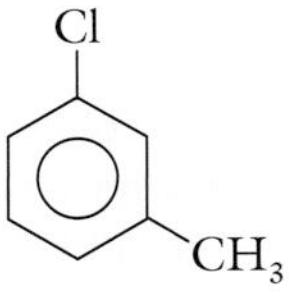

4.36 For each of the following pairs of structural formulas, indicate whether the pair represents (1) the same compound or (2) different compounds that are constitutional isomers or (3) different compounds that are cis-trans isomers or (4) different compounds that are not isomers.

(a) 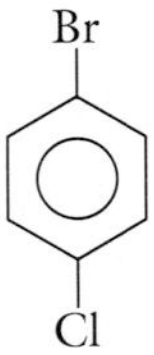 Br, Cl and 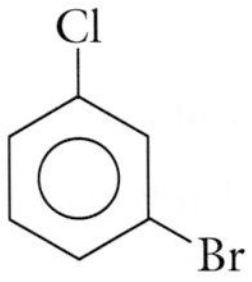 Cl, Br
(b) 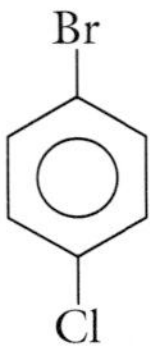 Br, Cl and 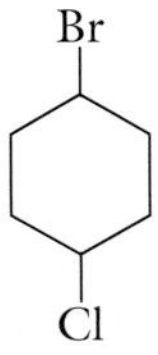 Br, Cl
(c)  Cl, Br and Cl, Br

4.37 Name each of the following compounds by the IUPAC system:

(a) 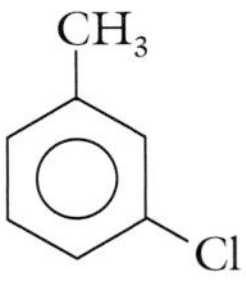 CH_3, Cl
(b) 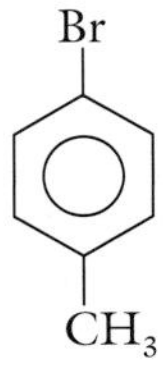 Br, CH_3
(c) 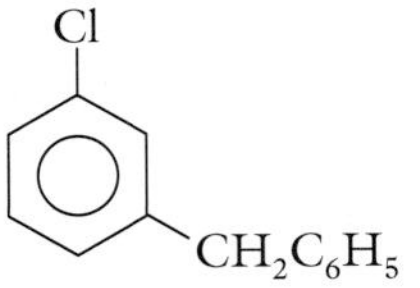 Cl, $CH_2C_6H_5$
(d) 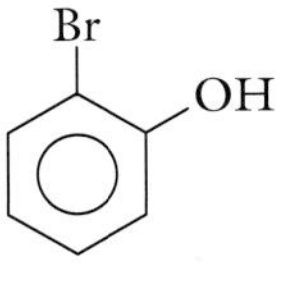 Br, OH
(e) 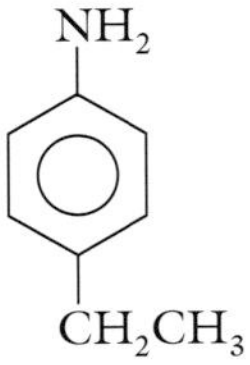 NH_2, CH_2CH_3
(f) 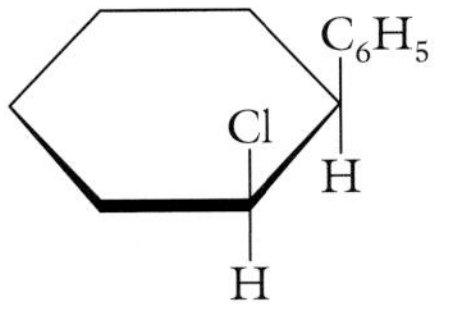 C_6H_5, H, Cl, H

4.38 Name each of the following compounds by the IUPAC system:

(a) 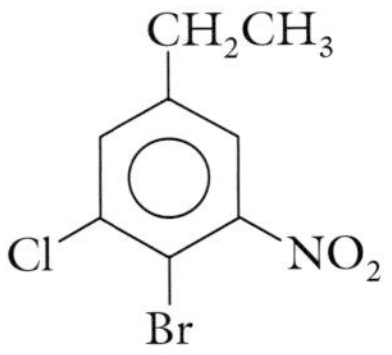 CH_2CH_3, Cl, NO_2, Br
(b) 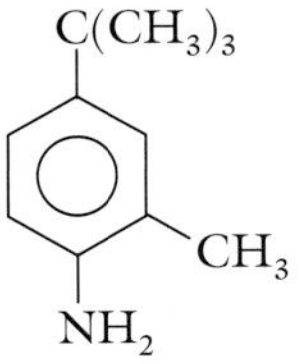 $C(CH_3)_3$, CH_3, NH_2
(c) 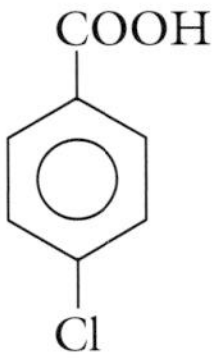 COOH, Cl
(d) 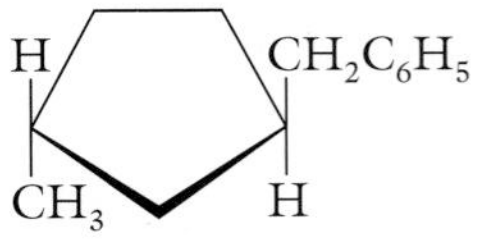 H, $CH_2C_6H_5$, CH_3, H

Reactions of Aromatics

4.39 Indicate whether a reaction takes place for each of the following reaction mixtures: (a) benzene + NaOH; (b) benzene + HBr; (c) benzene + SO_3 + H_2SO_4; (d) benzene + HNO_3 (with H_2SO_4 as catalyst); (e) combustion of benzene (give balanced equation).

If a reaction takes place, give the organic product(s) formed. If more than one product is formed, indicate only the major product(s). If no reaction takes place, write "no reaction."

4.40 Indicate whether a reaction takes place for each of the following reaction mixtures: (a) *t*-butylbenzene + hot acidic $K_2Cr_2O_7$; (b) benzene + H_2O; (c) benzene + $(CH_3)_2CHCl$ + $FeCl_3$; (d) combustion of ethylbenzene (give balanced equation); (e) benzene + Br_2 + $FeBr_3$.

If a reaction takes place, give the organic product(s) formed. If more than one product is formed, indicate only the major product(s). If no reaction takes place, write "no reaction."

Unclassified Exercises

4.41 Draw structural formulas of all possible isomers for compounds containing a benzene ring and having the molecular formula C_7H_7Br.

4.42 Draw structural formulas for each of the following compounds: (a) an alkene, C_6H_{12}, containing four methyl groups; (b) an alkyne, C_6H_{10}, containing a *t*-butyl group; (c) a cycloalkane, C_6H_{12}, containing one methyl group; (d) a cycloalkene, C_6H_{10}, containing no methyl groups; (e) a compound, C_8H_{10}, containing a benzene ring and two methyl groups; (f) a compound, C_8H_{10}, containing a benzene ring and one ethyl group.

4.43 Which of the following compounds exist as separate cis and trans isomers? Draw structural formulas of the cis and trans isomers where applicable. (a) 2-Hexene; (b) 1,2-dimethylbenzene; (c) 1-chloro-2-methylcyclohexane; (d) 1,2-dichlorocyclopentene; (e) 1-pentyne; (f) 2-pentyne.

4.44 Each of the following names is an incorrect IUPAC name. State why the name is incorrect. Indicate, if possible, what compound the namer of the compound had in mind and give the correct IUPAC name. (a) 4-Ethyl-1-pentene; (b) 5-*t*-butyl-1-isopropylcyclopentene; (c) 1,2,4-trimethyl-3-cyclohexene; (d) 4-butyl-1-methyl-2-hexene; (e) 1-propyl-5-methylbenzene; (f) *cis*-2-chloro-3-methyl-2-butene; (g) 2,4-pentadiene.

4.45 For each of the following pairs of structural formulas, indicate whether the pair represents (1) the same compound or (2) different compounds that are constitutional isomers or (3) different compounds that are cis-trans isomers or (4) different compounds that are not isomers.

(a) 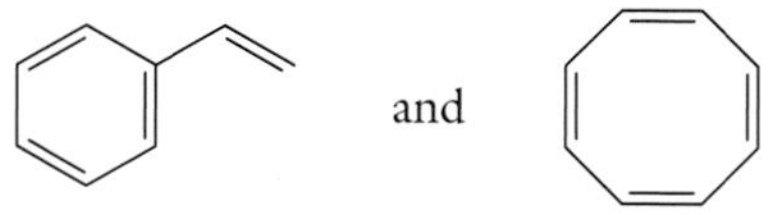 [structure] and [structure]

(b) 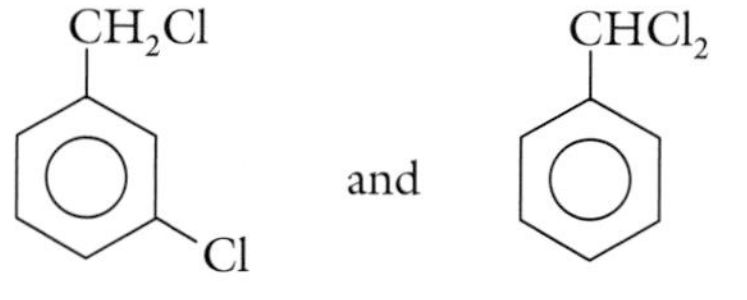 CH_2Cl (with Cl on ring) and $CHCl_2$

(c) 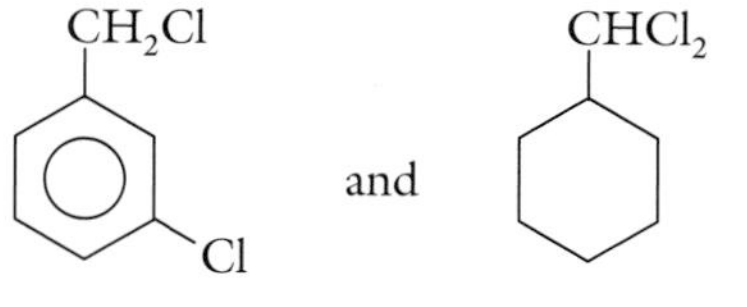 CH_2Cl (with Cl on ring) and $CHCl_2$

(d) 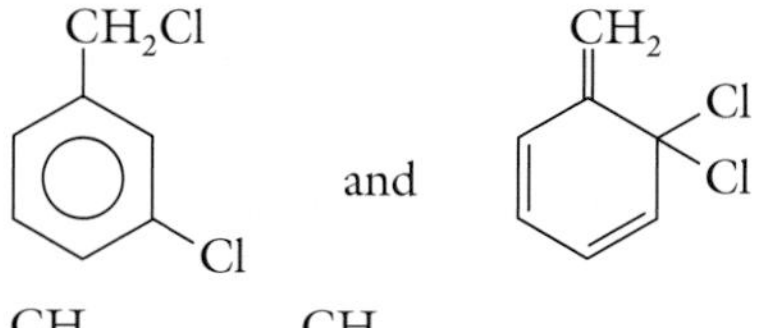 CH_2Cl (with Cl on ring) and CH_2 (with Cl, Cl)

(e) 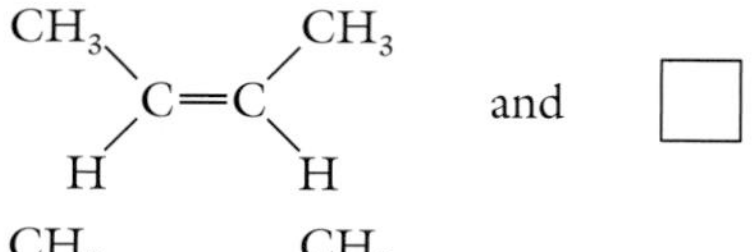 CH_3, CH_3, C=C, H, H and [structure]

(f) 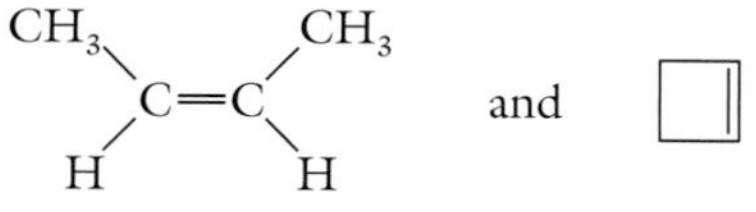 CH_3, CH_3, C=C, H, H and [structure]

4.46 For each of the following pairs of structural formulas, indicate whether the pair represents (1) the same compound or (2) different compounds that are constitutional isomers or (3) different compounds that are cis-trans isomers or (4) different compounds that are not isomers.

(a) 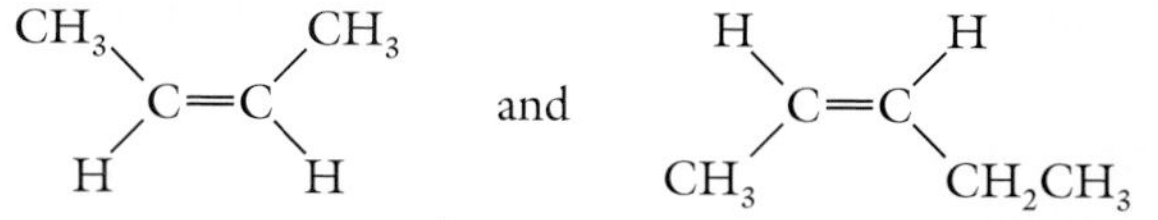 CH_3, CH_3, C=C, H, H and H, H, C=C, CH_3, CH_2CH_3

(b) [structure] and [structure]

(c) [structure] and [structure]

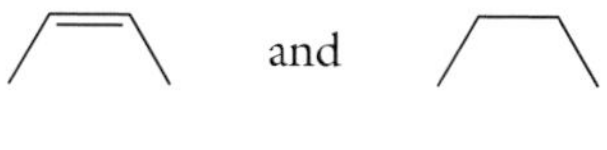

(d)  Cl, H, H, Cl and Cl, Cl, H, H

(e) $CH{\equiv}C{-}CH_2{-}CH_3$ and $CH_2{=}CH{-}CH{=}CH_2$

(f) $CH{\equiv}C{-}CH_2{-}CH_3$ and $CH_3{-}C{\equiv}C{-}CH_3$

4.47 Compare the boiling points and solubilities in water of cyclohexane, cyclohexene, and benzene.

4.48 Consider cyclohexane, cyclohexene, and benzene. Which will react with bromine under each of the following conditions? (a) Br_2; (b) Br_2 + high temperature or UV; (c) $Br_2 + FeBr_3$.

4.49 Consider cyclohexane, cyclohexene, and benzene. Which compound(s) will react with each of the following reagents? (a) Combustion (burning in air); (b) NaOH; (c) HCl; (d) H_2O/H^+; (e) hot acidic $KMnO_4$.

4.50 The addition of HCl to 1-pentene gives one major and one minor product, 2-chloropentane and 1-chloropentane, respectively. The addition of HCl to 2-pentene gives two products, 2-chloropentane and 3-chloropentane, in equal amounts. Explain the difference between the reactions of 1-pentene and 2-pentene.

4.51 What product is formed when 1,3-butadiene, $CH_2{=}CH{-}CH{=}CH_2$, reacts with excess Br_2?

4.52 Give the equation for the polymerization of each of the following monomers: (a) vinyl chloride (1-chloroethene); (b) isobutylene (2-methylpropene); (c) methyl methacrylate:

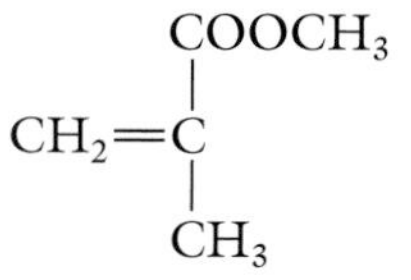

$$CH_2{=}C(COOCH_3)(CH_3)$$

4.53 What is the molecular formula of each of the following compounds?

(a) 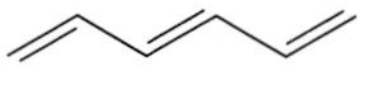 [structure] (b) 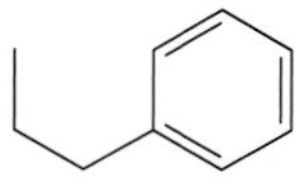 [structure]

(c) 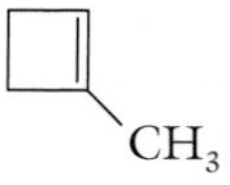 [structure with CH_3] (d) 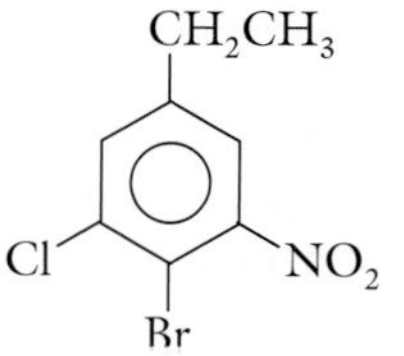 [structure with CH_2CH_3, Cl, NO_2, Br]

4.54 A chemist has two unknown samples, A and B. One of the samples is cyclohexene and the other is benzene, but the chemist does not know which sample is which. Both samples decolorize Br_2 when $FeBr_3$ is present, but only sample A decolorizes Br_2 in the absence of $FeBr_3$. Which sample is benzene and which sample is cyclohexene?

4.55 A chemist has two unknown samples, A and B. One of the samples is 1-hexene and the other is 1-hexyne, but the chemist does not know which sample is which. Both samples decolorize Br_2, but sample B decolorizes twice as much bromine as does sample A. Which sample is 1-hexene and which sample is 1-hexyne?

4.56 What is the hybridization of carbons 1 through 6 in the following compound? What are the bond angles about those carbons?

CH_3 (3)

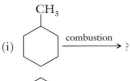

$CH_2-C=CH-$ (7, 2, 6, 1, 4, 5)

4.57 Give the organic product(s) formed in each of the following reactions. If no reaction takes place, write "no reaction." If more than one product is formed, indicate only the major product.

(a) $\xrightarrow{Br_2}$?

(b) $\xrightarrow{Br_2}$?

(c) $\xrightarrow[FeBr_3]{Br_2}$?

(d) $\xrightarrow{NaOH}$?

(e) $\xrightarrow{NaOH}$?

(f) CH_3 $\xrightarrow[\text{heat or light}]{Cl_2}$?

(g) CH_3 $\xrightarrow[FeCl_3]{Cl_2}$?

(h) CH_3 $\xrightarrow{\text{combustion}}$?

(i) CH_3 $\xrightarrow{\text{combustion}}$?

(j) $\xrightarrow{HCl}$?

(k) $\xrightarrow{HCl}$?

4.58 An unknown compound is hexane, methylcyclopentane, or 1-hexene. The purple color of permanganate is not decolorized when permanganate is added to the unknown. Identify the unknown.

4.59 An unknown compound is hexane, methylcyclopentane, or 1-hexene. The orange color of dichromate is replaced by a green-colored solution when dichromate is added to the unknown. Identify the unknown.

4.60 Which of the following compounds are aromatic?

(a) (b)

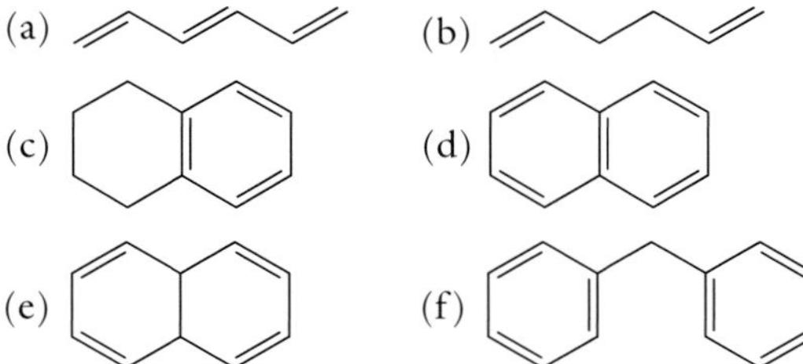

Expand Your Knowledge

Note: These icons denote exercises based on material in boxes.

4.61 The president of the ABC company wants to produce rubbing alcohol (2-propanol) by the acid-catalyzed addition of water to propene. One chemist in his employ, Bill Smith, says that the proposed reaction will not work, because the major product will be 1-propanol instead of 2-propanol. Another chemist, Mary Jones, disagrees and supports her viewpoint by invoking Markovnikov's rule. Who is correct?

$CH_3-CH_2-CH_2-OH$
1-Propanol

$CH_3-CH(OH)-CH_3$
2-Propanol

4.62 Compound A can exist as a pair of cis-trans isomers, but compound B cannot. Explain why.

CH_3 CH_3 CH_3 CH_3

A B

4.63 Geraniol and myrcene are extracted from the oils of roses and bay leaves, respectively. The structures of geraniol and myrcene are shown in Box 4.1. A perfume manufacturer has ordered some geraniol for use in formulating a perfume. The shipment arrives, but there is a suspicion that myrcene was delivered instead of geraniol. Suggest a simple chemical test to distinguish between geraniol and myrcene.

4.64 Box 4.9 shows the structure of phenanthrene, a carcinogen found in tobacco smoke and barbecued meats. What is the molecular formula of phenanthrene?

4.65 The medical community advises that we lower the proportion of our dietary fat intake that contains saturated fats, thereby increasing the proportion that contains unsaturated fats. Although we have not yet studied the structures of fats, what do you think are the differences between saturated and unsaturated fats?

4.66 2,4,6-Trinitrotoluene (also called TNT) is used as an explosive. It is produced from toluene ($C_6H_5—CH_3$). What reagent(s) and reaction conditions are needed to convert toluene into 2,4,6-trinitrotoluene?

4.67 The rate of addition of HCl to a series of alkenes is in the following order: 2-methylpropene > propene > ethene. Explain this order in relation to the mechanism for the addition reaction (see Boxes 4.4 and 4.5).

4.68 The addition of HCl to 2-methylpropene yields 2-chloro-2-methylpropane as the only product, whereas 2-pentene (both the cis and trans isomers) yields about equal amounts of two products: 2-chloropentane and 3-chloropentane. Explain the difference in relation to the mechanism for the Markovnikov rule (see Box 4.5).

4.69 Show the major product(s) formed when limonene (see Box 4.1) reacts with excess HCl.

4.70 Superglue is a polymer of methyl α-cyanoacrylate. Draw the structure of the polymer.

$COOCH_3$

$CH_2{=}C$

CN

Methyl α-cyanoacrylate

4.71 Terephthalic acid is one of the reactants used to produce the important synthetic fiber called poly(ethylene terephthalate) or polyester. Terephthalic acid is produced from *p*-dimethylbenzene. What reagent(s) and reaction condition are required for the conversion of *p*-dimethylbenzene into terephthalic acid?

COOH

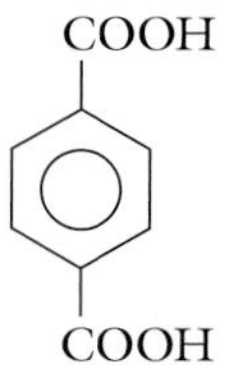

COOH

Terephthalic acid

4.72 Benzopyrene is one of the carcinogens in tobacco smoke and grilled meats (see Box 4.9). How many of the carbons in benzopyrene are bonded by π bonds? Draw a structure for benzopyrene to show all the π bonds.

4.73 There are two unlabeled vials, one containing vanillin (see Box 4.8) and the other limonene (see Box 4.1). Describe how a simple chemical test can be used to detect which vial contains vanillin and which contains limonene.

4.74 There are two different isomers of 1,4-dimethylcyclohexane but only one 1,4-dimethylbenzene. Explain.

4.75 An unknown is either *o*-, *p*-, or *m*-dimethylbenzene. Monobromination of the unknown with Br_2 and $FeBr_3$ yields only one product. Identify the unknown.

4.76 An unknown is either *o*-, *p*-, or *m*-chloromethylbenzene. Monobromination of the unknown with Br_2 and $FeBr_3$ yields two different isomeric products. Identify the unknown.

4.77 Aromatic substitution on monosubstituted benzenes follows a complex pattern because certain substituents when present on the benzene ring direct substitution to proceed in a selective manner. For example, a nitro group present on the benzene ring directs the next substitution to take place mostly at the meta position, whereas a methyl group present on the benzene ring directs the next substitution to take place mostly at the ortho and para positions. What sequence of reactions is needed to synthesize *m*-methylnitrobenzene from benzene? Show the required reagents and reaction conditions.

4.78 A sample of polypropylene has a molecular mass of 110,000. How many propylene units are linked together in the polypropylene molecule?

4.79 The addition of water to ethene does not take place unless a strong acid is present as a catalyst. Why?

4.80 Fused-ring compounds have rings that share two or more ring atoms (see Box 4.9). **Spiro-ring** compounds have rings that share one ring atom. Draw the structures of fused- and spiro-ring compounds containing two cyclohexane rings.

4.81 Explain why the oxidation state of the carbons of ethene is higher than that of the carbons of ethane. Oxidation state of carbon is defined in Section 3.10.

CHAPTER 5 ALCOHOLS, PHENOLS, ETHERS, AND THEIR SULFUR ANALOGUES

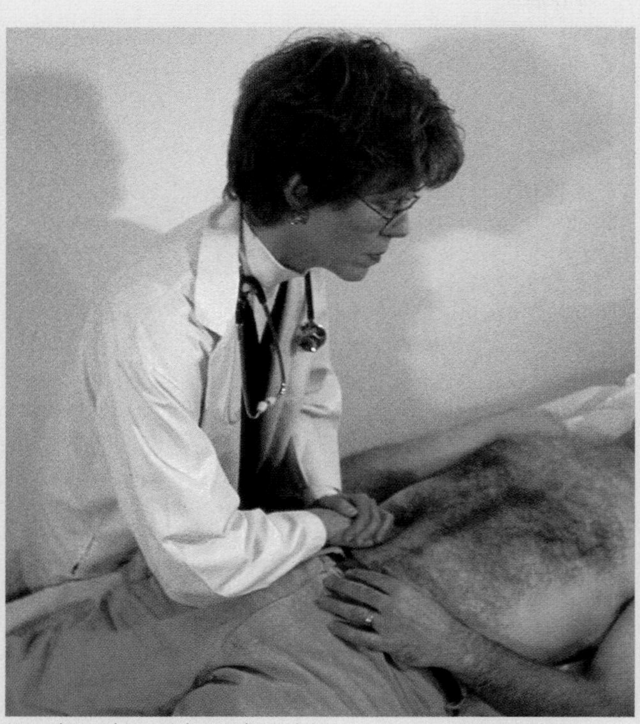
(David R. White, Richmond, Virginia.)

Chemistry in Your Future

The patient in bed 4 of your ward, suffering from alcoholic liver disease and, as of this morning, pneumonia, has rapidly gone from bad to worse. The liver can lose 75% of its tissue without ceasing to function, but it is no match for the toxic effects of long-term chronic abuse of alcohol. Why does this particular organic chemical pose such a risk to the body's largest organ? The answer is complex, but some aspects are explained in Chapter 5.

For more information on this topic and others in this chapter, go to www.whfreeman.com/bleiodian2e

Learning Objectives

- Draw and name alcohols.
- Describe the physical properties of alcohols.
- Describe and write equations for acid-base, dehydration, and oxidation reactions of alcohols.
- Draw and name phenols.
- Describe the properties of phenols.
- Draw and name ethers.
- Describe the properties of ethers.
- Draw and name thiols and disulfides.
- Describe the properties of thiols and disulfides.

Our study of organic compounds now turns to the many important compounds containing oxygen in addition to carbon and hydrogen. In this chapter, we will explore the chemistry of alcohols, phenols, and ethers—families in which oxygen is connected by single bonds to other atoms. We will also look briefly at some of the sulfur analogues of those three families. Then, in Chapters 6 through 8, we will cover the aldehydes, ketones, acids, esters, anhydrides, and amides—all families in which oxygen has a double bond to carbon.

—O—
Single bond
C=O
Double bond

We encounter alcohols in many aspects of our lives—in beer and other alcoholic beverages, in rubbing alcohol, as solvents for cosmetics and automotive antifreeze formulations, in cough medicines, and as starting materials for the synthesis of drugs and other widely used chemicals. Phenols find uses as antiseptics and disinfectants and as preservatives for food, gasoline, and other consumer goods. Ethers are employed as surgical anesthetics and in the manufacture of epoxy adhesives.

Many physiologically important compounds, including carbohydrates, proteins, and lipids, possess the same functional groups as those found in alcohols, phenols, ethers, and their sulfur analogues. Biological molecules tend to possess the functional groups of two or more families. For example, carbohydrates are both alcohols and aldehydes or ketones. Carbohydrates cannot be understood without first studying the alcohols, aldehydes, and ketones, because the functional groups of those families interreact in a unique manner.

5.1 STRUCTURAL RELATIONS OF ALCOHOLS, PHENOLS, AND ETHERS

Table 5.1 shows the general formulas, along with specific examples, of alcohols, phenols, and ethers. **Alcohols** and **phenols** contain an —OH group, a **hydroxyl group**, attached to a carbon atom. The carbon atom bearing the —OH in alcohols is a saturated carbon; that is, it is sp^3 hybridized and connected by single bonds to four adjacent atoms. In phenols, the —OH group is attached to one of the sp^2-hybridized carbons in a benzene ring. Note that the hydroxyl group of alcohols and phenols is not the hydroxide anion, OH^-, of strong bases such as KOH and NaOH.

Ethers, like alcohols and phenols, contain an oxygen that has bonds to two different atoms, but, in contrast with alcohols and phenols, neither of the bonds is to hydrogen. The oxygen of an ether has single bonds to two different carbons, which may be either aliphatic or aromatic carbons.

Alcohols, phenols, and ethers can be thought of as organic derivatives of water. Replacement of one of the hydrogens in water by an organic group

TABLE 5.1 Alcohols, Phenols, and Ethers

	General formula	Examples
alcohol	R—OH with R = aliphatic	CH_3—OH
phenol	R—OH with R = aromatic	OH attached to benzene ring
ether	R—O—R′ with R, R′ = same or different group, aliphatic or aromatic	CH_3—O—CH_3

yields an alcohol or phenol, depending on whether the organic group is aliphatic or aromatic. Replacement of the H in the —OH group of an alcohol or phenol (that is, replacement of the second H of water) by an organic group yields an ether:

$$\text{H—O—H} \xrightarrow{\text{replacement of H by R}} \underset{\substack{\textbf{Alcohol (R = aliphatic)}\\ \textbf{Phenol (R = aromatic)}}}{\text{R—O—H}} \xrightarrow{\text{replacement of H by R}} \underset{\textbf{Ether}}{\text{R—O—R}}$$

Example 5.1 Distinguishing among alcohols, phenols, and ethers

Which of the following structures is an alcohol, a phenol, an ether, or something else?

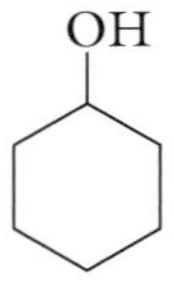

1: cyclohexane ring bearing OH

2: benzene ring bearing CH_3 and OH (meta)

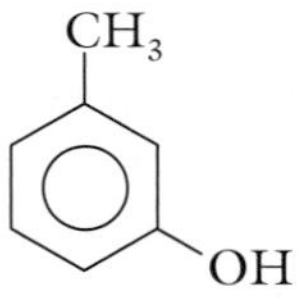

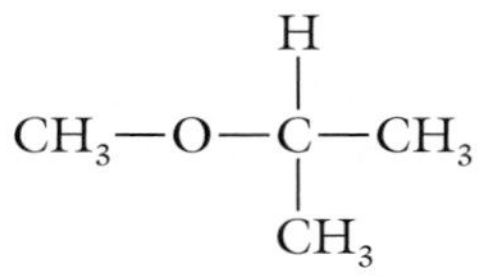

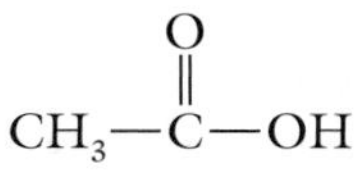

3: $CH_3—O—CH(CH_3)—CH_3$ (central C bearing H and CH_3)

4: $CH_3—C(=O)—OH$

Solution

Structure 1 is an alcohol because the —OH is attached to a saturated carbon. Structure 2 is a phenol because the —OH is attached to a benzene ring. Structure 3 is an ether because the oxygen atom is bonded to two different carbons. Structure 4 has a hydroxyl group, but it is neither an alcohol nor a phenol, because the carbon holding the —OH is neither saturated nor part of a benzene ring. Structure 4 is a carboxylic acid (Chapter 7).

Problem 5.1 Which of the following structures is an alcohol, a phenol, an ether, or something else?

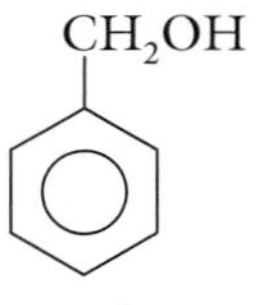

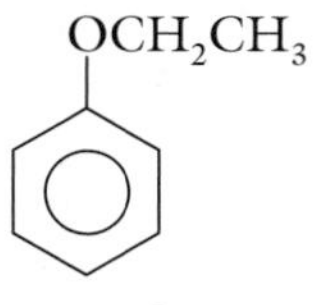

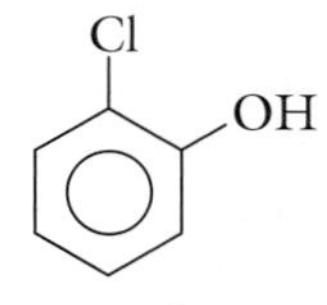

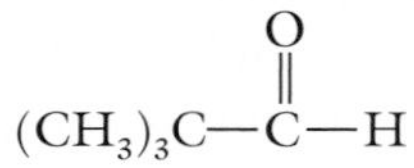

1: benzene ring bearing CH_2OH

2: benzene ring bearing OCH_2CH_3

3: benzene ring bearing Cl and OH (ortho)

4: $(CH_3)_3C—C(=O)—H$

The oxygen atoms in alcohols, phenols, and ethers use sp^3-hybrid orbitals in forming bonds, analogous to sp^3-hybridized carbons but with a difference. The oxygen atom has six outer-shell electrons, two more than carbon. Whereas carbon has four half-filled sp^3 orbitals and is tetravalent, oxygen has two half-filled and two filled sp^3 orbitals and is divalent: the pairs of electrons in the two filled sp^3 orbitals do not participate in bonding to other atoms and are called **nonbonding electrons.** The two half-filled orbitals bond oxygen to other atoms.

The nonbonding electrons are shown (by pairs of electron dots) in Lewis structures but not in structural formulas (Section 1.6). Whichever representation is used, always keep in mind that the nonbonding electrons are present. The oxygen atom in water, alcohols, phenols, and ethers has a complete octet of electrons.

$$\underset{\substack{\textbf{Lewis}\\ \textbf{structure}}}{\text{R—}\ddot{\underset{..}{\text{O}}}\text{H}} = \underset{\substack{\textbf{Structural}\\ \textbf{formula}}}{\text{R—OH}}$$

The bond angle about the sp^3-hybridized oxygen atom is 104.5° in water, distorted a few degrees from the tetrahedral bond angle of 109.5°, because the two nonbonded electron pairs repulse the adjacent bonded electron pairs and compress their bond angle (Section 1.7). The corresponding bond angles in alcohols, phenols, and ethers are much closer to the tetrahedral bond angle because the substituents attached to oxygen are larger than hydrogen (Figure 5.1). Repulsive interaction between the larger substituents counters the repulsions between nonbonded electron pairs.

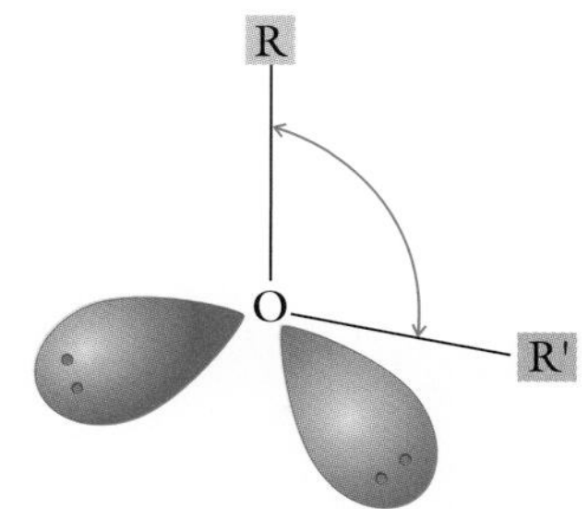

Figure 5.1 Bond angle for sp^3-hybridized oxygen in water (R = R′ = H), alcohols and phenols (R = aliphatic and aromatic, respectively, and R′ = H), and ethers (R, R′ = aliphatic or aromatic).

5.2 CONSTITUTIONAL ISOMERISM IN ALCOHOLS

Constitutional isomerism in alcohols, as in alkenes, arises out of two kinds of connectivity differences:

- Different carbon skeletons
- Different placement of the functional group (—OH) on a carbon skeleton

No isomeric alcohols are possible for the two simplest alcohols—methanol and ethanol—which contain one carbon and two carbons, respectively. There are, however, two C_3H_8O alcohols, owing to two different placements of the —OH group on the three-carbon chain:

$CH_3—OH$
Methanol

$CH_3—CH_2—OH$
Ethanol

$CH_3—CH_2—CH_2—OH$
1-Propanol

$CH_3—CH(OH)—CH_3$
2-Propanol

Example 5.2 Drawing constitutional isomers of alcohols

Both types of connectivity differences (different carbon skeletons and different placements of functional group) exist for $C_4H_{10}O$ alcohols, resulting in four constitutional isomers. Give structural formulas of the isomers.

Solution

Two different C_4 carbon skeletons are possible:

C—C—C—C
1

C—C(C)—C
2

Two isomers are possible for skeleton 1 because the —OH group can bond at either of two locations—on an end carbon or on a next-to-end carbon of the carbon chain:

$CH_3—CH_2—CH_2—CH_2—OH$
1a

$CH_3—CH(OH)—CH_2—CH_3$
1b

Be careful not to draw duplicate structural formulas of the same isomer. Placement of the —OH on the far-left carbon gives a duplicate of isomer 1a, whereas placement of the —OH on the second carbon from the right gives a duplicate of isomer 1b.

Two additional C_4 alcohols are created by attaching the —OH group at different locations on skeleton 2:

$CH_3—C(CH_3)(OH)—CH_3$
2a

$CH_3—CH(CH_3)—CH_2—OH$
2b

Problem 5.2 How many alcohol structures are possible for the molecular formula $C_5H_{12}O$? Draw one structural formula for each isomer.

The presence of an —OH group in a compound does not change the C:H ratio compared with the corresponding alkane, cycloalkane, or alkene. Therefore, the molecular formula tells us whether an alcohol has a ring or double bond:

- The C:H ratio is the same as that for an alkane, C_nH_{2n+2}, if the alcohol has no double bond or ring.
- The C:H ratio is the same as that for an alkene or cycloalkane, C_nH_{2n}, if the alcohol has a double bond or ring.

Example 5.3 Determining the general formula for an alcohol

Show that alcohols 1 and 2 follow the general formulas $C_nH_{2n+2}O$ and $C_nH_{2n}O$, respectively.

$CH_3—CH_2—CH_2—OH$ (1) $CH_2{=}CH—CH_2—OH$ (2)

Solution

The formula of alcohol 1 is C_3H_8O with $n = 3$ and $2n + 2 = 8$. The formula of alcohol 2 is C_3H_6O with $n = 3$ and $2n = 6$.

Problem 5.3 Show that alcohols 1 and 2 follow the general formulas $C_nH_{2n+2}O$ and $C_nH_{2n}O$, respectively.

$(CH_3)_3C—CH_2—OH$ (1)   OH on cyclopentane ring (2)

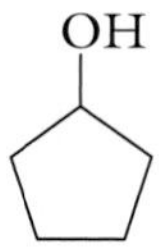

5.3 CLASSIFYING AND NAMING ALCOHOLS

Alcohols show differences in chemical reactivity, depending on their structural classification as **primary (1°), secondary (2°), or tertiary (3°) alcohols.** Alcohols are 1°, 2°, or 3° alcohols, depending on whether the carbon holding the —OH group is a 1°, 2°, or 3° carbon. Recall from Section 3.6 that a 1°, 2°, or 3° carbon is directly bonded to a total of one, two, or three other carbon atoms, respectively.

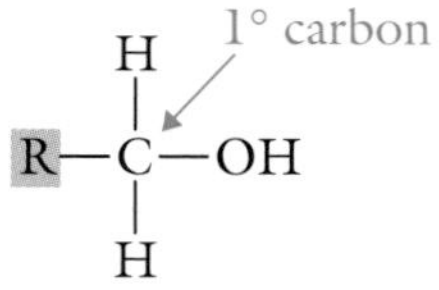

1° carbon	2° carbon	3° carbon
R—CH₂—OH (C bonded to R, H, H, OH)	R₂CH—OH (C bonded to R, R, H, OH)	R₃C—OH (C bonded to R, R, R, OH)
Primary alcohol	Secondary alcohol	Tertiary alcohol

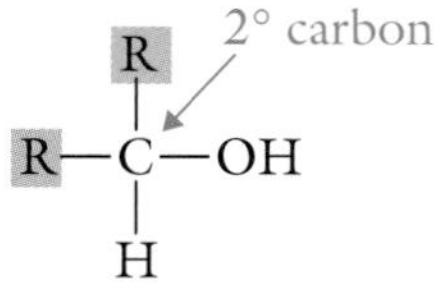

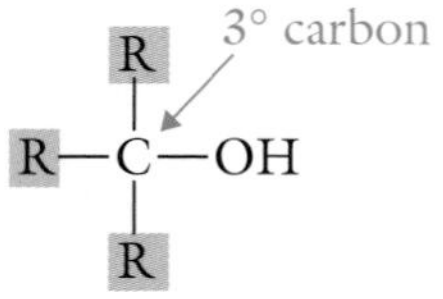

Example 5.4 Classification of alcohols

Classify the following alcohols as primary, secondary, or tertiary.

(a) $CH_3—CH_2—CH_2—OH$

(b) 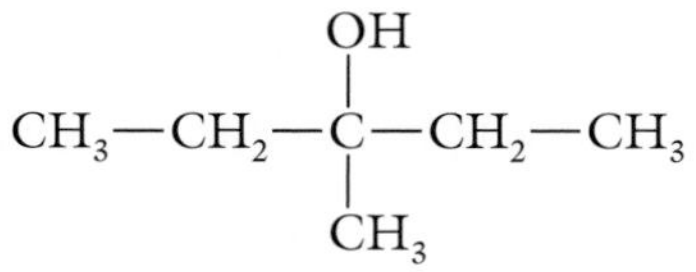 $CH_3—CH_2—C(OH)(CH_3)—CH_2—CH_3$

(c) $CH_3—CH(OH)—CH_2—CH_3$

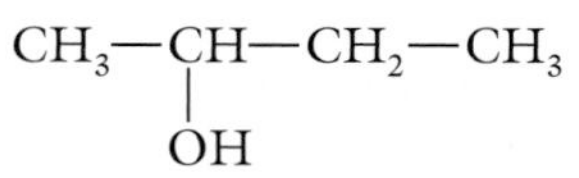

(d) 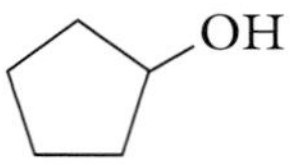 cyclopentane ring bearing OH

Solution

Locate the carbon bonded to the —OH group and then count the carbons directly attached to that carbon: (a) primary; (b) tertiary; (c) secondary; (d) secondary.

Problem 5.4 Classify the following alcohols as primary, secondary, or tertiary.

(a) 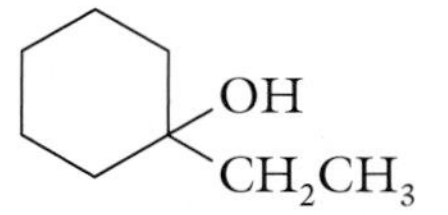 cyclohexane ring bearing OH and CH_2CH_3 on the same carbon

(b) $(CH_3)_3C—CH_2—OH$

(c) 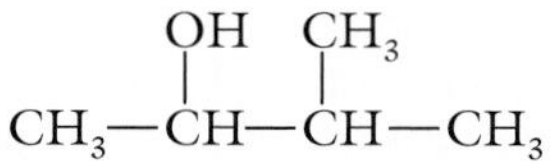 $CH_3—CH(OH)—CH(CH_3)—CH_3$

(d) 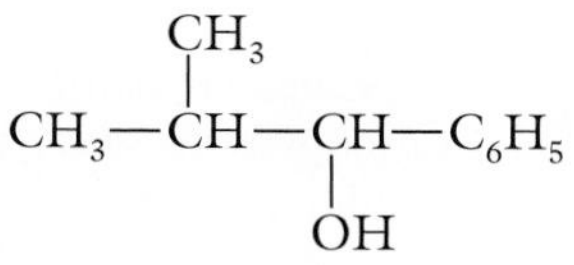 $CH_3—CH(CH_3)—CH(OH)—C_6H_5$

In the IUPAC nomenclature system, the rules for naming alcohols are a variation on the rules for naming alkanes:

Rules for naming alcohols

- Find the longest continuous chain that contains the carbon holding the —OH group.
- The ending **-e** of the alkane name is changed to **-ol.**
- The longest continuous chain is numbered from the end nearest the —OH group. The position of the —OH group is indicated by placing the number of the carbon holding the —OH in front of the base name.
- The names of substituents are added as prefixes, preceded by numbers indicating their positions on the longest chain.
- A cyclic compound with an —OH group attached to a ring carbon is named as a **cycloalkanol.** Numbering of the ring starts at the carbon holding the —OH and proceeds to other substituents on the ring in the direction that yields the lowest numbers for those substituents.
- A **polyfunctional alcohol** contains more than one —OH group in the molecule. Compounds with two and three —OH groups are called **diols** and **triols,** respectively. **Alkanediol** and **alkanetriol** form the basis of the IUPAC names of such compounds. Numbers are attached at the beginning of the names to specify the position of each —OH group (for example, 1,2-ethanediol, 1,2,3-propanetriol, and so forth).

Example 5.5 Naming alcohols by the IUPAC nomenclature system

Name the following compounds by the IUPAC system.

(a) 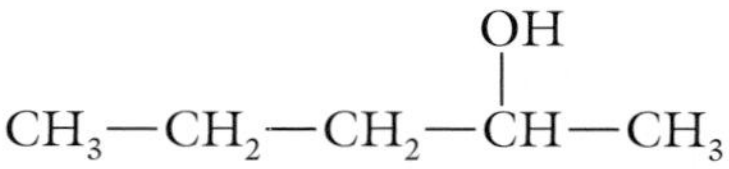 $CH_3—CH_2—CH_2—CH(OH)—CH_3$

(b) 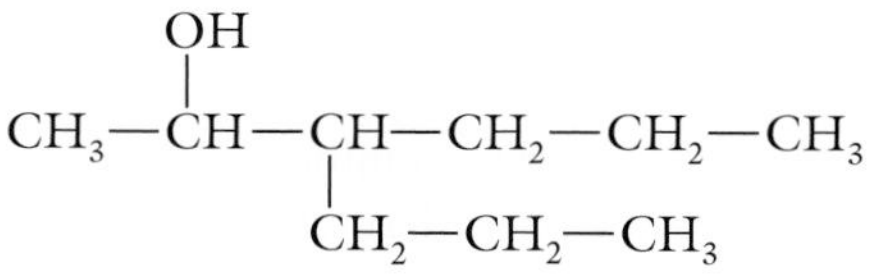 $CH_3—CH(OH)—CH(CH_2—CH_2—CH_3)—CH_2—CH_2—CH_3$

(c) $HO—CH_2—CH_2—CH_2—OH$

(d) 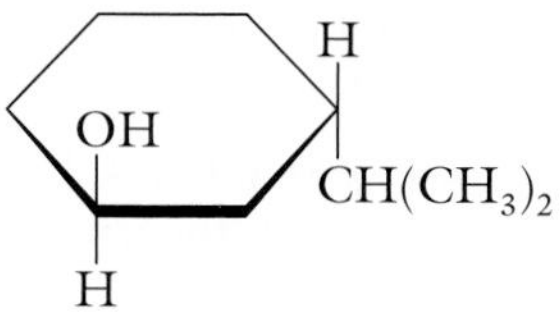 cyclohexane ring with OH and H on one carbon, and H and $CH(CH_3)_2$ on the adjacent carbon

5.1 CHEMISTRY AROUND US

Alcohols

Many alcohols, both naturally occurring and synthetic, have important industrial and everyday uses. Methanol (common name: methyl alcohol; see Table 5.2 for the structure of this alcohol and others) is used to manufacture formaldehyde (a raw material for certain plastics—see Box 6.2), as a solvent in which to carry out chemical reactions, and in Sterno heating pots.

Ethanol (common name: ethyl alcohol) comes from two sources: the fermentation of plant products and the hydration of ethene. Ethanol is the alcohol in alcoholic beverages. Its properties and effects as a beverage are described in Boxes 5.2 and 5.3. It is also used industrially and medically, in the production of plastics, drugs, and other organic chemicals; as a solvent for perfumes and varnishes; and as an antiseptic for sterilizing surgical instruments and the surface of a patient's body before surgery (as a 70% solution of ethanol in water).

2-Propanol (common name: isopropyl alcohol) is often called rubbing alcohol because it is applied to the skin to lower a fever. The isopropanol evaporates rapidly from the skin surface, removing heat and creating a cooling effect. 2-Propanol is also used as a solvent or carrier for cosmetics and skin creams and is used to manufacture acetone, an important industrial chemical (see Box 6.2).

Ethylene glycol (IUPAC name: 1,2-ethanediol) and propylene glycol (IUPAC name: 1,2-propanediol) are used in automotive cooling-system formulations. These mixtures of diol and water (typically combined in equal amounts) have higher boiling points and lower freezing points than those of water. They neither evaporate in summer nor freeze in winter and so prevent the damage that can happen to an engine when water alone is used to cool it. Another use of ethylene glycol is in the manufacture of poly(ethylene terephthalate), an important plastic (Mylar) and fiber (Dacron polyester) (Section 7.10).

1,2,3-Propanetriol (common names: glycerol, glycerin) is one of the products of our digestion of fats and oils. Unlike other alcohols, including the diols, glycerol is not toxic. Its lack of toxicity combined with its highly hydrophilic nature (a consequence of its three —OH groups) and sweet taste makes glycerol useful as a food additive to keep foods fresh by retention of moisture. It is also used in cosmetics, soaps, and skin creams. The reaction of glycerol with nitric acid produces nitroglycerin, a compound used both as a powerful explosive (employed in construction and, unfortunately, in warfare) and as a medicinal vasodilator. It is prescribed in the treatment of **angina pectoris,** a condition in which there is constriction of the blood vessels in heart muscle.

Menthol, obtained from peppermint oil, is used in cough drops and throat sprays to sooth the respiratory tract by increasing secretions from mucous membranes. Menthol is also an ingredient in shaving creams, toothpastes, and cigarettes.

CH_3 / OH / $CH(CH_3)_2$

Menthol

Solution

(a) The five-carbon chain holding the —OH group is numbered from the right side, the end nearest the —OH group, to yield the name 2-pentanol. Numbering from the other end is incorrect, because it yields 4-pentanol, containing a larger number for the position of the —OH group.

(b) The absolute longest chain contains seven carbons but does not contain the carbon holding the —OH. The longest chain holding the —OH group is the six-carbon chain shown here:

$$\overset{1}{CH_3}-\underset{2}{\overset{OH}{\overset{|}{CH}}}-\overset{3}{CH}(CH_2-CH_2-CH_3)-\overset{4}{CH_2}-\overset{5}{CH_2}-\overset{6}{CH_3}$$

The chain is numbered from the left side, the side nearest the —OH. Attached to the six-carbon chain is a propyl group at C3. The name is 3-propyl-2-hexanol.

(c) The compound contains a three-carbon chain with —OH groups at C1 and C3 and is named 1,3-propanediol.

(d) The —OH group is at C1 of the ring. Proceeding counterclockwise to the isopropyl group yields the name 3-isopropylcyclohexanol. Proceeding, clockwise is incorrect, because it yields a larger number in the name (4-isopropylcyclohexanol) for the position of the isopropyl group. The complete name is *trans*-3-isopropylcyclohexanol because the structure shown has the —OH and isopropyl groups projecting on different sides of the plane of the ring.

Problem 5.5 Name the following compounds by the IUPAC system.

(a) 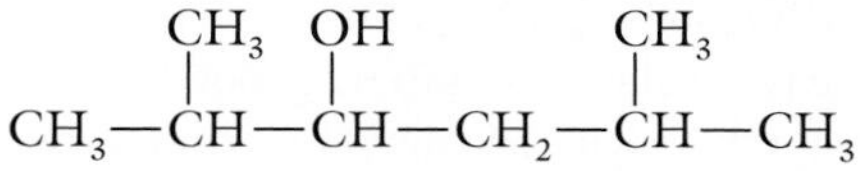 $CH_3-\overset{\overset{CH_3}{|}}{CH}-\overset{\overset{OH}{|}}{CH}-CH_2-\overset{\overset{CH_3}{|}}{CH}-CH_3$

(b) 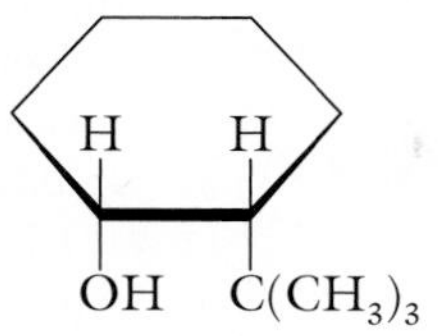 H H OH $C(CH_3)_3$

(c) $HO-CH_2-\overset{\overset{OH}{|}}{CH}-CH(CH_3)_2$

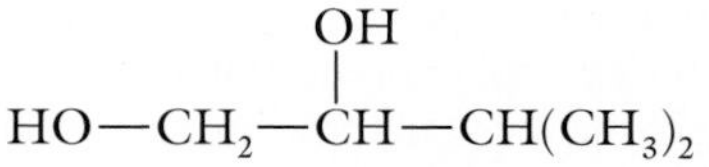

(d) 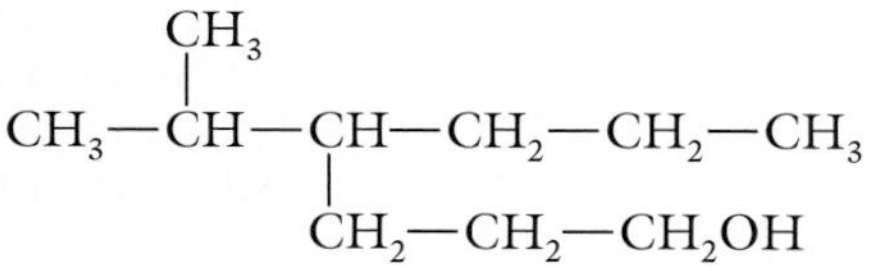 $CH_3-\overset{\overset{CH_3}{|}}{CH}-\underset{\underset{CH_2-CH_2-CH_2OH}{|}}{CH}-CH_2-CH_2-CH_3$

Simple alcohols are usually referred to by common names (Table 5.2), which consist of the name of the alkyl group attached to the —OH group followed by a space and then the word alcohol. Simple alcohols are those containing one of the simple alkyl groups: unbranched alkyl groups from methyl, ethyl, propyl, butyl through decyl; branched three- and four-carbon alkyl groups (isopropyl, isobutyl, *s*-butyl, *t*-butyl).

Box 5.1 describes many of the industrially important alcohols. Alcoholic beverages and their health aspects are described in Boxes 5.2 on page 158 and 5.3 on page 160.

TABLE 5.2 IUPAC and Common Names of Alcohols

Alcohol	IUPAC name	Common name
CH_3-OH	methanol	methyl alcohol
CH_3CH_2-OH	ethanol	ethyl alcohol
$CH_3CH_2CH_2-OH$	1-propanol	propyl alcohol
$CH_3\overset{\overset{OH}{|}}{C}HCH_3$	2-propanol	isopropyl alcohol
$CH_3CH_2CH_2CH_2-OH$	1-butanol	butyl alcohol
$CH_3\overset{\overset{OH}{|}}{C}HCH_2CH_3$	2-butanol	*s*-butyl alcohol
$CH_3\overset{\overset{CH_3}{|}}{C}HCH_2-OH$	2-methyl-1-propanol	isobutyl alcohol
$CH_3\overset{\overset{CH_3}{|}}{\underset{\underset{CH_3}{|}}{C}}-OH$	2-methyl-2-propanol	*t*-butyl alcohol
$HO-CH_2CH_2-OH$	1,2-ethanediol	ethylene glycol
$HO-CH_2\overset{\overset{OH}{|}}{C}HCH_3$	1,2-propanediol	propylene glycol
$HO-CH_2\overset{\overset{OH}{|}}{C}HCH_2-OH$	1,2,3-propanetriol	glycerol; glycerin

5.2 CHEMISTRY AROUND US

Types of Alcoholic Beverages

The alcohol in alcoholic beverages is ethanol. Although ethanol is synthesized in large amounts by the hydration of ethylene (Section 4.6), federal law in the United States requires that ethanol for consumption be produced by the fermentation of grains, fruits, or vegetables. Ethanol used for consumption is heavily taxed. Ethanol used for other purposes is not taxed but is carefully controlled by government statutes.

Fermentation is the anaerobic (in the absence of oxygen) conversion of the carbohydrates in a grain, fruit, or vegetable into ethanol. The process requires yeast, which supplies a mixture of enzymes needed to catalyze the process. Starch and sucrose (table sugar) in the grain, fruit, or vegetable are first hydrolyzed to glucose and fructose, and they in turn are converted into ethanol (Section 15.2):

$$\text{Starch, sucrose} \longrightarrow \text{glucose, fructose} \longrightarrow CH_3CH_2OH + CO_2$$

All alcoholic beverages consist almost entirely (>99%) of ethanol and water. These beverages differ from one another in the proportion of ethanol to water and in the chemicals that they contain other than ethanol and water. Beer contains from 3% to 8% v/v ethanol (lager beers contain less ethanol than do ale and stout beers). Wines contain from 10% to 14% v/v ethanol. Liquors have much higher ethanol contents, as much as 50% v/v ethanol. The ethanol content is usually expressed as **proof,** defined as twice the % v/v ethanol. Fermentation is limited to the production of beverages with contents no higher than 14% v/v ethanol, because yeast cannot survive at higher ethanol concentrations. Beers and wines are produced by fermentation followed by filtration, pasteurization, and carbonation. The higher ethanol content in liquors is produced by distillation of the fermentation mixture, taking advantage of the fact that ethanol has a lower boiling point than water.

Rows of distillation stills produce whiskey at the Glenfiddich Distillery in Dufftown, Scotland. (Paul Harris/Tony Stone.)

Ethanol itself has no taste. The individual tastes of different beers, wines, and liquors result from the presence of small amounts (<1%) of other substances, known as **congeners,** compounds specific to the source of starch and sucrose or to the manufacturing process. Beers in the United States are made from barley. Rice and sorghum seeds are used in Asia and Africa, respectively. Hops (flower clusters from the female of the vine *Humulus lupus*) are added to impart the bitter taste, whereas trace amounts of organic chemicals derived from the different grains are responsible for some of the more subtle flavors. Cask beers are aged in wooden casks and, in the process, acquire trace amounts of other congeners that originate in the wood. Light and dark beers are due to differences in the temperature (low and high, respectively) at which the grain is dried before fermentation. Most wines are produced from grapes, although other fruits or rice are sometimes used. Wine color and flavor are affected by the variety of grape and the geographical location in which the grapes are grown. Sparkling wine (champagne) contains carbon dioxide. Some wines, such as vermouth, are flavored with herbs and spices.

Liquors include the various whiskeys, rum, gin, vodka, and brandy. Whiskey is produced from grains: scotch whiskey from barley, bourbon from corn, and rye from rye. Rum is produced from molasses (sugar cane). Brandy is made by adding additional ethanol to wine. Vodka, made from potatoes and some grains, has almost no congeners. Gin, made from grain alcohol by a second or even third distillation, is second only to vodka in low congener content and is flavored with juniper berries.

5.4 PHYSICAL PROPERTIES OF ALCOHOLS

The physical properties of alcohols are very different from those of hydrocarbons (alkanes, alkenes, alkynes, and aromatics). Hydrocarbons, with their weak secondary attractive forces (Section 3.9), have low boiling and melting points and are insoluble in water. Alcohols have strong secondary attractive forces, and as a consequence are characterized by

- Much higher boiling and melting points than those of hydrocarbons
- Solubility in water

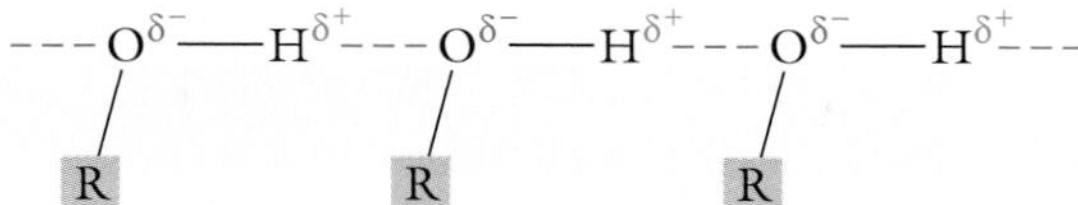

Figure 5.2 Hydrogen bonding in alcohols explains their high boiling points.

The physical properties of alcohols arise from the hydrogen bonds (Section 1.9) that result from the presence of the highly polar $^{\delta-}O—H^{\delta+}$ bonds. There are strong attractive forces between the partly positive hydrogen end of the O—H bond of one alcohol molecule and the partly negative oxygen end of the O—H bond of a neighboring alcohol molecule. This attractive force, depicted by the dashed line in Figure 5.2, must be broken for an alcohol to boil. Figure 5.2 is an oversimplification because any one alcohol molecule is surrounded (three-dimensionally) by hydrogen bonds to several neighboring alcohol molecules. Higher temperatures are required to boil alcohols compared with hydrocarbons because the hydrogen-bonding attractions in alcohols are much stronger than the London forces in hydrocarbons.

Table 5.3 shows that the difference in boiling points between hydrocarbons and alcohols is very large. There is a 154 Celsius degree difference in the boiling points of ethane (CH_3CH_3) and methanol (CH_3OH), compounds of comparable molecular mass (30 amu and 32 amu, respectively). The difference in boiling points between hydrocarbons and alcohols of comparable molecular size decreases, however, as the molecules increase in size. For example, the difference between 1-butanol ($CH_3CH_2CH_2CH_2OH$) and pentane ($CH_3CH_2CH_2CH_2CH_3$) is 81 Celsius degrees. The reason for the decreasing difference in boiling points is that, as alcohols become larger, the part of the molecule that participates in hydrogen bonding (the —OH group) becomes proportionately smaller, causing the alcohol molecule to increasingly resemble a nonpolar hydrocarbon.

TABLE 5.3 Physical Properties of Alkanes, Alcohols, and Ethers

Compound	Structure	Molecular mass (amu)	Boiling point (°C)	Solubility in water
ALCOHOLS				
methanol	CH_3OH	32	65	soluble
ethanol	CH_3CH_2OH	46	78	soluble
1-propanol	$CH_3CH_2CH_2OH$	60	97	soluble
1-butanol	$CH_3CH_2CH_2CH_2OH$	74	117	moderately soluble
1-pentanol	$CH_3CH_2CH_2CH_2CH_2OH$	88	138	slightly soluble
1-hexanol	$CH_3CH_2CH_2CH_2CH_2CH_2OH$	102	158	insoluble
1,2-ethanediol	$HOCH_2CH_2OH$	62	198	soluble
1,2,3-propanetriol	$HOCH_2CH(OH)CH_2OH$	92	290	soluble
ALKANES				
ethane	CH_3CH_3	30	−89	insoluble
propane	$CH_3CH_2CH_3$	44	−42	insoluble
butane	$CH_3CH_2CH_2CH_3$	58	−1	insoluble
pentane	$CH_3CH_2CH_2CH_2CH_3$	72	36	insoluble
ETHERS				
dimethyl ether	CH_3OCH_3	46	−23	soluble
diethyl ether	$CH_3CH_2OCH_2CH_3$	74	35	moderately soluble
WATER	H_2O	18	100	

5.3 CHEMISTRY WITHIN US

Health Aspects of Alcohol Consumption

Most people believe, erroneously, that ethanol is a stimulant. Because one of its initial effects is to reduce social inhibitions, it seems to stimulate conversation, relieve tension, and bolster confidence. However, ethanol is a powerful central-nervous-system depressant.

A low level of alcohol consumption, a drink or two, does most people no harm, psychologically or physically. There is even some evidence that drinking small amounts has cardiovascular benefits. However, there is no doubt about the negative consequences of excessive alcohol consumption.

Continued acute consumption of alcohol (consumption of large amounts in a short time) depresses the central nervous system more and more, causing increasing loss of inhibitions and control, false feelings of euphoria and superiority, and loss of motor skills and coordination. There are more than 40,000 traffic fatalities annually in the United States, about 40% of which are attributed to drivers whose judgments are impaired by alcohol consumption.

The point at which a person is legally considered in most localities to be drunk corresponds to a blood alcohol level of 0.08%. Acute consumption past this level can lead to blackouts (about 0.20% blood alcohol) and even death (0.30% and higher). Death occurs at those levels because ethanol shuts down the respiratory system by depressing the respiratory center in the brain—the person simply stops breathing.

Chronic alcohol abuse (habitual consumption of excessive amounts over a long period of time) may not result in the immediate illness or death brought on by acute consumption but, over time, also will result in major health problems. Slow liver destruction (cirrhosis) over many years has caused the deaths of many alcoholics.

There are more than 15,000 alcohol-related traffic fatalities annually in the United States. (Kolvoord/The Image Works.)

The consequences of uncontrolled alcohol consumption include not only the aforementioned traffic fatalities, but also domestic and neighborhood violence and lost productivity at home and in the workplace. The human devastation caused by alcohol consumption exceeds that attributed to hard drugs such as crack cocaine.

Pregnant women who drink alcohol (even modest social drinking) can give birth to babies with **fetal alcohol syndrome (FAS).** This syndrome consists of a nonreversible combination of mental retardation and birth defects. Among the specific problems are abnormal facial features, growth deficiences, and central nervous system problems. People with FAS have learning, memory, communication, hearing, vision, and attention-span deficits. Other terms—fetal alcohol effects, alcohol-related neurodevelopmental disorder, and alcohol-related birth defects—are used to describe children who have some, but not all, the clinical signs of FAS.

The deleterious effects of alcohol consumption are caused not directly by ethanol itself but by the product acetaldehyde formed when the liver attempts to detoxify the body of ethanol. The liver is able to detoxify the acetaldehyde by further oxidation and excretion of the resulting products when the ethanol load on the system is low. However, the liver is not able to keep up with a high ethanol load, and this failure results in a buildup of acetaldehyde. Highly reactive toward proteins, acetaldehyde denatures (Section 12.9) them so that they are no longer capable of performing their physiological functions.

Other alcohols are much more toxic than ethanol, and consumption of even small amounts can result in death. Desperate people have been known to drink methanol, for example, and methanol is exceptionally dangerous because its detoxification by the liver produces formaldehyde, which is more reactive than acetaldehyde toward proteins. Ethylene glycol, often found in households because of its common use as an antifreeze, also is highly toxic. Intravenous feeding of ethanol is the medical treatment for methanol and ethylene glycol poisoning. In this treatment, the ethanol is preferentially detoxified in the liver, whereas the methanol or ethylene glycol is eliminated in the urine without being "detoxified" into deleterious substances.

Chemical-grade ethanol, produced for nonbeverage uses, is generally denatured (rendered unfit for consumption) by the addition of toxic materials such as methanol and 2-propanol. This addition prevents chemical-grade ethanol from being diverted to use in alcoholic beverages. Certain industrial processes, however, require pure ethanol, devoid of denaturing agents. The sale and distribution of pure ethanol are very tightly controlled, with heavy penalties (prison and fines) for diversion to consumption.

INSIGHT INTO PROPERTIES

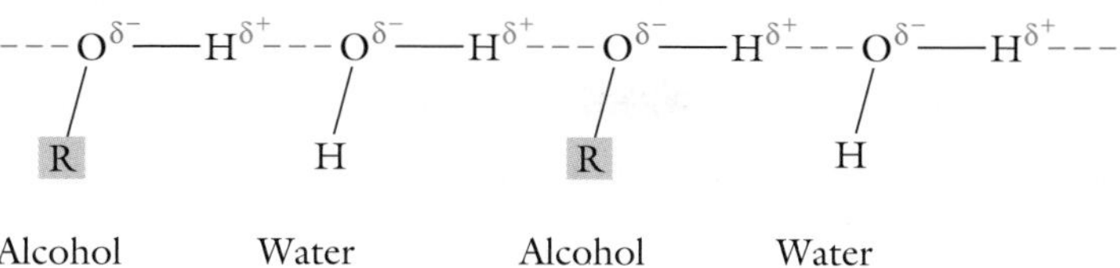

Figure 5.3 Hydrogen bonding between water and an alcohol explains why alcohols of three or fewer carbons are completely miscible with water.

Diols and triols, as might be expected, have especially high boiling points compared with alkanes of similar size. Alcohols containing more than one —OH group participate in more-extensive hydrogen bonding, which results in higher boiling points. For example, 1,2-ethanediol ($HOCH_2CH_2OH$) has a boiling point of 198°C compared with 97°C for 1-propanol ($CH_3CH_2CH_2OH$).

Hydrogen bonding also explains why alcohols of three or fewer carbons are completely miscible with water. Water alone and alcohol alone are extensively hydrogen bonded. The two compounds mix freely because alcohol molecules can hydrogen bond with water molecules. The O—H bonds attract one another irrespective of whether they reside in water or alcohol (Figure 5.3).

The amount of hydrogen bonding between alcohol and water decreases greatly as the number of carbons in the alcohol increases. The R part of an alcohol molecule is nonpolar, hydrophobic, and incapable of interaction (solvation) with the O—H bonds of water (see Table 5.3); only the hydrophilic —OH group can interact with water. Whereas alcohols with three or fewer carbons are completely soluble, the four-carbon alcohol is only moderately soluble in water, the five-carbon alcohol is slightly soluble, and alcohols of more than five carbons show negligible water solubility. An alcohol of more than five carbons becomes water soluble only if it contains more than one —OH group.

Many biochemical processes take advantage of the ability of —OH groups to water solubilize otherwise insoluble molecules. For example, one of the mechanisms by which the liver detoxifies toxic substances is to attach sufficient —OH groups to such molecules to render them water soluble (Section 18.4). Their solubility allows them to be transported in the blood and excreted in the urine.

Example 5.6 Explaining the physical properties of alcohols

Explain each of the following properties:

(a) The boiling point of water (100°C) is considerably higher than that of methanol (65°C) even though the molecular mass of methanol is almost twice that of water.

(b) 1-Hexanol is not water soluble, whereas 1,3-hexanediol is water soluble.

(c) Butene is not water soluble, whereas 1-propanol is water soluble.

Solution

(a) The extent of hydrogen bonding in water is much greater than in methanol because each water molecule has two O—H bonds, whereas methanol has one.

(b) The one —OH group in 1-hexanol cannot solubilize the six-carbon compound; the limit is three (possibly four) carbons per —OH group. The presence of two —OH groups in 1,3-hexanediol (that is, two —OH per six C, or a ratio of one —OH to three C) results in water solubilization.

(c) The "like dissolves like" rule applies. No attractive interactions are possible between butene and water, because butene is nonpolar. 1-Propanol, on the other hand, does hydrogen bond with water because, like water, it contains the —OH group.

Problem 5.6 Answer each of the following questions:
(a) Which has the higher boiling point, propanol or 1,2-ethanediol? Why?
(b) Why does butanol have a higher boiling point than *t*-butyl alcohol?
(c) Which is more soluble, propanol or butane, in a nonpolar solvent such as hexane?

5.5 THE ACIDITY AND BASICITY OF ALCOHOLS

Alcohols are usually considered neutral compounds. They neither turn blue litmus paper red nor turn red litmus paper blue. Aqueous solutions of alcohols have about the same pH value as that of water itself, pH = 7, a pH value often called neutral pH. However, alcohols are not neutral on an absolute scale. Whereas the hydrocarbon families of compounds show no acidic or basic behavior, alcohols resemble water in being very weakly amphoteric: they are both very weak acids and very weak bases (Section 2.4).

Figure 5.4 The weak acidity of an alcohol is expressed in its reaction with sodium metal to yield gaseous hydrogen (H_2). (Richard Megna/ Fundamental Photographs.)

The very weak basicity of an alcohol is expressed by its ability to accept a proton from strong acids such as sulfuric acid (H_2SO_4) and phosphoric acid (H_3PO_4):

$$\mathrm{R{-}O{-}H} + \mathrm{H_2SO_4} \rightleftharpoons \mathrm{R{-}\overset{\overset{\displaystyle H}{|}}{\underset{+}{O}}{-}H} + \mathrm{HSO_4^-}$$

Protonation takes place only in the presence of strong acids, and the equilibrium is far to the left. Less than 0.1% of the alcohol takes the protonated form, but this small concentration of protonated alcohol is important in dehydration reactions of alcohols (see Section 5.6 and Box 5.4).

The acidity of an alcohol is far too weak to be observed by reaction with NaOH or many other strong bases. However, the very weak acidity of an alcohol is expressed in its reaction with active metals such as sodium (which are much stronger bases than NaOH) to yield gaseous hydrogen (Figure 5.4):

$$2\ \mathrm{R{-}O{-}H} + 2\ \mathrm{Na} \longrightarrow 2\ \mathrm{R{-}ONa} + \mathrm{H_2}$$

5.6 THE DEHYDRATION OF ALCOHOLS TO ALKENES

Alcohols can be transformed into alkenes when heated in the presence of a catalytic amount of a strong acid such as sulfuric acid. An alcohol undergoes **intramolecular dehydration**—elimination of water—by losing an —OH group from a carbon atom and an H from an adjacent carbon atom:

Loss of H_2O

$$\mathrm{H{-}\underset{\underset{\displaystyle H}{|}}{\overset{\overset{\displaystyle H}{|}}{C}}{-}\underset{\underset{\displaystyle H}{|}}{\overset{\overset{\displaystyle OH}{|}}{C}}{-}H} \xrightarrow[\text{heat}]{H_2SO_4} \mathrm{H{-}\underset{\underset{\displaystyle H}{|}}{C}{=}\underset{\underset{\displaystyle H}{|}}{C}{-}H} + \mathrm{H_2O}$$

Note that dehydration cannot take place unless a strong acid catalyst is present.

All alcohols undergo intramolecular dehydration except alcohols without an H on a carbon atom adjacent to the carbon holding the —OH. Methanol is one such exception. Having only one carbon, it does not contain an adjacent carbon with an H.

5.4 CHEMISTRY IN DEPTH

Mechanism of Dehydration of Alcohols

The dehydration of alcohols does not take place in the absence of a strong acid catalyst, because the bond connecting carbon to oxygen is too strong to allow the —OH group to break away. The presence of a strong acid catalyst is the key to dehydration.

The first step in the reaction mechanism for dehydration is protonation of the oxygen of the alcohol by the strong acid catalyst, as illustrated on the right. Protonation is the bonding of H^+ to one of the oxygen's free electron pairs, a consequence of the basicity of alcohols. The resulting electron deficiency of oxygen, symbolized by its plus charge, weakens the carbon-to-oxygen bond, which breaks to eliminate water and form a carbocation. The carbocation attains stability by loss of a proton and formation of an alkene double bond.

The reactivity of alcohols in dehydration follows the order 3° > 2° > 1°, which corresponds to the order of stability of the carbocations that formed as intermediates from the respective alcohols.

$$\mathrm{H_3C{-}CH_2{-}OH} \underset{-H^+}{\overset{H^+}{\rightleftharpoons}} \mathrm{H_3C{-}CH_2{-}\overset{+}{O}H_2} \underset{H_2O}{\overset{-H_2O}{\rightleftharpoons}}$$

O is protonated — H_2O is eliminated

$$\mathrm{H_3C{-}\overset{+}{C}H_2} \underset{H^+}{\overset{-H^+}{\rightleftharpoons}} \mathrm{H_2C{=}CH_2}$$

e^- pair forms π bond — H^+ is eliminated

Example 5.7 Writing equations for the dehydration of alcohols: I

Which of the following alcohols undergo dehydration in the presence of a strong acid catalyst? If dehydration takes place, show the product. (a) 1-Propanol; (b) phenylmethanol; (c) cyclohexanol.

Solution

Find the carbon attached to the —OH group, and then look to see if there is a carbon adjacent to it that is carrying an H.

(a) $\mathrm{CH_3CH_2{-}CH_2{-}OH \xrightarrow[\text{heat}]{H_2SO_4} CH_3CH{=}CH_2 + H_2O}$

(b) $\mathrm{C_6H_5{-}CH_2{-}OH \xrightarrow[\text{heat}]{H_2SO_4}}$ no reaction

No H on adjacent C

(c) cyclohexanol (OH) $\xrightarrow[\text{heat}]{H_2SO_4}$ cyclohexene $+ H_2O$

Problem 5.7 Which of the following alcohols undergo dehydration in the presence of a strong acid catalyst? If dehydration takes place, show the product. (a) 2-Propanol; (b) 1-butanol; (c) *t*-butyl alcohol.

The dehydration of alcohols is the reverse of the hydration of alkenes. We can make the reaction proceed in either direction by putting Le Chatelier's principle to work (Section 2.3). When hydration is desired, we use a large excess of water relative to alkene to shift the equilibrium toward alcohol. When dehydration is desired, we shift the equilibrium toward alkene by taking advantage of the lower boiling point of the alkene compared with that of alcohol. We run the reaction at a temperature above the boiling point of alkene. Alkene distills out of the reaction vessel, and this distillation shifts the equilibrium to produce more alkene. Box 5.4 describes the mechanism of dehydration.

« The hydration of an alkene is the addition of water to the double bond, as described in Section 4.6.

The dehydration of many secondary and tertiary alcohols follows a more complicated course than that for primary alcohols. Most such alcohols have hydrogens attached to carbons on both sides of the carbon holding the —OH and will yield two different alkenes on dehydration if the alcohol is asymmetrical. For example, the dehydration of 2-butanol yields both 1-butene and 2-butene, by loss of hydrogens from C1 and C3, respectively:

Loss of H yields 2-butene

$$\underset{\uparrow}{CH_3}-\overset{OH}{\overset{|}{CH}}-\underset{\downarrow}{CH_2}-CH_3 \xrightarrow[-\,H_2O]{H_2SO_4} \underset{\substack{\textbf{1-Butene}\\(\text{Minor})}}{CH_2{=}CH-CH_2-CH_3} + \underset{\substack{\textbf{2-Butene}\\(\text{Major})}}{CH_3-CH{=}CH-CH_3}$$

Loss of H yields 1-butene

Dehydration is a selective process. The two alkenes are not formed in equal amounts. 2-Butene is the major product.

The relative stabilities of the different alkenes determine which alkene is the major product. The more stable the alkene, the higher its yield in the dehydration reaction. The stability of alkenes follows the order

Most stable **Least stable**

$$\underset{\textbf{Tetrasubstituted}}{R_2C{=}CR_2} > \overset{\textbf{Trisubstituted}}{R_2C{=}CHR} > R_2C{=}CH_2 \simeq \underset{\textbf{Disubstituted}}{RCH{=}CHR} > \overset{\textbf{Monosubstituted}}{RCH{=}CH_2} > \underset{\textbf{Unsubstituted}}{H_2C{=}CH_2}$$

Most substituted **Least substituted**

Stability increases with the degree of substitution of the double bond; that is, with increasing numbers of alkyl groups attached to the carbons of the double bond. The compound with the most substituted double bond is the major product. In the dehydration of 2-butanol, the major product (2-butene) has a disubstituted double bond, whereas the minor product (1-butene) has a monosubstituted double bond.

In actuality, the result of the dehydration of 2-butanol is more complicated than that just described. The 2-butene product is a mixture of *cis*-2-butene and *trans*-2-butene. The trans isomer is formed in greater abundance than the cis isomer because trans isomers are more stable than cis isomers.

Example 5.8 Writing equations for the dehydration of alcohols: II

Which of the following alcohols undergo dehydration in the presence of a strong acid catalyst? If dehydration takes place, show the product(s). If there is more than one product, indicate which is the major product. (a) 3-Hexanol; (b) 2-methyl-2-butanol.

Solution

(a) Look at the carbon atoms adjacent to the one bonded to the —OH.

Loss of H yields 2-hexene

$$\underset{\downarrow}{CH_3CH_2}-\overset{OH}{\overset{|}{CH}}-\underset{\uparrow}{CH_2}CH_2CH_3 \xrightarrow{-\,H_2O}$$

Loss of H yields 3-hexene

$$\underset{\textbf{2-Hexene}}{CH_3CH{=}CHCH_2CH_2CH_3} + \underset{\textbf{3-Hexene}}{CH_3CH_2CH{=}CHCH_2CH_3}$$

2-Hexene and 3-hexene are formed in equal amounts because each contains a disubstituted double bond. However, each is a mixture of cis and trans isomers, with the trans isomer being present in larger amounts.

(b) Loss of H yields 2-methyl-1-butene

$$CH_3-\overset{\displaystyle CH_3}{\underset{\displaystyle OH}{\overset{|}{\underset{|}{C}}}}-CH_2-CH_3 \xrightarrow{-\,H_2O}$$

Loss of H yields 2-methyl-2-butene

$$\underset{\substack{\textbf{2-Methyl-1-butene}\\(\text{Minor})}}{CH_2{=}\overset{\displaystyle CH_3}{\overset{|}{C}}-CH_2-CH_3} + \underset{\substack{\textbf{2-Methyl-2-butene}\\(\text{Major})}}{CH_3-\overset{\displaystyle CH_3}{\overset{|}{C}}{=}CH-CH_3}$$

2-Methyl-2-butene is the major product because it contains a trisubstituted double bond, whereas 2-methyl-1-butene contains a disubstituted double bond.

Problem 5.8 Which of the following alcohols undergo dehydration in the presence of a strong acid catalyst? If dehydration takes place, show the product(s). If there is more than one product, indicate which is the major product. (a) 3-Pentanol; (b) 1-methylcyclopentanol.

The dehydration of alcohols takes place in a number of physiological processes, including glycolysis (Section 15.1) and the citric acid cycle (Section 15.6).

›› Glycolysis is a metabolic process that generates energy and is considered in Chapter 15.

Concept check

✓ Intramolecular dehydration of an alcohol in the presence of an acid catalyst produces an alkene. When more than one alkene is possible, the major product is the most substituted alkene.

5.7 THE OXIDATION OF ALCOHOLS

Alcohols undergo two types of oxidation, depending on reaction conditions.

- **Combustion:** All carbon atoms in the organic compound are oxidized to their highest oxidation state.
- **Selective (mild) oxidation:** Only the carbon holding the —OH group is oxidized.

Alcohols, like hydrocarbons, undergo combustion to carbon dioxide and water. The balanced combustion equation for ethanol is

$$CH_3CH_2OH + 3\ O_2 \longrightarrow 2\ CO_2 + 3\ H_2O$$

Selective oxidations of organic compounds are oxidations at the weaker bonds in the molecules. Whereas combustion transforms an organic compound into carbon dioxide and water, selective oxidation transforms an organic compound into another organic compound, a member of a different family. Such transformations are important in biological systems as well as in industrial processes.

In an alcohol, the weaker bonds are the O—H bond and the adjacent C—H bond, and these weaker bonds are the bonds that undergo selective oxidation. Two of the most commonly used selective oxidizing agents are permanganate ion (MnO_4^-) and dichromate ion ($Cr_2O_7^{2-}$). Primary, secondary, and tertiary alcohols behave differently under selective oxidation conditions.

Primary alcohols are oxidized in two stages:

- Stage 1: Simultaneous loss of hydrogens (called **dehydrogenation**) from the —OH group and from the carbon holding the —OH produces a **carbonyl group** (C=O), as shown in the next equation. The resulting compound is classified as an **aldehyde** because an H is attached to the carbonyl carbon (Section 3.2).

›› Aldehydes and carboxylic acids are considered in Chapters 6 and 7, respectively.

- Stage 2: Oxidation of the H attached to the carbonyl carbon to —OH converts the carbonyl group into a **carboxyl group** (COOH). The resulting compound is a **carboxylic acid** (Section 3.2).

Loss of H

$$\mathrm{R{-}CH_2{-}O{-}H} \xrightarrow[\text{Stage 1}]{MnO_4^- \text{ or } Cr_2O_7^{2-}} \mathrm{R{-}C(H){=}O} \xrightarrow[\text{Stage 2}]{MnO_4^- \text{ or } Cr_2O_7^{2-}} \mathrm{R{-}C(OH){=}O}$$

Carbonyl group
Aldehyde

Carboxyl group
Carboxylic acid

Oxidation with permanganate or dichromate proceeds from primary alcohol all the way to carboxylic acid without stopping at the aldehyde stage because these oxidizing agents are too strong. For situations in which the aldehyde is the desired product, the reaction can be stopped at the aldehyde stage by using oxidizing agents that are milder than permanganate or dichromate.

Secondary alcohol oxidation follows a simpler course than the oxidation of primary alcohols. Oxidation cannot proceed past the carbonyl stage.

$$\mathrm{R{-}C(R')(H){-}O{-}H} \xrightarrow{MnO_4^- \text{ or } Cr_2O_7^{2-}} \mathrm{R{-}C(R'){=}O}$$

Loss of H

Carbonyl group
Ketone

›› Ketones are a subject of Chapter 6.

The product with the carbonyl group is a **ketone** (Section 3.2). (Note that ketones and aldehydes both have a carbonyl group. An aldehyde has one H and one R group attached to the carbonyl carbon, whereas a ketone has two R groups and no H attached to it. The R groups can be either aliphatic or aromatic.) Because a ketone lacks an H attached to the carbonyl carbon, there is no further oxidation under selective oxidation conditions. The H attached to the carbonyl carbon in an aldehyde is oxidized in the second stage.

Tertiary alcohols cannot undergo selective oxidation, because no H is attached to the carbon to which the —OH is bonded. The OH hydrogen cannot be lost by itself.

$$\mathrm{R{-}C(R')(R''){-}O{-}H}$$

C does not have H

The oxidation of an alcohol requires the loss of a pair of hydrogens, one from the —OH and one from the carbon holding the —OH.

Oxidations of alcohols are important in living systems, which use specific enzymes to control the reaction. Malate dehydrogenase catalyzes the conversion of malic acid into oxaloacetic acid (by oxidizing the —OH group of malic acid) in a key step of the citric acid cycle, an important energy-generating process (Section 15.5).

Formaldehyde and acetone, important industrial chemicals (Box 6.2), are produced by the oxidations of the corresponding alcohols, methanol and 2-propanol, respectively.

A PICTURE OF HEALTH

Examples of Alcohols, Phenols, Ethers, and Thiols

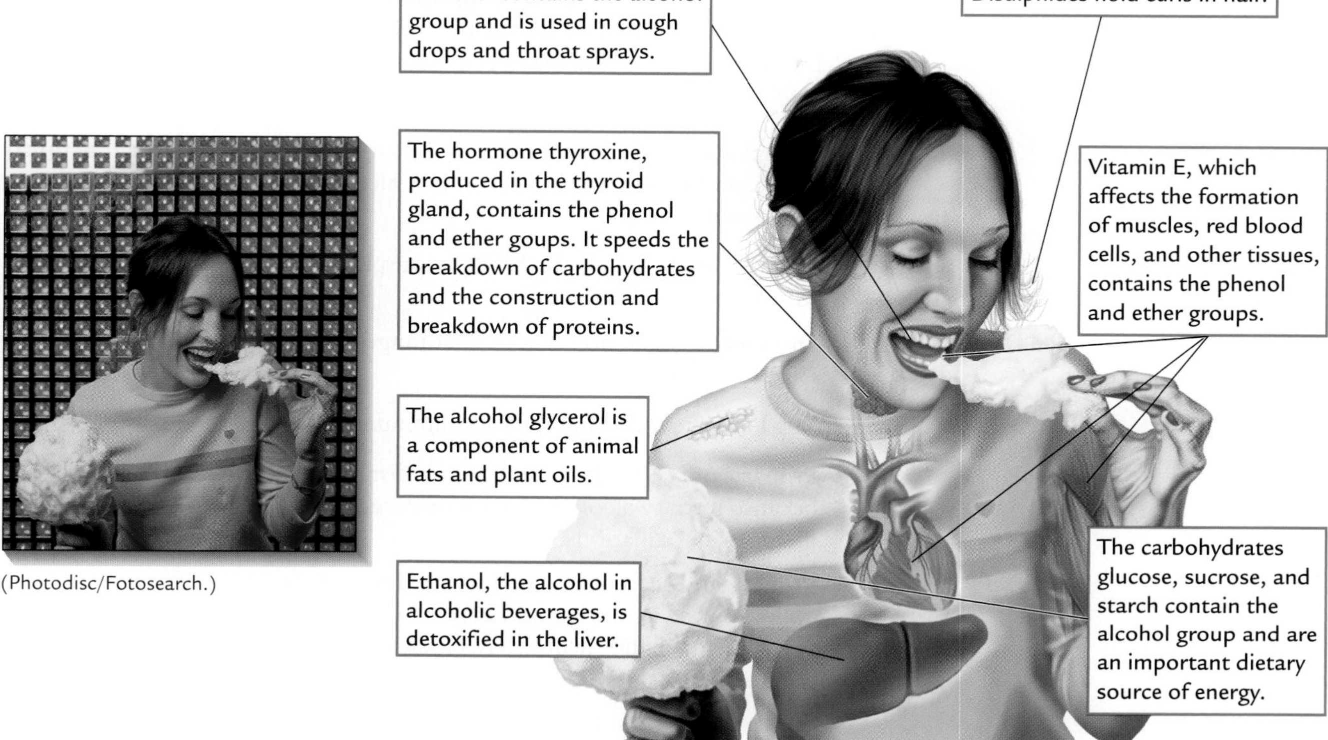
Menthol contains the alcohol group and is used in cough drops and throat sprays.
Disulphides hold curls in hair.
The hormone thyroxine, produced in the thyroid gland, contains the phenol and ether goups. It speeds the breakdown of carbohydrates and the construction and breakdown of proteins.
Vitamin E, which affects the formation of muscles, red blood cells, and other tissues, contains the phenol and ether groups.
The alcohol glycerol is a component of animal fats and plant oils.
Ethanol, the alcohol in alcoholic beverages, is detoxified in the liver.
The carbohydrates glucose, sucrose, and starch contain the alcohol group and are an important dietary source of energy.

(Photodisc/Fotosearch.)

Concept checklist

- ✓ Primary and secondary alcohols undergo selective oxidation by permanganate (MnO_4^-) and dichromate ($Cr_2O_7^{2-}$) ions.
- ✓ Primary alcohols are oxidized to aldehydes, which are subsequently oxidized to carboxylic acids. Secondary alcohols are oxidized to ketones with no further oxidation possible. Tertiary alcohols are not oxidized.

Example 5.9 Writing equations for the oxidation of alcohols

Which of the following alcohols undergo oxidation with dichromate or permanganate? If oxidation takes place, show the product. (a) 1-Propanol; (b) 1-methylcyclohexanol; (c) cyclohexanol.

Solution

(a) Primary alcohols undergo oxidation first to an aldehyde and then to a carboxylic acid:

$$CH_3CH_2CH_2{-}OH \xrightarrow[\text{or } MnO_4^-]{Cr_2O_7^{2-}} CH_3CH_2{-}\overset{\overset{\displaystyle O}{\|}}{C}{-}H \xrightarrow[\text{or } MnO_4^-]{Cr_2O_7^{2-}} CH_3CH_2{-}\overset{\overset{\displaystyle O}{\|}}{C}{-}OH$$

(b) Tertiary alcohols do not undergo selective oxidation:

No H

$$\text{1-methylcyclohexanol (OH, } CH_3\text{)} \xrightarrow{Cr_2O_7^{2-} \text{ or } MnO_4^-} \text{no reaction}$$

(c) Secondary alcohols undergo oxidation to ketones:

$$\text{cyclohexanol (C(OH)H)} \xrightarrow{Cr_2O_7^{2-} \text{ or } MnO_4^-} \text{cyclohexanone (C=O)}$$

Problem 5.9 Which of the following alcohols undergo oxidation with dichromate or permanganate? If oxidation takes place, show the product. (a) 2-Propanol; (b) phenylmethanol; (c) *t*-butyl alcohol.

Permanganate and dichromate oxidations are useful **simple chemical diagnostic tests** for primary and secondary alcohols. Each oxidation is accompanied by a visible change:

- Permanganate oxidation: MnO_4^- (purple solution) is converted into MnO_2 (brown precipitate).
- Dichromate oxidation: $Cr_2O_7^{2-}$ (orange solution) is converted into Cr^{3+} (green solution).

These tests have the following uses and limitations:

- Primary and secondary alcohols are distinguished from tertiary alcohols because tertiary alcohols do not undergo permanganate and dichromate oxidations.
- Primary and secondary alcohols are distinguished from alkanes, cycloalkanes, aromatics, and ethers, which do not undergo permanganate and dichromate oxidations (Sections 4.8 and 5.9).
- Positive tests with permanganate and dichromate do not distinguish primary and secondary alcohols from alkenes, alkynes, and phenols, because those compounds also are oxidized (Sections 4.8, 4.9, 5.8).

Dichromate oxidation forms the basis of an instrument used to monitor for alcohol abuse in several settings—for example, to monitor inmates in prisons, probationers, or patients at alcohol-rehabilitation clinics. There is a direct relation between the concentration of ethanol in the blood and that in exhaled breath. The test subject fills a balloon with breath, the filled balloon is attached to the instrument, and the sampled breath is allowed to flow through the dichromate solution for a set time period. The extent to which the dichromate solution changes color from orange to green indicates the blood-alcohol level.

Most instruments now used by law enforcement to test drivers for drunk driving are no longer based on dichromate oxidation. Different instruments, including those based on infrared spectroscopy (Box 6.3), fuel cells, and gas chromatography, have been developed and used.

Example 5.10 Identifying an unknown compound

An unknown compound is either 2-methyl-2-propanol or 1-butanol. When a few drops of permanganate solution are added to it, the purple color of the permanganate fades and a brown precipitate forms. What is the identity of the unknown? Explain your answer.

Solution

The unknown is 1-butanol. 1-Butanol, a primary alcohol, gives a positive test with permanganate, whereas 2-methyl-2-propanol, a tertiary alcohol, does not react with permanganate.

Problem 5.10 An unknown compound is either 2-butanol or 1-butanol. When a few drops of dichromate solution (orange) are added to it, the mixture turns green. What is the identity of the unknown? Explain your answer.

5.8 PHENOLS

A phenol is an organic compound in which an —OH group is attached to one of the sp^2-hybridized carbons of a benzene ring (see Table 5.1). The phenol family includes the parent compound phenol as well as a variety of compounds with other substituents attached to the benzene ring in addition to the —OH group. Many naturally occurring compounds contain the phenol structure (Box 5.5).

5.5 CHEMISTRY AROUND US

Phenols

Various phenols are used as antiseptics and disinfectants. Hexylresorcinol is commonly used in mouthwashes and throat lozenges, and *o*-phenylphenol serves as a disinfectant for walls, floors, furniture, and equipment both in hospitals and in homes.

OH, OH, $(CH_2)_5CH_3$
Hexylresorcinol

OH
***o*-Phenylphenol**

The active irritants in poison oak, poison ivy, and poison sumac belong to a subclass of phenols known as the urushiols. All urushiols have a benzene ring with two —OH groups and a 15-carbon unbranched chain but differ in the number of double bonds (from 0 to 3) in that chain.

OH, OH, $C_{15}H_{25-31}$
Urushiols

The drug marijuana, also called pot or grass, is obtained from the hemp plant *Cannabis sativa*. Leaves, flowers, and other parts of the plant are dried and smoked. The active ingredient tetrahydrocannabinol (THC), is a phenol. THC is a mild central-nervous-system depressant at low doses.

HO, $(CH_2)_4CH_3$, CH_3, O, CH_3 CH_3
Tetrahydrocannabinol

BHT (butylated hydroxytoluene) and BHA (butylated hydroxyanisole) are used as **antioxidant** additives to protect foods, gasoline, lubricating oils, plastics, and rubbers from degradation due to oxidation. Vitamin E has the same function in the body. A phenol protects other materials in its vicinity against oxidation by undergoing sacrificial oxidation itself.

OH, $(CH_3)_3C$, $C(CH_3)_3$, CH_3
2,6-Di-*t*-butyl-4-methylphenol (Butylated hydroxytoluene, BHT)

OH, $C(CH_3)_3$, OCH_3
2-*t*-Butyl-4-methoxyphenol (Butylated hydroxyanisole, BHA)

CH_3, HO, CH_3, O, CH_3, $(CH_2)_3CH(CH_2)_3CH(CH_2)_3CH—CH_3$ (with CH_3 branches), CH_3
Vitamin E

Note that some of the phenols illustrated here are members of other families as well. BHA, vitamin E, and THC are ethers. THC is also an alkene.

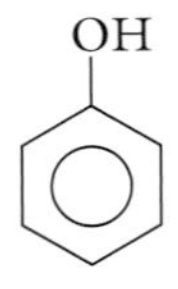

OH / $CH_2{=}CH$–OH

Phenol 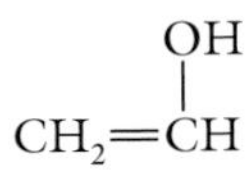 Enol

Enols constitute another organic family in which an —OH is attached to an *sp*²-hybridized carbon. They differ from phenols in that the —OH is attached not to a benzene ring but to a carbon–carbon double bond. Enols are generally unstable compounds that are encountered in very few systems. One exception is glycolysis, where an enol is an important intermediate in the process (Section 15.2).

In the IUPAC nomenclature system, substituted phenols are named as derivatives of the parent compound phenol.

Example 5.11 Naming phenols by the IUPAC nomenclature system

Name the following compounds by using the IUPAC system.

(a) HO, Cl, CH_3, CH_3 (b) $CH(CH_3)_2$, HO, NO_2

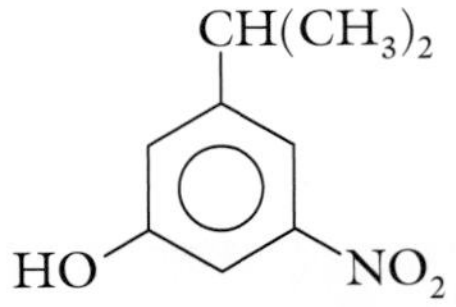

Solution

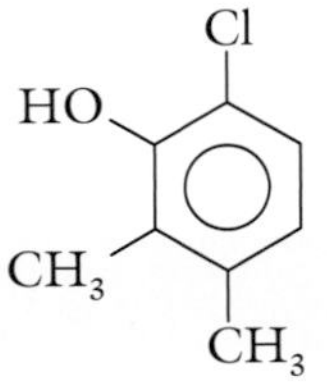

Carbon-1 is the carbon holding the —OH group. Number in the direction that assigns the set of lowest numbers to the other substituents on the benzene ring: (a) 6-chloro-2,3-dimethylphenol; (b) 3-isopropyl-5-nitrophenol.

Problem 5.11 Name the following compounds by using the IUPAC system.

(a) $CH_3CH_2CH_2CH_2$, $C(CH_3)_3$, OH (b) OH, CH_3, Br

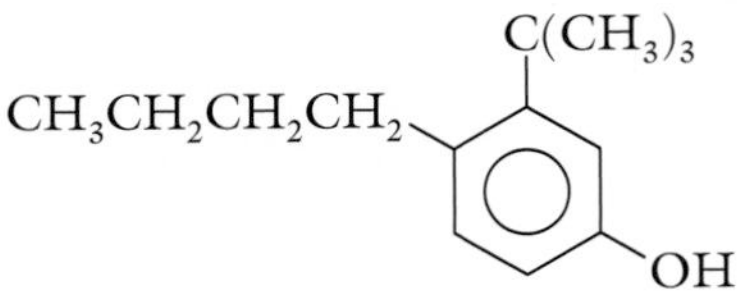

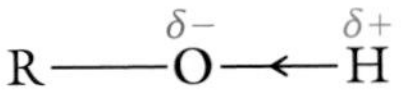

R—O←H ($\delta-$, $\delta+$)

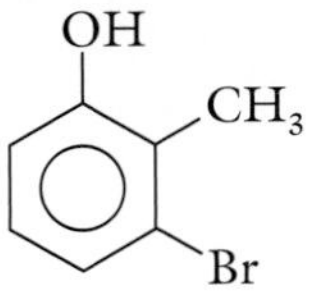

Alcohol (Less polar)

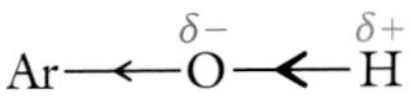

Ar←O←H ($\delta-$, $\delta+$)

Phenol (More polar)

The O—H bond in phenols is more polar than that in alcohols because, unlike an aliphatic group, the aromatic ring has an electron-withdrawing effect. All O—H bonds are polarized $^{\delta-}$O—H$^{\delta+}$ because oxygen is more electronegative than hydrogen. However, the electron-withdrawing effect of the aromatic ring (Ar) increases this polarization, causing the oxygen and hydrogen ends of the O—H bond to develop larger partial negative and partial positive charges, respectively, than are found on the O—H bond of an alcohol. The greater polarity of the O—H bond in phenols results in stronger hydrogen bonding and higher boiling and melting points as well as higher solubilities in water compared with alcohols. The boiling points of phenol and cyclohexanol, for example, are 182°C and 162°C, respectively. Their water solubilities are 9.3 g/100 mL and 3.6 g/100 mL, respectively, at 20°C.

Example 5.12 Identifying the hydrogen bonding of phenols

Which of the following representations correctly describe the hydrogen bonding that is responsible for the high boiling point of phenol?

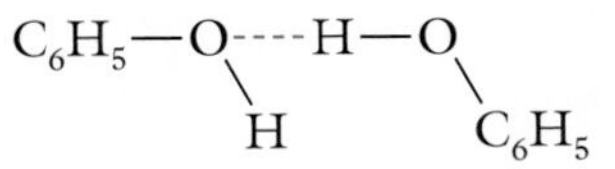

1: C_6H_5—O(H)----H—O—C_6H_5

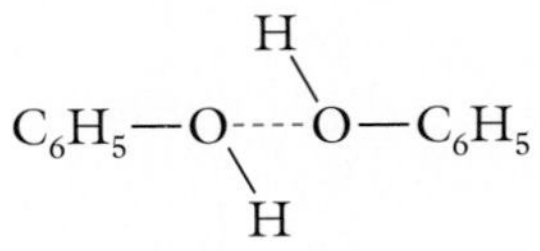

2: C_6H_5—O(H)----O(H)—C_6H_5

3: C_6H_5—O—H----C_6H_5—O—H

4: C_6H_5—O(H)----C_6H_5—OH

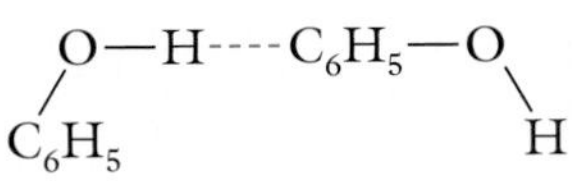

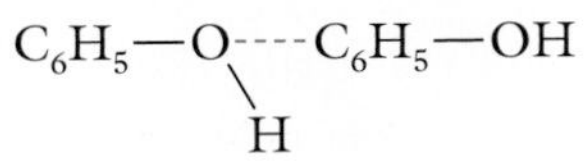

Solution

The correct representation of hydrogen bonding between phenol molecules is representation 1. There is an attraction between the oxygen of an O—H bond in one phenol molecule and the hydrogen of an O—H bond in another phenol molecule.

Problem 5.12 Which of the following representations correctly describe the hydrogen bonding that is responsible for the moderate solubility of phenol in water?

1: C_6H_5—O—H····H—O—H

2: H—O(—H)····H—OC_6H_5

3: C_6H_5—O(—H)····H—O—H

4: C_6H_5—O(—H)····O(—H)—H

Phenols are weak acids but much more acidic than alcohols. The extent of the ionization of phenols in water is much greater than that of alcohols.

$$C_6H_5{-}O{-}H + H_2O \rightleftharpoons C_6H_5{-}O^- + H_3O^+$$

For example, 0.1 molar aqueous solutions of cyclohexanol and phenol have pH values of about 7 and 5, respectively, corresponding to hydronium ion concentrations 100-fold higher in phenol than in cyclohexanol. The large difference in acidity between phenols and alcohols is easily tested by the use of an acid–base indicator such as litmus paper. Phenols turn blue litmus paper red, whereas, like water, alcohols show neutral behavior: blue litmus paper does not change color.

The greater acidity of phenols compared with alcohols also shows itself by the reaction of phenol with a strong base, such as the hydroxide ion:

$$C_6H_5{-}O{-}H + NaOH \longrightarrow C_6H_5{-}O^- + Na^+ + H_2O$$

Phenol is completely converted into phenoxide ion ($C_6H_5{-}O^-$) by reaction with the strong base: the equilibrium is completely to the right. The corresponding extent of reaction of an alcohol with NaOH is much less than 0.1%.

The higher acidity of phenols compared with that of alcohols is due to the electron-withdrawing effect of the phenyl ring compared with an R group. Electron withdrawal stabilizes the phenoxide anion by partly dispersing the negative charge around the phenyl ring, rather than leaving it entirely on the oxygen. This effect is called **charge dispersal:**

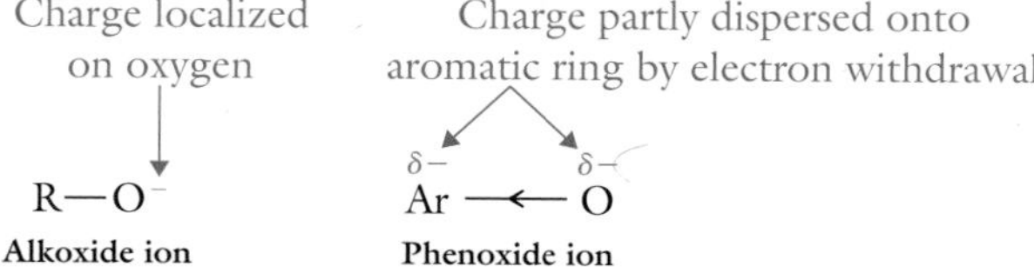

Its effect is to make the anion more stable and, therefore, it forms in higher concentration. The corresponding alkoxide anion is not stabilized by charge dispersal, because the R group is not electron withdrawing.

Phenols do not undergo dehydration, because dehydration would form a triple bond in the ring, resulting in destruction of the benzene structure. Phenols do undergo oxidation, however, even though it destroys the benzene ring. The oxidation reactions are quite complex and beyond the scope of this discussion. The oxidation of phenols forms the basis of their use as antioxidants, both in physiological systems and in commercial products (see Box 5.5).

5.9 ETHERS

Ethers, like alcohols and phenols, contain an oxygen that has bonds to two different atoms; but, in ethers, neither of those bonds is to hydrogen. Instead, the oxygen of an ether has single bonds to two different carbons, each of which may be either an aliphatic or an aromatic carbon (see Table 5.1). Review Example 5.1, Problem 5.1, and Exercise 5.1 if you are uncertain of the differences between ethers, alcohols, and phenols. Box 5.6 describes some important ethers.

A common nomenclature system is used for ethers containing simple groups. This system uses the names of the groups attached to the ether oxygen (in alphabetical order), followed by the word ether. There are spaces between the names of different groups and the word ether. The IUPAC nomenclature system is used for ethers containing other than the simple groups. In such cases, the ether is named as a member of some other family, with the more complex group determining the base name. The simpler group and the ether oxygen to which it is attached are treated as an alkoxy (RO) or aryloxy (ArO) group.

Example 5.13 Naming ethers

Name the following ethers:

(a) C_6H_5–OCH_2CH_3 (benzene ring bearing OCH_2CH_3)

(b) $CH_3{-}O{-}\underset{CH_3}{\overset{H}{C}}{-}CH_3$

(c) $CH_3{-}O{-}\underset{CH_3}{\overset{H}{C}}{-}CH_2{-}\overset{C_6H_5}{CH}{-}CH_3$

Solution

(a) Ethyl phenyl ether. (b) Isopropyl methyl ether. (c) The group on the right side of the oxygen does not have a simple name. The compound is named as a derivative of pentane. The pentane chain has a CH_3O (methoxy) group at C2 and a phenyl group at C4. Therefore, the name is 2-methoxy-4-phenylpentane.

Problem 5.13 Name the following ethers:

(a) $CH_3CH_2CH_2CH_2{-}O{-}\underset{CH_3}{\overset{CH_3}{C}}{-}CH_3$

(b) $CH_3CH_2CH_2CH_2{-}O{-}\underset{CH_3}{\overset{CH_3}{C}}{-}CH{=}CH_2$

(c) $(CH_3)_2CH{-}O{-}\overset{CH_3}{CH}{-}CH_2CH_3$

Common names are used for cyclic ethers:

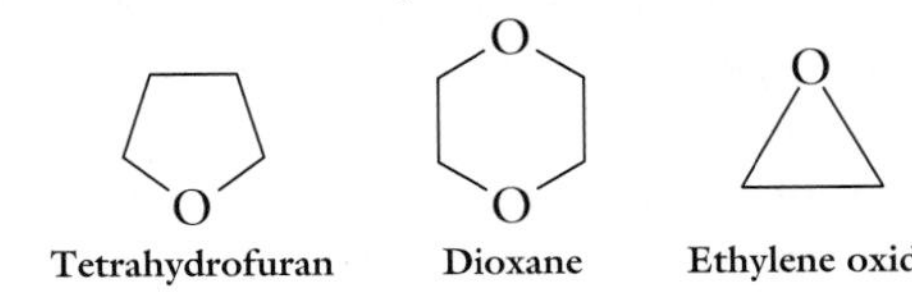

5.6 CHEMISTRY AROUND US

Ethers

Before the 1850s, few patients ever survived surgery, and physicians were reluctant to operate except in dire circumstances. Surgical patients died either from the shock of undergoing surgery without anesthesia or from bacterial infections (which were not understood at the time).

An American dentist, William Morton, tackled the first problem in the 1850s by introducing the use of diethyl ether as an anesthetic for surgery. English surgeon Joseph Lister addressed the other problem in 1865 by introducing the use of phenol as an antiseptic for sterilizing instruments and patients alike. Phenol itself is quite harsh in its effect and has long since been replaced by various substituted phenols (see Box 5.5). The use of anesthetics and antiseptics made modern surgery possible.

Diethyl ether and other ethers are general inhalation anesthetics. They produce both insensitivity to pain and unconsciousness, and they are introduced into the patient by inhalation. Most ethers have significant disadvantages, however. They are flammable and explosive, requiring great care in use, and they cause patients to suffer uncomfortable side effects, such as nausea, vomiting, and respiratory irritation. Fortunately, various halogenated ethers and halogenated alkanes with anesthetic properties have been found to be less volatile and less flammable and to have milder side effects. Examples are enflurane and fluothane.

$$H-CF_2-O-CF_2-CHF-Cl$$

Enflurane

$$CF_3-CH(Br)(Cl)$$

Fluothane

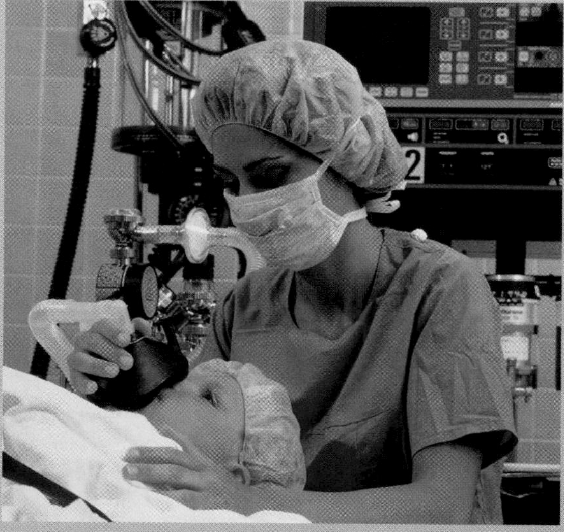

Administration of anesthesia to a child. (SIU/Photo Researchers.)

Ether functional groups are also found in many of the herbicides used to kill undesirable weeds, so as to improve the abundance of crops. Some of these herbicides were derived from polychlorinated phenols, a practice that has resulted in undesirable and unpredictable adverse side effects due to the presence of a certain class of impurities called **dioxins.** Dioxins are produced in minute amounts as by-products in the manufacture of herbicides. The most important dioxin is 2,3,7,8-tetrachlorodibenzo-*p*-dioxin (2,3,7,8-TCDD) because of its presence in the herbicide 2,4,5-trichlorophenoxyacetic acid (2,4,5-T). 2,4,5-T was banned in the United States in 1985 because of concerns about its adverse effects on humans. 2,4,5-T was one component of the defoliant **Agent Orange** used in Vietnam by the United States military to remove enemy cover and destroy their crops. Many Vietnamese and Americans exposed to Agent Orange have suffered from a variety of health problems. Note that 2,3,7,8-TCDD is simultaneously a cyclic diether, an aromatic compound, and a chlorinated compound.

2,3,7,8-Tetrachlorodibenzo-*p*-dioxin (2,3,7,8-TCDD)

$$Cl_2C_6H_2-OCH_2C(=O)-OH \text{ (with Cl at 2,4,5 positions)}$$

2,4,5-Trichlorophenoxyacetic acid (2,4,5-T)

Ethers in industrial use include tetrahydrofuran and dioxane, important solvents for carrying out chemical reactions. Unlike those two compounds and ethers in general, ethylene oxide is a reactive compound because of its highly strained bond angles (60° compared with the normal tetrahedral 109.5°). Ethylene oxide is used in the manufacture of various plastics, including epoxy adhesives. It has also been used for sterilizing surgical instruments in hospitals, as a fumigant, and as a sterilant for spices and cosmetics because it causes the death of microorganisms by reacting with their proteins and other biological molecules. The use of ethylene oxide is also hazardous to humans and must be carefully monitored and controlled.

t-Butyl methyl ether, usually incorrectly called methyl *t*-butyl ether or MTBE, has been added to gasoline to reduce hydrocarbon emissions (a result of incomplete combustion) by improving the efficiency of the combustion process. However, MTBE has found its way into ground and drinking waters in various locations because of leaking gasoline tanks. MTBE poses a sufficient health hazard that it is being phased out in most states, with replacement by ethanol.

$$(CH_3)_3C-O-CH_3$$

***t*-Butyl methyl ether**

$CH_3—O—CH_3$
Dimethyl ether

$CH_3CH_2—O—H$
Ethanol

Any ether is a constitutional isomer of an alcohol containing the same number of carbons. For example, dimethyl ether and ethanol have the same molecular formula, C_2H_6O, but differ in their structural formulas and possess very different properties.

Example 5.14 Drawing constitutional isomers of ethers

Draw structural formulas and name the different ethers of molecular formula $C_4H_{10}O$.

Solution

$CH_3—O—CH_2CH_2CH_3$
Methyl propyl ether

$CH_3—O—CH(CH_3)_2$
Isopropyl methyl ether

$CH_3CH_2—O—CH_2CH_3$
Diethyl ether

Problem 5.14 Draw structural formulas of the compounds (alcohols) that are isomeric with the ethers in Example 5.14.

Refer again to Table 5.3, which compares the boiling points and water solubilities of several alkanes, alcohols, and ethers. The lower-molecular-mass ether, dimethyl ether, has a higher boiling point than the corresponding alkane because of the polar $^{\delta-}O—C^{\delta+}$ bond. The resulting dipole–dipole secondary forces are stronger than the London secondary forces in alkanes. However, the effect drops off rapidly with increasing molecular mass, and diethyl ether has the same boiling point as pentane. Ethers have considerably lower boiling points than those of alcohols because the hydrogen-bond secondary forces of alcohols are stronger than the dipole–dipole secondary forces of ethers.

Ethers are almost as soluble in water as are alcohols. Hydrogen bonding with water is more limited because there are no —OH groups in ethers. Ethers cannot hydrogen bond with themselves, but they can hydrogen bond with water because the oxygen in an ether is a hydrogen-bond acceptor. Alcohols are more water soluble than ethers because alcohols are hydrogen-bond donors as well as hydrogen-bond acceptors.

Ethers are unreactive toward acids, bases, and oxidizing agents and, for this reason, are often used as solvents in which to carry out chemical reactions. Ethers, like alkanes, participate in combustion and halogenation reactions only.

Example 5.15 Identifying the hydrogen bonding of ethers

Which of the following representations correctly describe the hydrogen bonding between dimethyl ether and water?

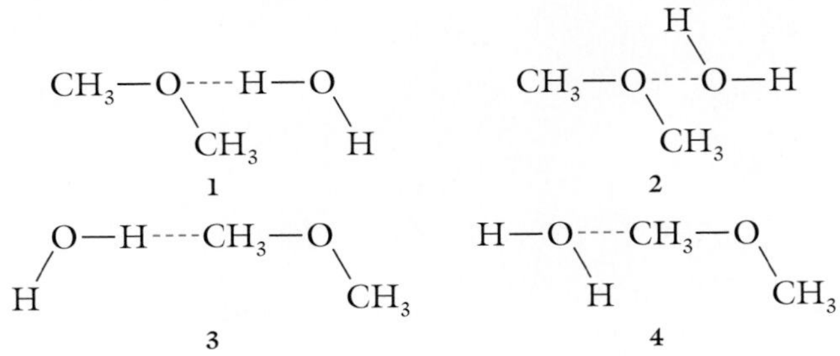

1: $CH_3—O(—CH_3)\cdots H—O—H$

2: $CH_3—O(—CH_3)\cdots O(—H)—H$

3: $H—O—H\cdots CH_3—O—CH_3$

4: $H—O(—H)\cdots CH_3—O—CH_3$

Solution

Hydrogen bonding is the attraction between the H of an O—H bond and the O of another bond (often but not limited to an O—H bond). Only representation 1 depicts hydrogen bonding.

Problem 5.15 Indicate which compound in each of the following pairs has the higher value of the specified property. If the two compounds in a pair have nearly the same value for the property, indicate that fact. Explain your conclusions.

(a) Methyl phenyl ether, *p*-methylphenol: boiling point.

(b) Dipropyl ether, 1-hexanol: solubility in heptane.

(c) Diethyl ether, 1-butanol: solubility in water.

5.10 THE FORMATION OF ETHERS BY DEHYDRATION OF ALCOHOLS

The discussion in Section 5.6 on the dehydration of alcohols to alkenes was an oversimplification for primary alcohols. Heating a primary alcohol in the presence of an acid catalyst produces not only an alkene, but also an ether:

$$CH_3CH_2CH_2{-}OH \xrightarrow{H_2SO_4} CH_3CH{=}CH_2 + H_2O \quad \text{(Major product at 180° C)}$$

$$CH_3CH_2CH_2{-}OH \xrightarrow{H_2SO_4} CH_3CH_2CH_2{-}O{-}CH_2CH_2CH_3 + H_2O \quad \text{(Major product at 140° C)}$$

Intramolecular dehydration to form the alkene competes with intermolecular dehydration to form the ether. In **intramolecular dehydration,** an —OH group and a hydrogen atom are removed from adjacent carbons in the alcohol molecule. In **intermolecular dehydration,** the —OH group and H atom come from two different alcohol molecules. The reaction can be directed toward the production of alkene or the production of ether by controlling the reaction temperature.

- Alkene is the major product at 180°C.
- Ether is the major product at 140°C.

Unlike primary alcohols, secondary and tertiary alcohols do not form ethers when heated with an acid catalyst. Intramolecular dehydration to alkenes is the major product at all temperatures. Secondary and tertiary carbons are too crowded because of the attached substituents to allow pairs of secondary or tertiary alcohol molecules to come sufficiently close together to undergo intermolecular dehydration. The effect is sometimes called **steric hindrance.**

Example 5.16 Writing equations for the dehydration of alcohols

Write an equation to show the product(s) formed when ethanol is heated at 140°C in the presence of an acid catalyst. Indicate both the major and the minor products if more than one product is formed.

Solution

Intermolecular dehydration is the major reaction at 140°C:

$$CH_3CH_2{-}O{-}H \xrightarrow[-H_2O]{H_2SO_4} \underset{\text{Major}}{CH_3CH_2{-}O{-}CH_2CH_3} + \underset{\text{Minor}}{CH_2{=}CH_2}$$

Problem 5.16 Give the product(s) formed when ethanol is heated at 180°C in the presence of an acid catalyst. Indicate both the major and the minor products if more than one product is formed.

5.11 THIOLS

Thiols, also called **mercaptans,** are the sulfur analogues of alcohols. The —SH group is called the **mercapto** or **sulfhydryl** group.

R—OH
Alcohol

R—SH
Thiol

In the IUPAC nomenclature system, the naming of thiols is similar to that of alcohols. The ending **-thiol** is added to the name of the alkane, but without dropping the -e from the alkane name.

Thiols have distinct, often disagreeable, odors and flavors. The offensive liquid squirted by skunks in their defense is due to a mixture of thiols, mostly 3-methyl-1-butanethiol. Natural gas is odorless, which would make it difficult to detect gas leaks. However, the gas companies deliberately add small amounts of ethanethiol, which gives natural gas an odor and allows its safe use. Ethanethiol is useful as a warning for a gas leak because our odor receptors are remarkably sensitive to low-molecular-mass thiols. We can smell one part of ethanethiol in 10 billion parts of air.

$(CH_3)_2CHCH_2CH_2—SH$
3-Methyl-1-butanethiol

$CH_3CH_2—SH$
Ethanethiol

Thiols differ from alcohols in several ways:

- Thiols have considerably lower boiling points than those of alcohols, even though thiols have higher molecular masses, because thiols do not possess the strong hydrogen bonding of alcohols.
- Thiols are weak acids but much stronger than alcohols, because the S—H bond is weaker than the O—H bond, a consequence of the larger size of sulfur compared with oxygen. However, thiols are weaker acids than phenols.
- Thiols are easily oxidized to disulfides by many oxidizing agents:

$$\underset{\text{Thiol}}{2\ \text{R—SH}} \xrightarrow{(\text{O})} \underset{\text{Disulfide}}{\text{R—S—S—R}}$$

  Disulfides are named by naming the R groups attached to the sulfur atoms followed by the word **disulfide.**
- Disulfides are easily reduced back to thiols by many reducing agents:

$$\text{R—S—S—R} \xrightarrow{(\text{H})} 2\ \text{R—SH}$$

  In the preceding equations, the abbreviations (O) and (H) indicate general conditions for selective oxidation and reduction, respectively, without specification of the exact oxidizing or reducing agent.
- Thiols react with heavy metal ions such as lead (Pb^{2+}) and mercury(II) (Hg^{2+}) to form insoluble salts:

$$2\ \text{R—SH} + \text{M}^{2+} \longrightarrow \text{R—S—M—S—R} + 2\ \text{H}^+$$

The reactions of thiols to form disulfides and insoluble heavy metal salts are of physiological importance (Sections 12.5 and 12.9). Many proteins contain thiol groups, and the formation of disulfides is often critical to their proper functioning. The formation of disulfides is also the basis for the permanent waving of hair (see Box 12.2). On the other hand, when such proteins form lead and mercury salts, the consequences to a living organism can be dire because the physiological functions of the proteins are altered. Brain development is retarded in infants who chew on paint that has peeled or flaked off the walls in old houses, because, until about 1980, house paint contained lead salts as the colorants. Inhalation of paint dust in such houses also is a problem. Lead also entered our environment through the use of tetraethyl lead as an additive to improve the performance of gasoline. Leaded gasoline has now been almost completely phased out.

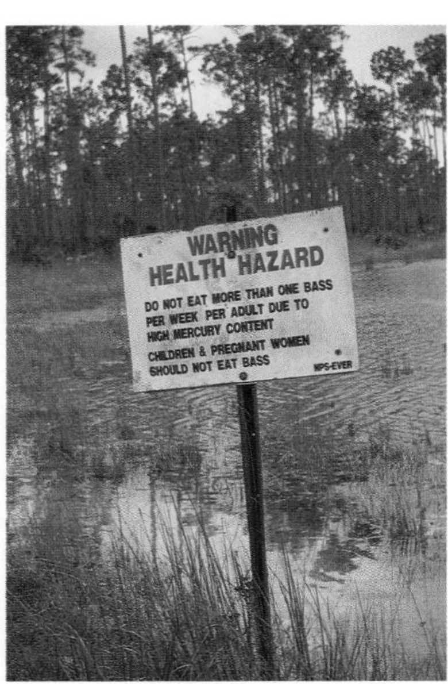

The sign warns that the fish in the lake are contaminated with mercury. (Karl H. Switak/Photo Researchers.)

People who eat mercury-contaminated fish caught in certain lake and ocean regions can suffer similar effects. Fish become contaminated with mercury because industrial wastes containing mercury were poured into various waters. Such waste disposal is no longer allowed, but the residual amounts of mercury compounds remaining in various waters create a long-lasting problem because decontamination is very difficult. The U.S. Food and Drug Administration has recommended that people, especially children and pregnant women, limit their intake of certain fish found to contain high levels of mercury. Swordfish, shark, tilefish, king mackerel, and tuna are among the species to be limited. Unfortunately, this limitation makes it much more difficult to follow the recommendation to incorporate more fish into the diet as a means of decreasing the amounts of meats such as beef that have high fat contents (Box 16.1). To follow both recommendations, people need to be selective about the fish species that they consume.

The aromas and flavors of garlic and onions are due largely to disulfides—diallyl disulfide and allyl propyl disulfide, respectively.

$$CH_2{=}CHCH_2{-}S{-}S{-}CH_2CH{=}CH_2$$

Diallyl disulfide

$$CH_2{=}CHCH_2{-}S{-}S{-}CH_2CH_2CH_3$$

Allyl propyl disulfide

Example 5.17 Writing equations for the reactions of thiols

Write equations for each of the following reactions: (a) ionization of methanethiol in water; (b) oxidation of ethanethiol; (c) reaction of methanethiol with Pb^{2+}.

Solution

(a) $CH_3SH + H_2O \longrightarrow CH_3S^- + H_3O^+$

(b) $CH_3CH_2SH \xrightarrow{(O)} CH_3CH_2{-}S{-}S{-}CH_2CH_3$

(c) $2\ CH_3SH + Pb^{2+} \longrightarrow CH_3S{-}Pb{-}SCH_3 + 2\ H^+$

Problem 5.17 Write equations for each of the following reactions: (a) neutralization of 1-propanethiol with NaOH; (b) reduction of diethyl disulfide; (c) reaction of 1-propanethiol with Hg^{2+}.

Summary

Structural Relations of Alcohols, Phenols, and Ethers

- Alcohols contain an —OH (hydroxyl) group attached to a saturated (sp^3) carbon.
- Phenols contain an —OH attached to one of the sp^2-hybridized carbons of a benzene ring.
- Ethers, like alcohols and phenols, contain an oxygen that has bonds to two different atoms but, in contrast with alcohols and phenols, neither of the bonds is to hydrogen.
- The oxygen of an ether has single bonds to two different carbons.

Constitutional Isomerism in Alcohols

- Constitutional isomerism in alcohols is due not only to different carbon skeletons but also to different placements of the —OH group within a carbon skeleton.
- The C:H ratio for an alcohol is the same as that for the corresponding hydrocarbon with the same number of double bonds and rings.

Classifying and Naming Alcohols

- Alcohols are 1°, 2°, or 3° alcohols, depending on whether the carbon holding the —OH group is a 1°, 2°, or 3° carbon.
- In the IUPAC system, alcohols are named by using the longest continuous carbon chain containing the —OH group.
- The base name is preceded by a number indicating the position of the —OH group, and the ending of the name is changed from -e to -ol.
- Prefixes with numbers are then added at the front to indicate substituents attached to the longest chain.

- Common names for simple alcohols are formed by using the name of the alkyl group attached to the —OH group followed by the word alcohol.

Physical Properties of Alcohols

- Alcohols participate in hydrogen bonding, which results in high boiling and melting points and solubility in water.

Acidity and Basicity of Alcohols

- Alcohols are very weak acids and bases.

Dehydration of Alcohols to Alkenes

- Intramolecular dehydration of an alcohol in the presence of an acid catalyst produces an alkene.
- When more than one alkene is possible, the major product is the most substituted alkene.
- A competing reaction for primary alcohols, but not secondary and tertiary alcohols, is intermolecular dehydration to an ether.
- At 180°C, the major product is alkene formed by intramolecular dehydration; at 140°C, the major product is ether formed by intermolecular dehydration.

Oxidation of Alcohols

- Primary and secondary alcohols undergo selective oxidation by permanganate (MnO_4^-) and dichromate ($Cr_2O_7^{2-}$) ions.
- Primary alcohols are oxidized to aldehydes, which are subsequently oxidized to carboxylic acids.
- Secondary alcohols are oxidized to ketones with no further oxidation possible.
- Tertiary alcohols are not oxidized.
- Permanganate and dichromate form the basis of simple chemical diagnostic tests for primary and secondary alcohols.

Phenols

- Substituted phenols are named by the IUPAC system as derivatives of the parent compound phenol.
- Phenols are more polar than alcohols and have higher boiling points and solubilities in water.
- Phenols are weak acids but much stronger than alcohols.
- Phenols undergo oxidation but not dehydration.

Ethers

- Ethers are isomeric with alcohols.
- The names of ethers containing simple groups consist of the names of the groups attached to the ether oxygen, followed by the word ether.
- Ethers have considerably lower boiling points than those of alcohols but are almost as soluble as alcohols in water.
- Ethers are unreactive toward acids, bases, and oxidizing agents.

Thiols

- Thiols contain the —SH group.
- They have lower boiling points than those of alcohols.
- Thiols are stronger acids than alcohols (but weaker than phenols), form insoluble products with heavy metal ions, and are oxidized to disulfides.
- Disulfides can be reduced back to thiols.

Summary of Reactions

ALCOHOLS

Acidity

$$2\ R{-}O{-}H + 2\ Na \longrightarrow 2\ R{-}O^- + Na^+ + H_2$$

Basicity

$$R{-}O{-}H + H_2SO_4 \rightleftharpoons R{-}\overset{+}{O}H_2 + HSO_4^-$$

Dehydration to alkene

$$-\overset{H}{\underset{|}{C}}-\overset{OH}{\underset{|}{C}}- \xrightarrow[\text{heat}]{H_2SO_4} -\underset{|}{C}{=}\underset{|}{C}- + H_2O$$

Most substituted double bond is major product

Dehydration to ether competes with alkene formation for 1° alcohols

$$2\ R{-}O{-}H \xrightarrow{H_2SO_4} R{-}O{-}R + H_2O$$

Ether is major product at 140°C
Alkene is major product at 180°C

Selective oxidation (3° alcohols do not oxidize)

$$R{-}CH_2{-}O{-}H \xrightarrow[Cr_2O_7^{2-}]{MnO_4^- \text{ or}} R{-}CH{=}O \xrightarrow[Cr_2O_7^{2-}]{MnO_4^- \text{ or}} R{-}C(OH){=}O$$

1° alcohol → Aldehyde → Carboxylic acid

$$R{-}CHR'{-}O{-}H \xrightarrow[Cr_2O_7^{2-}]{MnO_4^- \text{ or}} R{-}CR'{=}O$$

2° alcohol → Ketone

PHENOLS

Stronger acids compared with alcohols

$$C_6H_5{-}O{-}H + NaOH \longrightarrow C_6H_5{-}O^- + Na^+ + H_2O$$

ETHERS (UNREACTIVE)

THIOLS

Oxidation to disulfide

$$2\ R{-}SH \xrightarrow{(O)} R{-}S{-}S{-}R$$

Disulfides are reduced back to thiols

$$R{-}S{-}S{-}R \xrightarrow{(H)} 2\ R{-}SH$$

React with heavy metal ions (M = Pb, Hg)

$$2\ R{-}SH + M^{2+} \longrightarrow R{-}S{-}M{-}S{-}R + 2\ H^+$$

Precipitate

Key Words

alcohol, p. 151
dehydration, pp. 162, 175
dehydrogenation, p. 165
disulfide, p. 176
enol, p. 170
ether, pp. 151, 172
phenol, pp. 151, 169
primary alcohol, p. 154
secondary alcohol, p. 154
tertiary alcohol, p. 154
thiol, p. 176

Exercises

Structural Relation of Alcohols, Phenols, and Ethers

5.1 What structural features distinguish alcohols, phenols, and ethers?

5.2 Which of the following compounds is an alcohol, a phenol, an ether, or something else?

1. $(CH_3)_2CH{-}O{-}C_6H_5$
2. $C_6H_5{-}OCH_2CH_2OH$ (benzene ring bearing OCH_2CH_2OH)
3. $(CH_3)_3C{-}OH$
4. benzene ring bearing CH_2OH and OH (meta)

Constitutional Isomerism in Alcohols

5.3 Draw the structural formulas for (a) an alcohol, $C_5H_{12}O$, that contains one *t*-butyl group; (b) an alcohol, $C_4H_{10}O$, that contains one methyl group.

5.4 Draw the structural formulas for (a) a cyclic alcohol, $C_5H_{10}O$, that contains four CH_2 groups in the ring; (b) an alcohol, $C_6H_{14}O$, that contains four methyl groups but no *t*-butyl group.

5.5 For each of the following compounds, indicate whether the alcohol follows the general formula $C_nH_{2n+2}O$ or $C_nH_{2n}O$.

1. $CH_3{-}CH_2{-}C(OH)(CH_3){-}CH_2{-}CH_3$
2. cyclobutane ring bearing CH_2CH_3 and OH on adjacent carbons

5.6 Each of the following compounds is an alcohol. Indicate whether each alcohol possesses one or more double bonds or rings or both. (a) $C_4H_{10}O$; (b) C_4H_8O; (c) C_4H_6O.

Classifying and Naming Alcohols

5.7 Give the IUPAC name and classify the following alcohols as primary, secondary, or tertiary:

(a) $HO{-}CH_2{-}CH(CH_3){-}CH_2{-}CH_3$ (the CH_3 drawn above the CH)

(b) $CH_3{-}CH(CH_3){-}CH(OH){-}CH_3$ (the CH_3 drawn above the first CH, the OH below the second CH)

(c) [skeletal structure] OH

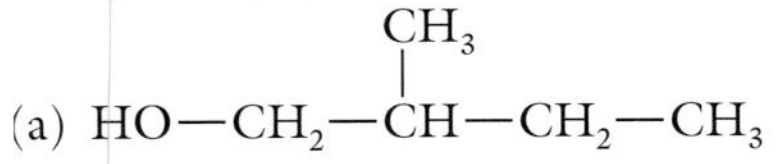

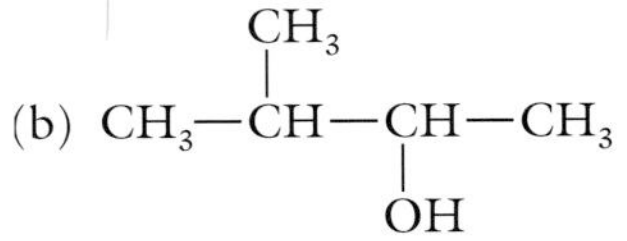

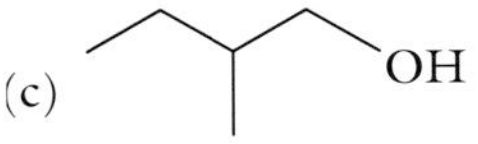

(d) $(CH_3)_3COH$

5.8 Draw the structural formula and classify the following alcohols as primary, secondary, or tertiary: (a) 5-chloro-4-methyl-2-hexanol; (b) 2,2-dimethylcyclobutanol; (c) 1,2,4-butanetriol; (d) *s*-butyl alcohol.

Physical Properties of Alcohols

5.9 Place the following compounds in order of boiling point: hexane, 1,4-butanediol, 1-pentanol, and 2-methyl-2-butanol. Explain the order.

5.10 Which of the following representations correctly describe the hydrogen bonding between methanol and water?

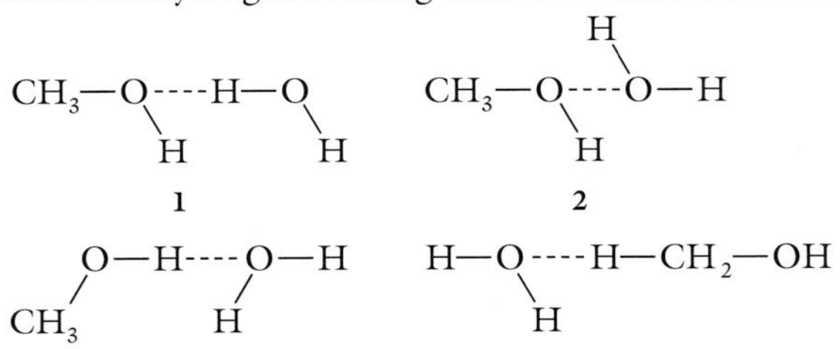

1: $CH_3{-}O(H)\cdots H{-}O{-}H$

2: $CH_3{-}O(H)\cdots O(H){-}H$

3: $CH_3{-}O{-}H\cdots O(H){-}H$

4: $H{-}O(H)\cdots H{-}CH_2{-}OH$

Acidity and Basicity of Alcohols

5.11 What is observed when a drop of ethanol is placed on blue litmus paper? Why?

5.12 What is observed when a drop of ethanol is placed on red litmus paper? Why?

5.13 Write the equation for the reaction of 1-butanol with HCl.

5.14 What occurs when 1-butanol is treated with each of the following reagents? If a reaction takes place, write an equation to show the product(s) formed. (a) NaOH; (b) Na.

Dehydration of Alcohols to Alkenes

5.15 Give the organic product(s) formed in each of the following reactions. If no reaction takes place, write "no reaction." If there is more than one product, indicate only the major product(s).

(a) $CH_3CH_2CH_2CH_2{-}OH \xrightarrow[\text{no heat}]{H_2SO_4} ?$

(b) $CH_3CH_2CH_2CH_2{-}OH \xrightarrow[\text{heat}]{} ?$

(c) $CH_3CH_2CH_2CH_2{-}OH \xrightarrow[\text{heat}]{H_2SO_4} ?$

5.16 Which of the following alcohols undergo dehydration in the presence of a strong acid catalyst? If dehydration takes place, show the product(s). If there is more than one product, indicate which is the major product. (a) 3-Methyl-2-butanol; (b) 2-methylcyclohexanol.

Oxidation of Alcohols

5.17 Which of the following alcohols undergo oxidation with dichromate or permanganate? If oxidation takes place, show the product. (a) 1-Pentanol; (b) 2-methyl-2-butanol.

5.18 Which of the following alcohols undergo oxidation with dichromate or permanganate? If oxidation takes place, show the product. (a) *t*-Butyl alcohol; (b) 3-methyl-2-butanol.

5.19 An unknown is either 1-butene or 1-butanol. Addition of a few drops of permanganate solution to the unknown results in decolorization of the purple color and a brown precipitate. What is the identity of the unknown? Explain your answer.

5.20 An unknown is either 2-methyl-2-propanol or 2-methyl-1-propanol. Addition of a few drops of dichromate solution (orange) to the unknown results in a green solution. What is the identity of the unknown? Explain your answer.

Phenols

5.21 Name the following compound by the IUPAC system.

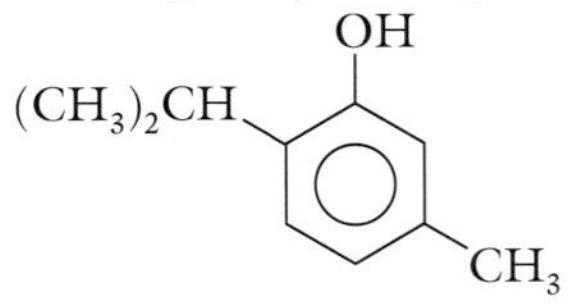

[benzene ring bearing OH, $(CH_3)_2CH$, and CH_3]

5.22 Draw the structure of 3-bromo-5-*t*-butylphenol.

5.23 Write an equation to show *p*-methylphenol acting as an acid when dissolved in water.

5.24 Place the following compounds in order of acidity: phenol, cyclohexanol, and HCl.

5.25 Indicate whether phenol reacts under each of the following conditions. If a reaction takes place, write the appropriate equation. (a) Combustion; (b) reaction with KOH.

5.26 Indicate whether phenol reacts under each of the following conditions. If a reaction takes place, write the appropriate equation. (a) Acid-catalyzed dehydration; (b) selective (mild) oxidation.

Ethers

5.27 Draw the structural formula for each of the following compounds: (a) 4-ethoxyphenol; (b) isobutyl propyl ether; (c) 3-isopropoxy-1-butanol.

5.28 Draw structural formulas of the ethers with molecular formula $C_5H_{12}O$. What other isomers exist for $C_5H_{12}O$?

5.29 Place the following compounds in order of boiling point: dimethyl ether, ethanol, and propane.

5.30 Which is more soluble in water, dimethyl ether or ethanol?

Formation of Ethers by Dehydration of Alcohols

5.31 What organic product(s) is formed when 1-methylcyclopentanol is heated at 180°C in the presence of an acid catalyst? What is the product at 140°C? Indicate only the major product(s) if more than one product is formed.

5.32 What alcohol(s) must be used to form each of the following ethers or alkenes on acid-catalyzed dehydration? Indicate the temperature that should be used to maximize the yield of the desired product. (a) Dibutyl ether; (b) 2-butene.

Thiols

5.33 Draw the structural formula for each of the following compounds: (a) 2-butanethiol; (b) dicyclopentyl disulfide.

5.34 Write equations for each of the following reactions: (a) neutralization of thiophenol (C_6H_5SH) with KOH; (b) oxidation of 2-propanethiol; (c) reaction of 1-propanethiol with Pb^{2+}; (d) reduction of diethyl disulfide.

Unclassified Exercises

5.35 Name the functional groups designated by A, B, C, D, and E.

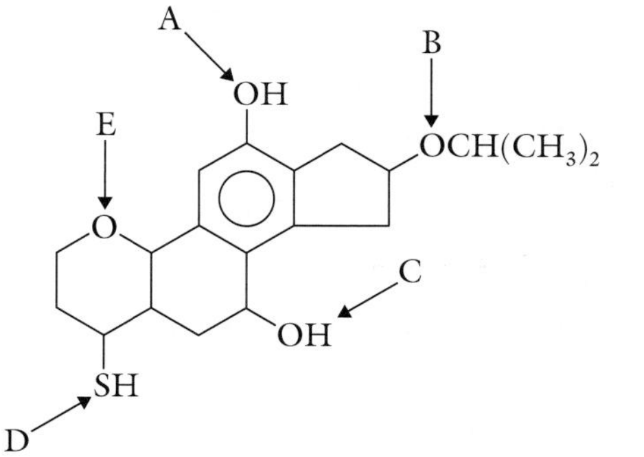

A → OH; B → $OCH(CH_3)_2$; C → OH; D → SH; E → O

5.36 Identify each of the following compounds as an alcohol, an ether, a phenol, a thiol, a disulfide, or something else.

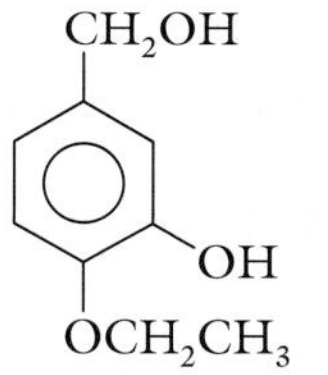

1: CH_2OH, OH, OCH_2CH_3 (benzene ring)

2: OH, CH_3, CH_2CH_3 (cyclohexane ring)

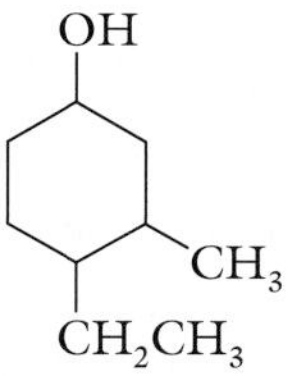

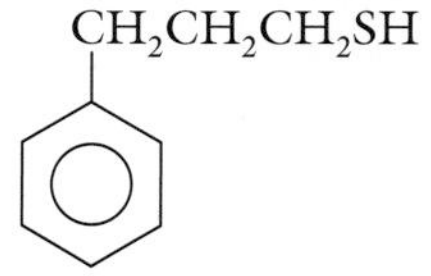

3: $CH_2CH_2CH_2SH$ (benzene ring)

4: $CH_3—\overset{\overset{\displaystyle O}{\|}}{C}—OH$

5.37 Identify each of the following compounds as an alcohol, an ether, a phenol, a thiol, a disulfide, or something else.

1: $CH_3CH_2—S—S—CH_2CH_3$

2: $CH_3—CH_2—C(OH)(CH_3)—CH_3$

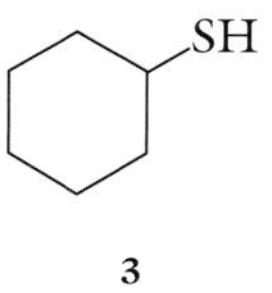

3: SH (cyclohexane ring)

4: $CH_3CH_2—C(OCH_3)(CH_3)—CH_3$

5.38 Draw the structural formulas for (a) a tertiary alcohol of molecular formula $C_5H_{12}O$; (b) a cyclic tertiary alcohol of five carbons and containing four carbons in the ring and one methyl group; (c) a phenol isomeric with *p*-chlorophenol; (d) an ether of molecular formula $C_6H_{14}O$ and containing a butyl group.

5.39 An unknown compound has the formula C_4H_8O. Is the unknown an ether or an alcohol? Does the unknown have a double bond or a ring?

5.40 For each of the following pairs of structural drawings, indicate whether the pair represents (1) the same compound or (2) different compounds that are constitutional isomers or (3) different compounds that are geometrical isomers or (4) different compounds that are not isomers.

(a) 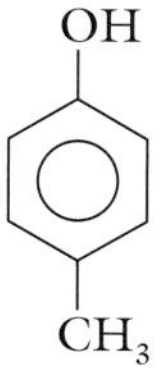 OH, CH_3 and CH_2OH

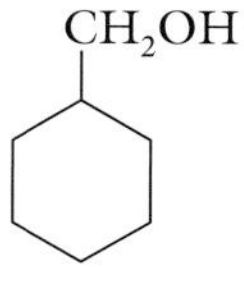

(b) $CH_3CH_2CH_2OCH(CH_3)_2$ and $CH_3CH_2CH_2CH_2CH_2OCH_3$

(c) $CH_3CH_2CH_2OCH(CH_3)_2$ and $CH_3CH_2CH_2CH_2CH_2CH_2OH$

(d) 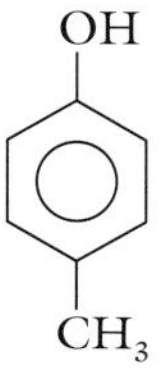 OH, CH_3 and CH_2OH

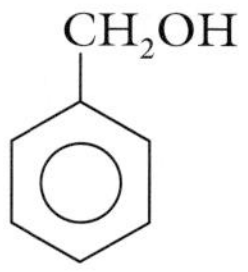

(e) $CH_3CH_2CH_2OCH(CH_3)_2$ and O

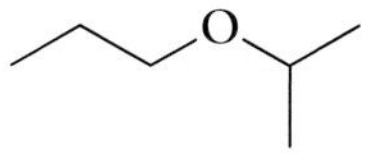

(f) $HO—CH_2—CH(CH_3)—CH_2—CH_3$ and $CH_3—CH(CH_3)—CH(OH)—CH_3$

(g) 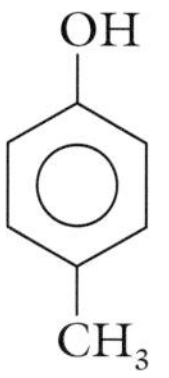 OH, CH_3 and OCH_3

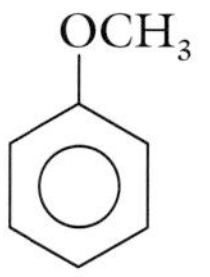

(h)  OH, CH_3 and CH_3, OH

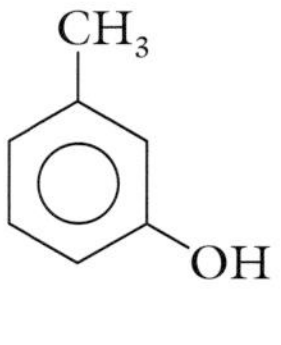

(i) 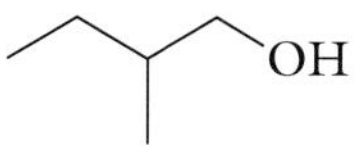 OH and $CH(OH)—CH_3$

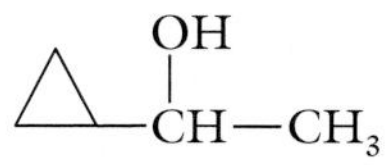

(j)  OH, CH_3 and CH_3, OH

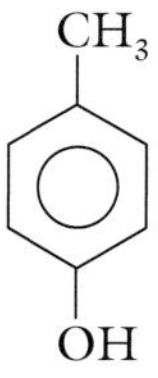

5.41 Draw the structural formula for each of the following compounds: (a) *s*-butyl propyl ether; (b) *m*-propylphenol; (c) *cis*-1,2-cyclohexanediol; (d) 2-pentanethiol; (e) 1-pentene-3-ol.

5.42 Place the following compounds in order of solubility in water: pentane, 1,4-propanediol, 1-butanol, and diethyl ether. Explain the order.

5.43 Indicate whether hydrogen bonding is present in the following compounds or mixtures. If present, draw a representation of the hydrogen bonding. (a) Ethanol; (b) propanol and water mixture; (c) diethyl ether; (d) dimethyl ether and methanol mixture; (e) phenol and water mixture.

5.44 Indicate which compound in each of the following pairs has the higher value of the specified property. If the two compounds in a pair have nearly the same value for the property, indicate that fact. Explain your conclusions.

(a) Dimethyl ether, propane: solubility in water.
(b) Cyclopropyl methyl ether, cyclobutanol: boiling point.
(c) Ethanol, pentane: solubility in water.
(d) 1-Butanol, 1-butanethiol: boiling point.

5.45 Place the following compounds in order of acidity: phenol, HCl, cyclohexanol, and cyclohexanethiol.

5.46 Give the organic product(s) formed in each of the following reactions. If no reaction takes place, write "no reaction." If there is more than one product, indicate only the major product(s).

(a) $CH_3CH_2CH_2CH_2{-}OH \xrightarrow[180°C]{H_2SO_4} ?$

(b) $CH_3CH_2CH_2CH_2{-}OH \xrightarrow[140°C]{H_2SO_4} ?$

(c) 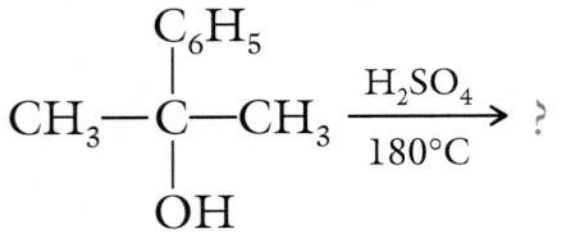 $CH_3{-}C(C_6H_5)(OH){-}CH_3 \xrightarrow[180°C]{H_2SO_4} ?$

(d) 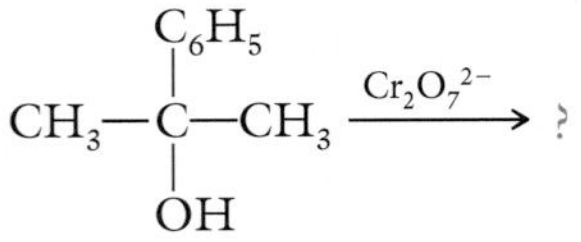 $CH_3{-}C(C_6H_5)(OH){-}CH_3 \xrightarrow{Cr_2O_7^{2-}} ?$

(e) $CH_3CH_2CH_2CH_2{-}OH \xrightarrow{MnO_4^-} ?$

(f) $CH_3CH_2OCH_2CH_3 \xrightarrow{MnO_4^-} ?$

5.47 Give the organic product(s) formed in each of the following reactions. If no reaction takes place, write "no reaction." If there is more than one product, indicate only the major product(s).

(a) 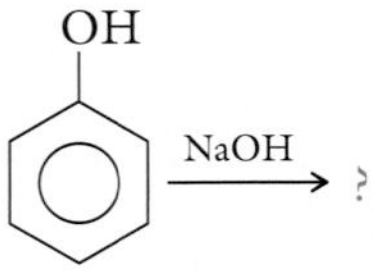 phenol (OH on benzene ring) $\xrightarrow{NaOH} ?$

(b) 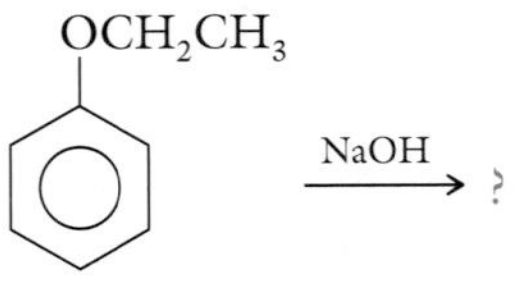 OCH_2CH_3 on benzene ring $\xrightarrow{NaOH} ?$

(c) 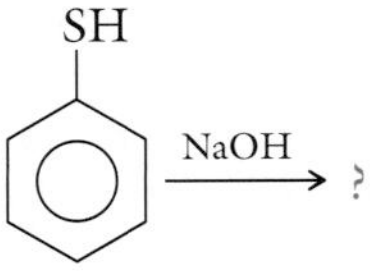 SH on benzene ring $\xrightarrow{NaOH} ?$

(d) 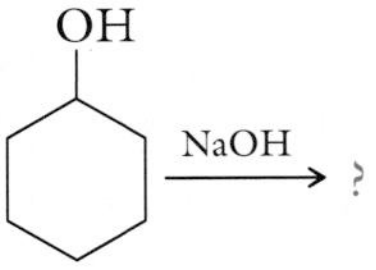 OH on cyclohexane ring $\xrightarrow{NaOH} ?$

(e) 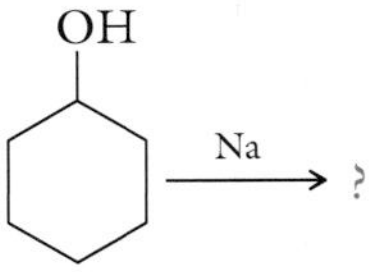 OH on cyclohexane ring $\xrightarrow{Na} ?$

(f) $CH_3CH_2CH_2CH_2{-}SH \xrightarrow{Pb^{2+}} ?$

(g) $CH_3CH_2CH_2CH_2{-}SH \xrightarrow{(O)} ?$

(h) $C_6H_5{-}S{-}S{-}C_6H_5 \xrightarrow{(H)} ?$

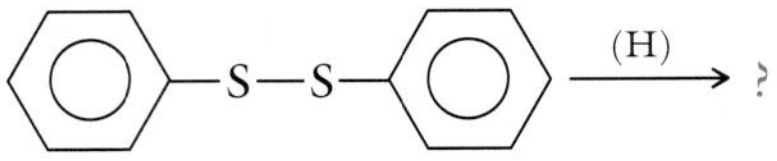

5.48 Write a balanced equation for the combustion of each of the following compounds: (a) 1-butanol; (b) diethyl ether.

5.49 For each of the following pairs of compounds, indicate which compound is more acidic: (a) 1-propanol and butane; (b) 1-propanol and HCl; (c) 1-propanol and water; (d) 1-propanol and phenol; (e) 1-propanol and 1-propanethiol.

5.50 An unknown is either *t*-butyl alcohol, ethyl methyl ether, pentane, or 1-butanol. Addition of a few drops of dichromate solution (orange) to the unknown results in a green solution. What is the identity of the unknown? Explain your answer.

5.51 An unknown is either *t*-butyl alcohol, ethyl methyl ether, pentane, or 1-butanol. The purple color of permanganate solution is not decolorized when permanganate is added to the unknown. What is the identity of the unknown? Explain your answer.

5.52 An unknown is either 1-butene or 1-butanol. Addition of a few drops of dichromate solution (orange) to the unknown results in a green solution. What is the identity of the unknown? Explain your answer.

5.53 An unknown is either cyclohexanol, toluene, or phenol. The unknown turns blue litmus paper red. What is the identity of the unknown? Explain your answer.

5.54 An unknown is either cyclohexanol, toluene, or phenol. The unknown is neutral to litmus paper: it does not turn blue litmus paper red or red litmus paper blue. What is the identity of the unknown? Explain your answer.

5.55 Give the bond angles for A, B, C, D, and E:

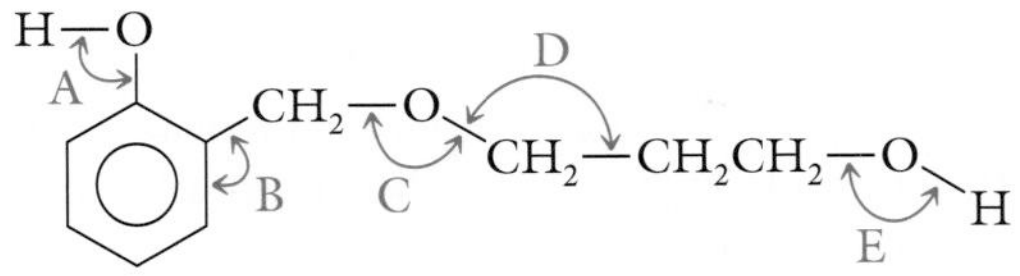

H–O (angle A) on benzene ring; ring–CH_2 (angle B) –O (angle C) –CH_2 (angle D) –CH_2CH_2–O–H (angle E)

5.56 What is the hybridization (sp^3 or sp^2) of atoms 1 through 6?

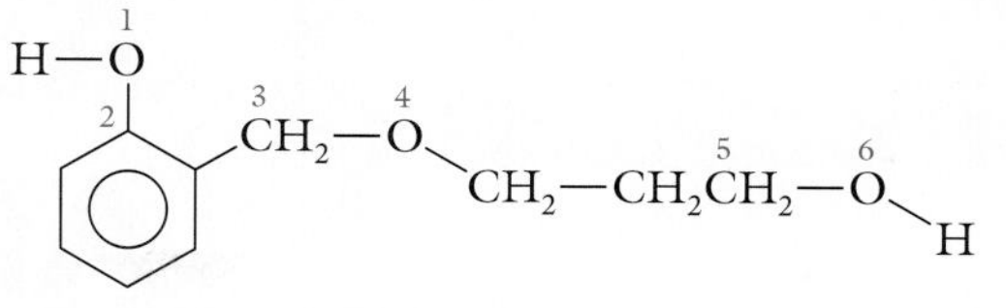

Expand Your Knowledge

Note: The icons denote exercises based on material in boxes.

5.57 The XYZ Cough Drop Company uses hexylresorcinol as an ingredient in its cough drops. The plant manager suspects that a shipment from the supplier of hexylresorcinol contains 3-hexyl-1,2-cyclohexanediol instead of hexylresorcinol. What simple chemical test distinguishes between the two compounds?

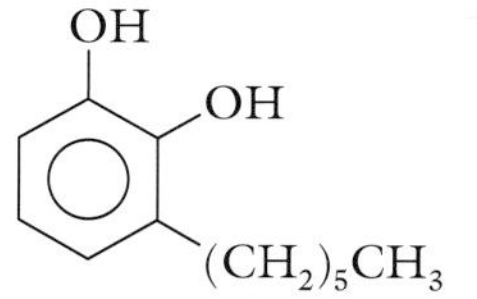

Hexylresorcinol 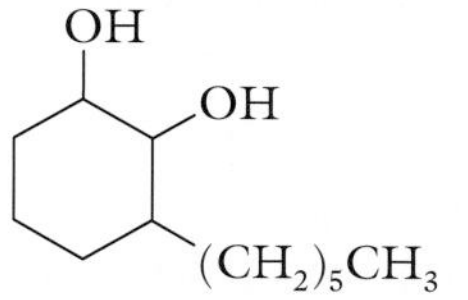 **3-Hexyl-1,2-cyclohexanediol**

5.1 5.2 **5.58** Ethanol has many important industrial uses in addition to its use in alcoholic beverages (see Box 5.1). Industrial ethanol is obtained from two sources: about 30% by fermentation of carbohydrates (see Box 5.2) and 70% by hydration of ethene. What reagent(s) and reaction conditions are required for producing ethanol by hydration of ethene?

5.1 **5.59** 2-Propanol (rubbing alcohol) is applied to the skin to lower a fever (see Box 5.1). Alcohols such as 1-decanol or 2-decanol are not useful for this purpose. How does 2-propanol lower a fever, and why are 1-decanol and 2-decanol not useful?

5.4 **5.60** Dehydration of alcohols requires a strong acid catalyst (see Box 5.4). Catalysts increase reaction rates by offering an alternate mechanism for reaction, one with a lower activation energy. What is the alternate mechanism for dehydration in the presence of an acid catalyst, and why does it take place more rapidly?

5.4 **5.61** 1-Propanol and 2-propanol yield the same product, propene, on dehydration (by heating in the presence of acid), but 2-propanol reacts faster than 1-propanol. Explain the difference in reactivity in reference to the mechanism of dehydration (see Box 5.4).

5.62 The cyclic ether ethylene oxide (Section 5.9) is an important industrial chemical. It is produced by selective oxidation of ethene and subsequently undergoes reaction with water to produce 1,2-ethanediol, the compound used as an antifreeze in automobile cooling systems. Write the equation for the conversion of ethylene oxide into 1,2-ethanediol. The reaction is classified as a **ring-opening addition** reaction and is catalyzed by strong acids.

3.2 **5.63** Whereas ethylene oxide readily undergoes ring-opening addition reactions such as that in Exercise 5.62, the cyclic ether dioxane is unreactive. Indicate the reason for the difference after referring to Box 3.2.

5.64 Lactic acid, produced in active skeletal muscles, is transported to the liver where it is eventually converted into glucose. The first step in this conversion produces pyruvic acid by oxidation of the hydroxyl group. Draw the structure of pyruvic acid.

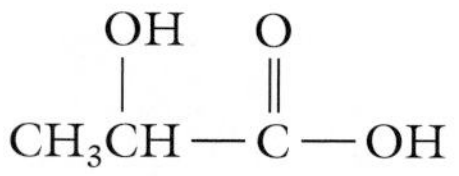

$CH_3CH(OH)—C(=O)—OH$

Lactic acid

5.65 Dehydration of 2-phosphoglyceric acid to phosphoenolpyruvic acid is an important step in the metabolism of glucose by glycolysis. Draw the structure of phosphoenolpyruvate.

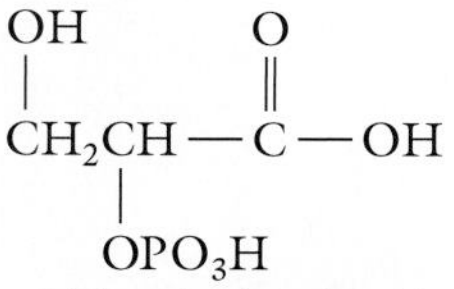

$CH_2(OH)CH(OPO_3H)—C(=O)—OH$

2-Phosphoglyceric acid

5.66 The XYZ Company needs 2-propanol but purchased 1-propanol by mistake. Jane Brown correctly suggests that the 1-propanol can be converted into 2-propanol by heating in the presence of a catalytic amount of a strong acid. What reactions are responsible for this conversion?

5.1 **5.67** Menthol is used in cough drops and throat sprays (see Box 5.1). Give the IUPAC name for menthol.

5.5 **5.68** Hexylresorcinol is used in mouthwashes (see Box 5.5). Give the IUPAC name for hexylresorcinol.

5.69 Explain why *p*-chlorophenol is more acidic than phenol.

5.70 Poly(vinyl alcohol) is useful in applications such as adhesives and coatings because of its solubility in water. Explain why this polymer is water soluble.

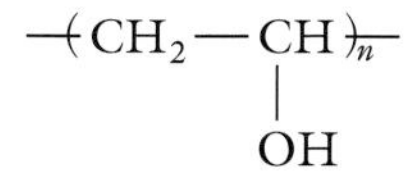

$—(CH_2—CH(OH))_n—$

Poly(vinyl alcohol)

5.71 Heating 1-butanol at 140°C in the presence of a strong acid produces dibutyl ether. Under the same conditions *t*-butyl alcohol produces only 2-methyl-1-propene; no di-*t*-butyl ether is produced. Why?

5.72 Diallyl disulfide, $(CH_2{=}CHCH_2S)_2$, is largely responsible for the aroma and flavor of garlic. Write an equation for the reaction of diallyl disulfide with (a) reducing agent; (b) Pb^{2+}. If no reaction takes place, write "no reaction."

5.73 Ethanol is added to gasoline to improve engine performance. Write the balanced equation for the combustion of ethanol.

5.74 The two carbons in ethanol have different oxidation states. Which carbon has the higher oxidation state? What is the basis of the assignment? The oxidation state of carbon is defined in Section 2.1.

CHAPTER 6 ALDEHYDES AND KETONES

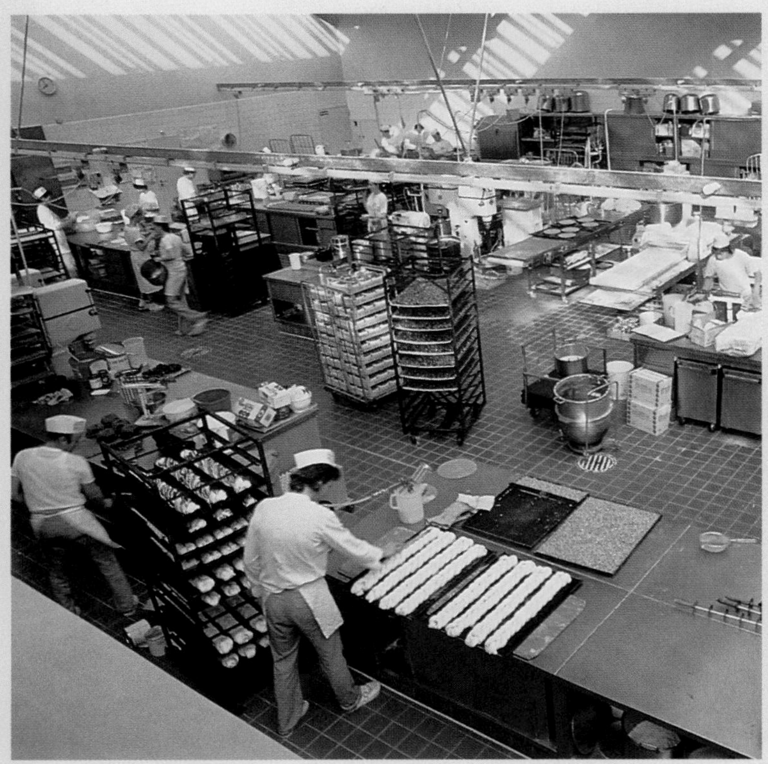

(Michael Rosenfeld/Tony Stone.)

Chemistry in Your Future

Your younger brother, a nursing student, enjoys visiting the kitchen under your management in a long-term convalescent care facility. Today, as he samples the pound cake that you've been baking, he complains about the difficulty of the organic and biochemistry course that he is taking, especially this week's topic—aldehydes and ketones. You surprise him by saying, "Those aldehydes and ketones are great stuff; I use them all the time in my recipes." You then take him on a short tour of the kitchen, pointing to sugar, butter, cornstarch, and various flavorings and spices (almond, cinnamon, vanilla, mint), all of which are aldehydes or ketones. Chapter 6 describes the common features and properties of this family of organic compounds.

For more information about this topic and others in the chapter, go to www.whfreeman.com/bleiodian2e

Learning Objectives

- Draw and name aldehydes and ketones.
- Describe the physical properties of aldehydes and ketones.
- Describe and write equations for oxidation and reduction reactions of aldehydes and ketones.
- Describe and write equations for hemiacetal and acetal formation by reaction of alcohols with aldehydes and ketones.
- Describe and write equations for the hydrolysis of acetals.

We now begin the study of oxygen-containing organic compounds in which the carbon and oxygen are connected by a double bond. In this chapter, we examine two closely related families of compounds: the aldehydes and ketones.

The aldehyde and ketone families include many biological molecules: carbohydrates such as the sugar glucose, an essential source of energy throughout the plant and animal world; lipids such as the male and female sex hormones testosterone and progesterone; and a variety of chemical intermediates that take part in biochemical pathways, as we will see in Chapters 14 through 17. An assortment of naturally occurring compounds with strong flavors and odors—for example, vanilla, almonds, mint, and butter—also belong to these families.

Aldehydes and ketones have a number of commercial uses as solvents and as starting materials for manufacturing many industrial chemicals, including plastics and adhesives.

6.1 THE STRUCTURE OF ALDEHYDES AND KETONES

The common structural feature of all **aldehydes** and **ketones** is the **carbonyl functional group,** a carbon–oxygen double bond (Section 3.2). Aldehydes are distinguished from ketones by the atoms directly attached to the carbon of the carbonyl group. This carbon is called the **carbonyl carbon.** In aldehydes, at least one hydrogen atom is directly attached to the carbonyl carbon, forming the **aldehyde group,** HC=O, often written as CHO. In ketones, only carbon atoms, no hydrogen atoms, are directly attached to the carbonyl carbon. The groups (represented by R and R′) attached to the carbonyl carbon can be either aliphatic or aromatic.

Other families also contain the carbonyl group (Table 6.1) but differ from aldehydes and ketones in having a **heteroatom**—any atom other than carbon or hydrogen—directly attached to the carbonyl carbon. In carboxylic acids and esters, the heteroatom is an oxygen; in amides, it is a nitrogen. Carboxylic acids, esters, and amides are the subject of Chapters 7 and 8.

INSIGHT INTO PROPERTIES

$$-\overset{\overset{\large O}{\|}}{C}-$$

Carbonyl group

Aldehyde

R = H, aliphatic or aromatic

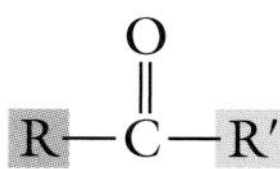

Ketone

R and R′ = aliphatic or aromatic

$$R-\overset{\overset{\large O}{\|}}{C}-OH$$

Carboxylic acid

$$R-\overset{\overset{\large O}{\|}}{C}-OR'$$

Ester

$$R-\overset{\overset{\large O}{\|}}{C}-NH_2$$

Amide

Example 6.1 Recognizing aldehydes and ketones

Which of the following compounds are aldehydes or ketones and which are not? Classify each compound.

$$\underset{1}{CH_3-\overset{\overset{\large O}{\|}}{C}-H} \qquad \underset{2}{CH_3-\overset{\overset{\large O}{\|}}{C}-OH} \qquad \underset{3}{CH_3-\overset{\overset{\large O}{\|}}{C}-NH_2} \qquad \underset{4}{CH_3-\overset{\overset{\large O}{\|}}{C}-CH_3}$$

TABLE 6.1 Families Containing the Carbonyl Group (C=O)

	Family	Atoms bonded to C=O	
		Y	Z
$Y-\overset{\overset{O}{\|}}{C}-Z$	aldehyde	H	H or C
	ketone	C	C
	carboxylic acid, ester	H or C	O
	amide	H or C	N

Solution

Compound 1 is an aldehyde because it has at least one H and no heteroatom attached to the carbonyl carbon. Compound 2 is a carboxylic acid because of the heteroatom oxygen attached to the carbonyl carbon. Compound 3 is an amide because there is a nitrogen attached to the carbonyl carbon. Compound 4 is a ketone because it has only carbons, no H and no heteroatom, attached to the carbonyl carbon.

Problem 6.1 Classify each of the following compounds as an aldehyde, ketone, carboxylic acid, ester, or amide.

$CH_3O{-}C({=}O){-}CH_2CH_2CH_3$ (1)

$H{-}C{=}O$ attached to a benzene ring (2)

Cyclohexane ring $=O$ (3)

$(CH_3)_3C{-}C({=}O){-}NH_2$ (4)

Aldehydes and ketones are related to alcohols in the same way that alkenes are related to alkanes. Loss of a hydrogen from each of two adjacent carbons of an alkane yields an alkene. Loss of the hydrogen from the —OH group of an alcohol, together with loss of a hydrogen from the carbon holding the OH, yields the carbonyl group:

$$\underset{\text{Alcohol}}{\text{H–O–}\overset{|}{\underset{|}{C}}\text{–H}} \xrightarrow{-2\,H} \underset{\text{Aldehyde or ketone}}{{-}C({=}O){-}} \quad \text{compared with} \quad \underset{\text{Alkane}}{\text{H–}\overset{|}{C}\text{–}\overset{|}{C}\text{–H}} \xrightarrow{-2\,H} \underset{\text{Alkene}}{{-}C{=}C{-}}$$

Recall that the presence of a double bond or ring lowers the C:H ratio in a compound by two hydrogens compared with alkanes. The presence of a C═O double bond has the same effect as that of a C═C double bond; aldehydes and ketones have the same C:H ratio as that of alkenes. The molecular formulas for the different families of organic compounds are shown in Table 6.2.

Another similarity between the aldehydes and ketones and the alkenes is in the bonding of the C═C and C═O double bonds. (Refer to Section 4.2 for a description of the carbon–carbon double bond.) Both the carbon and the oxygen atoms of the carbonyl group are sp^2 hybridized, and the carbonyl double bond contains one strong σ bond and one weak π bond. The bond angles about the carbonyl carbon are close to 120° (trigonal), as shown in Figure 6.1. Despite their structural similarities, the C═C and C═O double bonds do differ in important ways, as we shall soon see.

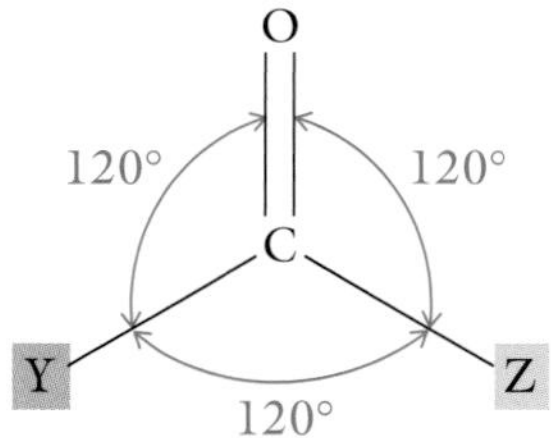

Figure 6.1 Bond angles in aldehydes (Y = H; Z = H, R) and ketones (Y = R; Z = R′).

Like other families of organic compounds, aldehydes and ketones can have constitutional isomers based on different carbon skeletons and different

TABLE 6.2 Molecular Formulas of Different Families

Family	Molecular formula
alkane	C_nH_{2n+2}
cycloalkane	C_nH_{2n}
alkene	C_nH_{2n}
alcohol, ether	$C_nH_{2n+2}O$
aldehyde, ketone	$C_nH_{2n}O$

placements of the carbonyl group. For example, both propanal and acetone have the molecular formula C_3H_6O:

$$\underset{\text{Propanal}}{CH_3CH_2-\overset{\overset{\displaystyle O}{\|}}{C}-H} \qquad \underset{\text{Acetone}}{CH_3-\overset{\overset{\displaystyle O}{\|}}{C}-CH_3}$$

Example 6.2 Drawing constitutional isomers of aldehydes and ketones

How many aldehydes and ketones are possible for the molecular formula C_4H_8O? Draw the structural formula for each isomer.

Solution

Two different C_4 carbon skeletons are possible:

$$\underset{1}{C-C-C-C} \qquad \underset{2}{C-\overset{\overset{\displaystyle C}{|}}{C}-C}$$

Both of these skeletons can yield an aldehyde, and skeleton 1 can also yield a ketone:

$$\underset{1a}{CH_3CH_2CH_2-\overset{\overset{\displaystyle O}{\|}}{C}-H} \qquad \underset{2a}{CH_3\overset{\overset{\displaystyle CH_3}{|}}{C}H-\overset{\overset{\displaystyle O}{\|}}{C}-H} \qquad \underset{1b}{CH_3-\overset{\overset{\displaystyle O}{\|}}{C}-CH_2CH_3}$$

No ketone is possible for skeleton 2, because the tertiary carbon already has three bonds to other carbon atoms and cannot also be a carbonyl carbon. We would have the impossibility of five bonds to carbon:

$$C-\underset{\underset{\displaystyle O}{\|}}{\overset{\overset{\displaystyle C}{|}}{C}}-C$$

C cannot have 5 bonds

Problem 6.2 Draw the structural formula for each aldehyde and ketone with the molecular formula $C_5H_{10}O$.

Some of the naturally occurring aldehydes and ketones are described in Box 6.1 on the following page. The commercial applications of formaldehyde and acetone, the simplest aldehyde and ketone, are described in Box 6.2 on page 191.

6.2 NAMING ALDEHYDES AND KETONES

In the IUPAC nomenclature system, aldehydes and ketones are named by using a variation of the rules for naming alkanes:

Rules for naming aldehydes and ketones

- Find the longest continuous chain that contains the carbonyl carbon.
- The ending **-e** of the alkane name is changed to **-al** for aldehydes and **-one** for ketones.
- The longest continuous chain is numbered from the end nearest the C=O group.
- For aldehydes, the carbonyl carbon is always C1, and the number 1 is not used in naming an aldehyde.
- The position of the carbonyl group is indicated for ketones by placing the number of the carbonyl carbon in front of the base name. The number 2 is not used for propanone and butanone, because only one ketone is possible for each.

6.1 CHEMISTRY AROUND US

Aldehydes and Ketones in Nature

Many of the aldehydes and ketones in nature are polyfunctional, such as the carbohydrate (sugar) glucose. Note that glucose contains an aldehyde group and five hydroxyl groups. Closely related is the ketone sugar fructose. Glucose is found in combined form as starch in potatoes, corn, rice, and peas. Glucose and fructose are found combined together as sucrose, the table sugar used to sweeten our baked goods, coffee, and tea.

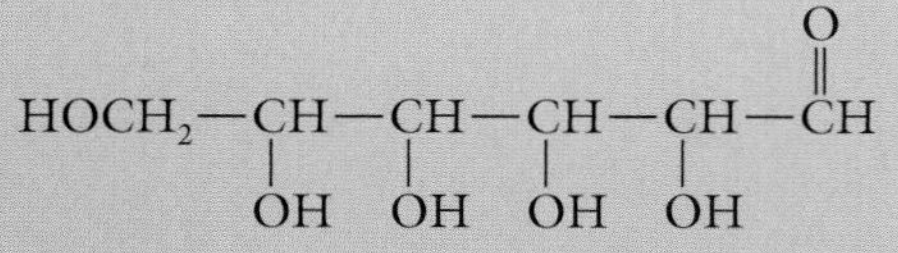

Glucose

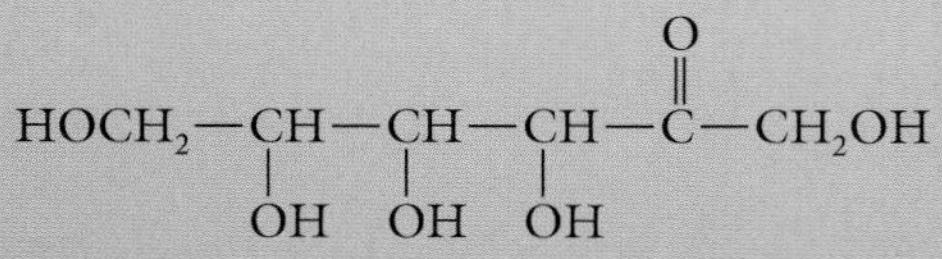

Fructose

The female and male sex hormones progesterone and testosterone, respectively, contain the ketone group. These hormones are members of the lipid family and belong to a class of lipids called steroids (Section 11.7).

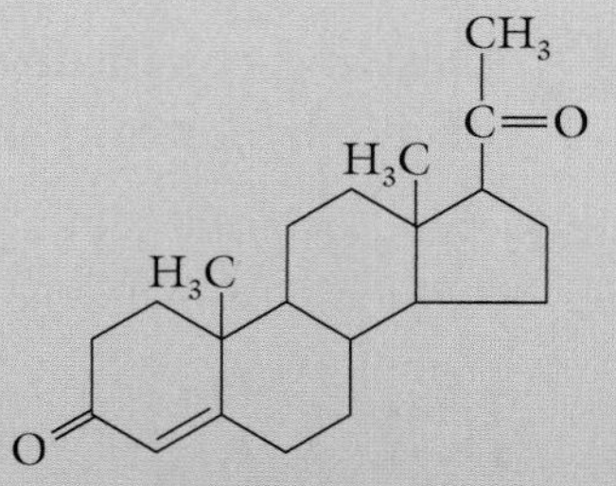

Progesterone

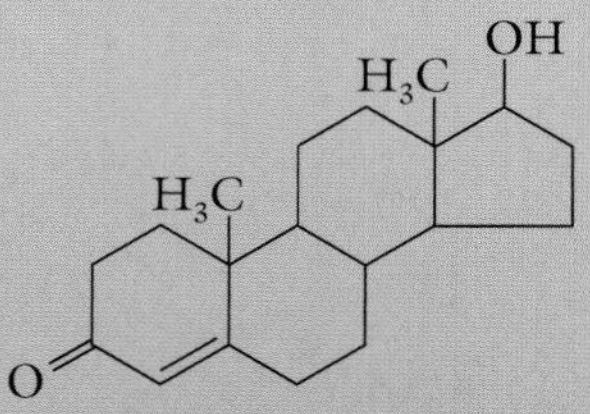

Testosterone

Aldehydes and ketones such as almond, cinnamon, and mint are used as flavorings in cooking. (*Left,* Inga Spence/Picture Cube; *center,* Ed Bock/The Stock Market; *right,* Wally Eberhart/Visuals Unlimited.)

- The names of substituents are used as prefixes, preceded by numbers indicating their positions on the longest chain.
- A cyclic compound containing a carbonyl carbon as part of the ring is named a **cycloalkanone.** Numbering of the ring starts at the carbonyl carbon and proceeds to other substituents on the ring in the direction that yields the lowest numbers for those substituents.
- C_6H_5–CHO is named benzaldehyde.

Example 6.3 Using the IUPAC nomenclature system

Name the following compounds by the IUPAC system:

(a) 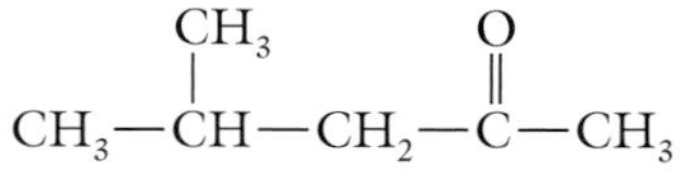

(b) 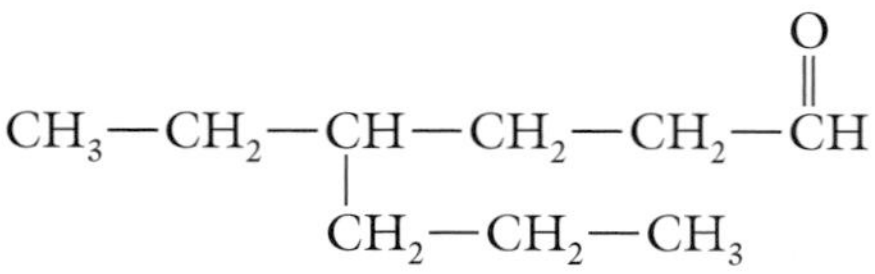

(c) 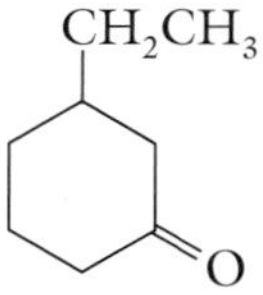

(d) 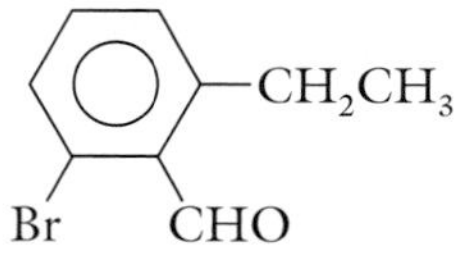

In Box 4.2, we learned that *cis*-11-retinal is essential to vision in most animals. *cis*-11-Retinal belongs to both the alkene and the aldehyde families, and both the C=C and the C=O functional groups are critical to its physiological function. The light-induced conversion of *cis*-11-retinal into *trans*-11-retinal, which triggers the electrical signal that the brain interprets as vision (Box 4.2), can take place only after *cis*-11-retinal reacts with the protein opsin to form rhodopsin. The presence of an aldehyde group in *cis*-11-retinal is critical to this event: rhodopsin is formed by a dehydration reaction between the carbonyl group of *cis*-11-retinal and an amine (NH_2) group of opsin.

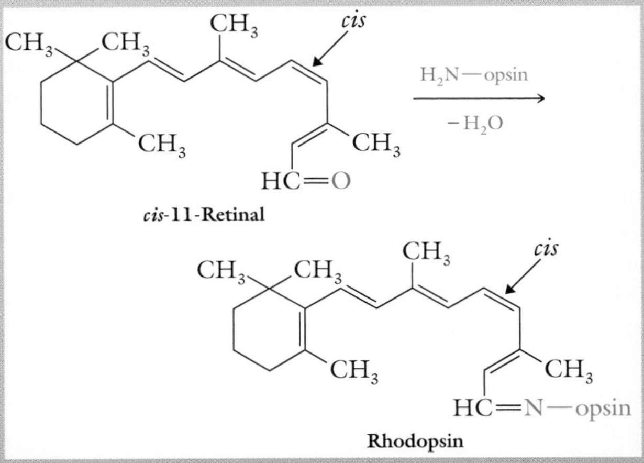

***cis*-11-Retinal**

Rhodopsin

Plants produce a variety of aldehydes and ketones that have long been used as flavorings or spices—for example, benzaldehyde (almond), cinnamaldehyde (cinnamon), vanillin (vanilla), and menthone (mint). Biacetyl, a diketone, is the compound responsible for the taste of butter and is often added to margarine and other shortenings to impart a buttery flavor.

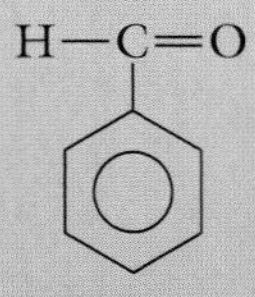

Benzaldehyde (Almond)

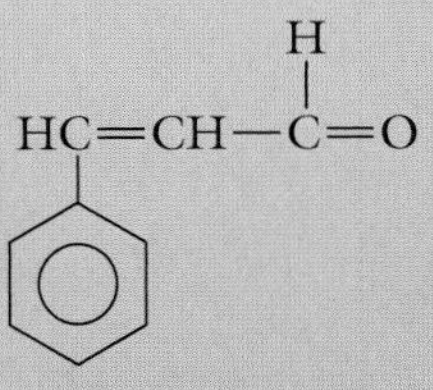

Cinnamaldehyde (Cinnamon)

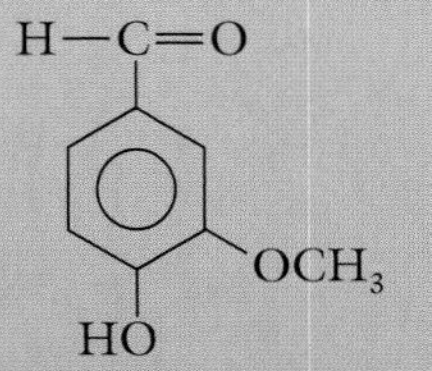

Vanillin (Vanilla)

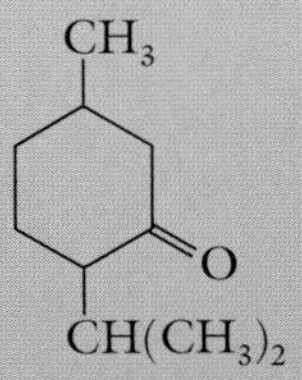

Menthone (Mint)

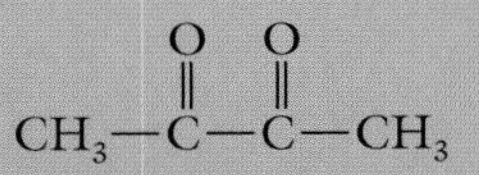

Biacetyl (Butter)

Solution

(a) This compound is a ketone. The longest continuous chain containing the carbonyl carbon is a five-carbon chain. The chain is numbered from the end nearest the carbonyl carbon to yield 2-pentanone as the base name. (Numbering the chain from the other end is incorrect, because it yields 4-pentanone, which has a higher number for the position of the carbonyl carbon.) The methyl group is at C4, which yields the name 4-methyl-2-pentanone.

(b) This compound is an aldehyde. The longest chain containing the carbonyl carbon is the seven-carbon chain:

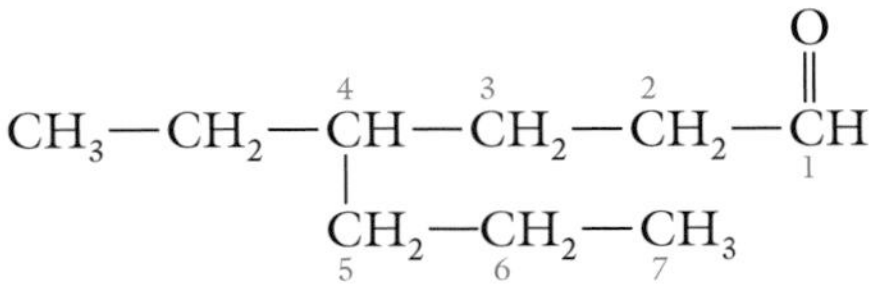

$$\overset{}{CH_3}-CH_2-\overset{4}{C}H(\underset{5}{C}H_2-\underset{6}{C}H_2-\underset{7}{C}H_3)-\overset{3}{C}H_2-\overset{2}{C}H_2-\underset{1}{C}H{=}O$$

The carbonyl carbon is C1, and there is an ethyl group at C4. The name is 4-ethylheptanal.

(c) In this cyclic ketone, the carbonyl carbon is C1, which places the ethyl group at C3. The name is 3-ethylcyclohexanone.

(d) This compound is named as a derivative of benzaldehyde, C_6H_5–CHO. The aldehyde group is attached at C1 of the ring. The two possible names are 2-bromo-6-ethylbenzaldehyde and 6-bromo-2-ethylbenzaldehyde, numbering in the clockwise and counterclockwise directions, respectively. Both names have the same set of numbers. The correct name is 2-bromo-6-ethylbenzaldehyde because the lower number is assigned to substituents in alphabetical order.

Problem 6.3 Name the following compounds by the IUPAC system:

(a) $CH_3CH(CH_3)—C(=O)—CH_2CH(CH_3)CH_3$ (b) $(CH_3)_3CCH_2CH_2—C(=O)H$

(c) 3-chlorocyclopentanone structure (Cl on ring, C=O on ring) (d) CH_3–(benzene ring, Br)–CHO

Unbranched acyclic aldehydes containing from one to four carbons are often referred to by common names. The common names are obtained by combining the prefixes **form-**, **acet-**, **propion-**, and **butyr-** with **-aldehyde** to give the names formaldehyde, acetaldehyde, propionaldehyde, and butyraldehyde (Table 6.3).

Common names for ketones are obtained by naming the groups attached to the carbonyl carbon followed by the word **ketone.** This method is used only when the groups attached to the carbonyl carbon are simple groups: the unbranched alkyl groups, from methyl, ethyl, propyl, butyl through decyl; the branched three- and four-carbon alkyl groups (isopropyl, isobutyl, *s*-butyl, *t*-butyl); and the unsubstituted cycloalkyl groups, such as cyclohexyl and cyclopentyl.

Note, however, that dimethyl ketone (IUPAC: propanone), an important industrial compound, is usually called acetone.

TABLE 6.3 IUPAC and Common Names of Aldehydes and Ketones

Compound	IUPAC name	Common name
$H—C(=O)—H$	methanal	formaldehyde
$CH_3—C(=O)—H$	ethanal	acetaldehyde
$CH_3CH_2—C(=O)—H$	propanal	propionaldehyde
$CH_3CH_2CH_2—C(=O)—H$	butanal	butyraldehyde
$C_6H_5—C(=O)—H$	benzaldehyde	benzaldehyde
$CH_3—C(=O)—CH_3$	propanone	acetone
$CH_3—C(=O)—CH_2CH_3$	butanone	ethyl methyl ketone

6.2 CHEMISTRY AROUND US

Important Aldehydes and Ketones

Formaldehyde (HCHO) is the simplest aldehyde, and acetone (CH_3COCH_3) is the simplest ketone. Both chemicals have a number of commercial applications.

Formaldehyde is used to preserve biological specimens and embalm cadavers, because it kills bacteria, viruses, and other microorganisms that would otherwise cause dead tissues to deteriorate. It kills the microorganisms by reacting with their proteins and destroying the physiological functions of these proteins. Similarly, it has been used to deactivate viruses for the preparation of vaccines. Formaldehyde has also been used for disinfecting buildings and ships, for tanning hides, and as a fungicide, but some of these uses have been curtailed because of the high toxicity of formaldehyde to humans. Formaldehyde is a gas at room temperature (boiling point = −21°C) and is generally used as an aqueous solution called **formalin.**

The extremely deleterious effects of drinking methanol noted in Box 5.3 are in fact due to the toxicity not of methanol itself but of the formaldehyde produced when the liver attempts to detoxify the body of methanol. Formaldehyde is toxic because it reacts with proteins and deactivates them—the same reaction that is useful for embalming purposes as well as the same reaction that forms rhodopsin from *cis*-11-retinal and opsin (see Box 6.1). The consumption of ethanol is less a problem because acetaldehyde, produced by the detoxification of ethanol, is less reactive in deactivating proteins. However, long-term consumption of large amounts of ethanol eventually takes its toll on the body and the mind.

Formaldehyde preserves biological specimens such as this grasshopper. (Richard Megna/Fundamental Photographs.)

Large amounts of formaldehyde are used in the manufacture of polymers. More than 8 billion pounds of polymers per year are produced in the United States by the reaction of formaldehyde with phenol, urea, and melamine. These polymers are used to manufacture plywood and particle board (for house construction, cabinets, and furniture), printed circuit boards, and the protective and decorative surfaces for kitchen and bathroom countertops and tables, as well as to impart crease resistance and wash-and-wear properties to cotton clothing and fabrics.

Phenol: C_6H_5OH Urea: $H_2N{-}C({=}O){-}NH_2$ Melamine: triazine ring with three NH_2 groups

Phenol **Urea** **Melamine**

Acetone is used in excess of 4 billion pounds per year in the United States. It and chemicals synthesized from it are useful as solvents for plastics, adhesives, pesticides, hydraulic fluids, and printing inks. Acetone is also used in nail-polish-remover formulations. Methyl methacrylate and bisphenol A are two important chemicals synthesized from acetone. Poly(methyl methacrylate) is manufactured from methyl methacrylate and is used for the manufacture of bathtubs, dentures, dental fillings, and windows for aircraft and factory buildings (Table 4.3). Bisphenol A is used to produce epoxy adhesives.

$$CH_2{=}C(CH_3){-}C({=}O){-}OCH_3$$

Methyl methacrylate

$$HO{-}C_6H_4{-}C(CH_3)_2{-}C_6H_4{-}OH$$

Bisphenol A

People with diabetes who are not receiving proper treatment may have a characteristic sweet odor of acetone on their breath. The diabetic does not adequately metabolize glucose, because of an insulin deficiency. This inadequacy causes a diversion of chemicals into other pathways, with the end result being an incomplete metabolism of fatty acids. Instead of conversion into carbon dioxide and water, the fatty acids are converted into acetone and other ketones called **ketone bodies** (Section 16.3) that are detected in the blood, urine, and breath. Section 18.6 describes the chemistry of diabetes in more detail.

Example 6.4 Using the common nomenclature system

Give the common name for each of the following ketones:

(a) 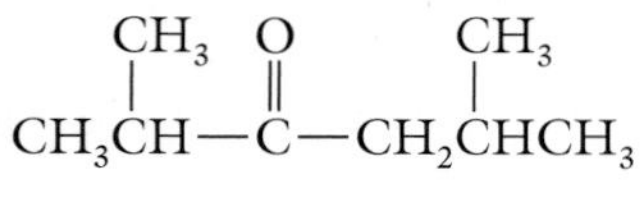 $CH_3CH(CH_3)-C(=O)-CH_2CH(CH_3)CH_3$ (b) 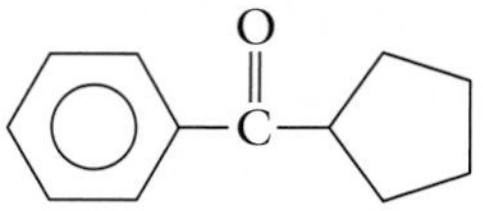 $C_6H_5-C(=O)-C_5H_9$

Solution

(a) The two groups attached to the carbonyl carbon are the isopropyl and isobutyl groups. The name is isobutyl isopropyl ketone because the groups are named in alphabetical order.

(b) The two groups attached to the carbonyl carbon are the phenyl and cyclopentyl groups. The name is cyclopentyl phenyl ketone.

Problem 6.4 Give the common name for each of the following ketones:

(a) 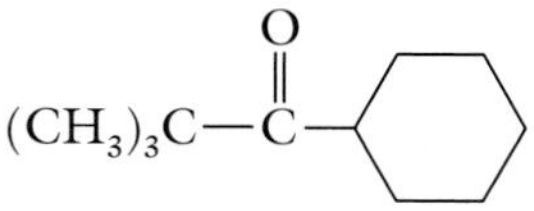 $(CH_3)_3C-C(=O)-C_6H_{11}$ (b) 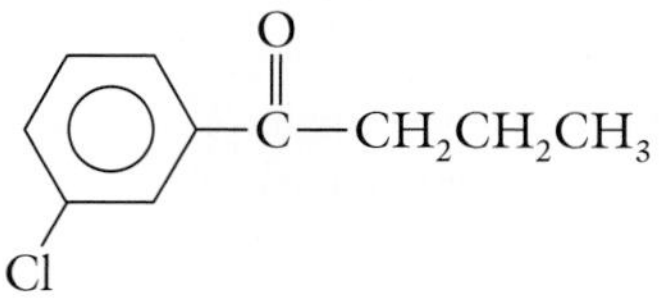 $ClC_6H_4-C(=O)-CH_2CH_2CH_3$

6.3 PHYSICAL PROPERTIES OF ALDEHYDES AND KETONES

Aldehydes and ketones are polar compounds because the carbonyl group is polar. The physical properties of aldehydes and ketones, the result of this polarity, are characterized by:

- Higher boiling and melting points than those of the corresponding hydrocarbons but lower than those of the corresponding alcohols
- Almost the same water solubility as that of alcohols

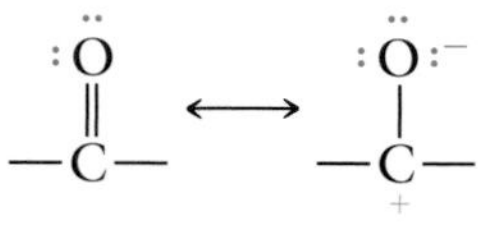

Unlike the carbon–carbon double bond of alkenes, the C=O bond is highly polarized. The weakly held electrons of the π bond are pulled closer to the oxygen of the carbonyl group than to the carbon because oxygen is much more electronegative than carbon. This polarization of the π-bond electrons is often expressed by using the ionic structure shown on the right-hand side in the margin. However, the π-bond electrons are not completely on oxygen; there are not full 1+ and 1− charges on carbon and oxygen, respectively. Neither the covalent structure nor the ionic structure accurately represents a carbonyl group. The π electrons are neither equally shared by carbon and oxygen nor completely on oxygen. The situation is in between those two extremes. The carbonyl group is a hybrid (blend) of the covalent and ionic structures, and we represent the situation by drawing the two structures separated by a two-headed arrow, as shown in the margin.

$$\underset{\text{Covalent}}{-\overset{\ddot{\text{O}}:}{\overset{\|}{\text{C}}}-} \longleftrightarrow \underset{\text{Ionic}}{-\overset{:\ddot{\text{O}}:^-}{\overset{|}{\text{C}^+}}-}$$

The carbonyl group is said to be a **resonance hybrid** of the two structures, but most of the time only one or the other of the two structural drawings is shown. Irrespective of which structure is drawn, the carbonyl group has partial charges, δ^+ on carbon and δ^- on oxygen, that account for the polar nature of the carbonyl group. An alternate representation of the carbonyl group is often used instead of having the covalent and ionic structures separated by the double-headed arrow:

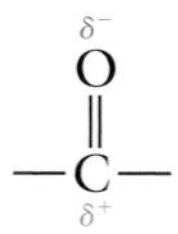

$$-\overset{\overset{\delta^-}{\text{O}}}{\overset{\|}{\underset{\delta^+}{\text{C}}}}-$$

INSIGHT INTO PROPERTIES

The polar carbonyl group is responsible for the strong secondary attractive forces in aldehydes and ketones. There are strong attractive forces between the partly positive carbon end of the C=O dipole of one molecule and the partly negative oxygen end of the C=O dipole of a neighboring molecule. These attractive forces are depicted for formaldehyde by the dashed lines in Figure 6.2.

Aldehydes and ketones have considerably higher boiling and melting points than those of the corresponding hydrocarbons because the dipole–dipole secondary forces in aldehydes and ketones are much stronger than the London forces in hydrocarbons. Higher temperatures are needed to overcome the attractive forces in aldehydes and ketones compared with those in hydrocarbons. For example, Table 6.4 shows the boiling points of butanal, 2-butanone, and pentane—all compounds having the same molecular mass. The boiling points of the aldehyde and ketone are considerably higher than the boiling point of the alkane (76°C and 80°C versus 36°C). However, aldehydes and ketones have boiling points considerably lower than those of alcohols of similar molecular mass. For example, the boiling point of 1-butanol is 117°C versus 76°C for butanal. These differences are also evident in Figure 6.3 and result from the differences in secondary attractive forces. Hydrogen-bonding attractive forces (alcohols) are stronger than dipole–dipole attractive forces (aldehydes, ketones), which are stronger than London forces (hydrocarbons), as illustrated in Figure 6.3 on the following page. The differences in boiling points between alkanes, aldehydes and ketones, and alcohols narrow with increasing molecular mass (Figure 6.4 on the following page). The effect of a C=O group in an aldehyde or ketone or an —OH group in an alcohol on the strength of the secondary attractive forces decreases as the fraction of the molecule that the C=O or —OH group constitutes becomes smaller.

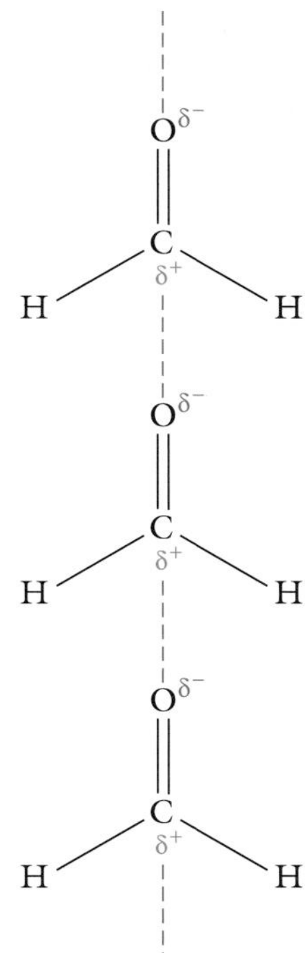

Figure 6.2 Dipole-dipole attractive forces in formaldehyde, which has a higher boiling point than that of ethane.

TABLE 6.4 Boiling Points of Aldehydes and Ketones

Compound	Structure	Molecular mass (amu)	Boiling point (°C)
ALDEHYDES AND KETONES			
formaldehyde	HCHO	30	−21
acetaldehyde	CH_3CHO	44	21
propanal	CH_3CH_2CHO	58	49
acetone	CH_3COCH_3	58	56
butanal	$CH_3CH_2CH_2CHO$	72	76
butanone	$CH_3COCH_2CH_3$	72	80
pentanal	$CH_3CH_2CH_2CH_2CHO$	86	103
2-pentanone	$CH_3COCH_2CH_2CH_3$	86	102
3-pentanone	$CH_3CH_2COCH_2CH_3$	86	102
ALKANE			
pentane	$CH_3CH_2CH_2CH_2CH_3$	72	36
ETHER			
diethyl ether	$CH_3CH_2OCH_2CH_3$	74	35
ALCOHOL			
1-butanol	$CH_3CH_2CH_2CH_2OH$	74	117

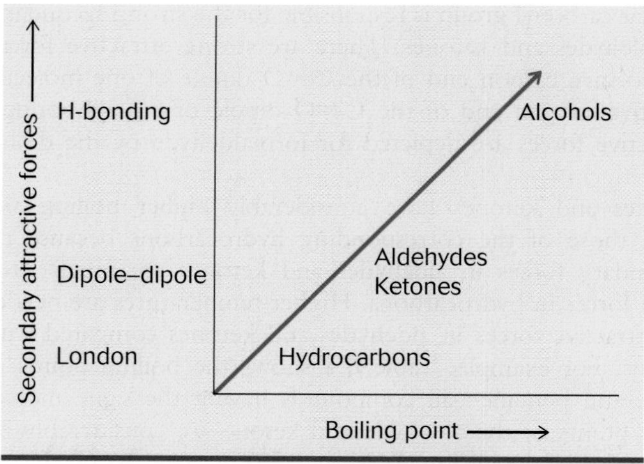

Figure 6.3 Relation between boiling point and strength of secondary attractive forces in hydrocarbons, aldehydes and ketones, and alcohols.

INSIGHT INTO PROPERTIES

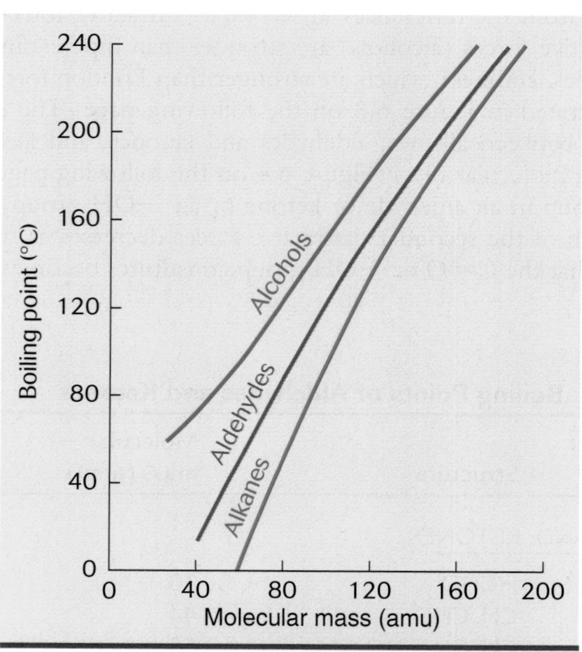

Figure 6.4 Boiling-point dependence on molecular mass for linear alkanes, aldehydes, and alcohols.

Although aldehydes and ketones have boiling points considerably lower than those of alcohols, they have almost the same water solubility as that of alcohols. The reason for this similarity is that aldehydes and ketones are hydrogen-bond acceptors and, therefore, they hydrogen bond with water even though they do not hydrogen bond with themselves.

There is little difference in physical properties between aldehydes and ketones. They have nearly the same boiling and melting points and water solubilities.

Example 6.5 Recognizing hydrogen bonding in aldehydes and ketones

Which of the following representations correctly describes the hydrogen bonding that takes place between acetaldehyde (ethanal) and water?

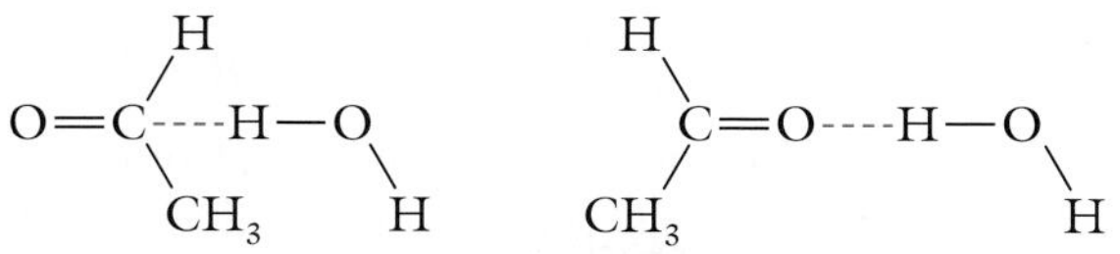

```
      H                        H
     /                          \
O=C----H—O                       C=O----H—O
    \      \                    /          \
    CH3     H               CH3             H
      1                            2

                    O                            O
                   //                           //
 O—H----CH3—C                 H—O----CH3—C
/                    \                \              \
H                    H                H              H
        3                               4
```

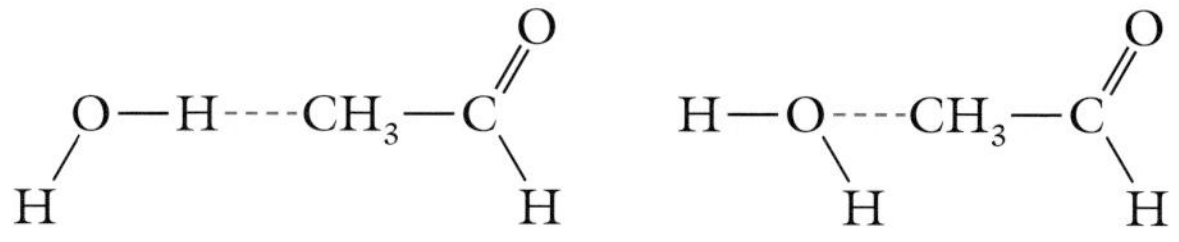

Solution

Only representation 2 describes hydrogen bonding. Hydrogen bonding here is the attraction between the H of an O—H bond of water and the O of the carbonyl group of the ketone. Hydrogen bonding does not take place between H and C, as described in representations 1 and 3, or between O and C, as described in representation 4.

Problem 6.5 Indicate which compound in each of the following pairs has the higher value of the specified property. If the two compounds in a pair have nearly the same value for the property, indicate that fact. Explain your conclusions.
(a) Butanal, hexanal: boiling water.
(b) Propanal, pentanal: solubility in water.
(c) Propanal, 2-propanol: boiling point.
(d) Butanal, 1-butanol: solubility in water.

6.4 THE OXIDATION OF ALDEHYDES AND KETONES

Aldehydes and ketones, like all organic compounds, undergo combustion to carbon dioxide and water. Selective (mild) oxidations take place with oxidizing agents such as permanganate ion (MnO_4^-) and dichromate ion ($Cr_2O_7^{2-}$):

- Aldehydes are oxidized to carboxylic acids.

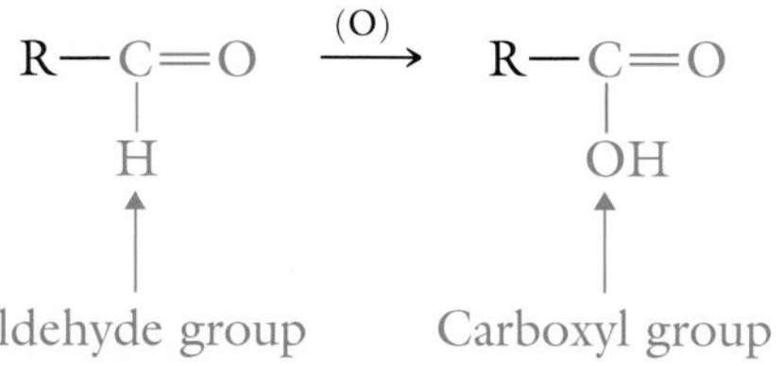

```
R—C=O   --(O)-->   R—C=O
  |                  |
  H                  OH
  ↑                  ↑
Aldehyde group     Carboxyl group
Aldehyde           Carboxylic acid
```

>> Carboxylic acids are a topic of Chapter 7.

- Ketones are not oxidized.

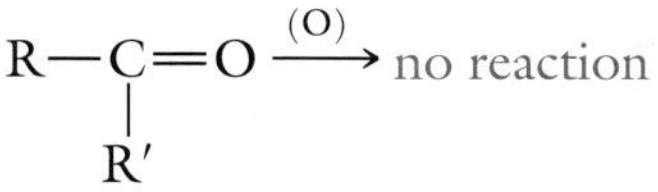

```
R—C=O  --(O)-->  no reaction
  |
  R'
Ketone
```

We encountered these same selective oxidations in our consideration of alcohols (Section 5.7). Aldehydes and ketones are the oxidation products of primary and secondary alcohols, respectively. Whereas aldehydes are oxidized further to carboxylic acids, ketones are inert to further oxidation under mild conditions.

The selective oxidation of aldehydes is useful for the industrial synthesis of carboxylic acids.

The oxidation of aldehydes to carboxylic acids also takes place slowly on exposure to air. Ketones are not affected.

Example 6.6 Recognizing the products of selective oxidation

Which of the following compounds undergo mild oxidation with dichromate or permanganate? If oxidation takes place, show the product. (a) Propanal; (b) acetone (propanone).

Solution

(a) Propanal, an aldehyde, is oxidized to a carboxylic acid.

$$CH_3CH_2-\overset{\overset{\displaystyle O}{\|}}{C}-H \xrightarrow{MnO_4^- \text{ or } Cr_2O_7^{2-}} CH_3CH_2-\overset{\overset{\displaystyle O}{\|}}{C}-OH$$

(b) Oxidation does not take place, because the compound is a ketone.

Problem 6.6 Which of the following compounds undergo mild oxidation with dichromate or permanganate? If oxidation takes place, show the product. (a) Cyclopentanone; (b) pentanal.

Dichromate and permanganate oxidations are useful simple chemical diagnostic tests for distinguishing aldehydes from ketones. Positive tests for an aldehyde are easily observed as color changes (Section 5.7). However, permanganate and dichromate oxidations cannot distinguish aldehydes from primary and secondary alcohols and alkenes, which also undergo oxidation under these conditions (Sections 4.8 and 5.7). Fortunately, there are other selective oxidizing reagents that are more selective and distinguish aldehydes not only from ketones, but also from primary and secondary alcohols and alkenes. Tollens's reagent is one such reagent. It oxidizes aldehydes but not the other compounds. **Tollens's reagent** is a colorless, alkaline solution of silver ion in ammonia, in which the silver ion is stabilized as its polyatomic ion, $[Ag(NH_3)_2]^+$. A positive test for an aldehyde is the formation of a shiny gray precipitate of silver metal, formed by reduction of silver ion. The silver usually coats the inside of the test tube and gives the appearance of a silver mirror.

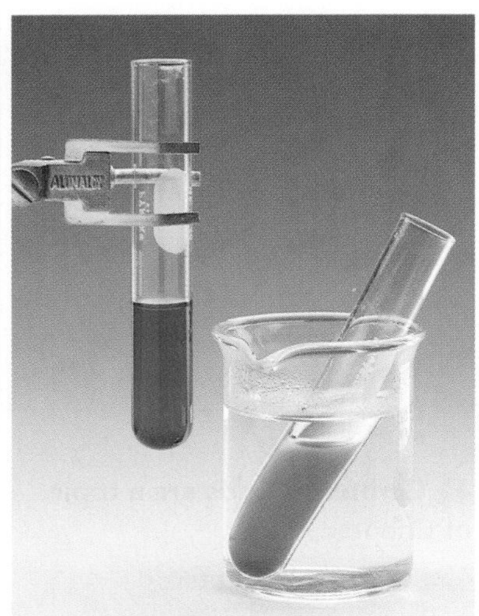

Figure 6.5 Benedict's reagent (blue Cu^{2+}) is used in a simple diagnostic chemical test for distinguishing α-hydroxy aldehydes and α-hydroxy ketones from simple aldehydes and ketones. In the presence of an α-hydroxy aldehyde or ketone (*right test tube*), the blue color disappears and a red precipitate (Cu_2O) forms. In the presence of a simple aldehyde or ketone (*left test tube*), no reaction takes place and the blue color of the reagent is unchanged. (Richard Megna/Fundamental Photographs.)

Benedict's reagent is another useful oxidizing agent, but its oxidizing action differs from that of Tollens's reagent. Benedict's reagent is a blue, alkaline solution of copper (II) ion (Cu^{2+}) stabilized as its polyatomic ion with citrate anion (citrate is the anion of citric acid). The blue color of copper (II) ion disappears and a red precipitate of Cu_2O forms in a positive test as copper (II) ion is reduced to copper (I) ion (Cu^+). These color changes can be seen in Figure 6.5. A positive test is one in which a substantial amount of red Cu_2O precipitate is rapidly formed. Simple aldehydes such as acetaldehyde and benzaldehyde, although oxidized by Tollens's reagent, do not give positive tests with Benedict's reagent. Some simple aldehydes give weak positive tests, with the red Cu_2O precipitate forming slowly or in only small amounts. Such results are considered to be negative tests, not positive tests. Only **α-hydroxy aldehydes** give positive tests with Benedict's reagent. The α notation refers to the carbon adjacent to the carbonyl carbon; so an α-hydroxyl group is an —OH group on an α-carbon:

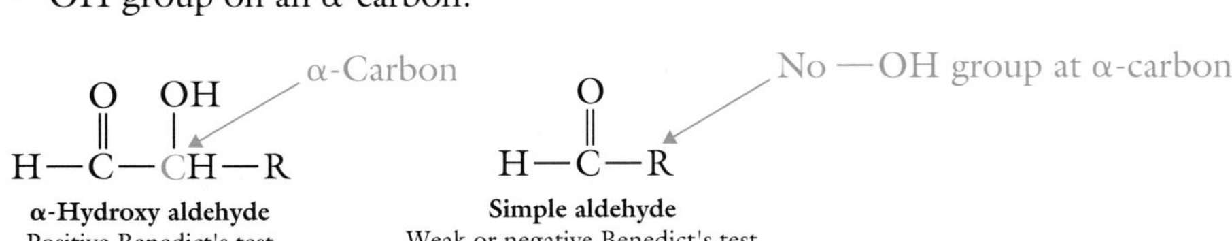

One group of ketones, **α-hydroxy ketones,** also gives a positive Benedict's test. α-Hydroxy ketones are ketones with an —OH group on a carbon atom adjacent to the carbonyl carbon:

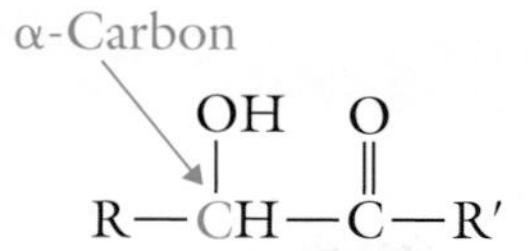

α-Carbon

$$R-\overset{\overset{\displaystyle OH}{|}}{C}H-\overset{\overset{\displaystyle O}{\|}}{C}-R'$$

α-Hydroxy ketone
Positive Benedict's test

Although the use of Benedict's reagent is limited to α-hydroxy aldehydes and α-hydroxy ketones, it is an important reagent in both chemical and clinical laboratory settings because it detects glucose and certain other carbohydrates (sugars). Glucose is not a simple aldehyde but a multifunctional molecule containing five hydroxyl groups as well as an α-hydroxy aldehyde group (see Box 6.1). Fructose (see Box 6.1), also an important carbohydrate, contains an α-hydroxy ketone group. Thus, glucose, fructose, and a number of other carbohydrates can be distinguished from simple aldehydes and ketones, alcohols, and other compounds by Benedict's test.

›› Carbohydrates are important biological compounds and the subject of Chapter 10.

Example 6.7 Recognizing compounds that react with Tollens's and Benedict's reagents

For each of the following compounds, indicate whether the compound gives a positive or negative test with (a) Tollens's reagent and (b) Benedict's reagent.

$$\underset{1}{H-\overset{\overset{\displaystyle O}{\|}}{C}-CH_2CH_2CH_2CH_3} \qquad \underset{2}{H-\overset{\overset{\displaystyle O}{\|}}{C}-\overset{\overset{\displaystyle OH}{|}}{C}H-CH_3}$$

$$\underset{3}{CH_3-\overset{\overset{\displaystyle O}{\|}}{C}-\overset{\overset{\displaystyle OH}{|}}{C}H-CH_3} \qquad \underset{4}{CH_3-\overset{\overset{\displaystyle O}{\|}}{C}-CH_2CH_3}$$

Solution

(a) Tollens's reagent gives positive tests with all aldehydes, but not with any ketones. Compounds 1 and 2 are aldehydes and give positive tests with Tollens's reagent. Compounds 3 and 4 are ketones and give negative tests.
(b) Only α-hydroxy aldehydes and α-hydroxy ketones give positive tests with Benedict's reagent. Compounds 2, an α-hydroxy aldehyde, and 3, an α-hydroxy ketone, give positive tests with Benedict's reagent. Compound 1, a simple aldehyde, gives a negative test. Compound 4, a simple ketone, gives a negative test.

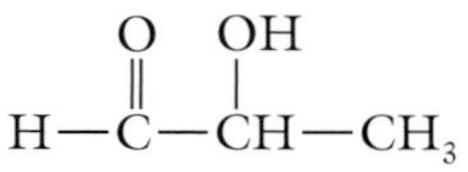

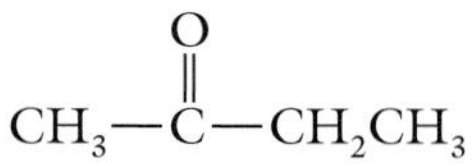

Problem 6.7 For each of the following compounds, indicate whether the compound gives a positive or negative test with (a) Tollens's reagent and (b) Benedict's reagent.

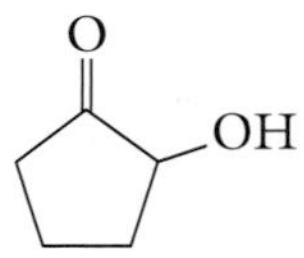

1 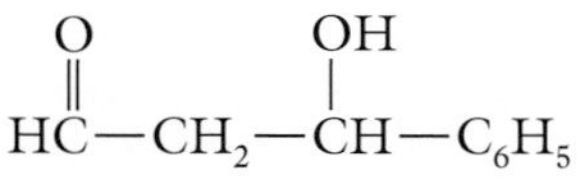 (2-hydroxycyclopentanone: cyclopentane ring with C=O and adjacent OH)

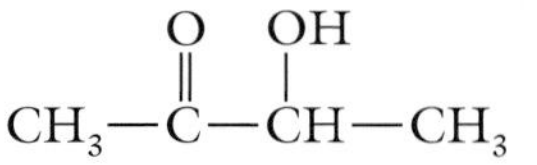

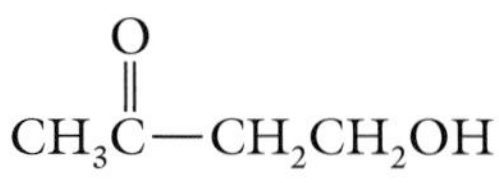

$$\underset{2}{H\overset{\overset{\displaystyle O}{\|}}{C}-CH_2-\overset{\overset{\displaystyle OH}{|}}{C}H-C_6H_5}$$

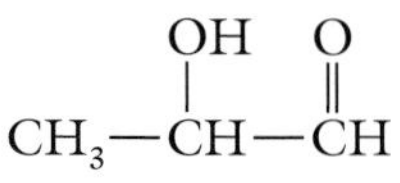

$$\underset{3}{CH_3\overset{\overset{\displaystyle O}{\|}}{C}-CH_2CH_2OH} \qquad \underset{4}{CH_3-\overset{\overset{\displaystyle OH}{|}}{C}H-\overset{\overset{\displaystyle O}{\|}}{C}H}$$

We have now considered the use of several simple chemical tests for analytical purposes—the bromine test for alkenes (Section 4.6) and various oxidation tests for alkenes, alcohols, and aldehydes (Sections 4.8, 5.7, and 6.4). Other much more powerful analytical tools are available for elucidating the structures of organic and biochemical molecules (Box 6.3 on the following page and Figures 6.6 and 6.7 on page 200).

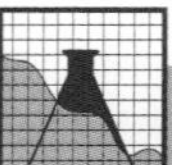

6.3 CHEMISTRY IN DEPTH

Absorption Spectroscopy: IR and NMR

Absorption spectroscopy entails the interaction of a compound with electromagnetic radiation and is useful for analyzing the chemical structure of unknown compounds. A molecule undergoes excitation by the absorption of energy, which raises the energy level of the molecule from a lower, ground state to a higher, excited state. A sample is subjected to incident radiation of different frequencies, and one detects the absorption of energy by comparing the transmitted frequencies with the incident frequencies. Missing frequencies correspond to frequencies that were absorbed by the sample. Each specific frequency (corresponding to a specific energy of radiation) that causes a molecular excitation yields information about the chemical structure of the compound.

Incident frequencies ⟶ Sample ⟶ Transmitted frequencies

Different types of spectroscopy, corresponding to different types of molecular excitations, are available for studying chemical structure.

Infrared (IR) Spectroscopy

Energies in the infrared region cause a variety of bond excitations, such as the stretching of bonds. The presence of specific bonds in a compound is detected by observing which frequencies of infrared radiation are absorbed by the compound. Different bonds require different energies for excitation because different bonds have different bond strengths. Thus, IR spectroscopy is useful for detecting the family of an unknown sample by detecting the characteristic frequencies that cause excitation of the characteristic bond(s) of the family (Table 6.5).

Figure 6.6 shows the IR spectra of 1-pentanol and 2-pentanone. Energy transmission is plotted against wave number (cm^{-1}), which is related to frequency. Energy transmission is the inverse of energy absorption. A downward-pointing peak corresponds to energy absorption. The IR spectra show many peaks because each compound has many bonds and each bond usually undergoes more than one kind of excitation. We search each spectrum for the peak(s) characteristic of the functional groups of different families:

- The O—H bond of an alcohol shows a peak in the 3200 to 3650 cm^{-1} region.
- The C=O bond of an aldehyde or ketone shows a peak in the 1690 to 1750 cm^{-1} region.

Alcohols are easily distinguished from aldehydes and ketones by looking for these characteristic excitation peaks

TABLE 6.5 Characteristic Excitation Frequencies of Different Bonds

Bond	Excitation frequency (cm^{-1})*
C—H	
alkane	2850–3000
alkene	3050–3150
alkyne	3250–3350
aromatic	3000–3050
C=C	
alkene	1620–1680
aromatic	1450–1600
O—H	
alcohol	3200–3650
phenol	3400–3600
carboxylic acid	2500–3300
C=O	
aldehyde, ketone	1690–1750
carboxylic acid	1710–1760
ester	1735–1750
amide	1630–1690
N—H	
amine, amide	3300–3500

*The frequency range for a bond is for compounds with the indicated bond. Different compounds with the same bond may have somewhat different excitation frequencies but in the indicated range for that bond.

6.5 THE REDUCTION OF ALDEHYDES AND KETONES

Aldehydes and ketones can be reduced to alcohols—aldehydes to primary alcohols, and ketones to secondary alcohols. The overall process consists of the addition of one hydrogen each to the carbon and oxygen of the carbonyl group. The reduction of aldehydes and ketones to alcohols is the reverse of the oxidation of alcohols to aldehydes and ketones described in Section 5.7.

in the IR spectrum. Thus, the IR spectrum of 1-pentanol (Figure 6.6) shows a peak for O—H at about 3350 cm^{-1}, but it shows no peak for C=O in the region from 1690 to 1750 cm^{-1}. The IR spectrum of 2-pentanone shows the C=O peak at about 1700 cm^{-1} but no peak for O—H in the region from 3200 to 3650 cm^{-1}.

Infrared spectroscopy is especially useful for detecting the family of an unknown sample. It is less easily used to differentiate between different members of the same family, although this differentiation can be done by an experienced infrared spectroscopist. Infrared spectroscopy is the basis of some instruments used by law enforcement to measure the ethanol content in the breath of persons suspected of drunk driving (Section 5.7).

Nuclear Magnetic Resonance (NMR) Spectroscopy

NMR spectroscopy is the most powerful of the different spectroscopic methods. Not only can it detect the family of an unknown sample, but it usually differentiates between different members of the same family. In NMR spectroscopy, a compound is placed in the magnetic field of a NMR spectrometer and then subjected to different frequencies of the radiofrequency region of electromagnetic radiation. The nuclei of certain atoms of a compound behave like tiny magnets—a property called **nuclear spin.** In the NMR spectrometer, there are two possible nuclear spin states—alignment of nuclear spin with or against the magnetic field of the spectrometer. Alignment with the magnetic field is lower in energy than alignment against the magnetic field. Radiofrequency energy causes the excitation of nuclear spins from alignment with the magnetic field to alignment against the magnetic field.

Proton NMR entails excitation of the nuclear spins of a compound's hydrogen atoms. Carbon-13 NMR entails excitation of the nuclear spins of a compound's C-13 carbons. (The carbon atoms in compounds consist of a mixture of C-12 and C-13 isotopes; only C-13 nuclei undergo the NMR excitation.) We will illustrate the great utility of NMR spectroscopy by showing how easily 2-pentanone and 3-pentanone can be distinguished from each other by C-13 NMR. Carbon-13 NMR spectroscopy has two important features:

- Each nonequivalent carbon in a compound produces a different peak.
- The frequency of a peak depends on both the type of carbon (for example, CH_2, CH_3, C=O) and the atoms or groups attached to the carbon.

Figure 6.7 shows the C-13 NMR spectra of the two ketones. Energy absorption is plotted against chemical shift (ppm), which is related to frequency. Absorption peaks point upward in NMR spectra instead of downward as in IR spectra. The C=O carbon peak is seen near 210 ppm. The trio of peaks near 80 ppm is due to the solvent used in the NMR experiment. 2-Pentanone has four additional peaks corresponding to carbons 1, 3, 4, and 5. 3-Pentanone has only two additional peaks because carbons 1 and 5 are equivalent and carbons 2 and 4 are equivalent. 2-Pentanone has a total of five peaks, whereas 3-pentanone has a total of three peaks. The two compounds are easily distinguished from each other.

This discussion only hints at the power of NMR spectroscopy. The method is enormously useful because information other than the number of different peaks is obtained. For both proton and C-13 NMR spectroscopy, the actual chemical shift for the same kind of atom differs, depending on the chemical environment near that atom. For example, in C-13 NMR, the peak for a CH_2 carbon is in the 20 to 30 ppm range when the groups attached to that carbon are alkane groups, but the signal is shifted to the 50 to 90 ppm range when a CH_2 carbon has an adjacent oxygen attached as in an alcohol or ether. NMR spectroscopy is also powerful because we can do both proton and C-13 NMR of the same compound to yield information on the compound's carbon and hydrogen structures. Furthermore, elements other than carbon and hydrogen can be examined, such as phosphorus and nitrogen and even some metals.

The NMR method has been developed as magnetic resonance imaging (MRI) for use in medical diagnosis. Magnetic resonance imaging produces images of tissues and organs by determining the different concentrations of hydrogens distributed throughout the tissues and organs. The hydrogens being detected are usually those of water in the tissues and organs, but hydrogens of carbohydrates, fats, and proteins also can be detected.

$$-\overset{\overset{\displaystyle O}{\|}}{C}- \underset{(O)}{\overset{(H)}{\rightleftharpoons}} -\underset{\underset{\displaystyle H}{|}}{\overset{\overset{\displaystyle OH}{|}}{C}}-$$

Recall that the abbreviations (O) and (H) indicate general conditions for selective oxidation and reduction, respectively, without specification of the exact oxidizing or reducing agent.

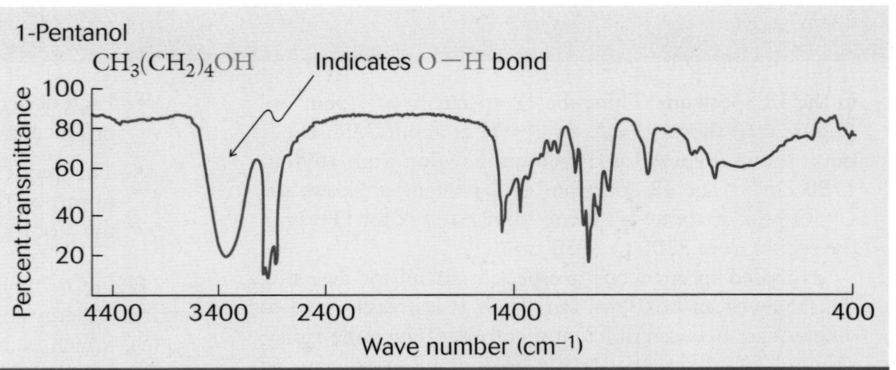

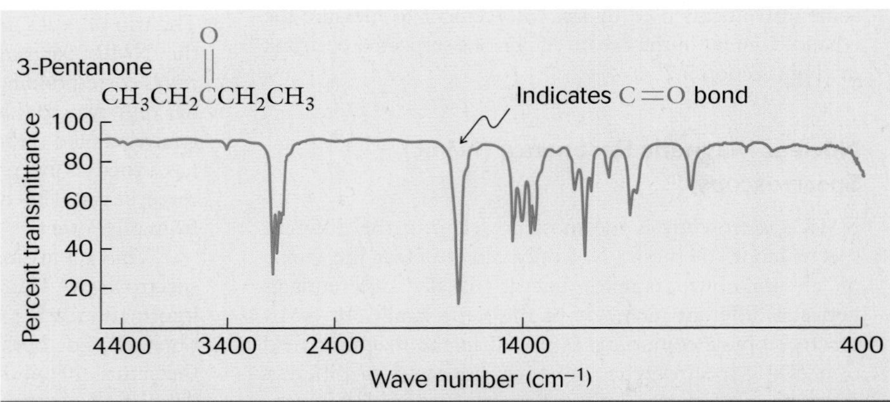

Figure 6.6 IR spectra of 1-pentanol and 3-pentanone.

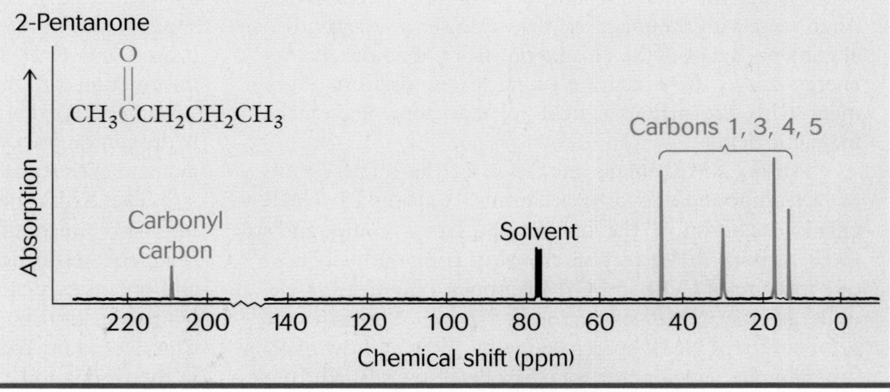

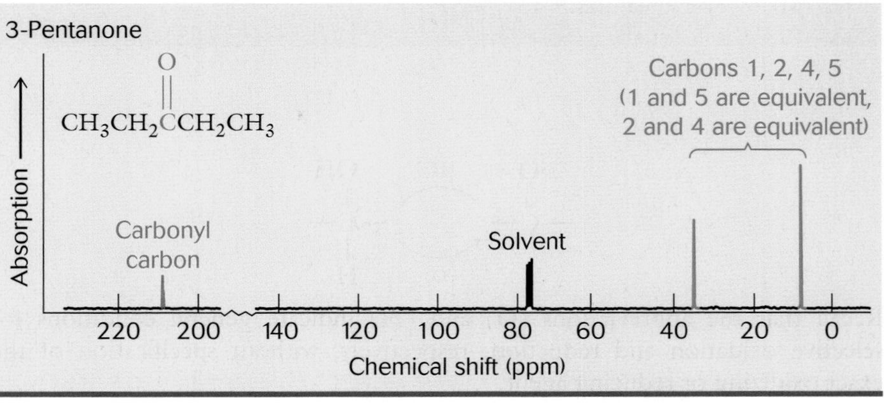

Figure 6.7 NMR spectra of 2-pentanone and 3-pentanone. The set of peaks at about 75 ppm is not for the compounds; it is for the solvent used to dissolve the sample before placement in the NMR instrument.

Two methods of reduction are useful for aldehydes and ketones:

- **Catalytic hydrogenation (catalytic reduction)** uses molecular hydrogen (H_2) as the reducing agent in the presence of a metal catalyst such as nickel or platinum:

$$-\overset{\overset{\displaystyle O}{\|}}{C}- \xrightarrow[Pt]{H_2} -\overset{\overset{\displaystyle OH}{|}}{\underset{\underset{\displaystyle H}{|}}{C}}-$$

  The π bond of the double bond breaks, which allows the addition of hydrogens to the carbon and oxygen of the carbonyl group. The reaction is similar to the catalytic hydrogenation of the carbon–carbon double bond (Section 4.6).

- **Hydride reduction** uses a metal hydride as the reducing agent—for example, lithium aluminum hydride ($LiAlH_4$). The aldehyde or ketone first undergoes reaction with the metal hydride followed by reaction with water:

$$-\overset{\overset{\displaystyle O}{\|}}{C}- \xrightarrow[2.\ H_2O]{1.\ LiAlH_4} -\overset{\overset{\displaystyle OH}{|}}{\underset{\underset{\displaystyle H}{|}}{C}}- \qquad (6.1)$$

To understand this reaction, we need to consider the nature of metal hydrides. Compounds containing H bonded to a more electronegative element (for example, Cl in HCl and O in H_2O or ROH) are sources of H^+ because H gives up the shared pair of electrons to the more electronegative element (Cl or O). Metal hydrides are different, being sources of hydride ion ($H{:}^-$, or simply H^-) because H is more electronegative than a metal.

$$H \rightarrow Cl \longrightarrow H^+ + Cl^- \qquad H \leftarrow \text{metal} \longrightarrow H^- + \text{metal}^+$$

Cl more electronegative than H **H more electronegative than metal**

Reduction of the carbonyl group by metal hydride proceeds because of the partly ionic character of the carbonyl group. In the first step, the $H{:}^-$ from the metal hydride bonds to the C^+ of the carbonyl group. In the second step, H^+ from H_2O bonds to the O^- of the carbonyl group:

$$-\overset{\overset{\displaystyle O^-}{|}}{\underset{+}{C}}- \xrightarrow[(H{:}^-)]{LiAlH_4} -\overset{\overset{\displaystyle O^-}{|}}{\underset{\underset{\displaystyle H}{|}}{C}}- \xrightarrow[(H^+)]{H_2O} -\overset{\overset{\displaystyle OH}{|}}{\underset{\underset{\displaystyle H}{|}}{C}}-$$

The reagents must be added sequentially, metal hydride followed by water, rather than simultaneously, to achieve reduction of the carbonyl group. In chemical equations, sequential addition is indicated by numbers in front of the reagents, as in Equation 6.1. Water must be added only after the aldehyde or ketone and metal hydride have been allowed an appropriate period of time to react with each other. The simultaneous addition of metal hydride and water results in reaction between hydride ion and water:

$$H{:}^- + H_2O \longrightarrow H_2 + HO^-$$

Reduction of the carbonyl group does not take place, because water is more reactive toward hydride than is the carbonyl group.

Hydride reduction is more selective than catalytic hydrogenation. Catalytic hydrogenation reduces both carbon–carbon and carbon–oxygen double bonds. Hydride reduction reduces only carbon–oxygen double bonds.

Hydride reductions of carbonyl groups are important in living cells. In the cell, however, these reactions do not use metal hydrides, because metal hydrides are not stable in aqueous environments. Metal hydrides react directly

›› NADH is an important biological reducing agent that will be encountered again in Chapters 14 through 18.

with water; hydride ion would not survive to reduce a carbonyl group. Instead, cells use a complex organic compound called **nicotinamide adenine dinucleotide (NADH),** to carry out hydride reductions. NADH is stable in aqueous environments. Figure 6.8 shows the structure of NADH and the location of its hydride ion. The structure of NADH is quite complicated. It is drawn in detail here to give you some idea of the complexity of biological molecules. We will refer to it simply as NADH.

An example of a hydride reduction in living cells is the reduction of pyruvic acid to lactic acid during strenuous exercise, such as running or bicycling. The reduction is a two-step process and is catalyzed by the enzyme alcohol dehydrogenase. It consists of the additions of H:$^-$ and H^+ to the carbonyl group. Supplied by NADH, H:$^-$ adds to the carbonyl carbon; this step is followed by the addition of H^+ from water to the carbonyl oxygen:

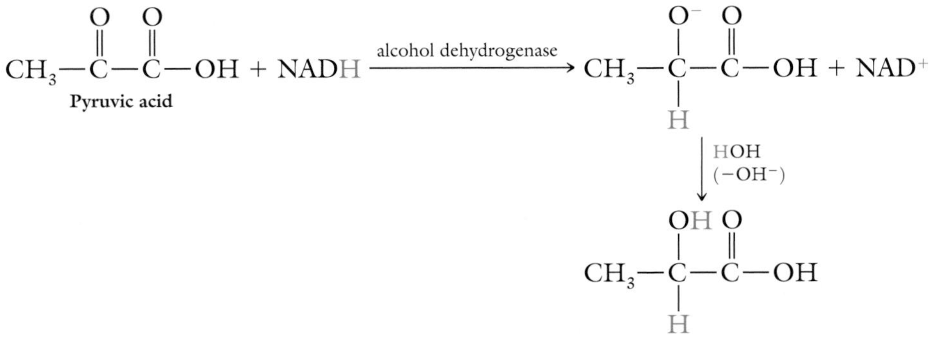

$$CH_3-\overset{O}{\overset{\|}{C}}-\overset{O}{\overset{\|}{C}}-OH + NADH \xrightarrow{\text{alcohol dehydrogenase}} CH_3-\underset{H}{\underset{|}{\overset{O^-}{\overset{|}{C}}}}-\overset{O}{\overset{\|}{C}}-OH + NAD^+$$

Pyruvic acid

$$\downarrow \text{HOH } (-OH^-)$$

$$CH_3-\underset{H}{\underset{|}{\overset{OH}{\overset{|}{C}}}}-\overset{O}{\overset{\|}{C}}-OH$$

Lactic acid

The oxidized form of NADH, after it has donated its hydride ion to another compound, is called NAD^+ and is used elsewhere in the organism as an oxidizing agent—for example, to oxidize ethanol to acetaldehyde. The role of NADH in this biological reduction and others is discussed further in Chapters 14 through 18.

Source of hydride ion

NADH $\longrightarrow$ NAD^+ + H:$^-$ (Hydride ion)

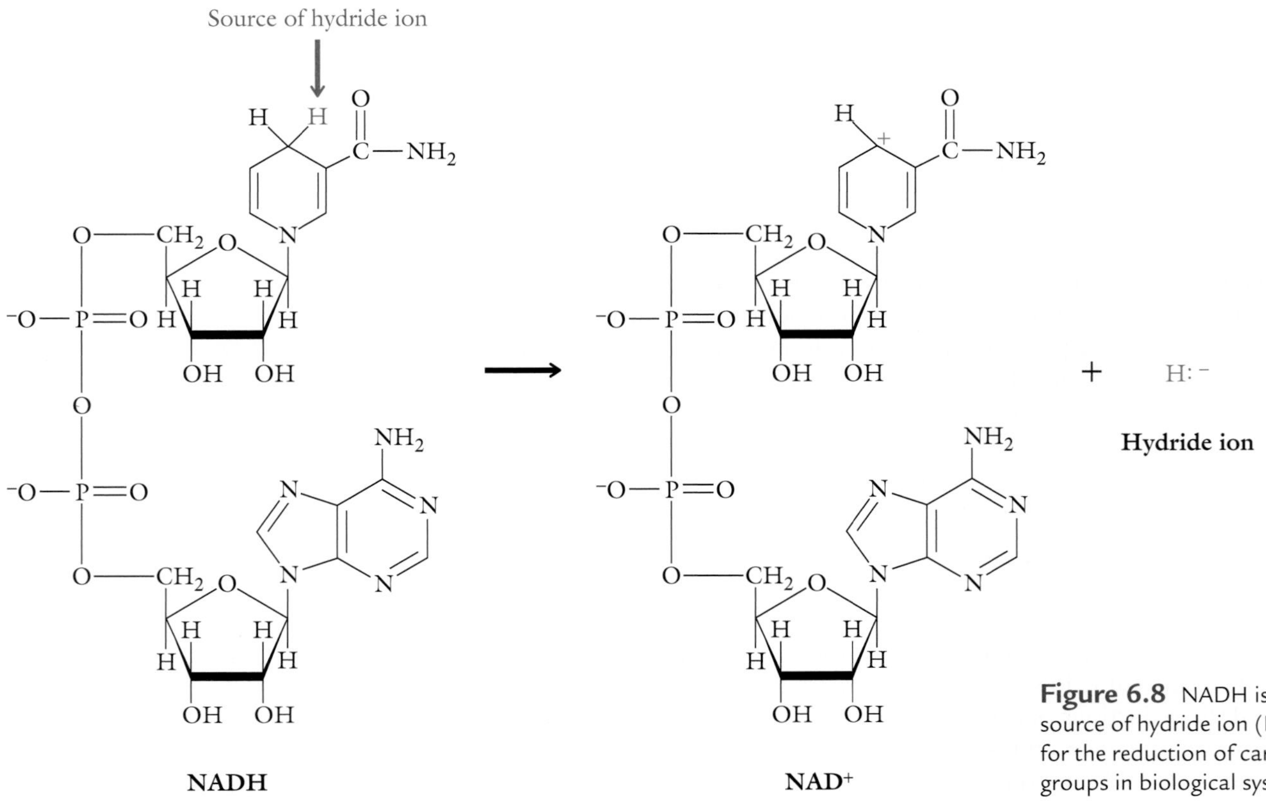

Figure 6.8 NADH is a source of hydride ion (H:$^-$) for the reduction of carbonyl groups in biological systems.

A PICTURE OF HEALTH
Examples of Aldehydes and Ketones

Our tongues recognize various aldehydes and ketones as the flavors of almond, cinnamon, mint, and vanilla among others.
cis-11-retinal, needed for the perception of vision, contains a carbonyl group.
The ketone acetone is found in the breath of diabetics who are not receiving proper treatment.
The carbohydrate glucose, an aldehyde, is a source of energy in all cells.
Ethanol is oxidized to acetaldehyde in the liver.
The female and male hormones progesterone and testosterone, produced in the ovaries and testes, respectively, are ketones.

(Rubberball/Fotosearch.)

Example 6.8 Writing equations for the reduction of aldehydes and ketones

Complete each of the following reactions by writing the structure of the organic product. If no reaction takes place, write "no reaction." If there is more than one product, show only the major product(s).

(a) $CH_3CH_2CH_2\overset{\overset{\Large O}{\|}}{C}H \xrightarrow[2.\ H_2O]{1.\ LiAlH_4} ?$

(b) $CH_3CH_2CH_2\overset{\overset{\Large O}{\|}}{C}H \xrightarrow[H_2O]{LiAlH_4} ?$

(c) $CH_3\overset{\overset{\Large O}{\|}}{C}CH_2CH_3 \xrightarrow[Ni]{H_2} ?$

(d) $CH_3\overset{\overset{\Large O}{\|}}{C}CH_2CH_3 \xrightarrow{H_2} ?$

(e) $CH_3CH_2CH_2\overset{\overset{\Large O}{\|}}{C}H \xrightarrow[2.\ H_2O]{1.\ NADH,\ \text{alcohol dehydrogenase}} ?$

Solution

(a) Sequential addition of metal hydride and water reduces the carbonyl group:

$$CH_3CH_2CH_2\overset{\overset{\displaystyle O}{\|}}{C}H \xrightarrow[2.\ H_2O]{1.\ LiAlH_4} CH_3CH_2CH_2\underset{\underset{\displaystyle H}{|}}{\overset{\overset{\displaystyle OH}{|}}{C}}H$$

(b) The aldehyde is not reduced, because $LiAlH_4$ and H_2O are not added sequentially. $LiAlH_4$ reacts with H_2O instead of with the carbonyl group.

(c) Catalytic hydrogenation reduces the ketone to a secondary alcohol:

$$CH_3\overset{\overset{\displaystyle O}{\|}}{C}CH_2CH_3 \xrightarrow[Ni]{H_2} CH_3\underset{\underset{\displaystyle H}{|}}{\overset{\overset{\displaystyle OH}{|}}{C}}CH_2CH_3$$

(d) Reduction does not take place, because the metal catalyst (Ni or Pt) is absent.

(e) Sequential reaction with NADH and water reduces the carbonyl group, exactly as in the metal hydride reduction with $LiAlH_4$:

$$CH_3CH_2CH_2\overset{\overset{\displaystyle O}{\|}}{C}H \xrightarrow[2.\ H_2O]{1.\ NADH,\ \text{alcohol dehydrogenase}} CH_3CH_2CH_2\underset{\underset{\displaystyle H}{|}}{\overset{\overset{\displaystyle OH}{|}}{C}}H$$

Problem 6.8 Write an equation for each of the following reactions: (a) reduction of cyclohexanone by using $LiAlH_4$; (b) catalytic reduction of benzaldehyde; (c) reduction of 2-pentanone with NADH.

6.6 HEMIACETAL AND ACETAL FORMATION BY REACTION WITH ALCOHOL

Aldehydes and ketones react with alcohols to form **hemiacetals** and **acetals.** Hemiacetals and acetals are of interest because they are important structures in carbohydrates.

The reaction of aldehydes and ketones with alcohols requires an acid catalyst and takes place in a two-step process. First, the alcohol adds to the carbonyl group to form a hemiacetal. Second, if excess alcohol is present, the hemiacetal reacts with a second molecule of alcohol to form an acetal. This second step is a dehydration reaction similar to the intermolecular dehydration of alcohols to form ethers (Section 5.10). The reaction is illustrated for propanal and methanol:

$$CH_3CH_2-\overset{\overset{\displaystyle O}{\|}}{C}-H \underset{H^+\ (-\ CH_3OH)}{\overset{CH_3OH,\ H^+}{\rightleftharpoons}} \underset{\text{Hemiacetal}}{CH_3CH_2-\underset{\underset{\displaystyle OCH_3}{|}}{\overset{\overset{\displaystyle OH}{|}}{C}}-H} \underset{H^+,\ H_2O\ (-\ CH_3OH)}{\overset{CH_3OH,\ H^+\ (-\ H_2O)}{\rightleftharpoons}} \underset{\text{Acetal}}{CH_3CH_2-\underset{\underset{\displaystyle OCH_3}{|}}{\overset{\overset{\displaystyle OCH_3}{|}}{C}}-H}$$

The reaction proceeds in a similar manner for ketones. The products formed from ketones were formerly called **hemiketals** and **ketals** to distinguish them from the products formed from aldehydes. However, the most recent conven-tion uses the names hemiacetals and acetals for the products derived from both aldehydes and ketones.

Most hemiacetals are unstable and cannot be isolated; that is, the equilibrium in the first reaction is far to the left. However, the equilibrium in the second reaction is far to the right, and the acetal product is isolated if excess alcohol is present. This process is an example of Le Chatelier's principle (Section 2.3): the favorable equilibrium in the second reaction shifts the first reaction, which has an unfavorable equilibrium, to the right.

Example 6.9 Drawing hemiacetal and acetal structures

Show the structures of the hemiacetal and acetal formed from the reaction of acetone (2-propanone) with excess ethanol.

Solution

To write the structure of the hemiacetal, first visualize the cleavage of the alcohol at the O–H bond to form RO^- and H^+ fragments:

$$CH_3CH_2OH \longrightarrow CH_3CH_2O^- + H^+$$

Next, visualize the carbonyl group of acetone as an ionic structure:

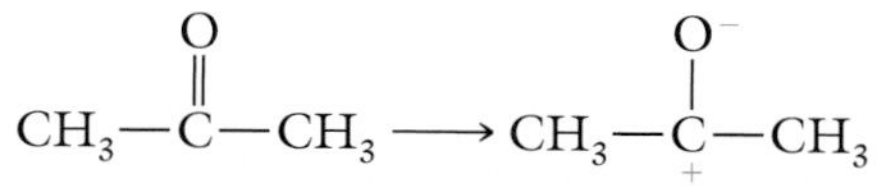

$$\underset{\textbf{Covalent}}{CH_3-\overset{\overset{\displaystyle O}{\|}}{C}-CH_3} \longrightarrow \underset{\textbf{Ionic}}{CH_3-\overset{\overset{\displaystyle O^-}{|}}{\underset{+}{C}}-CH_3}$$

The hemiacetal is formed by attachment of the RO^- and H^+ fragments to the oppositely charged ends of the carbonyl group:

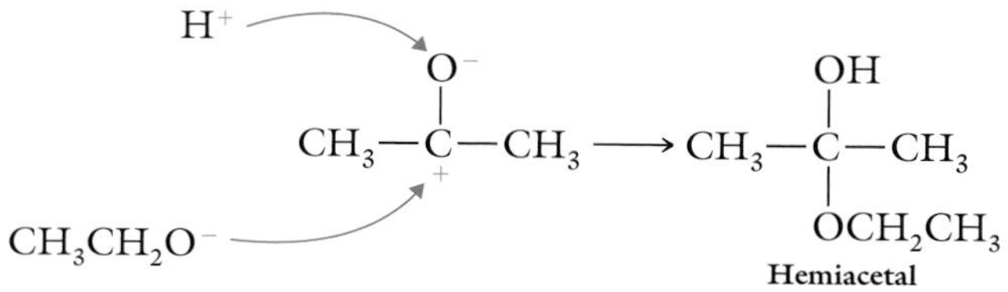

$$\overset{H^+}{} \quad CH_3-\overset{\overset{\displaystyle O^-}{|}}{\underset{+}{C}}-CH_3 \quad \underset{CH_3CH_2O^-}{} \longrightarrow \underset{\textbf{Hemiacetal}}{CH_3-\overset{\overset{\displaystyle OH}{|}}{\underset{\underset{\displaystyle OCH_2CH_3}{|}}{C}}-CH_3}$$

The acetal is formed by dehydration between the —OH group of the hemiacetal and the —OH group of the second molecule of ethanol:

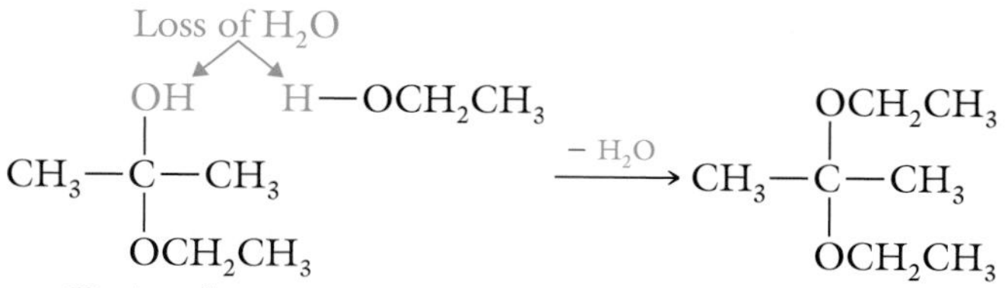

Loss of H_2O

$$\underset{\textbf{Hemiacetal}}{CH_3-\overset{\overset{\displaystyle OH}{|}}{\underset{\underset{\displaystyle OCH_2CH_3}{|}}{C}}-CH_3} \quad H-OCH_2CH_3 \xrightarrow{-H_2O} \underset{\textbf{Acetal}}{CH_3-\overset{\overset{\displaystyle OCH_2CH_3}{|}}{\underset{\underset{\displaystyle OCH_2CH_3}{|}}{C}}-CH_3}$$

Problem 6.9 Show the structures of the hemiacetal and acetal formed from the reaction of butanal with excess methanol.

Five- and six-membered cyclic hemiacetals are an important exception to the generalization that hemiacetals are unstable and cannot be isolated. Five- and six-membered cyclic hemiacetals are stable and can be isolated, although the reasons are not completely understood. Cyclic hemiacetals are formed when a compound possesses both alcohol and carbonyl groups. Cyclic hemiacetal structures are found extensively in carbohydrates (Section 10.4).

Example 6.10 Drawing the structure of a cyclic hemiacetal

Show the structure of the cyclic hemiacetal formed by intramolecular reaction between the alcohol group and the carbonyl group of 5-hydroxypentanal.

Solution

To write the structure of the hemiacetal, first visualize cleavage of the alcohol O–H bond to yield RO^- and H^+ fragments. Next, visualize the carbonyl group as an ionic structure. The hemiacetal is formed by attachment of the RO^- and H^+ fragments to the oppositely charged ends of the carbonyl group:

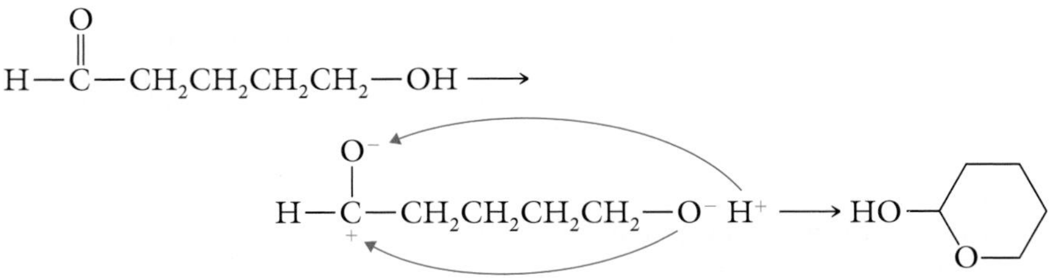

$$H-\overset{\overset{O}{\|}}{C}-CH_2CH_2CH_2CH_2-OH \longrightarrow$$

$$H-\overset{\overset{O^-}{|}}{\underset{+}{C}}-CH_2CH_2CH_2CH_2-O^-\ H^+ \longrightarrow HO-\text{(cyclic hemiacetal, six-membered ring containing O)}$$

Problem 6.10 Show the structure of the hemiacetal formed by intramolecular reaction between the —OH group and the carbonyl group of 6-hydroxyl-2-hexanone.

Recognition of hemiacetal and acetal structures is important because of their presence in carbohydrates. Both hemiacetals and acetals have two different oxygens bonded to the same carbon. In acetals, both oxygens are in —OR groups; in hemiacetals, one of the oxygens is in an —OH group and the other is in an —OR group:

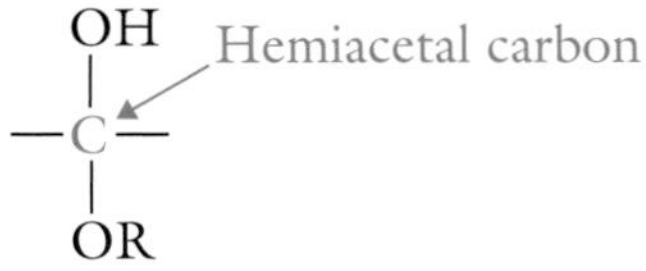

$$-\overset{\overset{OH}{|}}{\underset{\underset{OR}{|}}{C}}- \quad \text{Hemiacetal carbon} \qquad\qquad -\overset{\overset{OR}{|}}{\underset{\underset{OR}{|}}{C}}- \quad \text{Acetal carbon}$$

Hemiacetal 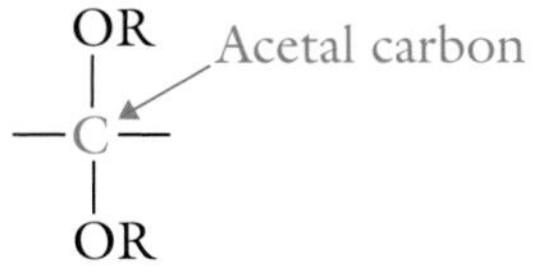   Acetal

Example 6.11 Recognizing hemiacetals and acetals

For each of the following compounds, indicate whether it is a hemiacetal, an acetal, or something different.

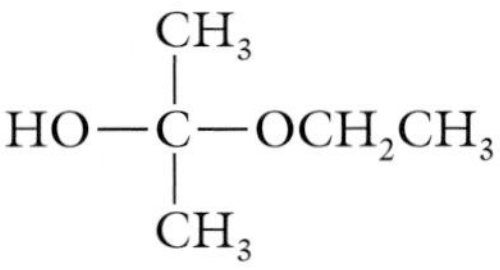

$$HO-\overset{\overset{CH_3}{|}}{\underset{\underset{CH_3}{|}}{C}}-OCH_2CH_3 \qquad HO-\overset{\overset{CH_3}{|}}{\underset{\underset{CH_3}{|}}{C}}-CH_2OCH_3$$

1 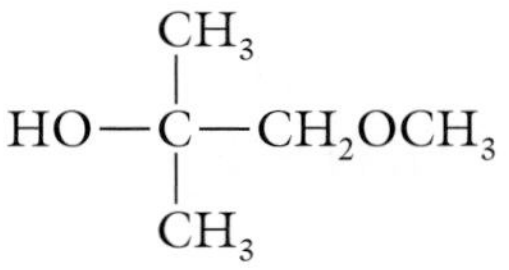   2

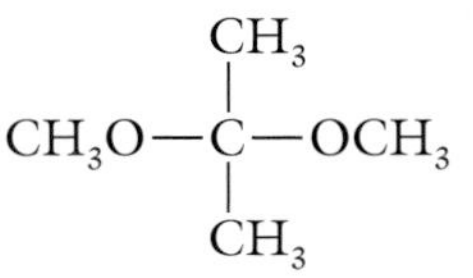

$$CH_3O-\overset{\overset{CH_3}{|}}{\underset{\underset{CH_3}{|}}{C}}-OCH_3$$

3

CH_3O, OC_6H_5 (substituents on adjacent carbons of a cyclohexane ring)

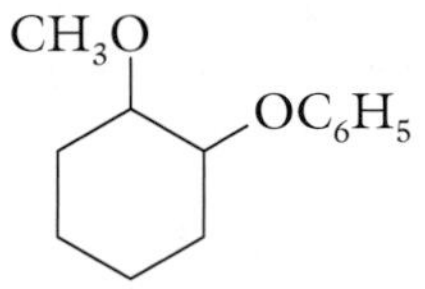

4

Solution

Compound 1 is a hemiacetal because it has a carbon with both an —OH and an —OR group attached:

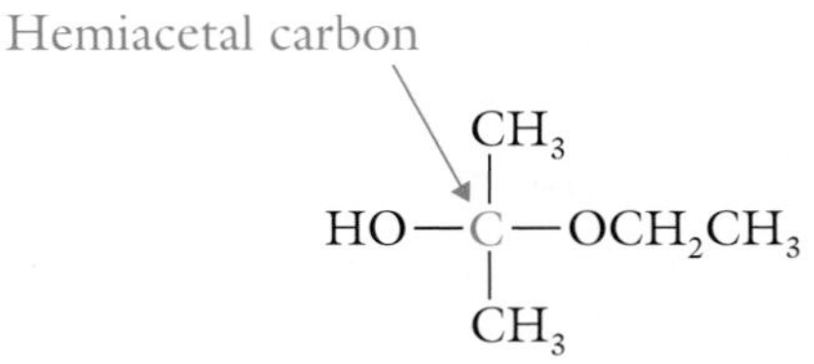

Hemiacetal carbon

$$HO-\overset{\overset{CH_3}{|}}{\underset{\underset{CH_3}{|}}{C}}-OCH_2CH_3$$

Compound 3 is an acetal because it has a carbon with two —OR groups attached:

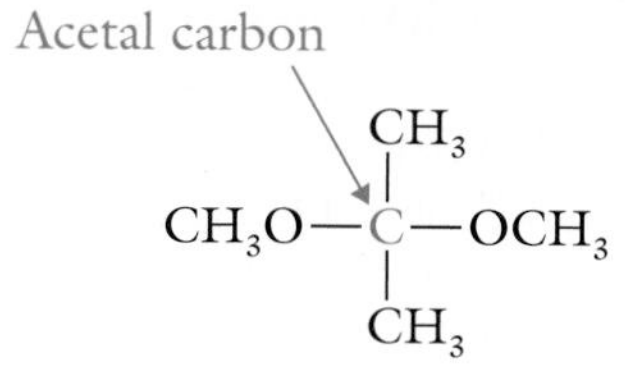

Acetal carbon

$$CH_3O—\overset{\overset{\displaystyle CH_3}{|}}{\underset{\underset{\displaystyle CH_3}{|}}{C}}—OCH_3$$

Compounds 2 and 4 are neither hemiacetals or acetals, because no carbon is bonded to two oxygens.

Problem 6.11 For each of the following compounds, indicate whether it is a hemiacetal, an acetal, or something different.

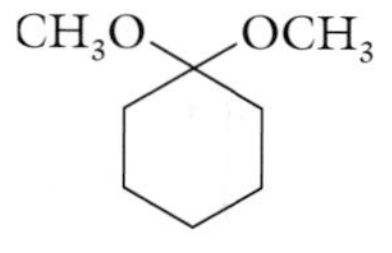

1: cyclohexane carbon bearing CH_3O and OCH_3

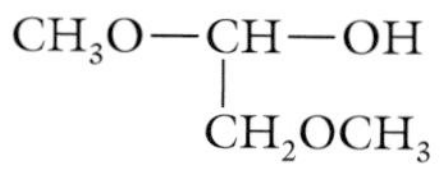

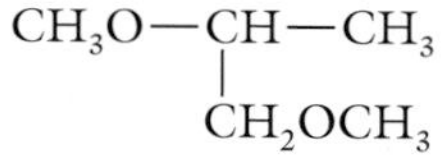

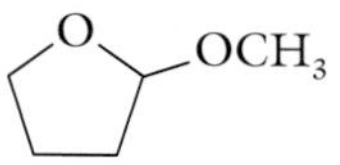

2: $CH_3O—\underset{\underset{\displaystyle CH_2OCH_3}{|}}{CH}—OH$

3: $CH_3O—\underset{\underset{\displaystyle CH_2OCH_3}{|}}{CH}—CH_3$

4: tetrahydrofuran ring (O) with OCH_3 on the carbon adjacent to O

Hemiacetal and acetal formations are reversible reactions. In the presence of excess water and an acid catalyst, an acetal is converted into the aldehyde or ketone and alcohol. This process is an example of a **hydrolysis** reaction—the reaction of an organic compound with water resulting in cleavage of the compound into two organic fragments each of which combines with a fragment (H^+ or OH^-) from water. Each of the organic products contains fewer carbons than did the original organic compound. The hydrolysis of an acetal proceeds through the intermediate formation of the hemiacetal, but the hemiacetal is not isolated, because of its instability. As mentioned earlier, Le Chatelier's principle determines the direction of the reaction. Acetal formation from an aldehyde or ketone requires an excess of alcohol, whereas the hydrolysis of an acetal requires an excess of water.

In animals, the hydrolysis of acetal groups is the first step in the digestive breakdown of starch and other carbohydrates in the digestive tract (Section 10.5).

Example 6.12 Writing the products of acetal or hemiacetal hydrolysis

Complete each of the following reactions by writing the structure of the product. If no reaction takes place, write "no reaction." If there is more than one product, show only the major product(s).

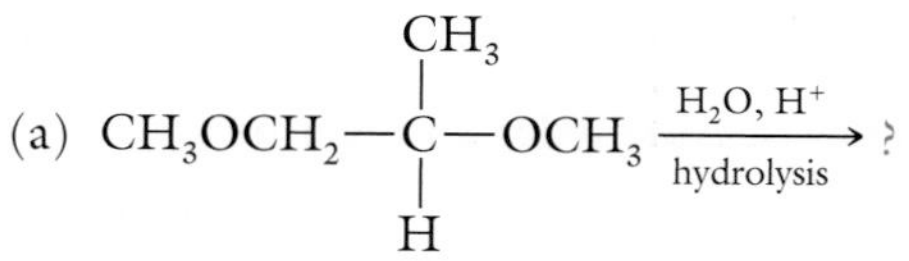

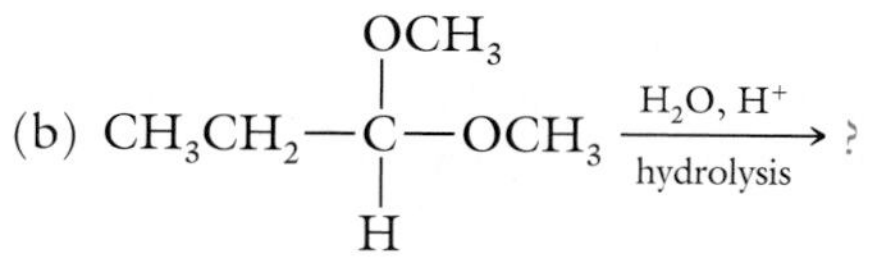

(a) $CH_3OCH_2—\overset{\overset{\displaystyle CH_3}{|}}{\underset{\underset{\displaystyle H}{|}}{C}}—OCH_3 \xrightarrow[\text{hydrolysis}]{H_2O,\ H^+} ?$

(b) $CH_3CH_2—\overset{\overset{\displaystyle OCH_3}{|}}{\underset{\underset{\displaystyle H}{|}}{C}}—OCH_3 \xrightarrow[\text{hydrolysis}]{H_2O,\ H^+} ?$

Solution

(a) No reaction. The compound is neither a hemiacetal nor an acetal. It is an ether (in fact, it has two ether groups), and ethers do not undergo hydrolysis.
(b) The compound is an acetal that undergoes hydrolysis. The hydrolysis of the acetal to the hemiacetal is the reverse of what takes place when a hemiacetal is converted into an acetal. One of the $—OCH_3$ groups leaves the acetal and combines with an H of water to form CH_3OH. The OH from water attaches to the carbon from which the $—OCH_3$ group left:

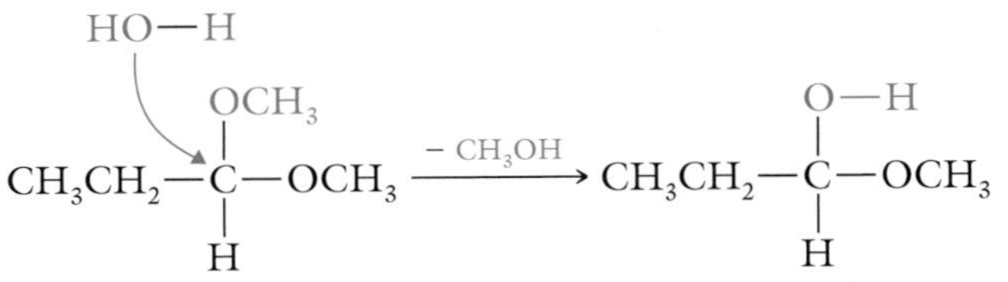

$$\begin{array}{c} HO—H \\ \searrow \end{array} \quad CH_3CH_2—\overset{\displaystyle OCH_3}{\underset{\displaystyle H}{\overset{|}{\underset{|}{C}}}}—OCH_3 \xrightarrow{-\,CH_3OH} CH_3CH_2—\overset{\displaystyle O—H}{\underset{\displaystyle H}{\overset{|}{\underset{|}{C}}}}—OCH_3$$

The hemiacetal is not isolated. Hemiacetals other than cyclic hemiacetals are unstable and are quickly converted into the aldehyde or ketone by loss of CH_3OH:

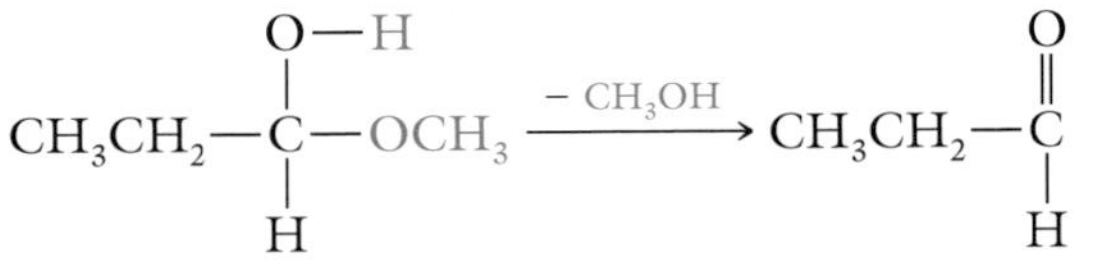

$$CH_3CH_2—\overset{\displaystyle O—H}{\underset{\displaystyle H}{\overset{|}{\underset{|}{C}}}}—OCH_3 \xrightarrow{-\,CH_3OH} CH_3CH_2—\overset{\displaystyle O}{\underset{\displaystyle H}{\overset{\|}{\underset{|}{C}}}}$$

The overall reaction consists of one molecule each of acetal and water reacting to yield one aldehyde and two alcohol molecules:

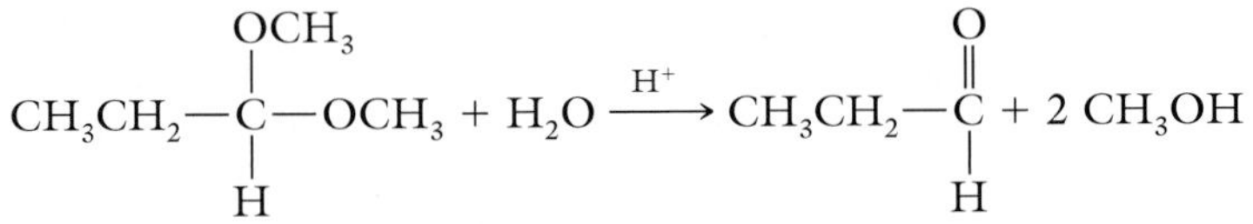

$$CH_3CH_2—\overset{\displaystyle OCH_3}{\underset{\displaystyle H}{\overset{|}{\underset{|}{C}}}}—OCH_3 + H_2O \xrightarrow{H^+} CH_3CH_2—\overset{\displaystyle O}{\underset{\displaystyle H}{\overset{\|}{\underset{|}{C}}}} + 2\ CH_3OH$$

Problem 6.12 Show the product(s) formed when each of the following compounds undergoes acid-catalyzed hydrolysis:

(a) 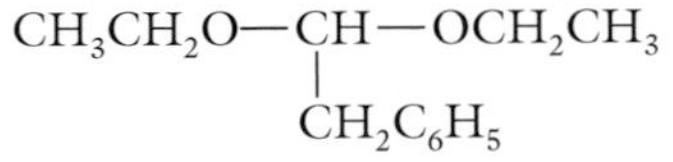 $CH_3CH_2O—\underset{\displaystyle CH_2C_6H_5}{\underset{|}{CH}}—OCH_2CH_3$ (b) 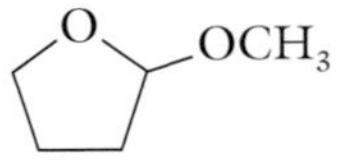 tetrahydrofuran ring (O in ring) bearing OCH_3 on the carbon adjacent to the ring O

Summary

Structural Features of Aldehydes and Ketones

- Aldehydes and ketones contain the carbonyl group ($C{=}O$).
- An aldehyde has at least one hydrogen atom directly attached to the carbon of the carbonyl group.
- A ketone has only carbon atoms, no hydrogen atoms, directly attached to the carbon of the carbonyl group.
- The C:H ratio for an aldehyde or ketone is two hydrogens fewer than the corresponding alcohol or alkane.

Naming Aldehydes and Ketones

- In the IUPAC system, aldehydes and ketones are named by identifying the longest continuous carbon chain containing the carbonyl group.
- Aldehydes are named as alkanals.
- Ketones are named as alkanones.
- For ketones, the position of the carbonyl carbon is indicated by a number in front of the name.
- For aldehydes, no number is needed, because the carbonyl carbon is always C1.
- Prefixes preceded by numbers indicate the substituents attached to the longest chain.
- C_6H_5–CHO is named benzaldehyde.
- Common names are used for some aldehydes and ketones.
- Common ketone names are obtained by naming the groups attached to the carbonyl carbon, followed by the word ketone.

Physical Properties of Aldehydes and Ketones

- Aldehydes and ketones have boiling and melting points intermediate between those of hydrocarbons and alcohols, a consequence of their dipole–dipole secondary forces being intermediate in strength between the London forces in hydrocarbons and the hydrogen-bond forces in alcohols.
- Aldehydes and ketones are only slightly less soluble than alcohols in water.

• Although aldehydes and ketones cannot hydrogen bond with themselves, they can hydrogen bond with water.

Chemical Reactions of Aldehydes and Ketones

• Aldehydes, but not ketones, undergo oxidation with several selective oxidizing agents—dichromate, permanganate, and Tollens's reagents.

• Dichromate and permanganate distinguish aldehydes from ketones but not from primary and secondary alcohols or from alkenes, which also give positive tests.

• Tollens's reagent is more selective than dichromate and permanganate, distinguishing aldehydes from ketones, alcohols, and alkenes.

• Benedict's reagent is useful for distinguishing α-hydroxy aldehydes and α-hydroxy ketones from other compounds.

• Aldehydes and ketones undergo addition reactions with hydrogen, hydrides, and alcohols.

• Aldehydes and ketones are reduced to primary and secondary alcohols, respectively.

• Both catalytic (H_2 with metal catalyst) and hydride (metal hydrides, NADH + enzyme) reductions are important for reducing aldehydes and ketones.

• Alcohol adds to the carbonyl group of aldehydes and ketones to form hemiacetals.

• Hemiacetals undergo dehydration with excess alcohol to form acetals.

• An acid catalyst is required for both hemiacetal and acetal formations.

• Acid-catalyzed hydrolysis reverses the reactions that form acetals and hemiacetals.

• Most hemiacetals are unstable and cannot be isolated, but cyclic hemiacetals are stable and can be isolated.

Summary of Reactions

SELECTIVE OXIDATION

$$\underset{\text{Aldehyde}}{\mathrm{R{-}\underset{|}{\overset{}{C}}(H){=}O}} \xrightarrow{(O)} \underset{\text{Carboxylic acid}}{\mathrm{R{-}C(OH){=}O}}$$

$$\underset{\text{Ketone}}{\mathrm{R{-}C(R'){=}O}} \xrightarrow{(O)} \text{no reaction}$$

- Primary and secondary alcohols, as well as alkenes, are oxidized by MnO_4^- and $Cr_2O_7^{2-}$.
- Tollens's reagent (Ag^+) oxidizes aldehydes but not ketones, alcohols, or alkenes.
- Benedict's reagent (Cu^{2+}) oxidizes α-hydroxy aldehydes and α-hydroxy ketones. Simple aldehydes give weak or negative results.

REDUCTION

$$\mathrm{{-}C({=}O){-}} \xrightarrow{(H)} \mathrm{{-}C(OH)(H){-}}$$

- Reduction is carried out with H_2 and Pt or with the sequential addition of hydride followed by water or, in biological systems, with NADH.

HEMIACETAL AND ACETAL FORMATION BY REACTION WITH ALCOHOL

$$\mathrm{R^2{-}C({=}O){-}R^1} \underset{H^+(-R^3OH)}{\overset{R^3OH,\ H^+}{\rightleftharpoons}} \underset{\text{Hemiacetal}}{\mathrm{R^2{-}C(OH)(OR^3){-}R^1}} \underset{H^+,\ H_2O\ (-R^3OH)}{\overset{R^3OH,\ H^+(-H_2O)}{\rightleftharpoons}} \underset{\text{Acetal}}{\mathrm{R^2{-}C(OR^3)(OR^3){-}R^1}}$$

- Reaction requires an acid catalyst.
- Reaction is reversible. Hydrolysis yields the aldehyde or ketone and alcohol.

Key Words

acetal, p. 204
aldehyde, p. 185
Benedict's reagent, p. 196
carbonyl group, p. 185
catalytic reduction, p.198
hemiacetal, p. 204
hydride reduction, p. 199
hydrolysis, p. 207
α-hydroxy aldehyde, p. 196
α-hydroxy ketone, p. 196
ketone, p. 185
oxidation, p. 195
Tollens's reagent, p. 196

Exercises

Structure of Aldehydes and Ketones

6.1 Does the molecular formula $C_5H_{12}O$ fit an alcohol, ether, aldehyde, or ketone?

6.2 Does the molecular formula $C_5H_{10}O$ fit an alcohol, ether, aldehyde, or ketone?

6.3 Draw structural formulas of all aldehydes having the molecular formula $C_6H_{12}O$.

6.4 Draw structural formulas of all ketones having the molecular formula $C_6H_{12}O$.

6.5 Which of the following compounds are aldehydes or ketones and which are not? Classify each compound. (Refer to Table 6.1 as needed.)

1: $H{-}\overset{O}{\overset{\|}{C}}{-}CH_2CH_2C_6H_5$

2: $CH_3{-}\overset{O}{\overset{\|}{C}}{-}OCH(CH_3)_2$

3: cyclohexanone (six-membered ring with C=O)

4: $CH_3{-}\overset{O}{\overset{\|}{C}}{-}NH_2$

6.6 For each of the following pairs of structural formulas, indicate whether the pair represents (1) the same compound or (2) different compounds that are constitutional isomers or (3) different compounds that are geometrical isomers or (4) different compounds that are not isomers.

(a) $CH_3{-}\overset{O}{\overset{\|}{C}}{-}CH_2CH_3$ and $CH_3CH_2CH_2{-}\overset{O}{\overset{\|}{C}}H$

(b) $CH_3{-}\overset{O}{\overset{\|}{C}}{-}CH_2CH_3$ and $CH_3{-}\overset{O}{\overset{\|}{C}}{-}CH_2CH_2CH_3$

(c) $CH_3{-}\overset{O}{\overset{\|}{C}}{-}CH_2CH_3$ and $CH_2{=}CHCH_2CH_2OH$

(d) $CH_3{-}\overset{O}{\overset{\|}{C}}{-}CH_2CH_2CH_3$ and $CH_3CH_2{-}\overset{O}{\overset{\|}{C}}{-}CH_2CH_3$

(e) $CH_3{-}\overset{O}{\overset{\|}{C}}{-}OCH(CH_3)_2$ and $CH_3CH_2{-}\overset{O}{\overset{\|}{C}}{-}CH_2CH_3$

(f) $CH_3{-}\overset{O}{\overset{\|}{C}}{-}CH_2CH_3$ and

6.7 What are the hybridizations of atoms a, b, c, d, and e in the following compound?

$$CH_3\overset{a}{C}H_2{-}\underset{b}{\overset{\overset{c}{O}}{\overset{\|}{C}}}{-}\overset{d}{C}H_2{-}\overset{e}{C}H_3$$

6.8 What are bond angles A, B, and C in the following compound?

$$CH_3CH_2{-}\overset{O}{\overset{\|}{C}}{-}CH_2{-}CH_3$$

(A: angle CH_2–C=O; B: angle C–CH_2–CH_3; C: angle CH_2–C–CH_2)

Naming Aldehydes and Ketones

6.9 Draw the structural formula for each of the following compounds: (a) 2,3-dimethylhexanal; (b) ethyl isopropyl ketone; (c) 3-*s*-butyl-4-ethylbenzaldehyde.

6.10 Draw the structural formula for each of the following compounds: (a) 4,5-dimethyl-2-hexanone; (b) 2-phenyl-3-heptanone; (c) 3-methylcyclohexanone.

6.11 Name each of the following compounds:

(a) $(CH_3)_2CH\underset{CH_2CH_3}{\underset{|}{C}H}{-}\overset{O}{\overset{\|}{C}}H$

(b) $C_6H_5{-}\overset{O}{\overset{\|}{C}}{-}C(CH_3)_3$

(c) cyclopentanone ring with C=O and adjacent CH_3

(d) $H\overset{O}{\overset{\|}{C}}CH_2\overset{Cl}{\overset{|}{C}}HC(CH_3)_3$

6.12 Name each of the following compounds:

(a) benzene ring bearing CH_3, Cl, and $-\overset{O}{\overset{\|}{C}}{-}H$

(b) $CH_3\overset{CH_3}{\overset{|}{C}}HCH_2\overset{O}{\overset{\|}{C}}CH_2\overset{CH_2CH_2CH_3}{\overset{|}{C}}HC(CH_3)_3$

(c)

Physical Properties of Aldehydes and Ketones

6.13 Indicate which compound in each of the following pairs has the higher value of the specified property. If the two compounds in a pair have nearly the same value for the property, indicate that fact. Explain your conclusions.

(a) 2-Butanone, 3-pentanone: boiling point.
(b) Butanal, 2-butanone: boiling point.
(c) Butanal, 2-butanone: solubility in water.
(d) Butanal, pentane: boiling point.
(e) Butanal, pentane: solubility in water.
(f) 2-Butanone, 3-pentanone: solubility in water.
(g) 2-Butanone, 2-butanol: boiling point.
(h) 2-Butanone, 2-butanol: solubility in water.

6.14 Indicate whether hydrogen bonding is present in each of the following compounds or mixtures. Draw a representation of the hydrogen bonding if present. (a) Ethanal; (b) water; (c) ethanal and water; (d) acetone (2-propanone) and ethanol.

The Oxidation of Aldehydes and Ketones

6.15 Which of the following compounds undergo oxidation with dichromate and permanganate reagents? If oxidation takes place, write the structure(s) of the product(s).

$$\underset{1}{CH_3CH_2-\overset{\overset{\displaystyle O}{\|}}{C}-CH_3} \qquad \underset{2}{CH_3CH_2-\overset{\overset{\displaystyle O}{\|}}{C}-H}$$

$$\underset{3}{CH_3CH_2CH_2CH_2OH}$$

6.16 Which of the compounds in Exercise 6.15 undergo oxidation with (a) Tollens's reagent and (b) Benedict's reagent? Briefly explain what is observed visually when there is a positive test.

6.17 For each of the following compounds, indicate whether the compound gives a positive test with (a) Tollens's reagent and (b) Benedict's reagent?

$$\underset{1}{CH_3CH_2\overset{\overset{\displaystyle O}{\|}}{C}-\overset{\overset{\displaystyle OH}{|}}{C}H-CH_3}$$

$$\underset{2}{HO-\overset{\overset{\displaystyle O}{\|}}{C}CH_2\overset{\overset{\displaystyle OH}{|}}{C}H-CH_3}$$

$$\underset{3}{CH_3CH_2-\overset{\overset{\displaystyle O}{\|}}{C}-CH_2-\overset{\overset{\displaystyle O}{\|}}{C}H}$$

$$\underset{4}{H-\overset{\overset{\displaystyle O}{\|}}{C}-\overset{\overset{\displaystyle OH}{|}}{C}H-CH_2CH_3}$$

For both Tollens's and Benedict's reagents, briefly explain what is observed visually when there is a positive test.

6.18 An unknown compound is either 2-pentanone or pentanal. The addition of a few drops of Tollens's reagent to the unknown results in the formation of a silver mirror. Identify the unknown.

6.19 An unknown compound is either 1-pentanol or pentanal. When a few drops of basic permanganate solution is added to the unknown, the purple color of the permanganate disappears and a brown precipitate forms. Identify the unknown.

6.20 An unknown compound is either 1-pentanol or pentanal. A shiny mirror forms when Tollens's reagent is added to the unknown. Identify the unknown.

The Reduction of Aldehydes and Ketones

6.21 Which of the following compounds undergo reduction with metal hydrides? If reduction takes place, give the product(s).

(a) $CH_3CH_2-\overset{\overset{\displaystyle O}{\|}}{C}-CH_3$

(b) $CH_3CH_2-\overset{\overset{\displaystyle O}{\|}}{C}-H$

(c) $CH_3CH_2CH_2CH_2OH$

6.22 Which of the compounds in Exercise 6.21 undergo reduction with H_2 and Pt? If reduction takes place, give the structure(s) of the product(s).

6.23 Write the equation showing hydride ion, $H{:}^-$, reacting with water. Is the hydride ion an acid or a base in this reaction?

6.24 For each of the following compounds, indicate whether it can be synthesized from an aldehyde or ketone by reduction. If yes, give the structure of the aldehyde or ketone.

(a) $CH_3-\overset{\overset{\displaystyle CH_3}{|}}{\underset{\underset{\displaystyle OH}{|}}{C}}-CH_3$

(b) $CH_3-\overset{\overset{\displaystyle CH_3}{|}}{C}H-CH_2-OH$

(c) OH (d) OH

Hemiacetal and Acetal Formation by Reaction with Alcohol

6.25 Write equations to show the reaction of benzaldehyde with excess methanol in the presence of an acid catalyst. Show the formation of the hemiacetal and acetal.

6.26 For each of the following compounds, indicate whether it is a hemiacetal, an acetal, or something different.

CH_3 O (1) OCH_3 O (2)

CH_3O OCH_3 (3) HO OCH_3 (4)

6.27 Give the structures of the aldehyde or ketone and the alcohol that are needed to synthesize each of the following hemiacetals or acetals:

(a) $C_6H_5CH(OCH_2CH_3)_2$

(b) (tetrahydrofuran ring with O and OCH_3 on the carbon next to the ring O)

(c) $HO-C(OCH_2CH_3)(CH_2CH_3)-CH_2CH_2CH_3$

(d) (cyclopentane ring)$-CH(OCH_3)_2$

6.28 Which of the following compounds undergo hydrolysis? Give the product(s) of hydrolysis for any compound that undergoes hydrolysis. Indicate "no reaction" if hydrolysis does not take place.

(a) $C_6H_5CH(OCH_2CH_3)_2$

(b) $HOCH_2CH_2OCH_3$

(c) $CH_3CH_2-C(=O)-CH_3$

(d) (tetrahydrofuran ring with O and OCH_3 on the carbon next to the ring O)

Unclassified Exercises

6.29 Compare the C=C and C=O double bonds with respect to (a) hybridization of the atoms; (b) bond angles about the atoms; (c) polarity; (d) chemical reactions.

6.30 Identify the families of functional groups 1–9 in the following compound:

(structure with numbered groups: 1 $HOCH_2$; 2 C=C in ring; 3 C=O; 4 ring O; 5 OCH_2CH_3; 6 benzene ring; 7 CH_2CHO; 8 ring O; 9 OH)

6.31 Draw structural formulas for each of the following compounds: (a) $C_5H_{10}O$ aldehyde containing one methyl group; (b) cyclic ketone of five carbons with all carbons in the ring; (c) C_8 ketone containing a benzene ring; (d) $C_6H_{12}O$ aldehyde containing a *t*-butyl group; (d) $C_6H_{12}O$ ketone containing a *t*-butyl group.

6.32 For each of the following pairs of structural formulas, indicate whether the pair represents (1) the same compound or (2) different compounds that are constitutional isomers or (3) different compounds that are geometrical isomers or (4) different compounds that are not isomers.

(a) $CH_3CH(CH_3)CH_2-C(=O)-H$ and $CH_3-C(=O)-CH(CH_3)_2$

(b) $CH_3CH(CH_3)CH_2-C(=O)-H$ and $CH_3-C(=O)-CH_2CH(CH_3)_2$

(c) $CH_3CH_2CH_2CH(OCH_3)_2$ and $CH_3CH(OCH_2CH_3)_2$

(d) $CH_3CH_2CH_2CH(OCH_3)_2$ and $CH_3C(OCH_2CH_3)_2CH_3$

(e) $CH_3C(OH)(OCH_2CH_3)CH_3$ and $CH_3CH(OH)CH_2CH_2OCH_3$

(f) $CH_3-C(=O)-CH_2CH_2CH_2CH_2-OH$ and (tetrahydropyran ring with CH_3 and HO on the carbon next to the ring O)

(g) CH_3O-(tetrahydropyran ring, on the carbon next to the ring O) and (tetrahydropyran ring with CH_3 and HO on the carbon next to the ring O)

6.33 Although the following names are incorrect, you can deduce the structure for each compound. Give the correct names. (a) 4-Methylcyclopentanone; (b) 2-propylbutanal; (c) 1-pentanone; (d) 3-chloro-4-pentanone.

6.34 A cyclic, nonaromatic compound containing an attached aldehyde group is named by assigning a name to the cyclic compound exclusive of the aldehyde group and then adding the suffix **-carbaldehyde.** Name each of the following compounds:

(a) (cyclopentane ring with H and $(CH_3)_2CH$ on one carbon and H and $C(=O)H$ on the adjacent carbon)

(b) (cyclohexane ring with H and CH_3 on one carbon and H and $C(=O)H$ on another carbon)

6.35 Indicate which compound in each of the following pairs has the higher value of the specified property. If the two compounds in a pair have nearly the same value for the property, indicate that fact. Explain your conclusions.
(a) Butanal, 1-butanol: boiling point.
(b) 1-Butanol, 2-butanone: solubility in water.
(c) Butanal, diethyl ether: boiling point.
(d) Butanal, diethyl ether: solubility in water.

6.36 Place the following compounds in order of increasing boiling point: pentane, butanal, 1-butanol, diethyl ether. Explain the order.

6.37 Place the following compounds in order of increasing solubility in water: pentane, butanal, 1-butanol, diethyl ether. Explain the order.

6.38 Indicate whether hydrogen bonding is present in each of the following compounds or mixtures. Draw a representation of the hydrogen bonding if present.
(a) Propanal; (b) propanal and propanol; (c) propanal and diethyl ether.

6.39 An unknown is either 2-pentanone or 2-methyl-2-butanol. Identify the unknown if it gives a negative test with Tollens's reagent.

6.40 The structure of glucose is shown here. Does glucose give a positive test with Benedict's and Tollens's reagents?

$$HOCH_2-CH(OH)-CH(OH)-CH(OH)-CH(OH)-CH(=O)$$

Glucose

6.41 Does glucose undergo reaction with H_2 and Pt? If reduction takes place, write the structure(s) of the product(s).

6.42 Show the structure of the cyclic hemiacetal formed by intramolecular reaction between the alcohol group and the carbonyl group of 5-hydroxyl-2-hexanone.

6.43 Write equations to show the reaction of 2-butanone with excess methanol. Show the formation of the hemiacetal and acetal.

6.44 Give the organic product(s) formed in each of the following reactions. If no reaction takes place, write "no reaction." If there is more than one product, indicate only the major product(s). If one molecule of the compound is capable of reacting with more than one molecule of reagent, assume that the reaction is carried out with excess reagent.

(a) $CH_3C(=O)CH_2CH_3 \xrightarrow[2.\ H_2O]{1.\ LiAlH_4} ?$

(b) $CH_3C(=O)CH_2CH_3 \xrightarrow[H_2O]{LiAlH_4} ?$

(c) cyclohexanone $\xrightarrow[H^+]{CH_3OH} ?$

(d) $CH_3-C(OCH_3)_2-C_6H_5 \xrightarrow{H^+,\ H_2O} ?$

(e) $CH_3-C(CH_2OCH_3)_2-C_6H_5 \xrightarrow{H^+,\ H_2O} ?$

(f) $CH_3CH_2CH_2C(=O)H \xrightarrow{MnO_4^-} ?$

(g) $CH_3CH_2CH_2CH_2OH \xrightarrow{MnO_4^-} ?$

(h) cyclohexanone $\xrightarrow[2.\ H_2O]{1.\ NADH,\ alcohol\ dehydrogenase} ?$

6.45 Give the organic product(s) formed in each of the following reactions. If no reaction takes place, write "no reaction." If there is more than one product, indicate only the major product(s). If one molecule of the compound is capable of reacting with more than one molecule of reagent, assume that the reaction is carried out with excess reagent.

(a) 3-methylcyclopentanone $\xrightarrow[H^+]{CH_3OH} ?$

(b) $CH_3-C(=O)-C_6H_5 \xrightarrow[Ni]{H_2} ?$

(c) $CH_3-C(OCH_3)_2-CH_3 \xrightarrow[Ni]{H_2} ?$

(d) $CH_3CH_2CH_2C(=O)H \xrightarrow{Tollens's\ reagent} ?$

(e) $CH_3-C(OCH_3)_2-C_6H_5 \xrightarrow{Benedict's\ reagent} ?$

(f) $H-C(=O)-CH_2C_6H_5 \xrightarrow[2.\ H_2O]{1.\ NADH,\ alcohol\ dehydrogenase} ?$

(g) 3-methylcyclopentanone $\xrightarrow[2.\ H_2O]{1.\ LiAlH_4} ?$

(h) 2-methoxytetrahydrofuran $\xrightarrow[H^+]{H_2O} ?$

6.46 An unknown compound C_3H_6O is converted into $C_3H_6O_2$ on reaction with dichromate. Identify the compound.

6.47 Identify the compound that yields 2-butanol on reduction with a metal hydride.

6.48 An unknown compound C_4H_8O, either an aldehyde or ketone, gives a negative test with Tollens's reagent but undergoes hydride reduction to form an alcohol. Identify the compound.

6.49 What aldehyde or ketone, if any, can be synthesized by oxidation of each of the following compounds?

(a) $CH_3-C(CH_3)(OH)-CH_3$

(b) $CH_3-CH(CH_3)-CH_2-OH$

(c) 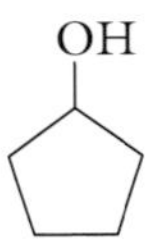 cyclopentanol (cyclopentane ring bearing OH)

(d) 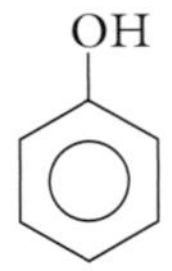 phenol (benzene ring bearing OH)

6.50 What aldehyde yields each of the following carboxylic acids by selective oxidation?

(a) $CH_3CH_2CH_2CH_2-\overset{O}{\overset{\|}{C}}-OH$

(b) 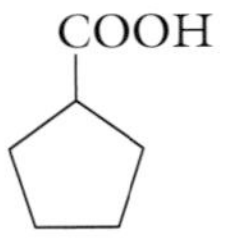 cyclopentane ring bearing COOH

(c) $(CH_3)_2CH\overset{O}{\overset{\|}{C}}-OH$

Expand Your Knowledge

Note: The icons [flask icons] denote exercises based on material in boxes.

6.51 [6.1] The structures of menthone (mint) and vanillin (vanilla) are shown in Box 6.1. What simple chemical tests distinguish the two compounds?

6.52 One of the industrial processes for producing acetone (2-propanone) uses 2-propanol as the starting material. What reaction is used to convert 2-propanol into acetone?

6.53 [6.1] The structure of cinnamaldehyde (cinnamon) is shown in Box 6.1. Cinnamaldehyde, C_9H_8O, can be reduced by both the catalytic hydrogenation and the hydride reduction methods. However, the products are different, having the molecular formulas $C_9H_{12}O$ and $C_9H_{10}O$, respectively. Explain the difference.

6.54 [6.2] Formaldehyde is useful for preserving biological specimens and embalming cadavers. It kills microorganisms (see Box 6.2) by inactivating their proteins by reacting with their amine groups ($-NH_2$). The reaction is a dehydration between the amine group and the carbonyl group with the formation of a C=N double bond. Write the equation for this reaction with the simple amine CH_3-NH_2.

6.55 Although Exercise 6.40 shows the structure of glucose as 2,3,4,5,6-pentahydroxyhexanal, less than 0.2% of that acyclic structure is in aqueous solutions of glucose. Glucose is more than 99.8% a mixture of the following two cyclic structures:

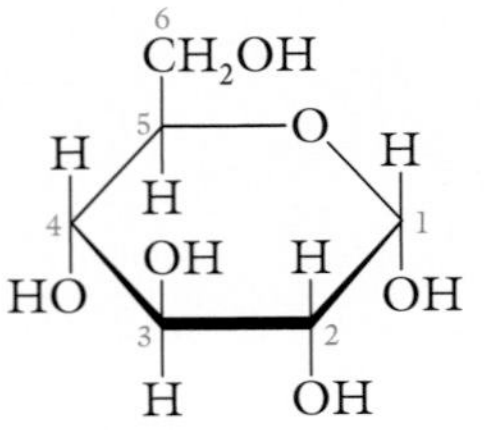

[Haworth projections: ring with C1–C5 and ring O; C5 bears $^6CH_2OH$; α-D-Glucose: C1 H (up), OH (down); β-D-Glucose: C1 OH (up), H (down); C2 H, OH; C3 OH, H; C4 H, HO]

α-D-Glucose 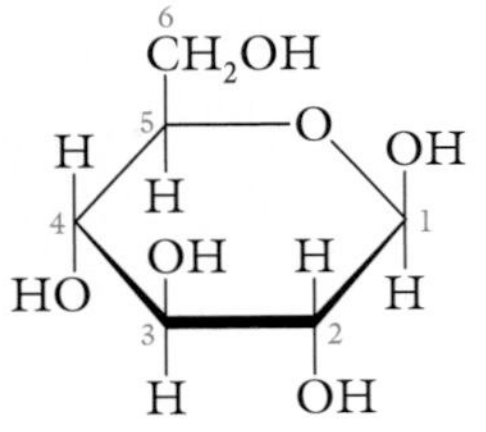   β-D-Glucose

What is the relation between these two cyclic structures and the acyclic structure shown in Exercise 6.40?

6.56 [6.1] The structures of benzaldehyde and cinnamaldehyde, responsible for the flavors of almond and cinnamon, respectively, are shown in Box 6.1. What simple chemical test distinguishes the two compounds?

6.57 [6.1] What is the molecular formula of the female sex hormone progesterone (see Box 6.1)?

6.58 Water adds to the carbonyl group of aldehydes and ketones to form compounds called **hydrates.** However, most hydrates are not stable, and the reaction reverses to give back the aldehyde or ketone and water. Chloral hydrate formed from chloral (trichloroethanal) is an exception. Chloral hydrate is a stable solid that has been used briefly in the past as a sedative but is no longer used, because it is a dangerous drug. It is also the drug referred to as "knockout drops" or "Mickey Finn" in some detective novels. Write the equation to show the formation of chloral hydrate from chloral.

6.59 Why is the number 2 required in the name 2-pentanone but not in the name butanone?

6.60 Write equations to show the sequence of reactions for the conversion of 1-butene into butanone.

6.61 [6.3] An unknown is either A, B, or C. The C-13 NMR spectrum shows peaks at 56, 114, 130, 132, 166, and 191 ppm, in addition to solvent peaks at 77 ppm (see Box 6.3). Identify the unknown.

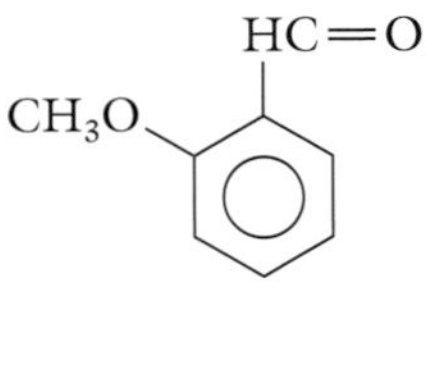

A: benzene ring with HC=O and ortho CH_3O

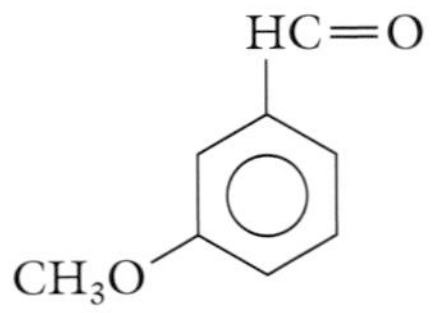

B: benzene ring with HC=O and para CH_3O

C: benzene ring with HC=O and meta CH_3O

6.3 **6.62** Is IR spectroscopy useful for differentiating between vanillin and menthone (see Box 6.3)? What peak(s) would be useful?

6.3 **6.63** Is IR spectroscopy useful for differentiating between progesterone and testosterone (see Box 6.3)? What peak(s) would be useful?

6.3 **6.64** Describe how C-13 NMR can be used to differentiate between 3-pentanone and pentanal (see Box 6.3).

6.65 Compound V (C_4H_8O) is reacted with H_2 and Pt to yield compound W ($C_4H_{10}O$). Compound W is heated in the presence of an acid catalyst to yield compound X (C_4H_8). Compound X is treated with H_2O and H^+ to yield compound Y ($C_4H_{10}O$). Compound Y is treated with $K_2Cr_2O_7$ and H^+ to yield compound Z (C_4H_8O). Compounds V and Z are isomers. Compounds W and Y are isomers. Identify V, W, X, Y, and Z.

6.66 Place the following compounds in order of increasing oxidation state for carbon (Sections 2.1 and 3.10): 1-butanol, butane, butanal.

6.67 Place numbers next to each carbon atom in hydroxypropanone to show the order of increasing oxidation state for carbon. Use 1 for the carbon atom having the highest oxidation state.

6.68 The compound shown here is used as the active ingredient in artificial tanning creams and lotions. The name dihydroxy-2-propanone is an incorrect IUPAC name. What is the correct IUPAC name?

$$HOCH_2-\overset{\overset{\displaystyle O}{\|}}{C}-CH_2OH$$

6.69 Why do biological systems use NADH instead of $LiAlH_4$ as the reducing agent for the conversion of a carbonyl group into an alcohol group—for example, in the conversion of pyruvic acid into lactic acid?

CHAPTER 7 CARBOXYLIC ACIDS, ESTERS, AND OTHER ACID DERIVATIVES

(David Joel/Tony Stone.)

Chemistry in Your Future

Because of his circulatory-system problems, your father has been advised by his physician to take an aspirin (acetylsalicylic acid) a day to reduce his risk of heart attack. Coincidentally, you have been reading about this risk reduction in one of your professional journals. As a quality-control manager in a food-processing plant, you are aware that some commonly used flavor additives for foods are salicylates and thus related to aspirin. A recent article suggested that the presence of these salicylates in foods may be responsible for the general decline in heart disease in the United States since World War II. However, the situation may not be so simple. Aspirin is both a carboxylic acid and an ester, but the salicylate flavor additives are only esters. Chapter 7 describes and compares these groups.

For more information about this topic and others in the chapter, go to www.whfreeman.com/bleiodian2e

Learning Objectives

- Draw and name carboxylic acids and their derivatives (esters, acid anhydrides, and acid halides).
- Describe the physical properties of carboxylic acids and their esters.
- Describe and write equations for the synthesis of carboxylic acids and their esters.
- Describe and write equations to demonstrate the acidity of carboxylic acids.
- Draw and name carboxylate salts and describe the cleaning action of soaps.
- Describe and write equations for the hydrolysis and saponification of esters.
- Draw and name phosphate esters and anhydrides.

Carboxylic acids and their derivatives are all around us. The tart taste of citrus fruits, vinegar, and rhubarb; the sharp sting of red ants; the unsavory smell and taste of rancid butter—all are due to carboxylic acids. Vitamin C, which is an essential part of our diet, is a carboxylic acid.

The pleasant odors and tastes of many fruits are due to carboxylic acid derivatives called esters, which are therefore commonly used in the manufacture of flavors. Esters are also used in the manufacture of plastics and textile fibers—not only for clothing and household goods, but also for medical and surgical applications—and in medicines such as aspirin, ibuprofen, and acetaminophen. Animal fats and plant oils are ester derivatives of carboxylic acids called fatty acids, and the sodium and potassium salts of these acids are used as soaps. The salts of other carboxylic acids are used as food preservatives. Proteins are amides, which are nitrogen-containing derivatives of carboxylic acids.

This chapter focuses mainly on carboxylic acids and their ester derivatives. Amides are so important to biological systems that a separate chapter (Chapter 8) deals with them. Also very important to living organisms are the phosphoric acid anhydrides and esters, which we consider at the end of this chapter and will meet again in our study of biochemistry. In contrast, the carboxylic acid halides and anhydrides are derivatives that are not present as such in living organisms and are discussed only briefly later in the chapter.

7.1 CARBOXYLIC ACIDS AND THEIR DERIVATIVES COMPARED

Carboxylic acids are another family of organic compounds that, like aldehydes and ketones, contain the carbonyl group, C═O. The identifying characteristic of carboxylic acids (often simply called organic acids) is that the C═O is present in a **carboxyl group,** a carbonyl group with an —OH attached. The carboxyl group is often abbreviated as —COOH or —CO_2H and carboxylic acids as RCOOH or RCO_2H.

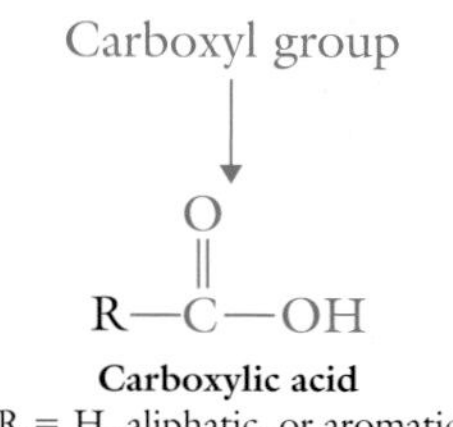

Carboxylic acid
(R = H, aliphatic, or aromatic)

The **carboxylic acid derivatives** are four additional families derived from the carboxylic acids—esters, acid anhydrides, amides, and acid halides:

$$R-\overset{\overset{O}{\|}}{C}-OR' \qquad R-\overset{\overset{O}{\|}}{C}-O-\overset{\overset{O}{\|}}{C}-R \qquad R-\overset{\overset{O}{\|}}{C}-NH_2 \qquad R-\overset{\overset{O}{\|}}{C}-X$$

Ester — **Acid anhydride** — **Amide** — **Acid halide** (X = halogen)

These derivatives differ from one another and from carboxylic acids in the group attached to the carbonyl carbon (Table 7.1).

TABLE 7.1 Carboxylic Acids and Their Derivatives

	Family	Z
$R-\overset{\overset{O}{\|}}{C}-Z$	acid	OH
R═H, R, Ar	ester	OR′ (R′═R′, Ar)
	acid anhydride	$O-\overset{\overset{O}{\|}}{C}-R$
	amide	NH_2, NHR, NRR′
	acid halide	X (halogen)

Esters and **anhydrides,** like carboxylic acids, have a second oxygen atom bonded to the carbonyl carbon, but the three families differ in the group that contains that oxygen: in esters, an —OR′ group; in anhydrides, an —O—CO—R group; and, in carboxylic acids, an —OH group. Anhydrides can also be defined as compounds with two different carbonyl carbons attached to the same oxygen.

Amides have a nitrogen attached in the form of an amino group—often the unsubstituted amino group $—NH_2$, but also the substituted amino groups —NHR and —NRR′. **Acid halides** have a halogen connected to the carbonyl carbon.

All four types of carboxylic acid derivatives can be synthesized from carboxylic acids and hydrolyzed back to them.

Example 7.1 Identifying carboxylic acids and their derivatives

Classify each of the following compounds as a carboxylic acid, ester, amide, anhydride, acid halide, or something else.

$$CH_3—\overset{O}{\overset{\|}{C}}—CH(CH_3)_2 \quad (CH_3)_2CH—\overset{O}{\overset{\|}{C}}—OH \quad C_6H_5—\overset{O}{\overset{\|}{C}}—NH_2$$

1 2 3

$$CH_3CH_2—\overset{O}{\overset{\|}{C}}—OCH_3 \quad CH_3—\overset{O}{\overset{\|}{C}}—Cl \quad CH_3CH_2—\overset{O}{\overset{\|}{C}}—O—\overset{O}{\overset{\|}{C}}—CH_2CH_3$$

4 5 6

Solution

Compound 1 is a ketone. It is neither a carboxylic acid nor a carboxylic acid derivative, because there are only carbons, no H and no heteroatom, attached to the carbonyl carbon. All of the other compounds are recognizable as acids or acid derivatives by the atom or group attached to the carbonyl carbon. Compound 2 is a carboxylic acid (—OH attached to the carbonyl carbon). Compound 3 is an amide (N attached to the carbonyl carbon). Compound 4 is an ester (—OR attached to the carbonyl carbon). Compound 5 is an acid halide (—Cl attached to the carbonyl carbon). Compound 6 is an acid anhydride (two carbonyl carbons attached to the same oxygen).

Problem 7.1 Classify each of the following compounds as a carboxylic acid, ester, amide, anhydride, or something else.

$$CH_3O—\overset{O}{\overset{\|}{C}}—CH_2CH_2CH_3 \quad C_6H_5—CH{=}O \quad (CH_3)_3C—\overset{O}{\overset{\|}{C}}—NH_2$$

1 2 3

$$CH_3—\overset{O}{\overset{\|}{C}}—Br \quad C_6H_5—\overset{O}{\overset{\|}{C}}—O—\overset{O}{\overset{\|}{C}}—C_6H_5 \quad HO—\overset{O}{\overset{\|}{C}}—CH_2CH_2CH_3$$

4 5 6

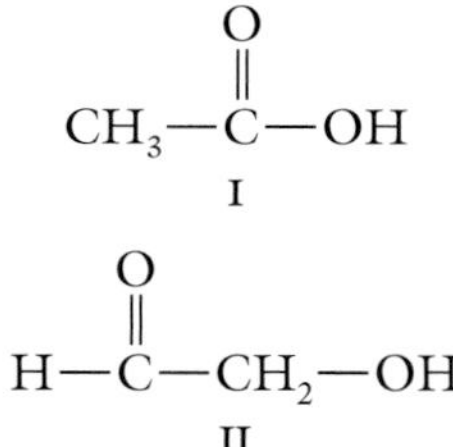

A compound is not a carboxylic acid, ester, amide, anhydride, or halide unless the —OH, —OR′, $—NH_2$, —O—CO—OR, or halogen group is directly attached to the carbonyl carbon. For example, consider isomeric compounds I and II shown in the margin. Compound I is a carboxylic acid. Compound II is not a carboxylic acid, and its properties are not those of a carboxylic acid, even though it contains both carbonyl and —OH groups. Instead, the properties of compound II are the sum of the properties of an aldehyde and an alcohol. The specific combination of the carbonyl and hydroxyl groups in the carboxyl group gives a carboxylic acid its unique properties.

As in other families of organic compounds, constitutional isomers (based on different carbon skeletons and different placements of the carboxyl group) are possible for carboxylic acids and acid derivatives. The orbital hybridization of the atoms in the C=O group and the bond angles about the C=O group in carboxylic acids and their derivatives are the same as those in aldehydes and ketones (sp^2 and 120°). The atoms (C, N, O) connected to the carbonyl carbon are all sp^3 hybridized with tetrahedral bond angles.

Note that, in the progression from saturated hydrocarbons to alcohols to aldehydes and ketones to carboxylic acids, we have studied organic compounds with carbon atoms of increasing oxidation state. The oxidation state of a carbon atom increases with an increasing number of bonds to oxygen or a decreasing number of bonds to hydrogen or both:

$$R{-}CH_2{-}H < R{-}CH_2{-}OH < R{-}\overset{O}{\overset{\|}{C}}{-}H < R{-}\overset{O}{\overset{\|}{C}}{-}OH$$

Lowest oxidation state — Highest oxidation state

7.2 THE SYNTHESIS OF CARBOXYLIC ACIDS

Methods of synthesizing carboxylic acids were encountered in preceding chapters, in the course of considering the reactions of other organic families. Benzoic acid, the simplest aromatic carboxylic acid, is synthesized from alkyl benzenes by the selective oxidation of the alkyl group to a carboxyl group (Section 4.4):

$$C_6H_5{-}CH_2{-}R \xrightarrow{MnO_4^- \text{ or } Cr_2O_7^{2-}} C_6H_5{-}C({=}O){-}OH + CO_2 + H_2O$$

« An aromatic ring is more stable to oxidation than an aliphatic group, as described in Section 4.4.

Carboxylic acids are the end products of the selective oxidation of primary alcohols and aldehydes (Sections 5.7 and 6.4):

$$R{-}CH_2{-}OH \xrightarrow{MnO_4^- \text{ or } Cr_2O_7^{2-}} R{-}\overset{H}{\overset{|}{C}}{=}O \xrightarrow{MnO_4^- \text{ or } Cr_2O_7^{2-}} R{-}\overset{OH}{\overset{|}{C}}{=}O$$

Example 7.2 Synthesizing carboxylic acids by selective oxidation

Write the equation for the selective oxidation of each of the following compounds to form a carboxylic acid: (a) toluene (methylbenzene); (b) 1-propanol; (c) propanal.

Solution

(a) $$C_6H_5{-}CH_3 \xrightarrow{MnO_4^- \text{ or } Cr_2O_7^{2-}} C_6H_5{-}C({=}O){-}OH$$

(b) $CH_3CH_2CH_2{-}OH \xrightarrow{MnO_4^- \text{ or } Cr_2O_7^{2-}} CH_3CH_2C(OH){=}O$

(c) $CH_3CH_2C(H){=}O \xrightarrow{MnO_4^- \text{ or } Cr_2O_7^{2-}} CH_3CH_2C(OH){=}O$

Problem 7.2 Write the equation for the selective oxidation of each of the following compounds: (a) ethylbenzene; (b) ethanol; (c) acetaldehyde (ethanal); (d) acetone (2-propanone).

7.3 NAMING CARBOXYLIC ACIDS

In the IUPAC nomenclature system, carboxylic acids are named according to the same rules as those for naming alkanes, with some modifications:

Rules for naming carboxylic acids

- Find the longest continuous chain that contains the carboxyl carbon. The carboxyl carbon is C1.
- The ending **-e** of the alkane name is changed to **-oic acid.**
- The names of substituents are used as prefixes, preceded by numbers to indicate their positions on the longest chain.
- A compound with two —COOH groups is named an **alkanedioic acid,** based on the longest continuous chain that contains both carboxyl carbons.
- The aromatic compound C_6H_5COOH is called **benzoic acid.** The ring carbon holding the —COOH is C1.

Many carboxylic acids are referred to by common names (Table 7.2). The prefixes used for the common names of carboxylic acids are the same as those used for the corresponding aldehydes; for example, formaldehyde and acetaldehyde (see Table 6.4) correspond to formic and acetic acids, respectively (Table 7.2).

TABLE 7.2 IUPAC and Common Names of Carboxylic Acids

Compound	IUPAC name	Common name	Melting point (°C)	Boiling point (°C)
H—COOH	methanoic acid	formic acid		101
CH_3—COOH	ethanoic acid	acetic acid		118
CH_3CH_2—COOH	propanoic acid	propionic acid		141
$CH_3CH_2CH_2$—COOH	butanoic acid	butyric acid		164
$CH_3(CH_2)_{10}$—COOH	dodecanoic acid	lauric acid	44	
$CH_3(CH_2)_{16}$—COOH	octadecanoic acid	stearic acid	71	
HOOC—COOH	ethanedioic acid	oxalic acid	190	
HOOC—CH_2—COOH	propanedioic acid	malonic acid	135	
HOOC—CH_2CH_2—COOH	butanedioic acid	succinic acid	188	
C_6H_5—COOH	benzoic acid	benzoic acid	122	249

Example 7.3 Naming carboxylic acids by the IUPAC system

Name each of the following compounds by the IUPAC system:

(a) 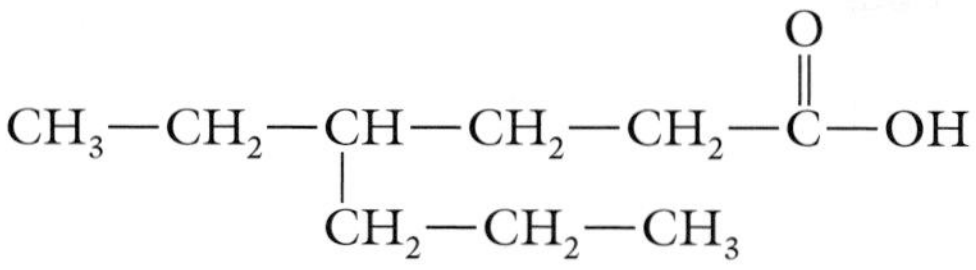 $CH_3-CH_2-CH(CH_2-CH_2-CH_3)-CH_2-CH_2-C(=O)-OH$

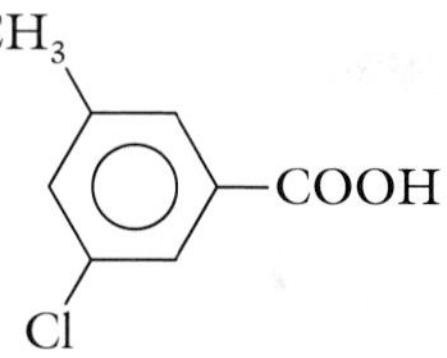

(b) benzene ring bearing CH_3, $-COOH$, Cl

(c) $HO-C(=O)-CH_2-CH(CH_2CH_3)-CH_2-C(=O)-OH$

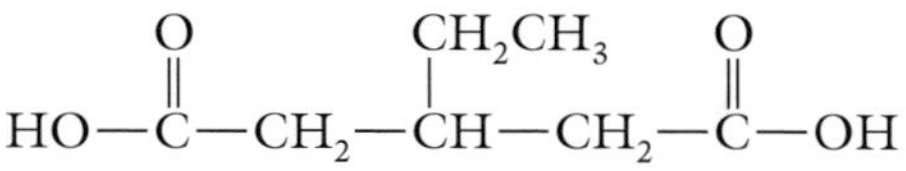

Solution

(a) The longest chain containing the COOH carbon is the seven-carbon chain. The COOH carbon is C1, and there is an ethyl group at C4. The name is 4-ethylheptanoic acid.

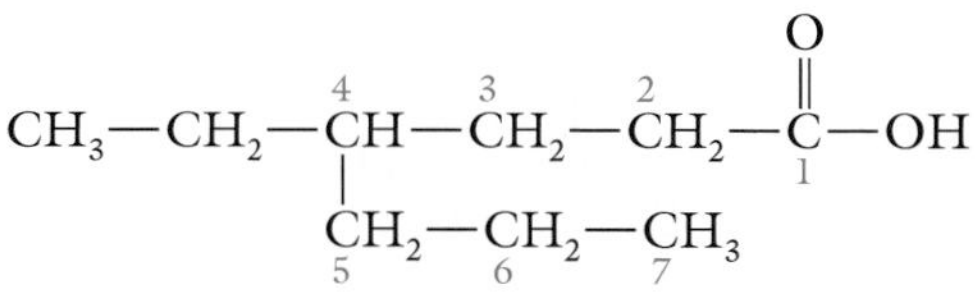

$$\overset{}{CH_3}-CH_2-\overset{4}{C}H(\underset{5}{C}H_2-\underset{6}{C}H_2-\underset{7}{C}H_3)-\overset{3}{C}H_2-\overset{2}{C}H_2-\underset{1}{C}(=O)-OH$$

(b) The —COOH group is at C1 of the benzene ring. The name is 3-chloro-5-methylbenzoic acid.

(c) This compound is an alkanedioic acid. The longest chain containing both —COOH groups has five carbons, with the —COOH groups at C1 and C5 and an ethyl group at C3. The name is 3-ethyl-pentanedioic acid.

Problem 7.3 Name each of the following compounds by the IUPAC system:

(a) 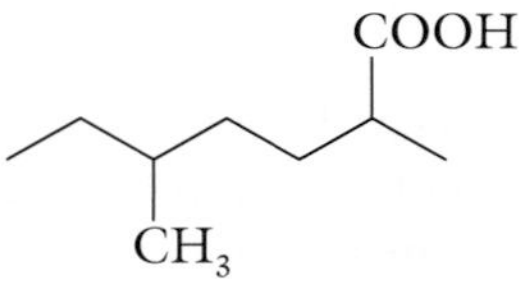 skeletal structure with COOH and CH_3 substituents

(b) 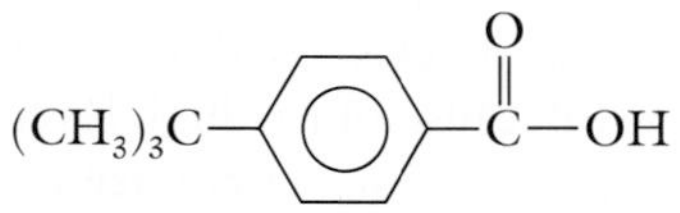 $(CH_3)_3C-C_6H_4-C(=O)-OH$

(c) $HOOC-CH_2CH_2CH(COOH)-CH_3$

Box 7.1 on page 223 describes some of the carboxylic acids that are commonly encountered in nature. Many of these carboxylic acids are polyfunctional compounds: either they contain two or more carboxyl groups or they contain other functional groups besides a carboxyl group. In the latter category are **hydroxy-**, **keto-**, and **aminocarboxylic acids**—carboxylic acids that contain hydroxyl, carbonyl, and amino groups, respectively.

7.4 PHYSICAL PROPERTIES OF CARBOXYLIC ACIDS

So far in our study of organic compounds, alcohols have been the family with the strongest secondary attractive forces and highest melting and boiling points, but carboxylic acids have even higher secondary attractive forces and even higher boiling and melting points than those of alcohols

TABLE 7.3 Comparison of Boiling Points of Compounds from Different Families

Compound	Structure	Molecular mass (amu)	Boiling point (°C)
pentane	$CH_3CH_2CH_2CH_2CH_3$	72	36
diethyl ether	$CH_3CH_2OCH_2CH_3$	74	35
methyl acetate	CH_3COOCH_3	74	57
butanal	$CH_3CH_2CH_2CHO$	72	76
butanone	$CH_3COCH_2CH_3$	72	80
1-butanol	$CH_3CH_2CH_2CH_2OH$	74	117
propanoic acid	CH_3CH_2COOH	74	141

(Table 7.3). The —OH and carbonyl groups of the carboxyl group are both polar:

The presence of the two polar groups results in very strong hydrogen bonding between pairs of carboxylic acid molecules, with strong attraction between the carbonyl oxygen and hydroxyl hydrogen:

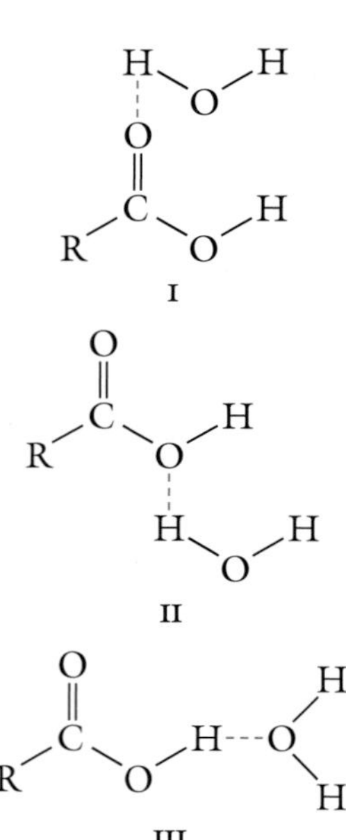

Thus, carboxylic acids exist as **dimers** (pairs of associated molecules) under most conditions. Because of the high level of attraction between the molecules in the dimer, high temperatures are needed to achieve melting and boiling.

Carboxylic acids show slightly higher solubility in water than do alcohols because of the three hydrogen-bond interactions with water shown in the margin: at the carbonyl oxygen (I), hydroxyl oxygen (II), and hydroxyl hydrogen (III).

The slightly greater water solubility of carboxylic acids compared with that of alcohols is observed as a more gradual decrease in solubility with increasing number of carbons. The water solubilities of 1-pentanol and pentanoic acid are 2.4 g/100 mL and 5.0 g/100 mL, respectively, at 20°C.

7.5 THE ACIDITY OF CARBOXYLIC ACIDS

Carboxylic acids are weak acids, undergoing ionization in water to form **carboxylate ion** and hydronium ion:

$$RC(=O){-}O{-}H + H_2O \rightleftharpoons RC(=O){-}O^- + H_3O^+$$

Carboxylic acid — Water — Carboxylate ion — Hydronium ion

The extent of ionization is less than about 1 to 2%; that is, the equilibrium is far to the left, as indicated by the unequal arrows. Carboxylic acids are much weaker acids than the strong inorganic acids such as HCl, H_2SO_4, HNO_3, and $HClO_4$, which are completely ionized in water. However, carboxylic acids are

7.1 CHEMISTRY AROUND US

Carboxylic Acids in Nature

Carboxylic acids are found in considerable abundance in nature. Most have sour (tart) tastes, and some have distinctive, often disagreeable, odors. Formic acid (IUPAC name: methanoic acid), the simplest of all carboxylic acids, is responsible for the irritation from the sting of red ants. Vinegar is a dilute (about 3–5%) aqueous solution of acetic acid (IUPAC name: ethanoic acid) formed by the enzymatic oxidation of ethanol in wine. Acetic acid is also an important industrial chemical, with large amounts used to manufacture cellulose acetate, which is used in fabrics and textiles, cigarette filters, and plastics applications. Propionic acid (IUPAC name: propanoic acid) is responsible for the odor and flavor of Swiss cheese, whereas the off-flavor and odor of rancid (spoiled) butter is due to butyric acid (IUPAC name: butanoic acid).

$$\mathrm{H{-}\overset{\overset{\displaystyle O}{\|}}{C}{-}OH}$$
Formic acid

$$\mathrm{CH_3{-}\overset{\overset{\displaystyle O}{\|}}{C}{-}OH}$$
Acetic acid

$$\mathrm{CH_3CH_2{-}\overset{\overset{\displaystyle O}{\|}}{C}{-}OH}$$
Propionic acid

$$\mathrm{CH_3CH_2CH_2{-}\overset{\overset{\displaystyle O}{\|}}{C}{-}OH}$$
Butyric acid

Many of the carboxylic acids in nature are polyfunctional—for example, oxalic, citric, lactic, and pyruvic acids.

Grapefruit, orange, and lemon contain citric acid. Spinach contains oxalic acid. Vinegar contains acetic acid. Swiss cheese contains propionic acid. (Richard Megna/Fundamental Photographs.)

Oxalic acid (IUPAC name: ethanedioic acid) is a toxic compound found in spinach and rhubarb. The concentration of oxalic acid in spinach leaves and rhubarb stalks (stems) is very low, well below levels that are toxic to humans. However, rhubarb leaves contain much higher concentrations, and ingestion of even a small amount of the rhubarb leaves can be dangerous.

Citric acid (IUPAC name: 3-hydroxy-1,3,5-propanetricarboxylic acid) is responsible for the tart taste of citrus fruits such as lemons, limes, oranges, and grapefruits.

$$\mathrm{HO{-}\overset{\overset{\displaystyle O}{\|}}{C}{-}\overset{\overset{\displaystyle O}{\|}}{C}{-}OH}$$
Oxalic acid

$$\mathrm{HO{-}\underset{\underset{\displaystyle CH_2COOH}{|}}{\overset{\overset{\displaystyle CH_2COOH}{|}}{C}}{-}COOH}$$
Citric acid

$$\mathrm{CH_3\overset{\overset{\displaystyle OH}{|}}{CH}{-}\overset{\overset{\displaystyle O}{\|}}{C}{-}OH}$$
Lactic acid

$$\mathrm{CH_3\overset{\overset{\displaystyle O}{\|}}{C}{-}\overset{\overset{\displaystyle O}{\|}}{C}{-}OH}$$
Pyruvic acid

Lemon juice is more than 5% citric acid. Lactic acid, produced by the bacterial fermentation of lactose (milk sugar), gives sour milk its disagreeable taste.

Lactic acid is used in cosmetic lotions and creams for treating dry, flaking, itchy skin and has been promoted as a beauty aid for removing wrinkles and moisturizing the skin. It is thought to hasten the sloughing off of dead cells from the skin. Such cosmetic preparations are often referred to as α-hydroxy acids, because a hydroxyl group is attached to the α-carbon (Section 6.4). Facial chemical peels, performed by a plastic surgeon, consist of much more aggressive treatment with trichloroacetic acid or phenol.

Our bodies produce certain carboxylic acids in the course of cellular metabolism. Lactic acid and pyruvic acid are produced in muscle tissue during vigorous exercise (Section 15.2). Citric and other acids are intermediates in the citric acid cycle (Section 15.5).

In Chapter 12, we will see how carboxylic acids containing α-amino (NH_2) groups, such as alanine, are used in nature as the building blocks for synthesizing proteins.

$$\mathrm{NH_2{-}\overset{\overset{\displaystyle CH_3}{|}}{CH}{-}COOH}$$
Alanine

stronger acids than phenols and much stronger acids than alcohols. Table 7.4 on the following page compares the $[H_3O^+]$, pH, and the ionization percentage of 0.1 M solutions of different compounds.

The carboxylic acids are more acidic than phenols because of the electron-withdrawing effect of the carbonyl group, a consequence of the carbonyl carbon's partial positive charge. Electron withdrawal stabilizes the carboxylate ion by charge dispersal: the negative charge does not reside entirely on the oxygen but is partly dispersed onto the carbonyl group. The extent of charge dispersal

TABLE 7.4 Relative Acidities of Organic Compounds

		Values for 0.10 *M* aqueous solution		
Compound	Structure	$[H_3O^+]$	pH	Ionization (%)
hydrochloric acid	HCl	0.10	1.00	100
benzoic acid	C_6H_5COOH	2.5×10^{-3}	2.60	2.5
acetic acid	CH_3COOH	1.3×10^{-3}	2.89	1.3
phenol	C_6H_5OH	3.3×10^{-6}	5.48	0.0033
ethanol	CH_3CH_2OH	1.0×10^{-7}	7.00	0.00010
water	H_2O	1.0×10^{-7}	7.00	0.00010

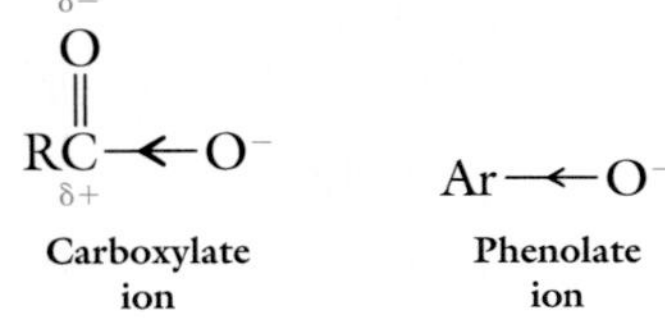

Carboxylate ion   Phenolate ion

is even greater for the carboxylate ion than for the phenolate ion because the carbonyl group is more strongly electron withdrawing than is a benzene ring.

Carboxylic acids are sufficiently acidic to turn blue litmus paper red—a simple chemical test that distinguishes carboxylic acids from alcohols (which do not react with the indicator dye in litmus paper) but not from phenols (which also turn blue litmus paper red).

Sodium hydroxide and sodium bicarbonate are sufficiently strong bases that their reactions with a carboxylic acid go to completion, forming the carboxylate salt:

$$\underset{\text{Carboxylic acid}}{\mathrm{RC(=O)-O-H}} + \underset{\text{Sodium hydroxide}}{\mathrm{NaOH}} \longrightarrow \underset{\text{Carboxylate salt}}{\mathrm{RC(=O)-O^-\ Na^+}} + \mathrm{H_2O}$$

$$\mathrm{RC(=O)-O-H} + \underset{\text{Sodium bicarbonate}}{\mathrm{NaHCO_3}} \longrightarrow \underset{\text{Carboxylate salt}}{\mathrm{RC(=O)-O^-\ Na^+}} + \mathrm{H_2CO_3}$$

$$\mathrm{H_2CO_3} \longrightarrow \mathrm{H_2O + CO_2}$$

Blue litmus paper turns red when treated with a carboxylic acid (*left*) but is unaffected by an alcohol (*right*). (Richard Megna/Fundamental Photographs.)

A solution of a carboxylic acid in pure water has a pH near 3, and only a small percentage of the carboxylic acid exists as ionized molecules (see Table 7.4). In more alkaline media (higher pH), the fraction of ionized, carboxylate ions increases. Almost all physiological fluids are maintained at near-neutral pH by the bicarbonate–carbonic acid buffer (Section 2.5). Blood pH is 7.35, and the pH in most cells ranges from 6.8 to 7.1. This physiological pH is 10,000 times as basic as pH 3. The greater basicity of physiological fluids relative to carboxylic acids results in neutralization of the carboxylic acids and conversion into carboxylate salts. Therefore, carboxylic acids, such as pyruvic, lactic, and citric acids (see Box 7.1), in living organisms are present almost entirely (>99%) in their carboxylate (ionized) form.

« pH = $-\log [H_3O^+]$ (Section 2.4)

Example 7.4 Ionization of carboxylic acids

Indicate whether a reaction takes place under each of the following conditions. If a reaction takes place, write the appropriate equation. (a) Benzoic acid + water; (b) benzoic acid + aqueous NaOH.

Solution

(a) Benzoic acid is only slightly ionized in water. The equilibrium is far to the left, as indicated by the unequal arrows:

$$\mathrm{C_6H_5C(=O)-O-H + H_2O} \rightleftharpoons \mathrm{C_6H_5C(=O)-O^- + H_3O^+}$$

(b) In the presence of a strong base such as NaOH, benzoic acid is almost completely ionized (indicated by the single arrow):

$$C_6H_5\overset{\overset{\displaystyle O}{\|}}{C}-O-H + NaOH \longrightarrow C_6H_5\overset{\overset{\displaystyle O}{\|}}{C}-O^- \; Na^+ + H_2O$$

Problem 7.4 Indicate whether a reaction takes place under the following conditions. If a reaction takes place, write the appropriate equation. (a) Propanoic acid + water; (b) propanoic acid + aqueous NaOH; (c) propanoic acid + aqueous $NaHCO_3$; (d) phenol + aqueous $NaHCO_3$.

Aromatic carboxylic acids are somewhat more acidic than aliphatic ones. Table 7.4 shows that $[H^+]$ for a benzoic acid solution is about twice that for an acetic acid solution. This difference is due to the electron-withdrawing characteristic of the benzene ring compared with an alkyl group. Greater electron withdrawal helps stabilize the aromatic carboxylate anion relative to the alkyl carboxylate anion.

7.6 CARBOXYLATE SALTS

A **carboxylate salt,** the product of reaction between a carboxylic acid and a strong base, is named as follows:

Rules for naming carboxylate salts

- The positive ion is named first.
- The carboxylate ion is named by changing the ending of the acid's name (either IUPAC or common) from **-ic acid** to **-ate.**

For example, the sodium salt of butanoic acid is named sodium butanoate.

Example 7.5 Naming carboxylate salts

Name the following carboxylate salts:

(a) $H-\overset{\overset{\displaystyle O}{\|}}{C}-O^- \; NH_4^+$ (b) $(C_6H_5-COO)_2Ca$

Solution

(a) This compound is the ammonium salt of methanoic (common name: formic) acid. The name is ammonium methanoate (common name: ammonium formate).

(b) This compound is the calcium salt of benzoic acid and is named calcium benzoate.

Problem 7.5 Name the following carboxylate salts:

(a) $(CH_3)_2CHCH_2\overset{\overset{\displaystyle O}{\|}}{C}ONa$ (b) $(CH_3COO^-)_3 \; Al^{3+}$

Carboxylate salts are ionic compounds and, therefore, possess much higher melting and boiling points than those of the corresponding carboxylic acids, which are covalent compounds. The attractive forces between ions in an ionic compound are stronger than any of the secondary attractive forces in covalent compounds (Sections 1.8 through 1.9). As a consequence, all carboxylate salts are solids at room temperature. For example, sodium formate is a solid with a melting point of 253°C, whereas formic acid is a liquid with a melting point of 8°C.

Carboxylate salts are much more soluble in water than the corresponding carboxylic acids. Carboxylic acids with more than 6 carbons are slightly soluble

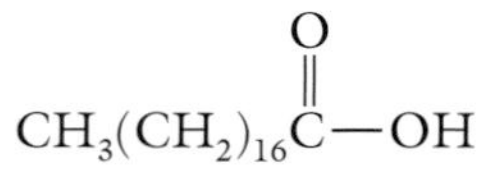

$$CH_3(CH_2)_{16}\overset{\overset{\displaystyle O}{\|}}{C}-OH$$

Stearic acid
Water insoluble

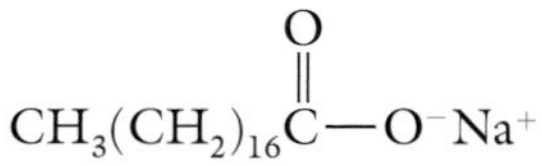

$$CH_3(CH_2)_{16}\overset{\overset{\displaystyle O}{\|}}{C}-O^-Na^+$$

Sodium stearate
Water soluble

or insoluble in water, but the salts are soluble up to a much larger number of carbons. For example, sodium stearate, with 18 carbons, is water soluble, whereas stearic acid is water insoluble. This increased solubility of carboxylate salts is due to the strong attractive forces between ions and water, much stronger than the hydrogen-bond attractions between carboxylic acids and water.

The solubility of a carboxylic acid is greatly increased in an environment of neutral or higher pH, because the carboxylic acid is converted into the carboxylate salt. This enhanced solubility is typical of physiological conditions, where carboxylic acids are solubilized by conversion into the carboxylate salt.

Medicines that contain carboxyl groups and need to be administered by injection are usually synthesized as their carboxylate salts. The carboxylate salt allows the preparation of higher concentrations of the medicine in aqueous solution, which results in faster absorption into the body on injection. Examples of such injectable medications are sodium ethacrynate and indomethacin sodium. Sodium ethacrynate is a diuretic used to treat edemas associated with congestive heart failure, cirrhosis of the liver, and renal disease. Indomethacin sodium is an anti-inflamatory, antipyretic, and analgesic used to treat gout and rheumatoid arthritis.

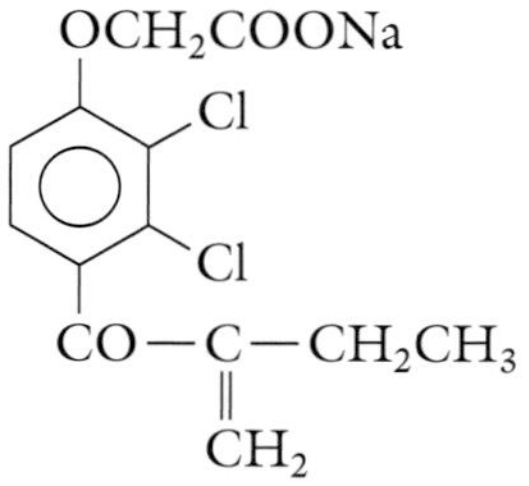

OCH_2COONa, Cl, Cl, $CO-\underset{\underset{\displaystyle CH_2}{\|}}{C}-CH_2CH_3$

Sodium ethacrynate

Cl, CO, N, CH_3, CH_3O, CH_2COONa

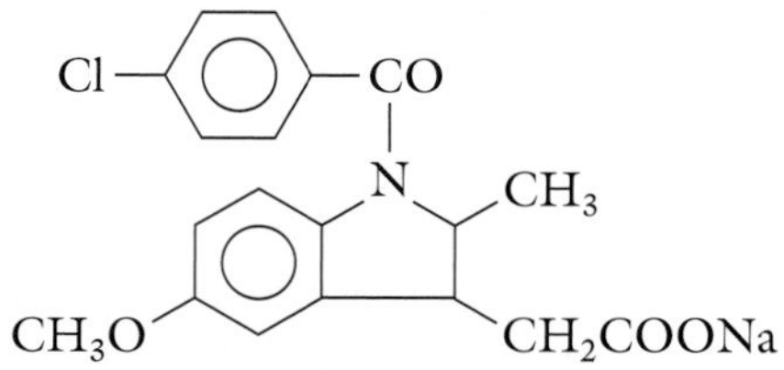

Indomethacin sodium

Example 7.6 Predicting the melting points of compounds

Place the following compounds in order of increasing melting point: sodium propanoate, 1-hexanol, pentanoic acid.

Solution

1-Hexanol $<$ pentanoic acid $\ll$ sodium propanoate. The ionic compound sodium propanoate has a much higher melting point than those of the two covalent compounds. Between the two covalent compounds of the same molecular mass, the carboxylic acid has the higher melting point because it has stronger hydrogen bonding (because of dimer formation) than the alcohol does.

Problem 7.6 Place the following compounds in order of increasing water solubility: $CH_3\,(CH_2)_{14}COOK$, $CH_3(CH_2)_{14}COOH$, $CH_3(CH_2)_{14}OH$.

A carboxylate salt, like the salt of any weak acid, is a weak base (proton acceptor) and undergoes acid–base reaction with water and strong acids (Section 2.4):

$$R\overset{\overset{\displaystyle O}{\|}}{C}-O^-K^+ + H_2O \rightleftharpoons R\overset{\overset{\displaystyle O}{\|}}{C}-OH + KOH$$

$$R\overset{\overset{\displaystyle O}{\|}}{C}-O^-K^+ + HCl \longrightarrow R\overset{\overset{\displaystyle O}{\|}}{C}-OH + KCl$$

Carboxylate salts are sufficiently basic to turn red litmus paper blue.

Some carboxylate salts have antibacterial and antifungal properties—they stop the growth of bacteria and fungi—and are used as food preservatives and in medications (Box 7.2).

7.2 CHEMISTRY AROUND US

Carboxylate Salts

Carboxylate salts are used in foods and other consumer products to stop or inhibit the growth of fungi (mold) and bacteria. Propionate salts, especially calcium and to a lesser extent sodium, are used as preservatives for baked goods, such as bread and cakes, and cheeses. Sodium benzoate exists naturally in cranberries, prunes, and other foods and is added as a food preservative to acidic food products such as carbonated beverages, ketchup, jams, jellies, and canned fruit. Sorbic acid (IUPAC name: 2,4-hexadienoic acid) and its sodium and potassium salts are used as preservatives for wines, carbonated beverages, pickled products, and fruit juices.

Two slices of moldy homemade wheat bread are at the left of commercial bread made with preservatives such as propionate salts. (Bill Aron/PhotoEdit.)

$$\left(CH_3CH_2-\overset{O}{\overset{\|}{C}}-O^-\right)_2Ca^{2+}$$

Calcium propionate

$$C_6H_5-\overset{O}{\overset{\|}{C}}-O^-Na^+$$

Sodium benzoate

$$CH_3CH{=}CHCH{=}CHCOOH$$

Sorbic acid

Monosodium glutamate (MSG) is used as a flavor enhancer in various foods. It is the monosodium salt of glutamic acid, one of the twenty amino acids used in synthesizing proteins in biological systems (Section 12.1).

$$H_2N-\underset{CH_2CH_2COOH}{\underset{|}{CH}}-\overset{O}{\overset{\|}{C}}-O^-Na^+$$

Monosodium glutamate (MSG)

Zinc 10-undecenoate is the active antifungal ingredient in athlete's-foot preparations. Aluminum acetate is used in creams and lotions for the treatment of diaper rash and acne.

$$[CH_2{=}CH(CH_2)_8COO^-]_2Zn^{2+}$$

Zinc 10-undecenoate

$$[CH_3COO^-]_3Al^{3+}$$

Aluminum acetate

7.7 SOAPS AND THEIR CLEANING ACTION

The **soaps** that we use to clean our bodies are the sodium or potassium carboxylate salts of long-chain carboxylic acids, called **fatty acids** and typically mixtures of unbranched aliphatic chains of 12 to 20 carbon atoms. The fatty acids are called fatty acids because they are obtained from animal fats and plant oils (for example, beef and pig fats, coconut oil). The carboxylate salts are derived from fats and oils by a process called saponification (Sections 7.3 and 11.4). An example of a soap molecule is sodium stearate, the sodium carboxylate salt of stearic acid, $CH_3(CH_2)_{16}COO^-Na^+$.

> **Fats and oils are lipids and the subject of Chapter 11.**

Dirt adheres to surfaces rather than dissolving in water, either because the dirt is hydrophobic (water hating) or because it is physically combined with body oils, cooking fats, greases, and other substances that are hydrophobic. Soap molecules in water allow greasy dirt to be solubilized because soap molecules are **amphipathic**; that is, they contain both hydrophobic and hydrophilic (water-loving) parts. The long hydrocarbon chain of the carboxylate salt is hydrophobic, and the ionic COO^-Na^+ end is hydrophilic. The hydrophobic and hydrophilic parts, called the hydrophobic tail and hydrophilic head, are represented by the wavy line and circle, respectively, in Figure 7.1.

The hydrophobic tail of a soap molecule has an affinity for the greasy dirt, whereas the hydrophilic head has an affinity for water, both consequences of the like-dissolves-like principle. Soap molecules, being attracted strongly to both dirt and water, break up the dirt particles into smaller particles. The

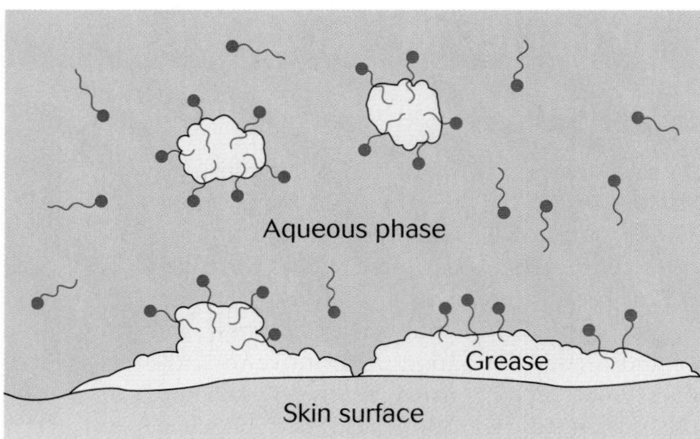

Figure 7.1 The cleaning action of soap. ~• = $R—COO^-Na^+$

surfaces of these particles are surrounded by soap molecules, which allows them to become solubilized in water.

The requirements for a good soap are an ionic group as the hydrophilic head and a long hydrocarbon group as the hydrophobic tail. A long-chain carboxylic acid, such as stearic acid, has only one of the requirements. It does not function well as a soap, because the carboxyl group is not nearly as hydrophilic as the carboxylate salt. Likewise, the carboxylate salt of a short-chain carboxylic acid, such as $CH_3(CH_2)_2COONa$, has only one of the requirements. It does not function well as a soap, because the small hydrocarbon tail is not sufficiently hydrophobic.

The soaps that we use contain small amounts of other materials such as perfumes, deodorants, skin moisturizers, and colorants. Some soaps have air blown into them so that they float on water. Antiseptics are added to make medicated soaps, and pumice is added to make scouring soaps.

Box 7.3 describes the special requirements of soaps used in regions with hard water.

7.3 CHEMISTRY AROUND US

Hard Water and Detergents

Soaps—sodium or potassium carboxylates—are one type of cleaning agent. Soaps work best in **soft water**—water that does not contain significant amounts of Ca^{2+}, Mg^{2+}, Fe^{2+}, and Fe^{3+}. When a soap is used in **hard water**—water containing these ions—the ions exchange with the Na^+ and K^+ of the soap to form insoluble calcium, magnesium, and iron carboxylate salts. For example:

$$2\ CH_3(CH_2)_{16}COO^-Na^+ + Ca^{2+} \longrightarrow [CH_3(CH_2)_{16}COO^-]_2\ Ca^{2+} + 2\ Na^+$$

Water-soluble soap → **Water-insoluble curd**

These salts precipitate from solution as **soap curd,** the scum responsible for the off-white or light-gray rings in bathtubs and sinks and the dull finish that sometimes appears on washed clothes. Only when the Ca^{2+}, Mg^{2+}, Fe^{2+}, and Fe^{3+} ions have been removed from hard water by the formation of curd can the remaining soap begin its cleaning action. Thus, in regions of hard water, much larger amounts of soap are required for cleaning.

One solution to the problem of hard water is to install a **water softener** in the water line in the house. This device converts hard water into soft water by replacing Ca^{2+}, Mg^{2+}, Fe^{2+}, and Fe^{3+} with Na^+ ions, a process called **ion exchange.** Another solution is to use a different type of cleaning agent—a **detergent,** such as sodium or potassium alkylbenzene sulfonates.

$$CH_3(CH_2)_{10}CH_2-C_6H_4-SO_2-O^-K^+$$

Alkylbenzene sulfonate

Like the carboxylate soaps, these compounds exchange their Na^+ and K^+ ions for the Ca^{2+}, Mg^{2+}, Fe^{2+}, and Fe^{3+} ions in the hard water, but the resulting calcium, magnesium, and iron alkylbenzene sulfonates are water soluble. Thus, they stay in solution and exert their cleaning action.

7.8 ESTERS FROM CARBOXYLIC ACIDS AND ALCOHOLS

Esters (RCOOR′) are formed from carboxylic acids by **esterification,** the reaction of an acid with an alcohol (R′ = aliphatic) or a phenol (R′ = aromatic). The two reactants are heated in the presence of a strong acid, such as sulfuric acid, as catalyst:

Bonds broken — Bond formed

$$\underset{\text{Acid}}{R{-}\overset{\overset{\displaystyle O}{\|}}{C}{-}OH} + \underset{\substack{\text{Alcohol}\\\text{or phenol}}}{H{-}O{-}R'} \xrightleftharpoons{H^+} \underset{\text{Ester}}{R{-}\overset{\overset{\displaystyle O}{\|}}{C}{-}O{-}R'} + H_2O$$

Esterification entails the breaking of two bonds: the O—H bond in the alcohol (or phenol) and the C—O bond between the hydroxyl oxygen and the carbonyl carbon of the carboxylic acid. Bond breakage is followed by an exchange of fragments and new bond formation. The OH from the carboxylic acid bonds to the H from the alcohol to form water. The alcohol oxygen bonds to the carbonyl carbon to form the ester.

Esterification is a reversible reaction. The yield of ester is increased by the removal of water to shift the equilibrium to the right side (Le Chatelier's principle). This removal is easily achieved in esterification because the reaction is usually run at a temperature considerably higher than the boiling point of water, causing water to distill out of the reaction vessel as it is formed.

Living organisms synthesize esters by reactions catalyzed by enzymes. These esters include a variety of plant and animal waxes, fats, and oils (Chapter 11). Esters are also synthesized from the reaction of an alcohol (or phenol) with a carboxylic acid anhydride or halide, rather than the carboxylic acid (Section 7.4).

Whereas most carboxylic acids have disagreeable odors, esters of carboxylic acids have pleasant odors and flavors. The lower-molecular-mass esters are responsible for the odors and flavors of many fruits. Synthetic esters are often added as flavoring agents to soft drinks, ice creams, yogurts, and other food products. For example, ethyl butanoate, isobutyl methanoate, and 3-methylbutyl ethanoate can be used to add pineapple, raspberry, and banana flavor and odor, respectively, to a food product. Various esters are also used to create the distinct fragrances of perfumes.

Esters are responsible for the pleasant odors and flavors of many fruits. (David Parker/SLP/Photo Researchers.)

One of the most widely used medicines in the world, aspirin, is an ester derived from *p*-hydroxybenzoic acid (common name: salicylic acid). Box 7.4 on the following page describes aspirin and related medicines.

Example 7.7 Writing an equation for the synthesis of an ester

Give the equation for ester formation between propanoic acid and 2-propanol.

Solution

$$CH_3CH_2{-}\overset{\overset{\displaystyle O}{\|}}{C}{-}OH + H{-}OCH(CH_3)_2 \xrightarrow[-H_2O]{H^+} CH_3CH_2{-}\overset{\overset{\displaystyle O}{\|}}{C}{-}OCH(CH_3)_2$$

Problem 7.7 Give the equation for ester formation between butanoic acid and phenol.

Thioesters are compounds formed by the esterification of carboxylic acids with thiols (Section 5.11):

$$\underset{\text{Acid}}{R{-}\overset{\overset{\displaystyle O}{\|}}{C}{-}OH} + \underset{\text{Thiol}}{H{-}S{-}R'} \xrightleftharpoons{H^+} \underset{\text{Thioester}}{R{-}\overset{\overset{\displaystyle O}{\|}}{C}{-}S{-}R'} + H_2O$$

7.4 CHEMISTRY AROUND US

Aspirin and Aspirin Substitutes

Aspirin (acetylsalicylic acid) is an **analgesic, antipyretic,** and **anti-inflammatory** medicine. That is, it relieves pain, such as the pain of headache, toothache, and sprains; it reduces fever but does not affect normal body temperature; and it reduces inflammation, such as the inflammation accompanying arthritis. Aspirin works by inhibiting the formation of prostaglandins (Section 11.8), a group of lipids responsible for the body's inflammatory and febrile responses and for the sensation and transmission of pain in the nervous system.

Acetylsalicylic acid, both a carboxylic acid and an ester, is synthesized by the acyl transfer reaction (Section 7.4) from acetic anhydride to the —OH group of salicylic acid:

$$\text{(Salicylic acid)} + CH_3-\overset{\overset{O}{\|}}{C}-O-\overset{\overset{O}{\|}}{C}-CH_3 \longrightarrow \text{(Acetylsalicylic acid)}$$

Salicylic acid

Acetylsalicylic acid

Aspirin can also have adverse effects in some people. It decreases the ability of blood to clot, which can result in gastrointestinal bleeding, general susceptibility to tissue bruising, and hemorrhagic stroke—that is, stroke resulting from excessive bleeding from ruptured blood vessels in the brain. The risk of gastrointestinal bleeding can be reduced by using enteric-coated aspirin—aspirin coated with a material that dissolves only in alkaline pH. Enteric-coated aspirin passes intact through the interior of the highly acidic stomach, and the coating does not dissolve until it reaches the alkaline contents of the intestine. Another potential adverse effect of aspirin is Reye's syndrome, a devastating and sometimes fatal illness including brain and liver dysfunction. It affects a small percentage of children who take aspirin in bouts of chickenpox or flu.

The anticlotting action of aspirin is used to protect people at risk for heart disease. Daily or every-other-day doses of aspirin prevent heart attacks by suppressing the formation of blood clots in clogged arteries feeding the heart.

Acetaminophen (an amide), ibuprofen (a carboxylic acid), and naproxen sodium are alternative medicines for people who cannot take aspirin. Acetaminophen (Tylenol, Panadol, Datril) is a useful alternative to aspirin as an analgesic and antipyretic, but it is not an anti-inflammatory drug. Ibuprofen (Advil, Motrin, Nuprin) is an anti-inflammatory drug, perhaps superior to aspirin, as well as an analgesic and antipyretic. Naproxen sodium (Aleve) is an anti-inflammatory drug, as well as an analgesic and antipyretic having a less-adverse effect on the gastrointestinal system than aspirin.

Ibuprofen

Acetaminophen

Naproxen sodium

These medicines should be used with caution and moderation. Indiscriminate use of acetaminophen has been linked to damage to the liver (especially when used with alcohol) and kidney. Long-term use of naproxen sodium also is implicated in kidney damage.

Thioesters derived from the thiol coenzyme A play important roles in several metabolic pathways, including the citric acid cycle (Section 15.5).

Thiol

Coenzyme A

7.9 NAMES AND PHYSICAL PROPERTIES OF ESTERS

The names of esters combine the names of the **alkoxy** or **aryloxy** (that is, the alcohol or phenol) part and the **acyl** (that is, the carboxylic acid) part of the compound.

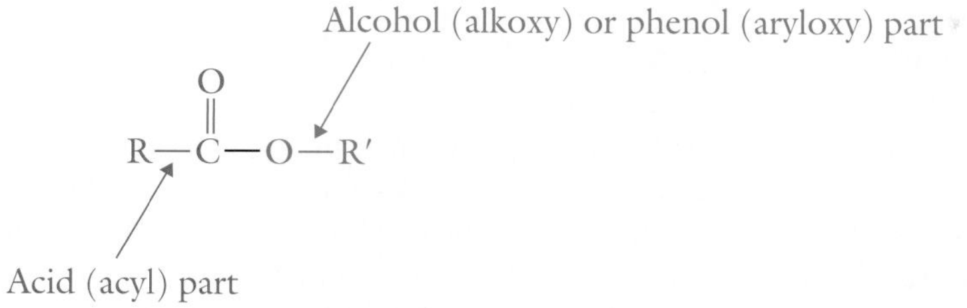

Alcohol (alkoxy) or phenol (aryloxy) part

$$R-\overset{\overset{\large O}{\|}}{C}-O-R'$$

Acid (acyl) part

Rules for naming esters

- The alkyl (or aryl) group of the alkoxy (or aryloxy) part of an ester is named first as a separate word.
- The acyl part is named by changing the ending of the name of the carboxylic acid (either IUPAC or common) from **-ic acid** to **-ate.**

Example 7.8 Naming esters by the IUPAC nomenclature system

Give the IUPAC names for the following compounds:

(a) $CH_3CH_2-\overset{\overset{O}{\|}}{C}-OCH(CH_3)_2$ (b) $CH_3CH_2O-\overset{\overset{O}{\|}}{C}-CH(CH_3)_2$

(c) 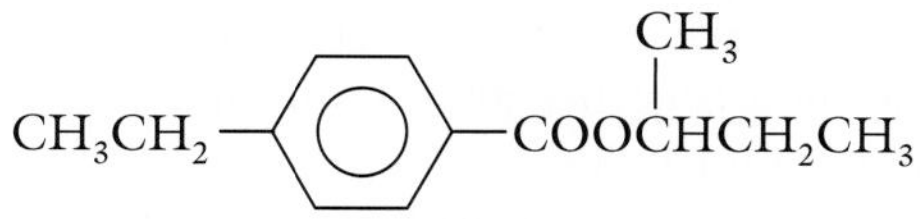 $CH_3CH_2-C_6H_4-COO\overset{\overset{CH_3}{|}}{C}HCH_2CH_3$

Solution

(a) The alkoxy part contains an isopropyl group. The acyl part comes from propanoic acid (common name: propionic acid). The IUPAC name of the ester is isopropyl propanoate (common name: isopropyl propionate).

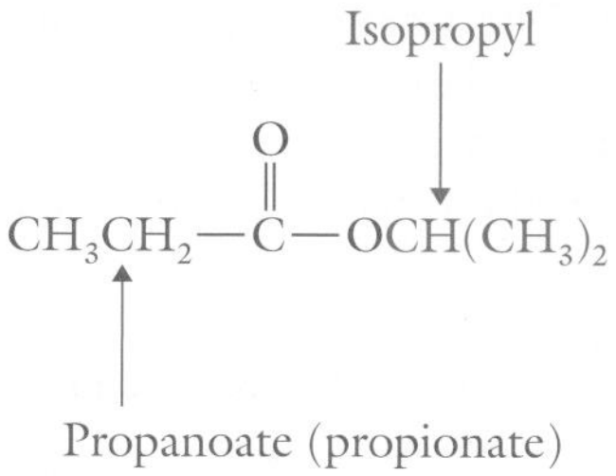

Isopropyl

$$CH_3CH_2-\overset{\overset{O}{\|}}{C}-OCH(CH_3)_2$$

Propanoate (propionate)

(b) Do not assume that the structure of an ester is always written with the acyl part on the left and the alkoxy part on the right. This ester is drawn in the opposite manner, acyl part on the right and alkoxy part on the left. Always remember that the acyl part does not contain an oxygen connected to the carbonyl carbon. The IUPAC name is ethyl 2-methylpropanoate.

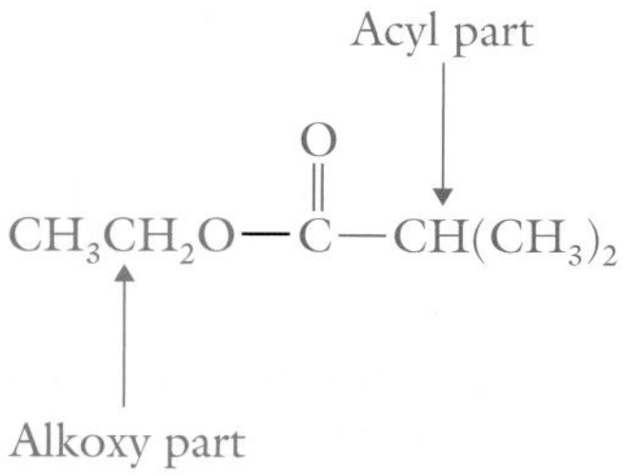

Acyl part

$$CH_3CH_2O-\overset{\overset{O}{\|}}{C}-CH(CH_3)_2$$

Alkoxy part

(c) The alkoxy part contains the *s*-butyl group, and the acyl part comes from *p*-ethylbenzoic acid. The ester is named *s*-butyl *p*-ethylbenzoate.

Problem 7.8 Give the IUPAC names of the following compounds:

(a) 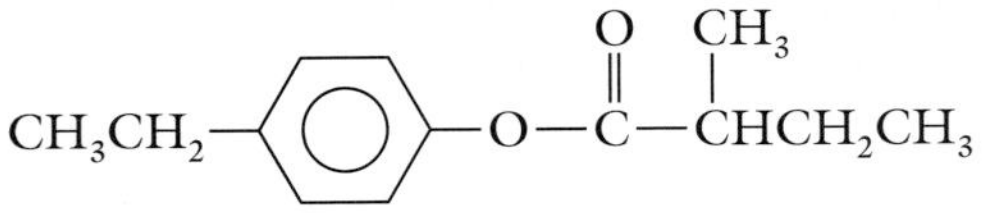 $CH_3CH_2-C_6H_4-O-\overset{O}{\overset{\|}{C}}-\underset{}{\overset{CH_3}{\overset{|}{C}}}HCH_2CH_3$

(b) $HCOOCH_2CH_2CH_2CH_2CH_3$ (c) $CH_3OOCCH_2CH_2CH_2CH_2CH_3$

Esters have much lower boiling and melting points than those of carboxylic acids because the secondary attractive forces in esters (dipole–dipole) are considerably lower than those in carboxylic acids (hydrogen bonding). For example, the boiling points of propanoic acid and methyl acetate are 141°C and 57°C, respectively (see Table 7.3). Carboxylic acids are more soluble in water than esters because carboxylic acids form strong hydrogen bonds with water. The attractive forces between esters and water are not as strong.

The polarity and secondary attractive forces in esters are lower than those in aldehydes and ketones, which results in lower melting and boiling points for esters relative to aldehydes and ketones. For example, the boiling points of methyl acetate and butanal are 57°C and 76°C, respectively (see Table 7.3). Esters have about the same water solubilities as those of aldehydes and ketones because they hydrogen bond equally well with water.

7.10 POLYESTER SYNTHESIS

The esterification reaction becomes a polymerization reaction if the reactants are **bifunctional;** that is, the reactants have two functional groups per molecule. They behave as **monomers,** joining together in large numbers (>50) to produce the much larger sized polymer molecules.

For example, the acid-catalyzed reaction of a dicarboxylic acid (which contains two —COOH groups) and a diol (which contains two —OH groups) produces a **polyester.** The reactants initially yield dimer molecules with —COOH at one end and —OH at the other end. The dimer molecules react to form a tetramer, which also contains a carboxyl group at one end and a hydroxyl at the other end. The tetramers react, and the process repeats itself over and over again to produce larger and larger molecules. When the molecular mass of the polyester is high enough to produce a solid material with the desired mechanical strength, the reaction can be stopped by cooling.

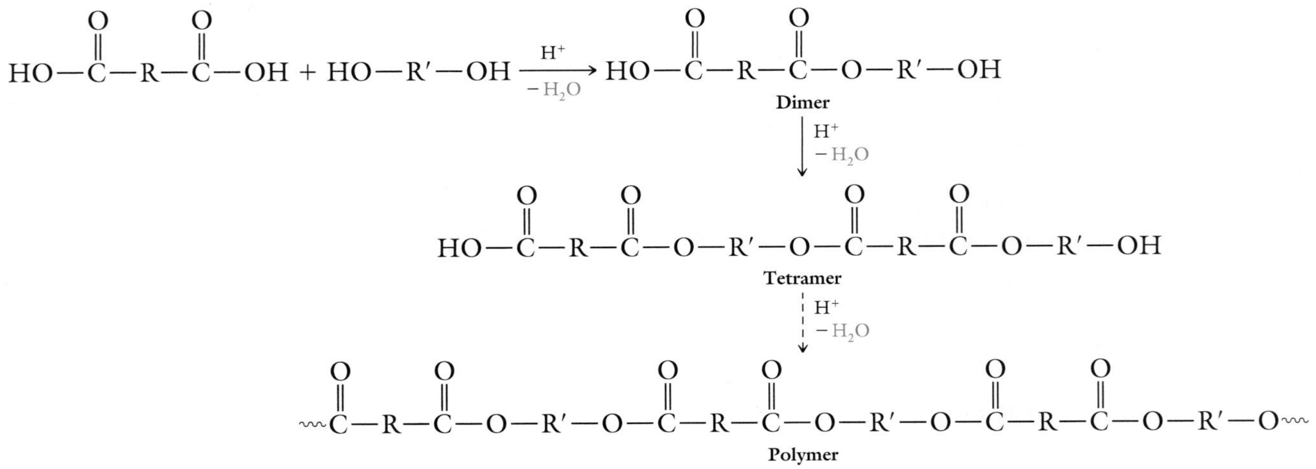

$$HO-\overset{O}{\overset{\|}{C}}-R-\overset{O}{\overset{\|}{C}}-OH + HO-R'-OH \xrightarrow[-H_2O]{H^+} \underset{\textbf{Dimer}}{HO-\overset{O}{\overset{\|}{C}}-R-\overset{O}{\overset{\|}{C}}-O-R'-OH}$$

$$\downarrow H^+, -H_2O$$

$$\underset{\textbf{Tetramer}}{HO-\overset{O}{\overset{\|}{C}}-R-\overset{O}{\overset{\|}{C}}-O-R'-O-\overset{O}{\overset{\|}{C}}-R-\overset{O}{\overset{\|}{C}}-O-R'-OH}$$

$$\downarrow H^+, -H_2O$$

$$\underset{\textbf{Polymer}}{\sim\overset{O}{\overset{\|}{C}}-R-\overset{O}{\overset{\|}{C}}-O-R'-O-\overset{O}{\overset{\|}{C}}-R-\overset{O}{\overset{\|}{C}}-O-R'-O-\overset{O}{\overset{\|}{C}}-R-\overset{O}{\overset{\|}{C}}-O-R'-O\sim}$$

The wavy lines in the polymer structure indicate that the structure goes on and on, with large numbers of monomer units linked together.

The polymer structure is usually drawn in much more abbreviated form, with brackets enclosing the **repeat unit:**

$$n\,\text{HO}-\overset{\displaystyle\text{O}}{\overset{\|}{\text{C}}}-\text{R}-\overset{\displaystyle\text{O}}{\overset{\|}{\text{C}}}-\text{OH} + n\,\text{HO}-\text{R}'-\text{OH} \xrightarrow[-\text{H}_2\text{O}]{\text{H}^+} \left(\!\!-\overset{\displaystyle\text{O}}{\overset{\|}{\text{C}}}-\text{R}-\overset{\displaystyle\text{O}}{\overset{\|}{\text{C}}}-\text{O}-\text{R}'-\text{O}-\!\!\right)_n$$

The subscript n indicates that the repeat unit repeats over and over many times (that is, n is a large number).

This type of polymerization reaction is called a **condensation polymerization** because a small molecule (water in this case) is a by-product of the process. Condensation polymerization is one of two types of polymerization reactions. The other type, in which no small by-product results, is called addition polymerization and is the type of polymerization that takes place with alkenes (Section 4.7 and Box 4.6).

Example 7.9 Writing an equation for a polyesterification

Write the equation to describe the polyesterification reaction between ethylene glycol ($HOCH_2CH_2OH$) and terephthalic acid (p-$HOOC—C_6H_4—COOH$).

Solution

$$n\,\text{HO}-\overset{\displaystyle\text{O}}{\overset{\|}{\text{C}}}-\text{C}_6\text{H}_4-\overset{\displaystyle\text{O}}{\overset{\|}{\text{C}}}-\text{OH} + n\,\text{HO}-\text{CH}_2\text{CH}_2-\text{OH} \xrightarrow[-\text{H}_2\text{O}]{\text{H}^+}$$

$$\left(\!\!-\overset{\displaystyle\text{O}}{\overset{\|}{\text{C}}}-\text{C}_6\text{H}_4-\overset{\displaystyle\text{O}}{\overset{\|}{\text{C}}}-\text{O}-\text{CH}_2\text{CH}_2-\text{O}-\!\!\right)_n$$

Problem 7.9 Each of the following pairs of reactants undergoes esterification. Which of the pairs produces a polymer? Write the equation for any reaction(s) that produce(s) a polymer.

(a) $\text{HO}-\overset{\displaystyle\text{O}}{\overset{\|}{\text{C}}}\text{CH}_2\text{CH}_2\text{CH}_2\text{CH}_2\overset{\displaystyle\text{O}}{\overset{\|}{\text{C}}}-\text{OH} + \text{HO}-\text{CH}_2\text{CH}_3$

(b) $\text{HOOCCH}_2\text{CH}_2\text{CH}_2\text{CH}_3 + \text{HOCH}_2\text{CH}_2\text{OH}$

(c) $\text{HO}-\overset{\displaystyle\text{O}}{\overset{\|}{\text{C}}}\text{CH}_2\text{CH}_2\text{CH}_2\text{CH}_2\overset{\displaystyle\text{O}}{\overset{\|}{\text{C}}}-\text{OH} + \text{HO}-\text{CH}_2\text{CH}_2-\text{OH}$

The polymer formed by the reaction in Example 7.9 is poly(ethylene terephthalate), or PET (also called Dacron, Mylar, or polyester). This commercially important material has an annual production of about 10 billion pounds in the United States. About 50% of the production goes into textile applications—including clothing, curtains, upholstery, and tire cord—and medical applications, such as the fabric used to repair blood vessels and the knitted tubing used in artificial heart valves. The remaining 50% of PET production goes into plastics applications. They include photographic, magnetic, and X-ray films and tapes and bottles for soft drinks, beer, and other food products.

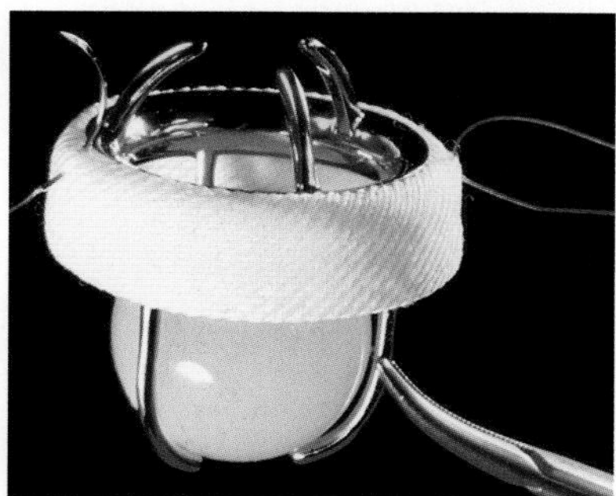

Polyester fabric is used in constructing artificial heart valves. (SIU/Visuals Unlimited.)

7.11 THE HYDROLYSIS OF ESTERS

Hydrolysis reactions were last encountered in Section 6.6, when we studied the hydrolysis of hemiacetals and acetals. Recall that hydrolysis is a reaction in which a molecule reacting with water is cleaved into two parts, each of which has fewer carbons than the original molecule. An ester hydrolyzes to the corresponding carboxylic acid and alcohol (or phenol) under either acidic or basic reaction conditions. Acidic hydrolysis requires the presence of a strong acid catalyst such as sulfuric acid:

$$\underset{\text{Ester}}{\text{R}-\overset{\overset{\text{O}}{\|}}{\text{C}}-\text{O}-\text{R}'} + \text{H}_2\text{O} \underset{}{\overset{\text{H}^+}{\rightleftharpoons}} \underset{\text{Acid}}{\text{R}-\overset{\overset{\text{O}}{\|}}{\text{C}}-\text{OH}} + \underset{\text{Alcohol or phenol}}{\text{H}-\text{O}-\text{R}'}$$

The acidic hydrolysis of an ester is the reverse of acid-catalyzed ester formation from a carboxylic acid and an alcohol (or phenol). Carrying out the reaction with a large excess of water shifts the equilibrium toward the hydrolysis products (Le Chatelier's principle). This condition is the reverse of those used to achieve esterification.

The hydrolysis and formation of esters take place continuously in living organisms. Triacylglycerols and many other lipids in our bodies and in our diets are esters. Digestion of these compounds is a simple matter of ester hydrolysis (Section 11.4). Our cells then synthesize new esters (to construct cell membranes, for example). These syntheses are catalyzed not by a strong acid, such as sulfuric acid, but by special proteins called enzymes.

Basic hydrolysis requires a strong base such as sodium or potassium hydroxide:

$$\underset{\text{Ester}}{\text{R}-\overset{\overset{\text{O}}{\|}}{\text{C}}-\text{O}-\text{R}'} + \text{NaOH} \xrightarrow{\text{H}_2\text{O}} \underset{\text{Carboxylate salt}}{\text{R}-\overset{\overset{\text{O}}{\|}}{\text{C}}-\text{ONa}} + \underset{\text{Alcohol or phenol}}{\text{H}-\text{O}-\text{R}'}$$

The basic hydrolysis of an ester produces the metal carboxylate salt instead of the carboxylic acid because the reaction is performed in the presence of the base. This reaction, also called **saponification,** is used to manufacture soaps from certain animal fats and vegetable oils (Section 11.4).

The basic hydrolysis and the acidic hydrolysis of esters differ in two important ways:

- Basic hydrolysis, unlike acidic hydrolysis, is not an equilibrium reaction. The formation of an ionic compound (the carboxylate salt) drives the equilibrium to the right, making the reaction irreversible. Basic hydrolysis is often preferred to acidic hydrolysis for this reason.
- The acid in acidic hydrolysis is a catalyst; it is not used up in the reaction and only small amounts are needed relative to the ester. The base in basic hydrolysis (saponification) is a reactant, needed in equimolar amounts relative to the ester, and is consumed in the reaction.

Example 7.10 Recognizing and writing esterification reactions

Complete the following equations. If no reaction takes place, write "no reaction."

(a) $\text{CH}_3\overset{\overset{\text{O}}{\|}}{\text{C}}-\text{OCH}_2\text{CH}_3 \xrightarrow{\text{H}_2\text{O}} ?$

(b) $\text{CH}_3\overset{\overset{\text{O}}{\|}}{\text{C}}-\text{OCH}_2\text{CH}_3 \xrightarrow[\text{H}^+]{\text{H}_2\text{O}} ?$

(c) $\text{CH}_3\overset{\overset{\text{O}}{\|}}{\text{C}}-\text{OCH}_2\text{CH}_3 \xrightarrow[\text{NaOH}]{\text{H}_2\text{O}} ?$

Solution

(a) No reaction, because no acid catalyst is present.

(b) Hydrolysis takes place because an acid catalyst is present. Two bonds break, an O—H of water and the C—O of the ester (as indicated below by the dashed line). Fragments from the ester and water bond together to form the products. The —OH from water attaches to the acyl part of the ester to form the carboxylic acid. The H from water attaches to the alkoxy part of the ester to form the alcohol. This reaction is the reverse of the esterification reaction described in Section 7.8.

$$CH_3\overset{O}{\overset{\|}{C}}\vdots OCH_2CH_3 \xrightarrow[H^+]{H_2O} CH_3\overset{O}{\overset{\|}{C}}-OH + H-OCH_2CH_3$$

$$HO\vdots H$$

(c) Saponification takes place when an ester is treated with aqueous strong base. The simplest way to determine the products is to write the equation for acidic hydrolysis and then have the carboxylic acid product react with base to convert it into the carboxylate salt:

$$CH_3\overset{O}{\overset{\|}{C}}\vdots OCH_2CH_3 \xrightarrow[NaOH]{H_2O} CH_3\overset{O}{\overset{\|}{C}}-OH + H-OCH_2CH_3$$

$$HO\vdots H$$

$$\downarrow NaOH$$

$$CH_3\overset{O}{\overset{\|}{C}}-ONa$$

The overall reaction is written in abbreviated form as

$$CH_3\overset{O}{\overset{\|}{C}}-OCH_2CH_3 \xrightarrow[NaOH]{H_2O} CH_3\overset{O}{\overset{\|}{C}}-ONa + H-OCH_2CH_3$$

Problem 7.10 Write the equation for each of the following reactions: (a) acidic hydrolysis of phenyl butanoate; (b) saponification of propyl propanoate with KOH.

7.12 CARBOXYLIC ACID ANHYDRIDES AND HALIDES

Carboxylic acid anhydrides are named by changing the end of the name of the corresponding carboxylic acid from **acid** to **anhydride.** Acid halides (also called acyl halides) are named by changing the ending from **-ic acid** to **-yl halide.**

$$R-\overset{O}{\overset{\|}{C}}-O-\overset{O}{\overset{\|}{C}}-R$$

Acid anhydride

$$R-\overset{O}{\overset{\|}{C}}-X$$

Acid halide
(X = halogen)

Example 7.11 Naming carboxylic acid anhydrides and halides

Name the following compounds:

(a) $CH_3CH_2-\overset{O}{\overset{\|}{C}}-Cl$

(b) $C_6H_5-\overset{O}{\overset{\|}{C}}-O-\overset{O}{\overset{\|}{C}}-C_6H_5$

Solution

(a) This compound is the acid chloride of propanoic acid (common name: propionic acid). The name is propanoyl chloride (common name: propionyl chloride).

(b) This compound is the anhydride of benzoic acid. The name is benzoic anhydride.

Problem 7.11 Give the structure of each of the following compounds: (a) butanoyl bromide; (b) acetic anhydride.

The term anhydride indicates a compound that is formed by dehydration. The anhydride group forms when two acid molecules are bonded together by dehydration. Section 7.5 will show that carboxylic acids are not the only acids to form anhydrides.

Acid anhydrides and halides are highly reactive with water and cannot exist in biological systems. They react with water to form the corresponding carboxylic acid:

$$\underset{\text{Acid anhydride}}{R{-}\overset{\overset{\displaystyle O}{\|}}{C}{-}O{-}\overset{\overset{\displaystyle O}{\|}}{C}{-}R} \xrightarrow{H_2O} \underset{\text{Carboxylic acid}}{2\ R{-}\overset{\overset{\displaystyle O}{\|}}{C}{-}OH}$$

$$\underset{\text{Acid halide}}{R{-}\overset{\overset{\displaystyle O}{\|}}{C}{-}Cl} \xrightarrow{H_2O} \underset{\text{Carboxylic acid}}{R{-}\overset{\overset{\displaystyle O}{\|}}{C}{-}OH} + HCl$$

Acid anhydrides are useful in the laboratory for synthesizing esters by reaction with alcohols (or phenols):

$$R{-}\overset{\overset{\displaystyle O}{\|}}{C}{-}O{-}\overset{\overset{\displaystyle O}{\|}}{C}{-}R + R'O{-}H \longrightarrow R{-}\overset{\overset{\displaystyle O}{\|}}{C}{-}OR' + R{-}\overset{\overset{\displaystyle O}{\|}}{C}{-}O{-}H$$

Acyl group

$$R{-}\overset{\overset{\displaystyle O}{\|}}{C}{-}Cl + R'O{-}H \longrightarrow R{-}\overset{\overset{\displaystyle O}{\|}}{C}{-}OR' + H{-}Cl$$

Acyl group

The synthesis of an ester from an acid anhydride or halide is more efficient than ester synthesis from the carboxylic acid. The industrial production of aspirin is carried out by using the reaction of acetic anhydride and salicylic acid (see Box 7.4).

Acid anhydrides and halides, as well as carboxylic acids, are often called **acyl transfer agents** because they transfer the **acyl group,** RCO, to the oxygen of an alcohol (or phenol). Reactions that include acyl transfer, called **acyl transfer reactions,** are important in biological systems. Protein synthesis takes place through acyl transfer reactions (Section 13.7). Thioesters derived from coenzyme A participate in metabolic processes through acyl transfer reactions (Chapters 15 and 16).

Example 7.12 Writing an equation for an acyl transfer reaction

Write the equation for the acyl transfer reaction between acetic anhydride and ethanol.

Solution

$$CH_3{-}\overset{\overset{\displaystyle O}{\|}}{C}{-}O{-}\overset{\overset{\displaystyle O}{\|}}{C}{-}CH_3 + CH_3CH_2O{-}H \longrightarrow$$

$$CH_3{-}\overset{\overset{\displaystyle O}{\|}}{C}{-}OCH_2CH_3 + CH_3{-}\overset{\overset{\displaystyle O}{\|}}{C}{-}O{-}H$$

Problem 7.12 Write the equation for the acyl transfer reaction between benzoyl chloride and methanol.

A PICTURE OF HEALTH
Examples of Carboxylic Acids, Esters, and Other Acid Derivatives

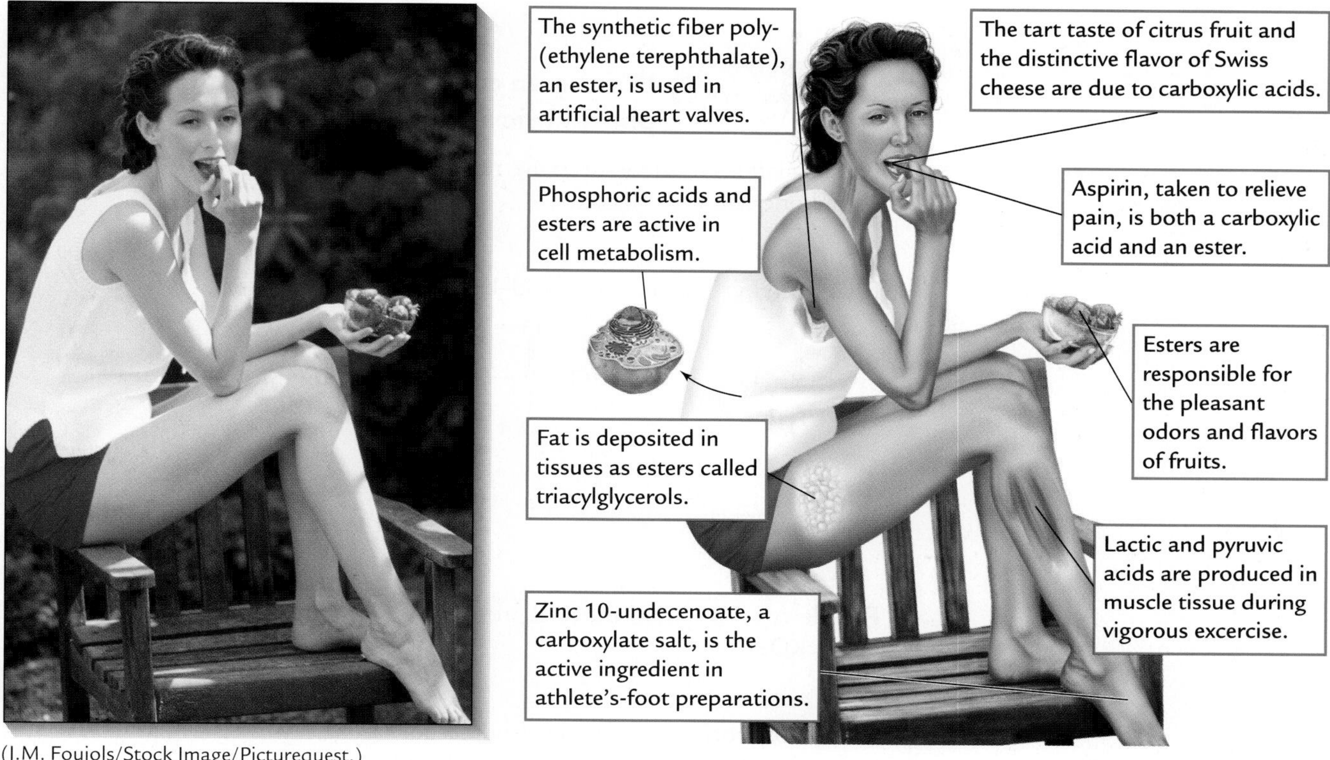

The synthetic fiber poly-(ethylene terephthalate), an ester, is used in artificial heart valves.
The tart taste of citrus fruit and the distinctive flavor of Swiss cheese are due to carboxylic acids.
Phosphoric acids and esters are active in cell metabolism.
Aspirin, taken to relieve pain, is both a carboxylic acid and an ester.
Esters are responsible for the pleasant odors and flavors of fruits.
Fat is deposited in tissues as esters called triacylglycerols.
Lactic and pyruvic acids are produced in muscle tissue during vigorous excercise.
Zinc 10-undecenoate, a carboxylate salt, is the active ingredient in athlete's-foot preparations.

(J.M. Foujols/Stock Image/Picturequest.)

7.13 PHOSPHORIC ACIDS AND THEIR DERIVATIVES

Although carboxylic acid anhydrides are not found in biological systems, other types of acid anhydrides—those derived from phosphoric acids—do play important roles in cell processes. Esters derived from phosphoric acids also are important in living systems—for example, phospholipids are used in constructing cell membranes.

›› Phospholipids are considered in Section 11.6.

The **phosphoric acids** are a group of three acids: phosphoric acid, diphosphoric acid, and triphosphoric acid. The simplest, **phosphoric acid,** undergoes intermolecular dehydration to form **diphosphoric acid:**

$$\underset{\text{Phosphoric acid}}{HO-\overset{\overset{\displaystyle O}{\|}}{\underset{\underset{\displaystyle OH}{|}}{P}}-OH} + \underset{\text{Phosphoric acid}}{HO-\overset{\overset{\displaystyle O}{\|}}{\underset{\underset{\displaystyle OH}{|}}{P}}-OH} \xrightarrow{-H_2O} \underset{\text{Diphosphoric acid}}{HO-\overset{\overset{\displaystyle O}{\|}}{\underset{\underset{\displaystyle OH}{|}}{P}}-O-\overset{\overset{\displaystyle O}{\|}}{\underset{\underset{\displaystyle OH}{|}}{P}}-OH}$$

Triphosphoric acid is formed by dehydration between diphosphoric and phosphoric acids:

$$HO-\overset{\overset{\displaystyle O}{\|}}{\underset{\underset{\displaystyle OH}{|}}{P}}-OH + HO-\overset{\overset{\displaystyle O}{\|}}{\underset{\underset{\displaystyle OH}{|}}{P}}-O-\overset{\overset{\displaystyle O}{\|}}{\underset{\underset{\displaystyle OH}{|}}{P}}-OH \xrightarrow{-H_2O} \underset{\text{Triphosphoric acid}}{HO-\overset{\overset{\displaystyle O}{\|}}{\underset{\underset{\displaystyle OH}{|}}{P}}-O-\overset{\overset{\displaystyle O}{\|}}{\underset{\underset{\displaystyle OH}{|}}{P}}-O-\overset{\overset{\displaystyle O}{\|}}{\underset{\underset{\displaystyle OH}{|}}{P}}-OH}$$

The HO—P=O groups in phosphoric, diphosphoric, and triphosphoric acids have acidic properties similar to those of the HO—C=O group in a carboxylic acid. The phosphoric acids are somewhat stronger acids than carboxylic acids but are still weak acids compared with the strong acids such as HCl, H_2SO_4, HNO_3, and $HClO_4$. Phosphoric, diphosphoric, and triphosphoric acids are polyprotic acids because each —OH group is acidic.

Example 7.13 Writing an equation for the reaction of phosphoric acid with strong base

Write equations to show successive reactions of each of the —OH groups in phosphoric acid with NaOH.

Solution

Each —OH group undergoes acid–base reaction in succession:

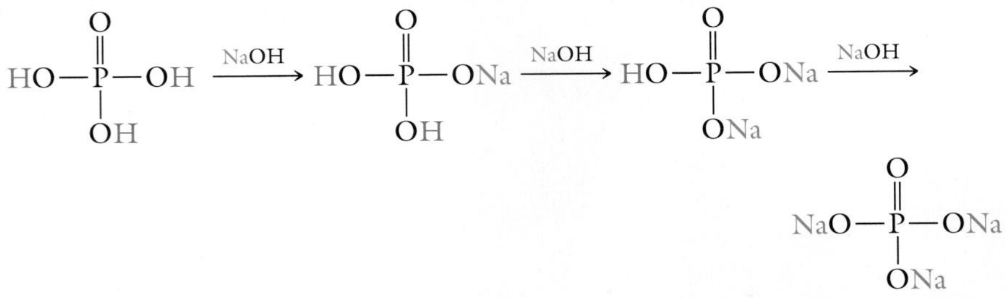

$$\mathrm{HO{-}P(=O)(OH){-}OH} \xrightarrow{\text{NaOH}} \mathrm{HO{-}P(=O)(OH){-}ONa} \xrightarrow{\text{NaOH}} \mathrm{HO{-}P(=O)(ONa){-}ONa} \xrightarrow{\text{NaOH}} \mathrm{NaO{-}P(=O)(ONa){-}ONa}$$

Problem 7.13 Show the final product after diphosphoric acid reacts with excess KOH.

Diphosphoric and triphosphoric acids are anhydrides as well as acids. Notice also that diphosphoric and triphosphoric acids have a close relation to carboxylic acid anhydrides:

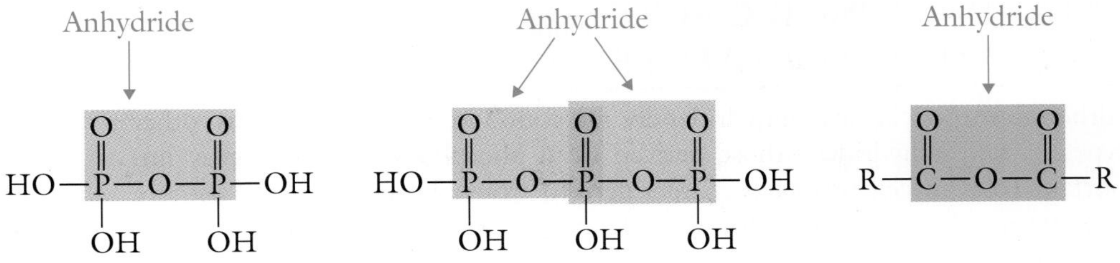

$$\mathrm{HO{-}P(=O)(OH){-}O{-}P(=O)(OH){-}OH}$$

Diphosphoric acid (Anhydride: P(=O)—O—P(=O))

$$\mathrm{HO{-}P(=O)(OH){-}O{-}P(=O)(OH){-}O{-}P(=O)(OH){-}OH}$$

Triphosphoric acid (Anhydride: two P(=O)—O—P(=O) linkages)

$$\mathrm{R{-}C(=O){-}O{-}C(=O){-}R}$$

Carboxylic acid anhydride (Anhydride: C(=O)—O—C(=O))

The HO—P=O groups in phosphoric, diphosphoric, and triphosphoric acids can undergo esterification with alcohols or phenols analogous to the esterification of the HO—C=O group of a carboxylic acid. The products are called **phosphoric acid esters** or **phosphate esters.** An example is the esterification of diphosphoric acid with ethanol:

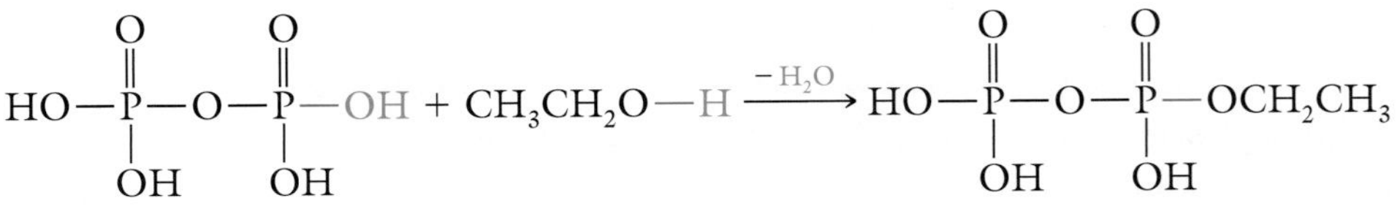

$$\underset{\text{Diphosphoric acid}}{\mathrm{HO{-}P(=O)(OH){-}O{-}P(=O)(OH){-}OH}} + \mathrm{CH_3CH_2O{-}H} \xrightarrow{-\mathrm{H_2O}} \underset{\text{Ethyl diphosphate}}{\mathrm{HO{-}P(=O)(OH){-}O{-}P(=O)(OH){-}OCH_2CH_3}}$$

Phosphate esters are named in the same way as carboxylic esters. The group derived from the alcohol (or phenol) is named followed by the name of the acid with the ending changed from **-ic acid** to **-ate.**

Example 7.14 Writing an equation for phosphate ester formation

Write the equation for the esterification of triphosphoric acid with 1-butanol. Name the ester.

Solution

$$\underset{\text{Triphosphoric acid}}{HO-\overset{O}{\overset{\|}{\underset{OH}{\underset{|}{P}}}}-O-\overset{O}{\overset{\|}{\underset{OH}{\underset{|}{P}}}}-O-\overset{O}{\overset{\|}{\underset{OH}{\underset{|}{P}}}}-OH} + H-OCH_2CH_2CH_2CH_3 \xrightarrow{-H_2O}$$

$$\underset{\text{Butyl triphosphate}}{HO-\overset{O}{\overset{\|}{\underset{OH}{\underset{|}{P}}}}-O-\overset{O}{\overset{\|}{\underset{OH}{\underset{|}{P}}}}-O-\overset{O}{\overset{\|}{\underset{OH}{\underset{|}{P}}}}-OCH_2CH_2CH_2CH_3}$$

Problem 7.14 Write the equation for the esterification of two of the —OH groups in phosphoric acid with methanol. Name the ester.

Phosphate esters are important in living systems. Many biological molecules that we will encounter in our study of biochemistry are phosphate esters: ATP, NADH and NAD^+, phospholipids, nucleic acids (DNA and RNA), and phosphate esters of carbohydrates, to name a few. Some of these compounds, such as ATP (Box 7.5) and NADH (see Figure 6.8), contain both phosphate ester and phosphate anhydride groups.

7.5 CHEMISTRY WITHIN US

Phosphate Esters in Biological Systems

Adenosine triphosphate (ATP) is the principal molecule in the process by which cells use energy. Acting as an energy-transferring intermediate, it makes the energy obtained from food available for all cell activities (see Section 14.9). Its role in energy transfer requires its hydrolysis to adenosine diphosphate (ADP), a process that releases energy. The phosphate —OH groups are shown here in ionized form because they are ionized under most physiological conditions (pH ~ 7).

Adenosine triphosphate (ATP)

Adenosine diphosphate (ADP)

Summary

Carboxylic Acids and Their Derivatives Compared

• Carboxylic acids contain the carboxyl group, —COOH, which is a carbonyl group with an —OH group attached to it.

• Carboxylic acids and their derivatives differ in the group attached to the carbonyl carbon.

$$\underset{\text{Acid}}{R-\overset{\overset{\displaystyle O}{\|}}{C}-OH} \qquad \underset{\text{Ester}}{R-\overset{\overset{\displaystyle O}{\|}}{C}-OR'}$$

$$\underset{\text{Acid anhydride}}{R-\overset{\overset{\displaystyle O}{\|}}{C}-O-\overset{\overset{\displaystyle O}{\|}}{C}-R} \qquad \underset{\text{Amide}}{R-\overset{\overset{\displaystyle O}{\|}}{C}-NH_2} \qquad \underset{\text{Acid halide}}{R-\overset{\overset{\displaystyle O}{\|}}{C}-X}$$

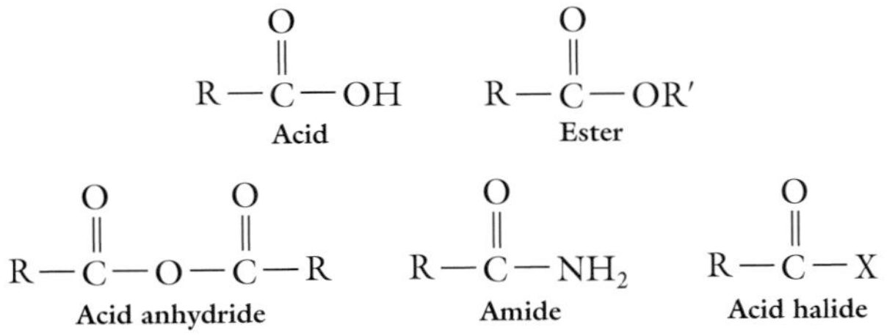

• Esters and anhydrides, like carboxylic acids, have an oxygen group bonded to the carbonyl carbon but differ in the type of oxygen group.

• Esters and anhydrides have —OR′ and —O—CO—R groups, respectively, whereas carboxylic acids have —OH.

• Amides have nitrogen (in an $-NH_2$, —NHR, or $-NR_2$ group), and acid halides have a halogen connected to the carbonyl carbon.

Carboxylic Acids

• Carboxylic acids are synthesized by the selective oxidation of alkylbenzenes, primary alcohols, and aldehydes.

• Carboxylic acids are named in the IUPAC system by naming the longest continuous carbon chain containing the carboxyl group.

• The ending of the corresponding alkane is changed from -e to -oic acid.

• The carboxyl carbon is C1.

• Prefixes preceded by numbers indicate substituents attached to the longest continuous chain.

• C_6H_5COOH is called benzoic acid.

• Carboxylic acids have considerably higher boiling and melting points than those of alcohols because of dimer formation by hydrogen bonding between pairs of acid molecules.

• Carboxylic acids have good water solubility, slightly higher than that of alcohols, owing to strong hydrogen-bond attractions with water.

• Carboxylic acids are weak acids but are stronger acids than phenols and much stronger acids than alcohols.

• Carboxylic acids and phenols can be distinguished from alcohols by litmus paper.

Carboxylate Salts

• Carboxylate salts, formed by the reaction of carboxylic acids with strong base, are named by the name of the positive ion followed by the name of the carboxylic acid, with the ending changed from -ic acid to -ate.

• Carboxylate salts are ionic compounds and, therefore, have much higher melting and boiling points and water solubility than those of carboxylic acids.

• Sodium and potassium carboxylate salts derived from unbranched long-chain acids with 12 to 20 carbons are used as soaps.

Carboxylic Acid Esters

• An ester is formed by acid-catalyzed dehydration between a carboxylic acid and an alcohol (or phenol).

• Esters are named by the name of the alkyl (or aryl) substituent of the alcohol (or phenol) followed by the name of the carboxylic acid, with the ending changed from -ic acid to -ate.

• Esters have slightly lower melting and boiling points than those of aldehydes and ketones and about the same water solubilities.

• Polyesters are formed by the reaction of bifunctional carboxylic acids and bifunctional alcohols (or phenols).

• An ester undergoes acidic hydrolysis to the corresponding carboxylic acid and alcohol (or phenol).

• Basic hydrolysis (saponification) produces the carboxylate salt instead of the carboxylic acid.

Carboxylic Acid Anhydrides and Halides

• Carboxylic acid anhydrides and halides undergo hydrolysis to form esters.

• They react with alcohols and phenols to form esters.

Phosphoric Acids and Their Derivatives

• Phosphoric, diphosphoric, and triphosphoric acids are stronger acids than carboxylic acids but are weaker than the strong acids, such as HCl, H_2SO_4, $HClO_4$, and HNO_3.

• Like carboxylic acids, phosphoric acids undergo esterification with alcohols or phenols.

Summary of Reactions

CARBOXYLIC ACIDS

Acidity

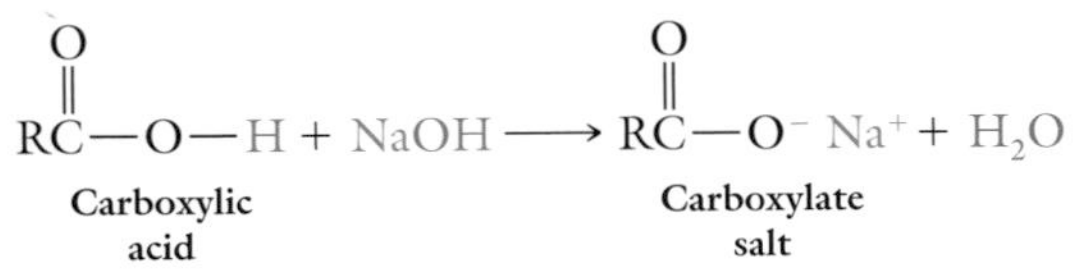

$$\underset{\text{Carboxylic acid}}{R\overset{\overset{\displaystyle O}{\|}}{C}-O-H} + NaOH \longrightarrow \underset{\text{Carboxylate salt}}{R\overset{\overset{\displaystyle O}{\|}}{C}-O^- \; Na^+} + H_2O$$

Esterification

$$\underset{\text{Acid}}{R-\overset{\overset{\displaystyle O}{\|}}{C}-OH} + \underset{\text{Alcohol or phenol}}{H-O-R'} \overset{H^+}{\rightleftharpoons} \underset{\text{Ester}}{R-\overset{\overset{\displaystyle O}{\|}}{C}-O-R'} + H_2O$$

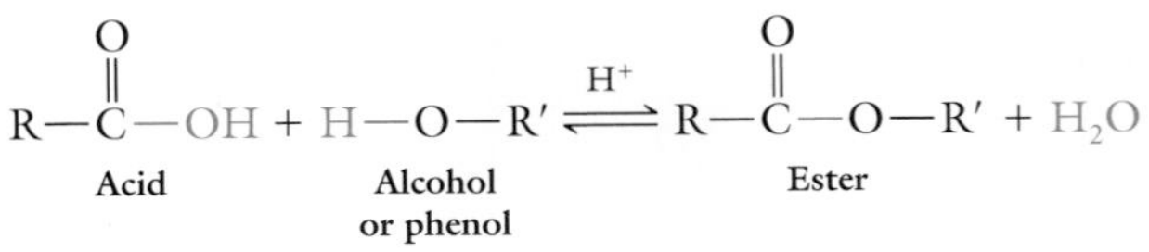

ESTERS

Acidic hydrolysis is the reverse of esterification:

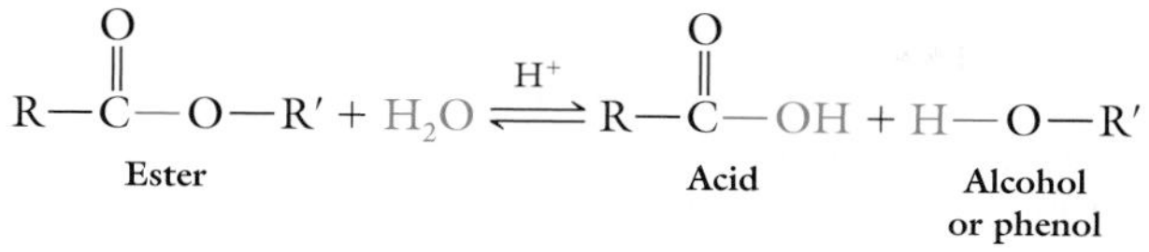

$$\underset{\text{Ester}}{R{-}\overset{O}{\overset{\|}{C}}{-}O{-}R'} + H_2O \overset{H^+}{\rightleftharpoons} \underset{\text{Acid}}{R{-}\overset{O}{\overset{\|}{C}}{-}OH} + \underset{\text{Alcohol or phenol}}{H{-}O{-}R'}$$

Saponification is hydrolysis with a base:

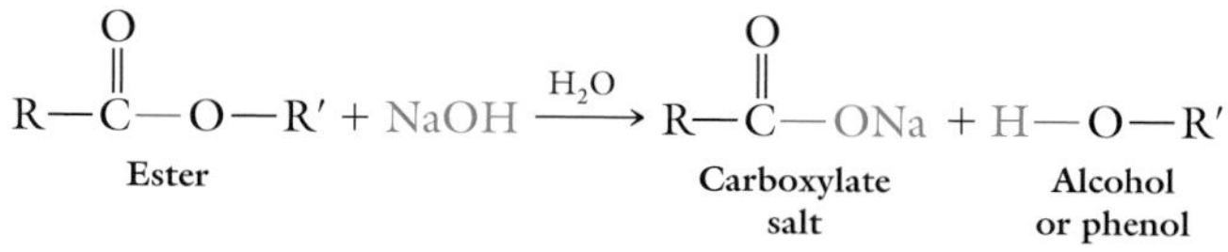

$$\underset{\text{Ester}}{R{-}\overset{O}{\overset{\|}{C}}{-}O{-}R'} + NaOH \xrightarrow{H_2O} \underset{\text{Carboxylate salt}}{R{-}\overset{O}{\overset{\|}{C}}{-}ONa} + \underset{\text{Alcohol or phenol}}{H{-}O{-}R'}$$

CONDENSATION POLYMERIZATION

$$n\,HO{-}\overset{O}{\overset{\|}{C}}{-}R{-}\overset{O}{\overset{\|}{C}}{-}OH + n\,HO{-}R'{-}OH \xrightarrow[-H_2O]{H^+} \left(\overset{O}{\overset{\|}{C}}{-}R{-}\overset{O}{\overset{\|}{C}}{-}O{-}R'{-}O\right)_n$$

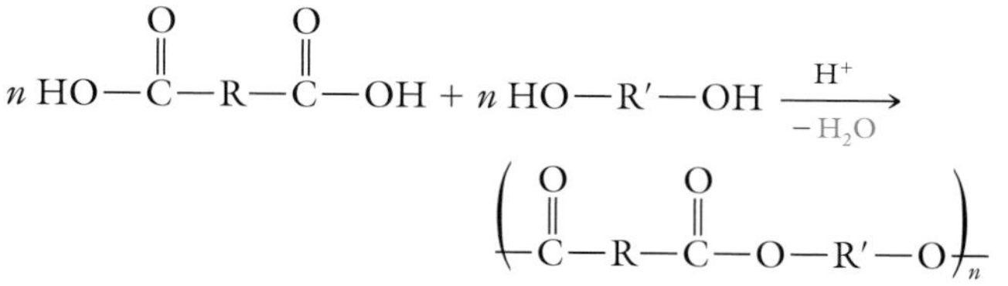

CARBOXYLIC ACID ANHYDRIDES (Z = —O—CO—R) AND HALIDES (Z = Cl, Br)

Hydrolysis

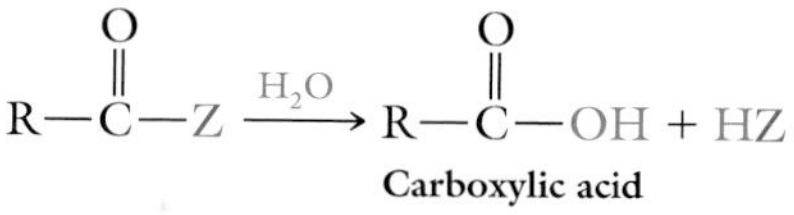

$$R{-}\overset{O}{\overset{\|}{C}}{-}Z \xrightarrow{H_2O} \underset{\text{Carboxylic acid}}{R{-}\overset{O}{\overset{\|}{C}}{-}OH} + HZ$$

Esterification

$$R{-}\overset{O}{\overset{\|}{C}}{-}Z \xrightarrow{R'OH} \underset{\text{Ester}}{R{-}\overset{O}{\overset{\|}{C}}{-}OR'} + HZ$$

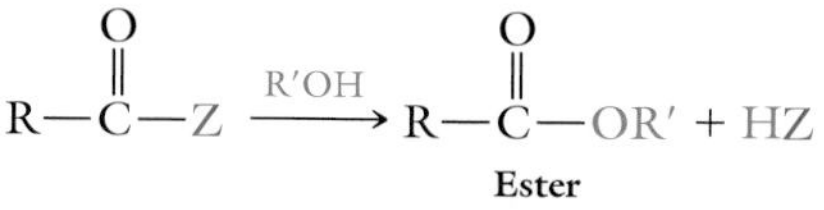

ESTERIFICATION OF PHOSPHORIC ACIDS

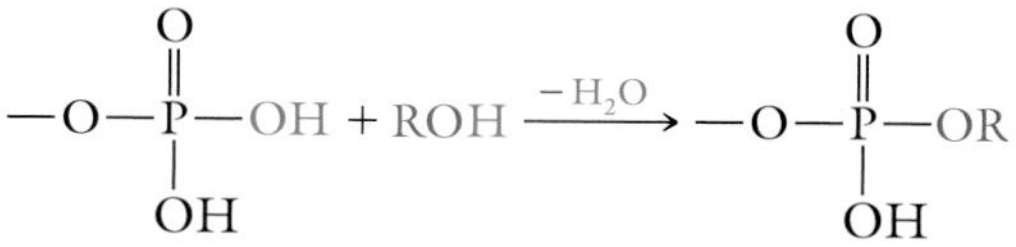

$$-O{-}\overset{O}{\overset{\|}{\underset{OH}{\underset{|}{P}}}}{-}OH + ROH \xrightarrow{-H_2O} -O{-}\overset{O}{\overset{\|}{\underset{OH}{\underset{|}{P}}}}{-}OR$$

Key Words

acid halide, pp. 218, 235
acyl group, p. 236
acyl transfer, p. 236
amide, p. 218
amphipathic, p. 227
anhydride, pp. 218, 235
bifunctional reactant, p. 232
carboxylate ion, p. 222
carboxyl group, p. 217
carboxylic acid, p. 217
condensation polymerization, p. 233
detergent, p. 228
diphosphoric acid, p. 237
ester, pp. 218, 229
esterification, p. 229
monomer, p. 232
phosphate ester, p. 238
phosphoric acid, p. 237
phosphoric acid anhydride, p. 238
saponification, p. 234
soap, p. 227
thioester, p. 229
triphosphoric acid, p. 237

Exercises

Comparison of Carboxylic Acids and Derivatives

7.1 Classify each of the following compounds as a carboxylic acid, ester, amide, anhydride, acid halide, or something else.

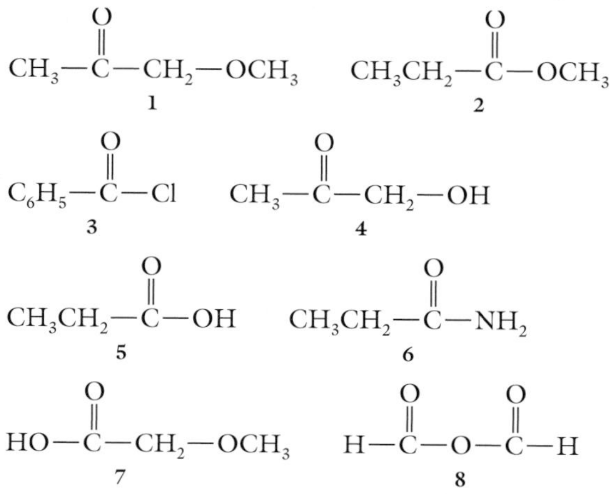

1. $CH_3{-}\overset{O}{\overset{\|}{C}}{-}CH_2{-}OCH_3$
2. $CH_3CH_2{-}\overset{O}{\overset{\|}{C}}{-}OCH_3$
3. $C_6H_5{-}\overset{O}{\overset{\|}{C}}{-}Cl$
4. $CH_3{-}\overset{O}{\overset{\|}{C}}{-}CH_2{-}OH$
5. $CH_3CH_2{-}\overset{O}{\overset{\|}{C}}{-}OH$
6. $CH_3CH_2{-}\overset{O}{\overset{\|}{C}}{-}NH_2$
7. $HO{-}\overset{O}{\overset{\|}{C}}{-}CH_2{-}OCH_3$
8. $H{-}\overset{O}{\overset{\|}{C}}{-}O{-}\overset{O}{\overset{\|}{C}}{-}H$

7.2 For each of the following pairs of structural formulas, indicate whether the pair represents (1) the same compound or (2) different compounds that are constitutional isomers or (3) different compounds that are geometrical isomers or (4) different compounds that are not isomers.

(a) 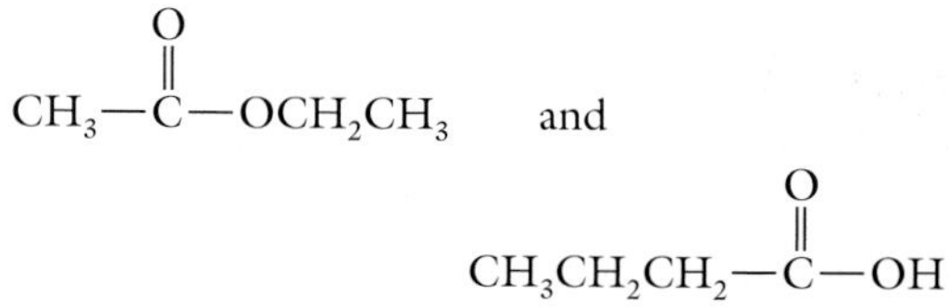 $CH_3{-}\overset{O}{\overset{\|}{C}}{-}OCH_2CH_3$ and $CH_3CH_2CH_2{-}\overset{O}{\overset{\|}{C}}{-}OH$

(b) $CH_3CH_2CH_2CH_2{-}\overset{O}{\overset{\|}{C}}{-}OH$ and $CH_3CH_2\overset{CH_3}{\overset{|}{C}H}{-}\overset{O}{\overset{\|}{C}}{-}OH$

(c) $CH_3CH_2CH_2CH_2{-}\overset{O}{\overset{\|}{C}}{-}OH$ and $CH_3CH_2\overset{CH_3}{\overset{|}{C}H}CH_2{-}\overset{O}{\overset{\|}{C}}{-}OH$

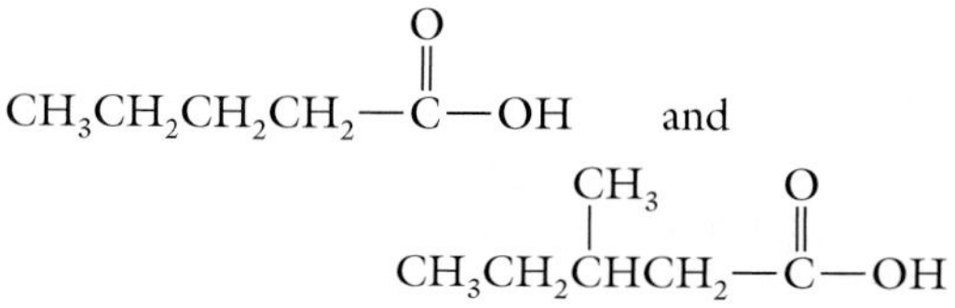

(d) 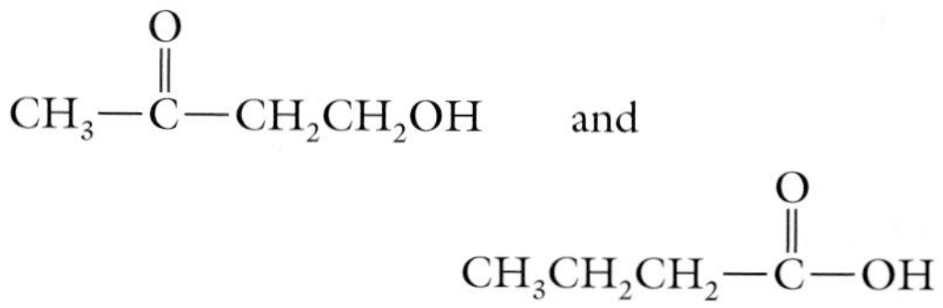 $CH_3{-}\overset{O}{\overset{\|}{C}}{-}CH_2CH_2OH$ and $CH_3CH_2CH_2{-}\overset{O}{\overset{\|}{C}}{-}OH$

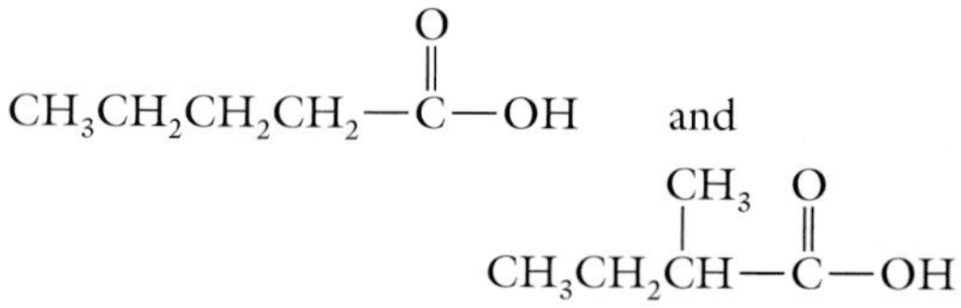

7.3 Draw a structural formula for each carboxylic acid with the molecular formula $C_5H_{10}O_2$.

7.4 Draw a structural formula for each ester with the molecular formula $C_4H_8O_2$.

Synthesis of Carboxylic Acids

7.5 Indicate which of the following compounds yield carboxylic acids on selective oxidation. For those that do, write an equation showing the conversion. Use permanganate or dichromate as the oxidizing agent. (a) Isopropylbenzene; (b) 1-pentanol; (c) 2-pentanol.

7.6 Indicate which of the following compounds yield carboxylic acids on selective oxidation. For those that do, write an equation showing the conversion. Use permanganate or dichromate as the oxidizing agent. (a) Butanal; (b) 2-butanone.

Naming Carboxylic Acids

7.7 Name each of the following carboxylic acids:

(a) 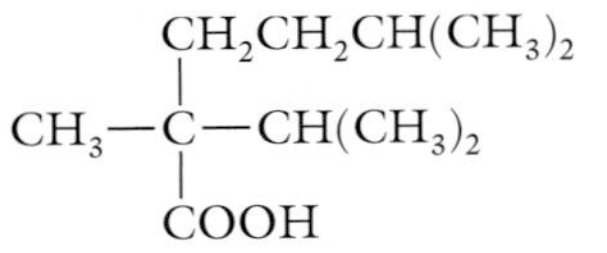 $CH_3-C(CH_2CH_2CH(CH_3)_2)(COOH)-CH(CH_3)_2$

(b) $(CH_3)_3CCH(Cl)CH_2COOH$

(c) $HO-C(=O)CH_2CH(CH_3)-C_6H_5$

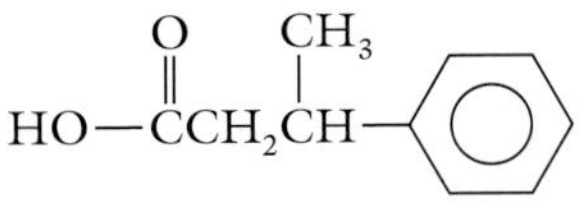

(d) 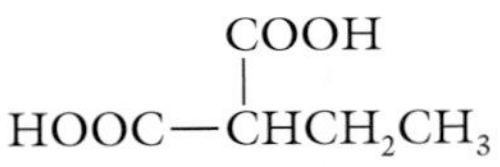 $HOOC-CH(COOH)CH_2CH_3$

7.8 Draw the structural formula for each of the following compounds: (a) 2-ethyl-4-methylbenzoic acid; (b) 2-isopropylbutanedioic acid.

Physical Properties of Carboxylic Acids

7.9 Draw a representation of the hydrogen bonding responsible for the high boiling point of propanoic acid.

7.10 Draw a representation of the hydrogen bonding present when propanoic acid dissolves in water.

7.11 Indicate which compound in each of the following pairs has the higher value of the specified property. If the two compounds in a pair have nearly the same value for the property, indicate that fact. Explain your conclusions.
(a) Ethanoic acid, 1-propanol: boiling point.
(b) Ethanoic acid, 1-propanol: solubility in water.

7.12 Indicate which compound in each of the following pairs has the higher value of the specified property. If the two compounds in a pair have nearly the same value for the property, indicate that fact. Explain your conclusions.
(a) Heptanoic acid, 1-octanol: solubility in water.
(b) Propanoic acid, butanal: boiling point.

Acidity of Carboxylic Acids

7.13 Isomers 1 and 2 both have —OH and C=O groups, but compound 1 is 10,000-fold more acidic than compound 2. Why?

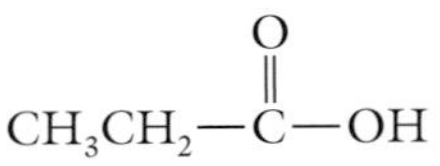

$CH_3CH_2-C(=O)-OH$ (1) $\qquad CH_3-C(=O)-CH_2OH$ (2)

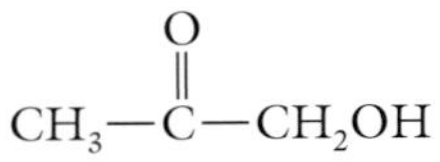

7.14 Place the following compounds in order of increasing acidity: phenol, propanol, propanoic acid. Explain the order of acidity in relation to the relative stability of the anion derived from each compound.

7.15 Indicate whether an acid–base reaction takes place under each of the conditions given. If reaction takes place, write the appropriate equation. (a) Ethanoic acid + water; (b) ethanoic acid + NaOH; (c) ethanoic acid + $NaHCO_3$; (d) phenol + water; (e) phenol + NaOH.

7.16 Show the reaction of hexanedioic acid, $HOOC(CH_2)_4COOH$, with excess NaOH.

7.17 What is the structure of the predominant chemical structure(s) present when propanoic acid is dissolved in water? What structure predominates when propanoic acid is added to a buffered solution at pH = 7 (similar to most physiological pH conditions)?

7.18 Compare the acidity of benzoic acid with that of hexanoic acid. What is the reason for the difference?

7.19 An unknown compound is either 1-propanol or propanoic acid. The unknown turns blue litmus paper red. Identify the unknown.

7.20 An unknown compound is either 1-propanol or propanoic acid. The unknown does not turn blue litmus paper red. Identify the unknown.

Carboxylate Salts

7.21 Draw the structural formula for each of the following compounds: (a) sodium hexanoate; (b) calcium benzoate.

7.22 Name the following carboxylate salts:

(a) 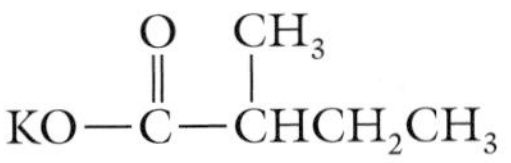 $KO-C(=O)-CH(CH_3)CH_2CH_3$

(b)  $NaO-C(=O)CH_2CH_2CH_2CH_2C(=O)-ONa$

7.23 Indicate which compound in each of the following pairs has the higher value of the specified property. If the two compounds in a pair have nearly the same value for the property, indicate that fact. Explain your conclusions.
(a) Sodium propanoate, propanoic acid: melting point.
(b) Sodium hexanoate, octanoic acid: solubility in water.
(c) Sodium hexanoate, octanoic acid: solubility in hexane.

7.24 Give the equations showing sodium propanoate acting as a base toward (a) water; (b) HCl.

7.25 An unknown is either butanoic acid or sodium butanoate. The unknown turns red litmus paper blue. Identify the unknown.

7.26 An unknown is either butanoic acid or sodium butanoate. The unknown turns blue litmus paper red. Identify the unknown.

Soaps and Their Cleaning Action

7.27 Which of the following compounds is a soap? Explain.

$$CH_3(CH_2)_2\overset{O}{\overset{\|}{C}}ONa \quad (1) \qquad CH_3(CH_2)_2\overset{O}{\overset{\|}{C}}OH \quad (2)$$

$$CH_3(CH_2)_{14}\overset{O}{\overset{\|}{C}}ONa \quad (3) \qquad CH_3(CH_2)_{14}\overset{O}{\overset{\|}{C}}OH \quad (4)$$

7.28 Explain how a soap removes greasy dirt from our bodies. Describe the process in relation to the secondary attractive forces that are operative.

Synthesis of Esters from Carboxylic Acids and Alcohols

7.29 Give the equation for ester formation between each of the following pairs of compounds: (a) pentanoic acid and ethanol; (b) benzoic acid and butanol.

7.30 Give the equation for ester formation between each of the following pairs of compounds: (a) 2-methylpropanoic acid and phenol; (b) pentanoic acid and ethanethiol.

7.31 What carboxylic acid and alcohol or phenol are needed to synthesize each of the following esters?

(a) $(CH_3)_3CCH_2CH_2{-}\overset{O}{\overset{\|}{C}}{-}OCH(CH_3)_2$

(b) 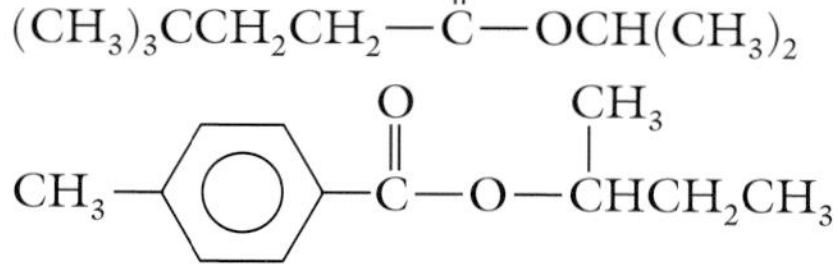 $CH_3{-}C_6H_4{-}\overset{O}{\overset{\|}{C}}{-}O{-}\overset{CH_3}{\overset{|}{C}}HCH_2CH_3$

(c) $CH_3(CH_2)_2\overset{O}{\overset{\|}{C}}{-}O{-}C_6H_4{-}CH_3$

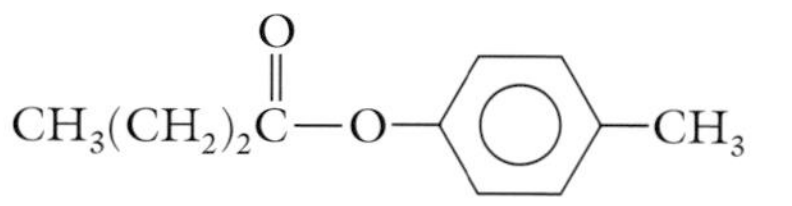

7.32 What carboxylic acid and alcohol or phenol are needed to synthesize each of the following esters?

(a) $CH_3CH_2O{-}\overset{O}{\overset{\|}{C}}{-}CH_2CH_2CH_2{-}\overset{O}{\overset{\|}{C}}{-}OCH_2CH_3$

(b) $CH_3CH_2\overset{O}{\overset{\|}{C}}{-}O{-}CH_2CH_2CH_2{-}O{-}\overset{O}{\overset{\|}{C}}CH_2CH_3$

Names and Physical Properties of Esters

7.33 Draw the structural formula for each of the following esters: (a) *t*-butyl butanoate; (b) cyclohexyl benzoate.

7.34 Name the following compounds:

(a) $CH_3CH_2CH_2CH_2CH_2\overset{O}{\overset{\|}{C}}{-}O{-}CH_2\overset{CH_3}{\overset{|}{C}}HCH_3$

(b) $(CH_3)_2CHO{-}\overset{O}{\overset{\|}{C}}{-}\overset{O}{\overset{\|}{C}}{-}OCH(CH_3)_2$

7.35 Indicate which compound in each of the following pairs has the higher value of the specified property. If the two compounds in a pair have nearly the same value for the property, indicate that fact. Explain your conclusions. (a) Methyl propanoate, pentanal: boiling point. (b) Methyl butanoate, hexanal: solubility in water. (c) Methyl propanoate, 1-pentanol: boiling point. (d) Methyl pentanoate, hexanoic acid: solubility in water. (e) Methyl propanoate, butanoic acid: boiling point.

7.36 Which of the following representations describe the hydrogen bonding present when methyl formate dissolves in water?

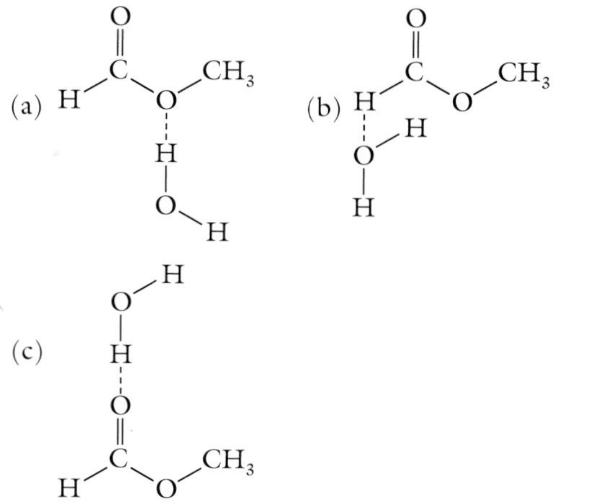

(a) $H{-}\overset{O}{\overset{\|}{C}}{-}O{-}CH_3$, with the ether O hydrogen-bonded (O···H) to H–O–H

(b) $H{-}\overset{O}{\overset{\|}{C}}{-}O{-}CH_3$, with the formyl H hydrogen-bonded (H···O) to H–O–H

(c) $H{-}\overset{O}{\overset{\|}{C}}{-}O{-}CH_3$, with the carbonyl O hydrogen-bonded (O···H) to H–O–H

Polyester Synthesis

7.37 What reactants are needed to synthesize the following polyester?

$$\left(\overset{O}{\overset{\|}{C}}{-}CH_2CH_2CH_2CH_2{-}\overset{O}{\overset{\|}{C}}{-}OCH_2CH_2CH_2O \right)_n$$

7.38 Draw the structure of the polyester formed from the reaction of butanedioic acid and 1,4-dihydroxybenzene.

Hydrolysis of Esters

7.39 Write the equation for each of the following reactions:
(a) acidic hydrolysis of propyl propanoate;
(b) saponification of phenyl butanoate with NaOH.

7.40 Give the product(s) formed in each of the following reactions. If no reaction takes place, write "no reaction." If more than one product is formed, indicate only the major product(s).

(a) 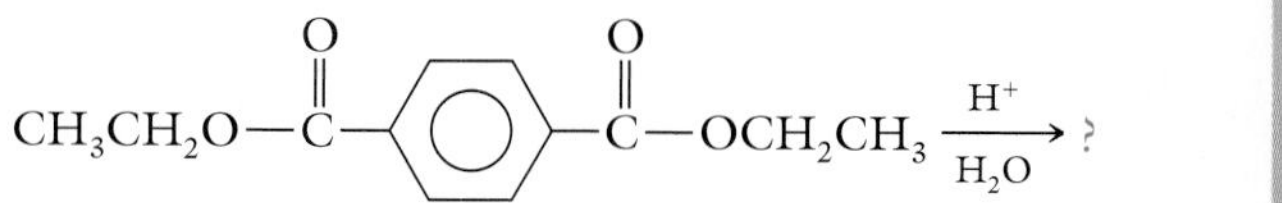 $CH_3CH_2O{-}\overset{O}{\overset{\|}{C}}{-}C_6H_4{-}\overset{O}{\overset{\|}{C}}{-}OCH_2CH_3 \xrightarrow[H_2O]{H^+} ?$

(b) $CH_3CH_2\overset{O}{\overset{\|}{C}}{-}O{-}CH_2CH_2CH_2{-}O{-}\overset{O}{\overset{\|}{C}}CH_2CH_3 \xrightarrow[H_2O]{NaOH} ?$

Carboxylic Acid Anhydrides and Halides

7.41 Draw the structural formula for each of the following compounds: (a) 2-methylpropanoyl chloride; (b) ethanoic anhydride.

7.42 Name the following compounds:

(a) $C_6H_5{-}\overset{O}{\overset{\|}{C}}{-}Br$

(b) $CH_3CH_2{-}\overset{O}{\overset{\|}{C}}{-}O{-}\overset{O}{\overset{\|}{C}}{-}CH_2CH_3$

7.43 Write an equation to show the reaction of benzoyl chloride with water.

7.44 Write an equation to show the reaction of benzoic anhydride with water.

Phosphoric Acids and Their Derivatives

7.45 Write the equation for the esterification of diphosphoric acid with an equimolar amount of 1-butanol. Name the ester.

7.46 Give the structural formula of isopropyl triphosphate.

7.47 Give the structural formula for the form of diphosphoric acid present at pH = 7.

7.48 Give the structural formula for the form of isopropyl triphosphate present at pH = 7.

Unclassified Exercises

7.49 A cyclic, nonaromatic compound containing an attached carboxyl group is named by naming the cyclic compound exclusive of the —COOH group and then adding the suffix **-carboxylic acid.** Draw the structure of *cis*-2-methylcyclopentanecarboxylic acid.

7.50 Carboxylic acids with a C=C in the longest chain containing the carboxyl group are named as alkenoic acids. Draw the structure of *trans*-2-pentenoic acid.

7.51 A compound that contains both hydroxyl and carboxyl groups undergoes intramolecular ester formation. The product is called a **lactone.** Show the lactone formed from 5-hydroxypentanoic acid.

7.52 Draw structural formulas for (a) an unbranched carboxylic acid of molecular formula $C_5H_{10}O_2$; (b) a branched carboxylic acid of molecular formula $C_5H_{10}O_2$ and containing an isopropyl group; (c) a lactone (cyclic ester) of four carbons with all carbons in the ring; (d) an ester of molecular formula $C_5H_{10}O_2$ and containing two ethyl groups; (e) a carboxylic acid of molecular formula $C_6H_{12}O_2$ and containing a *t*-butyl group; (f) an anhydride derived from a carboxylic acid of three carbons.

7.53 Draw the structural formula for each of the following compounds: (a) *trans*-3-chlorocyclohexanecarboxylic acid; (b) calcium acetate; (c) isopropyl benzoate; (d) pentanoic anhydride; (e) propanoyl chloride; (f) methyl diphosphate; (g) dimethyl phosphate.

7.54 Name the following compounds:

(a) $CH_3CH_2CH_2\overset{O}{\overset{\|}{C}}—OCH_2CH_2CH_3$

(b) $CH_3CH_2\underset{}{\overset{Cl}{\overset{|}{C}}}HCH_2CH_2\overset{O}{\overset{\|}{C}}—ONa$

(c) $HO—\underset{OH}{\underset{|}{\overset{O}{\overset{\|}{P}}}}—O—\underset{OCH_2CH_3}{\underset{|}{\overset{O}{\overset{\|}{P}}}}—OCH_2CH_3$

(d) 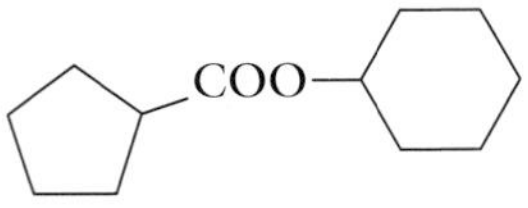 cyclopentyl—COO—cyclohexyl

7.55 What are the hybridizations of atoms a, b, c, and d in the following compound? What are bond angles A, B, and C?

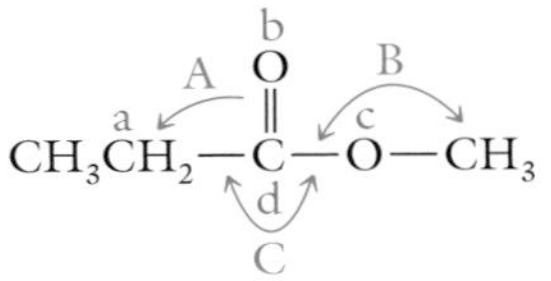

$CH_3CH_2—\overset{O}{\overset{\|}{C}}—O—CH_3$ (labels: a on CH_2 carbon, b on carbonyl O, c on ester O, d on carbonyl C; angles A, B, C)

7.56 Indicate which compound in each of the following pairs has the higher value of the specified property. If the two compounds in a pair have nearly the same value for the property, indicate that fact. Explain your conclusions.
(a) Methyl butanoate, hexanal: boiling point.
(b) Methyl butanoate, 2-hexanone: solubility in water.
(c) Methyl propanoate, 1-pentanol: boiling point.
(d) Methyl propanoate, butanoic acid: boiling point.

7.57 An unknown compound is either phenol, propanol, or propanoic acid. The unknown turns blue litmus paper red. Identify the unknown.

7.58 An unknown compound is either phenol, propanol, or propanoic acid. The unknown does not turn blue litmus paper red. Identify the unknown.

7.59 Give the equation for the esterification of acetic acid with 1-butanethiol.

7.60 Give the product formed when the ester group in the following lactone (cyclic ester) undergoes acid hydrolysis:

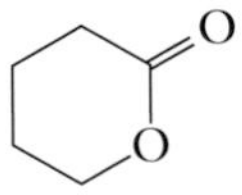

7.61 Give the product(s) formed in each of the following reactions. If no reaction takes place, write "no reaction." If more than one product is formed, indicate only the major product.

(a) $CH_3CH_2\overset{O}{\overset{\|}{C}}—OCH_2C_6H_5 \xrightarrow{H_2O} ?$

(b) $CH_3CH_2\overset{O}{\overset{\|}{C}}—OCH_2C_6H_5 \xrightarrow[H^+]{H_2O} ?$

(c) $CH_3CH_2\overset{O}{\overset{\|}{C}}—OCH_2C_6H_5 \xrightarrow[NaOH]{H_2O} ?$

(d) 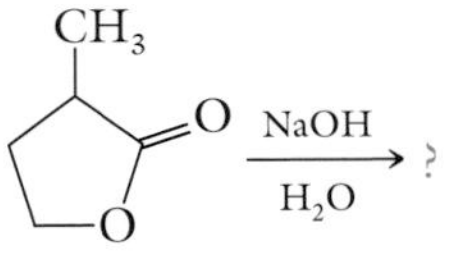 (five-membered lactone with CH_3 on the carbon adjacent to C=O) $\xrightarrow[H_2O]{NaOH} ?$

(e) $HO—\underset{OH}{\underset{|}{\overset{O}{\overset{\|}{P}}}}—O—\underset{OH}{\underset{|}{\overset{O}{\overset{\|}{P}}}}—O—\underset{OH}{\underset{|}{\overset{O}{\overset{\|}{P}}}}—OH \xrightarrow{CH_3CH_2CH_2OH} ?$

(f) $CH_3CH_2CH_2\overset{O}{\overset{\|}{C}}—OH \xrightarrow{NaHCO_3} ?$

(g) $(CH_3)_2CHCH_2CH_2\overset{O}{\overset{\|}{C}}—OH \xrightarrow[H^+]{C_6H_5OH} ?$

(h) $CH_3CH_2—\overset{O}{\overset{\|}{C}}—O—\overset{O}{\overset{\|}{C}}—CH_2CH_3 \xrightarrow{H_2O} ?$

(i) $CH_3CH_2—\overset{O}{\overset{\|}{C}}—Cl \xrightarrow{H_2O} ?$

(j) $CH_3CH_2\overset{O}{\overset{\|}{C}}—SCH_2C_6H_5 \xrightarrow[H^+]{H_2O} ?$

7.62 The ester $C_5H_{10}O_2$ yields ethanol on acid-catalyzed hydrolysis. Identify the other product(s) of hydrolysis.

7.63 An unknown compound is either an ester or a carboxylic acid. What simple chemical diagnostic test can be used to distinguish between the two possibilities?

Expand Your Knowledge

Note: The icons denote exercises based on material in boxes.

7.64 Low-molecular-mass carboxylic acids have distinct and often disagreeable odors. The neutralization of such acids with NaOH or other base results in a much less intense odor. Why?

[5.1] **7.65** Nitroglycerine is used as an explosive and to relieve chest pain due to angina pectoris (see Box 5.1). Nitroglycerine is the trinitroester of glycerol (also called glycerine), formed by dehydration between each of the —OH groups of glycerol with a molecule of nitric acid. Write an equation to show the synthesis of nitroglycerine from glycerol and nitric acid.

$$\begin{array}{ll} CH_2\text{—}OH & \\ | & \\ CH\text{—}OH & HO\text{—}NO_2 \\ | & \\ CH_2\text{—}OH & \\ \textbf{Glycerol} & \textbf{Nitric acid} \end{array}$$

7.66 Example 7.9 shows the conversion of terephthalic acid into the important plastic and fiber poly(ethylene terephthalate). Terephthalic acid is produced from *p*-xylene (*p*-dimethylbenzene). What reaction converts *p*-xylene into terephthalic acid?

7.67 By reference to Section 2.4, calculate the $[H_3O^+]$ and pH of a 1.00 *M* aqueous solution of acetic acid. K_a for acetic acid is 1.74×10^{-5}.

[7.4] **7.68** Two chemists at the XYZ Pharmaceutical Company are asked to devise synthetic strategies for the production of the active ingredient in aspirin formulations, acetylsalicylic acid (see Box 7.4). Betty Brown proposes esterification of salicylic acid by acetic anhydride, whereas John Smith proposes esterification of salicylic acid with methanol. John Smith's proposal is rejected because it produces the wrong ester. Betty Brown is placed in charge of the project to produce aspirin with the use of her proposed strategy. What is wrong with John Smith's strategy? Why is Betty Brown's strategy correct?

7.69 Why is a solution of sodium acetate a much stronger conductor of electricity than is a solution of acetic acid?

7.70 Why are carboxylic acid anhydrides and halides not found in biological systems?

7.71 The ester responsible for the distinctive odor and flavor of bananas is 3-methylbutyl ethanoate. Draw the structure of this ester and show its formation from the appropriate carboxylic acid and alcohol.

[7.4] **7.72** Jane Smith finds two bottles of aspirin (see Box 7.4) in her medicine cabinet. Opening both bottles, Jane notices an odor of vinegar (acetic acid) from one of the bottles. What should Jane do with that bottle? What does the odor of vinegar indicate?

7.73 Three different solutions (A, B, C) are prepared by dissolving the following compound(s) in 1 L each of water: A, 1 mol acetic acid; B, 1 mol each of acetic acid and HCl; C, 1 mol each of acetic acid and NaOH. Place the solutions in order of increasing concentration of acetate ion. Explain the order.

7.74 Apples contain hydroxybutanedioic acid (common name: malic acid). Draw the structure of this carboxylic acid.

7.75 Draw the structure of the carboxylic acid that is an isomer of hydroxypropanone.

7.76 The flavor of strawberries is due mostly to methyl thiobutanoate, synthesized from butanoic acid and methanethiol. Draw the structure of methyl thiobutanoate.

7.77 Starting only from compounds that are alcohols, write equations to show the synthesis of ethyl propanoate.

7.78 A carboxylic acid that is also both an aldehyde or ketone and an alkene is named as a ketoalkenoic acid. Draw the structure of 9-keto-*trans*-2-decenoic acid, a pheromone used by bees.

7.79 Draw the structure of benzocaine, a topical anesthetic synthesized from *p*-aminobenzoic acid and ethanol.

7.80 Compound X (C_2H_4O) is oxidized to compound Y ($C_2H_4O_2$). Compound Y reacts with isopropyl alcohol in the presence of acid to yield compound Z ($C_5H_{10}O_2$). Identify compounds X, Y, and Z.

7.81 Trichloroacetic acid, used in facial chemical peeling, is much more acidic than acetic acid. Explain the difference in acidity.

[6.3] **7.82** How would infrared spectroscopy be used to distinguish propanal from propanoic acid? Refer to Box 6.3 and Table 6.5.

7.83 The polymerization of lactic acid (IUPAC: 2-hydroxypropanoic acid) in the presence of an acid catalyst yields a polyester that is useful as absorbable suture material. Write an equation to show the polymerization.

7.84 Place the following compounds in order of increasing oxidation state for carbon: ethanoic acid, ethane, ethene, CO_2, ethanol, ethanal, methyl ethanoate.

7.85 Ester hydrolysis requires the presence of either a strong acid or a strong base. The roles of the acid and base are different. One is a catalyst, but the other is not. Describe the difference.

[7.4] **7.86** The ABC Company sells acetylsalicylic acid (aspirin) and ibuprofen (see Box 7.4). The labels have been lost from containers of the two compounds. Is there a simple chemical test that can be used to determine which compound is in which container? Explain.

[6.3] [7.4] **7.87** Can infrared spectroscopy (Box 6.3 and Table 6.5) be used to distinguish between acetylsalicylic acid (aspirin) and ibuprofen (see Box 7.4)? Explain.

[6.3] [7.4] **7.88** Can C-13 NMR spectroscopy (Box 6.3) be used to distinguish between acetylsalicylic acid (aspirin) and ibuprofen (see Box 7.4)? Explain.

CHAPTER 8 AMINES AND AMIDES

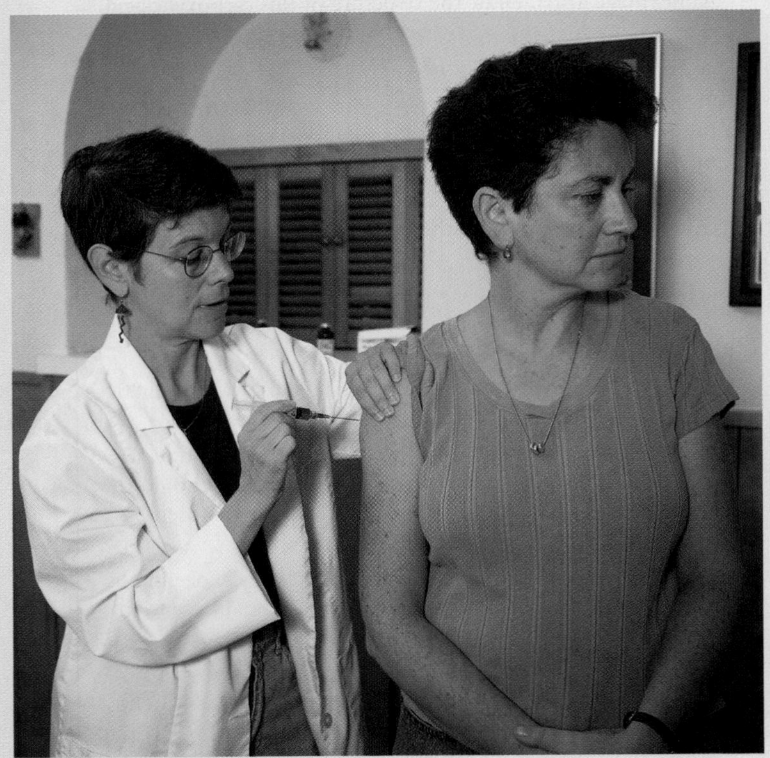
(Bill Aron/PhotoEdit.)

Chemistry in Your Future

Mrs. Carter is suffering greatly from hay fever this season, and her doctor has prescribed an injection of diphenhydramine hydrochloride (Benedryl). As you give her the injection, she asks if this drug is the same as the diphenhydramine ointment (Caladryl) that the doctor prescribed for her son's poison ivy last year. You tell her that, yes, the active ingredients in the injection and ointment are essentially the same and that you think that, when "hydrochloride" is attached to the name of a drug, it has something to do with making that particular compound more soluble in body fluids. Is this assumption of yours true, and, if so, to what kinds of compounds does it apply and why? The answer is contained in this chapter on amines and amides.

For more information about this topic and others in the chapter, go to www.whfreeman.com/bleiodian2e

Learning Objectives

- Draw and name amines.
- Describe the physical properties of amines.
- Explain and write equations to demonstrate the basicity of amines.
- Draw and name amine salts and describe their properties.
- Draw and name amides.
- Describe the physical properties of amides.
- Describe and write equations for the synthesis and hydrolysis of amides.

We now turn our attention from organic compounds containing oxygen to those containing nitrogen. These compounds are the amine and amide families—two families that play major roles in biological systems. Amino acids, the building blocks of proteins, are both amines and carboxylic acids; the proteins themselves are amides. The nucleic acids, deoxyribonucleic and ribonucleic acids (DNA and RNA, respectively), which are the molecules responsible for transmitting and expressing hereditary information, contain amine and amide groups (along with carbohydrate and phosphate ester groups). Some of the lipids that make up cell membranes contain amine groups. The polysaccharide chitin—the main structural component of the exoskeletons of crustaceans, insects, and spiders—contains amide groups.

A wide range of naturally occurring and synthetic amines and amides are physiologically active; that is, they affect the control and regulation of organs, tissues, or cells. Among them are the hormones adrenaline and melatonin; antibiotics, such as penicillin and tetracycline; and a vast number of substances used for medicinal and other purposes, such as caffeine, nicotine, narcotics, anesthetics, antidepressants, high-blood-pressure medications, and tranquilizers. An amide of enormous commercial importance is the textile fiber and plastic known as nylon.

8.1 AMINES AND AMIDES COMPARED

Amines and **amides** are organic derivatives of ammonia in which one or more of the hydrogens are replaced by organic groups (see below). The bonding of nitrogen directly to a carbonyl carbon forms the **amide group,** which distinguishes an amide from an amine and makes their properties quite different. In the amine structure, two of the R groups can be H; all three R groups can be H in the amide. The R groups can be aliphatic or aromatic in both amines and amides.

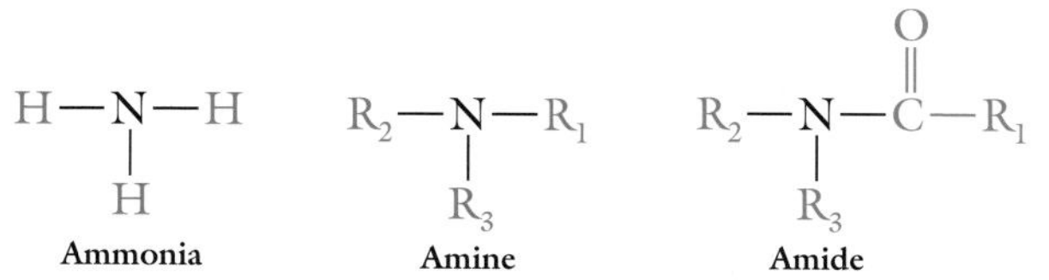

$$\underset{\text{Ammonia}}{\mathrm{H{-}\underset{\displaystyle H}{\underset{|}{N}}{-}H}} \qquad \underset{\text{Amine}}{\mathrm{R_2{-}\underset{\displaystyle R_3}{\underset{|}{N}}{-}R_1}} \qquad \underset{\text{Amide}}{\mathrm{R_2{-}\underset{\displaystyle R_3}{\underset{|}{N}}{-}\overset{\displaystyle O}{\overset{\|}{C}}{-}R_1}}$$

In forming the single bonds in ammonia, amines, and amides, nitrogen is sp^3 hybridized—like the oxygen in ethers and alcohols and the carbons in alkanes. However, the three elements differ in an important way. Nitrogen atoms contain one more electron than a carbon atom does and one less electron than an oxygen atom. Carbon is tetravalent with no nonbonding electrons, nitrogen is trivalent with one pair of nonbonding electrons, and oxygen is divalent with two pairs of nonbonding electrons. The nonbonded electron pairs of nitrogen and oxygen, shown below in red, are omitted from most structural drawings but are understood to be present. These nonbonded electrons contribute to the properties of amines and amides.

« sp^3 orbitals were described in Section 3.3.

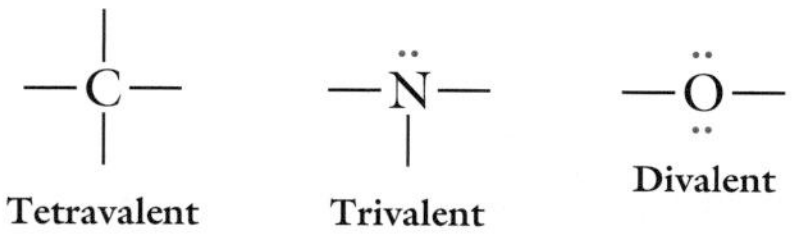

$$\underset{\text{Tetravalent}}{-\overset{|}{\underset{|}{C}}-} \qquad \underset{\text{Trivalent}}{-\overset{\cdot\cdot}{\underset{|}{N}}-} \qquad \underset{\text{Divalent}}{-\overset{\cdot\cdot}{\underset{\cdot\cdot}{O}}-}$$

The bond angles about the nitrogen atom in an amine are close to the tetrahedral bond angle of 109.5° (Section 1.7). The bond angles in amides differ from those in amines (Section 8.3).

8.2 CLASSIFYING AMINES

Amines are classified according to the number of carbons directly bonded to the nitrogen. **Primary, secondary,** and **tertiary amines** have one, two, and three carbons directly bonded to the nitrogen. Put another way, 1°, 2°, and 3° amines are derived from ammonia by the replacement of one, two, and three of the hydrogens by aliphatic or aryl groups.

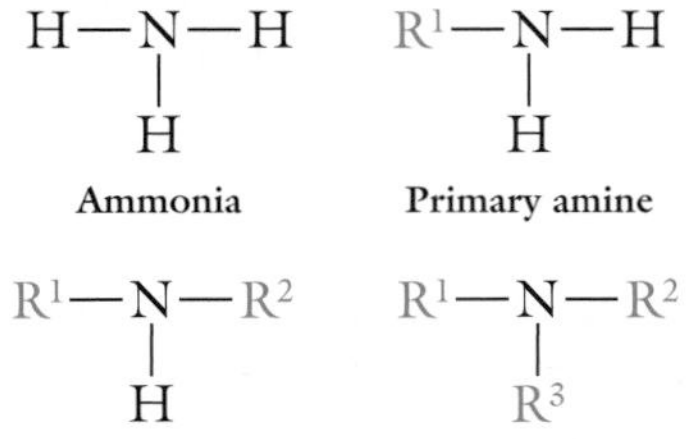

$$\underset{\text{Ammonia}}{H-\underset{\substack{|\\H}}{N}-H} \qquad \underset{\text{Primary amine}}{R^1-\underset{\substack{|\\H}}{N}-H}$$

$$\underset{\text{Secondary amine}}{R^1-\underset{\substack{|\\H}}{N}-R^2} \qquad \underset{\text{Tertiary amine}}{R^1-\underset{\substack{|\\R^3}}{N}-R^2}$$

Amines are also classified as aromatic (aryl) or aliphatic amines. An **aromatic (aryl) amine** is an amine in which at least one benzene or other aromatic group is directly bonded to nitrogen. An **aliphatic amine** is an amine in which no benzene or other aromatic groups are directly bonded to nitrogen.

Example 8.1 Classifying amines

Classify each of the following compounds as (a) amines, amides, or something else; (b) primary, secondary, or tertiary, if amines; (c) aliphatic or aromatic, if amines.

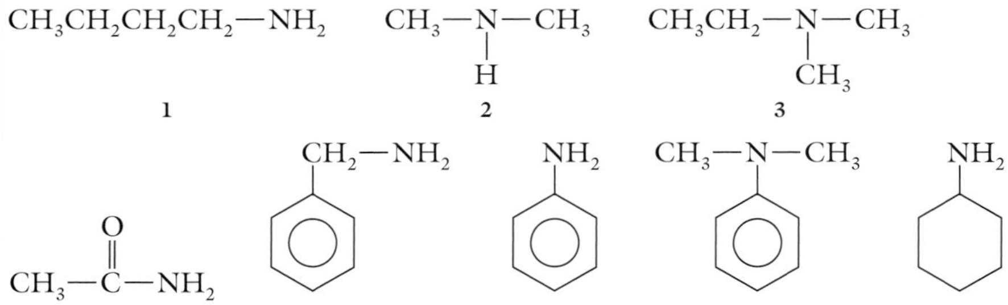

1: $CH_3CH_2CH_2CH_2-NH_2$

2: $CH_3-\underset{\substack{|\\H}}{N}-CH_3$

3: $CH_3CH_2-\underset{\substack{|\\CH_3}}{N}-CH_3$

4: $CH_3-\overset{\substack{O\\\|}}{C}-NH_2$

5: $C_6H_5-CH_2-NH_2$ (benzene ring bearing CH_2-NH_2)

6: $C_6H_5-NH_2$ (benzene ring bearing NH_2)

7: $C_6H_5-N(CH_3)_2$ (benzene ring bearing CH_3-N-CH_3)

8: $C_6H_{11}-NH_2$ (cyclohexane ring bearing NH_2)

Solution

(a) All compounds except compound 4 are amines because at least one organic group, but no carbonyl carbon, is attached to nitrogen. Compound 4 is an amide because a carbonyl carbon is attached to nitrogen.

(b) Compounds 1, 5, 6, and 8 are primary amines because only one carbon is directly attached to nitrogen. Compound 2 is a secondary amine because two carbons are directly attached to nitrogen. Compounds 3 and 7 are tertiary amines because three carbons are directly attached to nitrogen.

(c) Compounds 1, 2, 3, 5, and 8 are aliphatic amines because no aromatic group is bonded directly to nitrogen. Compounds 6 and 7 are aromatic amines because an aromatic group is directly bonded to nitrogen. Note that, in compound 7, both aliphatic and aromatic groups are bonded to nitrogen. Such amines are classified as aromatic amines. Compound 5 is an aliphatic amine because the aromatic group is not directly bonded to nitrogen. Compound 8 is not an aromatic amine, because the ring is an aliphatic ring, not an aromatic one.

Problem 8.1 Classify each of the following compounds as (a) amines, amides, or something else; (b) primary, secondary, or tertiary, if amines; (c) aliphatic or aromatic, if amines.

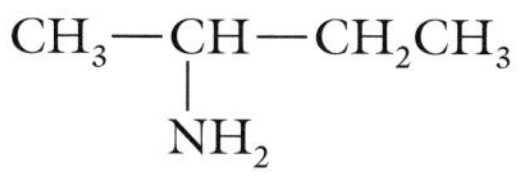

$$CH_3-\underset{\underset{NH_2}{|}}{CH}-CH_2CH_3$$
1

$N(CH_3)_2$ on cyclohexane ring
2

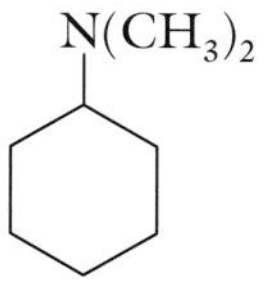

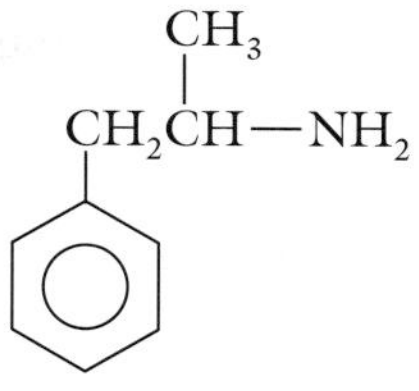

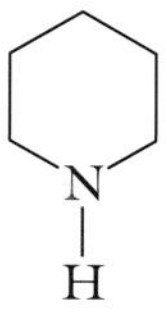

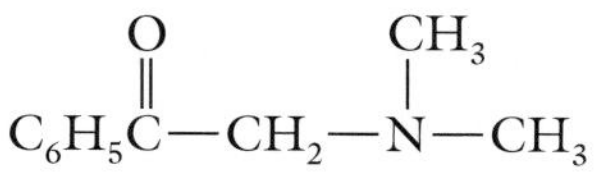

$$\overset{\overset{CH_3}{|}}{CH_2CH}-NH_2$$ on benzene ring
3

piperidine ring with N–H
4

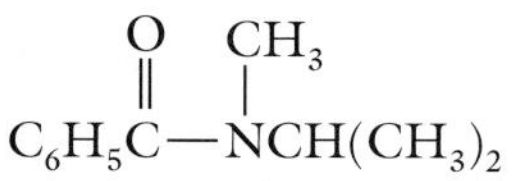

$$C_6H_5\overset{\overset{O}{\|}}{C}-\overset{\overset{CH_3}{|}}{N}CH(CH_3)_2$$
5

$$C_6H_5\overset{\overset{O}{\|}}{C}-CH_2-\overset{\overset{CH_3}{|}}{N}-CH_3$$
6

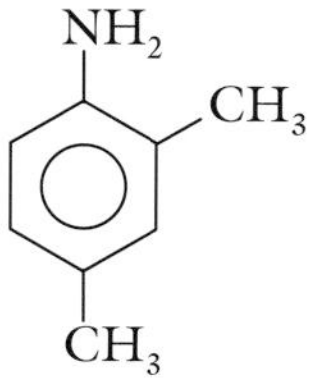

benzene ring with NH_2, CH_3, and CH_3
7

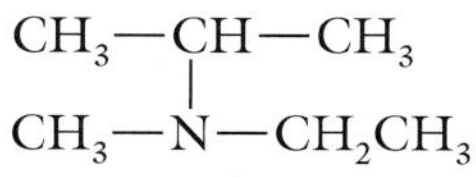

$$\underset{CH_3-N-CH_2CH_3}{CH_3-\underset{|}{CH}-CH_3}$$
8

Constitutional isomerism in amines, which is analogous to that in other families (Sections 4.3 and 5.2), is possible from two kinds of connectivity differences:

- Different carbon skeletons
- Different placements of the nitrogen on a carbon skeleton

Example 8.2 Drawing constitutional isomers

Draw the constitutional isomers of C_3H_9N.

Solution

The three carbons can make up one group, either a propyl or an isopropyl group, attached to nitrogen:

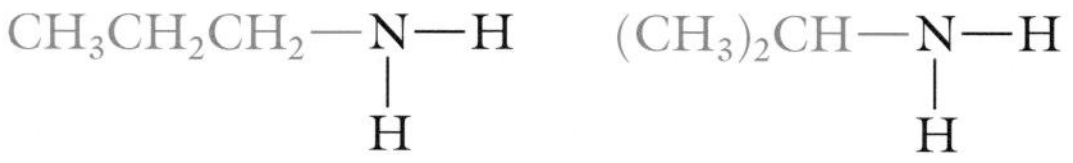

$$CH_3CH_2CH_2-\underset{\underset{H}{|}}{N}-H \qquad (CH_3)_2CH-\underset{\underset{H}{|}}{N}-H$$

Alternately, the three carbons need not be part of the same group. Either three methyl groups or one methyl and one ethyl group can be attached to nitrogen:

$$CH_3-\underset{\underset{CH_3}{|}}{N}-CH_3 \qquad CH_3CH_2-\underset{\underset{H}{|}}{N}-CH_3$$

Problem 8.2 Draw the constitutional isomers of $C_4H_{11}N$.

8.3 NAMING AMINES

The systems for naming amines are more varied than for any other family. Amines containing only the simple groups are best named by the common nomenclature system. The simple groups consist of the unbranched alkyl groups from methyl, ethyl, propyl, and butyl through decyl; the branched three- and four-carbon alkyl groups (isopropyl, isobutyl, *s*-butyl, *t*-butyl); and the unsubstituted cycloalkyl groups, such as cyclohexyl and cyclopentyl. In this common system, the groups (other than H) attached to nitrogen are named in alphabetical order followed without space by the suffix **-amine;** for example:

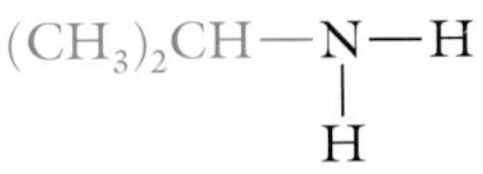

$(CH_3)_2CH-NH-H$ (N bonded to H below)

Isopropyl**amine**

(cyclohexyl)$-NH-CH_3$

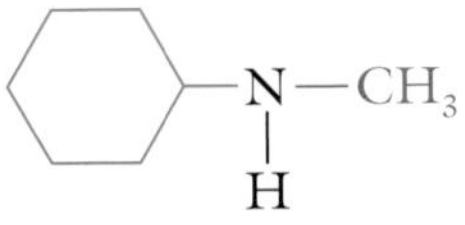

Cyclohexylmethyl**amine**

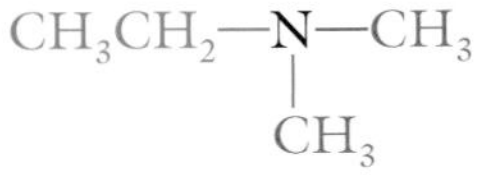

$CH_3CH_2-N(CH_3)-CH_3$

Ethyldimethyl**amine**

$C_6H_5NH_2$ is called **aniline,** and substituted anilines are named as derivatives of aniline. Substituents other than H on nitrogen are distinguished from substituents on the aromatic ring by using an *N*- prefix instead of a number or an *o*-, *m*-, or *p*- prefix; for example:

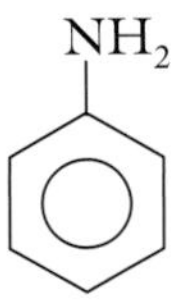

Aniline

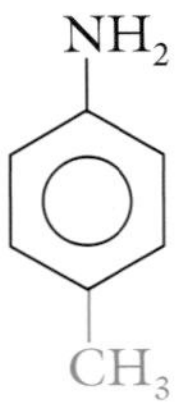

p-Methyl**aniline**
(4-Methylaniline)

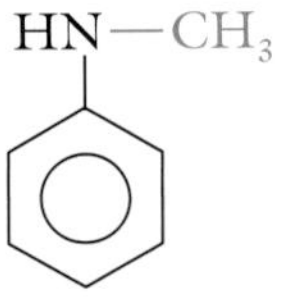

N-Methyl**aniline**

Because the IUPAC system for naming amines is not consistent with IUPAC nomenclature for other organic families, the most widely used system for naming amines is that recommended by ***Chemical Abstracts,*** *CA* for short. Published by the American Chemical Society, *CA* is a journal that abstracts the world's chemical literature. The *CA* rules for naming amines are similar to the IUPAC rules for naming alcohols:

Rules for naming amines

- Name primary amines on the basis of the longest continuous carbon chain that has the attached nitrogen.
- The amine is named as an **alkanamine** by changing the ending of the name of the alkane from **-e** to **-amine.** The position of attachment of nitrogen to the longest continuous carbon chain is indicated by a number in front of the name. The longest continuous carbon chain is numbered from the end nearest the nitrogen.
- Prefixes with numbers are used to indicate substituents attached to the longest continuous carbon chain.
- Compounds with two NH_2 groups are named as alkanediamines.
- A cyclic, nonaromatic compound containing an attached NH_2 group is named as a **cycloalkanamine.** The ring carbon holding the nitrogen is C1.
- Name secondary and tertiary amines on the basis of the group with the longest continuous carbon chain attached to nitrogen. The names of the other groups attached to nitrogen are placed in front of the alkanamine or cycloalkanamine name. *N*- and *N,N*- prefixes are used to indicate that the other groups are attached to nitrogen.

Although the *CA* nomenclature system uses the name benzenamine for $C_6H_5NH_2$, we will use the more widely accepted name aniline from the common nomenclature system.

Example 8.3 Naming amines

Give the common name or *CA* name, whichever is appropriate, for each of the following compounds:

(a) $CH_3CH_2CH_2CH_2{-}N(H){-}CH_3$ (b) $CH_3CH(CH_3){-}CH(CH_3)CH_2{-}N(H){-}CH_3$

(c) cyclohexane ring bearing NH_2 (d) benzene ring bearing $N(CH_3)_2$ (e) benzene ring bearing $HNCH_3$ and, at the meta position, CH_3

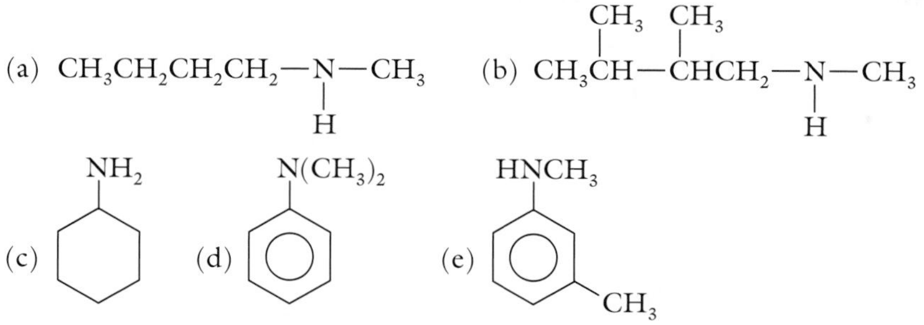

Solution

(a) Because only simple groups, butyl and methyl, are attached to nitrogen, the common name is used: butylmethylamine.

(b) The common nomenclature system is not applicable, because the larger group is not one of the simple groups. The *CA* nomenclature system must be used. The longest chain attached to the nitrogen is four carbons, with nitrogen attached at C1. The name is 2,3,*N*-trimethyl-1-butanamine.

(c) The common name, cyclohexylamine, is used because the cyclohexyl group is a simple group. (The *CA* name is cyclohexanamine.)

(d) The common name is used: *N,N*-dimethylaniline.

(e) The compound is a substituted aniline with one methyl group on nitrogen and one methyl substituted on the ring. The name is *m,N*-dimethylaniline (or 3,*N*-dimethylaniline).

Problem 8.3 Give the common name or *CA* name, whichever is appropriate, for each of the following compounds:

(a) $CH_3{-}CH_2{-}CH(N(CH_3)CH_2CH_3){-}CH_3$, i.e. $CH_3{-}CH_2{-}CH{-}CH_3$ with $CH_3{-}N{-}CH_2CH_3$ on the CH (b) $CH_3CH_2CH(NH_2)CH_2NH_2$

(c) cyclopentane ring bearing H and CH_3CH_2 on one carbon, and H and $N(CH_3)_2$ on an adjacent carbon (d) benzene ring bearing $HNCH_2CH_3$, with $CH(CH_3)_2$ at the meta position and CH_3 at the para position

(e) $(CH_3)_2CHCH_2CH(CH_2CH_3)CH_2CH(CH_3)_2$ with NH_2 on the first CH, i.e. $(CH_3)_2C(NH_2)HCH_2CH(CH_2CH_3)CH_2CH(CH_3)_2$

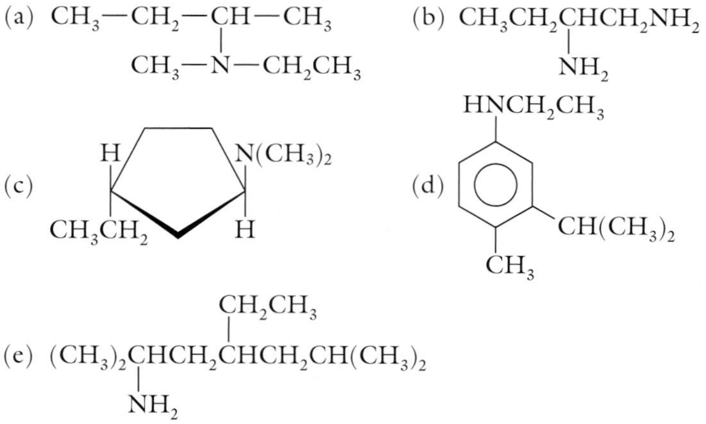

The groups $—NH_2$, $—NHR$, and $—NR_2$ are called amino, *N*-alkylamino, and *N,N*-dialkylamino groups, respectively. These names are used as prefixes in the names of compounds that contain both an amine group and an oxygen-containing functional group such as —COOH or —OH. Such compounds are named not as amines but as members of the family represented by the oxygen-containing group. In other words, the oxygen-containing functional groups have preference over nitrogen for nomenclature purposes. Examples are 4-aminobutanoic acid and 3-(*N,N*-dimethylamino)-1-hexanol.

$H_2N{-}CH_2CH_2CH_2C(=O){-}OH$

4-Aminobutanoic acid

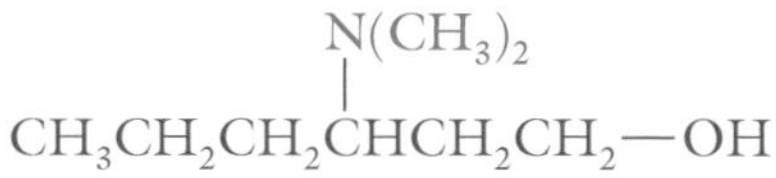

$CH_3CH_2CH_2CH(N(CH_3)_2)CH_2CH_2{-}OH$

3-(*N,N*-Dimethylamino)-1-hexanol

Cyclic compounds containing one or more nitrogen atoms in the ring (rather than attached to the ring) are called **nitrogen heterocyclics** or **heterocyclic amines** or **heterocyclic nitrogen bases.** Table 8.1 on the following page lists the most important types of nitrogen heterocyclics, many of which play important roles in biological systems.

TABLE 8.1 Common Names of Heterocyclic Structures

Name	Structure	Present in
imidazole	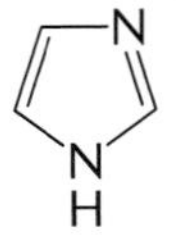	histidine (amino acid)
piperidine	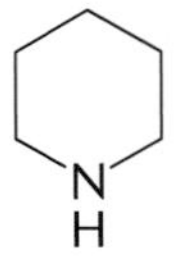	quinine and other drugs
purine	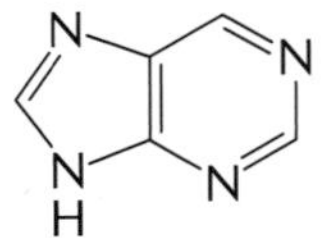	nucleic acids (DNA, RNA)
pyridine	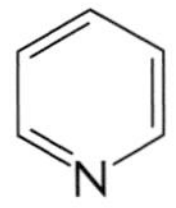	nicotine, vitamin B
pyrimidine	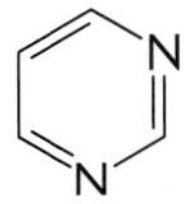	nucleic acids (DNA, RNA)
pyrrole		hemoglobin
pyrrolidine	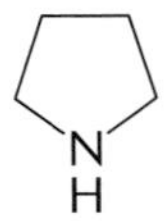	nicotine, proline (amino acid)

Alkaloids are amines produced by plants. They are usually heterocyclic amines, taste bitter, and have significant physiological effects. Alkaloids are used, either as the plant part itself or as the extracted compounds, for medicinal and other physiological purposes. Their effects can be put to good use or they can be disastrous and life threatening when used unwisely (Box 8.1).

Alkaloids are but one group of physiologically active amines. More physiologically active compounds, including synthetic and natural medications, belong to the amine family than to any other (Boxes 8.2 on page 255 and 8.3 on page 256). Some physiologically active compounds are amides, often in addition to being amines, and many also possess other functional groups such as ether, ester, or alcohol.

8.4 PHYSICAL PROPERTIES OF AMINES

The physical properties of an amine depend on whether it is a primary, secondary, or tertiary amine:

- Primary and secondary amines have melting and boiling points comparable to those of aldehydes and ketones (Table 8.2 on page 254).
- Tertiary amines have melting and boiling points considerably lower than those of primary and secondary amines and comparable to those of ethers and hydrocarbons.

8.1 CHEMISTRY WITHIN US

Opium Alkaloids

The **opium alkaloids,** also called **opiates,** include compounds obtained from the unripened seeds of the opium poppy *Papaver somniferum* (which is not the common poppy) as well as synthetic compounds of similar structure, all possessing similar physiological activity. Opiates are used medically as analgesics (pain relievers) and hypnotics (sleep inducers). They also induce a sense of euphoria (a "high"), which accounts for their illegal, nonmedical use. The downside of opium alkaloids is that they are addictive and induce confusion and lethargy. Moreover, they depress the respiratory system—that is, reduce its activity by affecting the brain centers that control respiration. Extreme depression of the brain's respiratory centers can result in death from respiratory arrest.

The raw **opium** extracted from the opium poppy contains almost two dozen different compounds. In the United States, raw opium was used in many patent medicines, such as Mrs. Winslow's Soothing Syrup, until the beginning of the twentieth century.

Morphine, the principal alkaloid in opium, was isolated from raw opium early in the nineteenth century. The invention of the syringe in the middle of the nineteenth century led to large-scale use of morphine for pain relief in the American Civil War (1861–1865). Direct injection of an aqueous solution of morphine (in the form of its amine salt) into the bloodstream allowed quick relief of pain from battle injuries but resulted in large numbers of soldiers addicted to morphine. Morphine and other narcotics became controlled substances under federal law in 1914. Physicians can still prescribe morphine for pain relief, but great care is taken to prevent prolonged use or high doses, either of which may lead to addiction.

Codeine, the monomethyl ether of morphine, is present in raw opium in much smaller amounts than is morphine. Codeine is usually produced commercially from morphine by converting one of the —OH groups into a methyl ether group. Codeine is similar to morphine in its physiological effects but is less potent and less addictive. It is used in prescription cough-suppressant syrups.

Heroin was first synthesized in the 1890s by the reaction of morphine with acetic anhydride to convert the —OH groups into acetate (ester) groups (an acyl transfer reaction; see Section 7.4). At that time, heroin was considered an antidote for morphine addiction, but nothing could have been farther from reality. Heroin produces a stronger high than morphine does and is much more addictive, and the addiction is harder to overcome. Heroin use is illegal, even by prescription, in the United States.

Opium poppies and seed pod. The seed pod is slit to collect the raw opium. (Rick Strange/Picture Cube.)

Morphine

Codeine

Heroin

If you compare the structures of morphine, codeine, and heroin, you see what appear to be modest differences, yet these differences result in significant differences in physiological activity. Opiates produce their physiological effects such as pain relief or euphoria by binding to receptor sites in the brain. Small differences in chemical structures result in different extents of binding at receptor sites and thus different physiological effects.

INSIGHT INTO PROPERTIES

$R_2N^{\delta-}$ --- $H^{\delta+}$ ··· $R_2N^{\delta-}$ --- $H^{\delta+}$ ··· $R_2N^{\delta-}$ --- $H^{\delta+}$

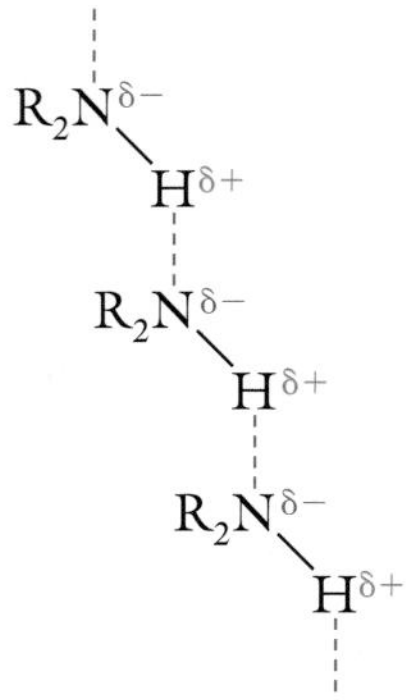

Figure 8.1 Hydrogen bonding in secondary amines.

TABLE 8.2 Comparison of Boiling Points of Amines and Compounds from Different Families			
Compound	Structure	Molecular mass (amu)	Boiling point (°C)
butane	$CH_3CH_2CH_2CH_3$	58	−1
2-methylpropane	$(CH_3)_3CH$	58	−12
trimethylamine (3°)	$(CH_3)_3N$	59	3
methylethylamine (2°)	$CH_3CH_2NHCH_3$	59	36
propylamine (1°)	$CH_3CH_2CH_2NH_2$	59	47
propanal	CH_3CH_2CHO	58	49
methyl formate	$HCOOCH_3$	60	32
propanone	CH_3COCH_3	58	56
1-propanol	$CH_3CH_2CH_2OH$	60	97
ethanoic acid	CH_3COOH	60	118
triethylamine (3°)	$(CH_3CH_2)_3N$	101	89
dipropyl ether	$(CH_3CH_2CH_2)_2O$	102	91
heptane	$CH_3(CH_2)_5 CH_3$	100	98

- All amines have water solubility comparable to those of aldehydes and ketones and slightly lower than that of alcohols.

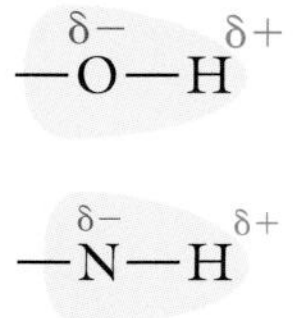

$\overset{\delta-}{-O}-\overset{\delta+}{H}$

$\overset{\delta-}{-N}-\overset{\delta+}{H}$

O–H bond is more polar than N–H bond

Primary and secondary amines, unlike tertiary amines, possess N–H bonds and participate in hydrogen bonding (Figure 8.1). This hydrogen bonding results in strong secondary forces and high melting and boiling points, slightly higher for primary amines than for secondary amines because primary amines have two N–H bonds, whereas secondary amines have only one. The N–H bond is less polar than the O–H bond, because nitrogen is less electronegative than oxygen and, therefore, the hydrogen bonding in primary and secondary amines is not as strong as that in alcohols. For this reason, alcohols have higher melting and boiling points than those of primary and secondary amines. The strength of secondary forces due to N–H hydrogen bonding is comparable to that of the dipole–dipole forces in aldehydes and ketones (see Table 8.2).

Tertiary amines have much weaker secondary forces than those of primary and secondary amines because tertiary amines possess only the weakly polar C–N bonds, no N–H bonds. The strength of these attractive forces is comparable to those in ethers and hydrocarbons.

Many amines have distinct and disagreeable fishy odors. The common names of some amines are based on their origin and odor; for example, putrescine (1,4-butanediamine) and cadaverine (1,5-pentanediamine) are found in rotting meat.

Amines of as many as three or four carbons are soluble in water in all proportions, with solubility dropping off as the number of carbons increases. Solubility is very low or negligible for amines with more than six or seven carbons. Tertiary amines are slightly less water soluble than primary and secondary amines because of differences in hydrogen bonding. The N–H bonds of primary and secondary amines participate in hydrogen-bonding attractions with water (Figure 8.2). In tertiary amines, hydrogen bonding with water is more limited because there are no N–H bonds. Hydrogen bonding takes place between the nitrogen of a tertiary amine and a hydrogen of water (Figure 8.3), analogous to the hydrogen bonding between an ether and water.

INSIGHT INTO PROPERTIES

$R_2N^{\delta-}$ --- $H^{\delta+}$ ··· $---\overset{\delta+}{H}-O^{\delta-}$ --- $H^{\delta+}$ ··· $R_2N^{\delta-}$ --- $H^{\delta+}$ ··· $---\overset{\delta+}{H}-O^{\delta-}$ --- $H^{\delta+}$

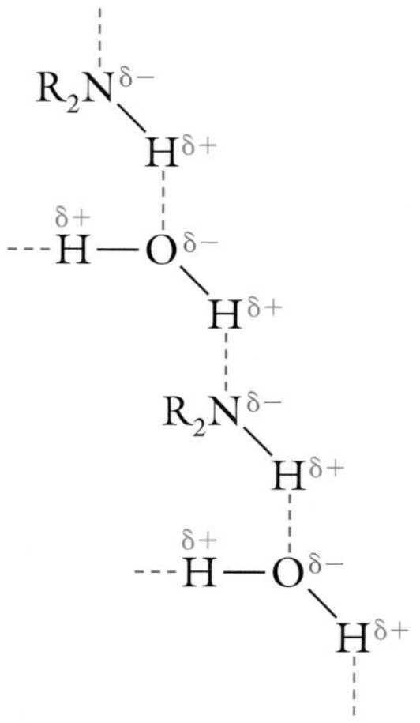

Figure 8.2 Hydrogen bonding between water and a secondary amine.

The slightly lower water solubility of amines compared with that of alcohol results from the slightly weaker attractive forces between water and the less-polar N–H bonds of primary and secondary amines or the nitrogen of tertiary

8.2 CHEMISTRY WITHIN US

Drugs for Controlling Blood Pressure

High blood pressure, often called the "silent killer" because it has no symptoms or very few of them in its early stages, has an unrelenting long-term adverse effect on the cardiovascular system. If undetected and untreated, the end result can be stroke or heart attack. Blood pressure is the pressure developed in blood vessels as the heart pumps blood through the blood vessels. A variety of drugs, mostly synthetic, are available to control high blood pressure, with an excellent prognosis for long-term health and quality of life.

Different types of drugs control blood pressure through different mechanisms. **Adrenergic inhibitors** lower blood pressure by interfering with the sympathetic nervous system, which controls the body's automatic response to stress, changes in blood pressure, heartbeat, and perspiration, for example. Adrenergic inhibitors include α-blockers such as terazosin (Hytrin) and β-blockers such as atenolol (Tenormin). **α-Blockers** relax blood vessels by blocking α-receptors that cause blood-vessel constriction. **β-Blockers** have a more general effect, acting on the β-receptors in heart, lung, and blood vessels that cause the heart to beat harder and faster and the blood vessels to constrict.

Diuretics such as furosemide (Lasix) cause the kidneys to produce more urine. The increased production of urine reduces blood pressure by lowering the volume of blood that the heart pumps to all parts of the body. **Calcium channel blockers** such as diltiazem (Cardizem) inhibit the entry of Ca^{2+}—an ion essential for muscle contraction—into the muscle cells of blood vessels. This inhibition results in the relaxation of blood vessel walls, which decreases blood pressure and heart rate. **Vasodilators** such as hydralazine (Apresoline) act on the walls and tissues of small arteries to decrease blood-vessel tension and resistance. **Angiotensin-converting enzyme (ACE)** inhibitors such as lisinopril (Prinivil) inhibit the conversion of angiotensin I into angiotensin II. Angiotensin II causes the constriction of blood vessels, and the inhibition of its formation results in a lowering of blood pressure.

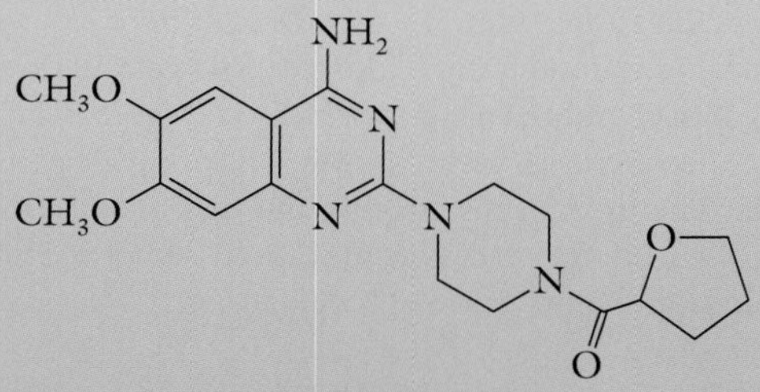

Terazosin (Hytrin)

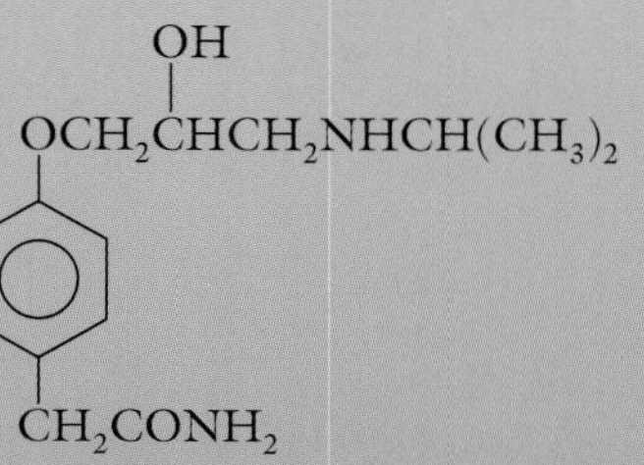

Atenolol (Tenormin)

amines. The attractive forces between water and the O–H of an alcohol are stronger. The water solubilities of amines are about the same as those of aldehydes and ketones, because the hydrogen-bond attractive forces between amines and water are comparable to those between water and the carbonyl oxygen of aldehydes or ketones.

INSIGHT INTO PROPERTIES

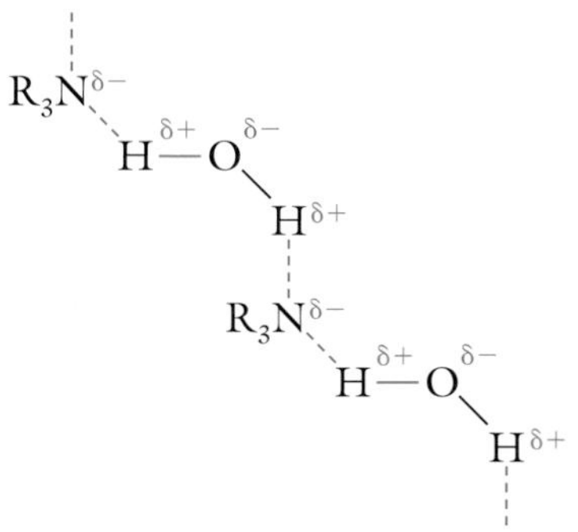

Figure 8.3 Hydrogen bonding between water and a tertiary amine.

8.5 THE BASICITY OF AMINES

Like ammonia (Section 2.4), amines are basic because of the ability of the nonbonded pair of electrons on nitrogen to accept (bond to) a proton from an acid such as water or HCl. The reaction is illustrated for methylamine reacting with water:

$$CH_3NH_2 + HOH \rightleftharpoons CH_3NH_3^+ + OH^-$$

Methylamine

Protonation of the nitrogen converts the amine into a positive ion because the nitrogen becomes electron deficient by sharing its pair of nonbonded electrons with the proton from water.

The reaction of an amine with water is an equilibrium reaction with the equilibrium far to the left. Methylamine is a typical aliphatic amine in this

8.3 CHEMISTRY WITHIN US

Other Amines and Amides with Physiological Activity

Caffeine is probably the most widely and often used mild central-nervous-system (CNS) stimulant. It increases the basal metabolism rate, the heart rate, the secretion of stomach acid, and urine production. Caffeine exists naturally in coffee and tea and is added to soft drinks. Most people claim to need a cup or two of coffee to get started in the morning or to stay awake if they are working late. Not surprisingly, caffeine is the main ingredient in stay-awake medications such as No Doz. There is evidence that caffeine is moderately addictive; so caffeine-free soft drinks, coffee, and tea are now readily available. Caffeine is also a component of many nonprescription drugs, including pain relievers, antihistamines, and cold medications. Its purpose is to counteract the unwanted sedative effects of the main ingredients of these medications.

Caffeine

Nicotine

Nicotine, a mild CNS stimulant found in tobacco, has been categorized by the U.S. surgeon general as an addictive drug. It is inhaled in tobacco smoke and absorbed orally when tobacco is chewed. Nicotine is highly toxic if injected directly into the bloodstream, but the amounts delivered into the bloodstream by smoking or chewing tobacco are far below the acute toxic dose. Nevertheless, regular tobacco use of any kind has long-term effects on the cardiovascular system, and smoking tobacco creates respiratory problems. Cigarette smoke contains significant amounts of carbon monoxide and fused-ring aromatic compounds. Carbon monoxide is absorbed by hemoglobin, interfering with the ability of blood to transport oxygen from the lungs to the rest of the body. Fused-ring aromatic compounds (see Box 4.9) are cancer-causing agents.

A woman is absorbing the central-nervous-system stimulants nicotine, by smoking, and caffeine, by drinking coffee. (Dion Ogust/Image Works.)

respect: only 6.6% of it is converted into the amine salt in water. Thus, amines are much weaker bases than strong inorganic bases such as NaOH and KOH.

Table 8.3 compares the $[OH^-]$, $[H_3O^+]$, pH, and ionization percentage of some inorganic and organic bases. Amines are considerably stronger bases than water. Amines are also stronger bases than alcohols and ethers, whose basicities are about the same as that of water. The oxygen atoms in water, alcohols, and ethers have the ability, like nitrogen, to bond to H^+ by using a

TABLE 8.3 Relative Basicities of Amines

	Values for 0.10 *M* aqueous solution			
Compound	$[OH^-]$	$[H_3O^+]$	pH	Ionization (%)
NaOH	0.10	1.0×10^{-13}	13.00	100
CH_3NH_2	6.8×10^{-3}	1.5×10^{-12}	11.82	6.8
$C_6H_5NH_2$	6.5×10^{-6}	2.4×10^{-9}	8.62	0.0065
H_2O	1.0×10^{-7}	1.0×10^{-7}	7.00	0.00010

Epinephrine (adrenaline), a hormone secreted by the adrenal glands when we are stressed or frightened, prepares the body to respond to a perceived threat. Several synthetic drugs called **amphetamines** are structurally related to adrenaline and mimic its stimulant effect. These drugs include isoproterenol, prescribed for emphysema and asthma, and methedrine, prescribed as an antidepressant and sometimes for weight control. Illegal use of amphetamines, called "uppers" or "pep pills" on the street, is a serious drug-abuse problem. Overuse can result in sleeplessness, paranoia, and hallucinations.

Epinephrine (Adrenalin), R = CH_3
Isoproterenol, R = $CH(CH_3)_2$

Methamphetamine (Methedrine)

Barbiturates, a class of amides that induce sedation and sleep, are called "downers" because of their mood-depressing effect. They are particularly dangerous when used in combination with alcohol because of a synergistic effect between the two in which their combined effect is greater than the effect of either drug alone. This combination is the agent of some suicides, suicide attempts, and accidental suicides. The barbiturate phenobarbital (Luminal) is used as an anticonvulsant agent for treating epilepsy and brain trauma.

Phenobarbital (Luminal)

Other synthetic drugs that contain amine or amide functional groups are antibiotics such as ampicillin, tranquilizers (antianxiety drugs) such as diazepam (Valium), antidepressants such as fluoxetine (Prozac), ulcer-healing medications such as cimetidine (Tagamet), and local anesthetics such as procaine (Novocain).

Ampicillin

Fluoxetine (Prozac)

Cimetidine (Tagamet)

pair of nonbonded electrons. However, oxygen is a poorer electron donor than nitrogen because oxygen, being more electronegative (Section 1.9), holds its electrons more tightly. Thus, the oxygen-containing organic compounds are weaker proton acceptors—weaker bases—than amines.

Amines are sufficiently basic to turn red litmus paper blue, which serves as a simple chemical diagnostic test for amines. The only other type of organic compound sufficiently basic to turn red litmus paper blue is a carboxylate salt, $RCOO^-$ (Section 7.6).

Even though an amine is a weak base, its reaction with an acid goes to completion with a strong acid such as HCl and the amine is converted into an **amine salt:**

$$CH_3{-}NH_2{:} + HCl \longrightarrow CH_3{-}NH_3^+\ Cl^-$$

Amine — Amine salt

The amine is often called **free amine** or **free base** to distinguish it from the amine salt.

Amines present in the body are almost entirely (>99%) in the amine salt (ionized) form. The pH of a solution of an amine in pure water is near 12 (see Table 8.3), and amines exist mostly in the un-ionized form at this high pH. However, most physiological fluids are buffered (by the bicarbonate–carbonic acid buffer) to a pH very near 7—five orders of magnitude more acidic than pH 12. At this higher acidity, amines exist as their amine salts.

Basic compounds such as amines as well as acidic compounds such as carboxylic acids (Section 7.5) are always present in the form of their neutralized, ionized forms under physiological conditions. Biological systems are buffered for purposes of neutralizing the bases and acids to maintain the system at constant pH. This constant pH is important because the physiological functions of most biological molecules such as proteins are very dependent on the pH. Biological systems typically operate within very narrow pH ranges. For example, blood pH is 7.4 (Section 2.5). Even small changes in pH—say, by 0.1 or 0.2 units—can alter the normal physiological functions of proteins (Section 12.9).

Example 8.4 Writing equations for the basicity of amines

Write the equation for each of the following reactions: (a) aniline is dissolved in water; (b) ethyldimethylamine is neutralized with HCl.

Solution

(a) Aniline ionizes weakly in water; the arrows show that the equilibrium is far to the left.

$$C_6H_5NH_2 + H_2O \rightleftharpoons C_6H_5NH_3^+ + OH^-$$

(b) This tertiary amine reacts with strong acid to form the amine salt; the equilibrium is completely to the right.

$$CH_3CH_2{-}\overset{\overset{\large CH_3}{|}}{\underset{\underset{\large CH_3}{|}}{N}}\!: + HCl \longrightarrow CH_3CH_2{-}\overset{\overset{\large CH_3}{|}}{\underset{\underset{\large CH_3}{|}}{N^+}}{-}H\ Cl^-$$

Problem 8.4 Write the equation for each of the following reactions: (a) aniline is neutralized with H_2SO_4; (b) diethylamine is dissolved in water.

$R{-}\ddot{N}H_2$
Alkylamine
Nonbonded electrons are more available

$Ar{\leftarrow}\ddot{N}H_2$
Arylamine
Nonbonded electrons are less available

Arylamines are much less basic than aliphatic amines. Table 8.3 shows that $[OH^-]$ is about 1000-fold greater for a methylamine solution than for an aniline solution. This difference is due to the electron-withdrawing effect of the benzene ring compared with an alkyl group. Greater electron withdrawal from nitrogen makes the nonbonded electron pair of nitrogen less available for bonding to a proton.

8.6 AMINE SALTS

An amine salt, the product of a reaction between an amine and a strong acid, is named by naming the positive ion first and then naming the negative ion. The positive ion is named differently, depending on whether the amine is aliphatic, aromatic, or heterocyclic:

Rules for naming amine salts

- For aromatic and heterocyclic amines, the final **-e** of the name of the amine is replaced by **-ium.**
- For aliphatic amines, the ending of the name of the amine is changed from **-amine** to **-ammonium.**

Pharmaceutical companies often use an alternate approach for naming amine salts used as medicines. The name of the amine is followed by the name of the acid used to synthesize the amine salt. Amine salts from HCl are called amine hydrochlorides. Those from H_2SO_4 are called amine hydrogen sulfates.

Example 8.5 Naming amine salts

Name the following amine salts:

(a) $CH_3CH_2{-}\overset{CH_3}{\underset{CH_3}{N^+}}{-}H\ Cl^-$ (b) $CH_3CH_2NH_2^+\ HSO_4^-$

(c) $CH_3CH_2{-}\overset{CH_3}{CH}CH_2{-}\overset{H}{\underset{H}{N^+}}{-}CH_3\ Br^-$

Solution

(a) The amine salt is derived from ethyldimethylamine and HCl. The name is ethyldimethylammonium chloride.

(b) The amine salt is derived from *N*-ethylaniline and H_2SO_4. The name is *N*-ethylanilinium hydrogen sulfate.

(c) The amine salt is derived from 2,*N*-dimethyl-1-butanamine and HBr. The name is 2,*N*-dimethyl-1-butanammonium bromide.

Problem 8.5 Draw the structures of the following amine salts: (a) cyclohexylammonium chloride; (b) *N,N*-dimethyl-1-butanammonium hydrogen sulfate; (c) pyridinium bromide (refer to Table 8.1).

Properties of Amine Salts

Amine salts are ionic and possess much higher secondary attractive forces than any covalent organic compounds do, even higher than carboxylic acids, which possess strong hydrogen-bond secondary attractions. The melting and boiling points of amine salts are much higher than those of the corresponding amines. All amine salts therefore have high melting points and are solids at room temperature. For example, methylammonium chloride is a solid with a melting point of 232°C and decomposes without boiling. Compare it with methylamine, which is a gas with a melting point of −94°C and a boiling point of −6°C. The difference in the melting and boiling points of amines and amine salts has had a significant and disturbing effect on cocaine use, increasing the number of people addicted to cocaine (Box 8.4 on the following page).

Amine salts are also much more water soluble than are amines. Amines of more than six carbons have very low or no solubility in water, whereas the solubility of amine salts extends to much larger numbers of carbons. This difference results from the much greater strength of the attractive forces between ions and water than that of the hydrogen-bond attractions between amines and water. Thus, the solubility of an amine is greatly increased in an environment of neutral or lower pH because the amine is converted into the amine salt. Amines such as hormones and neurotransmitters are solubilized in the body because they are in their salt form at physiological pH. The salt form allows them to be transported by the blood at the concentrations required for their specialized functions.

Amines that are used as medicines are almost always used in the form of their amine salts for two reasons. First, amines have a limited shelf life because they undergo slow oxidation by air. The amine salts have a much better shelf life because they are much more resistant to air oxidation. Second, water

8.4 CHEMISTRY WITHIN US

Cocaine: Free Base Versus Amine Salt

Cocaine is extracted from the leaves of the coca plant, which is found almost exclusively on the slopes of the Andes of South America. It has been used medically as a local anesthetic to relieve pain in severe injuries and in nasal surgery, but such use is restricted because of its highly addictive and destructive nature. Cocaine has become one of the most abused narcotics by so-called recreational users.

$\xrightleftharpoons[\text{NaOH}]{\text{HCl}}$

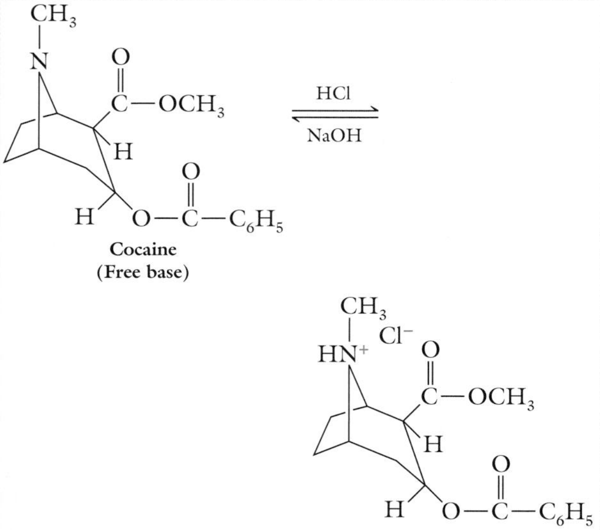

Cocaine (Free base)

Cocaine hydrochloride (Amine salt)

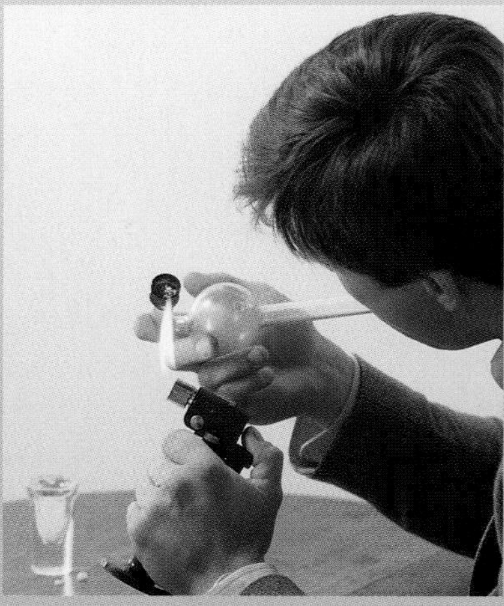

Crack and free-base cocaine are more dangerous than cocaine hydrochloride because crack and free-base cocaine can be smoked, which is easier than snorting or injecting cocaine hydrochloride. (Lawrence Migdale/Photo Researchers.)

Cocaine is a powerful central-nervous-system stimulant, speeding up the heart rate, raising the blood pressure, and stimulating the pleasure centers in the brain. Cocaine gives the user a great sense of euphoria, power, confidence, and increased stamina. The effect is short lived (30–60 min), however, and is followed by severe depression and a craving for more cocaine. The depression leads to repeated use, setting off a cycle of euphoria and depression that quickly results in psychological addiction.

solubility is required if a drug is to be administered intraveneously. Few medicines have enough water-solubilizing groups (such as hydroxyl or amino) relative to the number of carbons to allow their use as free amines. The amine salts are much more soluble in water. Water solubility is also important for applications such as cough syrups.

Example 8.6 Predicting the physical properties of amines and amine salts

Place the following compounds in order of increasing melting point: *N*-ethyl-*N*-methylaniline, anilinium chloride, *N*-propylaniline.

Solution

N-Ethyl-*N*-methylaniline $<$ *N*-propylaniline $\ll$ anilinium chloride.
The ionic compound anilinium chloride has a much higher melting point than those of the covalent compounds. Of the two covalent compounds of the same molecular mass, the secondary amine *N*-propylaniline has the higher melting point because it possesses hydrogen bonding. The tertiary amine *N*-ethyl-*N*-methylaniline has weaker secondary forces—dipole–dipole instead of hydrogen bonding—because it does not have N–H bonds.

Problem 8.6 Place the following compounds in order of increasing water solubility: $CH_3(CH_2)_6NH_2$, $CH_3(CH_2)_{14}NH_3{}^+Cl^-$, $CH_3(CH_2)_{14}NH_2$.

Long-term users often develop a paranoid psychosis and run a high risk of death from respiratory or heart failure.

Treating crushed coca leaves with aqueous HCl converts the cocaine into its amine salt **cocaine hydrochloride,** which dissolves in the aqueous solution. The crushed leaves are discarded, and the water is evaporated from the aqueous solution to yield solid cocaine hydrochloride, the original form of cocaine illegally sold on the streets. Cocaine hydrochloride is used in two ways: "snorting" and injection. Cocaine hydrochloride inhaled into the nasal passages dissolves in the watery mucous membranes of the nose and throat and enters the bloodstream. Long-term use results in the deterioration of the mucous membranes and nose cartilage and a chronic runny nose. Injection of an aqueous solution of cocaine hydrochloride directly into the blood stream gives a much faster and higher "high" than snorting does because the cocaine concentration in the bloodstream is much higher. The risks associated with injection, as for all injectable drugs, include hepatitis, AIDS, and other infections spread through the sharing of needles.

Today, much of the cocaine used illegally is cocaine, not cocaine hydrochloride. The cocaine is produced from cocaine hydrochloride by neutralization with a base. Two different forms of cocaine are produced, depending on the details of the neutralization process. **Free-base cocaine** is made by neutralizing the cocaine hydrochloride with aqueous base (NaOH) and then adding diethyl ether to extract the cocaine from the aqueous solution. Evaporation of the ether solvent yields free-base cocaine. This last step can be dangerous because diethyl ether is extremely flammable.

Crack cocaine, another form of free-base cocaine, is the street name for cocaine produced by a less-dangerous and simpler process. Cocaine hydrochloride is mixed with sodium bicarbonate (better known as baking soda) and water and then boiled. The sodium bicarbonate ($NaHCO_3$) reacts with the HCl of cocaine hydrochloride to form NaCl and carbonic acid (H_2CO_3). Carbonic acid decomposes to carbon dioxide and water with carbon dioxide volatilizing from the mixture. Heating is continued until a solid separates from the boiling mixture. The solid is from 75% to 90% cocaine, together with NaCl (and $NaHCO_3$ if too much was used in the recipe). The solid is removed, allowed to dry, and broken up into chunks.

Both free-base cocaine and crack cocaine can be smoked, which make them much more dangerous than cocaine hydrochloride. People are much less hesitant about smoking a drug than about injecting one. Cocaine can be smoked because it is a covalent compound. It has a lower melting point (98°C) and is much more volatile than the ionic compound cocaine hydrochloride. The latter cannot be smoked, because it does not melt and is not volatile. When smoked, cocaine is rapidly absorbed through the membranes of the respiratory system, giving an immediate high. The effect is the equivalent of injecting cocaine hydrochloride directly into the bloodstream, producing much faster and stronger euphoric effects than are achieved by snorting cocaine hydrochloride. The result has been a sharp rise in cocaine use and in the number of people addicted.

An amine salt, like any salt of a weak base, is a weak acid and undergoes acid-base reaction with water and strong bases:

$$CH_3NH_3^+\ Cl^- + H_2O \rightleftharpoons CH_3NH_2 + H_3O^+ + Cl^-$$

$$CH_3NH_3^+\ Cl^- + NaOH \longrightarrow CH_3NH_2 + H_2O + NaCl$$

Amine salts are sufficiently acidic to turn blue litmus paper red.

Quaternary Ammonium Salts

The ammonium salts just described have one, two, or three alkyl or aryl groups or combinations of alkyl and aryl groups, but it is also possible to synthesize amine salts having four groups. Such salts are called **quaternary ammonium salts** and are named in the same way as amine salts, taking into account the need to name four instead of three groups attached to nitrogen. Without a hydrogen attached to nitrogen, quaternary ammonium salts are not acidic, as are ammonium salts that contain one or more hydrogens. Quaternary ammonium salts do not turn blue litmus paper red.

$$R^1\text{—}\overset{\displaystyle R^2 \atop |}{\underset{| \atop \displaystyle R^3}{N^+}}\text{—}R^4\ Cl^-$$

Quaternary ammonium salt

Quaternary ammonium hydroxides are formed when a quaternary ammonium salt such as a halide undergoes reaction with strong base:

$$R_4N^+\ Cl^- + NaOH \longrightarrow R_4N^+\ OH^- + NaCl$$

Quaternary ammonium hydroxides are strong bases, as strong as NaOH and KOH.

$$CH_3\text{—}\overset{\displaystyle CH_3 \atop |}{\underset{| \atop \displaystyle CH_3}{N^+}}\text{—}CH_2CH_2\text{—}OH$$

Choline

Quaternary ammonium salts in which one of the alkyl groups is a long hydrocarbon chain (such as the chain $CH_3(CH_2)_{17}$—) are used as soaps

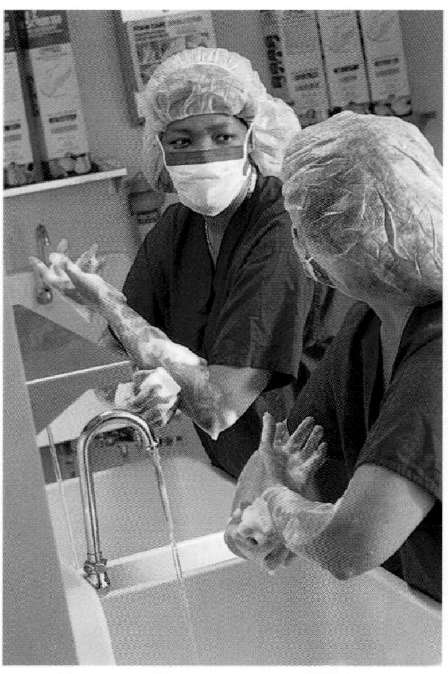

Surgical nurses scrub their hands with disinfectant before surgery. Many hospital disinfectants contain quaternary ammonium salts. (Charles Gupton/The Stock Market.)

» Neurotransmitters are chemicals that carry information from one nerve cell to another or to muscle, as shown in Section 18.5.

(Section 7.7), germicides, antiseptic soaps, and mouthwashes; they are used for disinfecting skin and hands preparatory to surgery and for sterilizing surgical instruments. The quaternary ammonium salt choline is a constituent of phospholipids used to construct cell membranes (Sections 11.6 and 11.10). It also functions as a neurotransmitter.

A PICTURE OF HEALTH

Examples of Amines and Amides

(Alamy.)

The hormone thyroxine, produced in the thyroid gland, contains an amine group.

Proteins such as muscle and hemoglobin are amides.

The hormone adrenalin, produced in the adrenal glands, contains the amine group.

Nicotine is an amine. Caffeine contains amine and amide groups.

Many medications contain amine or amide groups or both. The antibiotic ampicillin and the blood pressure medication terazosin contain both functional groups.

DNA and RNA contain amine and amide groups.

8.7 CLASSIFYING AMIDES

Amides contain the amide group—a carbonyl group attached directly to a nitrogen. A compound containing a nitrogen and a carbonyl is not an amide unless the two are bonded directly together.

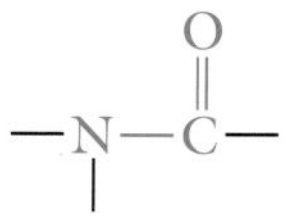

$$-\underset{|}{N}-\overset{O}{\overset{\|}{C}}-$$

Amide

$$-\underset{|}{N}-\underset{|}{\overset{|}{C}}-\overset{O}{\overset{\|}{C}}-$$

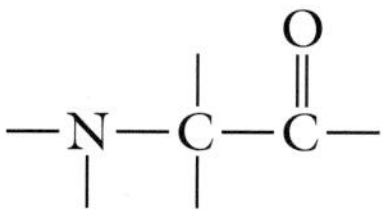

Not an amide

Amides are classified as primary, secondary, and tertiary amides when two, one, and zero hydrogens, respectively, are attached to nitrogen:

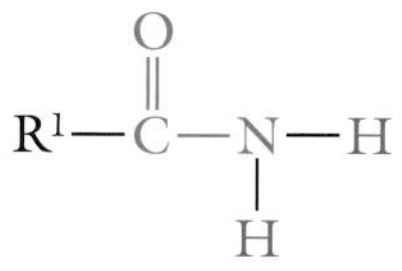

$$R^1-\overset{O}{\overset{\|}{C}}-\underset{H}{\underset{|}{N}}-H \qquad R^1-\overset{O}{\overset{\|}{C}}-\underset{H}{\underset{|}{N}}-R^2 \qquad R^1-\overset{O}{\overset{\|}{C}}-\underset{R^3}{\underset{|}{N}}-R^2$$

Primary 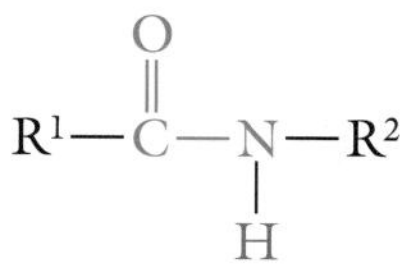 **Secondary** 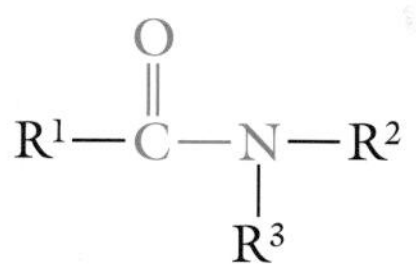 **Tertiary**

R^1, R^2, and R^3 can be aliphatic or aromatic.

Proteins are amides of central biological importance and will be considered in Chapter 12. Proteins have a variety of physiological functions in all living organisms. They are used in construction, protection, and movement (skin, bone, muscle); as catalysts in biochemical reactions (enzymes); in the regulation and control of physiological processes (hormones, neurotransmitters); and in the transport of oxygen throughout the body (hemoglobin).

Many physiologically active compounds are amides (see Boxes 8.2 and 8.3), including the sweetener aspartame (NutraSweet; Box 10.1).

Example 8.7 Classifying amides

All of the following compounds are constitutional isomers of C_4H_9NO. Which of the compounds are amides? Classify the amides as primary, secondary, or tertiary.

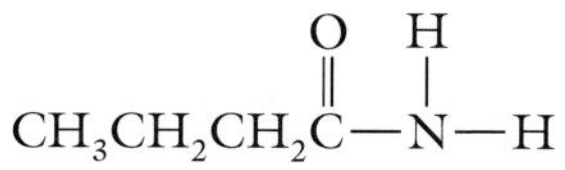

$$CH_3CH_2CH_2\overset{O}{\overset{\|}{C}}-\overset{H}{\overset{|}{N}}-H \quad (1) \qquad CH_3CH_2\overset{O}{\overset{\|}{C}}-\overset{CH_3}{\overset{|}{N}}-H \quad (2)$$

$$CH_3\overset{O}{\overset{\|}{C}}-\overset{CH_3}{\overset{|}{N}}-CH_3 \quad (3) \qquad CH_3CH_2\overset{O}{\overset{\|}{C}}-CH_2-\overset{H}{\overset{|}{N}}-H \quad (4)$$

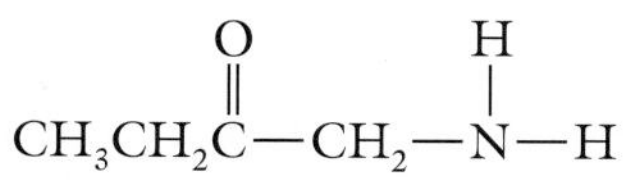

Solution

Compounds 1, 2, and 3 are amides because the nitrogen is directly bonded to the carbonyl carbon. Compound 4 is not an amide, because the nitrogen is not directly bonded to the carbonyl carbon. Compound 4 is an amine and a ketone. Compounds 1, 2, and 3 are primary, secondary, and tertiary amides, respectively, because there are two, one, and zero hydrogens bonded to nitrogen.

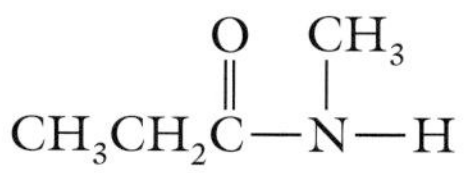

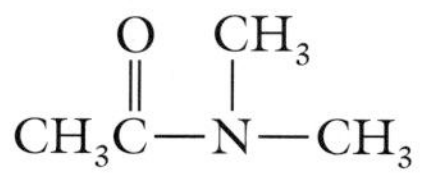

Problem 8.7 Which of the following compounds are amides? Classify the amides as primary, secondary, or tertiary.

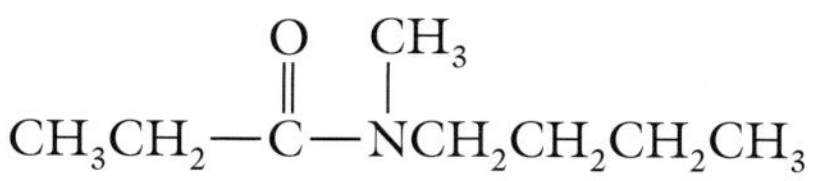

$$CH_3CH_2-\overset{O}{\overset{\|}{C}}-\overset{CH_3}{\overset{|}{N}}CH_2CH_2CH_2CH_3 \quad (1) \qquad C_6H_5-\overset{O}{\overset{\|}{C}}-\overset{H}{\overset{|}{N}}-C_6H_5 \quad (2)$$

$$CH_3\underset{NH_2}{\underset{|}{CH}}-\overset{O}{\overset{\|}{CH}} \quad (3) \qquad CH_3\underset{CH_3}{\underset{|}{CH}}\overset{O}{\overset{\|}{C}}-\overset{CH_3}{\overset{|}{N}}-CH_2C_6H_5 \quad (4)$$

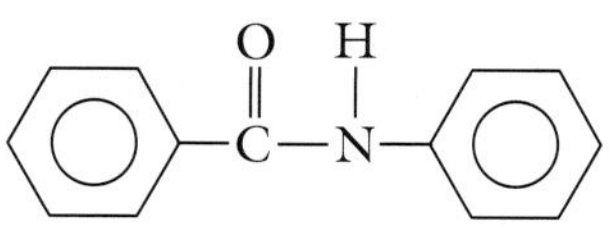

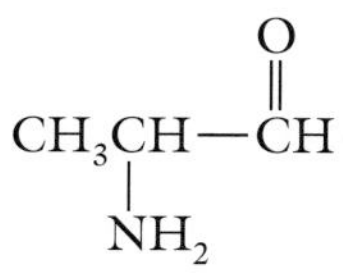

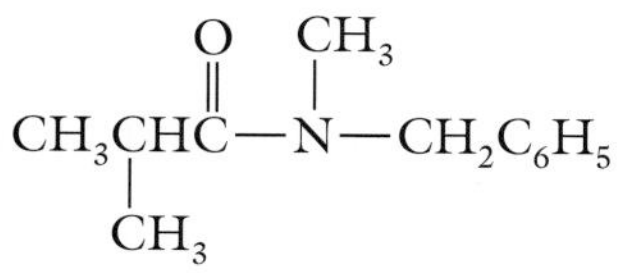

8.8 THE SYNTHESIS OF AMIDES

An amide is synthesized by the reaction of a carboxylic acid with ammonia or with a primary or secondary amine. However, a successful synthesis requires control of the reaction temperature. At ambient-to-moderate temperatures (about 20–50°C), there is no amide formation. There is only an acid–base reaction to form an ammonium carboxylate salt:

$$\underset{\text{Acid}}{R^1{-}\overset{\overset{\displaystyle O}{\|}}{C}{-}OH} + \underset{\text{Amine}}{H{-}\overset{\overset{\displaystyle R^3}{|}}{N}{-}R^2} \xrightleftharpoons{20\text{–}50°C} \underset{\text{Ammonium carboxylate}}{R^1{-}\overset{\overset{\displaystyle O}{\|}}{C}{-}O^-\ H{-}\underset{\underset{\displaystyle H}{|}}{\overset{\overset{\displaystyle R^3}{|}}{N^+}}{-}R^2}$$

R^1, R^2, and R^3 are hydrogen, aliphatic, or aryl.

When the reactions are carried out at elevated temperatures (>100°C), amides are formed in high yield by dehydration between the two reactants:

$$\underset{\text{Carboxylic acid}}{R^1{-}\overset{\overset{\displaystyle O}{\|}}{C}{-}OH} + \underset{\text{Ammonia}}{H{-}\overset{\overset{\displaystyle H}{|}}{N}{-}H} \xrightarrow{>100°C} \underset{\text{Primary amide}}{R^1{-}\overset{\overset{\displaystyle O}{\|}}{C}{-}\overset{\overset{\displaystyle H}{|}}{N}{-}H} + H_2O$$

$$\underset{\text{Carboxylic acid}}{R^1{-}\overset{\overset{\displaystyle O}{\|}}{C}{-}OH} + \underset{\text{Primary amine}}{H{-}\overset{\overset{\displaystyle H}{|}}{N}{-}R^2} \xrightarrow{>100°C} \underset{\text{Secondary amide}}{R^1{-}\overset{\overset{\displaystyle O}{\|}}{C}{-}\overset{\overset{\displaystyle H}{|}}{N}{-}R^2} + H_2O$$

$$\underset{\text{Carboxylic acid}}{R^1{-}\overset{\overset{\displaystyle O}{\|}}{C}{-}OH} + \underset{\text{Secondary amine}}{H{-}\overset{\overset{\displaystyle R^3}{|}}{N}{-}R^2} \xrightarrow{>100°C} \underset{\text{Tertiary amide}}{R^1{-}\overset{\overset{\displaystyle O}{\|}}{C}{-}\overset{\overset{\displaystyle R^3}{|}}{N}{-}R^2} + H_2O$$

The amide is also formed when the ammonium carboxylate is heated above 100°C after its formation at 20 to 50°C.

In amide formation, ammonia or an amine (minus one H) substitutes for (displaces) the —OH of the carboxylic acid. The reaction is very similar to esterification (Section 7.8). In both reactions, a carboxylic acid undergoes substitution with dehydration between the carboxylic acid and other reagent. The two reactions differ regarding the identity of the other reagent: an alcohol or phenol for esterification; ammonia or a primary or secondary amine for amide formation.

Amide formation is a reversible reaction. The yield of amide is increased by the removal of water, which shifts the equilibrium to the right side (Le Chatelier's principle). As in ester formation, this rightward shift is achieved by carrying out the reaction at a temperature higher than the boiling point of water. Water distills out of the reaction vessel as it is formed.

Note that tertiary amines cannot form amides, because they do not have the necessary hydrogen on the nitrogen to participate in the dehydration. At any temperature, a tertiary amine and a carboxylic acid undergo an acid–base reaction but not amide formation.

Like ester formation, amide formation is an acyl transfer reaction, with the carboxylic acid acting as an acyl transfer agent. We will consider acyl transfer reactions in detail when we study the biosynthesis of proteins in Chapter 13.

Example 8.8 Recognizing the conditions for amide formation

Complete each of the following reactions. If no reaction takes place, write "no reaction."

(a) $CH_3CH_2{-}\overset{\overset{\displaystyle O}{\|}}{C}{-}OH + HN(CH_3)_2 \xrightarrow{20-50°C} ?$

(b) $CH_3CH_2{-}\overset{\overset{\displaystyle O}{\|}}{C}{-}OH + HN(CH_3)_2 \xrightarrow{>100°C} ?$

(c) $CH_3CH_2{-}\overset{\overset{\displaystyle O}{\|}}{C}{-}OH + N(CH_3)_3 \xrightarrow{>100°C} ?$

Solution

Acid–base reactions take place at the lower temperatures. Amide formation requires temperatures above 100°C.

(a) $CH_3CH_2{-}\overset{\overset{\displaystyle O}{\|}}{C}{-}OH + HN(CH_3)_2 \xrightarrow{20-50°C} CH_3CH_2{-}\overset{\overset{\displaystyle O}{\|}}{C}{-}O^-\ H_2\overset{+}{N}(CH_3)_2$

(b) $CH_3CH_2{-}\overset{\overset{\displaystyle O}{\|}}{C}{-}OH + H{-}N(CH_3)_2 \xrightarrow[-H_2O]{>100°C} CH_3CH_2{-}\overset{\overset{\displaystyle O}{\|}}{C}{-}N(CH_3)$

(c) Amide formation does not take place with tertiary amines at any temperature. The only reaction is an acid–base reaction:

$$CH_3CH_2{-}\overset{\overset{\displaystyle O}{\|}}{C}{-}OH + N(CH_3)_3 \xrightarrow{>100°C} CH_3CH_2{-}\overset{\overset{\displaystyle O}{\|}}{C}{-}O^-\ H\overset{+}{N}(CH_3)_3$$

Problem 8.8 Complete each of the following reactions. If no reaction takes place, write "no reaction." (a) Benzoic acid + methylamine at 20–50°C; (b) acetic acid + pyridine at $>100°C$; (c) benzoic acid + dimethylamine at $>100°C$.

Acid halides and anhydrides are often used instead of carboxylic acids to form amides because the reactions proceed considerably faster, even at ambient temperatures. Moreover, the reactions easily proceed to completion because they are not reversible.

$$R^1{-}\overset{\overset{\displaystyle O}{\|}}{C}{-}Cl + H{-}\overset{\overset{\displaystyle R^3}{|}}{N}{-}R^2 \longrightarrow R^1{-}\overset{\overset{\displaystyle O}{\|}}{C}{-}\overset{\overset{\displaystyle R^3}{|}}{N}{-}R^2 + HCl$$

$$R^1{-}\overset{\overset{\displaystyle O}{\|}}{C}{-}O{-}\overset{\overset{\displaystyle O}{\|}}{C}{-}R^1 + H{-}\overset{\overset{\displaystyle R^3}{|}}{N}{-}R^2 \longrightarrow R^1{-}\overset{\overset{\displaystyle O}{\|}}{C}{-}\overset{\overset{\displaystyle R^3}{|}}{N}{-}R^2 + R^1{-}\overset{\overset{\displaystyle O}{\|}}{C}{-}O{-}H$$

Example 8.9 Writing the equation for amide synthesis from an acid halide

Give the equation for amide formation between propanoyl chloride and dimethylamine.

Solution

$$CH_3CH_2{-}\overset{\overset{\displaystyle O}{\|}}{C}{-}Cl + H{-}N(CH_3)_2 \xrightarrow{-HCl} CH_3CH_2{-}\overset{\overset{\displaystyle O}{\|}}{C}{-}N(CH_3)_2$$

Problem 8.9 Give the equation for amide formation between propanoic anhydride and dimethylamine.

8.9 POLYAMIDE SYNTHESIS

Polyamides, like polyesters (Section 7.10), are produced by the polymerization of bifunctional reagents (also called monomers). Whereas polyesters are produced from the reaction of dicarboxylic acids and diols, polyamide formation depends on the reaction of dicarboxylic acids and diamines. Large numbers (> 50) of monomer molecules react together to produce the much larger sized polymer:

$$n\,\mathrm{HO{-}\overset{\overset{\displaystyle O}{\|}}{C}{-}R{-}\overset{\overset{\displaystyle O}{\|}}{C}{-}OH} + n\,\mathrm{H_2N{-}R'{-}NH_2} \xrightarrow{-\mathrm{H_2O}} \left(\mathrm{\overset{\overset{\displaystyle O}{\|}}{C}{-}R{-}\overset{\overset{\displaystyle O}{\|}}{C}{-}\overset{\overset{\displaystyle H}{|}}{N}{-}R'{-}\overset{\overset{\displaystyle H}{|}}{N}}\right)_n$$

This polymerization, like polyesterification, is a condensation polymerization because water is a by-product of the reaction.

R and R′ in the preceding structures represent divalent units such as

$$\mathrm{{-}CH_2CH_2{-}} \qquad \mathrm{{-}CH_2CH_2CH_2CH_2{-}} \qquad {-}\mathrm{C_6H_4}{-}$$

As stated in Section 7.10, the repeat unit is the structure inside the brackets and the subscript n is used to indicate that the repeat unit repeats over and over many times; that is, n is a large number.

The polymerization reaction between 1,6-hexanediamine, $H_2N(CH_2)_6NH_2$, and hexanedioic acid, $HOOC(CH_2)_4COOH$, produces the important commercial material called poly(hexamethylene adipamide), or nylon-66:

$$\underset{\text{Hexanedioic acid}}{n\,\mathrm{HO{-}\overset{\overset{\displaystyle O}{\|}}{C}{-}(CH_2)_4{-}\overset{\overset{\displaystyle O}{\|}}{C}{-}OH}} + \underset{\text{1,6-Hexanediamine}}{n\,\mathrm{H_2N{-}(CH_2)_6{-}NH_2}} \xrightarrow{-\mathrm{H_2O}} \underset{\text{Nylon-66}}{\left(\mathrm{\overset{\overset{\displaystyle O}{\|}}{C}{-}(CH_2)_4{-}\overset{\overset{\displaystyle O}{\|}}{C}{-}\overset{\overset{\displaystyle H}{|}}{N}{-}(CH_2)_6{-}\overset{\overset{\displaystyle H}{|}}{N}}\right)_n}$$

The annual production of polyamides (also called nylon polymers) in the United States is more than 4 billion pounds. About two-thirds of this total is nylon-66. About 80% of polyamide production goes into textile applications such as wearing apparel, carpeting, automobile seat belts, tire cord, and parachutes. The remainder goes into plastics applications such as kitchen utensils, power-tool housings, automobile components (engine fans, brake and power-steering reservoirs, lamp housings), motor gears and bearings, ski boots, and tennis-racquet frames. Specialty polyamides are used in fire-resistant clothing for fire fighters and bulletproof vests for law-enforcement personnel. Nylon polymers are also used in surgical applications such as suture material.

The proteins present in biological systems contain polyamides called polypeptides (Chapter 12).

8.10 NAMING AMIDES

Like all derivatives of carboxylic acids, amide names are based on the name of the corresponding carboxylic acid:

Rules for naming amides

- The ending of the name of the corresponding carboxylic acid is changed from **-ic acid** (common) or **-oic acid** (IUPAC) to **-amide.**
- The names of groups attached to nitrogen are placed first, using an *N*- prefix for each group.

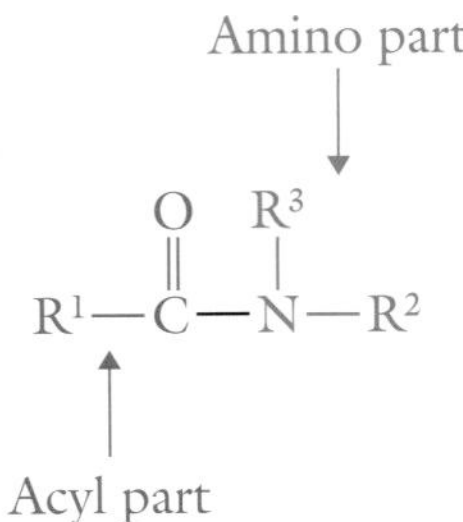

To name an amide, we need to distinguish the acyl and amino parts of the amide. The acyl and amino parts are those derived from the carboxylic acid and amine, respectively.

Example 8.10 Naming amides

Name the following compounds:

(a) $CH_3CH_2CH_2\overset{\overset{\displaystyle O}{\|}}{C}-\overset{\overset{\displaystyle CH_3}{|}}{N}-CH_2CH_3$ (b) $CH_3CH_2\overset{\overset{\displaystyle H}{|}}{N}-\overset{\overset{\displaystyle O}{\|}}{C}-CH_3$

(c) $CH_3CH_2\overset{\overset{\displaystyle O}{\|}}{C}-CH_2-\overset{\overset{\displaystyle H}{|}}{N}-H$

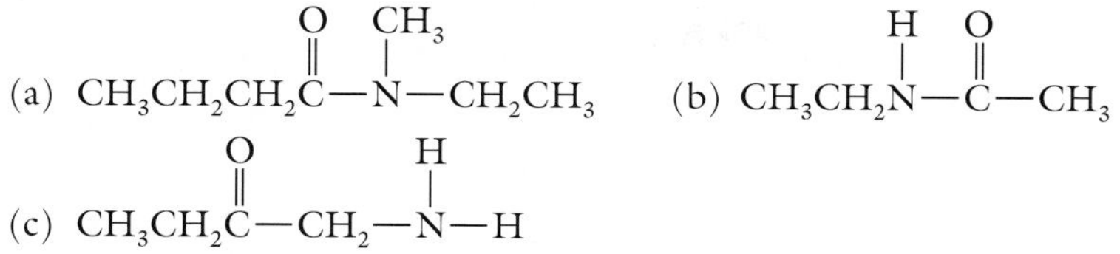

Solution

(a) The acyl and amino parts of the amide are:

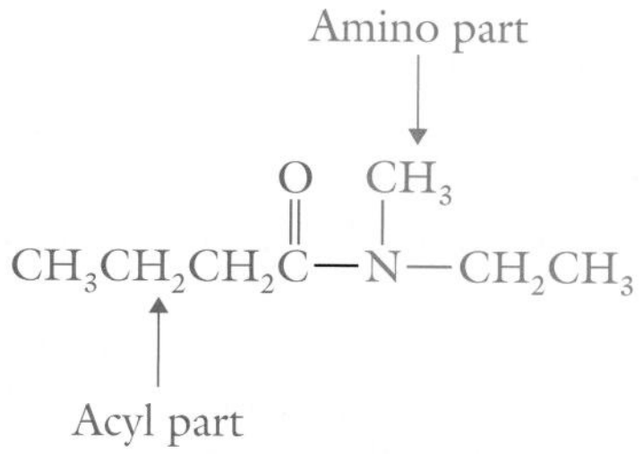

Amino part

$$CH_3CH_2CH_2\overset{\overset{\displaystyle O}{\|}}{C}-\overset{\overset{\displaystyle CH_3}{|}}{N}-CH_2CH_3$$

Acyl part

The carboxylic acid is butanoic acid (IUPAC name) or butyric acid (common name). One ethyl and one methyl group are attached to the nitrogen. The IUPAC name is *N*-ethyl-*N*-methylbutanamide; the common name is *N*-ethyl-*N*-methylbutyramide.

(b) This amide is not drawn in the usual manner with the acyl part on the left and the amino part on the right. It is drawn with the acyl part on the right and the amino part on the left. The carboxylic acid is ethanoic acid (IUPAC name) or acetic acid (common name). An ethyl group is attached to the nitrogen. The IUPAC name is *N*- ethylethanamide; the common name is *N*-ethylacetamide.

(c) The compound is not an amide. It is a ketone and an amine. It is named as a ketone because oxygen-containing functional groups have preference over amines (Section 8.3). The name is 1-aminobutanone.

Problem 8.10 Name the following compounds:

(a) $CH_3CH_2-\overset{\overset{\displaystyle O}{\|}}{C}-\overset{\overset{\displaystyle CH_3}{|}}{N}CH_2CH_2CH_2CH_3$ (b) $C_6H_5-\overset{\overset{\displaystyle O}{\|}}{C}-\overset{\overset{\displaystyle H}{|}}{N}-C_6H_5$

(c) $CH_3\underset{\underset{\displaystyle NH_2}{|}}{C}H-\overset{\overset{\displaystyle O}{\|}}{C}H$ (d) $CH_3\underset{\underset{\displaystyle CH_3}{|}}{C}H\overset{\overset{\displaystyle O}{\|}}{C}-\overset{\overset{\displaystyle CH_3}{|}}{N}-CH_2C_6H_5$

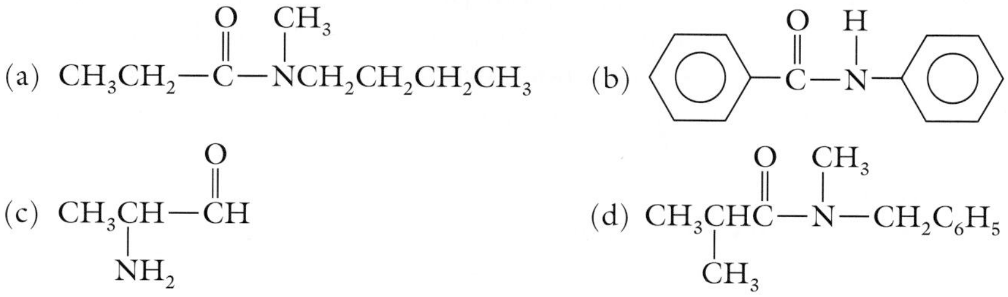

8.11 PHYSICAL AND BASICITY PROPERTIES OF AMIDES

The physical properties of amides are summarized by three observations:

- Amides have higher melting and boiling points than those of carboxylic acids (Table 8.4 on the following page).
- The order of melting and boiling points of amides is primary ≅ secondary > tertiary.
- Amides have slightly higher water solubility than that of carboxylic acids.

Amides have the strongest secondary attractive forces and the highest melting and boiling points of any covalent organic compounds. Recall that

TABLE 8.4 Comparison of Carboxylic Acids and Amides

Compound	Structure	Molecular mass (amu)	Melting point (°C)	Boiling point (°C)
acetic acid	CH_3COOH	60	17	118
acetamide	CH_3CONH_2	59	82	221
N-methylacetamide	$CH_3CONHCH_3$	73	28	204
N,N-dimethylacetamide	$CH_3CON(CH_3)_2$	87	−20	165

the melting and boiling points of carboxylic acids are much higher than those of other families (see Table 8.2), yet the amides have even higher melting and boiling points (see Table 8.4). For example, the melting and boiling points of acetamide are 82°C and 221°C, respectively, compared with 17°C and 118°C for acetic acid.

Amides possess very strong dipole–dipole attractive forces, stronger even than the very strong hydrogen-bond attractions in carboxylic acids. In an amide group, the polarized $^{\delta+}C{-}O^{\delta-}$ bond of the carbonyl group interacts strongly with the nonbonded electron pair of nitrogen. The electron pair of nitrogen is pulled toward the carbonyl carbon, and the amide group is best described as a resonance hybrid of the following structures:

$$R_2\ddot{N}{-}\overset{\overset{O}{\|}}{C}{-}R' \longleftrightarrow R_2\overset{+}{N}{=}\overset{\overset{O^-}{|}}{C}{-}R'$$

Dipolar ion

The structure on the right is called a **dipolar ion** because atoms bearing opposite charges are contained in the same molecule. This dipolar ion contains positive and negative charges on nitrogen and oxygen, respectively, separated by three atoms (instead of the usual two atoms as in the C=O group). The large charge separation results in greatly increased polarity and secondary attractive forces. In line with the N=C double-bond structure of the amide group, the bond angles about the nitrogen and carbon atoms are both close to 120°. This bond angle is about the same as the bond angles about the carbon atoms of the carbon–carbon double bonds in alkenes (Section 4.2). We will see that the dipolar ion structure of an amide is an important contribution to the structure and physiological functions of proteins (Section 12.5).

In primary and secondary amides, but not in tertiary amides, hydrogen-bond attractions are superimposed on the very strong dipole–dipole attractions. Thus, primary and secondary amides have higher secondary forces than those of tertiary amides, and this difference is apparent in the differences in their melting and boiling points (see Table 8.4). The melting and boiling points of primary and secondary amides are not only significantly higher than those of tertiary amides, but also much higher than those of carboxylic acids.

Amides are slightly more water soluble than carboxylic acids. Amides undergo very strong attractive interactions (both dipole–dipole and hydrogen bond) with water because of the high polarity of the $^+N{=}C{-}O^-$ structure.

Whereas amines are weak bases, amides are not more basic than water, ethers, and alcohols. The structural feature that results in the very high boiling and melting points of amides—the electron-withdrawing effect of the carbonyl group on the amide nitrogen—is also responsible for the lack of basicity in amides. This electron-withdrawing effect makes the nonbonded electron pair of nitrogen much less available for accepting a proton.

8.12 THE HYDROLYSIS OF AMIDES

Amides undergo hydrolysis (a cleavage reaction with water) less readily than do esters. In fact, the **amide bond** (the bond between the carbonyl carbon and the nitrogen) is very resistant to hydrolysis at neutral pH, a property of great importance in biological systems, given that the stability of proteins depends on the stability of the amide bond. However, there are two ways in which amides are made to hydrolyze to the corresponding carboxylic acid and ammonia or amine—by the use of a strong acid (acidic hydrolysis) or a strong base (basic hydrolysis). The reaction is similar to the hydrolysis of esters (Section 7.3).

$$-\overset{|}{N}-\overset{O}{\overset{\|}{C}}-$$

Amide group; Amide bond

In the absence of acid or base, the hydrolysis equilibrium is so unfavorable that we can say that hydrolysis does not take place. In the presence of strong acid or base, one of the products of hydrolysis is converted into a salt, and this conversion drives the equilibrium to the right. Acidic hydrolysis forms the amine salt; basic hydrolysis forms the carboxylate salt.

$$\underset{\text{Amide}}{R-\overset{O}{\overset{\|}{C}}-\overset{H}{\overset{|}{N}}-R'} + H_2O \rightleftharpoons \underset{\text{Acid}}{R-\overset{O}{\overset{\|}{C}}-OH} + \underset{\text{Amine}}{H-\overset{H}{\overset{|}{N}}-R'}$$

$$\xrightarrow{HCl} \underset{\text{Acid}}{R-\overset{O}{\overset{\|}{C}}-OH} + \underset{\text{Amine salt}}{R'\overset{H}{\overset{|}{N}}H_2^+\ Cl^-}$$

$$\xrightarrow{NaOH} \underset{\text{Carboxylate salt}}{R-\overset{O}{\overset{\|}{C}}-O^-\ Na^+} + \underset{\text{Amine}}{R'NH_2} + H_2O$$

After acidic or basic hydrolysis has taken place, the product can be neutralized by the addition of strong base or acid, respectively, to convert the amine salt or carboxylate salt into the corresponding amine or carboxylic acid. Note that the acid and base are not catalysts in the hydrolysis reactions. The acid or base is a reactant and is required in equimolar amounts relative to the amide.

Amide synthesis and acidic hydrolysis take place continuously in living organisms. Proteins are amides, and their digestion is simply the acidic hydrolysis of amide bonds (Section 12.4). Living organisms also need to synthesize various proteins (for example, bone, muscle, enzymes). Both the protein synthesis and the protein digestion reactions are catalyzed by enzymes.

» Enzymes are proteins that catalyze biological reactions, as described in Section 14.5.

Example 8.11 Writing equations for the hydrolysis of amides

Complete the following equations. If no reaction takes place, write "no reaction."

(a) $CH_3\overset{O}{\overset{\|}{C}}-\overset{H}{\overset{|}{N}}CH_2CH_3 \xrightarrow{H_2O}$?

(b) $CH_3\overset{O}{\overset{\|}{C}}-\overset{H}{\overset{|}{N}}CH_2CH_3 \xrightarrow[HCl]{H_2O}$?

(c) $CH_3\overset{O}{\overset{\|}{C}}-\overset{H}{\overset{|}{N}}CH_2CH_3 \xrightarrow[NaOH]{H_2O}$?

Solution

(a) No reaction takes place, because of the absence of either a strong acid or a strong base.

(b) Hydrolysis proceeds to completion because a strong acid is present.
Amide hydrolysis requires the breakage of one bond in each reactant, as indicated by the dotted line in the following reaction. The OH from water attaches to the acyl part of the amide to form the carboxylic acid. The H from water attaches to the nitrogen fragment from the amide to form the amine. The amine produced by hydrolysis is immediately converted into the amine salt by neutralization with HCl.

$$\mathrm{HO{\vdots}H}$$

$$\underset{N\text{-Ethylacetamide}}{\mathrm{CH_3\overset{\overset{O}{\|}}{C}\,\vdots\,\overset{\overset{H}{|}}{N}CH_2CH_3}} \xrightarrow{\mathrm{H_2O}} \underset{\text{Acetic acid}}{\mathrm{CH_3\overset{\overset{O}{\|}}{C}-OH}} + \underset{\text{Ethylamine}}{\mathrm{H-\overset{\overset{H}{|}}{N}CH_2CH_3}}$$

$$\text{Ethylamine} \xrightarrow{\mathrm{HCl}} \mathrm{H_2\underset{\underset{H}{|}}{\overset{+}{N}}CH_2CH_3\ Cl^-}$$

The overall reaction may be written in the following abbreviated form:

$$\mathrm{CH_3\overset{\overset{O}{\|}}{C}-\overset{\overset{H}{|}}{N}CH_2CH_3 \xrightarrow[\mathrm{HCl}]{\mathrm{H_2O}} CH_3\overset{\overset{O}{\|}}{C}-OH + H_3\overset{+}{N}CH_2CH_3\ Cl^-}$$

(c) Basic hydrolysis proceeds in a manner similar to acidic hydrolysis except that the presence of strong base converts the carboxylic acid product into the carboxylate salt:

$$\mathrm{CH_3\overset{\overset{O}{\|}}{C}\,\vdots\,\overset{\overset{H}{|}}{N}CH_2CH_3 \xrightarrow{\mathrm{H_2O}} CH_3\overset{\overset{O}{\|}}{C}-OH + H-\overset{\overset{H}{|}}{N}CH_2CH_3}$$

$$\mathrm{HO{\vdots}H}$$

$$\mathrm{CH_3\overset{\overset{O}{\|}}{C}-OH \xrightarrow[-\,H_2O]{NaOH} CH_3\overset{\overset{O}{\|}}{C}-O^-\ Na^+}$$

The overall reaction is written in the following abbreviated form:

$$\mathrm{CH_3\overset{\overset{O}{\|}}{C}-\overset{\overset{H}{|}}{N}CH_2CH_3 \xrightarrow{\mathrm{NaOH}} CH_3\overset{\overset{O}{\|}}{C}-O^-\ Na^+ + H_2NCH_2CH_3}$$

Problem 8.11 Complete the following equations. If no reaction takes place, write "no reaction."

(a) $$\mathrm{CH_3CH_2CH_2\overset{\overset{O}{\|}}{C}-\overset{\overset{H}{|}}{N}-C_6H_5 \xrightarrow[\mathrm{H_2O}]{\mathrm{HCl}} ?}$$

(b) $$\mathrm{CH_3CH_2\overset{\overset{O}{\|}}{C}-\overset{\overset{H}{|}}{N}-CH_2CH_2CH_3 \xrightarrow[\mathrm{H_2O}]{\mathrm{NaOH}} ?}$$

Summary

Amines and Amides Compared

• Amines and amides are organic derivatives of ammonia in which one or more of the hydrogens are replaced by organic groups:

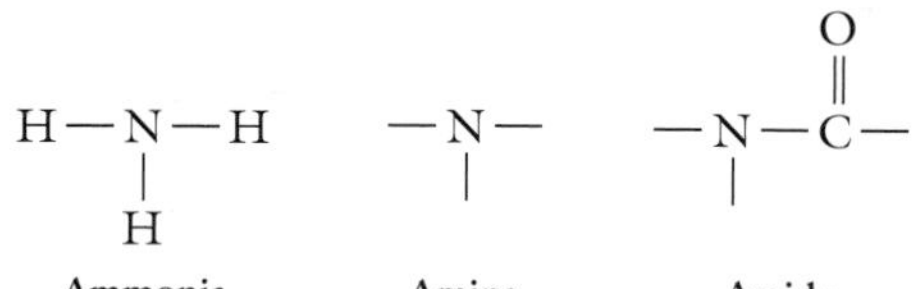

• A compound is an amide instead of an amine when it has a carbonyl carbon attached to nitrogen.

• Amines and amides are classified as primary, secondary, or tertiary, respectively, when one, two, or three of the hydrogens of ammonia are replaced by organic groups.

Naming Amines

• Amines containing simple groups are named by a common nomenclature system in which the names of the groups attached to nitrogen are followed without space by the suffix -amine.

• In the *CA* system, amine names are based on the longest continuous carbon chain attached to nitrogen.

• The ending of the corresponding alkane's name is changed from -e to -amine.

• The position of attachment of nitrogen to the carbon chain is indicated by a number in front of the name.

• The carbon chain is numbered from the end nearest the nitrogen.

• Prefixes preceded by numbers indicate groups attached to the longest chain.

• Prefixes preceded by *N*- and *N,N*- indicate other groups attached to nitrogen.

• A cyclic, nonaromatic compound containing an attached NH_2 group is named as a cycloalkanamine.

• The ring carbon holding the nitrogen is C1.

• $C_6H_5NH_2$ is called aniline, and substituted anilines are named as derivatives of aniline.

• Substituents on nitrogen are distinguished from substituents on the aromatic ring by using an *N*- prefix instead of a number (or the *o*-, *m*-, *p*- prefixes).

Physical Properties of Amines

• Primary and secondary amines, unlike tertiary amines, possess hydrogen-bonding capabilities and melting and boiling points comparable to those of aldehydes and ketones.

• Tertiary amines have melting and boiling points comparable to those of hydrocarbons and ethers.

• Amines of as many as three or four carbons are soluble in water in all proportions.

The Basicity of Amines

• Amines are weak bases because the nonbonded pair of electrons on nitrogen can accept (bond to) a proton from an acid.

• Aromatic amines are weaker bases than aliphatic amines.

Amine Salts

• Amine salts, formed by the reaction of amines with acids, are named by changing the ending of the name of the amine from -amine to -ammonium for aliphatic amines and from -e to -ium for aromatic and heterocyclic amines, followed by the name of the negative ion.

• Amine salts are ionic compounds with higher melting and boiling points and greater water solubilities than those of covalent organic compounds.

Synthesis and Naming of Amides

• Amides are formed by reactions of ammonia or primary or secondary amines with carboxylic acids, acid halides, or acid anhydrides.

• Amides are named by changing the ending of the name of the corresponding carboxylic acid from -ic acid or -oic acid to -amide.

• The names of groups attached to nitrogen are given first, using an *N*- prefix for each group.

• Polyamides are formed by reactions of bifunctional carboxylic acids and amines.

Physical Properties of Amides

• Amides have the highest melting and boiling points of all covalent organic compounds because of very strong dipole–dipole secondary attractions, a result of interaction between the carbonyl group and the nonbonded electron pair of nitrogen.

• Primary and secondary amides, which, unlike tertiary amides, have N—H bonds, also have hydrogen-bond secondary attractions and thus higher melting and boiling points than those of tertiary amides.

• Amides have slightly greater water solubility than that of carboxylic acids.

Chemical Properties of Amides

• Amides are far less basic than amines because the nonbonded electron pair of nitrogen is much less available, a consequence of electron withdrawal by the carbonyl group.

• Amides are no more basic than water or alcohols.

• Amides undergo acidic or basic hydrolysis to the corresponding carboxylic acid and amine or ammonia.

• The basic reaction produces the carboxylate salt, whereas the acidic reaction produces the amine salt.

Summary of Reactions

AMINES

Basicity

$$-\overset{|}{\underset{|}{N}}: + HCl \longrightarrow -\overset{|}{\underset{|}{N}}^{+}-H\ Cl^-$$

Amine → Amine salt

AMIDES

Formation

$$R^1-\overset{O}{\overset{\|}{C}}-OH + H-\overset{R^3}{\overset{|}{N}}-R^2 \xrightarrow{>100°C} R^1-\overset{O}{\overset{\|}{C}}-\overset{R^3}{\overset{|}{N}}-R^2 + H_2O$$

Carboxylic acid + Amine → Amide

Amide is primary if both R^2 and R^3 are H, secondary if only R^2 or R^3 is H, and tertiary if neither R^2 nor R^3 is H.

Polymerization

$$n\ HO-\overset{O}{\overset{\|}{C}}-R-\overset{O}{\overset{\|}{C}}-OH + n\ H_2N-R'-NH_2 \xrightarrow{-H_2O} \left(\overset{O}{\overset{\|}{C}}-R-\overset{O}{\overset{\|}{C}}-\overset{H}{\overset{|}{N}}-R'-\overset{H}{\overset{|}{N}}\right)_n$$

Hydrolysis

ACIDIC

$$R-\overset{O}{\overset{\|}{C}}-\overset{H}{\overset{|}{N}}-R' + H_2O \xrightarrow{HCl} R-\overset{O}{\overset{\|}{C}}-OH + R'NH_3^+\ Cl^-$$

Amide → Carboxylic acid + Amine salt

BASIC

$$R-\overset{O}{\overset{\|}{C}}-\overset{H}{\overset{|}{N}}-R' + H_2O \xrightarrow{NaOH} R-\overset{O}{\overset{\|}{C}}-O^-\ Na^+ + R'NH_2$$

Amide → Carboxylate salt + Amine

Key Words

aliphatic amine, p. 248
alkaloid, p. 252
amide, pp. 247, 263
amide bond, p. 269
amide group, pp. 247, 269
amine, p. 247
amine salt, pp. 257, 258
aromatic amine, p. 248
aryl amine, p. 248
dipolar ion, p. 268
heterocyclic amine, p. 251
heterocyclic nitrogen bases, p. 251
nitrogen heterocyclic, p. 251
polyamide, p. 266
quaternary ammonium salt, p. 261

Exercises

Classifying Amines

8.1 Draw structural formulas for tertiary amines with the molecular formula $C_5H_{13}N$.

8.2 Draw structural formulas for secondary amines with the molecular formula $C_5H_{13}N$.

8.3 Classify each of the following compounds as amines, amides, or something else; as primary, secondary, or tertiary, if amines; as aliphatic or aromatic, if amines.

1. $CH_3-CH(NH(CH_3))-CH(CH_3)_2$
2. 2-aminocyclohexanone (ring C=O with NH_2)
3. pyrimidine ring (N, N)
4. N-methyl lactam ring (CH_3, N, O)
5. $CH_3\overset{O}{\overset{\|}{C}}-N(CH_3)_2$
6. benzene ring with NO_2

8.4 Classify each of the following compounds as amines, amides, or something else; as primary, secondary, or tertiary, if amines; as aliphatic or aromatic, if amines.

1: cyclohexyl–N(H)–CH_3 2: NH_2 / benzene ring / CH_3 3: cyclic ring with O and NH

4: $C_6H_5C(=O)-CH_2N(CH_3)_2$ 5: NH_3 6: six-membered ring with N–H

Naming Amines

8.5 Draw the structure of each of the following compounds: (a) *N*-methyl-1-pentanamine; (b) isopropylmethylamine; (c) *cis*-4-ethyl-*N*-propylcyclohexanamine; (d) *N*-ethylaniline.

8.6 Draw the structure of each of the following compounds: (a) *t*-butylamine; (b) 1,5-pentanediamine; (c) *trans*-3-ethyl-*N,N*-dimethylcyclopentanamine; (d) *N*-isopropyl-3,4-dimethylaniline.

8.7 Give the common name or *CA* name, whichever is appropriate, for each of the following compounds:

(a) $CH_3CH_2-N(CH_3)-CH_3$ (b) cyclohexane ring carrying CH_3, H, H, and NH_2

(c) $(CH_3)_2CHCH_2CHCH_3$, with $CH_3-N-CH_2CH_3$ on the CH

(d) $CH_3CH_2-N(C_6H_5)(CH_3)$

8.8 Give the common name or *CA* name, whichever is appropriate, for each of the following compounds:

(a) $CH_3CH_2CH_2CH_2NHCH(CH_3)_2$

(b) benzene ring with NH_2, CH_3 (ortho to NH_2), and CH_3 (para to NH_2) (c) benzene ring–$CH_2CH(CH_2CH_3)-NH_2$

(d) five-membered ring with N–CH_3

Physical Properties of Amines

8.9 Draw structural representation(s) to describe the hydrogen bonding responsible for the high boiling point of ethylamine.

8.10 Draw structural representation(s) to describe the hydrogen bonding present when ethylamine dissolves in water.

8.11 Indicate which compound in each of the following pairs has the higher value of the specified property. If the two compounds in a pair have nearly the same value for the property, indicate that fact. Explain your conclusions.
(a) Butylamine, 1-butanol: boiling point.
(b) Butylamine, butanal: solubility in water.
(c) Butylamine, ethyldimethylamine: boiling point.
(d) Butylamine, diethylamine: solubility in water.

8.12 Indicate which compound in each of the following pairs has the higher value of the specified property. If the two compounds in a pair have nearly the same value for the property, indicate that fact. Explain your conclusions.
(a) Butylamine, butanal: boiling point.
(b) Butylamine, diethyl ether: boiling point.
(c) Butylamine, ethyldimethylamine: solubility in water.
(d) Butylamine, diethylamine: boiling point.

The Basicity of Amines

8.13 Give the equation to show propylamine acting as a base toward HCl.

8.14 Indicate whether any acid–base reaction takes place under each of the following conditions. If a reaction takes place, write the appropriate equation. Make sure to indicate whether the reaction goes to completion or is an equilibrium reaction. (a) Butylamine + water; (b) butylamine + H_2SO_4.

8.15 Place the following compounds in order of increasing basicity: propylamine, KOH, aniline, water, 1-propanol. Explain the order of basicity.

8.16 Which is the stronger base, aniline or cyclohexylamine? Why?

8.17 An unknown compound is either 1-propanol or propylamine. The unknown turns red litmus paper blue. Identify the unknown.

8.18 An unknown compound is either 1-propanol or propylamine. The unknown does not turn red litmus paper blue. Identify the unknown.

8.19 Show the reaction of 1,6-hexanediamine, $H_2N(CH_2)_6NH_2$, with excess HCl.

8.20 What is the structure of the predominant species present when ethylamine is dissolved in water? What structure predominates when ethylamine is added to a buffered solution at pH = 7 (similar to body pH)?

Amine Salts

8.21 Give the structural formula of each of the following compounds: (a) *N,N*-dimethylcyclopentanammonium chloride; (b) *N*-ethyl-*N*-methylpiperidinium sulfate; (c) ethylmethylpropylammonium bromide.

8.22 Name each of the following compounds:

(a) 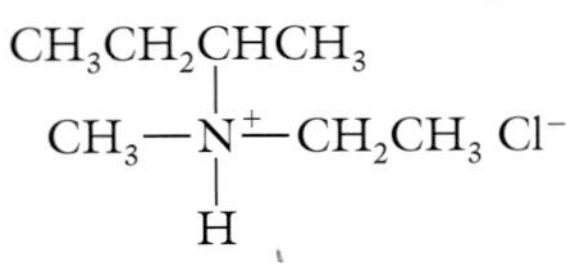 $CH_3-\overset{+}{N}(H)(CH_2CH_2CHCH_3\text{-attached } CH_3CH_2CHCH_3)-CH_2CH_3\ Cl^-$

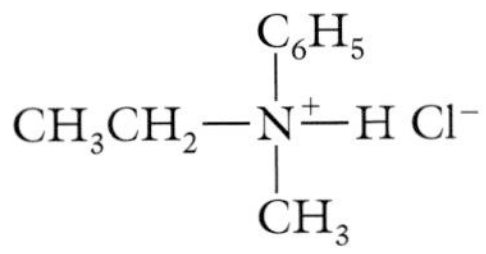

$$\begin{array}{c} CH_3CH_2CHCH_3 \\ | \\ CH_3-\overset{+}{N}-CH_2CH_3\ Cl^- \\ | \\ H \end{array}$$

(b)
$$\begin{array}{c} C_6H_5 \\ | \\ CH_3CH_2-\overset{+}{N}-H\ Cl^- \\ | \\ CH_3 \end{array}$$

(c)
$$\begin{array}{c} \quad CH_3 \qquad\quad H \\ \quad | \qquad\qquad | \\ (CH_3)_3C-CHCH_2-\overset{+}{N}-CH(CH_3)_2\ Br- \\ \qquad\qquad\qquad | \\ \qquad\qquad\qquad H \end{array}$$

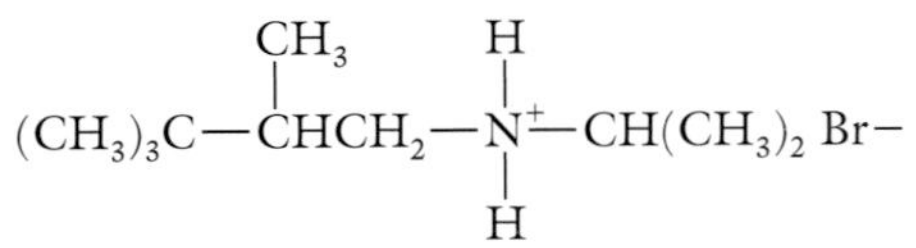

(d) 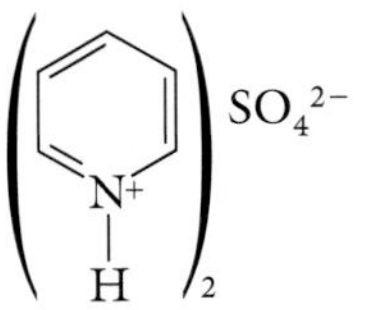 $\left(\text{pyridinium, } \overset{+}{N}-H\right)_2\ SO_4^{2-}$

8.23 Indicate which compound in each of the following pairs has the higher value of the specified property. If the two compounds in a pair have nearly the same value for the property, indicate that fact. Explain your conclusions.
(a) Trimethylammonium chloride, triethylamine: boiling point.
(b) Hexylamine, hexylammonium chloride: solubility in water.

8.24 Give the equations showing trimethylammonium chloride acting as an acid toward the following bases. Make sure to indicate whether the reaction goes to completion or is an equilibrium reaction. (a) Water; (b) NaOH.

8.25 An unknown is either butylamine or butylammonium chloride. The unknown turns red litmus paper blue. Identify the unknown.

8.26 An unknown is either butylamine or butylammonium chloride. The unknown turns blue litmus paper red. Identify the unknown.

Classifying Amides

8.27 Which of the following compounds are amides? Classify amides as primary, secondary, or tertiary.

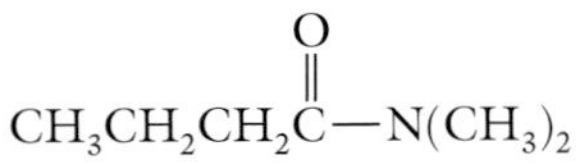

$$CH_3CH_2CH_2\overset{\overset{\large O}{\|}}{C}-N(CH_3)_2$$
1

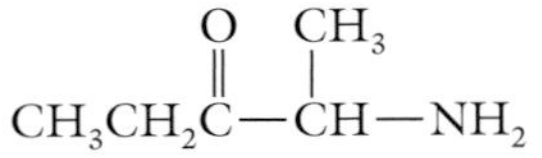

$$CH_3CH_2\overset{\overset{\large O}{\|}}{C}-\overset{\overset{\large CH_3}{|}}{C}H-NH_2$$
2

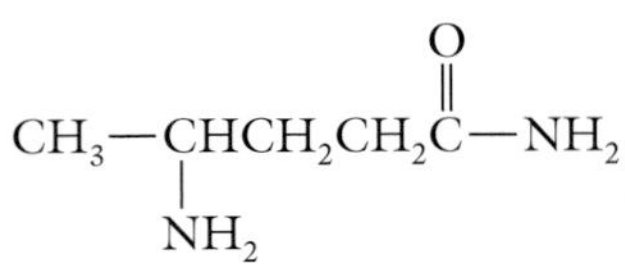

$$CH_3-\underset{\underset{\large NH_2}{|}}{C}HCH_2CH_2\overset{\overset{\large O}{\|}}{C}-NH_2$$
3

(2-pyrrolidinone: five-membered ring with $C{=}O$ and $N-H$)
4

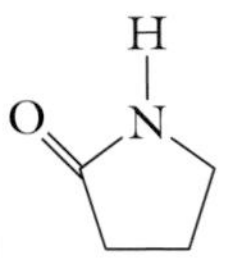

8.28 Draw the structures of all amides of formula C_3H_7NO.

Synthesis of Amides

8.29 Write the equation for the reaction that takes place between each of the following pairs of reactants: (a) benzoic acid and butylamine at 25°C; (b) benzoic acid and butylamine at > 100°C; (c) benzoyl chloride and ethylamine; (d) acetic anhydride and *N*-methylaniline.

8.30 Write the equation for the reaction that takes place between each of the following pairs of reactants: (a) acetic acid and trimethylamine at > 100°C; (b) butanoic acid and aniline at > 100°C; (c) acetyl bromide and aniline; (d) benzoic anhydride and ethylmethylamine.

8.31 What carboxylic acid and amine are needed to synthesize each of the following amides?

(a) 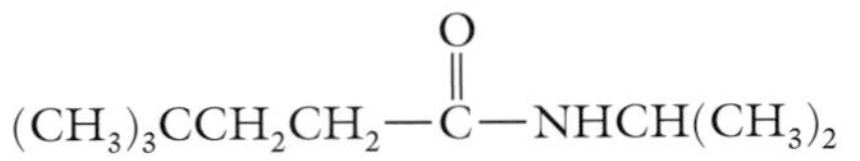 $(CH_3)_3CCH_2CH_2-\overset{\overset{\large O}{\|}}{C}-NHCH(CH_3)_2$

(b) 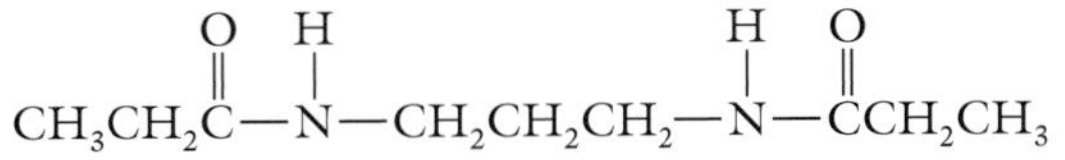 $CH_3CH_2\overset{\overset{\large O}{\|}}{C}-\overset{\overset{\large H}{|}}{N}-CH_2CH_2CH_2-\overset{\overset{\large H}{|}}{N}-\overset{\overset{\large O}{\|}}{C}CH_2CH_3$

8.32 What carboxylic acid and amine are needed to synthesize each of the following amides?

(a) 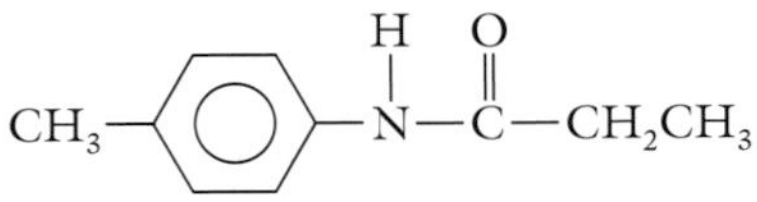 $CH_3-C_6H_4-\overset{\overset{\large H}{|}}{N}-\overset{\overset{\large O}{\|}}{C}-CH_2CH_3$

(b) 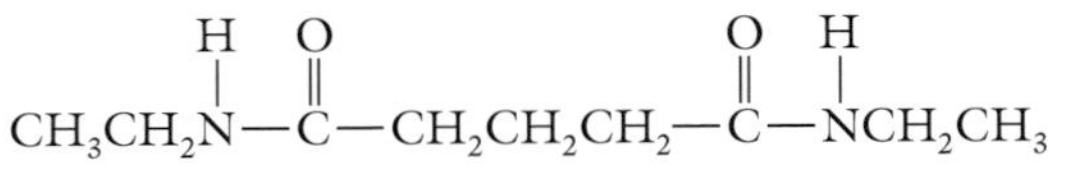 $CH_3CH_2\overset{\overset{\large H}{|}}{N}-\overset{\overset{\large O}{\|}}{C}-CH_2CH_2CH_2-\overset{\overset{\large O}{\|}}{C}-\overset{\overset{\large H}{|}}{N}CH_2CH_3$

Polyamide Synthesis

8.33 What diacid and diamine reactants are needed to synthesize the following polyamide?

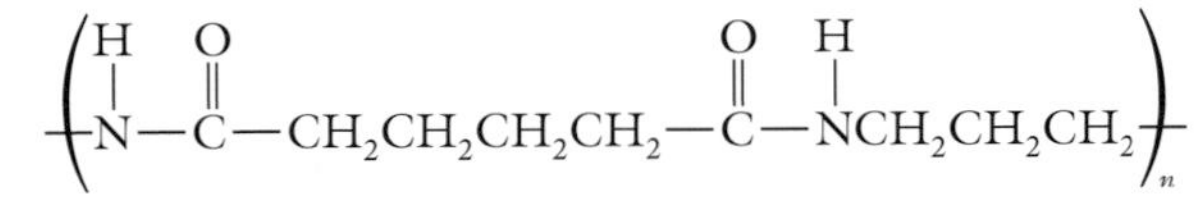

$$\left(\overset{\overset{\large H}{|}}{N}-\overset{\overset{\large O}{\|}}{C}-CH_2CH_2CH_2CH_2-\overset{\overset{\large O}{\|}}{C}-\overset{\overset{\large H}{|}}{N}CH_2CH_2CH_2\right)_n$$

8.34 Draw the structure of the polyamide formed from the reaction of hexanedioic acid and 1,4-benzenediamine.

Naming Amides

8.35 Give the structure of each of the following amides: (a) *N*-*t*-butylbutanamide. (b) *N*-phenyl-3-methylhexanamide.

8.36 Give the structure of each of the following amides:
(a) *N*-ethyl-*N*-propylbenzamide;
(b) 2,*N*-dimethyl-*N*-phenylpropanamide.

8.37 Name each of the following compounds:

(a) 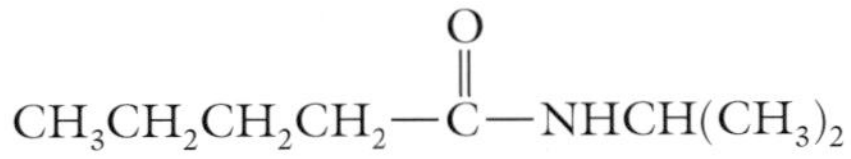 $CH_3CH_2CH_2CH_2-\overset{\overset{\large O}{\|}}{C}-NHCH(CH_3)_2$

(b) $C_6H_5-\overset{\overset{\large H}{|}}{N}-\overset{\overset{\large O}{\|}}{C}-CH_2CH_3$

(c) $(CH_3CH_2)_2N-\overset{\overset{\large O}{\|}}{C}-C_6H_5$

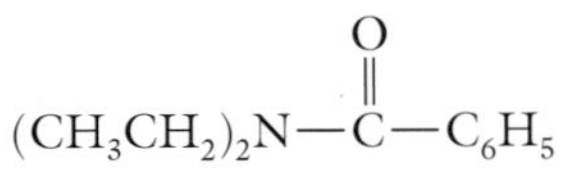

8.38 Name each of the following compounds:

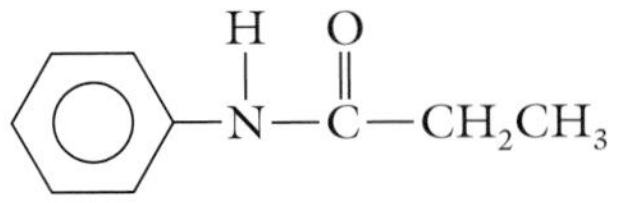

(a) 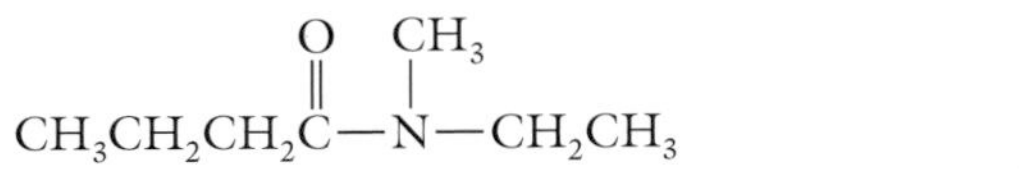 $CH_3CH_2CH_2\overset{\overset{\large O}{\|}}{C}-\overset{\overset{\large CH_3}{|}}{N}-CH_2CH_3$

(Exercise 8.38 continues on next page)

(b) 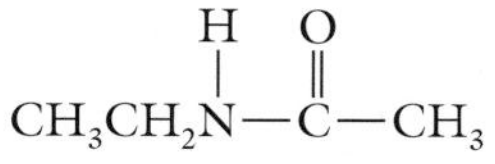

(c) 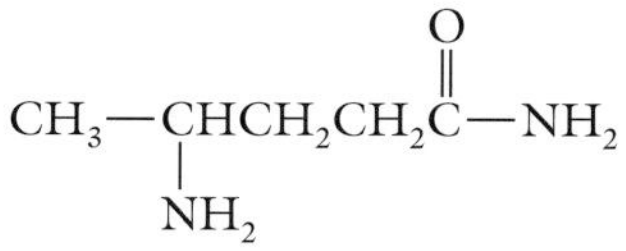

Physical and Basicity Properties of Amides

8.39 Indicate which compound in each of the following pairs has the higher value of the specified property. If the two compounds in a pair have nearly the same value for the property, indicate that fact. Explain your conclusions.
(a) Pentanamide, *N,N*-dimethylpropanamide: boiling point.
(b) Pentanamide, *N,N*-dimethylpropanamide: solubility in water. (c) Butanamide, butanoic acid: boiling point.
(d) Hexanamide, hexanoic acid: solubility in water.
(e) Butanamide, 1-pentanol: boiling point.

8.40 Which of the following representations correctly describe hydrogen bonding?

(a) 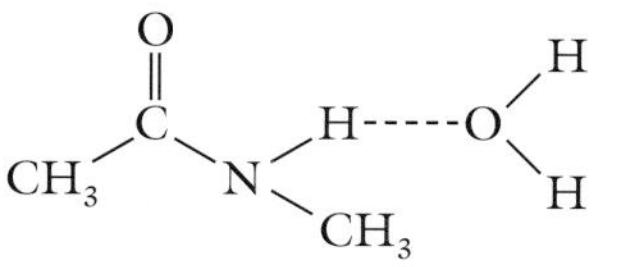

(b) 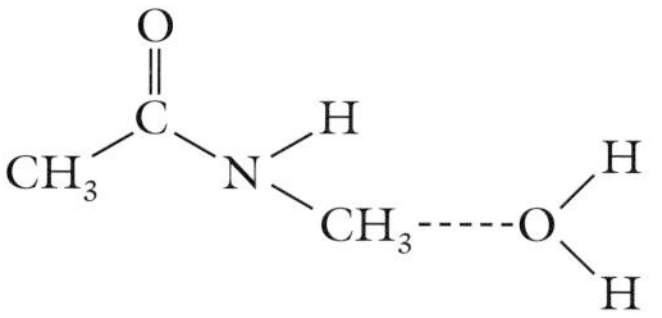

(c) 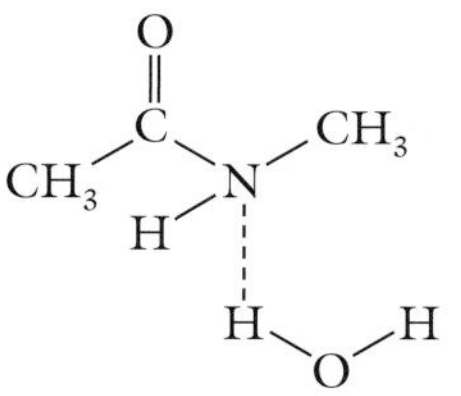 (d) 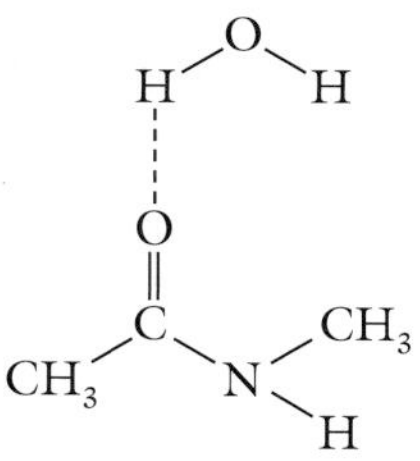

8.41 Place the following compounds in order of increasing basicity: ethanamide, propanamine, 1-propanol, NaOH, water. Explain the reasons for the order.

8.42 Isomers 1 and 2 both have NH_2 and C=O groups but isomer 1 is 10,000-fold less basic than isomer 2. What is the reason for the difference?

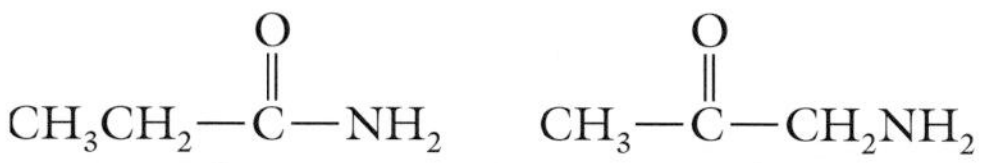

1 2

The Hydrolysis of Amides

8.43 Write the equation for (a) acidic hydrolysis of *N*-ethyl-*N*-methylpropanamide with HCl; (b) basic hydrolysis of *N*-phenylbenzamide with KOH.

8.44 Give the structure of the organic product in each of the following reactions. If no reaction takes place, write "no reaction." If there is more than one product, show only the major product(s).

(a) 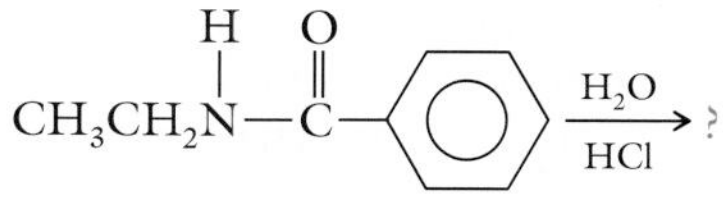

(b) 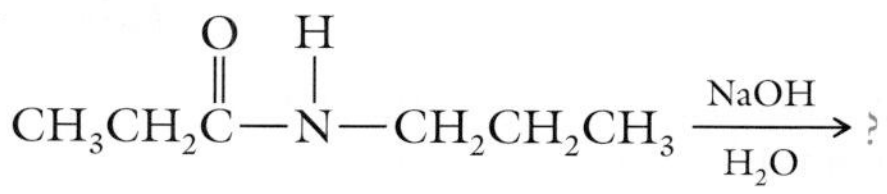

Unclassified Exercises

8.45 Draw structural formulas for each of the following compounds: (a) primary amide containing three carbons; (b) tertiary amide of four carbons and containing three methyl groups; (c) lactam (cyclic amide) of five carbons with all carbons in the ring; (d) secondary amine of molecular formula C_7H_9N and containing a benzene ring; (e) tertiary amine of molecular formula $C_6H_{15}N$ and containing a *t*-butyl group.

8.46 Give the structural formula of each of the following compounds: (a) *N*-isobutylaniline; (b) *p*-isobutylaniline; (c) *N,N*-dimethylpiperidinium chloride; (d) *N*-methyl-3-chlorobutanamide; (e) 3-aminopentanamide; (f) butylisopropylmethylamine.

8.47 Give the structural formula of each of the following compounds: (a) 5-amino-3-methyl-2-pentanone; (b) 4-pentene-1-amine.

8.48 What are the hybridizations of atoms a, b, c, and d in the following compound? What are bond angles A, B, and C?

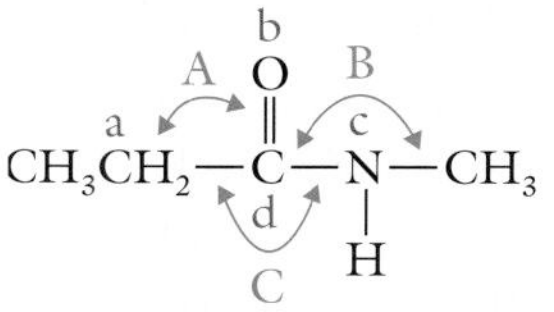

8.49 What are the hybridizations of atoms a, b, and c in the following compounds? What are bond angles A and B?

(a) 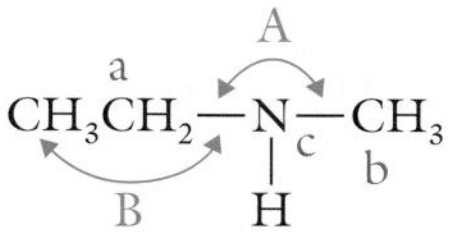 (b) 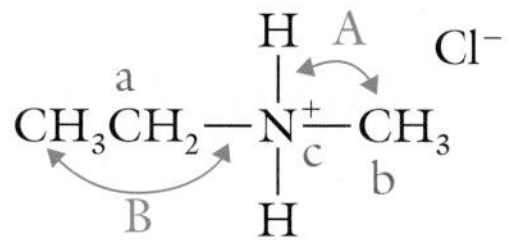

8.50 Identify amine and amide groups in the following compounds. Classify amines and amides as primary, secondary, or tertiary. Which amines are heterocyclic amines?

(a) 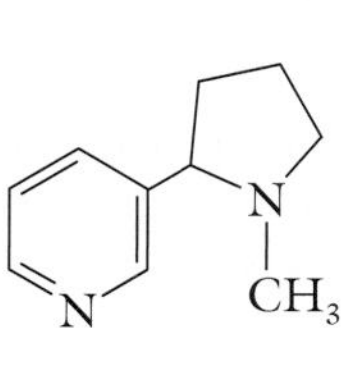

Nicotine

(b) 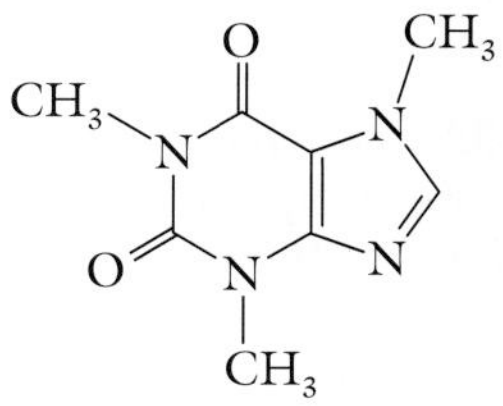

Caffeine

(c) 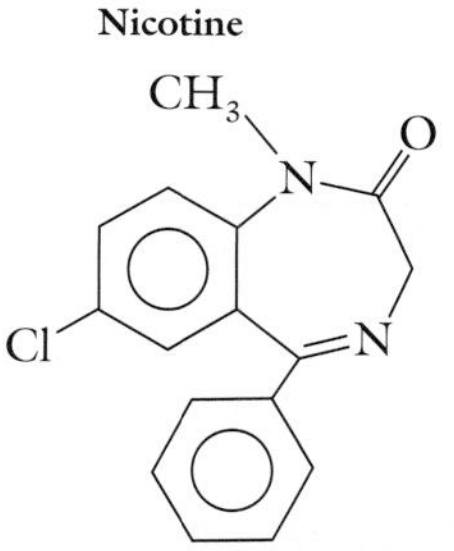

Diazepam (Valium)

8.51 Indicate which compound in each of the following pairs has the higher value of the specified property. If the two compounds in a pair have nearly the same value for the property, indicate that fact. Explain your conclusions.
(a) Butanamide, hexanamine: boiling point.
(b) Butanamide, hexanamine: solubility in water.
(c) Pentanamide, pentanoic acid: boiling point.
(d) Pentanamide, pentanoic acid: solubility in water.
(e) Trimethylammonium chloride, hexanamide: boiling point.
(f) Pentanamide, sodium butanoate: boiling point.
(g) Butanamine, 1,8-octanediamine: solubility in water.

8.52 Place the following compounds in order of increasing boiling point: methyl propanoate, 1-pentanol, pentylamine, butanoic acid, pentanal, trimethylammonium chloride, butanamide. Explain the order.

8.53 An unknown compound is either an amine or an amide. What simple chemical diagnostic test can be used to distinguish between the two possibilities?

8.54 An unknown compound is either propylamine, propanol, or ethanamide. The unknown turns red litmus paper blue. Identify the unknown.

8.55 An unknown compound is either propylamine, propanol, or ethanamide. The unknown does not turn blue litmus paper red. Identify the unknown.

8.56 Give the structure of the cyclic amide (lactam) formed from 5-aminopentanoic acid.

8.57 Each of the following pairs of reactants undergoes amide formation. Which of the pairs produces a polymer? Write the equation for the reaction(s) that produce(s) a polymer.

(a) $HO-\overset{O}{\overset{\|}{C}}CH_2CH_2CH_2CH_2\overset{O}{\overset{\|}{C}}-OH + H_2NCH_2CH_3$

(b) $HOOCCH_2CH_2CH_2CH_3 + H_2NCH_2CH_2NH_2$

(c) $HO-\overset{O}{\overset{\|}{C}}CH_2CH_2CH_2CH_2\overset{O}{\overset{\|}{C}}-OH + H_2NCH_2CH_2NH_2$

8.58 Give the structure of the organic product in each of the following reactions. If no reaction takes place, write "no reaction." If there is more than one product, show only the major product(s).

(a) $CH_3CH_2\overset{O}{\overset{\|}{C}}-NHCH_2C_6H_5 \xrightarrow[HCl]{H_2O} ?$

(b) $CH_3CH_2\overset{O}{\overset{\|}{C}}-NHCH_2C_6H_5 \xrightarrow[NaOH]{H_2O} ?$

(c) [five-membered lactam ring: N–H, C=O, with CH_3 on the carbon adjacent to C=O] $\xrightarrow[H_2O]{HCl} ?$

(d) $\left(\overset{H}{\overset{|}{N}}-\overset{O}{\overset{\|}{C}}(CH_2)_4\overset{O}{\overset{\|}{C}}-\overset{H}{\overset{|}{N}}(CH_2)_3\right)_n \xrightarrow{\text{hydrolysis}} ?$

(e) $CH_3CH_2CH_2CH_2\overset{O}{\overset{\|}{C}}-OH \xrightarrow[25°C]{C_6H_5NH_2} ?$

(f) $CH_3CH_2CH_2CH_2\overset{O}{\overset{\|}{C}}-OH \xrightarrow[>100°C]{C_6H_5NH_2} ?$

(g) $CH_3CH_2-\underset{H}{\underset{|}{\overset{C_6H_5}{\overset{|}{N}}}} \xrightarrow{H_2SO_4} ?$

(h) $CH_3CH_2CH_2CH_2\overset{O}{\overset{\|}{C}}-Cl \xrightarrow{CH_3NH_2} ?$

(i) $CH_3CH_2\overset{H}{\overset{|}{N}}-\overset{O}{\overset{\|}{C}}-C_6H_4-\overset{O}{\overset{\|}{C}}-\overset{H}{\overset{|}{N}}CH_2CH_3 \xrightarrow[HCl]{H_2O} ?$ (para-substituted benzene ring)

8.59 For each of the following pairs of structural formulas, indicate whether the pair represents (1) the same compound or (2) different compounds that are constitutional isomers or (3) different compounds that are geometrical isomers or (4) different compounds that are not isomers.

(a) $CH_3CH_2-\overset{O}{\overset{\|}{C}}-N(CH_3)_2$ or $CH_3CH_2-\overset{O}{\overset{\|}{C}}-CH_2N(CH_3)_2$

(b) $CH_3CH_2-\overset{O}{\overset{\|}{C}}-N(CH_3)_2$ or $H-\overset{O}{\overset{\|}{C}}-CH_2CH_2N(CH_3)_2$

(c) $(CH_3)_3N$ or $CH_3CH_2CH_2NH_2$

(d) [five-membered lactam ring: C=O, NH, with CH_3 on the carbon adjacent to NH] and [six-membered lactam ring: C=O, NH]

Expand Your Knowledge

Note: The icons denote exercises based on material in boxes.

8.60 The distinct and disagreeable odors sometimes associated with fish are due to low-molecular-mass amines formed by the degradation of proteins. Cooks often squeeze lemon juice, which contains citric acid (see Box 7.1), on the fish to cut down on the fishy odors resulting from such amines. How does the addition of citric acid result in a less-intense fishy odor?

8.61 Using the method described in Section 2.4, calculate the pH of a 1.00 *M* aqueous solution of methylamine. The value of K_b for methylamine is 4.59×10^{-4}.

8.62 Ephedrine is used as a bronchodilator and decongestant. Phenobarbital is used as an anticonvulsant and sedative. What simple chemical test distinguishes these two medicines?

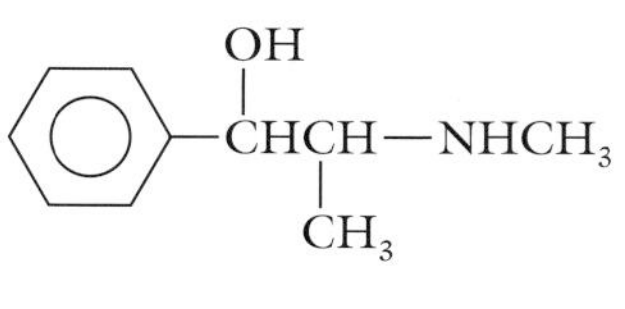

Ephedrine 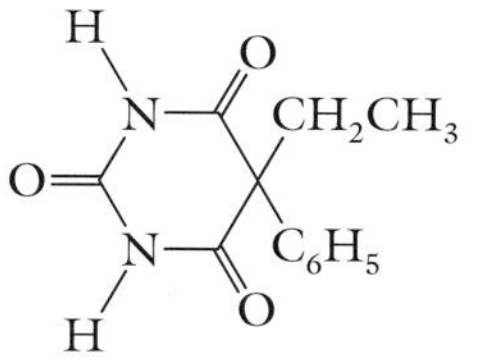 **Phenobarbital**

8.63 The amide $C_5H_{11}NO$ yields ethylamine on hydrolysis. Identify the other product(s) of hydrolysis.

8.64 1-Decanol and 1-decanamine are products marketed by the XYZ Chemical Company. A mistake in the plant results in the mixing of these two products. There is a need to quickly separate the mixture into the individual components because customers are awaiting delivery of the two chemicals. Distillation would be a useful method for the needed separation, but the distillation equipment is broken and will not be repaired for several days. Describe an alternate method of separating 1-decanol from 1-decanamine. Note that both 1-decanol and 1-decanamine are insoluble in water.

8.65 A 10% aqueous ammonia (NH_3) solution is useful as a reflex respiratory stimulant ("smelling salt"). Explain why a solution of ammonium chloride (NH_4Cl) is not useful for the same purpose.

8.66 Terazosin is an α-blocker used to treat high blood pressure (see Box 8.2). Identify amide and amine groups in terazosin and classify them as primary, secondary, and tertiary.

8.67 Three different solutions (A, B, C) are prepared by dissolving the following compound(s) in 1 L each of water: A, 1 mol of methylamine; B, 1 mol each of methylamine and HCl; C: 1 mol each of methylamine and NaOH. Place the solutions in order of increasing concentration of methylammonium ion. Explain the order.

8.68 Kevlar is a synthetic polyamide used to produce bulletproof vests for law-enforcement personnel and heat-resistant protective clothing for firefighters. It is produced by the reaction between terephthalic acid and 4-aminoaniline. Write the equation for the synthesis of Kevlar.

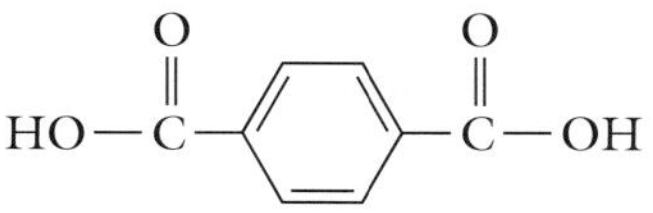

Terephthalic acid 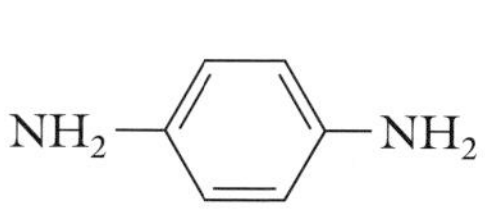 **4-Aminoaniline**

8.69 In a comparison of polymer samples of equal molecular mass, the polyamide called nylon-66 (Section 8.9) is physically stronger than polyethylene (Section 4.7). Why?

8.70 How many liters of 0.50 *M* HCl will be needed to neutralize 0.25 mol of ethylamine?

8.71 Which of the following compounds is a soap? Explain.

$CH_3(CH_2)_2N(CH_3)_3{}^+\ Cl^-$
1

$CH_3(CH_2)_{17}N(CH_3)_3{}^+\ Cl^-$
2

$CH_3(CH_2)_{17}N(CH_3)_2$
3

$CH_3(CH_2)_2N(CH_3)_2$
4

8.72 Benzalkonium chloride, used as a cleaning agent with antiseptic and disinfectant powers, contains benzyldimethyloctadecylammonium chloride. Draw the structure of this compound. Note: The prefix octadec- specifies C_{18}.

8.73 Crack cocaine, a free-base form of cocaine, can be smoked in a pipe, but the amine salt form of cocaine cannot (see Box 8.4). Why?

8.74 The ABC Company sells methamphetamine and phenobarbital (see Box 8.3). The labels have been lost from containers of the two compounds. What simple chemical test can be used to determine which compound is in which container?

8.75 How can infrared spectroscopy (Box 6.3 and Table 6.5) be used to distinguish between methamphetamine and phenobarbital (see Box 8.3)?

8.76 How can C-13 NMR spectroscopy (Box 6.3) be used to distinguish between methamphetamine and phenobarbital (see Box 8.3)?

8.77 A quaternary ammonium salt is synthesized by the reaction of a tertiary amine with an alkyl halide. Give the equation for the synthesis of tetramethylammonium chloride from trimethylamine and methyl chloride.

8.78 Hydrogen bonding in primary and secondary amines is weaker than hydrogen bonding in alcohols. Use the electronegativity values given in Figure 1.9 (Section 1.9) to show that the N—H bond is less polar than the O—H bond.

8.79 What is the molecular formula of caffeine (see Box 8.3)?

8.80 The C:H ratio for alkanes and alcohols is $n{:}2n + 2$ if no rings or double bonds are present. What is the C:H ratio for an amine containing only one nitrogen if no rings or double bonds are present? Why is the ratio different from that for an alcohol with no rings or double bonds?

8.81 Compound X (C_2H_4O) is oxidized to compound Y ($C_2H_4O_2$). Compound Y reacts with ethylamine in the presence of acid to yield compound Z (C_4H_9NO). Identify compounds X, Y, and Z.

8.82 The C:H ratio for a compound that contains a nitrogen and has no rings or double bonds is $n{:}2n + 3$. How many rings or double bonds or both are present in a compound with the molecular formula C_6H_7N?

CHAPTER 9 STEREOISOMERISM

(David Hanover/Tony Stone.)

Chemistry in Your Future

On his way to see the doctor for an examination, a patient pauses to discuss his high-blood-pressure drug with you. He says that he's been meaning to ask why his drug suddenly underwent an increase in price a few months ago. You explain that the FDA recently approved a "single enantiomer" version of the drug. The new version is purer and more effective and causes fewer side effects, and so the clinic thinks that it is worth the cost of the more expensive manufacturing process. "Purer in what way?" he asks. This chapter helps you to answer his question.

For more information about this topic and others in the chapter, go to www.whfreeman.com/bleiodian2e

Learning Objectives

- Distinguish between constitutional isomers and stereoisomers.
- Distinguish between chiral and achiral molecules.
- Identify tetrahedral stereocenters.
- Draw and name enantiomers of compounds with one tetrahedral stereocenter.
- Describe the optical activity of chiral compounds.
- Describe the process of chiral recognition.
- Draw enantiomers and diastereomers of compounds with two tetrahedral stereocenters.
- Draw enantiomers and diastereomers of cyclic compounds with tetrahedral stereocenters.

In Chapter 3, where we began our study of organic chemistry, two kinds of isomers were defined: constitutional isomers and stereoisomers. In reality, there are two classes of stereoisomers—enantiomers and diastereomers—and there are several subclasses of diastereomers. In Chapters 3 and 4, we studied one subclass of diastereomers: geometrical (also called cis-trans) isomers of alkenes and cycloalkanes.

Before beginning the part of this book on biochemistry, we need to introduce enantiomers and the remaining subclasses of diastereomers, because they are found in biological molecules (carbohydrates, lipids, proteins, and nucleic acids). Evolution has occurred in such a way that, for any molecule that can exist as two or more stereoisomers, virtually all known organisms, with the exception of bacteria such as *E. coli*, use only one of the stereoisomers. Thus, stereoisomerism is crucial to the selectivity of biochemical processes. We observed such selectivity in the effect of cis-trans isomerism on our perception of light and in the action of pheromones in insects (Boxes 4.2 and 4.3). In biochemical reactions, differences in stereoisomers can be as important as or more important than differences in functional groups.

Before considering the stereoisomers that we have not yet encountered, let's review what we have learned about isomers in earlier chapters.

9.1 REVIEW OF ISOMERISM

Isomers are different compounds that have the same molecular formula. Figure 9.1 shows that there are two different types of isomers: constitutional isomers and stereoisomers. **Constitutional isomers,** also called **structural isomers,** are different compounds having the same molecular formula but different **connectivity,** the order in which atoms are attached to one another (Sections 3.5, 4.3, 5.2). There are three types of connectivity differences, any one of which gives rise to constitutional isomers:

- Different carbon skeletons, as in butane and 2-methylpropane:

$$CH_3{-}CH_2{-}CH_2{-}CH_3 \qquad CH_3{-}\underset{\underset{\displaystyle CH_3}{|}}{CH}{-}CH_3$$

Butane **2-Methylpropane**

- Different placements of the functional group(s) on the same carbon skeleton, as in 1-propanol and 2-propanol:

$$CH_3{-}CH_2{-}CH_2{-}OH \qquad CH_3{-}\overset{\overset{\displaystyle OH}{|}}{CH}{-}CH_3$$

1-Propanol **2-Propanol**

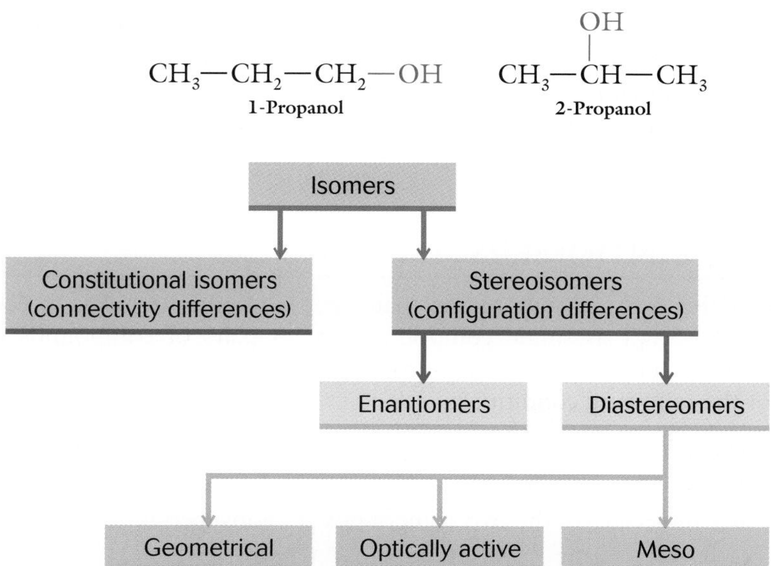

Figure 9.1 Different types of isomers.

- Different functional groups together with different carbon skeletons, as in ethanol and dimethyl ether:

$$CH_3CH_2{-}OH \qquad CH_3{-}O{-}CH_3$$

Ethanol **Dimethyl ether**

Stereoisomers have the same connectivity but differ in their configurations. **Configuration** describes the relative orientations in space of the atoms of a stereoisomer, independent of changes that occur by rotation about single bonds. (Configuration must not be confused with conformation. **Conformation** describes the relative orientations of the atoms of a compound caused by rotation about single bonds—Section 3.3.)

There are two classes of stereoisomers (see Figure 9.1):

- **Enantiomers** are stereoisomers that are nonsuperimposable mirror images of each other.
- **Diastereomers** are stereoisomers that are not enantiomers; that is, they are not mirror images of each other.

One type of diastereomer, **geometrical stereoisomers** (also called **cis-trans stereoisomers**), was discussed in Sections 3.8 and 4.5. Geometrical stereoisomers are found in alkenes and cyclic compounds when the two carbons of the double bond or any two carbons of a ring each contain two different substituents. Such carbon atoms are called **stereocenters** because a hypothetical exchange of the positions of any two substituents would convert one stereoisomer into the other. Examples of geometrical isomers are *cis*- and *trans*-2-butene and *cis*- and *trans*-1,2-dimethylcyclopropane. The cis isomer differs from the trans isomer in the configuration at a stereocenter.

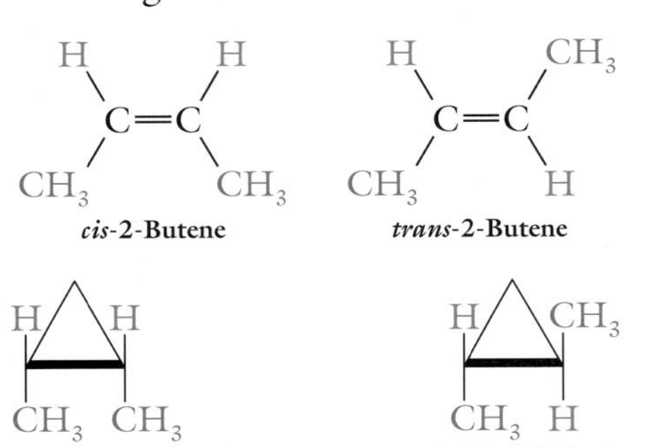

cis-2-Butene *trans*-2-Butene

cis-1,2-Dimethylcyclopropane *trans*-1,2-Dimethylcyclopropane

The stereocenters in alkenes are sometimes called **trigonal stereocenters** to distinguish them from the stereocenters in alkanes and cylcoalkanes, which are called **tetrahedral stereocenters.**

This chapter deals with enantiomers as well as those diastereomers not yet encountered.

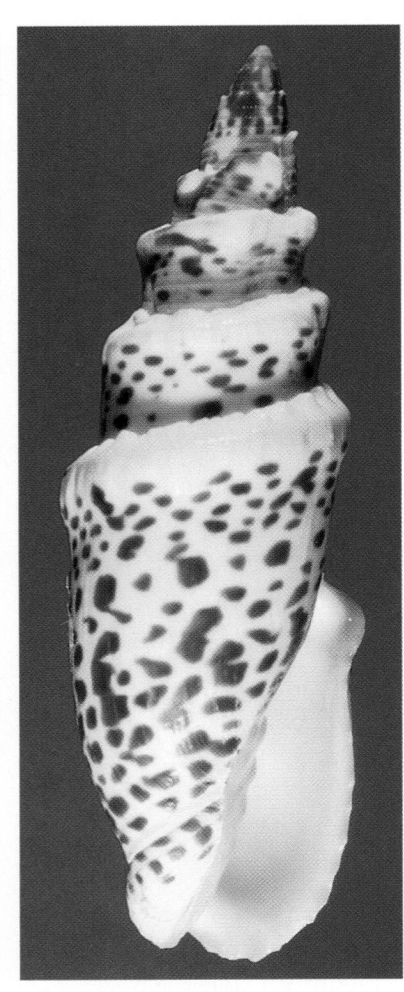

The spiraled seashell is chiral. (A. Kerstitou/Bruce Coleman.)

9.2 ENANTIOMERS

Many of the organic compounds that we have come across in preceding chapters exist not as single compounds but as pairs of compounds called **enantiomers.** Two compounds constitute a pair of enantiomers if they meet both of the following conditions:

- The molecules of the two compounds are mirror images of each other.
- The molecules of the two compounds are nonsuperimposable on each other.

How do we test two molecules to decide if they meet these two conditions?

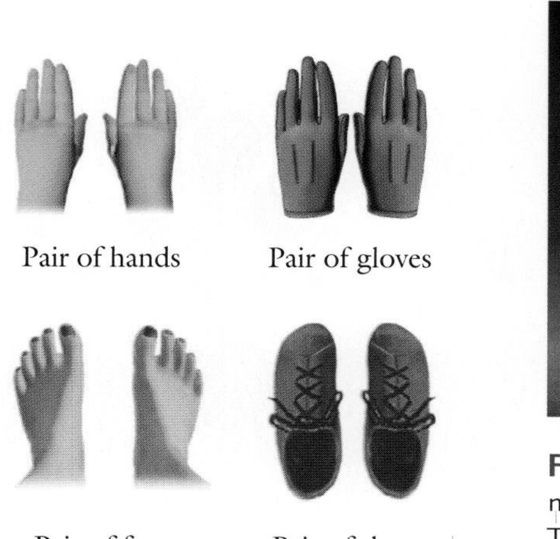

Figure 9.2 Chiral objects.

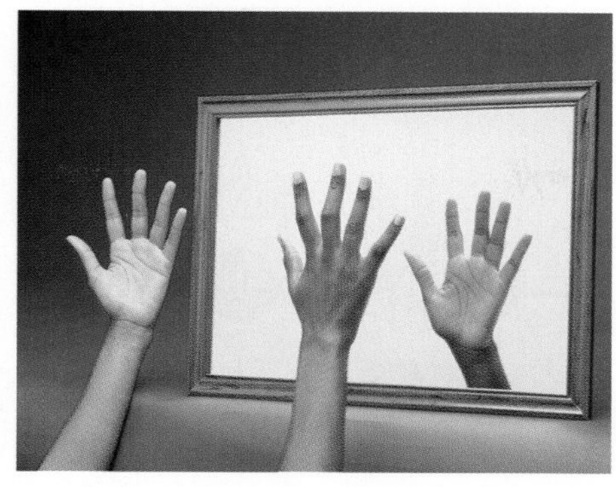

Figure 9.3 Our hands are chiral—nonsuperimposable, mirror-image objects. This photograph shows that the left hand is identical with the mirror image of the right hand. (Martin Bough/Fundamental Photographs.)

- **Mirror-image condition:** Imagine each molecule looking into a mirror. The two molecules are mirror images if each sees the other as its mirror image.
- **Nonsuperimposable condition:** Imagine the process of moving one molecule so that it merges into the other. The two molecules are **superimposable** if they become identical; that is, each and every part of one molecule coincides exactly with its corresponding part in the other molecule. The two molecules are **nonsuperimposable** if they do not become identical.

A molecule that is nonsuperimposable on its mirror-image molecule is said to be **chiral** or to have **chirality.** Molecules that do not possess chirality are **achiral.**

Chirality in molecules is easier to understand if we look at some chiral and achiral macroscopic objects in everyday life. Our hands and feet are chiral, as are a pair of shoes and a pair of gloves (Figure 9.2). Consider the relation between a pair of hands (Figure 9.3). A left hand is the nonsuperimposable mirror image of a right hand. By analogy, objects and molecules that are chiral are said to possess **handedness.** Our feet, too, are chiral; but most kinds of socks for our feet are not—they are achiral. Other familiar achiral objects are coffee cups and forks (Figure 9.4). These achiral objects are superimposable on their mirror-image objects. Some achiral objects become chiral objects when they are decorated. For example, a coffee cup without any design is achiral, but the same coffee cup with the name MARY on it is chiral.

An example of a chiral molecule is 2-butanol. Two 2-butanols exist, whose molecules are nonsuperimposable mirror images of each other (Figure 9.5).

Figure 9.4 Achiral objects.

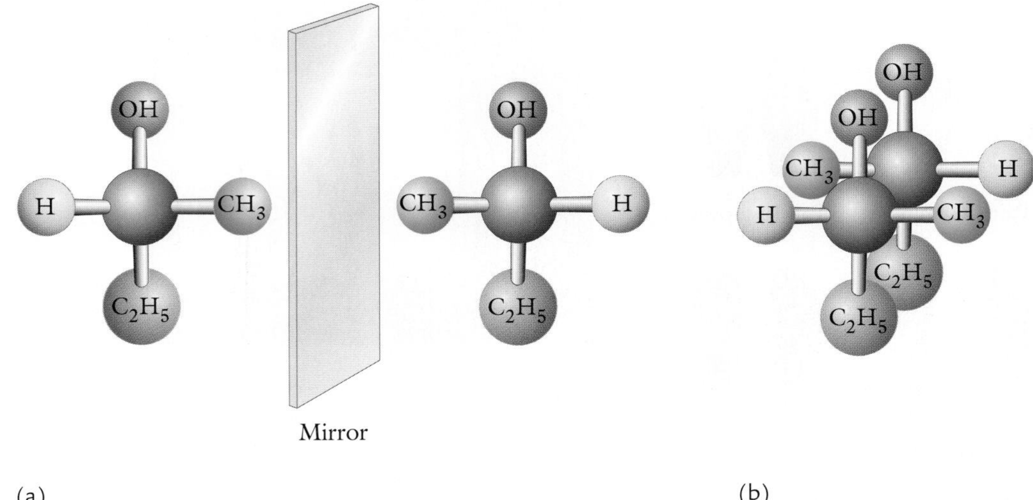

Figure 9.5 Enantiomers of 2-butanol. (a) The two compounds are nonsuperimposable mirror images. (b) The two compounds cannot be superimposed by sliding one over and merging it into the other, because the —H and —CH_3 substituents do not superimpose.

Sliding one molecule over and into the other superimposes the —OH and —C_2H_5 groups but not the —CH_3 and —H groups. Each of the molecules is chiral.

Figure 9.6 shows two alternate structural representations of the 2-butanol enantiomers, **wedge-bond** and **Fischer-projection** representations. The wedge-bond representations retain some three-dimensional perspective but not as much as the molecular models in Figure 9.5. These drawings use the following convention:

www.whfreeman.com/bleiodian2e

- The solid, horizontal wedge bonds (▬) extend in front of the plane of the paper.
- The dashed, vertical wedge bonds (┅) extend behind the plane of the paper.
- Both solid and dashed wedge bonds are oriented with their wider ends nearer the viewer.

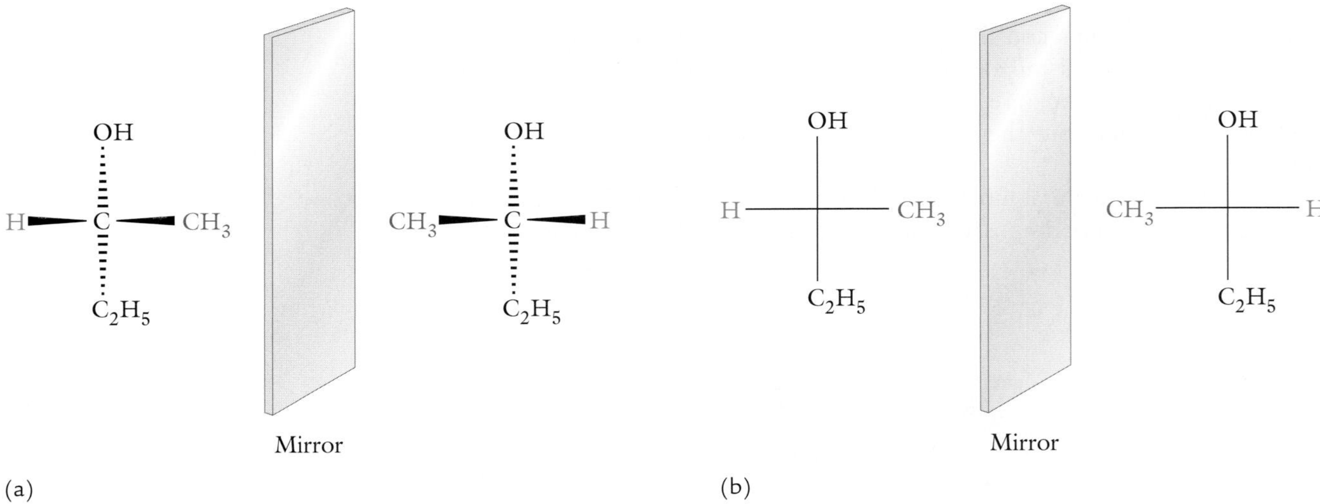

Figure 9.6 Alternate representations of the 2-butanol enantiomers: (a) wedge-bond structural drawing; (b) Fischer-projection structural drawing.

Most chemists prefer to use Fischer projections because they can be drawn much more quickly. The Fischer projections' lack of three-dimensional perspective would be a major drawback except that we rigidly adhere to the same convention used in the wedge-bond representations: horizontal bonds extend in front of the plane of the paper, and vertical bonds extend behind the plane of the paper. This convention is an absolute requirement for representing chiral molecules when we need to distinguish between the two molecules of an enantiomeric pair. It is not as critical for representing achiral molecules, because there is only one compound with that particular connectivity. It is also not critical when we do not need to distinguish between the two molecules of an enantiomeric pair, which is why the convention regarding horizontal and vertical bonds was not considered in earlier chapters.

The use of a molecular-model kit can be helpful if you have trouble visualizing three-dimensional structures and understanding the wedge-bond and Fischer-projection representations.

Figure 9.7 shows mirror-image molecules of 2-methyl-2-butanol. Unlike those of 2-butanol, the mirror-image molecules of 2-methyl-2-butanol are the same compound because they are superimposable on each other. 2-Methyl-2-butanol is not a chiral molecule and exists as one compound, not as a pair of enantiomers.

The ultimate test of whether some compound is one stereoisomer of a pair of enantiomers is to determine whether a molecule of the compound is superimposable on its mirror image. However, a simpler test can be applied:

- A pair of enantiomers is possible only when a compound contains a tetrahedral stereocenter. For acyclic compounds, a tetrahedral stereocenter is a carbon atom with four different substituents. (In older terminology, such stereocenters were called chiral centers.)

The C2 of 2-butanol is a tetrahedral stereocenter (see Figure 9.5). Four different substituents are attached to it: —H, —OH, —CH_3, and —C_2H_5. As mentioned earlier, a stereocenter is an atom situated in a compound in such a way that a hypothetical exchange of the positions of any two substituents would convert one stereoisomer into the other (Section 9.1). This is exactly the case for a tetrahedral stereocenter. Referring to Figure 9.5a, we can see that exchanging the positions of the —H and —CH_3 groups converts the stereoisomer on the left side into the stereoisomer on the right side, and vice

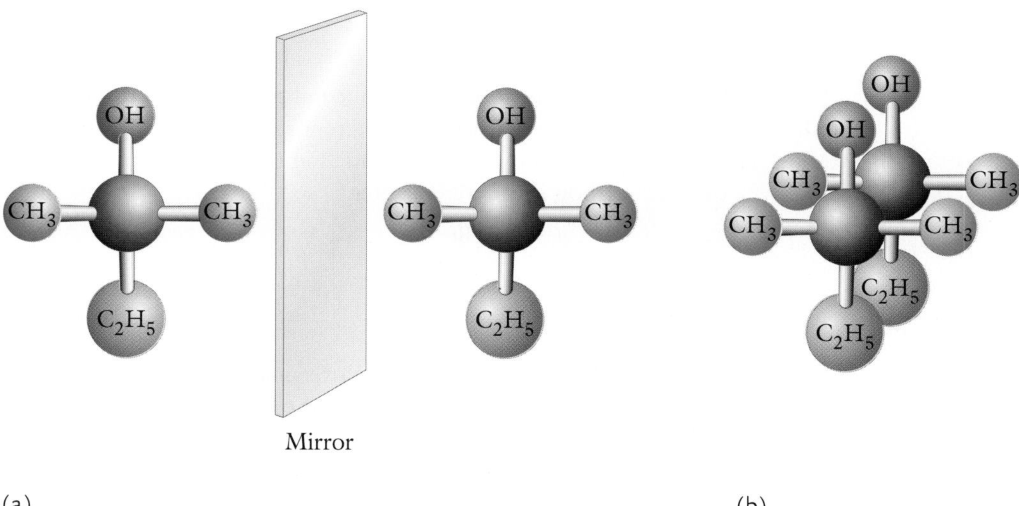

Figure 9.7 (a) Mirror-image molecules of 2-methyl-2-butanol. (b) The two molecules can be superimposed by sliding one over and merging it into the other.

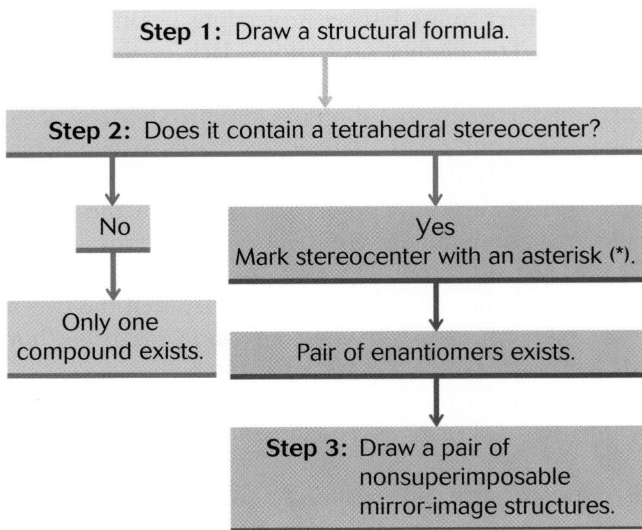

Figure 9.8 Procedure for determining whether a structural formula represents a single compound or a pair of enantiomers.

versa. The two enantiomers of 2-butanol differ in their configurations at the tetrahedral stereocenter.

None of the carbons in 2-methyl-2-butanol is a tetrahedral stereocenter (see Figure 9.7). For example, C2 does not have four different substituents; two of the substituents attached to C2 are the same (two CH_3 groups). 2-Methyl-2-butanol is achiral, as indicated earlier.

Figure 9.8 outlines a procedure for determining whether a compound is chiral. The procedure is illustrated in Example 9.1.

Example 9.1 Determining whether a compound is chiral

Which of the following compounds can exist as a pair of enantiomers and which cannot? If enantiomers are possible, draw structural formulas to represent the enantiomers. (a) 1-Bromo-1-chloroethane; (b) 1,1-dichloroethane.

Solution

Follow the procedure described in Figure 9.8.

(a) **Step 1.** Draw a structural formula of 1-bromo-1-chloroethane.

Br
H—C—CH_3
Cl

Step 2. Does the compound contain a tetrahedral stereocenter? Yes, C1 is a tetrahedral stereocenter because it has four different substituents (—H, —Br, —Cl, —CH_3). Marking the stereocenter with an asterisk is a useful way of keeping track of which carbon is the stereocenter.

Br
H—C*—CH_3
Cl

Step 3. Draw a pair of nonsuperimposable mirror-image structures. Show the tetrahedral stereocenter as a central carbon to which the four different substituents are attached.

Br H—C*—CH_3 Cl 1	Br CH_3—C*—H Cl 2

(b) **Step 1.** Draw a structural formula of 1,1-dichloroethane.

$$\begin{array}{c} Cl \\ | \\ H-C-CH_3 \\ | \\ Cl \end{array}$$

Step 2. Does the compound contain a tetrahedral stereocenter? No, neither carbon has four different substituents. C1 has two chlorines, whereas C2 has three hydrogens. There are no enantiomers of 1,1-dichloroethane. Only one compound is possible.

Problem 9.1 Which of the following compounds can exist as a pair of enantiomers and which cannot? If enantiomers are possible, draw structural formulas to represent the enantiomers. (a) 3-Hydroxypropanal; (b) 2-hydroxypropanal.

9.3 INTERPRETING STRUCTURAL FORMULAS OF ENANTIOMERS

In this section, we will develop some guidelines for interpreting structural drawings of enantiomers. Consider structures 1 and 2 in the Solution to Example 9.1, part a. These structures are one way of representing the pair of enantiomers of 1-bromo-1-chloroethane, but several other structural representations are possible (Figure 9.9). This single pair of enantiomers has several different representations because molecules, being three-dimensional, can be viewed from different directions: from the left or the right, from below or above, from the front or the back, and so forth. Any one pair of structural drawings, 1 and 2 or 3 and 4 or 5 and 6 or 7 and 8, is a valid representation of the two enantiomers, but only one pair of drawings should be used to represent them because there is only one pair of enantiomers.

There is no such ambiguity when we work with molecular models. Such models show the molecules as the three-dimensional structures that they are. The ambiguity arises only when we draw and observe three-dimensional structures on a two-dimensional (flat) surface.

Even experienced chemists can find it difficult to compare a pair of structural drawings and decide whether they represent a pair of enantiomers or the same compound (which may be an achiral compound or the same enantiomer of a pair of enantiomers). It is not easy, for example, to compare structures 1 and 4 or 2 and 7, and so forth, of Figure 9.9 and determine that they are the same molecule seen from different directions. We will limit ourselves in this book to comparing structural drawings originating from the same observation site. This limitation is important because it means that any two structural

1	2	3	4
Br (top), H (left), CH_3 (right), Cl (bottom)	Br (top), CH_3 (left), H (right), Cl (bottom)	Cl (top), CH_3 (left), H (right), Br (bottom)	Cl (top), H (left), CH_3 (right), Br (bottom)

5	6	7	8
Cl (top), Br (left), CH_3 (right), H (bottom)	Cl (top), CH_3 (left), Br (right), H (bottom)	Cl (top), H (left), Br (right), CH_3 (bottom)	Cl (top), Br (left), H (right), CH_3 (bottom)

Figure 9.9 Different representations of the one pair of enantiomers of 1-bromo-1-chloroethane: 1 = 3 = 5 = 7 and 2 = 4 = 6 = 8.

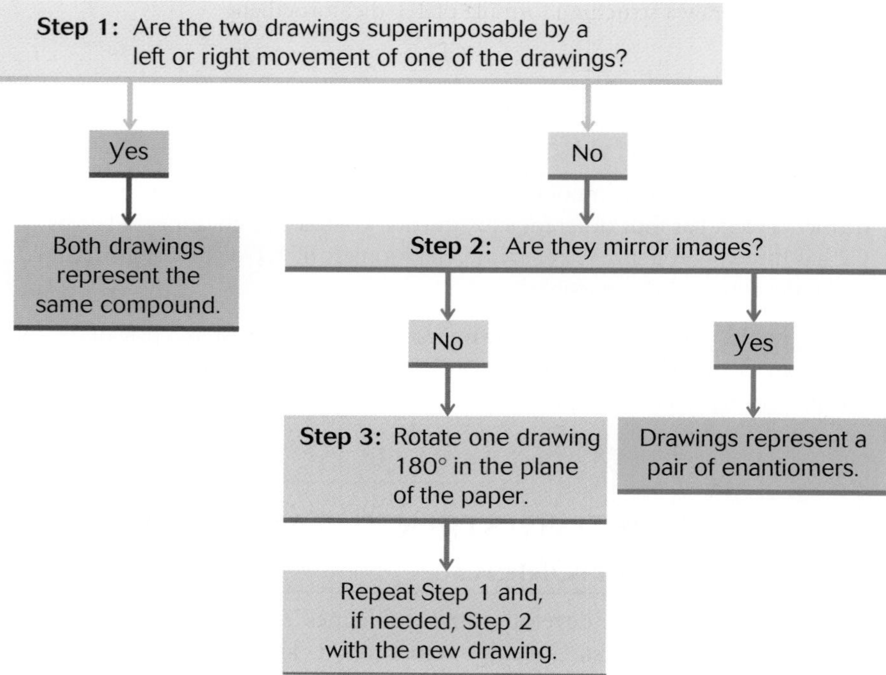

Figure 9.10 Procedure for determining whether two structural drawings represent the same compound or a pair of enantiomers.

drawings can be compared either as shown or after a simple rotation of one or the other drawing (but not both) 180° in the plane of the paper. No other rotations of the drawings will be allowed. Rotations of any degrees other than 180° in the plane of the paper and rotations of any kind out of the plane of the paper will not be allowed, because they would give structures that violate the convention of horizontal bonds extending forward and vertical bonds extending backward. The procedure for comparing two structural drawings is detailed in Figure 9.10 and applied in Example 9.2.

Example 9.2 Determining whether two structural drawings represent a pair of enantiomers or the same compound

Indicate whether each of the following pairs of structural drawings represents the same compound or a pair of enantiomers.

(a) Structure 1: top OH; H, C*, CH_3 (horizontal); bottom CH_2OH. Structure 2: top CH_2OH; CH_3, C*, H (horizontal); bottom OH.

(b) Structure 3: top CH_2OH; H, *, CH_3 (horizontal); bottom OH. Structure 4: top OH; H, *, CH_3 (horizontal); bottom CH_2OH.

Solution

(a) **Step 1.** Structures 1 and 2 are not superimposable.

Step 2. Structures 1 and 2 are not mirror images.

Step 3. Rotation of structure 1 by 180° in the plane of the paper yields structure 2 (that is, structures 1 and 2 are superimposable), which means that structures 1 and 2 represent the same compound.

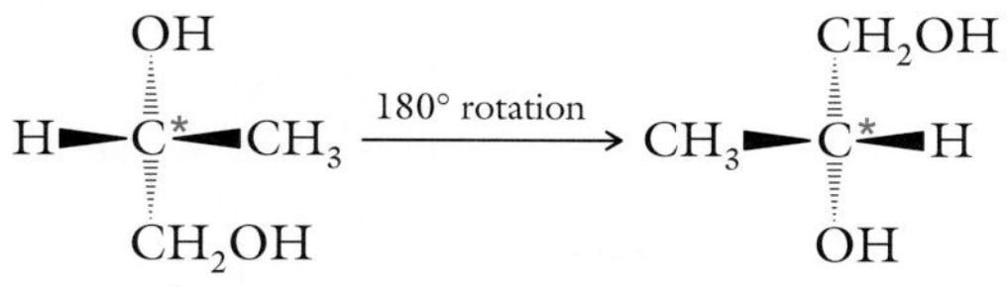

OH / H—C*—CH_3 / CH_2OH (1) $\xrightarrow{180° \text{ rotation}}$ CH_2OH / CH_3—C*—H / OH (2)

(b) **Step 1.** Structures 3 and 4 are not superimposable.

Step 2. Structures 3 and 4 are not mirror images.

Step 3. Rotation of structure 3 by 180° in the plane of the paper yields structure 3′, which is the nonsuperimposable mirror image of structure 4. Structures 3 and 4 represent enantiomers.

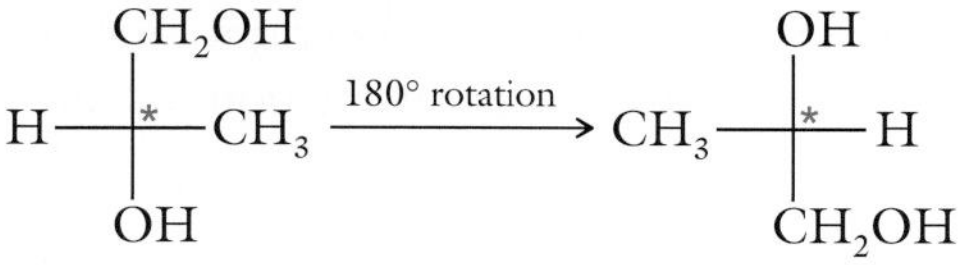

CH_2OH / H—*—CH_3 / OH (3) $\xrightarrow{180° \text{ rotation}}$ OH / CH_3—*—H / CH_2OH (3′)

Problem 9.2 Indicate whether each of the following pairs of structural drawings represents the same compound or a pair of enantiomers.

(a) 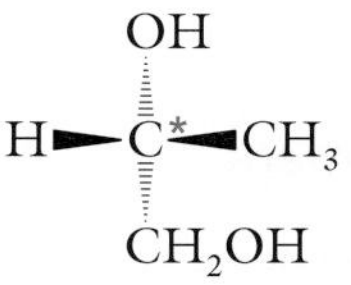 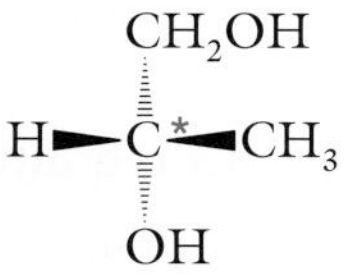 OH / H—C*—CH_3 / CH_2OH (1) CH_2OH / H—C*—CH_3 / OH (2)

(b) 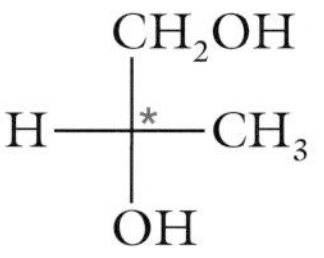 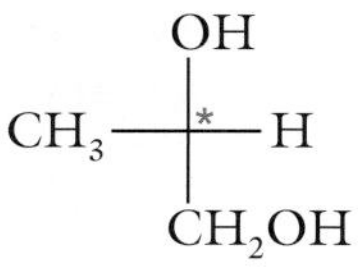 CH_2OH / H—*—CH_3 / OH (3)   OH / CH_3—*—H / CH_2OH (4)

9.4 NOMENCLATURE FOR ENANTIOMERS

One enantiomer of a pair of enantiomers is distinguished from the other enantiomer by adding a prefix before the name of the compound. The prefix indicates the configuration—the order of attachment of the four substituents—at the tetrahedral stereocenter. Before adding this prefix, chemists must experimentally determine the configuration at the tetrahedral stereocenter of each enantiomer by a technique called X-ray crystallography. In X-ray crystallography, the scattering of X-ray energy by the atomic nuclei of a molecule is used to determine the three-dimensional arrangement of the atoms of the compound.

There are two systems of nomenclature for enantiomers: the original nomenclature system uses the prefixes D- and L-; a more recent system uses the prefixes (*R*)- and (*S*)- (Box 9.1 on page 289). The D/L system is simpler and, though more limited in its application than the *R*/*S* system, is used extensively by biochemists and biologists in naming carbohydrates, amino acids, and other important biochemicals. We concentrate here on the D/L system.

The D/L system applies only when the tetrahedral stereocenter has the following substituents:

- Hydrogen
- A heteroatom substituent X (such as OH or NH_2) with the heteroatom bonded to the tetrahedral stereocenter
- Two different R substituents, R^1 and R^2, each of which has a carbon bonded to the tetrahedral stereocenter

Naming enantiomers by the D/L system requires that Fischer projections (or wedge-bond representations) of the enantiomers be drawn from an observation site such that the placement of the R^1 and R^2 substituents is vertical (they extend backward), whereas the placement of the H and heteroatom substituents is horizontal (they extend forward). The Fischer projections are then oriented so that R^1, the substituent having the most-substituted carbon atom (the carbon with the fewest hydrogens) attached to the tetrahedral stereocenter, is located upward and that R^2, the substituent having the least-substituted carbon atom (the carbon with the most hydrogens) attached to the tetrahedral stereocenter, is located downward. If necessary, the Fischer projection is rotated 180° in the plane of the paper to achieve this orientation. The enantiomer with the heteroatom bonded on the right side is the D-enantiomer, whereas the enantiomer with the heteroatom bonded on the left side is the L-enantiomer:

R¹ (top), H—C—X, R² (bottom): D

R¹ (top), X—C—H, R² (bottom): L

Example 9.3 Identifying the D- and L-enantiomers

Chemist Emil Fischer originated and first applied the D/L nomenclature system in 1891 for the two enantiomers of glyceraldehyde. Which of the following compounds is D-glyceraldehyde and which is L-glyceraldehyde?

1: HC=O (top), H—C—OH, CH_2OH (bottom)

2: HC=O (top), HO—C—H, CH_2OH (bottom)

Solution

Structures 1 and 2 are correctly oriented in space for application of the D/L nomenclature system: the carbon substituents are oriented vertically with the most (HC=O) and least (CH_2OH) substituted carbon atoms placed up and down, respectively. The enantiomer with the heteroatom (OH) bonded on the right side of the tetrahedral stereocenter, structure 1, is D-glyceraldehyde. The enantiomer with the OH on the left, structure 2, is L-glyceraldehyde.

Problem 9.3 Designate the following enantiomer of alanine, an amino acid used by nature in synthesizing proteins, as D- or L-.

COOH (top), NH_2—C—H, CH_3 (bottom)

9.5 PROPERTIES OF ENANTIOMERS

Chiral compounds differ from achiral compounds in one physical property (optical activity) and one chemical property (chiral recognition).

Optical Activity

Chiral compounds exhibit **optical activity,** and are said to be **optically active.** Achiral compounds are **optically inactive.** The enantiomers of an enantiomeric pair are identical in all their physical properties (such as boiling and melting points and solubility) except optical activity. Optical activity is the

9.1 CHEMISTRY AROUND US

The *R/S* Nomenclature System for Enantiomers

The D/L nomenclature system applies only to compounds in which the tetrahedral stereocenter is connected to H, a heteroatom substituent (such as OH or NH_2), and two different carbon substituents. Thus, it cannot be used to name a variety of enantiomers of compounds such as 2-methylbutanoic acid and the antihypertensive medication methyldopa (Aldomet). 2-Methylbutanoic acid has no heteroatom substituent connected to the tetrahedral stereocenter. Methyldopa has no H connected to the tetrahedral stereocenter.

2-Methylbutanoic acid: central carbon with CH_3 (top), H (left), COOH (right), C_2H_5 (bottom)

Methyldopa: central carbon with COOH (top), CH_3 (left), CH_2–(3,4-dihydroxyphenyl, OH, OH) (right), NH_2 (bottom)

2-Methylbutanoic acid **Methyldopa**

The *R/S* nomenclature system is used for compounds with any combination of substituents connected to the tetrahedral stereocenter. The *R/S* nomenclature system is the preferred nomenclature system for chemists and is becoming increasingly preferred by biochemists. For a structure such as methyldopa, the two enantiomers are designated (*R*)-methyldopa and (*S*)-methyldopa.

The (*R*) and (*S*) configurations are designated by using the following procedure:

1. Each substituent connected to the tetrahedral stereocenter is assigned a **priority** by a set of rules:

a. Priority increases with increasing atomic number of the atom directly connected to the tetrahedral stereocenter. The order of priority for atoms is

$$Br > Cl > S > F > O > N > C > H$$

b. For different substituents that have the same kind of atom connected to the tetrahedral stereocenter, priority is decided by the priority of the next atom or set of atoms. Continue to the next atoms or set of atoms, and the next after that, and so on, until there is a difference in priority of the atoms between the different substituents. The order of priority for some common substituents is

$$\overset{O}{\overset{\|}{C}}{-}OH > \overset{O}{\overset{\|}{C}}{-}NH_2 > \overset{O}{\overset{\|}{C}}{-}H > CH_2OH >$$

$$C(CH_3)_3 > CH(CH_3)_2 > CH_2CH_3 > CH_3 > H$$

A carbon–oxygen double bond has a higher priority than a carbon–oxygen single bond because, in the double bond, the carbon atom is considered to be the equivalent of a carbon atom bonded to two different oxygen atoms. A *t*-butyl group has higher priority than an isopropyl group because the *t*-butyl's carbon atom connected to the tetrahedral stereocenter is itself connected to three carbon atoms compared with two carbon atoms for the isopropyl group.

Application of these priority rules yields an order of priority for the four substituents connected to the tetrahedral stereocenter.

2. The stereoisomer is viewed from a position such that the substituent of lowest priority is farthest from the viewer. We view the stereoisomer on an imaginary line going through the tetrahedral stereocenter and then to the substituent of lowest priority.

3. The stereoisomer is assigned the (*R*) configuration if priority decreases in a clockwise direction for the other three substituents and the (*S*) configuration if the priority decreases in a counterclockwise direction for the other three substituents.

The procedure is illustrated for the enantiomer shown earlier for 2-methylbutanoic acid. First, we list the order of priority for the four substituents: COOH > C_2H_5 > CH_3 > H. Second, the structural drawing for the enantiomer is converted into

Structure: H–C with wedged COOH (up), dashed CH_3 (back), wedged C_2H_5 (down); Viewer on the right.

This conversion is not simple and is aided by the use of molecular models. From the specified viewing position, the order of priority decreases counterclockwise for the three substituents other than H. This enantiomer is (*S*)-2-methylbutanoic acid.

- When enantiomers contain the required substituents that allow the use of the D/L nomenclature system, the enantiomers can be named by both the *R/S* and D/L nomenclature systems. D- and L-enantiomers always translate into (*R*) and (*S*), respectively.
- Designation as (*R*) or (*S*), like the designation as D or L, does not specify the direction of optical rotation. (+) and (−) can be determined only by the polarimeter experiment. The configuration—whether (*R*) or (*S*)—is determined only by X-ray crystallography.

ability of a compound to rotate the plane of plane-polarized light. To understand what this ability means, we need to look briefly at the nature of light and the measurement of optical activity.

Optical activity is measured by an instrument called a **polarimeter** (Figure 9.11), composed of a light source, a polarizer, a sample cell (into which a solution of the chemical compound to be studied is placed), and an analyzer, together with the observer (human eye). A beam of ordinary light such as that from a light bulb contains electromagnetic energy vibrating in all angular orientations around the axis of the light beam. Light vibrating in only one angular orientation, called **plane-polarized light,** is isolated by passing ordinary light through a **polarizer.** Materials such as calcite and Polaroid filters (the same Polaroid filters used in sunglasses to reduce glare) are used as polarizers. The polarizer is set up so that the plane-polarized light emerges in a specified plane—for instance, the vertical plane as in Figure 9.11. The **analyzer,** made of the same material as the polarizer, allows light vibrating in only one direction to pass through it. The analyzer is initially set up to match the polarizer, allowing only light in the same plane—in this case, the vertical plane—to pass through.

When the sample cell contains a solution of an achiral compound (Figure 9.11a), there is no observed interaction of the compound with the plane-polarized light. The plane-polarized light passes through unchanged; that is,

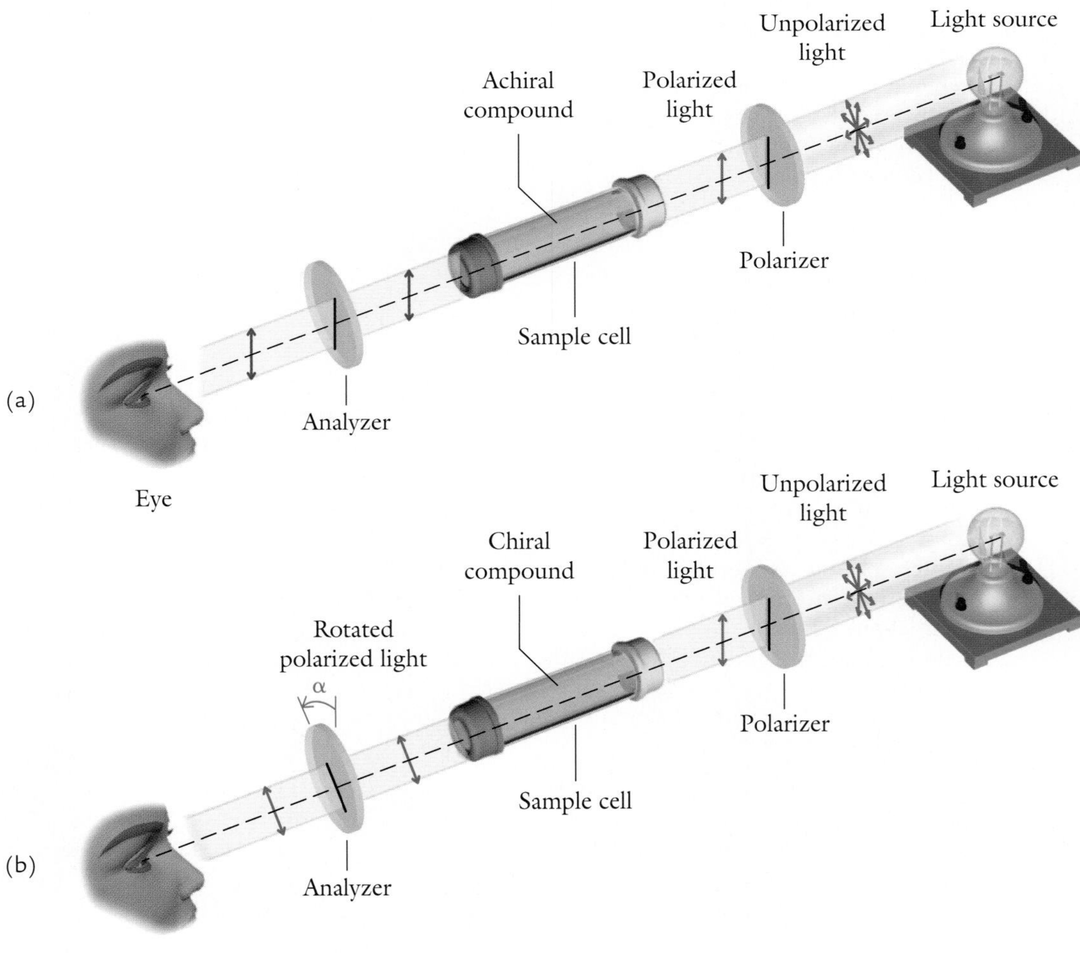

Figure 9.11 Polarimeter measurement of optical activity: (a) achiral compound; (b) chiral compound.

the light emerging from the sample cell is still polarized in the same plane. The light passes through the analyzer and can be observed.

When the sample cell contains a solution of a chiral compound (Figure 9.11b), the plane-polarized light interacts with the compound as it travels through the sample cell. The plane of the plane-polarized light that emerges is rotated at an angle α to the vertical plane. The rotated light cannot pass through the analyzer and cannot be observed. The experimenter then rotates the analyzer until light passes through and can be observed. This angle of rotation also is α, the amount to which the plane of the plane-polarized light has been rotated.

The two enantiomers of a pair of enantiomers rotate the plane of plane-polarized light by the same number of degrees but in opposite directions. One enantiomer rotates light in the clockwise direction and is called **dextrorotatory.** The other enantiomer rotates light in a counterclockwise direction and is called **levorotatory.** The direction of rotation of plane-polarized light is indicated in the names of enantiomers by placing (+)- and (−)- as prefixes to their names—for example, (+)-glyceraldehyde and (−)-glyceraldehyde.

It is important to keep in mind that the prefixes (+)- and (−)- (corresponding to dextrorotatory and levorotatory) are not related to the prefixes D- and L-. The direction of optical rotation—whether (+) or (−)—is determined by experiments with the polarimeter. The configuration at the tetrahedral stereocenter—whether D or L—is determined by X-ray crystallography. There is no general relation between configuration and direction of optical rotation. Some D-enantiomers rotate in a clockwise (dextrorotatory) direction; other D-enantiomers rotate in a counterclockwise (levorotatory) direction.

If both the configuration and the optical rotation of a chiral compound are known, both are indicated in the name. For example, the dextrorotatory enantiomer of glyceraldehyde is the D-enantiomer and the name is D-(+)-glyceraldehyde. The levorotatory enantiomer of fructose, an important carbohydrate, is the D-enantiomer and the name is D-(−)-fructose.

Specific Rotation

The degree of rotation α observed for a chiral compound in a polarimeter experiment depends on the number of chiral molecules encountered by the light beam as it passes through the sample tube. The value of α increases if the experimenter uses a more concentrated solution or a longer sample cell or both. To prevent ambiguities, chemists standardize the units of rotation by calculating a quantity called the **specific rotation,** $[\alpha]$, defined by

$$[\alpha] = \frac{\alpha}{CL}$$

where α is the observed rotation in degrees for a sample concentration of C grams per milliliter (g/mL) and sample-cell length of L decimeters (1 dm = 10 cm). (For pure liquids, C is the density of the liquid.)

Whereas α values for the same chiral compound will be different for experiments performed by different workers (using different concentrations and different sample-cell lengths), $[\alpha]$ is independent of the details of the particular polarimeter experiment. All polarimeter experiments with the same compound yield the same value of $[\alpha]$. Specific rotation is a physical property of the enantiomer and has a definite and constant value, just as boiling and melting points, density, color, and solubility have. The specific rotations of the enantiomers in an enantiomeric pair are the same in degrees but in the opposite direction—for example, $[\alpha] = +13.5°$ and $-13.5°$ for the enantiomers of glyceraldehyde. Like other physical properties, $[\alpha]$ is useful for identifying a compound.

The details of the measurement of specific rotation are more complicated than indicated because $[\alpha]$ varies somewhat, depending on the frequency of light, the solvent used, and the temperature of the experiment. To compare $[\alpha]$ values obtained by different workers, the values must have been obtained under exactly the same conditions: the same temperature, solvent, and frequency of light.

An equimolar (1 : 1) mixture of two enantiomers (of a pair of enantiomers) does not exhibit optical activity, because, in such a mixture, known as a **racemic mixture** or **racemate,** the rotation in one direction by one enantiomer is canceled by rotation in the opposite direction by the other.

Example 9.4 Calculating specific rotation, $[\alpha]$

A solution of L-valine at a concentration of 1.50 g/mL gives an optical rotation of $-20.3°$ when observed in the 5.00-cm-long sample cell of a polarimeter. Calculate the specific rotation of L-valine.

Solution

Substitute the values of α, C, and L into the equation $[\alpha] = \alpha/CL$, making sure that the correct units are used for C (g/mL) and L (dm). The concentration is given in the correct units, but the sample-cell length is not; 5.00 cm must be converted into 0.500 dm before substitution into the equation for $[\alpha]$.

$$[\alpha] = \frac{\alpha}{CL} = \frac{120.3°}{1.50 \text{ g/mL} \times 0.50 \text{ dm}} = \frac{127.1°}{\text{g/mL} \times \text{dm}} = 127.1°$$

Although the units of specific rotation are usually reported simply as degrees, it should always be understood that the actual units are degrees per gram per milliliter per decimeter.

Problem 9.4 D-Glucose has a specific rotation of $+53°$. Calculate the observed rotation of a solution of D-glucose at a concentration of 30 g/L when measured in a 15-cm-long sample cell.

Chemical Properties: Chiral Recognition

Whether the enantiomers of a chiral compound have the same or different chemical reactivity depends on the nature of the other reactants or the enzyme (biological catalyst) or both.

- When reacting with achiral compounds, enantiomers usually have the same reactivity.
- When reacting with chiral compounds or in reactions catalyzed by chiral enzymes, enantiomers usually have very large differences in reactivity.

These differences in the interaction of chiral compounds with achiral compounds and with other chiral compounds or enzymes can be understood, again, by considering feet, socks, and shoes (see Figures 9.2 and 9.4). Most socks are not chiral, though feet are. Either one of a pair of socks can be worn on either foot, but there are very strict requirements for wearing shoes. Only the left shoe fits well on the left foot, and only the right shoe fits well on the right foot.

Many reactions in living cells include one or more chiral molecules and are catalyzed by enzymes, which are chiral. Such reactions require a unique fit between two chiral species for reaction to take place. In enzyme-catalyzed reactions, one or more of the reactants bind to a site on the enzyme known as the **active site** (Section 2.2). For many enzymes, the active site is chiral. Figure 9.12 on page 294 illustrates the difference in fit between two enantiomers at the active site of a chiral enzyme. Only one enantiomer fits into (binds to) the active site, and reaction proceeds only with this enantiomer. This phenomenon, called **chiral recognition** or **chiral discrimination,** is one mechanism by which enzymes show discrimination (selectivity) in biological systems.

Not all biological discriminations are based on chirality. Some are based on geometrical isomerism and others on molecular (or ionic) size, shape, and polarity (or charge). These bases constitute the complementarity principle (Sections 2.2 and 14.5, Box 4.2). Boxes 4.2 and 4.3 describe the effects of geometrical isomerism on vision and pheromone action, respectively.

9.2 CHEMISTRY WITHIN US

Senses of Smell and Taste

The senses of taste **(gustation)** and smell **(olfaction)** depend on **chemoreceptors,** sensory cells that detect certain molecules or ions by binding to them through secondary attractive forces. Taste chemoreceptors are organized into **taste buds** on the upper surface of the tongue and, to a lesser extent, on the roof of the mouth. There are four primary taste perceptions—sweet, sour, salt, and bitter. Sour and salt tastes are associated with chemoreceptors sensitive to H_3O^+ and Na^+ (or other metal cations), respectively. Sweet and bitter tastes are associated with a variety of organic compounds. A number of different compounds bind to these chemoreceptors.

Binding takes place at specific proteins located on the cell membranes of chemoreceptor cells. The structural requirements for the binding of a molecule to a chemoreceptor include overall molecular size and shape, stereoisomerism, and secondary attractive forces in different parts of the molecule and the chemoreceptor protein. These requirements constitute the complementarity principle (Sections 2.2 and 14.5, Box 4.2). Binding is generally possible only between specific parts of the chemoreceptor protein and the molecule. Binding to the chemoreceptor cell generates an electrical impulse, which is transmitted to the end of the sensory cell and then from neuron (nerve cell) to neuron until it reaches the sensory centers in the brain. The brain interprets the electrical impulse as a specific taste.

Smell is perceived in a similar manner, through olfactory chemoreceptors lining the upper part of the nasal cavity.

There is considerable interaction between the sensations of taste and smell. Each reinforces the other. In actuality, a good deal of what we call taste is smell. For example, taste is sharply reduced when the nasal passages are blocked by a head cold. And foods often have more taste when hot because higher temperatures volatilize chemicals, which reach the olfactory chemoreceptors in the nasal passages.

The effect of stereoisomerism on taste and smell is often very striking, indicating that some chemoreceptor sites are themselves chiral and able to bind only one enantiomer of a pair of enantiomers. Consider the enantiomers of carvone. One enantiomer is the principal component of spearmint oil and the other the principal component of caraway seed oil, each with its characteristic taste and odor.

(−)-Carvone (Spearmint oil) | (+)-Carvone (Caraway seed oil)

The amino acid asparagine has a pair of enantiomers. D- and L-asparagines are found in the plants vetch and asparagus, respectively. D-Asparagine has a sweet taste. L-Asparagine is tasteless to some people but bitter tasting to others. The contradictory reports on the taste of L-asparagine illustrate the variations in the perceptions of taste. These variations arise from individual differences in the numbers and types of chemoreceptor cells as well as differences in the response of individual nervous systems (including the sensory centers in the brain).

D-Asparagine | L-Asparagine

A central taste bud is surrounded by many papillae in this electron micrograph of a part of the tongue surface. (SIU/Photo Researchers.)

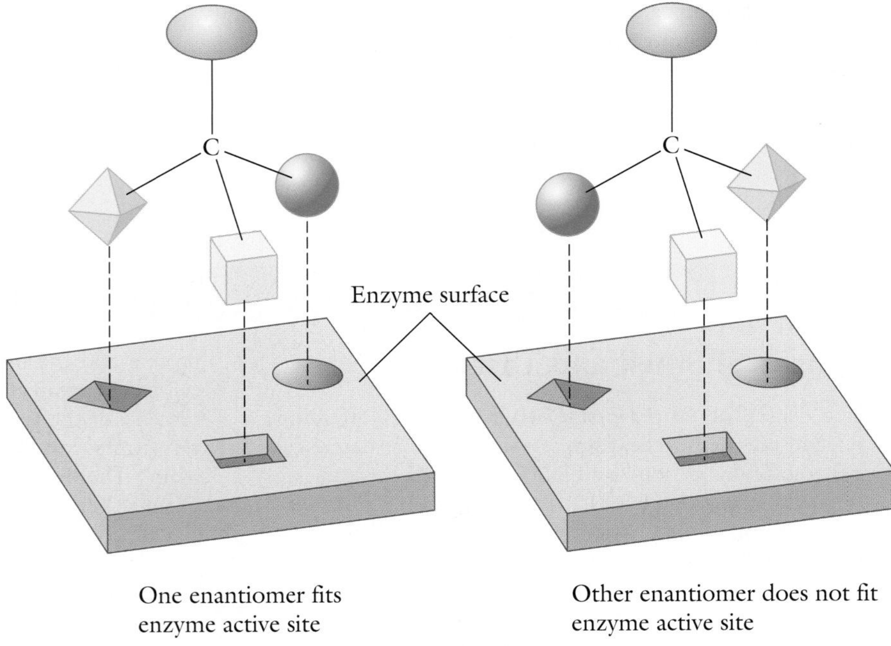

Figure 9.12 Chiral recognition of enantiomers by a chiral enzyme.

Chiral recognition happens in many physiological processes. For example, we ingest starch (a carbohydrate) and a variety of proteins and digest—that is, hydrolyze—them into simpler substances (glucose and amino acids, respectively). We then use these smaller compounds for a variety of purposes: to produce energy; to synthesize energy-storing materials such as glycogen; and to synthesize structural materials such as the components of bone, skin, and muscle. The enzymes used in these processes have evolved as highly selective chiral catalysts: only D-carbohydrates and L-proteins are used or synthesized in humans and most other organisms. In fact, evolution has produced a biological world in which, with very few exceptions, all carbohydrates are D-carbohydrates and all proteins are L-proteins (Sections 10.3 and 12.1). Whereas reactions in living cells produce only one enantiomer of a pair of enantiomers, synthetic reactions in the laboratory usually produce racemic mixtures.

›› Carbohydrates and proteins are the subjects of Chapters 10 and 12, respectively.

Chiral recognition in biological systems does not always mean that one enantiomer participates in a reaction and the other does not react at all. Different enantiomers may simply produce different effects, as in our sense of taste and smell (Box 9.2 on the previous page). Or the "inactive enantiomer," when it enters the body, may participate in a reaction that is undesirable and detrimental to the well-being of the organism. For example, synthetic drugs are produced in industrial reactions that often result in racemic mixtures, not a single enantiomer, and this result has had some important consequences (Box 9.3).

Concept checklist

- ✓ Enantiomers are a pair of stereoisomers whose molecules are nonsuperimposable mirror images.
- ✓ The only difference in physical properties of a pair of enantiomers is the direction of rotation of plane-polarized light.
- ✓ Enantiomers show no difference in chemical reactivity toward achiral compounds.
- ✓ Enantiomers show large differences in chemical reactivity toward chiral compounds or in reactions catalyzed by chiral enzymes.

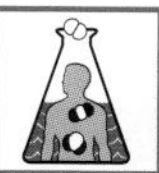

9.3 CHEMISTRY WITHIN US

Synthetic Chiral Drugs

About 40% to 50% of synthetic drugs consist of compounds containing one or more tetrahedral stereocenters, and the typical manufacturing process, unlike biological reactions, yields racemic mixtures (racemates) of enantiomers.

The body responds to a racemate in one of two ways:

- Both enantiomers react but in different processes, one of which may produce an undesired effect on the organism.
- One enantiomer reacts in a biological process and the other passes through the organism with no effect.

An example of the first response is in the treatment of people with Parkinson's disease. In this disease, the production of a neurotransmitter called dopamine is reduced. The consequence is an impaired ability to transmit information to the brain centers that control muscle movement, resulting in uncontrolled tremors, especially in the limbs, and difficulty in walking. The synthetic drug L-dopa (L-3,4-dihydroxyphenylalanine) is used to treat this condition. L-Dopa is actually L-(−)dopa and is called levodopa because it is levorotatory. The original medication used was racemic dopa, which showed some adverse side effects, notably granulocytopenia—a decreased leukocyte count in the blood. The D-dopa enantiomer was subsequently found to be both responsible for the adverse side effects and ineffective in treating Parkinson's disease.

L-Dopa is taken orally and subsequently absorbed into the bloodstream and transported to the brain, where it enters brain cells and is converted into dopamine. Dopamine itself cannot be used to treat the disease, because it cannot cross from the bloodstream into brain cells; it cannot cross the so-called blood–brain barrier. Unfortunately, L-dopa becomes progressively less effective with time because Parkinson's disease also causes degeneration of the dopamine receptors.

$CH_2CH_2NH_2$ / OH / OH — Dopamine

COOH / H—*—NH_2 / CH_2 / OH / OH — D-Dopa

COOH / NH_2—*—H / CH_2 / OH / OH — L-Dopa

An example of the second type of response to taking a drug as a racemate is provided by propranolol (Inderal), a β-blocker used as an antihypertensive, antiarrhythmic, and antianginal drug. The drug that is currently available is racemic propranolol, but only (−)-propranolol is effective. (+)-Propranolol is not effective but has no known adverse effects.

CH_2 / H—*—OH / $CH_2NHCH(CH_3)_2$ — D = Propranolol

CH_2 / HO—*—H / $CH_2NHCH(CH_3)_2$ — L = Propranolol

Drug companies have begun to synthesize and market **single-enantiomer drugs** in place of racemates to optimize the effectiveness of drugs even when the second enantiomer has no effect, whether desirable or adverse. The body is taxed by having to metabolize and excrete the unnecessary enantiomer, which may limit the dosage of the desirable enantiomer that can be prescribed. Among the single-enantiomer drugs now used are L-dopa, lisinopril (an ACE inhibitor for high blood pressure; Box 8.2), sertraline (an antidepressant), and naproxen sodium (an anti-inflammatory drug; Box 7.4).

The synthesis of a single-enantiomer drug requires special techniques. Biological reactions produce single-enantiomer compounds under the direction of chiral enzymes. Most syntheses carried out in the laboratory, though, result in racemates because the syntheses are not controlled by chiral catalysts. Most successful syntheses of single-enantiomer drugs have used one of two approaches:

- **Resolution:** The drug is produced as a racemate; then special separation techniques are used to separate the enantiomers.
- **Stereoselective synthesis:** Researchers devise a synthetic route that has a reaction step that can be controlled by an available chiral catalyst to produce only the desired enantiomer.

A PICTURE OF HEALTH
Examples of Stereoselective Molecules

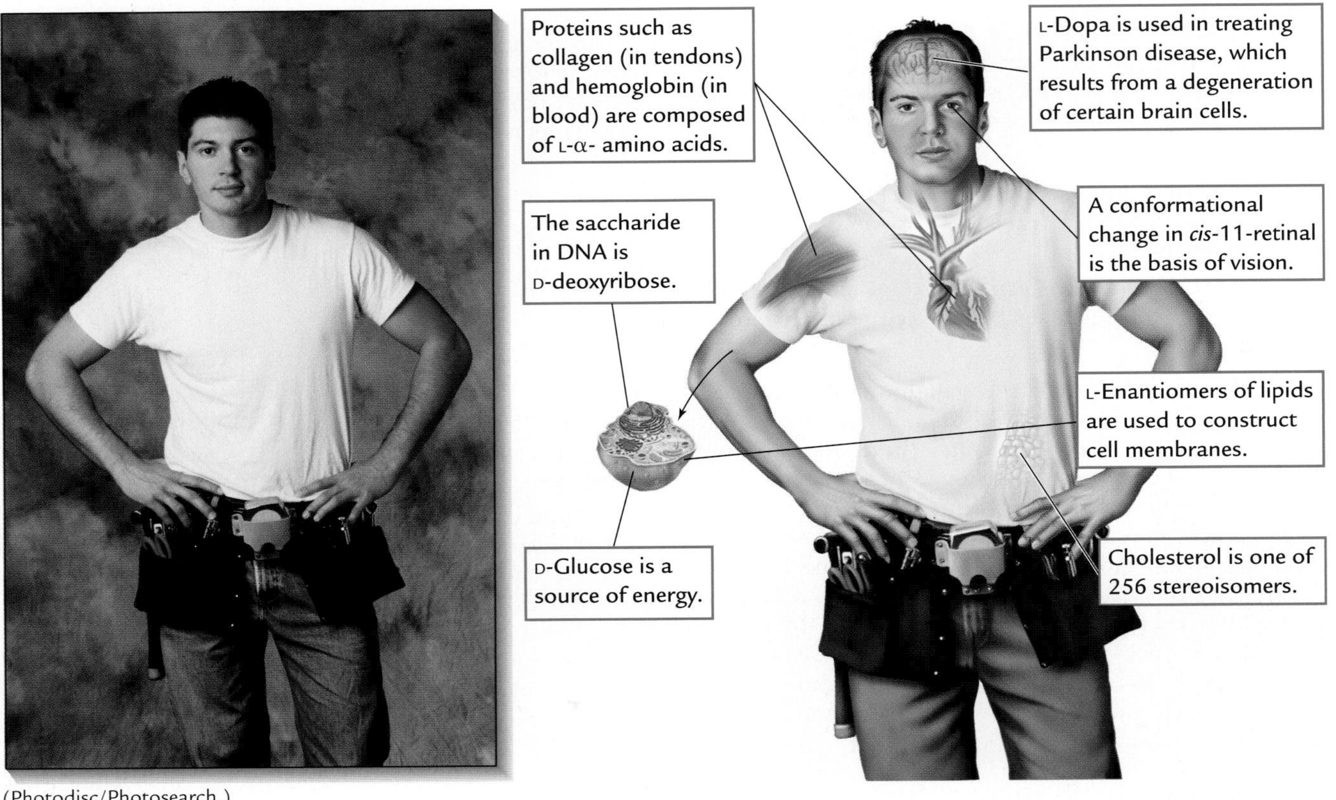
Proteins such as collagen (in tendons) and hemoglobin (in blood) are composed of L-α- amino acids.
L-Dopa is used in treating Parkinson disease, which results from a degeneration of certain brain cells.
The saccharide in DNA is D-deoxyribose.
A conformational change in *cis*-11-retinal is the basis of vision.
L-Enantiomers of lipids are used to construct cell membranes.
D-Glucose is a source of energy.
Cholesterol is one of 256 stereoisomers.
(Photodisc/Photosearch.)

9.6 COMPOUNDS CONTAINING TWO OR MORE TETRAHEDRAL STEREOCENTERS

So far, we have focused on compounds with one tetrahedral stereocenter, but many molecules (especially biologically important ones) contain more than one tetrahedral stereocenter. The number of possible stereoisomers increases as the number of tetrahedral stereocenters increases.

- The maximum number of stereoisomers possible for a compound with n tetrahedral stereocenters is 2^n.

Thus, as many as 4, 8, or 16 stereoisomers are possible when there are two, three, or four tetrahedral stereocenters, respectively.

It is important to recognize the presence of more than two stereocenters in a compound because many biologically important compounds contain many stereocenters. D-Glucose, the most plentiful carbohydrate and an energy source used by most plant and animal cells, contains four tetrahedral stereocenters (Section 10.2). Starch and cellulose, consisting of hundreds of glucose units bonded together (Section 10.6), contain many hundreds of tetrahedral stereocenters. Proteins, consisting of large numbers of combined amino acids, also contain a great many tetrahedral stereocenters. A biologically active molecule containing stereocenters is always one specific stereoisomer out of a large number of possibilities. D-Glucose is one stereoisomer out of a total of 16. Starch, cellulose, and proteins are each found in nature as one of an astronomical number of possible stereoisomers.

H—C=O
H—C*—OH
HO—C*—H
H—C*—OH
H—C*—OH
CH_2OH

D-Glucose

Compounds containing more than one tetrahedral stereocenter are too complex for us to consider in great detail. As we shall see, not all the stereoisomers of compounds with multiple tetrahedral stereocenters are enantiomers, and not all the stereoisomers are optically active. We will limit our detailed exploration to compounds with two tetrahedral stereocenters.

Two Tetrahedral Stereocenters with Different Sets of Substituents

Consider a compound with two tetrahedral stereocenters. The maximum number of four stereoisomers exists when the two tetrahedral stereocenters do not have the same set of four different substituents. In 2,3-pentanediol, for example, one tetrahedral stereocenter has —H, $—CH_3$, —OH, and $—CH(OH)C_2H_5$ and the other has —H, $—C_2H_5$, —OH, and $—CH(OH)CH_3$. There are $2^2 = 4$ stereoisomers of 2,3-pentanediol, differing from one another in the configurations at C2 and C3:

I	II	III	IV
CH_3	CH_3	CH_3	CH_3
H—*—OH	HO—*—H	HO—*—H	H—*—OH
HO—*—H	H—*—OH	HO—*—H	H—*—OH
C_2H_5	C_2H_5	C_2H_5	C_2H_5

I and II: Pair of enantiomers; III and IV: Pair of enantiomers

Diastereomers

Note that one of the vertical bonds in each of these structural drawings does not conform to the convention described in Section 9.2. The bonds from C1 to C2 and from C3 to C4 extend behind the plane of the paper, but the bond between C2 and C3 is in the plane of the paper.

The four stereoisomers consist of two pairs of enantiomers: I and II are one pair, and III and IV are the other pair. Neither I nor II is an enantiomer of III or IV, and none of these compounds are superimposable on each other—all four are different stereoisomers. The term **diastereomer** (Section 9.1) describes the relation of either compound from one enantomeric pair to either compound from the other enantiomeric pair. Enantiomers I and II are diastereomers of III and IV, and vice versa. Diastereomers are stereoisomers that are not enantiomers; that is, they are neither superimposable nor mirror images. (Recall that the term diastereomer also describes the relation between a pair of cis and trans stereoisomers; Sections 3.8 and 4.5.).

All diastereomers (except meso compounds, which are described in the next subsection, and cis and trans compounds) are optically active. However, unlike enantiomers, diastereomers have specific rotations that usually differ in their numerical value and may have the same or opposite direction of rotation. If the specific rotations are the same, it is entirely by chance. Even the direction of optical rotation can be the same or different for two diastereomers. Another difference from enantiomers is that diastereomers differ in other physical properties, such as melting and boiling points and solubility. Usually, diastereomers also differ in their chemical properties.

The artificial sweetener aspartame (NutraSweet), *N*-L-aspartyl-L-phenylalanine methyl ester, is an example of chiral recognition and of the differences in properties seen in enantiomers and diastereomers. This compound contains two tetrahedral stereocenters that do not have the same set of four different substituents. Of the four possible stereoisomers, only the L,L-stereoisomer (the one in which both tetrahedral stereocenters have the L configuration) is sweet—almost 200 times as sweet as sucrose (ordinary table sugar). The L,D-stereoisomer is not sweet; in fact, it is bitter tasting.

CH_2COOH
H—*—NH_2
C=O
NH
H—*—$COOCH_3$
$CH_2C_6H_5$

Aspartame

Two Tetrahedral Stereocenters with the Same Set of Substituents

Let's consider a compound with two tetrahedral stereocenters that have the same set of four different substituents. An example is 2,3-butanediol. Each stereocenter has —H, $—CH_3$, —OH, and $—CH(OH)CH_3$. We can refer to the structural drawings of the four stereoisomers of 2,3-pentanediol on page 297 to draw the 2,3-butanediols. All that is needed is to replace the ethyl groups with methyls:

CH_3; H—*—OH; HO—*—H; CH_3 (V)

CH_3; HO—*—H; H—*—OH; CH_3 (VI)

CH_3; HO—*—H; HO—*—H; CH_3 (VII) = CH_3; H—*—OH; H—*—OH; CH_3 (VIII)

V VI: Pair of enantiomers

VII: Meso

Diastereomers

Note the equal sign between drawings VII and VIII. These drawings represent the same compound, not different compounds. Drawings VII and VIII represent the same compound because a 180° rotation of either structure makes it superimposable with the other structure.

When the two tetrahedral stereocenters of a compound have the same set of four different substituents, there are only three, not four, stereoisomers. There is a pair of enantiomers (V and VI) and a lone, third stereoisomer (VII = VIII) called a **meso compound.** The meso compound is not an enantiomer of V or VI. It is neither a mirror image nor superimposable on V or VI. Therefore, V and VI are diastereomers of the meso compound, and vice versa. The meso stereoisomer can be shown as either VII or VIII but not both. If you draw both VII and VIII, you must place an equal sign between them. Meso compounds are achiral and optically inactive even though they contain two tetrahedral stereocenters.

Example 9.5 Determining the stereoisomers of compounds with two tetrahedral stereocenters

Draw structural formulas of all stereoisomers of 2-methyl-1,3-butanediol:

$HOCH_2CH(CH_3)CH(OH)CH_3$

Which stereoisomers are optically active? Which are enantiomers and which are diastereomers?

Solution

Because this compound has two tetrahedral stereocenters with different sets of four different substituents, four stereoisomers are possible:

1: CH_3; H—*—CH_2OH; HO—*—H; CH_3

2: CH_3; $HOCH_2$—*—H; H—*—OH; CH_3

3: CH_3; $HOCH_2$—*—H; HO—*—H; CH_3

4: CH_3; H—*—CH_2OH; H—*—OH; CH_3

All four compounds are optically active. Each stereoisomer is simultaneously both an enantiomer and a diastereomer. Stereoisomers 1 and 2 are enantiomers of each other, and each is a diastereomer of both 3 and 4. Stereoisomers 3 and 4 are enantiomers of each other, and each is a diastereomer of both 1 and 2.

Problem 9.5 Draw structural formulas of all stereoisomers of tartaric acid (2,3-dihydroxy-1,4-butanedioic acid):

```
         HO  OH
          |   |
HOOC—CHCH—COOH
```

Which stereoisomers are optically active? Which are enantiomers and which are diastereomers?

Concept checklist

- ✓ There are two pairs of enantiomers when the two tetrahedral stereocenters do not have the same set of four different substituents.
- ✓ There is one pair of enantiomers and a meso compound when the two tetrahedral stereocenters have the same set of four different substituents.

9.7 CYCLIC COMPOUNDS CONTAINING TETRAHEDRAL STEREOCENTERS

Cyclic compounds that have tetrahedral stereocenters also can exist as enantiomers. The basic criterion remains the same. Enantiomers are possible when a compound is nonsuperimposable on its mirror image.

A ring carbon is a tetrahedral stereocenter if

- the two nonring substituents are different and
- the ring is not symmetrical with respect to that carbon.

Consider methylcyclohexane (Figure 9.13a). It fulfills the first requirement: it has two different nonring substituents attached to C1. However, the second requirement is not fulfilled: the two halves of the ring on each side of C1 are the same ($CH_2CH_2CH_2$). There is no tetrahedral stereocenter, and the

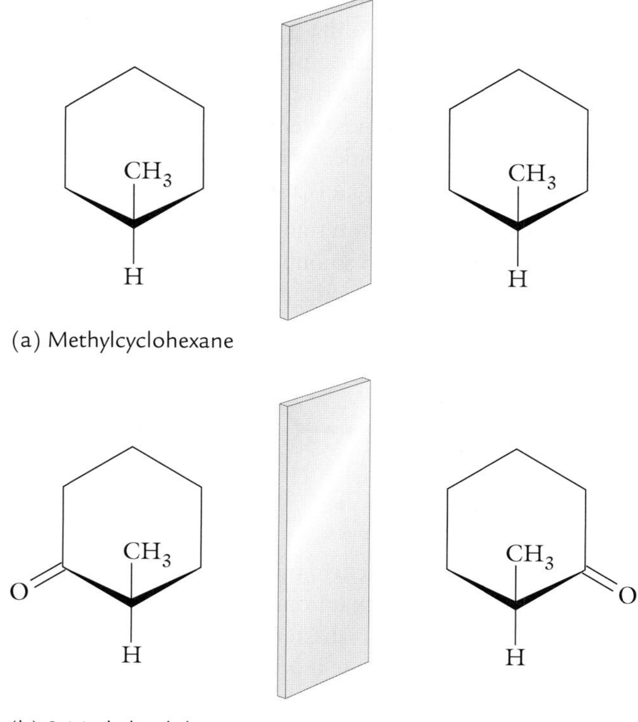

Figure 9.13 Mirror-image molecules of (a) methylcyclohexane (superimposable) and (b) 2-methylcyclohexanone (nonsuperimposable).

mirror-image molecules are superimposable. Thus, methylcyclohexane exists as a single compound.

2-Methylcyclohexanone, however, exists as a pair of enantiomers (Figure 9.13b). Carbon 2 is a tetrahedral stereocenter: the two nonring substituents are different, and the two halves of the ring on each side of C2 are different ($COCH_2CH_2$ versus $CH_2CH_2CH_2$). The mirror-image molecules are nonsuperimposable, and 2-methylcyclohexanone exists as a pair of enantiomers.

Example 9.6 Determining the stereoisomers of cyclic compounds

Draw the stereoisomers of (a) 1-chloro-2-methylcyclohexane; (b) 1,2-dimethylcyclohexane. Which stereoisomers are optically active? Which are enantiomers and which are diastereomers?

Solution

Both 1-chloro-2-methylcyclohexane and 1,2-dimethylcyclohexane possess two tetrahedral stereocenters. Each stereocenter has two different nonring substituents, and the two halves of the ring are different at both C1 and C2. The situation is similar to that for acyclic compounds with two tetrahedral stereocenters.

(a) Because the two tetrahedral stereocenters of 1-chloro-2-methylcyclohexane do not have the same set of four different substituents, the maximum number of four stereoisomers is possible. There are a trans pair of enantiomers and a cis pair of enantiomers:

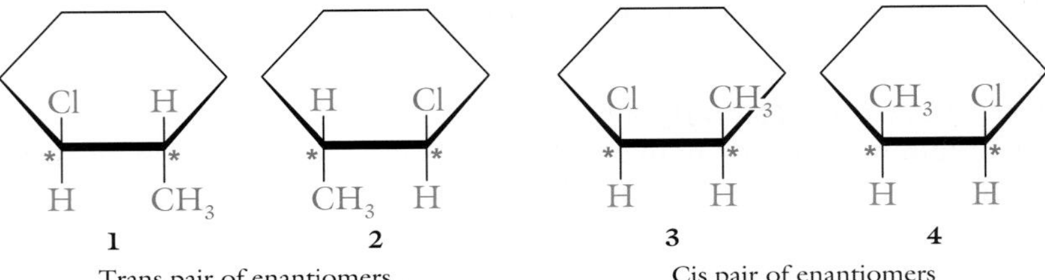

Trans pair of enantiomers Cis pair of enantiomers

The trans and cis designations refer to the placements of substituents on the opposite or same sides of the ring, respectively (Section 3.8).

All four compounds are optically active. Each stereoisomer is simultaneously both an enantiomer and a diastereomer. Stereoisomers 1 and 2 are enantiomers of each other, and each is a diastereomer of both 3 and 4. Stereoisomers 3 and 4 are enantiomers of each other, and each is a diastereomer of both 1 and 2.

(b) 1,2-Dimethylcyclohexane has two tetrahedral stereocenters with the same set of four different substituents, and so only three stereoisomers are possible. There are a trans pair of enantiomers and a cis meso compound:

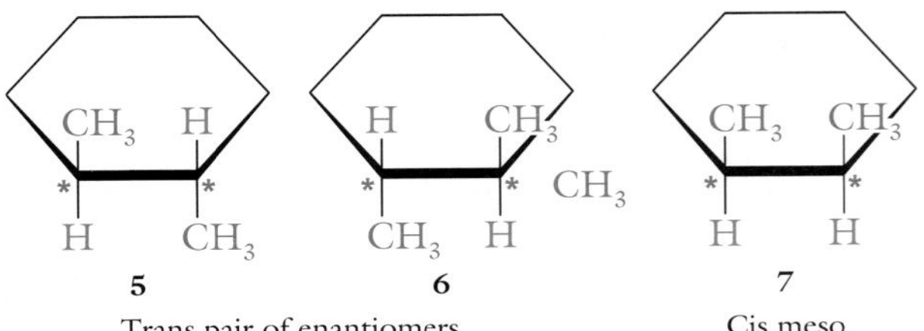

Trans pair of enantiomers Cis meso

The trans compounds are optically active, but the meso compound is optically inactive. Stereoisomers 5 and 6 are enantiomers of each other, and each is a diastereomer of 7. Stereoisomer 7 is a diastereomer of both 1 and 2. The cis meso compound is easy to spot because it is superimposable on its mirror image. A pair of cis enantiomers exists for 1-chloro-2-methylcyclohexane, but only one cis compound exists for 1,2-dimethylcyclohexane because the nonring substituents at the tetrahedral stereocenters are not the same for 1-chloro-2-methylcyclohexane, whereas they are the same for 1,2-dimethylcyclohexane.

Problem 9.6 Draw the stereoisomers of (a) chlorocyclohexane; (b) 1-chloro-3-methylcyclohexane. Which stereoisomers are optically active? Which are enantiomers and which are diastereomers?

Cyclic compounds with tetrahedral stereoisomers are common in nature. Cholesterol is such a compound. It has eight tetrahedral stereocenters, and there are 2^8, or 256, possible stereoisomers. Only one stereoisomer (the one called cholesterol) exists in nature. Cholesterol is an important component of cell membranes and the starting material for the biosynthesis of many important compounds, including vitamin D and the sex hormones progesterone and testosterone (Sections 11.7 and 11.9).

›› Cholesterol, progesterone, and testosterone are lipids and the subjects of Chapter 11.

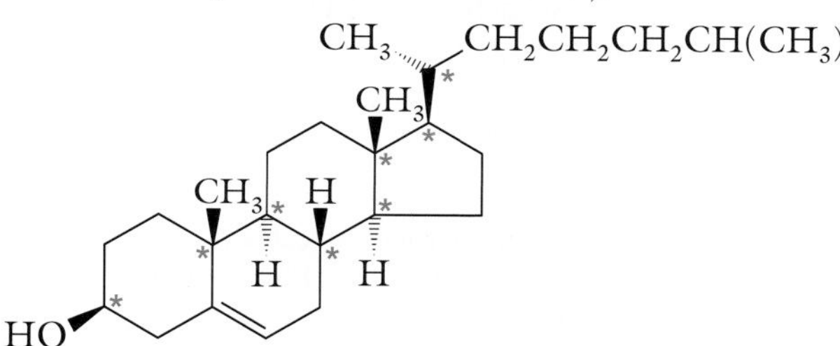

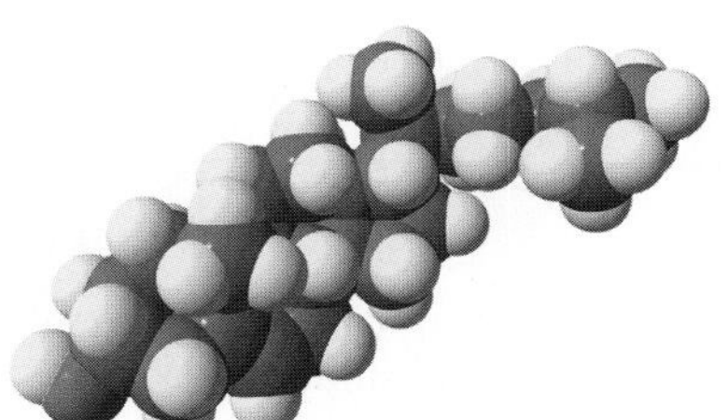

Cholesterol

Summary

Isomers

- Isomers are different compounds that have the same molecular formula.
- There are two types of isomers: constitutional isomers and stereoisomers.
- Constitutional (structural) isomers differ in connectivity—that is, in the order of attachment of atoms to one another.
- Stereoisomers have the same connectivity but differ in configuration—that is, the relative orientations in space of the atoms of a compound, independent of changes due to rotation about single bonds.
- Stereoisomers that are mirror images of each other are called enantiomers.
- Diastereomers are stereoisomers (including geometrical isomers) that are not enantiomers; that is, they are not mirror images of each other.

Enantiomers

- A pair of enantiomers—nonsuperimposable mirror-image compounds—exists when there is a tetrahedral stereocenter, a carbon with four different substituents.
- Such compounds are also called chiral compounds.
- Enantiomers can be differentiated by a naming system that uses the prefixes D- and L-.
- Chiral compounds are optically active: they rotate plane-polarized light, as measured by a polarimeter.
- Each enantiomer in a pair of enantiomers rotates plane-polarized light the same number of degrees but in opposite directions.
- A racemic mixture—an equimolar mixture of enantiomers—is optically inactive.
- Enantiomers have the same chemical reactivity when reacting with achiral reactants but often show very large differences in reactivity when reacting with chiral reactants or in reactions catalyzed by chiral enzymes, a phenomenon called chiral recognition or chiral discrimination.

Compounds Containing Two or More Tetrahedral Stereocenters

- The maximum number of stereoisomers possible for a compound with n tetrahedral stereocenters is 2^n.
- For a compound with two tetrahedral stereocenters that do not have the same four different substituents, two pairs of enantiomers are possible.
- Either compound from one pair of enantiomers is a diastereomer of either compound of the other pair of enantiomers.
- The maximum number of stereoisomers is not observed when the two tetrahedral stereocenters have the same set of four different substituents. In this case, there are three stereoisomers—a pair of enantiomers and a meso compound.
- Either compound of the pair of enantiomers is a diastereomer of the meso compound and vice versa.
- Meso compounds are achiral and not optically active.

Cyclic Compounds Containing Tetrahedral Stereocenters

- Enantiomers are also possible for cyclic compounds having one or more tetrahedral stereocenters.
- A ring carbon is a tetrahedral stereocenter if the two nonring substituents are different and the ring is not symmetrical with respect to that carbon.

Key Words

achiral, p. 281
chiral, p. 281
chiral discrimination (chiral recognition), p. 292
diastereomer, pp. 280, 297
enantiomer, p. 280
Fischer projection, p. 282
meso compound, p. 298
nonsuperimposable, p. 281
optical activity, p. 288
racemic mixture, p. 292
specific rotation, p. 291
stereocenter, p. 280
stereoisomer, p. 280
superimposable, p. 281
tetrahedral stereocenter, p. 280
trigonal stereocenter, p. 280
wedge bond, p. 282

Exercises

Review of Isomerism

9.1 Draw the structural formulas of constitutional isomers of C_4H_{10}.

9.2 Draw the structural formulas of constitutional isomers of C_5H_{12}.

9.3 Draw the structural formulas of ethers that are constitutional isomers of butanol.

9.4 Draw the structural formulas of ketones that are constitutional isomers of pentanal.

9.5 Draw the structural formula of the diastereomer of *cis*-2-butene.

9.6 Draw the structural formula of the diastereomer of *cis*-1,2-dimethylcyclobutane.

Enantiomers

9.7 Which of the following objects are chiral and which achiral? (a) A person; (b) an automobile; (c) a basketball without any design or name imprinted on it; (d) a basketball with the name MICHAEL JORDAN imprinted on it.

9.8 Which of the following objects are chiral and which achiral? (a) A clear glass coffee mug; (b) a coffee mug with your name imprinted on it; (c) a dog; (d) a television set.

9.9 Which of the following structures can exist as a pair of enantiomers and which cannot? Draw structural formulas of enantiomers, placing an asterisk next to each tetrahedral stereocenter.

(a) $CH_3CH_2CH(CH_3)CH_2CH_3$

(b) $CH_3CH(CH_3)CH_2CH_2CH_3$

(c) $HO—CH_2CH(NH_2)—CH_3$

(d) 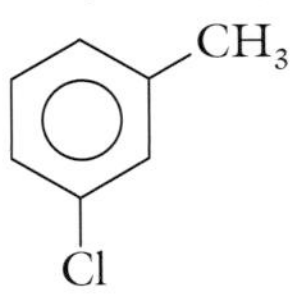

(e) 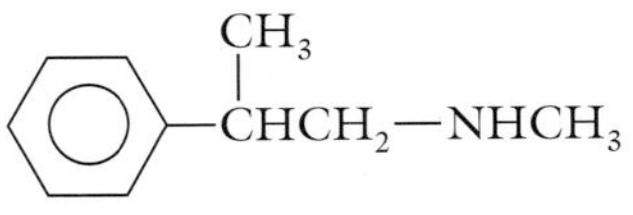

Methamphetamine

9.10 Which of the following structures can exist as a pair of enantiomers and which cannot? Draw structural formulas of enantiomers, placing an asterisk next to each tetrahedral stereocenter.

(a) $CH_3—CH(OH)—CH_2CH_3$

(b) $CH_3O—CH(CH_3)—CH_3$

(c) $HO—C(=O)—CH(CH_3)CH_2CH_3$

(d) $CH_2═CH—CH_2CH_2CH_3$

(e) 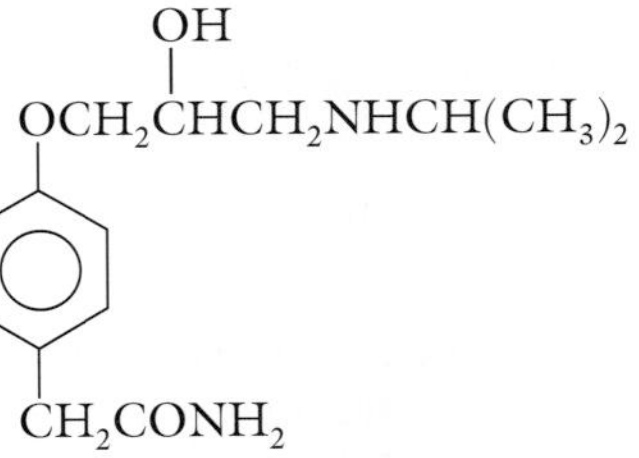

Metoprolol (Lopressor)

Interpreting Structural Formulas of Enantiomers

9.11 For each of the following pairs of structural formulas, indicate whether the pair represents (1) the same compound or (2) different compounds that are constitutional isomers or (3) different compounds that are enantiomers or (4) different compounds that are diastereomers or (5) different compounds that are not isomers.

(a) 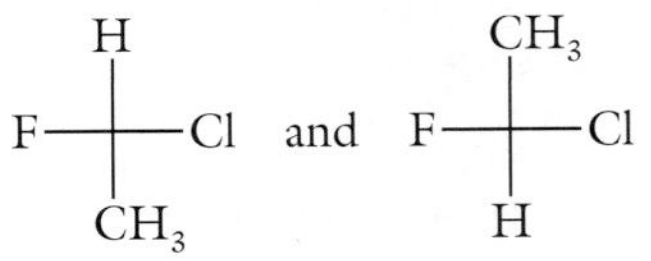

(b) 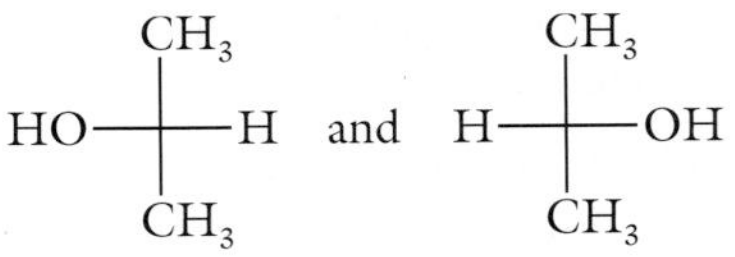

(c)

$$\begin{array}{c} HC{=}O \\ H{-}\!\!\!+\!\!\!{-}OH \\ CH_2OH \end{array} \quad \text{and} \quad \begin{array}{c} HC{=}O \\ H{-}\!\!\!+\!\!\!{-}OH \\ CH_2OH \end{array}$$

(d)

$$\begin{array}{c} COOH \\ CH_3{\blacktriangleright}C{\blacktriangleleft}H \\ OH \end{array} \quad \text{and} \quad \begin{array}{c} OH \\ CH_3{\blacktriangleright}C{\blacktriangleleft}H \\ COOH \end{array}$$

(e)

$$\begin{array}{c} Cl \\ H{-}\!\!\!+\!\!\!{-}CH_3 \\ Br \end{array} \quad \text{and} \quad \begin{array}{c} Cl \\ H{-}\!\!\!+\!\!\!{-}CH_3 \\ Br \end{array}$$

9.12 For each of the following pairs of structural formulas, indicate whether the pair represents (1) the same compound or (2) different compounds that are constitutional isomers or (3) different compounds that are enantiomers or (4) different compounds that are diasteromers or (5) different compounds that are not isomers.

(a)

$$\begin{array}{c} H \\ Cl{-}\!\!\!+\!\!\!{-}Cl \\ CH_3 \end{array} \quad \text{and} \quad \begin{array}{c} H \\ Cl{-}\!\!\!+\!\!\!{-}Cl \\ CH_3 \end{array}$$

(b)

$$\begin{array}{c} H \\ F{-}\!\!\!+\!\!\!{-}Cl \\ CH_3 \end{array} \quad \text{and} \quad \begin{array}{c} H \\ Cl{-}\!\!\!+\!\!\!{-}F \\ CH_3 \end{array}$$

(c)

$$\begin{array}{c} HC{=}O \\ H{-}\!\!\!+\!\!\!{-}OH \\ CH_2OH \end{array} \quad \text{and} \quad \begin{array}{c} CH_2OH \\ HO{-}\!\!\!+\!\!\!{-}H \\ HC{=}O \end{array}$$

(d)

$$\begin{array}{c} HC{=}O \\ H{-}\!\!\!+\!\!\!{-}OH \\ CH_3 \end{array} \quad \text{and} \quad \begin{array}{c} HC{=}O \\ HO{-}\!\!\!+\!\!\!{-}H \\ CH_3 \end{array}$$

(e)

$$\begin{array}{c} COOH \\ CH_3{\blacktriangleright}C{\blacktriangleleft}H \\ OH \end{array} \quad \text{and} \quad \begin{array}{c} OH \\ H{\blacktriangleright}C{\blacktriangleleft}CH_3 \\ COOH \end{array}$$

Nomenclature of Enantiomers

9.13 Identify each of the following stereoisomers as the D- or L-enantiomer. Check to make sure that the drawing is correctly oriented for making its assignment as D or L. If necessary, rotate the drawing by 180°.

$$\text{(a)} \begin{array}{c} HC{=}O \\ H{-}\!\!\!+\!\!\!{-}OH \\ CH_2OH \end{array} \qquad \text{(b)} \begin{array}{c} CH_2OH \\ H{-}\!\!\!+\!\!\!{-}NH_2 \\ HC{=}O \end{array}$$

$$\text{(c)} \begin{array}{c} HC{=}O \\ H{-}\!\!\!+\!\!\!{-}OH \\ CH_3 \end{array} \qquad \text{(d)} \begin{array}{c} CH_3 \\ H{\blacktriangleright}C{\blacktriangleleft}OH \\ COOH \end{array}$$

9.14 Identify each of the following stereoisomers as the D- or L-enantiomer. Check to make sure that the drawing is correctly oriented for making its assignment as D or L. If necessary, rotate the drawing by 180°.

$$\text{(a)} \begin{array}{c} CH_2OH \\ HO{-}\!\!\!+\!\!\!{-}H \\ HC{=}O \end{array} \qquad \text{(b)} \begin{array}{c} CH_2OH \\ H_2N{-}\!\!\!+\!\!\!{-}H \\ HC{=}O \end{array}$$

$$\text{(c)} \begin{array}{c} COOH \\ HO{\blacktriangleright}C{\blacktriangleleft}H \\ CH_3 \end{array} \qquad \text{(d)} \begin{array}{c} CH_3 \\ H{-}\!\!\!+\!\!\!{-}NH_2 \\ HC{=}O \end{array}$$

Properties of Enantiomers

9.15 Which of the following compounds rotate plane-polarized light? If so, in which direction? (a) Ethanol; (b) D-glucose; (c) (+)-phenylalanine; (d) racemic glutamic acid.

9.16 Which of the following compounds rotate plane-polarized light? If so, in which direction? (a) Pentane; (b) (−)-glucose; (c) L-phenylalanine; (d) racemic lactic acid.

9.17 D-Glutamic acid has a specific rotation of −31.5°. What is the specific rotation of L-glutamic acid?

9.18 Is D or L the correct designation for the enantiomer of lactic acid with $[\alpha] = -13.5°$?

9.19 The following compound is (−)-alanine. Draw (+)-alanine.

$$\begin{array}{c} COOH \\ H{-}\!\!\!+\!\!\!{-}NH_2 \\ CH_3 \end{array}$$

9.20 The following compound is L-valine. Draw D-valine.

$$\begin{array}{c} COOH \\ H_2N{-}\!\!\!+\!\!\!{-}H \\ CH(CH_3)_2 \end{array}$$

9.21 The water solubility of L-alanine is 127 g/L at 25°C. What is the water solubility of D-alanine?

9.22 The density of L-valine is 1.316 g/mL at 25°C. What is the density of D-valine at 25°C?

9.23 A solution of D-fructose (4.0 g/100 mL) gives an optical rotation of −3.6° when measured in a polarimeter sample cell 10.0 cm long. Calculate the specific rotation of D-fructose.

9.24 A solution of (+)-menthol (10.0 g/100 mL) gives an optical rotation of +2.6° when measured in a polarimeter sample cell 5.0 cm long. Calculate the specific rotation of (−)-menthol.

9.25 The specific rotation of (+)-penicillin V is +223°. What will be the observed rotation when using a solution concentration of 60.0 g per 1000 mL of solution and a polarimeter sample cell of 15.0-cm length?

9.26 The specific rotation of (−)-morphine is −132°. What will be the observed rotation when using a solution concentration of 10.0 g per 250 mL of solution and a polarimeter sample cell of 20.0-cm length?

Compounds Containing Two or More Tetrahedral Stereocenters

9.27 Use an asterisk to indicate each tetrahedral stereocenter (if any) in each of the following structures. Draw all possible stereoisomers for each structure. Which stereoisomers are optically active? Which stereoisomers are enantiomers and which are diastereomers?

(a) $CH_3CH_2—CH(CH_3)—CH(CH_2CH_3)—CH_2CH_3$

(b) $CH_3CH_2—CH(CH_3)—CH(CH_3)—CH_2CH_3$

(c) $CH_3CH_2—CH(CH_3)—CH(OH)—CH_2CH_3$

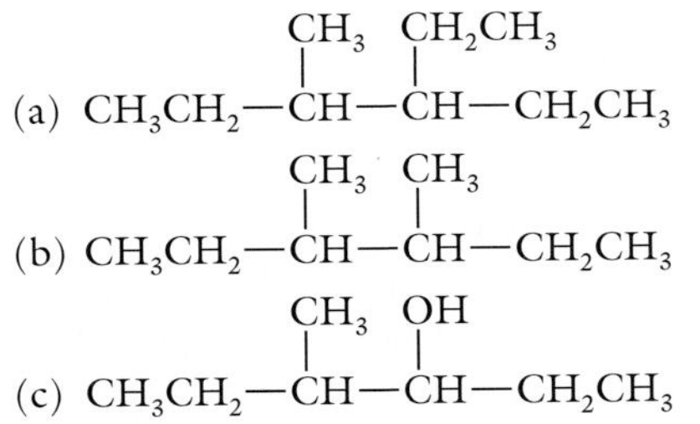

9.28 Use an asterisk to indicate each tetrahedral stereocenter (if any) in the following structures. Draw all possible stereoisomers for each structure. Which stereoisomers are optically active? Which stereoisomers are enantiomers and which are diastereomers?

(a) $HOOC—CH(CH_3)—CH(COOH)—COOH$

(b) $HOOC—CH(OH)—CH(OH)—COOH$

(c) $HOOC—CH(CH_3)—CH(OH)—COOH$

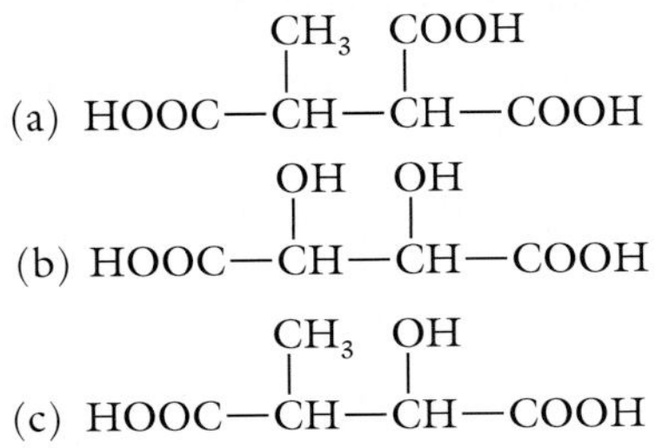

9.29 There are four stereoisomers (1, 2, 3, and 4) of 2-methyl-1,3-butanediol. The structures of two of the stereoisomers are

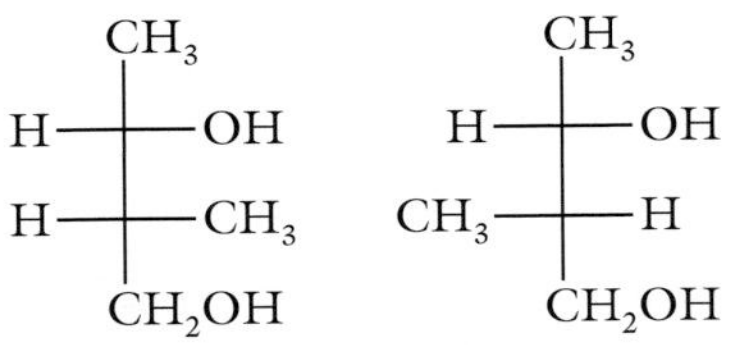

1: Fischer projection — top CH_3; upper carbon: H (left), OH (right); lower carbon: H (left), CH_3 (right); bottom CH_2OH

2: Fischer projection — top CH_3; upper carbon: H (left), OH (right); lower carbon: CH_3 (left), H (right); bottom CH_2OH

Stereoisomer 1 has a specific rotation of + 15° and a boiling point of 180–182°C. Stereoisomer 2 has a specific rotation of + 26° and a boiling point of 163–165°C. Draw the structures, and give the boiling points and specific rotations of stereoisomers 3 and 4.

9.30 There are three stereoisomers (1, 2, and 3) of tartaric acid. The structures of two of the stereoisomers are

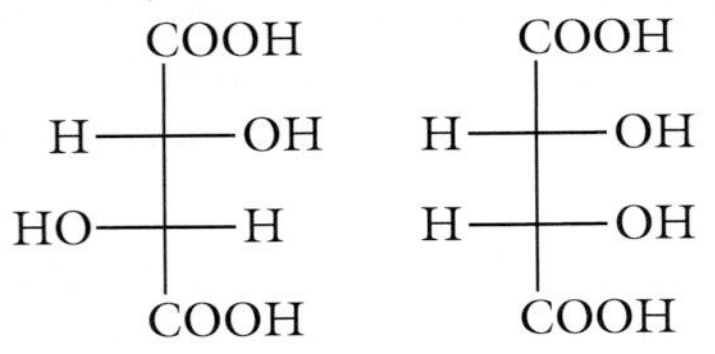

1: Fischer projection — top COOH; upper carbon: H (left), OH (right); lower carbon: HO (left), H (right); bottom COOH

2: Fischer projection — top COOH; upper carbon: H (left), OH (right); lower carbon: H (left), OH (right); bottom COOH

Stereoisomer 1 has a specific rotation of + 12.7° and a melting point of 172–174°C. Stereoisomer 2 is not optically active and has a melting point of 146–148°C. Draw the structure, and give the melting point and specific rotation of stereoisomer 3. Why is stereoisomer 2 not optically active?

9.31 Use an asterisk to indicate each tetrahedral stereocenter (if any) in each of the following structures. What is the maximum number of stereoisomers possible for each of these structures?

(a) $HC{=}O—HC(OH)—HC(OH)—HC(OH)—HC(OH)—H_2C—OH$ (vertical chain: HC=O; HC—OH; HC—OH; HC—OH; HC—OH; H_2C—OH)

(b) $H_2N—CH(CH_3)—C(=O)—N(H)—CH(CH_2OH)—C(=O)—N(H)—CH(CH_2COOH)—C(=O)—OH$

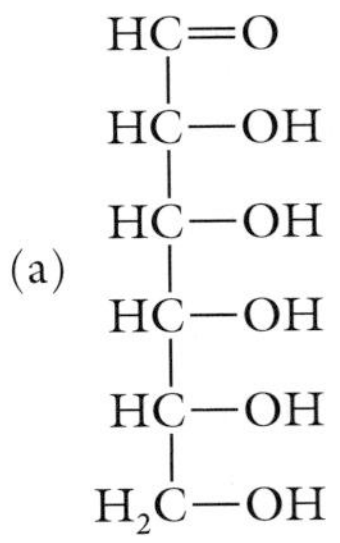

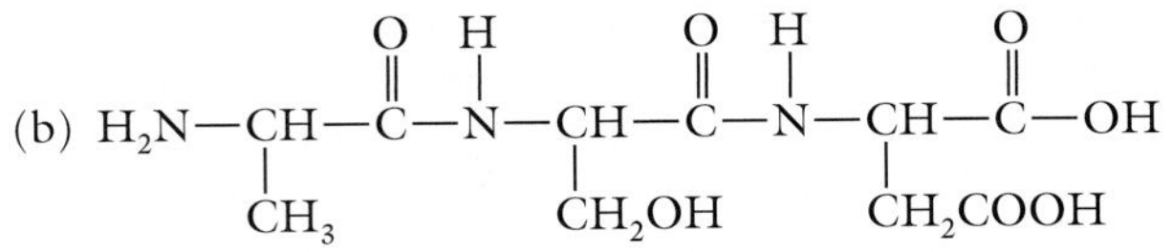

9.32 Use an asterisk to indicate each tetrahedral stereocenter (if any) in each of the following structures. What is the maximum number of stereoisomers possible for each of these structures?

(a) vertical chain: H_2C—OH; C=O; HC—OH; HC—OH; HC—OH; H_2C—OH

(b) $H_2N—CH(CH(CH_3)_2)—C(=O)—N(H)—CH(CH_2COOH)—C(=O)—N(H)—CH_2—C(=O)—OH$

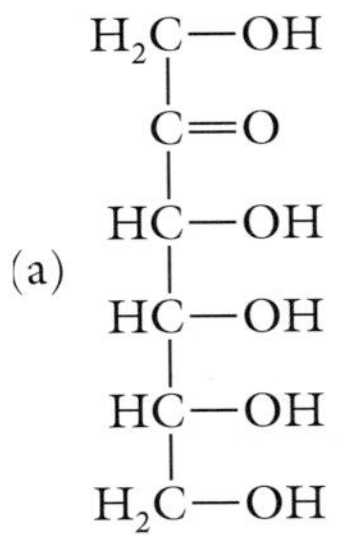

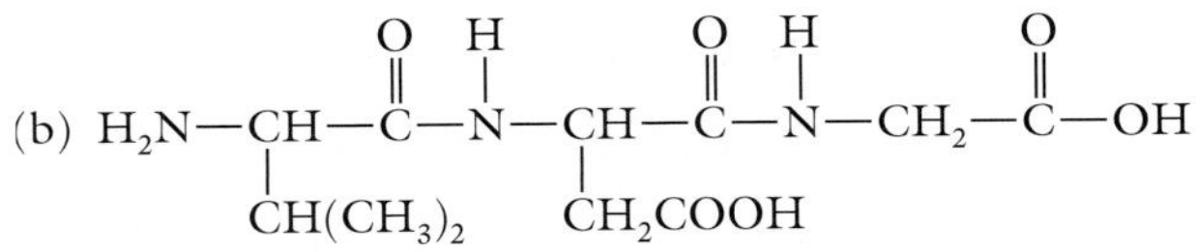

Cyclic Compounds Containing Tetrahedral Stereocenters

9.33 Draw the stereoisomers (if any) of each of the following compounds: (a) 3-methylcyclohexanone; (b) 1,1-dimethylcyclopentane; (c) 2-methyl-1,3-cyclohexanedione; (d) 1-hydroxy-2-methylcyclohexane; (e) 1,2-dihydroxycyclohexane. Which stereoisomers are optically active? Which stereoisomers are enantiomers and which are diastereomers?

9.34 Draw the stereoisomers (if any) of each of the following compounds: (a) chlorocyclopentane; (b) 1,1-dichloro-2-chlorocyclopentane; (c) 1,1,3,3-tetrachloro-2-methylcyclopentane; (d) 1-chloro-2-methylcyclopentane; (e) 1,2-dichlorocyclopentane. Which stereoisomers are optically active? Which stereoisomers are enantiomers and which are diastereomers?

9.35 Use an asterisk to indicate each tetrahedral stereocenter (if any) in each of the following structures. How many stereoisomers are possible for each of these structures?

(a) 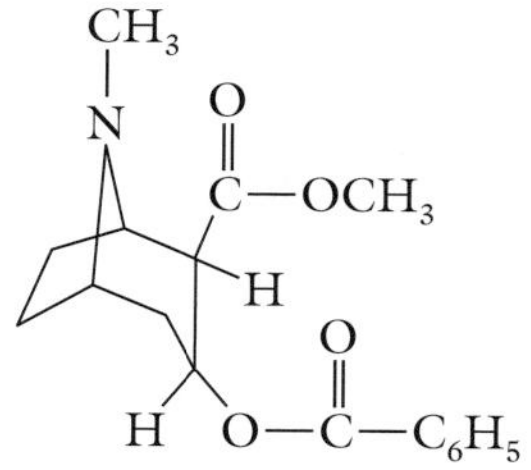

Cocaine

(b) 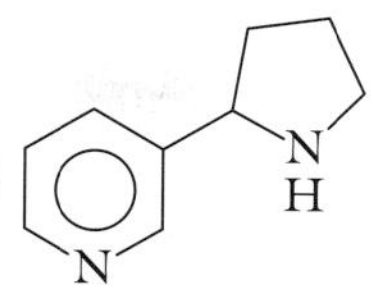

Nicotine

(c) 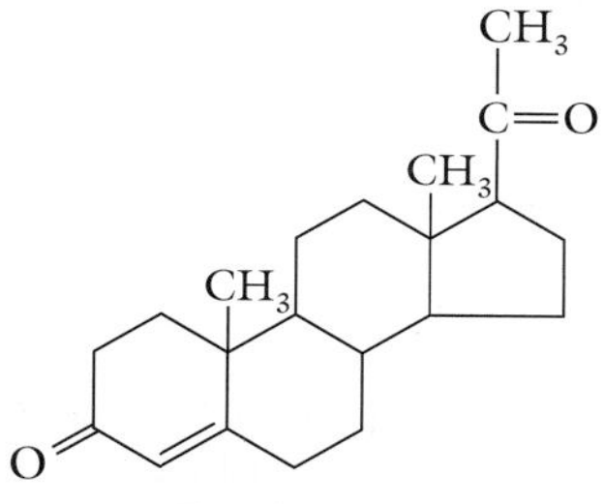

Progesterone

9.36 Use an asterisk to indicate each tetrahedral stereocenter (if any) in each of the following structures. How many stereoisomers are possible for each of these structures?

(a) 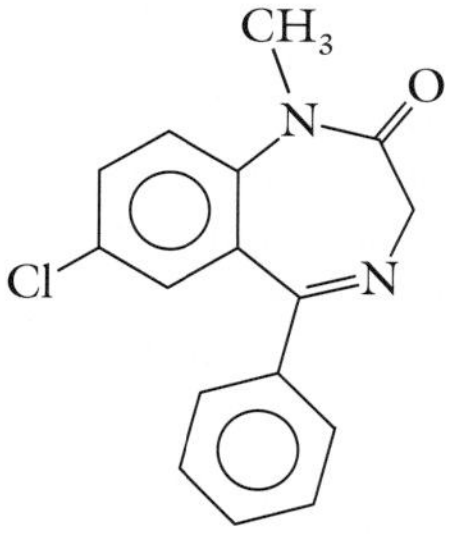

Diazepam (Valium)

(b) 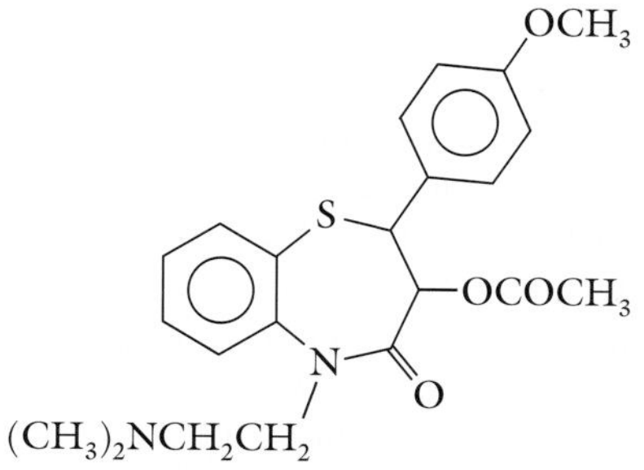

Diltiazem (Cardizem)

(c) 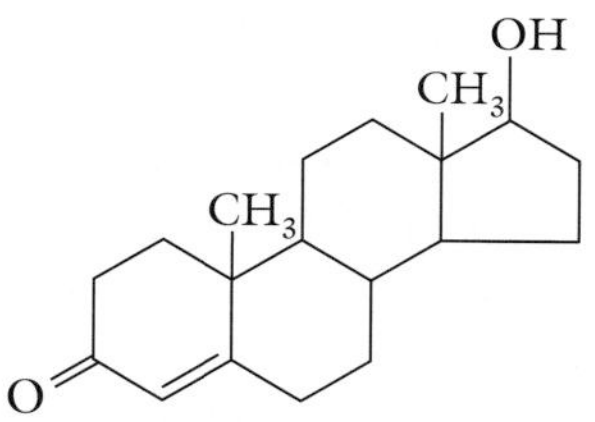

Testosterone

Unclassified Exercises

9.37 For each of the following pairs of structural formulas, indicate whether the pair represents (1) the same compound or (2) different compounds that are constitutional isomers or (3) different compounds that are enantiomers or (4) different compounds that are diasteromers or (5) different compounds that are not isomers.

(a) $CH_3CH_2CH_2OCH(CH_3)_2$ and $CH_3CH_2CH_2CH_2CH_2OCH_3$

(b) 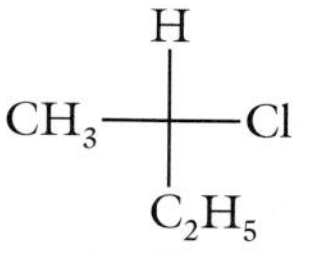 and 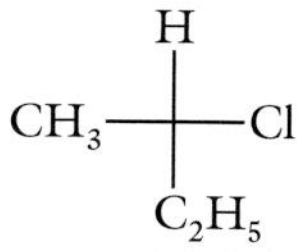

(c) 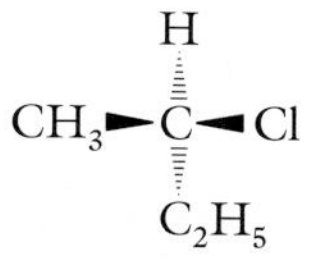 and 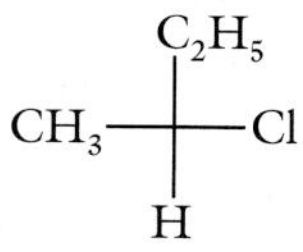

(d) $CH_3CH_2CH_2OCH(CH_3)_2$ and $CH_3CH_2CH_2CH_2CH_2CH_2OH$

(e) $CH_3CH_2CH_2OCH(CH_3)_2$ and 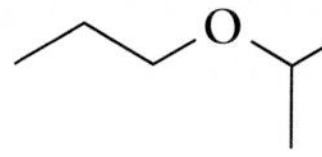

(f) 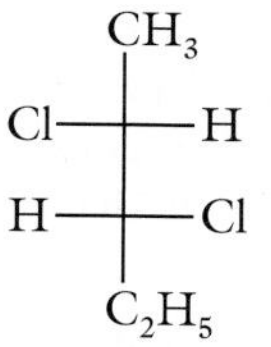 and 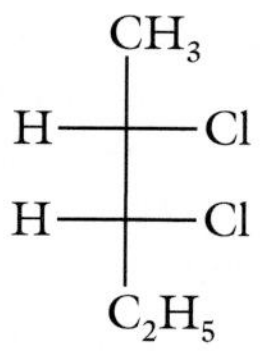

(g) 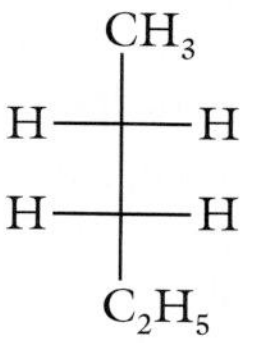 and 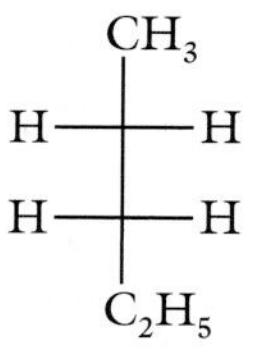

(h) 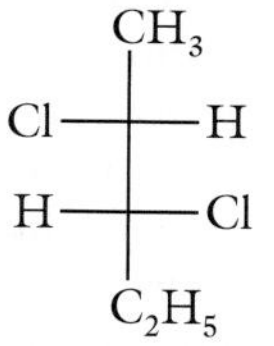 and 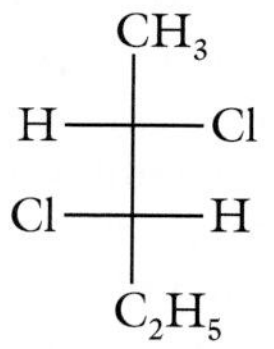

(i) 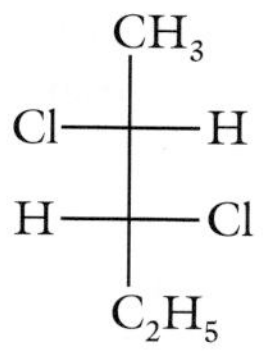 and 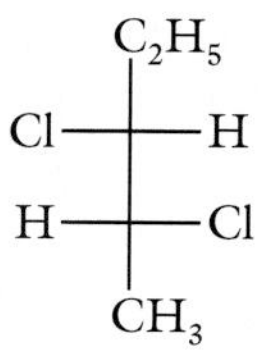

(j) 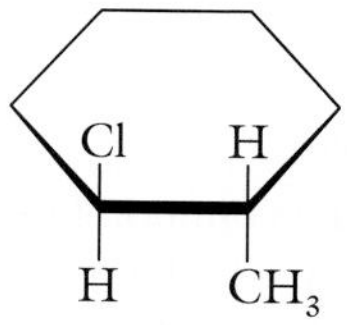 and 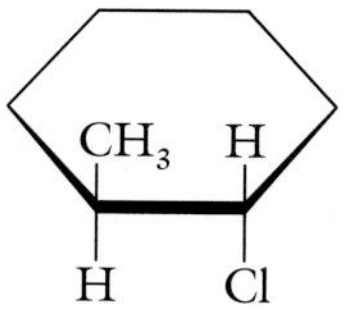

(k) 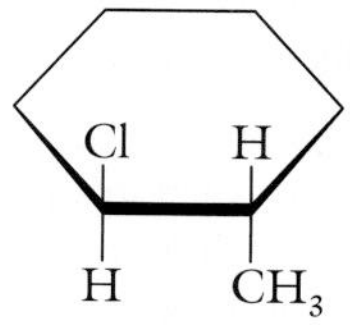 and 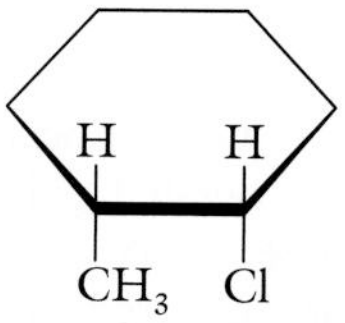

(l) 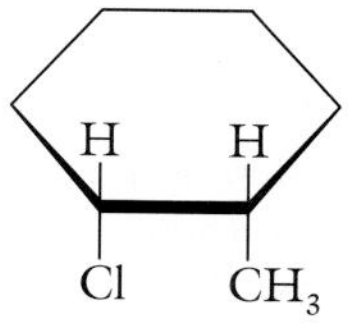 and 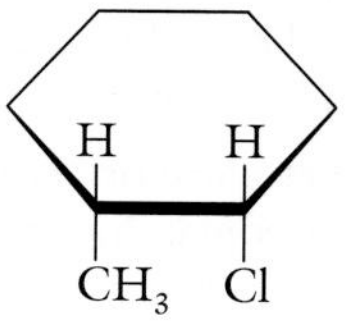

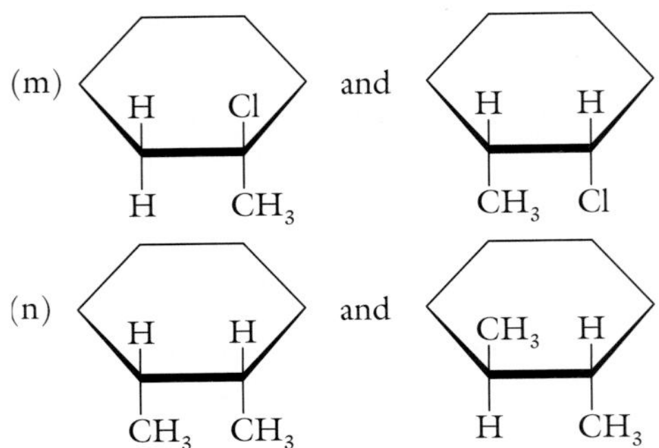

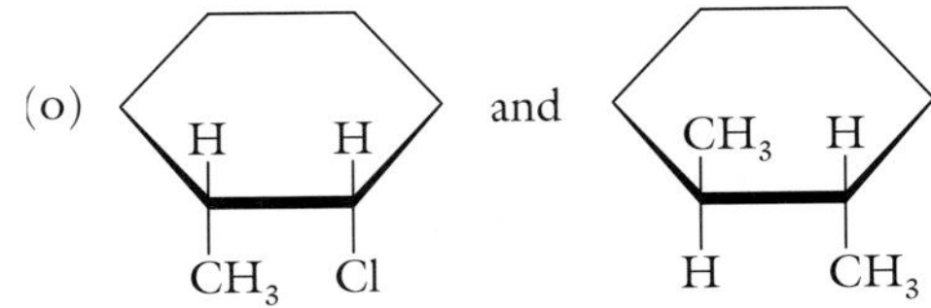

9.38 What is chiral recognition in biological systems?

9.39 Draw the four stereoisomers of 4-chloro-2-pentene. Use an asterisk to indicate each tetrahedral stereocenter.

Expand Your Knowledge

Note: The icons denote exercises based on material in boxes.

9.40 (+)-2-Methylbutanoic acid and (−)-2-methylbutanoic acid react at the same rate of esterification with ethanol. However, in esterification with (+)-*s*-butyl alcohol, (−)-2-methylbutanoic acid reacts much faster than does (+)-2-methylbutanoic acid. Describe the mechanism responsible for this behavior.

9.41 The addition of HCl to 1-butene produces optically inactive 2-chlorobutane. Why is the product optically inactive even though C2 of 2-chlorobutane is a tetrahedral stereocenter?

9.42 Racemic thalidomide was used as a sedative and antinausea agent by pregnant women in Germany and England in the late 1950s and early 1960s. Many of those women gave birth to babies with severely deformed or missing arms and legs and often with abnormalities of the eyes, ears, and digestive system. For many years, these birth defects were thought to have been caused solely by L-thalidomide. Subsequently, it was found that, although the D-enantiomer does not directly cause the birth defects, it does do so indirectly because it is converted under physiological conditions into the racemic mixture. The structure of L-thalidomide is shown belown. Draw D-thalidomide.

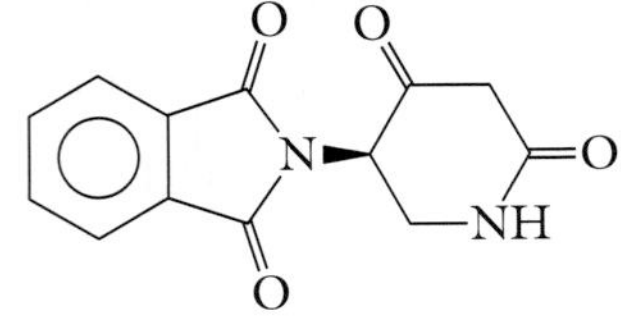

L-Thalidomide

9.43 The sex attractant for the common housefly is muscalure (*cis*-tricosene; see Box 4.3), the trans isomer being totally inactive. Is chiral recognition responsible for the large difference in physiological effect of the two isomers?

9.44 The absence or occurrence of stereoisomers for organic compounds containing single bonds to carbon supports the tetrahedral geometry of carbon atoms. How many stereoisomers are predicted for ethanol if the four bonds of carbon are directed to the four corners of a square? Draw all isomers. How many stereoisomers actually exist for ethanol?

9.45 The absence or occurrence of stereoisomers for organic compounds containing single bonds to carbon supports the tetrahedral geometry of carbon atoms. How many stereoisomers are predicted for CHClBrI if the four bonds of carbon are directed to the four corners of a square? Draw all isomers. How many stereoisomers actually exist for CHClBrI?

9.46 Which of the constitutional isomers of C_4H_9COOH possesses a tetrahedral stereocenter? What type(s) of stereoisomers are possible? Draw the stereoisomers.

9.47 The specific rotation of (+)-penicillin V is +223°. The ABC Company purchases an aqueous solution of (+)-penicillin V at a concentration of 0.100 g/mL. Mary Jones checks the concentration of (+)-penicillin V with a polarimeter by using a 20-cm-long sample cell and the aqueous solution of (+)-penicillin V at its delivered concentration. The optical rotation is found to be +22.3°. (a) Is the concentration of the delivered product 0.100 g/ml? (b) If not, what is the concentration?

9.48 The specific rotation of (−)-morphine is −132°. A sample of (−)-morphine has become contaminated with (+)-morphine. The sample contains 75% (−)-morphine and 25% (+)-morphine. What is the specific rotation of the contaminated sample?

9.49 (−)-Carvone (see Box 9.2), the compound that gives spearmint its taste and odor, cannot be designated as D- or L-. Why? Name (−)-carvone by the *R*/*S* nomenclature system (see Box 9.1).

9.50 L-Dopa is used to treat Parkinson's disease (see Box 9.3). Name L-dopa by the *R*/*S* nomenclature system (see Box 9.1).

9.51 The enantiomer of methyldopa shown in Box 9.1 is the effective enantiomer for treating hypertension. Name it by the *R*/*S* nomenclature system.

9.52 Monochlorination of pentane produces a mixture of three products (Section 3.10). (a) What are the products? (b) Although one of the products contains a tetrahedral stereocenter, the product is not optically active. Explain.

9.53 Naproxen sodium is an anti-inflammatory, analgesic, and antipyretic substitute for aspirin (see Box 7.4). It is sold as the (*S*)-enantiomer even though both enantiomers are effective. However, the (*S*)-enantiomer is much more effective than the (*R*)-enantiomer, and this allows the use of a minimal dosage, which minimizes the possibility of liver damage. Draw the structure of (*S*)-naproxen sodium.

PART 3

BIOCHEMISTRY

We now apply organic chemistry to biochemistry, a study of the structures and physiological functions of biochemicals—such as the physical processes and chemical reactions by which organisms extract, transform, and use energy and materials from their environment. Most biochemicals have complex structures that are uniquely designed for their physiological functions. An example is hemoglobin—the protein that enables oxygen to be transported throughout the body. Hemoglobin is composed of four polypeptide molecules, each of which has a heme molecule in its interior. Each heme has iron at its center, and this iron ion holds and stores oxygen.

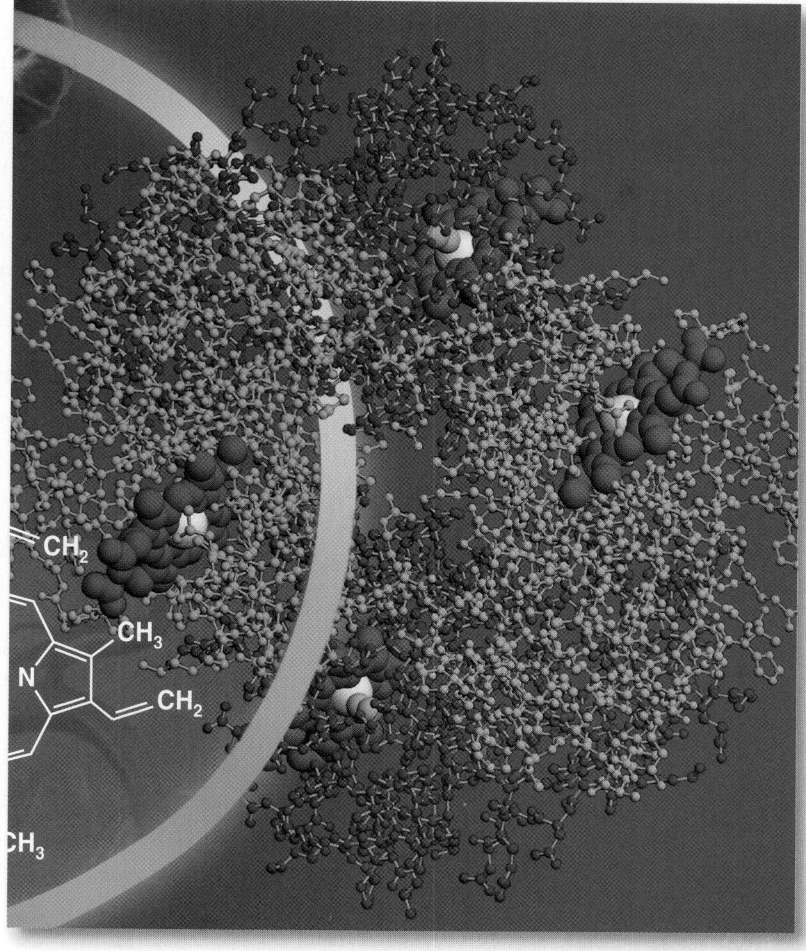

CHAPTER 10 CARBOHYDRATES

(Lynn M. Stone/Picture Cube.)

Chemistry in Your Future

Two children who are raising a heifer as a 4H project have brought her to the animal hospital where you work as a technician. They are worried that she is not gaining weight. As you weigh the animal, take her temperature, and ask questions about her eating habits, they ask you to explain why grass, which can't be digested by humans, is good food for a cow. You tell them that both the cellulose in grass and, say, the starch in the potatoes that humans eat are made from glucose. The difference is in the way in which the units of glucose are joined together; so grass and potatoes require different enzymes for digestion. Humans don't have the enzyme necessary for digesting cellulose. This chapter describes the structures of these carbohydrates and the differences in their digestibility.

For more information about this topic and others in the chapter, go to www.whfreeman.com/bleiodian2e

Learning Objectives

- Define the scope of biochemistry.
- Describe the biological roles of carbohydrates and distinguish between monosaccharides, oligosaccharides, and polysaccharides.
- Describe how monosaccharides are classified and named.
- Describe the characteristics of the D families of aldoses and ketoses.
- Draw the cyclic hemiacetal structures of saccharides.
- Describe and write equations for the mutarotation, oxidation, and acetal-formation reactions of monosaccharides.
- Describe the structures and functions of disaccharides.
- Describe the structures and functions of polysaccharides.
- Describe photosynthesis and the interdependence of plants and animals.

With this chapter, we begin our study of **biochemistry,** the study of the chemical activities within individual cells and among communities of cells. The relation between anatomical structure and physiological function is a powerful theme that is traditionally used to organize data and provide a context for biological information. For example, a description of the human kidney or heart would be incomplete if it did not couple the anatomical details with their physiological functions. Biochemistry extends the theme of structure and function downward from the cellular level to the subcellular and molecular levels of organization.

Life is not simple to define, and few of us would attempt to do it in one or two sentences. However, the functions of life are well recognized: to extract materials and energy from the environment; to use the materials and energy to maintain the structure of the organism and carry out essential processes; to compete with other life forms and survive to replicate and produce succeeding generations. Despite the enormous diversity among species, both the molecules of which they are composed (**biomolecules**) and the biochemical processes that they use to maintain and propagate life are remarkably similar. We begin our consideration of biochemistry with the study of biomolecules, and our presentation emphasizes three main aspects:

- **Structure:** The chemical structure of biomolecules; their organization into cellular components, cells, tissues, and macroscopic structures such as muscle and bone; and the relation of molecular structure to physiological function. Structure is explored in Chapters 10 through 21.
- **Transmission of information:** The molecular basis of heredity, the mechanism by which genetic information is stored and transmitted. Genetic transmission is the focus of Chapter 13.
- **Metabolism:** Collectively, the physical processes and chemical reactions by which organisms extract, transform, and use energy and materials from their environment. These subjects are treated in Chapters 14 through 18.

Organisms require both inorganic and organic materials to sustain life. The necessary inorganic substances are water and a host of inorganic ions such as Na^+, K^+, Fe^{2+}, Ca^{2+}, Cl^-, HPO_4^{2-}, and HCO_3^-. Many organic biomolecules, particularly proteins, function only in the presence of a specific inorganic ion. Sometimes, the ion is an integral part of the molecular structure. For example, hemoglobin, the protein that transports oxygen from the lungs to other tissues, contains Fe^{2+} (Chapter 12). Many physiological functions depend on the difference in concentrations of certain ions inside and outside cells, a difference maintained by the cell membrane (Chapter 11).

The four principal families of organic biomolecules are the carbohydrates, lipids, proteins, and nucleic acids. Proteins have many functions: they are the structural materials of muscle and bone; they are the enzymes that catalyze biochemical reactions; they transport oxygen through the blood to tissues; they protect an organism against viruses and bacteria; and they function as hormones to regulate physiological functions (Chapter 12). Lipids are the main components of the membranes that enclose all living cells; they are sources of energy; and some function as hormones to regulate physiological functions (Chapter 11). Nucleic acids direct and control the transmission of hereditary information and the synthesis of proteins and, ultimately, all cellular materials (Chapter 13).

Carbohydrates and their various functions are the subject of the remainder of this chapter.

10.1 INTRODUCTION TO CARBOHYDRATES

Carbohydrates, also called **saccharides,** are the single most abundant family of organic compounds to be found in nature. They have a variety of functions:

- The metabolism (breakdown) of the carbohydrate glucose generates the energy required for all life processes. Starch in plants and glycogen in animals are carbohydrates serving as storage forms of glucose.
- Carbohydrates serve as structural and protective materials. One example is cellulose, used in cell walls and in the extracellular structures of plants; another is chitin, which forms the exoskeletons of crustaceans and insects.
- Carbohydrates are precursors for the biosynthesis of proteins, lipids, and nucleic acids.
- Carbohydrates attached to proteins or lipids in cell membranes help cells recognize one another as well as specific molecules, triggering physiological processes such as fertilization, cell growth, and immune responses.
- The carbohydrates ribose and deoxyribose are components of ribonucleic acids (RNAs) and deoxyribonucleic acids (DNAs), respectively.

Carbohydrates are familiar in our lives. Starch is a major component of our diet—in potatoes, cereals, beans, peas, corn, rice, pasta, and bread. We use table sugar, or sucrose, to sweeten our baked goods, coffee, tea, and soft drinks. The wood used to build our houses, the paper on which this book is printed, and the cotton in our clothes are largely composed of cellulose.

Every carbohydrate is either a **polyhydroxyaldehyde** or a **polyhydroxyketone,** polyfunctional molecules containing an aldehyde or ketone group and two or more hydroxyl groups. Derivatives and polymers of these molecules also are considered carbohydrates. Carbohydrates are further classified as monosaccharides, oligosaccharides, and polysaccharides. **Monosaccharides,** the simplest saccharides, not only have important biological roles themselves but also serve as the building blocks (called **monomers, residues,** or **repeat units**) for synthesizing larger saccharides. Oligosaccharides and polysaccharides can be hydrolyzed to monosaccharides, but monosaccharides cannot be further hydrolyzed to still smaller saccharides. Glucose, fructose, and galactose are important monosaccharides.

Polysaccharides contain large numbers, usually hundreds or even thousands, of monosaccharide units bonded together. Important examples are starch, cellulose, glycogen, and chitin. The name **oligosaccharide** is loosely used for any saccharide larger than a monosaccharide but smaller than a polysaccharide. There is no well-defined demarcation between oligosaccharides and polysaccharides. Most biochemists consider oligosaccharides to contain as many as 10 to 20 monosaccharide units. More-specific terms such as **disaccharide, trisaccharide,** and **tetrasaccharide** are more useful. Sucrose and lactose (milk sugar) are important disaccharides. Monosaccharides and disaccharides are also called **sugars** because almost all of them taste sweet.

IUPAC names are not generally used for carbohydrates, because they are excessively long and cumbersome, and common names have become strongly entrenched in the chemical and biological literature. The common names of most, but not all, saccharides have the ending **-ose,** as in glucose, fructose, lactose, sucrose, and cellulose. Exceptions include most polysaccharides, such as starch, chitin, and glycogen.

Dietary carbohydrates are categorized as **simple** or **complex carbohydrates.** Simple carbohydrates are monosaccharides and disaccharides. Complex carbohydrates are polysaccharides.

10.2 MONOSACCHARIDES

Monosaccharides have one of the following structures:

$$
\begin{array}{c}
\mathrm{H{-}C{=}O} \\
| \\
(\mathrm{H{-}C{-}OH})_a \\
| \\
\mathrm{CH_2OH}
\end{array}
\qquad\qquad
\begin{array}{c}
\mathrm{CH_2OH} \\
| \\
\mathrm{C{=}O} \\
| \\
(\mathrm{H{-}C{-}OH})_b \\
| \\
\mathrm{CH_2OH}
\end{array}
$$

Aldose (Polyhydroxyaldehyde) | Ketose (Polyhydroxyketone)

For monosaccharides of biological importance, $a = 1$ to 4 and $b = 0$ to 3; that is, they are monosaccharides containing from three to six carbon atoms. The carbon chains are numbered from the end nearest the carbonyl carbon. In polyhydroxyaldehydes, the carbonyl carbon is C1. In polyhydroxyketones, the carbonyl carbon is C2.

Classification and Nomenclature

Monosaccharide names classify the compounds in two ways simultaneously, by combining two kinds of prefixes before **-ose:**

Rules for naming monosaccharides

- Monosaccharides with an aldehyde group are **aldoses;** those with a ketone group are **ketoses.**
- Monosaccharides with three, four, five, and six carbons are **trioses, tetroses, pentoses,** and **hexoses,** respectively.

For example, a five-carbon monosaccharide with a ketone group is called a ketopentose.

Example 10.1 Classifying monosaccharides

Classify each of the following monosaccharides to indicate both the type of carbonyl group and the number of carbons present:

$$
\begin{array}{c}
\mathrm{CH_2OH} \\
| \\
\mathrm{C{=}O} \\
| \\
\mathrm{H{-}C{-}OH} \\
| \\
\mathrm{H{-}C{-}OH} \\
| \\
\mathrm{CH_2OH} \\
\\
1
\end{array}
\qquad
\begin{array}{c}
\mathrm{H{-}C{=}O} \\
| \\
\mathrm{H{-}C{-}OH} \\
| \\
\mathrm{HO{-}C{-}H} \\
| \\
\mathrm{H{-}C{-}OH} \\
| \\
\mathrm{H{-}C{-}OH} \\
| \\
\mathrm{CH_2OH} \\
2
\end{array}
\qquad
\begin{array}{c}
\mathrm{CH_2OH} \\
| \\
\mathrm{C{=}O} \\
| \\
\mathrm{CH_2OH} \\
\\
3
\end{array}
\qquad
\begin{array}{c}
\mathrm{H{-}C{=}O} \\
\mathrm{H{-}{+}{-}OH} \\
\mathrm{HO{-}{+}{-}H} \\
\mathrm{H{-}{+}{-}OH} \\
\mathrm{H{-}{+}{-}OH} \\
\mathrm{CH_2OH} \\
4
\end{array}
$$

Solution

First determine whether the molecule is an aldehyde or ketone, and then count the number of carbons. Monosaccharide 1 is a ketopentose, 2 is an aldohexose, 3 is a ketotriose, 4 is an aldohexose. Note that monosaccharides 2 and 4 are identical. Monosaccharide 4 is the Fischer projection of 2. Fischer projections are used more often than not.

Problem 10.1 Classify each of the following monosaccharides to indicate both the type of carbonyl group and the number of carbons present:

1: H—C=O; H—C—OH; CH_2OH

2: CH_2OH; =O; HO—H; H—OH; H—OH; CH_2OH

3: H—C=O; H—C—OH; HO—C—H; H—C—OH; CH_2OH

4: H—C=O; H—OH; HO—H; CH_2OH

Stereoisomerism

Figures 10.1 and 10.2 show the D-aldoses and D-ketoses. Each figure starts with the triose at the top and proceeds downward through the tetroses, pentoses, and hexoses. The aldotriose D-glyceraldehyde (Figure 10.1) has one tetrahedral stereocenter, and each additional carbon in tetroses, pentoses, and hexoses adds another tetrahedral stereocenter. The ketotriose dihydroxyacetone

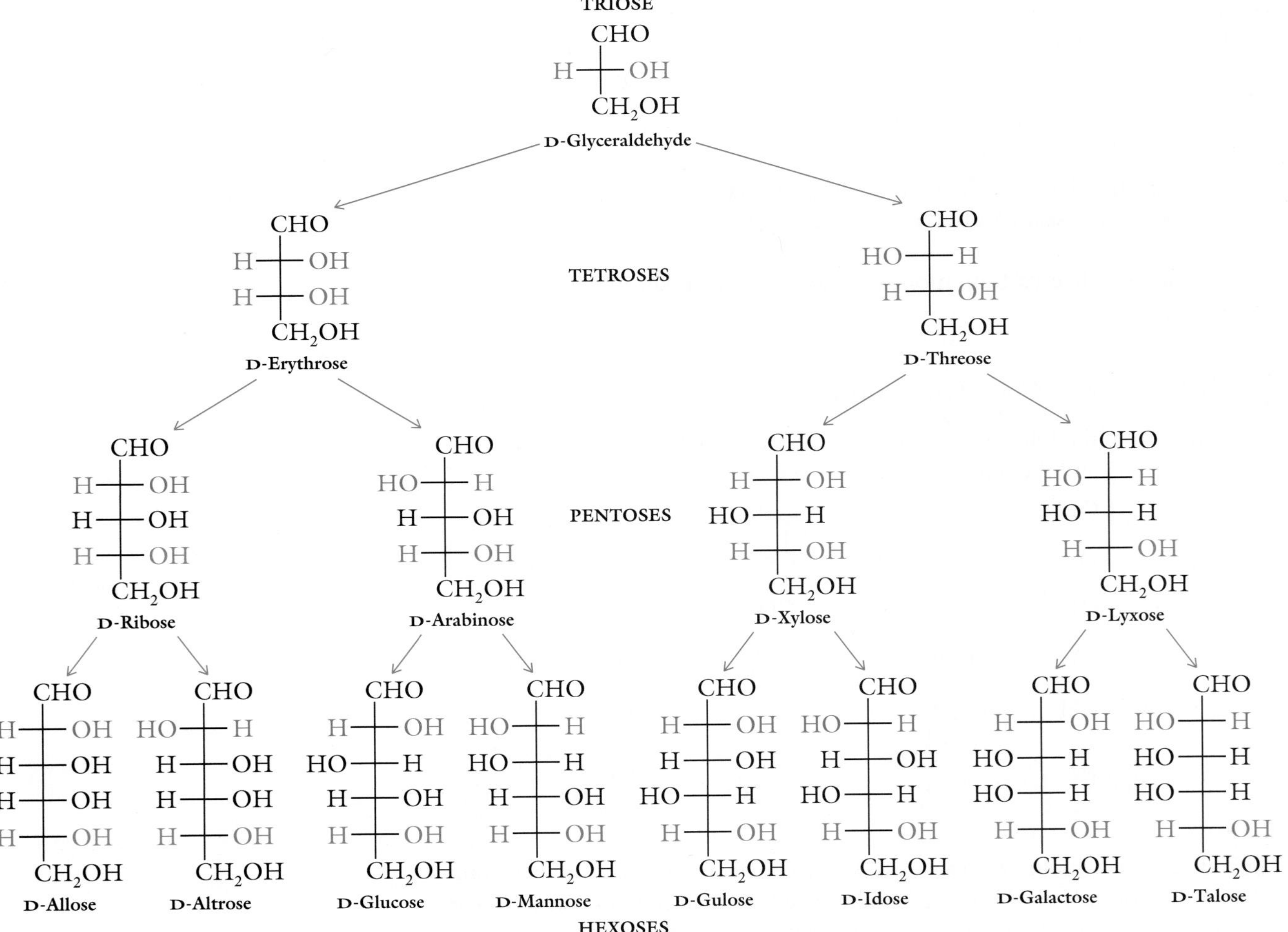

Figure 10.1 D-Aldoses. Each pair of arrows denotes a pair of diastereomers whose configurations are identical at all tetrahedral stereocenters except C2 (red). All D-aldoses have the same configuration at the stereocenter (blue) farthest from the carbonyl group.

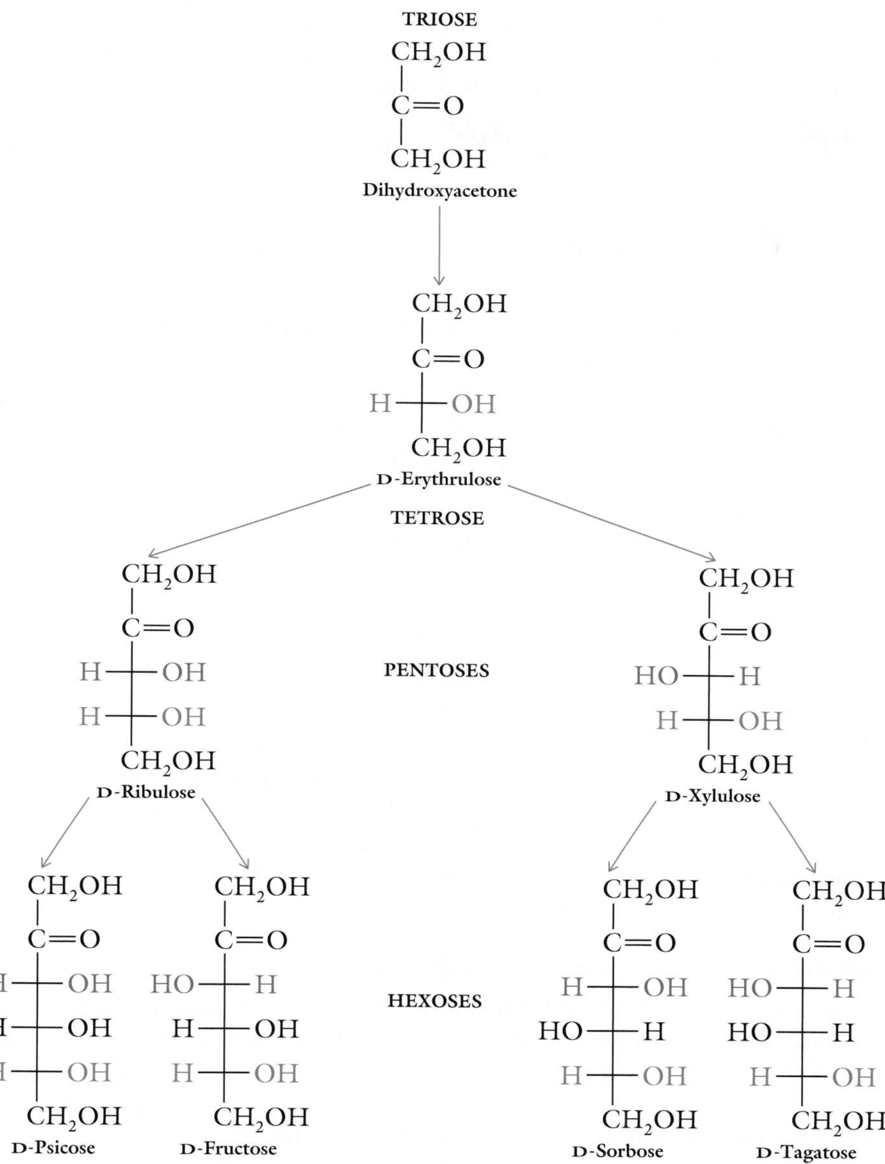

Figure 10.2 D-Ketoses. Each pair of arrows denotes a pair of diastereomers whose configurations are identical at all tetrahedral stereocenters except C3 (red). All D-ketoses have the same configuration at the stereocenter (blue) farthest from the carbonyl group.

(Figure 10.2) does not have a tetrahedral stereocenter. The ketotetrose D-erythrulose is the smallest ketose to possess a tetrahedral stereocenter. Each additional carbon adds a tetrahedral stereocenter.

A complete description of the stereoisomerism of an aldose or ketose requires a separate D or L designation (Section 9.4) for each tetrahedral stereocenter, but this designation is not necessary for our purposes. Our primary interest lies in the relations between the various aldoses and those between the various ketoses. All of the monosaccharides in Figures 10.1 and 10.2 have the same configuration at the tetrahedral stereocenter (shown in blue) farthest from the carbonyl carbon. That configuration is the D configuration—the —OH group is on the right side of the Fischer projection—and all of the monosaccharides are called D-saccharides to indicate this common structural feature.

In theory, each of the D-saccharides in Figures 10.1 and 10.2 has a corresponding mirror-image enantiomer, but these L-saccharides are rarely found in nature. The D-sugars predominate in nature because they are synthesized and used by stereoselective enzymes that function through the principle of chiral recognition (Section 9.5).

The honey produced by bees is a concentrated aqueous solution that is more than 80% carbohydrates, most of which are D-fructose and D-glucose. (Scott Camazine/Photo Researchers.)

In discussions of saccharides, the names of saccharides are often used without the prefix D-. Whenever the stereochemistry of a saccharide is not indicated, such as in "glucose" or "fructose," you should assume that the D-enantiomer is meant.

The saccharides in any horizontal grouping in Figure 10.1 or in Figure 10.2 are diastereomers of each other; that is, they are stereoisomers that are not enantiomers.

Example 10.2 Recognizing structural relations

For each of the following pairs of compounds, indicate whether the pair consists of different compounds that are (1) constitutional isomers or (2) stereoisomers that are enantiomers or (3) stereoisomers that are diasteromers or (4) not isomers. Refer to Figures 10.1 and 10.2 for the structures of the compounds. (a) D-Glucose and D-mannose; (b) D-ribose and D-xylulose; (c) D-fructose and D-arabinose; (d) D-sorbose and L-sorbose; (e) D-sorbose and D-fructose.

Solution

(a) Diastereomers; they are stereoisomers that are not enantiomers.

(b) Constitutional isomers; D-ribose is an aldopentose and D-xylulose is a ketopentose.

(c) Not isomers; D-fructose is a hexose and D-arabinose is a pentose.

(d) Enantiomers; they are nonsuperimposable mirror-image stereoisomers.

(e) Diastereomers; they are stereoisomers that are not enantiomers.

Problem 10.2 For each of the following pairs of compounds, indicate whether the pair consists of different compounds that are (1) constitutional isomers or (2) stereoisomers that are enantiomers or (3) stereoisomers that are diasteromers or (4) not isomers. Refer to Figures 10.1 and 10.2 for the structures of the compounds. (a) D-Glucose and D-talose; (b) D-ribose and D-sorbose; (c) D-fructose and L-fructose; (d) D-erythrose and L-erythrulose; (e) D-sorbose and D-psicose.

Some Important Monosaccharides

D-Glucose is overwhelmingly the most abundant monosaccharide in nature, existing mostly in combined form in starch, cellulose, glycogen, chitin, lactose, and sucrose. Dietary starch and stored glycogen (in the liver) are broken down to glucose, which is transported to cells and taken in for energy production. A 5% aqueous solution of D-glucose is used for the intravenous feeding of patients unable to take nourishment by mouth. Glucose is also known as **blood sugar** or **dextrose.**

Two other hexoses are very important. D-Fructose, also called **levulose** (because it is levorotatory; Section 9.5), is bonded to D-glucose in the disaccharide sucrose, the sugar in fruits and table sugar. D-Galactose is bonded to D-glucose in the disaccharide lactose, the sugar in mammalian milk. Although D-glucose is the most common component of polysaccharides, D-fructose and D-mannose also are present in some polysaccharides.

The high stereoselectivity in nature is evident in the disproportionately high use of D-glucose relative to the other aldohexoses. There are four tetrahedral stereocenters and 2^4, or 16, possible stereoisomers for D-glucose: the 8 D-aldohexoses of Figure 10.1 plus the corresponding mirror-image L-aldohexoses. But the amount of D-glucose found in nature far exceeds the total of all other aldohexoses, as well as all other monosaccharides.

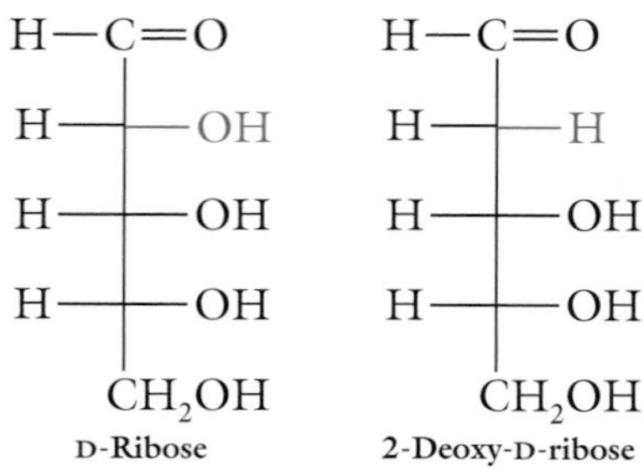

D-Ribose 2-Deoxy-D-ribose

D-Ribose and D-xylose are important pentoses. D-Xylose is a component of some plant polysaccharides. D-Ribose and 2-deoxy-D-ribose are components of ribonucleic and deoxyribonucleic acids, respectively. 2-Deoxy-D-ribose is a **deoxysaccharide,** or **deoxysugar,** having one less oxygen than its corresponding saccharide at C2: an —OH group of the saccharide is replaced by —H in the deoxysaccharide.

The trioses D-glyceraldehyde and dihydroxyacetone are important intermediates in metabolic processes (Sections 15.1 and 15.2).

10.3 CYCLIC HEMIACETAL STRUCTURES

The structural representations that we have used so far for pentoses and hexoses have not been entirely correct. These monosaccharides are actually polyhydroxyhemiacetals. Recall from Section 6.6 that the —OH group of an alcohol and the carbonyl group of an aldehyde or ketone undergo an acid-catalyzed addition reaction to form a hemiacetal:

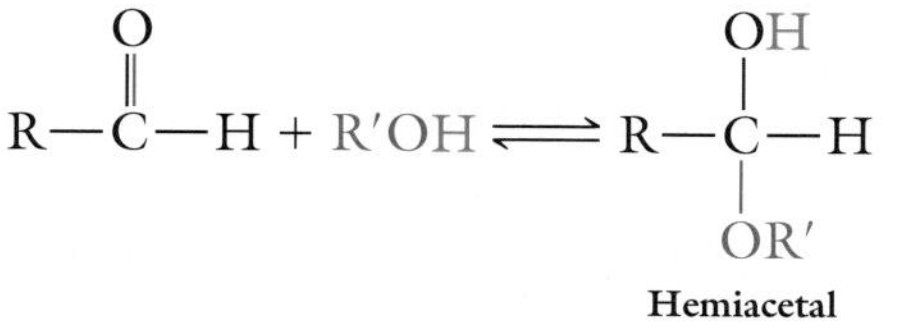

$$R-\overset{\overset{O}{\|}}{C}-H + R'OH \rightleftharpoons R-\underset{\underset{OR'}{|}}{\overset{\overset{OH}{|}}{C}}-H$$

Hemiacetal

Furthermore, we saw that, in a molecule with both an alcohol —OH group and an aldehyde or ketone carbonyl group, intramolecular hemiacetal formation produces a **cyclic hemiacetal.** For example, the addition of the —OH group at C5 of D-glucose to its own carbonyl group at C1 forms a six-membered cyclic hemiacetal, called a **pyranose ring** (Figure 10.3). Intramolecular hemiacetal formation is the key to understanding the structures and reactions of saccharides. Enzymes catalyze the formation of cyclic hemiacetals in biological systems.

We can visualize the conversion of the acyclic D-glucose structure into the cyclic hemiacetal by the following process:

- Draw the Fischer projection of the acyclic structure and then turn it sideways.
- Rotate the bond between C4 and C5 to bring the C5 —OH group close to the carbonyl group.
- Add the alcohol —OH to the carbonyl group: break the O–H bond, and then bond the H of the —OH to the carbonyl O and bond the O of the —OH to the carbonyl C.

INSIGHT INTO STRUCTURE

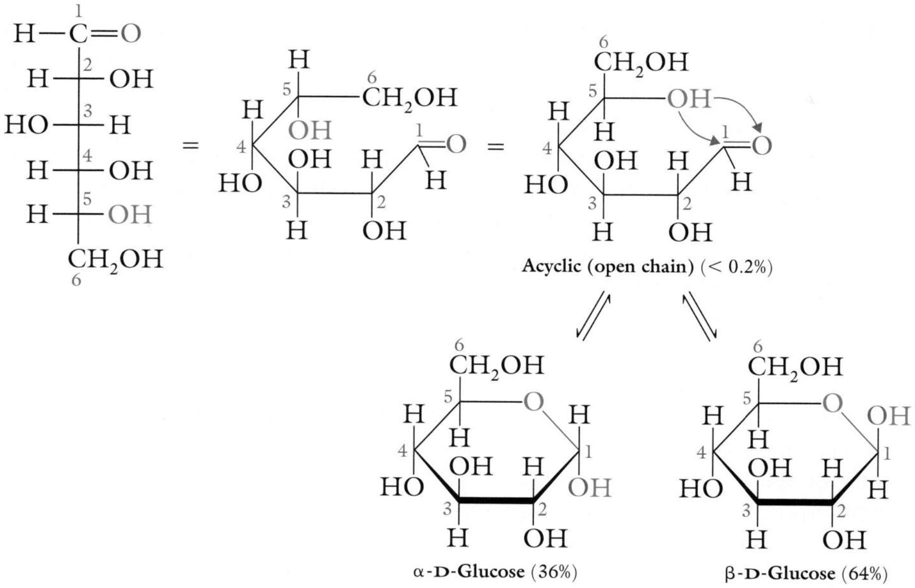

Acyclic (open chain) (< 0.2%)

α-D-Glucose (36%)

β-D-Glucose (64%)

Figure 10.3 Hemiacetal formation in D-glucose. The oxygen of the OH at C5 of the acyclic structure becomes the oxygen in the hemiacetal ring, whereas the carbonyl oxygen becomes the OH at C1.

Hemiacetal formation converts the carbonyl carbon into a new tetrahedral stereocenter. This new stereocenter is C1 of the cyclic hemiacetal in Figure 10.3 and is called the **hemiacetal carbon.** You can recognize the hemiacetal carbon as the carbon with two different oxygen groups attached: an alcohol —OH and an ether —OR.

Two stereoisomers of the cyclic hemiacetal, called the α and β configurations, are possible. These stereoisomers are diastereomers. Ring closure produces a mixture of the two diastereomers because neither configuration is excluded from forming in the reaction. Thus, D-glucose is a mixture of the two hemiacetals α-D-glucose and β-D-glucose. These diastereomers, differing only in the configuration at the hemiacetal carbon, are called **anomers.**

- The α-anomer of D-glucose has a trans relation between the —OH at the hemiacetal carbon (C1) and the $—CH_2OH$ at C5.
- The β-anomer of D-glucose has a cis relation between the —OH at the hemiacetal carbon (C1) and the $—CH_2OH$ at C5.

As we shall see later in the chapter, the difference between the α and β configurations at the hemiacetal carbon is highly significant to saccharide structure and function.

As noted in Section 6.6, hemiacetal formation between simple alcohols and aldehydes or ketones is not important, because the equilibrium is overwhelmingly to the left; the hemiacetal is very unstable. However, for cyclic hemiacetal formation in saccharides, the equilibrium is overwhelmingly in favor of the cyclic hemiacetal, which is a stable product. The acyclic structure, often called the **open-chain structure,** is not stable relative to the cyclic hemiacetal. In solutions of D-glucose, more than 99.8% of the D-glucose is present as the cyclic hemiacetal structure with less than 0.2% present as the acyclic structure. The β-anomer is about twice as abundant as the α-anomer.

Concept checklist

✓ Hexoses and pentoses exist predominantly as cyclic hemiacetals, not as acyclic compounds containing alcohol —OH and aldehyde or ketone C=O groups.

✓ Even when saccharides are represented by open-chain structures, the actual structure of the saccharide is predominantly the cyclic hemiacetal.

D-Fructose and other ketohexoses also exist as cyclic hemiacetals. The addition of the —OH at C5 to the carbonyl group forms a five-membered cyclic hemiacetal called a **furanose ring** (Figure 10.4). The hemiacetal carbon is C2. Again, a pair of anomers is formed. In α-D-fructose, the —OH group at C2 and the $—CH_2OH$ group at C5 have a trans relation. In β-D-fructose, the two groups have a cis relation.

The cyclic hemiacetal structures of saccharides shown in Figures 10.3 and 10.4 are called **Haworth structures** or **Haworth projections.** Such projections are drawn with bold bonds in the lower half of the ring to convey the three-dimensional aspects of the cyclic hemiacetals. We used the same convention described earlier for cyclic compounds (Section 3.8): the ring is perpendicular to the plane of the paper, and the groups attached to the carbons of the ring are parallel to the plane of the paper. Haworth projections are not always drawn with bold bonds.

The convention of depicting horizontal bonds extending forward and vertical bonds extending backward in Fischer projections should not be confused with the spatial orientations in Haworth projections. There is a simple translation between the two structures: groups appearing to the right in a Fischer projection extend downward from the ring in a Haworth projection; those to the left in a Fischer projection extend upward from the ring in a Haworth projection.

INSIGHT INTO STRUCTURE

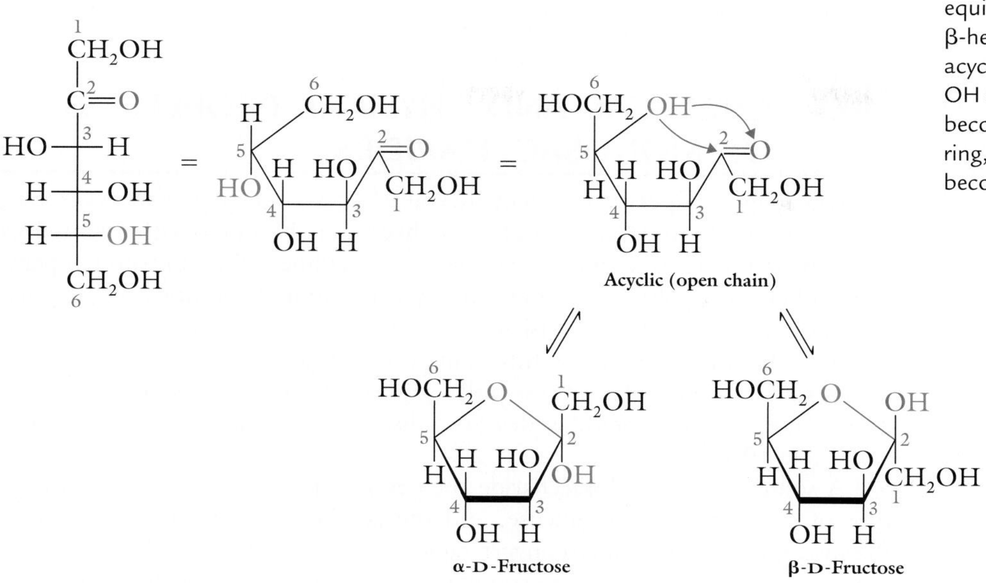

Figure 10.4 D-Fructose is an equilibrium mixture of α- and β-hemiacetal structures and an acyclic structure. The oxygen of the OH at C5 of the acyclic structure becomes the oxygen in the hemiacetal ring, whereas the carbonyl oxygen becomes the OH at C2.

Acyclic (open chain)

α-D-Fructose

β-D-Fructose

Example 10.3 Drawing the cyclic hemiacetal structure of a monosaccharide

Draw the cyclic structure of α-D-galactose, referring only to Figure 10.1. Indicate how this task is simplified if you are allowed to directly refer to both Figures 10.1 and 10.3.

Solution

Use the three-step process described near the beginning of Section 10.3 and illustrated in Figure 10.3 for D-glucose: draw the Fischer projection, turn it sideways, rotate the bond between C4 and C5, and add the alcohol O–H to the carbonyl group.

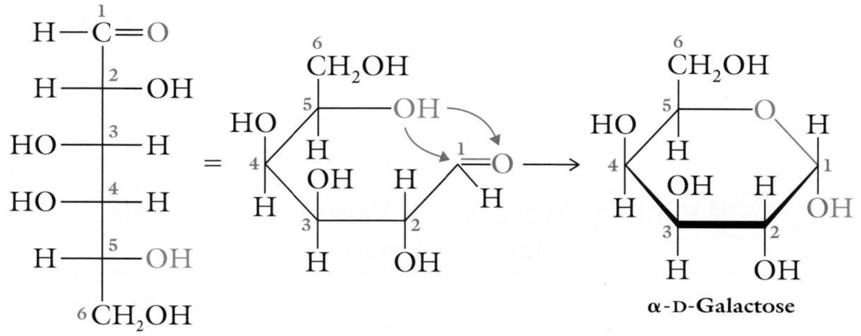

α-D-Galactose

Figure 10.1 shows that D-galactose differs from D-glucose only at the configuration at C4. If Figure 10.3 is available for reference, all that is needed is to take the structure of α-D-glucose and reverse the positions of the —H and —OH groups at C4:

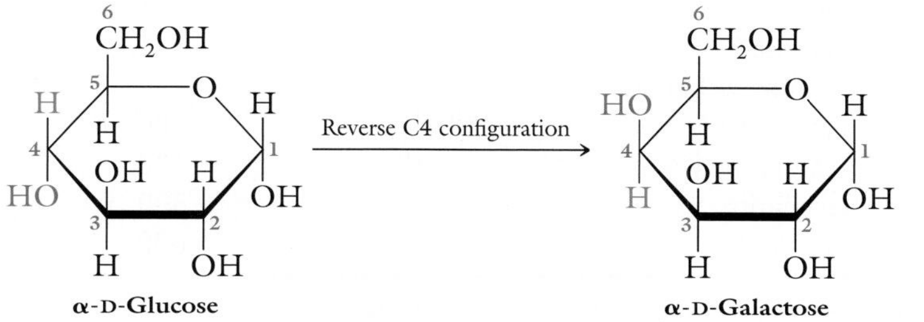

α-D-Glucose

Reverse C4 configuration

α-D-Galactose

Problem 10.3 Draw the cyclic structure of β-D-sorbose by reference to Figures 10.2 and 10.4.

10.4 CHEMICAL AND PHYSICAL PROPERTIES OF MONOSACCHARIDES

As you might expect of polyhydroxyaldehydes and polyhydroxyketones, monosaccharides undergo considerable hydrogen bonding with themselves and with water. Thus, monosaccharides are crystalline solids at room temperature and are very soluble in water. Highly concentrated solutions are very viscous liquids; think of the consistency of honey, maple syrup, and molasses. Monosaccharides are only slightly soluble in alcohols such as methanol and ethanol and are insoluble in less-polar solvents such as ethers and hydrocarbons. Almost all monosaccharides (and disaccharides) taste sweet (Box 10.1 and Table 10.1).

A solution of a monosaccharide such as D-glucose is an equilibrium mixture of α-D-glucose, β-D-glucose, and the acyclic, or open-chain, structure. Because the molecules interconvert rapidly from one structure into another, the properties of D-glucose are simultaneously those of an aldehyde with four —OH groups and those of a hemiacetal with five —OH groups. The presence of this equilibrium has been proved by the mutarotation and oxidation experiments described next.

Mutarotation

Pure crystalline samples of α-D-glucose or β-D-glucose can be obtained by crystallization from different solvents. But all aqueous solutions of D-glucose consist of the equilibrium mixture of about 36% α-D-glucose, 64% β-D-glucose, and less than 0.2% open-chain structure. This equilibrium is responsible for an optical rotation phenomenon called **mutarotation.** The specific rotations of α-D-glucose and β-D-glucose are +112.2° and +18.7°, respectively. These rotations can be measured immediately after dissolving each pure anomer separately in water. On standing, however, each solution undergoes a change in specific rotation until the equilibrium value of +52.7° is obtained. Each of the pure anomers undergoes equilibration to the same mixture of α-D-glucose, β-D-glucose, and open-chain compound.

TABLE 10.1 Sweetness of Various Compounds Relative to Sucrose

Compound	Sweetness*
lactose	0.16
galactose	0.32
maltose	0.33
glucose	0.74
sucrose	1.00
fructose	1.7
calcium cyclamate	30
aspartame	180
acesulfame K	200
saccharin	300
sucralose	600

*Relative to sucrose = 1.00 (on a weight basis).

10.1 CHEMISTRY WITHIN US

How Sweet Is It?

Sweet taste is one of the four primary taste perceptions, the others being sour, salty, and bitter (see Box 9.1). Almost all monosaccharides and disaccharides taste sweet, as do many other compounds with —OH groups on adjacent carbon atoms. For saccharides, sweetness depends on molecular structure, size, and stereoisomerism. The relative sweetnesses of various monosaccharides, disaccharides, and artificial sweeteners are listed in Table 10.1. Sweetness is measured relative to sucrose, set at 1.00 on a weight basis.

Excessive sugar intake resulting in obesity is a major health problem in the developed countries, and this problem has spurred the growth of the artificial sweetener industry. A number of synthetic organic compounds are far sweeter than sucrose, which allows their use as artificial sweeteners without significant caloric value because they are used in very small amounts. The greater sweetness of artificial sweeteners surprises many people. In fact, the artificial sweeteners are structurally very different from natural sugars. It clearly indicates a complex relation between the taste chemoreceptors and the structures that they respond to most strongly. Artificial sweeteners are also useful for diabetics who must restrict their intake of sugars (Box 12.3).

Saccharin and cyclamates were among the first artificial sweeteners. Cyclamates were banned in the United States in 1970 because the results of some studies showed that they cause cancer in laboratory animals. Saccharin was implicated in bladder cancer in laboratory animals in 1978 but was not banned, because legislators were reluctant to ban the only artificial sweetener available. Saccharin does, however, carry a warning that it may be a health hazard.

Newer artificial sweeteners include aspartame (Section 17.6), acesulfame K, and sucralose. Extensive testing suggests no major ill effects on health. Aspartame has largely replaced saccharin as an artificial sweetener. Acesulfame K and sucralose are useful in food products that require cooking, because higher temperatures degrade aspartame but not acesulfame K and sucralose.

Aspartame should not be used by people with phenylketonuria (PKU), a disorder of phenylalanine metabolism that can lead to mental retardation. After ingestion, aspartame is hydrolyzed to phenylalanine, aspartic acid, and methanol. Although methanol is toxic (Box 5.3), it is not dangerous at the minute concentrations formed from aspartame intake. However, even the small amounts of phenylalanine produced are a serious problem for phenylketonurics. Instead of transforming phenylalanine into tyrosine, people with PKU transform it into phenylpyruvate, which causes severe mental retardation.

✓ Each D-glucose anomer undergoes ring opening to the open-chain structure followed by ring closure to the mixture of α- and β-anomers.

Concept check

Monosaccharides—and dissacharides, as we shall see—are often drawn as either the α- or β-anomer to conserve space. This convention is not meant to denote the presence of only one type of anomer. Every solution of a saccharide with a hemiacetal group consists of an equilibrium mixture of the two anomers and the open-chain structure, whether the three structures are drawn or not.

The Oxidation of the Aldehyde Group

Mild oxidizing reagents such as Benedict's reagent (Cu^{2+} complexed with citrate ion in alkaline solution) oxidize the aldehyde group in an aldose to a carboxyl group (Section 6.4). Note that this oxidation is a reaction of an aldehyde (the open-chain structure of D-glucose); it is not a reaction of a hemiacetal (the ring structure of D-glucose):

Reducing agent is oxidized — Oxidizing agent is reduced

CH_2OH … $=O$ + Cu^{2+} (Blue) $\xrightarrow{Base}$ CH_2OH … $=O$ (OH) + Cu_2O (Red precipitate)

Oxidized product — **Reduced product**

(The oxidized product, D-gluconic acid, is shown with its carboxyl group as COOH but will actually exist in its ionized form COO^- under the basic conditions of the reaction.)

The oxidation reaction demonstates that α-D-glucose and β-D-glucose are in equilibrium with the open-chain aldehyde. The oxidation is a quantitative reaction: one mole of aldose is oxidized, even though, at equilibrium, less than 0.2% of the aldose is present as the aldehyde structure. As the small amount of aldehyde is oxidized, some α-D-glucose and β-D-glucose convert into the aldehyde, which is then oxidized; more α-D-glucose and β-D-glucose convert into the aldehyde, which is then oxidized; and so forth. Eventually, all of the α-D-glucose and β-D-glucose are converted into the aldehyde and oxidized, according to Le Chatelier's principle.

The oxidation of an aldose by Benedict's reagent is used in a clinical test for monosaccharides in urine. The blue color of the reagent is replaced by the red color of the Cu_2O precipitate. The intensity of the red color observed is directly proportional to the concentration of monosaccharide in the urine. The clinical test uses a paper strip impregnated with Benedict's reagent. The paper strip is dipped in the urine sample, and the resulting red color is compared with a standard color chart to determine the monosaccharide concentration.

As noted in Section 6.4, most ketones are not oxidized by Benedict's reagent, but α-hydroxy ketones such as D-fructose and other ketoses give positive tests because they are converted into aldoses by the alkaline conditions of Benedict's reagent. Aldoses and ketoses are called **reducing sugars** because the sugar is the reducing agent in the oxidation reaction with Benedict's reagent.

The **glucose oxidase test** is a more specific clinical test than Benedict's reagent. Whereas Benedict's reagent oxidizes all monosaccharides, glucose oxidase is specific for D-glucose. The enzyme glucose oxidase catalyzes the oxidation of D-glucose by molecular oxygen. D-Glucose is oxidized to D-gluconic acid with O_2 reduced to hydrogen peroxide (H_2O_2). *o*-Toluidine (*o*-methylaniline), present in the reaction mixture, is oxidized by H_2O_2 to colored products. The depth of the color gives a quantitative measure of the D-glucose level.

Acetal Formation: The Production of Glycosides

Monosaccharides, like all hemiacetals, are converted into acetals by the acid-catalyzed dehydration of the hemiacetal —OH group with an alcohol (Section 6.6). For example, the reaction of D-glucose with methanol yields a mixture of methyl α- and β-acetals. Dehydration takes place preferentially at the hemiacetal —OH group because it is the most reactive —OH group in the molecule.

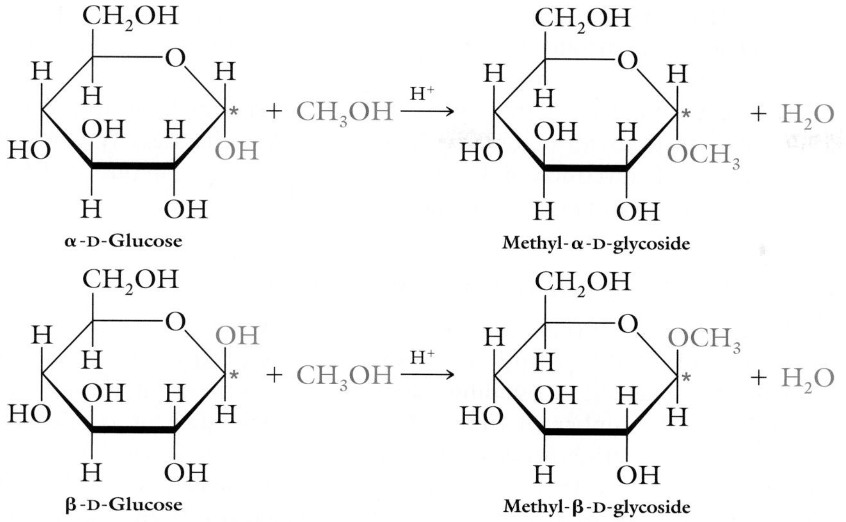

α-D-Glucose + CH_3OH $\xrightarrow{H^+}$ Methyl-α-D-glycoside + H_2O

β-D-Glucose + CH_3OH $\xrightarrow{H^+}$ Methyl-β-D-glycoside + H_2O

The hemiacetal and acetal carbons are marked with asterisks. You can recognize the acetal carbon by its attachment to two ether, —OR, groups.

Acetals of carbohydrates are called **glycosides,** and the acetal carbon and its two —OR groups form a **glycosidic linkage.** A glycosidic linkage forms as either an α- or a β-glycosidic linkage, depending on whether the α- or the β-acetal takes part in the reaction. In laboratory reactions catalyzed by acids such as sulfuric acid, a mixture of the two acetals is formed. In living systems, the formation of glycosidic linkages is catalyzed by enzymes stereoselectively—that is, one or the other glycosidic linkage is formed exclusively—which has important consequences for the physiological functioning of the acetal.

Example 10.4 Writing equations for reactions of monosaccharides

Show the formation of methyl α- and β-D-glycosides from the reaction of D-fructose with methanol. Place asterisks next to the hemiacetal and acetal carbons.

Solution

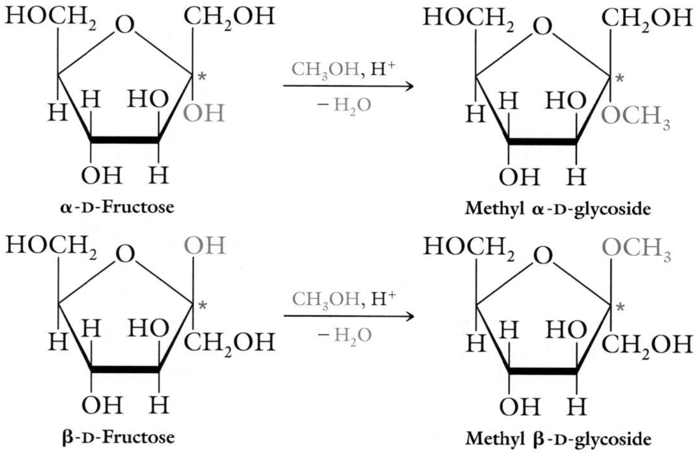

α-D-Fructose $\xrightarrow[-H_2O]{CH_3OH,\ H^+}$ Methyl α-D-glycoside

β-D-Fructose $\xrightarrow[-H_2O]{CH_3OH,\ H^+}$ Methyl β-D-glycoside

Problem 10.4 Show the product(s) formed when D-fructose is oxidized by Benedict's reagent.

Glycosides are not reducing sugars, because they are not in equilibrium with an open-chain compound, whether α-hydroxy aldehyde or ketone. For the same reason, they do not undergo mutarotation. Glycosides can be hydrolyzed to the saccharide and alcohol by reaction with water in the presence of an acid catalyst or appropriate enzyme.

The glycosidic linkage is the type of bond through which monosaccharide residues are linked to form disaccharides and polysaccharides. Whether the bond is an α- or β-glycosidic linkage is important to the structure and function of disaccharides and polysaccharides.

Other Derivatives of Monosaccharides

Phosphate esters are produced by dehydration between an —OH group of phosphoric acid, H_3PO_4, and an —OH of a saccharide (Section 7.5). D-Ribose-5-phosphate is a building block in the synthesis of nucleic acids (Section 13.1). A number of other phosphate esters, such as D-glucose-6-phosphate and D-fructose-6-phosphate, are intermediates in the metabolism of saccharides (Section 15.1). D-Glucuronic acid, in which the $—CH_2OH$ group of glucose has been oxidized to a carboxyl group, is an **acidic sugar.** D-Glucosamine, in which the —OH group at C2 is replaced by an amino (NH_2) group, is an **aminosugar.** (Note that the phosphate, carboxyl, and amino groups are shown in the ionized form in which they exist at physiological pH.)

N-Acetyl-D-glucosamine, derived from D-glucosamine, is the building block for chitin—the hard external covering (exoskeleton) of crustaceans. *N*-Acetyl-D-glucosamine and D-glucuronic acid are building blocks for hyaluronic acid, which is found in the synovial fluid of joints, where it acts as a shock absorber, and in the vitreous humor of the eye ball, where it serves to hold the retina in place.

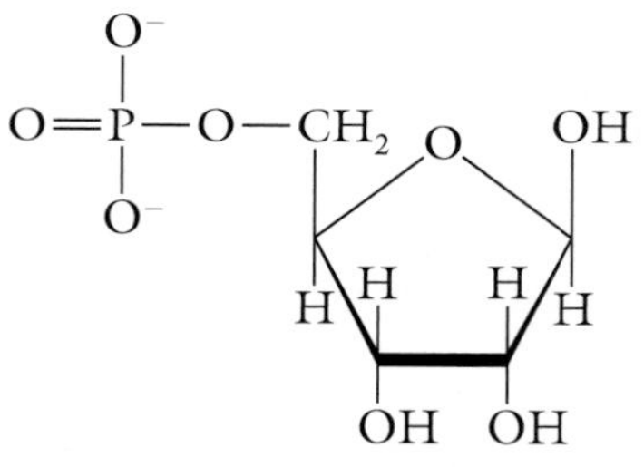

β-D-Ribose-5-phosphate

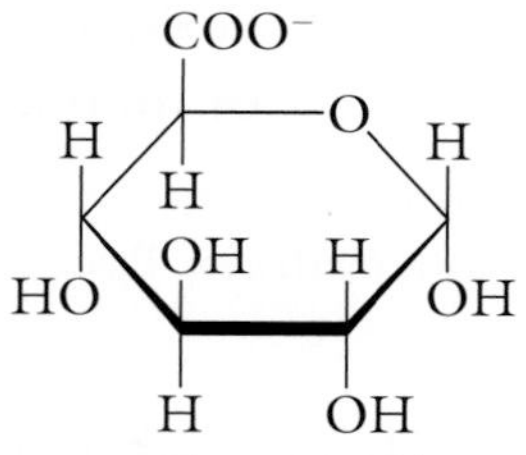

α-D-Glucuronic acid

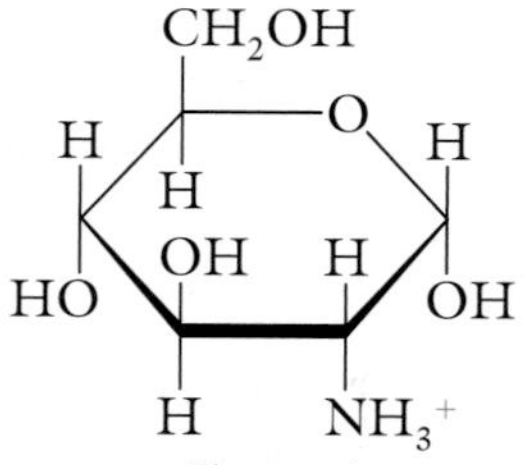

α-D-Glucosamine

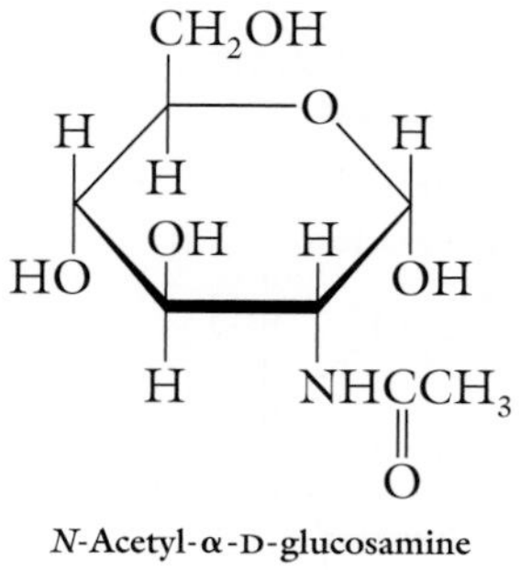

***N*-Acetyl-α-D-glucosamine**

α-D-Glucuronic acid-6-sulfate and *N*-sulfo-α-D-glucosamine-6-sulfate, derived from D-glucuronic acid and D-glucosamine, respectively, are building blocks for heparin, an anticoagulant that is normally present in many body tissues. Heparin binds strongly to the blood protein antiprothrombin and the resulting complex inhibits blood clotting. Heparin is used medicinally to reduce blood clotting in persons who are at risk of heart attack or who have undergone heart surgery. It is also added to blood during blood donation and to blood stored in blood banks.

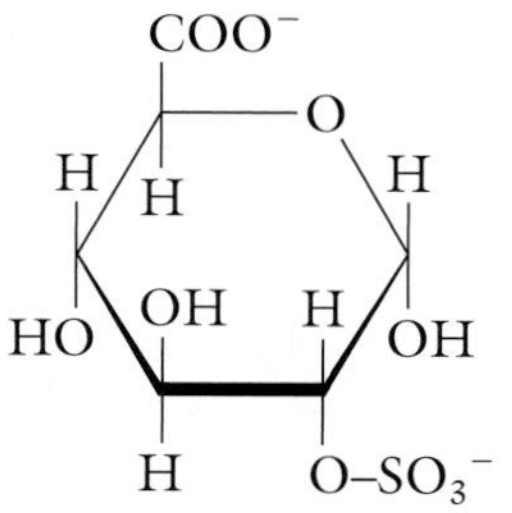

α-D-Glucuronic acid-2-sulfate

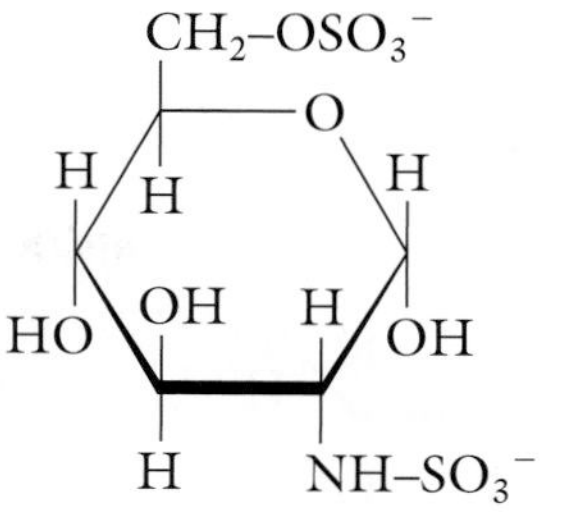

N-Sulfo-α-D-glucosamine-6-sulfate

10.5 DISACCHARIDES

Disaccharides consist of two monosaccharide residues, either the same or different monosaccharides, joined by a glycosidic linkage. We can visualize the formation of the glycosidic linkage as dehydration between the hemiacetal —OH of one monosaccharide and an —OH group (either the hemiacetal —OH or an alcohol —OH) of the other monosaccharide. This linkage is the same as that described in Section 10.4 for the reaction of a monosaccharide with an alcohol.

To characterize a disaccharide, we need to determine several structural features:

- What are the monosaccharides?
- Is the glycosidic linkage an α or a β linkage? Laboratory reactions produce mixtures of α- and β-glycosides. Nature, however, is stereoselective, forming exclusively one or the other.
- Do both monosaccharides link through their hemiacetal —OH groups? If not, for the monosaccharide that does not link through its hemiacetal —OH, which —OH group participates in the linkage? For aldohexoses such as D-glucose, it is usually (but not always) the —OH at C4.

We will consider four disaccharides: lactose and sucrose, the two most important naturally occurring disaccharides; and maltose and cellobiose, which are not present as such in nature but are produced in the breakdown of the naturally occurring polysaccharides starch and cellulose.

Maltose

Maltose, also called **malt sugar** or **corn sugar,** is produced by the partial digestion (hydrolysis) of starch by the enzyme **amylase,** which is secreted into the mouth by the salivary glands and into the small intestine by the pancreas. Plants also produce amylase, which is responsible for the formation of maltose when grains such as barley are allowed to soften in water and germinate. Subsequent fermentation by yeast produces alcoholic beverages (Box 5.2). Maltose is also produced in processed corn syrup, which is used in beverages such as malted milk. The hydrolysis of maltose, using either the enzyme **maltase** or an acid catalyst, shows that maltose is diglucose. The two D-glucose residues are linked together by an α(1 → 4)-glycosidic linkage. Glycosidic linkages are denoted by the following convention:

- The α in α(1 → 4)- indicates the anomeric configuration of the residue that is linked through its hemiacetal —OH.
- The first number in α(1 → 4)- indicates the position of the hemiacetal carbon in the same residue: C1 in aldoses; C2 in ketoses.
- The second number in α(1 → 4)- indicates the position of the carbon atom on the other residue to which it is attached.

The linkage between monosaccharide residues is most easily understood by visualizing the formation of D-maltose from two molecules of D-glucose by a

INSIGHT INTO STRUCTURE

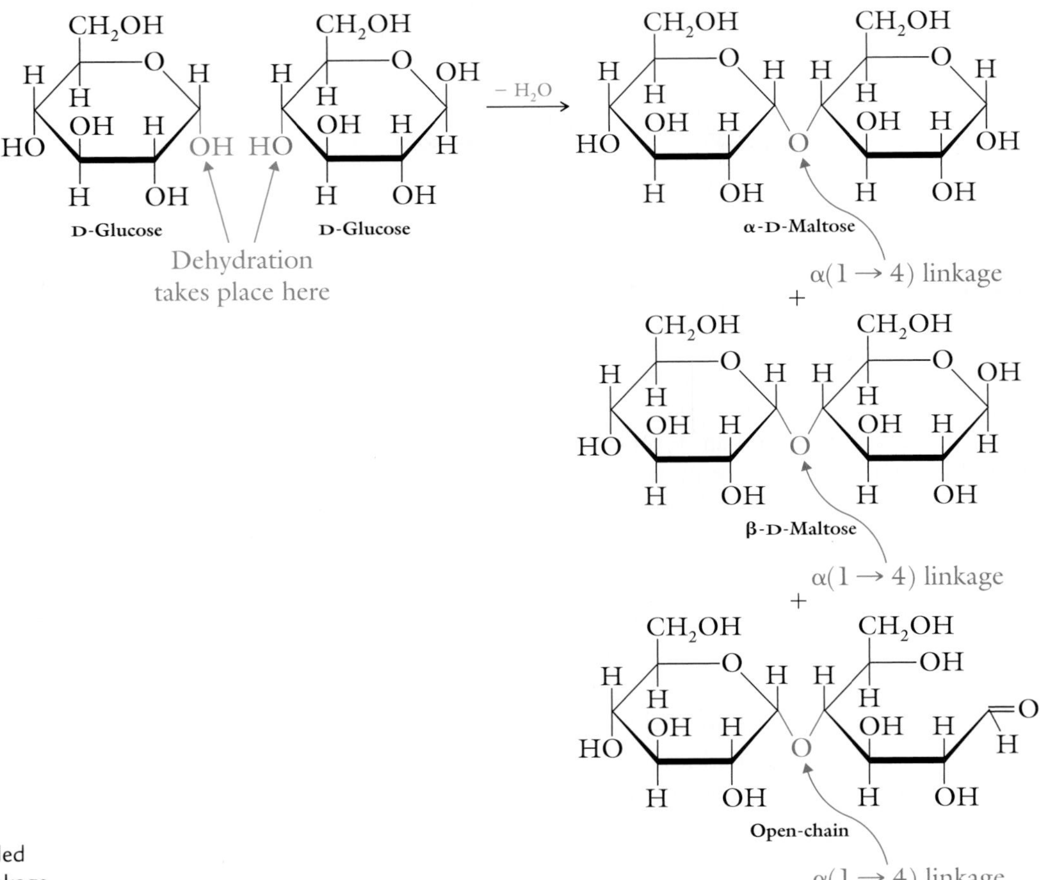

Figure 10.5 Maltose has two D-glucose residues bonded by an α(1 → 4)-glycosidic linkage.

dehydration reaction, as shown in Figure 10.5. (The linking together of monosaccharides in living cells is more complicated than a simple dehydration—Section 15.9.)

Because maltose has a free hemiacetal —OH on one of the two D-glucose residues (the one on the right side in Figure 10.5), it consists of an equilibrium mixture of α-maltose, β-maltose, and open-chain maltose. Thus, maltose is a reducing sugar, it gives a positive test with Benedict's reagent, and it undergoes mutarotation.

Do not confuse the α- in α-maltose with that in α(1 → 4)-. The α- in α-maltose refers to the α-anomer of maltose. The α- in α(1 → 4)- refers to the configuration of the glycosidic linkage between the two monosaccharide residues.

Note that we often omit the "D-" from the names of maltose and other saccharides, especially those larger than monosaccharides. Always remember that all saccharides are D-saccharides, unless otherwise noted.

Milk contains the disaccharide lactose; the cookies contain the disaccharide sucrose and the polysaccharide starch. (Ronnie Kaufman/The Stock Market.)

Cellobiose

Cellobiose is produced by the partial hydrolysis of cellulose, brought about by the enzyme **cellulase** or an acid catalyst. The hydrolysis of cellobiose by the enzyme **cellobiase** or an acid catalyst shows that cellobiose is diglucose. The only difference between maltose and cellobiose is the linkage between D-glucose residues: cellobiose has a β(1 → 4)-glycosidic linkage rather than an α(1 → 4)-glycosidic linkage (Figure 10.6). Like maltose, cellobiose is a reducing sugar and undergoes mutarotation.

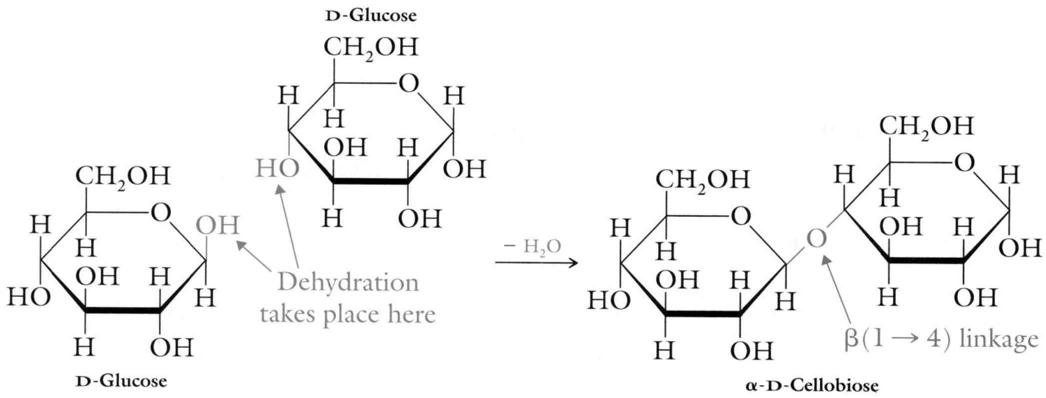

Figure 10.6 Cellobiose has two D-glucose residues bonded by a β(1 → 4)-glycosidic linkage. Cellobiose is an equilibrium mixture of α- and β-anomers and the open-chain (acyclic) structure. (Only the α-anomer is shown.)

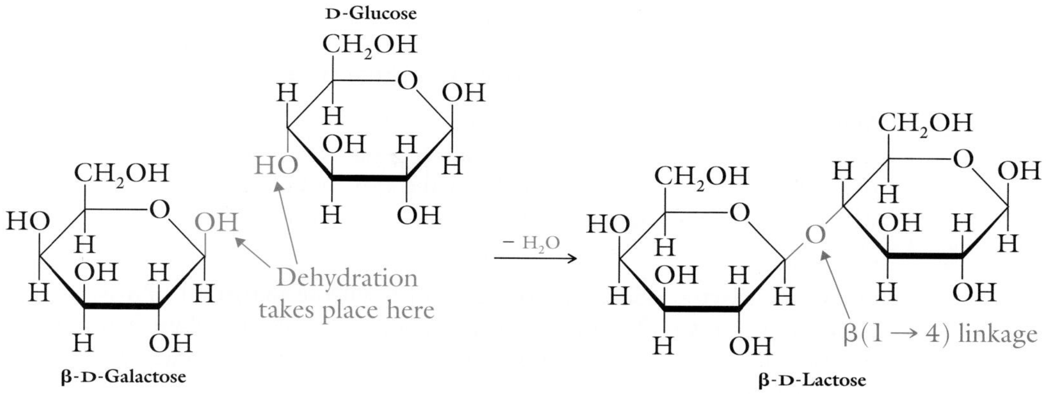

Figure 10.7 Lactose has D-galactose bonded to D-glucose by a β(1 → 4)-glycosidic linkage. Lactose is an equilibrium mixture of α- and β-anomers and the open-chain (acyclic) structure. (Only the β-anomer is shown.)

Lactose

Lactose, or **milk sugar,** constitutes from about 4% to 8% of mammalian milk and is a major energy source for nursing young. Lactose contains D-galactose and D-glucose joined by a β(1 → 4)-glycosidic linkage, from the hemiacetal —OH (at C1) of D-galactose to the C4 position of D-glucose (Figure 10.7). Because the hemiacetal —OH of D-glucose is retained in lactose, lactose is a reducing sugar and undergoes mutarotation.

Lactose intolerance and galactosemia are significant hereditary diseases (Box 10.2 on page 327 and Section 18.1).

Sucrose

Sucrose is the most abundant disaccharide in nature. It is found in the fruits and vegetables of our diet. Sugar cane and sugar beets are the commercial sources of sucrose used as table sugar to sweeten coffee, tea, soft drinks, ice cream, and baked goods. Sucrose is formed by acetal formation between the hemiacetal —OH groups of α-D-glucose and β-D-fructose (Figure 10.8 on the following page). The linkage between the monosaccharide residues is an α,β(1 → 2)-glycosidic linkage; that is, the linkage is an α-glycosidic linkage with respect to D-glucose but a β-glycosidic linkage with respect to D-fructose. Unlike the other disaccharides, sucrose does not contain a hemiacetal linkage

Sugar cane contains the disaccharide sucrose. (Arthur Morris/Visuals Unlimited.)

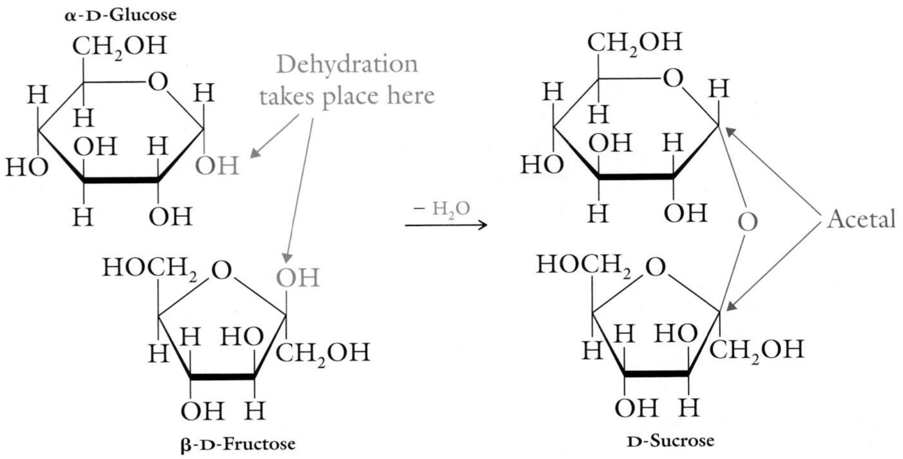

Figure 10.8 Sucrose has D-glucose and D-fructose bonded by an α,β(1 → 2)-glycosidic linkage. There are no anomers and no open-chain structure for sucrose.

α-D-Glucose

Dehydration takes place here

$-H_2O$

Acetal

β-D-Fructose

D-Sucrose

and is therefore not a reducing sugar; it does not give a positive Benedict's test and does not undergo mutarotation.

Example 10.5 Drawing the structure of a disaccharide

Melibiose, a disaccharide found in some plants, contains D-galactose linked to D-glucose by an α(1 → 6)-glycosidic linkage—an α-glycosidic linkage from C1 of D-galactose to C6 of D-glucose. Draw the structure of melibiose and indicate whether it is a reducing sugar. Does it undergo mutarotation?

Solution

Draw α-D-galactose above and to the left of D-glucose so that C1 of α-D-galactose is near C6 of D-glucose. The removal of a molecule of water from the hemiacetal —OH of α-D-galactose and the —OH at C6 of D-glucose gives the structure of melibiose.

α-D-Galactose

Dehydration takes place here

$-H_2O$

D-Glucose

β-D-Melibiose

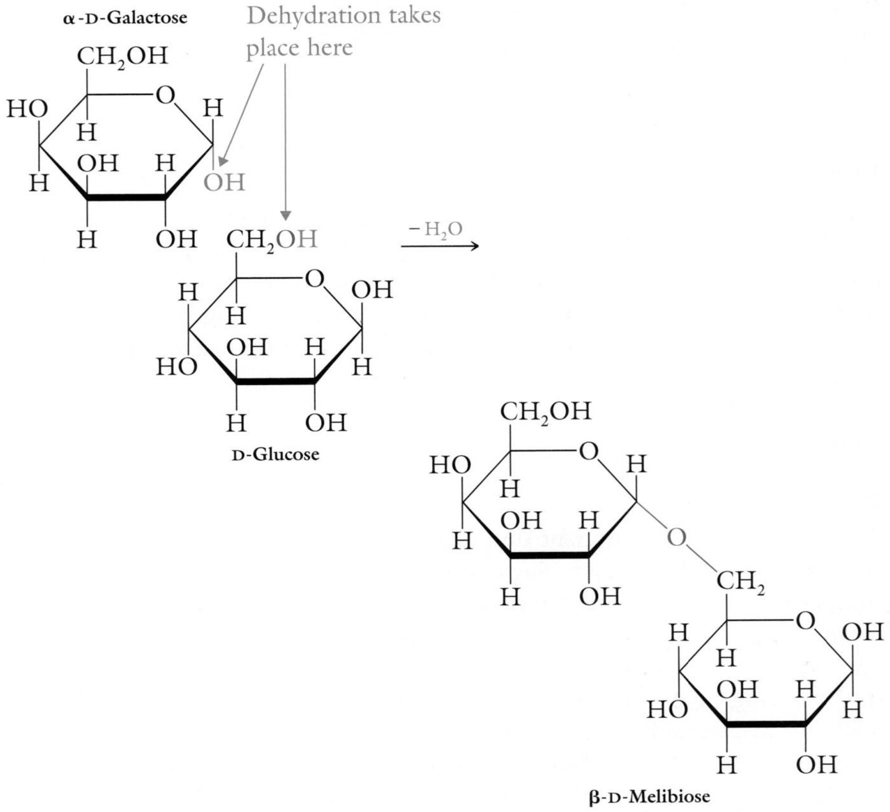

10.2 CHEMISTRY WITHIN US

Hereditary Problems of Lactose Use

Milk and milk products, such as cheese and ice cream, are not suitable foods for everyone. Some people are unable to use—in fact, unable to tolerate—lactose, the disaccharide present in milk. The digestion of lactose to D-galactose and D-glucose requires the enzyme lactase. Most infants and children produce sufficient lactase to digest lactose, but, in adulthood, many people develop a significant lactase deficiency that severely limits their ability to digest milk and milk products, a hereditary condition known as **lactose intolerance.** Unhydrolyzed lactose accumulates in the small intestine after ingestion because disaccharides are too large to pass through the intestinal membrane and enter the metabolic pathways. Two problems result: (1) osmotic pressure in the intestine increases, causing an influx of fluid into the intestine to dilute the lactose solution, and (2) bacterial fermentation of lactose in the colon produces large amounts of carbon dioxide and irritating acids. The consequences are abdominal distention and cramping, nausea, pain, and diarrhea.

People can manage their lactose intolerance by limiting the intake of milk and milk products, taking commercially prepared lactase orally with food, and using milk and milk products in which the lactose has been hydrolyzed enzymatically.

The incidence of lactose intolerance in population groups varies with geographical origins. Most Asians, almost 80% of Africans, and 20% of Caucasians (but less than 5% of Scandinavians) are lactose intolerant.

In another hereditary condition, **galactosemia,** lactose is hydrolyzed to D-galactose and D-glucose, but the enzymes that catalyze the conversion of D-galactose into D-glucose, a step necessary for normal metabolism, are defective. Infants with galactosemia fail to thrive on milk products. D-Galactose accumulates in the bloodstream and causes mental retardation, impaired liver function, cataracts, and even death. Galactosemia is treated only by a galactose-free diet. If the disorder is recognized in infancy, its effects can be eliminated by removing milk and milk products and other sources of D-galactose from the diet. Except for mental retardation, which can develop in a child before the disorder is diagnosed, most of the symptoms of galactosemia are minimized by controlling the diet. Galactosemia is generally not a problem in people other than infants, because the body eventually produces the required enzyme for galactose metabolism.

Melibiose has a hemiacetal group on the D-glucose residue and, therefore, is a reducing sugar and undergoes mutarotation.

Problem 10.5 What monosaccharides are obtained when the following disaccharide is hydrolyzed? (Refer to Figures 10.1 through 10.4 as needed.) Is the disaccharide a reducing sugar? Does it undergo mutarotation? Characterize the linkage between monosaccharide residues as α or β.

CH_2OH H O H H OH HO HO OCH_2 H H O CH_2OH H H HO OH OH H

The Digestion and Absorption of Carbohydrates

Maltose (from partial digestion of starch), lactose, and sucrose are part of the diet of humans and other mammals. However, these disaccharides cannot be absorbed as such from the intestinal tract (Section 18.1). Only monosaccharides are small enough to pass through the membranes of the intestinal cells and into the bloodstream. Larger saccharides must first be digested to monosaccharides, which is accomplished by the enzymes maltase, lactase, and sucrase, respectively.

Note that enzymes are generally named by using the prefix of the name of the compound undergoing reaction (digestion in this case) and adding **-ase** as a suffix.

10.6 POLYSACCHARIDES

Polysaccharides are polymers containing large numbers (usually hundreds or even thousands) of monosaccharide residues (repeat units) bonded together. They have a variety of structures and functions in living organisms. Some, such as starch and glycogen, serve as storage forms of D-glucose. Other polysaccharides, such as cellulose and chitin, serve as structural materials. Some contain only one type of monosaccharide residue; others contain more than one. Some oligosaccharides combine with proteins or lipids to form complex compounds that permit cells to recognize other cells or molecules in their surroundings.

Polysaccharides differ structurally in several ways:

- In the monosaccharide(s) that constitute(s) the residues.
- In the —OH group(s) that participate(s) in linking the monosaccharide residues.
- In the glycosidic linkage (α or β), when one of the —OH groups is the hemiacetal —OH (as is usual).
- In the presence or absence of branching in the polysaccharides (as described next).

Starch and Glycogen

Starch and glycogen are **storage (nutritional) polysaccharides:** D-glucose is stored as starch in plants and as glycogen in animals. **Starch** is deposited in plant cells as insoluble granules composed of two different types of polyglucose molecules, **amylose** and **amylopectin.** Most starches are from 10% to 30% amylose and from 70% to 90% amylopectin, the relative amounts depending on the plant species.

Amylose is a linear (unbranched) polymer of D-glucose residues linked through $\alpha(1 \rightarrow 4)$-glycosidic linkages. Starch contains a mixture of amylose molecules, ranging from a few thousand to one-half million atomic mass units in molecular mass. Figure 10.9 shows the structure of amylose.

Amylopectin, the major component of starch, is a branched polymer. Like amylose, it contains D-glucose residues linked through $\alpha(1 \rightarrow 4)$-glycosidic linkages but, in addition, it branches repeatedly through $\alpha(1 \rightarrow 6)$-glycosidic linkages (Figure 10.10). Not only are there branches, but there are branches on branches (Figure 10.11). Except at the branch points, the linkages between D-glucose residues are $\alpha(1 \rightarrow 4)$-glycosidic linkages. Amylopectins have molecular masses as high as 1 million amu or more.

Glycogen is similar to amylopectin but more extensively branched.

Recall that reducing sugars are those that have a hemiacetal group and thus an aldehyde group. One of the polymer chain ends of amylose—the chain end on the right side of the molecule described by Figure 10.9—is a hemiacetal and has reducing power, but the other end does not. Only one chain end of amylopectin is a hemiacetal. All other chain ends are acetals and have no reducing power. Even though both amylose and amylopectin have one reducing end, starch does not give a positive test with Benedict's reagent and does not undergo mutarotation. The reducing end of each amylose and amylopectin molecule constitutes less than 0.1% of each molecule because of their high

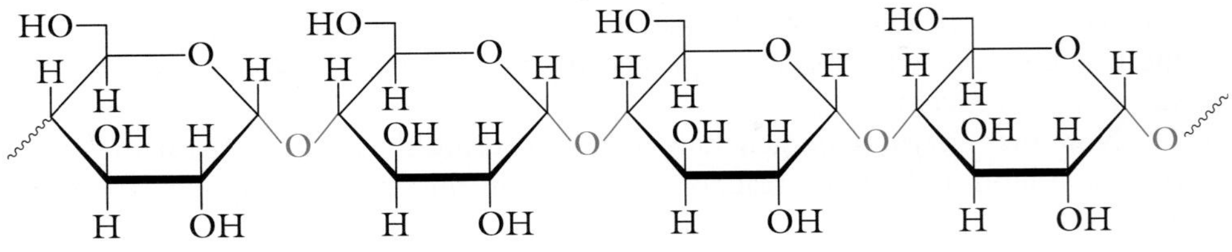

Figure 10.9 Amylose is a linear polymer of D-glucose residues bonded together by $\alpha(1 \rightarrow 4)$-glycosidic linkages.

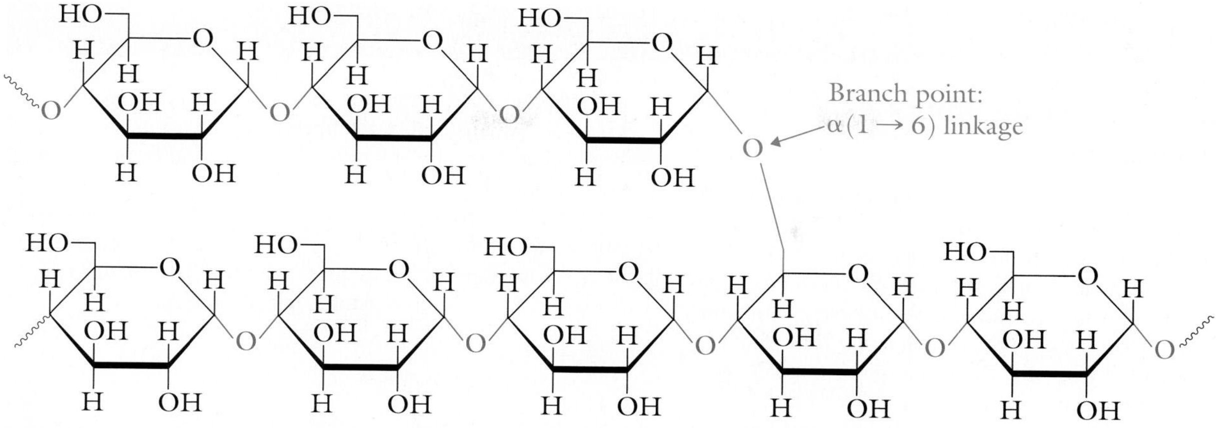

Figure 10.10 Amylopectin contains D-glucose residues bonded together by $\alpha(1 \rightarrow 4)$-glycosidic linkages (red) with branching through $\alpha(1 \rightarrow 6)$-glycosidic linkages (blue).

molecular masses. This minute amount of reducing function in starch is insufficient to give positive results in the Benedict's and mutarotation experiments.

Concept check

✓ All polysaccharides are nonreducing sugars; they do not give a positive Benedict's test, and they do not undergo mutarotation.

Starch and Glycogen in Digestion and Metabolism

Starch is the principal carbohydrate of our diet and is digested to D-glucose. Amylase in the digestive tract hydrolyzes the amylose and amylopectin to maltose. Maltase cleaves maltose to D-glucose (Section 10.5). Amylase and maltase are unable to hydrolyze $\alpha(1 \rightarrow 6)$-glycosidic linkages, however, and thus leave the parts of amylopectin at the branch points unhydrolyzed. This product, **dextrin,** is hydrolyzed by the enzyme **dextrinase** to D-glucose.

Some of the D-glucose produced by the digestion of starch is used immediately for energy production. Some of the excess D-glucose is polymerized into glycogen by a process called **glycogenesis** and stored in the liver and skeletal muscle. D-Glucose in excess of the amount needed to maintain the glycogen reserve is converted into lipids, which are deposited in fat tissue. When required for muscle action and other metabolic processes, D-glucose is cleaved from the chain ends of glycogen by a process called **glycogenolysis**. The highly branched glycogen structure is very efficient for fast generation of D-glucose because cleavage takes place at the multitude of chain ends simultaneously.

›› The roles of glycogenesis and glycogenolysis in utilizing D-glucose are considered in Chapter 15.

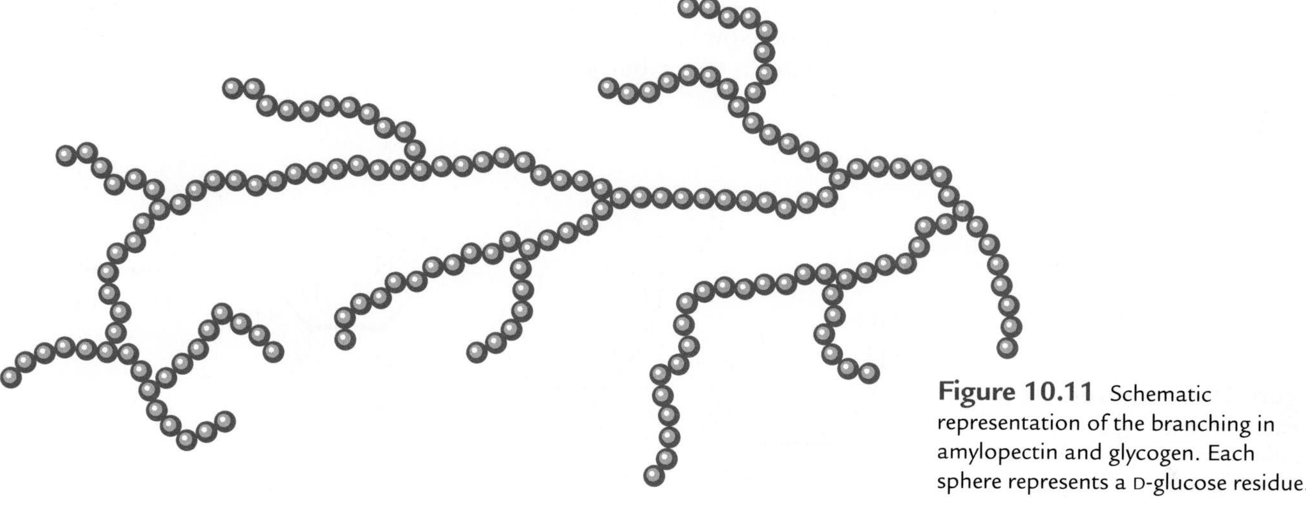

Figure 10.11 Schematic representation of the branching in amylopectin and glycogen. Each sphere represents a D-glucose residue.

10.3 CHEMISTRY AROUND US

Plastics and Textile Fibers from Cellulose

The high cellulose content of wood and cotton is responsible for the wide-ranging utility of both materials. The hairs on the seed of the cotton plant, for example, are more than 95% cellulose. Fabrics suitable for clothing are produced when the seed hairs are spun into thread and then woven. Not only do cotton fabrics hold up well to laundering, but they are comfortable on the body, the latter because of cotton's high moisture absorption and heat conduction. The annual worldwide production of cotton is more than 40 billion pounds.

The overall composition of wood is about 50% cellulose, with most of the remainder made up of **lignin,** an organic polymer containing aromatic groups. The cellulose fibers are imbedded in lignin, thus creating a composite material of high physical strength. (The same approach is used to strengthen concrete by reinforcing it with steel rods.) The composite structure gives large trees their enormous strength and ability to withstand their own weight as well as very strong winds.

Wood is processed into a variety of products. It is cut and used to construct furniture, housing, and other structures. It is converted into paper for books, newspapers, magazines, and cardboard boxes. Cellulose separated from wood is the starting material for various fiber (rayon) and plastic (cellophane, cellulose acetate) products. **Cellophane** is used in packaging wrap for food and tobacco products. **Rayon** is a widely used textile fiber, although not as strong as cotton fiber. The annual worldwide production is more than 4 billion pounds.

Cellulose acetate, produced by the reaction of cellulose with acetic anhydride, $(CH_3CO)_2O$, is used in cigarette filters, photographic film, lacquers and protective coatings for automobiles and furniture, and eyeglass frames.

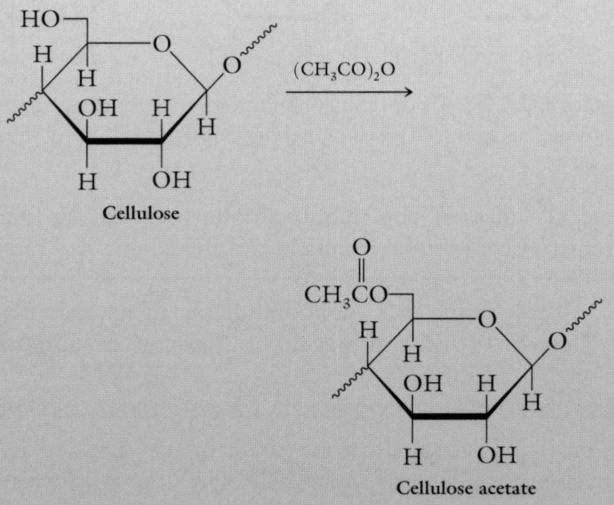

$(CH_3CO)_2O$

Cellulose

Cellulose acetate

Cellulose

Cellulose, a **structural polysaccharide,** is the most abundant organic compound in the biosphere, accounting for more than half of all organic carbon. It is a component of the construction material of plant-cell walls that gives plants their overall shape and physical strength. Cotton is almost pure cellulose. Cellulose is linear polyglucose with β(1 → 4)-glycosidic linkages between D-glucose residues (Figure 10.12), as described for cellobiose in Section 10.5.

Cellulose and starch—both polyglucose molecules produced by plants—are very different materials. Consider what happens when you add water to starch—say, corn starch. The starch becomes swollen by water and forms a paste, or colloidal dispersion. We use this property of starch to thicken gravies and sauces. Now consider adding water to cellulose—say, a piece of wood. The wood absorbs water but generally retains its shape and much of its physical

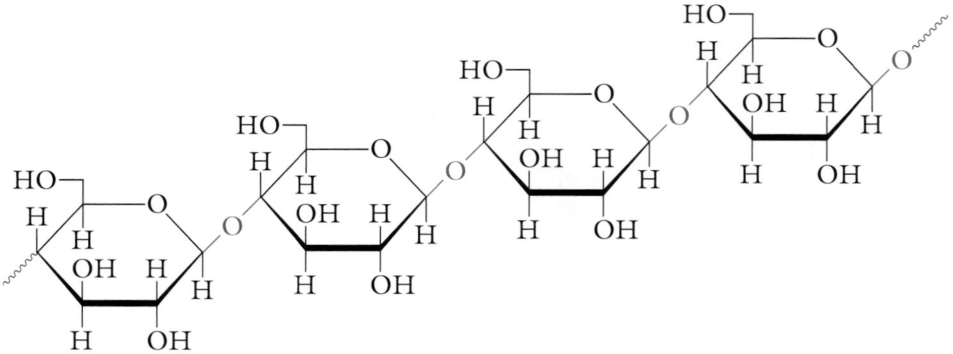

Figure 10.12 Cellulose is a linear polymer of D-glucose residues bonded by β(1 → 4)-glycosidic linkages.

strength. The difference in properties is a consequence of differences between the three-dimensional shapes, or conformations (Section 3.3), of the cellulose and starch molecules, which in turn depend on the difference in glycosidic linkages.

Many different conformations are possible for polymers because of rotations about the many single bonds in the polymer chain. We generally find that some feature of the polymer's chemical structure (the glycosidic linkage in this case) results in the polymer's existing in a specific conformation because that conformation is more stable than all others. Most importantly, the physiological functioning of biological polymers such as polysaccharides critically depends on conformation. (Chapter 12 contains a more detailed discussion of polymer conformation and its effect on physiological function.)

Cellulose molecules exist in extended chain conformations because of the β(1 → 4)-glycosidic linkages between glucose units. In the extended conformations, the chains form ribbons that pack side by side and one on top of the other and are stabilized by intramolecular and intermolecular hydrogen bonding (Figure 10.13). This property of cellulose makes it useful as a structural material in nature, as well as in the industrial manufacture of paper, cardboard, structural materials, textiles, and plastics (Box 10.3). The α(1 → 4)-glycosidic linkages of amylose and amylopectin, on the other hand, prevent the formation

Cellulose is a component of the construction materials that give this sunflower its overall shape and physical strength. (N/A Jasmin/PNI.)

INSIGHT INTO FUNCTION

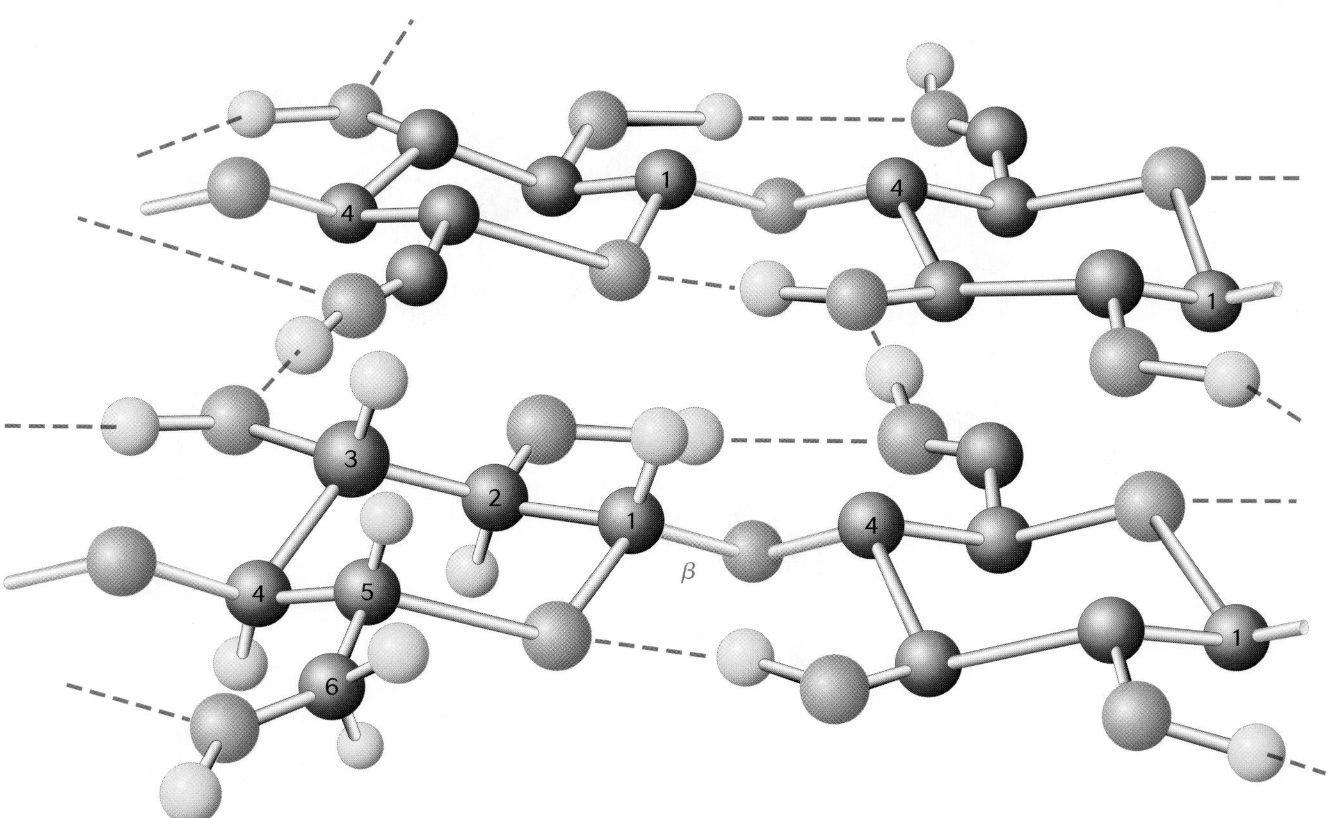

Figure 10.13 Extended chain conformations of parts of two different cellulose chains are shown held together by intermolecular hydrogen bonding. Hydrogen bonding continues in the same plane with other chains as well as in planes above and below this plane to form strong, fibrous bundles. For simplicity, only the glucose residue in the left forefront is shown with all hydrogens attached to carbons. (Adapted from C. K. Mathews and K. E. Van Holde, *Biochemistry,* 2d ed. Menlo Park, CA: Benjamin Cummings, 1996. ©Irving Geis.)

of extended chain conformations. Instead, amylose and amylopectin form helices stabilized by intramolecular hydrogen bonding (Figure 10.14).

Concept checklist

✓ Extensive intermolecular hydrogen bonding between adjacent molecules gives cellulose its great physical strength. Cellulose molecules are bound together in fibrous bundles that exclude large attractive interactions with water.

✓ The lack of intermolecular hydrogen bonding imparts only modest physical strength to starch. The amylose and amylopectin helices are easily solvated by water, which allows for rapid hydrolysis to glucose.

The difference in the glycosidic linkages of cellulose and starch have another important consequence: the enzymes that cleave $\alpha(1 \rightarrow 4)$ linkages do not cleave $\beta(1 \rightarrow 4)$ linkages and vice versa. Humans cannot digest cellulose, because we do not possess the enzyme cellulase, which cleaves $\beta(1 \rightarrow 4)$ linkages. However, cellulose is the main source of nutritional carbohydrate for grazing animals such as cows, sheep, horses, and deer and for insects such as termites. These species do not themselves possess cellulase, but their digestive

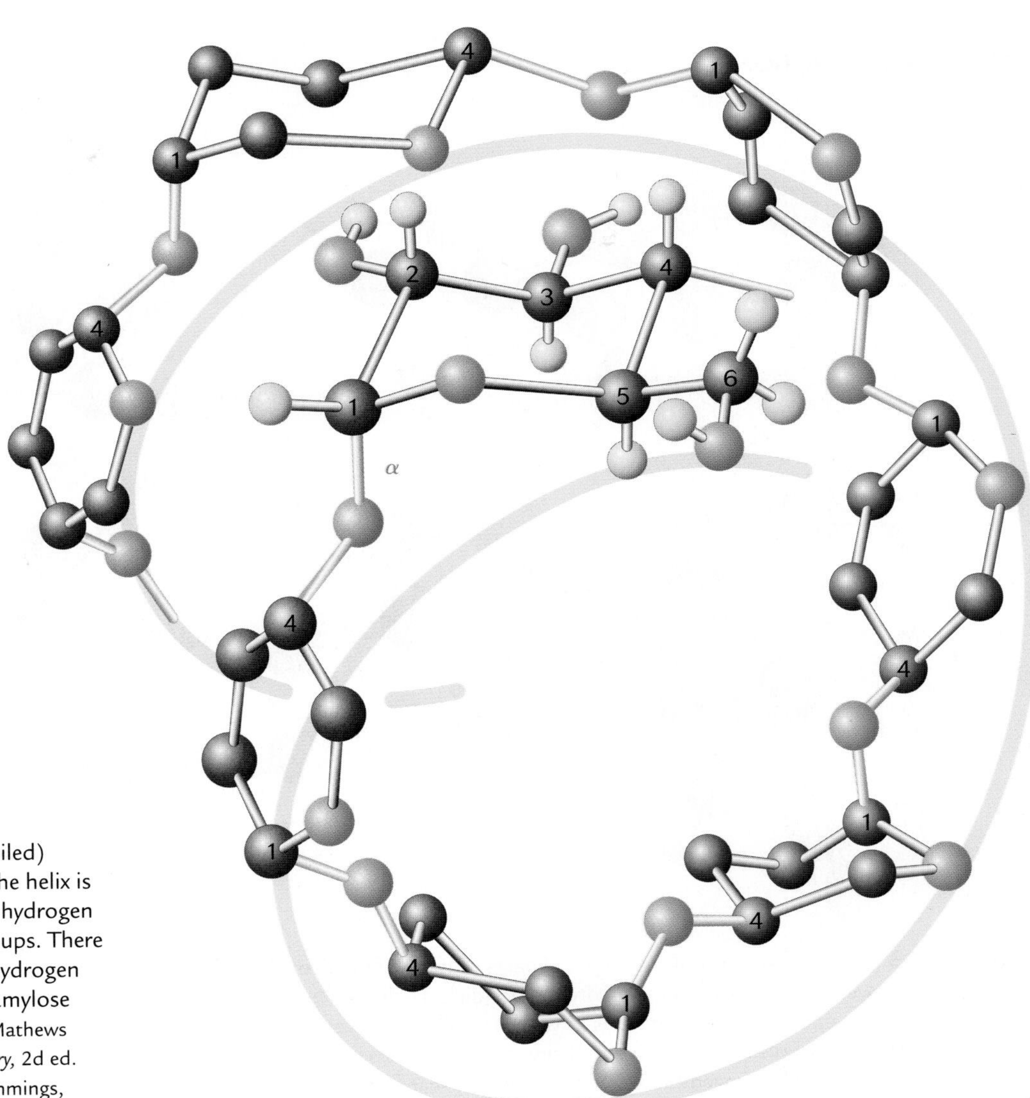

Figure 10.14 Helical (coiled) conformation of amylose. The helix is stabilized by intramolecular hydrogen bonding between —OH groups. There is very little intermolecular hydrogen bonding between different amylose helices. (Adapted from C. K. Mathews and K. E. Van Holde, *Biochemistry,* 2d ed. Menlo Park, CA: Benjamin Cummings, 1996. ©Irving Geis.)

10.4 CHEMISTRY WITHIN US

Dietary Fiber

Not all components of the foods that we ingest are digestible. Nondigestible polysaccharides called **dietary fiber, bulk,** or **roughage** are part of any diet that includes fruits, vegetables, and grains. These polysaccharides have beneficial effects even though they have no nutritional value. Dietary fiber softens stools by absorbing water and helps stimulate the peristaltic movements of the bowel—contractions of the muscles of the intestinal wall—thus facilitating the more rapid passage of material through the intestinal tract and improving the regularity of bowel movements. When stools move more quickly through the colon, intestinal bacteria have less time in which to produce harmful chemicals. An additional benefit of dietary fiber is control of appetite and weight by giving a feeling of fullness.

Dietary fiber can be insoluble, as is the cellulose present in vegetable stalks and leaves, wheat, bran, and brown rice, or it can be soluble, as is the pectin present in fruits, barley, and oats. **Pectin** is a polysaccharide containing mostly residues of D-galacturonic acid with α(1→4)-glycosidic linkages. Legumes such as beans and peas contain both insoluble and soluble fiber.

α-D-Galacturonic acid

Although the separate roles of insoluble and soluble fiber are not understood, there is speculation that soluble fiber is effective in absorbing harmful materials, including carcinogens and cholesterol.

Epidemiological evidence suggests that high-fiber diets reduce the incidence of colon and rectal cancer, hemorrhoids, diverticulitis, diabetes, and cardiovascular disease. On the other hand, the principal reason for the reduction may be that diets high in fiber are generally low in meats and fats.

tracts contain symbiotic microorganisms that secrete it. Although cellulose does not play a nutritional role for humans, dietary cellulose has some beneficial effects (Box 10.4).

Starch can be differentiated from cellulose not only by the difference in their physical properties but also by a simple chemical test. The addition of a few drops of iodine (I_2) solution to starch results in a dark blue color. The dark blue color is not observed when iodine is added to cellulose. The difference in conformation between starch and cellulose is responsible for the different results. Iodine is held tightly inside the amylose helix by strong attractive forces between iodine and amylose. This strong interaction between iodine and amylose is responsible for the dark blue color.

Note that the six-membered rings are shown in their chair conformations in Figures 10.13 and 10.14. Because of the tetrahedral bond angles for carbon atoms, the chair is the actual structure assumed by the six-membered ring (see Box 3.2). The chair structure and the type of glycosidic linkage are responsible for the formation of extended-chain or helical conformations of the different polysaccharides. For simplicity, however, we usually show the six-membered ring as a flat planar structure.

Cell Recognition: Glycolipids and Glycoproteins

Every multicellular organism has a variety of cell types, each performing specialized tasks. It is crucial that different kinds of cells recognize and interact properly with other cells and molecules and that an organism must recognize its own cells as distinct from foreign cells. In many cases, these tasks are accomplished by a process called **cell recognition,** whereby cells recognize one another because of saccharides attached to cell surfaces. The saccharides, usually oligosaccharides, are present as **glycolipids** or **glycoproteins,** molecules having saccharide parts bonded to lipid or protein molecules, respectively. The lipid or protein part of the glycolipid or glycoprotein is integrated into the cell-membrane structure, with the saccharide part located on the external membrane surface. The glycolipids and glycoproteins are actually part of the cell membrane.

» **Every cell is separated from its extracellular environment by a cell membrane, as considered in Section 11.10.**

TABLE 10.2 ABO Blood-Typing System

	Blood type			
Characteristic	A	B	AB	O
antigen on red-blood-cell surface	A	B	A, B	O
makes antibodies against antigen	anti-B	anti-A	none	anti-A, anti-B
can receive from blood type	O, A	O, B	O, A, B, AB	O
can donate to blood type	A, AB	B, AB	AB	O, A, B, AB

Cell recognition is usually through secondary attractive forces between the saccharide on the cell surface and a protein on another cell surface or in the extracellular matrix. An example of this phenomenon is fertilization, the union of an ovulated egg and a sperm. The cell membrane of an ovulated egg has oligosaccharide units projecting from its glycoproteins. Protein receptor sites on a sperm cell of the same species recognize and bind to these oligosaccharides. Binding is followed by the release of enzymes by the sperm cell. The enzymes dissolve the coat of the egg and allow entry of the sperm.

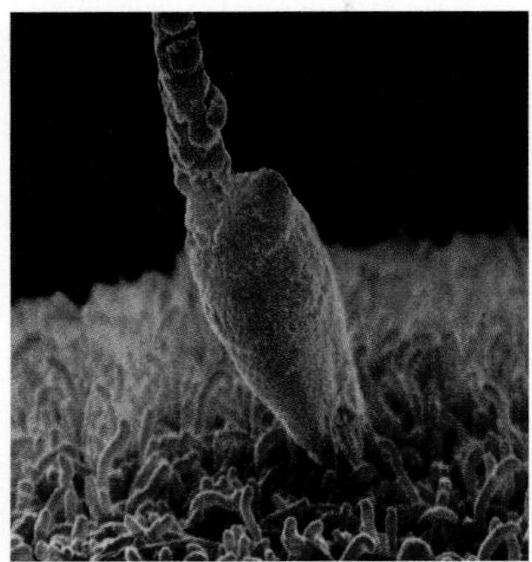

Scanning electron micrograph of human sperm and egg at the moment of penetration. (David M. Phillips/ Photo Researchers.)

Cell recognition is critical to success in performing blood transfusions. An animal produces a protein called an **antibody** (also called an **immunoglobulin**) in response to the entry of a foreign substance called an **antigen.** Antigens can be proteins, saccharides, or nucleic acids. Antibodies recognize and have a high affinity for the antigens that triggered their production. The four types of human red blood cells (A, B, AB, and O) have different oligosaccharides projecting from their cell membranes. The oligosaccharides act as antigens when a foreign red blood cell is transfused into a patient. The body produces antibodies that recognize and distinguish the oligosaccharide of its own blood type from those of other blood types. There are three types of antigens (A, B, and O), but only two types of antibodies (anti-A and anti-B). There are no antibodies against O antigens. The A and B oligosaccharide antigens contain six monosaccharide residues; the O antigen contains five monosaccharide residues.

Certain blood types are compatible for transfusion and others are not, depending on the antigen–antibody combination, as described in Table 10.2. For example, people with type O blood are universal donors (can donate blood to people of any type), whereas those with type AB blood are universal recipients (can receive blood from people of any type). Incompatible antigen–antibody combinations can result in death on transfusion. An example is the transfusion of type A blood into a type B recipient. The antibodies in the recipient's blood are anti-A and recognize the A antigens of the transfused blood cells as foreign and then bind to them through secondary attractive forces. This antibody–antigen binding results in the clumping together of the transfused blood cells and possible death due to the clogging of arteries.

Other processes that require cell recognition include:

- Inhibition of cell growth. Normal cells cease growing at the outer boundaries of organs. Cancerous cells do not possess this inhibition of growth.
- Infection of cells by bacteria and viruses.

10.7 PHOTOSYNTHESIS

Life as we know it, both plant and animal, is possible only because of **photosynthesis,** the process that plants use to trap solar energy and then use the energy to transform carbon from carbon dioxide into organic compounds, in the form of D-glucose and other saccharides. Before the evolution of photosynthetic organisms 3 billion years ago, Earth's atmosphere was lacking in O_2 but rich in CO_2, and primitive organisms were generating energy from molecules produced by nonbiological, nonrenewable means. Without photosynthesis, these energy sources would have eventually run out, and life would have ended. The new, photosynthetic organisms synthesized carbohydrates from carbon dioxide, water, and sunlight. In the process, they added oxygen to Earth's atmosphere and paved the way for oxygen-using, or aerobic, metabolism and the evolution of higher plants and animals. Photosynthesis is the ultimate source of energy and of all organic molecules used by all plant and animal life forms.

Photosynthesis, then, is the process of **carbon fixation**—the conversion of inorganic carbon (CO_2) into organic carbon in the form of carbohydrates, with the release of oxygen into the environment. It takes place in algae and higher plants. About 20% of Earth's photosynthesis takes place in land plants, and about 80% takes place in aquatic plants. The overall equation describing photosynthesis is deceptively simple:

$$x\,CO_2 + x\,H_2O + \text{sunlight} \xrightarrow{\text{chlorophyll}} (CH_2O)_x + x\,O_2$$

where $(CH_2O)_x$ represents monosaccharides with $x = 3$ to 6. The details of the process are quite complicated. The green pigment chlorophyll absorbs the energy of sunlight, which drives a series of reactions to produce the high-energy saccharides from low-energy carbon dioxide and water. The energy stored in saccharides is subsequently retrieved when the saccharides are metabolized by the plants or by animals, which are dependent on plants. The energy is retrieved in a form that is useful to the organisms through a complicated series of reactions that consume oxygen and release carbon dioxide and water.

Plants convert the D-glucose and other monosaccharides formed by photosynthesis into starch, cellulose, and other saccharides, which serve various nutritional and structural functions, as described in Section 10.6. Plants supply not only carbohydrates, but also many other organic biochemicals needed by animals. For example, plants convert carbohydrates into lipids, some of which are essential for animals (that is, lipids that the animals require but cannot themselves synthesize), and plants also convert carbohydrates into amino acids, proteins, nucleic acids, and other nitrogen-containing compounds, some of which are also essential in animal diets. The nitrogen required for these conversions is made available to plants by microorganisms living in and around their roots. Thus, plants are also the source of nitrogen for animals.

In summary, the source of all organic biomolecules for animals is plants. Herbivores obtain all of their carbon and nitrogen directly from plants. Carnivores obtain them from plants indirectly, from the flesh and viscera of the animals in their diets. Omnivores such as humans obtain them directly and indirectly from plants and animals.

Plants and animals are interdependent (Figure 10.15). Plants transform the inorganic carbon of carbon dioxide into organic carbon compounds—lipids, proteins, nucleic acids, and other biomolecules—by using the energy of the sun. In the process, they produce oxygen, which both plants and animals need to metabolize the organic carbon compounds so as to generate energy, synthesize other biomolecules, and carry out many life processes. In doing so, they return carbon dioxide along with water to the biosphere. Thus, there is a **carbon cycle,** in which carbons are shuffled back and forth between inorganic and organic forms, and an **oxygen cycle,** with both plants and animals using oxygen for energy-producing metabolism and plants regenerating oxygen by photosynthesis.

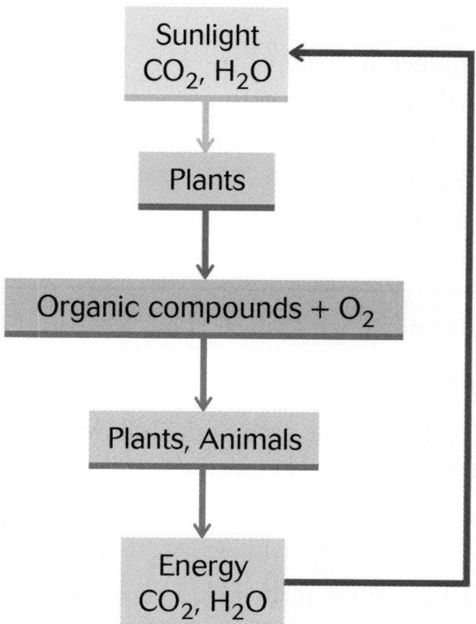

Figure 10.15 Photosynthesis.

Summary

Biochemistry

- Biochemistry is the study of the structure of biomolecules and of how they transmit genetic information and store and use energy.
- Carbohydrates, lipids, proteins, and nucleic acids are the principal families of organic biomolecules.

Monosaccharides

- Carbohydrates (saccharides), the most abundant of the organic biomolecules, are aldoses (polyhydroxyaldehydes) or ketoses (polyhydroxyketones) or compounds that can be hydrolyzed to aldoses or ketoses or are derived from them.
- Monosaccharides, the simplest saccharides, are the building blocks for producing larger saccharides, including disaccharides and polysaccharides.
- Monosaccharides of biological interest are those from three to six carbons.
- Monosaccharides are classified as aldoses or ketoses and by the number of carbon atoms in the chain containing the carbonyl group.
- Aldoses of three or more carbons and ketoses of four or more carbons contain tetrahedral stereocenters.
- Almost all monosaccharides of biological interest belong to the D family, those having the D configuration (the —OH group on the right side in the Fischer projection) at the tetrahedral stereocenter farthest from the carbonyl group.
- D-Glucose, D-fructose, D-glyceraldehyde, dihydroxyacetone, and D-galactose are important monosaccharides.
- D-Glucose is combined with D-fructose and D-galactose in sucrose and lactose, respectively.
- Starch, glycogen, and cellulose are polymers of D-glucose.
- A monosaccharide exists as cyclic hemiacetal α- and β-anomers in equilibrium with very low concentrations of the acyclic (open-chain) compound.
- Diastereomers are possible for monosaccharides containing more than one tetrahedral stereocenter. Anomers are diastereomers that differ in configuration only at the cyclic hemiacetal carbon.
- Monosaccharides undergo mutarotation, oxidation of the aldehyde group with Benedict's reagent, and acetal formation.

Disaccharides and Polysaccharides

- Disaccharides and polysaccharides are formed by glycosidic (acetal) linkages between monosaccharide residues.
- Disaccharides and polysaccharides are characterized by the monosaccharides from which they are produced, the positions of linkage between the monosaccharide rings, and the type of glycosidic linkage, α or β. Maltose, cellobiose, lactose, and sucrose are important disaccharides.
- Starch (composed of amylose and amylopectin), glycogen, cellulose, and chitin are important polysaccharides.
- Glycolipids and glycoproteins take part in cell-recognition processes.

Photosynthesis

- Photosynthesis, the process for carbon fixation, is the source of all carbohydrates.
- Plants use carbon dioxide, water, and sunlight to form monosaccharides, such as D-glucose, from which they synthesize starch, cellulose, and other saccharides, as well as lipids, proteins, and nucleic acids.
- Animals are dependent on plants as their source of organic compounds and energy.

Key Words

aldose, p. 311
anomer, p. 316
carbohydrate, p. 310
carbon fixation, p. 335
cell recognition, p. 333
glycoside, p. 321
glycosidic linkage, p. 321
Haworth projection, p. 316
hemiacetal carbon, p. 316
ketose, p. 311
mutarotation, p. 318
photosynthesis, p. 334
reducing sugar, p. 320
saccharide, p. 310
storage polysaccharide, p. 328
structural polysaccharide, p. 330
sugar, p. 310

Exercises

Monosaccharides

10.1 Define each of the following terms: (a) Aldose; (b) hexose; (c) ketopentose; (d) aldotetrose; (e) D family.

10.2 Define each of the following terms: (a) Ketose; (b) pentose; (c) aldohexose; (d) ketotetrose; (e) L family.

10.3 How many aldopentoses are possible? How many L-aldopentoses are possible?

10.4 How many ketohexoses are possible? How many D-ketohexoses are possible?

10.5 Why are there eight D-aldohexoses but only four D-ketohexoses?

10.6 Why are there four D-aldopentoses but only two D-ketopentoses?

10.7 For each of the following pairs of compounds, indicate whether the pair represents different compounds that are (1) constitutional isomers or (2) stereoisomers that are enantiomers or (3) stereoisomers that are diasteromers or (4) not isomers. Refer to Figures 10.1 through 10.4 as needed for the structures of the compounds. (a) D-Glucose and D-sorbose; (b) D-fructose and D-sorbose; (c) (+)-tagatose and (−)-tagatose; (d) D-sorbose and L-sorbose; (e) D-ribose and 2-deoxy-D-ribose.

10.8 For each of the following pairs of compounds, indicate whether the pair represents different compounds that are (1) constitutional isomers or (2) stereoisomers that are enantiomers or (3) stereoisomers that are diasteromers or (4) not isomers. Refer to Figures 10.1 through 10.4 as needed for the structures of the compounds. (a) D-Glucose and 3-deoxy-D-glucose; (b) D-ribose and D-glucose; (c) D-mannose and L-mannose; (d) D-galactose and D-mannose; (e) D-galactose and D-fructose.

Cyclic Hemiacetal Structures

10.9 When are Haworth projections used for monosaccharides?

10.10 When are Fischer projections used for monosaccharides?

10.11 What is a pyranose?

10.12 What is a furanose?

10.13 What structural feature distinguishes between the α- and the β-anomers of D-glucose?

10.14 What structural feature distinguishes between the α- and the β-anomers of D-fructose?

10.15 Draw the cyclic structure of α-D-allose. Refer to Figures 10.1 and 10.3.

10.16 Draw the cyclic structure of β-D-mannose. Refer to Figures 10.1 and 10.3.

10.17 Draw the cyclic structure of β-D-psicose. Refer to Figures 10.2 and 10.4.

10.18 Draw the cyclic structure of α-D-tagatose. Refer to Figures 10.2 and 10.4.

Chemical and Physical Properties of Monosaccharides

10.19 A solid sample of β-D-galactose is 100% β-D-galactose. What accounts for a solution of β-D-galactose giving a positive Benedict's test? What is experimentally observed in performing the Benedict's test?

10.20 A solid sample of α-D-fructose is 100% α-D-fructose. What accounts for a solution of α-D-fructose undergoing mutarotation? What is experimentally observed during mutarotation?

10.21 How many tetrahedral stereocenters are present in each of the following monosaccharides? (a) D-Glyceraldehyde; (b) D-sorbose; (c) L-mannose; (d) 2-deoxy-D-ribose.

10.22 How many tetrahedral stereocenters are present in each of the following monosaccharides? (a) Dihydroxyacetone; (b) D-ribose; (c) 3-deoxy-D-sorbose; (d) D-galactose.

10.23 Write an equation for the reaction of D-glucose with ethanol to form the ethyl β-D-glycoside.

10.24 Write an equation for the reaction of D-fructose with ethanol to form the ethyl α-D-glycoside.

10.25 Give the structure of the organic product produced when D-galactose reacts with Benedict's reagent.

10.26 Draw the structure of α-D-glucose-1-phosphate, the phosphate ester formed from α-D-glucose at C1.

10.27 Which of the following compounds are reducing sugars? Which undergo mutarotation? (a) D-Glucose; (b) the ethyl β-glycoside of D-glucose; (c) D-mannose; (d) L-glucose.

10.28 Which of the following compounds are reducing sugars? Which undergo mutarotation? (a) D-Sorbose; (b) D-fructose; (c) the methyl α-glycoside of D-fructose; (d) L-fructose.

Disaccharides

10.29 What is an α(1 → 4)-glycosidic linkage between two aldohexoses?

10.30 What is a β(1 → 4)-glycosidic linkage between two aldohexoses?

10.31 What are the products of human digestion of each of the following disaccharides? (a) Maltose; (b) lactose; (c) sucrose; (d) cellobiose.

10.32 What is obtained when each of the following disaccharides undergoes acid-catalyzed hydrolysis? (a) Maltose; (b) lactose; (c) sucrose; (d) cellobiose.

10.33 The disaccharide sucrose contains D-glucose linked to D-fructose by an α,β(1 → 2)-glycosidic linkage; that is, the linkage is an α-glycosidic linkage with respect to D-glucose but a β-glycosidic linkage with respect to D-fructose. Draw the structure of sucrose. Explain why sucrose is not a reducing sugar and does not undergo mutarotation.

10.34 The disaccharide lactose contains D-galactose linked by a β(1 → 4)-glycosidic linkage to D-glucose. Draw the structure of lactose. Explain why lactose is a reducing sugar and undergoes mutarotation.

10.35 Gentiobiose, a disaccharide found in some plants, consists of two D-glucose units linked together by a β(1 → 6)-glycosidic linkage. Draw the structure of gentiobiose. Is gentiobiose a reducing sugar? Does it undergo mutarotation?

10.36 Melibiose consists of D-galactose and D-glucose residues. There is an α(1 → 6)-glycosidic linkage from D-galactose to D-glucose; that is, D-glucose is linked through its C6, and its hemiacetal group is intact. Draw the structure of melibiose. Is melibiose a reducing sugar? Does it undergo mutarotation?

10.37 An unknown is either sucrose or lactose. The blue color of Benedict's reagent is not discharged when the reagent is added to the unknown. Identify the unknown and indicate how you arrive at your answer.

10.38 An unknown is either sucrose or lactose. The blue color of Benedict's reagent is discharged and a red precipitate forms when the reagent is added to the unknown. Identify the unknown and indicate how you arrive at your answer.

Polysaccharides

10.39 Compare the structures and functions of amylose and amylopectin.

10.40 Compare the structures and functions of glycogen and cellulose.

10.41 What are the products of human digestion of each of the following polysaccharides? What are the products of digestion of these polysaccharides by grazing animals such as cows? (a) Amylose; (b) amylopectin; (c) glycogen; (d) cellulose.

10.42 What is obtained when each of the following polysaccharides undergoes acid-catalyzed hydrolysis? (a) Amylose; (b) amylopectin; (c) glycogen; (d) cellulose.

10.43 Chitin, the second most abundant polysaccharide after cellulose, contains *N*-acetyl-D-glucosamine residues (Section 10.4) bonded through β(1 → 4)-glycosidic linkages. Draw the structure of chitin.

10.44 Dextran, the storage polysaccharide in yeasts and bacteria, contains D-glucose residues linked together by α(1 → 6)-glycosidic linkages. Draw the structure of dextran.

Photosynthesis

10.45 What are the immediate sources of organic compounds for animals? What is the ultimate source of organic compounds for animals?

10.46 What is the source of organic compounds for plants?

Unclassified Exercises

10.47 What structural feature(s) define(s) a carbohydrate?

10.48 What is the difference between monosaccharides, disaccharides, and polysaccharides?

10.49 Which carbohydrates are called sugars?

10.50 What structural characteristic in the Fischer projection defines a D-aldohexose rather than an L-aldohexose? in the Haworth representation? Refer to Figure 10.3 if needed.

10.51 What structural characteristic in the Fischer projection defines a D-ketohexose rather than an L-ketohexose? in the Haworth representation? Refer to Figure 10.4 if needed.

10.52 Why are there eight D-aldohexoses but only two D-ketopentoses?

10.53 Use the term monosaccharide, disaccharide, or polysaccharide to categorize each of the following compounds: (a) Sucrose; (b) galactose; (c) lactose; (d) amylose; (e) amylopectin; (f) fructose; (g) cellulose.

10.54 Which of the following compounds give a positive Benedict's test? Which undergo mutarotation? (a) Sorbose; (b) glycogen; (c) sucrose; (d) lactose; (e) starch; (f) talose; (g) cellulose.

10.55 For each of the following pairs of compounds, indicate whether the pair represents different compounds that are (1) constitutional isomers or (2) stereoisomers that are enantiomers or (3) stereoisomers that are diasteromers or (4) not isomers. Refer to Figures 10.1 through 10.4 as needed for the structures of the compounds. (a) α-D-Glucose and β-D-glucose; (b) (+)-galactose and (−)-fructose; (c) (+)-galactose and (−)-galactose; (d) maltose and cellobiose; (e) glucose and amylose; (f) glucose and maltose; (g) cellulose and amylose, both having the same molecular mass.

10.56 Draw the cyclic structure of β-D-idose. Refer to Figures 10.1 and 10.3.

10.57 What is the difference in function between structural and storage polysaccharides?

10.58 Why are humans unable to digest cellulose?

10.59 Why are grazing animals such as cows and sheep able to digest cellulose? Why are they also able to digest starch?

10.60 The structure of raffinose, a trisaccharide found in some plants, is shown here. Identify the monosaccharide residues in raffinose. Identify the linkages between monosaccharide residues as acetal or hemiacetal linkages. Is raffinose a reducing sugar? Does it undergo mutarotation?

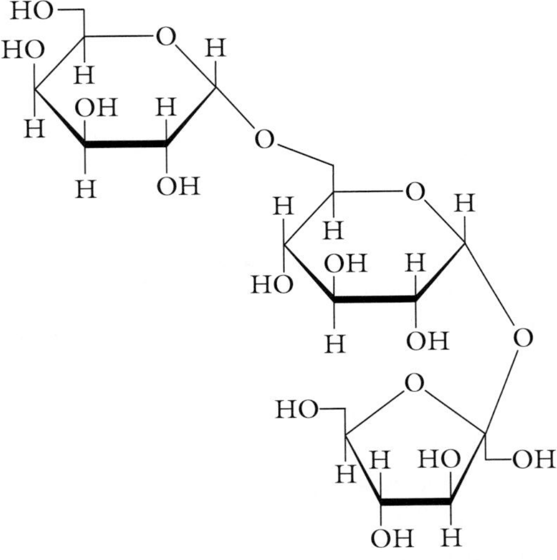

10.61 The hydrolysis of 1 mol of a carbohydrate yields 4 mol of monosaccharide. What is the structure of the carbohydrate?

10.62 Draw the cyclic structure of α-L-glucose. Refer to Figures 10.1 and 10.3 as needed.

10.63 Isomaltose is the disaccharide formed by the partial hydrolysis of dextran, a storage polysaccharide in yeasts and bacteria. Isomaltose is diglucose with an α(1 → 6)-glycosidic linkage. Draw the structure of isomaltose. Is isomaltose a reducing sugar? Does it undergo mutarotation?

10.64 Draw the structure of the polysaccharide that consists of D-fructose residues linked together by α(2 → 6)-glycosidic linkages.

10.65 Write the balanced overall equation for the photosynthesis of D-glucose.

10.66 Where is most of Earth's photosynthesis carried out?

10.67 Describe the oxygen cycle.

Expand Your Knowledge

Note: The icons [flask icons] denote exercises based on material in boxes.

10.68 D-Sorbitol, $C_6H_{14}O_6$, is found in apples, cherries, and other fruits. It can be synthesized by the reduction of D-glucose by the addition of H_2 in the presence of a catalyst such as Pt. Write an equation that shows this synthesis.

10.69 How many glucose residues are there in each molecule of amylose of molecular mass 162,000 amu? in cellulose of molecular mass 162,000 amu?

10.70 The polysaccharide hyaluronic acid is found in the synovial fluid of joints, where it acts as a shock absorber, and in the vitreous humor of the eye ball, where it serves to hold the retina in place. Hyaluronic acid contains alternating D-glucuronic acid and *N*-acetyl-D-glucosamine residues (Section 10.4). D-Glucuronic acid is connected to *N*-acetyl-D-glucosamine by a β(1 → 3)-glycosidic linkage; *N*-acetyl-D-glucosamine is connected to D-glucuronic acid by a β(1 → 4)-glycosidic linkage. Draw a part of the structure of hyaluronic acid that shows each of the two residues.

10.71 Both D-glucose and D-fructose are mixtures of the α- and β-anomers and the acyclic compound in aqueous solution. Sucrose is formed by the exclusive linking of α-D-glucose through C1 and β-D-fructose through C2. Speculate on the mechanism by which this exclusive link is formed.

10.72 The cell-surface oligosaccharides that determine blood typing contain L-fucose, one of the rare L-saccharides found in nature. L-Fucose is 6-deoxy-L-galactose. Draw the structure of L-fucose.

10.2 **10.73** Explain why galactosemia is a much more serious genetic disease than lactose intolerance (see Box 10.2).

10.74 D-Xylitol is used as the sweetener in many sugarless chewing gums and candies. It can be produced by the reduction of the carbonyl group in D-xylose. Write an equation for this reaction. Make sure to include the required reagent(s) and reaction conditions.

10.1 **10.75** An individual-sized package of saccharin contains about 35 mg of saccharin. The manufacturer indicates that one package of saccharin is equivalent to 2 teaspoons of sucrose. A teaspoon of sucrose contains 4.0 g of sucrose. Does this equivalence correspond to that expected from Table 10.1 (see Box 10.1)? If not, what explains the difference?

10.3 **10.76** The manufacturing of a plastic material into objects such as eyeglass frames usually includes melting of the plastic followed by forcing of the molten plastic into the desired shape and then cooling of the object to set its shape. Cellulose is difficult to process because its melting temperature is extremely high. Cellulose acetate (see Box 10.3) is much easier to process into plastic objects because cellulose acetate has a much lower melting temperature. Why does cellulose acetate have a much lower melting temperature than cellulose?

10.4 **10.77** Pectin, the soluble fiber present in fruits, barley, and oats, is a polysaccharide consisting mostly of D-galacturonic acid residues with α(1 → 4)-glycosidic linkages (see Box 10.4). Draw a two-residue-long segment of pectin.

10.78 Draw the structure of 3-deoxy-β-D-galactosamine.

10.79 Explain why people with type AB blood can be transfused with any type blood.

10.80 Explain why people with type A blood can be transfused with A and O blood types, but not with B and AB.

10.2 **10.81** The caloric values of sucrose and aspartame are about the same (4.0 kcal/g). What is the daily savings in kilocalories for someone who drinks five cups of coffee or tea daily if two teaspoons of sucrose per cup are replaced by the equivalent amount of aspartame (see Box 10.2)? A teaspoon of sucrose contains 4.0 g of sucrose.

10.82 A freshly made solution of sucrose has a specific rotation of +66.5°, and the specific rotation does not change with time. Why does sucrose not undergo mutarotation?

10.83 Although sucrose does not undergo mutarotation, treating a sucrose solution with either acid or sucrase results in a change in specific rotation. Explain.

7.1 10.1 **10.84** Recipes for some candy products describe the enhancement of sweetness by adding lemon juice to an aqueous sucrose solution followed by heating before using the sucrose solution. How does heating with lemon juice enhance sweetness (see Boxes 7.1 and 10.1)?

10.85 The anticoagulant heparin is a polysaccharide that contains alternating residues of α-D-glucuronic acid-6-sulfate and *N*-sulfo-α-D-glucosamine-6-sulfate connected by α(1 → 4)-glycosidic linkages. Draw a part of heparin that shows one each of the two residues.

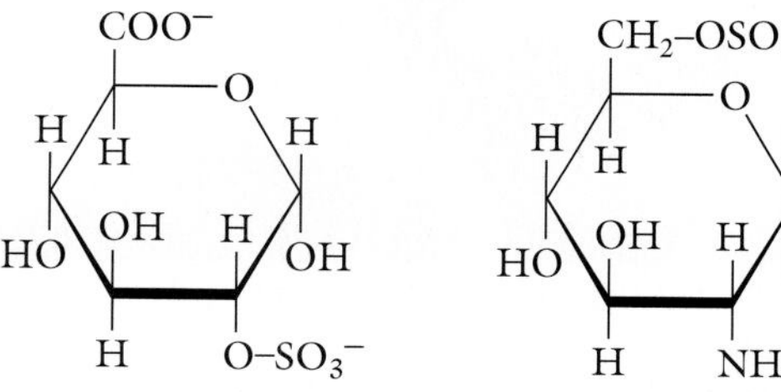

α-D-Glucuronic acid-2-sulfate **N-Sulfo-α-D-glucosamine-6-sulfate**

10.86 Carbon is conserved on our planet; that is, the total number of carbon atoms is constant, but the carbon atoms are cycled among different molecules. Explain.

10.87 Although both starch and cellulose are polyglucose, the two materials are quite different in their physical interaction with water. Describe the difference and explain the reason for the difference.

10.4 **10.88** Explain why pectin (soluble fiber) is soluble in water (see Box 10.4), whereas starch is only wetted by water.

10.89 Write a balanced equation for the combustion of (a) D-glucose; (b) D-fructose.

10.90 Write a balanced equation for the combustion of (a) starch; (b) cellulose.

10.91 What is the relation between the balanced equations for the photosynthesis and the combustion of D-glucose?

10.92 Explain how to differentiate between cellulose and starch by using a simple chemical test.

CHAPTER 11 LIPIDS

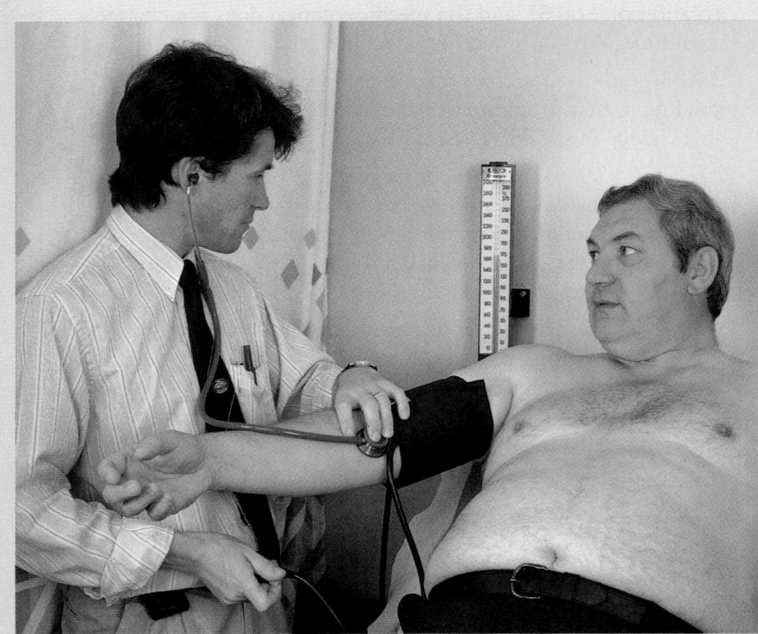
(Simon Fraser/Science Photo Library/Photo Researchers.)

Chemistry in Your Future

A month after coronary angioplasty, a patient visits the HMO for a follow-up exam. As you take his blood pressure, you ask if he has made any of the dietary changes that you recommended at the time of his discharge. Proudly, he announces that, instead of eggs, bacon, and potatoes every morning (20 grams of fat, at least), he is now enjoying a big bowl of granola. Unfortunately, you discover, he has been buying a brand loaded with palm and coconut oils, probably containing as much saturated fat as his former breakfast did. How can the fat from a plant oil, he asks, be as unhealthy as the fat from a pig? This chapter has some of the answers.

For more information about this topic and others in the chapter, go to www.whfreeman.com/bleiodian2e

Learning Objectives

- Draw the structures and describe the physical properties of saturated and unsaturated fatty acids.
- Draw the structures and describe the physical properties of triacylglycerols.
- Describe and write equations for the hydrolysis, catalytic hydrogenation, and rancidity reactions of triacylglycerols.
- Describe and draw the structures of waxes.
- Describe and draw the structures of glycerophospholipids and sphingolipids, and write equations for their hydrolysis.
- Describe and draw the structures of steroids, eicosanoids, and fat-soluble vitamins, and outline their functions.
- Describe the structure of biological membranes.
- Describe the processes by which molecules and ions are transported across biological membranes.

Lipids are the predominant and most efficient form of stored energy in animals. Together with carbohydrates and proteins, they are the major components of our diet. We obtain lipids from animal fats and plant oils and use them as a source of energy. Any excess lipids taken in are stored as fat. Among our dietary lipids are beef, pork, and poultry fats; fats in milk and milk products such as butter and cheese; fish oils; plant oils such as peanut, corn, and olive oils, and margarine manufactured from plant oils.

Lipids come in a variety of molecular structures with a broad array of functions:

- Triacylglycerols—animal fats and plant oils—are sources of energy and a storage form of energy not required for immediate use.
- Glycerophospholipids, sphingolipids, and cholesterol (together with proteins) are the primary structural components of the membranes that surround all cells and organelles.
- Steroid hormones, including the sex hormones, and other hormonelike lipids act as chemical messengers, initiating or altering activity in specific target cells.
- The lipid-soluble (fat-soluble) vitamins A, D, E, and K are required for a variety of physiological functions.
- Bile salts are needed for the digestion of lipids in the intestinal tract.

11.1 CLASSIFYING LIPIDS

Other families of organic compounds, including carbohydrates, proteins, and nucleic acids, are defined by the presence of a common functional group. This is not true of lipids, a family that includes compounds with diverse functional groups: ester, amide, alcohol, phosphate, acetal, and others.

The classification of compounds as lipids is based on their solubility behavior. To test for solubility, animal or plant matter is crushed into small pieces or powder and then mixed with a nonpolar or low-polarity solvent, such as toluene, dioxane, or carbon tetrachloride. Those compounds that dissolve in the solvent are classified as lipids.

Concept check

✓ Lipids are compounds, present in plants and animals, that are soluble in nonpolar or low-polarity solvents.

Lipids are nonpolar compounds or compounds of low polarity. They are insoluble (or only very slightly soluble, at most) in water. Most lipids other than triacylglycerols are **amphipathic:** one part of the molecule is hydrophobic and another part is hydrophilic (Section 7.7). The reason that lipids have little or no solubility in water is that the hydrophobic part is much larger than the hydrophilic part.

Lipids are classified as hydrolyzable or nonhydrolyzable. **Hydrolyzable lipids** undergo hydrolytic cleavage into two or more considerably smaller compounds in the presence of an acid, a base, or a digestive enzyme. All hydrolyzable lipids contain at least one ester group. Some also contain an amide, phosphate, or acetal group. **Nonhydrolyzable lipids** do not undergo hydrolytic cleavage into considerably smaller compounds, because they do not contain ester, amide, phosphate, or acetal groups. The terms **saponifiable** and **nonsaponifiable lipids** also are used. Saponification is hydrolysis with a base (Section 7.3). Hydrolyzable lipids include the triacylglycerols, waxes, glycerophospholipids, and sphingolipids (Figure 11.1 on the following page). Nonhydrolyzable lipids include steroids, eicosanoids, and fat-soluble vitamins.

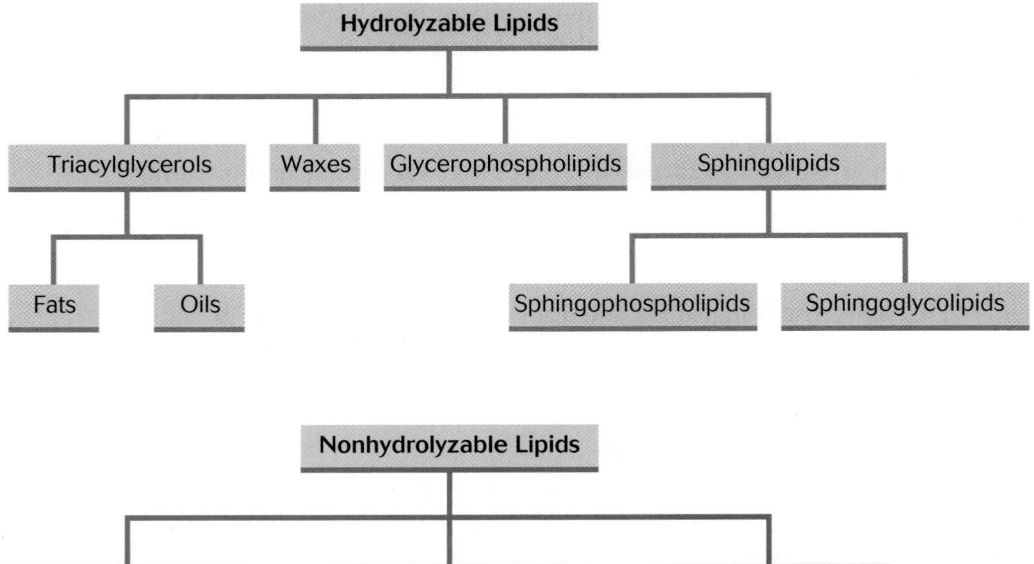

Figure 11.1 Classification of lipids.

In this chapter, we describe the different lipids and their functions, as well as the structure of biological membranes and how they serve as selective membranes.

11.2 FATTY ACIDS

All hydrolyzable lipids contain one or more **fatty acids**—carboxylic acids with long hydrocarbon chains (Section 7.7). Almost all fatty acids are found in nature in combined form. Only very small amounts are found as uncombined, free acids. The fatty acids used in nature to construct hydrolyzable lipids are almost exclusively the linear (unbranched) acids with an even number of carbons listed in Table 11.1.

Fatty acids may be either saturated or unsaturated. **Saturated fatty acids** contain no C=C double bonds. **Unsaturated fatty acids** contain one or more C=C double bonds. They are the **monounsaturated** and **polyunsaturated fatty acids.** In naturally occurring fatty acids, almost all double bonds have the cis configuration. Palmitic and stearic acids are the most abundant saturated fatty acids. Oleic acid is the most abundant unsaturated fatty acid. Two unsaturated fatty acids, linoleic and linolenic, are **essential fatty acids**—fatty acids that are synthesized only by plants but are required by animals and must be included in their diets. All other fatty acids are **nonessential fatty acids** that animals synthesize either from other fatty acids or from other precursors (derived from other foodstuffs).

The position of the double bond nearest the methyl group in unsaturated fatty acids is often indicated by an **omega number** (ω-). The ω-1 carbon is the methyl carbon in the fatty acid chain; that is, the carbon atom farthest from the carboxyl group. The omega number of the fatty acid is the number of the first carbon of the first double bond, counting from the ω-1 carbon. (This numbering method is the opposite of that of the IUPAC nomenclature system in which the counting begins at the carboxyl carbon, as C1.) Linoleic and linolenic acids are ω-6 and ω-3 fatty acids, respectively.

TABLE 11.1 Common Saturated and Unsaturated Fatty Acids

Name of acid	Number of double bonds	Number of carbon atoms	Structure	Melting point (°C)
SATURATED FATTY ACIDS				
lauric	0	12	$CH_3(CH_2)_{10}COOH$	44
myristic	0	14	$CH_3(CH_2)_{12}COOH$	54
palmitic	0	16	$CH_3(CH_2)_{14}COOH$	63
stearic	0	18	$CH_3(CH_2)_{16}COOH$	69
arachidic	0	20	$CH_3(CH_2)_{18}COOH$	77
behenic	0	22	$CH_3(CH_2)_{20}COOH$	81
lignoceric	0	24	$CH_3(CH_2)_{22}COOH$	84
MONOUNSATURATED FATTY ACIDS				
palmitoleic	1	16	$CH_3(CH_2)_5CH{=}CH(CH_2)_7COOH$	1
oleic	1	18	$CH_3(CH_2)_7CH{=}CH(CH_2)_7COOH$	13
nervonic	1	24	$CH_3(CH_2)_7CH{=}CH(CH_2)_{13}COOH$	42
POLYUNSATURATED FATTY ACIDS				
linoleic	2	18	$CH_3(CH_2)_4(CH{=}CHCH_2)_2(CH_2)_6COOH$	−5
linolenic	3	18	$CH_3CH_2(CH{=}CHCH_2)_3(CH_2)_6COOH$	−11
arachidonic	4	20	$CH_3(CH_2)_4(CH{=}CHCH_2)_4(CH_2)_2COOH$	−49
eicosapentaenoic	5	20	$CH_3CH_2(CH{=}CHCH_2)_5(CH_2)_2COOH$	−54

Example 11.1 Recognizing naturally occurring fatty acids

Which of the following carboxylic acids are present in animal and plant lipids?

$$\underset{1}{CH_3(CH_2)_{13}COOH} \qquad \underset{2}{CH_3\overset{\overset{\large CH_3}{|}}{C}HCH_2(CH_2)_{10}COOH}$$

$$\underset{3}{trans\text{-}CH_3(CH_2)_5CH{=}CH(CH_2)_7COOH}$$

Solution

None of these carboxylic acids are components of animal and plant lipids. Acid 1 has an odd number of carbon atoms. Acid 2 both has an odd number of carbons and is branched. Acid 3 has a trans double bond. The fatty acids of animal and plant lipids are unbranched, with an even number of carbons. Double bonds, when present, have a cis configuration.

Problem 11.1 Which of the following carboxylic acids are present in animal and plant lipids?

$$\underset{1}{CH_3(CH_2)_{14}COOH} \qquad \underset{2}{CH_3\overset{\overset{\large CH_3}{|}}{C}HCH_2(CH_2)_{11}COOH}$$

$$\underset{3}{cis\text{-}CH_3(CH_2)_5CH{=}CH(CH_2)_7COOH}$$

The physical properties of fatty acids depend on the number of carbons and double bonds. Fatty acids have very low water solubility, and solubility decreases as the number of carbons increases. The water solubilities of lauric and stearic acids at 30°C are 0.063 g/L and 0.0034 g/L, respectively. Although the carboxyl group of a fatty acid can hydrogen bond with water, the much larger, nonpolar, hydrocarbon part cannot. The overall nonpolar nature of fatty acids is apparent when we compare the water solubilities of D-glucose and lauric acid, which have similar molecular masses. D-Glucose is much more soluble (1100 g/L) than lauric acid (0.063 g/L). The large hydrocarbon part of a fatty acid is responsible for its solubility in nonpolar solvents such as benzene. The solubilities of lauric and stearic acids in benzene are 124 and 2600 g/L, respectively.

The melting points of fatty acids (and lipids that contain them) increase as the number of carbons increases and decrease as the number of double bonds increases (see Table 11.1). The effect of double bonds on melting point is large. Stearic acid, a saturated fatty acid, has a melting point of 70°C. The melting point is lowered to 13, −5, and −11°C with the presence of one, two, and three double bonds, respectively, in oleic, linoleic, and linolenic acids.

Cis double bonds decrease the melting points by altering molecular shape, thus affecting the ability of the molecules to pack tightly together. Figure 11.2 shows the shapes of stearic, oleic, and linoleic acids. Saturated fatty acid molecules, with elongated conformations, pack closely together, with strong secondary attractive forces and, thus, high melting points. The cis double bonds in unsaturated fatty acids create kinks, or bends, in the hydrocarbon chain, and these kinks prevent tight molecular packing. With weaker secondary attractive forces, lower temperatures are needed to melt the compound.

Trans double bonds also lower the melting points of fatty acids, but much less than do cis double bonds. For example, the melting point of elaidic acid, the *trans*-isomer of oleic acid, is 45°C, lower than that of stearic acid (69°C) but higher than that of oleic acid (13°C). The molecular kinks placed in molecules by trans double bonds do not distort the elongated conformation of the hydrocarbon chain nearly as much as cis double bonds do.

INSIGHT INTO PROPERTIES

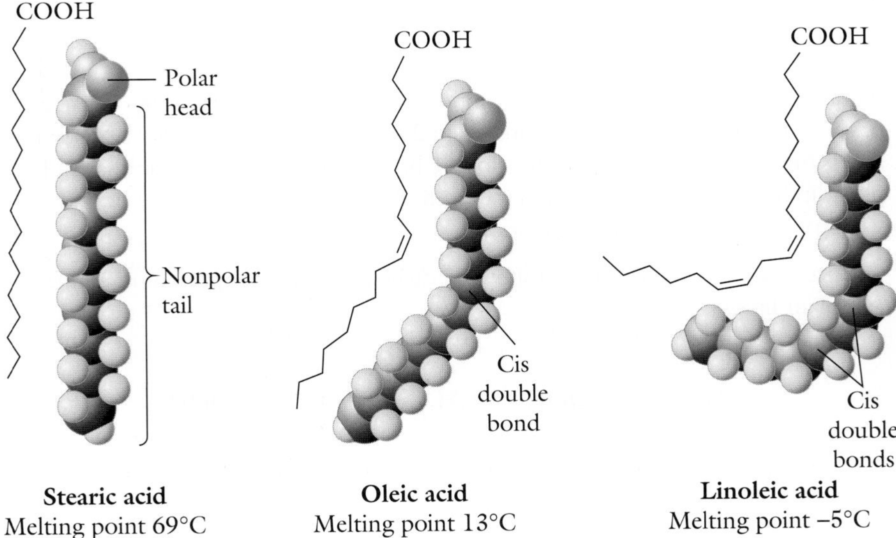

Figure 11.2 Molecular shapes of saturated and unsaturated fatty acids. These differences in shape cause melting-point differences.

11.3 THE STRUCTURE AND PHYSICAL PROPERTIES OF TRIACYLGLYCEROLS

Triacylglycerols (TAGs), also called **triglycerides,** make up about 90% of our dietary lipid intake. All **animal fats** such as beef, butter, pork, and poultry fats and all **plant oils** such as corn, olive, peanut, and soybean oils are triacylglycerols. They are esters (Section 7.8)—in fact, triesters—of glycerol. We can visualize their formation by the esterification of glycerol (1,2,3-propanetriol) with fatty acids:

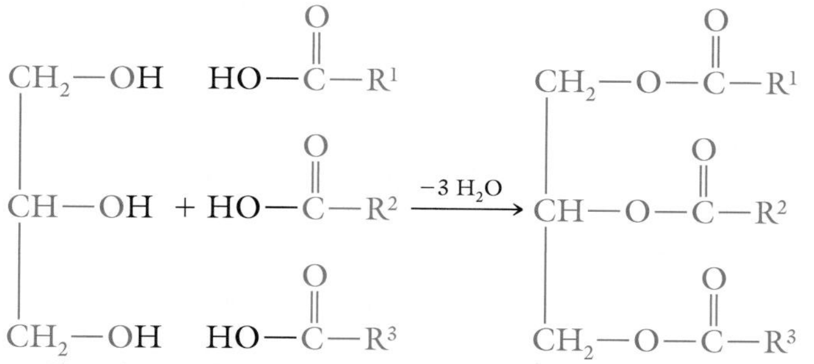

$$\begin{array}{lll} CH_2{-}OH & HO{-}\overset{\overset{\displaystyle O}{\|}}{C}{-}R^1 & & CH_2{-}O{-}\overset{\overset{\displaystyle O}{\|}}{C}{-}R^1 \\ | & & & | \\ CH{-}OH & +\ HO{-}\overset{\overset{\displaystyle O}{\|}}{C}{-}R^2 & \xrightarrow{-3\,H_2O} & CH{-}O{-}\overset{\overset{\displaystyle O}{\|}}{C}{-}R^2 \\ | & & & | \\ CH_2{-}OH & HO{-}\overset{\overset{\displaystyle O}{\|}}{C}{-}R^3 & & CH_2{-}O{-}\overset{\overset{\displaystyle O}{\|}}{C}{-}R^3 \\ \text{Glycerol} & \text{Fatty acids} & & \text{Triacylglycerol} \end{array}$$

In **simple triacylglycerols,** $R^1 = R^2 = R^3$. In **complex,** or **mixed, triacylglycerols,** R^1, R^2, and R^3 are different.

Example 11.2 Drawing the structures of triacylglycerols

Draw the structure of the triacylglycerol containing equimolar amounts of myristic, palmitic, and oleic acids.

Solution

Draw glycerol. Draw the three fatty acids, each lined up with its carboxyl group alongside an —OH group of glycerol. Form the ester group at each of the three carbons of glycerol by carrying out the dehydrations between glycerol and fatty acid —OH groups indicated in blue.

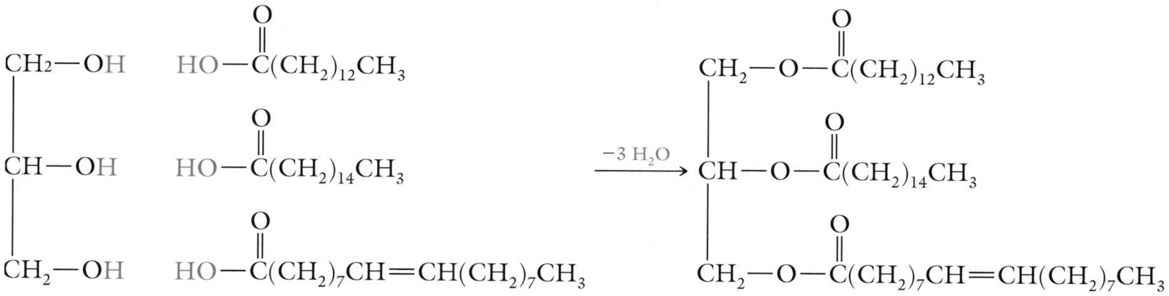

$$\begin{array}{llll} CH_2{-}OH & HO{-}\overset{\overset{\displaystyle O}{\|}}{C}(CH_2)_{12}CH_3 & & CH_2{-}O{-}\overset{\overset{\displaystyle O}{\|}}{C}(CH_2)_{12}CH_3 \\ | & & & | \\ CH{-}OH & HO{-}\overset{\overset{\displaystyle O}{\|}}{C}(CH_2)_{14}CH_3 & \xrightarrow{-3\,H_2O} & CH{-}O{-}\overset{\overset{\displaystyle O}{\|}}{C}(CH_2)_{14}CH_3 \\ | & & & | \\ CH_2{-}OH & HO{-}\overset{\overset{\displaystyle O}{\|}}{C}(CH_2)_7CH{=}CH(CH_2)_7CH_3 & & CH_2{-}O{-}\overset{\overset{\displaystyle O}{\|}}{C}(CH_2)_7CH{=}CH(CH_2)_7CH_3 \end{array}$$

Problem 11.2 Draw the structure of the triacylglycerol containing equimolar amounts of lauric, palmitoleic, and linoleic acids.

Naturally occurring triacylglycerols contain a variety of fatty acid components. Moreover, any fat or oil is a mixture of many different triacylglycerols, very few of which are simple triacylglycerols. Table 11.2 on the following page gives the approximate fatty acid composition of various animal and plant triacylglycerols. Triacylglycerols from animal sources other than fish contain a lower percentage of unsaturated fatty acids than do most plant triacylglycerols, from about 40% to 60% versus 85% to 90%. The terms saturated and unsaturated are used to indicate the different types of fatty acid content.

TABLE 11.2 Fatty Acid Composition of Triacylglycerols

Source of fat or oil	Percent fatty acid								
	Saturated				Monounsaturated			Polyunsaturated	
	$<C_{14}$	C_{14}	C_{16}	C_{18}	C_{14}	C_{16}	C_{18}	C_{18}	$>C_{18}$
ANIMAL									
beef (tallow)		6	27	14			50	3	
butter	14	11	29	9		5	27	4	1
human		3	24	8		5	47	10	3
pork (lard)		1	28	12		3	48	6	2
herring		7	13			5		21	53
sardine		5	15	3	15	12	←18*→		32
PLANT									
corn		1	10	3		2	34	50	
olive			7	2			85	3	
peanut			8	3			56	26	7
soybean			10	2			29	58	1
coconut	60	19	11	2			8		
palm		1	40	6			43	10	

*Total of monounsaturated and polyunsaturated C_{18} fatty acids.

Fish triacylglycerols differ considerably in composition from those of other animals: from 75% to 80% of the fatty acids are unsaturated, compared with 40% to 60% in other animals, and a much greater proportion of the unsaturated component comprises polyunsaturated fatty acids.

A few plants have triacylglycerol compositions much like those of animals other than fish. Examples are palm and coconut oils. Coconut triacylglycerol is extremely atypical of plant triacylglycerols; it has a saturated fatty acid content of 92%, far exceeding that of any animal triacylglycerol.

The melting points of plant triacylglycerols are lower than those of the animal triacylglycerols because of the higher content of unsaturated fatty acids in the plant lipids. Cis double bonds lower secondary attractive forces in triacylglycerols, similar to their effect in fatty acids (Section 11.2). Plant triacylglycerols are liquid, whereas animal triacylglycerols are solid at ambient temperatures (about 20°C). The terms **fats** and **oils** are used to indicate solid and liquid triacylglycerols, respectively.

All triacylglycerols, irrespective of the source, are water insoluble. They are even less soluble than the corresponding fatty acids, because of the low polarity of the ester groups.

11.4 CHEMICAL REACTIONS OF TRIACYLGLYCEROLS

Triacylglycerols undergo hydrolysis, catalytic hydrogenation, and rancidity reactions.

Hydrolysis

Triacylglycerols undergo hydrolysis at the ester groups (Section 7.3) to produce glycerol and fatty acids. Hydrolysis in the laboratory requires an acid or base: acidic hydrolysis yields the fatty acids and basic hydrolysis yields the fatty acid carboxylate salts:

$$\underset{\text{Triacylglycerol}}{\begin{array}{l} CH_2{-}O{-}\overset{O}{\overset{\|}{C}}{-}R^1 \\ | \\ CH{-}O{-}\overset{O}{\overset{\|}{C}}{-}R^2 \\ | \\ CH_2{-}O{-}\overset{O}{\overset{\|}{C}}{-}R^3 \end{array}} \xrightarrow{H_2O,\ H^+} \underset{\text{Glycerol}}{\begin{array}{l} CH_2{-}OH \\ | \\ CH{-}OH \\ | \\ CH_2{-}OH \end{array}} + \underset{\text{Fatty acids}}{\begin{array}{l} HO{-}\overset{O}{\overset{\|}{C}}{-}R^1 \\ HO{-}\overset{O}{\overset{\|}{C}}{-}R^2 \\ HO{-}\overset{O}{\overset{\|}{C}}{-}R^3 \end{array}}$$

$$\text{Triacylglycerol} \xrightarrow{H_2O,\ NaOH} \underset{\text{Glycerol}}{\begin{array}{l} CH_2{-}OH \\ | \\ CH{-}OH \\ | \\ CH_2{-}OH \end{array}} + \underset{\text{Fatty acid salts}}{\begin{array}{l} NaO{-}\overset{O}{\overset{\|}{C}}{-}R^1 \\ NaO{-}\overset{O}{\overset{\|}{C}}{-}R^2 \\ NaO{-}\overset{O}{\overset{\|}{C}}{-}R^3 \end{array}}$$

Basic hydrolysis, or **saponification,** of animal fats and some plant oils produces soaps, which are the mixtures of fatty acid sodium (or potassium) salts that remain after the removal of glycerol. Various additives are used to enhance the cleaning power of commercial soaps or to give them other properties desirable to the consumer (Section 7.7).

°Example 11.3 Writing equations for the hydrolysis of triacylglycerols

Write the equation for the acid-catalyzed hydrolysis of the triacylglycerol containing equimolar amounts of myristic, stearic, and palmitoleic acids.

Solution

Hydrolyze each ester group to an alcohol and carboxylic acid:

$$\begin{array}{l} CH_2{-}O{-}\overset{O}{\overset{\|}{C}}(CH_2)_{12}CH_3 \\ | \\ CH{-}O{-}\overset{O}{\overset{\|}{C}}(CH_2)_{16}CH_3 \\ | \\ CH_2{-}O{-}\overset{O}{\overset{\|}{C}}(CH_2)_7CH{=}CH(CH_2)_5CH_3 \end{array} \xrightarrow[H^+]{H_2O} \begin{array}{l} CH_2{-}OH \\ | \\ CH{-}OH \\ | \\ CH_2{-}OH \end{array} + \begin{array}{l} HO{-}\overset{O}{\overset{\|}{C}}(CH_2)_{12}CH_3 \\ HO{-}\overset{O}{\overset{\|}{C}}(CH_2)_{16}CH_3 \\ HO{-}\overset{O}{\overset{\|}{C}}(CH_2)_7CH{=}CH(CH_2)_5CH_3 \end{array}$$

Problem 11.3 Write the equation for the saponification with NaOH of the triacylglycerol containing equimolar amounts of myristic, stearic, and palmitoleic acids.

The Digestion and Storage of Triacylglycerols

The digestion (hydrolysis) of dietary triacylglycerols to smaller molecules is necessary for their use because triacylglycerols are too large to diffuse through the intestinal membranes. This digestion is carried out with enzymes called

lipases, with the help of bile salts (Section 11.7), and takes place in the small intestine, which has a basic pH. Digestion is more complicated than hydrolysis in the laboratory (Section 18.1). The digestion of triacylglycerols produces a mixture of products, most of which are monoacylglycerols and fatty acids and some of which are diacylglycerols and glycerol. The cells of the intestinal lining then rebuild triacylglycerols from their smaller components and package them, with proteins, into lipoprotein particles called **chylomicrons,** which are transported through the lymphatic system to the bloodstream. The blood carries the chylomicrons to various tissues where the triacylglycerol is separated from the protein and hydrolyzed to fatty acids and glycerol, which are then metabolized to generate energy. Fatty acids not needed for immediate energy generation are reconverted into triacylglycerols and stored as fat droplets in **adipose cells,** also called **adipocytes** or **fat cells.** Such cells make up most of the fatty, or adipose, tissue in animals.

» The structure and properties of chylomicrons are considered in Section 18.3.

11.1 CHEMISTRY WITHIN US

Noncaloric Fat

Artificial sweeteners (see Box 10.1) have been used for many years to limit caloric intake without eliminating pastries and other sweet foods from the diet. There has been a corresponding effort to produce artificial fats. The result has been Olestra—an ester of sucrose. The sucrose ester used is a sucrose hexa-, hepta-, or octaester or a mixture of the different esters, produced by the esterification of hydroxyl groups of sucrose with fatty acids derived from edible fats and oils (soybean, corn, cottonseed).

Frying foods in sucrose ester gives them the flavor of foods fried in fat. Unlike other fats used for frying, however, sucrose ester is noncaloric. The sucrose ester cannot be digested even though the sucrose and fatty acids from which it is made can be digested. The lipase enzymes that hydrolyze the ester groups in triacylglycerols and other hydrolyzable lipids cannot hydrolyze the ester groups of sucrose ester, because the ester is too large and bulky to bind to the active sites of the enzymes. It is also too large to enter the cells lining the intestine. Sucrose ester therefore passes through the intestinal tract and is excreted in the feces.

There are problems associated with the ingestion of sucrose ester, and so it should not be consumed in large amounts. It can cause gastrointestinal cramping, flatulence, anal leakage, and loose bowels in some people, especially if they consume too much of it. It also interferes with the absorption of fat-soluble vitamins (Section 11.9) and other nutrients. In 1996, the FDA approved sucrose ester for use, but only in prepackaged, ready-to-eat salty snacks such as corn and potato chips. Sucrose ester is not available for consumers to use directly in frying foods, because such use would result in excessive consumption with the resultant problems just described.

Potato chips made from Olestra are free of calories from fat. (Richard Megna/Fundamental Photographs.)

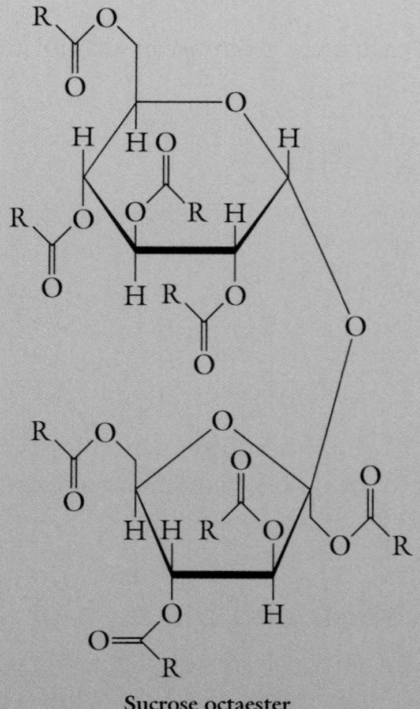

Sucrose octaester

Triacylglycerols are the main storage form of energy in animals. Recall that animals also store energy as glycogen (Section 10.6), but triacylglycerols are a far more efficient form of energy storage: the energy produced by metabolic oxidation of triacylglycerols is 9.2 kcal/g (dry weight) versus 4.0 kcal/g for glycogen (dry weight) (1 cal = 4.184 J). The reason for this difference in the amount of energy produced is that the carbon atoms of triacylglycerols are more reduced than those of glycogen; that is, they contain more hydrogen and less oxygen. Furthermore, triacylglycerols are stored in highly anhydrous form, whereas glycogen is stored in hydrated form. One gram of stored triacylglycerol is almost pure triacylglycerol, but one gram of stored glycogen is two-thirds water. If we take the difference in hydration into account, 1 g of stored triacylglycerol yields about seven times the energy of 1 g of stored glycogen.

A bear and her cub hibernating in the winter. Fat reserves are used for energy, and they insulate the bear from coldness. (Lynn Rogers/Peter Arnold, Inc./Alamy.)

The glycogen stored in your liver and muscle is enough to fill your energy needs for about 12 h. The triacylglycerol stored in adipose cells can fill your energy needs for several weeks to a couple of months. Hibernating animals accumulate sufficient triacylglycerol reserves before hibernation to last through a winter season. Plants store triacylglycerols in their seeds to provide the energy required for germination. Layers of adipose tissue under the skin also serve as thermal insulation against very low temperatures for penguins, seals, walruses, and other warm-blooded polar animals.

As described in Box 10.1, the increase in obesity in Western developed countries led to the development of artificial sweeteners. More recently, the use of artificial, or noncaloric, fats was approved in the United States (Box 11.1).

Catalytic Hydrogenation

Catalytic hydrogenation is the addition of hydrogen to alkene double bonds in the presence of a metal catalyst such as nickel (Ni) or platinum (Pt) (Section 4.6). This process is the basis for the conversion of plant oils into margarine and other products.

Shortening and margarine are produced by the hydrogenation of vegetable oils. (Richard Megna/Fundamental Photographs.)

Butter and other animal triacylglycerols are implicated in the development of atherosclerosis because of their high content of saturated triacylglycerols and cholesterol (Box 16.1). Vegetable oils such as corn oil are preferable dietary lipids because of their higher content of unsaturated triacylglycerols and almost total absence of cholesterol. However, most people would find it unpalatable to spread corn oil on their toast or baked potatoes. An alternative is margarine, produced by the **hardening** of corn or other vegetable oils by partial hydrogenation in which some but not all double bonds are hydrogenated. Enough of the double bonds are hydrogenated to convert the oil into a semisolid or solid product acceptable to our palates. Milk and artificial coloring added to the product give it an attractive flavor, color, and consistency. Partial hydrogenation is also used to produce cooking oils such as Crisco and Spry from vegetable oils. Other partially hydrogenated oils are used in a variety of baked goods, candy bars, cookies, crackers, and breakfast cereals. Decreasing the unsaturated fatty acid content of these products also imparts longer shelf life because double bonds undergo oxidation reactions (see following page under Rancidity).

For some time, margarine has been considered a more healthy dietary fat than butter, but recent evidence shows that the situation is complicated. In the partial-hydrogenation process, some of the cis double bonds undergo conversion into trans double bonds, a process called **isomerization.** The hydrogenation catalyst fosters hydrogenation by an easy breaking of the π bond of the double bond, which is followed by the addition of hydrogen. However, the re-forming of the π bond competes with hydrogen addition, and this competition is the mechanism for isomerization. The re-forming of the double bond yields predominantly trans double bonds rather than cis double bonds because trans double bonds are more stable than cis (Section 5.6).

Fatty acids with trans double bonds have physical properties much more like those of saturated fatty acids than like cis unsaturated fatty acids (Section 11.1). Furthermore, there is evidence that trans unsaturated fatty acids raise blood cholesterol levels more than cis unsaturated fatty acids do, possibly even more than saturated fatty acids. Trans unsaturated fatty acids also raise low-density-lipoprotein levels and decrease high-density-lipoprotein levels (Box 16.1).

The U.S. Food and Drug Administration (FDA) has mandated that, by January 2006, the labels on food products must include the trans unsaturated fatty acid content in addition to saturated and total unsaturated fatty acid, cholesterol, and carbohydrate contents. Concern about trans unsaturated fatty acids has led manufacturers to find alternate methods of decreasing the unsaturated fatty acid content without increasing the trans unsaturated fatty acid content. The mixing together of natural saturated and unsaturated triacylglycerols allows the production of products with any desired unsaturated fatty acid content without increasing the trans unsaturated fatty acid content.

Example 11.4 Writing equations for the hydrogenation of triacylglycerols

Write the equation for the Pt-catalyzed hydrogenation of the triacylglycerol containing equimolar amounts of lauric, palmitoleic, and linoleic acids. Assume complete hydrogenation of all double bonds unless otherwise indicated.

Solution

Hydrogen is added to each C–C double bond.

$$\begin{array}{l} CH_2{-}O{-}\overset{O}{\overset{\|}{C}}(CH_2)_{10}CH_3 \\ | \\ CH{-}O{-}\overset{O}{\overset{\|}{C}}(CH_2)_7CH{=}CH(CH_2)_5CH_3 \\ | \\ CH_2{-}O{-}\overset{O}{\overset{\|}{C}}(CH_2)_6(CH_2CH{=}CH)_2(CH_2)_4CH_3 \end{array} \xrightarrow[Pt]{H_2} \begin{array}{l} CH_2{-}O{-}\overset{O}{\overset{\|}{C}}(CH_2)_{10}CH_3 \\ | \\ CH{-}O{-}\overset{O}{\overset{\|}{C}}(CH_2)_{14}CH_3 \\ | \\ CH_2{-}O{-}\overset{O}{\overset{\|}{C}}(CH_2)_{16}CH_3 \end{array}$$

Problem 11.4 Write the equation for the Pt-catalyzed hydrogenation of the triacylglycerol that contains equimolar amounts of palmitic, oleic, and linoleic acids.

Rancidity

Butter, salad and cooking oils, mayonnaise, and fatty meats can become **rancid,** developing unpleasant odors and flavors. Rancidity is due to a combination of two reactions:

- Bacterial hydrolysis of ester groups
- Air oxidation of alkene double bonds

In fats containing triacylglycerols with some low-molecular-mass carboxylic acids, hydrolysis by airborne bacteria under moist, warm conditions is directly responsible for rancid odors and flavors. For example, butter contains triacylglycerols that are from 3% to 4% butanoic acid and from 1% to 2% hexanoic acid. Bacterial hydrolysis (digestion) of butter releases these volatile, malodorous, and off-flavor acids.

Foul-tasting and malodorous carboxylic acids are also generated by air oxidation of double bonds in fats and oils or in the unsaturated acids released by bacterial hydrolysis of fats and oils. Oxidation cleaves double bonds, with each carbon of the double bond being converted into a —COOH group. The end result is the creation of low-molecular-mass, volatile, offensive-smelling carboxylic and dicarboxylic acids.

The rancidity reactions are slowed by refrigeration. Fats and oils stored at room temperature, such as salad oil, should be tightly capped to impede the entry of moisture, bacteria, and air—all of which are required for the hydrolysis and oxidation reactions. Some manufacturers of vegetable oils and other products add substituted phenols such as BHA (butylated hydroxyanisole) and BHT (butylated hydroxytoluene) as antioxidants to inhibit the oxidation reaction (see Box 5.5).

Example 11.5 Writing equations for rancidity reactions

A triacylglycerol undergoes hydrolysis by bacteria to yield a mixture of lauric, palmitic, and palmitoleic acids. Write the equation for the subsequent air oxidation of palmitoleic acid. Why don't lauric and palmitic acids undergo air oxidation?

Solution

$$CH_3(CH_2)_5CH{=}CH(CH_2)_7\overset{\overset{\displaystyle O}{\|}}{C}{-}OH \xrightarrow{(O)} CH_3(CH_2)_5\overset{\overset{\displaystyle O}{\|}}{C}{-}OH + HO{-}\overset{\overset{\displaystyle O}{\|}}{C}(CH_2)_7\overset{\overset{\displaystyle O}{\|}}{C}{-}OH$$

Lauric and palmitic acids do not contain double bonds. Oxidation takes place only at double bonds.

Problem 11.5 Write the equation for the air oxidation of oleic acid.

Air oxidation of C=C double bonds is due to both oxygen (O_2) and ozone (O_3). Ozone is present at much lower concentrations than oxygen but is a more powerful oxidizing agent. The oxidation of C=C double bonds is largely responsible for the respiratory distress in the elderly and others with compromised lung function during periods of "environmental ozone alerts." During periods in which the ozone concentrations increase many times and temperatures are high, the unsaturated amphipathic lipids that are a main component of the cell membranes that line the air passages and lungs (Section 11.10) undergo considerable oxidative damage and loss of physiological function.

11.5 WAXES

Waxes have a variety of protective functions in plants and animals. Many fruits, vegetables, and plant leaves are coated with waxes that protect against parasites and mechanical damage and prevent excessive water loss. The skin glands of many vertebrates secrete waxes that protect hair, furs, feathers, and skin, keeping them lubricated, pliable, and waterproof. Waterfowl depend on waxes to make their feathers water repellent and to prevent them from becoming water-soaked and less buoyant. Bees secrete wax to construct the honeycomb in which eggs are laid and new bees develop. In some marine organisms, notably the plankton, waxes are used instead of triacylglycerols for energy storage.

Waxes coat the duck's body and feathers and enable it to float effortlessly in the water. (Ron Sanford/The Stock Market.)

Most waxes are mixtures of esters of fatty acids with long-chain alcohols having one —OH group. The acids and alcohols usually contain an even number of carbons, in the range of 14 to 36, and are unbranched. For example, the major component of beeswax is melissyl palmitate, the ester derived from palmitic acid and melissyl alcohol:

$$\underbrace{CH_3(CH_2)_{14}{-}\overset{\overset{\displaystyle O}{\|}}{C}}_{\text{Palmitate}}{-}\underbrace{O{-}(CH_2)_{29}CH_3}_{\text{Melissyl}}$$

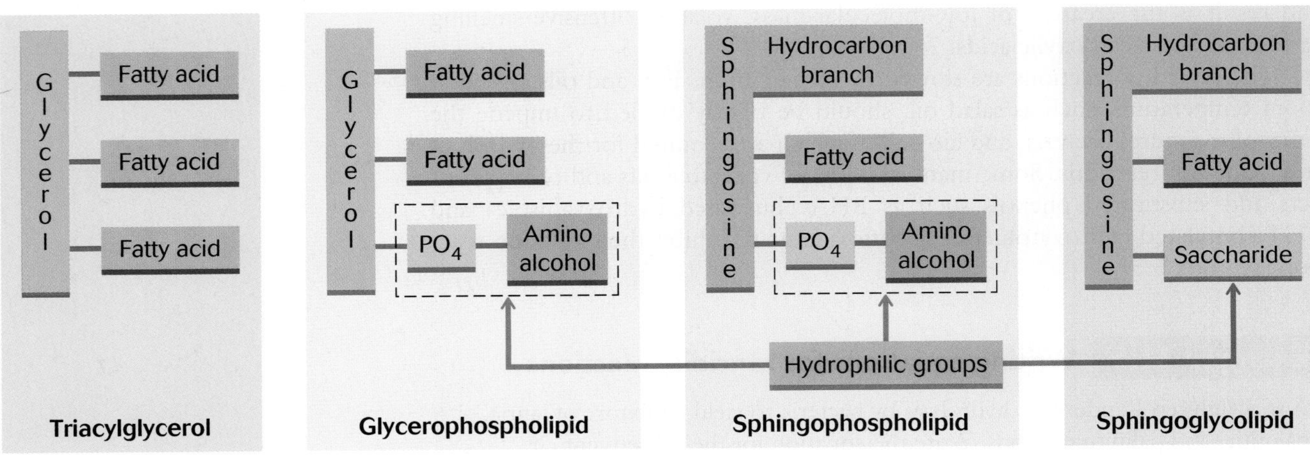

Figure 11.3 Structural comparison of different hydrolyzable lipids.

Waxes have many commercial uses. Carnauba wax, from the leaves of Brazilian palm trees, is used for high-gloss, hard finishes for automobiles, boats, and floors. Lanolin, a component of wool wax, is used in cosmetic and pharmaceutical creams, lotions, and other products. Spermaceti, a component of the oil obtained from the head of a sperm whale, was also used in these products until such use was banned in the late 1970s.

11.6 AMPHIPATHIC HYDROLYZABLE LIPIDS

The predominant lipids of cell membranes are the amphipathic hydrolyzable lipids. There are two broad groups: those based on glycerol, the **glycerophospholipids (phosphoglycerides),** and those based on sphingosine, the **sphingolipids.** Unlike triacylglycerols, which have only nonpolar groups, the glycerophospholipids and sphingolipids have one highly hydrophilic group (Figure 11.3). The hydrophilic group is responsible for the amphipathic nature of these lipids, which allows their assembly into cell membranes (Section 11.10).

Glycerophospholipids

Like triacylglycerols, glycerophospholipids have fatty acid ester groups at two of the carbons of glycerol, but, at the third carbon, they have a phosphodiester group. The esterification of glycerol with two fatty acids and one phosphoric acid forms a **phosphatidic acid.** One of the ester groups is a phosphate ester (Section 7.5). Further esterification of the phosphate ester with an alcohol produces a glycerophospholipid by the formation of a phosphodiester group:

$$
\begin{array}{llcllcl}
\textbf{Glycerol} & \textbf{Fatty acids} & & \textbf{Phosphatidic acid} & & \textbf{Glycerophospholipid} \\
CH_2{-}OH & HO{-}\overset{O}{\overset{\|}{C}}{-}R^3 & & CH_2{-}O{-}\overset{O}{\overset{\|}{C}}{-}R^3 & & CH_2{-}O{-}\overset{O}{\overset{\|}{C}}{-}R^3 \\
| & & & | & & | \\
CH{-}OH\ + & HO{-}\overset{O}{\overset{\|}{C}}{-}R^2 & \xrightarrow{-3\,H_2O} & CH{-}O{-}\overset{O}{\overset{\|}{C}}{-}R^2 & & CH{-}O{-}\overset{O}{\overset{\|}{C}}{-}R^2 \\
| & & & | & & | \\
CH_2{-}OH & HO{-}\underset{O^-}{\underset{|}{\overset{O}{\overset{\|}{P}}}}{-}OH & & CH_2{-}O{-}\underset{O^-}{\underset{|}{\overset{O}{\overset{\|}{P}}}}{-}OH & \xrightarrow[-H_2O]{HOR^1} & CH_2{-}O{-}\underset{O^-}{\underset{|}{\overset{O}{\overset{\|}{P}}}}{-}OR^1 \\
 & \textbf{Phosphoric Acid} & & & & \text{Phosphodiester group}
\end{array}
$$

TABLE 11.3 Alcohols in Glycerophospholipids

Alcohol (R^1OH)	Structure of R^1	Alcohol (R^1OH)	Structure of R^1
ethanolamine	$-CH_2CH_2\overset{+}{N}H_3$	inositol	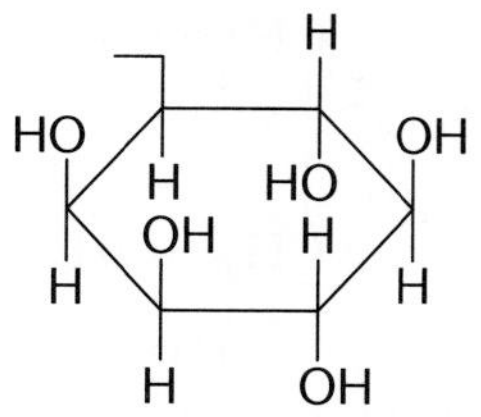
choline	$-CH_2CH_2\overset{+}{N}(CH_3)_3$		
serine	$-CH_2CH(COO^-)\overset{+}{N}H_3$		

Although all P—OH groups actually exist as P—O$^-$ groups at physiological pH (pH ~ 7), the phosphoric and phosphatidic acids, respectively, are shown with one and two P—OH groups in un-ionized form. This form allows us to more easily visualize the formations of the phosphatidic acid and glycerophospholipid by dehydration reactions.

Table 11.3 lists the alcohols (R^1OH in the preceding reaction) found in the phosphodiester group of various glycerophospholipids. Glycerophospholipids are often named by placing **phosphatidyl-** before the name of the alcohol—for example, phosphatidylcholine when the alcohol is choline. However, this nomenclature does not specify the identities of the two fatty acid units. Some glycerophospholipids have other common names. Those containing choline and ethanolamine are called **lecithins** and **cephalins,** respectively.

The digestion or laboratory hydrolysis of a glycerophospholipid yields glycerol, fatty acids, phosphoric acid, and alcohol.

Example 11.6 Drawing the structures of glycerophospholipids

Draw the glycerophospholipid whose hydrolysis yields equimolar amounts of glycerol, palmitic acid, oleic acid, phosphoric acid, and ethanolamine.

Solution

Draw glycerol. Draw palmitic, oleic, and phosphoric acids alongside the —OH groups of glycerol. Draw ethanolamine alongside phosphoric acid. Carry out the dehydrations indicated in blue.

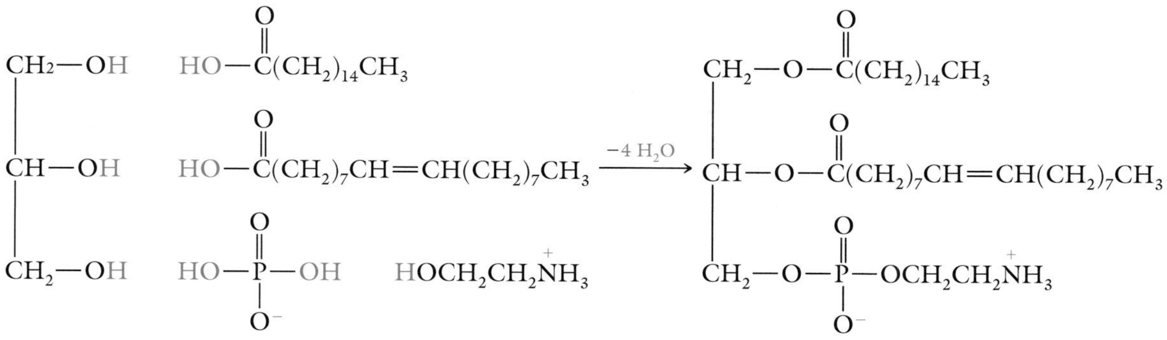

$$\begin{array}{lll} CH_2{-}OH & HO{-}\overset{O}{\overset{\|}{C}}(CH_2)_{14}CH_3 & \\ | & & \\ CH{-}OH & HO{-}\overset{O}{\overset{\|}{C}}(CH_2)_7CH{=}CH(CH_2)_7CH_3 & \\ | & & \\ CH_2{-}OH & HO{-}\overset{O}{\overset{\|}{P}}{-}OH & HOCH_2CH_2\overset{+}{N}H_3 \\ & \quad\;\; | & \\ & \quad\;\; O^- & \end{array} \xrightarrow{-4\,H_2O} \begin{array}{l} CH_2{-}O{-}\overset{O}{\overset{\|}{C}}(CH_2)_{14}CH_3 \\ | \\ CH{-}O{-}\overset{O}{\overset{\|}{C}}(CH_2)_7CH{=}CH(CH_2)_7CH_3 \\ | \\ CH_2{-}O{-}\overset{O}{\overset{\|}{P}}{-}OCH_2CH_2\overset{+}{N}H_3 \\ \qquad\quad\;\; | \\ \qquad\quad\;\; O^- \end{array}$$

Problem 11.6 Draw the glycerophospholipid whose hydrolysis yields equimolar amounts of glycerol, stearic acid, linoleic acid, phosphoric acid, and choline.

Sphingolipids

Sphingolipids are amphipathic hydrolyzable lipids based on sphingosine instead of glycerol. Sphingosine is a long-chain, unsaturated amino alcohol. Sphingolipids have an amide group at C2, formed by the reaction of a fatty acid with the amine group. The two types of sphingolipids—sphingophospholipids and

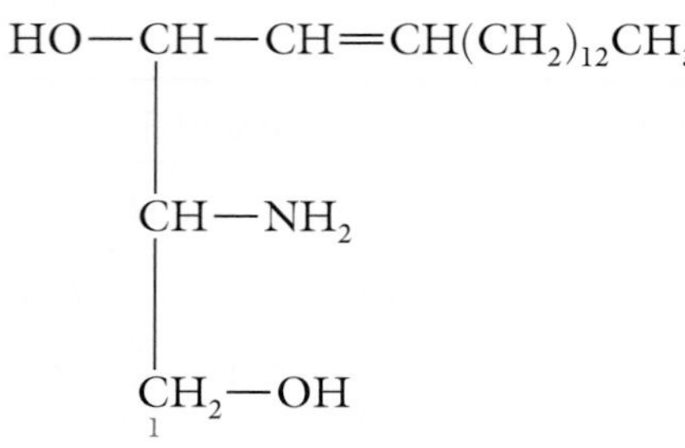

$$HO-CH-CH=CH(CH_2)_{12}CH_3$$
$$|$$
$$CH-NH_2$$
$$|$$
$$\underset{1}{C}H_2-OH$$

Sphingosine

sphingoglycolipids—differ in the structural unit at C1, which is the carbon atom at the bottom of the sphingosine structure shown in the margin.

Like glycerophospholipids, **sphingophospholipids** have a phosphodiester group at C1. For this reason, glycerophospholipids and sphingophospholipids are often classified together as **phospholipids.** In the following structure, $—R^1$ represents one of the alcohol units in Table 11.3; that is, the same alcohols are found in both glycerophospholipids and sphingophospholipids. **Sphingoglycolipids,** or **glycolipids** for short, contain a saccharide unit at C1, joined through an acetal linkage. Some sphingoglycolipids contain monosaccharide units. Others contain oligosaccharide units.

Sphingophospholipid

$$HO-CH-CH=CH(CH_2)_{12}CH_3$$
$$CH-N(H)-C(=O)-R^2$$
$$CH_2-O-P(=O)(O^-)-OR^1$$

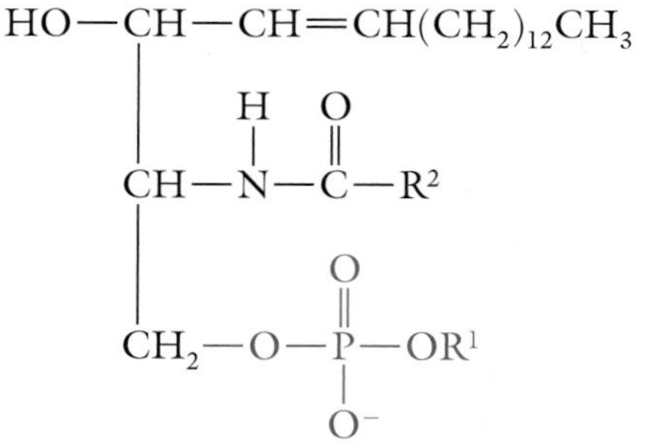

Phosphodiester group

Sphingoglycolipid

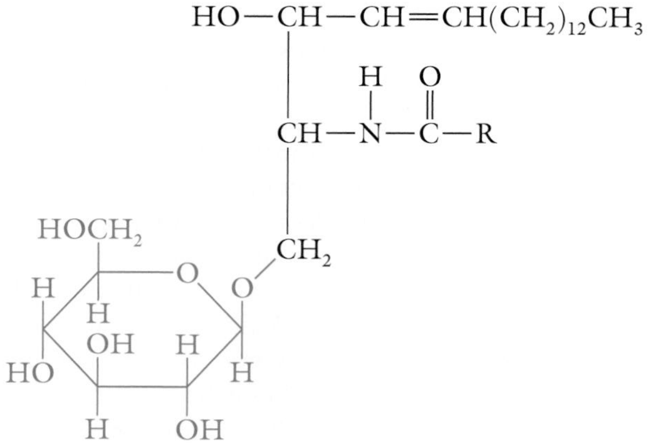

$$HO-CH-CH=CH(CH_2)_{12}CH_3$$
$$CH-N(H)-C(=O)-R$$
$$CH_2-O-(\text{saccharide})$$

HOCH2, H, H, OH, H, HO, H, H, OH, O, O

Saccharide group

Example 11.7 Drawing the structures of sphingolipids

Draw the sphingolipid whose hydrolysis yields equimolar amounts of sphingosine, palmitoleic and phosphoric acids, and choline.

Solution

Draw sphingosine. Draw palmitoleic acid alongside the $—NH_2$ group and phosphoric acid alongside the $—OH$ group. Draw choline alongside phosphoric acid. Carry out the dehydrations indicated in blue.

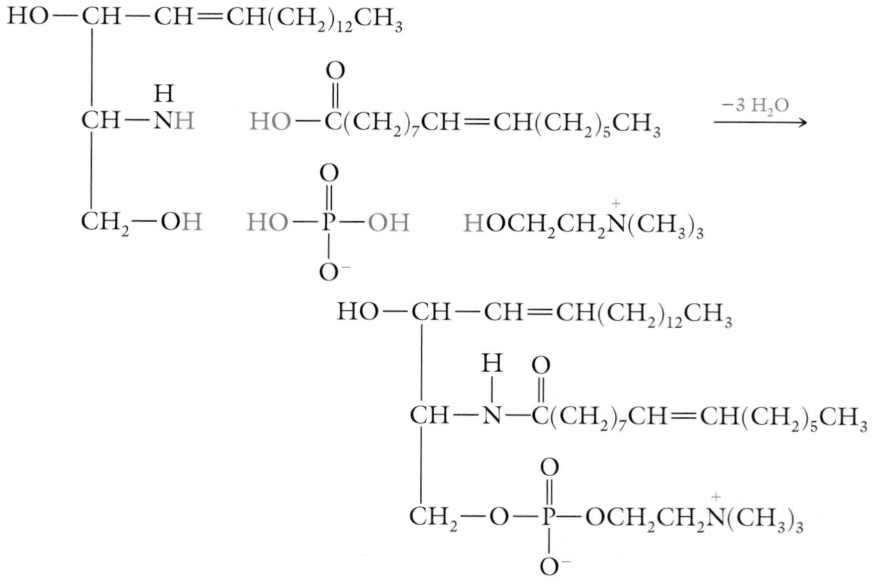

$$HO-CH-CH=CH(CH_2)_{12}CH_3$$
$$CH-NH(H) \quad HO-C(=O)(CH_2)_7CH=CH(CH_2)_5CH_3 \quad \xrightarrow{-3\,H_2O}$$
$$CH_2-OH \quad HO-P(=O)(O^-)-OH \quad HOCH_2CH_2\overset{+}{N}(CH_3)_3$$

$$HO-CH-CH=CH(CH_2)_{12}CH_3$$
$$CH-N(H)-C(=O)(CH_2)_7CH=CH(CH_2)_5CH_3$$
$$CH_2-O-P(=O)(O^-)-OCH_2CH_2\overset{+}{N}(CH_3)_3$$

Problem 11.7 Draw the sphingolipid whose hydrolysis yields equimolar amounts of sphingosine, oleic acid, and β-D-glucose.

Cell membranes contain mixtures of different glycerophospholipids and sphingolipids (as well as cholesterol). Because different cell membranes have different functions, the relative amounts of the different amphipathic hydrolyzable lipids vary from one type of cell to another. For example, the membranes of the myelin sheath that surrounds and electrically insulates many nerve-cell axons are rich in sphingophospholipids. Sphingoglycolipids containing oligosaccharides are usually found at cell surfaces where they perform cell-recognition functions (Section 10.6). Many illnesses arise from inadequate metabolism and utilization of glycerophospholipids and sphingolipids—for example, Niemann-Pick, Tay-Sachs, Gaucher's, and Fabry's diseases (Section 16.6).

11.7 STEROIDS: CHOLESTEROL, STEROID HORMONES, AND BILE SALTS

Steroids are nonhydrolyzable lipids that contain the **steroid ring structure,** which consists of four fused rings, three of them six-membered rings and one of them a five-membered ring. This group includes cholesterol, adrenocortical and sex hormones, and bile salts.

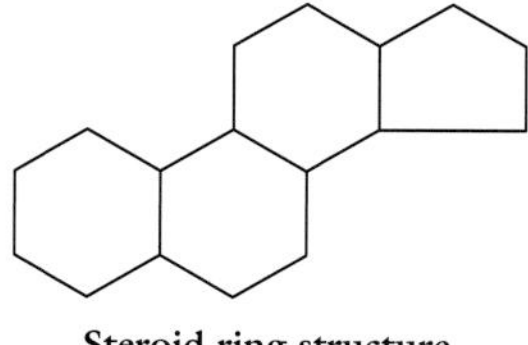
Steroid ring structure

Cholesterol is the major steroid in animals. Plants contain very little cholesterol but do contain related compounds in their membranes. The cholesterol molecule contains eight tetrahedral stereocenters (indicated by asterisks) and thus $2^8 = 256$ stereoisomers (128 pairs of enantiomers), but it exists in nature as only one stereoisomer—the one shown here:

Cholesterol

« Figure 1.6 shows a space-filling molecular model of cholesterol.

This stereoisomer is another example of the remarkable stereoselectivity of biological systems.

We now know that cholesterol plays an important role in cardiovascular disease and are advised to lower our intake of this lipid (Box 16.1). Nevertheless, we must keep in mind that cholesterol is critical to many physiological functions. In particular, it serves as a component of membranes (Section 11.10) and as a precursor for all other animal steroids, including the adrenocortical and sex hormones and the bile salts (Figure 11.4 on the following page). The body synthesizes all the cholesterol that it needs if cholesterol is excluded from the diet.

Steroid Hormones

The steroid hormones are just one group of **hormones,** substances synthesized in the endocrine glands from which they are excreted and then transported in the bloodstream to target tissues, where they regulate a variety of cell functions. Hormones are often called chemical messengers because they relay messages between different parts of the body. They are effective at very low concentrations, often in the picomolar to nanomolar range (10^{-12}–10^{-9} M).

The **adrenocortical hormones,** produced in the adrenal glands, located at the top of the kidneys, include cortisol and aldosterone. Cortisol helps the

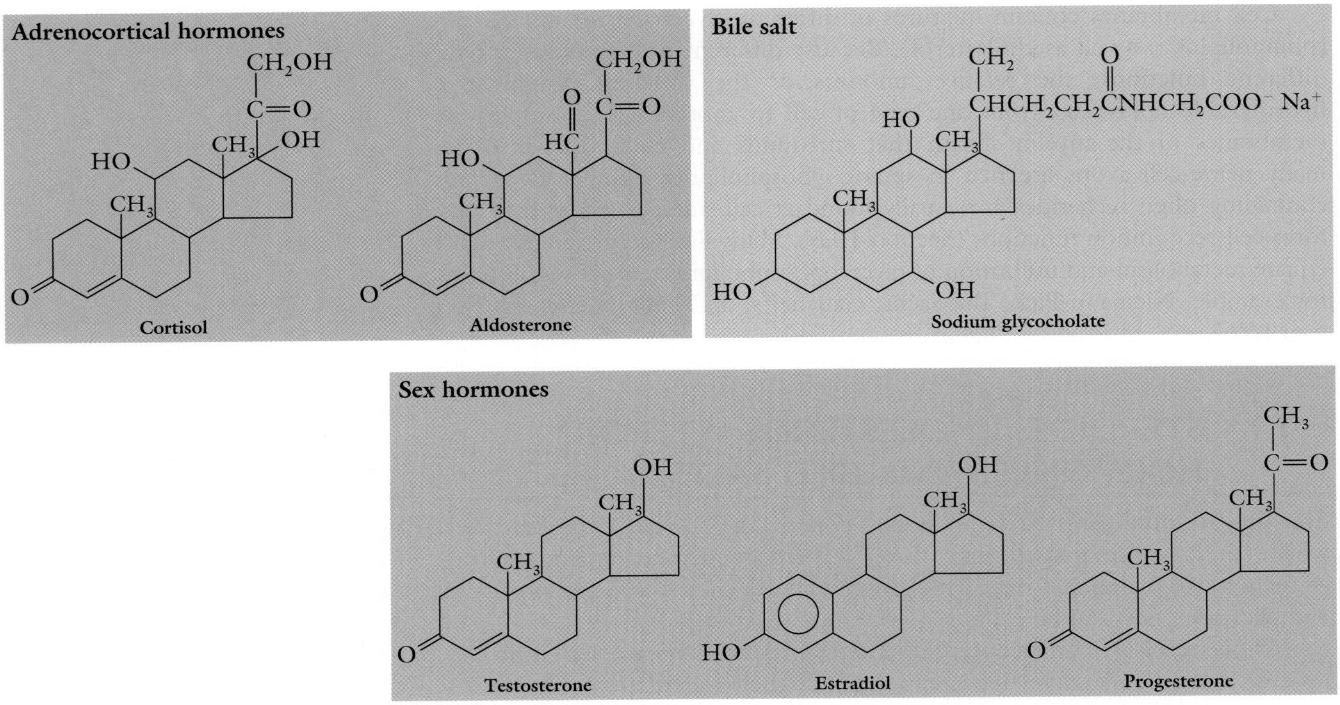

Figure 11.4 Steroids.

body respond to short-term stress. Because a high blood-glucose level is needed for your brain to function, cortisol stimulates other cells to decrease their use of glucose and, instead, metabolize fats and proteins for energy. Cortisol also blocks the immune system, because a stress situation is not the time to feel sick or have an allergic reaction. For this reason, cortisol is used as an anti-inflammatory and antiallergy medication. However, long-term use of cortisol can be damaging to the body because of its effect on the immune system.

Aldosterone takes part in the regulation of electrolyte concentrations in tissues. It stimulates Na^+, K^+, and water reabsorption by the kidney, helping to maintain blood volume and blood pressure.

The **sex hormones** are produced in the gonads: ovaries in females and testes in males. There are two kinds of female sex hormones—the **estrogens** and **progestins**—and one kind of male sex hormone—the **androgens.** The ovaries produce mostly estrogens, such as estradiol. They also produce small quantities of androgens. The testes produce mostly androgens, such as testosterone. They also produce small quantities of estrogens. Progestins are female hormones produced only in the ovaries. The most abundant progestin is progesterone, which, together with estradiol, regulates the menstrual cycle. Box 11.2 describes the hormonal changes in the menstrual cycle and the use of hormones in contraceptive drugs.

The androgens and estrogens regulate the development of the sex organs, the production of sperm and ova, and the development of secondary sex characteristics: lack of facial hair, increased breast size, and high voice in women; facial hair, increased musculature, and deep voice in men. Synthetic androgens are used by some athletes to promote muscle growth (Box 11.3 on page 358).

11.2 CHEMISTRY WITHIN US

The Menstrual Cycle and Contraceptive Drugs

After puberty, a woman secretes sex hormones in a repeating pattern, called the **menstrual cycle,** that causes periodic changes in her reproductive organs. The most important changes are the development and release of a single egg cell (ovum) and the growth of a uterine lining (endometrium), which prepares the uterus to receive a fertilized ovum.

A cycle, lasting 28 days on average, begins on the first day of menstruation (day 1). The levels of estrogens and progestins, mainly estradiol and progesterone, are low at this time, which triggers the hypothalamus to release gonadotropin-releasing hormone (GnRH). GnRH travels in the bloodstream to the pituitary gland, where it stimulates the release of follicle-stimulating hormone (FSH) and luteinizing hormone (LH) into the bloodstream.

FSH and LH stimulate ovum development in an ovarian structure called a **follicle.** Ovulation—the release of an ovum into one of the fallopian tubes where it can be fertilized by a sperm—takes place on or close to day 14. The ovum travels through the fallopian tube to the uterus. The follicle that released the ovum becomes a corpus luteum—an endocrine gland that secretes estrogen and progestin. These hormones control the development and maintenance of the uterine endometrium for receiving a fertilized ovum.

Negative feedback by estradiol and progesterone to the hypothalamus stops the release of GnRH, which in turn halts the secretion of FSH and LH. In a menstrual cycle, when the ovum is unfertilized, the corpus luteum shrinks, secretion of estradiol and progesterone decreases, and the endometrium begins to slough off. The negative feedback on the hypothalamus is relieved, and the pituitary gland increases its secretion of FSH and LH. Thus the cycle begins again.

If the ovum is fertilized as it travels through the fallopian tube, the corpus luteum continues to secrete estradiol and progesterone, promoting implantation of the fertilized egg in the uterine wall and the subsequent development of an embryo.

Oral contraceptives (birth-control pills) are combinations of a synthetic estrogen, such as ethynyl estradiol, and a synthetic progestin, such as norethindrone, that act by preventing ovulation. Natural estrogens and progestins cannot be used, because they are broken down if taken orally. This combination of hormones, taken for days 5 through 24 of the menstrual cycle, mimics the hormone levels present in early pregnancy and tricks the body into thinking that it is pregnant. Negative feedback inhibits GnRH, FSH, and LH secretion, and thus ovulation does not take place.

The negative side effects of oral contraceptives—chiefly an increased risk of cardiovascular complications such as blood clots, hypertension, stroke, and heart attack—are mostly associated with the high hormone levels used in early contraceptive pills. Modern contraceptive pills carry very low risks, except for women older than 35 who smoke or have hypertension.

OH, H_3C, $C{\equiv}CH$, HO

Ethynyl estradiol

OH, H_3C, $C{\equiv}CH$, O

Norethindrone

Mifepristone (RU-486) is a postcoital contraceptive pill. Often called the **morning-after pill,** it is almost completely effective when taken within 72 h after intercourse. Mifepristone works by blocking the progesterone receptors in the uterine lining, thus preventing progesterone from exerting its endometrium-maintenance effects. The endometrium sloughs off, as in menstruation. The effectiveness of mifepristone is enhanced by a dose of a prostaglandin taken 2 days later to induce uterine contractions. Mifepristone, together with a prostaglandin, causes a spontaneous abortion (miscarriage) and is used for that purpose, as an alternative to aspiration abortion, as late as 1 to 2 months after the last menstrual period.

H_2N, OH, H_3C, $C{\equiv}CCH_3$, O

Mifepristone (RU-486)

Mifepristone is used widely in Europe, especially France, and in East Asia. It was approved by the FDA in September 2000 for use in the United States.

11.3 CHEMISTRY WITHIN US

Anabolic Steroids

Anabolic steroids are steroids that stimulate the biosynthesis of proteins, the primary constituent of muscle. Some athletes use them to enhance their performance, especially in sports where increased size, strength, and stamina are advantageous: football, weight lifting, bodybuilding, and track and field events such as the discus and hammer throw and the shot put. The enhanced performance is achieved through both increased muscle development and increased aggressiveness.

Testosterone is a natural anabolic steroid. When taken orally or by intramuscular injection, testosterone is not very effective in enhancing athletic performance, because most of it undergoes metabolic breakdown. Laboratory modification of testosterone has produced synthetic anabolic steroids, such as nandrolone decanoate and stanozolol, that are more resistant to metabolic breakdown and more effective than testosterone.

Participants in a drug-free bodybuilding championship in Austin, Texas. (Daemmrich/The Image Works.)

$O-C(=O)-(CH_2)_8CH_3$, H_3C, O

Nandrolone decanoate

OH, H_3C, CH_3, H_3C, HN, N

Stanozolol

The use of anabolic steroids is banned by all sports organizations, but there is great variability in testing procedures. At one end of the spectrum are the Olympics. Urine testing of athletes is routinely required before and after athletic events. Athletes with positive test results have been disqualified and even banned from future competition. Baseball is at the other end of the spectrum, with weaker testing and penalty policies. Even when athletes are tested, it is not a certainty that cheaters will be found out. New synthetic anabolic steroids are always being developed, and testing methods may not keep ahead of these developments.

Anabolic steroids are controlled substances in the United States, under the Anabolic Steroids Act of 1990. However, some are approved for therapeutic uses: testosterone replacement in hypogonadal men, treatment of endometriosis and fibrocystic disease of the breast in women, and rehabilitation of patients with muscle atrophy resulting from severe injury. Therapeutic uses of anabolic steroids involve far lower doses than those typically used by athletes for performance enhancement.

Anabolic steroids can have severe side effects, ranging from temporary acne to increased liver and cardiovascular disease and including psychological effects such as mood swings and antisocial and violent behavior. Men may experience testicular atrophy, temporary infertility, and accelerated male-pattern baldness. Women may experience altered menstruation, clitoral enlargement, deepened voice, increased facial and body hair, and male-pattern baldness.

Bile Salts

Bile salts, also called **bile acids,** have a carboxylate salt group attached to the steroid ring structure. They play a crucial role in lipid digestion. Bile salts are synthesized from cholesterol in the liver and stored in the gall bladder as a solution called **bile,** which is a mixture of bile salts, cholesterol, and pigments from the breakdown of red blood cells. Sodium glycocholate (see Figure 11.4) is a bile salt.

Bile is secreted into the small intestine after a fatty meal and participates in the digestion of hydrolyzable lipids. Dietary lipid arrives in the intestine in the form of insoluble lipid globules. Hydrolysis by lipases can take place only at the surfaces of the globules. The larger the globules, the smaller the total surface area of lipid available for digestion and the slower the rate of digestion by the lipases.

The structure and function of bile salts resemble those of soaps (Section 7.7). Bile salts contain a large hydrophobic part (steroid ring system) and a small hydrophilic part (carboxylate). Bile salts break up large lipid globules into much smaller ones, greatly increasing the surface area available to lipases, and this increases the rate of digestion.

The conversion of cholesterol into bile salts, as well as the solubilization of cholesterol by bile salts in the bile, is the way in which excess cholesterol is eliminated through the intestinal tract. When the cholesterol concentration in bile is too high, however, cholesterol precipitates in the form of gallstones. Some gallstones are small enough to pass through the bile duct to the intestine, causing considerable pain at times but no long-term consequences. Larger gallstones may become stuck in the bile duct, however, and the consequences are severe. The duct blockage prevents the person from digesting fats, because bile cannot enter the intestine. The resulting abdominal pain and nausea do not subside. The skin becomes yellow—a condition called jaundice—as bile pigments enter the blood, and the stools lose their brown color. The dissolution of gallstones with medication is possible in some cases but can be very slow. Surgical removal of the gallbladder is the more common treatment.

Bile salts are also needed for the efficient intestinal absorption of the fat-soluble vitamins (A, D, E, and K; Section 11.9).

11.8 EICOSANOIDS

The **eicosanoids** are nonhydrolyzable lipids derived from the polyunsaturated C_{20} fatty acid called arachidonic acid (Figure 11.5). There are three groups of eicosanoids: the leukotrienes, prostaglandins, and thromboxanes. In a **leukotriene,** the 20-carbon chain of arachidonic acid and its carboxyl group are unchanged. A **prostaglandin** is similar to a leukotriene but has a cyclopentane ring formed by bond formation between C8 and C12 of the 20-carbon chain. A **thromboxane** is similar to a prostaglandin but, instead of the cyclopentane ring, C8 through C12 form a six-membered cyclic ether structure. As shown in Figure 11.5, individual leukotrienes, prostaglandins, and thromboxanes are distinguished by a letter and subscript number.

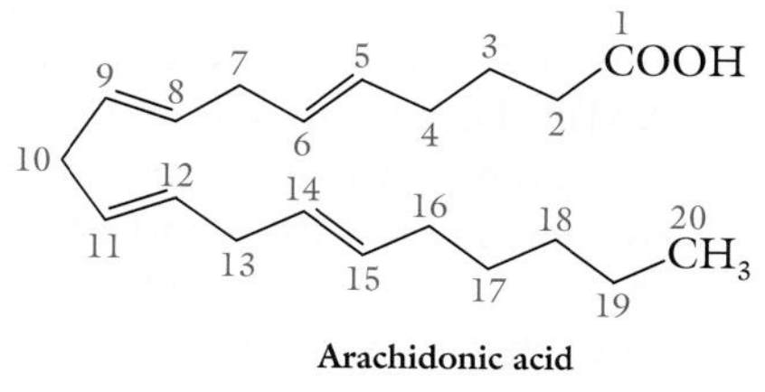

Figure 11.5 Examples of the three categories of eicosanoids, derived from arachidonic acid.

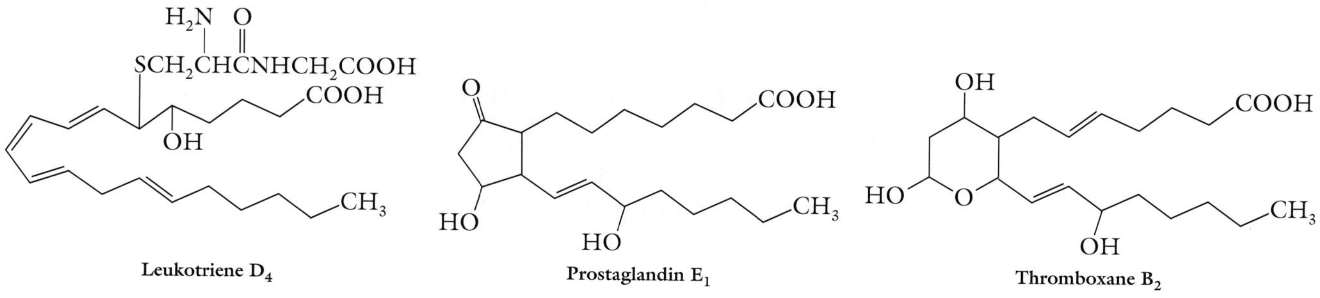

Eicosanoids have hormonelike physiological functions. Like hormones, eicosanoids produce their regulatory effects at low concentrations. Unlike hormones, eicosanoids are not transported in the bloodstream from their sites of synthesis to their sites of action. Eicosanoids are **local hormones** or **local mediators,** acting in the same tissues in which they are synthesized. They are produced in most tissues. Eicosanoids play roles in

- the inflammatory response in joints (rheumatoid arthritis), skin (psoriasis), muscle (overexertion), and eyes;
- the production of pain and fever in disease and injury;
- the regulation of blood pressure;
- blood clotting;
- the induction of labor;
- the regulation of the wake–sleep cycle; and
- allergic and asthmatic reactions.

The pain and swelling caused by arthritis and related illnesses results from the production of prostaglandins. Various anti-inflammatory drugs prevent the synthesis of prostaglandins by reacting with the enzyme prostaglandin H_2 synthase (PGHS) and inhibiting its activity. PGHS catalyzes the synthesis of an important intermediate compound en route to the prostaglandins. Steroidal anti-inflammatory drugs such as cortisol are effective but of limited use for long-term use because of adverse effects on the immune system (Section 11.7). Nonsteroidal anti-inflammatory drugs (NSAIDs) are more useful because their side reactions are generally less severe. Among the NSAIDs are aspirin, ibuprofen, acetaminophen, naproxen sodium, and indomethacin sodium (Box 7.4 and Section 7.6). The most recent NSAIDs are the COX-2 inhibitors celecoxib (Celebrex), rofecoxib (Vioxx), and valdecoxib (Bextra).

There are two forms of prostaglandin H_2 synthase, cyclooxygenase-1 (COX-1) and cyclooxygenase-2 (COX-2). COX-1 is generated in most tissues in the body and is responsible for maintaining the normal health of organs and tissues as well as responding to inflammatory stimuli. In the stomach, COX-1 is responsible for maintaining the health of the mucous lining. COX-2 is much more specific in its actions compared with COX-1. COX-2 is not produced in the stomach and is generally produced only in response to inflammatory stimuli. Aspirin inhibits the activity of both COX-1 and COX-2. This inhibition reduces the inflammatory response but also risks gastric irritation, which can lead to bleeding and ulcers. Ibuprofen, acetaminophen, and naproxen sodium have other side reactions (Box 7.4).

The COX-2 inhibitors were heavily prescribed as safer alternatives to the other NSAIDs because they inhibited COX-2 but not COX-1. Thus COX-2 inhibitors prevented the inflammatory response without risking gastric irritation. However, in late 2004 and early 2005, respectively, Vioxx and Bextra were withdrawn from the market by their manufacturers. This action resulted from studies that showed an increased risk of heart attack and stroke. There are now concerns about Celebrex as well as some of the other NSAIDs.

11.9 FAT-SOLUBLE VITAMINS

Vitamins are organic compounds required in trace amounts for normal metabolism but not synthesized by the organism that requires them. They must be included in the diet. Vitamins are classified as water soluble or fat soluble, on the basis of their solubility characteristics. The **water-soluble**

A PICTURE OF HEALTH
Examples of Lipids in the Body

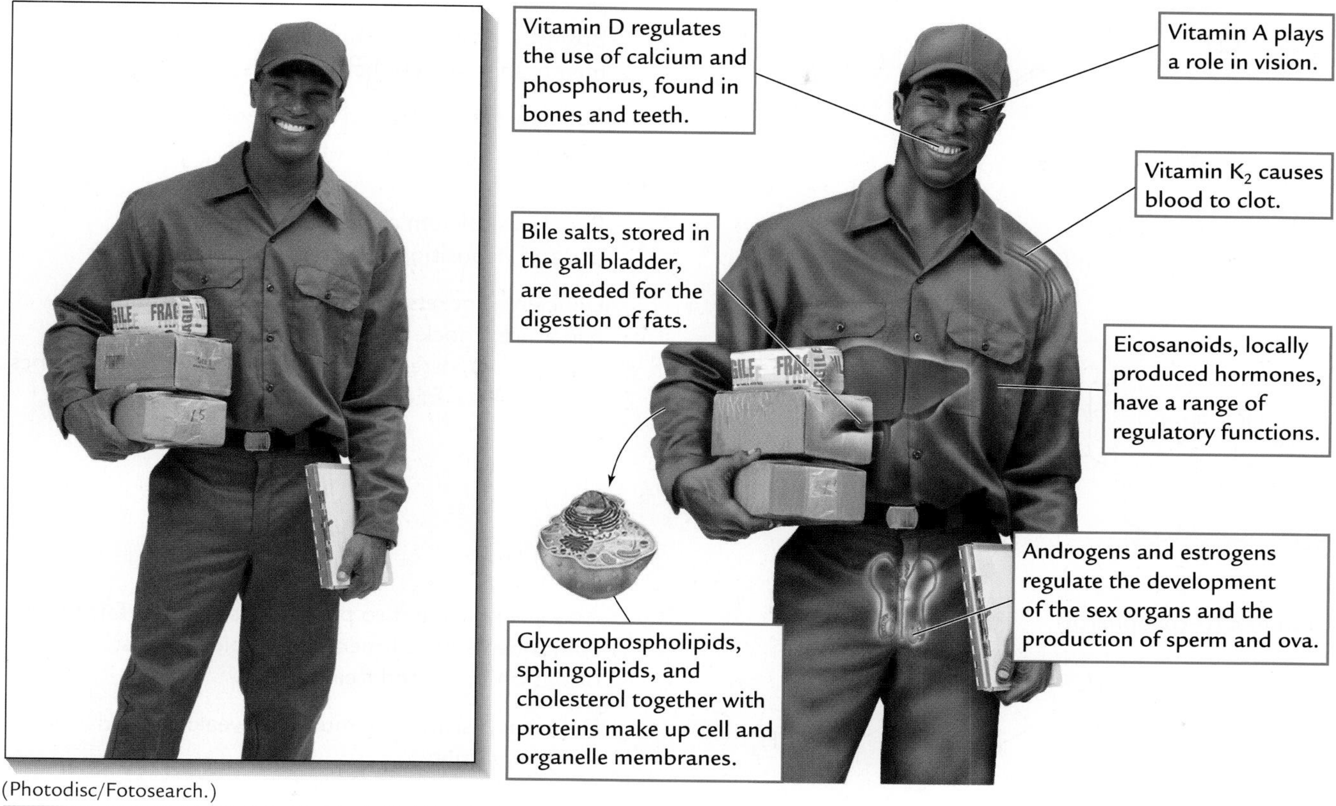

Vitamin D regulates the use of calcium and phosphorus, found in bones and teeth.
Vitamin A plays a role in vision.
Vitamin K_2 causes blood to clot.
Bile salts, stored in the gall bladder, are needed for the digestion of fats.
Eicosanoids, locally produced hormones, have a range of regulatory functions.
Androgens and estrogens regulate the development of the sex organs and the production of sperm and ova.
Glycerophospholipids, sphingolipids, and cholesterol together with proteins make up cell and organelle membranes.

(Photodisc/Fotosearch.)

vitamins, which are the B and C vitamins, contain sufficient polar functional groups such as hydroxyl and amine groups to make them water soluble. The **fat-soluble vitamins,** which are the A, D, E, and K vitamins, are a group of nonhydrolyzable lipids (Table 11.4 on the following page). Most vitamins consist of a closely related group of compounds.

›› Water-soluble vitamins are considered in Chapter 18.

Strictly speaking, vitamin D is a steroid hormone and not a vitamin, because it can be synthesized from cholesterol. Recall that we do not depend on our diet for cholesterol. Vitamin D has been traditionally classified as a vitamin because, for many years, scientists did not know that it is synthesized from cholesterol.

Fat-soluble vitamins are obtained mostly from vegetables and fruits, fish, liver, and dairy products. Deficiences of these vitamins cause serious health problems, as indicated in Table 11.4. However, excess fat-soluble lipids are stored in fat tissues, and so they need not be ingested daily. In fact, the consumption of megadoses of fat-soluble vitamins can produce problems because of their accumulation in the body. For example, excess vitamin D causes the deposition of abnormal amounts of calcium in bone, resulting in bone pain. It also causes calcification of soft tissue such as the kidneys and brain, resulting in impaired kidney function and mental retardation. High consumption of water-soluble vitamins presents much less risk, because the excess is excreted in the urine.

TABLE 11.4 Fat-Soluble Vitamins

Name and structure	Function
vitamin A (*trans*-retinol) 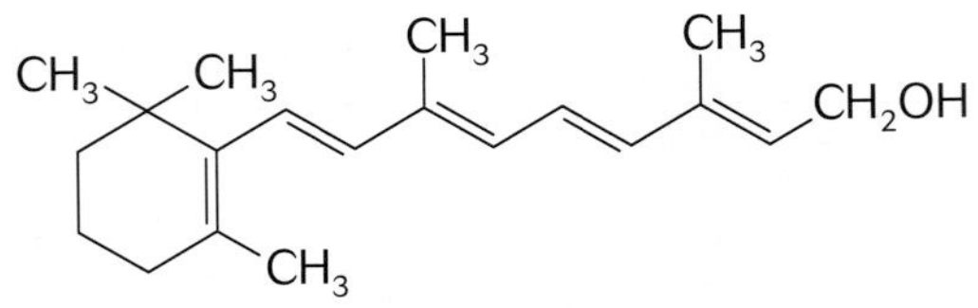	Plays key role in vision by its conversion into *cis*-11-retinal and subsequently into rhodopsin (see Box 4.2). Aids proper functioning of mucous membranes and epithelial tissues. Deficiency: dry eyes and skin, sterility in males, night blindness
1,25-dihydroxyvitamin D_3 (active form of Vitamin D) 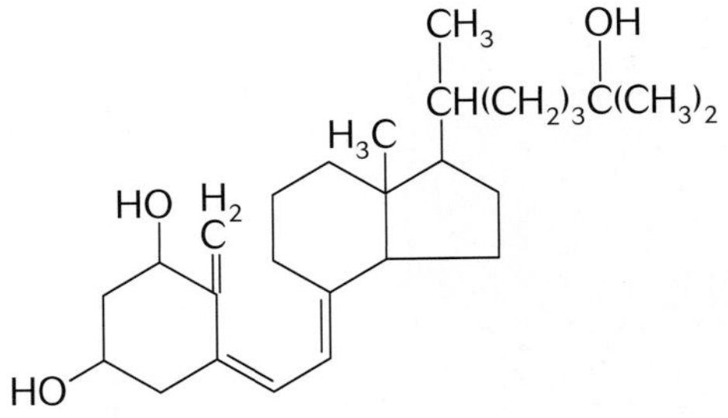	Regulates calcium and phosphate use and deposition in bone and cartilage. Deficiency: rickets in children (bowlegs, spinal curvature, knock-knees, pelvic and thoracic deformities); osteomalacia in adults (weakened bones susceptible to fracture)
vitamin E (α-tocopherol) 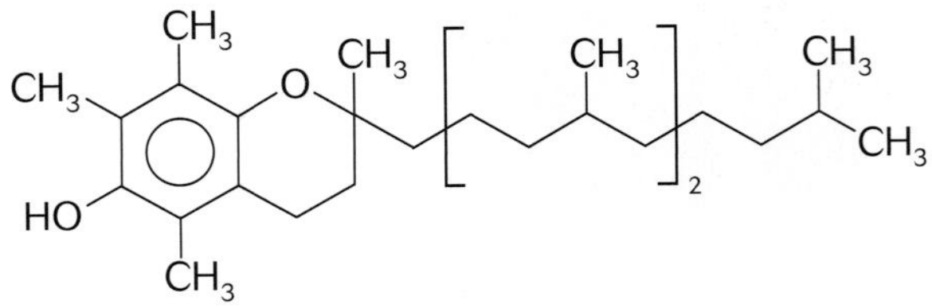	Acts as antioxidant to protect unsaturated fatty acid components of cell-membrane lipids against oxidation by air and free radicals. Deficiency: scaly skin, muscular weakness and atrophy, sterility
vitamin K_2 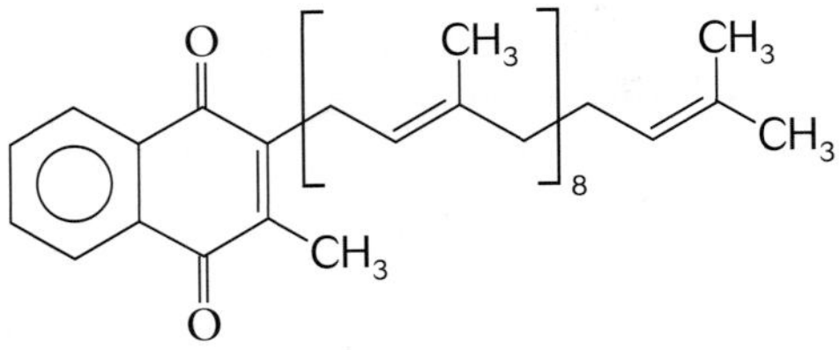	Regulates formation of prothrombin, needed for blood clotting. Deficiency: increased time for blood clotting, a serious problem when a person is bruised, wounded, or undergoing surgery

11.10 BIOLOGICAL MEMBRANES

Every cell, whether prokaryotic (bacteria) or eukaryotic (higher organisms), is separated from its extracellular environment by a cell membrane (Section 14.1). Eukaryotic cells are more highly organized into separate membrane-enclosed **organelles** (see Figure 14.1). Each of the organelles performs a specialized function. The nucleus synthesizes nucleic acids and stores the cell's genetic information. Lysosomes digest macromolecules into their smaller components, which are then recycled by cells. Mitochondria metabolize carbohydrates and lipids to generate energy. The endoplasmic reticulum synthesizes proteins and lipids. The Golgi apparatus synthesizes oligosaccharides and adds to them lipids and proteins to form glycolipids and glycoproteins.

Biological membranes, both cell and organelle membranes, are not inert barriers. They are highly selective permeability barriers that regulate the molecular and ionic composition within cells and organelles:

- Membranes control the entry into cells and organelles of the materials required for synthesizing biomolecules, the metabolic fuels for generating energy, and other materials to be modified or broken down. Membranes also control the exit of products from organelles for use elsewhere in the cell and from cells for use elsewhere in the organism and the exit of waste products of cellular processes.
- Membranes play the central role in cell recognition (Section 10.6).
- Membranes take part in cell communication. Receptor molecules in membranes receive information from other cells—in the form of hormones, for example—and translate it into molecular changes within the cell.
- Membranes maintain different concentrations of ions on each side of the membrane barrier. These concentration gradients are used to generate chemical or electrical signals for the transmission of nerve impulses and muscle action and to transport other ions across the membrane.

Membrane Structure

Biological membranes consist mostly of lipids and proteins. The relative amounts vary considerably from one type of membrane to another. At one extreme, the myelin sheath—multiple layers of membranes that insulate nerve cells in some parts of the nervous system—is about 80% lipid and 20% protein by weight. At the other extreme, the inner membrane of the mitochondrion is about 20% lipid and 80% protein. Most membranes, such as the plasma membrane of human erythrocytes (red blood cells), contain about equal amounts of lipid and protein. The typical membrane is also from about 2% to 10% carbohydrate.

The lipids of membranes are the amphipathic lipids—glycerophospholipids, sphingolipids, and cholesterol—all of which have

- a long hydrophobic (nonpolar) part, or **tail**—the two hydrocarbon chains in glycerophospholipids and sphingolipids or the steroid ring structure with its attached alkyl group in cholesterol; and
- a short hydrophilic (polar or ionic) part, or **head**—the phosphodiester group in glycerophospholipids and sphingophospholipids, the saccharide in sphingoglycolipids, or the hydroxyl group in cholesterol.

Figure 11.6 on the next page shows the **fluid-mosaic model** that biochemists use to conceptualize biological membranes. The membrane lipids are organized into a **lipid bilayer.** That is, two lipid layers, each having a hydrophobic side and a hydrophilic side, are arranged with the hydrophobic sides in contact and the hydrophilic sides forming the inner and outer surfaces of the membrane, in contact with the internal and external aqueous environments. The driving force for the formation of this lipid bilayer comprises secondary attractive and repulsive forces. The orientation of the tails toward the membrane interior and of the heads toward the inner and outer membrane surfaces

- prevents the repulsive interactions between the hydrophobic tails and the intracellular and extracellular aqueous environments;
- maximizes the attractive forces between the hydrophobic tails, both within each layer and between the two layers of the bilayer; and
- maximizes the attractive forces of the hydrophilic heads with each other, as well as between the hydrophilic heads and the intracellular and extracellular aqueous environments.

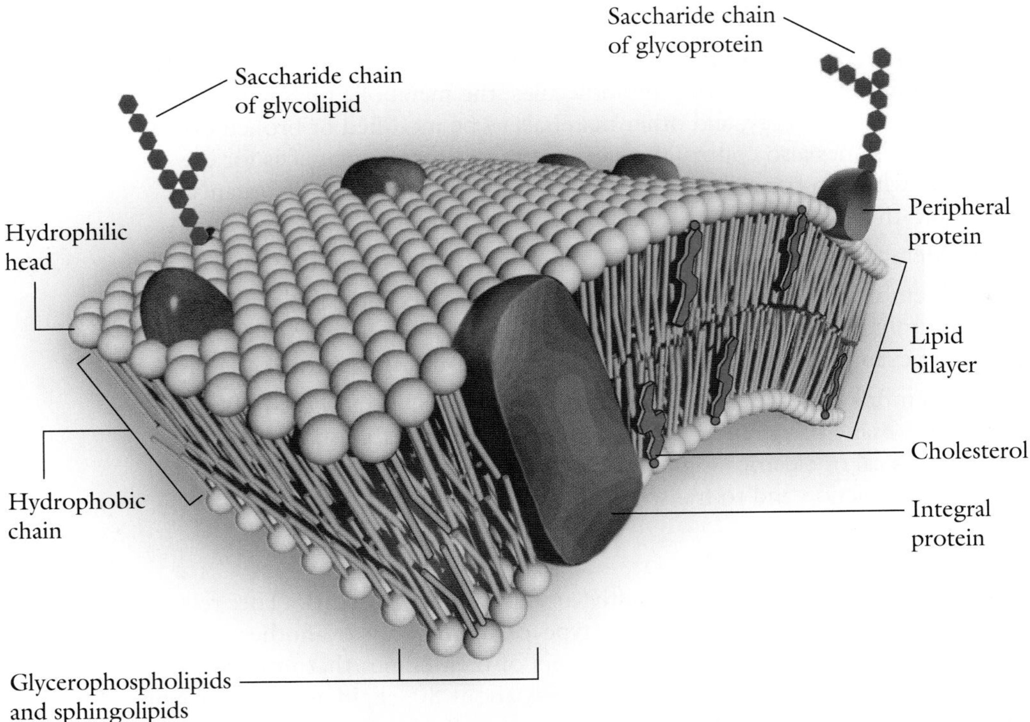

Figure 11.6 Fluid-mosaic model of the cell membrane.

» Membrane-lipid diseases are considered in Section 16.6.

Stereoselectivity of the amphipathic lipids is important in their ability to aggregate together into the lipid bilayer. All of the glycerophospholipids and sphingolipids are L-enantiomers (C2 in the glycerol and sphingosine components is a tetrahedral stereocenter). As stated earlier, cholesterol is one specific stereoisomer of a possible 256 stereoisomers (Section 9.7).

Membrane lipids establish the compartments in which the cell's metabolic processes take place, but membrane proteins play the crucial role in mediating those processes by functioning as enzymes and as transporters of molecules and ions through the membrane. There are two general types of membrane proteins: integral and peripheral proteins. **Integral,** or **intrinsic, proteins** are embedded in the membrane. Some penetrate only part way through the lipid bilayer, on one side or the other of the membrane. Others penetrate completely from one side to the other. **Peripheral,** or **extrinsic, proteins** are located on the membrane surfaces and do not penetrate into the membrane. Membrane proteins associate with the lipid bilayer through secondary attractive forces. Peripheral proteins undergo ionic and hydrogen-bond interactions with the lipid heads at the membrane surface. Integral proteins have both hydrophobic parts, which allow them to penetrate the hydrophobic interior of the lipid bilayer, and hydrophilic parts, which prefer contact with the aqueous medium.

The carbohydrates of membranes are on the extracellular hydrophilic surfaces. They are covalently bound to membrane lipids as glycolipids and to proteins as glycoproteins. The carbohydrate parts of glycolipids and glycoproteins serve as the receptor sites in cell-recognition and cell-communication processes.

The fluid-mosaic model of membrane structure seen in Figure 11.6 is aptly named. The lipid bilayer is a mosaic—a complex pattern of different lipids and proteins. Membranes are structurally and functionally asymmetrical. The outer and inner surfaces have different components and functions, and each surface is itself asymmetrical, with different components and functions at different locations. The lipid bilayer is fluid because the individual lipid and protein molecules have lateral mobility: they can diffuse sideways in the plane of each layer of the bilayer. This mobility is possible because the lipids and proteins are held together by secondary attractive forces, not covalent bonds.

The fluidity of membranes varies with membrane composition and temperature:

- Shorter and unsaturated fatty acid components in the glycerophospholipids and sphingolipids reduce secondary attractive forces and increase fluidity.
- Lower temperatures decrease fluidity through decreased motion of the lipid tails.
- Cholesterol modulates the fluidity of the membrane. Low cholesterol levels strengthen the membrane, but higher levels make the membrane too rigid.

The precise fluidity needed for a membrane depends on its function. The relative amounts of cholesterol, saturated chains, and unsaturated chains vary among species of organisms, among cell and organelle membranes in the same organism, and even in the same type of membrane over time with changes in temperature and dietary lipid composition.

Transport Through Membranes

Small molecules such as water, inorganic ions, simple carbohydrates, amino acids, and lipids pass through cell and organelle membranes by three processes: simple transport, facilitated transport, and active transport (Figure 11.7).

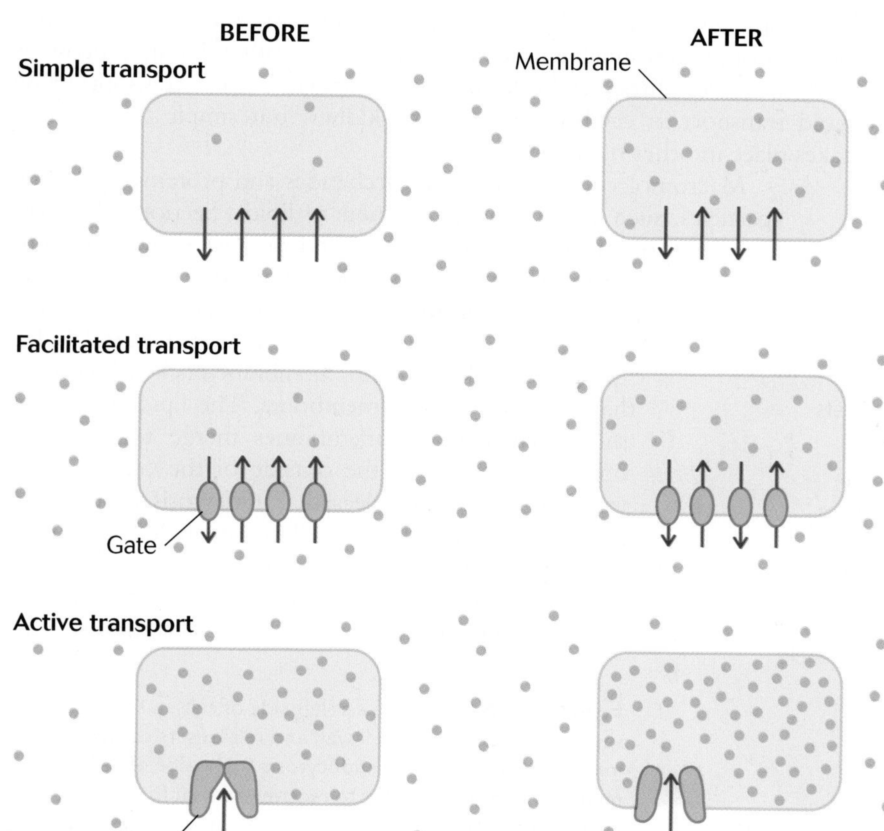

Figure 11.7 Transport across a biological membrane. Both simple and facilitated transport equalize solute concentrations across a membrane. Active transport proceeds against a concentration gradient and either maintains or increases the concentration gradient.

Simple transport (diffusion) is the passage of a solute across a membrane along a concentration gradient from the high-concentration side to the low-concentration side. Simple transport accommodates only a few small molecules such as O_2, N_2, H_2O, urea, and ethanol. For ions and larger polar molecules, simple transport is prevented by the large secondary repulsive forces encountered in the hydrophobic regions of the lipid bilayer. Transport of these solutes requires facilitated or active transport.

Facilitated (or **passive**) **transport** also is transport along a concentration gradient but, in contrast with simple transport, specific integral proteins—called **transporters** or **permeases**—facilitate and speed up the transport of solute. Transporters act as selective **gates,** or **channels,** for the binding and transport of specific solutes. Chloride ion (Cl^-) and bicarbonate ion (HCO_3^-), for example, cross membranes by facilitated transport.

Active transport is transport across a membrane against a concentration gradient, from the low-concentration side to the high-concentration side. Like facilitated transport, it requires specific transporters. In active transport, the transporters act as **pumps** to drive solute in the energetically unfavorable direction against the concentration gradient.

Active transport requires energy, which is supplied by coupling the transport process to a metabolic process that is energetically favorable. For example, nerve-impulse transmission and muscle contraction require nerve and muscle cells to maintain a lower concentration of Na^+ and a higher concentration of K^+ in the cell than in the extracellular fluid. The energetically unfavorable transport of Na^+ out and K^+ in (called the **Na^+/K^+ pump**) is coupled with and balanced by the energetically favorable hydrolysis of adenosine triphosphate (ATP) to adenosine diphosphate (ADP). Many other ions and molecules, including H^+, Ca^{2+}, and amino acids pass through cell and organelle membranes by active transport.

›› The function of the Na^+/K^+ pump in nerve-cell membranes is described in Section 18.5.

Some solutes are transported by more than one mechanism, depending on the cell or organelle. For example, glucose is cotransported with Na^+ by active transport through intestinal epithelial cells. On the other hand, glucose is transported into erythrocytes by facilitated transport. Urea is transported by facilitated transport in erythrocytes and the kidney, but simple transport of urea takes place in other tissues.

Macromolecules, such as polysaccharides and proteins, and large particles, such as lipoproteins (carriers of lipids; Section 18.3), are transported into and out of cells and organelles by a different mechanism. The process is called **exocytosis** for the export of material and **endocytosis** for the import of material (Figure 11.8). Macromolecules and other materials for export are packaged into membrane-enclosed **vesicles** that move to the cell membrane. The lipid bilayers of the vesicle and cell membranes merge together, an opening develops to the outside of the cell, and the vesicle contents are released. Endocytosis proceeds in the reverse manner. The cell or organelle membrane forms a pocket that receives material from the environment. The pocket develops into a vesicle that moves into the cell or organelle and releases its contents.

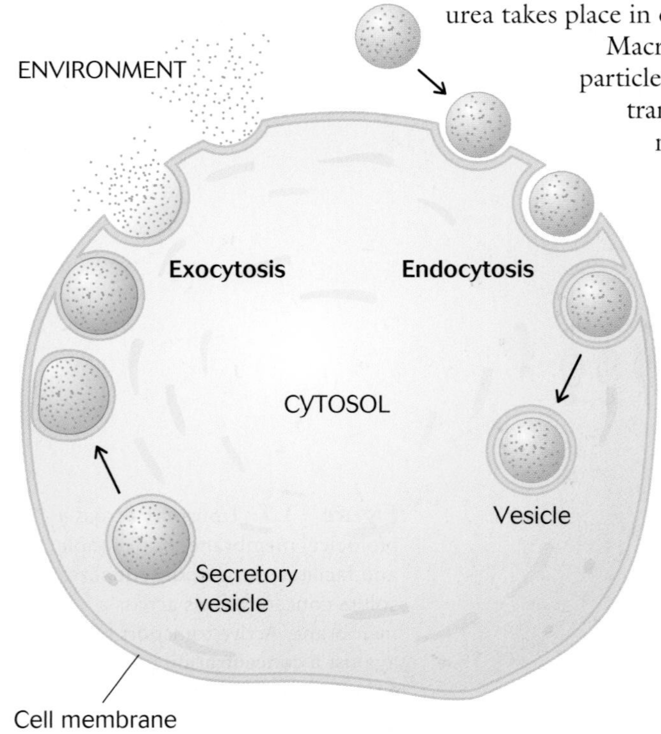

Figure 11.8 In exocytosis (left side of cell), a secretory vesicle merges with the cell wall and releases its contents into the environment. In endocytosis (right side of cell), the membrane forms a pocket to receive material from the environment. The pocket closes to become a vesicle that moves into the cell interior and then releases its contents.

Summary

Classifying Lipids

- The lipids are biomolecules soluble in nonpolar or low-polarity solvents.
- Lipids are classified as hydrolyzable or nonhydrolyzable, depending on the presence or absence of hydrolyzable groups.
- Triacylglycerols, waxes, glycerophospholipids, and sphingolipids are the hydrolyzable lipids; steroids, eicosanoids, and fat-soluble vitamins are the nonhydrolyzable lipids.

Fatty Acids

- Fatty acids are the carboxylic acid components of hydrolyzable lipids and usually contain from 12 to 24 carbons.
- Fatty acids are almost exclusively linear (unbranched) acids with an even number of carbons.
- Saturated fatty acids contain no C=C double bonds.
- Unsaturated fatty acids contain one or more cis C=C double bonds resulting in the monounsaturated and polyunsaturated fatty acids.

Triacylglycerols

- Animal fats and plant oils are triacylglycerols, the triesters formed by the esterification of glycerol with fatty acids.
- Animal fats are generally richer in saturated fatty acids, and plant oils are richer in unsaturated fatty acids.
- These lipids undergo hydrolysis (including saponification and digestion), catalytic hydrogenation, and rancidity reactions.
- Triacylglycerols are the main storage form of energy in animals.
- Energy is produced more efficiently by triacylglycerol metabolism than by glycogen metabolism.

Waxes

- Waxes, the esters of long-chain monohydric alcohols and fatty acids, perform a variety of protective functions in plants and animals.

Amphipathic Hydrolyzable Lipids

- Glycerophospholipids, which are amphipathic lipids containing a highly hydrophilic ionic group, are the major component of biological membranes.
- Sphingolipids, which are amphipathic lipids based on sphingosine instead of glycerol, also are components of biological membranes.

Steroids

- Steroids are nonhydrolyzable lipids that contain the steroid ring structure, consisting of four fused rings, of which three are six-membered rings and one is a five-membered ring.
- Cholesterol, steroid hormones, and bile salts are steroids.
- Cholesterol is a component of biological membranes and is the precursor to all other steroids.
- The adrenocortical and sex hormones are synthesized in endocrine glands and transported in the bloodstream to target tissues where they regulate various functions ranging from the control of metabolism to the development of the sex organs and secondary sexual characteristics.
- Hormones are effective at very low concentrations.
- Bile salts are needed for the efficient digestion of hydrolyzable lipids and absorption of fat-soluble vitamins.

Eicosanoids

- The eicosanoids—consisting of prostaglandins, leukotrienes, and thromboxanes—act as local hormones.
- Unlike hormones, eicosanoids regulate functions in the tissues in which they are produced.
- Eicosanoids mediate such functions as the inflammatory response, muscle contraction, and blood clotting.

Fat-Soluble Vitamins

- Vitamins are organic compounds required in trace amounts for an organism's normal metabolism, but they cannot be synthesized by the organism and must be included in the diet.
- The fat-soluble vitamins are A, D, E, and K; the water-soluble vitamins are B and C.

Biological Membranes

- Biological membranes consist mostly of lipids and proteins.
- The lipids used for membranes are the amphipathic lipids—glycerophospholipids, sphingolipids, and cholesterol.
- Membrane lipids are organized into a bilayer in which membrane proteins are embedded.
- The proteins function as permeases, carrying solutes through the membrane.
- There are three types of transport mechanisms for small-sized solutes: simple transport, facilitated transport, and active transport.
- Macromolecules and large particles are transported into the out of cells and organelles by endocytosis and exocytosis.

Key Words

Exercises

Fatty Acids

11.1 Which of the following carboxylic acids are present in animal and plant lipids?

$CH_3(CH_2)_{16}COOH$
1

$CH_3CH(CH_3)CH_2(CH_2)_{13}COOH$
2

cis-$CH_3(CH_2)_5CH{=}CH(CH_2)_8COOH$
3

11.2 Which of the following carboxylic acids are present in animal and plant lipids?

$CH_3(CH_2)_{13}COOH$
1

$CH_3(CH_2)_{14}COOH$
2

trans-$CH_3(CH_2)_5CH{=}CH(CH_2)_7COOH$
3

11.3 What structural features distinguish between saturated, monounsaturated, and polyunsaturated fatty acids?

11.4 Categorize the following fatty acids as saturated, monounsaturated, polyunsaturated, ω-3, ω-6, and ω-9:
(a) $CH_3(CH_2)_{16}COOH$;
(b) $CH_3(CH_2)_4(CH{=}CHCH_2)_2(CH_2)_6COOH$;
(c) $CH_3CH_2(CH{=}CHCH_2)_3(CH_2)_6COOH$;
(d) $CH_3(CH_2)_7CH{=}CH(CH_2)_7COOH$.

11.5 Which of the following compounds has the higher melting point and why?

cis-$CH_3(CH_2)_5CH{=}CH(CH_2)_7COOH$
1

$CH_3(CH_2)_{14}COOH$
2

11.6 Which of the following compounds has the higher melting point and why?

trans-$CH_3(CH_2)_5CH{=}CH(CH_2)_7COOH$
1

cis-$CH_3(CH_2)_5CH{=}CH(CH_2)_7COOH$
2

The Structure and Physical Properties of Triacylglycerols

11.7 Draw the structure of the triacylglycerol containing equimolar amounts of lauric, myristic, and oleic acids.

11.8 Draw the structure of the triacylglycerol containing equimolar amounts of palmitic, stearic, and palmitoleic acids.

11.9 Consider triacylglycerols with the following fatty acid compositions: (a) palmitic, stearic, stearic; (b) palmitic, oleic, linoleic; (c) palmitic, stearic, oleic. Place the triacylglycerols in order of melting points. Explain the order.

11.10 Explain why fish triacylglycerols are liquid, whereas beef triacylglycerols are solid.

Chemical Reactions of Triacylglycerols

11.11 Consider the following triacylglycerol:

$$\begin{array}{l} CH_2{-}O{-}C(=O)(CH_2)_{12}CH_3 \\ | \\ CH{-}O{-}C(=O)(CH_2)_{14}CH_3 \\ | \\ CH_2{-}O{-}C(=O)(CH_2)_7CH{=}CH(CH_2)_7CH_3 \end{array}$$

Write equations to describe the following reactions of this compound: (a) saponification (hydrolysis) with NaOH; (b) catalytic hydrogenation (H_2/Pt).

11.12 Write the equation for the acid-catalyzed hydrolysis of the triacylglycerol in Exercise 11.11. How does digestion with lipase differ from the acid-catalyzed hydrolysis?

11.13 Explain why bacterial hydrolysis of the following triacylglycerol produces a malodorous mixture. Write the equation for the reaction.

$$\begin{array}{l} CH_2{-}O{-}C(=O)(CH_2)_2CH_3 \\ | \\ CH{-}O{-}C(=O)(CH_2)_{14}CH_3 \\ | \\ CH_2{-}O{-}C(=O)(CH_2)_7CH{=}CH(CH_2)_5CH_3 \end{array}$$

11.14 Explain why air oxidation of the triacylglycerol in Exercise 11.13 produces a malodorous mixture. Write the equation for the reaction.

11.15 What reaction is used to harden corn oil in the manufacture of margarine?

11.16 What reaction is used to produce a soap from beef fat?

Waxes

11.17 One component of the wax obtained from the leaves of a South American tree has the formula $C_{40}H_{80}O_2$. What is the structure of this wax component: 1, 2, or 3?

$CH_3CH_2CH_2COO(CH_2)_{35}CH_3$
1

$CH_3(CH_2)_{16}COO(CH_2)_{21}CH_3$
2

$CH_3(CH_2)_{15}COO(CH_2)_{22}CH_3$
3

11.18 Spermaceti, a fragrant mixture of lipids isolated from sperm whales, was used in formulating cosmetics until banned in the late 1970s. A major component of spermaceti is the wax cetyl palmitate formed from palmitic acid and cetyl alcohol, $CH_3(CH_2)_{15}OH$. Draw the structure of cetyl palmitate.

Amphipathic Hydrolyzable Lipids

11.19 What structural feature distinguishes glycerophospholipids from triacylglycerols?

11.20 What structural feature distinguishes sphingolipids from triacylglycerols?

11.21 Draw the structure of the glycerophospholipid whose hydrolysis yields equimolar amounts of glycerol, myristic acid, oleic acid, phosphoric acid, and serine.

11.22 Draw the structure of the glycerophospholipid whose hydrolysis yields glycerol, lauric acid, linolenic acid, phosphoric acid, and choline.

11.23 What are the products when the following lipid is hydrolyzed in the presence of an acid?

$$
\begin{array}{l}
CH_2{-}O{-}\overset{\overset{\displaystyle O}{\|}}{C}(CH_2)_{14}CH_3 \\
|\\
CH{-}O{-}\overset{\overset{\displaystyle O}{\|}}{C}(CH_2)_{16}CH_3 \\
|\\
CH_2{-}O{-}\underset{\underset{\displaystyle O^-}{|}}{\overset{\overset{\displaystyle O}{\|}}{P}}{-}OCH_2CH_2\overset{+}{N}(CH_3)_3
\end{array}
$$

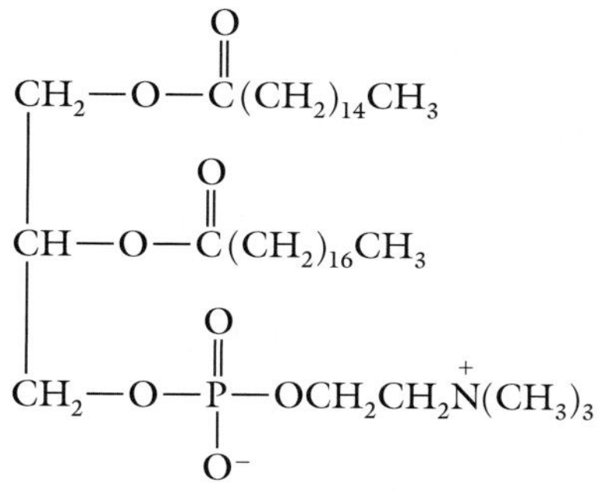

11.24 What are the products when the lipid in Exercise 11.23 undergoes saponification with KOH?

11.25 Draw the structure of the sphingolipid containing linoleic acid, phosphoric acid, and choline.

11.26 Draw the structure of the sphingolipid containing palmitic acid and α-D-glucose.

11.27 What are the products when the following lipid is hydrolyzed in the presence of an acid?

$$
\begin{array}{l}
HO{-}CH{-}CH{=}CH(CH_2)_{12}CH_3 \\
\quad\;\; |\\
\quad\;\; CH{-}\overset{\overset{\displaystyle H}{|}}{N}{-}\overset{\overset{\displaystyle O}{\|}}{C}(CH_2)_7CH{=}CH(CH_2)_7CH_3 \\
\quad\;\; |\\
\quad\;\; CH_2{-}O{-}\underset{\underset{\displaystyle O^-}{|}}{\overset{\overset{\displaystyle O}{\|}}{P}}{-}OCH_2CH_2\overset{+}{N}(CH_3)_3
\end{array}
$$

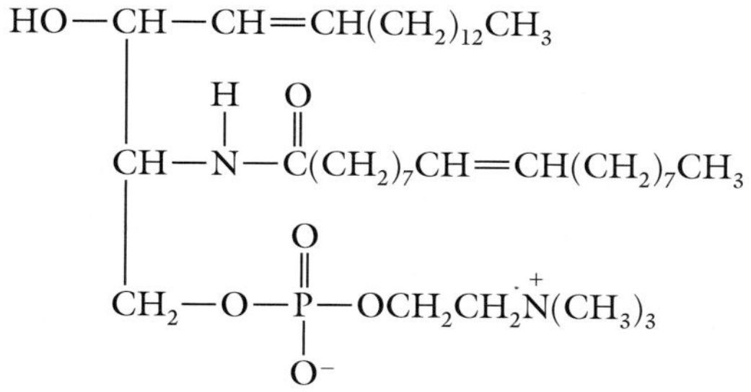

11.28 What are the products when the lipid in Exercise 11.27 undergoes saponification with NaOH?

Steroids

11.29 Why are steroids not classified as hydrolyzable lipids?

11.30 What structural feature is characteristic of all compounds classified as steroids?

11.31 Testosterone is one of many stereoisomers. Draw the structure of testosterone (Figure 11.4) and place an asterisk next to each tetrahedral stereocenter. How many stereoisomers are possible for testosterone?

11.32 Estradiol is one of many stereoisomers. Draw the structure of estradiol (Figure 11.4) and place an asterisk next to each tetrahedral stereocenter. How many stereoisomers are possible for estradiol?

11.33 What are the physiological functions of the female and male sex hormones?

11.34 What are the physiological functions of the adrenocortical hormones?

11.35 What compound is the precursor to the steroid hormones and bile salts?

11.36 What structural differences are there between testosterone and estradiol (Figure 11.4)?

11.37 What structural feature of bile salts distinguishes them from other steroids?

11.38 How do bile salts aid in the digestion of lipids?

Eicosanoids

11.39 Even though leukotriene D_4 (Figure 11.5) has an amide group that undergoes hydrolysis, it is not classified as a hydrolyzable lipid. Why?

11.40 What feature is characteristic of all compounds classified as eicosanoids?

11.41 What structural feature distinguishes prostaglandins from leukotrienes?

11.42 What structural feature distinguishes leukotrienes from thromboxanes?

11.43 What is a hormone? Why are eicosanoids called local hormones, whereas steroid hormones are called hormones?

11.44 What are the physiological functions of eicosanoids?

Fat-Soluble Vitamins

11.45 Define vitamin.

11.46 What distinguishes vitamins from hormones?

11.47 What property distinguishes fat-soluble from water-soluble vitamins?

11.48 Why is there a much greater danger of overdosing on fat-soluble vitamins than on water-soluble vitamins?

11.49 Describe the physiological functions of vitamins A, D, E, and K.

11.50 Some textbooks classify vitamin D as a vitamin; others classify it as a steroid hormone. Why?

Biological Membranes

11.51 Where are membranes found in the body?

11.52 What are the functions of biological membranes?

11.53 What lipids are present in biological membranes? What is their function?

11.54 What nonlipids are present in biological membranes? What is their function?

11.55 Why does the flexibility of cell membranes increase with increasing content of unsaturated fatty acids in the glycerophospholipids and sphingolipids?

11.56 Why does the flexibility of cell membranes decrease with increasing content of saturated fatty acids in the glycerophospholipids and sphingolipids?

11.57 Why do ions such as Na^+ and HCO_3^- and molecules such as D-glucose not pass through membranes by simple transport? Such ions and molecules pass through membranes only by facilitated transport or active transport.

11.58 What is the difference between simple transport and facilitated transport?

11.59 Why does active transport require energy?

11.60 Distinguish between integral and peripheral proteins.

11.61 Lactose is transported into cells through integral proteins under conditions in which the intracellular concentration of lactose exceeds the extracellular concentration. Does this transport constitute simple transport, facilitated transport, or active transport?

11.62 Glucose is transported into cells through integral proteins under conditions in which the extracellular concentration of glucose exceeds the intracellular concentration. Does this transport constitute simple transport, facilitated transport, or active transport?

Unclassified Exercises

11.63 Which solvent would be least effective in dissolving lipids? $CH_3(CH_2)_4CH_3$; CH_3OH; CCl_4; $C_2H_5OC_2H_5$.

11.64 What is the difference between essential and nonessential fatty acids?

11.65 Give the ω number for each of the following fatty acids:

(a) $CH_3(CH_2)_{14}COOH$;
(b) $CH_3(CH_2)_5CH{=}CH(CH_2)_7COOH$;
(c) $CH_3(CH_2)_4(CH{=}CHCH_2)_2(CH_2)_6COOH$.

11.66 Consider the following triacylglycerol:

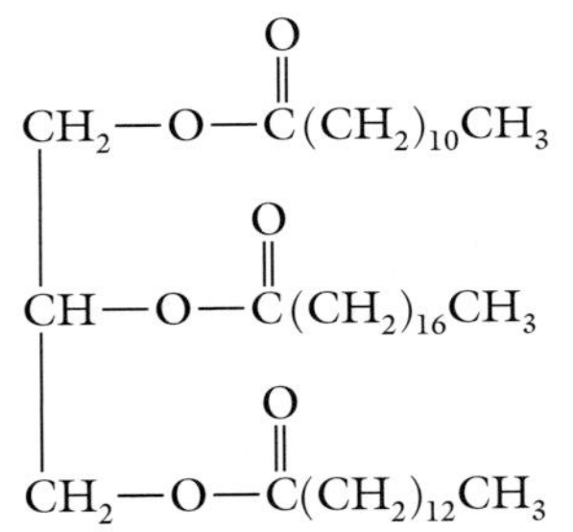

$$\begin{array}{l} CH_2{-}O{-}\overset{O}{\overset{\|}{C}}(CH_2)_{10}CH_3 \\ | \\ CH{-}O{-}\overset{O}{\overset{\|}{C}}(CH_2)_{16}CH_3 \\ | \\ CH_2{-}O{-}\overset{O}{\overset{\|}{C}}(CH_2)_{12}CH_3 \end{array}$$

What are the products in each of the following reactions of this compound? (a) Acid-catalyzed hydrolysis; (b) saponification with NaOH; (c) catalytic hydrogenation (H_2/Pt).

11.67 What are the physiological functions of waxes?

11.68 Consider the following glycerophospholipid:

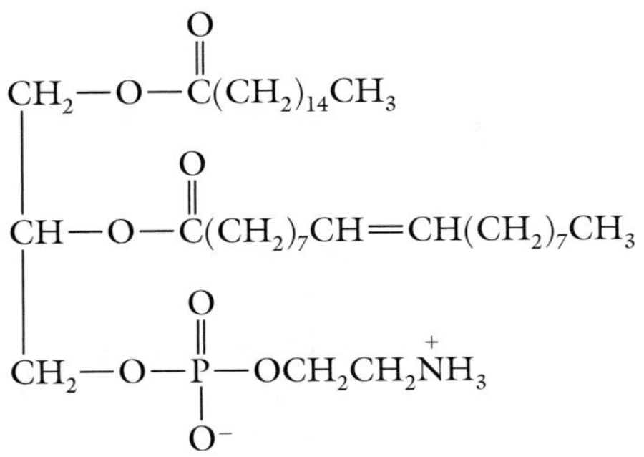

$$\begin{array}{l} CH_2{-}O{-}\overset{O}{\overset{\|}{C}}(CH_2)_{14}CH_3 \\ | \\ CH{-}O{-}\overset{O}{\overset{\|}{C}}(CH_2)_7CH{=}CH(CH_2)_7CH_3 \\ | \\ CH_2{-}O{-}\overset{O}{\overset{\|}{P}}{-}OCH_2CH_2\overset{+}{N}H_3 \\ \qquad\quad\; | \\ \qquad\quad\; O^- \end{array}$$

What are the products in each of the following reactions of this compound? (a) Acid-catalyzed hydrolysis; (b) saponification with KOH; (c) catalytic hydrogenation (H_2/Pt).

11.69 How does the digestion of triacylglycerols and other hydrolyzable lipids differ from hydrolysis in the laboratory with the use of an acid or base?

11.70 Which of the following lipids are based on glycerol? Which are based on sphingosine? (a) Sphingolipids; (b) leukotrienes; (c) glycerophospholipids; (d) prostaglandins; (e) steroids; (f) triacylglycerols; (g) waxes; (h) sex hormones.

11.71 Which of the following lipids are hydrolyzable? (a) Sphingolipids; (b) leukotrienes; (c) glycerophospholipids; (d) prostaglandins; (e) steroids; (f) triacylglycerols; (g) waxes; (h) sex hormones.

11.72 Which of the following lipids are components of biological membranes? (a) Sphingolipids; (b) leukotrienes; (c) glycerophospholipids; (d) prostaglandins; (e) cholesterol; (f) triacylglycerols; (g) waxes; (h) sex hormones.

11.73 Which of the following lipids undergo saponification? (a) Sphingolipids; (b) leukotrienes; (c) glycerophospholipids; (d) prostaglandins; (e) steroids; (f) triacylglycerols; (g) waxes; (h) sex hormones.

11.74 Write the equation for the reaction of the lipid in Exercise 11.68 with Br_2.

11.75 A triacylglycerol contains equimolar amounts of myristic, palmitic, and stearic acids. Three constitutional isomers exist for this composition. Draw the isomers.

11.76 What reactions are responsible for rancidity in butter and other fats and oils?

11.77 Triacylglycerol A contains equimolar amounts of lauric, stearic, and myristic acids, whereas triacylglycerol B contains equimolar amounts of oleic, stearic, and myristic acids. Air oxidation produces a malodorous mixture from triacylglycerol B but not from triacylglycerol A. Explain.

11.78 What important compounds are synthesized in the body from cholesterol? What other purpose does cholesterol serve?

11.79 Draw the structure of progesterone (Figure 11.4) and place an asterisk next to each tetrahedral stereocenter. How many stereoisomers are possible for progesterone?

11.80 What is the common feature of all lipids in cell membranes?

11.81 Why does the lipid bilayer not include triacylglycerols?

11.82 Oxygen is transported into cells without the aid of proteins under conditions in which the extracellular concentration of oxygen exceeds the intracellular concentration. Does this transport constitute simple transport, facilitated transport, or active transport?

11.83 Certain cells that line the stomach pump out protons against a concentration gradient of more than 1 million to 1. Does this transport constitute simple transport, facilitated transport, or active transport?

Expand Your Knowledge

Note: The icons denote exercises based on material in boxes.

11.84 (a) Compare the energies generated by the oxidation (metabolism) of 100 g each of dry triacylglycerol and dry glycogen. (Refer to Section 11.4.) (b) In the body, triacylglycerol is stored in anhydrous form but glycogen is hydrated (67% water). Compare the energies generated by the oxidation of 100 g each of dry triacylglycerol and hydrated glycogen.

11.85 Coconut oil is an atypical plant triacylglycerol. Explain why coconut oil is a liquid instead of a solid in spite of its very high saturated fatty acid composition (Table 11.2).

11.86 Triacylglycerol A contains equimolar amounts of lauric, stearic, and myristic acids, whereas triacylglycerol B contains equimolar amounts of butanoic, stearic, and myristic acids. Bacterial hydrolysis produces a malodorous mixture from triacylglycerol B but not from triacylglycerol A. Explain.

11.87 One of the cholesteryl esters used to transport cholesterol in the blood is cholesteryl linoleate, the ester of cholesterol and linoleic acid. Draw the structure of cholesteryl linoleate.

11.88 Membrane compositions of fish and other cold-blooded animals change when their environmental temperature is lowered. The unsaturated fatty acid content of the lipids in the cell membranes increases when the organism becomes adapted to the lower temperature. What is the purpose of this increase?

11.89 A clogged kitchen sink is often due to the buildup of fats in the trap. Explain how the addition of an aqueous NaOH solution eliminates these clogs.

11.90 Punctures and other mechanical disruptions in cell membranes are quickly resealed. What mechanism is responsible for the self-sealing property of cell membranes?

11.91 Imagine the planet Gibo in a far-off galaxy where the life forms have a similar appearance to those on Earth. However, there is one huge difference: heptane on Gibo has the role played by water on Earth. If the molecules that form cell membranes are the same as those on Earth, what is the major difference in the construction of cell membranes on Gibo compared with Earth?

11.92 How many constitutional isomers are possible for a triacylglycerol containing one each of myristic, palmitic, and oleic acids? Show structural representations of each constitutional isomer. Are enantiomers possible for any of the constitutional isomers?

11.93 Bile salts such as sodium glycocholate (Figure 11.4) function similarly to carboxylate soaps (Section 7.7). They break up larger-sized hydrophobic materials into smaller-sized ones to make them more compatible with an aqueous medium. What are the common structural features of bile salts and carboxylate soaps?

11.94 A triacylglycerol containing one each of lauric, palmitic, and stearic acids has the molecular formula $C_{49}H_{94}O_6$. Write the balanced equation for the combustion of this triacylglycerol.

11.95 Hexane is soluble in nonpolar solvents but is not classified as a lipid. Why?

11.96 Double bonds in biological systems are predominantly cis double bonds even though they are less stable than trans double bonds. Speculate on the mechanism responsible for the synthesis of the less-stable double bonds.

11.97 The unsaturated fatty acid components of triacylglycerols contain predominantly cis double bonds. Hydrogenation of triacylglycerols (such as those in corn oil) with H_2 in the presence of a metal catalyst such as Pt decreases the cis-double-bond content but increases the trans-double-bond content. What is the mechanism responsible for the increase in trans-double-bond content?

11.2 **11.98** An unknown sample is either ethynyl estradiol or norethindrone (see Box 11.2). What simple chemical test can be used to identify the sample?

6.3 **11.99** How can infrared spectroscopy (see Box 6.3) be used to distinguish between testosterone and progesterone?

11.100 How are polysaccharides transported into and out of cells and organelles?

11.101 A triacylglycerol has the molecular formula $C_{49}H_{88}O_6$. How many carbon–carbon double bonds are present in a molecule of this triacylglycerol?

11.102 The fatty acid content of triacylglycerol A is 70% stearic acid and 30% oleic acid. That of triacylglycerol B is 10% butanoic acid, 60% stearic acid, and 30% oleic acid. Which triacylglycerol has the lower melting temperature? Explain.

11.103 A triacylglycerol contains three fatty acid components. However, the hydrolysis of a fat or oil usually yields a mixture of more than three different fatty acids. For example, a mixture of myristic, palmitic, stearic, oleic, linoleic, and linolenic acids is obtained from the hydrolysis of a particular fat. What does this mixture tell us about the triacylglycerol composition of a fat or oil?

11.1 **11.104** Sucrose esters (see Box 11.1) are manufactured entirely from naturally occurring components—specifically, sucrose and fatty acids derived from triacylglycrols. However, sucrose esters are not digested by either the enzyme sucrase (Section 10.5), which hydrolyzes sucrose, or the lipases that hydrolyze triacylglycerols. Why?

11.105 The hydrogenation of triacylglycerols (such as those in corn oil) with H_2 proceeds very slowly unless a metal catalyst such as Pt is present. Is the activation energy E_a of the catalyzed reaction larger, smaller, or the same as that of the uncatalyzed reaction? Is the heat of reaction ΔH of the catalyzed reaction larger, smaller, or the same as that of the uncatalyzed reaction? Refer to Section 2.2.

CHAPTER 12 PROTEINS

(Mark Richards/PhotoEdit.)

Chemistry in Your Future

As a USDA meat inspector, you are often asked about the outbreaks of bovine spongiform encephalopathy ("mad cow disease") in the United Kingdom. The pathogen is neither a virus nor a bacterium. One day, you read about the 1997 Nobel Prize in medicine awarded to Stanley Prusiner, who believes that the disease is caused by a prion, a rogue form of a protein that is normally present in mammals. The protein can exist in two conformations: the normal, harmless conformation and an improperly folded one that causes mad cow disease. Evidence that humans might contract a version of the disease by ingesting beef from diseased cattle created something of a panic in the United Kingdom and concern in other countries. This chapter describes the importance of protein conformation and folding and their effects on physiological functions.

For more information about this topic and others in the chapter, go to www.whfreeman.com/bleiodian2e

Learning Objectives

- Draw and categorize the α-amino acids.
- Describe the effect of pH changes on α-amino acid structure.
- Describe the structures of peptides, and write equations for their synthesis from α-amino acids.
- Describe and write equations for the reactions of peptides.
- Distinguish between the primary, secondary, tertiary, and quaternary structures of proteins.
- Describe the forces that stabilize proteins.
- Describe the differences between fibrous and globular proteins.
- Describe the structures and functions of some proteins (α-keratin, collagen, silk fibroin, myoglobin, hemoglobin, and lysozyme).
- Describe mutations and how they affect living organisms.
- Describe protein denaturation by environmental changes.

The primary functions of the biomolecules that you have studied thus far—the carbohydrates and lipids—are to provide an organism with energy, with precursors to other biomolecules, and with molecules with which to construct cell membranes. The primary functions of proteins are the building and maintenance of the organism. Proteins are the most plentiful organic chemicals in the body, making up more than half of its dry weight. The protein family comprises about 100,000 different compounds or more—a far greater number than that of any other family of biomolecules. And proteins are responsible for the greatest range of functions. We can summarize the functions of proteins as follows:

- **Catalytic proteins,** or **enzymes,** catalyze the synthesis and utilization of proteins (including enzymes themselves), carbohydrates, lipids, nucleic acids, and almost all other biomolecules. The different enzymes that an organism produces are determined by heredity and are in fact what distinguishes one species from another and one individual member of a species from another. In Chapter 13, we shall see how deoxyribonucleic acids (DNAs) direct the synthesis of enzymes and all other proteins.
- **Transport proteins** bind and carry specific molecules or ions from place to place. Hemoglobin transports oxygen from the lungs to other tissues (Section 18.3). Integral membrane proteins transport molecules and ions across cell and organelle membranes (Section 11.10).

›› Enzymes are considered in Section 14.5.

A PICTURE OF HEALTH
Examples of Proteins in the Body

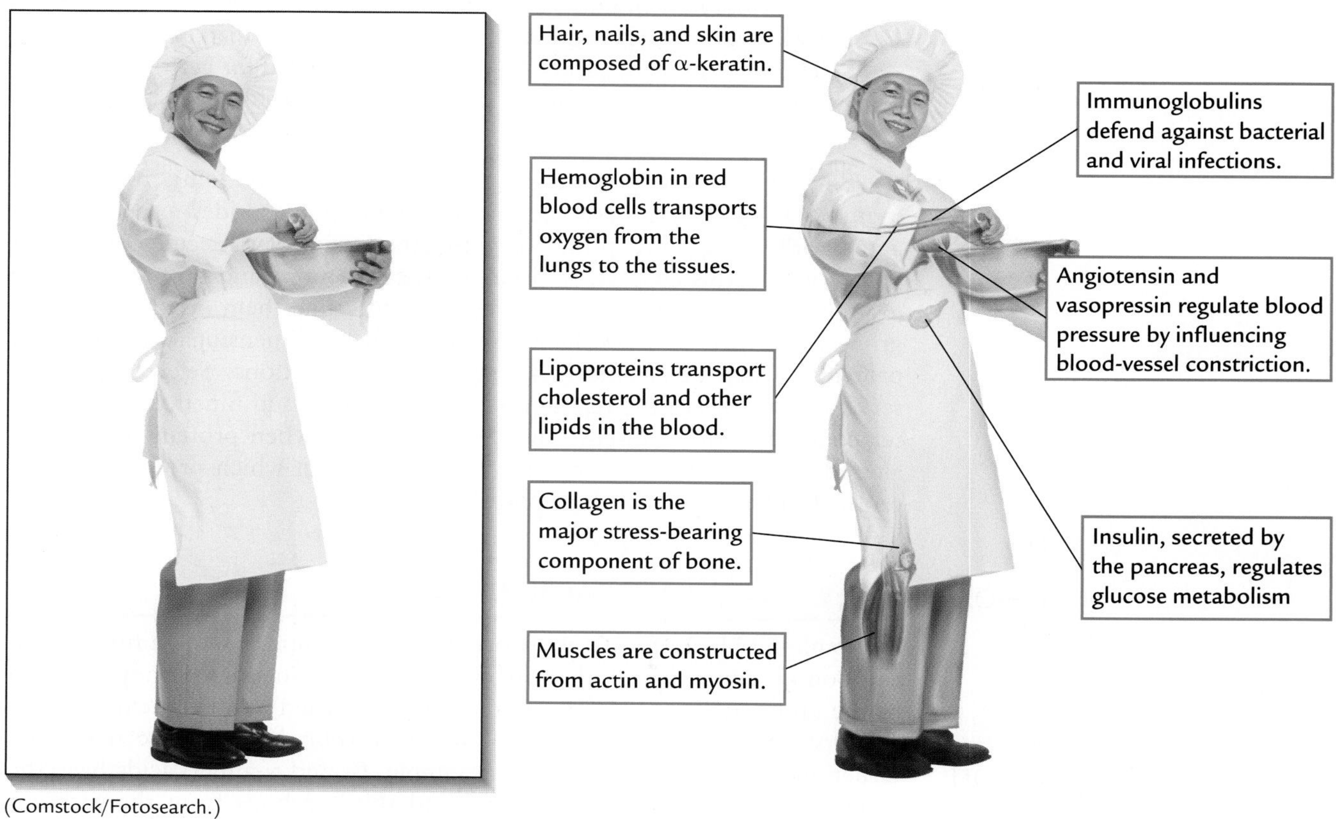
Hair, nails, and skin are composed of α-keratin.
Immunoglobulins defend against bacterial and viral infections.
Hemoglobin in red blood cells transports oxygen from the lungs to the tissues.
Angiotensin and vasopressin regulate blood pressure by influencing blood-vessel constriction.
Lipoproteins transport cholesterol and other lipids in the blood.
Collagen is the major stress-bearing component of bone.
Insulin, secreted by the pancreas, regulates glucose metabolism
Muscles are constructed from actin and myosin.

(Comstock/Fotosearch.)

- **Regulatory proteins** control cellular activity. In addition to lipid hormones (considered in Section 11.7), there are protein hormones (for example, insulin, which regulates glucose metabolism; Section 18.6). Some neurotransmitters, which relay information within the nervous system, are proteins. Proteins called **inducers** and **repressors** control gene expression—the way in which information in DNA is translated into the synthesis of proteins.
- **Structural proteins** give the physical shape, and the strength to maintain it, to structures in animals—the role played by cellulose in plants. The protein collagen is the major component of bone, tendons, and cartilage. Hair, fingernails, feathers, and horns consist mostly of the protein α-keratin.
- **Contractile proteins** provide cells and organisms with the ability to change shape and move. Muscles contract and cells change shape through the actions of actin and myosin. A sperm cell is propelled by its flagellum, composed of the protein tubulin, and an ovum passes through a fallopian tube assisted by movements of cilia, also composed of tubulin.
- **Protective proteins** defend against invaders and prevent or minimize damage after injury. Immunoglobulins, or antibodies, recognize and inactivate invading bacteria, viruses, and foreign proteins. The blood-clotting proteins fibrinogen and thrombin prevent the loss of blood when the vascular system is damaged. Toxins and venoms serve defensive or predatory functions for some plants and animals.
- **Storage proteins** provide a reservoir of nitrogen and other nutrients, especially when external sources are low or absent. Ovalbumin, the protein of egg white, supplies the developing bird embryo with nutrients during its isolation in the egg. Casein in milk serves the same function for mammalian infants. The seeds of many plants contain proteins needed for germination and early plant development. Ferritin serves as a storage form of iron.

Polypeptides are polymers constructed by the bonding together of α-amino acids. Proteins are naturally occurring polypeptides, usually aggregated with other polypeptides or with other types of molecules or ions or with both. In this chapter, we examine the structures and properties of the α-amino acids and the polypeptides formed from them. With this background, we turn to a consideration of the three-dimensional structures of proteins, which determine their physiological functions. Selected proteins serve as examples of the relation between structure and function. The chapter concludes with a discussion of what happens when protein structure is altered because of mutations in genes and ways in which protein function can be lost through environmental changes.

α-Carbon

$$NH_2{-}\overset{\overset{R}{|}}{C}H{-}\overset{\overset{O}{\|}}{C}{-}OH$$

α-Amino acid

α-Amino group

$$\underset{\gamma}{C}{-}\underset{\beta}{C}{-}\underset{\alpha}{C}{-}\overset{\overset{O}{\|}}{C}{-}OH$$

12.1 α-AMINO ACIDS

The building blocks for naturally occurring polypeptides are **α-amino acids,** compounds containing both a carboxyl group and an amino group attached to the same carbon atom. The amino acids are called α-amino acids because the first carbon attached to the carboxyl carbon is called the α-carbon. Other amino acids are known. For example, β- and γ-amino acids have the amino group located on the second and third carbons from the carboxyl carbon, respectively (see structures in margin on the facing page). Nearly all polypeptides of all species of plants and animals, however, are constructed from the 20 α-amino acids shown in Table 12.1 on the next two pages. Whenever we use the term amino acid in discussing polypeptides and proteins, we mean α-amino acid, unless otherwise noted.

Mammals require all 20 α-amino acids for protein synthesis but are unable to synthesize half of them. These 10 **essential α-amino acids** must be obtained from the diet (see Table 12.1). Diets containing animal proteins supply all of the essential α-amino acids. Vegetarians must exercise extra care to obtain sufficient amounts of all of the essential α-amino acids (Box 12.1 on page 378).

The α-amino acids are categorized according to the R group, or **side group,** on the α-carbon:

- **Nonpolar neutral** α-amino acids have neutral hydrophobic side groups.
- **Polar neutral** α-amino acids have neutral hydrophilic side groups.
- **Polar acidic** α-amino acids have acidic hydrophilic side groups.
- **Polar basic** α-amino acids have basic hydrophilic side groups.

The structures and functions of proteins strongly depend on the side groups of their constituent α-amino acids.

Proline differs from all other α-amino acids in having the nitrogen atom of its α-amino group as part of a ring structure. The α-amino group of proline is thus a secondary amine, whereas the α-amino groups of all other α-amino acids are primary amines. Note that, in Table 12.1, the structure shown for proline in the "side group (R)" column is the complete structure of proline. For all other α-amino acids, only the structure of the side group is given.

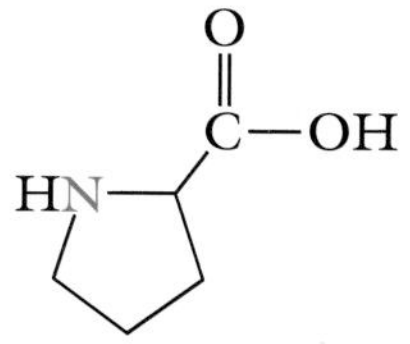

O
C—OH
HN
Proline

The common names of the α-amino acids are also the accepted IUPAC names. Each α-amino acid has a standard three-letter and one-letter abbreviation as well. The three-letter abbreviations are the first three letters of the names of all except four α-amino acids. For those four α-amino acids, the first letter of the name is combined with two other letters in the name. The three-letter abbreviations are more widely used and easier to remember than the one-letter abbreviations and will be used in this book. Biochemists use the one-letter symbols mainly for denoting the often very long sequences of α-amino acids in polypeptides and proteins.

The α-carbon of all α-amino acids except glycine is a tetrahedral stereocenter. The four different groups attached to the tetrahedral stereocenter are the amino group, carboxyl group, H, and R. The α-carbon of glycine is not a tetrahedral stereocenter, because two of the attached groups are the same (hydrogen). Thus, all α-amino acids except glycine can exist as a pair of enantiomers, nonsuperimposable mirror-image molecules (Section 9.2). Figure 12.1 shows D-alanine and L-alanine. The D- and L-designations are based on the convention for D- and L-glyceraldehyde, as described in Section 9.4 and shown in Figure 12.1. The carbon substituents, COOH and R, are placed vertically, with the more highly oxidized substituent (COOH) placed at the top. In the D-enantiomers, the heteroatom substituent—NH_2 for an amino acid and OH for glyceraldehyde—is on the right; in the L-enantiomers, it is on the left.

With very rare exceptions, only the L-α-amino acids exist in the proteins of plants and animals. Some bacterial cell walls contain D-alanine and D-glutamic acid. Unless indicated otherwise, the term amino acid in this book refers to the L-α-enantiomer.

INSIGHT INTO PROPERTIES

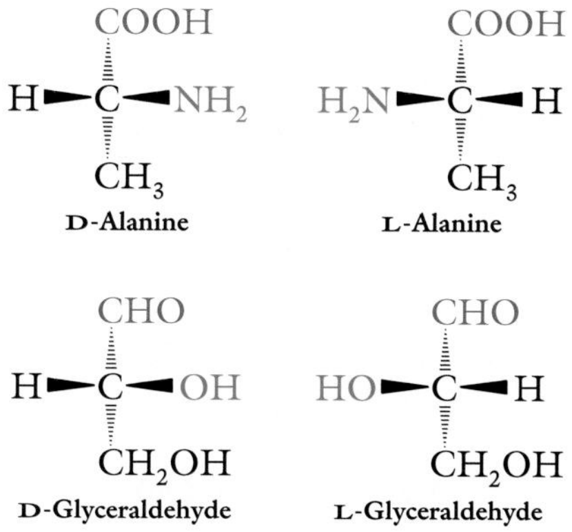

COOH
H—C—NH_2
CH_3
D-Alanine

COOH
H_2N—C—H
CH_3
L-Alanine

CHO
H—C—OH
CH_2OH
D-Glyceraldehyde

CHO
HO—C—H
CH_2OH
L-Glyceraldehyde

Figure 12.1 The D- and L-enantiomers of alanine and their relation to D- and L-glyceraldehyde.

TABLE 12.1 α-Amino Acids

Name	Abbreviations		Side group (R) of $NH_2-CH(R)-C(=O)-OH$	Isoelectric point (pI)
NONPOLAR NEUTRAL				
glycine	Gly	G	$H-$	5.97
alanine	Ala	A	CH_3-	6.01
valine*	Val	V	$(CH_3)_2CH-$	5.96
leucine*	Leu	L	$CH_3CH(CH_3)CH_2-$	5.98
isoleucine*	Ile	I	$CH_3CH_2CH(CH_3)-$	6.02
phenylalanine*	Phe	F	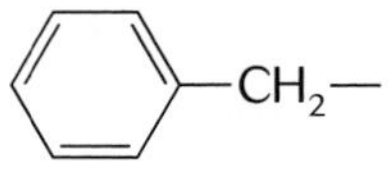 $C_6H_5-CH_2-$	5.48
methionine*	Met	M	$CH_3SCH_2CH_2-$	5.74
proline†	Pro	P	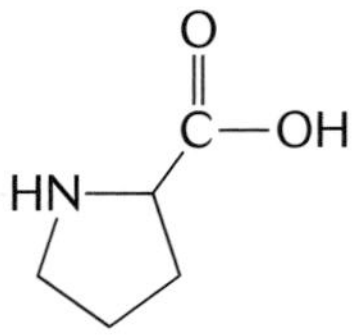 (pyrrolidine ring: HN– ring –C(=O)–OH)	6.30
tryptophan*	Trp	W	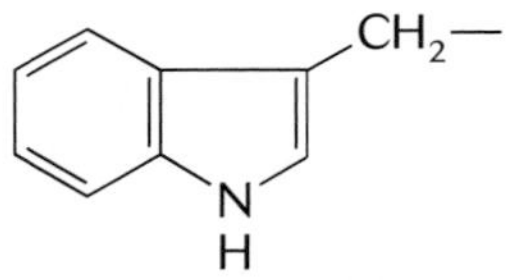 (indole ring, N–H)$-CH_2-$	5.88

(*Continued at the top of the following page.*)

12.2 THE ZWITTERIONIC STRUCTURE OF α-AMINO ACIDS

The structures representing α-amino acids to this point are not completely correct. An α-amino acid exists not as an uncharged molecule but as a **zwitterion,** or **dipolar ion,** which we can visualize to be formed by intramolecular proton transfer between the acidic —COOH and basic $-NH_2$ groups:

$$\underset{\substack{\textbf{Uncharged}\\ \text{(Nonexistent)}}}{NH_2-\overset{R}{\overset{|}{C}H}-\overset{O}{\overset{\|}{C}}-OH} \longrightarrow \underset{\substack{\textbf{Zwitterion}\\ \text{(Exists in both solid and solution)}}}{{}^{+}NH_3-\overset{R}{\overset{|}{C}H}-\overset{O}{\overset{\|}{C}}-O^-}$$

$${}^{+}NH_3-\overset{CH_3}{\overset{|}{C}H}-COO^-$$
L-Alanine

$$HO-\overset{CH_3}{\overset{|}{C}H}-COOH$$
L-Lactic acid

A zwitterion contains one negative (carboxylate) and one positive (ammonium) charge center and has a net charge of zero. The zwitterion structure of α-amino acids is not unexpected: an acid–base reaction takes place whenever a carboxylic acid and an amine are mixed together (Section 8.5). In this case, the carboxyl and amino groups are part of the same molecule.

Strong secondary attractive forces between the negative and the positive charge centers of zwitterions result in high melting points. For example, the

TABLE 12.1 (*continued*)

Name	Abbreviations		Side group (R)	Isoelectric point (pI)
POLAR NEUTRAL				
cysteine	Cys	C	$HSCH_2$—	5.05
serine	Ser	S	$HOCH_2$—	5.68
threonine*	Thr	T	$CH_3CH(OH)$—	5.60
asparagine	Asn	N	$H_2NC(=O)CH_2$—	5.41
glutamine	Gln	Q	$H_2NC(=O)CH_2CH_2$—	5.65
tyrosine	Tyr	Y	$HO-C_6H_4-CH_2$—	5.66
POLAR ACIDIC				
aspartic acid	Asp	D	$HOC(=O)CH_2$—	2.77
glutamic acid	Glu	E	$HOC(=O)CH_2CH_2$—	3.22
POLAR BASIC				
lysine*	Lys	K	$H_2NCH_2CH_2CH_2CH_2$—	9.74
arginine*	Arg	R	$H_2NC(=NH)NHCH_2CH_2CH_2$—	10.76
histidine*	His	H	(imidazol-4-yl)$-CH_2$—	7.59

*Essential for mammals.

†Complete structure of proline is shown.

melting point of L-alanine (314°C) is much higher than that of L-lactic acid (53°C), a polar compound of similar molecular mass. Strong secondary attractive forces between zwitterion charge centers and water results in water solubility for α-amino acids.

Concept check

✓ The zwitterion structures of α-amino acids are responsible for their physical properties: high melting point and significant solubility in water.

α-Amino acids exist exclusively as zwitterions in the solid state. The structure in aqueous solution is more complicated. There is an equilibrium between three species—zwitterion, cation, and anion:

$$\underset{\substack{\textbf{Cation}\\ \text{High }[H^+]\\ \text{Low pH }(<1)}}{{}^+NH_3-CH(R)-COOH} \underset{HO^-}{\overset{H^+}{\rightleftharpoons}} \underset{\substack{\textbf{Zwitterion}\\ \text{pH} = \text{pI}}}{{}^+NH_3-CH(R)-COO^-} \underset{H^+}{\overset{HO^-}{\rightleftharpoons}} \underset{\substack{\textbf{Anion}\\ \text{Low }[H^+]\\ \text{High pH }(>12)}}{NH_2-CH(R)-COO^-}$$

12.1 CHEMISTRY WITHIN US

Proteins in the Diet

Our bodies store significant amounts of lipids (mainly as triacylglycerols) and carbohydrates (as glycogen) to supply energy but only minimal reserves of α-amino acids for the production of proteins. This situation necessitates an almost daily intake of the requisite amounts of the essential α-amino acids. Protein synthesis can proceed only when all 12 α-amino acids are present. It also calls for the specific synthesis of nonessential α-amino acids from the essential α-amino acids or other precursors.

Dietary proteins are classified as **complete** or **incomplete proteins,** depending on their α-amino acid content. Complete proteins supply all of the essential α-amino acids in the amounts needed for protein synthesis. Incomplete proteins do not. Animal proteins are complete proteins. For example, human milk protein contains all of the α-amino acids, both essential and nonessential, in the proportions needed by growing infants. Animal proteins such as beef, pork, poultry, seafood, cow's milk, and hen's eggs have very nearly the same nutritional value as that of human milk protein.

Plant proteins are incomplete proteins, varying in their nutritional value. All essential α-amino acids can be obtained from plant sources, but no plant provides all of them in sufficient amounts: different plant proteins are deficient in different α-amino acids. Proteins from grains such as corn, oats, rice, and wheat are low in lysine, and some are also low in tryptophan. Proteins from legumes such as beans and peas are low in methionine. An all-grain or all-legume diet does not supply all essential α-amino acids in the required amounts.

A vegetarian diet must contain a complementary mixture of plant proteins to supply all essential α-amino acids. This mixture can be accomplished by combining a grain with a legume. The legume is low in methionine but high in lysine and tryptophan. The grain is high in methionine but low in lysine (and some are also low in tryptophan). Nuts such as almonds and walnuts have protein compositions similar to that of grains and can be used to complement legumes. However, nuts are high in lipids.

Much of the world's population is vegetarian. Some people follow a vegetarian diet for religious or ethical convictions; others do so for health reasons (to restrict cholesterol, saturated fat, and total fat intake). But most do so for economic reasons. Animal proteins are more expensive to produce than plant proteins. Many cultures have discovered specific dietary combinations of plant proteins that supply all the α-amino acids required for normal human growth: in Mexico, corn tortillas and beans; in Japan, rice and soybean curd (tofu); in the U.S. South, rice and black-eyed peas. Each of these diets provides a balance of the required α-amino acids.

The protein in steak is complete. Although the proteins in tofu and rice are individually incomplete, the combination of tofu and rice supplies all the essential α-amino acids. (Roy Morsch (*left*), Michael A. Keller (*center*), Don Mason (*right*)/The Stock Market.)

The cation is formed from the zwitterion by protonation of the carboxylate group. The anion is formed by ionization (deprotonation) of the ammonium group. Compounds with this dual ability to act as both base (proton acceptor) and acid (proton donor) are called **amphoteric** compounds.

The equilibrium between the three species (zwitterion, cation, and anion) varies with pH. Each α-amino acid has a pH value, called the **isoelectric point** or **pI,** at which almost all molecules (>99.9%) are present as the zwitterion with no net electrical charge. Table 12.1 lists the pI values of the α-amino acids. As pH falls below pI, the concentration of cation increases and the concentration of zwitterion decreases. As pH rises above pI, the concentration of anion increases and the concentration of zwitterion decreases.

The acid–base behavior of neutral (nonpolar and polar) α-amino acids is quantitatively different from that of the acidic or basic α-amino acids. Neutral α-amino acids act as moderately strong buffers, resisting changes in ionization state with changes in pH. Calculations using the Henderson–Hasselbalch equation (Section 2.5) show that more than 97% of a neutral α-amino acid is still in its zwitterion form when the pH is within two pH units above or below pI. The cation and anion forms become the dominant forms (>90%) only in highly acidic (pH < 1) and highly basic (pH > 12) media, respectively. The pI values for the neutral α-amino acids fall in the relatively narrow range of 5.05 to 6.30. The carboxyl and amino groups of these α-amino acids are charged at physiological pH because physiological pH is near neutral (7.35 for blood and 6.8–7.1 for most cells), which is within two pH units of pI.

Acidic α-amino acids have pI values below (on the acidic side of) the pI of the neutral α-amino acids because of the acidic carboxyl group in the R side chain. The pI values of aspartic and glutamic acids are 2.77 and 3.22, respectively.

Basic α-amino acids have pI values above (on the basic side of) the pI of the neutral α-amino acids because of the basic amino side group in the R side chain. The pI values of lysine, arginine, and histidine are 9.74, 10.76, and 7.59, respectively.

Concept check

✓ All carboxyl and amino groups, including those in side groups, of all α-amino acids are charged at physiological pH. The charge on carboxyl groups is negative. The charge on amino groups is positive.

Whereas the neutral α-amino acids exist as zwitterions with zero overall charge at physiological pH, the acidic and basic α-amino acids have overall charges of 1− and 1+, respectively:

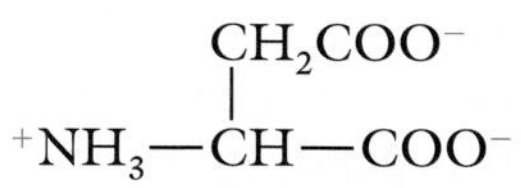

$$\underset{\text{Aspartic acid }(1-)}{{}^{+}NH_3-\underset{}{\overset{CH_2COO^-}{\overset{|}{C}H}}-COO^-} \qquad \underset{\text{Lysine }(1+)}{{}^{+}NH_3-\overset{(CH_2)_4\overset{+}{N}H_3}{\overset{|}{C}H}-COO^-}$$

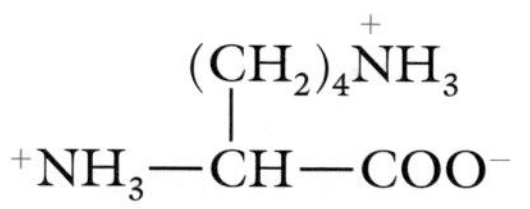

Example 12.1 Determining the ionization state of α-amino acids in media of different pH values

Show the predominant structure of valine at its isoelectric point (5.96), at physiological pH (7), at pH < 1, and at pH > 12.

Solution

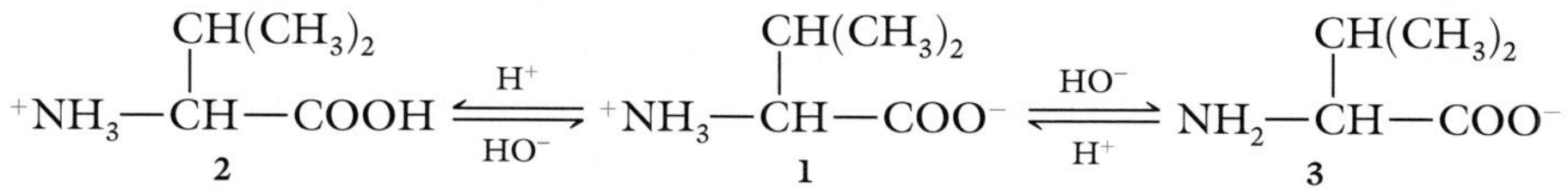

$$\underset{\substack{\mathbf{2}\\ pH<1}}{{}^{+}NH_3-\overset{CH(CH_3)_2}{\overset{|}{C}H}-COOH} \underset{HO^-}{\overset{H^+}{\rightleftharpoons}} \underset{\substack{\mathbf{1}\\ pH = 5.96 \text{ and } 7}}{{}^{+}NH_3-\overset{CH(CH_3)_2}{\overset{|}{C}H}-COO^-} \underset{H^+}{\overset{HO^-}{\rightleftharpoons}} \underset{\substack{\mathbf{3}\\ pH>12}}{NH_2-\overset{CH(CH_3)_2}{\overset{|}{C}H}-COO^-}$$

The zwitterion (structure 1) exists at pI. The zwitterion structure also predominates at physiological pH, which is within two pH units of pI. When conditions are much more acidic than pI, at pH < 1, the carboxylate group is protonated to form species 2, with an overall charge of 1+. When conditions are much more basic than pI, at pH > 12, the ammonium group loses a proton to form species 3, with an overall charge of 1−.

Problem 12.1 Show the predominant structure of phenylalanine at its isoelectric point (5.96), at physiological pH (7), at pH < 1, and at pH > 12.

The water solubility of α-amino acids varies with pH. Although α-amino acids are fairly soluble in water in general, they are least soluble at the isoelectric

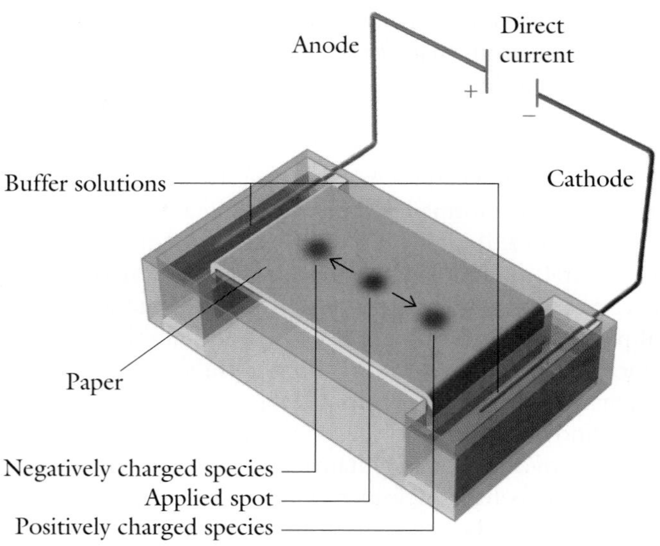

Figure 12.2 Electrophoresis. A strip of paper, saturated with a buffer solution, is positioned with its ends in buffer reservoirs that hold the electrodes. The researcher places a drop of sample solution in the middle of the paper and then turns on the electric current (direct current). After some period of time, the researcher turns off the current, removes and dries the paper strip, and stains it with a dye specific for the substance(s) under study.

point. Zwitterions aggregate together: the positive end of one zwitterion associates with the negative end of another through strong secondary attractive forces, minimizing their ability to associate with water. At pH above or below pI, α-amino acids are cations or anions, and intermolecular associations between them are much weaker. Strong attractive interactions with water result in increased solubility.

Electrophoresis is useful for the analysis of mixtures of α-amino acids (and peptides, as described later). One type of electrophoresis, called **paper electrophoresis,** is shown in Figure 12.2. This technique identifies substances in an electrical field by separation on the basis of their overall electrical charge. Different α-amino acids show different migration behaviors in the electrical field because of charge differences, which depend on structure and pH. An α-amino acid does not migrate in electrophoresis at pH equal to its pI, because it has zero charge. Electrophoresis at physiological pH distinguishes between neutral, acidic, and basic α-amino acids: a neutral α-amino acid has zero charge and does not migrate, an acidic α-amino acid is charged 1− and migrates to the anode (positive electrode), and a basic α-amino acid is charged 1+ and migrates to the cathode (negative electrode).

12.3 PEPTIDES

Peptides are polyamides formed by α-amino acids reacting with one another. The details of the reaction are complex and will be considered in Section 13.7. Overall, the reaction is viewed, at least conceptually, as a dehydration between the carboxyl and amino groups of different α-amino acid molecules:

$$H_3\overset{+}{N}-\underset{}{\overset{R^1}{\overset{|}{C}H}}-\overset{O}{\overset{\|}{C}}-O^- + H_3\overset{+}{N}-\overset{R^2}{\overset{|}{C}H}-\overset{O}{\overset{\|}{C}}-O^- \xrightarrow{-H_2O} H_3\overset{+}{N}-\overset{R^1}{\overset{|}{C}H}-\overset{O}{\overset{\|}{C}}-\overset{H}{\overset{|}{N}}-\overset{R^2}{\overset{|}{C}H}-\overset{O}{\overset{\|}{C}}-O^-$$

Dehydration takes place here — Peptide group — Peptide bond

The α-amino acids in the peptide, called **amino acid residues** or **monomers,** are linked together by a **peptide bond,** the bond between the carbonyl carbon and the nitrogen. The functional group formed by this reaction is an amide group and is called the **peptide group** (—CO—NH— or —CONH—). In the laboratory, amides are formed by the reaction of carboxyl and amino groups at elevated temperatures (Section 8.8). In biological systems, peptide formation is catalyzed by an enzyme.

Peptides are called **dipeptides, tripeptides, tetrapeptides, pentapeptides, hexapeptides,** and so forth, depending on the number of amino acid residues. The term **polypeptide** refers to peptides containing many amino acid residues. Biochemists do not agree on the exact size of peptide at which to use the term polypeptide. Most biochemists use polypeptide when the size exceeds from 10 to 20 amino acid residues. **Oligopeptide** is a loosely defined term used to refer to peptides smaller than polypeptides. And biochemists often use the term peptide loosely to refer to peptides of any size.

Amino Acid Sequences and Constitutional Isomers

Different constitutional (structural) isomers are possible whenever α-amino acids combine to form peptides. For example, two different dipeptides are possible from the reaction of glycine and alanine, depending on which amino group reacts with which carboxyl group:

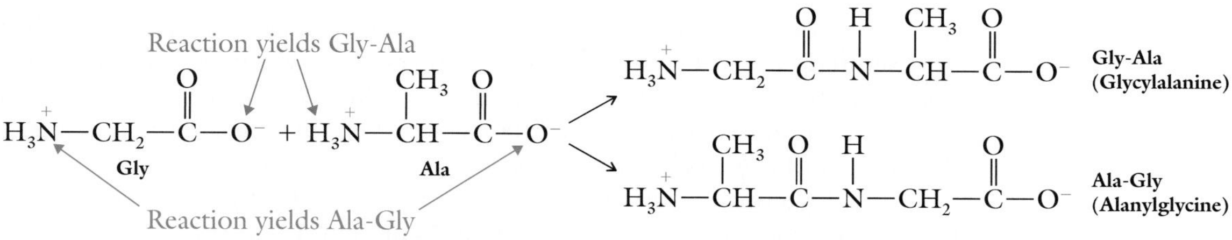

Reaction yields Gly-Ala

$$\underset{\text{Gly}}{H_3\overset{+}{N}-CH_2-\overset{O}{\overset{\|}{C}}-O^-} + \underset{\text{Ala}}{H_3\overset{+}{N}-\overset{CH_3}{\overset{|}{C}H}-\overset{O}{\overset{\|}{C}}-O^-}$$

Reaction yields Ala-Gly

$$H_3\overset{+}{N}-CH_2-\overset{O}{\overset{\|}{C}}-\overset{H}{\overset{|}{N}}-\overset{CH_3}{\overset{|}{C}H}-\overset{O}{\overset{\|}{C}}-O^- \quad \textbf{Gly-Ala (Glycylalanine)}$$

$$H_3\overset{+}{N}-\overset{CH_3}{\overset{|}{C}H}-\overset{O}{\overset{\|}{C}}-\overset{H}{\overset{|}{N}}-CH_2-\overset{O}{\overset{\|}{C}}-O^- \quad \textbf{Ala-Gly (Alanylglycine)}$$

All peptides contain an α-amino group at one end and an α-carboxyl group at the other end. The α-amino acid residues containing the amino and carboxyl groups are the **N-terminal (amino-terminal)** and **C-terminal (carboxyl-terminal) residues,** respectively. The convention for drawing peptides is to place the N-terminal residue at the left and the C-terminal residue at the right. Peptides are named as follows:

Rules for naming peptides

- The C-terminal residue, located at the far right, keeps its amino acid name.
- For each of the other amino acid residues in a peptide, the **-ine** or **-ic acid** ending of the amino acid name is replaced by **-yl,** except for tryptophan, for which **-yl** is added to the name.
- Naming begins at the N-terminal residue.

For example, the two dipeptides formed from glycine and alanine are glycylalanine and alanylglycine, as shown in the preceding reaction. More often, however, the three-letter abbreviations of the α-amino acid residues are used, such as Gly-Ala and Ala-Gly, rather than the full name.

Recall that all α-amino acids used in plants and animals are the L-enantiomers. The complete name for a peptide would include L- before the name of each residue—for example, L-Gly-L-Ala. The inclusion of L is usually not done, it being understood that only L-enantiomers are present (unless otherwise stated).

Example 12.2 Drawing and naming peptides

Show the formation of Val-Ser-Asp from the individual α-amino acids by using complete structural formulas. Give the full name for this tripeptide. Show all amino and carboxyl groups in their ionic forms.

Solution

Draw the α-amino acids in the order Val-Ser-Asp from left to right, each amino acid drawn with its α-amino group left and carboxyl group right. Carry out dehydrations between adjacent carboxyl and amino groups.

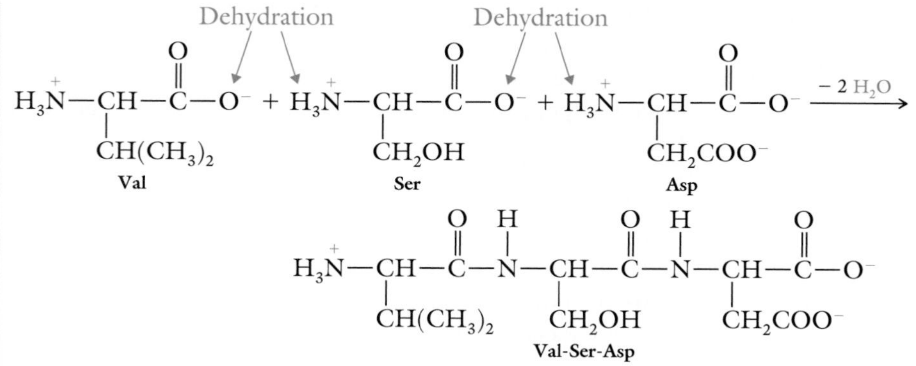

$$\overset{+}{H_3N}-\underset{\underset{\textbf{Val}}{CH(CH_3)_2}}{CH}-\overset{O}{\overset{\|}{C}}-O^- + \overset{+}{H_3N}-\underset{\underset{\textbf{Ser}}{CH_2OH}}{CH}-\overset{O}{\overset{\|}{C}}-O^- + \overset{+}{H_3N}-\underset{\underset{\textbf{Asp}}{CH_2COO^-}}{CH}-\overset{O}{\overset{\|}{C}}-O^- \xrightarrow{-2\,H_2O}$$

(Dehydration arrows indicate the bonds formed between each carboxylate and the next amino group.)

$$\overset{+}{H_3N}-\underset{CH(CH_3)_2}{CH}-\overset{O}{\overset{\|}{C}}-\overset{H}{\overset{|}{N}}-\underset{CH_2OH}{CH}-\overset{O}{\overset{\|}{C}}-\overset{H}{\overset{|}{N}}-\underset{CH_2COO^-}{CH}-\overset{O}{\overset{\|}{C}}-O^-$$

Val-Ser-Asp

The full name of the peptide is valylserylaspartic acid.

Problem 12.2 Show the formation of Ala-Lys-Phe from the individual α-amino acids by using complete structural formulas. Give the full name for this tripeptide. Show all amino and carboxyl groups in their ionic forms.

The physiological function of a peptide is determined largely by its **amino acid sequence,** the sequence of amino acid residues, which we list by convention from the N-terminal to the C-terminal residue. As already noted, peptides with the same amino acids but in different sequences are constitutional isomers: the number of isomers is the number of possible amino acid sequences.

The ***n* factorial** ($n!$) **rule** gives the number of constitutional isomers for peptides containing one each of n different amino acid residues: the number of isomers is $n!$. For example, the number of constitutional isomers for a decapeptide containing ten different amino acid residues is

$$10! = 10 \times 9 \times 8 \times 7 \times 6 \times 5 \times 4 \times 3 \times 2 \times 1 = 3{,}628{,}800$$

The peptides that make up proteins are much larger than decapeptides, often containing hundreds and sometimes thousands of amino acid residues. Any polypeptide is but one of an astronomical number of possible constitutional isomers, another example of the high selectivity of biological systems.

Example 12.3 Calculating the number of constitutional isomers of peptides

How many different constitutional isomers are possible for a tripeptide containing tyrosine, histidine, and proline? Give the abbreviated names for the different amino acid sequences.

Solution

There are three different amino acids and thus $3! = 3 \times 2 \times 1 = 6$ isomers. The isomers are simply all the possible sequences:

Tyr-His-Pro	His-Pro-Tyr	Pro-Tyr-His
Tyr-Pro-His	His-Tyr-Pro	Pro-His-Tyr

Problem 12.3 How many different constitutional isomers are possible for tripeptides containing glutamic acid, isoleucine, and lysine? Give the abbreviated names for the different amino acid sequences.

The $n!$ rule applies only when a peptide contains only one each of all of its different amino acid residues. No simple rule applies when a peptide contains more than one of any of its amino acid residues. In the latter case, the number of constitutional isomers is less than that given by the $n!$ rule. For example, there are only three tripeptides with two Tyr residues and one His residue: Tyr-Tyr-His, His-Tyr-Tyr, and Tyr-His-Tyr.

The Peptide Bond

The peptide C–N bond, although drawn as a single bond, does not have the properties of a single bond. There is no free rotation about this bond, and its length is shorter than expected for a single bond. The peptide bond, like all amide bonds (Section 8.3), has considerable double-bond character. This double-bond character results from resonance interaction between the π electrons of the carbonyl group and the nonbonded electron pair of nitrogen:

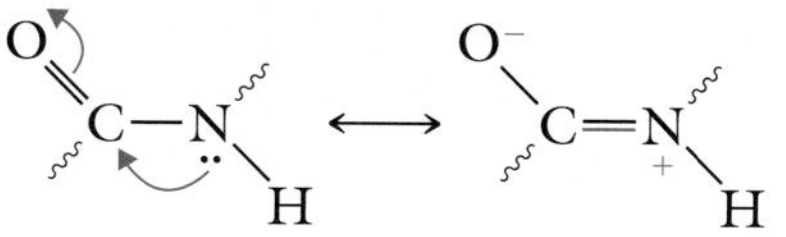

$$\mathrm{O{=}C(\sim)-\ddot{N}(\sim)H \longleftrightarrow {}^{-}O-C(\sim){=}\overset{+}{N}(\sim)H}$$

The peptide-chain segments, shown as wavy lines, are trans to each other on the C=N bond. This configuration is more stable than the cis configuration, in which the peptide-chain segments would sterically (spatially) interfere with each other. The atoms of the double bond and the atoms directly attached to the double bond are coplanar: they lie in the same plane.

In simple amides, the double-bond character of the peptide bond is responsible for very high melting and boiling points and a lack of basicity (Section 8.3). In polypeptides, it plays a role in determining three-dimensional structure and function (Section 12.5).

The Ionization of Peptides

A peptide, like the amino acids from which it is constructed, has an isoelectric point—the pH at which the peptide has an overall zero charge and does not migrate in electrophoresis.

- The pI value of a peptide containing only neutral α-amino acid residues or equal numbers of acidic and basic residues or both is in the range of pI values for neutral α-amino acids (pH = 5.05–6.30).
- The pI of a peptide containing acidic and basic α-amino acid residues is on the acidic side (lower than 5.05–6.30) if there is an excess of acidic residues and on the basic side (higher than 5.05–6.30) if there is an excess of basic residues.

The most important thing to keep in mind about the ionization state of peptides in biological systems is:

- All amino and carboxyl groups, including those on side groups of acidic and basic α-amino acid residues, are charged at physiological pH (Section 12.2).

Example 12.4 Writing the structures of peptides at different pH values

Show the structure of Asp-Lys-Gly at physiological pH. Does the peptide migrate in electrophoresis at physiological pH? If it migrates, to which electrode? Is the pI value for the peptide on the acidic or basic side of the pI values for polypeptides containing only neutral amino acid residues?

Solution

Asp-Lys-Gly has one carboxyl-containing side group and one amino-containing side group. All amino and carboxyl groups are charged at physiological pH and the resulting structure is

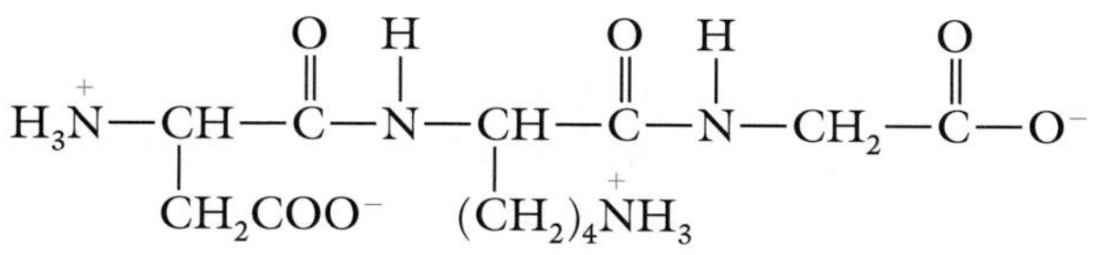

$$\mathrm{H_3\overset{+}{N}-\underset{\displaystyle CH_2COO^-}{CH}-\overset{\displaystyle O}{\overset{\|}{C}}-\overset{\displaystyle H}{N}-\underset{\displaystyle (CH_2)_4\overset{+}{N}H_3}{CH}-\overset{\displaystyle O}{\overset{\|}{C}}-\overset{\displaystyle H}{N}-CH_2-\overset{\displaystyle O}{\overset{\|}{C}}-O^-}$$

Asp-Lys-Gly has two positive charges and two negative charges, with an overall charge of zero, and does not migrate in electrophoresis at physiological pH.

The pI value is about the same as that for peptides containing only neutral amino acid residues because there are equal numbers of acidic and basic side groups.

Problem 12.4 Show the structure of the following peptides at physiological pH. (a) Ala-Lys-Ala; (b) Asp-Lys-Asp. Does each peptide migrate in electrophoresis at physiological pH? If it migrates, to which electrode? Is the pI value for each peptide on the acidic or basic side of the pI values for polypeptides containing only neutral amino acid residues?

Like α-amino acids, peptides have solubility and electrophoresis properties that are pH dependent (Section 12.2). A peptide shows minimum solubility at its pI. Solubility increases at higher and lower pH values. Electrophoresis (Section 12.2) is useful for the identification and analysis of peptides. For example, the two peptides Asp-Lys-Gly and Asp-Lys-Asp have net charges of 0 and 1−, respectively, at physiological pH. The two tripeptides can be distinguished because, at physiological pH, Asp-Lys-Gly does not migrate in an electrophoresis experiment, whereas Asp-Lys-Asp migrates to the anode.

12.4 CHEMICAL REACTIONS OF PEPTIDES

Cysteine is the only α-amino acid that contains the **sulfhydryl** group, —SH (Section 5.3). Thus equipped, the cysteine residues of a peptide often have the function of linking together two peptides or different parts of the same peptide through formation of a **disulfide bridge.** Disulfide bridges are partly responsible for the three-dimensional structures (and resulting biological functions) of proteins (Section 12.5).

Disulfide bridges are formed by selective oxidation—that is, a loss of hydrogens from the sulfhydryl groups of a pair of cysteine residues:

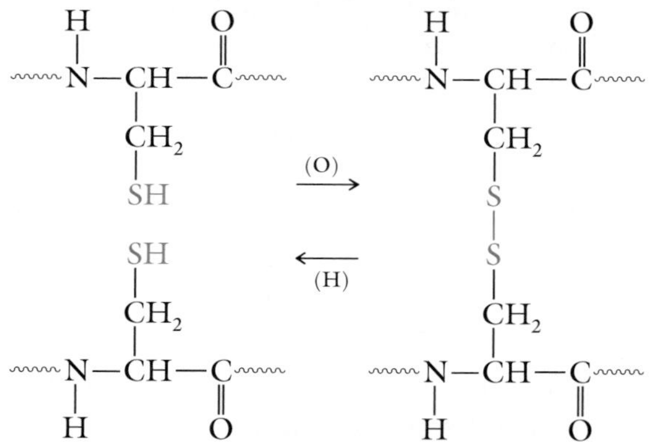

$$\begin{array}{c} \sim NH-CH(CH_2SH)-C(=O)\sim \\ \sim NH-CH(CH_2SH)-C(=O)\sim \end{array} \underset{(H)}{\overset{(O)}{\rightleftharpoons}} \begin{array}{c} \sim NH-CH-C(=O)\sim \\ |\;CH_2-S-S-CH_2\;| \\ \sim NH-CH-C(=O)\sim \end{array}$$

The reverse reaction, the selective reduction of disulfide bridges to form sulfhydryl groups, also takes place in biological systems. These reactions can be carried out in the laboratory by selective oxidation and reduction.

›› The digestion of polypeptides is considered in Section 18.1.

In the digestion of a polypeptide, which takes place in the stomach and intestines, enzymes hydrolyze peptide bonds—the reverse of peptide formation. The digestive process is complex, but the end result is essentially the same as that in laboratory hydrolysis by acid or base (Section 8.4): α-amino acids. Disulfide bridges are not cleaved in digestion and hydrolysis. Only reduction cleaves disulfides. The reduction of disulfide bridges takes place in the liver subsequent to digestion.

Consider a hypothetical hexapeptide derived by disulfide-bridge formation between Ala-Cys-Ser peptides. Digestion (as well as laboratory hydrolysis) would break all peptide bonds but not the disulfide bridge:

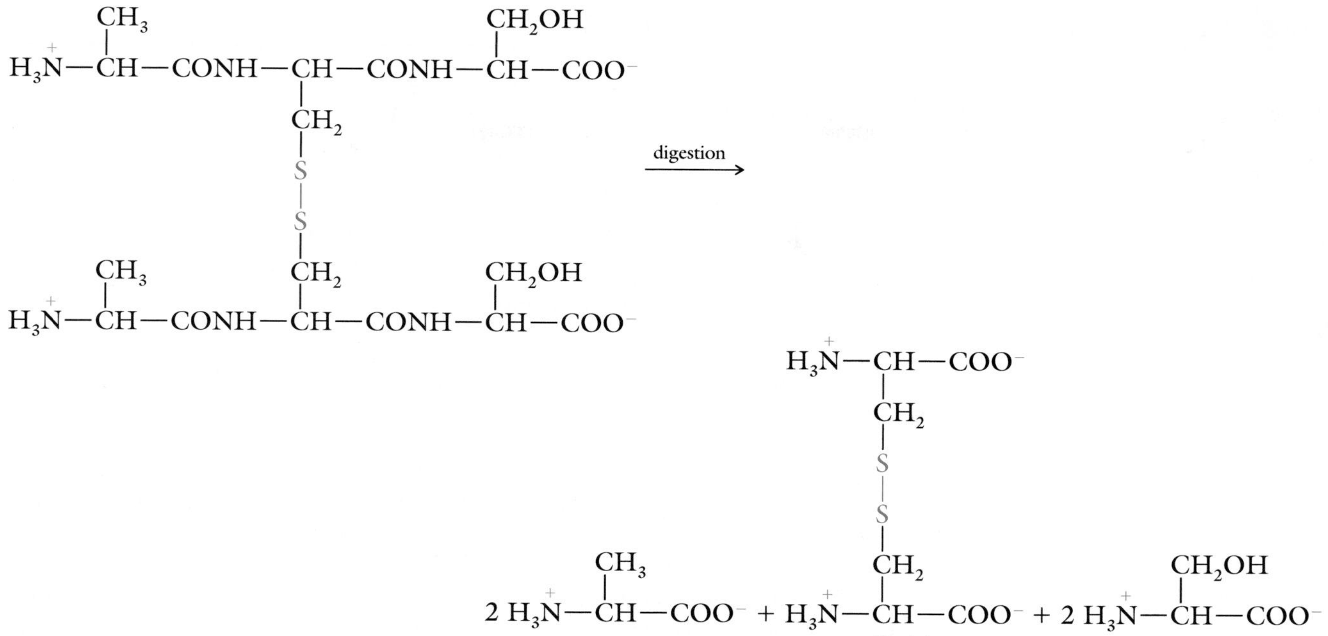

$$
\begin{array}{l}
H_3\overset{+}{N}-\underset{}{\overset{CH_3}{CH}}-CONH-\underset{\underset{\underset{\underset{S}{|}}{S}}{|}}{\underset{CH_2}{CH}}-CONH-\overset{CH_2OH}{CH}-COO^- \\
H_3\overset{+}{N}-\overset{CH_3}{CH}-CONH-\overset{CH_2}{CH}-CONH-\overset{CH_2OH}{CH}-COO^-
\end{array}
\xrightarrow{\text{digestion}}
$$

$$
2\,H_3\overset{+}{N}-\overset{CH_3}{CH}-COO^- + H_3\overset{+}{N}-CH(COO^-)-CH_2-S-S-CH_2-CH(COO^-)-\overset{+}{N}H_3 + 2\,H_3\overset{+}{N}-\overset{CH_2OH}{CH}-COO^-
$$

Cystine

The two cysteine residues joined by a disulfide bridge make up **cystine.** A reduction reaction cleaves cystine to form two cysteines. This type of cleavage takes place in the liver:

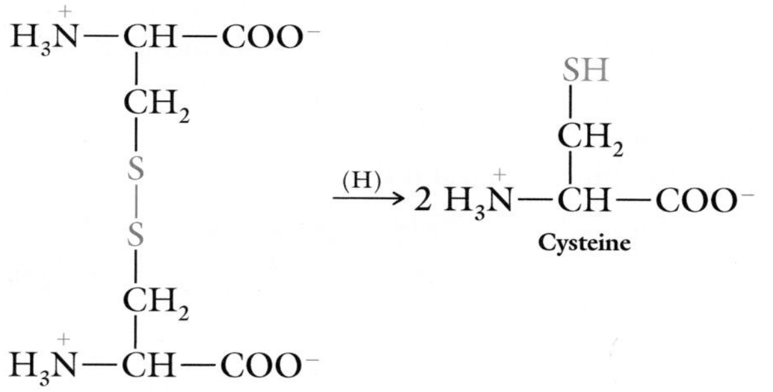

$$
H_3\overset{+}{N}-CH(COO^-)-CH_2-S-S-CH_2-CH(COO^-)-\overset{+}{N}H_3 \xrightarrow{(H)} 2\,H_3\overset{+}{N}-\overset{\overset{SH}{|}}{\overset{CH_2}{CH}}-COO^-
$$

Cystine **Cysteine**

Peptide structures in which the three-letter abbreviations for amino acid residues are used can be written for peptides containing disulfide bridges by using —S—S— to bridge two cysteine residues, as shown in the margin.

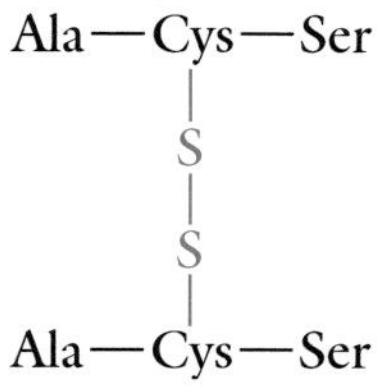

Ala—Cys—Ser
|
S
|
S
|
Ala—Cys—Ser

12.5 THE THREE-DIMENSIONAL STRUCTURE OF PROTEINS

As mentioned earlier, biologically active peptides vary greatly in size, from a few α-amino acid residues to hundreds or thousands of residues. Not all are called proteins. Biochemists often distinguish between **peptides** and **proteins** on the basis of the number of α-amino acid residues. The term protein is generally not used for peptides containing fewer than 50 residues even when such peptides have physiological functions similar to those of proteins (Table 12.2 on the following page). They are referred to simply as peptides.

Polypeptides containing more than 50 residues are called proteins if they have physiological functions as individual polypeptide molecules. However, the typical protein is not a single polypeptide molecule. It is an aggregation of two or more identical or different polypeptides that forms a three-dimensional structure with a specific function. In many cases, the aggregation of polypeptides is associated with ions or molecules other than polypeptides.

TABLE 12.2 Physiological Functions of Some Peptides

Name	Number of α-amino acid residues	Function
angiotensin	8	regulates blood pressure by constricting arteries
enkephalin	5	reduces pain sensation by binding to brain receptors
gastrin	17	aids digestion by stimulating HCl and pepsinogen secretion in stomach
glutathione	3	maintains cysteine and iron in hemoglobin in reduced states by scavenging oxidizing agents
oxytocin	9	induces labor by contracting uterine muscles
vasopressin	9	regulates blood pressure by stimulating excretion of water by kidneys

For example, hemoglobin, the carrier of oxygen in the blood, contains four polypeptide molecules (two each of two different polypeptides) and four heme molecules (Section 12.7).

Proteins containing only polypeptide molecules are called **simple proteins.** Those also containing nonpolypeptide molecules or ions are called **conjugated proteins:** the polypeptide part is the **apoprotein;** the nonpolypeptide molecules and ions, such as heme, are the **prosthetic groups.** Table 12.3 lists the major classes of conjugated proteins, with the prosthetic group of each class.

As stated earlier, there are more than 100,000 different proteins, each with a unique physiological function. This enormous diversity in function begins with the diversity of α-amino acid sequences of polypeptide chains. Each sequence produces a unique three-dimensional structure, which in turn determines the unique physiological function of the protein.

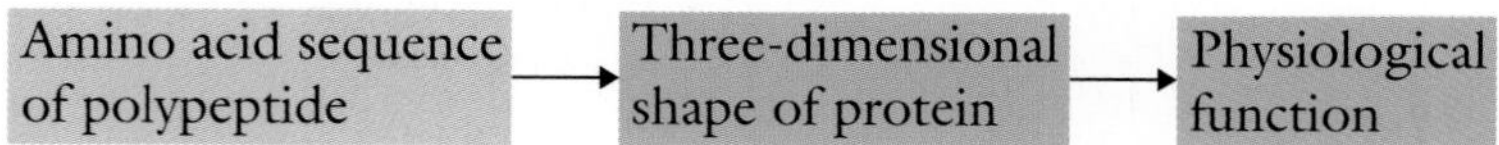

TABLE 12.3 Classification of Conjugated Proteins

Class	Prosthetic group	Example
glycoprotein	saccharide	immunoglobulin (antibody); interferon (antiviral agent); mucin (food lubricant in saliva)
hemoprotein	heme	hemoglobin (O_2 carrier in blood); myoglobin (O_2 storage in muscle)
lipoprotein	lipid	chylomicron, VLDL, LDL, HDL (lipid carriers)
metalloprotein	metal ion	Ca^{2+} in calmodulin (muscle contraction); Fe^{2+} in hemoglobin and myoglobin; Fe^{2+} in ferritin (Fe^{2+} storage); Zn^{2+} in carboxypeptidase (protein digestive enzyme)
nucleoprotein	nucleic acid	RNA-bound protein (protein synthesis in ribosome)
phosphoprotein	phosphate ester	casein (milk protein)

Abbreviations: VLDL, very low density lipoprotein; LDL, low-density lipoprotein; HDL, high-density lipoprotein; RNA, ribonucleic acid.

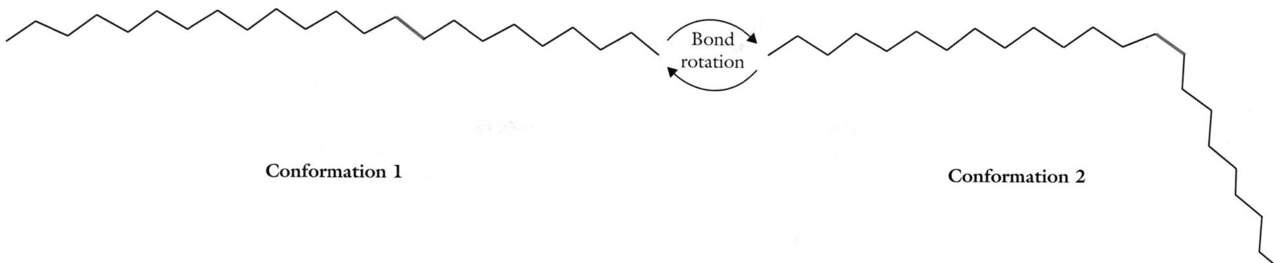

Figure 12.3 Conformations are interconvertible through rotation about single bonds in a polypeptide chain. The bond undergoing rotation in this hypothetical polypeptide is shown in red.

The conformation of a polypeptide or protein determines its three-dimensional structure. The conformation of a protein containing more than one polypeptide is the sum of the conformations of each of its constituent polypeptides. Theoretically, a polypeptide can have many different conformations owing to rotations about the various single bonds in the polymer chain. The formation of polypeptide conformations by rotations about single bonds is referred to as the **folding** of the polypeptide. Figure 12.3 shows how rotation about one bond changes the conformation of a molecule. Rotations about many different bonds in a polypeptide result in an even greater number of possible conformations. Nevertheless, a polypeptide normally exists as a single specific conformation: the unique α-amino acid sequence of a polypeptide stabilizes one specific conformation more than any of the others. This one specific conformation is essential to the molecule's physiological function under normal physiological conditions.

Protein structure is described at four levels:

- **Primary structure** is the α-amino acid sequence of a polypeptide.
- **Secondary structure** is the conformation in a local region of a polypeptide molecule. The conformations are the same in different regions of the molecule for some polypeptides but are different in different regions for other polypeptides.
- **Tertiary structure** exists when the polypeptide has different secondary structures in different local regions. Tertiary structure describes the three-dimensional relation among the different secondary structures in different regions.
- **Quaternary structure** exists only in proteins in which two or more polypeptide molecules aggregate together. It describes the three-dimensional relation among the different polypeptides.

Determinants of Protein Conformation

Two aspects of polypeptide structure determine a polypeptide's (and protein's) most stable conformation: (1) the bonds in the linear chain and (2) the interactions of the side groups. First, the allowed conformations are limited by the requirements that all bonds in the polymer chain (particularly the peptide bond with its double-bond nature; Section 8.3) maintain their normal bond angles and bond lengths and that all amino acid residues are the L-enantiomers. These structural features disallow a significant fraction of the possible conformations that would distort and destabilize the molecule. Then, of the very large number of possible conformations remaining, the most stable one is determined by the second aspect: the way in which the amino acid residues interact with each other and with the aqueous environment. One conformation is stable relative to other conformations for the following reasons:

1. Shielding of nonpolar α-amino acid residues from water. Most polypeptides fold into a conformation that buries the side groups of nonpolar α-amino acid residues in the interior of the polypeptide molecule or in the interior of an aggregate of polypeptide molecules. This **hydrophobic effect** prevents the highly destabilizing contact between the nonpolar residues and the highly polar aqueous environment. (The same driving force is responsible for the lipid bilayer of cell and organelle membranes; Section 11.10.)

2. Hydrogen bonding between peptide groups. The peptide group (—CO—NH—) can hydrogen bond with water or with other peptide groups. The peptide group prefers not to hydrogen bond with water, because such bonding does not minimize the exposure of nonpolar α-amino acid residues to water. Hydrogen bonding between different peptide groups predominates because it produces conformations with maximum shielding of nonpolar residues from water. The secondary structure of polypeptides is determined mainly by hydrogen bonding between different peptide groups.

3. Attractive interactions between side groups of α-amino acid residues. Folding proceeds with the placement of appropriate α-amino acid residues near each other to minimize repulsive interactions and maximize attractive interactions. Three types of attractive interactions take place (Figure 12.4) and have an important effect on the tertiary and quaternary structures of proteins:

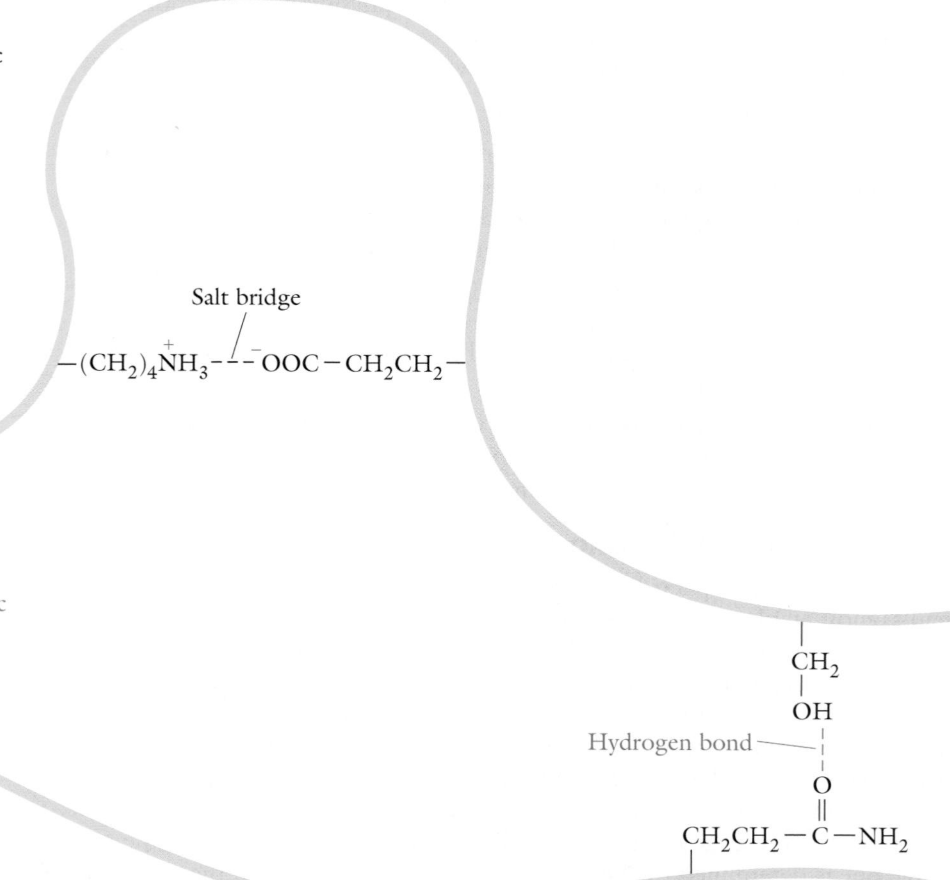

Figure 12.4 Noncovalent secondary attractive interactions of amino acid side groups: hydrophobic attraction between Phe and Ala; salt bridge between Lys and Glu; hydrogen bonding between Ser and Gln.

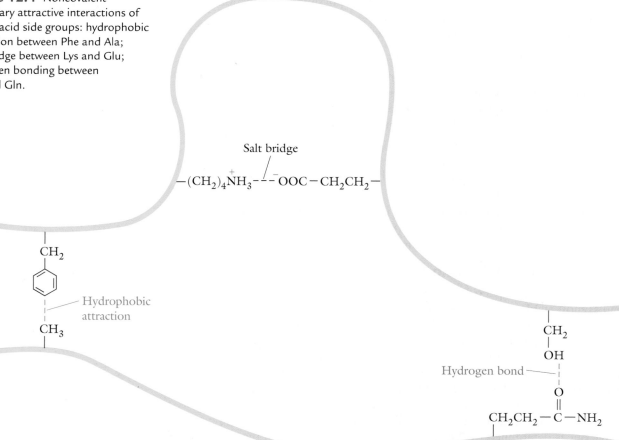

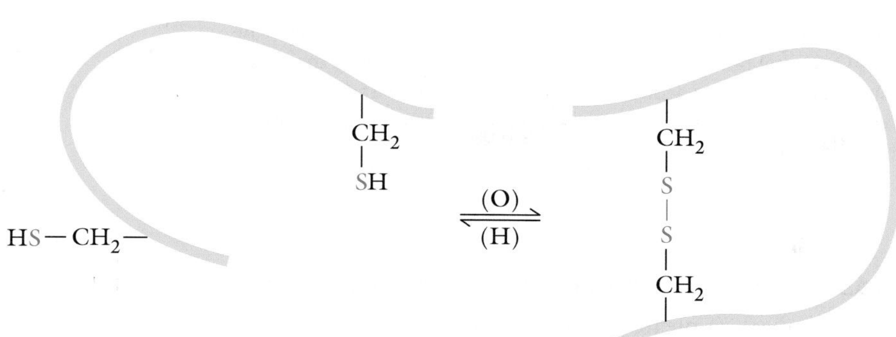

Figure 12.5 Formation of disulfide bridge between Cys residues.

a. Nonpolar residues are placed near each other and participate in **hydrophobic attractions.** Hydrophobic attractions are not the same as the hydrophobic effect, but there is a relation between the two: the hydrophobic effect shields nonpolar residues from the aqueous environment and places them near each other, which results in hydrophobic attractions.

b. Polar, neutral residues are placed near each other and participate in **hydrogen-bonding attractions.**

c. Polar, acidic residues (Asp, Glu) are placed near polar, basic (Arg, His, Lys) residues and participate in **salt-bridge (ionic) attractions.** (Recall that all acidic and basic side groups are charged at physiological pH.)

4. Attractive interactions of side groups of polar α-amino acid residues with water. Some proteins fold so that polar residues, both charged and uncharged, are on the protein surface, where they undergo attractive interactions with water. **Globular proteins** (Section 12.7), such as hemoglobin, whose physiological functions require solubility in the aqueous environment, are folded in this way. The functions of **fibrous proteins** (Section 12.6), such as α-keratin of hair, nails, and skin, require water insolubility. These molecules have far fewer polar residues on their surfaces.

5. Disulfide bridges formed in the folding process. Disulfide bridges are the only covalent bonds contributing to the stabilization of protein conformation through side-group interactions. The disulfide bridge in Figure 12.5 is an intramolecular disulfide bridge (formed between different parts of the same molecule). Intermolecular disulfide bridges, between different polypeptide molecules, also exist. Disulfide bridges are important for determining tertiary and quaternary structures.

The **native conformation** of a protein is its conformation under normal physiological conditions and is the result of the various secondary interactions and disulfide bridges just described. Proteins in their native conformations are called **native proteins.** The folding of proteins into their native conformations is aided by proteins called **chaperones.**

Concept checklist

- ✓ Hydrogen bonding between peptide groups is mainly responsible for the secondary structure of polypeptides.
- ✓ Disulfide bridges and secondary interactions between side groups are mostly responsible for the tertiary and quaternary structures of proteins.
- ✓ Disulfide bridges are more resistant to environmental changes such as those in temperature and pH than are the various secondary attractive forces.

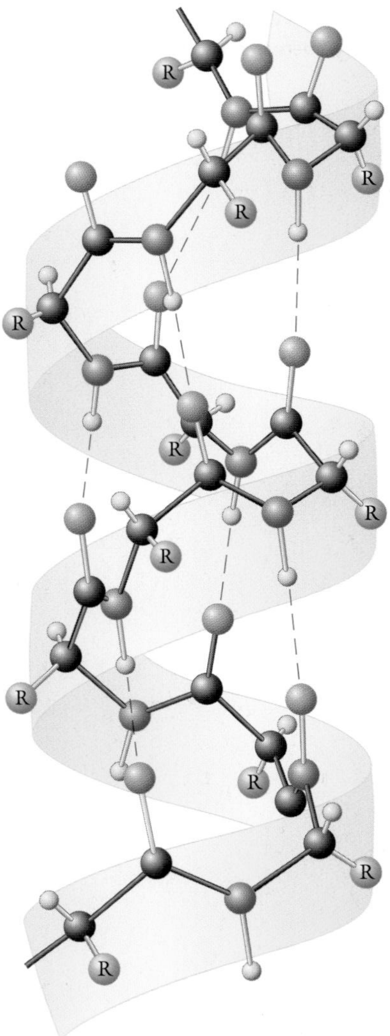

Figure 12.6 α-Helical conformation, a secondary structure. The C=O double bond is shown as a single bond to simplify the drawing. Nitrogen and oxygen atoms are represented by blue and red spheres, respectively. Side groups of α-amino acid residues are represented by green spheres.

Basic Patterns of Protein Conformation

The α-helical, β-pleated-sheet, β-turn, and loop conformations are the most important secondary structures in naturally occurring polypeptides and proteins. The **α-helix** has the overall shape of a coiled spring (Figure 12.6). This helical (spiral) folding of the polypeptide chain is due to stabilization by hydrogen bonding between the oxygen of each peptide group's C=O and the hydrogen of the peptide group's N—H of the fourth residue farther along the polypeptide chain. The side groups of the residues protrude outward at right angles to the long axis (length) of the α-helix. The α-helix is a right-handed helix, a clockwise movement in either direction being required in a procession along the length of the helix.

In the **β-pleated sheet,** the polypeptide chains are not coiled but are in extended conformations with side-by-side alignment of adjacent chains, in either a parallel or an antiparallel arrangement. The antiparallel arrangement (Figure 12.7) is the more common one: adjacent chains run in opposite directions; that is, the N-terminal- to C-terminal-residue orientation of adjacent chains is in opposite directions. Hydrogen bonding between adjacent chains—that is, between the peptide-group oxygen of one chain and the peptide-group hydrogen of an adjacent chain—stabilizes the β-pleated sheet. Each polypeptide chain, except for residues at the edges of the sheet, is hydrogen bonded to adjacent chains on each side. The sheets are pleated, not flat, because of the planar peptide groups and the tetrahedral bond angles about the other atoms in the polymer chains. Side groups of α-amino acid residues alternately protrude above and below the plane of the sheets. The side-by-side alignment is not always between segments of different polypeptide molecules. In many polypeptides, the molecules turn back on themselves, in U-turn style; so different segments of the same chain are aligned side by side. Thus, in some polypeptides, the β-pleated-sheet conformation is intramolecular. In others, it is intermolecular.

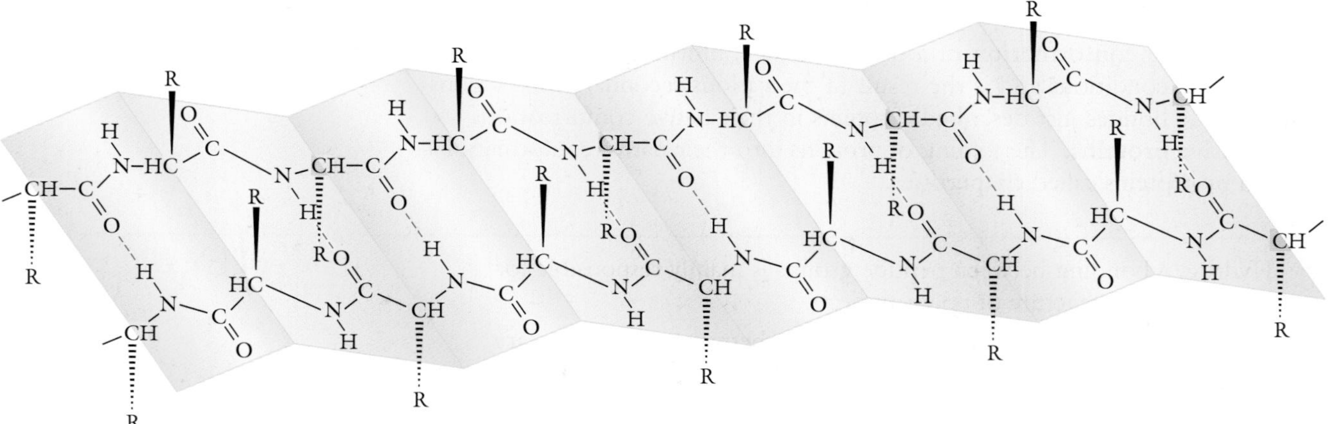

Figure 12.7 β-Pleated-sheet conformation, a secondary structure. Side groups of α-amino acid residues protrude alternately above (solid wedge bonds) and below (dashed wedge bonds) the plane of the sheet.

The **β-turn,** also called the **β-bend,** is the part of a polypeptide chain where the chain abruptly changes direction. Such turns often connect adjacent intramolecular segments of β-pleated sheets in a polypeptide chain. β-Turns are less ordered than α-helices and β-sheets.

Loop conformations are segments of polypeptide chains that are less ordered than β-turns and much less ordered than α-helices and β-pleated sheets.

Specific conformations are favored or disfavored by certain α-amino acid residues:

- The β-pleated sheet is favored only for polypeptides with small side groups—specifically, for polypeptides with a high content of Gly, with most other residues being Ala and Ser. Residues with larger side groups are generally not accommodated by the β-pleated-sheet conformation. Large side groups spatially interfere with each other: polypeptide chains cannot pack sufficiently close to allow attractive interactions.
- The α-helix can better accommodate large side groups than can the β-pleated sheet, but certain residues disfavor the α-helix. For example, proline is called a **helix breaker** because its presence prevents the formation of the α-helix: proline's rigid cyclic structure does not fit into the helical conformation. Several consecutive residues with like-charged side groups (for example, Asp or Lys) or large side groups (for example, Trp) also disfavor the α-helix because of repulsive interactions between side groups (electrostatic and spatial, respectively).

The effects of specific residues on conformation is a highly complex topic that we will not pursue further.

Concept checklist

✓ α-Helical, β-pleated-sheet, β-turn, and loop conformations are secondary structures.

12.6 FIBROUS PROTEINS

Fibrous proteins—water-insoluble proteins with elongated shapes having one dimension much longer than the others—serve as structural and contractile proteins. A characteristic of the secondary structure of their polypeptides is the repetition of a single conformational pattern throughout all or almost all of the chain. For this reason, the elongated chains, whether α-helix or β-pleated sheet, aggregate tightly into fibers or sheets with strong secondary attractive forces, producing macroscopic structures with high strength.

Fibrous proteins are usually described as containing no tertiary structure, because the polypeptide chains have only one conformational pattern. On the other hand, fibrous proteins almost always have quaternary structure, because they are composed of two or more polypeptide molecules aggregated together into a specific conformational pattern.

α-Keratins

α-Keratins are the structural components of hair, horn, hoofs, nails, skin, and wool. Electron microscopy and X-ray diffraction show that these materials have hierarchical structures. An example is the structure of hair (Figure 12.8 on the following page). Its individual polypeptide chains are almost entirely in the α-helical conformation, with very short disordered regions separating much longer α-helical regions. A pair of right-handed α-helices coil around each other in a left-handed manner to form a double-stranded helical coil (a **double helix**) called a **supercoil** or **superhelix.** A pair of supercoils coil around each other to form a **protofibril.** Protofibrils coil together to form **microfibrils,** which in turn coil into **macrofibrils,** which are further

The horns of this bighorn ram are constructed of α-keratin proteins. (Leonard Lee Rue III/Photo Researchers.)

INSIGHT INTO FUNCTION

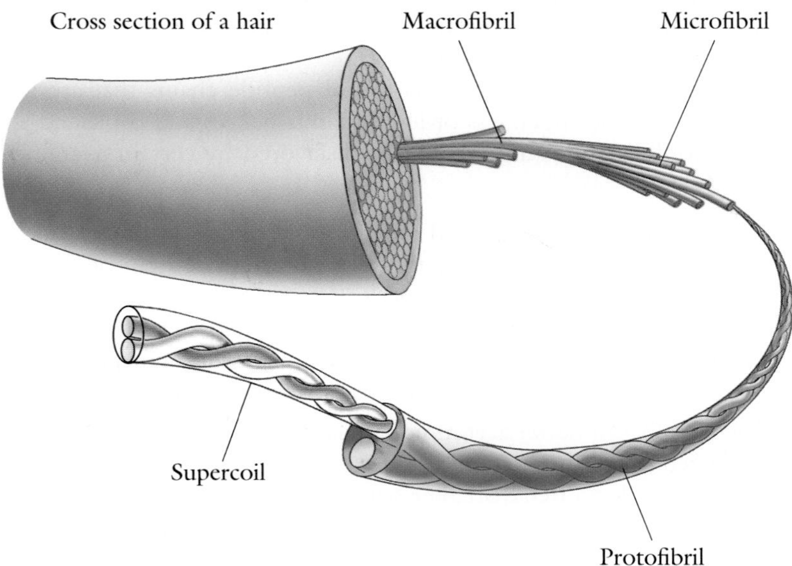

Figure 12.8 The hierarchical structure of a strand of hair explains many of the hair's properties.

assembled into a hair. Coiling at higher and higher levels is a mechanism for enhancing physical strength. At each level—superhelix, protofibril, microfibril, and macrofibril—structures are both packed side by side and staggered lengthwise to produce a final biological structure that is much longer and thicker than any of the individual elements. The springiness of hair and wool results from the tendency of the coiled structure to untwist when stretched and to recover its original conformation when released.

This packing into coiled conformations at higher and higher levels of organization is stabilized by secondary attractive forces and disulfide bridges between different polypeptide molecules. Disulfide bridges are more important than secondary interactions in imparting insolubility, strength, and resistance to stretching. Figure 12.9 shows disulfide bridges between the polypeptide chains in a protofibril—between chains of the same superhelix and between polypeptide chains of different superhelices. These disulfide bridges are intermolecular, which distinguishes them from the intramolecular disulfide bridges illustrated in Figure 12.5. Intermolecular bridges are also called **cross-links.**

α-Keratins are hard or soft, depending on the cysteine content and thus the number of disulfide bridges. Hard α-keratins (hair, horn, nail) have higher cysteine contents and are harder and less pliable than soft α-keratins (skin, callus). Permanent waving of hair is the deliberate rearrangement of the hair's disulfide bonds (Box 12.2 and Figure 12.10).

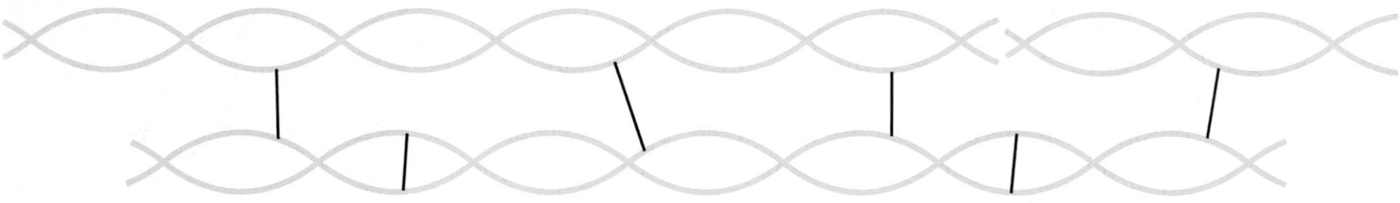

Figure 12.9 Disulfide cross-links between polypeptide chains in a protofibril. Cross-links are represented by straight lines between helices. The representation of disulfide cross-links is simplified by not showing the two superhelices coiled around each other.

Reduction → Curl → Oxidation

Figure 12.10 Permanent waving of hair.

Collagen

The most abundant protein in vertebrates is **collagen,** the major stress-bearing component of connective tissues such as bone, cartilage, cornea, ligament, teeth, tendons, and the fibrous matrices of skin and blood vessels. Collagen, like the α-keratins, has a hierarchical structure based on helical polypeptide chains—but with some important differences. Collagen contains much more

12.2 CHEMISTRY WITHIN US

Permanent Waving of Hair

Whether your hair is straight or curly and, if curly, whether loosely or tightly curled are hereditary traits. Heredity determines the primary structure of the α-keratin of your hair, and the amount and location of cysteine residues in that structure determine the pattern of disulfide bridges between polypeptide chains, which determines the shape of your hair. However, straight hair can be changed to curly hair or the size of the curl in curly hair can be changed by permanent waving.

First, the hair is treated with a reducing agent to break the disulfide bridges responsible for holding the hair in its present shape (see Figure 12.10). This treatment converts the disulfide bridges into sulfhydryl groups and imparts some conformational mobility to the α-helical polypeptide molecules of α-keratin. Next, a new shape is imposed on the hair through a process that adjusts the relative placement of the polypeptide chains—shaping the hair around curlers of the appropriate diameter and then setting the shape with an oxidizing agent to convert the sulfhydryl groups into a new pattern of disulfide bridges. When the oxidizing agent is rinsed out and the curlers are removed, the hair keeps the new shape. A permanent wave is not truly permanent, because new hair will grow in its genetically determined shape. The style can be maintained only by periodically repeating the permanent-waving process.

In the last step of the permanent waving of hair, an oxidizing agent is applied to form disulfide bridges between cysteine residues in the proteins of the hair. (Brent T. Madison, Shizuoka.Ken, Japan.)

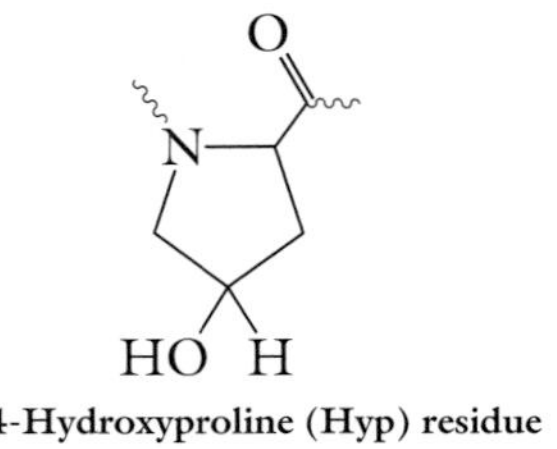

4-Hydroxyproline (Hyp) residue

glycine and proline than does α-keratin, and a considerable proportion of the proline residues are converted into 4-hydroxyproline (Hyp) residues in the biosynthesis of collagen. Additionally, collagen contains very little cysteine.

This difference in amino acid content and thus primary structure results in a different type of helical structure. The collagen polypeptide forms a left-handed helix, more elongated than the α-helix. Three collagen polypeptides wind around one another with a long right-handed twist to form a right-handed superhelix called a **triple helix** or **tropocollagen** (Figure 12.11). These triple helices are further organized into fibrils and higher-level structures. Coiling from one level to the next is in opposite directions. The packing at each level is very tight, and, along with the opposite-direction coiling, this packing produces a very strong structure. The same mechanism is used to produce the cables that support suspension bridges.

Hydrogen bonding and cross-linking between polypeptide chains contribute considerably to the strength of collagen. Most of the hydrogen bonding between peptide groups is not within the same polypeptide chain but is between different chains—both within and between triple helices. The hydroxyl groups of 4-hydroxyproline residues also participate in hydrogen bonding. In collagen, cross-linking is not through disulfide bridges, because collagen is almost devoid of cysteine. Cross-linking is accomplished through a series of complex reactions between lysine and histidine residues.

In bone and teeth, collagen fibrils are embedded in **hydroxyapatite,** $Ca_5(PO_4)_3(OH)$, an inorganic calcium phosphate polymer. This composite structure has very high physical strength. The same strength-enhancing principle applies to concrete structures that contain reinforcing metal rods as well as to wood that contains cellulose imbedded in lignin (Box 10.3).

INSIGHT INTO FUNCTION

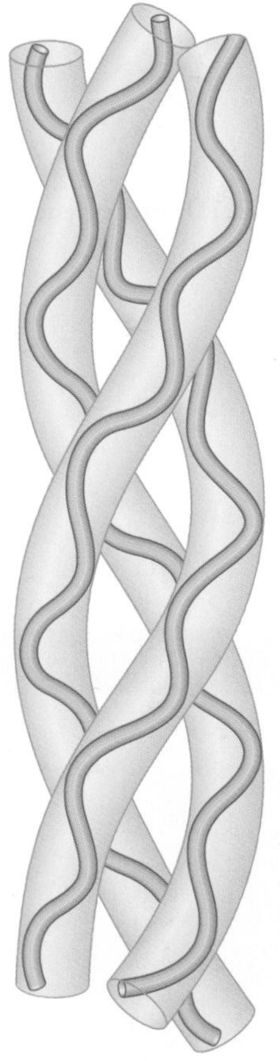

Figure 12.11 The triple helix of collagen explains that substance's flexibility and strength.

β-Keratins and Silk Fibroins

Among the few proteins having conformations that are almost completely β-pleated sheets are the β-keratins, the proteins in bird feathers and reptile scales, and **silk fibroin,** the protein of the silks secreted by many insects to fabricate cocoons, webs, and nests. In most silks, a gummy protein called **serican** cements the silk fibroin fibers together.

12.7 GLOBULAR PROTEINS

Globular proteins, unlike fibrous proteins, do not aggregate into macroscopic structures. Whereas fibrous proteins form the structural elements of an organism, the globular proteins do most of the metabolic work—catalysis, transport, regulation, protection—which requires solubility in blood and the other aqueous media of cells and tissues.

How are such large molecules made water soluble? In globular proteins, the primary structures are folded and organized into globular (globelike) conformations with a preponderance of hydrophilic α-amino acid residues on their outer surfaces. Globular proteins do not aggregate into macroscopic structures, because there are negligible secondary attractive forces between them. (Recall that highly branched molecules have lower boiling points than do unbranched molecules because their globular shape does not allow significant secondary attractive forces between molecules; Section 3.9.)

Globular proteins are solubilized because the hydrophilic nature of their surfaces allows strong hydrogen-bond attractions with water. Unlike fibrous proteins, a typical globular protein possesses tertiary structure—different conformational structures in different segments of its polypeptide chains, with α-helical or β-pleated-sheet segments or both types of segments connected by β-turn or loop segments or both.

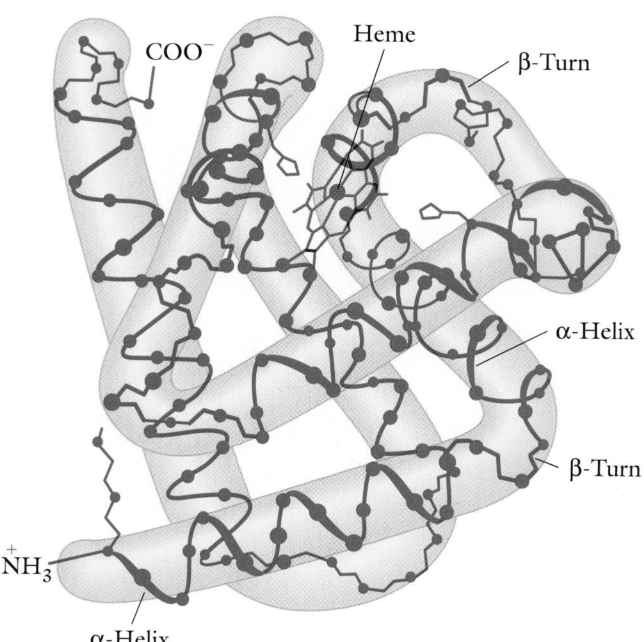

Figure 12.12 Myoglobin. α-Helical sections are separated by β-turns.

Myoglobin and Hemoglobin

The hemoglobin in red blood cells picks up oxygen in the lungs and transports it in the bloodstream to tissues throughout the body. Myoglobin, a similar protein present in muscle tissues, has a higher affinity for oxygen than does hemoglobin. It picks up oxygen from hemoglobin and stores it as a reserve for times when a working muscle's demand for oxygen is too high to be met by hemoglobin.

Myoglobin consists of one polypeptide containing 153 α-amino acid residues. Three-fourths of the residues are folded into eight α-helical sections (shown as straight tubular sections in Figure 12.12). The α-helical sections are connected by short β-turns containing mostly proline, the α-helix breaker (Section 12.5). The folding of the polypeptide chain places nonpolar residues together in the interior, where they are shielded from water. All polar residues except two histidines are located on the exterior surfaces, where they can hydrogen bond with water. The prosthetic group **heme,** an organic molecule with an Fe^{2+} at its center, is held in a cavity inside the polypeptide molecule by hydrophobic attractive forces. The Fe^{2+} is held in the center of heme by the free electron pairs of the four heme nitrogens and by ionic attractions with an interior histidine residue. The Fe^{2+} ion is the site where oxygen is bound when myoglobin takes on oxygen and stores it.

CH_3, CH_2, CH_3, CH_3, CH_2, N, N, N, N, Fe^{2+}, $HOOC{-}CH_2CH_2$, $HOOC{-}CH_2CH_2$, CH_3

Heme

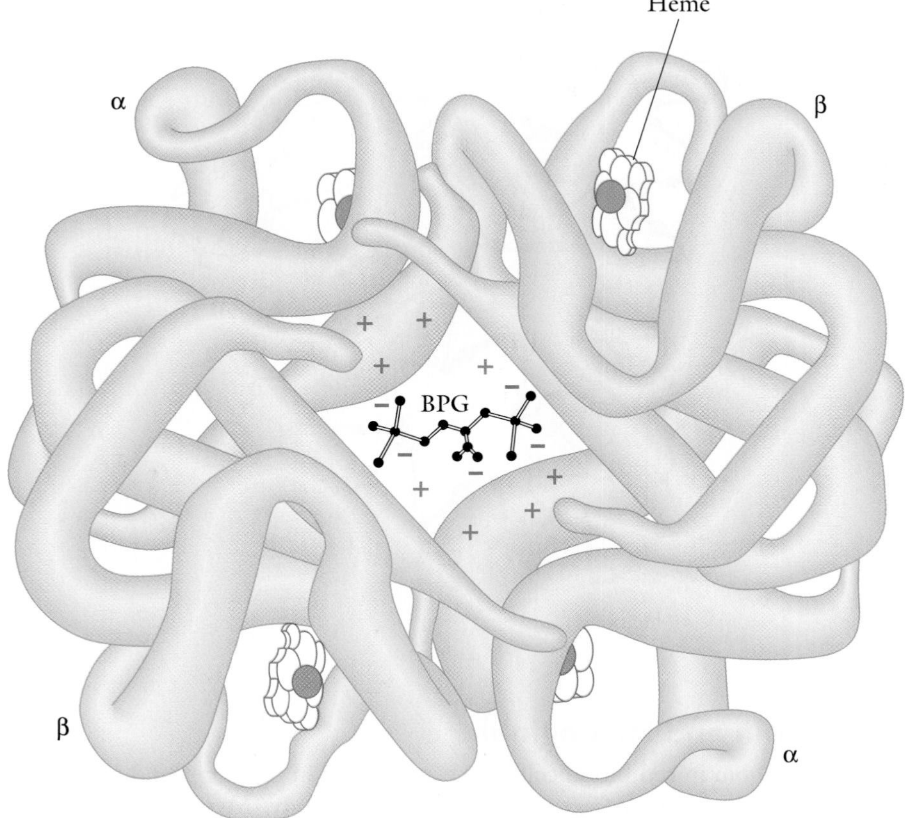

Figure 12.13 Hemoglobin contains four polypeptide chains (two α and two β) held together by secondary attractive forces. Each polypeptide chain contains a heme with its Fe^{2+} ion. 2,3-Bisphosphoglycerate (BPG) is located in the central cavity between the four polypeptide chains.

Hemoglobin has four polypeptide chains: a pair of α-chains and a pair of β-chains containing 141 and 146 α-amino acid residues, respectively (Figure 12.13). The primary structures of the α- and β-chains are similar to that of myoglobin and thus have similar secondary and tertiary structures, with α-helical sections separated by β-turns. However, the α- and β-chains of hemoglobin have hydrophobic α-amino acid residues at several positions where myoglobin has hydrophilic residues. These hydrophobic residues do not fold into the interior of the α- and β-chains but remain on their surfaces. Hydrophobic interactions among these nonpolar residues, avoiding repulsive interactions with water, serve to hold the four polypeptide chains together in the quaternary structure.

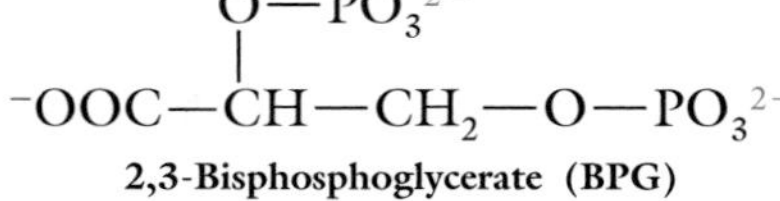

2,3-Bisphosphoglycerate (BPG)

›› Section 18.3 describes the transport of oxygen by hemoglobin in more detail.

Each polypeptide of hemoglobin carries a heme with its Fe^{2+} ion. In deoxygenated hemoglobin—that is, hemoglobin that is not carrying any oxygen—a 2,3-bisphosphoglycerate ion (BPG) is held in a central cavity between the four polypeptide chains by interaction with four positively charged groups on each of the two β-chains. BPG moves out of hemoglobin as hemoglobin picks up oxygen in the lungs. BPG is again picked up by hemoglobin after oxygen has been unloaded in peripheral tissues. BPG and environmental conditions (pH, CO_2 concentration) regulate the oxygen affinity of hemoglobin and are responsible for hemoglobin having a lower oxygen affinity than does myoglobin.

Carbon monoxide is a poison because it binds much more strongly than oxygen to the Fe^{2+} centers of heme, preventing oxygen from being picked up in the lungs and thus preventing the transport of oxygen to tissues, with the consequent cessation of all metabolic processes. Death is the result unless oxygen is administered to reverse the binding of carbon monoxide to hemoglobin. Heavy smokers of tobacco and other plants have a significant fraction of their hemoglobin tied up by carbon monoxide (produced by incomplete combustion) at all times, causing their shortness of breath.

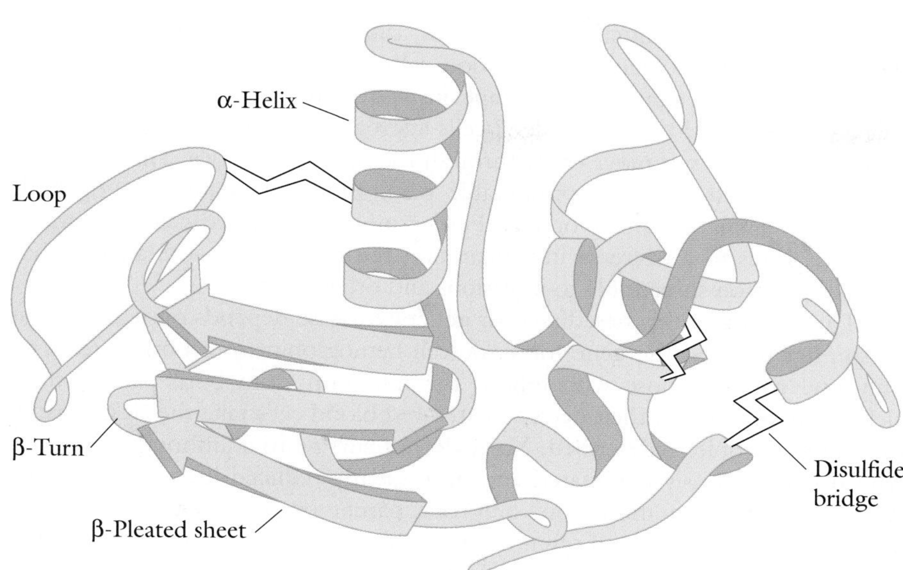

Figure 12.14 Hen egg-white lysozyme contains all four types of secondary structures (α-helix, β-pleated sheet, β-turn, and loop).

Lysozyme

Lysozyme, an enzyme in the cells and secretions of vertebrates, hydrolyzes bacterial cell walls. This enzyme does not kill bacteria but helps dispose of bacteria after they have been killed by other means. Figure 12.14 shows the structure of hen egg-white lysozyme, a polypeptide of 129 α-amino acid residues. Lysozyme illustrates a protein that contains all four types of secondary structures that we have considered—α-helix, β-pleated sheet, β-turn, and loop. The β-pleated sheets are shown as flat arrows to indicate the direction of the polypeptide chain. The arrowhead points in the direction of the C-terminal residue. Lysozyme has intramolecular alignment of β-pleated sheets in contrast with the β-keratins and silk fibroin, which have intermolecular alignments of β-pleated sheets.

12.8 MUTATIONS: SICKLE-CELL HEMOGLOBIN

Protein function, as we have seen, ultimately depends on primary structure. Primary structure is, as we shall see in Section 13.7, determined by the structure of the DNA (deoxyribonucleic acid) of the gene(s) that directs the protein's synthesis. A **genetic mutation** is an alteration in the DNA structure of a gene that may in turn produce a change in the primary structure of a protein. Secondary, tertiary, and quaternary structures can change as a result, which could have consequences for the protein's function. However, not all alterations in the primary structure produce significantly altered function.

›› Mutations are considered further in Section 13.8.

- The substitution of an α-amino acid residue similar in size, charge, and polarity often has minimal effect on three-dimensional structure and may have no effect on function. Examples are the substitution of Ile for Val, of Asp for Glu, and of Thr for Ser.
- The substitution of an α-amino acid residue very different in size, charge, and polarity may result in a large change in three-dimensional structure. Examples are Trp for Gly, Leu for Glu, and Ser for Ala. Substitutions of this type will most likely result in significant alterations in function unless the structural change is in a part of the protein without function.

Many mutant variations of human hemoglobin have been reported, of which a significant number have deleterious effects. **Sickle-cell hemoglobin,**

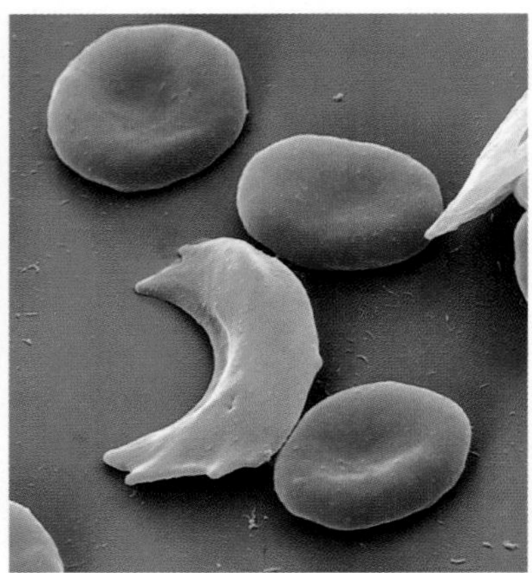

Figure 12.15 Electron micrograph of normal (circular) and sickled red blood cells. (Meckes/Ottawa/Photo Researchers.)

the most harmful of mutant hemoglobins, causes misery and early death for many humans. Deoxygenated red blood cells containing sickle-cell hemoglobin take on an elongated sickle shape instead of the normal biconcave disk shape (Figure 12.15). The "sickled" cells aggregate together into long rodlike structures that do not move easily through the blood capillaries. The capillaries become inflamed, causing considerable pain. Blood circulation is impaired, leading to damage to many organs. Furthermore, the sickled cells are fragile. Breakage leads to a decrease in functioning red blood cells—the anemia that leaves the person susceptible to infections and other diseases.

The severity of this hereditary, or genetic, disease depends on whether the person has received the gene for sickle-cell hemoglobin from both parents or only one parent. Those who inherit the mutation from both parents have **sickle-cell anemia:** all their deoxygenated red blood cells take the sickle shape, causing the problems described. Many do not survive to adulthood, and those who do are seriously debilitated and have greatly reduced life spans. Persons who inherit the mutation from only one parent produce both normal and sickle-cell hemoglobins and often do not suffer severe symptoms except under conditions of severe oxygen deprivation, such as at high altitudes or after strenuous aerobic activity. This condition is called **sickle-cell trait.**

Sickle-cell hemoglobin and normal hemoglobin differ in only one α-amino acid residue in each of the two β-polypeptide chains. Valine substitutes for the glutamic acid residue at position 6:

Normal hemoglobin	Val-	His-	Leu-	Thr-	Pro-	Glu-	Glu-	Lys~~~~
Sickle-cell hemoglobin	Val-	His-	Leu-	Thr-	Pro-	Val-	Glu-	Lys~~~~
β	1	2	3	4	5	6	7	8

Although a hydrophobic residue replaces a charged residue, sickle-cell hemoglobin does not differ greatly from normal hemoglobin in its ability to pick up oxygen, because the regions critical to the picking up of oxygen (the heme and BPG cavities) are not altered. However, residue 6 of each of the β-polypeptide chains is at the surface of hemoglobin, and the substitution results in a disastrous interaction between different hemoglobins. A hydrophobic pocket is exposed at the surface of hemoglobin, both normal and sickle cell, when hemoglobin is deoxygenated. The valine residue at position 6 of one sickle-cell hemoglobin fits into and sticks (by hydrophobic attractions) to the hydrophobic pocket of the next hemoglobin (Figure 12.16). Sickling is the sequential aggregation of many such hemoglobins. It does not occur with normal hemoglobin, because there are no attractive interactions between the hydrophobic pocket and residue 6 when that residue is a charged glutamic acid residue. Moreover, sickling occurs only when there is a high concentration of deoxygenated hemoglobin. The hydrophobic pocket of hemoglobin is not exposed at the surface in oxygenated hemoglobin, because of conformational differences between oxygenated and deoxygenated hemoglobins.

INSIGHT INTO FUNCTION

Val at residue 6 of β-chain

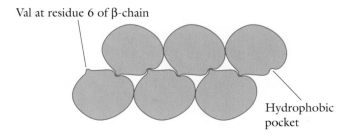

Hydrophobic pocket

Figure 12.16 Formation of rodlike structures by the sequential aggregation of many deoxygenated sickle-cell hemoglobin entities. This phenomenon explains the painful and dangerous symptoms of sickle-cell anemia.

Sickle-cell anemia is found almost exclusively in populations arising in tropical regions of the world, notably central and western Africa. Why should a genetic disease be restricted geographically? An answer is suggested by the high incidence of malaria, a tropical disease, found in the same geographical regions. The malarial parasite spends much of its life cycle in red blood cells, but sickled cells are a less-hospitable environment for the malarial parasite than normal red blood cells. This built-in protection against malaria gives little advantage to persons with sickle-cell anemia, who already have limited life spans. But those with sickle-cell trait have a survival advantage against malaria, a higher likelihood of surviving to adulthood to pass on the trait to their offspring. Unfortunately, there is a 25% chance, based on the principles of genetics, that an offspring from two parents with sickle-cell trait will have sickle-cell anemia.

Some relief for sickle-cell anemia is achieved with hydroxyurea, which stimulates the production of fetal hemoglobin. Fetal hemoglobin does not contain the sickle mutation, because it contains only the α-polypeptide chains and no β-polypeptide chains. There is also an experimental treatment in which normal bone marrow is transplanted into patients with sickle-cell anemia to stimulate the production of normal hemoglobin.

The dependence of physiological function on protein structure is often subtle and not easy to predict. The substitution of valine for glutamic acid at position 6 of the β-polypeptide chain of hemoglobin results in a terrible genetic disease. However, replacement of the same glutamic acid residue by a different nonpolar residue, alanine, instead of valine, a mutation called hemoglobin G Makassar, has no significant effect on the function of hemoglobin. The methyl side group of alanine apparently does not interact strongly enough with the hydrophobic pocket of a neighboring hemoglobin. Other genetic diseases are described in Section 13.8.

The subtleness of the interrelation of primary, secondary, tertiary, and quaternary structures and physiological functions is also evident in the use of bovine and pig insulins as substitutes for human insulin in the treatment of diabetes (Box 12.3 on the following page).

12.9 DENATURATION

Denaturation is a loss of native conformation brought about by a change in environmental conditions, resulting in loss of physiological function. The protein is then called a **denatured protein.**

Denaturation alters secondary, tertiary, and quaternary structures through conformational changes. Denaturation should not be confused with digestion or laboratory hydrolysis, which affect the primary structure of a protein. Denaturation takes place under milder conditions than those of digestion or hydrolysis; the secondary, tertiary, and quaternary structures are held together mainly by noncovalent attractions, which are much weaker than covalent forces and more easily disrupted. Disulfide bridges impart considerable resistance to denaturation because they are stronger than the noncovalent interactions.

The whipping of egg whites into a frothy solid for meringues and soufflés is achieved through a change in the conformational structure of the polypeptide chain. (David Hundley/ The Stock Market.)

Globular proteins possess weaker secondary attractive forces and denature more readily than fibrous proteins. The denaturation of globular proteins usually results in an insoluble or much less soluble protein compared with the native protein as polypeptide chains unfold from the secondary, tertiary, and quaternary structures of the native protein and become tangled together in random conformations, with nonpolar residues no longer folded inside and protected from the aqueous environment. When we cook or whip egg white, the visible changes are due to the denaturation of albumin.

12.3 CHEMISTRY WITHIN US

Diabetes Mellitus and Insulin

Insulin, a peptide hormone secreted by the pancreas, is essential to the regulation of glucose metabolism. Abnormalities in insulin action cause two types of the disorder known as diabetes mellitus. Insufficient production of insulin by the pancreas is the cause of **type I diabetes mellitus,** also called **insulin-dependent diabetes** or **juvenile-onset diabetes.** Insufficient insulin results in defects not only in carbohydrate metabolism, but also in lipid and protein metabolism because of the interrelations between metabolic pathways (Section 18.6). Insulin-dependent diabetes usually develops before the age of 40, often in adolescence. The consequences of untreated diabetes are severe: increased risk of athereosclerosis (Box 16.1); peripheral neuropathy (a nerve disorder); diabetic retinopathy, an eye disorder often resulting in blindness; and chronic kidney failure.

Insulin-dependent diabetes is treated by subcutaneous injection of replacement insulin. Both bovine (beef) and porcine (pig) insulin have been used for this purpose. These insulins are not identical with human insulin, however. Insulin is composed of two polypeptide chains, A and B, covalently linked by disulfide bridges. Chain A has 21 and chain B has 30 α-amino acid residues. Human and pig insulins differ in only one residue, whereas human and beef insulins differ in three residues:

	Chain A			Chain B
Position of residue	8	9	10	30
human insulin	Thr	Ser	Ile	Thr
pig insulin	Thr	Ser	Ile	Ala
beef insulin	Ala	Ser	Val	Ala

These differences in primary structure do not cause large differences in the physiological functions of the three insulins. Either the three-dimensional structures are the same or any differences that do exist are not in functionally important regions of the molecule, which is not necessarily what we would predict. Pig insulin has a nonpolar alanine residue instead of the human insulin's polar threonine residue at one position. Beef insulin has nonpolar alanine residues instead of human insulin's polar threonine residues at two positions and valine instead of isoleucine at a third position. Apparently, the polarity of the side group is not important. All of the variant residues have a common feature—the side groups are not too different in size.

Treatment with pig and beef insulins allows a reasonable quality of life for diabetics but does not completely solve the problem caused by the disease. On average, the life span is still decreased, perhaps because treatment does not maintain the ideal insulin concentration in the bloodstream at all times or because pig and beef insulins are good but not perfect replacements for human insulin. Many diabetics now test themselves for glucose blood levels throughout the day and adjust the amount of insulin that they use accordingly. The equivalent of human insulin is now available, synthesized by recombinant DNA technology (Section 13.3). This synthetic insulin may offer significant improvement over pig and beef insulins in the treatment of this form of diabetes.

About 90% of all diabetics have **type II diabetes mellitus,** called **non-insulin-dependent diabetes** or **adult-onset diabetes.** This condition generally develops in people older than 40 who are obese, physically inactive, and have carbohydrate-rich diets. The pancreas secretes sufficient insulin, but cells throughout the body do not properly respond to it. Non-insulin-dependent diabetes is much less severe than insulin-dependent diabetes, with most patients being able to manage their glucose blood levels by diet and exercise alone.

A variety of denaturing conditions or agents lead to protein denaturation:

1. Increased temperature (increased thermal energy) increases the thermal motions of proteins, disrupting the noncovalent attractions responsible for the secondary, tertiary, and quaternary structures of the native protein. We take advantage of these effects to kill microorganisms during the sterilization of surgical and laboratory instruments and for food-canning and food-preparation equipment, as well as in the sealing and cauterization of small blood vessels in surgery. Cooking fish, meats, and poultry both kills microorganisms and makes the foods more digestible by unraveling the native conformations.

Microwave radiation also disrupts native conformations by increasing the thermal motions of proteins. Microwave radiation heats the proteins by an indirect effect. Microwaves heat up the water molecules, which then transfer thermal energy to the protein molecules. This process is the same as that used when food is cooked in a microwave oven.

2. Ultraviolet and ionizing radiations disrupt native conformations by causing chemical reactions in the polypeptide chains.

3. Mechanical energy has similar effects. Violent mixing, as in the whipping or shaking of egg white into a frothy solid for meringues and souffles causes polypeptide chains to unfold and become tangled up in a random fashion.

4. Changes in pH from normal physiological pH alter protein conformations by interfering with salt-bridge interactions. A fraction of the positively charged (basic) side groups become uncharged when the pH increases above 7, and a fraction of the negatively charged (acidic) side groups become uncharged when the pH decreases below 7. The extent and types of changes depend on the extent of the pH changes.

When bacteria cause milk to sour, the protein casein precipitates. This precipitation is the basis for cheese production. The pH of milk is normally about 6.5. Casein has an isoelectric point of 4.7, which means that it contains an excess of acidic side groups and has an overall negative charge at pH 6.5. Bacterial metabolism of milk produces lactic acid, lowering the pH. The carboxylate side groups of casein become protonated and the protein denatures.

Acid denaturation of proteins in the stomach aids in the digestive process. The low pH (1.5–2.0) of the stomach unfolds protein structures, making peptide bonds more accessible to enzymatic hydrolysis.

5. Organic chemicals such as soaps and detergents, alcohols, and urea (NH_2CONH_2) denature proteins by interrupting interactions between side groups of α-amino acid residues. This type of denaturation is the basis for using 70% alcohol as an antiseptic agent, for example, to sterilize skin before an injection. Alcohol passes through the bacterial cell walls and denatures bacterial proteins.

6. Salts of heavy metals such as Pb^{2+}, Hg^{2+}, and Ag^{+} react with sulfhydryl groups of cysteine residues to form metal disulfide bridges that prevent the formation of native conformations by the normal disulfide bridges.

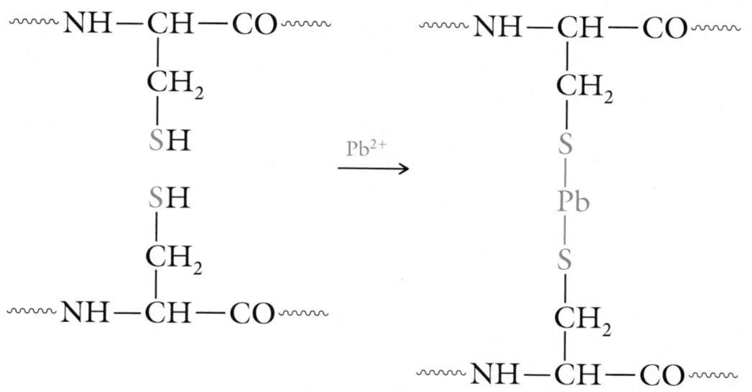

$$\begin{array}{c} \sim NH-CH-CO\sim \\ | \\ CH_2 \\ | \\ SH \\ \\ SH \\ | \\ CH_2 \\ | \\ \sim NH-CH-CO\sim \end{array} \xrightarrow{Pb^{2+}} \begin{array}{c} \sim NH-CH-CO\sim \\ | \\ CH_2 \\ | \\ S \\ | \\ Pb \\ | \\ S \\ | \\ CH_2 \\ | \\ \sim NH-CH-CO\sim \end{array}$$

Young children who ingest flaked-off paint chips in very old buildings, where lead-based paint still remains on walls and woodwork, can suffer from lead poisoning. Lead poisoning can also result when acidic foods leach out lead pigments from the glazes used in some countries to manufacture ceramic dinnerware. Mercury poisoning results from ingesting too much fish taken from waters into which chemical wastes containing mercury salts have been dumped.

7. Oxidizing and reducing agents alter native conformations, the first by forming disulfide bridges and the second by cleaving them.

Various neurodegenerative diseases such as mad cow and Alzheimer's diseases are caused by conformational changes (Box 12.4 on the next page).

12.4 CHEMISTRY WITHIN US

Conformational Diseases: Prion and Alzheimer's Diseases

Native conformations of proteins are critical to physiological function. If they undergo denaturation, native conformations are either reestablished through chaperone proteins or the denatured proteins are degraded by hydrolysis followed by the synthesis of new proteins. However, some diseases are caused by the formation of denatured (misfolded) proteins that are neither renatured nor degraded. Instead, the denatured proteins are deposited extracellularly in various tissues and organs as insoluble material called **amyloid plaque.** The amyloid plaque interferes with the normal physiological functions of cells, tissues, and organs. Cell death, organ failure, and death follow. Among the diseases caused by the extracellular deposition of misfolded proteins are Alzheimer's disease and prion diseases such as mad cow disease and Creutzfeldt–Jakob disease.

Prion Diseases

Prion diseases cause the mammalian central nervous system to undergo neurodegeneration (nerve-cell death) and the eventual death of the organism. These diseases are called **transmissible spongiform encephalopathies (TSEs)** because they are infectious diseases and the brain tissues develop a spongelike appearance owing to the deposition of amyloid plaques surrounded by dead and dying neurons. Among the prion diseases are **scrapie** in sheep and goats, **mad cow disease (bovine spongiform encephalopathy,** or **BSE)** in cattle, and **Creutzfeldt–Jakob (CJD) disease** in humans. These diseases are characterized by a change in the conformation of a protein called a **prion** (which stands for **proteinaceous infectious agent).**

The normal prion protein **(PrP^C)** is located mainly on the outer surfaces of neurons and its conformational structure has a high (about 40%) α-helix content with almost no β-pleated sheets. The altered prion protein **(PrP^{Sc})** has a lower α-helix content but a high (almost 50%) β-pleated-sheet content. Although the physiological function of PrP^C has not been established, its change into PrP^{Sc} results in neurodegeneration and eventual death. The mechanism of the conformational change of PrP^C into PrP^{Sc} is not well understood, but PrP^{Sc} appears to catalyze the change. PrP^{Sc} is the infectious agent for the conformational change. PrP^{Sc} is a rare infectious agent because it contains no nucleic acids (DNA, RNA; Chapter 13). The common infectious agents—bacteria and viruses—contain nucleic acids.

PrP^{Sc} is transmitted by the ingestion of nerve tissue, including brain and spinal cord, as well as bone meal from diseased mammals. PrP^{Sc} is not present in the meat or milk of the animals. However, contamination of the meat of animals with nerve tissue and bone meal can happen when the animals are slaughtered, because the safe slaughter of diseased animals to retain their meat is difficult. Thus, all diseased cattle were slaughtered and disposed of during the outbreak of mad cow disease in the United Kingdom in the early 1990s.

Mad cow disease is assumed to have arisen because cattle were fed meat-and-bone meal from scrapie-infected sheep and the nerve tissue and bone remnants left over from the slaughter of BSE-infected cattle. The transmission of BSE to humans has not been firmly established.

Creutzfeldt–Jakob disease is a rare TSE disease that arises in humans late in life in a sporadic and apparently spontaneous manner. (This manner of onset is indicative of a spontaneous mutation—Section 13.8.) Its symptoms and medical progression are similar to those of the other TSEs. However, in 1994, more than 100 teenagers and young adults in the United Kingdom developed a new variation of CJD, different from the sporadic CJD, which indicates transmission by ingestion of BSE-infected cattle.

Stanley Prusiner received the 1997 Nobel Prize in medicine for his work elucidating our understanding of the role of prions in TSEs. However, there is much that we do not understand, including the mechanism by which prions survive the digestive processes in the intestinal tract and enter the bloodstream.

Alzheimer's Disease

Alzheimer's disease (AD) is a neurodegenerative disease that causes a slow loss of memory, judgment, the ability to function socially, and eventual death. The progression of the disease often takes a decade or so, although it can be significantly shorter or longer. Death results from pneumonia, malnutrition, or general body wasting.

AD is not a simple disease but a group of closely related diseases. They can be divided into early- and late-onset variations of AD, depending on whether symptoms begin before or after age 65. Some of the variations of AD (about 25%) are inherited and run in families; that is, mutated genes are passed to offspring in a family. Other variations are sporadic, which means that other blood relatives are not affected. Some of these cases are due to spontaneous mutations over a person's lifetime. However, there appears to be a genetic link for other cases, with certain genes indicating a tendency toward spontaneous mutation to AD. Irrespective of the specific variation of Alzheimer's disease, the brain tissues of all AD patients show the deposition of amyloid plaques surrounded by dead and dying neurons—similar to the brain tissue of patients with prion diseases.

Summary

Proteins

- Proteins have a wider range of functions than any other family of biomolecules: enzymatic, transport, regulatory, structural, contractile, protective, and storage.
- Polypeptides are polymers constructed by the bonding together of α-amino acids.
- Proteins are naturally occurring polypeptides, usually aggregated with other polypeptides or with other types of molecules or ions or both.

α-Amino Acids

- An α-amino acid contains a carboxyl group and an amino group attached to the same carbon atom.
- α-Amino acids are categorized as nonpolar neutral, polar neutral, polar acidic, or polar basic, depending on the side group attached to the α-carbon.
- Only L-α-amino acids are used to synthesize the polypeptides of plants and animals.
- Ten of the 20 α-amino acids needed for polypeptide synthesis are not synthesized by mammals. These essential amino acids must be obtained from the diet.

Zwitterionic Structure of α-Amino Acids

- α-Amino acids exist as zwitterions (dipolar ions) in equilibrium with cation and anion forms.
- The isoelectric point, pI, of an α-amino acid is the pH at which the α-amino acid is present almost exclusively in its zwitterion form.
- Neutral α-amino acids have pI values ranging from 5.05 to 6.30.
- Acidic and basic α-amino acids have pI values on the acidic and basic sides, respectively, of this range.
- All amino and carboxyl groups of α-amino acids are charged at physiological pH, which is also the case for the α-amino acid residues of peptides and proteins.

Peptides

- Peptides are formed by dehydration between carboxyl and amino groups of α-amino acids to form peptide bonds.
- Peptides are named by naming the α-amino acid residues in sequence, beginning at the end with the free α-amino group (the N-terminal residue) and proceeding to the end with the free carboxyl group (the C-terminal residue).

Chemical Reactions of Peptides

- Peptides undergo disulfide-bridge formation by the oxidation of sulfhydryl groups on pairs of cysteine residues.
- Both intramolecular and intermolecular disulfide bridges exist in proteins.
- The digestion of peptides and proteins in the stomach and intestine consists of the hydrolysis of peptide bonds.
- Disulfide bonds are cleaved by reduction in the liver.

The Three-Dimensional Structure of Proteins

- The primary structure of a peptide—the sequence of α-amino acid residues—determines its three-dimensional structure (secondary, tertiary, and quaternary), which in turn determines its physiological function.
- Secondary structure is the conformation present in a local region of a polypeptide.
- The important secondary structures are the α-helix, β-pleated sheet, β-turn, and loop.
- Tertiary structure exists when the polypeptide has different secondary structures in different local regions.
- Tertiary structure describes the three-dimensional relation among the different secondary structures in different regions.
- Quaternary structure exists only in proteins with two or more polypeptide molecules. It describes the three-dimensional relation among the different polypeptides of a protein.
- Many factors determine the three-dimensional structure of a polypeptide or protein: the shielding of nonpolar α-amino acid residues from water; hydrogen bonding between peptide groups; hydrophobic, hydrogen-bonding, and salt-bridge attractions between α-amino acid side groups; the interaction of polar side groups with water; and disulfide bridges.

Fibrous Proteins

- Fibrous proteins are water-insoluble proteins with elongated shapes.
- The proteins aggregate tightly together into fibers or sheets with extensive intermolecular secondary forces to build macroscopic structures.
- These structures are the structural and contractile proteins of bone, hair, skin, and muscle.

Globular Proteins

- Globular proteins do not aggregate into macroscopic structures.
- They perform a wide range of functions (enzymatic, transport, regulatory, protective) that usually require solubility in blood and the other aqueous media of cells and tissues.
- Solubility depends on primary structures that fold and organize into globular conformations with hydrophilic residues on their outer surfaces.

Mutations

- A genetic mutation—an alteration in DNA structure—can result in the production of a protein with an abnormal α-amino acid sequence.
- Some mutations result in proteins with drastically altered physiological function. Sickle-cell anemia is one such mutation.

Denaturation

- Denaturation is the loss of the native conformation of a protein brought about by a change in environmental conditions.
- Denaturing conditions or agents include heat; microwave, ultraviolet, and ionizing radiations; mechanical energy; pH changes; organic chemicals; heavy metal salts; and oxidizing and reducing agents.

Key Words

α-amino acid, p. 374
α-helix, p. 390
amino acid residue, p. 380
amino acid sequence, p. 382
β-pleated sheet, p. 390
β-turn, p. 391
denaturation, p. 399
disulfide bridge, p. 384
electrophoresis, p. 380
fibrous protein, p. 389
folding, p. 387
globular protein, p. 389
heme, p. 395
isoelectric point, p. 378
mutation, p. 397
native conformation, p. 389
peptide, p. 380
peptide group, p. 380
polypeptide, p. 381
primary structure, p. 387
prion, p. 402
protein, pp. 373, 385
quaternary structure, p. 387
salt bridge, p. 389
secondary structure, p. 387
sickle-cell anemia, p. 398
sickle-cell trait, p. 398
side group, p. 375
tertiary structure, p. 387
zwitterion, p. 376

Exercises

α-Amino Acids

12.1 Classify the following amino acids as α-, β-, or γ-amino acids.

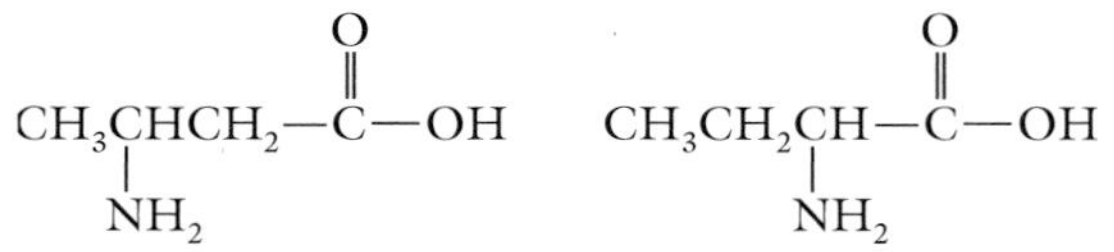

1: $CH_3CH(NH_2)CH_2—C(=O)—OH$

2: $CH_3CH_2CH(NH_2)—C(=O)—OH$

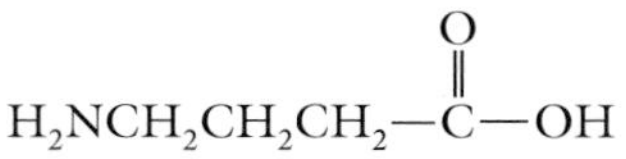

3: $H_2NCH_2CH_2CH_2—C(=O)—OH$

12.2 Classify the following amino acids as D or L.

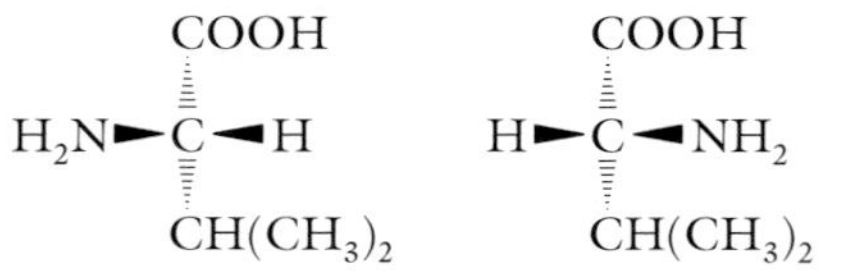

1: COOH above, H_2N ◄ C ► H, $CH(CH_3)_2$ below

2: COOH above, H ◄ C ► NH_2, $CH(CH_3)_2$ below

12.3 By reference to Table 12.1, use the three-letter abbreviations to name the α-amino acid(s) having (a) the smallest side group; (b) the functional group of the alcohol family; (c) a benzene ring; (d) two tetrahedral stereocenters; (e) a heterocyclic side group; (f) sulfur; (g) an acidic side group; (h) an α-amino group in a ring.

12.4 By reference to Table 12.1, use the three-letter abbreviations to name the α-amino acid(s) having (a) the largest side group; (b) the functional group of the amide family; (c) a heterocyclic structure; (d) no tetrahedral stereocenter; (e) four nitrogen atoms; (f) a basic side group; (g) an —SH group; (h) a secondary α-amino group.

Zwitterionic Structure of α-Amino Acids

12.5 Why does an α-amino acid exist as a zwitterion, both in the solid state and when dissolved in water?

12.6 Define isoelectric point.

12.7 (a) Show the structures of alanine at its isoelectric point, at physiological (7) pH, at very low (<1) pH, and at very high (>12) pH. (b) At what pH does alanine have its lowest solubility in water? its highest solubility? (c) To which electrode does alanine migrate in electrophoresis at its pI? at physiological (7) pH?

12.8 (a) Show the structure of glutamic acid at physiological (7) pH. (b) At what pH will glutamic acid have its lowest solubility in water? (c) To which electrode does glutamic acid migrate in electrophoresis at pH = pI? at physiological (7) pH?

Peptides

12.9 Draw the structure of Ser-Met at physiological pH and give its full name.

12.10 Draw the structure of Cys-Phe at physiological pH and give its full name.

12.11 Draw the structure of Thr-Ala-Asp at physiological pH and give its full name.

12.12 Draw the structure of Leu-Val-Lys at physiological pH and give its full name.

12.13 Draw the structure of Gly-Pro-Ala at physiological pH and give its full name.

12.14 Draw the structure of Ser-Ala-Pro at physiological pH and give its full name.

12.15 How many constitutional isomers are possible for tripeptides containing threonine, cysteine, and leucine? Give abbreviated names for the different amino acid sequences.

12.16 How many different constitutional isomers are possible for tetrapeptides containing alanine, glutamic acid, tyrosine, and valine? Give abbreviated names for the different amino acid sequences for those tetrapeptides that have valine as the C-terminal residue.

12.17 There are six possible stereoisomers of the type based on tetrahedral stereocenters for Lys-Ala-Glu. For each stereoisomer, give an abbreviated name that indicates the configuration at each of the stereocenters. Which stereoisomer would be found in nature?

12.18 Cis-trans isomers of peptides are possible because of the double-bond character of the peptide group. The peptide groups of peptides found in nature possess the trans configuration. Draw Gly-Ala in the trans configuration.

12.19 Draw the structure of Asp-Ala-Asp at physiological pH. To which electrode will the tripeptide move in electrophoresis at physiological pH?

12.20 Draw the structure of Lys-Ala-Lys at physiological pH. To which electrode will the tripeptide move in electrophoresis at physiological pH?

Chemical Reactions of Peptides

12.21 Give the names of the products of digestion of Phe-Asp-Lys-Gly. What products are formed if the same peptide is subjected to acidic hydrolysis? basic hydrolysis?

12.22 Give the names of the products of digestion of Pro-Ala-Thr-Lys. What products are formed if the same peptide is subjected to acidic hydrolysis? basic hydrolysis?

12.23 Hydrolysis of a tripeptide yields equimolar amounts of Ala, Gly, and Lys. What is the amino acid sequence of the tripeptide?

12.24 Compare the products of digestion of the two pentapeptides Ala-Gly-Asp-Gly-Phe and Gly-Asp-Gly-Phe-Ala.

12.25 Consider the following hexapeptide:

$$\begin{array}{l} H_3\overset{+}{N}-\underset{\displaystyle CH_3}{\underset{|}{CH}}-CO-NH-\underset{\displaystyle \begin{array}{c}CH_2\\|\\S\\|\\S\\|\\CH_2\end{array}}{\underset{|}{CH}}-CO-NH-\underset{\displaystyle CH_2C_6H_5}{\underset{|}{CH}}-COO^- \\ H_3\overset{+}{N}-\overset{\displaystyle CH_3}{\overset{|}{CH}}-CO-NH-\overset{|}{CH}-CO-NH-\overset{\displaystyle CH_2C_6H_5}{\overset{|}{CH}}-COO^- \end{array}$$

Write the structures of the products obtained, if any, when the tripeptide is (a) digested; (b) subjected to selective oxidizing conditions; (c) subjected to selective reducing conditions.

12.26 Consider the following tripeptide:

$$H_3\overset{+}{N}-CH_2-CONH-\underset{\displaystyle CH_2SH}{\underset{|}{CH}}-CONH-\overset{\displaystyle CH(CH_3)_2}{\overset{|}{CH}}-COO^-$$

Write the structures of the products obtained, if any, when the tripeptide is (a) digested; (b) subjected to selective oxidizing conditions; (c) subjected to selective reducing conditions.

Three-Dimensional Structure of Proteins

12.27 Distinguish between conjugated and simple proteins.

12.28 Distinguish between an apoprotein and a prosthetic group.

12.29 What takes place in the conversion of one conformation of a molecule into a different conformation?

12.30 Which of the following structural levels of proteins are the result of conformational differences? (a) Primary; (b) secondary; (c) tertiary; (d) quaternary.

12.31 Which of the following interactions are most important in determining the secondary structure of polypeptides? (a) Hydrogen bonding between peptide groups; (b) attractive interactions between side groups of residues.

12.32 Which of the following interactions are most important in determining the tertiary and quaternary structure of polypeptides? (a) Hydrogen bonding between peptide groups; (b) attractive interactions between side groups of residues.

12.33 What attractive interaction (hydrogen bond, salt bridge, hydrophobic, or none) can take place between side groups for each of the following pairs of α-amino acid residues under physiological conditions? (a) Pro and His; (b) Ser and Tyr; (c) Pro and Phe; (d) Lys and Arg; (e) Lys and Glu; (f) Ser and Val.

12.34 What attractive interaction (hydrogen bond, salt bridge, hydrophobic, or none) can take place between side groups for each of the following pairs of α-amino acid residues under physiological conditions? (a) Met and Arg; (b) Met and Ile; (c) Thr and Tyr; (d) Asp and Glu; (e) Thr and Phe; (f) Arg and Asp.

12.35 What secondary forces other than those between side groups of pairs of α-amino acid residues are important in determining protein conformations?

12.36 What covalent bond is important in stabilizing protein conformations?

12.37 What structural feature dictates the secondary, tertiary, and quaternary structures of proteins?

12.38 Is intermolecular or intramolecular hydrogen bonding between peptide groups more important for the α-helix? for the β-pleated sheet?

Fibrous Proteins

12.39 Why are fibrous protein structures required for proteins that function as structural and contractile proteins?

12.40 Which of the following structural features contribute to the high physical strength of fibrous proteins? (a) Double helix (for example, α-keratin); (b) triple helix (for example, collagen); (c) hierarchy of increasing levels of coiling; (d) intermolecular disulfide bridges; (e) composite construction (for example, bone).

12.41 Briefly describe the difference between a double helix and a triple helix.

12.42 Does the hardness of α-keratins increase or decrease with increasing cysteine content?

12.43 Where are α-keratins found? What is their physiological function?

12.44 Where are collagens found? What is their physiological function?

Globular Proteins

12.45 Why are globular protein structures required for proteins that function as catalytic, transport, regulatory, and protective proteins?

12.46 Why does a description of hemoglobin include quaternary structure but that of myoglobin does not?

12.47 Why does a description of myoglobin include tertiary structure but that of an α-keratin does not?

12.48 Classify hemoglobin and myoglobin as conjugated or simple proteins.

12.49 Which is the apoprotein and which the prosthetic group in myoglobin?

12.50 How do the physiological functions of hemoglobin and myoglobin differ?

12.51 What is the function of heme in hemoglobin and myoglobin?

12.52 Draw the abbreviated structure when the following polypeptide forms an intramolecular disulfide bridge.

Gly-Ala-Cys-His-Phe-Asp-Cys-Met-Gly-Thr

12.53 Which of the following α-amino acid residues are most likely to be in the exterior of a globular protein: Asp, Ile, Glu, His, Lys, Phe, Trp, Val? Explain.

12.54 Which of the following peptides will be more soluble in water at physiological pH? (a) Ala-Glu-Phe-Gly-Leu-Val-Ala-Phe; (b) Ala-Glu-Phe-Glu-Leu-Val-Ala-Glu. Explain.

Mutations

12.55 Define mutation.

12.56 Distinguish between the sickle-cell trait and sickle-cell anemia.

12.57 A hemoglobin mutation known as hemoglobin Hammersmith has Ser instead of Phe at position 42 of the β-polypeptide chains. This substitution is in the pocket of the polypeptide chain that holds heme. The overall result is a decrease in the picking up of oxygen by hemoglobin because hemoglobin tends to lose heme. A replacement of Phe by Ile at the same position has much less effect on the ability of hemoglobin to pick up oxygen. Why?

12.58 What is the consequence of a mutation that changes a hydrophilic residue into a hydrophobic residue on the surface of a globular protein?

12.59 What is the consequence of a mutation that changes a hydrophobic residue into a hydrophilic residue in the interior of a globular protein?

12.60 Explain each of the following observations regarding the effect of mutations on physiological function: (a) replacement of Trp often has a large effect; (b) replacement of Asp by Glu often has little effect; (c) replacement of Ile by Leu often has little effect; (d) replacement of Lys by Glu often has a large effect.

Denaturation

12.61 What is a native protein?

12.62 What is denaturation?

12.63 Compare protein digestion and denaturation, indicating the difference in the products.

12.64 Proteins with higher cysteine contents are more resistant to denaturation than those with lower cysteine contents. Why?

12.65 Drops of dilute silver nitrate solution placed in the eyes of newborn babies protect against gonorrhea by killing the gonorrhea organism. What mechanism is responsible for the protective effect?

12.66 Which of the following side-group attractive forces are most affected by pH changes? (a) Hydrophobic; (b) salt bridge; (c) hydrogen bond.

12.67 Polypeptides with low contents of acidic and basic residues are less sensitive to pH changes than those with high contents. Why?

12.68 What is the mechanism responsible for sterilization by heat?

Unclassified Exercises

12.69 Draw the structure of L-glutamic acid.

12.70 Place the following compounds in order of increasing melting point: glycine, butylamine, propanoic acid. Explain the order.

12.71 (a) Show the structure of methionine at its isoelectric point, at physiological pH, at very low (<1) pH, and at very high (>12) pH. (b) At what pH will methionine have the lowest solubility in water? its highest solubility in water? (c) To which electrode does methionine migrate in electrophoresis at its pI? at physiological pH? at pH = 1? at pH = 12?

12.72 Show the structure of lysine at physiological pH and indicate to which electrode it migrates.

12.73 How many different constitutional isomers are possible for pentapeptides containing one each of phenylalanine, lysine, threonine, proline, and valine? Give the sequence by using abbreviated names for the different amino acid sequences for those pentapeptides that have valine as the N-terminal residue, proline as the C-terminal residue, and lysine as the middle residue.

12.74 Draw Ser-Val in the trans configuration.

12.75 How many stereoisomers of the type based on tetrahedral carbons are possible for Lys-Gly-Ala? Give abbreviated names. Which is (are) probably present in nature?

12.76 Which would have higher solubility in aqueous solution—a peptide with a high lysine content or one with a high valine content?

12.77 Draw the structure of Met-Ser-Ile-Gly-Glu at physiological pH and give its full name.

12.78 Draw the structure of Pro-Gly-Pro at physiological pH and give its full name.

12.79 Show the structure of Asp-Lys-Lys at physiological pH and indicate to which electrode it migrates in electrophoresis at physiological pH. Is the pI for the peptide on the acidic or basic side of the pI for polypeptides containing only neutral amino acid residues?

12.80 To which electrode will Lys-Gly-Ile move in electrophoresis at physiological pH?

12.81 What is the net charge on Phe-Asp-Leu-Glu-Thr at physiological pH?

12.82 To which electrode will the following peptide migrate at physiological pH?

Phe-Lys-Leu-His-Thr-Glu-Met-Asp-Ser-Glu

12.83 Draw the abbreviated structure(s) of the product(s) formed when the following peptide is subjected to selective reduction.

```
          Gly—Ala—Cys—Gly—Lys
                   |        |
                   S       Phe
                   |        |
                   S       His
                   |        |
Asp—Gly—Met—Cys—His—Glu
```

12.84 The isoelectric point of the protein lysozyme is 11.0. Which of the following α-amino acid residues are present in larger amounts: Asp, Glu, Gly, Lys, Ser, Trp, Val? Explain.

12.85 The isoelectric point of albumin is 4.9. (a) The water solubility of albumin will be minimum at which of the following pH values: <4.9, 4.9, or >4.9? (b) To which electrode will albumin migrate in electrophoresis at physiological pH?

12.86 An unknown protein sample is either albumin ($pI = 4.9$) or lysozyme ($pI = 11.0$). In electrophoresis at physiological pH, migration of protein is only to the cathode. Identify the unknown.

12.87 An unknown protein sample is either pepsin ($pI = 1.0$) or human growth hormone ($pI = 6.9$). In electrophoresis at physiological pH, migration of protein is only to the anode. Identify the unknown.

12.88 The primary structures of the α- and β-polypeptides of hemoglobin are very similar to that of the polypeptide of myoglobin except that several hydrophilic residues on the exterior surface of myoglobin are replaced by hydrophobic residues on the surfaces of the α- and β-polypeptides of hemoglobin. Reconcile this difference with the fact that a hemoglobin has four polypeptides and myoglobin has only one.

12.89 Which proteins have more polar residues on their surfaces—fibrous or globular proteins? Why?

12.90 Parts A and B of a polypeptide have different amino acid compositions. Part A contains mostly Gly, Ala, and Ser; part B contains only small amounts of Pro, Gly, Ala, Ser, Glu, Asp, Lys, Arg, and His. The secondary structure of one part is α-helix, and the other is β-pleated sheet. Which part is α-helix and which β-pleated sheet?

12.91 Part X of a polypeptide contains considerable amounts of Pro and Glu and very little Gly, Ala, and Ser. What is the secondary structure of part X? Explain.

12.92 Which of the following functions are performed by globular proteins and which by fibrous proteins? Catalytic; structural; transport; regulatory; contractile; protective.

12.93 Explain each of the following observations regarding the effect of mutations on function: (a) replacement of Gly by Trp often has a large effect; (b) replacement of Arg by Lys often has little effect; (c) replacement of Thr by Ser often has little effect; (d) replacement of Arg by Asp often has a large effect.

12.94 What mechanism is responsible for the denaturation of a protein by each of the following changes: (a) decrease pH to 2.0; (b) microwave energy; (c) reducing agent; (d) mercury compounds.

Expand Your Knowledge

Note: These icons denote exercises based on material in boxes.

12.95 The following amino acid residues are found in certain proteins: residues 1 and 2 in collagen and residue 3 in prothrombin (a protein in blood-clot formation). Each of these residues is formed by modification of one of the 20 amino acids after a peptide has been synthesized. Which amino acid is modified to produce each residue?

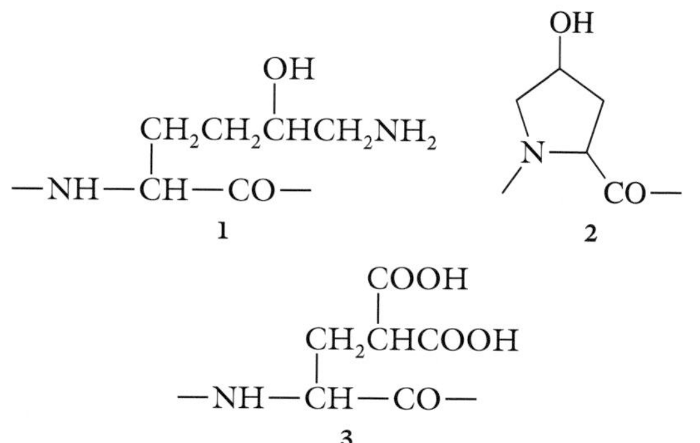

12.96 Glutathione (γ-glutamylcysteinylglycine) functions as a scavenger for oxidizing agents in red blood cells and other cells. It is an atypical peptide in that the γ-carboxyl group, not the α-carboxyl group, of glutamic acid is used in forming the peptide bond with the α-amino group of cysteine. Draw the structure of glutathione at physiological pH.

12.97 Consider an integral protein that spans the lipid bilayer of a cell membrane and protrudes on both the extracellular and the intracellular sides. Of the α-amino acid residues Glu, Leu, Lys, Phe, Ser, and Val, predict which are in contact with (a) the interior of the bilayer, (b) the extracellular environment, and (c) the intracellular environment. Explain.

12.2 **12.98** Describe the chemical reactions taking place in the permanent waving of hair (see Box 12.2).

12.3 **12.99** The defect in sickle-cell hemoglobin is caused by the replacement of only one α-amino acid residue: the glutamic acid residue at position 6 of the β-polypeptide chains is replaced by valine. On the other hand, beef insulin is useful for treating diabetics, even though it differs from human insulin in three residues (see Box 12.3). Explain the difference.

12.100 α-Keratins are the structural components of hair and nails. Hair splits most easily along its longest dimension, whereas fingernails tend to split across the finger direction not along the finger direction. What differences most likely exist in the fibrous structures of hair and nails?

12.101 The lipids and proteins of cell membranes have lateral mobility. Explain why the lateral motion of proteins is much slower than that of lipids (Section 11.10).

12.1 **12.102** What dietary problem results when a combination of rice and corn is used to supply the proteins in a diet (see Box 12.1)?

12.1 **12.103** One measure of the economic advancement of a society is the amount of animal protein consumed relative to vegetable protein (see Box 12.1). Explain.

12.104 How can electrophoresis be used to distinguish normal hemoglobin from sickle-cell hemoglobin?

12.2 **12.105** Helen White was born with straight hair, but likes curly hair. She is able to transform her straight hair into curly hair by permanent waving (see Box 12.2). Why is it necessary to repeat the process periodically?

12.106 A complete turn of a typical α-helix contains 3.6 amino acid residues and is 5.4 Å in length. How many amino acid residues are contained in a 60-Å-long α-helical segment of a polypeptide?

12.107 Each amino acid residue in a typical β-pleated sheet has a length of 3.5 Å. What is the length of a β-pleated-sheet segment that contains 20 residues?

12.3 **12.108** Compare the causes and treatments for types I and II diabetes mellitus (see Box 12.3).

12.3 **12.109** Diabetics take replacement insulin by injection instead of by mouth (see Box 12.3). Why?

12.110 The n! rule applies only when all amino acid residues in a peptide are different. When some of the amino acid residues are the same in a peptide, the number of peptide constitutional isomers is less than n factorial. Show that the number is less by comparing the number of peptide constitutional isomers for tripeptide A, which contains one each of Thr, Phe, and Gly, with the number of peptide constitutional isomers for tripeptide B, which contains two Thr residues and one Phe residue.

12.111 Pepsin, albumin, and lysozyme have pI values of 1.0, 4.9, and 11.1, respectively. Compare the amino acid compositions of the three proteins with respect to their relative amounts of acidic and basic amino acid residues.

12.112 Partial hydrolysis of a peptide is one technique used to determine the amino acid sequence in a peptide. Consider the two tripeptides Ala-Glu-Phe and Ala-Phe-Glu. Describe the difference in the dipeptides formed by partial hydrolysis of these two tripeptides.

12.113 The pI of albumin is 4.9. (a) At what pH will there be no migration of albumin in electrophoresis? (b) At what pH will albumin have a negative charge? (c) At what pH will all acidic and basic groups be uncharged?

12.114 Why are globular proteins with higher cysteine contents more resistant than globular proteins with lower cysteine contents against denaturation by heating?

12.115 The difference in hardness of the α-keratins that make up tortoise shells and human skin is due to the difference in the cysteine contents of the α-keratins. Which has the higher cysteine content?

9.1 **12.116** Explain why all L-amino acids except L-cysteine are (S)-amino acids (see Box 9.1), but L-cysteine is (R)-cysteine.

CHAPTER 13 NUCLEIC ACIDS

(Charles Gupton/Stock Boston.)

Chemistry in Your Future

You are a genetic counselor talking to a young married couple. The wife is pregnant, and you have the DNA results from the amniocentesis that was performed on her fetus. The couple are relieved to learn that their baby will be born free of sickle-cell anemia. Sickle-cell anemia develops when both parents pass to the fetus a mutated gene for the β-polypeptide chain of hemoglobin. The mutation is a point mutation at one of the 438 nucleotide bases of the gene. The messenger RNA produced from the gene for the β-polypeptide chain has a total of 146 codons, and the point mutation results in the wrong base triplet at codon 6. The β-polypeptide produced from the mutated gene has the wrong α-amino acid at position 6. What is a point mutation? What is messenger RNA? What is a codon? How does the mutation result in the synthesis of the wrong β-polypeptide? And why does a mutation at only one codon have such a large effect on a child's health? The answers are in this chapter.

For more information about this topic and others in the chapter, go to www.whfreeman.com/bleiodian2e

Learning Objectives

- Describe and draw the structures of nucleotides.
- Describe and draw the structures of nucleic acids, and write equations for their synthesis from nucleotides.
- Describe the three-dimensional structures of DNA and RNA.
- Describe the replication of DNA.
- Describe the formation of rRNA, tRNA, and mRNA.
- Describe the genetic code for polypeptide synthesis.
- Describe the synthesis of polypeptides.
- Describe mutations and how they affect living organisms.
- Describe how antibiotics fight infections.
- Describe viruses and vaccines.
- Describe recombinant DNA technology and some of its uses.

We learned in Chapter 12 that each biological species is different from every other because of structural differences in their proteins. Humans, elephants, salmon, shrimp, string beans, and roses each have a different (though not necessarily an entirely different) set of proteins. Proteins also differ somewhat between individual members of the same species, but the differences are much smaller than the differences that distinguish species. The proteins of the offspring of the same parents are much more alike than those of unrelated members of the same species. Thus, Mary Kelley and her brother Tom have many more traits in common with each other and with one or the other of their parents than they do with their friend Jeffrey Rogers and his sister Gloria.

Chromosomes, located in the nuclei of cells, contain the hereditary information that directs the synthesis of the approximately 100,000 proteins unique to a human being. The sum of all human proteins is called the **human proteome.** All human cells except germ cells (sperm and ova) are **diploid,** which means that they contain two copies of each chromosome: a total of 23 pairs of chromosomes. Germ cells are **haploid** cells, which means that they contain only one copy of each chromosome.

Every chromosome contains large numbers of **genes,** the fundamental units of heredity. Genes are responsible both for the traits common to a species and for the unique traits of an individual member of that species. Each gene carries the information for synthesizing one or more polypeptides, which are responsible for those hereditary traits.

At the molecular level, the growth and reproduction of organisms are directed and carried out by two types of **nucleic acids—ribonucleic acids (RNAs)** and **deoxyribonucleic acids (DNAs).** Chromosomes contain DNA molecules. Each gene is just a part of a DNA molecule. DNA contains the hereditary information and directs its own reproduction and the synthesis of RNA. RNA molecules leave the cell nucleus and direct the synthesis of proteins in ribosomes, which are organelles in the cytosol. The cytosol is the region of the cell outside of all the organelles in the cell (see Figure 14.1).

13.1 NUCLEOTIDES

Nucleic acids are polymers, as are the carbohydrates and polypeptides that you studied in Chapters 10 and 12. Their building blocks (monomers) are called **nucleotides.** Unlike the building blocks for carbohydrates and polypeptides, however, nucleotides are themselves hydrolyzable into three components—phosphoric acid, a pentose sugar, and a heterocyclic nitrogen base (Figures 13.1

The DNA of all species is based on the same four nucleotides. The differences between this man and the rhinoceroses reside in the differences in the nucleotide sequences of their DNA. (David Weintraub/Photo Researchers.)

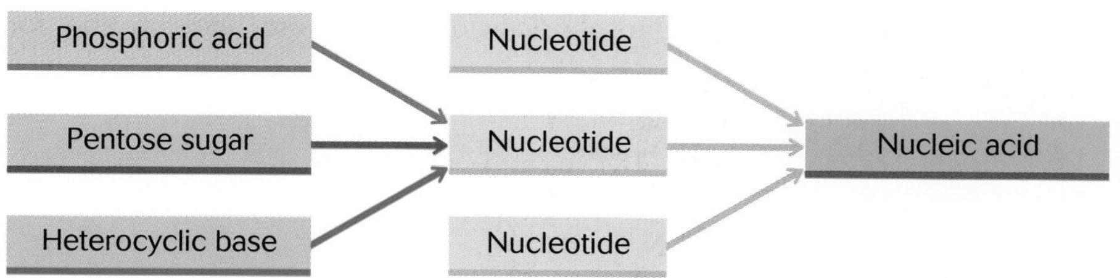

Figure 13.1 Relation of components of nucleotides to nucleic acids.

and 13.2). The pentose sugar found in RNA is D-ribose. The pentose sugar in DNA is 2-deoxy-D-ribose. The sugars are usually referred to simply as ribose and deoxyribose.

« The —OH group at C2 in D-ribose is replaced by —H in 2-deoxy-D-ribose, as described in Section 10.2.

RNA and DNA also differ in the composition of their heterocyclic nitrogen bases. A total of five different heterocyclic bases are used: three pyrimidine and two purine bases. A pyrimidine base contains one ring, whereas a purine base has a fused ring structure. Each of the five bases is symbolized by the first letter of its name—C, T, U, A, and G for cytosine, thymine, uracil, adenine, and guanine, respectively. A, G, and C are used in synthesizing both DNA and RNA nucleotides. T is used only for DNA, whereas U is used only for RNA.

Nucleotide synthesis is a complex process requiring metabolic energy, but the end result is easily understood by visualizing two dehydrations among the three components: one dehydration between phosphoric acid and the pentose sugar and the other between the pentose sugar and the heterocyclic base. The

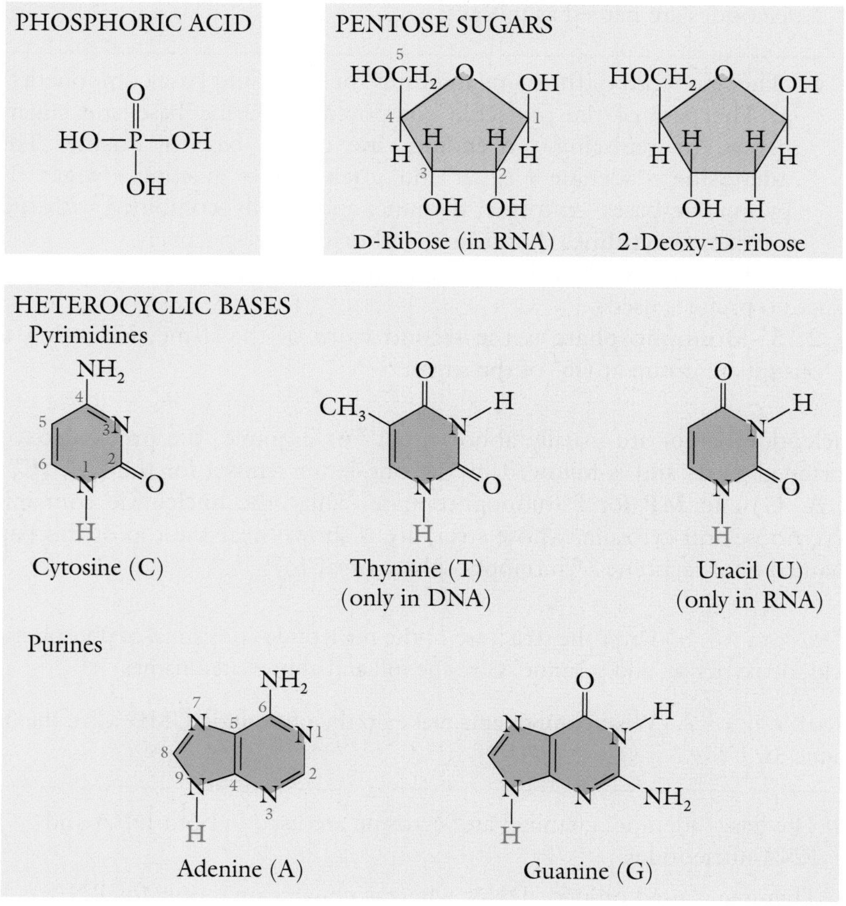

PHOSPHORIC ACID

PENTOSE SUGARS

D-Ribose (in RNA)

2-Deoxy-D-ribose

HETEROCYCLIC BASES

Pyrimidines

Cytosine (C)

Thymine (T) (only in DNA)

Uracil (U) (only in RNA)

Purines

Adenine (A)

Guanine (G)

« The heterocyclic bases are heterocyclic compounds with a basic nitrogen atom in the ring, as described in Section 8.3.

Figure 13.2 Individual components of nucleotides. The blue hydrogen is lost when a heterocyclic base is bonded to a sugar.

synthesis of a nucleotide from phosphoric acid, deoxyribose, and cytosine, for example, proceeds as follows:

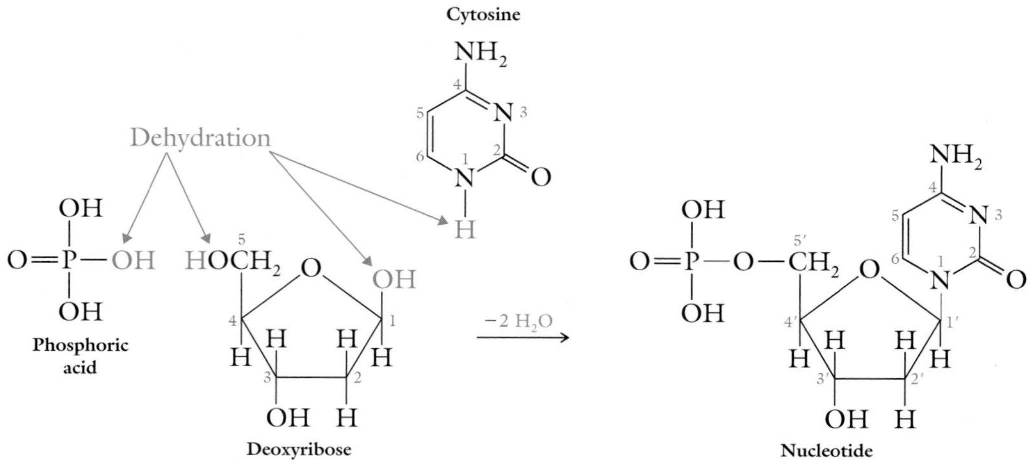

Cytosine

Dehydration

Phosphoric acid

Deoxyribose

$-2\ H_2O$

Nucleotide

Dehydration between phosphoric acid and deoxyribose takes place at C5 of the pentose ring. Dehydration between deoxyribose and cytosine takes place at C1 of the pentose ring and at an N—H hydrogen of the heterocyclic base. In pyrimidine bases, it is the hydrogen at position 1. In purine bases, it is the hydrogen at position 9. Note the use of primed numbers in depictions of the structures of nucleotides; primed numbers distinguish ring positions in the sugar from ring positions in the heterocyclic base.

Nucleotides derived from ribose and deoxyribose are called **ribonucleotides** and **deoxyribonucleotides,** respectively.

Nucleotides are named as follows:

Rules for naming nucleotides

1. The first word of the name indicates the sugar and base components.
   a. The part of the molecule composed of purine base and sugar is named by replacing the ending **-ine** of the base by **-osine.** Thus, **adenosine** = adenine + sugar and **guanosine** = guanine + sugar. The pyrimidine bases—cytosine, thymine, and uracil—combined with sugar are named **cytidine, thymidine,** and **uridine,** respectively.
   b. When deoxyribose is the sugar, the prefix **deoxy-** is used. Otherwise no prefix is used.
2. **5′-Monophosphate** is the second word of the name, indicating the phosphate group at C5′ of the sugar.

Nucleotide names are usually abbreviated. For example, the prefix deoxy- is shortened to **d-** and is followed by the one-letter symbol for the base **(C, T, U, A, G)** and **MP** for 5′-monophosphate. Thus, the nucleotide containing deoxyribose and cytosine, whose structure is shown near the top of this page, is named deoxycytidine 5′-monophosphate or dCMP.

Problem 13.1 Draw the structure of the nucleotide consisting of phosphoric acid, deoxyribose, and guanine. Give the full and abbreviated names.

Problem 13.2 What components make up the nucleotide UMP? Give the full name for UMP.

Concept checklist

✓ The bases adenine, guanine, and cytosine are used in both DNA and RNA nucleotides.

✓ Thymine is used only for DNA, whereas uracil is used only for RNA.

13.2 NUCLEIC ACID FORMATION FROM NUCLEOTIDES

As noted earlier, nucleic acids are polynucleotides. RNA is formed from ribonucleotides, and DNA is formed from deoxyribonucleotides. Like that of nucleotides, the synthesis of nucleic acids is a complex process. The products of nucleic acid synthesis are those that would be produced by a simple dehydration reaction between the —OH of the phosphate group at C5′ of one nucleotide molecule and the —OH at C3′ of another nucleotide molecule. The group formed by this reaction, called a **phosphodiester group,** joins one nucleotide residue to another. For example, the dinucleotide UMP-CMP is produced by the reaction of UMP with CMP through dehydration between the —OH group at C3′ of UMP and the phosphate group at C5′ of CMP:

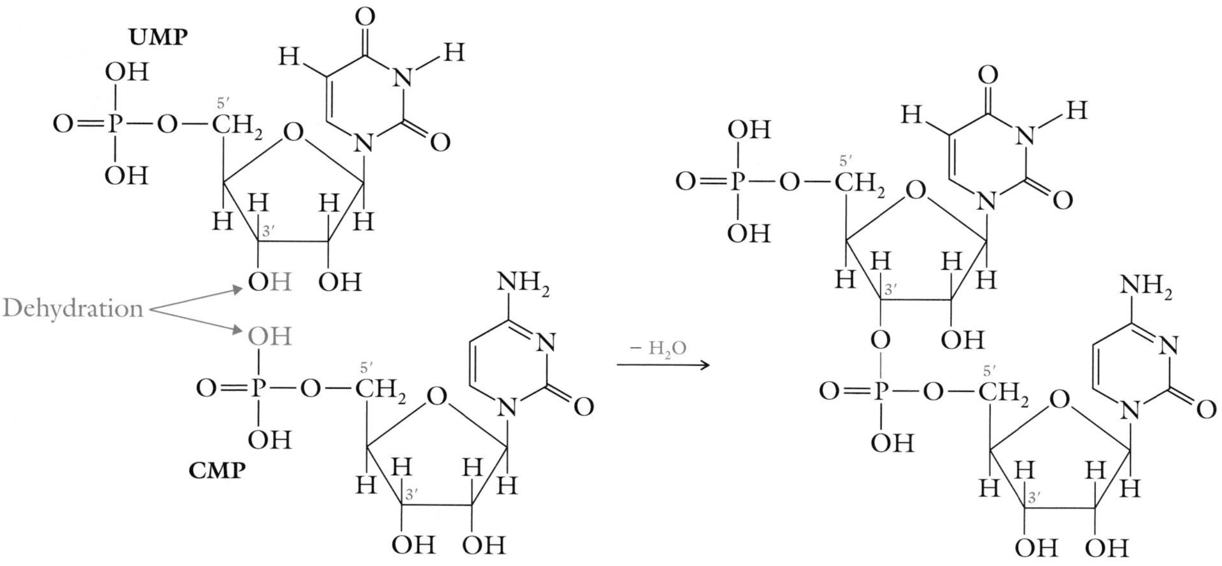

The formation of a three-nucleotide sequence, dCMP-dAMP-dTMP, is shown in Figure 13.3 on the following page. By convention, nucleotide sequences are named in the 5′ ⟶ 3′ direction. A nucleic acid has one 5′-end and one 3′-end. The **5′-end** has a phosphate group at C5′ that is attached to only one pentose ring. All other phosphate groups are attached to two different pentose rings. The **3′-end** has a pentose ring with an unreacted —OH group at C3′. The nucleotide residues in a nucleic acid are named by proceeding from the 5′-end to the 3′-end.

When the ends of the nucleic acid are not given, an alternate method for describing the 5′ ⟶ 3′ convention is used: Proceed from the 5′-carbon of a pentose ring to the 3′-carbon of the same pentose ring and then move to succeeding pentose rings. Do not proceed from the 5′-carbon of one pentose ring to the 3′-carbon of the adjacent pentose ring.

The importance attached to the 5′ ⟶ 3′ convention stems from the fact that the three-nucleotide sequence dCMP-dAMP-dTMP is only one of a total of six constitutional isomers—corresponding to six different ways that three nucleotides can be joined together. The other five constitutional isomers are:

dTMP-dAMP-dCMP
dCMP-dTMP-dAMP
dAMP-dTMP-dCMP
dAMP-dCMP-dTMP
dTMP-dCMP-dAMP

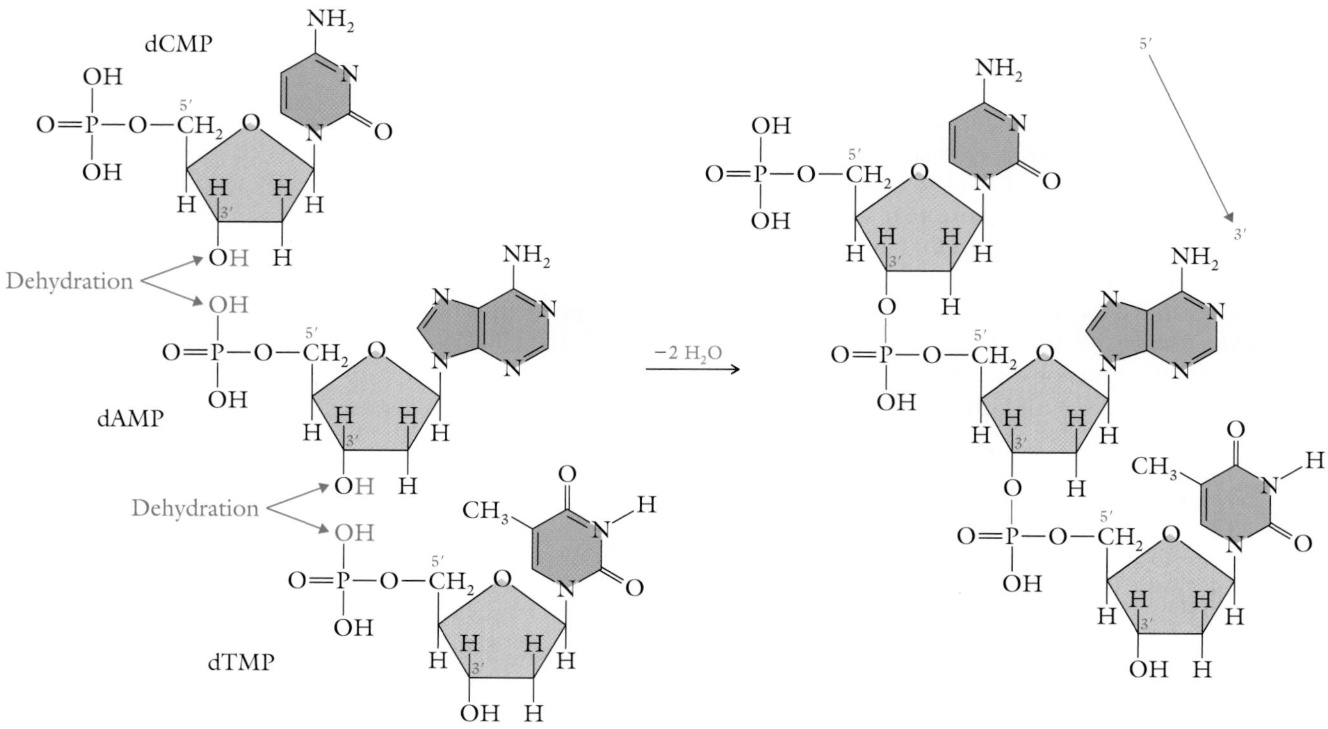

Figure 13.3 Synthesis of dCMP-dAMP-dTMP.

The nucleotide sequence in a specific nucleic acid is critical to its function. We will see in Section 13.7 that the nucleotide sequence in DNA is what determines the sequence of α-amino acid residues in a polypeptide.

Drawing the nucleic acid structures for sequences of more than two or three nucleotide residues becomes cumbersome. Figure 13.4 shows various abbreviated representations of a tetranucleotide sequence. Many structural details are left out of these representations so that, for example, instead of looking at the sugar to find out if a sequence represents DNA or RNA, we must look at other available information, such as whether T or U is present. By convention, the $5' \longrightarrow 3'$ direction proceeds either from top to bottom or from left to right. For sequences of more than a dozen nucleotide residues, the

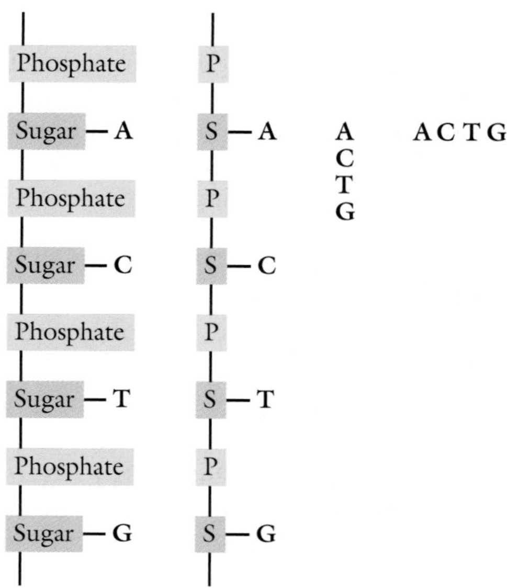

Figure 13.4 Different representations of deoxyribonucleic acid containing the nucleotide sequence dAMP-dCMP-dTMP-dGMP.

use of any representation other than the one on the far right in Figure 13.4, written left to right instead of top to bottom, is inconvenient.

All of these representations emphasize the key feature of nucleic acids: the sequence of bases present as side groups on a sugar–phosphate polymer chain. Note the analogy to polypeptides, where there is a sequence of different side groups on a polyamide chain.

✓ Nucleic acids are named in the $5' \longrightarrow 3'$ direction.

Concept check

Nucleotides and nucleic acids are acidic because of the presence of P–OH groups. Although the structures of nucleotides and nucleic acids are shown with the P–OH in nonionized form, it should be understood that they are ionized to $P{-}O^-$ at physiological pH.

Problem 13.3 Draw the complete structure of GMP-UMP. Indicate the $5' \longrightarrow 3'$ direction.

13.3 THE THREE-DIMENSIONAL STRUCTURE OF NUCLEIC ACIDS

The primary structure of a nucleic acid—the sequence of bases along the sugar–phosphate polymer chain—determines its higher-level conformational structure (secondary, tertiary, quaternary). The situation is analogous to that of proteins except that there are only a few different types of three-dimensional structures for nucleic acids—corresponding to their limited (but enormously critical) range of functions.

Deoxyribonucleic Acids

The three-dimensional structure of DNA was deduced by James Watson and Francis Crick in 1953 on the basis of two key pieces of data. First, before 1953, Erwin Chargoff showed that, although the base compositions of the DNA in different species of plants and animals varies considerably, the molar amounts of A and T are equal and the molar amounts of G and C are equal for each species. Second, Maurice Wilkins and Rosalind Franklin obtained the X-ray diffraction structure of DNA in 1953. The 1962 Nobel Prize in medicine or physiology was awarded to Crick, Watson, and Wilkins for their work in elucidating the structure of DNA. Franklin deserved to be included in the award but lost out on two counts—the Nobel Prize regulations limit the number of co-awardees to three persons and, more important in this instance, the prize cannot be awarded posthumously (Franklin died in 1958).

The Watson–Crick model for DNA is the **DNA double helix.** DNA molecules do not exist as individual polynucleotide molecules. Instead, two paired DNA molecules (often called **DNA strands**) fold or coil around each other to form a right-handed double helix **(duplex),** as shown in Figure 13.5 on the next page. The two DNA strands of the DNA double helix run in opposite directions, one in the $5' \longrightarrow 3'$ direction and the other in the $3' \longrightarrow 5'$ direction. The phosphate groups are located on the outer surface of the double helix, and the heterocyclic bases are inside the double helix. This interior placement of heterocyclic bases is critical to the formation of the double helix.

The orientation of the heterocyclic base on each nucleotide residue is perpendicular to the axis of the double helix; that is, the bases are in the horizontal plane when the double helix is viewed in the vertical plane. Bases stack on top of one another in the vertical plane, like the steps in a spiral staircase.

Two types of attractive secondary forces stabilize the DNA double helix—hydrogen-bonding and hydrophobic attractions. There are strong hydrogen-bonding attractive forces between a base on one DNA strand and a base on the

INSIGHT INTO FUNCTION

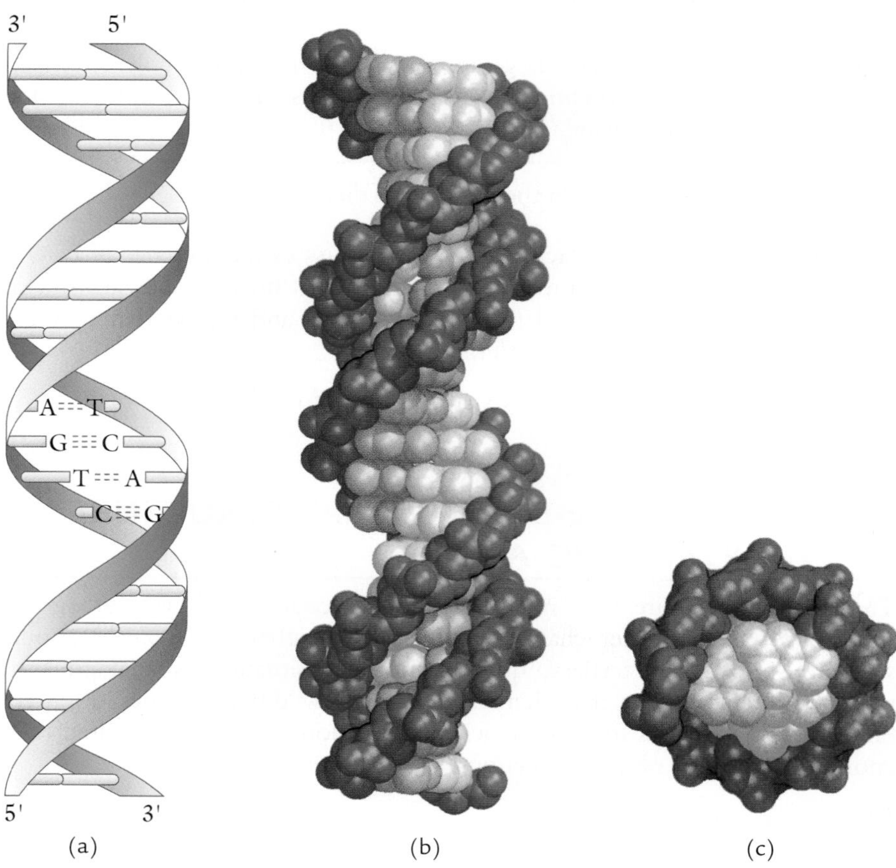

Figure 13.5 DNA double helix: (a) simple representation with hydrogen bonding shown between base pairs; (b and c) axial and radial views of space-filling models. One DNA strand is green and the other is red. The bases are shown in lighter shades of green and red. We will see how this structure provides the basis for biological growth and development, reproduction, and evolution.

other DNA strand. This attraction is called **base-pairing,** and the bases pair in a **complementary** manner—adenine pairs only with thymine, and cytosine pairs only with guanine. Adenine and thymine pairing is accomplished through two hydrogen bonds; cytosine and guanine pairing is accomplished through three hydrogen bonds (Figure 13.6). The two DNA strands in the double helix are **complementary strands** in the sense that each heterocyclic base is located opposite its complementary base (A opposite T; G opposite C). Base-pairing to form a DNA double helix with maximum hydrogen bonding (and maximum stability) requires that the two strands run in opposite directions. If the strands ran in the same direction, the bases would not be in the correct proximity to each other to form the maximum number of hydrogen bonds.

Hydrophobic attractions between bases contribute—perhaps even more than hydrogen bonding—to the stabilization of the DNA duplex. Each heterocyclic base in a base pair has hydrophobic attractive interactions with the bases directly above and below it.

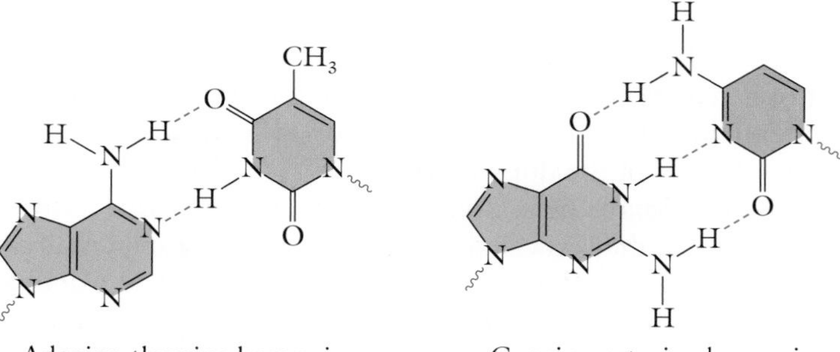

Figure 13.6 Complementary base-pairing in A-T and G-C. A-T and G-C pairings yield the same cross-sectional dimension throughout the DNA double helix.

Concept check

✓ The double helix is stabilized by hydrogen bonding and hydrophobic attractions between heterocyclic bases.

Why is base-pairing only between A and T and between C and G but not between A and C, G and T, C and T, or A and G? As Watson and Crick were the first to realize, the A-T and C-G base pairs have the same dimension (see Figure 13.6), which results in a uniform cross-sectional dimension throughout the entire DNA double helix. Other combinations of bases have different dimensions and would decrease the stability of the DNA duplex by causing variations in the cross-sectional dimension and a consequent decrease in hydrogen bonding and hydrophobic attractions between bases.

Example 13.1 Determining the base sequence in a DNA strand from the sequence in its complementary strand

If the base sequence in a section of one DNA strand is ACGTAG, reading in the $5' \longrightarrow 3'$ direction, what is the sequence, reading in the $5' \longrightarrow 3'$ direction, for the corresponding section of the complementary DNA strand?

Solution

The complementary strand has A opposite T in the given strand, T opposite A, C opposite G, and G opposite C.

5′ ACGTAG 3′ **Given strand**

3′ TGCATC 5′ **Complementary strand**

Reading in the $5' \longrightarrow 3'$ direction, the sequence for the complementary strand is CTACGT.

Instead of writing a base sequence as "ACGTAG, reading in the $5' \longrightarrow 3'$ direction," we can write 5′-ACGTAG-3′ as an abbreviation or just simply ACGTAG because the $5' \longrightarrow 3'$ direction is the convention unless otherwise indicated.

Problem 13.4 If the base sequence in a section of one DNA strand is 5′-GGCTAT-3′, what is the sequence for the corresponding section of the complementary DNA strand?

DNA strands are enormously large molecules, with molecular masses estimated to be from a few billion to as high as 100 billion. Considering that the human cell nucleus has a diameter of about 5 micrometers ($1\ \mu m = 10^{-6}$ m), whereas the total DNA (called the **human genome**) contained in its 23 pairs of chromosomes has 3.2 billion base pairs (and a length of about 2 m if stretched end to end), the DNA duplex must be highly compacted to fit into the nucleus (Figure 13.7 on the following page). This compaction is achieved by the double helix being folded around structures called **nucleosome cores.** Each nucleosome core is composed of two pairs each of four different proteins called **histones.** The DNA duplex wraps around the nucleosome cores to form a chain of **nucleosomes,** with about 150 to 200 base pairs per nucleosome. Large numbers of nucleosome cores are needed to compact each DNA duplex. Chromosomes contain about equal masses of DNA and histones. The chain of nucleosomes is coiled to higher and higher levels to form a highly compact, highly supercoiled **chromatin fiber.**

INSIGHT INTO PROPERTIES

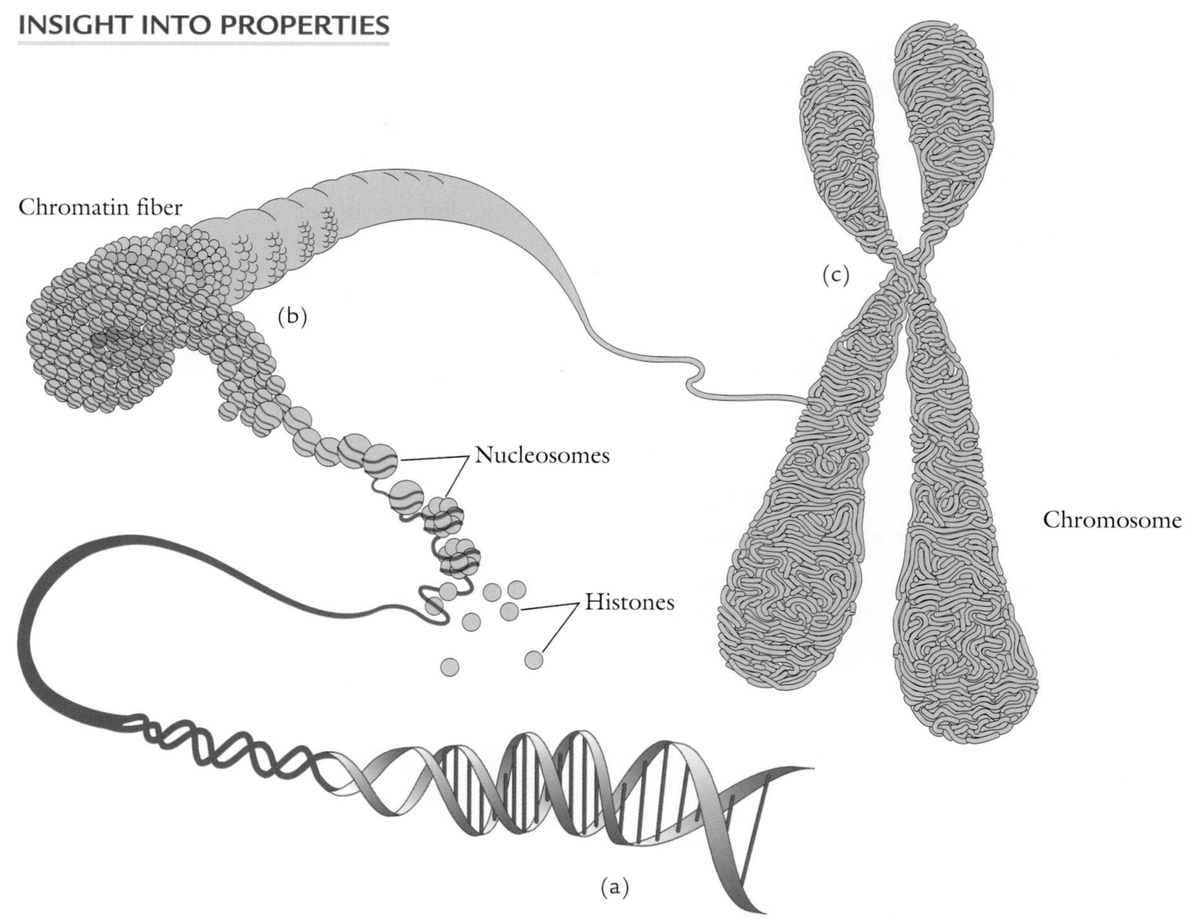

Figure 13.7 Different levels of DNA structure: (a) double helix (hydrogen bonding between base pairs is represented by straight lines between the two DNA strands); (b) highly supercoiled chromatin fiber; (c) chromosome. The compaction and supercoiling illustrated here are necessary to fit the enormously long DNA duplexes into the nuclei of human cells.

Ribonucleic Acids

Ribonulceic acids exist as single-strand molecules, not as double helices as does DNA. The spatial bulk of the —OH group at C2′ of the ribose units in RNA (in place of the —H of deoxyribose in DNA) prevents the formation of long stretches of double-helix conformation. Several different types of RNA molecules exist, each with its own three-dimensional structure. Figure 13.8 shows one type of RNA molecule, called **transfer RNA** or **tRNA.** The overall shape of tRNA resembles a cloverleaf, especially in the two-dimensional representation.

The secondary and tertiary structures of tRNA result from the intramolecular base-pairing of nucleotide residues. The base-pairing scheme is the same as that for DNA except that uracil is used in RNA instead of thymine. Adenine pairs with uracil; cytosine pairs with guanine. Roughly half the nucleotide bases in a tRNA molecule participate in base-pairing. Base-pairing takes place not in one long continuous sequence but in short sequences. Sections of the tRNA molecule with approximately helical conformation are separated by nonhelical sections without base-pairing. The nonhelical sections, except for the linear section at the 3′-end of the tRNA chain, are curved and called **loops.** The linear section at the 3′-end is called the **acceptor stem,** and the loop farthest from it is called the **anticodon.** There is no base-pairing in the loop sections, because complementary bases are not opposite each other or, if opposite each other, are not within hydrogen-bonding distance. The acceptor stem and anticodon play key roles in polypeptide synthesis (Section 13.7).

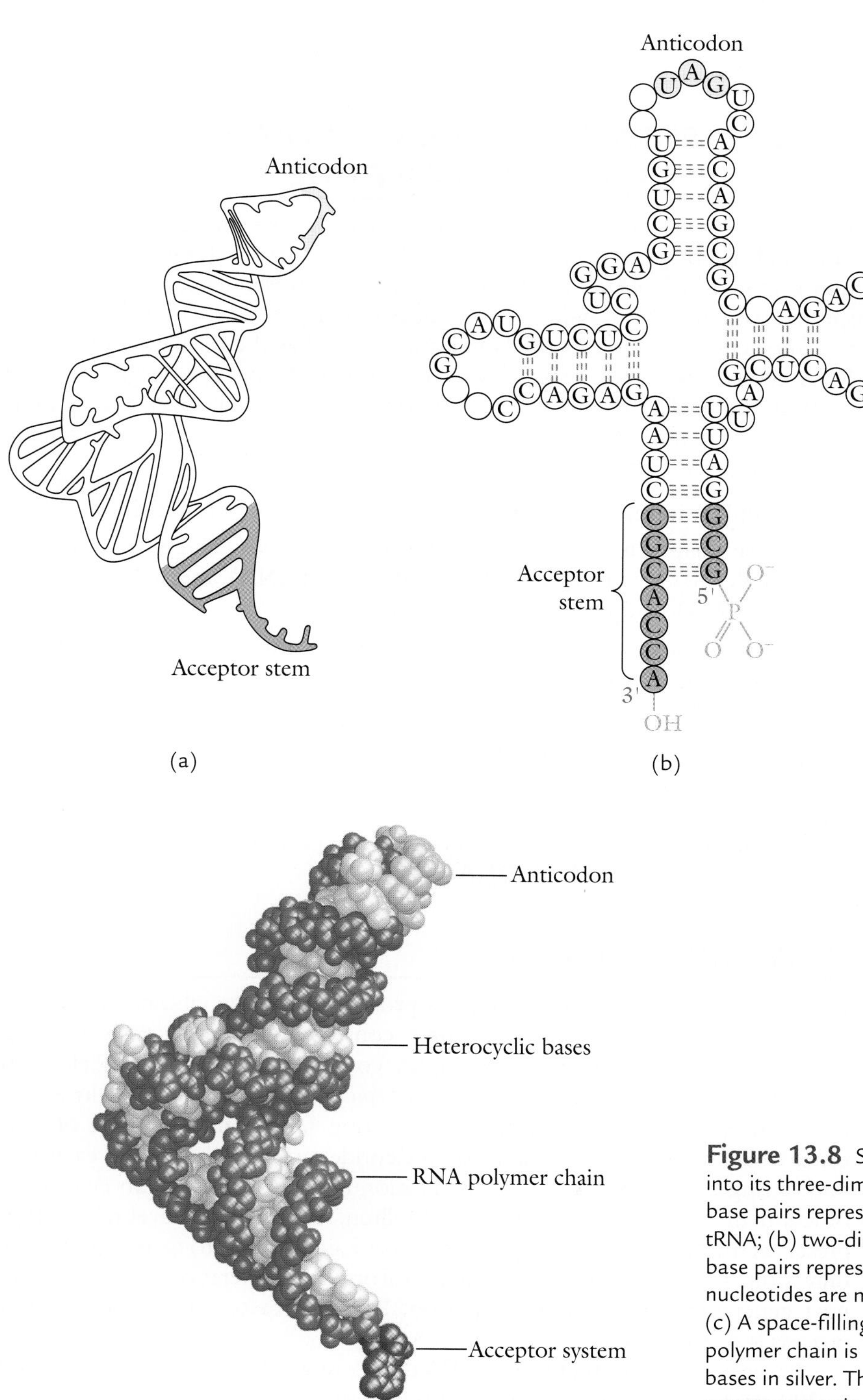

Figure 13.8 Structure of a tRNA molecule: (a) folded into its three-dimensional structure with hydrogen-bonded base pairs represented by straight bars connecting parts of tRNA; (b) two-dimensional structure with hydrogen-bonded base pairs represented by dashed lines. The unspecified nucleotides are modified analogues of A, C, G, or U. (c) A space-filling model of tRNA in which the RNA polymer chain is shown in dark blue and the heterocyclic bases in silver. The anticodon is shown in yellow and the acceptor stem in red. (Part c from *Biochemistry,* 4th ed., by Lubert Stryer. © 1975, 1981, 1988, 1995 by Lubert Stryer.)

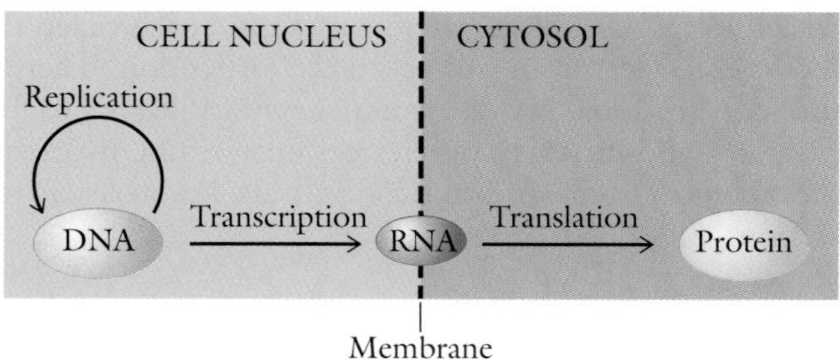

Figure 13.9 Overview of molecular genetics. Inside the nucleus, DNA directs its own replication and the synthesis of different RNAs. The RNAs travel outside the nucleus and carry out protein synthesis in ribosomes.

13.4 INFORMATION FLOW FROM DNA TO RNA TO POLYPEPTIDE

Figure 13.9 gives an overview of molecular genetics—the storage, transmission, and use of genetic information at the molecular level. Genetic information is stored in DNA as a sequence of bases. These sequences correspond to specific sequences of α-amino acid residues in polypeptides. DNA is the "master" molecule, directing its own replication as well as the formation of various RNA molecules. DNA remains in the nucleus, being too large to move through the nuclear membrane into the cytosol. The much smaller RNA molecules go out into the cytosol to carry out protein synthesis. The genetic process consists of three parts:

- **Replication** is the copying of DNA in the course of cell division.
- **Transcription** is the synthesis of RNA from DNA. Three types of RNA molecules are produced: **ribosomal RNA (rRNA), messenger RNA (mRNA),** and **transfer RNA (tRNA).**
- **Translation** is the synthesis of polypeptides through the combined efforts of rRNA, mRNA, and tRNA.

Each sequence of bases in DNA that results in the formation of one or more polypeptides is a **gene.** Overall, about 2% of the nucleotides in DNA result in the formation of polypeptides. These regions of DNA are called **coding regions,** and we often say that their messages (the sequence of bases) are **expressed.** The remaining 98% of nucleotides appear not to have any function and are sometimes called **junk DNA.**

13.5 REPLICATION

Replication is an extremely complex process, requiring about two dozen enzymes. Copying the human genome consists of simultaneously copying 23 pairs of DNA duplexes, each duplex containing an average of close to 70 million base pairs. Cell division of a typical human cell takes a little less than a day, with DNA replication accounting for about one-third of that time, which means that thousands of nucleotides are joined together each second during replication. All of this replication is carried out with an error level of less than 1 wrong nucleotide per 10 billion. This extreme level of accuracy preserves the integrity of the genome from generation to generation—not only from parents to offspring, but also from one generation of cells to the next generation within the same person. It is the reason that the replication process is so complex.

In replication, the strands of a DNA duplex unwind over a short span of 150 to 200 nucleotide residues, forming a **replication bubble** with a **replication fork** at each end (Figure 13.10a). Each strand acts as a **template strand** on which a complementary strand is synthesized. Enzymes catalyze the unwinding process by breaking the hydrogen bonding between base pairs. Other proteins stabilize the unwound DNA strands by hydrogen bonding with the bases. The enzyme **DNA polymerase** catalyzes the addition of nucleotides to growing DNA strands.

The two new strands grow from replication forks at opposite sides of the bubble. They grow in opposite directions because new bases are always added in the $5' \longrightarrow 3'$ direction. As DNA synthesis proceeds, the DNA duplex unwinds further, to elongate the replication bubble. One of the new DNA strands, the **leading strand,** is synthesized continuously. The other DNA strand, the **lagging strand,** is synthesized in a discontinuous manner as a series of smaller DNA fragments, called **Okazaki fragments.** These fragments are later linked together by the enzyme **DNA ligase.**

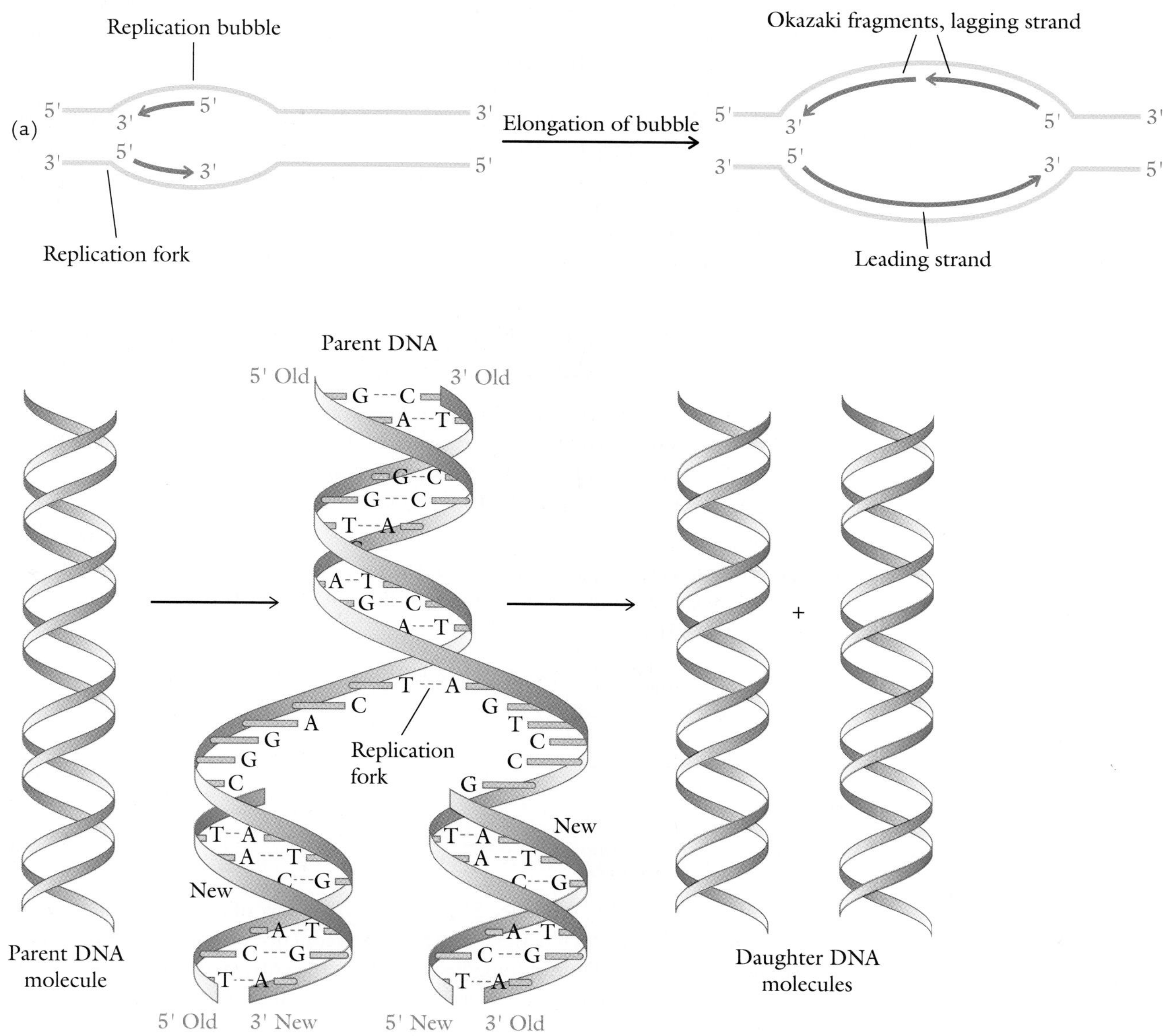

Figure 13.10 Replication of DNA duplex: (a) DNA synthesis at a pair of replication forks proceeds with elongation of the replication bubble (the figure is simplified by not showing the DNA strands in helical form); (b) semiconservative replication (old strands are red, new strands are green).

Replication proceeds simultaneously at thousands of replication bubbles along each DNA strand, completing replication of this enormously large molecule in a reasonable amount of time (about 10 min). The DNA fragments synthesized in different replication bubbles are bonded together by DNA ligase to produce the final new DNA strands.

DNA polymerase, with a molecular mass of one-half million, directs the synthesis process by simultaneous interaction (through secondary forces) with the template DNA strand, the new growing DNA strand, and nucleotides. The pairing of bases and the action of DNA polymerase together ensure the accuracy of replication. Thus, if a template strand has T at a certain position, only a dAMP nucleotide will hydrogen bond and be incorporated into the new strand at that point. After the addition of a nucleotide, DNA polymerase moves to the next nucleotide residue in the template strand and ensures that only the correct, complementary nucleotide will be added to the chain. DNA polymerase not only catalyzes the addition of nucleotides to growing DNA chains but also has a role, along with other enzymes, in proofreading and correcting errors, thus ensuring that replication is carried out with an error level of less than 1 wrong nucleotide per 10 billion.

Through replication, a single **parent** DNA duplex is transformed into two **daughter** DNA duplexes, each containing one old and one new DNA strand. Replication is therefore described as **semiconservative:** one parental strand is conserved intact within each daughter DNA duplex. The new DNA strand of a daughter DNA duplex is identical with the old DNA strand of the other daughter duplex (Figure 13.10b).

Example 13.2 Determining the base sequences in parent and daughter DNA strands

The sequence 5′-GCGTAA-3′ is part of one strand of a parent DNA double helix. What is the corresponding sequence on the complementary strand of the parent DNA double helix? What is the corresponding sequence on the new daughter strand made from the parent strand during replication?

Solution

The complementary parent strand and the new daughter strand are identical (5′-TTACGC-3′) and complementary to the 5′-GCGTAA-3′ sequence of the parent DNA strand:

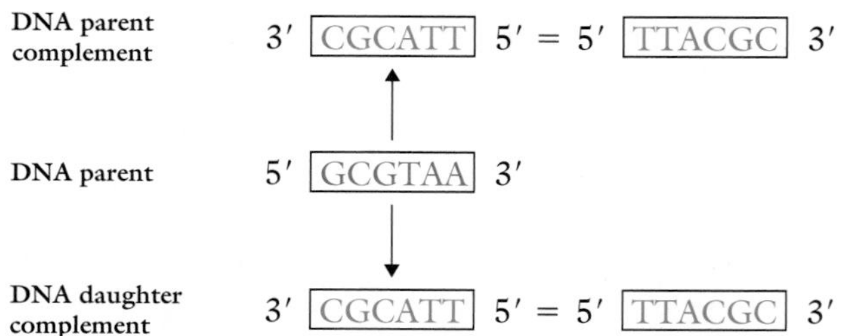

Problem 13.5 The sequence 5′-ACGTGC-3′ is part of one strand of a parent DNA double helix. What is the corresponding sequence on the complementary strand of the parent DNA double helix? What is the corresponding sequence on the new daughter strand made from this parent strand during replication?

13.6 TRANSCRIPTION

Transcription, the synthesis of rRNA, tRNA, and mRNA with the use of information from DNA, proceeds similarly to replication (Figure 13.11). The DNA double helix unwinds to form a **transcription bubble.** Only one of

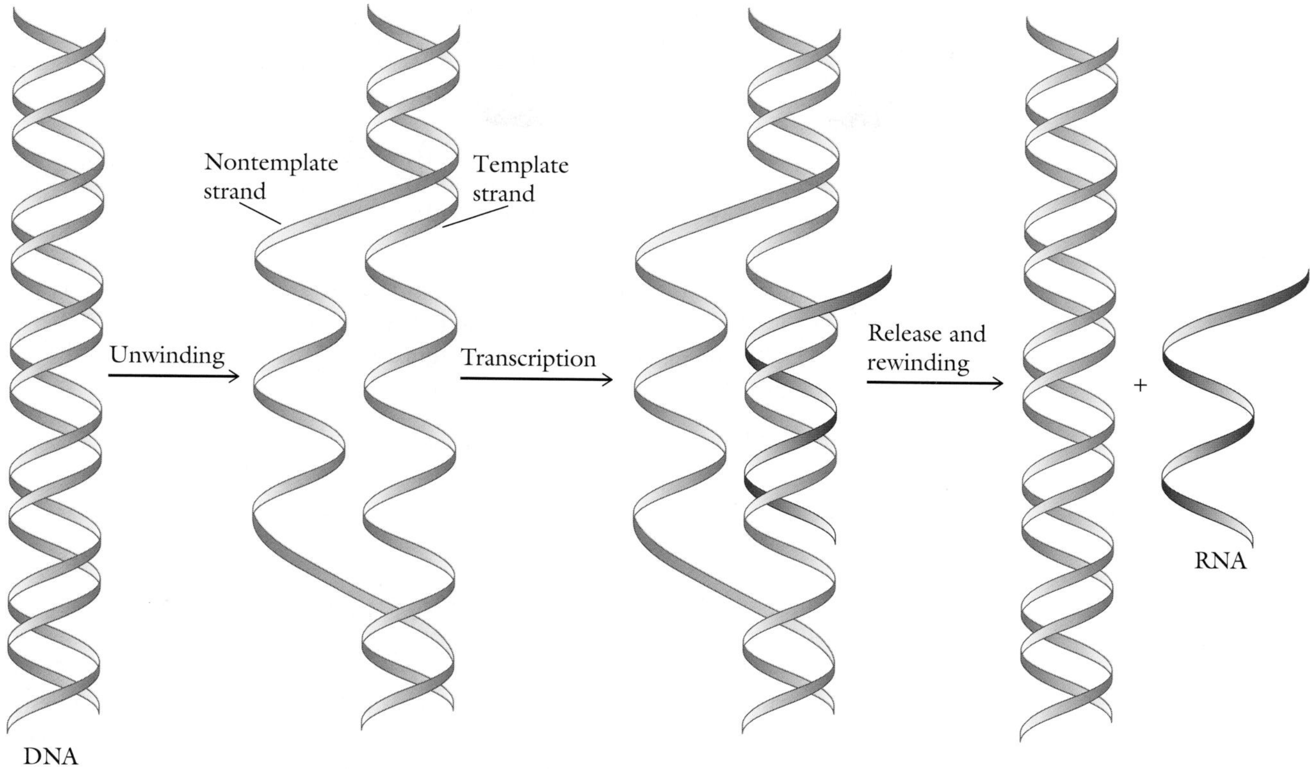

Figure 13.11 Transcription of DNA to form RNA. The template DNA strand is unwound from the nontemplate strand and used to synthesize RNA.

the two DNA strands in the bubble, called the **template strand,** is used as the template for RNA synthesis. The other DNA strand, called the **nontemplate strand,** is not used for RNA synthesis.

Transcription proceeds under the control of **RNA polymerase,** which catalyzes the joining together of nucleotides. Synthesis begins at an **initiation site** whose base sequence is recognized by RNA polymerase as a start signal. Synthesis stops when RNA polymerase encounters a **termination site** whose base sequence is recognized as a stop signal. The RNA polymerase and the newly synthesized RNA are released, and the template and nontemplate DNA strands are rewound into the DNA duplex.

Unlike DNA polymerase, RNA polymerase has no proofreading function, with the result that the error level in RNA synthesis, less than 1 base in 10,000 to 100,000 bases, is not nearly as low as that in DNA synthesis. The higher error level in RNA synthesis compared with that in DNA synthesis is not a serious problem, because RNA synthesis does not have a role in preserving the integrity of the genome from generation to generation.

As in replication, base-pairing determines the base sequence of RNA synthesized from a DNA template. A, G, C, and T in the template strand of DNA result in the incorporation of U, C, G, and A, respectively, in the RNA strand. As mentioned earlier, uracil is used in place of thymine in RNA. Thus, an A in DNA has U as its complementary base in RNA.

Concept check

✓ The RNA synthesized from a DNA template strand has the same base sequence as that of the DNA nontemplate strand (except that RNA has U in place of T).

Example 13.3 Determining the base sequences in DNA nontemplate and RNA strands from the sequence in the DNA template strand

For the sequence 5′-ACATGC-3′ in a DNA template strand, what is the base sequence in the DNA nontemplate strand? What is the base sequence in the RNA synthesized from the DNA template strand?

Solution

The sequences in the DNA nontemplate and RNA strands are complementary to the sequence in the DNA template strand. The base sequence is 5′-GCATGT-3′ in the DNA nontemplate strand, and its equivalent, 5′-GCAUGU-3′, is in the RNA strand:

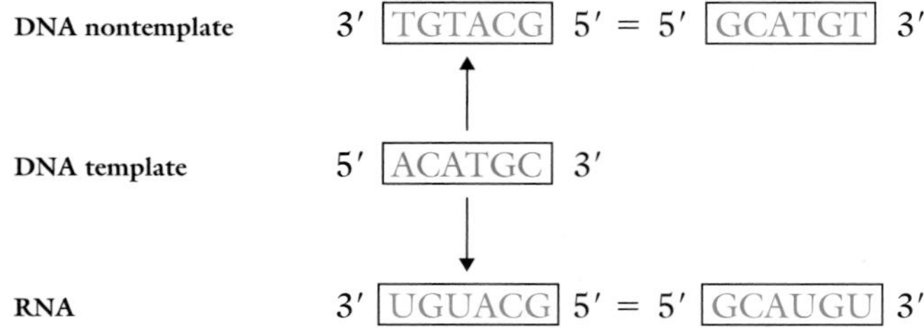

Note that the direction of the RNA strand, like that of the DNA nontemplate strand, is opposite that of the DNA template strand. However, the convention for reading and writing base sequences of all nucleic acids is in the 5′ ⟶ 3′ direction unless otherwise stated.

Problem 13.6 For the sequence 5′-TAAGTCAAC-3′ in a DNA template strand, what is the base sequence for the DNA nontemplate strand? What is the base sequence for the synthesized RNA?

A ribonucleic acid is called a **primary transcript RNA (ptRNA)** when initially formed. Primary transcript RNA is modified by **posttranscriptional processing** to produce messenger, ribosomal, and transfer ribonucleic acids. Some of this processing takes place in the cell nucleus and some in the cytosol. The purpose of posttranscriptional processing is to stabilize and optimize a specific RNA molecule for its function in polypeptide synthesis. **End capping** adds certain nucleotide units to the 5′- and 3′-ends of the ribonucleic acid. **Base modification** alters certain bases. **Splicing** deletes certain parts of the ribonucleic acid. The parts deleted are called **introns.** The parts retained are called **exons.** Figure 13.12 shows the splicing process for the conversion of a primary transcript into mRNA. The introns have no well-established physiological functions. The exons are bonded together after excision of the introns to form the mRNA. The base sequence in these exons is what defines the amino acid sequence in the polypeptide whose synthesis they will subsequently direct.

The diversity in the human proteome is five times as great as the diversity in the human genome. The number of genes in the human genome is close to 21,000 (Section 13.11), but the number of proteins in the human proteome is estimated at 100,000. Thus the average gene is responsible for producing several proteins. Posttranscriptional processing is partly responsible for this phenomenon. (Posttranslational processing also is responsible—Section 13.7.) Different mRNAs can be made from the same gene by alternate splicing mechanisms. The same ptRNA is spliced differently in different cells and tissues to produce different mRNAs, which then produce different polypeptides. For example, there are five different forms of the contractile protein tropomyosin

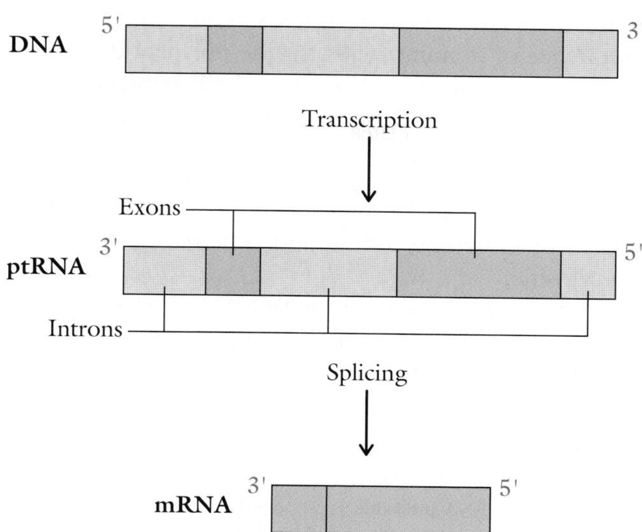

Figure 13.12 Posttranscriptional processing of ptRNA into mRNA by splicing. Introns are gray; exons are orange.

in these tissues: skeleton muscle, smooth muscle, fibroblast, liver, and brain. The ptRNA is the same for all five proteins, but the excision of different combinations of one or more exons produces five different mRNAs. Each mRNA produces a different form of tropomyosin.

The various RNAs are much smaller than DNA. Most rRNAs have molecular masses ranging from 0.5 million to 1 million. mRNAs have molecular masses ranging from 100,000 to 1 or 2 million. tRNAs are much smaller, with molecular masses below 50,000.

Every step in the expression of the genetic message, from transcription through translation, requires the coordinated effort of large numbers of enzymes. Furthermore, superimposed on the overall process are higher-level, regulatory mechanisms that determine which of all possible genetic events take place and when they take place within each individual cell. This level of control is necessary because an organism's entire genome—the complete blueprint for the production of all the proteins required by all cells—is found in every cell of the body. For example, heart cells contain the DNA not only for heart polypeptides and proteins, but also for bone, liver, skin, and other cells. Therefore, regulatory proteins are present in each cell to prevent the transcription of most of the DNA messages contained in its chromosomes: only a fraction of the coding regions of the DNA in any given cell is actually expressed. Heart cells do not need and must not synthesize the proteins found in liver, skin, and bone cells, and so forth. **Repressor proteins** turn off the synthesis of all proteins except those needed for the particular cell. Simultaneously, **inducer proteins** turn on the synthesis of proteins needed by the cell.

13.7 TRANSLATION

Translation is the process by which rRNA, mRNA, and tRNA work together outside the nucleus to carry out polypeptide synthesis. Polypeptides are produced at structures called **ribosomes** located in the cytosol. Ribosomes have two subunits, one large and one small, and are composed of rRNA (about 60%) and protein (most of which comprises the enzymes required for protein synthesis). mRNA, which carries the genetic message encoding the amino acid sequence of the polypeptide to be synthesized in the ribosome, binds to the ribosome and acts as the template for polypeptide synthesis. α-Amino acids are supplied for polypeptide synthesis when tRNAs transport them into the ribosome.

The **genetic message** carried by mRNA is its sequence of bases, which specifies the sequence of α-amino acids for the polypeptide to be synthesized. Each successive set of three consecutive bases (**base triplet**) in mRNA is called a **codon** because it codes for, or specifies, one specific tRNA that carries one specific α-amino acid. Mutual recognition between a codon of mRNA and its specified tRNA is accomplished through a base triplet, the **anticodon,** on the tRNA molecule (see Figure 13.8). By convention, the one-letter abbreviations for the bases are used in writing codons and anticodons; thus, UGC stands for uracil-guanine-cytosine.

The three-dimensional structures of tRNA and mRNA are such that a tRNA molecule hydrogen bonds (that is, base pairs) with an mRNA molecule only when the tRNA anticodon is complementary to the mRNA codon. For example, only the tRNA with anticodon 3′-ACG-5′ (5′-GCA-3′) base pairs to codon 5′-UGC-3′ of mRNA:

mRNA codon	5′-UGC-3′
tRNA anticodon	3′-ACG-5′

That particular tRNA carries the α-amino acid Cys.

The **genetic code** is the complete list of mRNA codons and the α-amino acids and other instructions that the codons specify (Table 13.1). Again, by convention, the codon sequences are always written in the 5′ ⟶ 3′ direction. Table 13.1 is used as follows to find the α-amino acid or instruction specified by a codon:

- Find the first base of the codon in the far-left vertical column.
- Move horizontally to the right of the first base to the vertical column headed by the second base of the codon. You are now in a section of the table with four entries.

TABLE 13.1 Genetic Code: Codon Assignments

First base (5′ end)	Second base				Third base (3′ end)
	U	C	A	G	
U	Phe	Ser	Tyr	Cys	U
	Phe	Ser	Tyr	Cys	C
	Leu	Ser	Stop	Stop	A
	Leu	Ser	Stop	Trp	G
C	Leu	Pro	His	Arg	U
	Leu	Pro	His	Arg	C
	Leu	Pro	Gln	Arg	A
	Leu	Pro	Gln	Arg	G
A	Ile	Thr	Asn	Ser	U
	Ile	Thr	Asn	Ser	C
	Ile	Thr	Lys	Arg	A
	Met*	Thr	Lys	Arg	G
G	Val	Ala	Asp	Gly	U
	Val	Ala	Asp	Gly	C
	Val	Ala	Glu	Gly	A
	Val	Ala	Glu	Gly	G

*Codon AUG for Met also codes for the initiation (start) of polypeptide synthesis.

- Of those four entries, the specified α-amino acid or instruction is the one on the same horizontal line as the third base, which is listed in the far-right vertical column.

Problem 13.7 For each of the following codons, what is the complementary tRNA anticodon, and what α-amino acid is specified by the codon? (a) UCC; (b) CAG; (c) AGG; (d) GCU.

The genetic code is "degenerate" because many of the 64 codons are redundant. Most of the α-amino acids are coded by two, three, or four codons, and Arg, Leu, and Ser are each coded by six codons. Only Met and Trp are coded by a single codon. This degeneracy acts as a protective mechanism against mutations (Section 13.8). The codon for Met, which is AUG, codes simultaneously for Met and the initiation of synthesis. This simultaneous coding means that the first codon transcribed in all mRNAs is AUG, and the first α-amino acid in all initially formed polypeptides is Met. Three of the codons (UAA, UAG, UGA) are "Stop" codons, coding for the termination of polypeptide synthesis because no tRNAs recognize and hydrogen bond with them.

- All mRNAs begin with AUG (for the initiation of polymer synthesis) and end with one of the Stop codons.
- Additional AUG codons after the initial AUG codon code for methionine (Met).

The genetic code is nearly universal. It applies to all organisms, whether plant or animal, with very few exceptions.

Example 13.4 Determining the polypeptide structure coded by a template DNA strand

Sections *X*, *Y*, and *Z* of the following part of a template DNA strand are exons of a gene:

DNA 5′ | TGT | GCG | GTG TAT | TGC AAC | CAT | 3′
X Y Z

What are the structures of the primary transcript and mRNA made from this template DNA? What polypeptide sequence will be synthesized?

Solution

The base sequence in the primary transcript is the complement of that in template DNA, except that U is used in RNA instead of T. The direction of RNA synthesis is opposite the direction of the template DNA strand. In the splicing process, introns are removed and the result is mRNA:

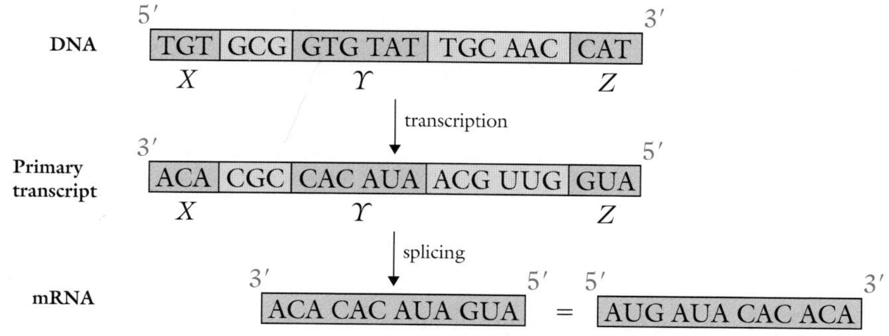

The codons in mRNA must be read in the $5' \longrightarrow 3'$ direction to correctly use the genetic code (see Table 13.1) to determine the sequence of α-amino acids in the polypeptide. The mRNA sequence AUG-AUA-CAC-ACA codes for the polypeptide sequence Met-Ile-His-Thr.

Problem 13.8 Sections *X*, *Y*, and *Z* of the following part of a template DNA strand are exons of a gene:

	5′							3′
DNA	CAC	CAC	ACC	GTA	TGT	GGA	CAT	
	X		*Y*				*Z*	

What are the structures of the primary transcript and mRNA made from this template DNA? What polypeptide sequence will be synthesized?

Each tRNA is designed to carry a specific α-amino acid, and the reaction that bonds the two together is catalyzed by the enzyme **aminoacyl-tRNA synthetase.** There is at least one synthetase for each α-amino acid. Each aminoacyl-tRNA synthetase has recognition sites specific to a particular α-amino acid and its corresponding tRNA. The synthetase causes the α-amino acid to bond to the acceptor stem (see Figure 13.8) at the 3′-end of tRNA by the formation of an ester linkage between the carboxyl group of the α-amino acid and the —OH group on C3′ (in some cases, the C2′) of ribose (Figure 13.13).

« The formation of an ester by reaction between —COOH and —OH groups was described in Section 7.8.

Polypeptide synthesis of the sequence Met-Gly-Ile is shown in Figure 13.14. It takes place at **peptidyl (P)** and **aminoacyl (A) sites** in the ribosome, located on the left and right sides of the ribosome structures in Figure 13.14. The P site holds the growing polypeptide chain, and the A site holds the tRNA with its α-amino acid. Initiation starts with Met, followed by the sequential addition of Gly and Ile. The process begins when the 5′-end of the mRNA enters the smaller subunit of the ribosome, which prepares it to receive the correct tRNAs.

In Figure 13.14a, the tRNAs for Met and Gly have already picked up Met and Gly from the cytosol, entered the ribosome, and lined up at the P and A sites, respectively (because of base-pairing between the mRNA codons and tRNA anticodons). The enzyme **peptidyl transferase,** contained in the large subunit of the ribosome, then catalyzes the transfer of Met from site P and the bonding of Met to Gly at site A. The reaction, called **acyl transfer,** yields a dipeptide at site A and an empty tRNA at site P (Figure 13.14b). The empty

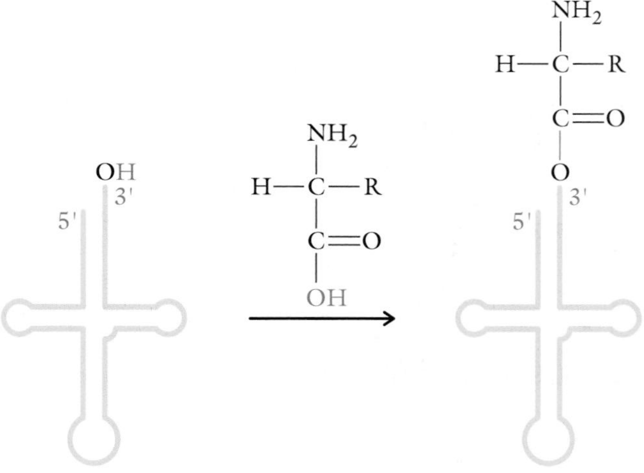

Figure 13.13 Picking up an α-amino acid at the 3′-end of tRNA.

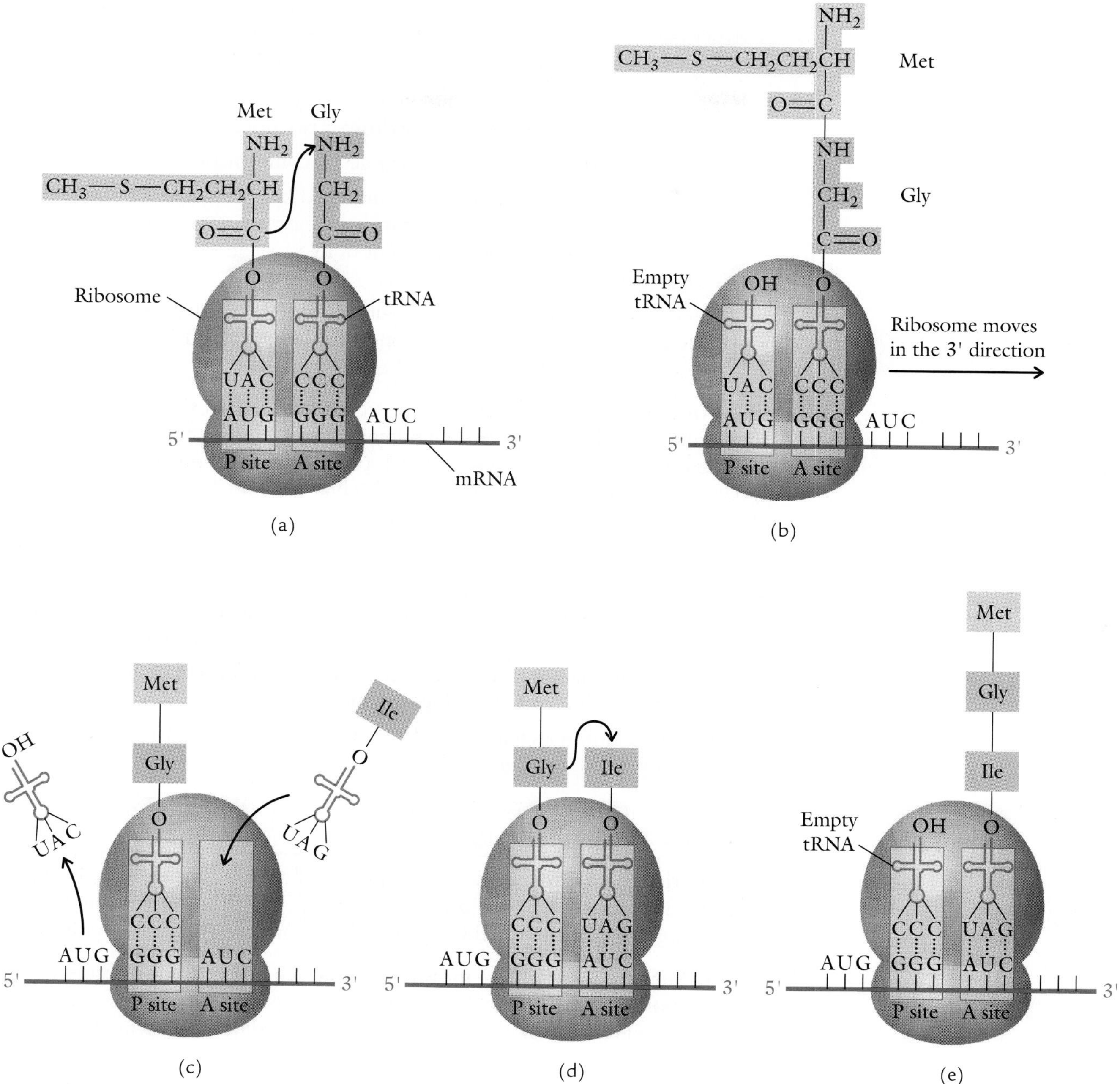

Figure 13.14 Synthesis of polypeptide. (a) Met and Gly tRNAs line up at mRNA; (b) Met transfers and bonds to Gly; (c) empty tRNA is released; (d) Ile tRNA lines up at mRNA; (e) Met-Gly transfers and bonds to Ile.

tRNA is then released from the ribosome (becoming available to transport another Met), and the ribosome moves along the mRNA in the $5' \longrightarrow 3'$ direction. The dipeptide is now at the P site (Figure 13.14c). A tRNA carrying Ile now base pairs at the empty A site (Figure 13.14d), after which the Met-Gly dipeptide is transferred and bonded to Ile to form the Met-Gly-Ile tripeptide at site A, leaving an empty tRNA at site P (Figure 13.14e). Synthesis proceeds in this manner over and over again, with the ribosome moving toward the 3′-end of the mRNA.

Synthesis terminates when a Stop codon is read on mRNA, and the polypeptide is released from the ribosome. The polypeptide then undergoes **posttranslational processing,** which brings it to its final, functional form:

- The methionine residue that initiated polypeptide synthesis is cleaved from most polypeptides by a hydrolysis reaction.
- Polypeptides are folded into their active conformations.
- Some α-amino acid residues are modified by chemical reactions; examples are disulfide-bridge formation at cysteine residues (Section 12.4) and the attachment of hydroxyl groups to proline residues (Section 12.6).
- For proteins with quaternary structure, different polypeptides and nonpolypeptide molecules or ions must be assembled together and folded into their active (native) conformations.

Posttranslational processing is the second mechanism that rationalizes the difference in diversity between the human genome and human proteome. The first mechanism is posttranscriptional processing (Section 13.6). Different polypeptides and proteins are produced when the initially produced polypeptide is processed differently in posttranslational processing by certain reactions:

- Proteolysis is the cleavage of a polypeptide by the hydrolysis of one or more of its amide linkages. In some cases, a piece of the polypeptide is cut off and discarded to alter physiological function. In other cases, a polypeptide is cut up to form two or more functional polypeptides.
- Phosphorylation adds phosphate groups to a polypeptide to alter its conformation and physiological function.
- Glycosylation attaches carbohydrates to a polypeptide to alter cell-recognition functions.

13.8 MUTATIONS

A **mutation** is an error in the base sequence of a gene. The end result can be the alteration or cessation of a polypeptide's or protein's functioning because of a change in its α-amino acid sequence. There are two types of mutations:

- **Substitution (point) mutations,** in which one base substitutes for another in the normal base sequence: one purine for another, one pyrimidine for another, a purine for a pyrimidine, or a pyrimidine for a purine.
- **Frameshift mutations,** in which a base is inserted into the normal base sequence or is deleted from it.

A substitution mutation changes one codon in the DNA base sequence. A frameshift mutation, although rarer than a substitution mutation, has a much larger effect because it changes all codons that follow the insertion or deletion.

Mutations that arise from random errors in replication are called **spontaneous mutations.** Superimposed on spontaneous mutations are those caused by **mutagens,** agents that produce mutation. Various chemicals and types of radiation act as mutagens by modifiying the nucleotide bases. Sodium nitrite ($NaNO_2$), used as a preservative for frankfurters, bologna, and other processed meats, is a chemical mutagen. It converts cytosine, which base pairs with guanine, into uracil, which then base pairs with adenine. Thus, it affects DNA structure by converting a C-G base pair into U-A. The overall danger is thought to be extremely low because of the very small amounts of nitrite used in meat preservation. However, it is probably prudent to limit the intake of such processed foods. Stopping the use of nitrite is not a practical alternative, because it would increase the incidence of botulism—food poisoning by the neurotoxins secreted by the airborne bacterium *Clostridium botulinum*. Viruses also may cause mutations (Section 13.10).

Frameshift mutations can occur when certain planar organic compounds such as benzopyrene and other fused-ring aromatic compounds (see Box 4.9) undergo reactions that subsequently result in their insertion between DNA base pairs. The insertion of these compounds distorts the DNA double helix and can result in the deletion or insertion of a base pair. Benzopyrene and similar compounds are found in automobile exhaust, tobacco smoke, and burnt (especially barbecued) meats. Here is one plausible mechanistic connection between smoking and cancer.

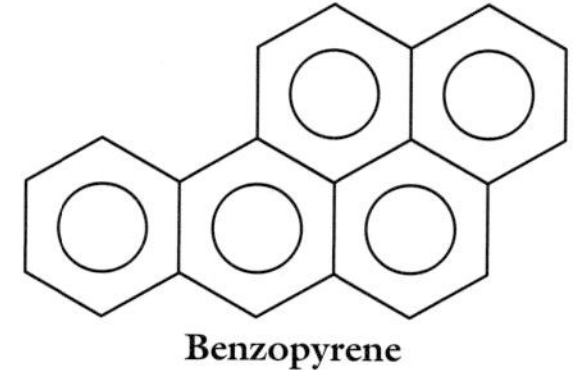
Benzopyrene

We are exposed to various types of radiation: ultraviolet light and cosmic rays (α- and β-particles, neutrons) from outer space, α- and β-particles and neutrons from the radioisotopes in the ground beneath us, and X-rays and radioisotopic radiation from medical imaging equipment. When the energy of radiation is absorbed by DNA, bonds may be broken and bases and base sequences modified.

Radiation may also lead to oxidative damage, perhaps the most significant source of mutations. Excited-oxygen species such as hydroxyl (HO·) and superoxide (O_2^-·) free radicals are formed from the effects of radiation on living cells, as well as in the course of normal aerobic metabolism. Because of their unpaired-electron structure, free radicals are extremely reactive species that can alter base structures and DNA base sequences. Fortunately, cells have elaborate antioxidative mechanisms for destroying these reactive species. Vitamins C and E are important antioxidants.

Preservation of the base sequence in an organism's genome is the key to the survival of the organism and the survival of its offspring. Two factors preserve the genome with great fidelity. First, replication errors are very rare because of the proofreading function of DNA polymerase. Second, several repair systems correct errors, both spontaneous mutations and mutations caused by mutagens.

Silent mutations are base-sequence errors that do not affect the functioning of an organism. Several factors are responsible for silent mutations:

- The redundancy of the genetic code makes the genome somewhat mutation resistant. Many base changes have no effect on the α-amino acid sequence in the polypeptide or protein synthesized from DNA (through RNA). Say, for example, that the codon CCU in mRNA has been mutated to CCC or CCA or CCG (see Table 13.1). The same α-amino acid, Pro, is encoded by all four codons. On the other hand, mutation of the first or second base in the CCU codon does result in a change in the encoded α-amino acid.
- Substitution of an α-amino acid residue for one similar in size, charge, and polarity often has no effect on function, because the three-dimensional structure of the polypeptide or protein is unchanged (Section 12.8).
- Substitution of an α-amino acid residue for a dissimilar residue may have little or no effect if the residue is located in a region of the polypeptide or protein that is unimportant for its function.
- Mutations, even frameshift mutations, in introns are constrained within the introns and have no effect on the base sequences in exons, because the introns are excised.
- Many genes are present in two or more copies. Unless all copies of a gene undergo mutation, an improbable occurrence, the organism still produces the required polypeptides and proteins.

Other mutations do have negative consequences, because they interfere with proper functioning:

- Substitution of an α-amino acid residue for a dissimilar residue in a region of the polypeptide or protein that is important for its function usually results in nonfunctional polypeptides and proteins.
- Changes that convert a codon for an α-amino acid into a Stop codon that terminates polypeptide synthesis usually result in nonfunctional polypeptides and proteins.
- A frameshift mutation in an exon almost always results in a nonfunctional polypeptide and protein because the deletion or insertion changes all codons in the gene from that point on.
- The polypeptides or proteins produced by mutant DNA may be toxic.

Mutations in **somatic cells** (cells other than egg and sperm) affect the functioning of an individual organism. Mutations in **germ cells** (egg and sperm cells) have greater significance because they are passed to offspring and subsequent generations. Thus, germ-cell mutations are the cause of **genetic (hereditary) diseases** that can be passed to offspring. Some of the more common human genetic diseases are described in Table 13.2. Diabetes mellitus (Box 12.3 and Section 18.6), familial hypercholesterolemia (Box 16.1), galactosemia and lactose intolerance (Box 10.2), and sickle-cell anemia (Section 12.8) are described elsewhere.

Although the fidelity of replication is high and the repair systems are very efficient, mutations do accumulate in an individual organism's lifetime. There is probably a link between the accumulation of a lifetime of mutations and the changes seen in cancer (uncontrolled growth of cells) and aging. A mutagen giving rise to **cancer** is called a **carcinogen.** Many mutagens, but not all, are carcinogens. The ultimate effects of carcinogens depend on the quantity of accumulated mutations and the particular functions affected. Box 13.1 on page 434 considers cancer and cancer therapy.

Not only are mutations inevitable, they are the basis for the evolution of a species. Sometimes, but rarely, a mutation may occur that gives some members

TABLE 13.2 Hereditary Diseases

Disease	Affected physiological function
albinism	Tyrosinase is deficient and melanin is not produced; melanin protects skin and the iris of the eye against UV damage and skin cancer.
cystic fibrosis	Viscous mucus in bronchial passages leads to pulmonary bacterial infections. Inadequate lipase and bile salts produced by the pancreas result in maladsorption of lipids and fat-soluble vitamins.
diabetes mellitus	Decreased insulin production results in poor glucose metabolism and the accumulation of glucose in the blood and urine, atherosclerosis, visual problems, and circulatory problems in legs and feet.
galactosemia	Defective transferase results in poor metabolism of galactose; accumulation of galactose in cells causes cataracts and mental retardation.
hemophilia	Defective blood-clotting factors result in poor clotting, excess bleeding, and internal hemorrhaging.
phenylketonuria (PKU)	Owing to lack of phenylalanine hydroxylase for converting Phe into Tyr, Phe is converted into phenylpyruvic acid, which causes brain damage.
sickle cell	Defective hemoglobin; sickled red blood cells aggregate to cause anemia, plugged capillaries, low oxygen pressure in tissues.
Tay-Sachs	Defective hexosaminidase A causes accumulation of glycolipids in brain and eyes; mental retardation and loss of motor control.
thalassemia	Defective hemoglobin.

of a species an advantage over other members in the face of certain adverse environmental conditions. The population containing that mutation survives and reproduces in greater numbers, passing its altered genome along to subsequent generations.

Example 13.5 Determining the effect of a mutation on the α-amino acid sequence of a polypeptide

Imagine the following mutations in the template DNA strand in Example 13.4: (a) TAT of the *Υ* exon is mutated to GAT; (b) TAT of the *Υ* exon is mutated to TAA; (c) G is inserted in the middle of the second intron. What effect would each of these mutations have on the codon of the resulting mRNA? What effect would they have on the α-amino acid sequence of the resulting polypeptide?

Solution

(a) 5′-GAT-3′ in DNA results in 3′-CUA-5′ in mRNA. The codon is 5′-AUC-3′, which codes for Ile. There would be no change in the α-amino acid sequence, because Ile was the residue encoded by the codon before mutation.

(b) 5′-TAA-3′ in DNA results in 3′-AUU-5′ in mRNA. The codon is 5′-UUA-3′, which codes for Leu. The α-amino acid sequence would be altered, with Ile changed to Leu. The polypeptide would be Met-Leu-His-Thr instead of Met-Ile-His-Thr.

(c) There would be no effects on either the mRNA or the polypeptide. A mutation in an intron is constrained within the intron and has no effect on the base sequence in the exons, because the intron is excised.

Problem 13.9 Imagine the following mutations in the template DNA strand in Problem 13.8: (a) GTA of the *Υ* exon is mutated to GAA; (b) GTA of the *Υ* exon is mutated to TTA; (c) TGT of the second intron is mutated to AGT. What effect would each of these mutations have on the codon of the resulting mRNA? What effect would they have on the α-amino acid sequence of the resulting polypeptide?

As mentioned in Section 13.6, the higher error level in RNA synthesis (less than one base in 10,000 to 100,000) compared with that in DNA synthesis (less than one base in 10 billion) is not a serious problem. Why is this the case? The function of RNA is the synthesis of polypeptides and proteins. The error level in RNA leads to a small number of defective copies of RNA, less than 1 in 10,000 to 100,000. An organism is not affected adversely, because many correct copies of RNA are produced (at least 9,999 to 99,999, depending on the exact error level). The overall result is that from 99.99% to 99.999% of the polypeptides and proteins produced have the correct amino acid sequence and are available to be used. The few defective copies produced are not used. On the other hand, an error in DNA synthesis in the course of replication results in at least one of the two cells produced in cell division having a mutated chromosome. All of that cell's progeny also will carry the mutated chromosome. All of the polypeptides and proteins produced in those cells will have the incorrect amino acid sequence. No copies of the correct polypeptide and protein will be produced.

13.9 ANTIBIOTICS

Antibiotics are chemicals, usually organic compounds, used to fight infections by microorganisms (bacterium, mold, yeast), especially bacterial infections. Bacteria enter the body but usually not the cells of the organisms that they infect. They harm the host through the toxicity of their metabolic products. Before the discovery and use of antibiotics in the 1930s, infectious diseases such as diphtheria, gonorrhea, pneumonia, tuberculosis, and syphilis were major

13.1 CHEMISTRY WITHIN US

Cancer and Cancer Therapy

We are made up of many different types of cells—brain, heart, lung, muscle, bone, skin, hair, and so forth—all dividing at different times and at different rates. Some cells, such as brain and other nerve cells, do not divide at all after reaching mature development. Cell division is a tightly regulated process in a healthy person, constrained at boundaries between different types of cells, tissues, and organs and controlled by regulatory proteins called **growth factors,** which circulate in the bloodstream. Each type of cell is regulated by its own growth factor and a corresponding **receptor protein** on its membrane. The genes for the growth factors and receptors, responsible for normal cell function, are called **proto-oncogenes.**

The feature common to all cancers and distinct from other illnesses is the uncontrolled growth and division of cells to form large masses called **tumors.** A **benign tumor,** which is not a cancer, is a limited growth that remains in its initial location. Cancers are **malignant tumors**—uncontrolled growths that consume the body's resources and invade and crush nearby structures (such as when a tumor invades and compromises the lungs), destroying their functions. Even worse, cancers often shed cells that spread to new sites in the body and colonize these sites—a process called **metastasis.**

Cancer begins when mutations convert proto-oncogenes into **oncogenes,** genes that no longer code for the correct regulatory proteins. Cell division goes out of control in the absence of the normal regulatory mechanism. Cancer-causing mutations occur through replication errors, the effects of chemical and radiation carcinogens, and some viruses (Sections 13.8 and 13.10).

The United Nations World Health Organization estimates that environmental and life-style factors account for more than 80% of all cancers, with the remainder being caused by genetic factors and viruses. The environmental and life-style factors include cigarette smoking, air and water pollution, diet, and occupational hazards.

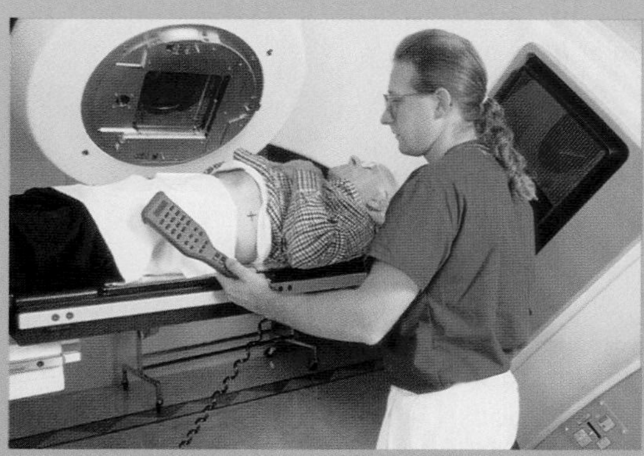

A patient is positioned before receiving radiation therapy to combat cancer. (L. Steinmark/Custom Medical Stock Photo.)

The main strategies for treating cancers are surgery, radiotherapy, and chemotherapy. The surgical removal of cancerous tissues is an option only if the patient can survive without the affected tissues. It requires knowledge of the cancer's exact location because the cancer will rebound after surgery if all cancerous cells are not removed. The outlook after surgical treatment is often enhanced by some radiotherapy or chemotherapy to ensure that no cancerous cells remain.

Radiotherapy is the irradiation of cancers with ionizing radiations. γ-Rays from radioisotopic sources such as cobalt-60 and cesium-137 are those most often used. X-rays also are used. Irradiation stops the growth of cancer cells through mutation of the cells' DNA. Hence, mutation, the mechanism that initiates cancer, may also be the means for stopping it. Radiotherapy damages healthy tissue along with the cancer cells but, because cancer cells grow more rapidly than normal cells do, they are more sensitive to mutation than the healthy tissues are. The objective of radiotherapy, therefore, is to kill the cancerous cells while keeping the damage to healthy tissue at a minimum. The total radiation delivered to a patient is held to the lowest effective level, and collimated (narrowed and focused) radiation beams are used to confine the radiation as much as possible to the cancerous tissues.

Chemotherapy is the use of chemicals to interfere with the replication of cancer cells. Among the many useful chemicals are 5-fluorouracil and cisplatin. Cisplatin binds with DNA and prevents its replication. 5-Fluorouracil deactivates one of the enzymes responsible for producing the nucleotide dUMP. Other chemicals used in chemotherapy interfere with replication by substituting for one of the heterocyclic bases, resulting in an incorrect base sequence. As with radiotherapy, the key to chemotherapy is that cancerous cells are more susceptible to disruption than are healthy cells. There is a continuing effort to find chemicals that are more selective in disrupting the replication of cancerous cells than that of healthy cells.

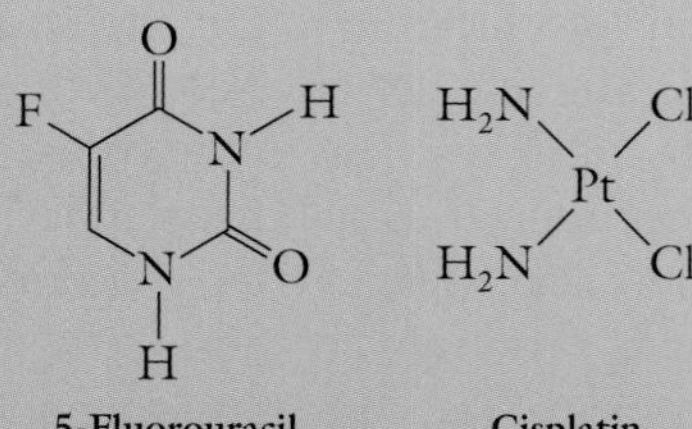

A recent trend in cancer treatment is to combine different kinds of therapy—radiotherapy with chemotherapy, surgery with either radiotherapy or chemotherapy, or chemotherapy with a combination of different drugs. The fine-tuning of combination therapies results in more complete destruction of cancerous cells with minimal damage to healthy cells.

TABLE 13.3 Antibiotic Inhibition of Protein Synthesis in Bacteria

Antibiotic	Mechanism for inhibition
chloramphenicol	inhibits peptide-bond formation by interfering with peptidyl transferase
erythromycin	prevents translocation of ribosome along mRNA
penicillin	inhibits formation of enzyme needed for cell-wall formation
puromycin	causes premature termination
streptomycin	inhibits initiation; causes misreading of mRNA
tetracycline	prevents binding of tRNA to mRNA

causes of death. Although most of the original antibiotics were isolated from microorganisms, most present-day antibiotics are synthesized in the laboratory.

Antibiotics fight bacterial infections by interfering with the genetic apparatus in bacteria. Protein synthesis is blocked, usually by preventing translation, and this stops bacterial reproduction (Table 13.3). Large-scale use of antibiotics has been highly successful, but some bacteria have "fought back" by spontaneously mutating into antibiotic-resistant strains. Any population of bacteria (or other organisms) undergoes spontaneous mutations. There is always the possibility that one of the spontaneous mutations imparts protection to the bacteria against a specific antibiotic. An antibiotic-resistant strain of bacteria is produced when those mutant bacteria survive a treatment with an antibiotic and propagate. The tendency for an antibiotic-resistant strain to arise increases with increased use of the antibiotic. Scientists attempt to keep ahead of the bacteria by synthesizing new antibiotics effective against the mutant strains, but it is not always easy to do so.

To be useful, an antibiotic must be specific for bacteria, with minimal effect on protein synthesis in humans and other animals. There are probably no antibiotics that completely meet this criterion. The typical antibiotic affects the host but less so than the bacteria. For this reason, antibiotics are prescribed for a short period only—say, 2 weeks—and then halted. Prolonged antibiotic use fosters the production of mutant strains of bacteria and may adversely affect the host. At the same time, patients are always instructed to complete the entire course of antibiotic treatment and not to stop before finishing should they feel better after a few days. If treatment is stopped too soon, the bacterial population may not have dropped sufficiently low for the patient's immune system to keep it under control, and the bacterial infection may rebound.

13.10 VIRUSES

Viruses are infectious, parasitic particles, usually smaller than bacteria, that consist of either DNA or RNA (but not both). **DNA viruses** contain only DNA. **RNA viruses** contain only RNA. The DNA or RNA is encapsulated by a protein coat called a **capsid.** The typical virus also has an additional protein protective coat called an **envelope** (or **overcoat**) whose surface contains glycoproteins. Viruses lack some or most of the cellular apparatus (α-amino acids, nucleotides, enzymes) needed for replication, transcription, and translation. Thus, when they are outside host cells, they show no signs of life: they do not generate energy, reproduce, or synthesize proteins. To reproduce, viruses must enter a cell and take control of its metabolic machinery. Viruses are responsible for many diseases—AIDS, chicken pox, the common cold, hepatitis, herpes, influenza, some leukemias, mononucleosis, poliomyelitis, rabies, shingles, smallpox, and various tumors (including some cancers).

13.2 CHEMISTRY WITHIN US

HIV and AIDS

The human immunodeficiency virus (HIV), the cause of acquired immune deficiency syndrome (AIDS), has a very specific target. It infects only the T cells of the human immune (lymphatic) system. Ironically, the normal function of these cells is to identify and destroy invaders in the body. The HIV count in the blood soars at the onset of the infection but usually drops to a very low number as the immune system fights back with an increased production of T cells. A struggle ensues between HIV and T cells but, as long as the immune system stays ahead of the virus, the infected person remains symptom free.

From several months to several years after infection, the immune system usually begins to lose the struggle, and symptoms appear. These symptoms include swollen lymph nodes and fungal and herpes simplex infections. Most people with these symptoms progress to the stage at which their immune systems become compromised. People do not die of AIDS directly but become easy prey to opportunistic infections—such as pneumonia, toxoplasmosis, and various cancers—that do not normally affect healthy persons.

AIDS is primarily a sexually transmitted disease. The virus is spread from one person to another through blood, semen, and vaginal fluid. It passes from the fluid of an infected person to the noninfected person's bloodstream through the thin linings of the noninfected person's anus, mouth, penis, vagina, or uterus. Torn or damaged tissues in these organs greatly ease the entry of HIV. Intravenous drug users are at high risk of contracting AIDS because they often share needles. HIV-infected mothers transmit the virus to their infants during birth or subsequently through breast feeding. Before 1985, blood transfusions also resulted in some transmission of HIV, but blood banks now test donors' blood for HIV before accepting blood donations.

The patterns of AIDS infection differ throughout the world. In the developed Western countries, AIDS is most prevalent in homosexual males and intravenous drug users, although there are indications that the incidence may be rising in the heterosexual population. In the less-developed countries of Africa and Asia, AIDS is mostly a heterosexual disease equally divided between males and females. More than an estimated 20 million people died from AIDS worldwide from 1981 to 2003. About 0.5 million of these deaths were in the United States. The United Nations' estimates for the number of persons infected with HIV as of the end of 2004 are 40 million worldwide, with 1 million in North America, 25 million in Sub-Saharan Africa, 7 million in South and Southeast Asia, 2 million in South and Central America and the Caribbean, and 2 million in Europe and Central Asia.

The spread of HIV has slowed down in the United States, probably a result of education about the mechanism of transmission. However, an increasing proportion of new infections, about 50% in 2003, are found among persons younger than 24. The epidemic rages on in the less-developed countries, where the number of people infected with HIV continues to soar.

There is an enormous worldwide effort to find a cure for AIDS. Efforts to find a vaccine to prevent infection by HIV are not encouraging, because the virus is constantly mutating, and vaccines with promise against one strain offer no hope against new strains.

However, there is consensus that ways can be found to control the reproduction of HIV within infected persons and delay the onset of AIDS by blocking one or more steps in the virus's life cycle (see Figure 13.15). New therapeutic drugs have been tested against almost all of the steps. To be useful, the drugs must slow or stop HIV reproduction without doing too much damage to the patient. No one drug has been found to be highly effective, although a number of drugs have shown promise. The problem in every instance has been that prolonged use of the drug results in resistant mutant strains. Yet, there is considerable optimism that a combination of effective drugs may work well, because the virus is less

Each kind of virus infects only specific types of cells in specific organisms. For example, there are large numbers of plant viruses, none of which invade animal cells. The infection of a host cell by a virus occurs when the virus adheres to the cell. Enzymes in the virus envelope enable the virus to penetrate the cell membrane and insert itself or its nucleic acid into the host cell.

When inside, the virus diverts the host's cellular machinery (enzymes, α-amino acids, nucleotides) from its normal role and puts it to work replicating new daughter viruses (using the viral nucleic acid as a blueprint). Hundreds of progeny viruses result from the entry of one virus into a host cell. The progeny then escape from the host cell and attack other host cells. Some viruses exit the host cell by breaking open its membrane, killing the cell instantly. Other viruses escape through the cell membrane without breaking it open. The host cell dies nevertheless, because its normal cell processes have been compromised.

The reproduction of RNA and DNA viruses follows different paths:

likely to undergo several simultaneous mutations that would circumvent all of the drugs.

The present therapy is not a cure, but it can stop the progression of the disease and improve both the quality of life and the longevity of infected patients. This therapy, called **combination therapy,** usually consists of a combination of three or more drugs that have to be taken together for the rest of a person's life. Among these drugs are a variety of **reverse transcriptase inhibitors (RTIs)** and **protease inhibitors (PIs).** A combination of AZT, 3TC, and Indinavir is often used as the initial combination for new patients. The reverse transcriptase inhibitors block the transcription of viral RNA to viral DNA (step 3 in Figure 13.15). Reverse transcriptase cannot discriminate AZT and 3TC from the normal nucleotide containing deoxyribose and thymidine, but AZT and 3TC differ from the normal thymidine nucleotide in not having an —OH group at the 3′-position of deoxyribose. Because of the absence of the —OH group, DNA synthesis terminates whenever AZT or 3TC is incorporated into the growing DNA strands in the course of reverse transcription. Protease inhibitors block step 7, the cutting up of the proteins produced by the translation of viral RNA, by inactivating the enzyme protease. The components needed to assemble the virus are then not available, and new HIVs are not produced.

New types of drugs are in development. One of them is the **fusion** or **entry inhibitor** enfuvirtide. Enfuvirtide is a polypeptide that binds to the protein envelope of the HIV virus and prevents its fusion with the cell membranes of T cells. Thus, the virus is prevented from entering the T cell. Future combination therapies may include a fusion inhibitor together with RTI or PI or both drugs.

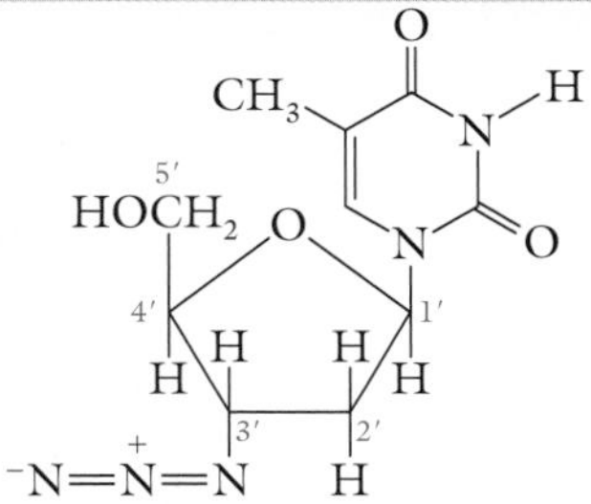

3′-Azido-2′,3′-dideoxythymidine (AZT)

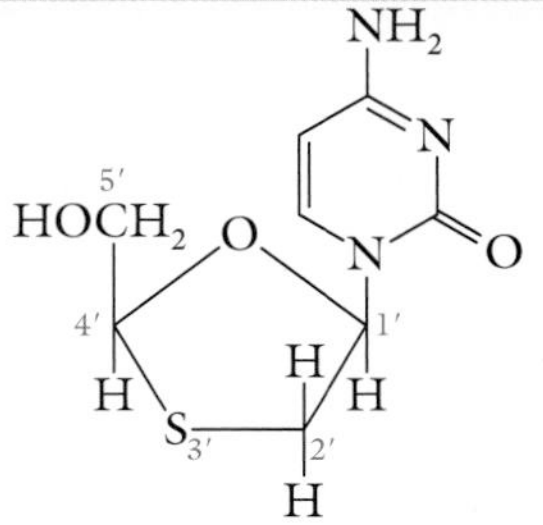

3′-Thia-2′,3′-dideoxycytidine (3TC)

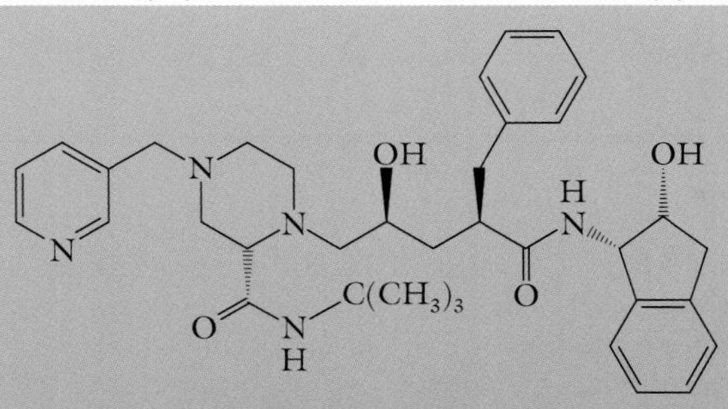

Protease inhibitor (Indinavir)

- DNA viruses enter the host nucleus and integrate themselves into the host genome. The modified DNA is treated by the host in the same manner as it treats its own (unmodified) DNA.
- RNA viruses must replicate RNA from RNA. The host cells do not have the enzyme, **RNA replicase,** required for this process, but the RNA virus contains information for synthesizing it. RNA replicase is synthesized in the host cell, and it directs the reproduction of viral RNA.
- **Retroviruses** are RNA viruses that reproduce through the intermediate formation of DNA. The retrovirus contains the enzyme **reverse transcriptase,** which directs the synthesis of viral DNA from viral RNA, a direction contrary to the normal pathway. The viral DNA becomes integrated into the host genome, and then the host reproduces the viral RNA by using the viral DNA as a template. The retroviruses include HIV and cancer-causing viruses **(oncogenic viruses).** Box 13.2 describes AIDS and, together with Figure 13.15, the life cycle of the HIV virus.

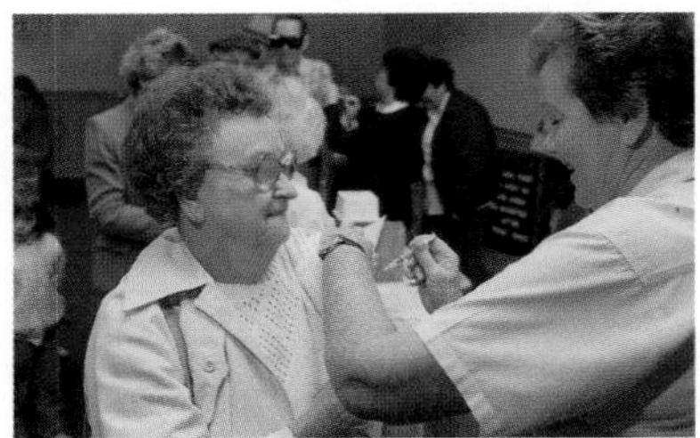

A nurse administers a vaccine to a senior citizen to protect her against the flu. (Kenneth Murray/Photo Researchers.)

1. Retrovirus attaches to host cell at membrane protein CD4 receptor.

2. The viral core is uncoated as it enters the host cell.

3. Viral RNA uses reverse transcriptase to make complementary DNA.

4. Single-stranded reverse transcript synthesizes second complementary DNA strand to form viral DNA duplex.

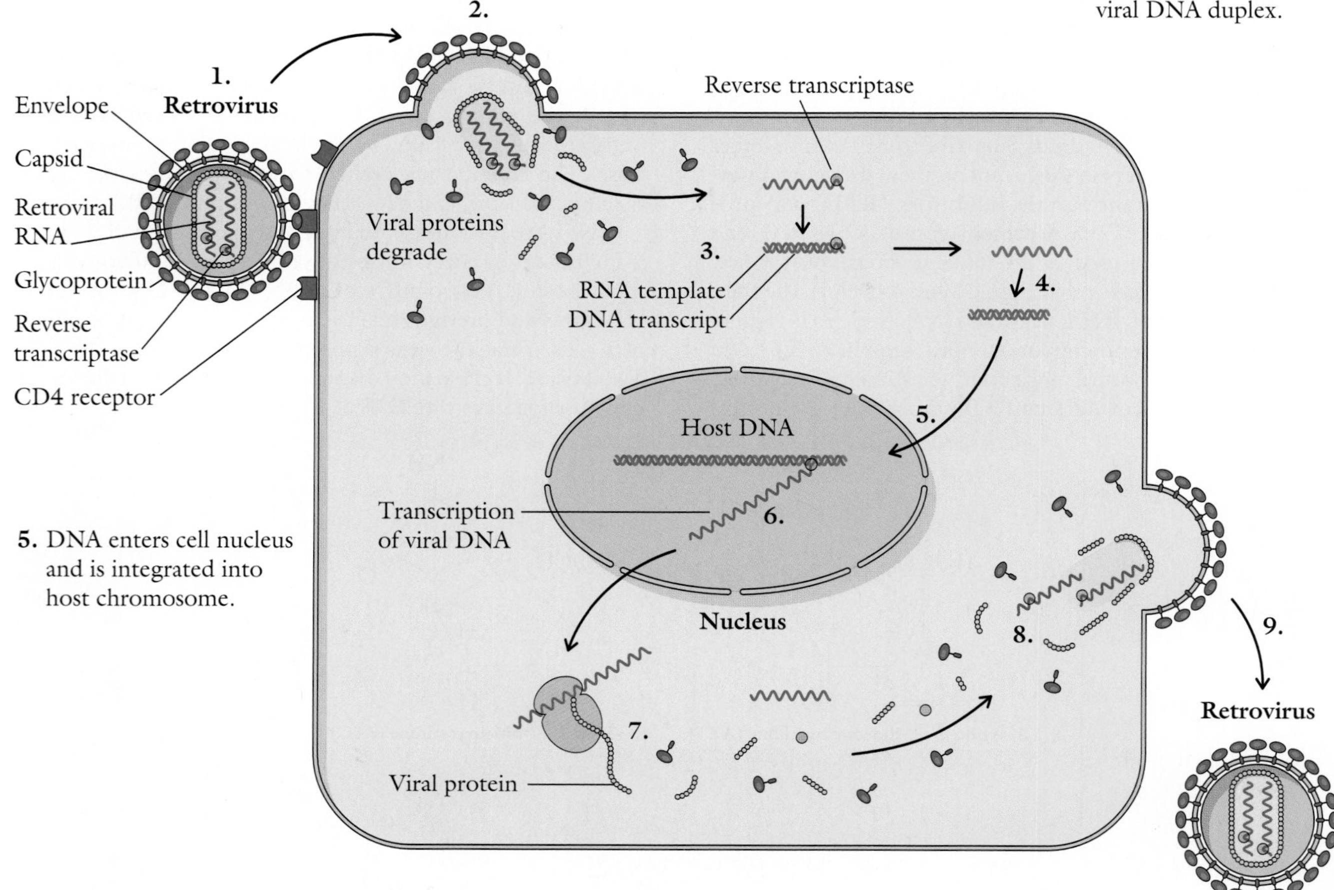

6. Viral DNA transcribes viral RNA, which is exported to cytoplasm.

7. Viral RNA is translated. Protease cuts viral proteins into smaller pieces for incorporation into new viruses.

8. Viral proteins, capsids, and envelopes are assembled.

9. Assembled virus emerges from the cell membrane.

Figure 13.15 Reproduction of HIV in a T cell.

Viral infections are generally more difficult to treat than bacterial infections because viruses, unlike bacteria, replicate inside host cells and spend little time outside of them. It is, therefore, much more difficult to design a drug that prevents the reproduction of the virus without compromising the host cell's normal activities.

The most successful approach to fighting viral infections is the preemptive use of vaccines. A **vaccine** is a preparation containing the weakened virus or its proteins. Its presence in the body stimulates the immune system to generate antibodies (immunoglobulins) that recognize the virus's antigens (Section 10.6). Any later entry of the virus into the host is thus quickly recognized by the host, which produces lymphocytes and antibodies to fight the invasion. Vaccines are also useful for protection against bacterial infections—for example, diphtheria and whooping cough.

« Immunoglobulins are proteins that have a protective function, described in Chapter 12.

13.11 RECOMBINANT DNA TECHNOLOGY

Recombinant DNA technology (also called **genetic engineering**) is used to alter the genome of an organism by transplanting DNA into it, either synthetic DNA or DNA from another organism (either the same or a different species). Research in recombinant DNA technology dates only to the mid-1970s, but we have already seen or are on the verge of seeing benefits:

- Organisms with altered DNA producing therapeutic drugs
- Agricultural improvements in animal and plant stocks
- Gene therapy to alleviate genetic diseases

The Production of Therapeutic Proteins

The first application of recombinant DNA technology was the use of an organism with altered DNA to produce human insulin for diabetics to use in place of bovine and ovine insulin. Both yeast and bacteria have been useful as the vehicle for this purpose. Figure 13.16 outlines the process for producing human insulin

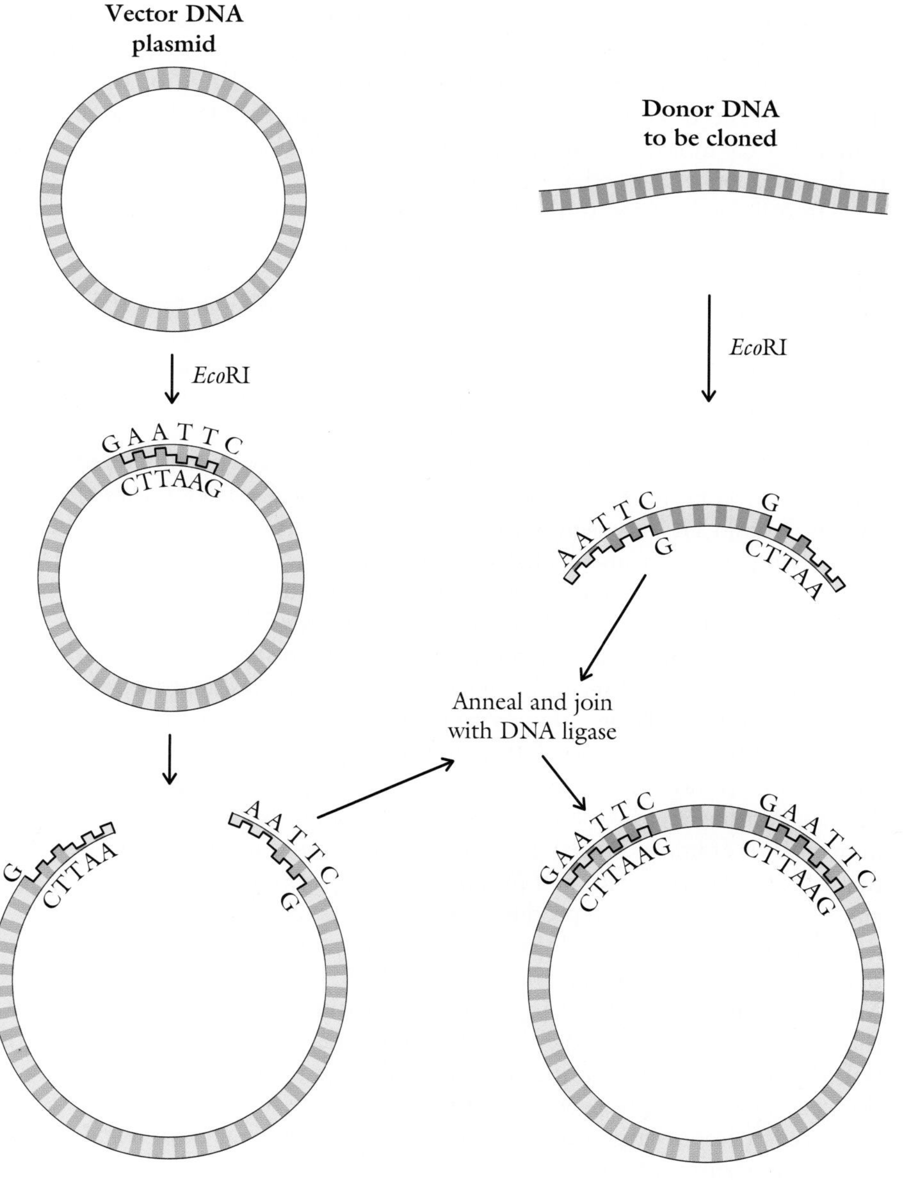

Figure 13.16 Formation of recombinant DNA with the use of *E. coli* plasmids. Vector DNA (plasmid) is cleaved open by a restriction enzyme (*Eco*RI). Donor DNA is cleaved at its ends by *Eco*RI. Vector and donor DNAs are joined together by DNA ligase.

by altering the DNA of *Escherichia coli* bacteria. *E. coli* contains DNA in two forms, a single chromosome and a large number of circular, double-stranded DNAs called **plasmids.** The more-plentiful plasmids are more useful than the single chromosome for genetic engineering.

The process begins after scientists have identified the gene encoding the synthesis of the desired polypeptide. The next step is to produce and splice this gene, called **donor DNA,** into the plasmid **(vector DNA).** The vector DNA is the DNA that is to be altered—that is, altered by the introduction of the donor DNA. *E. coli* cells are broken up, and their plasmids are isolated by extraction and centrifugation.

The vector DNA and donor DNA are mixed together. A **restriction enzyme** (*Eco*RI) is then added to the mixture, and this enzyme cleaves open the plasmid at one site containing a specific sequence of nucleotides. The cut is uneven, leaving overlapping, complementary, single strands at each end of the now linear DNA. For example, when *Eco*RI is used to cleave plasmids, one single-strand end may be TTAA and the other AATT.

The donor DNA must have uneven ends complementary to those on the vector DNA because the two are to be spliced (bonded) together. The donor DNA is prepared either by using the same restriction enzyme on the DNA from an organism that contains the gene of interest or by in vitro DNA synthesis. In vitro synthesis of donor DNA requires knowledge of the primary structure of the DNA that codes for the desired polypeptide. To use recombinant DNA technology to make human insulin, the donor DNA must contain the base sequence that codes for human insulin.

DNA ligase is added to the mixture to join together the vector and donor DNAs. Base-pairing of the complementary uneven ends ensures the proper alignment of vector and donor DNAs before covalent bonding by DNA ligase. The altered plasmid DNA is called **recombinant DNA.** This altered DNA is introduced into *E. coli* by any of several treatments that make bacterial walls transiently permeable to plasmids. One such treatment is **chemical shock**—*E. coli* are placed in aqueous calcium chloride at low temperature and then incubated with the plasmids at higher temperature.

The altered *E. coli* now contain both normal and recombinant DNA. When they replicate, they reproduce many identical copies **(clones)** of the altered plasmid DNA, which is why recombinant DNA technology, or genetic engineering, is often also called **cloning.** When the altered plasmid DNAs undergo transcription and translation, they produce not only the normal *E. coli* polypeptides, but also the polypeptide coded by the donor DNA—human insulin in this case. The human insulin is then separated and purified.

Dolly was the world's first clone of an adult mammal. (AP/Wide World Photos.)

Human insulin produced by genetic engineering was approved for use by the U.S. Food and Drug Administration in 1982 and is now the replacement insulin used to treat diabetics. Other substances produced by recombinant DNA technology are now in use or in various stages of development (Table 13.4). The applications are diverse and include use in animals and plants as well as in humans. For example, bovine growth hormone is used to increase milk production in cows. Sheep, goats, and pigs have recently been used in place of bacteria and yeast as the vehicles to produce various protein-based drugs in their milk.

Microinjection and viral vectors are techniques other than chemical shock that are useful for introducing foreign DNA into an organism's genome. **Microinjection** is the direct injection of DNA into the nucleus of a cell with the use of a very fine needle. This technique has a high success rate for the skilled practitioner but cannot be used to alter large numbers of cells, because the cells must be treated one at a time. It is well suited to the alteration of germ cells because large numbers of cells need not be altered. **Viral vectors** are altered viruses, often retroviruses, whose DNA has been engineered in the laboratory to include the desired DNA sequence and to lack the genes for viral

TABLE 13.4 Products of Recombinant DNA Technology

Product	Use
blood-clotting factor	Promotes blood clotting for treating hemophilia; eliminates risk of treatment by blood transfusion.
erythropoietin	Treats anemia resulting from kidney disease; stimulates erythrocyte production.
human growth factor (somatotropin)	Treats dwarfism in children.
interferons	Prevent viral reproduction by inhibiting protein synthesis in infected cells.
interleukins	Stimulate production of T leukocytes in immune disorders.
monoclonal antibodies	Treat cancer by binding and transporting drug, toxin, or radioactive compound to tumor.
tissue plasmologen activator (TPA)	Activates plasmin (enzyme); dissolves clots in heart attack.
vaccines	Like the traditional weakened viruses, viral coat proteins act as vaccines but are safer.

reproduction. When they enter a host cell, they incorporate the desirable DNA into the host's genome. Because these viruses are altered to prevent their replication, they do no harm to the host cell.

The Transgenic Breeding of Animals and Plants

Traditional practices for improving animal and plant stocks include selective breeding and selective feeding or fertilizing programs. **Transgenic breeding** is the selective production of organisms whose genes have been permanently altered by recombinant DNA technology. Transgenic breeding offers the potential for faster, more accurate, and more varied improvement in stocks. Accurate control of transplanted DNA allows desirable traits to be introduced into the transgenic animal or plant without the simultaneous introduction of too many other, possibly undesirable traits. The desirable traits can be introduced in one generation, compared with the several generations needed with traditional breeding. Furthermore, traits not normally available in a given species may be introduced into it through the transportation of DNA from a different species.

The objective of transgenic breeding is to produce stock that is larger and leaner, grows faster, uses feed more efficiently, and is more disease resistant. Many experimental efforts have reported success, although few transgenic animals have yet reached the market. Transplantation of the growth hormone gene from the sockeye salmon into the coho salmon has accelerated the coho's growth so that it is ready for market sooner than the standard farmed coho salmon, from about 1 to $1\frac{1}{2}$ years instead of 2 to 3 years. Sheep that grow superior wool without needing the standard dietary supplements and faster-growing and leaner pigs and cattle are other examples of genetic engineering.

A number of transgenic plants have been developed, including herbicide-resistant cotton, soybean, and rapeseed (canola), insect-resistant corn and potato, and delayed-ripening tomato. Transgenic crops have been grown on about 200 million acres worldwide in close to 20 countries. About 60% of the total acreage is in the United States.

Herbicide-resistant crops allow a more efficient use of herbicides to control weeds. The DNA of these plants is genetically engineered to contain a **tolerance gene** that protects them from certain broad-spectrum herbicides. These broad-spectrum herbicides kill nearly all plants except those containing the tolerance gene. The overall result is improved crop yield and quality. An additional benefit is that these herbicides break down quickly in the soil. The farmer does not need to limit the subsequent planting season's herbicide treatment because of a buildup of herbicide residue in the soil.

The DNA of insect-resistant plants is modified to contain a **toxin gene** that produces a protein toxin that is fatal to many insect larvae such as the corn borer and cotton bollworm. The plant produces the toxin, which is ingested when the insect bores into the plant or chomps on it. The insect dies within a few days.

Tomatoes are normally picked in the hard, green state so that they can be shipped long distances without rotting. At their final destination, they are treated with ethene (Box 4.1), which triggers the ripening process. Transgenic tomatoes, designed to ripen much more slowly, can be harvested later in the ripening process (red instead of green) and still reach market without losses caused by fast ripening during shipment.

The Human Genome Project

The Human Genome Project (HGP) is a project funded by the U.S. Department of Energy and National Institutes of Health with additional contributions from private industry as well as other countries. The project's objective was to identify the chromosomal locations and base sequences of all the genes in the human genome. The driving force for the HGP was the thought that information on the human genome would be critical in battling a variety of genetic diseases, including some cancers. The Human Genome Project was initiated in 1990 and completed in 2003.

Some of the achievements of the HGP are as follows:

- The base sequences of the 3.2 billion base pairs in the human genome have been determined.
- Only about 2% of the base pairs code for polypeptides. The remaining 98% of base pairs appear not to have any function.
- The 2% of base pairs that are in coding regions are contained in 21,000 genes, which is one-fifth the total number of proteins.
- 99.9% of the genome is the same for all people. Differences among people come from differences in base sequences in 0.1% of the genome, corresponding to differences in slightly more than 3 million base pairs.
- The locations of specific genes on specific chromosomes have been determined.

This last achievement is of great importance because it gives precise information on a variety of inherited diseases such as cystic fibrosis, Huntington's, sickle-cell anemia, Tay-Sachs, amylotrophic lateral sclerosis, muscular dystrophy, Down syndrome, certain cancers, Alzheimer's, retinoblastoma, phenylketonuria, and hemochromatosis. We now know which gene on which chromosome is responsible for each inherited disease.

We know the base sequence for each gene. More specifically, the HGP examined the base sequences in "normal" genes for people without inherited diseases and compared that information with the base sequences in the mutated genes present in people with an inherited disease. Data of this kind have the potential for yielding better treatments for diseases.

Gene Therapy

When the base sequence is known for a gene that is missing or defective in certain people, recombinant DNA technology offers two approaches to treatment. One approach is the production of therapeutic drugs, as described earlier, to be taken orally or by injection. However, most genetic diseases do not respond to this approach. The other approach is **gene therapy**—the use of recombinant DNA technology to modify a person's genome directly by inserting the gene that corrects the genetic disease. Research efforts are underway to furnish genes for missing or flawed genes to persons with cystic fibrosis, hemophilia, muscular dystrophy, and other genetic diseases.

The critical step that must be solved before gene therapy is successful is the delivery of recombinant DNA to the appropriate cells in the patient. The common cold virus, adenovirus, is the most-studied gene-delivery system. Researchers modify the adenovirus in two ways. First, certain viral genes are excised to prevent the virus from reproducing in cells and causing illness. Second, the excised genes are replaced with the DNA sequence whose absence is responsible for the genetic disease. A number of techniques are under study for delivery of the altered viral DNA to the patient's cells: aerosol spraying into the air passages; direct injection into the bloodstream; and the removal of cells from the patient followed by incubation with the altered viral DNA and the return of the treated cells to the patient.

Gene therapy is currently only an experimental treatment. There has been very little success in clinical trials. Several factors have prevented gene therapy from achieving success at present:

- Gene therapy alters only a portion of the patient's cells and the altered cells may not be stable. Some patients showed initial positive responses to gene therapy, but these responses were short-lived.
- Gene therapy can trigger a severe response by the patient's immune system. Some deaths have resulted from extremely severe responses.
- Many commonly occurring diseases, such as arthritis, diabetes, and Alzheimer's disease, are caused by simultaneous defects in several genes. The treatment of such diseases by gene therapy would be especially difficult.
- Viral vectors are not easy to control so as to prevent them from introducing problems for the patient.

Gene Testing

The HGP identified the genes responsible for a variety of inherited diseases. This identification allows **gene testing**—the analysis of the base sequences in a person's DNA to identify whether that person possesses mutated genes for inherited diseases. There are now more than 1000 specific tests available for inherited diseases. Such tests enable the screening and counseling of couples who are planning to have children.

For example, consider sickle-cell anemia, a serious and debilitating disease that is inherited when both parents have the mutated gene (Section 12.8). If only one parent has the mutated gene, the offspring will not have sickle-cell anemia. The offspring will have sickle-cell trait, which has much milder symptoms and is manageable.

Gene testing also enables prenatal and newborn screening; presymptomatic testing for adult-onset diseases, such as certain cancers and Alzheimer's and Huntington's diseases; and confirmation of a diagnosis for a symptomatic person.

DNA Fingerprinting

In DNA fingerprinting, DNA analysis is used for convicting felons and exonerating the innocent, establishing paternity and maternity of children, proving family relationships, identifying soldiers killed in war, and identifying people killed in natural disasters.

DNA fingerprinting does not analyze the complete 3.2 billion base pairs of a person's DNA, which would be very costly and time consuming. A shorter method based on the presence of repeated sequences of base pairs in the introns is used. These sequences are called **variable number of tandem repeats (VNTR).** VNTRs for different people differ in their base sequences, numbers of repeated sequences, and lengths of the repeated sequences. These differences are revealed in DNA fingerprinting through the identification of the VNTRs after their isolation from the complete DNA by digestion (hydrolysis) with specific restriction enzymes.

Social, Ethical, and Other Considerations

Although major new technologies have many predictable social effects, history teaches that many consequences of progress will come as a surprise. What will be the effect of recombinant DNA technology? Gene testing allows the testing of people for genetic diseases and for predispositions to conditions such as alcoholism, Alzheimer's disease, and hypercholesterolemia. The positive results include better and earlier treatments and genetic counseling for couples who are considering having children. On the other hand, the same information may be used to restrict people's access to health and life insurances and to employment.

Similarly, human gene therapy, which offers great potential for treating genetic diseases, poses equally great problems. Few would want to deny life-saving treatment to anyone, but some treatments may be prohibitively expensive. If gene therapy can eliminate a genetic disease, can it also enhance human intelligence, physical strength, and athletic ability? What about treating germ cells to improve the genome of the offspring? Who will benefit, and who might suffer as a result? How will these questions be debated and resolved in different cultures and societies throughout the world? A fraction (3–5%) of the federal funds spent on the HGP is allocated for studies of these ethical issues.

Summary

Nucleic Acids

- Genes, the fundamental units of heredity, are responsible for the traits of species and of individual members of a species.
- Each gene is a part of deoxyribonucleic acid and carries the information for synthesizing one or more polypeptides.
- Deoxyribonucleic and ribonucleic acids are responsible for carrying out the hereditary process.

Nucleotides

- Nucleotides are the building blocks for nucleic acids.
- Nucleotides are composed of phosphoric acid, a pentose sugar (either ribose or deoxyribose), and a heterocyclic nitrogen base.
- Adenine, guanine, cytosine, and thymine (A, G, C, and T) are the bases for DNA; uracil (U) replaces T for RNA.

Nucleic Acid Formation from Nucleotides

- Nucleic acids are polymers formed by dehydration between the —OH of the phosphate group at C5′ of one nucleotide residue and the —OH at C3′ of the next nucleotide residue.
- Nucleic acids are named in the $5' \longrightarrow 3'$ direction.

The Three-Dimensional Structure of Nucleic Acids

- The three-dimensional structure of a nucleic acid is determined by its primary structure, the sequence of bases along the polymer chain.
- DNA molecules exist not as individual molecules but as complementary, paired DNA molecules that fold around each other to form a right-handed double helix (duplex).
- The double helix is stabilized by hydrogen bonding and hydrophobic attractions between heterocyclic bases.
- RNA molecules exist as individual molecules with three-dimensional structures determined by intramolecular hydrogen bonding between bases.

Replication, Transcription, and Translation

- Replication is the copying of DNA in the course of cell division and is directed by the DNA itself.
- Transcription is the synthesis of ribosomal RNA, messenger RNA, and transfer RNA from DNA.
- Translation is the synthesis of polypeptides through the combined efforts of rRNA, mRNA, and tRNA.
- The genetic code is the complete description of the α-amino acids specified by various base triplets (codons) on mRNA.
- The genetic code is degenerate and nearly universal.

Mutations

- A mutation is an error in the DNA base sequence of a gene.
- It can yield altered or no biological activity for a polypeptide or protein, because of a change in the α-amino acid sequence.

- A substitution mutation consists of the substitution of one base for another; a frameshift mutation consists of a base being inserted into the normal base sequence or deleted from it.
- Spontaneous mutations are those resulting from random errors in replication.
- Other mutations are produced by radiation and chemical mutagens.
- Mutations in the DNA of germ cells may result in genetic (hereditary) diseases.

Antibiotics

- Antibiotics are organic compounds used to fight infections by bacteria, molds, and yeasts.

Viruses

- Viruses are infectious, parasitic particles, smaller than bacteria, composed of DNA or RNA.
- Viruses carry out life processes only within host cells, where they take over the host's cellular apparatus.
- The most successful approach to fighting viral infections is the preemptive use of vaccines.
- Vaccines contain the weakened virus or its proteins.

Recombinant DNA Technology

- Recombinant DNA technology (genetic engineering, or cloning) alters the genome of an organism by transplanting into it synthetic DNA or DNA from another organism (of the same or a different species).
- This technology allows the production of proteins and therapeutic drugs from altered organisms, agricultural improvements through transgenic breeding, and gene therapy to alleviate genetic diseases.
- The Human Genome Project identified the chromosomal locations and base sequences of all the genes in the human genome. Applications include various recombinant DNA technologies as well as gene therapy, gene testing, and DNA fingerprinting.

Key Words

antibiotic, p. 433
anticodon, p. 419
base-pairing, p. 416
base triplet, p. 426
cancer, p. 432
chromosome, p. 410
cloning, p. 440
codon, p. 426
deoxyribonucleic acid, p. 410
DNA double helix, p. 415
DNA fingerprinting, p. 443
gene, p. 410
gene testing, p. 443
gene therapy, p. 442
genetic code, p. 426
genetic engineering, p. 439
human genome, p. 417
Human Genome Project, p. 442
human proteome, p. 410
mutation, p. 430
nucleic acid, p. 410
nucleotide, p. 410
primary transcript, p. 424
recombinant DNA technology, p. 439
replication, p. 420
ribonucleic acid, p. 410
transcription, p. 420
translation, p. 420
vaccine, p. 438
virus, p. 435

Exercises

Nucleotides

13.1 What sugar is used to synthesize ribonucleotides? deoxyribonucleotides?

13.2 What bases are used to synthesize ribonucleotides? deoxyribonucleotides?

13.3 Draw the structure and give the abbreviated name of the deoxyribonucleotide whose components are phosphoric acid, deoxyribose, and thymine.

13.4 Draw the structure and give the full name of AMP.

13.5 What are the products of the hydrolysis of dGMP?

13.6 What are the products of the hydrolysis of UMP?

Nucleic Acid Formation from Nucleotides

13.7 Where in the cell is DNA found?

13.8 Where in the cell is RNA found?

13.9 Give the abbreviated name for the following dinucleotide:

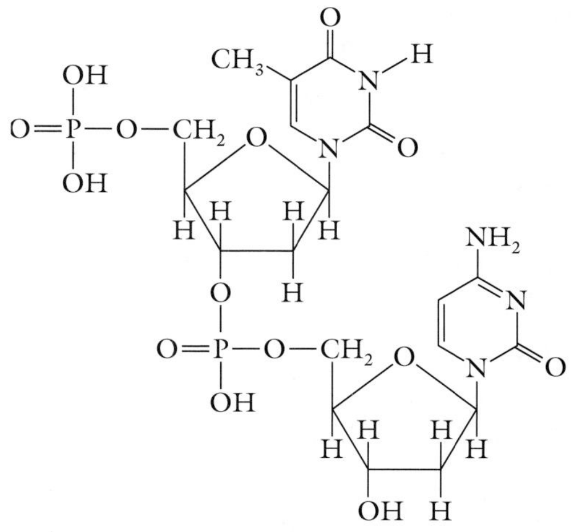

13.10 What is the relation between UMP-CMP and CMP-UMP?

13.11 Why does the sequence dUMP-dGMP not exist in nature?

13.12 Why does the sequence dGMP-CMP not exist in nature?

13.13 Draw the complete structure of dGMP-dTMP. Indicate the $5' \longrightarrow 3'$ direction.

13.14 Draw the complete structure of AMP-UMP. Indicate the $5' \longrightarrow 3'$ direction.

13.15 What nucleotides are obtained by the hydrolysis of dAMP-dTMP-dTMP-dGMP-dCMP?

13.16 What bases are obtained by the complete hydrolysis of dAMP-dTMP-dTMP-dGMP-dCMP? in what relative amounts?

The Three-Dimensional Structure of Nucleic Acids

13.17 What are histones and what is their function?

13.18 What is a DNA duplex?

13.19 What attractive forces result in base-pairing?

13.20 Which are the complementary base pairs?

13.21 Which of the following is a variable part of a specific DNA double helix? (a) Ratio of moles of adenine to moles of thymine present; (b) ratio of moles of sugar to moles of bases present; (c) sequence of sugars attached to the basic units; (d) sequence of bases attached to the sugar units.

13.22 What is the difference, if any, between chromosome, DNA double helix, and gene?

13.23 The DNA contained in the 23 pairs of human chromosomes is 30.4 mol % adenine and 19.6 mol % cytosine. What are the amounts of thymine and guanine?

13.24 One DNA strand in a DNA duplex is 34 mol % A, 28 mol % T, 16 mol % C, and 22 mol % G. What is the composition of the complementary strand?

13.25 The base sequence in a section of one DNA strand is 5′-TAACCG-3′. What is the sequence in the corresponding section of the complementary DNA strand?

13.26 The base composition of one of the DNA chains of a DNA double helix is 18 mol % A, 35 mol % T, 26 mol % C, and 21 mol % G.

(a) What is the base composition of the complementary DNA chain?

(b) Is the total amount of purine bases (A and G) equal to the total amount of pyrimidine bases (C and T) for the DNA double helix?

Information Flow from DNA to RNA to Polypeptide

13.27 Where is genetic information stored?

13.28 Distinguish among replication, transcription, and translation.

Replication

13.29 What is produced by replication?

13.30 Where does replication take place?

13.31 Define or describe the roles of the following elements in replication: (a) replication bubble; (b) leading and lagging strands; (c) Okazaki fragments.

13.32 What are the roles of the following enzymes in replication? (a) DNA polymerase; (b) DNA ligase.

13.33 Distinguish between parent and daughter DNA.

13.34 Why is DNA replication called semiconservative replication?

13.35 The sequence 5′-TTAGCG-3′ is contained in a part of the new DNA strand of a daughter DNA double helix.

(a) What is the corresponding sequence in the old DNA chain of the same daughter DNA double helix?

(b) What is the corresponding sequence in the old DNA chain in the other daughter DNA double helix?

13.36 The sequence 5′-CCTATC-3′ is contained in a part of one strand of a parent DNA double helix.

(a) What is the corresponding sequence in the complementary strand of the parent DNA double helix?

(b) What is the corresponding sequence in the new daughter strand made from this parent strand during replication?

Transcription

13.37 What is produced by transcription?

13.38 Where does transcription take place?

13.39 Which DNA strand is used in transcription, the template or nontemplate DNA strand?

13.40 What is the function of the enzyme RNA polymerase?

13.41 What is the purpose of posttranscriptional processing?

13.42 What chemical reactions take place in posttranscriptional processing?

13.43 Compare the molecular masses of DNA, rRNA, mRNA, and tRNA.

13.44 Define and distinguish between introns and exons. What are their functions?

13.45 The base composition of a DNA template strand for replication is 15 mol % A, 25 mol % C, 20 mol % G, 40 mol % T. What base composition is expected for the RNA synthesized from this template strand?

13.46 Consider the sequence 5′-ACAGGTTAC-3′ in a DNA template strand.

(a) What is the base sequence for the DNA nontemplate strand?

(b) What is the base sequence for the synthesized RNA?

Translation

13.47 What is produced by translation?

13.48 Where does translation take place?

13.49 What are the roles of rRNA, mRNA, and tRNA in translation?

13.50 Distinguish between codon and anticodon, indicating where each is located and their functions.

13.51 What constitutes the genetic message and what is it a message for?

13.52 What is the genetic code?

13.53 What is meant by the statement that the genetic code is nearly universal?

13.54 What is meant by the statement that the genetic code is degenerate?

13.55 What are the roles of aminoacyl-tRNA synthetase, the aminoacyl site, the peptidyl site, and acyl transfer in the translation process?

13.56 What are the functions of inducer and repressor proteins?

13.57 Define base triplet.

13.58 Consider each of the following codons. What is the anticodon on the tRNA complementary to the codon? What α-amino acid is specified by the codon? (a) UCC; (b) CAG; (c) AGG; (d) GCU.

13.59 Sections *X, Y,* and *Z* of the following part of a template DNA strand at its 3′-end are exons of a gene:

What are the structures of the primary transcript and mRNA made from this template DNA? What polypeptide sequence will be synthesized?

13.60 Sections *X, Y,* and *Z* of the following part of a template DNA strand at its 3′-end are exons of a gene:

5′ TAC | AGC | GTA ACC | GAT CCT | CAT 3′
X *Y* *Z*

What are the structures of the primary transcript and mRNA made from this template DNA? What polypeptide sequence will be synthesized?

Mutations

13.61 What is a mutation?

13.62 What is meant by the statement that the genetic code is very nearly mutation resistant?

13.63 Describe the difference between substitution and frameshift mutations.

13.64 Which has the greater potential for harm—insertion, deletion, or substitution mutations?

13.65 What is a spontaneous mutation?

13.66 What is a silent mutation?

13.67 Distinguish between germ and somatic cells.

13.68 What is a genetic (hereditary) disease?

13.69 What is cancer?

13.70 What is a mutagen? a carcinogen?

13.71 Which of the following substitution mutations is likely to be more harmful? Why? (a) Valine is substituted for glutamic acid; (b) aspartic acid is substituted for glutamic acid.

13.72 Which of the following substitution mutations is likely to be more harmful? Why? (a) Lysine is substituted for glutamic acid; (b) lysine is substituted for histidine.

13.73 The following diagram shows part of a template DNA strand at its 3′-end, with sections *Y* and *Z* being the exons of a gene:

(a) What polypeptide sequence will be synthesized?
(b) What polypeptide sequence will be synthesized if the GTC triplet of exon *Y* is mutated to CTC?
(c) What polypeptide sequence will be synthesized if the GTC triplet of exon *Y* is mutated to ATC?
(d) What polypeptide sequence will be synthesized if C is inserted between the two triplets of the intron between exons *Y* and *Z*?

13.74 The following diagram shows part of a template DNA strand at its 3′-end, with sections *Y* and *Z* being the exons of a gene:

(a) What polypeptide sequence will be synthesized?
(b) What polypeptide sequence will be synthesized if the ACT triplet of exon *Y* is mutated to AAT?
(c) What polypeptide sequence will be synthesized if the ACT triplet of exon *Y* is mutated to GCT?
(d) What polypeptide sequence will be synthesized if C is inserted between the two triplets of exon *Y*?

Antibiotics

13.75 What is an antibiotic, and what is its purpose?

13.76 What is the mechanism by which an antibiotic works?

Viruses

13.77 What is a virus?

13.78 How does a virus reproduce?

13.79 What is a vaccine and what is its purpose?

13.80 What is the mechanism by which a vaccine works?

Recombinant DNA Technology

13.81 What is the principle on which recombinant DNA technology is based?

13.82 What benefits are possible through recombinant DNA technology?

13.83 Distinguish among vector DNA, donor DNA, and recombinant DNA in recombinant DNA technology.

13.84 What is the role of *E. coli* plasmids in the production of human insulin?

Unclassified Exercises

13.85 What is the relation between nucleotides and nucleic acids?

13.86 What component besides a sugar and a base is contained in nucleotides?

13.87 Distinguish between the 5′-end and the 3′-end of a nucleic acid.

13.88 Why does the sequence CMP-TMP-AMP not exist in nature?

13.89 Why does the sequence dGMP-dCMP-dUMP not exist in nature?

13.90 Why is base-pairing only between adenine and thymine and between guanine and cytosine in DNA? Why is base-pairing not between adenine and guanine and between thymine and cytosine?

13.91 What nucleotides are obtained by the hydrolysis of dGMP-dTMP-dAMP-dTMP-dTMP-dGMP-dCMP?

13.92 What are the major differences in the primary structures of DNA and RNA?

13.93 What is the major difference in the three-dimensional structures of DNA and RNA?

13.94 The total DNA in the chromosomes of carrots is 36 mol % adenine and 14 mol % cytosine. What are the amounts of thymine and guanine?

13.95 Strand 1 of a DNA double helix is 31 mol % G and 14 mol % T; strand 2 of the double helix contains 17 mol % G and 38 mol % T. Answer the following questions.

(a) What are the amounts of C and A in strand 2?
(b) What are the amounts of C and A in strand 1?

13.96 If one particular mRNA is 26 mol % U and 33 mol % G, what are the amounts of A and C?

13.97 Sections *X*, *Y*, and *Z* of the following part of a template DNA strand at its 3′-end are exons of a gene:

5′ | ACT | CAC | AAA TTG | TGT GGT | CAT | 3′

ACT	CAC	AAA TTG	TGT GGT	CAT
X		*Y*		*Z*

What mRNA is made from this template DNA? What polypeptide sequence will be synthesized?

13.98 Which, if any, of the following statements is (are) true? (a) All mutagens are carcinogens; (b) all carcinogens are mutagens.

13.99 What is a transgenic plant or animal?

13.100 What is the objective of the Human Genome Project?

13.101 What is the objective of gene therapy?

13.102 What do transgenic breeding and gene therapy have in common?

Expand Your Knowledge

Note: These icons denote exercises based on material in boxes.

13.103 At high temperatures, deoxyribonucleic acids become denatured: they unwind from double helices into disordered single strands. Account for the fact that, the higher the content of guanine–cytosine base pairs relative to adenine–thymine base pairs, the higher the temperature required to denature a DNA double helix.

13.104 A mutant hemoglobin has aspartic acid at position 5 of the α-polypeptide chains instead of alanine. This mutation occurred by substitution of a single base in the normal codon for the alanine residue. Give the codons for the normal and mutant residues.

13.105 Explain why the ratio of guanine to cytosine is 1 : 1 in DNA, but it is usually not 1 : 1 in RNA.

13.106 Erwin Chargoff showed that the base compositions of the DNA in different species of plants and animals vary considerably. If two species are found with identical base compositions, does this finding necessarily indicate that the two species have identical DNA? Explain.

13.107 What is the minimum number of nucleotide bases in the gene for the β-polypeptides of human hemoglobin (146 amino acid residues)? Why is the number of nucleotide bases in the gene likely to be much larger?

13.108 The error level in DNA synthesis is much lower than that in RNA synthesis, less than 1 base in 10 billion compared with less than 1 base in 10,000 to 100,000. This difference is due to the fact that DNA polymerases can proofread and correct errors, but RNA polymerases cannot. Because an error of a single base in either replication or transcription can lead to an error in protein synthesis, what is the biological explanation for this difference in error levels between DNA and RNA syntheses?

13.109 Jerry North reads the patient information sheet that came with the antibiotic levofloxacin prescribed for his bacterial infection. There is a precautionary statement that "overuse can lead to the antibiotic's decreased effectiveness." What is the meaning of this precaution?

13.110 Explain why radiotherapy and chemotherapy are effective for treating many cancers even though both therapies damage healthy cells as well as cancer cells (see Box 13.1).

13.111 The human genome consists of 3.2 billion base pairs contained in its 46 chromosomes. Calculate the average molecular mass of a DNA strand, assuming that the average molecular mass of a nucleotide is 340.

13.112 Explain why a genetic code based on codons consisting of only two bases instead of three bases would be inadequate.

13.113 If five different bases were used, would a genetic code based on codons consisting of two bases be adequate to code for the 20 amino acids and the initiation and Stop signals? What major deficiency would such a genetic code have compared with the existing genetic code?

13.114 Distinguish between benign and malignant tumors (see Box 13.1).

13.115 What is the relation between proto-oncogenes and oncogenes (see Box 13.1)?

13.116 Distinguish between HIV and AIDS (see Box 13.2).

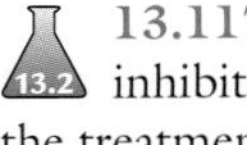

13.117 Describe the functions of reverse transcriptase inhibitors, protease inhibitors, and fusion inhibitors in the treatment of HIV (see Box 13.2).

ENZYMES AND METABOLISM

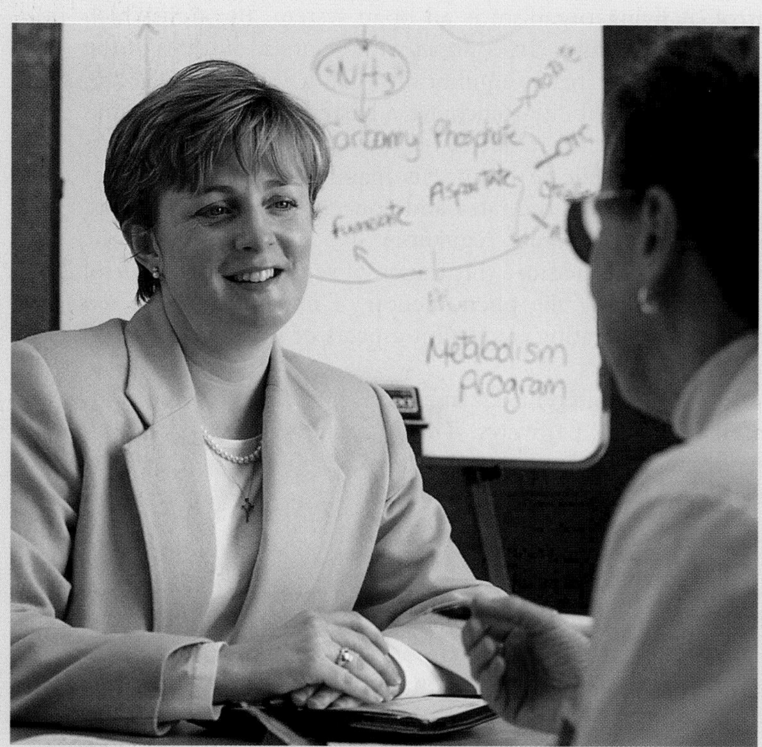

(John Coletti, Boston.)

Chemistry in Your Future

You have applied for a job in a major hospital's inborn errors of metabolism program, and, before your first interview, you decide to brush up on what you learned in nursing school about inherited metabolic disorders. You locate an impressive book titled *Inborn Errors of Metabolism* and, scanning its table of contents, find lengthy lists of disorders. Some you have heard of, and you know that they are rare and difficult to manage, although the high mortality rates of a generation ago are gradually going down. Each of these disorders is caused by the lack of some functional enzyme. What is so special about the role of enzymes in metabolism, anyway? This chapter throws some light on the question.

For more information on this topic and others in this chapter, go to www.whfreeman.com/bleiodian2e

Learning Objectives

- Compare and contrast the characteristics of eukaryotic and prokaryotic cells.
- Describe the sense in which metabolism is a balance between catabolic and anabolic processes.
- Describe how adenosine triphosphate links catabolism to anabolism.
- Explain how enzymes and their modifiable activities make metabolism possible.
- Describe the role of high-energy compounds in metabolism, and give an example of how they work.

The preceding four chapters considered the structures and properties of the four principal families of organic biomolecules: the carbohydrates, lipids, proteins, and nucleic acids. In the body, these biomolecules combine to form structures (such as ribosomes, lysosomes, the Golgi apparatus, and mitochondria) whose conjoint task is to extract energy from the environment and use it to sustain life. Chapters 14 through 18 examine these molecular and subcellular structures for the purpose of understanding how they perform their biochemical functions—the functions known collectively as metabolism.

A key theme in any discussion of metabolism is the central importance of regulation and control. A living organism is an open system, through which mass and energy flow at varying rates. In contrast with chemical systems in the laboratory, which will eventually reach equilibrium, living systems never do. The life of each cell rests on the ability to achieve a balance between degradative processes that break tissues and molecules down and synthetic processes that build new ones up. This delicate balance is maintained by a variety of regulatory mechanisms that control the rates and concentrations of enzymes. We will not explore the details of every regulatory step in all the metabolic processes considered herein. But, because this theme of regulation and control is so important for understanding the phenomenon of metabolism, selected examples of regulation, demonstrating important aspects of molecular control mechanisms, will be presented.

« **The principles underlying enzyme structure are covered in Chapter 12; the DNA-directed synthesis of enzymes is described in Chapter 13.**

« **The basic concepts of biochemical catalysis are introduced in Section 2.2.**

Metabolism requires many different enzymes and their catalyzed reactions but only a few central metabolic pathways. These central pathways, virtually identical in all forms of life, will be the principal topic of this chapter and the following three chapters.

14.1 CELL STRUCTURE

In chemistry laboratories and in industrial manufacturing facilities, temperatures of hundreds of degrees, pressures of tens of atmospheres, organic solvents, and extremes of pH are commonly required to induce chemical compounds to react. Moreover, these reactions take place in vessels millions of times as large as living cells. Chemical reactions in cells, on the other hand, take place in aqueous media at neutral pH and at low and constant temperatures. The cell itself is an extraordinarily fragile vessel of extremely small dimensions. Let us start with a brief survey of its interior.

For our purposes, the animal world can be divided into two types of organisms, called **eukaryotes** and **prokaryotes.** The biochemistry of the two groups is strikingly similar. Moreover, in both groups, all cells are enclosed within a complex membrane capable of exquisite control over the entry and exit of nutrients and waste products (Section 11.10). The differences between eukaryotes and prokaryotes lie in their internal organization and modes of reproduction. Eukaryotes include all cells of multicellular organisms, such as the vertebrates, and many single-celled organisms, such as the yeasts and the protists *Euglena* and *Paramecium*. Almost all prokaryotes are bacteria.

In eukaryotic cells, many of the cellular macromolecules are packaged into **organelles,** subcellular structures surrounded by their own membranes. Organelles form separate functional compartments within the cell. The internal landscape of prokaryotes, as we shall soon see, is quite different.

Figure 14.1 is a sketch of a typical animal cell, identifying its principal compartments and structures, and Table 14.1 lists their major functions. The cell is surrounded by a **cell membrane** responsible for recognizing both small and large molecules and for the transport of molecules and ions into and out of the cell. The area outside of all the compartments in the cell, filled with a jellylike background substance, is called the **cytosol.**

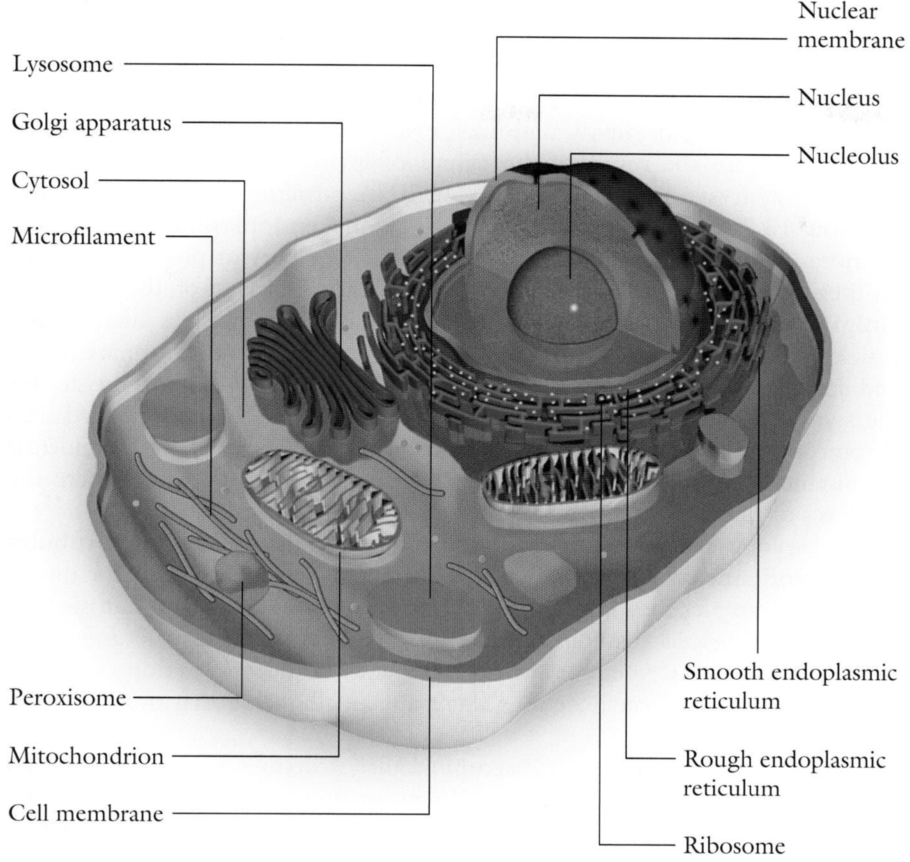

Figure 14.1 Drawing of a typical animal cell identifying its principal compartments and structures.

Biochemists have adopted the words aerobic and anaerobic from bacteriologists, who denote cell growth in the presence or absence of oxygen, respectively. Some anaerobic bacteria can grow only in the absence of oxygen and are called obligate anaerobes. Others can grow in the absence or presence of oxygen and are called facultative anaerobes. The biochemical use of the word **aerobic** describes a chemical reaction that requires oxygen. Biochemically, the word **anaerobic** describes a chemical reaction in which oxygen is not a reactant, but the reaction can take place in the presence of oxygen. In fact, oxygen is present in the cytosol, and anaerobic metabolic processes take place there. Aerobic metabolic processes, those requiring oxygen as a reactant, take place

TABLE 14.1 Animal-Cell Compartments and Their Major Functions

Compartment	Functions
cell membrane	transport of ions and molecules; receptors for biomolecules
nucleus	DNA synthesis and repair; RNA synthesis
nucleolus	ribosome synthesis
endoplasmic reticulum	synthesis of proteins and the lipids used for the manufacture of cell organelles; lipid synthesis
ribosome	protein synthesis; usually attached to the endoplasmic reticulum
Golgi apparatus	modification of proteins both for export and for incorporation into organelles
mitochondrion	cellular respiration, oxidation of lipids and carbohydrates; conservation of energy; urea synthesis
peroxisome	reactions with oxygen, decomposition of hydrogen peroxide
lysosome	hydrolysis of carbohydrates, proteins, lipids, and nucleic acids
microfilament	cell cytoskeleton; intracellular movements
cytosol	metabolism of carbohydrates, lipids, amino acids, and nucleotides; protein synthesis

chiefly in **mitochondria.** Mitochondria are subcellular organelles that possess both an outer and an inner membrane specialized for specific transport of molecules and electrons.

Lysosomes are the cell's digestive system, responsible for the hydrolysis of carbohydates, proteins, lipids, and nucleic acids. In Section 16.7, we list some diseases that are the result of the failure of lysosomes to hydrolyze certain lipids. Protein synthesis takes place at RNA structures called **ribosomes** (Section 13.7). Ribosomes are attached to the **endoplasmic reticulum**—membranous structures that, in addition to taking part in protein and lipid synthesis, help to transport macromolecules into and out of the cell. The **Golgi apparatus** is a network of flattened, smooth membranes surrounding vesicles in which proteins synthesized on the endoplasmic reticulum are modified and packaged for transport out of the cell. It is also an important site for the synthesis of new membrane material. Eukaryotic DNA is organized into chromosomes located within the **nucleus,** which is enclosed by the nuclear membrane. Within the nucleus is a structure called the **nucleolus,** where the synthesis of ribosomes takes place.

Eukaryotic cells also possess a cytoskeleton consisting of **microtubules** and **microfilaments.** They give cells a shape and provide a means for movement; for example, they impel the movement of chromosomes in mitosis (cell division) and meiosis (haploid germ-cell formation). At small structures called **peroxisomes,** direct reactions with oxygen and the decomposition of hydrogen peroxide take place. Reactions there are usually those that detoxify toxic substances.

In prokaryotes, subcellular macromolecular structures either are anchored directly to the inside of the cell membrane or move about freely within the cell. The DNA is organized into a single chromosome located in a **nuclear zone,** which is not surrounded by a membrane. Prokaryotic cells have little internal membranous structure, but their internal protein concentration is very high, about 20%. This high concentration leads to a very large internal osmotic pressure, perhaps from 2 to 3 atm. No lipid-based cell membrane could withstand such high internal pressure, but prokaryotic cells possess, in addition to the cell membrane, a strong, rigid polysaccharide framework called a cell wall. The cell wall is strong enough to resist the internal osmotic pressures typical of prokaryotes.

14.2 GENERAL FEATURES OF METABOLISM

Metabolism in animal cells has four functions: (1) to obtain energy in a chemical form by the degradation of nutrients; (2) to convert a wide variety of nutrient molecules into the few precursor molecules used to build proteins, carbohydrates, nucleic acids, lipids, and other cell molecules; (3) to synthesize cell molecules; and (4) to produce or modify the biomolecules necessary for specific functions in specialized cells or both.

Metabolism takes place through the interaction of two processes: catabolism and anabolism. **Catabolism** is the biochemical degradation of energy-containing compounds, leading to the capture of that energy in new chemical forms that the cell can use for biosynthesis, movement, and secretion. As a result of catabolic reactions, the many different types of energy-containing molecules are converted into a limited number of simpler molecules, which are then used for the synthesis of components vital to cell structure and function.

Anabolism is the biochemical synthesis of biomolecules. A few simple molecules serve as building blocks to create a large variety of macromolecules and complex cellular components. These simple, building-block molecules have other uses as well. Amino acids are also the precursors for hormones, alkaloids, and porphyrins, and they serve as neurotransmitters. Nucleotides are also precursors of energy carriers and coenzymes. Anabolism includes the work necessary to transport ions and biomolecules across cell membranes.

Stoichiometrically, the oxidation of D-glucose in the body is identical with its combustion under commercial or laboratory conditions; that is

$$C_6H_{12}O_6 + 6\ O_2 \longrightarrow 6\ CO_2 + 6\ H_2O$$

In those combustions, the energy contained in D-glucose is converted into heat. However, in cellular metabolism, some of the reaction energy is captured. The metabolic process consists of sequences of reactions in which nutrients are broken down to simpler molecules while, at the same time, substances that fuel biosynthetic processes are synthesized. These substances, called high-energy compounds, are discussed in Section 14.9. It is in this sense that metabolism extracts energy from the environment and uses it to sustain life.

14.3 STAGES OF CATABOLISM

Catabolism, the degradation of the major energy-yielding compounds ingested by cells, proceeds in the sequence of steps shown in Figure 14.2. These steps can be grouped into three major stages.

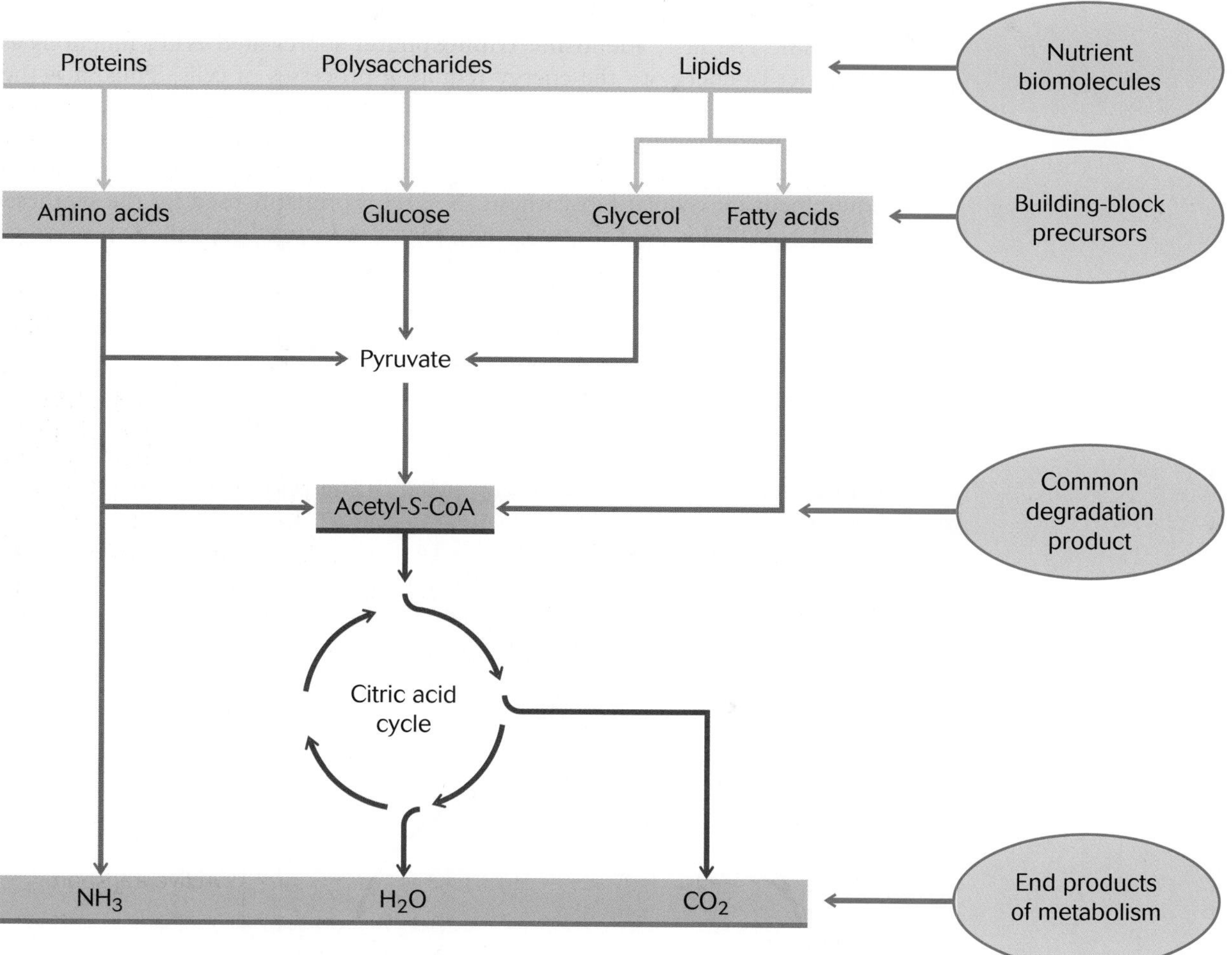

Figure 14.2 Catabolism of the major energy-yielding compounds of cells proceeds in a sequence of steps that consists of three major stages: (1) nutrient biomolecules are degraded to monosaccharides, amino acids, glycerol, and fatty acids; (2) these molecules are converted into a common degradation product, acetyl-*S*-CoA, which is (3) degraded into the end products carbon dioxide and water.

In the first stage, nutrient molecules are degraded to their lower-molecular-mass components: α-amino acids, monosaccharides, fatty acids, and glycerol.

The next stage converts these different products into one simple molecule—**acetyl-*S*-coenzyme A** (acetyl-*S*-CoA). The collected monosaccharides are first converted into a three-carbon intermediate, pyruvate, and then into the two-carbon derivative acetyl-*S*-CoA (Section 14.9). The carbon chains of fatty acids and most of the amino acids also are converted into the same two-carbon derivative. The second stage of catabolism ends with the production of this one molecule from the myriad types entering the degradative process.

In the final stage of catabolism, acetyl-*S*-CoA enters the **citric acid cycle.** In this part of the pathway, acetyl-*S*-CoA is oxidized to carbon dioxide and water.

Concept check

✓ Catabolism is characterized by the conversion of the many different types of nutrient molecules into a final, common end product.

14.4 THE TRANSFORMATION OF NUTRIENT CHEMICAL ENERGY INTO NEW FORMS

Much of the energy of nutrients captured in the second and third stages of catabolism is used to synthesize three important compounds that fuel anabolism. The first, **adenosine triphosphate,** abbreviated **ATP,** functions as the carrier of energy to the energy-requiring processes of cells. Thus, it is the link between catabolism and anabolism. The second and third products, **NADH (nicotinamide adenine dinucleotide)** and **NADPH (nicotinamide adenine dinucleotide phosphate)** capture the reducing power that is a key requirement for cellular biosynthesis. NADH is primarily used for the synthesis of ATP. NADPH is used almost exclusively for reductive biosynthetic processes.

Figure 14.3 is a representation of how ATP links the processes of catabolism to those of anabolism. In general, as ATP yields its energy to drive an anabolic reaction, it is broken into its hydrolysis products, adenosine diphosphate (ADP) and inorganic phosphate (see Box 7.5). (Inorganic phosphate, consisting of all species of phosphate, HPO_4^{2-}, $H_2PO_4^-$, and PO_4^{3-}, is represented by the symbol P_i for convenience in describing phosphorylation reactions in metabolism. Phosphorylation describes a reaction in which the hydroxyl group of an alcohol is converted into a phosphate ester group; see Section 7.5.)

The ATP molecule, seen in Figure 14.4, is composed of the nitrogen base adenine, ribose, and phosphates esterified to carbon 5′ of the ribose group. Figure 14.4 indicates the cleavage point for the conversion of ATP into ADP. Note that, at cellular pH, the phosphate groups of both ATP and ADP are ionized.

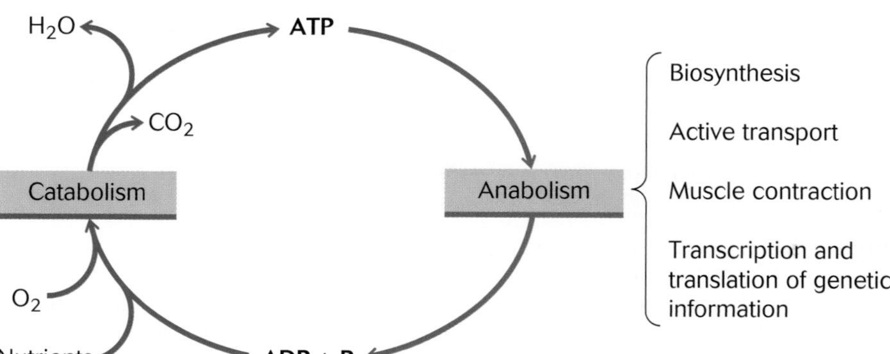

Figure 14.3 ATP is the principal link between the processes of catabolism and anabolism.

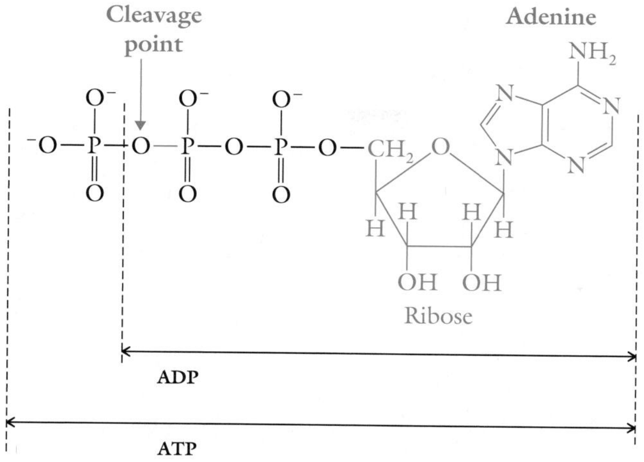

Cleavage point
Adenine
ADP
ATP
Ribose

Figure 14.4 The structure of ATP is composed of the nitrogen base adenine, ribose, and phosphates esterified to carbon 5′ of the ribose group. Note the cleavage point for the hydrolysis of ATP into ADP.

ATP synthesis takes place in two ways. In one way, ADP reacts directly with P_i. This synthesis of ATP is associated with the simultaneous oxidation of NADH in a sequence of reactions within mitochondria called the electron-transport chain. This form of ATP synthesis is called **oxidative phosphorylation** and will be considered in detail in Section 15.7. In the other way in which ATP is synthesized, called **substrate-level phosphorylation,** a phosphate group is transferred to ADP directly from another phosphorylated compound. (These phosphorylated compounds, called high-energy compounds, are considered in more detail in Section 14.9.)

NAD^+ (Figure 14.5) is one of a class of substances that link metabolic oxidations to metabolic reductions. Metabolic oxidations, like any redox reaction (Section 2.1), consist of an oxidation half-reaction and a reduction half-reaction. For the most part, metabolic oxidations are not direct reactions with oxygen but consist of transfers of hydrogen atoms and electrons and are better described as **dehydrogenations.** In a metabolic dehydrogenation,

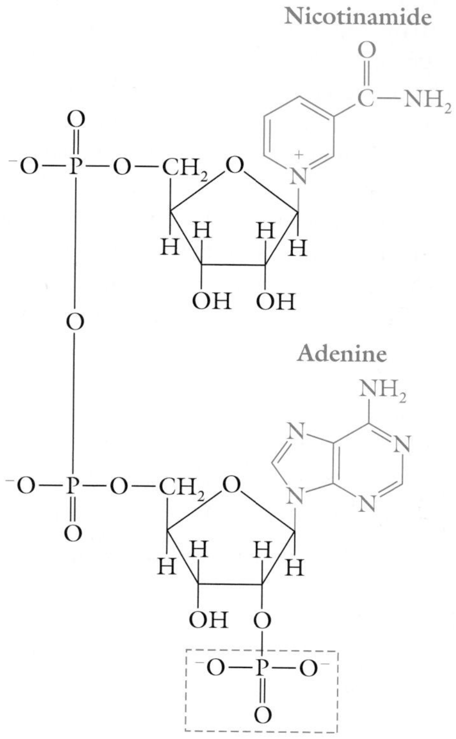

Nicotinamide
Adenine

Figure 14.5 The structure of NAD^+ (and of $NADP^+$) shows that, in the oxidized state, the nicotinamide ring has a positive charge. In the reduced state of NAD(P)H, the positive charge is neutralized when the hydride ion is transferred to the nicotinamide ring. The phosphate group outlined by red dashed lines, esterified to the 2′-OH of the ribose group of NAD^+, transforms the NAD^+ into $NADP^+$, nicotinamide adenine dinucleotide phosphate.

Figure 14.6 The structure of flavin adenine dinucleotide (FAD). The arrows point to the nitrogen atoms that become hydrogenated when FAD is reduced to form $FADH_2$.

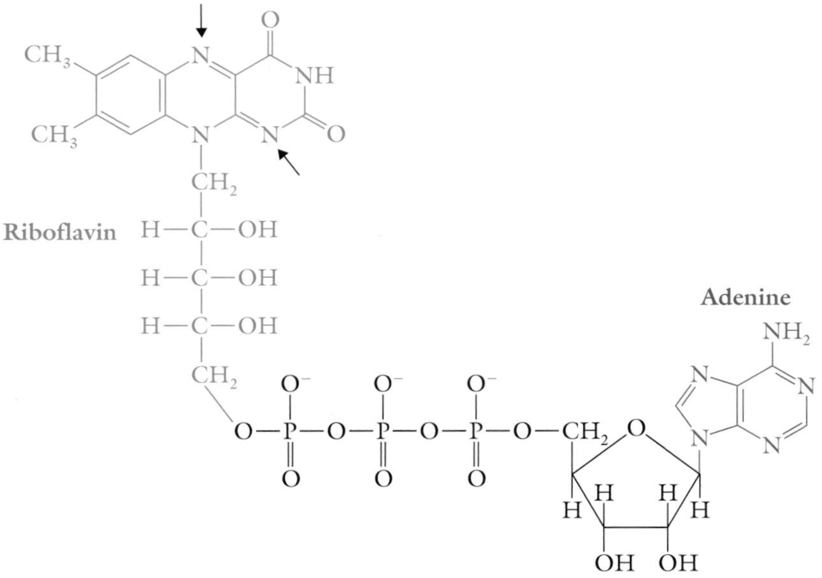

Riboflavin

Adenine

Flavin adenine dinucleotide (FAD)

two hydrogen atoms and two electrons are transferred to an acceptor. Although comparatively few, direct reactions with oxygen do take place, and they are called oxygenations.

In catabolic dehydrogenations, NAD^+ receives electrons and a hydrogen atom in the form of a hydride ion ($H:^-$) to form NADH (see Figure 6.8). NADH can then be oxidized in subsequent reactions through which hydrogen atoms and electrons are eventually transferred to oxygen to form water. NADPH is chiefly responsible for providing the electrons and hydrogen atoms in anabolic syntheses—for example, the hydrogenation of double bonds in the formation of steroids.

The structures of NAD^+ and $NADP^+$ are identical except for the phosphate group, outlined by red dashed lines in Figure 14.5, esterified to the 2′-OH of the ribose group of NAD^+. The addition of that phosphate group transforms the NAD^+ into **$NADP^+$** (nicotinamide adenine dinucleotide phosphate). The complex adenine nucleotide part of the molecule does not participate in redox reactions. Its function is to provide the structural properties that allow interaction with a specific enzyme.

Note that, in the oxidized state, the nicotinamide ring has a positive charge, and, in the reduced state (NADH), the positive charge is neutralized when the hydrogen atom and its electrons are transferred to the nicotinamide ring. Because NAD^+ and $NADP^+$ undergo the same reduction reaction, biochemists symbolize both of them by the formula $NAD(P)^+$. When $NAD(P)^+$ participates in the dehydrogenation of reduced nutrients, one hydrogen atom with two electrons is transferred as a hydride ion ($H:^-$) to the nicotinamide ring, whereas the second hydrogen atom enters the reaction medium as a proton (H^+). With AH_2 representing the reduced nutrient, the typical reaction is

$$AH_2 + NAD(P)^+ \longrightarrow A + NAD(P)H + H^+$$

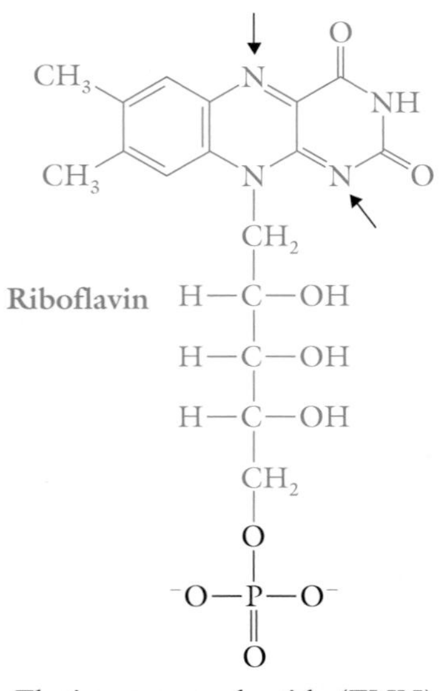

Flavin mononucleotide (FMN)

Figure 14.7 The structure of flavin mononucleotide (FMN). FMN is essentially phosphorylated riboflavin. Reduction (hydrogen addition) takes place at the nitrogen atoms identified by the arrows.

The other important substances that link metabolic oxidations to metabolic reductions are the flavin nucleotides, FAD (flavin adenine dinucleotide; Figure 14.6) and FMN (flavin mononucleotide; Figure 14.7). In contrast with the pyridine nucleotides [$NAD(P)^+$], both hydrogen atoms are incorporated into a flavin nucleotide when it is reduced:

$$AH_2 + FAD \longrightarrow A + FADH_2$$

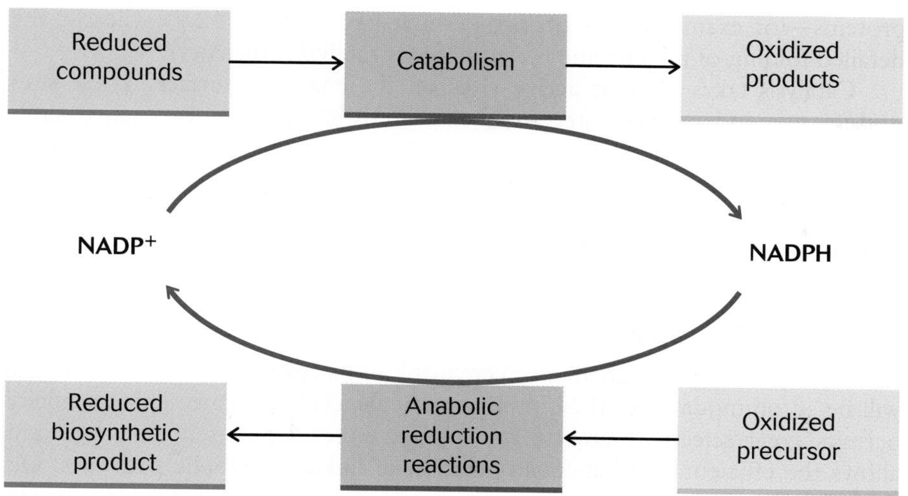

Figure 14.8 $NADP^+$ links the oxidation reactions of catabolism to the reduction reactions of anabolism. NADPH is primarily used for reductive biosynthetic reactions.

To function as hydrogen acceptors or donors, the pyridine and flavin nucleotides must be bound to enzymes specific for the particular reaction taking place. $NAD(P)^+$ is loosely bound to its enzyme, but FAD is so tightly bound that its combination with its enzyme is called a flavoprotein. As we shall see, they are cofactors (Section 14.5) in enzymatic dehydrogenations. The linking of the oxidizing reactions of catabolism and the reducing requirements of anabolism is illustrated in Figure 14.8.

Concept checklist

- ✓ The chemical energy obtained from nutrients during the second and third stages of catabolism is used to build the major products, ATP and NAD(P)H.
- ✓ Adenosine triphosphate functions as the carrier of energy to the energy-requiring processes of cells.
- ✓ NAD(P)H provides the reducing power required for cellular biosynthesis.

14.5 ENZYMES

A single living cell is a far more efficient chemical factory than has ever been devised by chemical engineers. Multicellular organisms synthesize thousands of different types of proteins, as well as other molecules, all within a single cell having a volume of about 1×10^{-12} cm^3.

The cell is able to channel biomolecules into the construction of specific products in precisely the required amounts because its reactions are linked into specific sequences that can be controlled in a number of ways. The key to this control is the existence of unique catalysts called enzymes. Enzymes as catalysts were first considered in Section 2.2. The amounts and activities of the enzymes catalyzing each step of each reaction sequence can be modified, allowing the cell to carry out its metabolic functions in an integrated, harmonious way.

Enzymes Are Proteins

Although recent work has demonstrated modest catalytic activity in some nucleic acids, the overwhelming evidence is that enzymes are protein molecules possessing unique catalytic properties that depend on their structural integrity.

When an enzyme is hydrolyzed by the digestive enzyme trypsin, its catalytic activity disappears completely. Thus, the primary structure (the backbone) of the enzyme is necessary for the enzyme to function. Catalytic activity will also be lost if the enzyme is subjected to conditions that denature proteins—for example, heat or low or high pH (Section 12.9). Clearly, the detailed folding of the enzyme is central to its catalytic efficiency.

Catalysis takes place at **active sites** on the enzyme's surface. These sites possess two independent structural characteristics: part of the active site's structure provides the catalytic activity, and part functions as a binding site. A binding site is usually a cleft or indentation occupying a very small part of the enzyme's surface. At that location, noncovalent or secondary forces bring **substrate** molecules—the substance on which the enzyme acts—into close contact with the chemical groups responsible for catalytic activity.

Enzymes Are Specific

« » Specific binding sites are also important in cell recognition (Section 10.6), lipid transport (Section 18.3), and nerve transmission (Section 18.5).

Many enzymes possess great specificity: only molecules of a specific structure will be accommodated at the binding site of a specific enzyme. This specificity permits great selectivity among the myriad molecules present in a cell and allows the efficient simultaneous operation of linked metabolic pathways. On the other hand, there are enzymes acting outside of cells that have a broad specificity and act on many different compounds that have a common structural feature. Typical enzymes of this type are the digestive enzymes, each of which can catalyze the hydrolysis of broad classes of molecules.

The binding site accepts its substrate in much the way that a lock accepts its key: the substrate's structure complements the structure of the binding site. This **complementarity** of structure includes stereochemistry and shape (Section 9.5 and Figure 9.2) as well as electrical charge and hydrophobic–hydrophilic considerations. In other words, the shape and charge of a molecule must be complementary to the shape and charge of the binding site. Moreover, the affinity of a hydrophilic molecule for a binding site is enhanced if that substrate also has some hydrophobic character that is matched by that of the binding site. In certain cases, the shape of an enzyme's binding site changes on binding of substrate. This phenomenon, called **inducible fit** because of the flexible nature of the protein, enhances the complementary character of the binding site as it binds substrate.

There are substances that can change the activity of an enzyme by changing the way in which a substrate is bound to the enzyme. Such substances are known as **inhibitors.** Substances that structurally resemble the substrate but undergo no reaction or very slow reaction are known as **competitive inhibitors.** They bind to the active site and can be displaced by an increase in the concentration of the normal substrate; so, the inhibition is reversible. For example, methanol, CH_3OH, a toxic alcohol, competes with ethanol, CH_3CH_2OH, the alcohol found in ordinary alcoholic beverages, for the active site on the enzyme alcohol dehydrogenase. One of the products of methanol dehydrogenation is the principal toxin. Therefore, if methanol can be prevented from binding to the enzyme, it can't be oxidized, and, because it is water soluble, in due time it will be excreted without doing damage. A patient who has ingested methanol can be treated by having the patient drink copious quantities of ethanol, which successfully competes for the enzyme's active site and displaces methanol. The same prescription is used for accidental ingestion of ethylene glycol, $HO{-}CH_2{-}CH_2{-}OH$.

Other substances not structurally related to substrate also can reduce an enzyme's activity. They do so by binding to a site distant from the catalytic site, which alters the protein's conformational structure. This change in conformational structure, in turn, alters the structure of the catalytic site and changes the enzyme's activity. This type of inhibition is called **noncompetitive inhibition,** and it cannot be reversed by increasing the concentration of substrate. As an example, some enzymes possess sulfhydryl (—SH) groups that are distant from the active sites of enzymes, but are important for maintaining the integrity of their conformational structures. Heavy metals, such as Ag^+, or

TABLE 14.2 **Inorganic Cofactors and Their Enzymes**

Cofactor	Enzyme
Cu^{2+}	cytochrome oxidase
Fe^{2+}, Fe^{3+}	catalase; cytochrome oxidase; peroxidase
K^{+}	pyruvate kinase (also needs Mg^{2+})
Mg^{2+}	hexokinase; glucose 6-phosphatase
Mn^{2+}	arginase
Ni^{2+}	urease
Se^{2+}	glutathione peroxidase

Hg^{2+}, will bind to those sites and modify the conformational structure so that enzyme activity is reduced (Section 12.9).

Enzyme Cofactors and Coenzymes

Most metabolic reactions are catalyzed by proteins that are combined with specialized small molecules called cofactors or coenzymes. Cofactors or coenzymes combine transiently with a noncatalytic protein, called an **apoenzyme,** to form the catalytic **holoenzyme.** The purpose of a **cofactor** is to maintain the protein conformation appropriate for the binding of substrate to form the activated complex. A list of some cofactors can be found in Table 14.2. Note that they are usually specific ions. Although acid- or base-catalyzed metabolic reactions—that is, hydrolyses—can be carried out by proteins alone, the catalysis of oxidations and reductions or group transfers cannot be carried out by any of the amino acid residues found in proteins. These functions are effected by small molecules called **coenzymes** that are transiently bound to apoenzymes. A list of coenzymes is given in Table 14.3. Many of the dietary vitamins, such as the B vitamins, become coenzymes and serve as temporary carriers of atoms or functional groups in oxidation–reduction and group-transfer reactions. The pyridine and flavin nucleotide cofactors are derived from the vitamins nicotinamide and riboflavin, respectively.

14.6 ENZYME CLASSIFICATION

The early phases of every science are characterized by much naming but only rudimentary understanding. Thus, enzymes were named as they were discovered, before very much was known about their structure or the way in which they worked. In many cases, chemists simply added the suffix -**ase** to the name of the compound acted on by the enzyme, as in hexokinase, urease, and cytochrome oxidase. In other cases, the names (trypsin or pepsin, for example) do not indicate the substrates. Although the historical names are still in

TABLE 14.3 **Vitamins and Corresponding Coenzymes Serving as Group-Transfer Carriers**

Vitamin	Coenzyme	Group transferred
biotin	biocytin	carbon dioxide
folic acid	tetrahydrofolate	other one-carbon groups
pantothenic acid	coenzyme A	acyl groups
cobalamin (vitamin B_{12})	5′-deoxyadenosylcobalamine	alkyl groups, hydrogen atoms
riboflavin	flavin adenine dinucleotide	hydrogen atoms
niacin (nicotinic acid)	nicotinamide adenine dinucleotide	hydride ion ($H:^-$)
pyridoxine (vitamin B_6)	pyridoxal phosphate	amino groups
thiamine	thiamine pyrophosphate	aldehydes

TABLE 14.4 Enzyme Classification Based on Catalyzed Reactions

Class	Reaction type
oxidoreductase	electron transfer
transferase	group-transfer reactions
hydrolase	hydrolysis reactions
lyase	addition or removal of groups to or from double bonds
isomerase	conversion of one isomer into another by group transfer within a molecule
ligase	formation of C–C, C–S, C–O, and C–N bonds by condensation reactions coupled to ATP cleavage

common use, a new system of nomenclature has been developed that eliminates ambiguity in naming and classifying enzymes. This system places an enzyme in one of six major classes, listed in Table 14.4, with subclasses based on the reactions catalyzed.

Hexokinase is the common name for the enzyme catalyzing the reaction between adenosine triphosphate and hexoses such as D-glucose:

$$\text{D-Glucose} + \text{ATP} \rightleftharpoons \text{D-glucose-6-}PO_4 + \text{ADP}$$

The formal systematic name for the enzyme is ATP: hexose phosphotransferase. This name indicates that the enzyme catalyzes the transfer of a phosphate group from ATP to any hexose. The phosphotransferase specific for glucose is called glucokinase. Although the formal nomenclature eliminates ambiguity, its use is often quite cumbersome. Most writers and workers in the field use the simpler historically derived common or trivial names of enzymes unless the situation warrants formal nomenclature, and that will be our policy in this book. As stated in Section 10.2, the D form of monosaccharides predominates in nature. As a consequence, whenever the stereochemistry of a saccharide is not indicated, such as in "glucose" or "fructose," you should assume that the D-enantiomer is meant. The enantiomer will be identified in chemical equations.

14.7 ENZYME ACTIVITY

When an enzyme is discovered, one of the first aspects to be studied is its effect on the rate of its catalyzed reaction: how many moles of substrate are converted into product per unit time? To answer this question, the experimenter adds increasing concentrations of substrate to a fixed concentration of enzyme. For example, the experimenter might prepare a rack of test tubes each containing 0.01 mM of enzyme solution (Appendix 4) and then add a different amount of substrate to each one (Figure 14.9). After the initial rate of reaction in each tube has been measured and plotted as a function of substrate concentration, many enzymes (for example, pepsin or salivary amylase) produce curves resembling the one in Figure 14.9. The rate of the enzymatically catalyzed reaction is called the **enzyme activity.**

The graph in Figure 14.9 shows that the rate of this enzyme-catalyzed reaction reaches a maximum value when the substrate concentration exceeds the enzyme concentration. At this point, the enzyme has reached its capacity to bind substrate and has become "saturated." A fixed concentration of enzyme means a fixed and limited number of active sites. The rate of reaction at each site is constant, so an increase in the substrate concentration past the enzyme saturation point cannot increase the rate of reaction. When the enzyme is saturated with substrate, the enzyme activity is described by the **turnover number**: the number of substrate molecules turned into product by

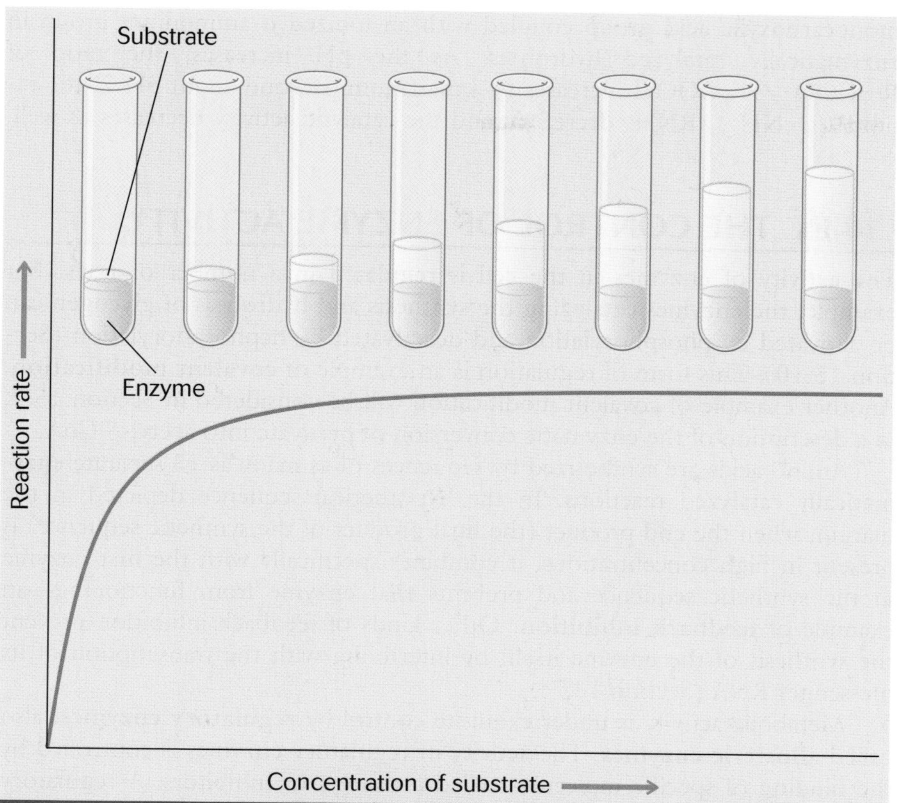

Figure 14.9 The rates of an enzyme-catalyzed reaction are measured in a series of test tubes in which the enzyme concentration is kept constant and the substrate concentration is continually increased. The rate of the enzyme-catalyzed reaction increases and reaches a limit as the substrate concentration increases.

one molecule of enzyme in unit time. The added substrate must wait its turn to enter the active site, just as any passenger must do at a crowded bus stop.

When the reaction rate becomes constant, it is possible to define the enzyme activity in terms of the ratio of catalyzed rate to uncatalyzed rate. For example, the enzyme carbonic anhydrase catalyzes the formation of bicarbonate ion in the reaction of carbon dioxide with water. Found in high concentrations in the human erythrocyte, it is among the most active of all enzymes. The ratio of enzyme-catalyzed rate to the uncatalyzed rate of hydration of CO_2 is: $\text{rate}_{\text{catalyzed}}/\text{rate}_{\text{uncatalyzed}} = 1 \times 10^7$. In other words, the catalyzed reaction takes place 10 million times as fast as the uncatalyzed reaction.

Because enzymes are proteins, they can be denatured at extremes of pH and temperature. However, changes in temperature that do not cause denaturation will result in changes in reaction rates that are reversible. Studies of the variation in reaction rates with variation in temperature have provided researchers with activation energies for detailed descriptions of mechanisms. The effect of pH on reaction rates can reveal the identity of acid–base groups in catalysis. Figure 14.10 is a typical bell-shaped curve implicating an ionized

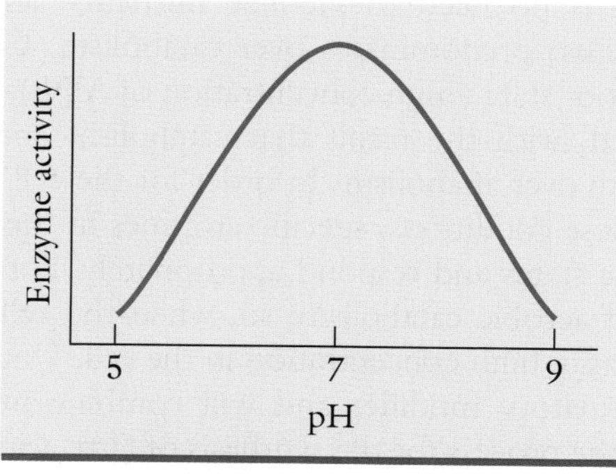

Figure 14.10 The increase and decrease in enzyme activity as the pH increases from pH 5.0 to pH 9.0.

monocarboxylic acid group coupled with an ionized α-ammonium group in enzymatically catalyzed hydrolysis. As the pH increases, the ratio of $R{-}COO^-/R{-}COOH$ increases to a maximum. Beyond about pH 7, the ratio of $R{-}NH_3^+/RNH_2$ decreases, and the catalytic activity decreases as well.

14.8 THE CONTROL OF ENZYME ACTIVITY

$Substrate_1$
$Enzyme_1$
$Substrate_2$
$Enzyme_2$
$Substrate_3$
$Enzyme_3$
Product

The activity of enzymes in the cell is regulated in a number of ways. For example, the enzymes catalyzing the synthesis and hydrolysis of glycogen can be activated by phosphorylation and deactivated by dephosphorylation (Section 15.10). This form of regulation is an example of **covalent modification.** Another example of covalent modification will be considered in Section 15.2, in a description of the enzymatic conversion of pyruvate into acetyl-*S*-CoA.

Amino acids are synthesized by sequences of as many as 13 separate enzymatically catalyzed reactions. In the hypothetical sequence depicted in the margin, when the end product (the final product of the synthetic sequence) is present in high concentrations, it combines specifically with the first enzyme in the synthetic sequence and prevents that enzyme from functioning—an example of **feedback inhibition.** Other kinds of feedback inhibition prevent the synthesis of the enzyme itself, by interfering with the transcription of its messenger RNA (Section 13.7).

Metabolic activity is under exquisite control by **regulatory enzymes,** also called **allosteric enzymes.** The activity of regulatory enzymes is controlled by the binding of specific molecules called activators or inhibitors. A regulatory modifier of an allosteric enzyme combines not with the enzyme's active site but with another specific binding site of the enzyme that is complementary to the modifier's structure. The binding of the modifier molecule at the nonenzymatic site, as illustrated in Figure 14.11, in turn modifies the enzyme's conformational structure (Section 12.5), causing changes in its catalytic activity. Those changes can be either positive, leading to increased activity, or negative, leading to inhibition or decreased activity.

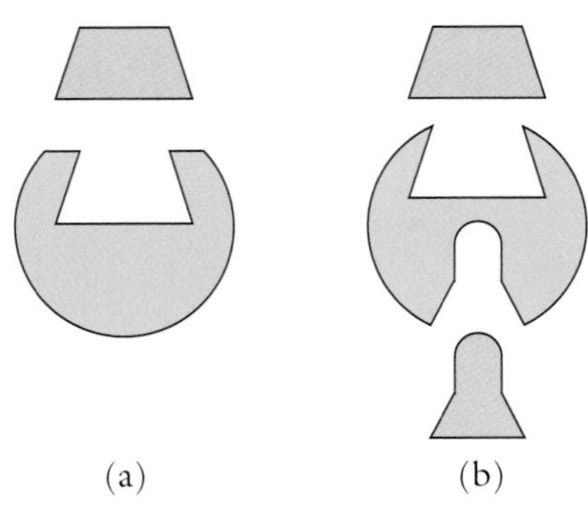

Figure 14.11 Representations of (a) a nonregulatory enzyme as described in Section 2.2 and (b) a regulatory enzyme. The difference between the two is that the regulatory enzyme possesses a binding site distant and, more importantly, different from the substrate-binding site. The specifically structured modifier molecule binds to this distant site and modifies the enzyme's activity.

A different kind of regulation is effected by the covalent modification of enzymes by other enzymes. The first of this type to be elucidated was the enzyme phosphorylase, which catalyzes the release of free glucose from glycogen, its storage form. The inactive form of this enzyme is converted into its active form by the catalytic addition of phosphate to certain serine residues, called phosphorylation, and is converted back into its inactive form by the catalytic removal of those groups, called dephosphorylation. The catalysis is carried out by different enzymes. We will consider these conversions in detail in Section 15.10. It is important to note that, since the discovery of this phenomenon, it has been found to be at the heart of a wide variety of metabolic controls over cellular activities such as intra- and intercellular communication.

Regulatory enzymes control the balance between energy-rich and energy-poor states of the cell. When the cell is in an energy-rich state (high concentration of ATP), there is no need to sacrifice nutrients, and the result is that anabolism (biosynthesis) predominates over catabolism. Conversely, when the cell is in an energy-poor state (high concentration of ADP), the synthesis of biomolecules is curtailed, with the result that catabolism and the production of ATP will predominate over anabolism. In order for the cell's metabolic machinery to respond to these conditions, specific enzymes in the metabolic pathways must recognize those states and respond appropriately. For example, citrate is a major component of aerobic catabolism; so, when the cell is in a high-energy state, citrate is present in high concentration in the cell. Under those conditions, citrate acts as a regulatory modifier and will combine with and activate the enzyme that begins the process for the synthesis of fatty acids. (Section 16.6.)

Concept check

✓ Regulatory enzyme activity is modified by specific, low-molecular-mass activator or inhibitor molecules that bind to an enzyme and alter its conformation.

Some enzymes are present in the cell as individual catalytic molecules. Others combine into structural complexes consisting of many different enzymes. The enzymes responsible for the synthesis of fatty acids in human cells, for example, are present in the cell cytosol in the form of a single multifunctional protein that catalyzes seven different reactions. The enzymes of aerobic respiration are arranged in close contact with one another within the membranes of mitochondria. The rates at which sequences of metabolic reactions take place within such complexes are markedly enhanced by the enzymes' organization into closely arranged, functionally integrated structures.

14.9 HIGH-ENERGY COMPOUNDS

Remember that catalysts cannot alter the equilibrium of a chemical reaction; they only increase the rate of approach to equilibrium. Many metabolic reactions have small equilibrium constants; that is, they are unfavorable and yield small quantities of product. They could not take place within cells without the presence of ATP and a small number of other **high-energy compounds** to drive them forward. The mechanism for achieving this impetus was detailed in Example 2.8, which showed how a reaction with a small equilibrium constant can be coupled to a favorable reaction through a common intermediate to produce a large overall equilibrium constant.

High-energy compounds tend to contain groups such as phosphate and acetate, whose transfer to water is characterized by very large equilibrium constants. These compounds turn up repeatedly in each metabolic pathway, serving as common intermediates in reactions whose equilibrium constants are small. To illustrate again how the strategy works, we will examine the first reaction of anaerobic metabolism, in which glucose is phosphorylated by ATP.

The reaction equation is

$$\text{D-Glucose} + \text{ATP} \xrightleftharpoons{\text{hexokinase}} \text{D-glucose-6-PO}_4 + \text{ADP}$$

It is critical to note that this reaction, like all others of metabolism, requires a specific enzyme to proceed at a rate appropriate to the time requirements of metabolism. The enzyme is specific in two key respects: (1) it transfers a phosphate group (all enzymes that catalyze the transfer of a phosphate group are called kinases) and (2) it is structurally adapted to hexose sugars.

In the enzyme-catalyzed reaction, ATP reacts directly with glucose. However, if we picture this single-step catalyzed reaction as taking place in two sequential steps with a common intermediate, the role of ATP in this reaction and similar ones can be made more clear. The two sequential steps will be (1) the reaction between glucose and inorganic phosphate and (2) the reaction of ATP with water. The first step, the direct reaction of glucose with inorganic phosphate, is characterized by a small equilibrium constant, whereas the second step, the hydrolysis of ATP, has a very large equilibrium constant:

1. $\text{D-Glucose} + \text{P}_i \rightleftharpoons \text{D-glucose-6-PO}_4 + \text{H}_2\text{O}$ $\quad K_1 = 3.8 \times 10^{-3}$
2. $\text{ATP} + \text{H}_2\text{O} \rightleftharpoons \text{ADP} + \text{P}_i$ $\quad K_2 = 2.26 \times 10^{6}$

When the two equations are added together, the P_i and H_2O on each side cancel, resulting in the equation describing the one-step process:

3. $\text{D-Glucose} + \text{ATP} \rightleftharpoons \text{D-glucose-6-PO}_4$

« Section 2.3 presents an example of how to calculate the overall equilibrium constant for a sequence of reactions.

The equilibrium constant for the one-step process is obtained by multiplying the equilibrium constants of reactions 1 and 2:

$$K_1 \times K_2 = K_3 = 8.6 \times 10^3$$

The overall equilibrium constant now favors the formation of D-glucose-6-PO_4.

High-energy compounds are compounds that have large equilibrium constants in certain reactions, such as the transfer of phosphate or other groups to water (HOH) and other hydroxyl-bearing (ROH) compounds. However, keep in mind that a favorable equilibrium constant does not always mean that a product is formed quickly. It turns out that the activation energies for direct transfer to water are quite high, and the rates of hydrolysis for these compounds would be very slow if it were not for the presence of transferase enzymes such as hexokinase, which accelerate the reaction rate.

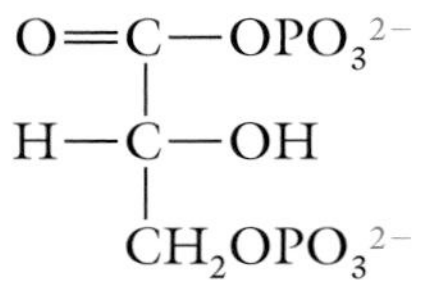

There are other high-energy compounds in metabolism that have even larger equilibrium constants for phosphate transfer than does ATP. As a consequence, the ATP–ADP system can operate as a phosphate shuttle between the high-energy compounds with large equilibrium constants for phosphate transfer and others with lower equilibrium constants for phosphate transfer. The high-energy compounds can transfer, for example, a phosphate group to ADP, which then can transfer that phosphate group to glucose. Reactions in which parts of molecules such as phosphate groups ($—PO_4$), amino groups ($—NH_2$), or methyl groups ($—CH_3$), are transferred from one molecule to another are called group-transfer reactions.

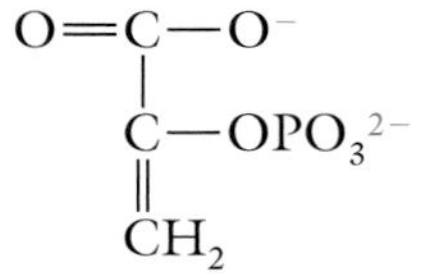

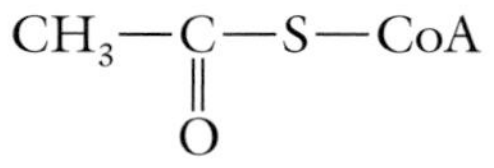

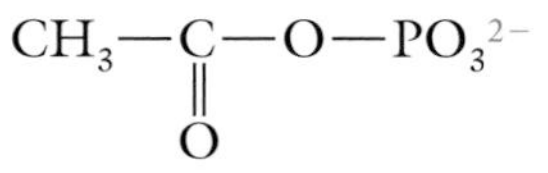

$O{=}C—OPO_3^{2-}$
$H—C—OH$
$CH_2OPO_3^{2-}$

1,3-Bisphosphoglycerate

$O{=}C—O^-$
$C—OPO_3^{2-}$
$\|$
CH_2

Phosphoenol pyruvate

$CH_3—C({=}O)—S—CoA$

Acetyl-*S*-CoA

$CH_3—C({=}O)—O—PO_3^{2-}$

Acetyl phosphate

Two such compounds are **1,3-bisphosphoglycerate** and **phosphoenolpyruvate.** Both are synthesized in the initial stage of glucose catabolism and in turn are used to synthesize ATP by substrate-level phosphorylation; that is, direct phosphate-group transfer from a phosphorylated high-energy compound to a phosphate-accepting compound. Other high-energy compounds include the **thioesters,** such as **acetyl-*S*-CoA,** and **acid anhydrides,** such as **acetyl phosphate.**

Concept check

✓ Reactions that include high-energy compounds enable the chemically unfavorable reactions of metabolism (that is, reactions characterized by small equilibrium constants) to produce adequate quantities of desired products.

Full details of the biochemical processes of catabolism and anabolism will be presented in subsequent chapters, beginning in Chapter 15 with a discussion of carbohydrate metabolism.

Summary

General Features of Metabolism

- Metabolism is the sum of the physical and chemical processes with which an organism extracts energy from the environment and uses that energy for sustaining life.
- Living systems are never at equilibrium: they are open systems that are balanced between degradative processes, called catabolism, and synthetic processes, called anabolism.
- The balance between catabolism and anabolism is maintained by control over the activity of enzymes and their concentrations.

Stages of Metabolism

- In the first stage of catabolism, nutrient macromolecules are degraded to their building blocks: amino acids, monosaccharides, fatty acids, and glycerol.

• In the next stage, hexoses are converted into acetyl-*S*-coenzyme A. The carbon chains of fatty acids and most of the amino acids also are converted into that same two-carbon derivative.

• In the next stage of catabolism, acetyl-*S*-CoA enters the citric acid cycle and is oxidized to carbon dioxide.

• The connection between catabolism and anabolism is made by the conversion of the chemical energy of nutrients chiefly into adenosine triphosphate (ATP) and reducing power in the form of NADH and NADPH.

Enzymes, Cofactors, and Coenzymes

• Enzymes—protein catalysts—are central to the integrated operation of all the metabolic pathways within cells.

• Enzymes permit chemical reactions to take place rapidly at low temperatures in aqueous environments and at neutral pH in cells.

• They are structurally specific for their substrates, and many of them require the presence of nonprotein molecules (cofactors and coenzymes) to function.

• The activity of allosteric or regulatory enzymes is modified by molecules that are not their substrates.

• Other controls of enzyme activity act by covalent modification and by feedback inhibition.

High-Energy Compounds

• High-energy compounds act as common intermediates in group-transfer reactions and cause the overall equilibrium constants for metabolic reactions to be highly favorable.

Key Words

Exercises

Cell Structure

14.1 In what way is biochemistry a new dimension of biology?

14.2 What are the two fundamental types of cells?

14.3 What is the chief difference between the two fundamental types of cells?

14.4 What is the cytosol?

14.5 Where do oxidative processes take place in animal cells?

14.6 Where is DNA located in animal cells?

14.7 Where does protein synthesis take place in animal cells?

14.8 What is the endoplasmic reticulum?

14.9 Where is DNA located in bacterial cells?

14.10 Why do bacteria possess cell walls in addition to cell membranes?

General Features of Metabolism

14.11 What are the functions of metabolism?

14.12 Define catabolism.

14.13 Define anabolism.

14.14 In what general way does the oxidation of glucose in the test tube differ from the oxidation of glucose in the body?

14.15 What is the function of the first stage of catabolism?

14.16 What takes place in the second stage of catabolism?

14.17 What takes place in the final stage of catabolism?

14.18 What are the two major products derived from the energy in nutrients?

14.19 What is the most fundamental role played by ATP in metabolism?

14.20 What chemical process does ATP undergo when its energy is used in a metabolic process?

14.21 What are the two processes by which ATP is synthesized?

14.22 Write the reaction describing the fate of hydrogen when NAD^+ or $NADP^+$ takes part in a dehydrogenation.

14.23 What functions differentiate NAD^+ from $NADP^+$?

Enzymes, Cofactors, and High-Energy Compounds

14.24 To what family of biomolecules do enzymes belong, and what role do they play in the cell?

14.25 What is an enzyme's active site, and what functions does the active site serve?

14.26 What is complementarity?

14.27 What is meant by inducible fit?

14.28 What are coenzymes, and why are they necessary?

14.29 What factors in our diets are an important source of coenzymes?

14.30 What is the meaning of enzyme activity?

14.31 Define turnover number.

14.32 What is feedback inhibition?

14.33 What is a regulatory enzyme?

14.34 True or false: Enzymes always operate as individual molecules. Explain your answer.

14.35 What is the chemical role of high-energy compounds?

Expand Your Knowledge

14.36 Are the biochemical processes of living cells at chemical equilibrium?

14.37 How is the metabolic machinery of cells able to handle the many different types of molecules used as nutrients?

14.38 Is the energy of nutrients used directly and without modification in the metabolic activity of cells? Explain.

14.39 Do all enzymes operate at full activity at all times?

14.40 How are biochemically unfavorable reactions able to produce adequate amounts of products?

14.41 Is it possible for an enzyme to require both a cofactor and a coenzyme to be catalytically active? Explain your answer.

14.42 Explain how a molecule can serve as both an inhibitor and a substrate in an enzyme system characterized by feedback inhibition.

14.43 This diagram represents a feedback-inhibition system. Each arrow with its number represents an enzyme-catalyzed reaction. Each letter represents a substrate molecule. Which substrate is the inhibitor, and, if it can inhibit only one reaction, which one must it inhibit to prevent the synthesis of the final product?

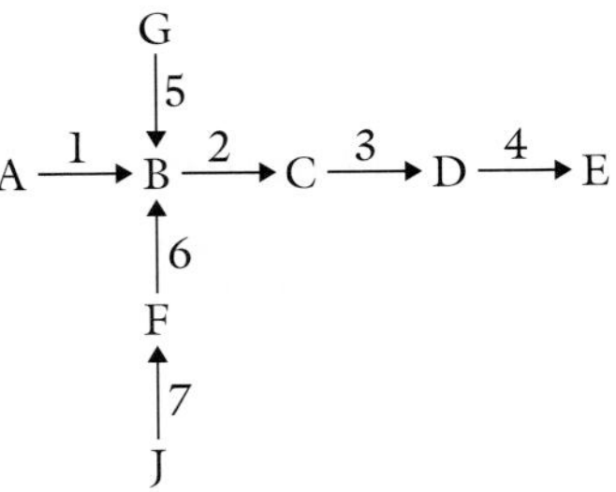

14.44 Propose a model for the mechanism of mercury toxicity in the body.

14.45 How does competitive enzyme inhibition differ from noncompetitive enzyme inhibition?

14.46 What would be the effect of a high fever on the body's metabolic activity?

14.47 Is all enzyme inhibition reversible? Explain your answer.

14.48 What is the meaning of enzyme control by covalent modification?

14.49 True or false: Enzymes reduce the heat of reaction of a chemical process. Explain your answer.

14.50 Can acid–base enzyme catalysis be detected by variation in temperature?

CHAPTER 15 CARBOHYDRATE METABOLISM

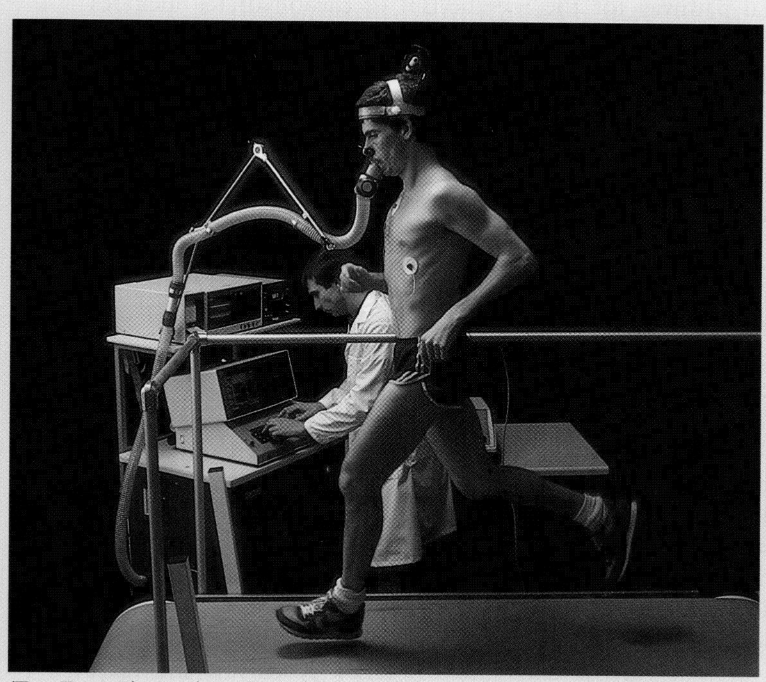

(Tom Tracy/The Stock Market.)

Chemistry in Your Future

An increasingly busy sports physician has asked you, her assistant, to write a list of general guidelines for clients to follow whenever they are preparing for a demanding athletic event. She's provided some notes to get you started, the first of which reads, "Increase carbohydrate consumption, for efficient ATP synthesis." How can you reword this guideline in a way that will motivate your clients? This chapter should give you some ideas.

For more information on this topic and others in this chapter, go to www.whfreeman.com/bleiodian2e

Learning Objectives

- Describe the first stage of carbohydrate catabolism, glycolysis, which takes place in the cell cytosol.
- Describe the second stage of carbohydrate catabolism, the citric acid cycle, which takes place in the cell's mitochondria.
- Describe gluconeogenesis, the synthesis of glucose from the product of glycolysis.
- Describe the role of hormones in the synthesis and degradation of glycogen.
- Describe how ATP is synthesized through the operation of the electron-transport chain.
- Describe the selective permeability of mitochondria.
- Explain the comparative efficiencies of glycolysis and the citric acid cycle.

In Chapter 14, you learned that catabolism has three stages. In the first stage, nutrients are degraded to monosaccharides, α-amino acids, and fatty acids. In the second stage, hexoses are converted into pyruvate, a three-carbon intermediate. Pyruvate, the carbon chains of fatty acids, and many of the α-amino acids are then converted into a unique two-carbon derivative, acetyl-*S*-coenzyme A (acetyl-*S*-CoA). In the final stage of catabolism, acetyl-*S*-CoA enters a pathway in which it is oxidized to carbon dioxide and water. Two other end products—ammonia and urea—are subsequently produced by other pathways.

The biochemical pathway for the metabolism of carbohydrates includes the catabolic pathway through which α-amino acids and fatty acids also are oxidized. Our examination of carbohydrate metabolism here, in Chapter 15, prepares the way for a fruitful consideration of the metabolism of α-amino acids and fatty acids in Chapters 16 and 17.

15.1 GLYCOLYSIS

Glucose is the principal nutrient that fuels metabolism. Like most saccharides in biological systems, the glucose in our bodies consists of the D-enantiomer (Section 10.3). Whenever the stereochemistry of a biological saccharide or its derivatives is not indicated, such as in "glucose" or "fructose," you should assume that the D-enantiomer is meant.

As pointed out in Section 3.10, we do not regularly balance equations describing organic chemical reactions. The reasons were spelled out in that chapter and will not be repeated here. What is important to note is that the equations describing biochemical reactions will often be treated the same way. We wish to bring attention to the transformations of one substance into another, rather than focusing on considerations of mass or charge balance. Where such considerations are of primary importance, they will be presented in detail, as in Section 15.7.

«« Major cell structures are described in Section 14.1.

The first pathway that glucose encounters is called **glycolysis** and takes place in the cell cytosol. In glycolysis, the six-carbon glucose molecule is converted in a sequence of nine enzymatically catalyzed steps into two molecules of pyruvate, a three-carbon molecule. In the process, ATP and NADH are produced. The pyruvate produced in glycolysis proceeds to be oxidized, first, to acetyl-*S*-CoA and, then, to carbon dioxide and water in the next pathway of metabolism, the citric acid cycle (Section 15.5).

Glycolysis itself can be seen as consisting of two major stages. In the first stage, glucose and other hexoses, such as fructose and galactose, are converted into the three-carbon product glyceraldehyde-3-phosphate. In the second stage, glyceraldehyde-3-phosphate undergoes oxidation and then rearrangement into two high-energy phosphorylated compounds that are used to phosphorylate ADP. Although two ATPs are necessary to start the process, four ATPs are produced by it in phosphorylations of ADP during the process. Both NAD^+ and inorganic phosphate (P_i) are required.

The stoichiometry of the process, including priming by ATP, is

$$\underset{\text{Glucose}}{C_6H_{12}O_6} + 2\ ATP + 2\ NAD^+ + 2\ ADP + 2\ P_i \longrightarrow$$
$$2\ \underset{\text{Pyruvate}}{C_3H_4O_3} + 4\ ATP + 2\ NADH + 2\ H^+ + 2\ H_2O$$

By subtracting the priming ATP from both sides, we find that the net result is

$$C_6H_{12}O_6 + 2\ NAD^+ + 2\ ADP + 2\ P_i \longrightarrow$$
$$2\ C_3H_4O_3 + 2\ ATP + 2\ NADH + 2\ H^+ + 2\ H_2O$$

Figure 15.1 diagrams that glycolytic pathway.

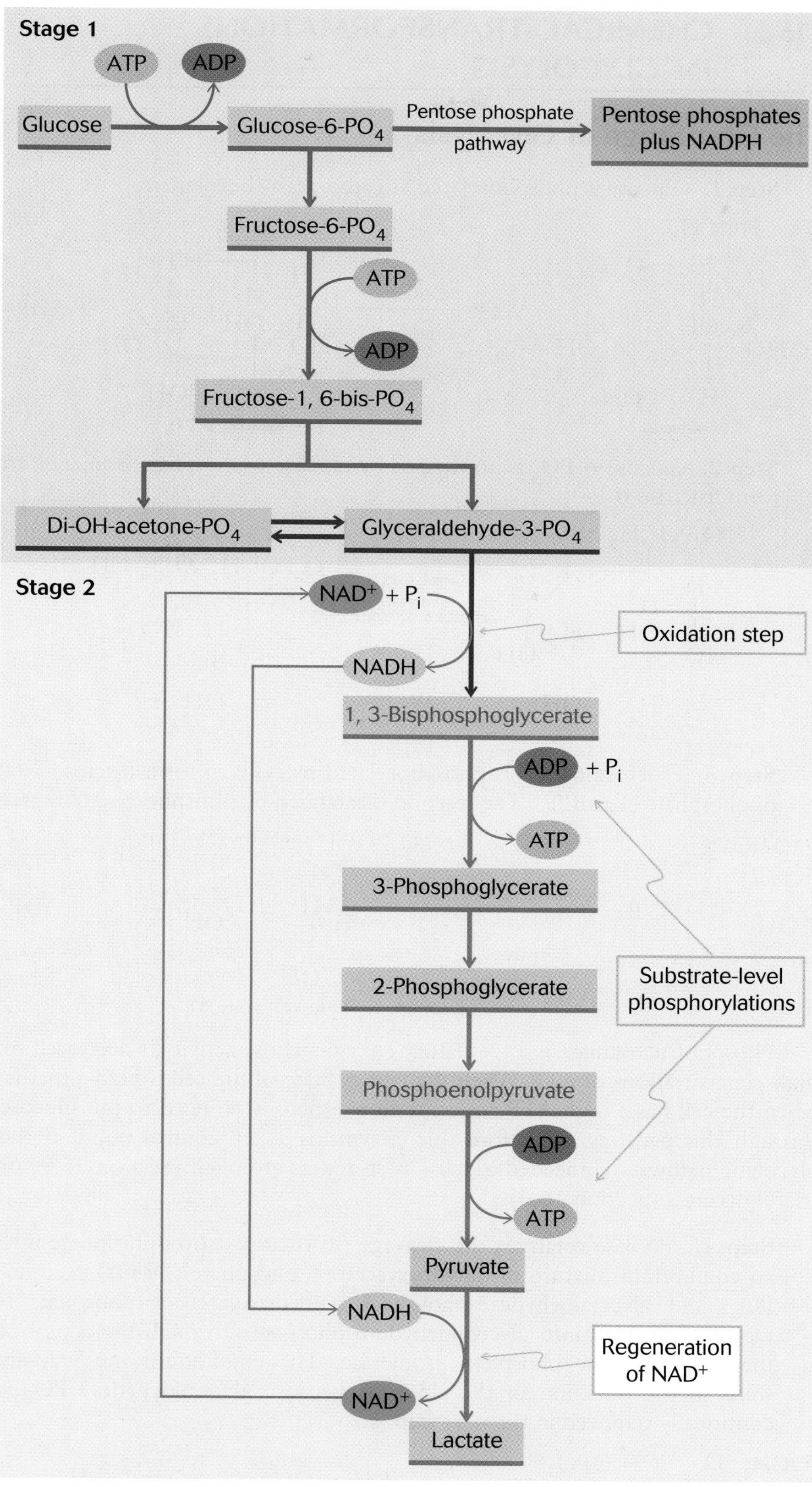

Figure 15.1 The glycolytic pathway. Note the steps for the production of the high-energy compounds 1,3-bisphosphoglycerate and phosphoenolpyruvate, the two substrate-level phosphorylations of ADP, and the regeneration of NAD^+. Keep in mind that 2 mol of glyceraldehyde-3-PO_4 are produced for each mole of glucose.

15.2 CHEMICAL TRANSFORMATIONS IN GLYCOLYSIS

The First Stage of Glycolysis

Step 1. Glucose is phosphorylated at carbon 6 by hexokinase.

Glucose + ATP $\xrightarrow{\text{hexokinase}}$ Glucose-6-PO$_4$ + ADP

Glucose **Glucose-6-PO$_4$**

Step 2. Glucose-6-PO$_4$ is isomerized by glucose-6-phosphate isomerase to form fructose-6-PO$_4$.

Glucose-6-PO$_4$ $\underset{\text{isomerase}}{\overset{\text{glucose-6-PO}_4}{\rightleftharpoons}}$ Fructose-6-PO$_4$

Glucose-6-PO$_4$ **Fructose-6-PO$_4$**

Step 3. Fructose-6-PO$_4$ is phosphorylated by ATP to form fructose-1,6-bisphosphate (1,6-BPF). The reaction is catalyzed by **phosphofructokinase.**

Fructose-6-PO$_4$ + ATP $\xrightarrow{\text{phosphofructokinase}}$ Fructose-1,6-bis-PO$_4$ + ADP

Fructose-6-PO$_4$ **Fructose-1,6-bis-PO$_4$**

Phosphofructokinase is a regulatory enzyme whose activity is decreased by high concentrations of ATP. When the energy state of the cell is high—that is, when the cell has a high ATP concentration—there is no need to put glucose through this pathway. Therefore this enzyme is a key control point in the glycolytic pathway. Unneeded glucose is stored as glycogen (Section 15.9) or triacylglycerols (Section 16.6).

Step 4. Aldolase catalyzes the cleavage of fructose-1,6-bisphosphate into an equilibrium mixture of dihydroxyacetone phosphate (Di-OH-acetone-PO$_4$) and glyceraldehyde-3-phosphate. Dihydroxyacetone phosphate is rapidly converted into glyceraldehyde-3-phosphate through the action of the enzyme triose phosphate isomerase. The equilibrium then rapidly shifts in the direction of the aldehyde because glyceraldehyde-3-PO$_4$ is continually removed in the next step (step 5).

Fructose-1,6-bis-PO$_4$ $\overset{\text{aldolase}}{\rightleftharpoons}$ Dihydroxyacetone-PO$_4$ + Glyceraldehyde-3-PO$_4$

Fructose-1,6-bis-PO$_4$ **Dihydroxyacetone-PO$_4$** **Glyceraldehyde-3-PO$_4$**

In effect, 1 mol of glucose has been transformed into 2 mol of glyceraldehyde-3-PO_4. In the next steps, we will follow the fate of 1 mol of this three-carbon molecule, but keep in mind that two such molecules have been formed from 1 mol of glucose.

The Second Stage of Glycolysis

Step 5. The aldehyde group of glyceraldehyde-3-PO_4 (3-PGAL) is oxidized to a carboxyl group and esterified with inorganic phosphate to form the high-energy compound 1,3-bisphosphoglycerate (1,3-BPG). The reaction is catalyzed by glyceraldehyde-3-PO_4 dehydrogenase, in which NAD^+ is reduced to NADH.

$$\begin{array}{c} O{=}C{-}H \\ | \\ H{-}C{-}OH \\ | \\ CH_2OPO_3^{2-} \end{array} + NAD^+ + P_i \xrightleftharpoons[\text{dehydrogenase}]{\text{glyceraldehyde-3-}PO_4} \begin{array}{c} O{=}C{-}OPO_3^{2-} \\ | \\ H{-}C{-}OH \\ | \\ CH_2OPO_3^{2-} \end{array} + NADH + H^+$$

3-PGAL → **1,3-Bisphosphoglycerate**

Step 6. Phosphoglycerate kinase catalyzes the phosphorylation of ADP by the high-energy compound 1,3-bisphosphoglycerate to form ATP and 3-phosphoglycerate. Phosphorylation of ADP to form ATP by reaction with a phosphorylated high-energy compound is called substrate-level phosphorylation.

$$\begin{array}{c} O{=}C{-}OPO_3^{2-} \\ | \\ H{-}C{-}OH \\ | \\ CH_2OPO_3^{2-} \end{array} + ADP \xrightleftharpoons[\text{kinase}]{\text{phosphoglycerate}} \begin{array}{c} O{=}C{-}O^- \\ | \\ H{-}C{-}OH \\ | \\ CH_2OPO_3^{2-} \end{array} + ATP$$

1,3-BPG → **3-Phosphoglycerate**

Step 7. Phosphoglyceromutase catalyzes the rearrangement of 3-phosphoglycerate into 2-phosphoglycerate.

$$\begin{array}{c} O{=}C{-}O^- \\ | \\ H{-}C{-}OH \\ | \\ H{-}C{-}OPO_3^{2-} \\ | \\ H \end{array} \xrightleftharpoons{\text{phosphoglyceromutase}} \begin{array}{c} O{=}C{-}O^- \\ | \\ H{-}C{-}OPO_3^{2-} \\ | \\ H{-}C{-}OH \\ | \\ H \end{array}$$

3-Phosphoglycerate → **2-Phosphoglycerate**

Step 8. Enolase then catalyzes the dehydration of 2-phosphoglycerate to yield the phosphorylated enol form of pyruvate, phosphoenolpyruvate (PEP), the second high-energy compound produced by the glycolytic process.

$$\begin{array}{c} O{=}C{-}O^- \\ | \\ H{-}C{-}OPO_3^{2-} \\ | \\ HO{-}C{-}H \\ | \\ H \end{array} \xrightleftharpoons{\text{enolase}} \begin{array}{c} O{=}C{-}O^- \\ | \\ C{-}OPO_3^{2-} \\ \| \\ CH_2 \end{array} + H_2O$$

2-Phosphoglycerate → **Phosphoenolpyruvate**

Step 9. ADP is then enzymatically phosphorylated by pyruvate kinase, and the enol form of pyruvate rapidly rearranges nonenzymatically to the keto form of pyruvate.

$$\begin{array}{c} O{=}C{-}O^- \\ | \\ C{-}OPO_3^{2-} \\ \| \\ CH_2 \end{array} + ADP \xrightleftharpoons{\text{pyruvate kinase}} \begin{array}{c} O{=}C{-}O^- \\ | \\ C{=}O \\ | \\ CH_3 \end{array} + ATP$$

PEP → **Pyruvate**

Glycolysis can continue only if NAD^+ can be regenerated from its reduced form, NADH, produced in step 5. In active skeletal muscle, this regeneration is accomplished through the next step (step 10), which results in the production of large quantities of lactate. The lactate diffuses from the muscle cells, enters the circulation, and is transported to the liver, where it is converted back into glucose in a process called gluconeogenesis (Section 15.8).

Step 10. NAD^+ is regenerated by the reduction of pyruvate to lactate, catalyzed by lactic dehydrogenase.

$$\begin{array}{c} O{=}C{-}O^- \\ | \\ C{=}O \\ | \\ CH_3 \\ \textbf{Pyruvate} \end{array} + NADH + H^+ \underset{}{\overset{\text{lactic dehydrogenase}}{\rightleftharpoons}} \begin{array}{c} O{=}C{-}O^- \\ | \\ H{-}C{-}OH \\ | \\ CH_3 \\ \textbf{Lactate} \end{array} + NAD^+$$

The overall result of steps 1 through 10 is

$$\text{Glucose} + 2\ \text{ADP} + 2\ P_i \longrightarrow 2\ \text{lactate} + 2\ \text{ATP}$$

A much greater amount of ATP per mole of glucose is produced in the part of the carbohydrate catabolic pathway that follows glycolysis—the citric acid cycle. However, the rate at which ATP is produced in glycolysis is far greater than that in the citric acid cycle. For this reason, glycolysis is the principal metabolic pathway for glucose in active skeletal muscle, where large and rapidly available quantities of ATP are required.

In other tissues, such as brain or kidney, where a high rate of ATP production is not needed, pyruvate becomes the final product of the glycolytic breakdown of glucose, and this process can be represented by steps 1 through 9:

$$C_6H_{12}O_6 + 2\ NAD^+ + 2\ \text{ADP} + 2\ P_i \longrightarrow 2\ \text{pyruvate} + 2\ \text{ATP} + 2\ \text{NADH} + 2\ H^+$$

The NADH generated by glycolysis is oxidized by transferring its reducing power from the cytosol to the electron-transport system in the mitochondria, where the electrons and hydrogen are eventually passed on to molecular oxygen to form water (Section 15.11):

$$2\ \text{NADH} + 2\ H^+ + O_2 \longrightarrow 2\ NAD^+ + 2\ H_2O$$

Some microorganisms such as yeast (*Saccharomyces cerevisiae*) regenerate NAD^+ by the conversion of pyruvate into ethanol and carbon dioxide. The reaction takes place in two steps: first, the decarboxylation of pyruvate to form acetaldehyde and, then, the reduction of the acetaldehyde by the use of NADH to form ethanol and the regenerated NAD^+:

$$\begin{array}{c} O{=}C{-}O^- \\ | \\ C{=}O \\ | \\ CH_3 \\ \textbf{Pyruvate} \end{array} \xrightarrow[\searrow CO_2]{\text{pyruvate decarboxylase}} \begin{array}{c} H \quad O \\ \diagdown \;\, /\!\!/ \\ C \\ | \\ CH_3 \\ \textbf{Acetaldehyde} \end{array} + NADH + H^+ \xrightarrow{\text{alcohol dehydrogenase}} \begin{array}{c} H \\ | \\ H{-}C{-}OH \\ | \\ CH_3 \\ \textbf{Ethanol} \end{array} + NAD^+$$

15.3 THE PENTOSE PHOSPHATE PATHWAY

Pentoses and NADPH (the reduced cofactor, Section 14.4) are synthesized in a glucose-requiring metabolic sequence called the **pentose phosphate pathway.** In the initial step, 1 mol of glucose-6-PO_4 is oxidized, resulting in the formation of 2 mol of NADPH, 1 mol of ribose-5-PO_4 (Section 10.4), and 1 mol of CO_2:

$$\text{Glucose-6-}PO_4 + 2\ NADP^+ \longrightarrow \text{ribose-5-}PO_4 + 2\ \text{NADPH} + 2\ H^+ + CO_2$$

This step is an important source of reducing power in the biosynthesis of lipids (Section 16.5), and the pathway is quite active in tissue such as mammary glands. The pathway includes a set of enzymes that allow the interconversion of hexoses and pentoses so that, in combination with glycolysis, the pentose phosphate pathway can supply the cell with any of four alternatives: (1) both NADPH and pentoses, (2) only NADPH when no pentoses are needed, (3) pentoses (for nucleic acid or cofactor synthesis) when no NADPH is needed, and (4) both ATP (from glycolysis) and NADPH for situations in which both are needed.

15.4 THE FORMATION OF ACETYL-*S*-COENZYME A

After glycolysis, the catabolism of glucose continues with the entry of pyruvate into the mitochondria. There the pyruvate is oxidatively transformed into acetyl-*S*-CoA, which is subsequently oxidized in the citric acid cycle (Section 15.5).

The overall reaction producing acetyl-*S*-CoA from pyruvate is

$$\text{Pyruvate} + \text{NAD}^+ + \text{CoA-SH} \longrightarrow \text{acetyl-}S\text{-CoA} + \text{NADH} + \text{H}^+ + \text{CO}_2$$

(**Coenzyme A,** abbreviated here as **CoA-SH,** is a complex thioalcohol containing the vitamin pantothenic acid; see Figure 15.2.) The requisite dehydrogenation and decarboxylation of pyruvate is carried out in a series of steps by a complex of enzymes called the pyruvate dehydrogenase multienzyme complex, located in the mitochondria of eukaryotic cells. The details of this complicated process illustrate the economy and efficiency that result when related enzyme activities are organized into one macromolecular complex. They also demonstrate the role of vitamins as cofactors in enzyme reactions and they show how a regulatory enzyme responds both allosterically and by covalent modification to the energy needs of the cell.

›› The body's daily vitamin requirements are presented in Section 18.2.

The Reaction Steps

The pyruvate dehydrogenase multienzyme complex consists of three different enzymes and five cofactors, assembled together in such a way that the reactions are carried out sequentially. All of these cofactors, except lipoic acid, are derived from vitamins required for human nutrition. The cofactors are

1. **Thiamine** (vitamin B_1) in thiamine pyrophosphate (TPP). The combination of this cofactor with its apoenzyme is abbreviated as Enz_1-TPP.
2. **Lipoic acid,** a growth factor that can be synthesized by vertebrates. The combination of lipoic acid with its apoenzyme is abbreviated as shown below.

$$\text{Enz}_2\text{—R}\langle{}^{\text{S}}_{\text{S}}|$$

3. **Pantothenic acid** in coenzyme A.
4. **Riboflavin** in flavin adenine dinucleotide (FAD; see Figure 14.6). The combination of FAD with its apoenzyme is abbreviated in this case as Enz_3-FAD.
5. **Nicotinic acid** in nicotinamide adenine dinucleotide (NAD; see Figure 14.5.)

The structures and names of the first three cofactors are shown in Figure 15.2 on the following page. Table 15.1 lists the three enzymes and their cofactors that take part in the following sequence of steps of the overall reaction. Although the details of the reaction are complex, we will present the process in outline in five steps.

TABLE 15.1 Enzymes and Cofactors in the Conversion of Pyruvate into Acetyl-*S*-CoA

Enzyme	Cofactor
pyruvate dehydrogenase	thiamine pyrophosphate
dihydrolipoyl transacetylase	lipoic acid, CoA-SH
dihydrolipoyl dehydrogenase	flavin adenine dinucleotide, nicotinamide adenine dinucleotide

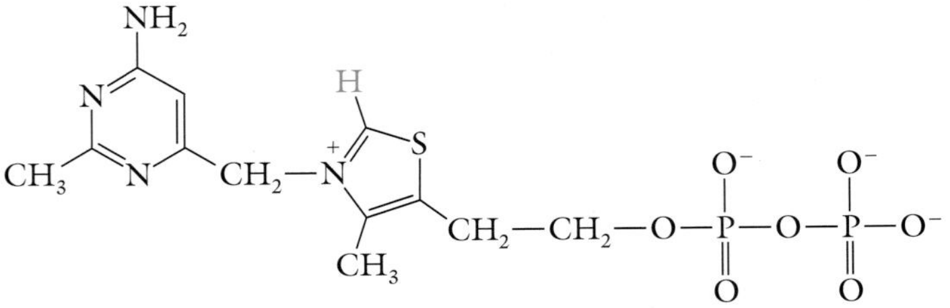

Figure 15.2 Three of the cofactors of pyruvate dehydrogenase required for the conversion of pyruvate into acetyl-S-CoA. Note that, in coenzyme A, the β-mercaptoethylamine group becomes acetylated at the SH group.

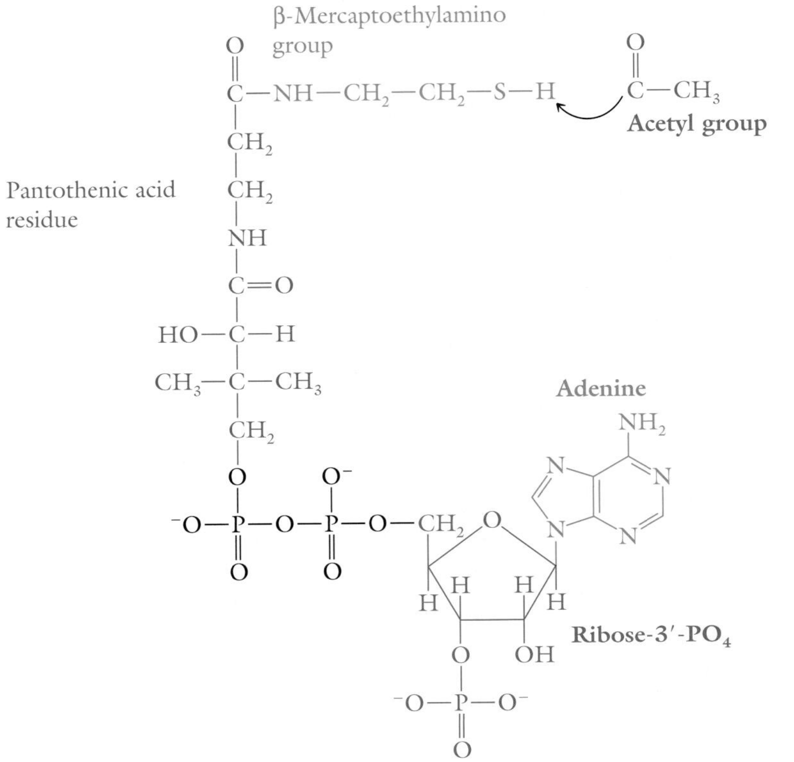

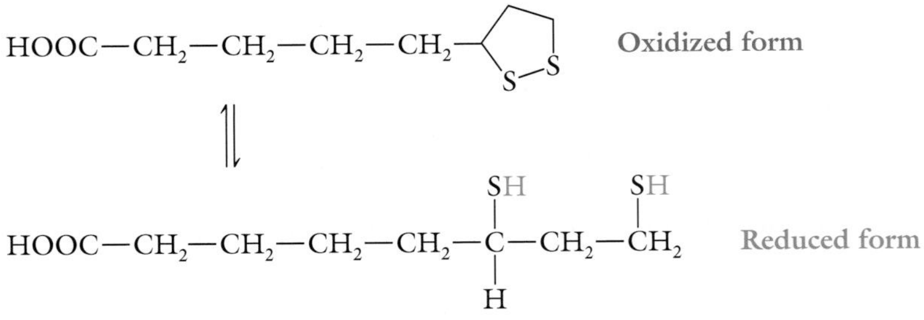

In the first step, pyruvate replaces the TPP hydrogen shown in blue in Figure 15.2, with the result that carbon dioxide is lost and thiamine pyrophosphate is converted into a hydroxyethyl derivative of the cofactor bound to enzyme 1.

$$\textbf{Step 1.}\quad Enz_1—TPP + CH_3—C(=O)—COO^- \longrightarrow Enz_1—TPP—C(CH_3)(OH)—H + CO_2$$

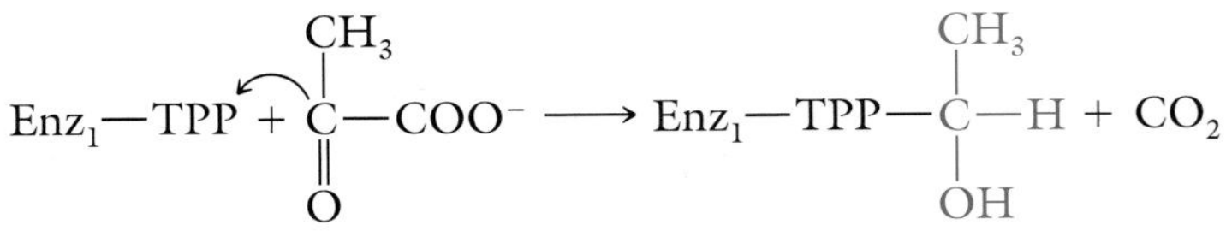

In step 2, the hydroxyethyl group is transferred enzymatically to the oxidized form of the lipoic acid cofactor of enzyme 2 to form an acetyl thioester along with a sulfhydryl group.

$$\textbf{Step 2.}\quad \text{Enz}_1\text{—TPP—}\underset{\text{OH}}{\overset{\text{CH}_3}{\text{C}}}\text{—H} + \text{Enz}_2\text{—R}\langle{}^{\text{S}}_{\text{S}}| \rightleftharpoons \text{Enz}_1\text{—TPP} + \text{Enz}_2\text{—}\underset{\text{SH}}{\text{R}}\text{—S—}\underset{\text{O}}{\overset{}{\text{C}}}\text{—CH}_3$$

In step 3, acetyl-*S*-CoA is formed as the thioester is transferred to CoA-SH, and the lipoic acid becomes fully reduced.

$$\textbf{Step 3.}\quad \text{Enz}_2\text{—}\underset{\text{SH}}{\text{R}}\text{—S—}\underset{\text{O}}{\text{C}}\text{—CH}_3 + \text{CoA-SH} \rightleftharpoons \text{Enz}_2\text{—R}\langle{}^{\text{SH}}_{\text{SH}} + \text{CoA—S—}\underset{\text{O}}{\text{C}}\text{—CH}_3$$

Steps 4 and 5 consist of the transfer of reducing power from reduced lipoic acid, first to FAD and then to NAD^+ to form NADH.

$$\textbf{Step 4.}\quad \text{Enz}_2\text{—R}\langle{}^{\text{SH}}_{\text{SH}} + \text{Enz}_3\text{—FAD} \rightleftharpoons \text{Enz}_2\text{—R}\langle{}^{\text{S}}_{\text{S}}| + \text{Enz}_3\text{—FADH}_2$$

$$\textbf{Step 5.}\quad \text{Enz}_3\text{—FADH}_2 + \text{NAD}^+ \rightleftharpoons \text{Enz}_3\text{—FAD} + \text{NADH} + \text{H}^+$$

The NADH then yields its electrons to the mitochondrial electronic-transport system.

Adding the five reactions and canceling common terms, we find the overall process to be

$$\text{Pyruvate} + \text{NAD}^+ + \text{CoA-SH} \longrightarrow \text{acetyl-}S\text{-CoA} + \text{NADH} + \text{H}^+ + \text{CO}_2$$

Steps 2 through 5 are reversible; however, Step 1 is irreversible. Thus, pyruvate cannot be synthesized by the addition of carbon dioxide to acetyl-*S*-CoA.

Because the five cofactors and three enzymes are all located within the same complex: (1) diffusion distances are minimized, enhancing the rates of reaction and increasing efficiency; (2) side reactions are minimized, increasing economy; and (3) control mechanisms can be integrated and coordinated.

Concept check

✓ A multienzyme complex that carries out a sequence of metabolic reactions is a common structural theme of cellular metabolism.

The Control Mechanisms

The activity of the pyruvate dehydrogenase complex is controlled by both product inhibition and covalent modification.

Product inhibition is simply Le Chatelier's principle at work. When acetyl-*S*-CoA and NADH are present in high concentrations, steps 3 and 5 are driven in the reverse direction. Enzyme 1 cannot deliver its hydroxyethyl group to enzyme 2, and the decarboxylation of pyruvate is inhibited.

Within the multienzyme complex, two allosterically controlled enzymes—pyruvate dehydrogenase kinase and pyruvate dehydrogenase phosphatase (abbreviated PDK and PDP)—covalently modify enzyme 1, causing its activation or inactivation. PDK inactivates enzyme 1 by phosphorylation. PDK itself is allosterically activated by acetyl-*S*-CoA and NADH and inactivated by pyruvate, ADP, and Ca^{2+}. PDP removes the phosphate group by hydrolysis and reactivates enzyme 1. It has no inactivators and is active only in the presence of high concentrations of Ca^{2+} and Mg^{2+}. Table 15.2 on the following page summarizes the allosteric control of these two enzymes. There are many examples of this kind of control in cellular metabolism. We will see it again in glycogenolysis (Section 15.10).

TABLE 15.2 Allosteric Control of Pyruvate Dehydrogenase Kinase (PDK) and Pyruvate Dehydrogenase Phosphatase (PDP)

Enzyme	Activator	Inhibitor
PDK	acetyl-*S*-CoA NADH	pyruvate ADP Ca^{2+}
PDP	Ca^{2+} Mg^{2+}	none

Concept check

✓ Covalent modification of an enzyme by allosterically controlled phosphorylation and dephosphorylation is an important molecular theme of cellular metabolism.

15.5 THE CITRIC ACID CYCLE

In contrast with the glycolytic pathway, which is a linear sequence of enzymatically catalyzed reactions, the aerobic phase of glucose catabolism is cyclic. It begins with the formation of citric acid—hence its name, the **citric acid cycle.** It is also called the tricarboxylic acid cycle (TCA cycle) or the Krebs cycle (in honor of its discoverer, Hans Krebs).

The cyclic nature of the process is illustrated in Figure 15.3. The cycle begins when acetyl-*S*-CoA donates an acetyl group to the four-carbon

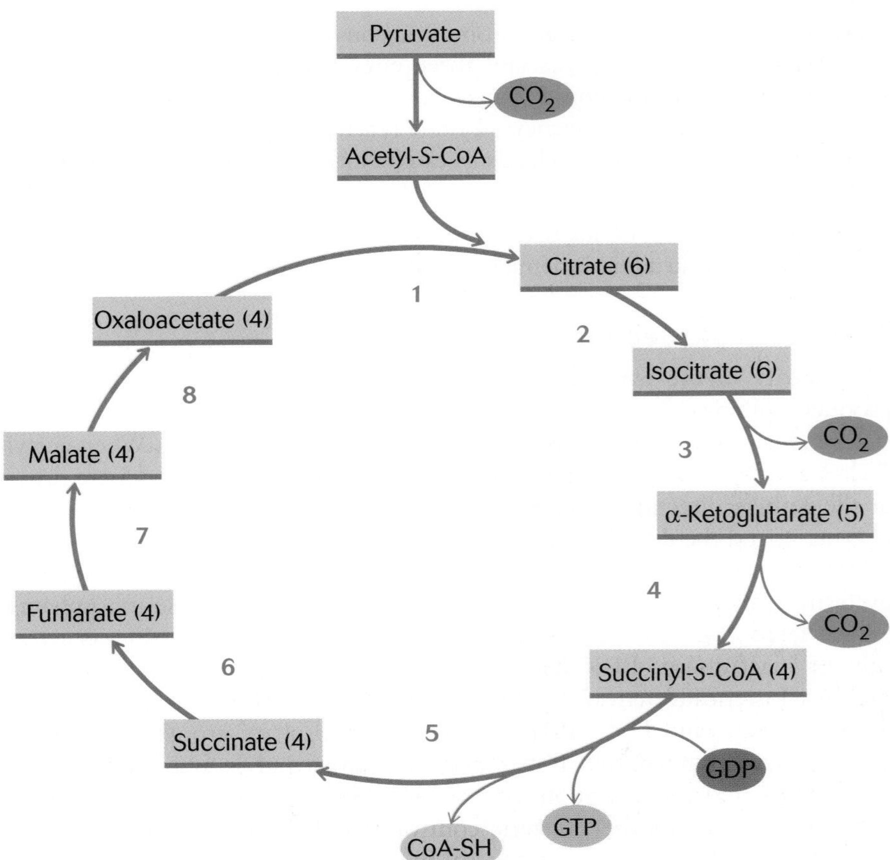

Figure 15.3 An outline of the citric acid cycle identifying components by name and the number of carbon atoms (given in parentheses) in the intermediate.

TABLE 15.3 Mitochondrial Location of Citric Acid Cycle Enzymes

Inner membrane	Matrix space
aconitase	citrate synthase
succinate dehydrogenase	isocitrate dehydrogenase
electron-transport chain	α-ketoglutarate dehydrogenase
succinyl-*S*-CoA synthetase	
fumarase	
malate dehydrogenase	
pyruvate dehydrogenase	

dicarboxylic acid oxaloacetate to form the six-carbon tricarboxylic acid citrate. Citrate is then converted into isocitrate, which, in turn, is oxidatively decarboxylated with the loss of CO_2 and the production of NADH to form the five-carbon α-ketoglutarate. This compound undergoes an oxidative dehydrogenation and a series of transformations to yield a second molecule of CO_2 and NADH and emerges as the four-carbon succinate. The enzyme system that decarboxylates α-ketoglutarate is almost identical with the pyruvate complex except that it has no regulatory components. Succinate is then converted in three steps into the four-carbon oxaloacetate, which begins the cycle again.

The result is that two carbons enter the cycle in the form of acetyl-*S*-CoA, two carbons leave as CO_2, and oxaloacetate is regenerated to begin the cycle again. No oxaloacetate is lost, and therefore one molecule of it can effect the oxidation of limitless numbers of acetyl groups.

All of the enzymes of the citric acid cycle, as well as those of the electron-transport system, are located in the mitochondria of animal cells. Their precise locations are listed in Table 15.3.

15.6 REACTIONS OF THE CITRIC ACID CYCLE

Let's now look more closely at the citric acid cycle's eight successive reactions. The first reaction, acetyl-*S*-CoA with oxaloacetate, is catalyzed by citrate synthase. The methyl carbon of acetyl-*S*-CoA reacts with the carbonyl group of oxaloacetate, and, at the same time, the thioester bond is hydrolyzed to produce free CoA-SH.

$$\mathrm{O{=}C(COO^-)(CH_2COO^-)} + \mathrm{CH_3{-}C({=}O){-}S{-}CoA} \xrightleftharpoons[\text{citrate synthase}]{+\,H_2O} \mathrm{HO{-}C(COO^-)(CH_2COO^-)(CH_2COO^-)} + \text{CoA-SH}$$

Citrate

CoA-SH is now free to participate in another round of the pyruvate dehydrogenation reaction. Citrate synthase cannot bind acetyl-*S*-CoA unless oxaloacetate is first bound to the enzyme.

In the next step of the cycle, the enzyme aconitase catalyzes the transformation of citrate into isocitrate. It does so through the reversible dehydration–hydration of the double bond of an enzyme-bound intermediate called *cis*-aconitate.

$$\begin{array}{c} COO^- \\ | \\ H-C-H \\ | \\ HO-C-COO^- \\ | \\ CH_2 \\ | \\ COO^- \\ \textbf{Citrate} \end{array} \underset{\text{aconitase}}{\overset{-\,H_2O}{\rightleftharpoons}} \begin{array}{c} COO^- \\ | \\ C-H \\ \| \\ C-COO^- \\ | \\ CH_2 \\ | \\ COO^- \\ \textbf{\textit{cis}-Aconitate} \\ \textbf{(enzyme bound)} \end{array} \underset{\text{aconitase}}{\overset{+\,H_2O}{\rightleftharpoons}} \begin{array}{c} COO^- \\ | \\ HO-C-H \\ | \\ H-C-COO^- \\ | \\ CH_2 \\ | \\ COO^- \\ \textbf{Isocitrate} \end{array}$$

CH_2

In the next step, the six-carbon isocitrate is dehydrogenated and decarboxylated by the enzyme isocitrate dehydrogenase to form the five-carbon α-ketoglutarate. This reaction requires NAD^+.

$$\begin{array}{c} COO^- \\ | \\ HO-C-H \\ | \\ H-C-COO^- \\ | \\ CH_2 \\ | \\ COO^- \\ \textbf{Isocitrate} \end{array} + NAD^+ \xrightarrow[\text{dehydrogenase}]{\text{isocitrate}} \begin{array}{c} COO^- \\ | \\ C=O \\ | \\ CH_2 \\ | \\ CH_2 \\ | \\ COO^- \\ \boldsymbol{\alpha}\textbf{-Ketoglutarate} \end{array} + CO_2 + NADH + H^+$$

In the following step, α-ketoglutarate is acted on by the isocitrate dehydrogenase complex, which catalyzes the formation of succinyl-*S*-CoA with the loss of CO_2. The reaction is

$$\begin{array}{c} COO^- \\ | \\ C=O \\ | \\ CH_2 \\ | \\ CH_2 \\ | \\ COO^- \\ \boldsymbol{\alpha}\textbf{-Ketoglutarate} \end{array} + \text{CoA-SH} + NAD^+ \xrightarrow[\text{dehydrogenase}]{\alpha\text{-ketoglutarate}} \begin{array}{c} S-CoA \\ | \\ C=O \\ | \\ CH_2 \\ | \\ CH_2 \\ | \\ COO^- \\ \textbf{Succinyl-}\textbf{\textit{S}}\textbf{-CoA} \end{array} + CO_2 + NADH + H^+$$

The α-ketoglutarate complex is virtually identical in structure and function with the pyruvate dehydrogenase complex. However, the α-ketoglutarate complex does not possess the regulatory properties of the pyruvate dehydrogenase complex.

« **High-energy compounds are the subject of Section 14.9.**

Like acetyl-*S*-CoA, succinyl-*S*-CoA is a high-energy compound that undergoes an energy-conserving reaction in which the compound GDP is phosphorylated to form GTP. GDP and GTP have the same structures, respectively, as ADP and ATP except that adenine (A) is replaced by guanine (G; Section 13.1). Hence GDP and GTP are guanosine diphosphate and guanosine triphosphate, respectively. GTP can phosphorylate ADP and takes part in a variety of membrane processes. The reaction is

$$\begin{array}{c} S-CoA \\ | \\ C=O \\ | \\ CH_2 \\ | \\ CH_2 \\ | \\ COO^- \\ \textbf{Succinyl-}\textbf{\textit{S}}\textbf{-CoA} \end{array} + GDP + P_i \underset{\text{synthetase}}{\overset{\text{succinyl-}S\text{-CoA}}{\rightleftharpoons}} \begin{array}{c} COO^- \\ | \\ CH_2 \\ | \\ CH_2 \\ | \\ COO^- \\ \textbf{Succinate} \end{array} + GTP + \text{CoA-SH}$$

Succinyl-*S*-CoA is the main building block of porphyrins, compounds that are key components of the hemes of hemoglobin and myoglobin and the cytochromes of the electron-transport chain (Section 15.11).

In the next step, succinate is dehydrogenated by the enzyme succinic dehydrogenase to form the trans isomer fumarate. Succinic dehydrogenase is a component of the inner mitochondrial membrane (Section 15.12). The hydrogen acceptor here is FAD (see Figure 14.6) represented as E-FAD. Dehydrogenases that contain FMN or FAD are called **flavoproteins** because these cofactors are so tightly bound that they are considered to be part of the protein molecule; E-FAD is therefore a flavoprotein.

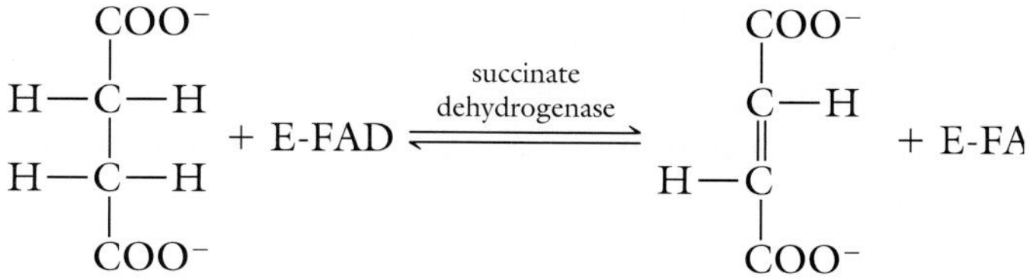

$$\begin{array}{c} COO^- \\ | \\ H-C-H \\ | \\ H-C-H \\ | \\ COO^- \\ \textbf{Succinate} \end{array} + \text{E-FAD} \xrightleftharpoons{\text{succinate dehydrogenase}} \begin{array}{c} COO^- \\ | \\ C-H \\ \| \\ H-C \\ | \\ COO^- \\ \textbf{Fumarate} \end{array} + \text{E-FA}$$

In the next reaction, fumarate is reversibly hydrated by the enzyme fumarate hydratase, also called fumarase. The product is the chiral L-malate. The hydration is specific for the trans dicarboxylic acid; fumarase will not hydrate maleate, the cis isomer of fumarate.

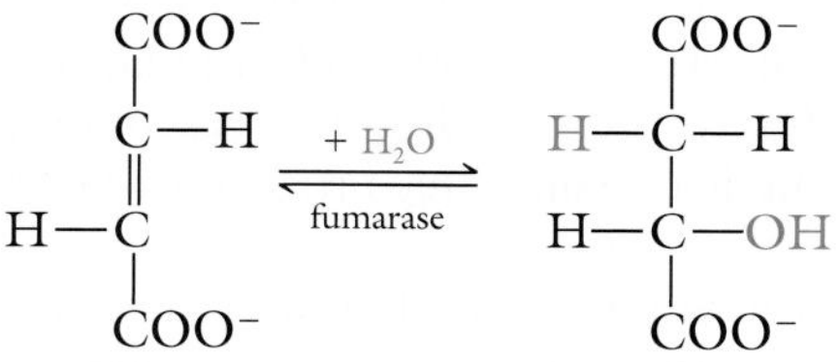

$$\begin{array}{c} COO^- \\ | \\ C-H \\ \| \\ H-C \\ | \\ COO^- \\ \textbf{Fumarate} \end{array} \underset{\text{fumarase}}{\overset{+\,H_2O}{\rightleftharpoons}} \begin{array}{c} COO^- \\ | \\ H-C-H \\ | \\ H-C-OH \\ | \\ COO^- \\ \textbf{L-Malate} \end{array}$$

The last step in the citric acid cycle is the dehydrogenation of L-malate to form oxaloacetate. The enzyme catalyzing this step is NAD^+-linked malate dehydrogenase.

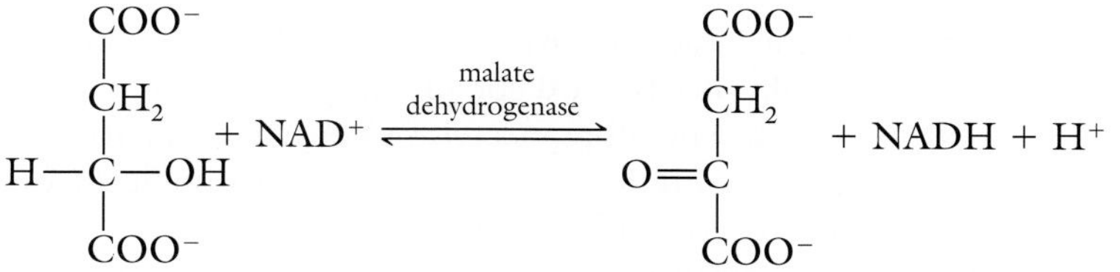

$$\begin{array}{c} COO^- \\ | \\ CH_2 \\ | \\ H-C-OH \\ | \\ COO^- \\ \textbf{L-Malate} \end{array} + NAD^+ \xrightleftharpoons{\text{malate dehydrogenase}} \begin{array}{c} COO^- \\ | \\ CH_2 \\ | \\ O=C \\ | \\ COO^- \\ \textbf{Oxaloacetate} \end{array} + NADH + H^+$$

Oxaloacetate is again available to react with a new incoming acetyl-*S*-CoA to begin a new cycle. The equilibrium constant for this final reaction is 6.0×10^{-6}, which means that the cellular concentration of oxaloacetate is very low at all times. Therefore, slight changes in the concentration of oxaloacetate have a great effect on the overall rate of the cycle.

The net reaction of the citric acid cycle is

$$\text{Acetyl-}S\text{-CoA} + 3\,NAD^+ + FAD + GDP + P_i \longrightarrow 2\,CO_2 + 3\,NADH + FADH_2 + GTP + \text{CoA-SH}$$

This equation should not be taken to convey that, after a single turn of the cycle, the two carbon atoms of acetyl-*S*-CoA are the same two carbon atoms that emerge as CO_2. They are not. They will emerge as CO_2 in the next cycle. One ATP in the equivalent form of GTP is generated. The hydrogens removed from four intermediates emerge as three molecules of NADH and one molecule of $FADH_2$. We shall examine their fates when we consider the electron-transport system in Section 15.11.

15.7 THE REPLENISHMENT OF CYCLE INTERMEDIATES

Some of the intermediates of the citric acid cycle are important components in other metabolic processes. For example, when α-ketoglutarate, succinate, and oxaloacetate are drained away to be converted into α-amino acids, their concentrations in the cycle must be replenished. Some of the intermediates can be replenished from the breakdown of α-amino acids and carbohydrates.

Oxaloacetate is replaced through the action of the enzyme pyruvate carboxylase, which catalyzes the following reversible reaction:

$$\text{Pyruvate} + CO_2 + \text{ATP} + H_2O \rightleftharpoons \text{oxaloacetate} + \text{ADP} + P_i + 2\ H^+$$

This complex reaction is described in detail next, in the section on gluconeogenesis, because the concentration of oxaloacetate is also a key regulatory aspect of that process.

15.8 GLUCONEOGENESIS

Gluconeogenesis is a process that takes place in the liver; in this process, glucose is resynthesized from lactate, the chief product of glycolysis in active muscle. The pathway is largely but not quite the reverse of glycolysis: three irreversible steps in glycolysis must be bypassed, which is accomplished by their replacement by different reactions and different enzymes. All the other, reversible steps in the synthesis utilize the reactions and enzymes of the glycolytic pathway. The irreversible steps that must be bypassed are indicated in Figure 15.4.

The first bypass reaction is required because there is no enzymatic mechanism for the direct synthesis of the high-energy compound phosphoenolpyruvate from pyruvate. Phosphoenolpyruvate synthesis in gluconeogenesis begins in the mitochondria. There, oxaloacetate is synthesized from pyruvate by reaction with carbon dioxide:

$$\text{Pyruvate} + CO_2 + \text{ATP} \longrightarrow \text{oxaloacetate} + \text{ADP}$$

The enzyme catalyzing this reaction, pyruvate carboxylase, is a mitochondrial regulatory enzyme virtually inactive except in the presence of acetyl-*S*-CoA, which is a specific activator. Therefore, acetyl-*S*-CoA is the key to turning on this phase of carbohydrate synthesis, and it is present in significant quantities only when the cell is in an energy-rich phase.

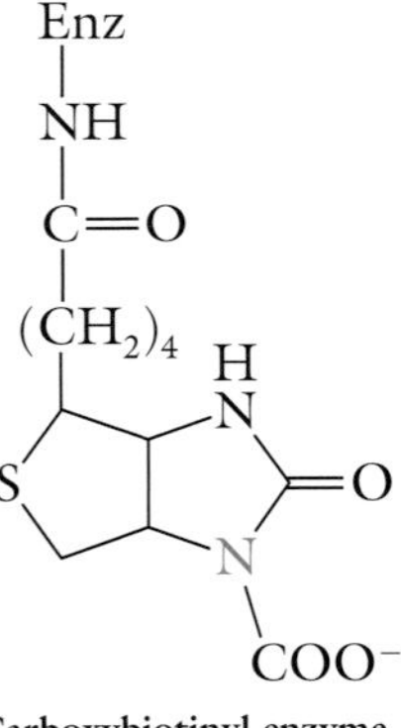

Carboxybiotinyl enzyme

The cofactor in carboxylations is biotin. It acts as a carrier of carboxyl ($—COO^-$) groups in enzymatically catalyzed carboxylation reactions in which ATP also is required. The structure in the margin shows that biotin is covalently attached to its carrier protein, indicated by Enz, and the transient carboxyl group is attached to the nitrogen atom shown in blue.

Remember that glycolysis takes place in the cytosol. Gluconeogenesis also must take place there. Oxaloacetate, which is synthesized in the mitochondria, cannot traverse the mitochondrial membrane into the cytosol, but malate can. So the next step is a reduction of mitochondrial oxaloacetate to malate by mitochondrial malate dehydrogenase with the use of NADH. This step is followed by the transport of malate into the cytosol, where it is reoxidized to oxaloacetate by cytosolic malate dehydrogenase.

Cytosolic phosphoenolpyruvate is synthesized from the extramitochondrial oxaloacetate in a reaction requiring GTP and catalyzed by the enzyme phosphoenolpyruvate carboxykinase. This reaction is the first one to bypass the enzymes of the glycolytic sequence:

$$\text{Oxaloacetate} + \text{GTP} \longrightarrow \text{phosphoenolpyruvate} + CO_2 + \text{GDP}$$

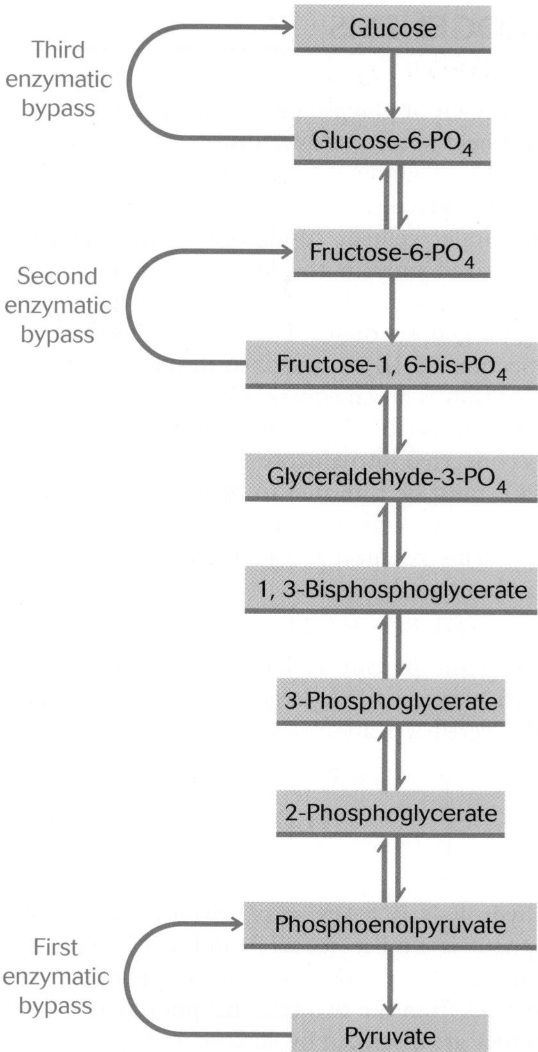

Figure 15.4 Gluconeogenesis is the reversal of glycolysis with the exception of three reactions that bypass the three irreversible reactions in the glycolytic pathway. These bypass reactions employ three new enzymes, providing alternative reaction pathways.

Note that this reaction takes place only under conditions of high ATP and NADH concentrations, which also drive the next several reversible steps to produce fructose-1,6-bisphosphate.

The formation of fructose-1,6-bisphosphate from fructose-6-PO_4 in glycolysis is irreversible. Therefore, in a second bypass reaction, the gluconeogenesis pathway utilizes a new enzyme, fructose bisphosphatase, to irreversibly hydrolyze the bisphosphate to fructose-6-PO_4:

$$\text{Fructose-1,6-bis-}PO_4 + H_2O \longrightarrow \text{fructose-6-}PO_4 + P_i$$

The third bypass reaction accomplishes the hydrolysis of glucose-6-PO_4 to glucose, which then can leave the liver and enter the blood for distribution to other tissues. This reaction requires a third new enzyme, glucose-6-phosphatase:

$$\text{Glucose-6-}PO_4 + H_2O \longrightarrow \text{glucose} + P_i$$

In vertebrates, glucose-6-phosphatase is located only in the liver and is not found in any other tissue. Thus, only the liver can furnish free glucose to the blood, and therefore the liver is the control center that regulates the glucose concentration of the blood.

15.9 GLYCOGENESIS

Glucose can be converted into a polysaccharide called **glycogen** (Section 10.6), which consists of a chain of glucose molecules linked through $\alpha(1 \rightarrow 4)$ bonds (Section 10.7) and held in reserve as a future energy source. The human body stores about half of its daily energy supply as glycogen. The process for the synthesis of glycogen, **glycogenesis,** requires an entirely new set of enzymes that operate only under cellular conditions of high concentrations of ATP. Glycogen is synthesized only in the liver and in muscle cells and is stored as glycogen granules. Insulin also plays a role in glycogenesis by increasing the activity of glycogen synthase. The role of insulin in glycogenesis and glycogenolysis (Section 15.10) is described in more detail in Chapter 18. The first step in its synthesis is an enzymatic isomerization in which the phosphate group attached to carbon 6 of glucose is transferred to carbon 1; that is, glucose-6-PO_4 is converted into glucose-1-PO_4.

Glucose-1-PO_4 must be activated to be incorporated into glycogen, a step accomplished by reaction with uridine triphosphate (UTP) to form uridine diphosphoglucose. This reaction is catalyzed by the specific enzyme UDP-glucose pyrophosphorylase:

$$\text{Glucose-1-PO}_4 + \text{UTP} \longrightarrow \text{UDP-glucose} + \text{PP}_i$$

PP_i stands for **inorganic pyrophosphate,** $P_2O_7^{4-}$, which undergoes an enzymatic hydrolysis characterized by a large equilibrium constant. That secondary process drives the overall activation reaction:

$$^{-}\text{O}-\underset{\underset{\text{O}}{\|}}{\overset{\overset{\text{O}^-}{|}}{\text{P}}}-\text{O}-\underset{\underset{\text{O}}{\|}}{\overset{\overset{\text{O}^-}{|}}{\text{P}}}-\text{O}^- + \text{H}_2\text{O} \xrightarrow{\text{pyrophosphatase}} 2\ \text{HO}-\underset{\underset{\text{O}^-}{|}}{\overset{\overset{\text{O}}{\|}}{\text{P}}}-\text{O}^-$$

Pyrophosphate hydrolysis is a key step driving many biosynthetic reactions. Uridine triphosphate is similar to ATP and GTP, with the substitution of the nitrogen base uracil for either adenine or guanine (Section 13.1). The pyrophosphate derives from the two terminal phosphate groups that are cleaved from UTP when they are displaced by glucose-1-PO_4 to form UDP-glucose.

The activated glucose becomes the substrate for glycogen synthase, a regulatory enzyme that is activated by glucose-6-PO_4. This enzyme catalyzes the addition of the activated glucose to the nonreducing end (Section 10.6) of a growing chain of glycogen:

$$\text{UDP-glucose} + (\text{glucose})_n \longrightarrow (\text{glucose})_{n+1} + \text{UDP}$$

Glycogen does not consist solely of straight chains of linkages. It is, in fact, a highly branched chain structure. Branching makes more chain ends available, which maximizes the rate of glucose release and glycogen synthesis. If glycogen had only one end, both glucose release and glycogen formation would be too slow to support life. Branching is accomplished by the presence of a branching enzyme, which breaks the $\alpha(1 \longrightarrow 4)$ linkage and replaces it with an $\alpha(1 \longrightarrow 6)$ link. The branching enzyme is very demanding. When the linear chain is in the process of elongation and is at least 11 units long, 7 of those units are transferred as a chain to a site closer to the interior of the glycogen molecule. Furthermore, the new branch site cannot be closer than 4 units from another branch point.

15.10 GLYCOGENOLYSIS

Glycogen in the liver is degraded to form glucose whenever the blood's concentration of glucose drops below normal levels, about 5 m*M*. This process, called **glycogenolysis,** proceeds by a route different from that of the synthesis of glycogen. The glycogen's chain-terminal glucose is removed through phosphorolysis by the enzyme phosphorylase in its active form, phosphorylase$_a$. Phosphorolysis accomplishes a result similar to that of hydrolysis except that the elements of phosphoric acid instead of water are used to cleave the acetal group between the glucose units.

$$(\text{Glucose})_n + P_i \xrightarrow{\text{phosphorylase}_a} (\text{glucose})_{n-1} + \text{glucose-1-PO}_4$$

Structurally, phosphate cleaves the $\alpha(1 \rightarrow 4)$ bond of glycogen at the nonreducing end of the polyglucose chain. In so doing, it (1) releases glucose-1-PO_4 and (2) shortens the polyglucose chain by one monomer. Schematically,

HPO_4^{2-} + HO–[glucose]–O–[glucose]–O–[glucose]–O~ $\xrightarrow{\text{phosphorylase}_a}$

HO–[glucose]–OPO_3^{2-} + HO–[glucose]–O–[glucose]–O~

Phosphorylase is ordinarily present in its inactive form, phosphorylase$_b$. It must be activated through a series of steps initiated in the liver by the hormone glucagon and in both the liver and muscle by the hormone epinephrine. **Glucagon** is a polypeptide hormone synthesized by pancreatic α cells, and **epinephrine** (Box 8.3) is synthesized in the adrenal cortex from the α-amino acid tyrosine. The details of the process are presented in Figure 15.5 on the following page.

›› The role of hormones in metabolism is considered in Section 18.6.

The hormone epinephrine is released very quickly when any sudden environmental change is sensed by the central nervous system, a response known as the fight-or-flight reflex. It stimulates glycogenolysis primarily in muscle cells. Epinephrine is also under the control of cells in the hypothalamus of the brain that are sensitive to the blood concentration of glucose. Lowered glucose concentration will raise the epinephrine level. Epinephrine also increases blood pressure and heart rate.

The mechanism of action of the hormone glucagon is similar to that of epinephrine, but its role is to regulate the blood-glucose concentration over the long term. Therefore, it is active during fasting periods and between meals. Its secretion into the blood is directly controlled by the blood-glucose concentration. When that concentration falls below 80 to 100 mg/dL, the concentration optimum for brain function, the pancreatic α cells secrete glucagon. The hormone then stimulates glycogenolysis in both liver and muscle cells, which require glucose for energy, although, as you will recall, only liver cells, not muscle, can contribute glucose to the blood. Unlike epinephrine, glucagon has no effect on blood pressure or heart rate.

Hormones are classified as either water soluble or water insoluble. Water-insoluble hormones, such as steroids (Section 11.7), penetrate cell membranes, enter the cells, and directly affect their metabolic targets. Water-soluble hormones, such as epinephrine and glucagon, do not penetrate the cell membrane but bind directly to specific receptors on the outer surface of the membrane. Each hormone that binds to the cell membrane has a unique receptor, but each receptor, when occupied by its hormone, ultimately activates the same membrane-bound enzyme, adenylyl cyclase. Activation takes place in three steps. First, hormone bound to its cell-surface receptor causes a conformational change in the receptor protein. Second, this conformational change causes a second membrane protein, called a GTP-binding protein (abbreviated G protein) to bind GTP. In its inactive state, the G protein binds GDP. When the G protein is stimulated by the hormone–receptor complex, the GDP dissociates, GTP is bound in its place, and, in the last of the three steps, the GTP-protein then activates adenylyl cyclase.

Adenylyl cyclase catalyzes the formation of the cyclic form of AMP (c-AMP), which is called the **second messenger** (signaling within a cell), the hormone being the **first messenger** (signaling from cell to cell).

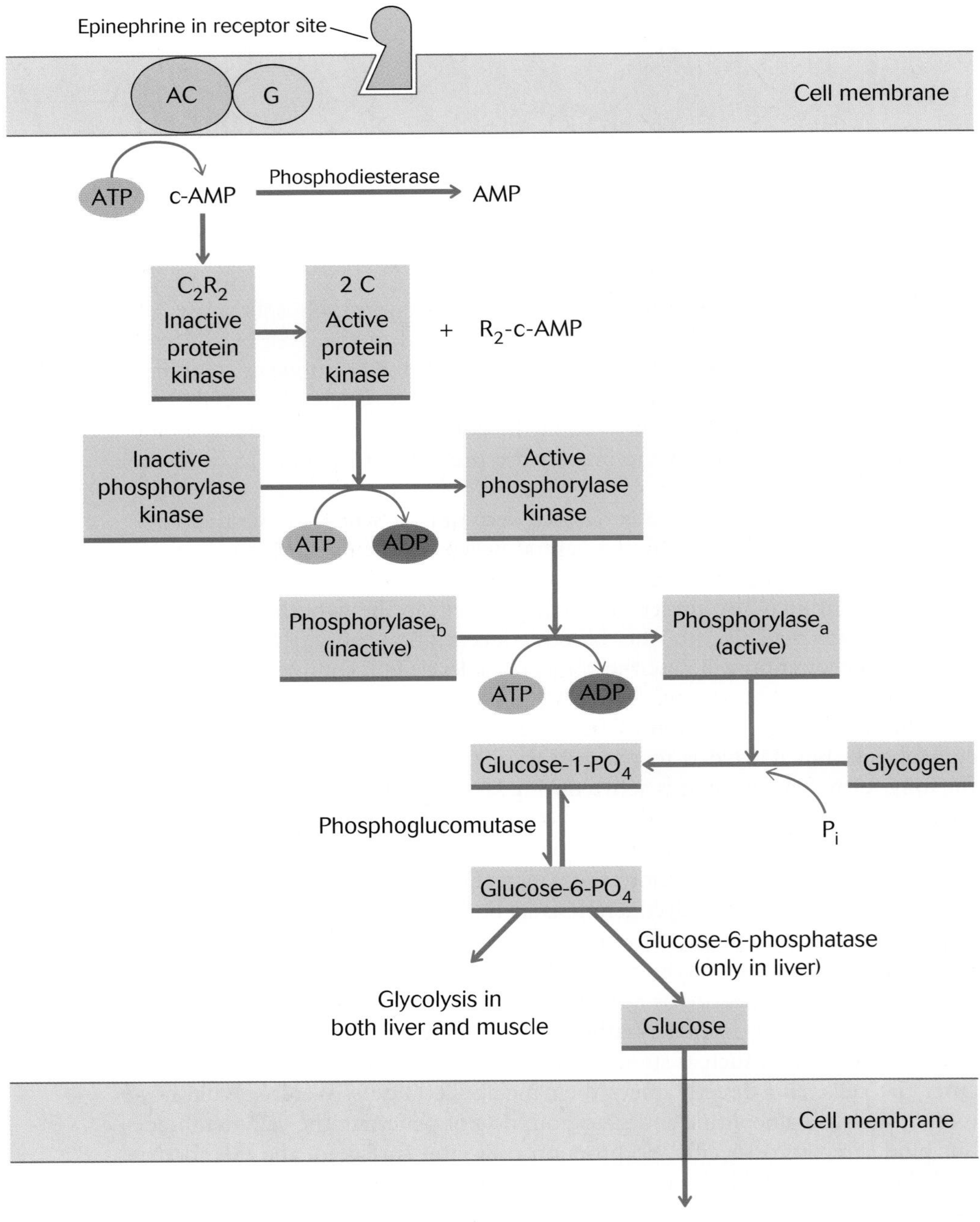

Figure 15.5 The enzymatic cascade in glycogenolysis is initiated by the action of the hormone epinephrine in the liver and in skeletal muscle cells. Abbreviations: G, GTP-binding protein; AC, active adenylyl cyclase. The pancreatic hormone glucagon causes the same result in the liver.

$$\text{ATP} \xrightarrow{\text{adenylyl cyclase}} \text{c-AMP} + PP_i$$

A second enzyme of the glycogenolysis pathway is ordinarily in its inactive form because it binds a regulatory inhibitor, R, to form an inactive C_2R_2 complex. When cyclic AMP binds to the R subunit, the C_2R_2 complex dissociates, thus freeing C, the active form of the second enzyme. The reaction is

$$C_2R_2 + 2\ \text{c-AMP} \longrightarrow 2\ \text{C} + 2\ \text{R-c-AMP}$$

In its active form, C catalyzes the phosphorylation and activation of a third enzyme. The third enzyme catalyzes the phosphorylation of inactive phosphorylase, phosphorylase$_b$, to yield the fourth enzyme, the active form of the phosphorylase, phosphorylase$_a$. Each phosphorylation requires ATP. Phosphorylase$_a$ in turn cleaves glucose from glycogen by phosphorolysis to yield glucose-1-PO_4. Glucose-1-PO_4 is converted into glucose-6-PO_4 by phosphoglucomutase.

Up to that point, the reactions in both liver and muscle cells are identical. However, glucose-6-phosphatase is present only in liver cells. Therefore, only liver cells can produce free glucose from glucose-6-PO_4, and, consequently, only liver cells can increase the blood-glucose concentration. The result of the process in muscle cells is a great increase in the rate of glycolysis, accompanied by increased availability of ATP.

In glycogenolysis, enzymes sequentially activate a series of other enzymes. The action of a single enzyme can be likened to opening a door through which other molecules may emerge. If each of the other molecules also is an enzyme, the number of catalytic units multiplies rapidly. Figure 15.6 depicts each enzyme as activating three other enzymes. In the cell, however, one enzyme

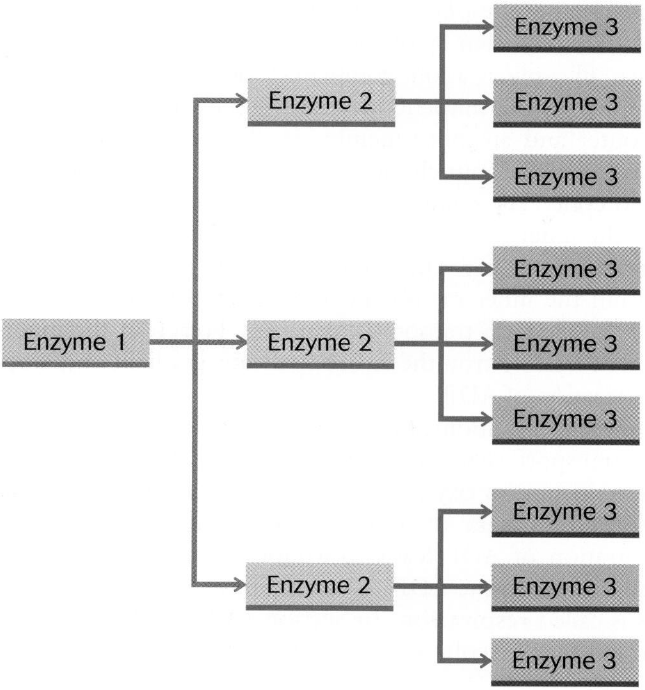

Figure 15.6 When an enzyme activates a series of enzymes of a pathway, the original signal that turned on enzyme 1 is amplified manyfold.

may activate thousands of others. The original signal becomes greatly amplified, depending on how many activation steps take place, and, in this case, has been estimated to be as great as 2.5×10^7. Just a few molecules of epinephrine suffice to release many grams of glucose into the blood.

A fifth enzyme, phosphodiesterase, is active at all times in both muscle and liver cells. It catalyzes the hydrolysis of c-AMP to AMP and quickly puts an end to the enzyme cascade leading to glycogenolysis. This enzyme's presence ensures that glycogenolysis will take place only as long as either epinephrine or glucagon continue to be present at the cell membrane. After all, you can't fight or flee all the time. An interesting sidelight here is that phosphodiesterase is inhibited by caffeine and theobromine, the alkaloids in coffee and tea, respectively. Their effect is to prolong the active lifetime of c-AMP, as though epinephrine were constantly being delivered to cell membranes.

15.11 THE ELECTRON-TRANSPORT CHAIN

So far in our consideration of glucose metabolism, energy extracted from nutrients in the form of ATP has been obtained by substrate-level phosphorylations. In this section, we shall see how the oxidation of NADH also results in the formation of ATP. This NADH-dependent synthesis of ATP is called **oxidative phosphorylation** (Section 14.4). Stored electrons (in the form of NADH or $FADH_2$) pass through a highly organized sequence of enzymes and cofactors, called the **electron-transport chain,** in the inner mitochondrial membrane, to finally react with molecular oxygen to form water.

In Chapter 2, weak acids were arranged in an order based on their tendencies to lose a proton. The strongest lost protons more readily than the weakest, and the conjugate base of the weakest gained protons more easily than the strongest. An analogous ordering among partners in redox reactions is seen in the organization of the electron-transport chain: NADH tends to reduce components of its dehydrogenase, and that dehydrogenase tends to reduce components immediately adjacent to it, and so forth. Each component is alternately reduced and oxidized as electrons move along to the final electron acceptor, oxygen.

During electron transport, a proton gradient forms across the inner mitochondrial membrane: the pH outside that membrane is lower than the inside pH. ATP synthesis is driven by this proton gradient.

Oxidative phosphorylation completely depends on the structural and osmotic integrity of the mitochondria. A mitochondrion possesses two membranes, an outer and an inner membrane. The space enclosed by the inner membrane is called the mitochondrial matrix, and it contains the enzymes of the citric acid cycle. The components of the electron-transport chain are integral parts of the inner membrane, which is repeatedly folded so as to produce a very large surface area. The combination of the special organization of enzymes within the inner membrane and the large surface area enhances the efficiency of the electron-transport chain. Any breach in the inner mitochondrial membrane will destroy the hydronium ion gradient and stop the oxidative phosphorylation of ADP.

Under normal conditions, the movement of electrons and hydrogen along the electron-transport chain to oxygen is coupled to the phosphorylation of ADP: if the reduction of oxygen is halted, the synthesis of ATP also stops; at the same time, the uptake of oxygen depends on the availability of ADP. A high concentration of ADP leads to a rapid uptake of oxygen, and oxygen uptake is reduced when the ADP concentration is lowered. The uptake of oxygen by cells is called **respiration.** In Section 15.7, we shall see that respiration is more efficient than glycolysis at producing ATP from glucose.

15.12 ENZYMES OF THE ELECTRON-TRANSPORT CHAIN

The electron-transport chain has been shown to consist of four distinct enzyme complexes. Although the complexes are integral structures of the mitochondrial membrane, they do not seem to be fixed in position but are free to diffuse within the membrane. Figure 15.7 depicts the mitochondrial membrane containing the four enzyme complexes, with arrows indicating the flow of electrons. Complex I is essentially NADH dehydrogenase. Here, the reducing power in the form of electrons and hydrogen from NADH is transferred to flavin mononucleotide, FMN (see Figure 14.7). Remember that dehydrogenases that contain FMN or FAD are called flavoproteins. Also present is iron in the form of iron–sulfur centers incorporated into the protein. The iron in these centers undergoes reversible oxidation–reduction between Fe^{2+} and Fe^{3+} states during electron transport.

Electrons from complex I are transported to the next complex by a lipid-soluble molecule called ubiquinone (UQ) because of its ubiquitous nature, being found in virtually all forms of life. It is also called coenzyme Q (CoQ). UQ, shown in the margin, is free to diffuse within the lipophilic environment of the inner mitochondrial membrane. It undergoes reduction in two steps, one electron at each step.

Ubiquinone (UQ)
(R = hydrocarbon chain)

$\xrightarrow{e^-, H^+}$ $\xrightarrow{e^-, H^+}$ **UQH_2**

The oxidation of one NADH and the reduction of one UQ by complex I is accompanied by the net transport of protons from the matrix side of the mitochondrial membrane to the cytosolic side. The flow of electrons provides the energy to drive this proton transport, an example of active transport (Section 11.10). Approximately 3 mol of ATP are synthesized by the oxidation of 1 mol of NADH. Details about the stoichiometry of ATP production are presented in Section 15.7.

UQ receives electrons from three other sources. One of these sources is complex II. Complex II, the flavoprotein complex succinic dehydrogenase, is a member of the electron-transport chain in the mitochondrial membrane and, at the same time, is an integral member of the citric acid cycle. Succinic dehydrogenase contains an FAD covalently bound to its protein. The oxidation of

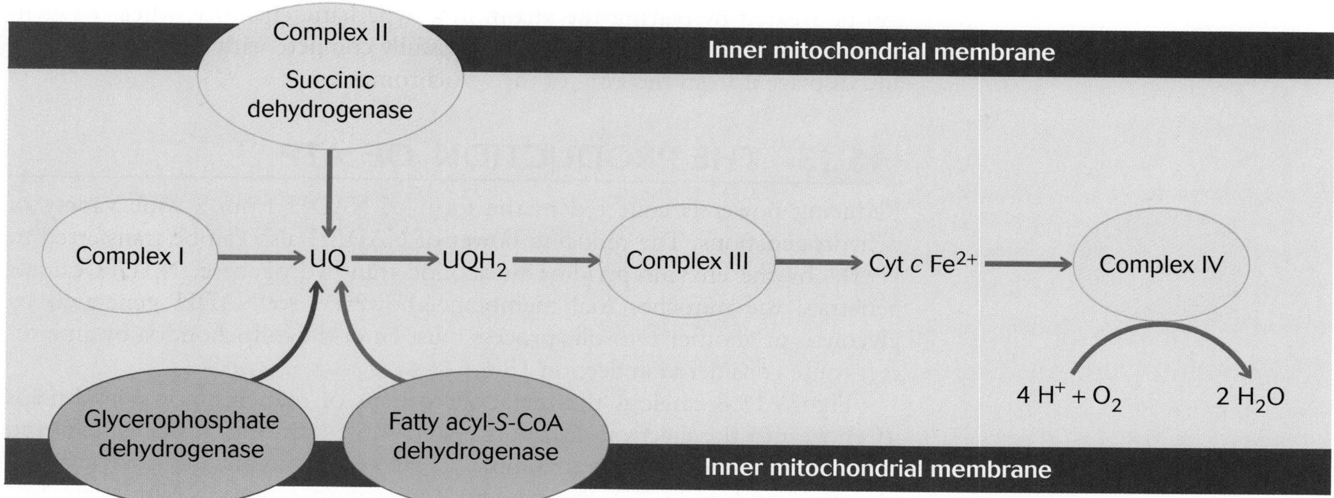

Figure 15.7 The enzyme complexes of the electron-transport chain in the mitochondrial membrane. Complex II, glycerophosphate dehydrogenase, and fatty acyl-*S*-CoA dehydrogenase are incorporated into the inner membrane. Abbreviations: UQ, ubiquinone; UQH_2, reduced ubiquinone; Cyt *c*, cytochrome *c*.

succinate to fumarate results in the formation of $FADH_2$, which transfers its electrons to UQ. The oxidation of 1 mol of $FADH_2$ eventually results in the formation of 2 mol of ATP.

UQ is also reduced by electrons from the oxidation of fatty acids by an electron-transferring flavoprotein (Section 16.4) and from a reaction catalyzed by glycerophosphate dehydrogenase.

Reduced ubiquinone (UQH_2) delivers its reducing power to complex III. Complex III contains a family of iron-containing proteins called cytochromes—specifically cytochromes *b* and c_1—which undergo reversible oxidation–reductions. The cytochromes possess the same type of iron-containing heme as that found in hemoglobin, but, in the cytochromes, heme combines with protein in such a way as to prevent its iron center from combining with molecular oxygen. Iron–sulfur centers also are present. The passage of electrons through complex III also is accompanied by the transport of protons to the cytosolic side of the inner membrane. Electrons from this complex are conducted to the next complex by a freely diffusing protein called cytochrome *c* (Cyt *c*).

Complex IV is called cytochrome oxidase because it transfers electrons from the electron-carrier cytochrome *c* to molecular oxygen to form water. The complex contains cytochromes *a* and a_3 and two copper ions that shuttle between the Cu^{2+} and Cu^{+} states. The reaction of protons, electrons, and oxygen to form water is

$$4\ \text{Cyt}\ c(\text{Fe}^{2+}) + 4\ \text{H}^+ + \text{O}_2 \xrightarrow{\text{complex IV}} 4\ \text{Cyt}\ c(\text{Fe}^{3+}) + 2\ \text{H}_2\text{O}$$

The reduction of oxygen in this reaction is accompanied by proton transport across the inner membrane. This is the final reaction in which electrons are derived from reduced nutrient molecules.

Carbon monoxide or cyanide poisoning prevents the combination of hydrogen and electrons with oxygen and shuts down the electron-transport chain at complex IV. Cyanide combines with Fe^{3+}, and carbon monoxide combines with Fe^{2+} of the cytochromes. Clinical treatment of cyanide poisoning includes the introduction of nitrites, which convert the Fe^{2+} of hemoglobin in the blood into the Fe^{3+} of methemoglobin. Blood methemoglobin can compete with cellular cytochrome Fe^{3+} for the cyanide and reverse the symptoms. Cyanide can then be transformed into the much less toxic cyanate by the therapeutic administration of thiosulfate. Carbon monoxide toxicity can be treated by placing the victim in a hyperbaric chamber, where oxygen, supplied at elevated pressures, can successfully compete with carbon monoxide and displace it from the Fe^{2+} of the cytochromes.

15.13 THE PRODUCTION OF ATP

Reducing power is collected in the form of NADH from a wide variety of dehydrogenations. The reducing power of NADPH also can be transferred to NAD^+ by the enzyme pyridine nucleotide transhydrogenase. NADH cannot penetrate the mitochondrial membrane, however; so NADH generated by glycolysis or another cytosolic process must enter the mitochondria by an indirect route considered in Section 15.6.

Figure 15.8 catalogs the metabolic origins of reducing power and maps its entry into the electron-transport chain. Most reducing power is delivered directly to the flavoprotein at complex I. However, succinic dehydrogenase of the citric acid cycle is a flavoprotein and cannot reduce NADH dehydrogenase. This flavoprotein and others deliver reducing power directly to UQ and thence to complex III. Because it bypasses complex I, reducing power from the flavoprotein succinate dehydrogenase produces fewer moles of ATP. The origin of reducing power from fatty acids and α-amino acids will be taken up in Chapters 16 and 17.

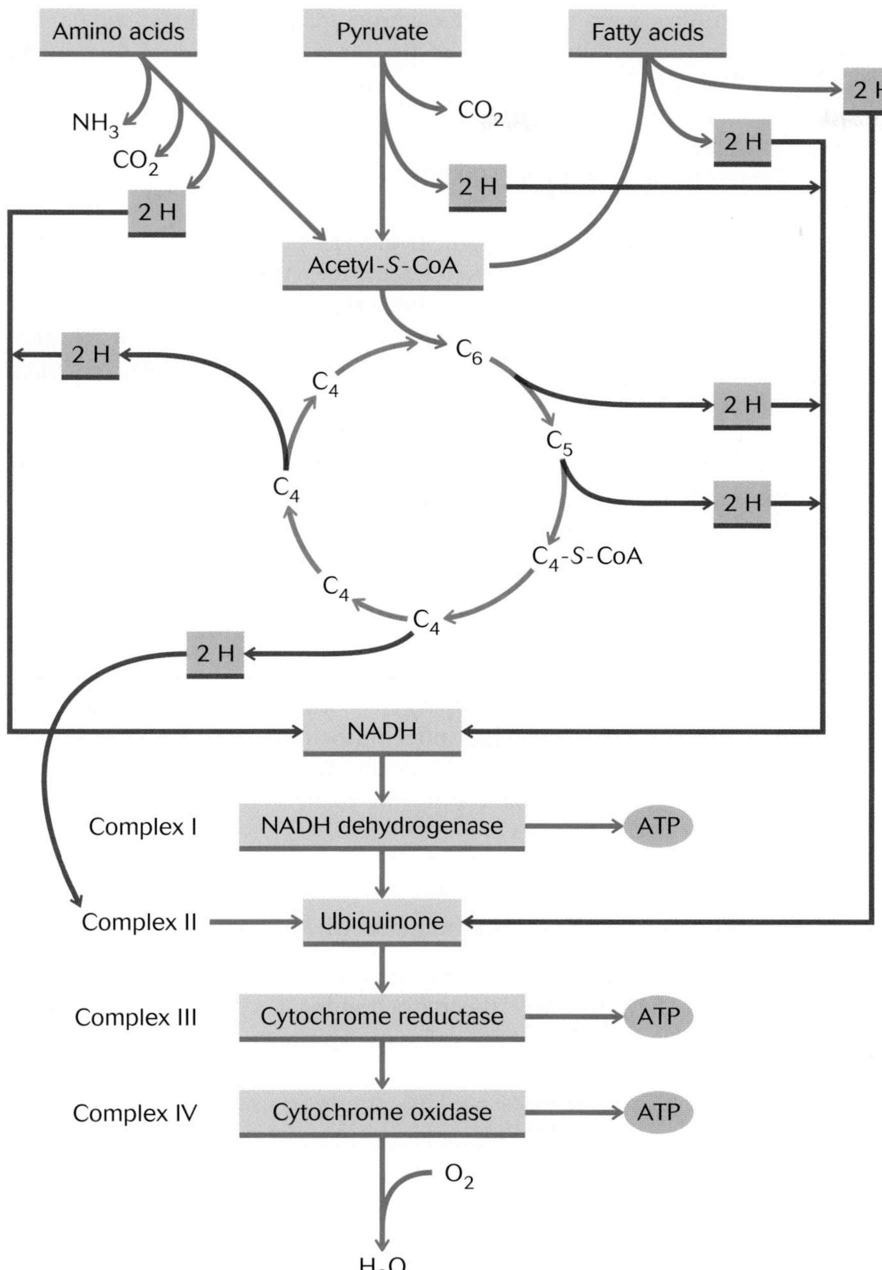

Figure 15.8 The origins and fate of reducing power extracted from nutrients in catabolism. Subscript numbers identify the components of the cycle: for example, C_6, citrate; C_5, α-ketoglutarate.

As electrons move through the electron-transport chain during respiration, hydrogen ions are transported from the inner matrix to the intermembrane space of the mitochondrion. As a result, a large diffusion gradient of protons forms, which provides the driving force for the formation of ATP. This process is called the **chemiosmotic mechanism** because it effects chemical change through an osmotic gradient. The concept was developed by Peter Mitchell, who was awarded a Nobel Prize in 1969 for this powerful idea.

The inner membrane is the location of a complex enzyme that has been shown to possess two functions: (1) a proton-transport channel and (2) an ATPase. This complex enzyme, called ATP synthase, has been extracted from mitochondria and dissociated into several structural components. The functional properties of the parts have been established and the parts have been reassembled into a working enzyme.

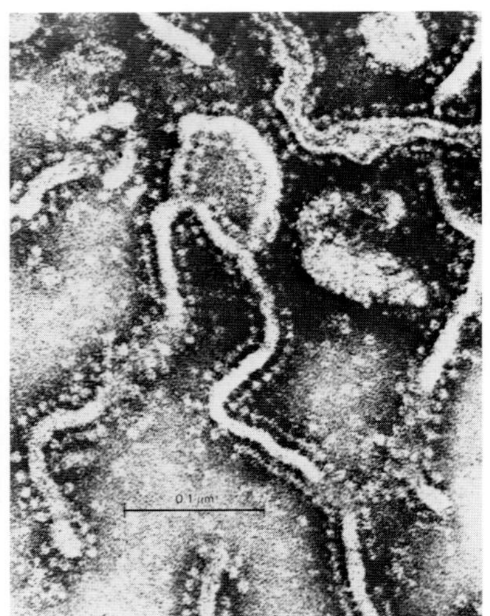

Figure 15.9 Electron micrograph of a part of the inner mitochondrial membrane showing the lollypop structure of ATP synthase. (Courtesy of Dr. Donald F. Parsons.)

Structural studies with the use of electron microscopy have shown the enzyme to be an integral component of the inner mitochondrial membrane and to have a "lollypop" appearance with a stalk and head protruding into the mitochondrial matrix. Figure 15.9 contains one such image of a part of the inner mitochondrial membrane, showing the lollypop structure of ATP synthase. There can be as many as 100,000 of these ATP-synthesizing "factories" in a single mitochondrion.

As the transmembrane proton gradient grows, the electrical charge of the intermembrane space becomes increasingly positive. At some point, the transport of electrons along the electron-transport chain tends to stop because of this buildup of positive charge. The stoppage is relieved, however, when protons begin to flow inward through the proton-transport channel. As a result of this inward flow, ATP synthase catalyzes the formation of a phosphoanhydride bond between ADP and P_i, which allows electron transport to resume. The process is diagrammed in Figure 15.10. Calculations show that the transmembrane proton gradient is great enough at three locations along the electron-transport chain to account for the synthesis of 1 mol of ATP at each location.

15.14 MITOCHONDRIAL MEMBRANE SELECTIVITY

The outer mitochondrial membrane is freely permeable to almost all low-molecular-mass solutes. However, the inner mitochondrial membrane is permeable only to solutes corresponding to specific transport systems embedded in that membrane. It is impermeable not only to H^+, but to OH^-, K^+, and Cl^- ions as well.

Adenine Nucleotide and Phosphate Translocases

Two of the most important inner-membrane-specific transport systems are

1. **The adenine nucleotide translocase:** This transport system allows ADP and phosphate to enter the mitochondria and ATP to leave. It is specific for ADP and ATP and will not transport AMP or any other nucleotide such as GTP or GDP.
2. **The phosphate translocase:** This transport system promotes the simultaneous transport of $H_2PO_4^-$ and H^+ into the mitochondria.

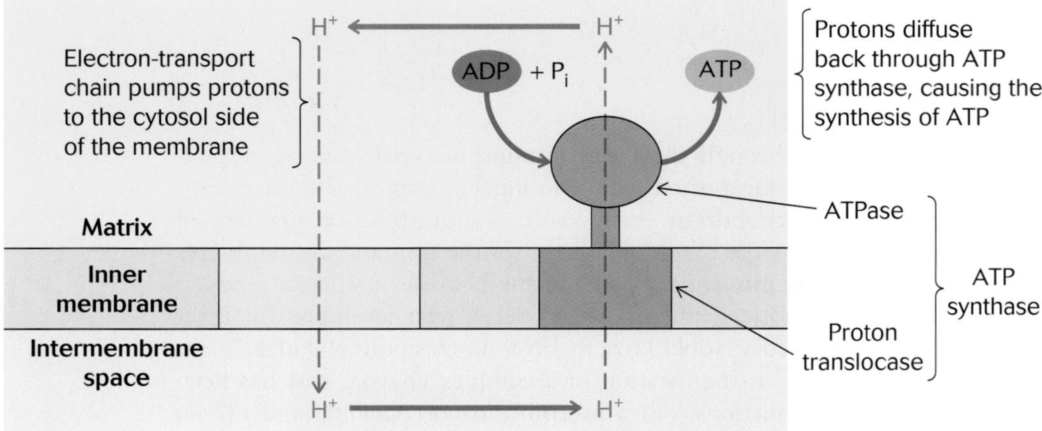

Figure 15.10 Formation of ATP as a result of the movement of protons along the proton gradient generated during electron transport in mitochondria, showing the "lollypop" structure of ATP synthase.

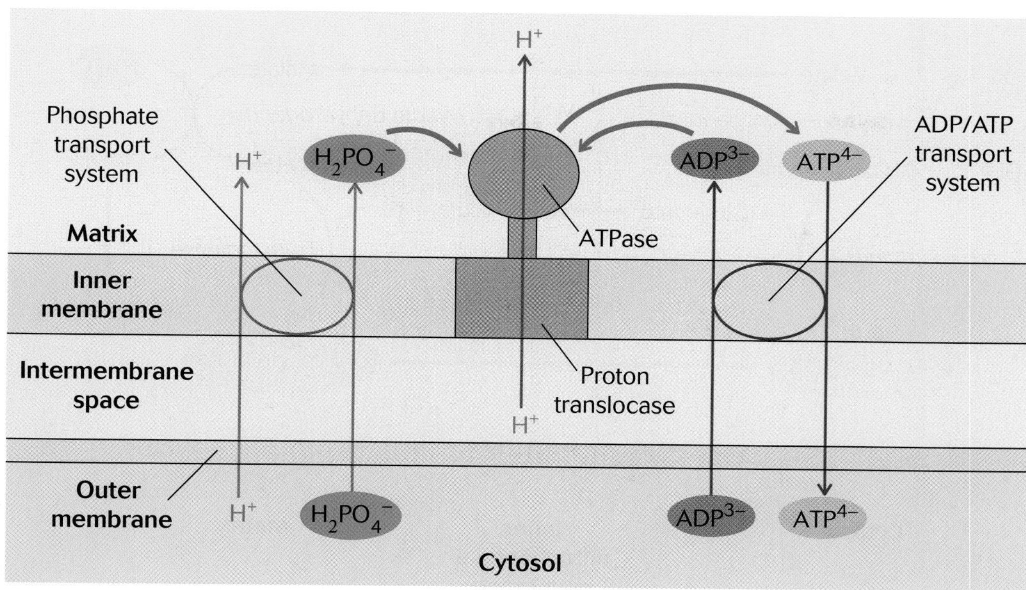

Figure 15.11 The coordinated transport of P_i and ADP into the mitochondria and ATP out of it.

The coordinated operation of these two transport systems brings ADP and P_i into the mitochondrion from the cytosol and sends the newly synthesized ATP out to the cytosol to energize the cell's activities. These coordinated systems are illustrated in Figure 15.11.

Other specific transport systems of the inner mitochondrial membrane are: a system for pyruvate, which is formed in the cytosol and must enter the mitochondrion to enter the citric acid cycle; a dicarboxylate transport system for malate and succinate; a tricarboxylate transport system for citrate and isocitrate; and two other systems that allow the reducing power of cytosolic NADH to enter the mitochondrion for oxidation.

The Impermeability of the Mitochondrial Membrane to NAD^+ and NADH

NADH cannot pass through the inner mitochondrial membrane. It can deliver its reducing power to the electron-transport chain by one of two indirect routes. The first route, the malate–aspartate shuttle (shuttle A) is active in heart, liver, and kidney mitochondria, where it delivers its reducing equivalents to complex I, producing 3 mol of ATP per mole of NADH. The second route is the glycerol phosphate shuttle (shuttle B) active in brain and muscle mitochondria, where it delivers its electrons directly to complex II to produce 2 mol of ATP per mole of NADH.

The malate–aspartate shuttle is illustrated in Figure 15.12 on the next page. NADH from glycolysis is used to reduce cytosolic oxaloacetate to form malate. The reaction is catalyzed by cytosolic malate dehydrogenase. Malate is then transported into the mitochondrial matrix by the dicarboxylate transport system (M in Figure 15.12), where it is in turn oxidized by mitochondrial malate dehydrogenase to form mitochondrial oxaloacetate. The other product, mitochondrial NADH, delivers its reducing equivalents to complex I of the electron-transport chain. Cytosolic oxaloacetate must now be regenerated to continue the cycle. Mitochondrial oxaloacetate cannot pass through the membrane, but it is converted into aspartate by transamination with glutamate (see Section 17.2). Both aspartate and glutamate can penetrate the membrane through an α-amino acid transport system (G in Figure 15.12).

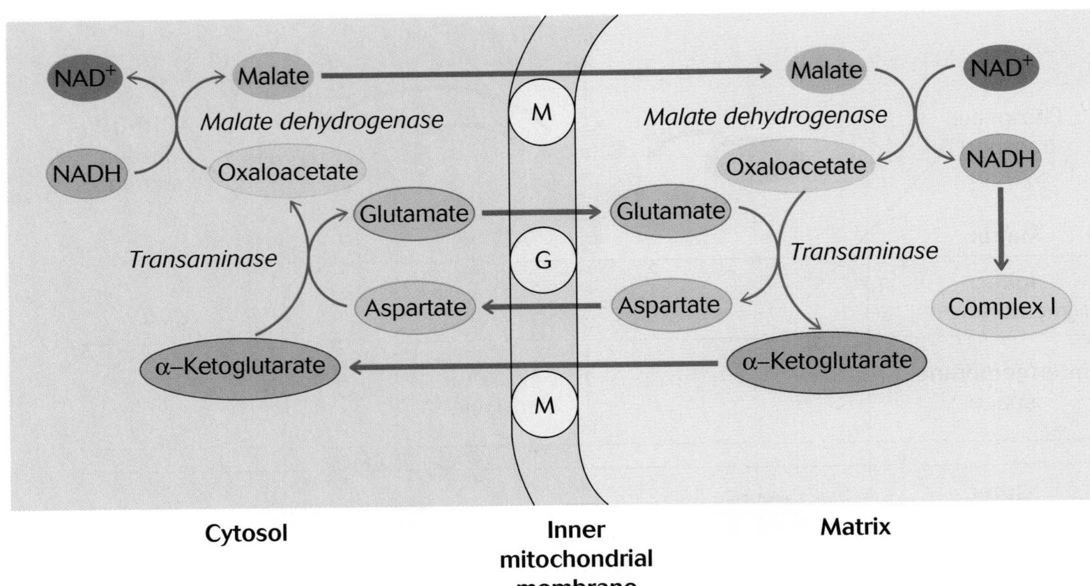

Figure 15.12 The malate–aspartate shuttle brings reducing equivalents generated in the cytosol into the mitochondria. Abbreviations: M, dicarboxylate acid transport system; G, α-amino acid transport system.

The α-ketoglutarate generated in the transaminase reaction producing mitochondrial aspartate passes through the membrane by means of the same dicarboxylate transport system used by malate.

The glycerol phosphate shuttle begins in the cytosol with the reduction of dihydroxyacetone phosphate by NADH to form 3-phosphoglycerol:

$$\begin{array}{c} CH_2OPO_3^{2-} \\ | \\ C{=}O \\ | \\ H{-}C{-}OH \\ | \\ H \end{array} + NADH + H^+ \xrightleftharpoons[\text{dehydrogenase}]{\text{3-phosphoglycerol}} \begin{array}{c} CH_2OPO_3^{2-} \\ | \\ H{-}C{-}OH \\ | \\ H{-}C{-}OH \\ | \\ H \end{array} + NAD^+$$

The product then enters the intermembrane space of the mitochondrion and interacts with a flavoprotein at the surface of the mitochondrial inner membrane. There it is oxidized back to dihydroxyacetone phosphate, and the reducing power, in the form of $FADH_2$, is passed on to ubiquinone and then to complex III of the electron-transport chain to result in the synthesis of 2 mol of ATP.

15.15 ENERGY YIELD FROM CARBOHYDRATE CATABOLISM

Table 15.4 presents the energy yield of carbohydrate metabolism in moles of NADH and $FADH_2$ generated at each step of the citric acid cycle and in glycolysis. Glycolysis produces a net of 2 mol of ATP per mole of glucose. The oxidation of 1 mol each of NADH and $FADH_2$ results in the synthesis of 3 mol of ATP and 2 mol of ATP, respectively. Table 15.4 uses these values to calculate the overall yield of ATP per mole of glucose.

The total number of moles of ATP produced by both glycolysis and the citric acid cycle is 38 or 36 mol of ATP per mole of glucose, depending on which shuttle is used to transfer reducing power from the cytosol into the mitochondria (Section 15.14). The citric acid cycle can be seen to produce at least 18 times as much ATP from glucose as that produced by glycolysis.

TABLE 15.4 An Accounting of the ATP Yield from Carbohydrate Catabolism

Process	NADH	$FADH_2$	ATP
GLYCOLYSIS			
glucose → glucose-6-PO_4			−1
fructose-6-PO_4 → 1,6-BPF			−1
2 3-PGAL → 2 1,3-BPG	2		2
2 PEP → 2 pyruvate			2
CITRIC ACID CYCLE			
2 pyruvate → 2 acetyl-*S*-CoA + 2 CO_2	2		
2 isocitrate → 2 α-ketoglutarate + 2 CO_2	2		
2 α-ketoglutarate → 2 succinate + 2 CO_2	2		
2 succinyl-*S*-CoA + 2 GDP →			2
2 succinate → 2 fumarate		2	
2 malate → 2 oxaloacetate	2		
Sum of moles per mole of glucose	10	2	4

	Sum of ATP yields	
	Shuttle A	Shuttle B
ATP from glycolysis	2	2
NADH from glycolysis	6	4
NADH from citric acid cycle	24	24
$FADH_2$ from citric acid cycle	4	4
ATP from GTP	2	2
Total ATP* per mole of glucose	38	36

*3 ATP/NADH, 2 ATP/$FADH_2$

The overall result of glycolysis is

$$\text{Glucose} + 2\ \text{ADP} + 2\ \text{P}_i + 2\ \text{NAD}^+ \longrightarrow 2\ \text{pyruvate} + 2\ \text{ATP} + 2\ \text{NADH} + 2\ \text{H}^+$$

The net result of optimal oxidation of glucose is

$$\text{Glucose} + 38\ \text{ADP} + 38\ \text{P}_i + 6\ \text{O}_2 \longrightarrow 6\ \text{CO}_2 + 38\ \text{ATP} + 44\ \text{H}_2\text{O}$$

Although mitochondrial oxidative phosphorylation produces more than 18 times as much ATP per mole of glucose as glycolysis produces, the rate of glycolysis is greater than that of the oxidative process. This greater rate is primarily because the concentration of the glycolytic enzymes is much greater than that of the enzymes in the citric acid cycle. This greater concentration allows anaerobic processes, primarily intense muscular activity, to take place at adaptively useful rates.

If all the potential energy contained in the glucose molecule could be harnessed in the form of ATP, it would generate about 94 mol of ATP per mole of glucose. Because 38 mol of ATP are optimally formed, the efficiency of the metabolic process can be calculated to be about 40%. This efficiency significantly exceeds the typical efficiencies of mechanical engines, about 10%, and is close to the efficiency of electrochemical cells that directly transform the oxidation of hydrocarbons into electricity.

Summary

Glycolysis

- In glycolysis, a linear enzymatic pathway located in the cell cytosol, glucose is converted into two molecules of pyruvate. In the process, ATP and NADH are produced.
- Pyruvate enters the citric acid cycle to be oxidized, first, to acetyl-*S*-CoA and, then, to carbon dioxide and water.
- In active muscle, pyruvate is reduced to lactate. This reduction regenerates NAD^+; so glycolysis can take place at required rates.

The Citric Acid Cycle

- The aerobic phase of glucose catabolism is a cyclic enzymatic pathway called the citric acid cycle, located in the mitochondria of animal cells.
- A turn of the cycle begins with the condensation of an acetyl group of acetyl-*S*-CoA with oxaloacetate to form citrate.
- In seven succeeding enzymatically catalyzed steps, oxaloacetate is regenerated. This process results in the loss of two molecules of CO_2 and in the production of two molecules of NADH, one molecule of $FADH_2$, and one molecule of GTP.
- The NADH and $FADH_2$ produced as a result of the oxidations is used to generate ATP.

Gluconeogenesis

- Glucose is resynthesized in the liver from the products of glycolysis.
- Three irreversible steps in glycolysis must be bypassed to synthesize glucose from pyruvate.
- These irreversible steps are replaced by different reactions and enzymes.
- All other steps in the synthesis use the reversible reactions and enzymes of the glycolytic pathway.

Glycogenesis

- Glucose is stored as a polysaccharide called glycogen.
- Glycogen is synthesized only in liver and muscle cells and only when these cells contain high concentrations of ATP. It is stored as glycogen granules.

Glycogenolysis

- The chain-terminal glucose residue of glycogen is released by phosphorolysis.
- Phosphorolysis is accomplished by the active form of the enzyme phosphorylase.
- Phosphorylase is activated in the liver through the action of the hormone glucagon and in both liver and muscle by the hormone epinephrine.

The Electron-Transport Chain

- ATP synthesis takes place in the inner mitochondrial membrane and is catalyzed by a sequence of enzymes called the electron-transport chain.
- The reduced cofactors generated in catabolism are oxidized to form water, and that oxidation is coupled to the simultaneous synthesis of ATP.
- The synthesis of ATP by the electron-transport chain is called oxidative phosphorylation.
- Oxidative phosphorylation is driven by a diffusion gradient of protons established across the inner mitochondrial membrane and is called the chemiosmotic mechanism of ATP synthesis.
- Specific transport systems in the inner mitochondrial membrane allow the permeability only of solutes that take part in oxidative phosphorylation.

Energy Yield from Carbohydrate Catabolism

- Glycolysis produces a net of 2 mol of ATP per mole of glucose.
- The total ATP produced by the citric acid cycle is from 36 to 38 mol per mole of glucose.
- Because it takes place at a much greater rate than that of oxidative phosphorylation, glycolysis is the principal catabolic route in active skeletal muscle, where ATP is used at high rates.

Key Words

chemiosmotic mechanism, p. 489
citric acid cycle, p. 476
electron-transport chain, p. 486
flavoprotein, p. 479
gluconeogenesis, p. 480
glycogenesis, p. 482
glycogenolysis, p. 482
glycolysis, p. 468
hormone, p. 483
oxidative phosphorylation, p. 486
substrate-level phosphorylation, p. 471

Exercises

Glycolysis

15.1 What nutrient is the principal source of energy for metabolism?

15.2 Does the first metabolic pathway require the presence of oxygen?

15.3 What are the most important products of glycolysis?

15.4 Are all the steps of the glycolytic pathway reversible? Explain your answer.

15.5 Are there any important regulatory control points in glycolysis?

15.6 Identify the oxidation step in glycolysis.

15.7 Identify the steps in glycolysis in which ATP is synthesized.

15.8 What is substrate-level phosphorylation?

15.9 How is NAD^+ regenerated in glycolysis in animal cells?

15.10 How is NAD^+ regenerated in glycolysis in yeast cells?

15.11 What would be the effect on glycolysis if inorganic phosphate were unavailable?

Formation of Acetyl-*S*-CoA

15.12 What reaction does pyruvate undergo when oxygen is available within a cell?

15.13 Where does the reaction in Exercise 15.12 take place?

15.14 What are the names of the four vitamins having roles in the oxidation and decarboxylation of pyruvate?

15.15 Is the pyruvate dehydrogenase complex under regulatory control? Explain your answer.

Citric Acid Cycle

15.16 What is the first reaction of the citric acid cycle?

15.17 In which of the reactions of the citric acid cycle is CO_2 generated?

15.18 Does a substrate-level phosphorylation take place in the citric acid cycle? Explain your answer.

15.19 Why is the citric acid cycle called a cycle rather than a pathway?

15.20 Is there an enzyme complex in the citric acid cycle similar to the pyruvate dehydrogenase complex? Explain your answer.

15.21 Explain why the citric acid cycle is sometimes referred to as a catalytic cycle.

15.22 Does the concentration of oxaloacetate have an effect on the overall rate of the citric acid cycle? Explain your answer.

15.23 How is oxaloacetate replaced after its concentration has been reduced by conversion into other biomolecules?

Gluconeogenesis

15.24 Is gluconeogenesis simply the reverse of glycolysis? Explain your answer.

15.25 How is phosphoenolpyruvate synthesized?

15.26 What is the cofactor in enzymatic carboxylation reactions?

15.27 How does oxaloacetate, synthesized in mitochondria, enter the cytosol?

15.28 Can muscle or brain supply free glucose to the blood? Explain your answer.

Glycogenesis

15.29 What is glycogen?

15.30 Can glycogen be synthesized by brain cells? Explain your answer.

15.31 What is the important metabolic function of pyrophosphate, $P_2O_7^{4-}$?

15.32 In what activated form is glucose added to a growing glycogen chain?

Glycogenolysis

15.33 Glucose is removed from a glycogen polymer by phosphorolysis. Compare that process with hydrolysis.

15.34 Is the enzyme that catalyzes glycogenolysis always present in its active form?

15.35 What are glucagon and epinephrine?

15.36 Explain the roles of glucagon and epinephrine in glycogenolysis.

15.37 Does either glucagon or epinephrine have any function other than in glycogenolysis?

15.38 What are the first and second messengers in glycogenolysis?

15.39 Why are liver cells the only cells that can produce free glucose from glucose-6-PO_4?

Electron-Transport Chain

15.40 What is meant by oxidative phosphorylation?

15.41 At how many points along the electron-transport chain can ATP be synthesized?

15.42 What are the cytochromes?

15.43 Can ATP synthesis take place in mitochondria whose inner membranes have been mechanically disrupted?

15.44 What fundamental process taking place in mitochondria provides the driving force for the synthesis of ATP?

15.45 Is it possible for any cellular substance to diffuse into the inner mitochondrial space? Explain your answer.

15.46 ATP is synthesized inside the mitochondria but is used outside the mitochondria. How does the cell accomplish this?

Energy Yield from Carbohydrate Catabolism

15.47 What is the total amount of ATP produced under aerobic conditions per mole of glucose?

15.48 Compare the energy yields of glycolysis and the citric acid cycle.

15.49 Why is glycolysis the preferred carbohydrate catabolic pathway in active skeletal muscle cells?

Expand Your Knowledge

15.50 Is it possible for a liver cell to synthesize glucose by reversing glycolysis?

15.51 What are the two critical conditions that must be fulfilled if muscle cells are to produce ATP during intense activity?

15.52 True or false: When a cell is in a high-energy state, pyruvate dehydrogenase becomes active. Explain your answer.

15.53 If NADH cannot penetrate the inner mitochondrial membrane, explain how it can deliver its reducing power to the electron-transport chain.

15.54 Why does the dehydrogenation of succinate lead to the synthesis of only 2 mol of ATP?

15.55 What is the major difference between glycolysis and the citric acid cycle?

15.56 How can you tell if a metabolite is present in a cell at high or low concentrations? Give an example.

15.57 What are possible consequences of a mutation causing the loss of the glycogen branching enzyme?

15.58 What is a likely consequence of a mutation leading to the loss of glucose-6-phosphatase?

15.59 Describe a metabolic process that is, in effect, "controlled" by the Le Chatelier effect.

15.60 What is the effect of phosphorylation on protein kinase?

15.61 A glycogen molecule of 18,000 residues is branched every 11 residues. How many reducing ends does it have? (Consult Section 10.6.)

CHAPTER 16 FATTY ACID METABOLISM

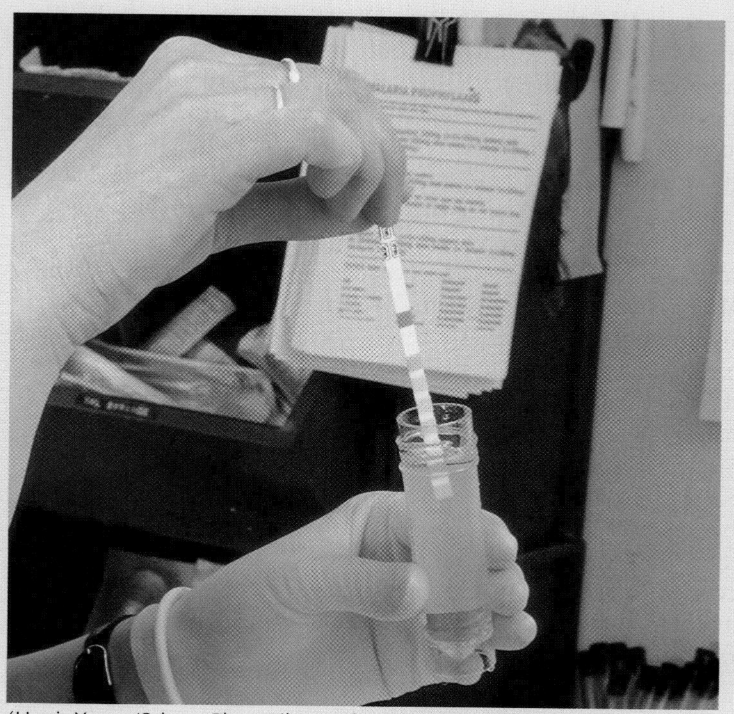
(Hattie Young/Science Photo Library/Photo Researchers.)

Chemistry in Your Future

A patient admitted yesterday evening in a diabetic coma has recovered quite rapidly, and this morning his levels of blood glucose, urinary glucose, and urinary ketones, as well as his mental status, are almost normal. Nevertheless, you continue to monitor him carefully. Most of the urine and blood analyses are performed in the hospital lab, but you have been doing spot tests for ketonuria (ketones in the urine) by immersing a reagent test strip into a urine sample and observing the extent of the reagent's color change. In presenting the metabolism of fatty acids, this chapter describes ketone bodies and explains why they are sometimes excreted in the urine.

For more information on this topic and others in this chapter, go to www.whfreeman.com/bleiodian2e

Learning Objectives

- Describe how fatty acids are activated and transported from the cytosol into the mitochondria.
- Describe the fundamental steps by which a fatty acid chain is shortened by two carbon atoms.
- Describe ketone bodies and how they are used in cellular metabolism.
- Describe the circumstances under which fatty acid biosynthesis is initiated.
- List the sequence of reactions by which a fatty acid chain is elongated by two carbon atoms.
- Show how triacylglycerols are synthesized and describe their principal cellular function.
- Describe the biosynthesis of membrane lipids.

In this chapter, we will explore how fatty acids (Section 11.2) function as a major energy source for the cell, how they are synthesized, and how they serve as the building blocks for the biosynthesis of triacylglycerols, membrane lipids, and steroids. In addition, we will look at some health problems related to the metabolism of membrane lipids.

Both fatty acids, stored as triacylglycerols (Section 11.3) in adipose cells, and glucose, stored as glycogen (Section 10.6) in the liver and in muscle, serve as sources of energy. In addition to serving as a source of energy, glycogen in the liver is used to maintain a constant blood concentration of glucose. The amount of glycogen stored in the liver and muscle of a 70-kg man is about 400 g. His body needs about 2400 kcal/day but, at 4 kcal/gram, 400 g of glycogen cannot supply all of that need. Both the need for energy and the need to spare glycogen for blood-glucose maintenance are satisfied by the use of stored triacylglycerols.

When we eat more carbohydrate than is required to satisfy our immediate energy needs, the liver and muscle cells convert the excess into glycogen. When the capacity to store glycogen is reached, any remaining carbohydrate is transformed into fatty acids and then stored as triacylglycerol. In contrast with the limitations of glycogen storage, there is no apparent limit to the amount of triacylglycerols that can be stored, and, as a result, the biosynthesis of triacylglycerols in animals, including humans, proceeds very actively.

To understand how good an energy source triacylglycerols are, consider that the average male human stores about 17% of body weight as triacylglycerols. The oxidation of triacylglycerols yields about 9 kcal of energy per gram; therefore the stored lipid provides a 6-week supply of energy for a 70-kg man who consumes about 2400 kcal/day. The need for triacylglycerols arises between meals, when the hormone glucagon is released in response to the lowered glucose concentration in the blood (Section 15.6). We begin our presentation of fatty acid metabolism with a consideration of the release of fatty acids from storage.

16.1 FATTY ACID MOBILIZATION

Triacylglycerols, commonly called triglycerides (Section 11.3), are the fats stored in fat cells. The mobilization of triacylglycerols from storage takes place principally through the action of the hormone glucagon. Between meals, glucagon is released in response to lowered blood-glucose concentration and initiates a sequence of enzymatic reactions that activate cellular triacylglycerol lipase. This enzyme hydrolyzes the triacylglycerols and frees fatty acids, along with glycerol, for oxidative processes.

The mechanism of action of glucagon in fatty acid mobilization is similar to that which results in the activation of phosphorylase in glycogenolysis

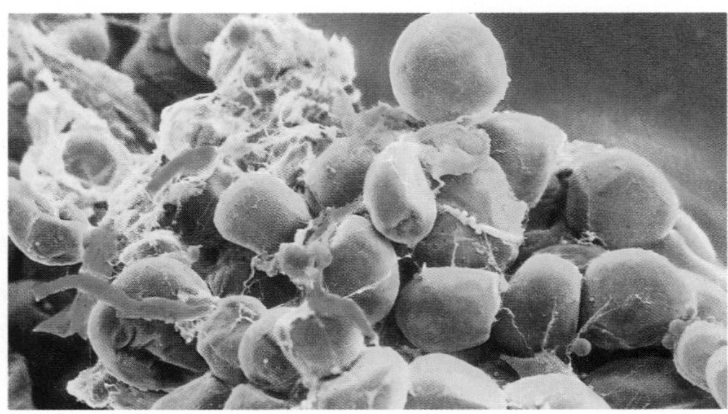

Adipose tissue. Colored scanning electron micrograph of adipocytes (yellow), which build up the adipose connective tissue. Connective-tissue fibers at top left support the fat cells. Almost the entire volume of each adipocyte is occupied by a single lipid droplet of triacylglycerols. Fat to be stored arrives at these cells through small capillaries, seen here as blue tubules. (P. Motta, Department of Anatomy, University La Sapienza, Rome/ Science Photo Library/Photo Researchers.)

(Section 15.10). The free fatty acids leave the adipose-tissue cells and become bound to serum albumin. Water-soluble glycerol and the protein-bound fatty acids then circulate to other organs to be utilized as energy sources.

Glycerol is processed through the glycolytic pathway. It is first phosphorylated by ATP, then oxidized to dihydroxyacetone phosphate, and subsequently isomerized into glyceraldehyde-3-phosphate to form pyruvate. Pyruvate can then be oxidized through the citric acid cycle or converted into glucose through gluconeogenesis.

The fatty acids take a more complex route. Fatty acids can be oxidized only when in the form of an acyl-*S*-CoA thioester. This transformation, which takes place in the cytosol, requires the use of 2 mol of ATP. The fatty acid used in the following discussion is palmitic acid, a 16-carbon fatty acid. At the pH of the cell, the acid exists as palmitate and the activation reaction is

$$\underset{\textbf{Palmitate}}{CH_3(CH_2)_{14}COO^-} + ATP + CoA{-}SH \longrightarrow \underset{\textbf{Palmitoyl-}S\textbf{-CoA}}{CH_3(CH_2)_{14}CO{-}S{-}CoA} + AMP + PP_i$$

As mentioned in Chapter 15, PP_i represents pyrophosphate, $P_2O_7^{4-}$, derived from the hydrolysis of ATP. We have seen this type of process in the synthesis of glycogen (Section 15.9). The subsequent hydrolysis of pyrophosphate catalyzed by the enzyme pyrophosphatase is characterized by a very large equilibrium constant, thus providing the driving force for the overall reaction.

As described in the preceding equation, the activation process appears to require only 1 mol of ATP, when, in fact, 2 mol of ATP are consumed. The reason is that AMP (adenosine monophosphate) cannot be phosphorylated by either substrate-level or oxidative phosphorylation. It can, however, react with another mole of ATP, catalyzed by the enzyme adenylate kinase, to form 2 mol of ADP:

$$AMP + ATP \longrightarrow 2\ ADP$$

The ADP thus generated now can be phosphorylated by substrate-level or oxidative phosphorylation to form ATP. The result of adding the two equations is:

$$\underset{\textbf{Palmitate}}{CH_3(CH_2)_{14}COO^-} + 2\ ATP + CoA{-}SH \longrightarrow \underset{\textbf{Palmitoyl-}S\textbf{-CoA}}{CH_3(CH_2)_{14}CO{-}S{-}CoA} + 2\ ADP + 2\ P_i$$

The oxidation of the fatty acid must take place in mitochondria, but the acyl-*S*-CoA derivative cannot penetrate the inner mitochondrial membrane. Therefore, in the next step, the CoA derivative exchanges its CoA part for another substance that confers membrane (lipid) solubility on the activated fatty acid. The substance is **carnitine.** In a reaction catalyzed by the enzyme carnitine transferase, the thioacyl ester is exchanged for a carboxylic ester at the OH group of carnitine:

$$CH_3(CH_2)_{14}CO{-}S{-}CoA + \underset{\textbf{Carnitine}}{H_3C{-}\overset{\overset{CH_3}{|}}{\underset{\underset{CH_3}{|}}{N^+}}{-}CH_2{-}\overset{\overset{H}{|}}{\underset{\underset{OH}{|}}{C}}{-}CH_2{-}COO^-} \xrightleftharpoons{\text{carnitine transferase}}$$

$$\underset{\textbf{Acylcarnitine}}{H_3C{-}\overset{\overset{CH_3}{|}}{\underset{\underset{CH_3}{|}}{N^+}}{-}CH_2{-}\overset{\overset{H}{|}}{\underset{|}{C}}{-}CH_2{-}COO^-} \quad \text{with } O{-}C(=O){-}(CH_2)_{14}{-}CH_3 \text{ (Palmitoyl group) on C} \quad + CoA{-}SH$$

The acylcarnitine molecule passes through the inner mitochondrial membrane into the matrix. Once there, it reacts with mitochondrial CoA-SH, catalyzed by a mitochondrial form of carnitine acyltransferase, to reform the acyl-*S*-CoA derivative in the reverse of the preceding reaction. The oxidation process can now take place.

The pools of cytosolic and intramitochondrial CoA-SH are kept separated, as are those of ATP and NAD^+. Compartmentalization keeps regulatory control of anabolic processes separate from the regulatory controls of catabolic processes and independent of them.

Concept checklist

- ✓ Fatty acids released from adipose tissue are transported in the blood to other tissues as fatty acid–serum albumin complexes.
- ✓ Before they can be oxidized, fatty acids must be converted into acetyl-*S*-CoA derivatives, and this process takes place in the cytosol.
- ✓ To leave the cytosol and enter mitochondria where oxidation takes place, the acetyl-*S*-CoA derivatives must be transformed into carnitine derivatives.

16.2 FATTY ACID OXIDATION

In stage 1 of fatty acid oxidation, which consists of four steps, the products of all fatty acids are acetyl-*S*-CoA, NADH, and $FADH_2$. In stage 2, those products enter the citric acid cycle and the electron-transport chain for ultimate oxidation to CO_2 and H_2O.

Step 1. The first step in the oxidation of the 16-carbon fatty acid palmitate is a dehydrogenation between carbons 2 and 3 of the acyl-*S*-CoA derivative to produce the trans unsaturated acyl-*S*-CoA derivative of palmitate. This process is also called **β-oxidation** because the dehydrogenation takes place between carbons 2 and 3 (the α and β carbon atoms). In the following reaction equation, the double bond between carbons 2 and 3 is indicated by the symbol Δ^2, in which Δ signifies a double bond and the superscript specifies that the double bond is located between carbons 2 and 3, counting from the carboxyl carbon. The coenzyme reduced in this reaction is FAD (Section 14.4 and Figure 14.6), and the enzyme is acyl-*S*-CoA dehydrogenase. With carbon atoms 2 and 3 identified, the reaction is

$$\underset{\text{Acyl-}S\text{-CoA}}{C_{12}H_{25}{-}CH_2{-}\overset{3}{C}H_2{-}\overset{2}{C}H_2{-}\overset{O}{\overset{\|}{C}}{-}S{-}CoA} \xrightarrow[\text{E—FAD} \;\; \text{E—FADH}_2]{} \underset{trans\text{-}\Delta^2\text{-Enoyl-}S\text{-CoA}}{C_{12}H_{25}{-}CH_2{-}\underset{H}{\overset{3}{C}}{=}\overset{H}{\underset{2}{C}}{-}\overset{O}{\overset{\|}{C}}{-}S{-}CoA}$$

The E-$FADH_2$ delivers its reducing power to a flavoprotein, which passes it along to ubiquinone from which it moves on to complex II of the electron-transport chain (Section 15.12). This sequence of reactions leads to the production of 2 mol of ATP.

Step 2. The next step is an enzymatic hydration of the trans double bond to form 3-hydroxyacyl-*S*-CoA. The enzyme, enoyl-*S*-CoA hydratase, catalyzes the hydration of the double bond between carbon atoms 2 and 3 of this fatty acid derivative, leading to the L-hydroxy configuration. The reaction is

$$\underset{trans\text{-}\Delta^2\text{-Enoyl-}S\text{-CoA}}{C_{12}H_{25}{-}CH_2{-}\underset{H}{\overset{3}{C}}{=}\overset{H}{\underset{2}{C}}{-}\overset{O}{\overset{\|}{C}}{-}S{-}CoA} \xrightarrow{H_2O} \underset{\text{L-3-Hydroxyacyl-}S\text{-CoA}}{C_{12}H_{25}{-}CH_2{-}\underset{H}{\overset{OH}{\overset{3|}{C}}}{-}\underset{H}{\overset{H}{\overset{|2}{C}}}{-}\overset{O}{\overset{\|}{C}}{-}S{-}CoA}$$

The synthesis of fatty acids takes place in the cytosol and is thus separated from their oxidation in the mitochondria. In the synthesis, the catabolic sequence is largely reversed (Section 16.4) Here it is important to note why we stress the stereochemical L configuration of the 3-hydroxyacyl-*S*-CoA product of the oxidative pathway. The use of a different reductase in the cytosolic synthetic process that yields the D-hydroxy configuration also controls and separates the synthetic and oxidative pathways.

Step 3. The next step is the dehydrogenation of L-3-hydroxyacyl-*S*-CoA by L-3-hydroxyacyl-*S*-CoA dehydrogenase, with NAD^+ as cofactor, to produce the 3-ketoacyl-*S*-CoA. The reaction is

$$C_{12}H_{25}{-}CH_2{-}\overset{\overset{\displaystyle OH}{|}}{\underset{\underset{\displaystyle H}{|}}{\overset{3\;\;\;2}{C}}}{-}CH_2{-}\overset{\overset{\displaystyle O}{\|}}{C}{-}S{-}CoA \xrightarrow[NAD^+\;\;NADH]{} C_{12}H_{25}{-}CH_2{-}\overset{\overset{\displaystyle O}{\|}}{\underset{3}{C}}{-}\overset{2}{C}H_2{-}\overset{\overset{\displaystyle O}{\|}}{C}{-}S{-}CoA$$

L-3-Hydroxyacyl-*S*-CoA **3-Ketoacyl-*S*-CoA**

The dehydrogenase delivers its reducing power to complex I of the electron-transport chain (Section 15.12), thereby producing 3 mol of ATP per mole of NADH.

Step 4. At the fourth and last step of the oxidation cycle, an enzymatic cleavage takes place that requires a molecule of CoA-SH. The enzyme catalyzing this reaction is 3-ketothiolase. The result is the production of a molecule of acetyl-*S*-CoA derived from carbons 1 and 2 of the original fatty acid, along with another fatty acyl-*S*-CoA molecule having two fewer carbon atoms than the original acyl-*S*-CoA derivative. The reaction is

$$C_{12}H_{25}{-}CH_2{-}\overset{\overset{\displaystyle O}{\|}}{\underset{3}{C}}{-}\underset{2}{CH_2}{-}\overset{\overset{\displaystyle O}{\|}}{C}{-}S{-}CoA \xrightarrow{CoA{-}SH} C_{12}H_{25}{-}CH_2{-}\overset{\overset{\displaystyle O}{\|}}{\underset{3}{C}}{-}S{-}CoA + CH_3{-}\overset{\overset{\displaystyle O}{\|}}{\underset{2}{C}}{-}S{-}CoA$$

3-Keto(C_{16})acyl-*S*-CoA **(C_{14})Acyl-*S*-CoA** **Acetyl-*S*-CoA**

The new, shortened acyl-*S*-CoA undergoes the same sequence of reactions, ending again with the production of a second molecule of acetyl-*S*-CoA and another acyl-*S*-CoA derivative with two fewer carbon atoms in the chain. The reaction equation for one oxidation cycle consisting of steps 1 through 4 and beginning with palmitic acid (C_{16}) thioester is

$$C_{16}\text{-}S\text{-CoA} + \text{CoA-SH} + \text{FAD} + \text{NAD}^+ + H_2O \longrightarrow$$
$$C_{14}\text{-}S\text{-CoA} + \text{acetyl-}S\text{-CoA} + \text{FADH}_2 + \text{NADH} + H^+$$

At each passage through this sequence of reactions, the fatty acid chain loses two carbon atoms as acetyl-*S*-CoA; so, if we start with palmitoyl-*S*-CoA, a 16-carbon chain acid, seven of these sequences will take place. Furthermore, because the cycle begins with a CoA derivative, only seven more molecules of CoA-SH are required to produce overall eight molecules of acetyl-*S*-CoA, as indicated in the following sketch:

$$\overbrace{C{-}C}^{8}\overbrace{C{-}C}^{7}\overbrace{C{-}C}^{6}\overbrace{C{-}C}^{5}\overbrace{C{-}C}^{4}\overbrace{C{-}C}^{3}\overbrace{C{-}C}^{2}\overbrace{C{-}C}^{1}{-}S{-}CoA$$

In addition, 14 pairs of hydrogen atoms are removed from palmitoyl-*S*-CoA, of which 7 emerge as $FADH_2$ and 7 as NADH + H^+. The equation for seven cycles that reduce the C_{16} acid to eight molecules of acetyl-*S*-CoA is

$$\text{Palmitoyl-}S\text{-CoA} + 7\ \text{CoA-SH} + 7\ \text{FAD} + 7\ \text{NAD}^+ + 7\ H_2O \longrightarrow$$
$$8\ \text{acetyl-}S\text{-CoA} + 7\ \text{FADH}_2 + 7\ \text{NADH} + 7\ H^+$$

Each $FADH_2$ produces 2 ATP, and each NADH produces 3 ATP; so the dehydrogenations leading to the formation of acetyl-*S*-CoA, the first stage of fatty acid oxidation, result in the production of 35 ATP. The oxidation of unsaturated fatty acids skips step 1 and proceeds through steps 2 through 4.

The second stage of fatty acid oxidation is the disposal of the acetyl-*S*-CoA produced in the first stage. That acetyl-*S*-CoA is identical with the acetyl-*S*-CoA produced by the dehydrogenation of pyruvate (Section 15.4). Much of it enters the citric acid cycle and is oxidized to CO_2 and water (but see Section 16.3).

Table 15.4 shows that the oxidation of 1 mol of acetyl-*S*-CoA results in the production of 12 ATP. Therefore 8 mol of acetyl-*S*-CoA yield 96 ATP. The net production of ATP from the first and second stages of fatty acid oxidation is 35 ATP + 96 ATP − 2 ATP (activation step) = 129 ATP per mole of palmitate.

16.3 KETONE BODIES AND CHOLESTEROL

In the mitochondria of the human liver, acetyl-*S*-CoA is transformed in two ways: oxidation by the citric acid cycle as indicated in Section 16.2 and, in a second pathway, the formation of **ketone bodies,** a group of compounds consisting of **acetoacetate,** D-3-hydroxybutyrate, and acetone. The ketone bodies arise as an overflow from fatty acid oxidation. They are normally present and make up about 10% of the metabolic energy supply. However, they are present in high concentrations in untreated diabetes mellitus (Section 18.6 and Box 12.3). In diabetes mellitus, peripheral tissues fail to oxidize glucose, because it cannot penetrate into cells, and so fatty acids must be oxidized to compensate for the unavailable energy. Under those circumstances, the excess acetyl-*S*-CoA is converted into ketone bodies, a condition known as **ketosis.** Starvation is another condition that leads to ketosis. In starvation, oxaloacetate is necessarily very low in concentration. Therefore, the acetyl-*S*-CoA produced from fatty acid oxidation cannot react with oxaloacetate to form citric acid and so must go the route of ketone-body formation.

The formation of acetoacetate takes place in liver mitochondria and consists of the reaction of 2 mol of acetyl-*S*-CoA to form acetoacetyl-*S*-CoA, catalyzed by the enzyme thiolase:

$$\underset{\text{Acetyl-}S\text{-CoA}}{CH_3{-}\overset{\overset{\displaystyle O}{\|}}{C}{-}S{-}CoA} + \underset{\text{Acetyl-}S\text{-CoA}}{CH_3{-}\overset{\overset{\displaystyle O}{\|}}{C}{-}S{-}CoA} \rightleftharpoons \underset{\text{Acetoacetyl-}S\text{-CoA}}{CH_3{-}\overset{\overset{\displaystyle O}{\|}}{C}{-}CH_2{-}\overset{\overset{\displaystyle O}{\|}}{C}{-}S{-}CoA} + CoA{-}SH$$

Next, to form acetoacetate, the acetoacetyl-*S*-CoA must lose its CoA component. This loss takes place in two steps, the first being the reaction of acetoacetyl-*S*-CoA with acetyl-*S*-CoA to form 3-hydroxy-3-methylglutaryl-*S*-CoA (HMG-*S*-CoA).

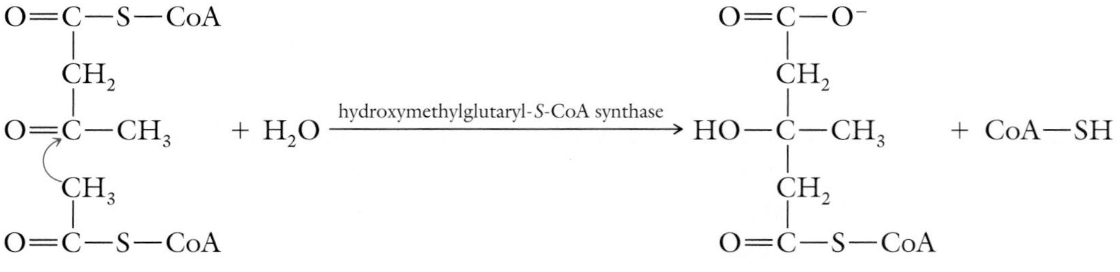

$$\begin{array}{c} O{=}C{-}S{-}CoA \\ | \\ CH_2 \\ | \\ O{=}C{-}CH_3 \\ \\ CH_3 \\ | \\ O{=}C{-}S{-}CoA \end{array} + H_2O \xrightarrow{\text{hydroxymethylglutaryl-}S\text{-CoA synthase}} \begin{array}{c} O{=}C{-}O^- \\ | \\ CH_2 \\ | \\ HO{-}C{-}CH_3 \\ | \\ CH_2 \\ | \\ O{=}C{-}S{-}CoA \\ \text{HMG-}S\text{-CoA} \end{array} + CoA{-}SH$$

In the next step, acetoacetate is cleaved from HMG-*S*-CoA by hydroxymethylglutaryl-*S*-CoA lyase:

$$\begin{array}{c} O{=}C{-}O^- \\ | \\ CH_2 \\ | \\ HO{-}C{-}CH_3 \\ | \\ CH_2 \\ | \\ O{=}C{-}S{-}CoA \end{array} \xrightarrow{\text{hydroxymethylglutaryl-}S\text{-CoA lyase}} \begin{array}{c} \textbf{Acetoacetate} \\ O{=}C{-}O^- \\ | \\ CH_2 \\ | \\ O{=}C{-}CH_3 \\ + \\ CH_3 \\ | \\ O{=}C{-}S{-}CoA \end{array}$$

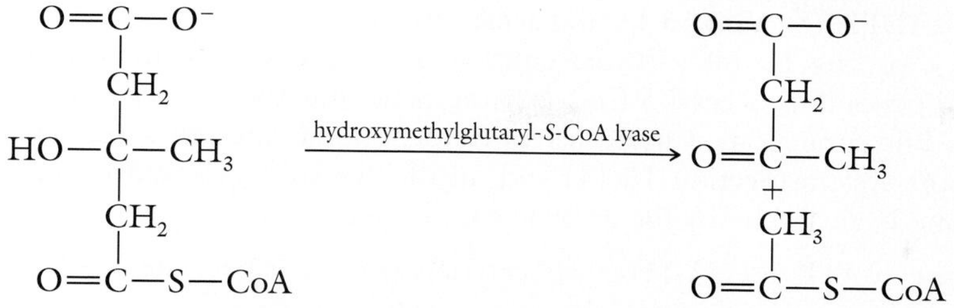

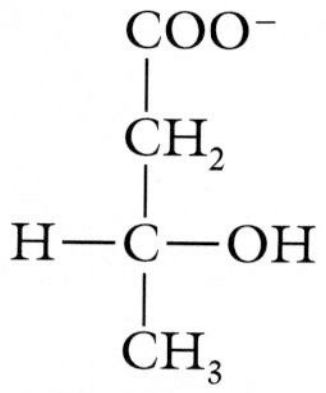

Acetoacetate is reduced by D-3-hydroxybutyrate dehydrogenase to produce D-3-hydroxybutyrate, which is transported to peripheral tissues. There, by another complex series of reactions, D-3-hydroxybutyrate is converted into acetyl-*S*-CoA, which is oxidized in the citric acid cycle, supplying cardiac and skeletal muscle with as much as 10% of their daily energy requirements. In starvation, D-3-hydroxybutyrate can supply as much as 75% of the brain's energy requirement. Acetoacetate is unstable and can spontaneously or enzymatically lose its carboxyl group to form acetone, a volatile substance that can be detected in the breath of diabetics as a sweet odor.

$$\begin{array}{c} COO^- \\ | \\ CH_2 \\ | \\ H{-}C{-}OH \\ | \\ CH_3 \end{array}$$

D-3-Hydroxybutyrate

3-Hydroxy-3-methylglutaryl-*S*-CoA is also synthesized in the cytosol by a different group of enzymes. As the precursor molecule central to the synthesis of many important biomolecules such as cholesterol and vitamins A, E, and K, HMG-*S*-CoA plays a central role in lipid metabolism. The biosynthesis of cholesterol, for example, begins with the reduction of 3-hydroxy-3-methylglutaryl-*S*-CoA to form mevalonic acid catalyzed by HMG-*S*-CoA reductase. That reduction step controls the overall rate of cholesterol biosynthesis, because the reductase is a regulatory enzyme that exists in both phosphorylated (inactive) and dephosphorylated (active) forms. In cases of uncontrolled high concentrations of cholesterol, drugs such as lovastatin (Box 16.1 on page 506) inhibit the reductase and block cholesterol's synthesis.

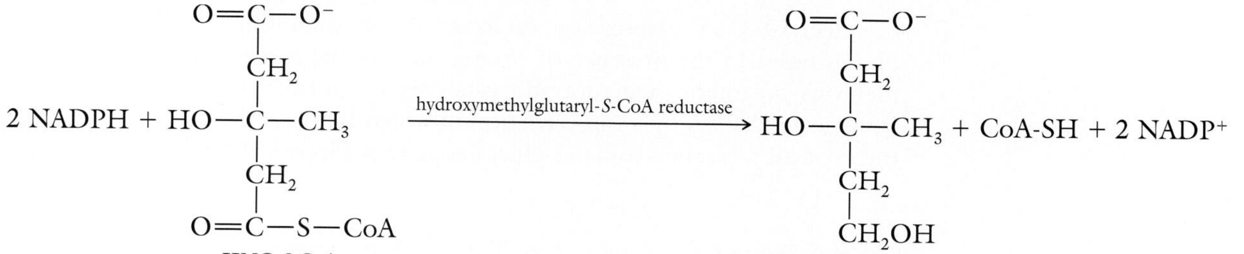

$$2\ NADPH + \begin{array}{c} O{=}C{-}O^- \\ | \\ CH_2 \\ | \\ HO{-}C{-}CH_3 \\ | \\ CH_2 \\ | \\ O{=}C{-}S{-}CoA \\ \textbf{HMG-}S\textbf{-CoA} \end{array} \xrightarrow{\text{hydroxymethylglutaryl-}S\text{-CoA reductase}} \begin{array}{c} O{=}C{-}O^- \\ | \\ CH_2 \\ | \\ HO{-}C{-}CH_3 \\ | \\ CH_2 \\ | \\ CH_2OH \\ \textbf{Mevalonate} \end{array} + CoA\text{-}SH + 2\ NADP^+$$

The mevalonate is converted by a complex series of reactions into Δ^3-isopentenylpyrophosphate. Six moles of this compound are polymerized and cyclized to form lanosterol, the precursor of cholesterol. Cholesterol (Section 11.7) is the precursor of the steroid hormones, bile salts, and vitamin D. Excessive cholesterol production leads to the formation of deposits, called plaque, on the internal surfaces of arteries and is thought to be a major factor in heart disease (see Box 16.1). Δ^3-Isopentenylpyrophosphate is the basic building block of vitamins A, E, and K and of natural rubber. In polymerized form, it forms the long hydrocarbon chain of ubiquinone (Section 15.12).

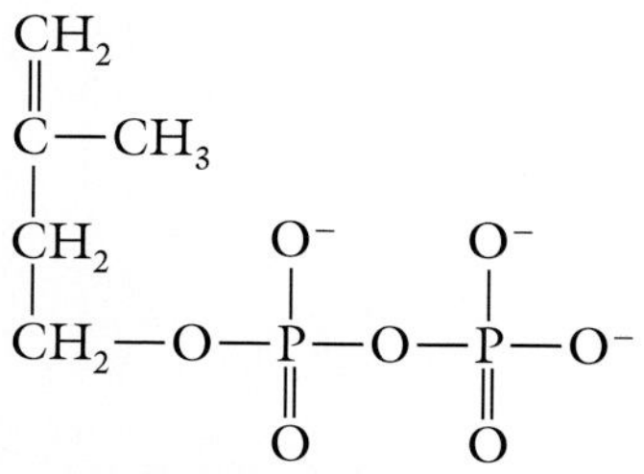

$$\begin{array}{l} CH_2 \\ \| \\ C{-}CH_3 \\ | \\ CH_2 \\ | \\ CH_2{-}O{-}P(O^-)(=O){-}O{-}P(O^-)(=O){-}O^- \end{array}$$

Δ^3-Isopentenylpyrophosphate

16.4 THE BIOSYNTHESIS OF FATTY ACIDS

The synthesis of fatty acids essentially utilizes the same chemical reactions as those of fatty acid catabolism. However, fatty acid synthesis takes place in the cytosol rather than in mitochondria and it requires a different activation mechanism and different enzymes and coenzymes. Fatty acids are synthesized when the cell is in an energy-rich state; that is, when there is an abundance of ATP, NADPH (Section 15.1), and acetyl-*S*-CoA.

The process begins with the entry of acetyl-*S*-CoA into the cytosol from the mitochondria. Acetyl-*S*-CoA cannot penetrate the mitochondrial membrane, but citrate can. Citrate leaves mitochondria through a tricarboxylate-transport system (Section 15.14) and, in the cytosol, is cleaved into acetyl-*S*-CoA and oxaloacetate by the enzyme citrate lyase:

$$\text{Citrate} + \text{ATP} + \text{CoA-SH} \longrightarrow \text{acetyl-}S\text{-CoA} + \text{oxaloacetate} + \text{ADP} + P_i$$

Citrate is also an allosteric signal for the activation of the next enzyme in the overall process, acetyl-*S*-CoA carboxylase.

The oxaloacetate produced by the cleavage of citrate is transformed into pyruvate by the following cytosolic reactions. The pyruvate can then enter the mitochondria for reconversion into oxaloacetate.

$$\underset{\textbf{Oxaloacetate}}{^{-}OOC{-}CH_2{-}C({=}O){-}COO^{-}} + \text{NADH} \xrightleftharpoons{\text{malate dehydrogenase}} \underset{\textbf{L-Malate}}{^{-}OOC{-}CH_2{-}CH(OH){-}COO^{-}} + \text{NAD}^{+}$$

$$\underset{\textbf{L-Malate}}{^{-}OOC{-}CH_2{-}CH(OH){-}COO^{-}} + \text{NADP}^{+} + H^{+} \xrightleftharpoons{\text{malic enzyme}} \underset{\textbf{Pyruvate}}{^{-}O{-}C({=}O){-}C({=}O){-}CH_3} + \text{NADPH} + CO_2$$

Acetyl-*S*-CoA carboxylase catalyzes the rate-limiting step in fatty acid biosynthesis. In the presence of citrate, the enzyme forms a highly active, filamentous polymeric structure and catalyzes the formation of malonyl-*S*-CoA from acetyl-*S*-CoA by a carboxylation in which biotin is the cofactor. A similar carboxylation reaction initiates gluconeogenesis (Section 14.8).

$$\underset{\textbf{Acetyl-}S\textbf{-CoA}}{CH_3{-}C({=}O){-}S{-}CoA} + CO_2 + \text{ATP} + H_2O \xrightarrow{\text{acetyl-}S\text{-CoA carboxylase}} \underset{\textbf{Malonyl-}S\textbf{-CoA}}{^{-}OOC{-}CH_2{-}C({=}O){-}S{-}CoA} + \text{ADP} + P_i + H^{+}$$

After its formation, malonyl-*S*-CoA then reacts with the fatty acid synthase complex.

« **The advantage of a multifunctional enzyme complex is described in Section 14.8.**

In bacteria, the fatty acid synthase complex consists of seven different enzymes. In animal cells, however, these enzymatic activities take place at specialized regions, or catalytic domains, of a single multifunctional polypeptide chain. Acyl groups are esterified to CoA-SH in the oxidation of fatty acids, but, in fatty acid biosynthesis, acyl groups are esterified to —SH groups on the different protein domains of the synthase complex. One of the —SH groups is on an enzyme called β-keto-synthase or the **condensing enzyme,** abbreviated **CE.** The other is on a protein called the **acyl carrier protein,** abbreviated **ACP.** The —SH group on the condensing enzyme is part of a cysteine residue (Section 12.4) at the active site of the enzyme. We will call that —SH group the **α-SH** group. The second —SH group will be designated the **β-SH** group.

It is located on a pantotheine residue, as in CoA-SH (see Figure 15.2), but is attached directly to the protein rather than to adenine. The two —SH groups have different functions.

The chain-lengthening reaction begins with the reaction of acetyl-*S*-CoA with the α-SH group, catalyzed by acetyl-ACP transferase. In a companion reaction, malonyl-*S*-CoA reacts with the β-SH group, catalyzed by malonyl-ACP transferase.

$$\underset{\text{Acetyl-}S\text{-CoA}}{CH_3-\overset{O}{\overset{\|}{C}}-S-CoA} + HS_\alpha\text{-(CE/ACP)-}S_\beta H + \underset{\text{Malonyl-}S\text{-CoA}}{{}^-OOC-CH_2-\overset{O}{\overset{\|}{C}}-S-CoA} \longrightarrow CH_3-\overset{O}{\overset{\|}{C}}-S_\alpha\text{-(CE/ACP)-}S_\beta-\overset{O}{\overset{\|}{C}}-CH_2-COO^- + 2\ CoA-SH$$

The next reaction is the first of four steps in the lengthening of the fatty acid chain. The acetyl group bound to the α-SH group is transferred to the malonyl group at the β-SH site to form the acetoacetyl derivative. At the same time, CO_2 is displaced and lost as bicarbonate.

$$CH_3-\overset{O}{\overset{\|}{C}}-S_\alpha\text{-(CE/ACP)-}S_\beta-\overset{O}{\overset{\|}{C}}-CH_2-COO^- + H_2O \longrightarrow HS_\alpha\text{-(CE/ACP)-}S_\beta-\overset{O}{\overset{\|}{C}}-CH_2-\overset{O}{\overset{\|}{C}}-CH_3 + HCO_3^-$$

The loss of CO_2, as HCO_3^-, has a very large equilibrium constant; so the reaction is irreversible and provides the driving force for this step in the chain-lengthening process. The acetyl group on the α-SH group supplies the methyl terminal group of the growing chain.

The next reactions affect the four-carbon derivative covalently bound at the β-SH site. In the first of these reactions, the acetoacetyl-S_β-ACP-CE-S_αH derivative is reduced to form the D-3-hydroxybutyryl derivative. The cofactor is NADPH, and the enzyme is 3-ketoacyl-ACP reductase. Note that the reductase uses NADPH rather than the NADH used in fatty acid oxidation. An additional contrasting feature is that the product is the D-derivative rather than the L-derivative seen in the fatty acid oxidation pathway.

$$\underset{\text{Acetoacetyl-}S_\beta\text{-ACP-CE-}S_\alpha H}{CH_3-\overset{O}{\overset{\|}{C}}-CH_2-\overset{O}{\overset{\|}{C}}-S_\beta-ACP-CE-S_\alpha H} + NADPH + H^+ \longrightarrow \underset{\text{D-3-Hydroxyacyl-}S_\beta\text{-ACP-CE-}S_\alpha H}{CH_3-\overset{OH}{\overset{|}{\underset{H}{\underset{|}{C}}}}-CH_2-\overset{O}{\overset{\|}{C}}-S_\beta-ACP-CE-S_\alpha H} + NADP^+$$

The next step is a dehydration catalyzed by 3-hydroxyacyl-ACP dehydratase, which produces *trans*-Δ^2-enoyl-S_β-ACP-CE-S_αH:

$$\underset{\text{D-3-Hydroxyacyl-}S_\beta\text{-ACP-CE-}S_\alpha H}{CH_3-\overset{OH}{\overset{|}{\underset{H}{\underset{|}{C}}}}-CH_2-\overset{O}{\overset{\|}{C}}-S_\beta-ACP-CE-S_\alpha H} \xrightarrow{-H_2O} \underset{trans\text{-}\Delta^2\text{-Enoyl-}S_\beta\text{-ACP-CE-}S_\alpha H}{CH_3-\overset{H}{\overset{|}{C}}=\underset{H}{\underset{|}{C}}-\overset{O}{\overset{\|}{C}}-S_\beta-ACP-CE-S_\alpha H}$$

16.1 CHEMISTRY WITHIN US

Atherosclerosis

Cardiovascular (blood-vessel) disease is the single largest killer in the developed Western world, being responsible for about half the deaths each year. Death occurs by heart attack or stroke, the end result of atherosclerosis (hardening of the arteries). Atherosclerosis is called the silent killer because it begins and progresses for many years before symptoms are present.

Atherosclerosis is a buildup of cholesterol-rich, fatty deposits called **plaque** on the interior endothelial lining of arteries. Cholesterol is cleared from the bloodstream when the lipid-transport system (Section 18.3) is working properly and the dietary intake of cholesterol is not excessive. Plaque develops in response to damaged arterial lining resulting from chronic high blood pressure **(hypertension),** stress, smoking, a high-fat and high-cholesterol diet, and genetic and other factors. White blood cells migrate into damaged endothelial tissue in an attempt to repair the lesion. Growth factors are released that cause the proliferation of tissue at the site of damage, and this proliferation is followed by the deposition of lipids—mostly cholesteryl esters. Fibrous connective tissue including calcium deposits cap off the thickened area to yield narrowed and roughened arterial walls that are much less elastic.

Plaque buildup promotes the formation of a blood clot **(thrombus).** Blood flow becomes turbulent when blood flows over plaque. The rough plaque damages blood platelets and acts as the nucleus for clot formation, which further narrows the artery. The coronary arteries—the arteries that feed the heart—are especially prone to narrowing and clot formation. A **heart attack (myocardial infarction)** with death of heart tissue results when a clot in a coronary artery **(coronary thrombosis)** stops blood flow. In a **stroke (cerebral infarction),** arteries in the brain are blocked (and often ruptured). Death of brain tissue results in muscle, sensory, or language impairment, depending on the part of the brain damaged. Strokes as well as damage to other tissues and organs can also occur when a piece of a coronary thrombus breaks loose, travels through the bloodstream, and lodges in a narrower artery. Plaque buildup also damages the cardiovascular system by further elevating the blood pressure. Hypertension weakens heart muscle (which thickens because of the strain) and decreases the efficiency of the heart's pumping action. Blood backs up into the heart and lungs, a condition called **congestive heart failure** that is often fatal.

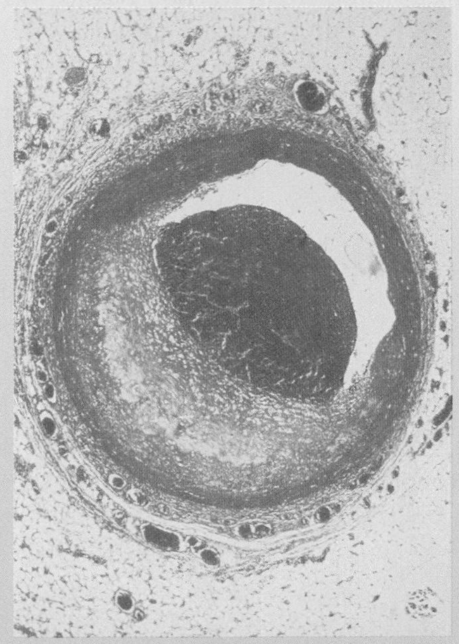

Light micrograph showing atherosclerosis in coronary artery. Most of the artery is blocked by plaque deposits. (Biophoto Associates/Photo Researchers.)

Many epidemiological studies indicate a strong correlation between elevated cholesterol and lipoprotein levels in the blood and the incidence of cardiovascular disease. Various medical organizations and journals such as the American Heart Association and the *New England Journal of Medicine* have published guidelines for blood cholesterol and lipoproteins (Table 18.11). Experts believe that adults should maintain cholesterol levels below 200 mg/dL to minimize the risk of cardiovascular disease. **Low-density lipoprotein (LDL),** the main carrier of cholesterol in the blood, is sometimes called **"bad cholesterol."** LDL levels below 130 mg/dL are recommended. However, the relation between blood cholesterol and cardiovascular disease is very complex because **high-density lipoprotein (HDL),** sometimes called **"good cholesterol,"** protects against cardiovascular disease by returning cholesterol to the liver.

Heredity, diet, and life style have roles in the development of atherosclerosis. The importance of heredity is clearly evident, given that there are people who consume a diet high in cholesterol and yet have blood-cholesterol

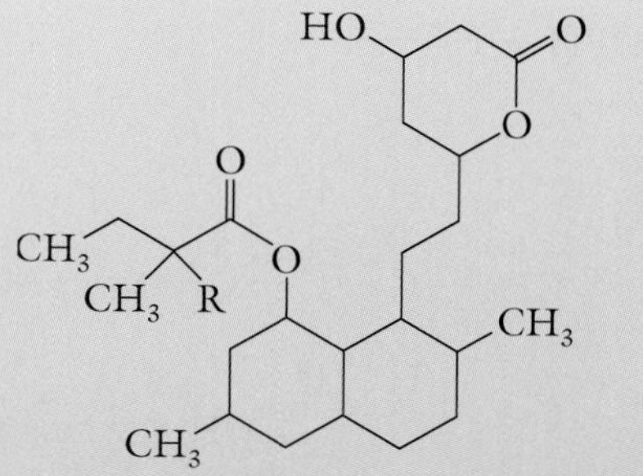

Lovastatin (R = H)
Simvastatin (R = CH_3)

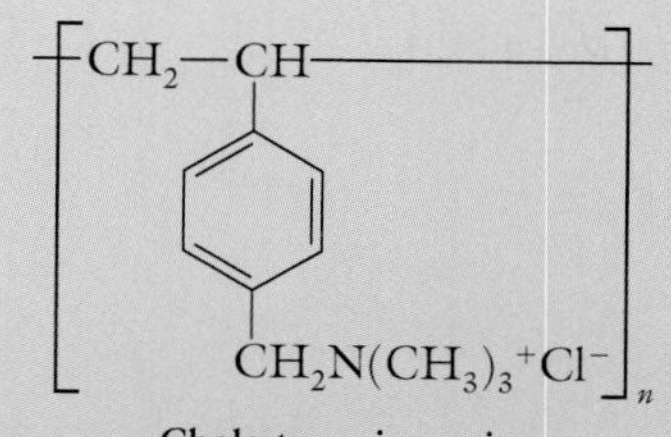

Cholestyramine resin

levels below 200 mg/dL. Other people have high blood-cholesterol levels in spite of diets low in cholesterol. **Familial hypercholesterolemia** is a hereditary condition in which the blood-cholesterol level is grossly elevated (as high as 500 mg/dL and higher). Michael Brown and Joseph Goldstein were awarded the 1985 Nobel Prize in physiology or medicine for their study of this disorder. People with familial hypercholesterolemia have defective LDL receptors or an insufficient number of LDL receptors. Cholesterol is not taken up by cells, the feedback mechanism for shutting down the endogenous synthesis of cholesterol does not function, and the cholesterol level in the blood becomes elevated. When dietary control is ineffective, the drugs lovastatin and simvastatin are useful for lowering cholesterol by inhibiting endogenous cholesterol synthesis through the inhibition of the enzyme HMG-*S*-CoA reductase (Section 16.3). Also useful is cholestyramine resin (a synthetic polymer; Section 4.7), which binds bile salts in the intestine and thus promotes the conversion of more cholesterol into bile salts in the liver.

The extent to which blood cholesterol can be lowered by diet has limits, because cholesterol is essential for life and the body synthesizes cholesterol. However, diet clearly has an effect on the blood-cholesterol level. Prudent dietary recommendations are based on epidemiological correlations between diet and cardiovascular disease:

- Consume fewer animal products (meat and dairy products) relative to plant products (fruits, vegetables, and grains).
- Consume more chicken and fish relative to red meat.

These recommendations have two objectives. First, the total intake of lipids is lowered because meat, especially red meat, has a higher lipid content than do fruits, vegetables, and grains. Second, the amount of saturated fatty acids and cholesterol relative to unsaturated fatty acids is lowered because fruits and vegetables have no cholesterol and are lower in saturated fatty acids compared with meats. The amount of saturated fatty acids is also lowered because fish and chicken are lower in saturated fats relative to red meat.

Decreased lipid intake not only decreases the blood-cholesterol level, but also tends to reduce weight because lipids generate more than twice the energy per gram than does protein or carbohydrate. Lower lipid intake tends to translate into less excess (unmetabolized) lipid that is stored as fat (adipose) tissue. Decreased weight lowers the load on the heart and the rest of the cardiovascular system, decreasing the tendency for hypertension, lesions, and atherosclerosis. The specific effects of fatty acids are not simple. Saturated fatty acids increase cholesterol and LDL levels, whereas unsaturated fatty acids lower cholesterol and LDL levels. Monounsaturated fatty acids (olive oil) are more effective than polyunsaturated fatty acids in lowering LDL levels. Dietary fiber (Box 10.4) also lowers cholesterol levels.

Many other, poorly understood factors are important. Smoking and a stressful life style correlate with a higher incidence of atherosclerosis, perhaps by increasing arterial lesions. Aerobic exercise raises the level of HDL, the good cholesterol. Small amounts of alcohol, especially red wine, appear to decrease the tendency of cholesterol to adhere to lesions.

The dehydration reaction is followed by a reduction at the double bond to form the saturated acyl derivative. The enzyme is enoyl-*S*-ACP reductase, and the cofactor is NADPH.

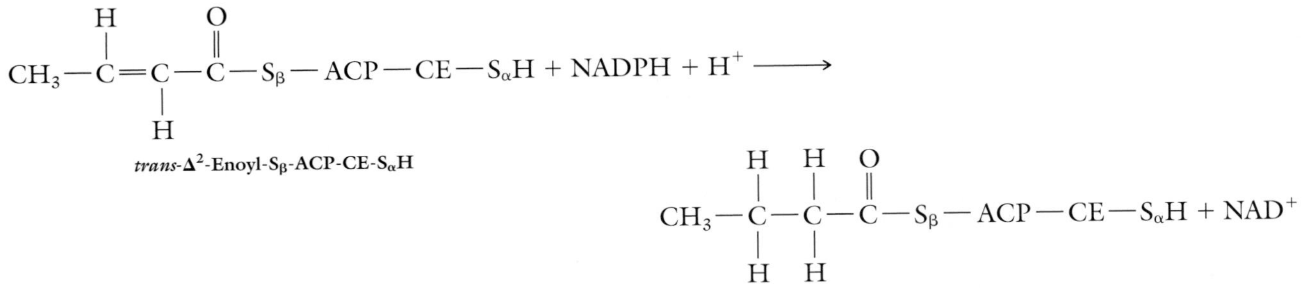

$$\underset{\textit{trans}\text{-}\Delta^2\text{-Enoyl-S}_\beta\text{-ACP-CE-S}_\alpha\text{H}}{CH_3{-}\overset{\overset{H}{|}}{C}{=}\underset{\underset{H}{|}}{C}{-}\overset{\overset{O}{\|}}{C}{-}S_\beta{-}ACP{-}CE{-}S_\alpha H} + NADPH + H^+ \longrightarrow$$

$$\underset{\text{Acyl-S}_\beta\text{-ACP-CE-S}_\alpha\text{H}}{CH_3{-}\overset{\overset{H}{|}}{\underset{\underset{H}{|}}{C}}{-}\overset{\overset{H}{|}}{\underset{\underset{H}{|}}{C}}{-}\overset{\overset{O}{\|}}{C}{-}S_\beta{-}ACP{-}CE{-}S_\alpha H} + NAD^+$$

With the completion of this first round of reactions, the newly formed four-carbon saturated chain is transferred from the β-SH site to the α-SH site formerly occupied by acetyl-*S*-CoA.

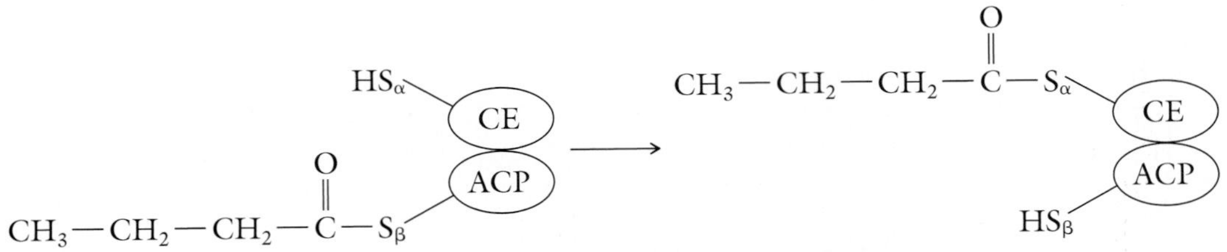

$$HS_\alpha{-}CE/ACP{-}S_\beta{-}\overset{\overset{O}{\|}}{C}{-}CH_2{-}CH_2{-}CH_3 \longrightarrow CH_3{-}CH_2{-}CH_2{-}\overset{\overset{O}{\|}}{C}{-}S_\alpha{-}CE/ACP{-}HS_\beta$$

The next round of reactions begins with the addition of another malonyl-*S*-CoA to the now empty β-SH site. This reaction is followed by the transfer of the four-carbon chain at the α-SH site to the malonyl group, with the loss of CO_2 as before. We now have a six-carbon chain, and it will undergo the four reactions just detailed to yield a saturated chain lengthened by two carbons.

After seven rounds of addition of malonyl-*S*-CoA to the first molecule of acetyl-*S*-CoA, a molecule of palmitoyl-S_β-ACP-S_αH is produced. At this point, the chain-lengthening process ceases, and palmitate is freed from the enzyme by the action of the hydrolytic enzyme palmitoyl thioesterase.

Because ionic bicarbonate appears as a product, the stoichiometry of palmitate synthesis, including charge balance, is

$$\text{Acetyl-}S\text{-CoA} + 7\ \text{malonyl-}S\text{-CoA}^- + 14\ \text{NADPH} + 14\ \text{H}^+ \longrightarrow \text{palmitoyl-}S\text{-CoA} + 7\ \text{HCO}_3^- + 14\ \text{NADP}^+ + 7\ \text{CoA-SH}$$

When the ATPs required for the synthesis of seven molecules of malonyl-*S*-CoA derived from acetyl-*S*-CoA are taken into account,

$$7\ \text{Acetyl-}S\text{-CoA} + 7\ \text{HCO}_3^- + 7\ \text{ATP}^{4-} \longrightarrow 7\ \text{malonyl-}S\text{-CoA}^- + 7\ \text{ADP}^{3-} + 7\ \text{P}_i^{2-} + 7\ \text{H}^+$$

the overall stoichiometry for the synthesis of palmitoyl-*S*-CoA from acetyl-*S*-CoA is

$$8\ \text{Acetyl-}S\text{-CoA} + 7\ \text{ATP}^{4-} + 14\ \text{NADPH} + 7\ \text{H}^+ \longrightarrow \text{palmitoyl-}S\text{-CoA} + 14\ \text{NADP}^+ + 7\ \text{CoA-SH} + 7\ \text{ADP}^{3-} + 7\ \text{P}_i^{2-}$$

Hormonal Regulation

>> The roles of insulin and glucagon in the regulation of metabolic processes are summarized in Section 18.6.

The two hormones that regulate fatty acid biosynthesis are glucagon and insulin. Glucagon initiates a sequence of enzymatic reactions that phosphorylate and inactivate acetyl-*S*-CoA carboxylase. The mechanism is similar to the one that activates phosphorylase in glycogenolysis (Section 15.10). In its phosphorylated form, acetyl-*S*-CoA carboxylase is active only in the presence of high concentrations of citrate; so fatty acid biosynthesis virtually ceases. At the same time, triacylglycerol lipase is activated by phosphorylation and frees fatty acids for oxidative processes. Insulin, on the other hand, activates the phosphodiesterase that hydrolyzes cyclic AMP to noncyclic AMP, ending the cascade initiated by glucagon and allowing fatty acids to be synthesized and stored.

Further Processing of Palmitate

Chains longer than 16 carbon atoms—for example, the 18-carbon-atom stearate—are made from palmitate by special elongation reactions in mitochondria or at the surface of the endoplasmic reticulum. Shorter chains are made when the growing fatty acid chain is released from the synthase complex before reaching its maximum 16-carbon length.

Eukaryotes can introduce a single cis double bond into stearate. It is a complex dehydrogenation requiring molecular oxygen and NADH, and it takes place at the surface of the endoplasmic reticulum with the formation of oleyl-*S*-CoA from stearoyl-*S*-CoA. Oleate is the only unsaturated fatty acid with the cis double bond as close to the methyl end (the 9,10 position of stearoyl-*S*-CoA) that animals can synthesize; however, animals require polyunsaturated fatty acids with double bonds even closer to the methyl end. These fatty acids are linoleic and linolenic acids, and they must be obtained through the diet. Such fatty acids are therefore called essential (Section 11.2). Mammals do, however, have the ability to introduce additional double bonds into the carbon chains of linoleic and linolenic acids between carbon 9 and the carboxyl carbon. In this way, mammals can synthesize arachidonic acid, a precursor of eicosanoids, a group of compounds derived from fatty acids that induce a wide variety of physiological responses (Section 11.8).

16.5 THE BIOSYNTHESIS OF TRIACYLGLYCEROLS

The biosynthesis of cellular structures containing lipids begins with the fatty acids synthesized in the cytosol. The structures and properties of many of these lipids were considered in Sections 11.3, 11.4, and 11.6. In this section, we consider the biosynthesis of triacylglycerols.

Triacylglycerols, as already noted, are the principal storage forms of fatty acids. Glycerol and fatty acids are the building blocks for synthesizing triacylglycerols. When biosynthesis and oxidation take place at about equal rates, the amount of fat in the body remains quite constant. But, if the intake of carbohydrate, fat, or protein is in excess of required amounts, the excess calories are stored as triacylglycerols—in other words, as additional deposits of fat.

The rate of the formation of triacylglycerols is strongly affected by the hormone insulin, which promotes the formation of triacylglycerols from carbohydrates (Section 16.4). Patients who suffer from diabetes mellitus cannot utilize glucose properly and are therefore unable to synthesize fatty acids and triacylglycerols from carbohydrates.

The precursors of triacylglycerols are fatty acyl-*S*-CoA and glycerol-3-phosphate (Section 15.14). The biosynthesis of triacylglycerols begins with the acylation of the free hydroxy groups of glycerol-3-PO_4 with 2 mol of fatty acyl-*S*-CoA:

$$2 \text{ Fatty acyl-}S\text{-CoA} + \text{glycerol-3-PO}_4 \longrightarrow \text{diacylglycerol-3-PO}_4 + 2 \text{ CoA-SH}$$

The product diacylglycerol-3-PO_4, also known as **phosphatidic acid** (Section 11.6), exists as **phosphatidate** at cellular pH and is a key intermediate in membrane-lipid biosynthesis.

In the next stage, phosphatidate is hydrolyzed to the diacylglycerol, which can then be acylated by a third fatty acyl-*S*-CoA to form the final product, triacylglycerol. Seven moles of ATP are required for this synthesis: 1 mol for the phosphorylation of glycerol, and 2 mol for the activation of each of three fatty acids to the fatty acyl-*S*-CoA form. The number of moles of ATP required for the total synthesis of tripalmitoylglycerol is

$$\begin{aligned} \text{15 ATP/mole of palmitate} \times \text{3 mol of palmitate} &= \text{45 ATP} \\ \text{ATP/mole for tripalmitoylglycerol synthesis} &= \underline{\text{7 ATP}} \\ \text{Total moles of ATP per mole of tripalmitoylglycerol} &= \text{52 ATP} \end{aligned}$$

Although the synthesis of triacylglycerol entails a significant cost in ATP, the energy yield of triacylglycerol far outweighs the investment in its formation. To place the relative costs in perspective, let us calculate the ATP yield per mole of tripalmitoylglycerol oxidized. We can begin by noting that, as a result of oxidation, there are 8 mol of acetyl-*S*-CoA, 7 mol of NADH, and 7 mol of $FADH_2$ generated per mole of palmitate. The following table shows the ATP yield per mole of palmitate.

		ATP yield
fatty acid activation	−2 ATP	−2
from electron transport	7 NADH	21
	7 $FADH_2$	14
from oxidation of 8 acetyl CoA	24 NADH	72
	8 $FADH_2$	16
from succinyl-CoA synthetase	8 GTP	8
Total ATP per mole of palmitate		129

Furthermore, the glycerol from triacylglycerol enters the glycolytic pathway after having been primed by ATP and therefore yields an additional 20 mol of ATP through the glycerol phosphate shuttle (or 22 mol ATP through the malate–aspartate shuttle). The total yield of ATP per mole of tripalmitoylglycerol oxidized is

$$\begin{aligned} \text{Fatty acid} \quad 3 \times 129 &= 387 \\ \text{Glycerol} \quad &= \underline{\ 20} \\ \text{Sum of ATP per mole} &= 407 \end{aligned}$$

Now subtract the 2 mol of ATP required for the activation of each fatty acid to form the CoA derivative:

$$\begin{aligned} 3 \times 2 &= \underline{\ \ 6} \\ \text{Total ATP yield per mole of tripalmitoylglycerol oxidized} &= 401 \end{aligned}$$

This yield is almost eight times the ATP cost of synthesizing the tripalmitoylglycerol.

It is interesting to compare our biochemical estimates of the energy available in lipids and carbohydrates with measured nutritional values (9.13 kcal/g for palmitic acid and 3.81 kcal/g for glucose). We can do so by comparing two ratios: (1) ATP yield per gram of tripalmitoylglycerol to ATP yield per gram of glucose and (2) the ratio of the measured nutritional energy of lipid to the measured nutritional energy of carbohydrate.

First, we calculate the ATP yield per gram of each compound.

$$\text{Formula mass of tripalmitoylglycerol: 807.3 g/mol}$$
$$\text{407 mol ATP/807.3 g} = \text{0.504 mol ATP/g}$$

$$\text{Formula mass of glucose: 180.2 g/mol}$$
$$\text{38 mol ATP/180.2 g} = \text{0.211 mol ATP/g}$$

Then we calculate the ratio of those two values:

$$\left(\frac{\dfrac{\text{mol ATP}}{\text{g TPG}}}{\dfrac{\text{mol ATP}}{\text{g glucose}}} \right) = \frac{0.504}{0.211} = 2.39$$

Finally, we calculate the ratio of nutritional energy values:

$$\frac{9.13 \text{ kcal/g palmitate}}{3.81 \text{ kcal/g glucose}} = 2.40$$

The closeness (less than 1% difference) between measured nutritional energy and that calculated from ATP yield suggests that metabolic energy can indeed be estimated by using the ATP yields from the stoichiometry of metabolic reactions.

16.6 THE BIOSYNTHESIS OF MEMBRANE LIPIDS

The structures of membrane lipids were presented in Section 11.6. They will not be described here, except to repeat that it is useful to think of these lipids in the general sense illustrated in Figure 11.6—that is, as possessing a strongly polar or ionic head group, represented by a sphere, and large hydrophobic tail groups, represented by long rods. The long rods represent fatty acid chains with varying degrees of unsaturation. The polar head groups vary from the small ethanolamine head group in phosphatidylethanolamine to the complex carbohydrates in the sphingoglycolipids.

The biosynthesis of membrane lipids begins with diacylglycerol. The head groups are added in an activated form much as glucose, in the form of a uridine diphosphoglucose (UDP-glucose) derivative, is added to glycogen. In the formation of the membrane lipid phosphatidyl ethanolamine, the activated head group is in the form of **cytidine diphosphoethanolamine (CDP-ethanolamine).**

The formation of the activated head group begins with the phosphorylation of ethanolamine by ATP catalyzed by the enzyme ethanolamine kinase:

$$\text{Ethanolamine} + \text{ATP} \longrightarrow \text{phosphoethanolamine} + \text{ADP}$$

Phosphoethanolamine then reacts with cytidine triphosphate (CTP). The enzyme in this step is phosphoethanolamine cytidylyl transferase.

$$\text{Phosphoethanolamine} + \text{CTP} \longrightarrow \text{CDP-ethanolamine} + \text{PP}_i$$

This activation reaction is driven by the favorable hydrolysis of pyrophosphate. Finally, the activated head group is added to diacylglycerol by the enzyme ethanolaminephosphotransferase:

$$\text{Diacylglycerol} + \text{CDP-ethanolamine} \longrightarrow \text{phosphatidylethanolamine} + \text{CMP}$$

In similar reaction pathways, CDP derivatives react with diacylglycerol to produce other membrane phospholipids such as phosphatidylcholine, phosphatidylserine, phosphatidylinositol, and cardiolipin.

As noted earlier, the nucleotide adenine forms the core of the carrier of activated phosphate groups and the nucleotide uridine forms the core of the carrier for glucosyl groups. We can now add cytidine to our list of nucleotides that act as carriers of a specific chemical group in cellular metabolism.

Membrane lipids undergo very rapid turnover in the cell, but a critical balance between synthesis and degradation keeps their concentrations steady. Imbalances between the two processes are the causes of many early-childhood diseases.

Membrane lipids as well as other biomolecules are degraded within **lysosomes,** membrane-bounded structures in the cytosol that contain a large array of hydrolytic enzymes of great specificity. Normally, membrane lipids find their way into a lysosome and are degraded into soluble components that then leave the lysosome to be salvaged and incorporated into newly synthesized biomolecules. However, certain recessive human mutations result in the absence of specific lysosomal hydrolytic enzymes. In consequence,

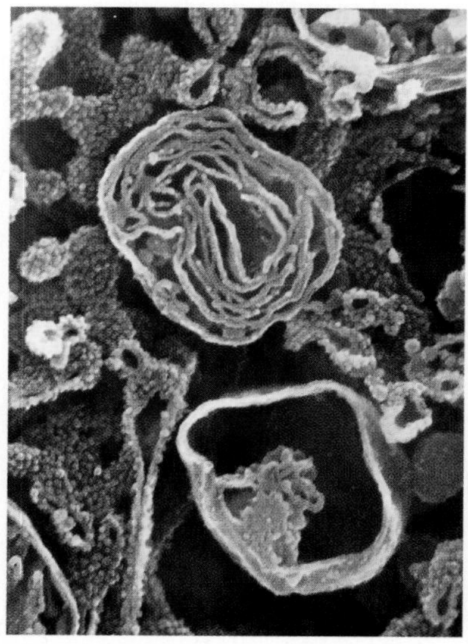

Lysosomes. Colored high-resolution scanning electron micrograph of two lysosomes in a pancreatic acinar cell, which is a secretory cell surrounding a cavity. Lysosomes (rounded, center) are bounded by a single membrane. Old or nonfunctioning organelles become enclosed in the lysosome and are digested by a concentrated mixture of digestive enzymes. Partly digested cell organelles can be seen in each lysosome. (P. Motta and T. Naguro/ Science Photo Library/Photo Researchers.)

membrane lipids that enter lysosomes do not leave, and so the lysosomes swell, causing functional disruption and, ultimately, cell death. This cell death causes mental retardation in many cases and death at an early age. There are no treatments for these lysosomal diseases at present, but genetic engineering offers hope for the future. In the interim, amniocentesis and genetic counseling have reduced the incidence of these diseases. Table 16.1 lists some of these lysosomal diseases and the enzyme that is defective or absent in each disease.

TABLE 16.1 Lysosomal Diseases Due to Faulty Degradation of Membrane Lipids

Disease	Defective enzyme
Gaucher's disease	glucocerebrosidase
Tay-Sachs disease	*N*-acetylhexosaminidase
Fabry's disease	trihexosylceramide galactosyl hydrolase
Hurler's syndrome	α-L-iduronidase
Krabbe's disease	galactosylceramide β-galactosylhydrolase
Niemann-Pick disease	sphingomyelinase

Summary

Fatty Acid Mobilization

• The glycogen storage capacity in humans can support about 12 hours of metabolic energy requirements. The deficit is made up by stored triacylglycerols.

• Fatty acids derived from stored triacylglycerols are activated in the cytosol to acyl-*S*-CoA thioesters and are then transported into the mitochondria in the form of carnitine complexes.

Fatty Acid Oxidation

• Fatty acid oxidation begins with a dehydrogenation between carbons 2 and 3 of the acyl CoA derivative.

• Next, the trans double bond produced in the first step undergoes enzymatic hydration.

• The derivative produced in the second step then undergoes dehydrogenation.

• There are two final products: (1) a molecule of acetyl-*S*-CoA from carbons 1 and 2 of the original fatty acid and (2) another fatty acyl-*S*-CoA molecule having two fewer carbon atoms than does the original acyl-*S*-CoA derivative.

• The acetyl-*S*-CoA formed in this process enters the citric acid cycle and is oxidized to CO_2 and water.

Ketone Bodies and Cholesterol

• Ketone bodies are formed in the mitochondria of the human liver. They consist of acetoacetate, D-3-hydroxybutyrate, and acetone.

• A key intermediate in their biosynthesis is also used to synthesize many important biomolecules such as cholesterol and vitamins A, E, and K.

The Biosynthesis of Fatty Acids

• The biosynthesis of fatty acids begins when citrate leaves the mitochondria and enters the cytosol where the fatty acid synthase complex lies.

• Citrate is both the allosteric signal for the initiation of fatty acid biosynthesis and the origin of acetyl-*S*-CoA, which is the substrate for the first reaction at the fatty acid synthase complex—the formation of malonyl-*S*-CoA.

• The central feature of the fatty acid synthase complex is the presence of two unique —SH groups to which growing fatty acid chains are esterified.

• In the first chain-lengthening step, acetyl-*S*-CoA is transferred to the —SH group on the condensing enzyme, and malonyl-*S*-CoA is transferred to the —SH group on the acetyl carrier protein.

• The acetyl group is then condensed with the malonyl group to form the acetoacetyl derivative. At the same time, CO_2 is displaced and lost.

• The acetyl group becomes the methyl terminal group of the growing chain.

• In the next three reactions, the β-keto group is reduced, dehydrated, and finally reduced to form an aliphatic chain increased by two carbon atoms.

• The process is repeated until a 16-carbon chain is formed and released by hydrolysis as palmitoyl-*S*-CoA.

The Biosynthesis of Triacylglycerols

• The biosynthesis of triacylglycerols begins with the formation of diacylglycerol-3-PO_4, also known as phosphatidic acid.

• Phosphatidate is hydrolyzed to diacyl glycerol, which is acylated by a third fatty acyl-*S*-CoA to form the final product, triacylglycerol.

The Biosynthesis of Membrane Lipids

• A membrane lipid is composed of a large hydrophobic end group and an ionic or strongly polar head group.

• The biosynthesis of membrane lipids begins with diacylglycerol.

• The head groups are added in an activated form—for example, cytidine diphosphoethanolamine in the formation of phosphatidyl ethanolamine.

• A number of recessive human mutations result in the loss of hydrolytic enzymes, creating an imbalance between the synthesis and the degradation of membrane lipids and causing childhood diseases such as Tay-Sachs disease.

Key Words

α-SH, p. 504
acetoacetate, p. 503
acyl carrier protein (ACP), p. 504
β-oxidation, p. 500
β-SH, p. 505
carnitine, p. 499
CDP-ethanolamine (cytidine diphosphoethanolamine), p. 511
condensing enzyme (CE), p. 504
ketone body, p. 502
ketosis, p. 502
lysosome, p. 511
phosphatidic acid, p. 509

Exercises

16.1 Can glycogen supply all of the body's daily energy requirements? Explain your answer.

16.2 What is the principal source of stored energy in humans?

16.3 Under what conditions is stored fat mobilized for cellular oxidation?

16.4 What is the first step in the mobilization of the fat stored in animals?

16.5 Show the fatty acid activation reaction that takes place in the cytosol.

16.6 How many moles of ATP are required for the activation of one mole of fatty acid? Explain.

Fatty Acid Transport into Mitochondria

16.7 Can activated fatty acids directly enter mitochondria? Explain your answer.

16.8 What is carnitine and what is its role in fatty acid oxidation?

Oxidation in Mitochondria

16.9 What is the first product of palmitate oxidation in mitochondria?

16.10 What is the coenzyme in the first oxidation step of palmitate, and what is the fate of its reducing power?

16.11 What is the second product of palmitate oxidation in mitochondria?

16.12 What is unique about the structure of the product in Exercise 16.11?

16.13 What is the third product of palmitate oxidation in mitochondria?

16.14 What is the coenzyme in the third step of the mitochondrial oxidation of palmitate, and what is the fate of its reducing power?

16.15 What are the reactants in the fourth step of the oxidation of palmitate?

16.16 Describe the products of the fourth step of the oxidation of palmitate.

Ketone Bodies

16.17 Name the ketone bodies synthesized in the liver.

16.18 What is the word used to indicate that ketone bodies are in abnormal excess?

16.19 What are two physiological conditions leading to an abnormal excess of ketone bodies?

16.20 Do ketone bodies serve any normal metabolic role?

Fatty Acid Biosynthesis

16.21 What physical method is employed by animal cells to separate the fatty acid biosynthetic apparatus from the oxidative process?

16.22 What biochemical method is employed by animal cells to separate the fatty acid biosynthetic apparatus from the oxidative process?

16.23 What hormone is responsible for initiating fatty acid biosynthesis?

16.24 What hormone is responsible for inhibiting fatty acid biosynthesis?

16.25 What is the principal source of acetate in fatty acid biosynthesis?

16.26 What is the rate-limiting step in fatty acid biosynthesis?

16.27 Describe the steps that alter the fatty acid synthase molecule before chain lengthening.

16.28 Describe the process that leads to fatty acid chain lengthening.

16.29 What are the roles of CO_2 and biotin in the process that leads to fatty acid chain lengthening?

16.30 Describe a biochemical synthesis other than fatty acid chain lengthening in which CO_2 and biotin are required.

16.31 Describe conditions at the α-SH and β-SH sites of the fatty acid synthase at the start of chain lengthening.

16.32 Describe conditions at the α-SH and β-SH sites of the fatty acid synthase when chain lengthening is complete.

The Biosynthesis of Triacylglycerols

16.33 What is the first step in the biosynthesis of triacylglycerols?

16.34 What is the role of phosphatidic acid in lipid metabolism?

16.35 Is the return—that is, the yield of stored energy—worth the investment of energy required to synthesize triacylglycerols? Explain.

16.36 Is it legitimate to express metabolic energy as quantities of ATP? Explain your answer.

The Biosynthesis of Membrane Lipids

16.37 In what form is ethanolamine added to diacylglycerol?

16.38 Name the nucleotides that act as specific chemical-group carriers in cell metabolism.

16.39 What is the chemical basis of many lysosomal diseases?

16.40 What is the genetic basis of the lysosomal diseases?

Unclassified Exercises

16.41 What are some of the features differentiating the biosynthetic from the catabolic reactions of the fatty acids?

16.42 What is a central structural feature of the fatty acid synthase complex in animal cells?

16.43 What is the overall stoichiometry for the synthesis of palmitate from acetyl-*S*-CoA?

16.44 Criticize this statement: Ketone bodies are synthesized from cytosolic 3-hydroxy-3-methylglutaryl-*S*-CoA.

16.45 Can the oxidation of fatty acids generate ATP if the resultant acetyl-*S*-CoA is not oxidized? Explain.

Expand Your Knowledge

Note: The icons  denote exercises based on material in boxes.

16.46 What physiological condition arises when oxaloacetate is absent from animal cells?

16.47 True or false: The presence of high concentrations of citrate in cells triggers the biosynthesis of fatty acids. Explain your answer.

16.48 An unconscious older man was brought to the emergency room. The attending physician smelled the man's breath and immediately ordered intravenous infusion of glucose. What led him to order this treatment?

16.49 How do fatty acids arise in adipose-tissue cells and move from those cells to where they are needed for energy?

16.50 True or false: The activation of fatty acids for oxidation requires 1 mol of ATP. Explain your answer.

16.51 List two ways by which citrate stimulates fatty acid synthesis.

16.52 Account for the fact that citrate yields CO_2 and NADPH in the cytosol.

16.53 Sam adheres strictly to a fat-free diet yet finds the increase in his body fat out of control. Is there a biochemical basis for his problem?

16.54 Suppose a person has a genetic condition in which CTP (cytidine triphosphate) is present in an abnormally low concentration. How will this concentration affect lipid metabolism?

16.55 What is the physiological consequence of Tay-Sachs disease?

16.56 What is the medical treatment for Gaucher's disease?

16.57 When does diet have a minimal effect on lowering blood cholesterol? (See Box 16.1.)

CHAPTER 17 AMINO ACID METABOLISM

(Bruce Ayres/Tony Stone.)

Chemistry in Your Future

You have enjoyed the experience of interviewing Americans from all walks of life for the USDA's nationwide food-consumption survey, and it has made you more aware of your own eating habits. For example, because you worry about gaining weight, you have always tried not to eat foods that are high in fat. As a consequence, you almost never eat meats and dairy products. Now you realize that your diet includes few other sources of protein and that you are eating far less than the recommended daily amount. This chapter on amino acid metabolism will tell you why it is important to maintain a consistent level of protein intake.

For more information on this topic and others in this chapter, go to www.whfreeman.com/bleiodian2e

Learning Objectives

- Describe transamination and oxidative deamination.
- Describe how amino groups are collected in the liver in the form of glutamate.
- Describe how ammonia is transported through the blood to the liver.
- Describe the ways in which ammonia is transformed into urea, and indicate where this transformation takes place in the cell.
- Describe the end products of the catabolism of an amino acid's carbon skeleton.
- Identify and describe some of the heritable genetic defects of amino acid metabolism.
- Learn why some amino acids are called "essential" and others "nonessential."
- Describe the biosynthesis of nonessential amino acids.
- Describe how feedback inhibition affects amino acid biosynthesis.

Amino acids, largely derived from dietary proteins, are chiefly used to synthesize proteins. A smaller proportion is used to synthesize specialized biomolecules such as the hormones epinephrine and norepinephrine, neurotransmitters, and the precursors of purines and pyrimidines. Whenever we use the term amino acid in considering polypeptides and proteins, we always mean α-amino acids, whether or not the α prefix is shown. In addition, whether or not indicated and regardless of how drawn, all amino acids are L-amino acids unless otherwise noted (Section 12.1). The structures of the amino acids found in proteins are given in Table 12.1. Unlike carbohydrates and lipids, they cannot be stored in the body for later use; so any amino acids not required for immediate biosynthetic needs are either degraded, supplying energy in the process, or converted into acetyl-*S*-CoA and then into fatty acids (and thus triacylglycerol stores).

When carbohydrates are not available—or cannot be used properly, as in diabetes mellitus—amino acids become a primary source of energy. Thus, when starvation, fasting, or the result of some other condition depletes all carbohydrate supplies, the body's own proteins, especially muscle proteins, are broken down and used for energy.

›› This effect and other effects of diabetes are explained in Section 18.6.

In this chapter, we shall explore the catabolic pathways by which amino acids are degraded and then consider some aspects of their biosynthesis. The discussion of catabolism will focus, first, on the production and fate of ammonia and, then, on the breakdown of the α-keto acids formed when the α-amino group is removed from the α-amino acids.

17.1 AN OVERVIEW OF AMINO ACID METABOLISM

Amino acids are produced from the breakdown of dietary protein and from the breakdown, through normal metabolic turnover, of the body's proteins. Enzymes and muscle proteins have a particularly rapid turnover.

The catabolism of amino acids takes place in three stages: (1) removal of the amino group, leaving the carbon skeleton of the amino acid; (2) breakdown of the carbon skeleton to an intermediate of the glycolytic pathway or citric acid cycle, or to acetyl-*S*-CoA; and (3) oxidation of these intermediates to carbon dioxide and water with the production of ATP. This third stage takes place in the pathways considered in Chapter 15.

Unique to amino acid catabolism is the extraction of the amino group ($—NH_2$) from the rest of the amino acid molecule, a process that must be accomplished without producing toxic levels of ammonia (NH_3) in blood and tissues. At cellular pH, NH_3 exists as the NH_4^+ ion; so, when ammonia appears in cells, it is immediately converted into ammonium ion, which is the form in which it participates in almost all biochemical reactions. The normal concentration of ammonium ion in the blood is in the range of 3.0×10^{-5} to 6.0×10^{-5} *M*. Above these concentrations (hyperammonemia), coma may result. The reasons for the toxicity of ammonium ion are not entirely clear. One possibility is the reaction of the ammonium ion with α-ketoglutarate to form glutamate (Section 17.3). α-Ketoglutarate is usually present at very low concentrations in cells, particularly brain cells, and any further reduction in concentration by conversion into glutamate could result in severe metabolic stress.

The pathway for the amino groups extracted from amino acids consists of

- the conversion of amino groups from all amino acids into a single product, glutamate, by transamination;
- the conversion of glutamate into α-ketoglutarate by oxidative deamination, releasing NH_4^+; and

- the conversion of NH_4^+ into the nontoxic compound urea, which the blood carries to the kidneys for excretion.

These processes take place in the liver. Amino groups and ammonium ion that must be conveyed from other tissues to the liver are transported in the form of glutamine or alanine, as we shall see. The amino acids glutamate and glutamine play central roles in amino acid catabolism, as well as in biosynthesis.

❮❮ **The structures and properties of the amino acids are described in Sections 12.1 and 12.2.**

17.2 TRANSAMINATION AND OXIDATIVE DEAMINATION

After the digestion of a protein-rich meal, amino acids are absorbed from the intestine into the bloodstream, and those not used in biosynthetic reactions undergo degradation in the liver. The degradation is accomplished through two kinds of reaction: transamination and oxidative deamination.

In **transamination** reactions, the α-amino group of an amino acid is transferred to the carbon atom of the carbonyl group of an α-keto acid, which is then converted into the corresponding amino acid. The original amino acid is converted into its corresponding α-keto acid. That is,

$$\text{Amino acid}_1 + \alpha\text{-keto acid}_2 \longrightarrow \alpha\text{-keto acid}_1 + \text{amino acid}_2$$

(Recall that α-keto acids are components of the citric acid cycle, undergoing the catabolic degradation described in Section 15.5.) In essence, transamination is the exchange of an amino group for a carbonyl group. It can take place not only in liver cells but in all cells.

Transamination reactions are catalyzed by enzymes called amino transferases or transaminases. Most transaminases are specific for α-ketoglutarate but are less specific for the amino acid. For this reason, the principal product of transamination from a wide array of amino acids is glutamate. Therefore, the oxidative pathways for amino acids tend to converge to form a single product. An important exception is a group of transaminases in muscle cells that use pyruvate as the amino-group acceptor. For this reason, alanine is the principal product of amino acid catabolism in muscle cells. Both types of transaminase are of central importance in the transport systems considered in detail in Section 17.3.

The enzymes are named for the amino acid that donates the amino group: for example, alanine transaminase, aspartate transaminase, and leucine transaminase. The transamination reaction between aspartate and α-ketoglutarate, catalyzed by asparatate transaminase, is

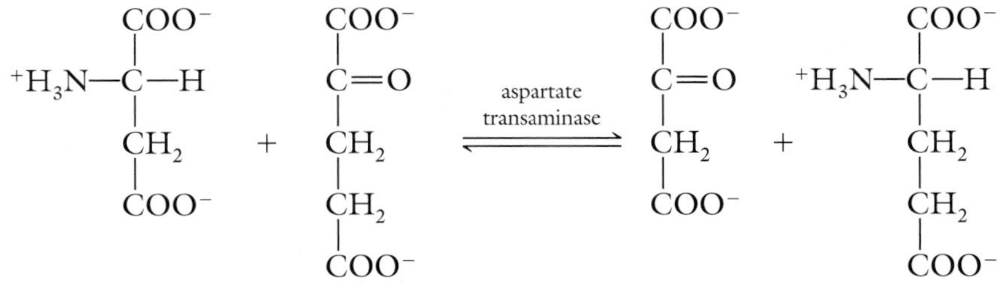

$$\underset{\text{L-Aspartate}}{{}^+H_3N{-}CH(CH_2COO^-){-}COO^-} + \underset{\alpha\text{-Ketoglutarate}}{{}^-OOC{-}C(=O){-}CH_2{-}CH_2{-}COO^-} \xrightleftharpoons{\text{aspartate transaminase}} \underset{\text{Oxaloacetate}}{{}^-OOC{-}C(=O){-}CH_2{-}COO^-} + \underset{\text{L-Glutamate}}{{}^+H_3N{-}CH(CH_2CH_2COO^-){-}COO^-}$$

All transaminases utilize pyridoxine (vitamin B_6) in the form of **pyridoxal phosphate** as the coenzyme in the transamination reaction.

$$H{-}C{=}O,\ {}^{2-}O_3PO{-}CH_2,\ OH,\ CH_3 \text{ (substituents on pyridine ring)}$$

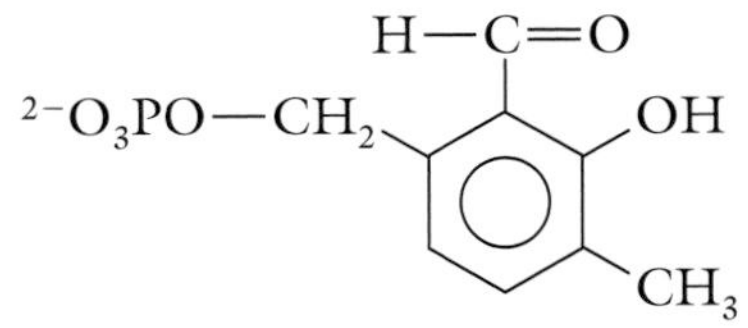

Pyridoxal phosphate

Measurements of the levels of two transaminases in the blood—alanine transaminase, also called glutamate:pyruvate transaminase (GPT), and aspartate transaminase, also called glutamate:oxaloacetate transaminase (GOT)—are used in the diagnosis of liver disease or toxicity. These conditions can result in tissue damage, and, as a consequence, transaminases, along with other enzymes, will leak into the bloodstream. The severity and stage of liver damage can be assessed by measuring the concentrations of these enzymes in the blood serum, measurements known as the serum GPT and GOT or, more commonly, SGPT and SGOT tests.

Glutamate produced from transamination reactions can undergo **oxidative deamination,** releasing ammonium ion and producing α-ketoglutarate. The reaction is catalyzed by glutamate dehydrogenase, with NAD^+ as coenzyme.

$$\underset{\textbf{L-Glutamate}}{\begin{array}{c} COO^- \\ | \\ {}^+H_3N-C-H \\ | \\ CH_2 \\ | \\ CH_2 \\ | \\ COO^- \end{array}} + NAD^+ + H_2O \xrightleftharpoons{\text{glutamate dehydrogenase}} \underset{\boldsymbol{\alpha}\textbf{-Ketoglutarate}}{\begin{array}{c} COO^- \\ | \\ C=O \\ | \\ CH_2 \\ | \\ CH_2 \\ | \\ COO^- \end{array}} + NH_4^+ + NADH + H^+$$

Because it is reversible, this reaction also affords a mechanism both for effectively assimilating ammonium ions into metabolic pathways in the human liver and kidney and for lowering toxic levels of ammonium ions in all cells.

In extrahepatic tissues, amino acids are produced from the metabolic turnover of proteins. These amino acids, if not reused, must be degraded. As in the liver, the amino acids are removed in transamination reactions to form glutamate. The glutamate cannot pass through the cell membrane; as shown in the next section, it must be converted into another compound for passage into the bloodstream.

17.3 AMINO-GROUP AND AMMONIA TRANSPORT

The amino groups collected in the form of glutamate in extrahepatic tissues, as well as NH_4^+ formed in those tissues from the breakdown of amino acids and other nitrogen-containing biomolecules, are packaged in a nontoxic form that can leave the cell, enter the circulation, and travel to the liver. The conversion of NH_4^+ into nontoxic form takes place by two processes, resulting in two different transport forms:

- In most cell types, the production of glutamine
- In muscle cells, the production of alanine

The enzyme glutamine synthetase, in almost all cell types, catalyzes the formation of glutamine from glutamate.

$$\underset{\textbf{L-Glutamate}}{\begin{array}{c} COO^- \\ | \\ {}^+H_3N-C-H \\ | \\ CH_2 \\ | \\ CH_2 \\ | \\ COO^- \end{array}} + NH_4^+ + ATP \longrightarrow \underset{\textbf{L-Glutamine}}{\begin{array}{c} COO^- \\ | \\ {}^+H_3N-C-H \\ | \\ CH_2 \\ | \\ CH_2 \\ | \\ C=O \\ | \\ NH_2 \end{array}} + ADP + P_i + H^+$$

The amination of the γ-glutamyl carboxyl group converts the negatively charged glutamate into glutamine, which has a net charge of zero and can pass through cell membranes into the blood.

The amide group of glutamine is the source of amino groups in many biosynthetic reactions, including amination of α-keto acids in amino acid synthesis (Section 18.4). The concentration of glutamine in blood is significantly higher than that of any other amino acid.

Concept check

✓ Glutamine is a nontoxic transport form of NH_4^+ and a temporary storage form of amino groups in the body.

On reaching the liver, glutamine can be deaminated in another reaction by the enzyme glutaminase:

$$\text{Glutamine} + H_2O \longrightarrow \text{glutamate} + NH_4^+$$

Some of the glutamate can undergo oxidative deamination, as described earlier, releasing another molecule of NH_4^+.

In active muscle cells, there is considerable turnover of amino acids and nucleotides, which gives rise to large quantities of ammonium ion. The ammonium ion is transported from muscle to the liver in the form of the amino acid alanine through the action of the **glucose–alanine cycle** (Figure 17.1).

The cycle begins in the mitochondria of muscle cells with a process called **reductive amination,** in which α-ketoglutarate reacts with NH_4^+ to form glutamate, a reaction catalyzed by glutamate dehydrogenase—in this case, an irreversible reaction in which NADPH is the cofactor:

$$NH_4^+ + \alpha\text{-ketoglutarate} + \text{NADPH} \longrightarrow \text{glutamate} + \text{NADP}^+ + H_2O$$

This reaction also constitutes another way in which ammonium ion can be incorporated to form an amino acid.

The glutamate then moves into the cytosol where it undergoes transamination with pyruvate catalyzed by alanine transaminase:

$$\text{Glutamate} + \text{pyruvate} \longrightarrow \text{alanine} + \alpha\text{-ketoglutarate}$$

Pyruvate is readily available in actively contracting muscle, where the glycolytic pathway is extremely active. Alanine has no net charge at physiological pH and can pass through cell membranes to enter the circulation and thence the liver.

After alanine enters the liver, transamination in the cytosol transfers alanine's amino group to α-ketoglutarate, forming glutamate and pyruvate. The glutamate gives up its NH_4^+ by oxidative deamination catalyzed by glutamate dehydrogenase (Section 18.4) or, as we shall see in Section 17.4, by transamination with oxaloacetate to form aspartate. Pyruvate is converted into glucose by gluconeogenesis, and glucose returns through the circulation to muscle cells.

As we can see, the glucose–alanine cycle kills two birds with one stone: pyruvate produced in muscle cells is converted into glucose in the liver and returned to muscle; the waste product ammonium ion is removed from muscle cells and converted into urea in the liver. The urea is then excreted by the kidneys.

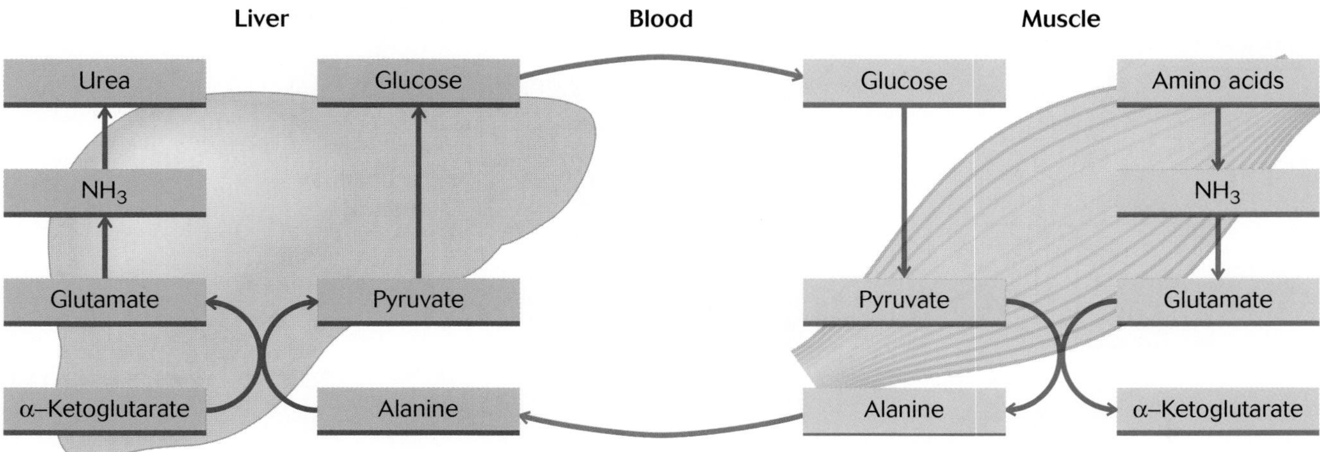

Figure 17.1 The glucose–alanine cycle by which ammonium ion is transported from muscle cells to the liver in the form of an amino acid.

17.4 THE UREA CYCLE

Having considered the ways in which amino groups and NH_4^+ are transported to the liver and how NH_4^+ is released in liver cells, we now look at the way in which this toxic compound is converted into a nontoxic, excretable form. Humans as well as most other terrestrial animals (land-dwelling animals rather than those that live in water) transform ammonium ion into a water-soluble, nonionic compound called urea. Such organisms are called **ureotelic.** Urea excretion is advantageous for two reasons. First, the water solubility of ammonium ion makes keeping it in the urine formed in the kidney problematic because the ammonium ion can easily pass back into the blood. Second, excretion of its water-soluble form, NH_4^+, requires the simultaneous excretion of an oppositely charged ion (a counterion). This requirement would lead to significant loss of metabolically important anionic counterions—for example, phosphate and bicarbonate. The conversion of ammonium ion into nontoxic, nonionic form nullifies these problems.

The biosynthesis of urea is a complex process that takes place only in the liver. It is a cyclic process, one that takes place partly in mitochondria and partly in the cytosol. Two intermediates of the cycle are amino acids not found in proteins: ornithine and citrulline. The other amino acid in the cycle is arginine, a common component of proteins. In the **urea cycle,** the biosynthesis of 1 mol of urea fixes 2 mol of ammonium ion. The steps of this cycle are outlined in Figure 17.2.

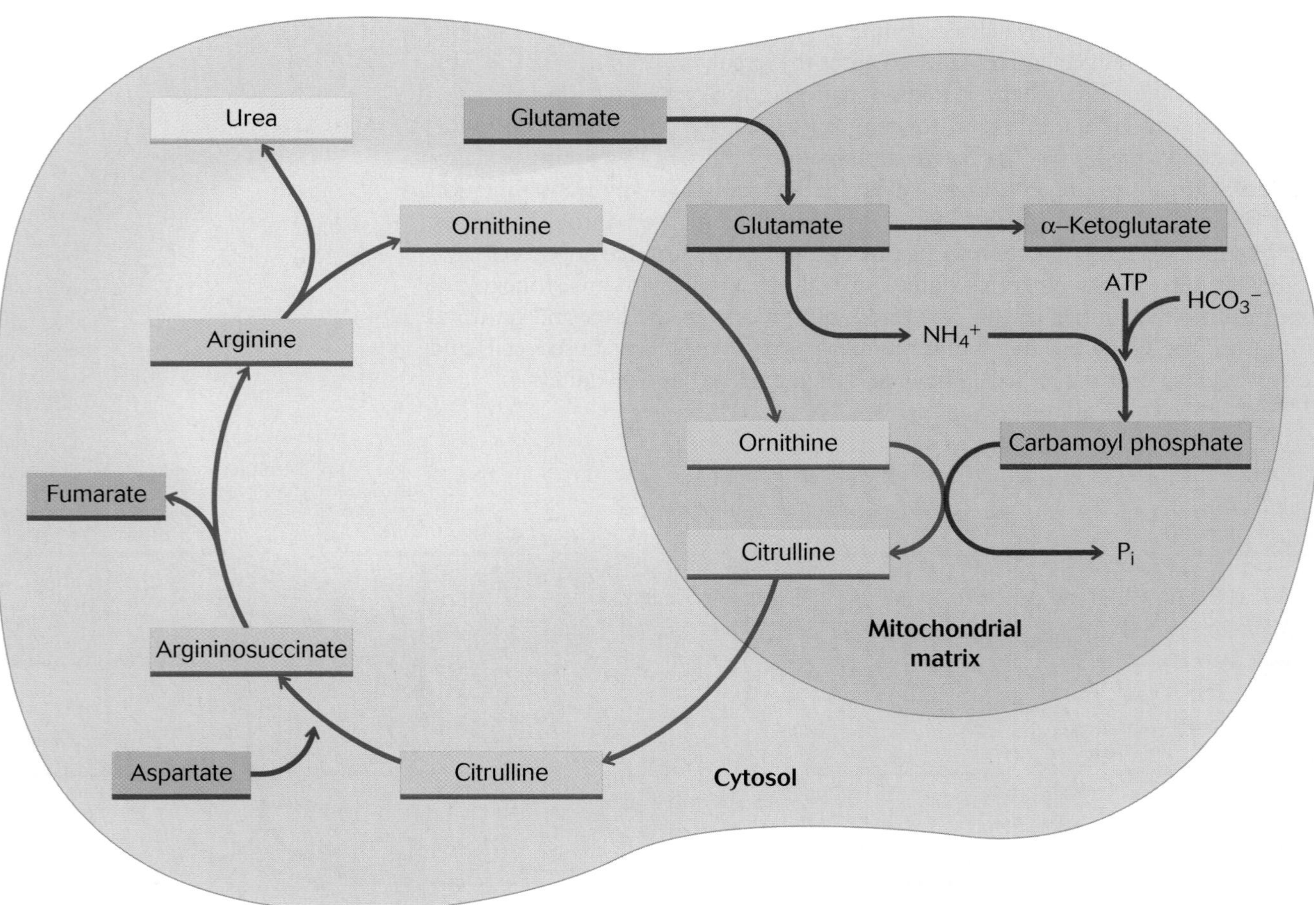

Figure 17.2 The urea cycle consists of reactions within the cytosol and reactions within mitochondria.

Reactions in the Mitochondria

In liver mitochondria, NH_4^+ reacts with bicarbonate and two molecules of ATP to produce carbamoyl phosphate. The reaction is catalyzed by carbamoyl phosphate synthetase I (as distinct from the cytosolic carbamoyl phosphate synthetase II, an enzyme of nucleotide synthesis). This reaction is an activation reaction in which carbamoyl phosphate is an activated form of the carbamoyl group, which then enters the urea cycle.

$$HCO_3^- + NH_4^+ + 2\ ATP^{4-} \longrightarrow H_2N{-}\overset{\overset{\displaystyle O}{\|}}{C}{-}OPO_3^{2-} + 2\ ADP^{3-} + P_i^{2-} + 2\ H^+$$

Carbamoyl phosphate

Step 1. Carbamoyl phosphate reacts with ornithine to form citrulline, a reaction catalyzed by ornithine transcarbamoylase.

$$\underset{\text{Ornithine}}{{}^{+}H_3N{-}CH(COO^-){-}CH_2{-}CH_2{-}CH_2{-}NH_3^+} + \underset{\text{Carbamoyl phosphate}}{H_2N{-}C({=}O){-}O{-}PO_3^{2-}} \longrightarrow \underset{\text{Citrulline}}{{}^{+}H_3N{-}CH(COO^-){-}CH_2{-}CH_2{-}CH_2{-}NH{-}C({=}O){-}NH_2} + P_i$$

Citrulline then leaves the mitochondrion and enters the cytosol.

In subsequent reaction equations, the ornithine part of citrulline will be represented by Orn because it undergoes no fundamental structural changes and emerges unchanged at the end of the process to begin a new cycle. Thus,

$$\underset{\text{Citrulline}}{{}^{+}H_3N{-}CH(COO^-){-}CH_2{-}CH_2{-}CH_2{-}NH{-}C({=}O){-}NH_2} = H_2N{-}C({=}O){-}Orn$$

Reactions in the Cytosol

In the cytosol of liver cells, glutamate, the carrier of amino groups from amino acids, undergoes transamination with oxaloacetate to form aspartate. Aspartate carries the second molecule of ammonium ion into the urea cycle.

Step 2. Aspartate reacts with citrulline to form argininosuccinate. In this reaction, requiring ATP and catalyzed by argininosuccinate synthetase, the two terminal phosphates are split off as pyrophosphate ($P_2O_7^{4-}$), which is subsequently hydrolyzed to 2 P_i. As we have seen, this hydrolysis has a very large equilibrium constant (Section 15.9) and is the driving force for the synthesis of argininosuccinate.

Citrulline

$$\text{Citrulline: } H_2N{-}C({=}O){-}\text{Orn} \;+\; \text{Aspartate: } {}^{-}OOC{-}CH_2{-}CH(NH_3^+){-}COO^- \;+\; \text{ATP} \xrightarrow{\text{argininosuccinate synthetase}} {}^{+}H_2N{=}C(\text{Orn}){-}NH{-}CH(COO^-){-}CH_2{-}COO^- \;+\; \text{AMP} + PP_i$$

Aspartate **Argininosuccinate**

Step 3. In the next step, the argininosuccinate is cleaved into arginine and fumarate by the enzyme argininosuccinate lyase.

$$\text{Argininosuccinate: } {}^{+}H_2N{=}C(\text{Orn}){-}NH{-}CH(COO^-){-}CH_2{-}COO^- \xrightarrow{\text{argininosuccinate lyase}} \text{Arginine: } {}^{+}H_2N{=}C(\text{Orn}){-}NH_2 \;+\; \text{Fumarate: } H{-}C(COO^-){=}C(H){-}COO^-$$

Argininosuccinate **Arginine** **Fumarate**

Fumarate enters the citric acid cycle.

Step 4. The enzyme arginase catalyzes the cleavage of arginine into urea. Birds, reptiles, and bony fishes do not possess this enzyme. It is found only in the liver of ureotelic organisms.

$$ {}^{+}H_2N{=}C(\text{Orn}){-}NH_2 \xrightarrow[H_2O]{\text{arginase}} \text{Ornithine} + H_2N{-}C({=}O){-}NH_2 \;(\text{Urea})$$

Ornithine is now available for another round of the urea cycle. Urea passes from liver cells into the blood and is excreted by the kidneys.

A significant amount of energy goes into urea synthesis: 2 ATP for the synthesis of carbamoyl phosphate, 2 ATP for the synthesis of argininosuccinate, 1 ATP converted into AMP in the synthetase reaction, and 1 ATP required for the conversion of AMP into ADP (Section 16.1). Thus the total energy cost for the synthesis of a mole of urea is 4 mol of ATP. The price for the detoxification of ammonium ion is about 20% (see Table 15.4) of the available energy in the amino acids oxidized.

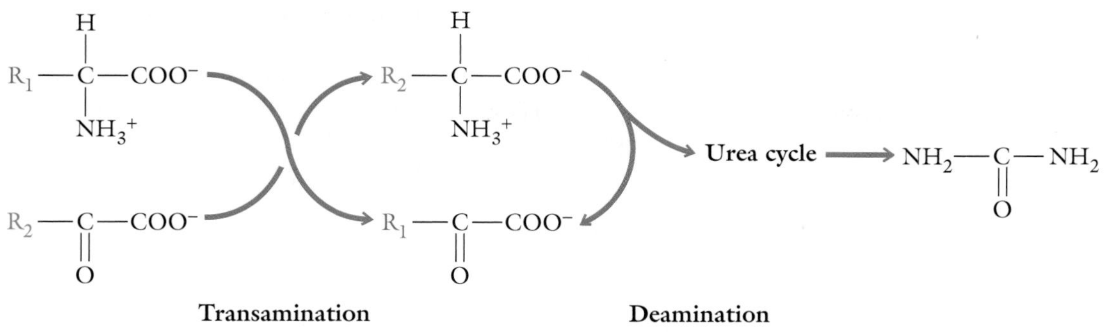

$$R_1-\underset{NH_3^+}{\overset{H}{C}}-COO^- + R_2-\underset{O}{\overset{\|}{C}}-COO^- \xrightarrow{\text{Transamination}} R_2-\underset{NH_3^+}{\overset{H}{C}}-COO^- + R_1-\underset{O}{\overset{\|}{C}}-COO^-$$

$$R_2-\underset{NH_3^+}{\overset{H}{C}}-COO^- \xrightarrow{\text{Deamination}} \text{Urea cycle} \longrightarrow NH_2-\underset{O}{\overset{\|}{C}}-NH_2$$

Figure 17.3 Outline of the processes through which amino groups of amino acids are converted into urea.

Figure 17.3 summarizes the processes by which ammonium ion is removed from amino acids and converted into urea.

17.5 THE OXIDATION OF THE CARBON SKELETON

We now turn to the second stage of amino acid catabolism: the breakdown of the carbon skeletons of the deaminated α-keto acids.

The α-keto acids of many amino acids are the same as those found in the glycolytic pathway and the citric acid cycle. Therefore they undergo the oxidative degradation described in Chapter 15. These α-keto acids thus replenish the citric acid cycle intermediates, either directly or through the formation of pyruvate and its carboxylation to oxaloacetate (Section 15.7).

Other α-keto acids produced from amino acids are broken down to acetyl-*S*-CoA or acetoacetyl-*S*-CoA or both and thus can enter the citric acid cycle or be converted into ketone bodies (Section 16.3). These α-keto acids—just like fatty acids, which produce acetyl-*S*-CoA—do not replenish citric acid cycle intermediates.

Every amino acid has a different degradative pathway; however, we will not consider these pathways in detail. All pathways lead to intermediates that find their way into the citric acid cycle for complete oxidation to CO_2 and H_2O. Six amino acids are degraded to pyruvate, seven to acetyl-*S*-CoA, four to α-ketoglutarate, three to succinyl-*S*-CoA, two to fumarate, and two to oxaloacetate (Figure 17.4 on the following page).

The amino acids that can be converted into pyruvate, α-ketoglutarate, succinyl-*S*-CoA, fumarate, and oxaloacetate can give rise to glucose by gluconeogenesis (Section 15.8), and these amino acids are said to be **glucogenic.** Seven amino acids are converted into acetyl-*S*-CoA or acetoacetyl-*S*-CoA or both and can thus yield ketone bodies in the liver (Section 16.3); they are the **ketogenic** amino acids. Glucose cannot be produced from amino acids that form only acetyl-*S*-CoA or acetoacetyl-*S*-CoA.

Concept check

✓ Only amino acids that replenish oxaloacetate can give rise to glucose by gluconeogenesis.

Large amounts of ketone bodies are produced from amino acids, as well as from lipids (Section 16.3), in untreated diabetes mellitus. In starvation and fasting, as well as in diabetes, large amounts of glucose also are produced from

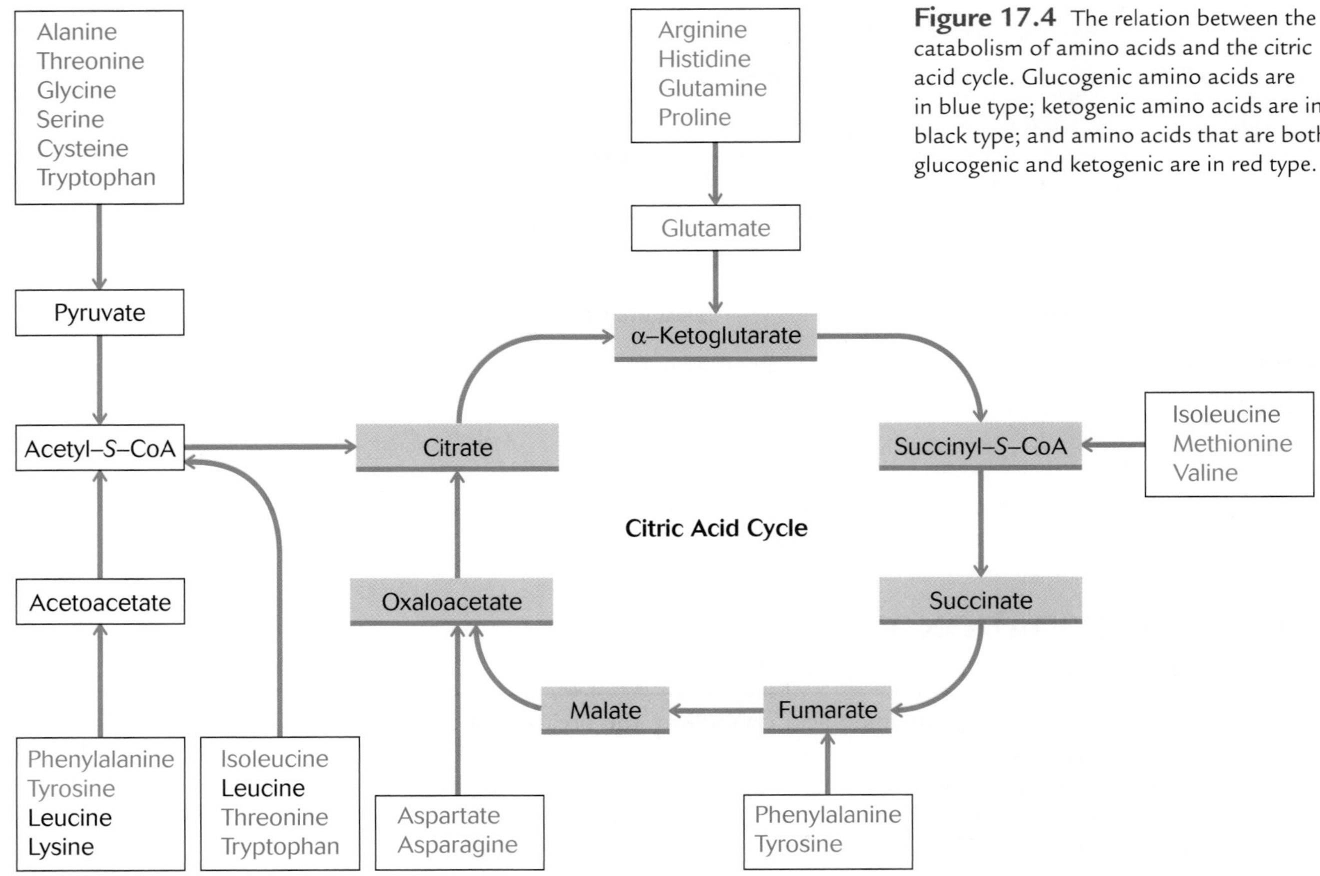

Figure 17.4 The relation between the catabolism of amino acids and the citric acid cycle. Glucogenic amino acids are in blue type; ketogenic amino acids are in black type; and amino acids that are both glucogenic and ketogenic are in red type.

amino acids. Phenylalanine, tyrosine, and isoleucine are both ketogenic and glucogenic: their degradation yields both glucose-forming and ketone-body-forming precursors (see Figure 17.4). Table 17.1 lists the glucogenic and ketogenic amino acids. Note that only leucine and lysine cannot act as precursors for glucose synthesis.

TABLE 17.1 Glucogenic and Ketogenic Amino Acids

Glucogenic only	Glucogenic and ketogenic	Ketogenic only
alanine	phenylalanine	leucine
arginine	isoleucine	lysine
asparagine	tyrosine	
aspartate		
cysteine		
glutamate		
glutamine		
glycine		
histadine		
methionine		
proline		
serine		
threonine		
tryptophan		
valine		

17.6 HERITABLE DEFECTS IN AMINO ACID METABOLISM

A number of heritable diseases result from defects in enzymes catalyzing steps in amino acid catabolism. Such a defect arises from genetic mutations that cause errors in the amino acid sequence of the primary peptide chain with consequent loss in function of the enzyme (Section 12.5). Because there are usually several enzymatically catalyzed steps in a catabolic sequence, normal intermediate compounds are not processed; they therefore accumulate and often lead to negative physiological consequences. Defects in the degradative pathway for the amino acid phenylalanine are the cause of several heritable diseases.

« The causes of genetic mutations are discussed in Section 13.8.

About 25% of the normal concentration of phenylalanine goes into the synthesis of proteins, and the rest is used to synthesize tyrosine. Aside from incorporation into proteins, tyrosine is the precursor of the hormone thyroxine. The first step in the conversion of phenylalanine into tyrosine is the introduction of a hydroxyl group in the aromatic ring of phenylalanine. In the following structural diagram, this step is represented in the upper sequence. If that step cannot take place (crossed-out arrow), the amino group of phenylalanine can be removed by transamination to form phenylpyruvate. In **phenylketonuria (PKU),** one of the first human defects of metabolism to be discovered, the enzyme catalyzing this step, phenylalanine hydroxylase, is defective, and phenylalanine accumulates. In phenylketonurics, the plasma levels of phenylalanine can be as much as 20 times its normal concentration. Because the normal degradative pathway to tyrosine is blocked, phenylalanine reacts (by transamination) with pyruvate to form phenylpyruvate.

Phenylalanine **Tyrosine**

$$C_6H_5-CH_2-CH(NH_3^+)-COO^- \not\longrightarrow HO-C_6H_4-CH_2-CH(NH_3^+)-COO^-$$

pyruvate → alanine

$$C_6H_5-CH_2-C(=O)-COO^-$$

Phenylpyruvate

Phenylpyruvate is not further metabolized. It accumulates in tissue and blood, and, because it is soluble, it is excreted along with phenylalanine in the urine. Although the biochemical basis for the consequences of the accumulation of phenylalanine in the early stages of life remains unknown, the most serious of these consequences is the impairment of normal development of the brain, causing mental retardation and severely shortened life expectancy.

Phenylalanine and phenylpyruvate are easily detected in blood and urine, and, in the United States and many other countries, newborns are now routinely tested for PKU. When PKU is diagnosed early, mental retardation can be reduced by modifying the infant's diet. Because phenylalanine is an essential amino acid, it cannot be completely eliminated from the diet, but it can be reduced. For example, casein from milk can be hydrolyzed and most of the phenylalanine removed. This special low-phenylalanine diet is particularly effective in childhood. After the nervous system is fully developed, such prescriptions have little effect on the outcome of the disease. However, certain precautions are still useful—for example, the avoidance of foods and drinks sweetened with aspartame (Section 9.6). Aspartame is the methyl ester of the dipeptide aspartylphenylalanine, and one of the products of its digestion is phenylalanine.

Another disorder of phenylalanine catabolism leads to the accumulation of the intermediate homogentisate, a nontoxic water-soluble compound excreted in the urine. Rapid air oxidation of this compound produces an intensely black pigment, which gave this disorder the name "black urine" disease, or **alkaptonuria.** The disease has no negative physiological consequences but, as might be imagined, led to some psychological and social problems before it was understood as a benign genetic defect.

A group of related conditions called maple syrup urine disease are caused by deficiencies in the catabolism of the branched-chain amino acids such as isoleucine. The disease is usually detected because it results in acidosis in newborns and young children. The most common metabolic defect lies in the inability to oxidize the α-keto acids resulting from transamination. All patients with this disease excrete α-keto acids and other side products. An unidentified product gives rise to the characteristic odor that lends its name to this group of diseases—maple syrup urine. Although some cases respond to dietary intervention, most cases result in mental retardation and early death.

A brief listing of several other genetic defects in amino acid metabolism includes a deficiency in mitochondrial ornithine transaminase. This deficiency results in a progressive loss of vision caused by atrophy of the retina. Plasma levels of ornithine, a critical component of the urea cycle, can be elevated as a result of a deficiency in an aminotransferase that catalyzes the first step in the conversion of ornithine into glutamate. Another disorder of ornithine metabolism has been found to be caused by the defective transport of ornithine into the mitochondria. A deficiency or absence of cytosolic tyrosine transaminase results in a disease characterized by skin and eye lesions, often accompanied by mental retardation. The enzyme tyrosinase catalyzes the formation of a precursor of melanin through the oxidation of tyrosine. Melanin is a high-molecular-mass polymer that is insoluble and very dark in color. Its absence leads to the condition known as albinism—a lack of skin and hair color.

17.7 THE BIOSYNTHESIS OF AMINO ACIDS

Human adults can synthesize 10 of the 20 amino acids required for protein synthesis. These amino acids are classified as **nonessential**—meaning nonessential in the diet (Section 12.1). Those amino acids that humans cannot synthesize are classified as **essential** and must be obtained by dietary intake (Box 12.1). We consider, here, two of the more central features of the biosynthesis of the nonessential amino acids.

Amino-Group Donation

Glutamate and glutamine play essential roles as amino-group donors in amino acid synthesis. Reductive amination, producing glutamate, takes place in all cells; the reaction is catalyzed by glutamate dehydrogenase, with NADPH as cofactor:

$$NH_4^+ + \alpha\text{-ketoglutarate} + NADPH \xrightarrow[\text{dehydrogenase}]{\text{glutamate}} \text{glutamate} + NADP^+ + H_2O$$

Glutamate is used to synthesize amino acids, by transamination, from appropriate precursors—primarily α-keto acids of the citric acid cycle and glycolysis or α-keto acids derived from intermediates in these two processes. The amino acids synthesized directly from citric acid cycle intermediates by transamination or reductive amination are alanine, aspartate, asparagine, glutamate, and glutamine. The biosynthetic pathways for the other amino acids are complex and consist of many enzymatically catalyzed steps and many types of intermediate compounds.

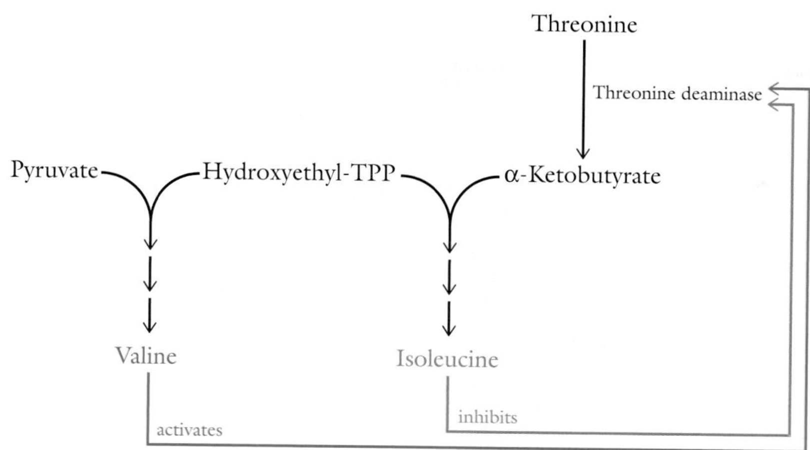

Figure 17.5 Biosynthetic feedback control of the balance of concentrations of valine and isoleucine.

Feedback Control

In Chapter 14, we considered feedback inhibition as a control mechanism for enzyme activity. In fact, it is the chief mechanism for the control of amino acid biosynthesis. An interesting example of how it works is the control of the biosynthesis of isoleucine by the concentration of valine. Figure 17.5 illustrates the general features of the process.

The synthesis of both valine and isoleucine begins with a common intermediate, hydroxyethyl-TPP (HE-TPP), that you first encountered in Section 15.4. After HE-TPP reacts with pyruvate, valine is produced after a number of intervening enzyme-catalyzed steps. The synthesis of isoleucine begins with a product of the deamination of another amino acid, threonine. The product is α-ketobutyrate, and the catalysis is carried out by the enzyme threonine deaminase. After HE-TPP reacts with α-ketobutyrate, isoleucine emerges after a few catalyzed reaction steps. As Figure 17.5 indicates, isoleucine allosterically inhibits the enzyme that initiates its synthesis, typical of feedback control. But, in addition, the same enzyme is allosterically activated by the product of a competitive pathway, valine. The result is that the concentrations of valine and isoleucine are kept in balance for protein biosynthesis. This method of balancing amino acid concentrations is common to many of the amino acid biosynthetic pathways.

Summary

Transamination and Oxidative Deamination

- Amino acids that are not required for the synthesis of proteins or other biomolecules undergo catabolic degradation.
- The amino groups of all amino acids are collected in the liver by transamination with α-ketoglutarate to form glutamate and the α-keto acid of the amino acid.
- In other cells, ammonium ion is removed from amino acids by oxidative deamination.
- The α-keto acids then undergo catabolic degradation.
- The ammonium ion is converted into urea and excreted.

Amino-Group and Ammonia Transport

- Two processes convert ammonium ion produced in extrahepatic tissues into a nontoxic, transportable form.
- The first process is the synthesis of glutamine from glutamate, which takes place in most tissues.
- Glutamine leaves cells and travels to the liver, where the ammonium ion is released and converted into urea.
- In the second process, ammonium ion produced in muscle cells is first transferred to glutarate by reductive amination and then to pyruvate by transamination to form alanine, which is then transported to the liver.

• On reaching the liver, alanine's amine group is converted into urea, and the resulting pyruvate is converted into glucose by gluconeogenesis.

• Glucose circulates back to muscle to begin the cycle again.

The Urea Cycle

• Humans and most other terrestrial animals package ammonium ion into water-soluble, nonionic, nontoxic urea, which is excreted by the kidneys.

• The biosynthesis of urea requires about 20% of the energy available in the amino acids.

• The urea cycle takes place only in liver cells, and its intermediates include arginine and two amino acids not found in proteins—ornithine and citrulline.

• Urea is split from arginine by the enzyme arginase, unique to urotelic (urea-forming) organisms.

• The regenerated ornithine begins a new turn of the cycle.

The Oxidation of the Carbon Skeleton

• All amino acid carbon skeletons are oxidized to carbon dioxide and water in the citric acid cycle.

• Seventeen amino acids can be converted into glucose and are called glucogenic.

• Seven amino acids are converted into acetyl-*S*-CoA or acetoacetyl-*S*-CoA or both, yield ketone bodies in the liver, and are called ketogenic.

• Phenylalanine, tyrosine, and isoleucine are both ketogenic and glucogenic. Their degradation yields both glucose-forming and ketone-body-forming precursors.

Heritable Defects in Amino Acid Metabolism

• A number of diseases characterized by the incomplete catabolic degradation of amino acids are caused by heritable mutations.

• In these diseases, normal intermediates are not processed past some point in the catabolic sequence. As a result they accumulate in tissues and cause a variety of physiological abnormalities.

• Two of these diseases are phenylketonuria and alkaptonuria. Both are caused by defects in the degradative pathway of phenylalanine.

• Although they can lead to severe mental retardation if untreated, the effects can be alleviated by detection in newborns followed by strict adherence to a low-phenylalanine diet.

The Biosynthesis of Amino Acids

• Most amino acids are synthesized by the transamination of α-keto acids by glutamate, which is, in turn, synthesized by reductive amination.

• Adult humans can synthesize 10 of the 20 amino acids required for protein synthesis from intermediates of glycolysis and the citric acid cycle.

• Those amino acids that humans cannot synthesize must be obtained through the diet.

Key Words

alkaptonuria, p. 526
essential amino acid, p. 526
glucose–alanine cycle, p. 519
glucogenic amino acid, p. 523
ketogenic amino acid, p. 523
nonessential amino acid, p. 526
oxidative deamination, p. 518
phenylketonuria (PKU), p. 525
pyridoxal phosphate, p. 517
reductive amination, p. 519
transamination, p. 517
urea cycle, p. 520
ureotelic animal, p. 520

Exercises

Amino Acid Metabolism

17.1 Describe the process called transamination. Give an example.

17.2 What vitamin forms the coenzyme used in transamination?

17.3 What happens to dietary amino acids in excess of those required for biosynthesis?

17.4 What happens to proteins of the body if carbohydrates are not available in the diet?

Transamination and Oxidative Deamination

17.5 Describe a reaction that is a first step in the catabolism of ingested amino acids.

17.6 Describe an alternative first step in the catabolism of ingested amino acids that takes place in the liver.

17.7 How is ammonium ion that is generated in the liver rendered nontoxic in humans?

17.8 How is ammonium ion that is generated in cells other than the liver rendered nontoxic in humans?

17.9 What is a ureotelic animal?

17.10 What unique enzyme is possessed by ureotelic animals?

17.11 What is the likely mechanism of ammonium ion toxicity in cells?

17.12 Describe the method used by all cells of the body to render ammonium ion harmless for transport through the blood to the liver.

17.13 What is the method used by muscle cells to render ammonium ion harmless for transport through the blood to the liver?

The Urea Cycle

17.14 In what organ is urea synthesized?

17.15 Do the reactions of the urea cycle take place in the cytosol or in mitochondria? Explain your answer.

17.16 Name and draw the structure of an amino acid that is not used to synthesize proteins but participates in urea's synthesis.

17.17 Name and draw the structure of an amino acid other than that named in Exercise 17.16 that also is not used to synthesize proteins but participates in urea's synthesis.

17.18 List the urea-cycle reactions that require ATP.

17.19 List the urea-cycle reactions that do not require ATP.

17.20 What is the total energy cost for the synthesis of 1 mol of urea in moles of ATP?

17.21 What percentage of the energy available in amino acids is used to synthesize urea?

The Oxidation of the Carbon Skeleton

17.22 Name two glucogenic amino acids.

17.23 Why are the amino acids named in Exercise 17.22 called glucogenic?

17.24 Name two ketogenic amino acids.

17.25 Why are the amino acids named in Exercise 17.24 called ketogenic?

17.26 Name two products of the catabolism of tyrosine that cause that amino acid to be called both glucogenic and ketogenic (see Figure 17.4).

17.27 Name two products of the catabolism of phenylalanine that cause that amino acid to be called both glucogenic and ketogenic (see Figure 17.4).

Heritable Defects in Amino Acid Metabolism

17.28 What is the origin of heritable metabolic disorders of amino acid metabolism?

17.29 How are heritable metabolic defects in amino acid metabolism manifested?

17.30 What is the name of one of the heritable diseases caused by defects in phenylalanine catabolism?

17.31 What is the result of the disease named in Exercise 17.30 if untreated, and how can it be treated?

The Biosynthesis of Amino Acids

17.32 What is meant by an essential amino acid?

17.33 What is meant by a nonessential amino acid?

Unclassified Exercises

17.34 True or false: The citric acid cycle is a source of amino acids. Explain your answer.

17.35 Aside from preventing its toxicity, is there any other advantage for converting ammonia into urea?

17.36 True or false: Dietary amino acids cannot be used to maintain normal blood-glucose concentrations.

17.37 Can arginase be extracted from the liver cells of a fish?

Expand Your Knowledge

17.38 True or false: Ornithine is both a substrate and a product of the urea cycle. Explain your answer.

17.39 Aspartame, a nonnutritional sweetener, is L-aspartyl-L-phenylalanine methyl ester. Why do products containing it warn phenylketonurics not to use the product?

17.40 Explain how reductive amination allows a citric acid cycle intermediate to replenish supplies of amino acids.

17.41 True or false: Oxidative deamination of aspartate leads to the formation of pyruvate and ammonia. Explain your answer.

17.42 Humans can synthesize arginine but not in sufficient amounts. Is this amino acid considered essential or nonessential? Explain your answer.

17.43 Consider the following branched pathway for the biosynthesis of amino acids Y and Z. Propose two control mechanisms for maintaining the concentration of Y equal to that of Z.

$$A \longrightarrow B \longrightarrow C \begin{cases} \nearrow D \longrightarrow E \longrightarrow F \longrightarrow Y \\ \searrow G \longrightarrow H \longrightarrow I \longrightarrow Z \end{cases}$$

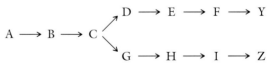

17.44 True or false: The molecular mechanism underlying feedback control of amino acid biosynthesis is competitive inhibition. Explain your answer.

17.45 True or false: The catabolism of leucine gives rise to glucose. Explain your answer.

17.46 Can a person be cured of a heritable amino acid metabolic defect?

CHAPTER 18 NUTRITION, NUTRIENT TRANSPORT, AND METABOLIC REGULATION

(Charles Gupton/The Stock Market.)

Chemistry in Your Future

Every August, you work as resident nurse at a summer camp near your favorite mountain resort. The first days are busy, as the children wrestle with homesickness and as their bodies adjust to the moderately high altitude. The camp staff are well trained, however, and do not allow the children to overexert themselves. Soon the complaints of headache, nausea, sleeplessness, and malaise—whether due to homesickness or lack of oxygen—subside, leaving you time to enjoy your beautiful surroundings. The transport of oxygen through the blood (and the body's adjustment to breathing at high altitudes) is one of the topics in this chapter.

For more information on this topic and others in this chapter, go to www.whfreeman.com/bleiodian2e

Learning Objectives

- Describe digestion and how its products constitute nutrition.
- Describe blood and its role in the transport of nutrients, oxygen, and carbon dioxide.
- Describe the chief metabolic requirements of the major organs.
- Describe the special role of the liver in metabolic regulation.
- Explain how hormones and the nervous system regulate metabolism.
- Describe the biochemical and hormonal characteristics of the absorptive and postabsorptive states.
- Compare and contrast starvation and diabetes with the absorptive and postabsorptive states.

So far, we have looked at the details of metabolism within a cell. We have studied how cells use glucose, fatty acids, and amino acids to extract energy and synthesize biomolecules. However, we do not consist of single cells. Furthermore, our bodies are not merely a collection of different organs made up of single cells.

Organs such as the heart and the liver are themselves organized into a system that is far from the sum of the metabolism of individual cells or organs. This larger system requires the organs of the body to communicate and interact with one another. The balance of all this communication and interaction keeps the body in a healthy steady state called **homeostasis.**

The purpose of this chapter is to present the problems that stand in the way of homeostasis and to illustrate how homeostasis is achieved by describing how the body responds to physiological stress. To do so, we'll first study digestion—that is, the conversion of foodstuffs into small molecules or nutrients. Next, we'll examine the properties of nutrients needed to sustain metabolism. We'll then explore how nutrients and waste products are transported to and from all the cells of the body. This topic will be followed by an inventory of the metabolic needs of individual organs. We'll then examine how the nervous system and its hormones regulate metabolism. Finally, we will illustrate metabolic regulation by describing how the body responds to physiological stress.

18.1 DIGESTIVE PROCESSES

In digestion, foods are enzymatically degraded to low-molecular-mass components to prepare them for absorption into the cells lining the gut (through which they pass into the bloodstream). This enzymatic degradation is necessary because the cells lining the intestine can absorb only small molecules, and most nutrients are ingested in the form of biopolymers—that is, proteins and carbohydrates.

The reduction of the major components of food—proteins, carbohydrates, and triacylglycerols—to low-molecular-mass components begins in the mouth, with the mechanical action of chewing and the secretion of amylase (a starch-degrading enzyme) in the saliva. The next stage takes place in the stomach, where the secretion of the hormone gastrin is stimulated by the entry of protein into the stomach. A summary of the secretions of the human digestive system can be found in Table 18.1.

TABLE 18.1 Secretions of the Human Digestive System

Location	Proenzyme (zymogen)	Enzyme	Hormone	Other
mouth		amylase		
stomach	pepsinogen	pepsin	gastrin	HCl
intestine		enterokinase carboxypeptidase aminopeptidase	secretin	
within intestinal cells		nucleotidases disaccharidases		
pancreas	trypsinogen chymotrypsinogen prolipase	trypsin chymotrypsin lipase amylase		HCO_3^-
liver		colipase		bile salts

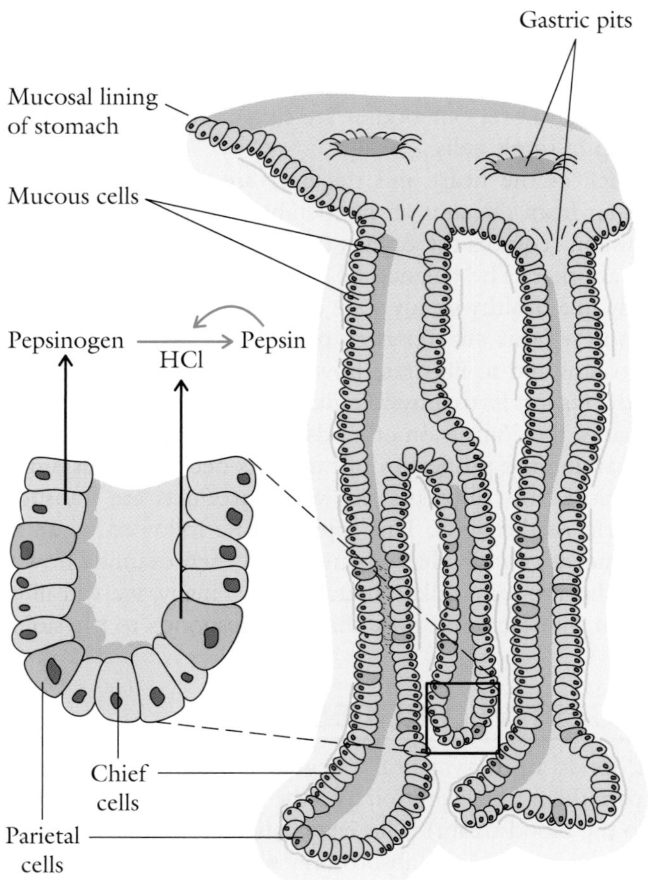

Figure 18.1 Gastric glands are located within gastric pits in the stomach lining. These glands contain both chief cells, which secrete pepsinogen, and parietal cells, which secrete HCl. (Adapted from Figure 37.12, page 802, N. A. Campbell, *Biology,* 3d ed., Redwood City, CA, Benjamin Cummings, 1993.)

Gastrin stimulates the gastric glands in the stomach's lining to secrete pepsinogen and the parietal cells in the stomach's lining to secrete HCl (Figure 18.1). The HCl brings the pH of the stomach to between 1.5 and 2.0, a level of acidity that denatures, or unfolds, proteins, making their internal peptide bonds accessible to enzymatic hydrolysis.

Pepsinogen is a **zymogen,** or inactive enzyme precursor. The HCl in the stomach converts some of the pepsinogen into the active enzyme, pepsin, by removing a small terminal peptide. The pepsin thus activated becomes the catalyst for the rapid conversion of the remaining pepsinogen into pepsin. Pepsin, which attacks the peptide bonds of amino acids possessing hydrophobic side groups (leucine and phenylalanine, for example), reduces large proteins to mixtures of smaller peptides. Other enzymatic hydrolases secreted as zymogens (proenzymes) and activated by similar processing are found in the small intestine.

As the stomach contents pass into the small intestine, their low pH stimulates the secretion of the hormone secretin by duodenal cells. When this hormone reaches the pancreas, it stimulates that organ to secrete bicarbonate into the gut. Because intestinal digestive enzymes function at or near neutral pH, the secretion of bicarbonate is necessary to neutralize the low pH of the entering stomach contents.

As amino acids enter the small intestine, they stimulate intestinal-cell secretion of the specialized enzyme enterokinase. This enzyme converts the zymogen trypsinogen, secreted by the pancreas, into the active proteolytic enzyme trypsin. Trypsin then converts pancreatic chymotrypsinogen into the active chymotrypsin, and the two enzymes proceed to hydrolyze polypeptides to small peptides. The further hydrolysis of the small peptides to their constituent amino acids is accomplished by two other enzymes secreted by the pancreas and intestinal cells: carboxypeptidase, which attacks peptides at the free carboxyl end, and aminopeptidase, which attacks peptides at the free amino end. The mixture of amino acids is then transported across the intestinal cells and into the blood, which carries it directly to the liver.

The principal carbohydrates in food are starch, cellulose (from plants), and glycogen (from animals). Mammals lack enzymes capable of hydrolyzing the $\beta(1 \rightarrow 4)$ linkages of cellulose, which therefore passes through and out of the human gut in the form of roughage (Box 10.4). The hydrolysis of starch and glycogen at the $\alpha(1 \rightarrow 4)$ linkage, begun in the mouth, is completed in the small intestine chiefly by the action of pancreatic amylase. Disaccharides derived directly from dietary sources (such as sucrose from fruits and lactose from milk) are hydrolyzed to a mixture of monosaccharides by enzymes located within cells lining the small intestine; the monosaccharides are then absorbed into the blood and brought directly to the liver. However, many people lack the enzyme lactase, which hydrolyzes lactose (Box 10.2). In these persons, lactose remains in the gut, where it is partly fermented by intestinal microorganisms, a process resulting in diarrhea and the formation of gases. This condition is known as lactose intolerance.

Small amounts of nucleic acids (DNA and RNA) are present in food. They are hydrolyzed to nucleotides by pancreatic enzymes. Enzymes from the epithelial cells lining the intestine break the nucleotides down to free heterocyclic bases (Section 13.1) and monosaccharides.

The digestion of triacylglycerols begins in the small intestine. The pancreas secretes a zymogen called prolipase, which is converted into the active lipase by intestinal proteases. In the presence of a special protein called colipase and bile salts synthesized in the liver and stored in the gall bladder, lipase begins the hydrolysis of fatty acids from triacylglycerols. The bile salts are emulsifying agents that, in combination with the churning action of the intestine, produce a suspension of triacylglycerol droplets with sufficiently large total surface area to permit efficient enzymatic hydrolysis (Section 11.7).

Generally, only one or two fatty acid chains of a triacylglycerol are released by hydrolysis. The product is therefore a mixture of sodium and potassium salts of fatty acids along with monoacylglycerols and diacylglycerols. The mixture is absorbed by intestinal cells and reassembled within them into new

People with lactose intolerance can substitute a variety of soy-based nutritional supplements and drinks for lactose-containing milk. (Keith/Custom Medical Stock.)

» Chylomicrons are considered further in Section 18.3.

triacylglycerols. These reassembled triacylglycerols combine with protein to form **chylomicrons** (Section 11.4), droplets of triacylglycerols surrounded by mono- and diacylglycerols covered by a surface layer of protein. The chylomicrons leave the intestinal cells by exocytosis (Section 11.10, Figure 11.8) to enter the interstitial space between the intestinal cells and the vascular system.

The chylomicrons do not penetrate the capillaries to be carried directly to the portal circulation of the liver, as do amino acids and monosaccharides. Instead, they pass into the lymphatic system by means of small lymph vessels called lacteals in the intestinal walls. The lymphatic system eventually connects with the vascular system at the thoracic duct, where the contents of the lymphatic system empty into the subclavian vein and enter the blood. Lipids absorbed into the blood combine with proteins produced by the liver to form **lipoproteins.** The details of lipoprotein structure, transport, and interaction with cells are described in Section 18.3.

18.2 NUTRITION

The end product of digestion is a complex mixture of biomolecules that must be sufficient to fulfill an organism's requirements for biosynthesis, motion (muscle contraction), ion transport, and secretion. To be adequate, it must contain five basic classes of nutrients: energy sources, essential amino acids, vitamins, minerals, and essential fatty acids. The components of each class are listed in Table 18.2.

TABLE 18.2 Nutrients Required by Humans

ENERGY SOURCES		
carbohydrates		
fats		
proteins		
ESSENTIAL AMINO ACIDS		
arginine	lysine	threonine
histidine	methionine	tryptophan
isoleucine	phenylalanine	valine
leucine		
ESSENTIAL FATTY ACIDS		
linoleic acid		
linolenic acid		
VITAMINS		
thiamine	pantothenic acid	vitamin B_{12}
riboflavin	folic acid	ascorbic acid
niacin	biotin	vitamins A, D, E, K
pyridoxine		
MINERALS		
arsenic	iron	selenium
calcium	magnesium	silicon
chlorine	nickel	sodium
chromium	molybdenum	tin
copper	phosphorus	vanadium
fluorine	potassium	zinc
iodine		

TABLE 18.3 Energy Equivalents of Nutrients

Nutrient	Energy content (kcal/g)*
carbohydrates	4.0
fats	9.2
proteins	4.2

*Per gram of dry weight.

The energy content of foodstuffs varies. The approximate caloric values of generic carbohydrates, fatty acids, and proteins listed in Table 18.3 are based on a varied diet in which all nutritional components are present. (Recall, for example, that humans cannot synthesize glucose from fatty acids. If carbohydrate is absent from the diet, fatty acid metabolism becomes inefficient, and the caloric value of fatty acids decreases.)

« Categories of amino acids are summarized in Table 12.1.

The energy requirement of a body at complete rest 12 h after eating is called the **basal metabolic rate,** which is considered the energy required to maintain the body's basic "housekeeping" functions. The basal metabolic rates for men and women in their early 20s are 1800 and 1300 kcal/day, respectively. However, a person's daily activity controls his or her total caloric requirement. In general, variations depend on the extent of muscular activity, body weight, age, and sex. Table 18.4 lists the recommended caloric intake as a function of sex and age.

Proteins

A diet of only glucose would be adequate to fulfill all the body's carbohydrate requirements. However, the requirements for amino acids and fatty acids are more complicated, because certain amino acids and fatty acids are considered essential and others nonessential from a dietary point of view. Of the 20 amino acids required for protein synthesis, there are 10 that humans either cannot synthesize at all or cannot synthesize in sufficient quantities. Those ten essential amino acids must be obtained from the diet.

TABLE 18.4 Recommended Daily Energy Allowances

	Age (years)	Weight (kg)	Energy (kcal)
infants	0.0–0.5	6	650
	0.5–1.0	9	970
children	1–3	13	1300
	4–6	20	1700
	7–10	28	2400
females	11–14	46	2200
	15–18	55	2100
	19–22	55	2100
	23–50	55	2000
	50+	55	1800
males	11–14	45	2700
	15–18	66	2800
	19–22	70	2900
	23–50	70	2700
	50+	70	2400

TABLE 18.5 **Chemical Scores and Biological Values of Some Food Proteins**

Protein source	Chemical score	Biological value
human milk	100	95
beefsteak	98	93
whole egg	100	87
cow's milk	95	81
corn	49	36
polished rice	67	63
whole wheat bread	47	30

Proteins are required not for their caloric value but for their content of amino acids. The biosynthesis of specific proteins demands that each required amino acid be present at the synthesis site or synthesis will cease. If even one amino acid needed for the protein is not present, synthesis will stop, and the previously synthesized nascent polypeptide chain will be dismantled. Experimental animals fed a synthetic diet of amino acids do not grow if one essential amino acid is omitted from the diet. However, when that amino acid is added, growth begins within hours.

In evaluating the protein content of a meal, we must ask two questions: Does a food protein contain the correct types of amino acids? And how accessible are the amino acids—that is, how digestible is the food? These qualities of a dietary protein are expressed as its **biological value.** For example, if a given protein provides all the required amino acids in the proper proportions and all are released on digestion and absorbed, the protein is said to have a biological value of 100. When the biological value of a protein is high, only small daily amounts of that protein are required to keep a person in nitrogen balance. **Nitrogen balance** means that the body's intake of protein nitrogen is equal to the nitrogen excreted in urine and feces.

A protein's biological value will be less than 100 if (1) it is incompletely digestible, as is true of keratin; (2) it is a protein of plant origin surrounded by cellulosic husks, as is true of cereal grains; (3) it is deficient in one or more essential amino acids. In the last case, large quantities of the protein would have to be ingested to obtain enough of the essential amino acid, whereas the amino acids in abundance would be used calorically (Box 12.1).

A somewhat different but experimentally useful protein classification is called the **chemical score,** obtained by completely hydrolyzing the protein and comparing its amino acid composition with that of human milk. The biological values and chemical scores of some food proteins are listed in Table 18.5.

Fatty Acids

Fatty acids containing more than one unsaturated bond past carbon 9 of a saturated chain, counting from the carboxyl end, cannot be synthesized by humans. Therefore, two polyunsaturated fatty acids of plant origin, linoleic acid and linolenic acid, are essential to human nutrition. They are used by mammals to synthesize arachidonic acid, which in turn is used to synthesize leucotrienes, thromboxanes, and prostaglandins, a family of lipid-soluble organic acids that have hormonelike physiological functions (Section 11.8 and Figure 11.5). Deficiencies in these fatty acids are rare, because they are present in abundance in edible plants and in fowl and fish.

Although the caloric values of saturated and unsaturated fatty acids are comparable, the proportion of saturated to unsaturated fatty acids in the diet has significant physiological consequences. A great deal of evidence has accumulated over many years that correlates decreased concentrations of high-density lipoproteins, increased concentrations of low-density lipoproteins, and total blood cholesterol with diets that are rich in saturated fatty acids. The studies also relate such diets to a predisposition to develop coronary artery disease. For this reason, experts are urging the people of developed countries (where foods tend to be high in saturated fatty acids) to increase the proportion of unsaturated fatty acids in the diet (Box 16.1). The typical compositions of various plant oils and animal fats are listed in Table 11.2.

British sailors came to be called "limeys" because they consumed limes to prevent scurvy. (Culver Pictures.)

Vitamins

Many vitamins—vitamin C (ascorbic acid), for example—were discovered when a disorder caused by their absence from the diet was cured by their addition to the diet. One of the earliest documentations of a vitamin deficiency appears in the journals of Jacques Cartier, who explored North America in 1535. He described a disease that came to be known as scurvy, which was manifested in his sailors as terrible skin disorders accompanied by tooth loss. Two hundred years later, a British physician found that he could cure scurvy by adding citrus fruits such as lemons and limes to the diet and, 200 years after that, in 1932, the antiscurvy vitamin was isolated and given the name vitamin C.

Another deficiency disease, known as beriberi, is characterized by neurological disorders. Originally thought to be an infectious disease, beriberi was unknown until the early nineteenth century, when rice-polishing machines were invented to remove the brown outer hull of the rice seed. The cure of beriberi was discovered when the addition of the outer hull of rice to a victim's diet completely reversed the symptoms. The critical dietary component found in rice hulls is thiamine, the coenzyme in decarboxylations. The blood of people with a thiamine deficiency contains elevated levels of pyruvate, which must be decarboxylated before it can enter the tricarboxylic acid cycle.

Other vitamins are known as growth factors, because experiments showed that test animals do not grow if certain substances other than carbohydrates, triacylglycerols, and proteins are omitted from the diet. Vitamins were chemically analyzed, and, in many cases, their metabolic roles as enzyme cofactors were identified. In other cases, the vitamin's precise biochemical function is yet to be made clear.

Vitamins, both water soluble and fat soluble (Section 11.9), can be divided into two groups that depend on the effects of their deficiencies. The first group includes thiamine, riboflavin, niacin, ascorbic acid, and folic acid. In affluent countries, marginal deficiencies of these vitamins are common but, in many parts of the world, the deficiencies are great enough to be life threatening. The other group of vitamins includes pyridoxine, pantothenic acid, biotin, vitamin B_{12}, and the fat-soluble vitamins A, D, E, and K. Deficiencies of these vitamins are rare. However, people suffering from fat-absorption disorders are deficient in the fat-soluble vitamins; vitamin A deficiency is the cause of xerophthalmia, or night blindness, owing to an insufficient synthesis of the visual pigment rhodopsin. Pyridoxine is needed for transaminations; so the body's requirement depends on the quantity of protein in the diet—the more dietary protein, the greater the need for pyridoxine. Biotin, pantothenic acid, and vitamin B_{12} are not ordinarily required in the diet, because they are usually synthesized in adequate amounts by intestinal bacteria. Nevertheless, a diet rich in egg-white protein can cause a serious biotin deficiency, because egg white contains the protein avidin, which binds very strongly to biotin to form an avidin–biotin complex that cannot be absorbed. Vitamin B_{12} deficiency, resulting in pernicious anemia, does occasionally occur, and its likely cause is the absence of **intrinsic factor,** a glycoprotein synthesized by the stomach. Vitamin B_{12} must be transported across the intestinal-cell membrane as a complex with intrinsic factor. In people who cannot synthesize this protein, the vitamin must be administered by injection directly into the bloodstream.

TABLE 18.6 **Vitamin Needs of Men from 23 to 50 Years of Age**

Vitamin	Coenzyme form, where known	Metabolic role or associated deficiency disease or both, where known	Recommended daily allowance
thiamine	thiamine pyrophosphate	decarboxylation coenzyme; deficiency causes beriberi	1.5 mg
niacin	NAD^+	dehydrogenase coenzyme; deficiency causes pellagra	19 mg
ascorbic acid	unknown	unknown; deficiency causes scurvy	60 mg
riboflavin	FAD	dehydrogenase coenzyme	1.7 mg
pyridoxine	pyridoxal phosphate	transamination coenzyme	2.2 mg
folic acid	tetrahydrofolate	one-carbon-group transfer; deficiency causes anemia	400 μg
pantothenic acid	coenzyme A	fatty acid oxidation	5–10 mg
biotin	biocytin	CO_2-transferring enzymes	150 μg
vitamin B_{12}	deoxyadenosylcobalamine	odd-numbered fatty acid oxidation; deficiency causes pernicious anemia	3 μg
vitamin A_1	unknown	visual-cycle intermediate; deficiency causes night blindness (xerophthalmia)	1 mg
vitamin D_3	1,25-dihydroxycholecalciferol	hormone controlling calcium and phosphate metabolism; deficiency causes rickets	10 μg
vitamin E	unknown	protects against damage to membranes by oxygen; deficiency causes liver degeneration	10 mg
vitamin K_1	unknown	activation of prothrombin; deficiency causes disorders in blood clotting	1 mg

The fat-soluble vitamins are stored in body fat and therefore need not be ingested daily. However, the water-soluble vitamins are excreted or destroyed in the course of metabolic turnover and must be replaced by regular ingestion. Table 18.6 lists the known essential vitamins (both the fat soluble and the water soluble), their coenzyme forms if known, their physiological functions, and the recommended daily allowance (RDA) for men between 23 and 50 years of age.

Minerals

Carbohydrates, proteins, triacylglycerols, and nucleic acids are composed of six elements: carbon, hydrogen, nitrogen, oxygen, phosphorus, and sulfur. In addition to these elements, many other minerals are required for experimental mammals and presumed to be required for humans. These minerals are divided into two groups—bulk and trace minerals—and are presented in Table 18.7. Table 18.8 lists some minerals whose functions are known or whose deficiencies result in well-recognized symptoms.

TABLE 18.7 Minerals Required by Humans

BULK ELEMENTS*		
calcium	magnesium	potassium
chlorine	phosphorus	sodium
TRACE ELEMENTS†		
copper	molybdenum	nickel‡
fluorine	selenium	silicon‡
iodine	zinc	tin‡
iron	arsenic‡	vanadium‡
manganese	chromium‡	

*These minerals are required in doses higher than 100 mg/day.
†These minerals are required in doses of 1 to 3 mg/day.
‡These minerals are known to be required in test animals and are likely to be required in humans.

TABLE 18.8 Minerals and Their Nutritional Functions

Element	Nutritional function
calcium	bones, teeth
phosphorus	bones, teeth
magnesium	cofactor for many enzymes
potassium	water, electrolyte, acid–base balance; location is intracellular
sodium	water, electrolyte, acid–base balance; location is extracellular
iron	iron porphyrin proteins
copper	iron porphyrin synthesis and cytochrome oxidase
iodine	synthesis of thyroxin; lack leads to goiter
fluorine	forms fluoroapatite, strengthens bones and teeth
zinc	cofactor for many enzymes
tin	growth factor for mammals grown in ultraclean conditions*
nickel	growth factor for mammals grown in ultraclean conditions*
vanadium	growth factor for mammals grown in ultraclean conditions*
chromium	growth factor for mammals grown in ultraclean conditions*
silicon	growth factor for mammals grown in ultraclean conditions*
selenium	component of the enzyme glutathione peroxidase
molybdenum	component of the enzymes xanthine and aldehyde oxidases

*Probably required under normal conditions.

18.3 NUTRIENT TRANSPORT

All the cells of the human body depend on the blood to bring them oxygen and nutrients and to carry CO_2, the principal waste product of metabolism, away. As the chief transport system of the body, the blood is also responsible for the overall coordination of metabolic processes in the various organs and tissues.

The Composition of Blood

The blood's composition is extremely complex, owing to its role in the coordinated operation and integration of all the body's organ systems. For this reason, routine blood tests provide many clues that enable physicians to diagnose pathological conditions.

TABLE 18.9 **Components of Human Blood Plasma**

Inorganic components	Organic metabolites	Plasma proteins
buffers { $NaHCO_3$, Na_2HPO_4 }	glucose	serum albumin
	amino acids	VLDLs
NaCl	lactate	LDLs
$CaCl_2$	pyruvate	HDLs
$MgCl_2$	ketone bodies	immunoglobulins
KCl	citrate	fibrinogen
Na_2SO_4	urea	prothrombin
	uric acid	specialized transport proteins
	creatinine	

Abbreviations: VLDL, very low density lipoprotein; LDL, low-density lipoprotein; HDL, high-density lipoprotein.

Whole blood consists of about a 50:50 volume ratio of cells to a liquid fraction called plasma. The cells are mostly erythrocytes (red blood cells), leukocytes (white blood cells), and platelets (clotting cells). Although the plasma is not as complex as the blood's cellular contents, it contains a wide variety of dissolved inorganic components, organic metabolites, waste products, and so-called plasma proteins.

Plasma is 90% water and 10% solutes. The composition of the solute fraction is 70% plasma proteins, 20% organic metabolites, and 10% inorganic salts. The major components of each of these categories are listed in Table 18.9. One class of plasma proteins, the immunoglobulins, is a group of structurally related proteins that may consist of hundreds of different types. Transferrin, a substance that transports iron, is an example of a number of specialized transport proteins.

The total plasma protein concentration is between 8.6 and 11.7 g/dL. The major protein components of plasma, their normal concentrations, and their chief functions are listed in Table 18.10, and the concentrations of the

TABLE 18.10 **Principal Protein Fractions of Blood Plasma**

Protein component of plasma	Normal concentration range (mg/dL)	Function
serum albumin	3500–4500	osmotic regulation of blood volume; transport of fatty acids
α_1-globulins	300–600	transport of lipids, thyroxine, and ACTH
α_2-globulins	400–900	transport of lipids and copper
β-globulins	600–1100	transport of lipids, iron, and hemes; antibody activity
γ-globulins	700–1500	almost all circulating antibodies
fibrinogen	3000	fibrin precursor for blood clotting
prothrombin	100	precursor of thrombin required for blood clotting

Abbreviation: ACTH, adrenocorticotrophic hormone.

TABLE 18.11 Normal Concentrations of Organic Substances in Blood Plasma

Nonprotein plasma component	Normal concentration range (mg/dL)
CARBOHYDRATES	
glucose	70–90
fructose	6–8
ORGANIC ACIDS	
lactate	8–17
pyruvate	0.4–2.5
ketone bodies	1–4
citrate	1.5–3.0
NITROGENOUS COMPOUNDS	
amino acids	35–65
urea	20–30
uric acid	2–6
creatinine	1–2
LIPIDS*	
total lipids	300–700
triacylglycerols	80–240
cholesterol and esters	130–240
phospholipids	160–270

*All lipids in blood plasma are bound to proteins.

nonprotein components of plasma are listed in Table 18.11. Among the latter are urea, uric acid, and creatinine, the chief nitrogenous waste products of metabolism excreted by the kidneys.

The blood communicates with the external environment through the kidneys and the lungs. Nonvolatile substances are excreted through the kidneys, and volatile substances and gases are excreted through the lungs.

The Transport of Lipids

The hydrophobic lipids (cholesterol and triacylglycerols) are insoluble in blood and must be packed inside of soluble, spherical **lipoprotein particles** to be transported from one tissue to another through the bloodstream. The lipoprotein particles' structure is reminiscent of the structure of the micelles formed when soap combines with hydrophobic substances. Each lipoprotein particle consists of a core of hydrophobic lipids surrounded by a shell of amphipathic lipids (phospholipids and cholesterol, mostly in the form of cholesteryl esters) and proteins (Figure 18.2 on the following page). The amphipathic lipids and proteins have their hydrophobic ends oriented inward toward the hydrophobic lipids and their hydrophilic ends oriented outward toward the aqueous plasma.

The lipoproteins differ in the relative amounts of lipid and protein that they contain, as well as in the relative amounts of different lipids

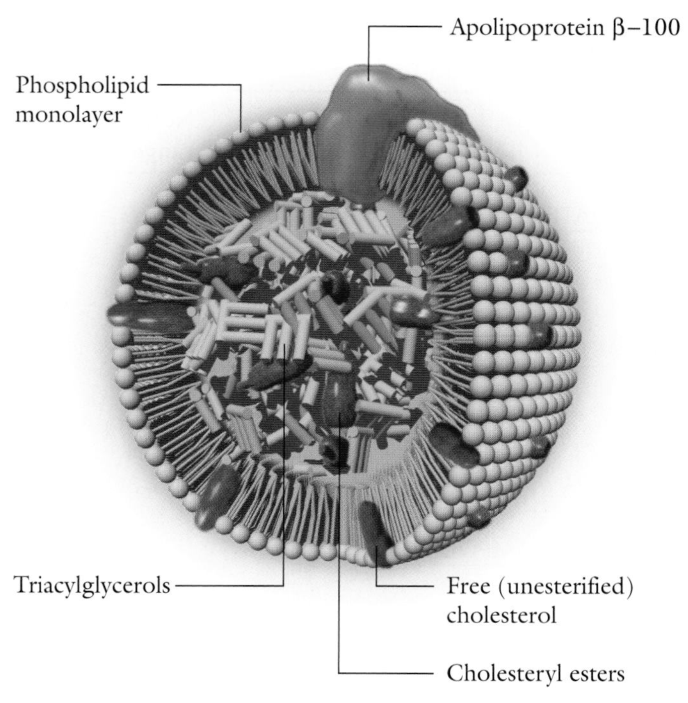

Figure 18.2 Structure of low-density lipoprotein (LDL). The protein is apolipoprotein β-100, which recognizes receptor sites on adipocytes.

(Table 18.12). These differences result in differences in density, by which the lipoproteins are classified. The least-dense lipoproteins are the **chylomicrons,** followed by **very low density lipoproteins (VLDLs), low-density lipoproteins (LDLs),** and **high-density lipoproteins (HDLs).** A higher protein content results in higher density, because protein is more dense than lipid. Normal, borderline high, and high levels for blood lipids are listed in Table 18.13.

Each lipoprotein has a specific function based on its point of synthesis, lipid composition, and protein composition. The proteins in lipoproteins have several roles: solubilization of lipids, targeting of specific lipoproteins to specific tissues (by cell recognition between those proteins and the glycoproteins or glycolipids of cell membranes), and activation of enzymes that hydrolyze and unload lipids from lipoproteins. As lipoproteins circulate through the liver and other tissues in the body, cell recognition between proteins of the lipoproteins and receptors on the capillary membranes results in the binding of the lipoproteins to the cells. The binding of lipoproteins is followed by the transfer of lipid between lipoproteins and cells.

Chylomicrons, which transport dietary (exogenous) triacylglycerols from the intestine to adipose tissue and cholesterol to the liver, are assembled in

TABLE 18.12 Density and Composition of Lipoproteins

	Chylomicron	VLDL	LDL	HDL
Density (g/mL)	<0.95	0.95–1.006	1.006–1.063	1.063–1.210
Composition (wt %)				
cholesterol	1	7	8	2
cholesteryl ester	3	13	38	15
triacylglycerol	85	51	10	4
phospholipid	9	19	20	24
protein	2	10	24	55

TABLE 18.13 **Lipid Levels in Blood**

	Concentration (mg/dL)		
	Desirable	Borderline high	High
cholesterol*	160	200–240	>240
low-density lipoprotein	<130	130–160	>160
high-density lipoprotein	>35		
triacylglycerol	<200	200–400	>400

*Includes cholesteryl esters.

intestinal cells and carried through the lymph system to the bloodstream. The enzyme lipoprotein lipase, located on the extracellular sides of the membranes of muscle and adipose cells, is activated by proteins on the chylomicrons. This enzyme hydrolyzes the triacylglycerols to fatty acids and glycerol, which are then taken up by the cells in the target tissues. (The fatty acids not oxidized for energy are reesterified for storage as triacylglycerols in adipose cells.) Devoid of triacylglycerols but still rich in cholesterol, the chylomicrons, now called **chylomicron remnants,** are taken up by the liver.

Very low density lipoproteins transport cholesterol and triacylglycerols synthesized in the liver to other tissues. Cholesterol delivered to the liver by chylomicron remnants also is packed into VLDLs. The VLDLs deliver triacylglycerols to extrahepatic tissue (other-than-liver tissue) by using the same mechanisms as chylomicrons do. As the lipoproteins unload their triacylglycerols, lipoprotein density increases, and the VLDLs become low-density lipoproteins. LDLs transport cholesterol from the liver to other cells for the construction of cell membranes and the synthesis of steroid hormones. The LDLs then return to the liver for reprocessing. Recognition between proteins on LDLs and LDL receptors on cell membranes is critical to the uptake of LDLs and the unloading of cholesterol in liver cells, as well as all other cells.

High-density lipoproteins, synthesized in the liver, remove cholesterol from dying cells and from membranes undergoing turnover in cells outside the liver. The cholesterol is returned to the liver, both directly by HDLs and through transfer to LDLs. Excess cholesterol is converted into bile acids, which are secreted into the intestine and then excreted in the feces.

The cholesterol level in the blood results from the interaction of the different types of lipoproteins with one another and with liver and other cells and from the regulation of the biosynthesis of cholesterol in the liver. Cholesterol synthesis in cells is regulated through a feedback mechanism: cholesterol uptake from the bloodstream shuts down the synthesis of the enzyme HMG-*S*-CoA reductase required for cholesterol biosynthesis (Section 16.4).

Fatty acids liberated from adipose-tissue cells by glucagon-sensitive cellular lipase are transported through the blood in the form of serum albumin–fatty acid complexes to other tissues for oxidation.

The Transport of Oxygen

Oxygen is required by respiring cells, and carbon dioxide is generated by those cells. Simple diffusion would not allow tissues buried deep within a multicellular organism to obtain atmospheric oxygen or to dissipate carbon dioxide at a rate sufficient to sustain metabolism. Circulatory systems and respiratory proteins such as hemoglobin solved the diffusion problem. Now we will see how they work.

In terrestrial animals, oxygen enters the circulatory system through the respiratory membranes of the lungs. It diffuses across the membranes passively,

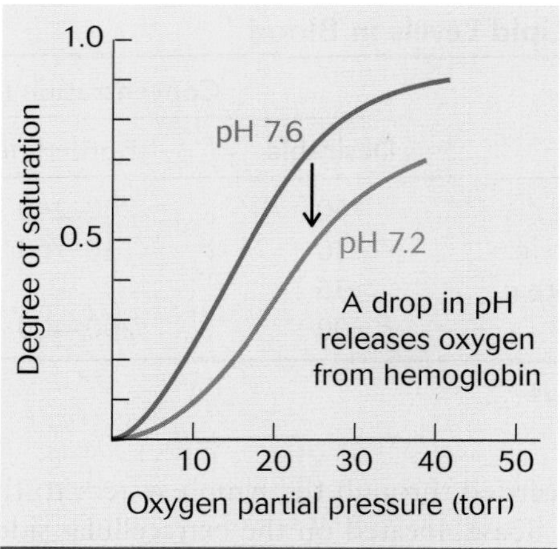

Figure 18.3 Oxygen-saturation curves for hemoglobin at two values of pH.

under a gradient in concentration; that is, the concentration of oxygen in the alveolar spaces of the lung is greater than that in the blood plasma. The solubility of oxygen in plasma is too low for the plasma alone to absorb enough oxygen to fuel aerobic metabolism. However, sufficient oxygen is carried to the tissue level by another mechanism.

Oxygen is transported to the tissues not by plasma but by the erythrocytes of the blood—the red blood cells. The blood of an adult human has a volume of 5 to 6 L. About half of this volume is composed of erythrocytes, a mass equivalent to that of the liver. Erythrocytes are very small, degenerate cells containing no nuclei, mitochondria, or any other subcellular organelle. Their only metabolic fuel is glucose, and their metabolic engine consists only of glycolysis, which is their sole source of ATP. Their chief function is to transport oxygen from the lungs to the tissues and carbon dioxide from the tissues to the lungs. They contain large quantities of the iron-containing protein hemoglobin, which is approximately 90% of the erythrocyte's protein content.

Oxygen combines with hemoglobin. In doing so, the oxygen is not dissolving into the blood but instead is complexed reversibly with the Fe(II) of the heme in the hemoglobin protein (Section 12.7). This complex increases the blood concentration of oxygen to such an extent that 100 mL of whole blood carries about 21 mL of oxygen, about 50 times the amount that dissolves in the plasma.

The oxygen-binding curves for hemoglobin at two values of pH are presented in Figure 18.3. The y-axis denotes the degree of hemoglobin's saturation with oxygen. If we designate hemoglobin as Hb and oxygenated hemoglobin as HbO_2, the percentage of saturation of hemoglobin with oxygen is defined as

$$\text{Percent saturation} = \left(\frac{[HbO_2]}{[Hb + HbO_2]}\right) \times 100\%$$

The x-axis shows the partial pressure of oxygen. The oxygen partial pressure at 50% saturation is known as the $\boldsymbol{P_{50}}$.

The curves in Figure 18.3 have a characteristic S-shape, known as a **sigmoid shape,** that has an important physiological significance. Each molecule of hemoglobin consists of four polypeptide subunits, and each subunit has an iron-bearing heme molecule that binds one molecule of oxygen (Section 12.7). The sigmoid shape indicates that the oxygen molecule bound to the first subunit increases the binding affinity of the remaining subunits for the next molecules of oxygen to be bound. In other words, the first oxygen bound increases the capacity of hemoglobin to bind more oxygen: this characteristic is called **cooperative binding.** The extent of binding increases as the partial pressure of oxygen increases until a maximum is reached and no more can be bound. That maximum partial pressure is about equal to the oxygen tension at the lungs.

Concept check

✓ The steepness of the sigmoid oxygen-binding curve shows that oxygen can be loaded (bound) and unloaded (dissociated) over a narrow range of oxygen tensions.

The amount of oxygen that combines with hemoglobin depends not only on the partial pressure of the oxygen present, but also on the pH, the partial pressure of CO_2, and the presence of a compound, 2,3-bisphosphoglycerate, generated by the glycolytic apparatus of the erythrocyte. 2,3-Bisphosphoglycerate lowers the affinity of hemoglobin for oxygen (Section 12.7). In solution, the P_{50} for hemoglobin is 1 torr. In the presence of 2,3-bisphosphoglycerate, the P_{50} increases to 26 torr. This increase is helpful in situations where the external oxygen concentration is low. For example, at 15,000 feet, the altitude of Lake Titicaca in Peru, hemoglobin cannot be fully saturated. Therefore, at the partial pressure of oxygen at the tissue level, only a small part of the bound oxygen could be unloaded into the cells. The erythrocytes of people living at that altitude contain increased levels of 2,3-bisphosphoglycerate, lowering the oxygen affinity of hemoglobin enough so that a significant fraction of its oxygen can be unloaded at the tissue level. When people move from low to high altitudes, a hormone called erythropoietin is synthesized by the kidneys. This hormone initiates the synthesis of erythrocytes, thus increasing the oxygen-carrying capacity of the blood. The administration of erythropoietin is also used to alleviate clinical states of anemia.

The pH of the blood has an additional significant effect on the oxygen-carrying properties of hemoglobin. If we use the abbreviation **HHb** to represent the deoxygenated form of hemoglobin, a Brønsted–Lowry acid, and HbO_2^- to represent the oxygenated form, a Brønsted–Lowry base—and if we print hydrogen ions in color—the oxygenation of hemoglobin can be represented by the following equilibrium:

$$HHb + O_2 \rightleftharpoons HbO_2^- + H^+$$

The reversibility of this system means that, if the pH is raised, more oxygen will be bound. If the pH is lowered, oxygenated hemoglobin will give up its oxygen. This effect is known as the **Bohr effect,** named after its discoverer, Christian Bohr, the father of Niels Bohr. The effect of pH on the oxygen-binding curves of hemoglobin in Figure 18.3 can be understood if we follow

This child is acclimatized to living at the high altitude of Lake Titicaca in Peru. (Kenneth Murray/Photo Researchers.)

the series of events taking place from the time of loading in the lungs to the unloading of oxygen at the tissue level.

Oxygen is bound, or loaded, by hemoglobin in the lungs at an oxygen tension of about 100 torr. There the CO_2 concentration is low and, consequently, the pH is relatively high, about 7.6. At that pressure, the hemoglobin becomes almost completely saturated. The oxygen is then carried to the tissues, where the CO_2 concentration is high, and the pH is lowered to about 7.2 to 7.3. The oxygen concentration is also much lower in the tissues (an oxygen tension from about 25 to 40 torr) than in the lungs; so the unloading of oxygen (dissociation from hemoglobin) is favored. At the cell level, the partial pressure of oxygen is about 40 torr, and the binding curve is steep. Therefore, over a narrow range of oxygen concentrations, hemoglobin cycles between about 60% and 95% saturation. Thus, the acid–base properties of hemoglobin allow a very efficient delivery of oxygen at the tissue level.

The fact that dissociation of hydrogen ion is caused by the oxygenation of hemoglobin greatly affects events within the lungs. When CO_2 arrives at the lungs, it is in the form of HCO_3^-; the H^+ from the dissociation of H_2CO_3 in the erythrocytes, is bound to hemoglobin. As O_2 is loaded, H^+ ions are released from the newly oxygenated hemoglobin, and the following reaction takes place:

$$HCO_3^- + H^+ \rightleftharpoons H_2CO_3 \rightleftharpoons CO_2 + H_2O$$

Because the oxygenation of hemoglobin makes hydrogen ion available, the binding of oxygen at the lungs increases the efficiency of the release of CO_2 to the atmosphere.

The effect of CO_2 on the oxygenation equilibrium of hemoglobin will be considered next.

The Transport of Carbon Dioxide

The CO_2 generated in the tissues diffuses under a concentration gradient into the **interstitial space**—the space between cells—and then across the capillary walls into the blood, where it dissolves in the plasma. The bulk of the CO_2 produced by cellular metabolism arrives at the red blood cell in the dissolved state, $CO_2(aq)$. Erythrocytes possess the enzyme carbonic anhydrase, which rapidly catalyzes the formation of HCO_3^- from the incoming CO_2. Only about 0.5% of the CO_2 dissolved in plasma is in the form of H_2CO_3, and, of this amount, only a small proportion forms plasma HCO_3^- and H^+. The H^+ thus formed in the plasma is bound by the plasma proteins, and so there is no shift in blood pH.

Carbonic anhydrase, utilizing zinc ion as a cofactor, catalyzes the reaction between CO_2 and an OH^- ion from water to form HCO_3^- and leave an H^+ ion:

$$CO_2(aq) + OH^-(aq) + H^+(aq) \rightleftharpoons HCO_3^-(aq) + H^+(aq)$$

We write the equation with H^+ ion on both sides to emphasize the fact that CO_2 does not react directly with water but with one of the dissociation products of water, OH^-, and leaves the other product, H^+, behind. The enzyme increases the rate of HCO_3^- formation by a factor of 1×10^7 (Section 14.3). At body temperature, the solubility of $NaHCO_3$ in water is about 109 g/L, compared with a solubility of only about 0.12 g/L for CO_2 at its tissue partial pressure of 40 torr. The body's strategy of converting most of the CO_2 in the blood into HCO_3^- and transporting it in that form greatly increases the amount of CO_2 that the blood is able to carry.

The bicarbonate ion concentration is greater in the red blood cell than in the plasma because of its continuous high rate of formation in red blood cells. Therefore bicarbonate tends to diffuse out of the red blood cell into the plasma. Under most circumstances, the HCO_3^- anion would diffuse across the membrane only if a cation accompanied it (to maintain the balance of charge on both sides of the membrane); however, the red-blood-cell membrane is not permeable to K^+, the cation found within these cells. An equivalent alternative mechanism, known as the **chloride shift,** is used instead, in which an HCO_3^- anion is exchanged across the cell membrane with a Cl^- anion that diffuses into the cell from the plasma.

Now let us turn our attention to the proton generated by the formation of HCO_3^- in the erythrocyte. After its oxygenation in the lungs, hemoglobin is carried to the tissues in the form of HbO_2^-. Within the red blood cell, its negative charge is balanced by the K^+. When the red blood cells begin to absorb CO_2 at the tissues, the proton generated by the consequent formation of HCO_3^- reacts with the HbO_2^- formed at the lungs:

$$H^+ + HbO_2^- \rightleftharpoons HHb + O_2$$

In this way, the protons generated at the tissues enhance the unloading there of oxygen from the oxygenated hemoglobin formed in the lungs.

The carbon dioxide generated at the tissue level also has an effect on the unloading of oxygenated hemoglobin. A significant amount—about 20%—of the CO_2 in blood is carried in **carbamate** compounds formed by the reaction of CO_2 with the amino groups of hemoglobin. The general reaction of CO_2 with any $—NH_2$ group can be represented as

$$R—NH_2 + CO_2 \rightleftharpoons \underset{\text{Carbamate}}{R—NH—COO^-} + H^+$$

When CO_2 reacts in this way with the amino groups of oxygenated hemoglobin, it causes the dissociation of the oxygen:

$$CO_2 + HbO_2^- \rightleftharpoons HbCO_2^- + O_2$$

Therefore, the direct loading of CO_2 as a carbamate compound on hemoglobin results in the additional unloading of O_2 at the tissue level.

Finally, the equilibrium between oxygenated and deoxygenated hemoglobin is sensitive to pH. When the pH decreases at the cell level because of the high concentration of CO_2, O_2 leaves its bound form and can diffuse into the tissue under its concentration gradient.

The processes that take place in the red blood cell at the tissue level are reversed at the lungs, as shown in the following reaction equations. In these sequences of equations, protons generated in the red blood cell at the tissue level and eliminated in the lungs are identified by color. Remember that the pH is about 7.2 at the tissue level and about 7.6 at the lungs.

$$\begin{aligned} \text{Tissue level:}\quad & CO_2 + H_2O \rightleftharpoons H^+ + HCO_3^- \\ & H^+ + HbO_2^- \rightleftharpoons HHb + O_2 \\ \text{Lungs:}\quad & HHb + O_2 \rightleftharpoons HbO_2^- + H^+ \\ & H^+ + HCO_3^- \rightleftharpoons CO_2 + H_2O \end{aligned}$$

The flow of CO_2 represented in these equations—from the tissues to the external environment—is entirely through diffusion. No active-transport mechanism is required. The direction and rate of flow are completely dependent on the difference in concentrations of CO_2 between the internal and the external environment.

The HCO_3^- ion is one-half of the Brønsted–Lowry conjugate acid–base pair H_2CO_3–HCO_3^-. However, the red blood cell contains very little H_2CO_3, because of the action of carbonic anhydrase. As a consequence, the concentration of HCO_3^- in the plasma depends on dissolved CO_2, not H_2CO_3. The reactions of this conjugate pair at the tissue level and at the lungs constitute the bicarbonate buffer of the blood (Section 2.5):

$$\begin{aligned} \text{Tissue level:}\quad & CO_2 + H_2O \rightleftharpoons H^+ + HCO_3^- \\ \text{Lungs:}\quad & H^+ + HCO_3^- \rightleftharpoons CO_2 + H_2O \end{aligned}$$

Remember that both reactions are catalyzed by carbonic anhydrase. If they were not, then the $CO_2 + H_2O$ appearing in both equations could be replaced by H_2CO_3, and the more familiar Brønsted–Lowry conjugate acid–base pair would be present. Dissolved CO_2 takes the place of H_2CO_3 in the Brønsted–Lowry scheme (Section 2.4). The reactions in oxygen and carbon dioxide transport and the anatomical locations of these reactions are summarized in Figure 18.4.

An important point is that acid generated at the tissue level is eliminated by the exhalation of CO_2. If CO_2 is not removed from the lungs rapidly enough, the system "backs up": hydrogen ion will not be removed by reaction with HCO_3^-, and the blood pH will fall, a condition known as **respiratory acidosis.** When other physiological events cause a lowering of blood pH, the condition is called **metabolic acidosis.** The healthy body's response to acidosis is to increase the rate of breathing (hyperventilation). Chemoreceptors in the carotid artery, sensing a rise in levels of CO_2 and H^+ ion, send a message to increase the activity of the respiratory center of the brain.

The reverse problem, in which blood pH is excessively high, is called alkalosis. Victims of emphysema run the risk that their rapid, shallow breathing will eliminate so much CO_2, and therefore H^+ ion, that they will begin to suffer from **respiratory alkalosis.** When other physiological conditions are responsible, the condition is called **metabolic alkalosis.** One example is excessive vomiting. The loss of H^+ from the stomach causes a flow of replacement H^+ from the plasma, with a consequent rise in blood pH. The healthy body's response to such a rise in blood pH is to retain as much CO_2 as possible by reducing the breathing rate, or hypoventilating.

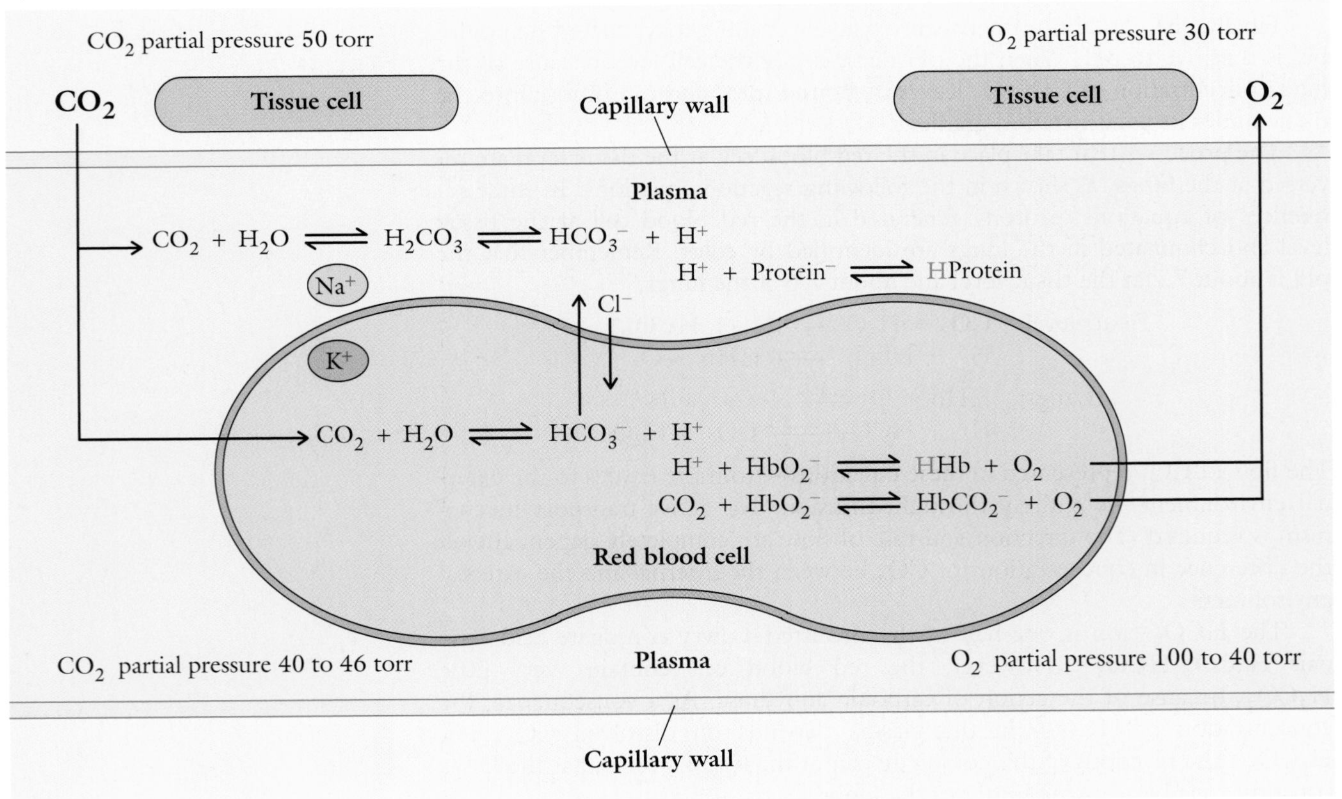

Figure 18.4 The reactions responsible for the transport of oxygen and carbon dioxide in the blood.

18.4 METABOLIC CHARACTERISTICS OF THE MAJOR ORGANS AND TISSUES

The cells of different organs have different metabolic requirements. Some use all three principal nutrients—glucose, fatty acids, and amino acids. Others may use only glucose or principally fatty acids. A good first step toward learning how metabolism is coordinated is to take a metabolic inventory of the major organ systems and tissues.

The Heart

The heart contains almost no stored energy (glycogen or triacylglycerol) and, although it can use a variety of metabolic fuels (glucose, fatty acids, lactate, and ketone bodies), most of its energy is derived from the oxidation of fatty acids. As much as 50% of heart-muscle-cell volume is taken up by mitochondria carrying out oxidative processes that lead to the production of ATP. Because heart muscle, unlike skeletal muscle, cannot function anaerobically for a brief period, a lack of oxygen results immediately in cell death. When oxygenated blood is prevented from reaching the heart because of blockage in blood vessels, a process called **myocardial infarction** (heart attack) occurs (Box 16.1).

Skeletal Muscle

Muscle can utilize a variety of fuels, but the choice depends on the muscle's degree of activity. The principal energy source of resting muscle is fatty acids. When exertion begins, glycogen reserves are mobilized to provide glucose in the form of glucose-6-phosphate. (About 75% of the body's glycogen supply is located in muscle, but, because muscle cells do not possess glucose-6-phosphatase, they cannot provide free glucose to other organs.) The rate of glycolysis, which produces lactate, is much greater than the rate of the citric acid cycle, which breaks it down; so lactate accumulates and is released into the blood. Alanine produced from pyruvate by transamination also is released (see Section 17.3), and both products are transported to the liver to be converted into glucose through gluconeogenesis. The glucose from liver is then transported back to the muscle as well as other tissues.

In addition to glycogen, muscle possesses a second energy-storage depot, **creatine phosphate.** This high-energy compound can phosphorylate ADP to produce ATP and thus supply energy for a short time in periods of extreme exertion. Its breakdown product, creatinine, is a normal component of urine.

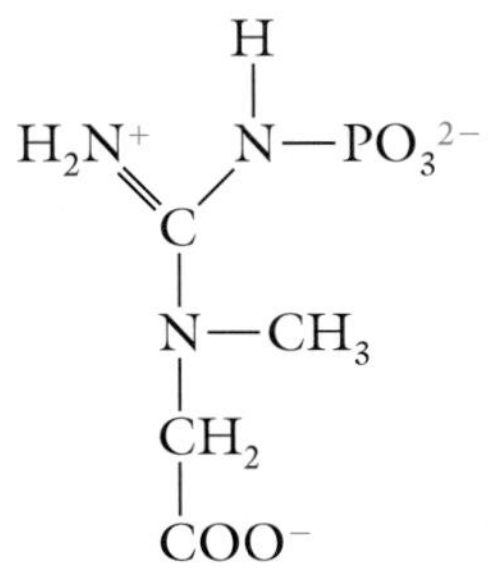

$$\mathrm{H_2N^+{=}C(-NH{-}PO_3^{2-})-N(CH_3)-CH_2-COO^-}$$

Creatine phosphate

Adipose Tissue

Adipose tissue consists of cells called **adipocytes.** About 17% of an average human male weighing 70 kg consists of triacylglycerol stored in adipose tissue. This stored triacylglycerol represents about 110,000 kcal (450,000 kJ) of stored energy—enough to sustain life for a few months.

When chylomicrons from intestinal absorption reach adipocytes, they are acted on by the lipoprotein lipase on the cell surface. The fatty acids freed by this action are either complexed with serum albumin for transport to other tissues for oxidation or absorbed into the adipocyte for storage.

The breakdown of triacylglycerols within adipocytes depends on the activity of a hormone-sensitive lipase. This internal lipase is activated by glucagon, which activates a phosphorylation cascade, as in glycogenolysis. The hormone insulin reverses this stimulation. As a result, when glucose levels are high, the rate of synthesis of triacylglycerols exceeds the rate of breakdown; however, when the glucose level falls and glucagon levels increase, triacylglycerol breakdown exceeds synthesis and fatty acids are released from the cell. These free fatty acids combine with the blood protein serum albumin to form soluble complexes that are thus able to travel through the blood to other tissues. The glycerol from the hydrolysis of triacylglycerols also enters the blood and travels to the liver, where it is used to produce glucose through gluconeogenesis.

The synthesis of triacylglycerols requires the presence of glycerol-3-phosphate as well as fatty acids. Glycerol-3-phosphate is produced by the reduction of dihydroxyacetone phosphate from glycolysis. The supply of glycerol-3-phosphate is controlled by the cell's concentration of glucose. Insulin stimulates the uptake of glucose in adipose tissue, and triacylglycerol synthesis will take place as long as the glucose supply is adequate. Lipid transport and the role of insulin are considered further in Section 18.6.

Kidneys

About 80% of the oxygen consumed by the kidneys generates the ATP used to pump Na^+ into the interstitial space surrounding the kidney tubules. The high ion concentration thus created outside the tubules establishes a strong osmotic gradient that withdraws water from the ultrafiltrate flowing through the tubules and therefore concentrates the urine. At the same time, other important substances, such as glucose, are actively transported from the ultrafiltrate back into the blood.

Like the heart, the kidneys work continuously and have a very active aerobic metabolism. They can use glucose, fatty acids, ketone bodies, and amino acids as metabolic fuels. All of these substances are ultimately degraded through the citric acid cycle to produce by oxidative phosphorylation the required amounts of ATP.

The kidney is an important component of the body's pH control system. For example, in starvation and in diabetes, large amounts of organic acids are produced, causing the blood's pH to decrease significantly—perhaps to a greater degree than the blood's bicarbonate buffer system could handle alone. Fortunately, the kidney has its own buffer system, consisting principally of ammonia and ammonium ion. The ammonia is generated by the deamination of amino acids in kidney cells and transported into the tubules. There it combines with excess hydronium ion to form ammonium ion. A significant amount of the ionic form of ammonia does not reenter the kidney from the tubules, and hydronium ion is therefore excreted. The body's reservoir of hydrogen ion also can be conserved. This conservation is accomplished by reducing the extent of the deamination of amino acids within kidney cells, thereby reducing the concentration of ammonia in the kidney tubules.

The Liver

With the exception of triacylglycerols, all nutrients absorbed by the intestinal tract are transported directly to the liver. There they are processed and distributed to the other organs and tissues.

Glucose Hexokinase can phosphorylate cellular glucose at the concentrations of glucose normally found in the blood—about 5 m*M*. In contrast, after a meal, the concentrations of glucose in the portal circulation of the liver can rise to about two to three times the normal concentrations. At those concentrations,

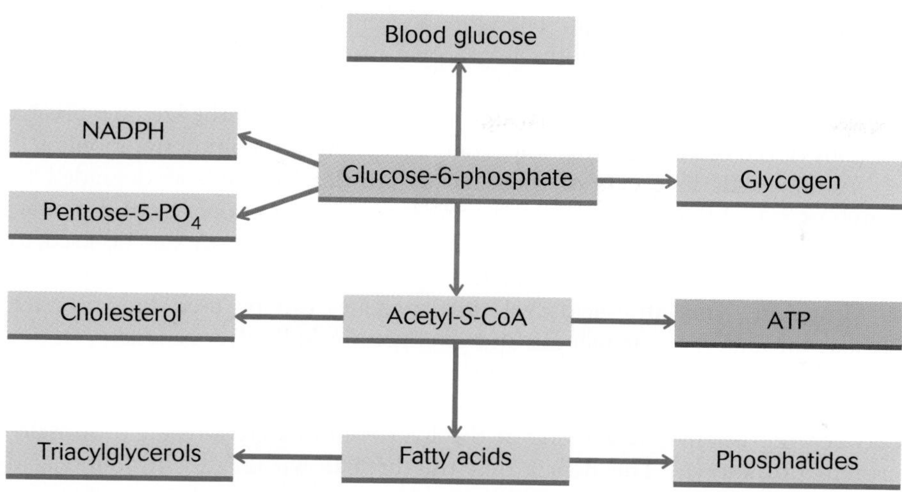

Figure 18.5 Metabolic pathways for glucose in the liver.

hexokinase cannot phosphorylate all the incoming glucose. However, liver cells possess a phosphorylating enzyme, **glucokinase,** that is not found in any other organ and is adapted to these high concentrations. Therefore all incoming dietary glucose is converted into glucose-6-phosphate by the liver.

Glucose-6-phosphate stands at the crossroads of the needs of all the body's organs. Liver cells contain glucose phosphatase—an enzyme not found in any other tissue. Therefore, when the concentration of blood glucose falls, the liver can dephosphorylate glucose-6-phosphate and supply free glucose to the blood. Glucose-6-phosphate not needed for the maintenance of blood glucose is stored as glycogen (Section 14.5).

Glucose-6-phosphate in excess of these two needs is degraded to acetyl-*S*-CoA, which is converted into malonyl-*S*-CoA for the synthesis of fatty acids and cholesterol (Section 16.5). The fatty acids are used to synthesize triacylglycerols and phospholipids. Some of the triacylglycerols and phosphatides are exported to other organs. The cholesterol is converted in part into bile salts stored in the gall bladder.

Some of the acetyl-*S*-CoA from glycolysis can be used to produce ATP through the citric acid cycle, but, normally, fatty acids are the principal oxidative fuel used by the liver. Finally, the reductive power used in fatty acid synthesis, NADPH, is produced in the liver by oxidation of glucose-6-phosphate through the pentose phosphate pathway (Section 15.1). These pathways are illustrated in Figure 18.5.

Amino acids Amino acids entering the liver after absorption in the intestines follow a number of metabolic pathways (Figure 18.6). The rate of catabolism

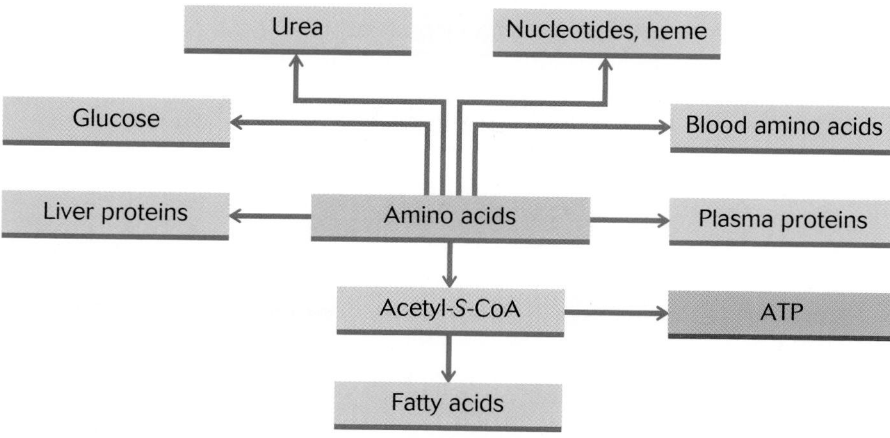

Figure 18.6 Metabolic pathways for amino acids in the liver.

of the liver's proteins is high. Therefore a significant portion of the entering amino acids is used to renew the liver's own proteins. Another portion enters the outgoing blood and travels to the other organs for biosynthesis into tissue proteins. Yet another portion is used by the liver to synthesize the plasma proteins of the blood, with the important exception of the immunoglobulins.

Amino acids in excess of those needs are deaminated and degraded to acetyl-*S*-CoA and citric acid cycle intermediates. The ammonia from deamination is converted into urea (Section 17.4). The acetyl-*S*-CoA can be used to synthesize fatty acids or can be further oxidized through the citric acid cycle to form ATP. The citric acid cycle intermediates can be used to synthesize glucose through gluconeogenesis and glycogen through glycogenesis.

The liver is also a participant in the glucose–alanine cycle (Section 17.3), which is an aid in maintaining the blood-glucose concentration in periods between meals. Alanine arriving at the liver is ultimately a result of muscle-protein breakdown. The deficit that this breakdown creates in amino acid concentration in muscle cells is corrected by nutrients from the next meal.

Amino acids are used as precursors of a variety of nitrogen-containing biomolecules, such as the heme of hemoglobin, peptide hormones, and the nucleotides used to synthesize ATP, RNA, and DNA. The degradation of the nitrogen heterocyclic bases (adenine, cytidine, and so forth) of nucleotides results in the production of uric acid.

« Nucleotide and nucleic acid structures are presented in Sections 13.1 through 13.3.

Fatty acids Fatty acids are the chief oxidative fuel of the liver (Figure 18.7). Their oxidation (Section 16.3) produces acetyl-*S*-CoA, which enters the citric acid cycle to yield ATP by oxidative phosphorylation. Acetyl-*S*-CoA in excess of that required for basic energy needs is converted into the ketone bodies acetoacetate and 3-hydroxybutyrate (Section 16.3). These ketone bodies enter the circulation to supply energy to peripheral tissues through the citric acid cycle. Ketone bodies can supply as much as one-third of the energy required by the heart. In starvation, the brain adapts to use 3-hydroxybutyrate in addition to glucose.

The bile salts necessary for fat digestion (Section 11.7) are synthesized from cholesterol, which is synthesized in the liver from acetyl-*S*-CoA derived from fatty acids and glucose. Fatty acids are also incorporated into the lipid parts of the plasma lipoproteins synthesized in the liver. The lipoproteins are critical in the transport of dietary lipids to adipose tissue for storage. Free fatty acids form complexes with serum albumin in the plasma to be transported to peripheral tissue for use as oxidative fuel.

« Box 5.3 describes some consequences of the liver's attempt to detoxify alcohol.

Detoxification The liver possesses oxidative enzyme systems specialized to detoxify nonphysiological or foreign organic substances such as drugs (including caffeine), food additives, paint-thinner vapors, and so forth. The general

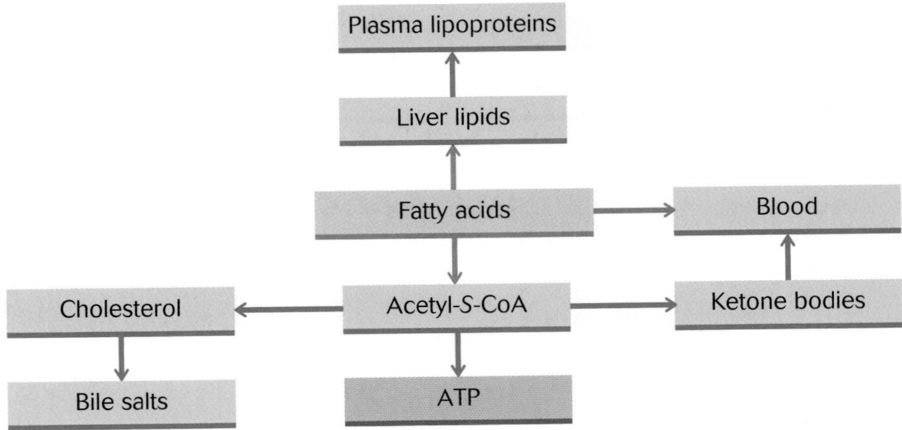

Figure 18.7 Metabolic pathways for fatty acids in the liver.

result of the oxidation is to introduce hydroxyl groups into the foreign organic molecule. The hydroxyl group is then esterified to an amphipathic molecule such as an amino acid, significantly increasing the water solubility of the foreign substance and thus enhancing the ability of the kidneys to excrete it.

18.5 CELLULAR COMMUNICATION

Hormones

Hormones can be low-molecular-mass substances (such as epinephrine), peptides (such as oxytocin), steroids (such as cortisol, Section 11.7 and Figure 11.4), or proteins (such as insulin). Many hormonal systems are governed by a part of the brain called the hypothalamus, but others, such as the digestive-system hormones, are subject to regulation within the target organs themselves. A **hormone** is secreted into the bloodstream in response to a neural or chemical signal, and it acts at an organ or tissue other than its place of origin.

When a hormone reaches its target organ, it alters some metabolic process in one of two general ways. One type of hormone—for example, water-soluble epinephrine—interacts with receptors on the surface of a cell. This interaction causes the production of a different molecule, a so-called second messenger, inside the cell. The second messenger modifies the activity of an enzyme or enzyme system within the cell. These hormones are rapidly inactivated; so their effects are short-lived. The other type of hormone, represented by the steroid hormones estradiol and testosterone, is carried through the bloodstream to its target cell, where, after entry, it forms a complex with a specific receptor protein. This intracellular hormone–receptor complex then interacts with the cell's DNA. The result is a modification of the level, or concentration, of a specific enzyme coded by the DNA. The effects of this type of hormone are generally long-lived. Both types act at very low concentrations, below micromolar—and often as low as nanomolar to picomolar—amounts.

The Brain and the Nervous System

The brain is the master control unit for metabolic integration. For this reason, most of our attention will be on its unique characteristics, but let's first consider its metabolic needs.

The brain contains no stored energy sources and uses only glucose for its energy needs, which amount to about 60% of the total resting human glucose consumption. Its very active aerobic metabolism uses about 20% of the total oxygen consumed by the body at rest. Even more interesting, the actual amounts of oxygen consumed in liters per minute remain constant whether a person is asleep or actively thinking. Because the brain stores no glycogen or triacylglycerol, it depends critically on the constant availability of glucose from the circulating blood. This dependency means that our blood glucose must be maintained at a constant concentration at all times (see Table 18.11). Brain function can undergo significant and irreversible damage if blood glucose should fall below critical levels for even quite short periods of time. Although it cannot use fatty acids, the brain can adapt to use 3-hydroxybutyrate (Section 16.4) as an energy source under certain circumstances, such as starvation. The transport of glucose into brain cells is noninsulin dependent (Section 18.6); so, provided that glucose is above minimal blood levels, brain function in diabetic patients is unaffected.

Electrical activity of the brain The special function of the brain rests on its ability to generate a large electrical potential across its cell membranes and to transmit an electrical signal from cell to cell. The anatomical features of nerve cells necessary to this function are presented in Box 18.1 on the following page.

18.1 CHEMISTRY WITHIN US

Nerve Anatomy

Nerve cells, or **neurons,** have a unique architecture. Short structures called dendrites emerge from one end of the neuronal cell body, and a single, long extension called the axon emerges from the other end. The axon may divide into many special branches called **synaptic terminals.**

There is no direct contact between neurons. They are separated by a space called the **synaptic cleft.** When a signal is transmitted from one neuron to the next, the end of the neuron that is sending the signal is called the **presynaptic ending,** and the part of the membrane of the adjacent neuron receiving the signal is called the **postsynaptic terminal.** All these functional components in the nerve junction constitute the **synapse.** The signal is transmitted across the synaptic cleft by chemicals called **neurotransmitters.** Neurotransmitters are synthesized in nerve-cell bodies and stored in secretory vesicles located in the presynaptic ending.

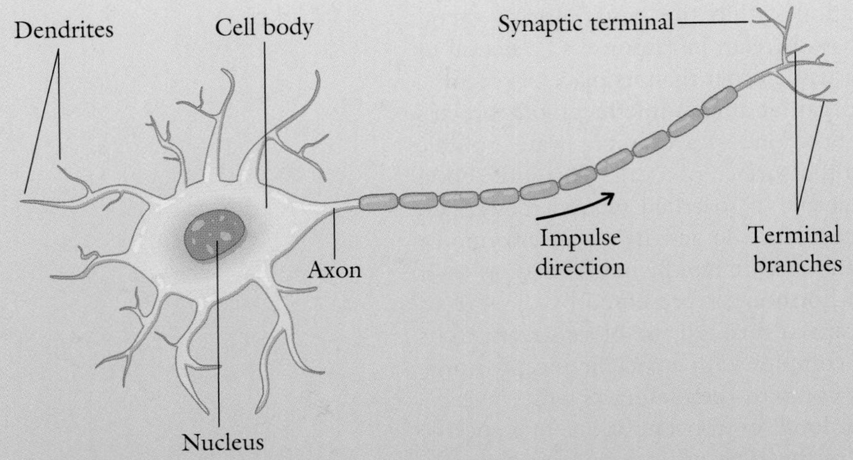

The structure of a typical vertebrate neuron. Two kinds of structures extend from the cell body: dendrites and an axon. The dendrites receive signals from other neurons, and the axon conveys signals away from the cell body. Synaptic terminals at the end of the axon make connections to other neurons.

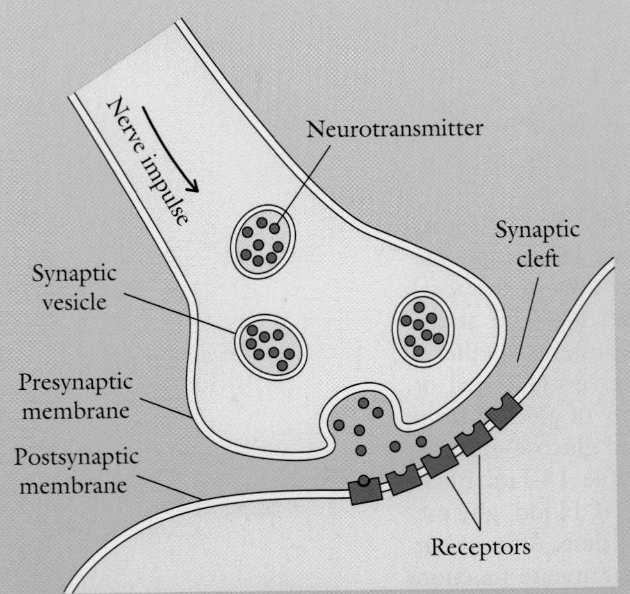

A synapse. An action potential causes synaptic vesicles in the presynaptic cell to fuse with the presynaptic membrane. Neurotransmitters are released into the synaptic cleft and diffuse to the postsynaptic membrane. There they bind to receptors that modify the permeability of the membrane, which in turn alters the membrane potential.

The electrical potential across nerve-cell membranes is the result of an asymmetrical distribution of sodium and potassium ions across nerve-cell membranes. The concentration of potassium ions within nerve cells is about 125 mM, and the concentration outside is about 10 mM. The concentration of sodium ions outside is about 150 mM; inside it is about 15 mM. These concentration asymmetries are created by a unique membrane active-transport enzyme system called the **sodium potassium ATPase (Na^+/K^+ ATPase,** sometimes called the Na^+/K^+ pump; Section 15.14). Utilizing a continuous supply of ATP, this enzyme system literally pumps three Na^+ ions out of nerve cells while simultaneously pumping two K^+ ions into nerve cells for every molecule of ATP hydrolyzed. The resulting asymmetrical distribution generates an electrical potential of about -60 to -70 millivolts (mV) with respect to the outside of the membrane. (The inside of the cell is negatively charged with respect to the outside of the cell.) The Na^+/K^+ ATPase is found in many other cells as well, notably intestinal epithelium and the cells lining the kidney tubule.

When an incoming signal sufficiently reverses this polarization at a synaptic contact (described in Box 18.1), it causes membrane depolarization immediately adjacent to it, which is repeated again and again along the membrane. The depolarization travels down the axon membrane and is called an **action potential.** The axon therefore functions to transmit a signal rapidly, as fast as 90 m/s, over long distances. For example, the sciatic nerve extends from the lower end of your spinal cord to your foot.

When an action potential reaches a presynaptic terminal, neurotransmitters are released into the synaptic cleft, diffuse to the postsynaptic terminal, bind to receptor sites there, and affect the membrane potential. In this way, a nerve impulse is transmitted from one cell to another. The neurotransmitters must then be rapidly inactivated or the next incoming signal will not be detected. Inactivation of the neurotransmitter is accomplished either enzymatically or by its reabsorption into the presynaptic cell.

Sensory neurons, or **receptor cells,** are neurons that are specialized to detect particular environmental changes, reacting, for example, to temperature, touch, or particular chemicals. They respond to changes in a given variable by becoming depolarized to varying extents, depending on the intensity of the environmental change. Sensory neurons generate action potentials spontaneously and continuously. The different levels of depolarization cause the action potentials to be generated at different frequencies—for example, 5/s, 15/s, or 35/s. Consequently, the intensity of environmental change is signaled by the frequency of action potentials, called the **nerve discharge.** The signal must be conveyed to some effector cell or organ in order for a response to be made to the receptor cell's output. Effector cells and organs are specialized to carry out functions appropriate to the incoming signal—for example, to cause muscle contraction or chemical secretion.

Many neurotransmitter and inhibitor substances function in the brain, between neurons outside the brain, and between neurons and effector organs such as muscles. Among them are aspartate, γ-aminobutyrate (GABA), glutamate, glutamine, glycine, and acetylcholine. Most of them and a number of other peptides and amino acid derivatives act specifically in particular regions of the brain.

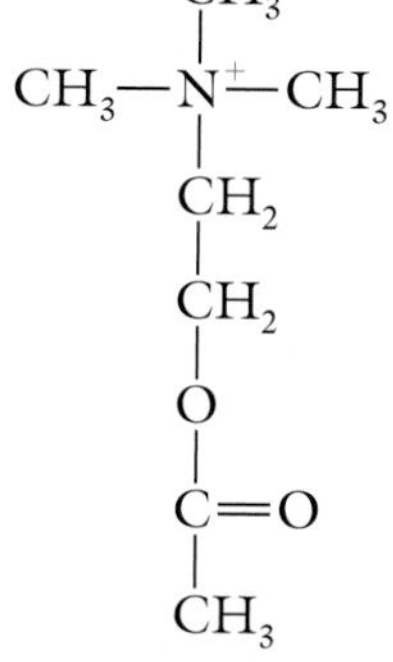

Acetylcholine functions as a neurotransmitter both in the brain and at neuromuscular junctions—that is, at the junction between nerve and muscle. There it is inactivated by the enzyme acetylcholine esterase, which hydrolyzes acetylcholine to acetate and choline. These components must then be reabsorbed, resynthesized into acetylcholine, and stored in presynaptic vesicles. All these processes require considerable amounts of ATP and account for the large consumption of oxygen by nerve tissue.

$$\begin{array}{c} CH_3 \\ | \\ CH_3{-}\overset{+}{N}{-}CH_3 \\ | \\ CH_2 \\ | \\ CH_2 \\ | \\ O \\ | \\ C{=}O \\ | \\ CH_3 \end{array}$$

Acetylcholine

Hormones of the brain Among its other brain functions, the hypothalamus is the site of synthesis of a large number of hormones and hormone-releasing agents. These substances can affect organ systems directly or cause other endocrine glands to secrete their hormones. For example, the hypothalamus secretes a specific **hypothalamic releasing hormone (HRH)** that causes

TABLE 18.14 Some Hypothalamic Releasing Hormones

Releasing hormone	Anterior pituitary hormone effect
thyrotropic-releasing hormone	thyrotropin released
gonadotropin-releasing hormone	lactogenic hormone and follicle-stimulating hormone (FSH) released
gonadotropin release inhibiting factor	lactogenic hormone and FSH release inhibited
corticotropin-releasing hormone (CRH)	adrenocorticotrophic hormone (ACTH) released
somatocrinin (growth-releasing hormone, GRH)	growth hormone (GH) released
somatostatin (growth-inhibiting hormone, GIH)	GH release inhibited

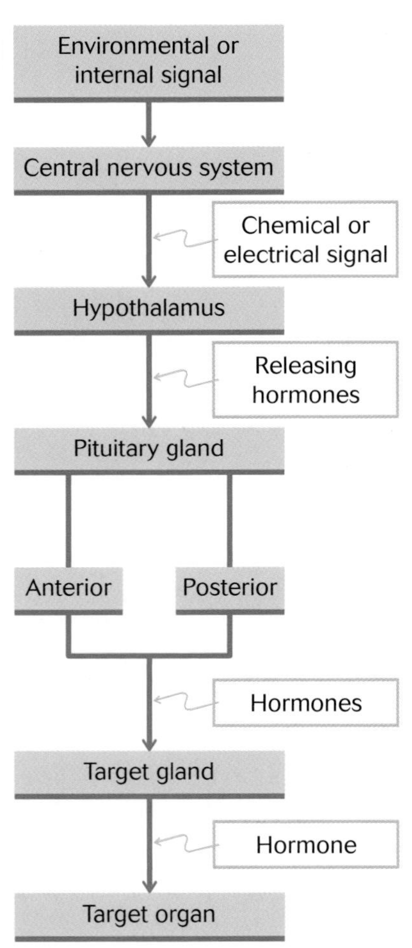

Figure 18.8 The signal pathway for hypothalamic releasing hormones in higher animals. Signals originate either internally or in the outside environment. The signal is then transmitted, either chemically or electrically (by specific neurons), first to the hypothalamus, then to the pituitary, and from there to a target gland, which secretes the final hormone that causes systemic events.

the anterior pituitary gland to secrete thyrotropic hormone, which, in turn, stimulates the thyroid gland to secrete thyroxin. Another HRH stimulates the anterior pituitary to secrete β-corticotropin, which stimulates the adrenal cortex to secrete cortisol (Figure 11.4). Yet another HRH causes the anterior pituitary to secrete prolactin, which acts directly on the mammary glands to stimulate milk production.

The hypothalamus releases hormones in response to input from the central nervous system. In higher animals, the central nervous system responds either to signals that are generated internally or to signals that originate in the outside environment, as illustrated in Figure 18.8. After a signal has been transmitted either chemically or electrically to the hypothalamus, the signal is relayed to the pituitary and from there to a target gland, which secretes the final hormone that causes systemic events. Examples of target glands are the thyroid gland—which produces thyroxin, a hormone that influences cellular oxidations—and the adrenal gland—which produces cortisol, a steroid hormone that mobilizes the body's defenses against stress. As the signal progresses along the pathway, each subsequent factor or hormone is released in a larger amount. Therefore the original signal becomes amplified in intensity, and the amplification is often referred to as a hormonal cascade. The releasing hormones are usually secreted in nanogram amounts, the hormones released from the anterior pituitary may be in microgram amounts, and the concentration of the ultimate hormone released from the target—for example, cortisol from the adrenal gland—can be in milligram amounts.

The posterior pituitary does not synthesize hormones but stores oxytocin and vasopressin and releases them in response to specific releasing hormones from the hypothalamus. Table 18.14 is a list of some hypothalamic releasing hormones and the anterior pituitary hormones that they affect.

18.6 METABOLIC RESPONSES TO PHYSIOLOGICAL STRESS

When our bodies are subjected to physiological stress, they respond so as to counteract the stress and return to their normal steady state, or homeostasis. A study of these responses reveals the mechanisms underlying the coordinated control of metabolism. We will explore these mechanisms by first examining the conditions existing (1) just after a meal—the absorptive, or "fed," state—when there is a transitory increase in the blood concentrations of nutrients and (2) some time after a meal—the postabsorptive state—when the concentration of nutrients must be maintained above a certain minimal level. With that preparation, we will then examine the body's responses under two abnormal stressful conditions: starvation and diabetes mellitus.

The **absorptive state** describes conditions in the body when nutrients are entering the blood—that is, during the ingestion of a meal, which ordinarily consists of carbohydrate, protein, and lipid. The **postabsorptive state** describes conditions some time after a meal, when the gastrointestinal tract is empty and the body must subsist on stored forms of energy.

Absorptive and Postabsorptive States

Between meals, the glucose concentration of the blood is normally in the range of 70 to 90 mg/dL. After a meal, it may rise to a range between 120 and 150 mg/dL. Blood rich in glucose flows through the pancreas and stimulates the β cells to secrete the protein hormone insulin into the blood. The absorptive state is therefore characterized by high concentrations of insulin in the blood.

Insulin acts on all tissues of the body, with the exception of the brain and erythrocytes, and sends the metabolic machinery into an anabolic mode. It acts on a cell by adsorbing to receptors on the cell's surface. The mechanism of this membrane interaction is not clear—the second messenger (Section 15.6) has not been determined—but, in every case, the insulin alters either (1) membrane transport or (2) cellular enzyme activity. It enhances the facilitated transport (Section 11.10) of glucose and amino acids into most cells, promotes the storage of lipids and glycogen, and increases the biosynthesis of proteins and nucleic acids.

More specifically, insulin enhances the entry of glucose into muscle and adipose tissue, activates glycolysis in the liver (thus providing glycerol phosphate for triacylglycerol synthesis), increases the rate of fatty acid and triacylglycerol synthesis in both the liver and adipose tissue, inhibits gluconeogenesis in the liver, increases glycogenesis in the liver and in muscle, increases amino acid uptake in muscle, leading to protein synthesis, and, at the same time, inhibits muscle-protein degradation.

In the absorptive state, glucose is the principal fuel used by all cells. Because it is usually present in excess of what is required for basic energy needs, most glucose is converted into storage molecules. The glycogen-storage capacity of the liver is rapidly filled, and the leftover glucose there is converted into fatty acids and subsequently into triacylglycerols. Amino acids not needed for protein synthesis also are converted into lipid for storage. Glucose is also transported to adipose tissue and converted into triacylglycerols, as well as to muscle cells to be stored as glycogen.

Concept check

✓ The absorptive state is signaled by the presence of insulin and characterized by the preponderance of anabolic processes over catabolic processes.

About 4 to 5 h after a meal, the blood-glucose concentration falls to its normal level of about 70 to 90 mg/dL. It is prevented from dropping below this value by the response of another set of hormone-producing cells in the pancreas, the α cells. These cells, like the β cells, are sensitive to the blood-glucose concentration. When that concentration falls below about 80 mg/dL, the α cells secrete the protein hormone glucagon into the blood.

Glucagon essentially reverses the effects of insulin, shifting the metabolic machinery toward catabolism. Its primary target is the liver, and its chief physiological role is to keep the blood-glucose level constant by increasing the cyclic AMP levels within the cells. This enhances glycogenolysis in the liver and, through the action of liver glucose-6-phosphatase, releases glucose to the blood. Blood-glucose levels are also kept constant because, in this state, the insulin levels are significantly reduced. This reduction in insulin means that glucose is primarily diverted to brain cells, and all other cells must switch to other metabolic fuels. Other fuel is available because glucagon also acts on adipose tissue. It increases the activity of the internal lipase of adipose tissue, thereby releasing fatty acids to the blood for transport to all tissues. The availability of fatty acids for oxidation spares the glucose concentration so that glucose can be directed principally to the brain.

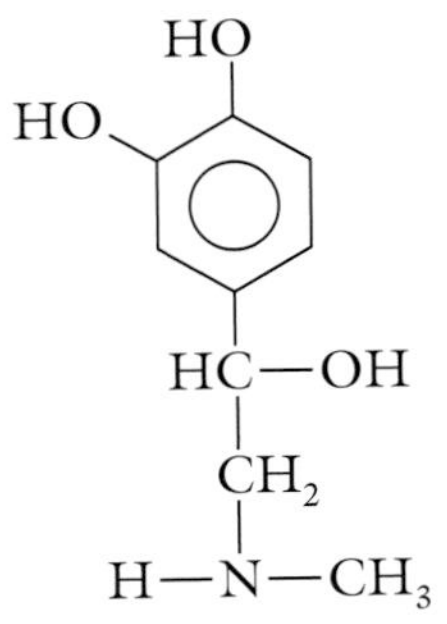

HO, HO, HC—OH, CH_2, H—N—CH_3

Epinephrine

The sympathetic nervous system also is capable of shifting the metabolic processes from an anabolic to a catabolic state. These shifts are short-lived and are caused by the presence of the neurotransmitter **epinephrine.** Epinephrine is synthesized by the adrenal gland and released into the blood in response to anger or fear signaled by the sympathetic nervous system—the "fight or flight" reflex. There are no epinephrine receptors in the liver; instead, the neurotransmitter's effects are directed toward muscle and adipose-tissue cells, where they cause rapid mobilization of stored energy resources—glycogenolysis in muscle and lipolysis in adipocytes.

Concept check

✓ The postabsorptive state is characterized by a shift to catabolic processes caused by the presence of the protein hormone glucagon.

Starvation

Starvation can be viewed as an unrelieved continuation of the postabsorptive state. The body must exist solely on endogenous supplies of energy. The question to be explored is: What happens when the postabsorptive state is extended past the usual fasting period that stretches from dinnertime to breakfast?

First, the mobilization of fat stores for their energy continues, but the problem of maintaining the required continuous glucose supply to the brain becomes more acute. Normally, glucose is supplied by glycogenolysis in the liver. But, after fasting, the liver glycogen supply becomes depleted and unavailable. The hydrolysis of triacylglycerols in both the liver and adipocytes will produce glycerol phosphate, which can be converted into glucose through gluconeogenesis; however, the amounts produced cannot equal those produced through glycogenolysis. In starvation, the glucose–alanine cycle in the liver that ordinarily provides some of the raw material for liver glucogenesis becomes a major route to gluconeogenesis. This route, combined with the fact that muscle protein constitutes about 65% of the body's mass, allows the catabolism of glucogenic amino acids to proceed at a rate that maintains a constant blood-glucose concentration. As a result, protein breakdown speeds up in the initial stages of starvation. Aside from the shortage of water and vitamins, continuous catabolism of the body's protein mass is the cause of the most serious consequences of starvation.

Large amounts of ketone bodies are present in the blood during starvation, because the body's energy is coming from the oxidation of fat, and the required amounts of oxaloacetate are not available for acetyl-*S*-CoA to enter the citric acid cycle (Section 16.3). Within days of the initiation of starvation, the brain adapts to the use of ketone bodies. As a result, the rate of protein catabolism can be reduced, and the reservoir of the body's protein can be "spared." Within a week, the brain derives about a third of its energy needs from acetoacetate and its reduction product 3-hydroxybutyrate, and protein catabolism has slowed significantly. By the end of 6 weeks, the brain takes about two-thirds of its energy needs from the ketone-body pool. As the brain uses more and more of the ketone bodies as fuel, the catabolism of muscle protein decreases by a factor of four from the first week to the sixth week. Although our ability to respond to physiological stresses beyond those imposed by starvation are severely compromised, depending on the size of the adipose tissue depot, life can continue for long periods of time. However, to do so requires the presence of adequate amounts of water, without which death will occur within days, not weeks.

Concept check

✓ Starvation is a continuation of the postabsorptive state accompanied by significant loss of protein mass and the presence of ketone bodies in the blood and urine.

Diabetes Mellitus

Diabetes mellitus, the third leading cause of death in the United States, is due to a functional deficiency of insulin. The disease is seen in two forms: type 1, or juvenile-onset diabetes, which is insulin dependent (meaning insulin must be administered externally); and type 2, or adult-onset, diabetes, which is not insulin dependent. Type 1 may afflict people at any age but usually appears early in life, and type 2 usually affects people over the age of 40. The type 1 disease is characterized by an inability to produce insulin and is believed to be caused by the destruction of the β cells in the pancreas by either a viral infection or an autoimmune disorder. About 10% of all cases of diabetes are type 1. The other 90% of cases are type 2 disease, in which insulin is produced but is unable to effect the passage of glucose across cell membranes. The chief cause is believed to be a lack of insulin membrane receptors. It is often accompanied by obesity and can be controlled by diet and drugs.

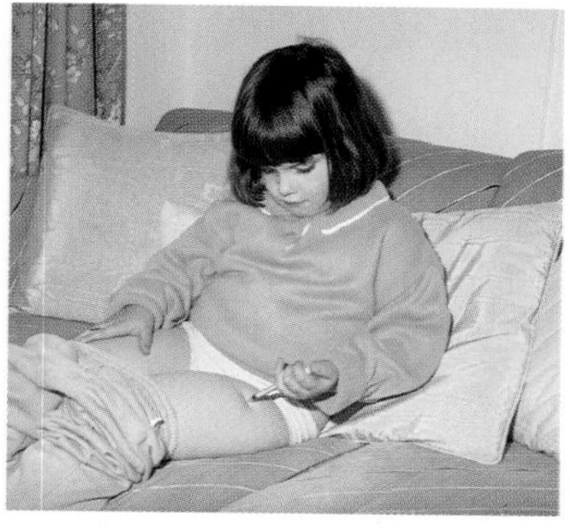

Young diabetic girl injecting herself with insulin. Diabetes mellitus is a disorder of carbohydrate metabolism due to a lack of the pancreatic hormone insulin. Treatment requires the use of daily insulin injections. (Science Photo Library/Photo Researchers.)

Those tending toward the development of type 2 diabetes can be detected by measurement of the blood-glucose concentration after a period of fasting. Table 18.15 indicates three categories of people: normal, impaired, and high. Those in the normal category are not diabetics and probably never will be. Those in the high category are considered to be type 2 diabetics. About 7 out of 100 people in the impaired category develop diabetes per year. Those in the impaired category must lose weight and engage in an exercise regimen if they are to prevent the development of diabetes.

The disease can have a genetic origin (Box 12.3). Just as some enzymes are synthesized as zymogens, the hormone insulin is synthesized as proinsulin. For proinsulin to be converted into insulin, a peptide of 33 amino acids must be excised from the proinsulin polypeptide. Mutations can cause defects preventing this conversion. Even if the conversion takes place properly, other mutations can render the insulin inactive. Other genetic defects affect cell-surface receptors or the internal mechanisms through which the insulin manifests its activity or both.

The physiological situation created by diabetes mellitus is somewhat paradoxical. Because no insulin is available, glucose cannot enter any cells except brain tissue and erythrocytes, which means that most cells are starving in the midst of plenty. The cells of the body of a diabetic operate in a metabolic state somewhere between the postabsorptive state and starvation. The liver attempts to increase its glycogen supplies by speeding up gluconeogenesis. This process requires glucogenic amino acids, which must be supplied from muscle protein. For this reason, uncontrolled diabetes was called "the wasting disease." The catabolism of amino acids leads to greatly increased urea concentrations in the urine, as much as five times the normal amounts, indicative of the severe imbalance between the daily intake and the breakdown of protein.

TABLE 18.15 Blood-Glucose Concentrations after a 10-Hour Fast on Two Separate Occasions

Normal	Impaired	High
below 110 mg/dL	110–126 mg/dL	above 126 mg/dL

As in starvation, adipose-tissue lipids are mobilized through the action of glucagon, and fatty acid oxidation becomes the principal source of ATP. The result is an elevation of the concentrations of ketone bodies and organic acids (acetoacetate and β-hydroxybutyrate) in the blood. The normal pH of the blood is 7.4, but, in severe diabetics, the pH can drop as low as 6.8. In uncontrolled diabetes, this decrease in pH can cause unconsciousness. An unfortunate side effect of the low pH is that the unstable acetoacetate is decarboxylated nonenzymatically to form acetone. The sweet smell of acetone on the breath, accompanied by unconsciousness, is symptomatic of a life-threatening diabetic coma, yet all too often is mistaken for a drunken stupor.

In paradoxical contrast with the concurrent starvation of body cells, the blood-glucose concentration in diabetes can exceed 200 mg/dL. At levels above 180 mg/dL, the kidneys can no longer reabsorb the glucose in the tubule ultrafiltrate. The glucose then appears in the urine in large quantities (hence the name mellitus—from the Latin "like honey"). The high concentrations of glucose in the ultrafiltrate create a large osmotic effect that keeps water in the tubules and prevents it from being reabsorbed. Another typical symptom of this disease, therefore, is frequent and excessive urination, from 2 to 10 L/day, accompanied by extreme thirst. The initial diagnosis of diabetes is often based on these effects.

Type 1, or insulin-dependent, diabetes, requiring daily injections of insulin, is the more difficult form to control. Insulin cannot be administered by mouth, because, as a protein, it is rapidly hydrolyzed in the gastrointestinal tract. The less-severe, adult-onset type of diabetes can often be controlled much more simply by diet and drugs. Until recently, the insulin administered to insulin-dependent diabetics was extracted from animals. Today, however, this insulin source has largely been replaced by recombinant DNA techniques (Section 13.11), which are used to induce *E. coli* bacteria to produce human insulin.

Summary

Digestion and Nutrition

- In digestion, high-molecular-mass foodstuffs are reduced to low-molecular-mass molecules.
- Nutrition describes the body's needs for low-molecular-mass molecules to sustain continuous metabolism.

Nutrient Transport

- Hydrophobic lipids are transported through the blood in the form of lipoprotein particles.
- Lipoproteins differ in their relative amounts of lipid and protein and in relative amounts of different lipids.
- The hemoglobin of erythrocytes combines reversibly with oxygen to increase the blood's oxygen concentration to about 21 mL/100 mL.
- Hemoglobin's release of oxygen at the tissue level is enhanced by a decrease in pH and an increase in carbon dioxide.
- Most of the CO_2 carried by the blood is in the form of bicarbonate ion.

Metabolic Characteristics of the Major Organs and Tissues

- The cells of different organs have different metabolic requirements.
- The brain uses only glucose for its energy needs.
- Heart muscle utilizes fatty acids as its primary energy source and operates aerobically at all times.
- The principal energy source of resting muscle is fatty acids.
- When exertion begins, the glycogen in muscle, about 75% of the body's glycogen supply, is mobilized to provide glucose.
- The breakdown of triacylglycerols within adipocytes is activated by epinephrine and reversed by insulin.
- The intense aerobic metabolism of the kidneys is supported by glucose, fatty acids, ketone bodies, and amino acids.
- All nutrients absorbed by the intestinal tract, except triacylglycerols, are transported directly to the liver, where they are processed and distributed to the other organs and tissues.

Cellular Communication

- The coordinated control of the operation of tissues and organs is accomplished by intercommunication through hormones.
- The brain is the center of the synthesis of a large number of hormones and hormone-releasing agents.
- Rapid intercellular communication is carried out by the nervous system because of its ability to transmit electrical signals from cell to cell.
- This signal transmission from cell to cell is accomplished by chemical substances called neurotransmitters.

Metabolic Responses to Physiological Stress

- The absorptive state describes metabolic conditions during the ingestion of a meal.
- This state is signaled by the presence of insulin and characterized by the predominance of anabolic over catabolic processes.
- The postabsorptive state describes conditions some time after a meal, when the gastrointestinal tract is empty and the body must subsist on stored forms of energy.
- The postabsorptive state is characterized by a shift to catabolic processes caused by the presence of the protein hormone glucagon.
- Starvation can be viewed as an unrelieved continuation of the postabsorptive state.
- Diabetes mellitus is caused by a functional deficiency of insulin, in consequence of which glucose cannot enter any cells except brain tissue.
- All the cells of the body of a diabetic operate in a metabolic state somewhere between the postabsorptive state and starvation.

Key Words

absorptive state, p. 557
acidosis, p. 548
action potential, p. 555
alkalosis, p. 548
chylomicron, p. 542
diabetes mellitus, p. 559
homeostasis, p. 531
hormone, p. 553
lipoprotein, p. 542
neuron, p. 554
postabsorptive state, p. 557
zymogen, p. 532

Exercises

Digestion

18.1 What is the outcome of the digestive process?

18.2 Why is the digestive process necessary?

18.3 What causes the secretion of gastrin in the stomach?

18.4 How is the low pH of the stomach contents raised to neutral pH in the small intestine?

18.5 What is a zymogen?

18.6 Are all digestive enzymes secreted as zymogens? Explain.

18.7 How do the digestive products of carbohydrates and proteins reach the liver?

18.8 How do the digestive products of triacyglycerols reach the liver?

18.9 Describe the role of gastrin in digestion.

18.10 Describe the role of secretin in digestion.

Nutrition

18.11 What is the meaning of the biological value of a protein?

18.12 How is the chemical score of a protein determined?

18.13 Why are the polyunsaturated fatty acids linoleic acid and linolenic acid considered essential?

18.14 Why is methionine considered an essential amino acid?

18.15 What is meant by a dietary-deficiency disease?

18.16 What is an example of a dietary-deficiency disease and its cure by replacement?

Nutrient Transport

18.17 Describe the cellular composition of blood.

18.18 Describe the composition of blood plasma.

18.19 Describe the structure of a typical lipoprotein.

18.20 List the various classes of lipoproteins and the chief property used in differentiating them.

18.21 What is the role of the protein components of lipoproteins?

18.22 What happens to a triacylglycerol at the surface of an adipocyte?

18.23 What characteristic of the blood increases its oxygen-carrying capacity 50-fold over the oxygen solubility of plasma alone?

18.24 What biochemical characteristic makes hemoglobin efficient in unloading oxygen at the low oxygen tension at the location of respiring cells?

18.25 What is the function of 2,3-bisphosphoglycerate?

18.26 How does blood pH effect the unloading of oxygen at respiring cells and the lungs?

18.27 The major form in which CO_2 is carried in the blood is bicarbonate ion. How is it formed?

18.28 What is the chloride shift?

18.29 How does CO_2 affect oxygenated hemoglobin when hemoglobin reaches respiring cells?

18.30 How does the formation of carbamate hemoglobin affect the unloading of oxygen at respiring cells?

18.31 What is respiratory alkalosis?

18.32 What is respiratory acidosis?

Metabolic Characteristics of the Major Organs and Tissues

18.33 What are the nutrients used for energy production by brain cells?

18.34 What effect does blood composition have on brain function?

18.35 What are the forms of energy storage in heart cells?

18.36 What is the principal fuel of heart cells?

18.37 What is the biochemical consequence of a heart cell's preference for its principal energy source?

18.38 What is the physiological consequence of a heart cell's preference for its principal energy source?

18.39 What is the principal energy source of active muscle?

18.40 What is the role of muscle cells in the maintenance of the blood-glucose concentration?

18.41 How is the triacylglycerol content of chylomicrons in the blood transferred into adipocytes?

18.42 Is all of the triacylglycerol content of chylomicrons transferred into adipocytes? Explain.

18.43 How do adipocytes respond to the hormone glucagon?

18.44 How do adipocytes respond to the hormone insulin?

18.45 What becomes of the glycerol from the hydrolysis of triacylglycerols in adipocytes?

18.46 How does glycolysis contribute to the synthesis of triacylglycerols in adipocytes?

18.47 Most of the respiration of kidney cells is put to what purpose?

18.48 How do the kidneys concentrate urine?

18.49 Give an example of how the kidneys can help in excreting excessive hydrogen ion from the body.

18.50 Give an example of how the kidneys can correct the pH of the body when it rises above normal levels.

18.51 How are the high concentrations of glucose in the liver after a meal handled by that organ?

18.52 How does the liver take part in the maintenance of a supply of free glucose in the blood?

18.53 How is the ammonia derived from the catabolism of amino acids in the liver detoxified?

18.54 How is the ammonia derived from the catabolism of amino acids detoxified in muscle cells?

18.55 What are ketone bodies?

18.56 What are the origin and source of most of the ketone bodies in circulating blood?

18.57 How does the liver contribute to the intestinal digestion of dietary lipids?

18.58 How does the liver contribute to the biosynthesis of the steroid hormones?

Cellular Communication

18.59 Are all hormones low-molecular-mass substances?

18.60 True or false: A hormone acts at the organ in which it is synthesized.

18.61 Do all hormones interact with receptors at the surface of a cell?

18.62 What is a second messenger?

18.63 When answering these questions, does your brain require more oxygen than when you are asleep?

18.64 What special property of nerve cells differentiates the brain from other organs?

18.65 True or false: Neurotransmitters have a long life.

18.66 In what way does the central nervous system respond to signals from other organs?

18.67 How does the central nervous system respond to the body's need for thyroxin?

18.68 What is a hormonal cascade?

18.69 What special physiological property is associated with brain cells?

18.70 What is the origin of the property referred to in Exercise 18.69?

18.71 Describe the quantitative distribution of sodium ions and potassium ions between the inside and the outside of nerve cells.

18.72 How is the distribution of sodium ions and potassium ions across nerve membranes achieved?

18.73 Describe the role of the brain in effecting changes in organ systems that are remote from the brain and its nerve connections.

18.74 Give a specific example of how the brain can affect the physiological behavior of an organ such as the kidney or the thyroid gland.

Metabolic Responses to Physiological Stress

18.75 Define the absorptive state.

18.76 Define the postabsorptive state.

18.77 What hormone is present in high concentrations in the absorptive state?

18.78 Describe the actions of the hormone given in answer to Exercise 18.77.

18.79 The presence of what hormone in the blood characterizes the postabsorptive state?

18.80 Describe the actions of the hormone given in answer to Exercise 18.79.

18.81 What is the major difference between the body's responses to the postabsorptive state and its responses to the initial stages of starvation?

18.82 What is the major difference between the body's initial response to starvation and its responses to the later stages?

18.83 What are the characteristics of juvenile-onset diabetes?

18.84 What are the characteristics of adult-onset diabetes?

18.85 How does the liver maintain its glycogen supplies in diabetes?

18.86 What is the explanation for the presence of high concentrations of urea in the urine of diabetics?

18.87 What is the explanation for the presence of high concentrations of ketone bodies in the blood of diabetics?

18.88 Explain the fact that the pH of the blood of diabetics is often lower than normal.

Unclassified Exercises

18.89 What is the meaning of nitrogen balance?

18.90 What are the consequences of a high proportion of saturated to unsaturated fatty acids in the diet?

18.91 Why is vitamin B_{12} not ordinarily required in the diet but must sometimes be supplemented?

18.92 What differentiates bulk elements from trace elements?

18.93 Describe how the liver treats glucose in excess of what is needed for the maintenance of blood-glucose level and storage.

18.94 What is the source of reducing power for biosynthesis in the liver?

Expand Your Knowledge

Note: The icons  denote exercises based on material in boxes.

18.95 What is the biological logic behind the pancreatic synthesis of chymotrypsinogen—a zymogen—rather than chymotrypsin?

18.96 Can an injection of insulin improve the intellectual capabilities of a type 1 diabetic?

18.97 How does dietary lipid enter adipose-tissue cells?

18.98 How would an abnormally low concentration of chloride ion in the blood affect the blood's pH?

18.99 How can a person suffer from protein deficiency when her diet consists of wheat products high in protein concentration?

18.100 How are nerve impulses propagated among neurons (see Box 18.1)?

18.101 What is the distinguishing characteristic of sensory neurons (see Box 18.1)?

APPENDIX 1
Nomenclature: Naming Chemical Compounds

Ionic compounds and molecular compounds are named according to different sets of rules.

In the names of **binary ionic compounds** (ionic compounds made up of two elements), the metal's name is given first and is unchanged; it is followed by the nonmetal element's name combined with the suffix **-ide.** Examples include potassium chloride (KCl), aluminum oxide (Al_2O_3), and calcium fluoride (CaF_2).

This simple rule applies to elements contained within the first three periods of the periodic table. However, many elements of period 4 and higher periods form ions of more than one charge type. For example, iron can form ions with charges of 2+ or 3+, and copper can form ions with charges of 1+ or 2+. Therefore, the chlorides of these metals can have the compositions $FeCl_2$, $FeCl_3$, CuCl, and $CuCl_2$. The older, and sometimes commonly used, names of these compounds are ferrous chloride, ferric chloride, cuprous chloride, and cupric chloride, respectively; but a modern, more systematic method for naming them is becoming more widely used.

The modern method of nomenclature used for ions of more than one charge type is called the **Stock system.** In this system, the charge of the cation is indicated with a roman numeral: iron(II) chloride, iron(III) chloride, copper(I) chloride, and copper(II) chloride. The combining ratios (the formulas) of such compounds can be deduced from their names and the charge of the anion. To illustrate, the Stock system names will be established for the following compounds: (a) $TiCl_3$; (b) $FeCl_3$; (c) CuCl; (d) $CoCl_2$.

For each of these compounds, the chloride ion possesses a stable electrical charge of 1−. Therefore, Ti must be Ti^{3+}, Fe must be Fe^{3+}, Cu must be Cu^{+}, and Co must be Co^{2+}. The compound names are therefore (a) titanium(III) chloride; (b) iron(III) chloride; (c) copper(I) chloride, and (d) cobalt(II) chloride.

Polyatomic ions consist of two or more atoms combined into a single charged unit. The names and formulas of some of the more important ones are listed in Table A1.1 on the following page. As with other ions, when a polyatomic ion is present in a compound, the sum of positive and negative charges of the compound must be zero. If two or more of a given polyatomic ion are present in a formula, the formula is written with the polyatomic ion enclosed in parentheses followed by a subscript indicating the number of those ions in the formula. For example, aluminum carbonate is $Al_2(CO_3)_3$, and calcium phosphate is $Ca_3(PO_4)_2$. If only one polyatomic ion is present, no parentheses (and subscript) are necessary. For example, potassium carbonate is K_2CO_3, and sodium phosphate is Na_3PO_4. Finally, by convention, in every ionic compound, the cation comes first, followed by the anion.

Many of the compounds that are referred to in this book—such as ammonia, carbon dioxide, and water—have common names with which you are already familiar, names adopted before the development of systematic methods of nomenclature, and those are the names that we use for them here. However, should an unfamiliar substance be referred to, a systematic method is available.

TABLE A1.1 Some Important Polyatomic Ions

Ion	Name	Ion	Name
NH_4^+	ammonium	CO_3^{2-}	carbonate
NO_3^-	nitrate	HCO_3^-	hydrogen carbonate
NO_2^-	nitrite	PO_4^{3-}	phosphate
OH^-	hydroxide	HPO_4^{2-}	hydrogen phosphate
$C_2H_3O_2^-$	acetate	$H_2PO_4^-$	dihydrogen phosphate
CN^-	cyanide	SO_4^{2-}	sulfate
HSO_4^-	hydrogen sulfate	HSO_3^-	hydrogen sulfite

Greek prefixes in chemical nomenclature

1	mono-
2	di-
3	tri-
4	tetra-
5	penta-
6	hexa-
7	hepta-
8	octa-
9	nona-
10	deca-

The systematic method for naming simple, **binary molecular compounds** is based on the fact that some pairs of elements can form more than one covalently bonded compound. Two such pairs, for example, are (1) nitrogen and oxygen and (2) carbon and oxygen. The system adopted for naming their compounds uses Greek prefixes (see the table in the margin) and is illustrated in Table A1.2.

A covalent compound's name comes from the order in which its elements appear in the chemical formula: the first element in the formula is named first; the second element is named second, but altered to accommodate the suffix **-ide** (as in the naming of simple ionic compounds). The Greek prefixes are added to the elements' names to indicate the numbers of different kinds of atoms in each molecule of the compound. Note, however, that NO is called nitrogen monoxide, not mononitrogen monoxide, and CO is called carbon monoxide, not monocarbon monoxide. The prefix **mono-** is never used for naming the first element. The order of elements in the formula is based on a consideration of electronegativity: in naming the compound, the less-electronegative element is followed by the more-electronegative element. For example, CCl_4 is carbon tetrachloride, PCl_3 is phosphorus trichloride, PCl_5 is phosphorus pentachloride, and SO_2 is sulfur dioxide.

TABLE A1.2 Systematic Names of Covalent Compounds

Nitrogen and oxygen compound	Systematic name	Carbon and oxygen compound	Systematic name
N_2O	dinitrogen monoxide	CO	carbon monoxide
NO	nitrogen monoxide	C_2O_3	dicarbon trioxide
NO_2	nitrogen dioxide	CO_2	carbon dioxide
N_2O_3	dinitrogen trioxide		
N_2O_4	dinitrogen tetroxide		
N_2O_5	dinitrogen pentoxide		

APPENDIX 2

Significant Figures

A series of measurements of the same property of the same object made by one or more persons are unlikely to result in precisely the same value. This inevitable variability is not the result of mistakes or negligence. No matter how carefully each measurement is made, there is no way to prevent small differences between measurements. These differences arise because no matter how fine the divisions of a measuring device may be, when a measure falls between two of them, an estimate or "best guess" must be made. This unavoidable estimate is called the **uncertainty** or **variability.** All measurements are made with the assumption that there is a correct or true value for the quantity being measured. The difference between that true value and the measured value is called the **error.**

The variability of a measurement can be reported as an average amount greater than or smaller than the measured value, such as 4 ± 0.01 g. This notation means that, every time a mass of 4 g is placed on a balance, the readings will be slightly different but will probably fall within 0.01 g of the actual mass (no higher than 4.01 g and no lower than 3.99 g). An alternative to this notation is the use of **significant figures.** This system indicates the uncertainty by the number of digits instead.

$4 \pm 1\ \text{g} = \underline{4}\ \text{g}$	One significant figure
$4 \pm 0.1\ \text{g} = \underline{4.0}\ \text{g}$	Two significant figures
$4 \pm 0.001\ \text{g} = \underline{4.000}\ \text{g}$	Four significant figures
$4 \pm 0.0001\ \text{g} = \underline{4.0000}\ \text{g}$	Five significant figures

For the purpose of counting significant figures, zero can have different meanings depending on its location within a number:

- A trailing zero is significant; therefore the number 4.130 has four significant figures.
- A zero within a number is significant; therefore the number 35.06 has four significant figures.
- A zero before a digit is not significant; therefore the number 0.082 has only two significant figures.

A report such as 20 cm is ambiguous. It could be interpreted as meaning "approximately 20" (say, 20 ± 10) or it might be understood as 20 ± 1. However, it might mean 20 cm exactly. The number of significant figures in 20 cm is unclear.

- A number ending in zero with no decimal point, such as 20, is ambiguous.

Ambiguities of this last type can be prevented by the use of exponential, or scientific, notation (Appendix 3).

APPENDIX 3
Scientific Notation

Scientific notation, or **exponential notation,** is a convenient method for preventing ambiguity in the reporting of measurements and for simplifying the manipulation of very large and very small numbers. To express a number such as 233 in scientific notation, we write it as a number between 1 and 10 multiplied by 10 raised to a whole-number power: 2.33×10^2. The number between 1 and 10 (in our example, the number 2.33) is called the **coefficient,** and the whole-number exponent of 10 (in our example, 10^2) is called the **exponential factor.** Remember in using scientific notation that any number raised to the zero power is equal to 1. Thus, $10^0 = 1$.

For any number greater than 1, the procedure is:

Step 1 Move the decimal point to the left to create a coefficient between 1 and 10.

Step 2 Create an exponential factor whose power is equal to the number of places that the decimal point was moved to the left.

To express a number smaller than 1 (decimal number) in scientific notation—for example, 0.2—transform it into a whole-number coefficient between 1 and 10, multiplied by an exponential factor that is decreased by the same power of ten. The number 0.365 is therefore written 3.65×10^{-1}. The numbers 0.046 and 0.00753 are written 4.6×10^{-2} and 7.53×10^{-3}, respectively.

For any number smaller than 1, the procedure is:

Step 1 Move the decimal point to the right to create a coefficient between 1 and 10.

Step 2 Create an exponential factor with a negative power that is equal to the number of places that the decimal point was moved to the right.

At the end of Appendix 2, a number ending in zero with no decimal point in the number (21,600, for example) was said to be ambiguous. Scientific notation provides a way to express the number without ambiguity. If the last zero in 21,600 is significant, the number has five significant figures and should be written 2.1600×10^4. If neither zero is significant, the number has three significant figures and should be written 2.16×10^4. If a number containing zeroes loses those zeroes when the number is expressed in scientific notation, they were not significant.

The results of multiplications and divisions of measured quantities are reported in accord with the following rule: The number of significant figures in a number resulting from multiplication or division may not exceed the number of significant figures in the least well known value used in the calculation. For example, the area of a square whose dimensions have been measured as 8.5 in. on a side is:

$$\text{Area of a square} = \text{side} \times \text{side} = 8.5 \text{ in.} \times 8.5 \text{ in.}$$
$$= 72.25 \text{ in.}^2$$

Multiplication of two two-digit numbers always yields a number with more than two digits. The length of the side is known to only two significant figures; therefore the area of the square (length $\times$ length) cannot be known with any greater accuracy. After rounding, the final value is reported as 72 in.^{2}.

A somewhat different approach is required for addition and subtraction. In both these situations, the number of figures after the decimal point decides the final answer. The final sum or difference cannot have more figures after the decimal point than are contained in the least well known quantity in the calculation. All significant figures are retained while doing the calculation, and the final result is rounded. For example, consider the addition of the measured quantities 24.62 g, 3.7 g, and 93.835 g. The least well-known quantity has only one significant figure after the decimal place, and so the final sum cannot have more than that. We add all the values to give 122.155 g and then round off to yield the final sum as 122.2 g.

APPENDIX 4

Molarity

Molarity is a measure of concentration that is based on chemical mass units. For liquid solutions—that is, gases, liquids, or solids dissolved in liquids—molarity is defined as moles of solute per liter of solution. Concentrations described in this way are called **molar** and are denoted by the symbol ***M***. For example, a 2.0 *M* solution is a 2.0 molar solution, a solution whose molarity is 2.0 mol/L.

One context in which molarity is used as a measure of concentration is the preparation of a solution of a specific molarity. Specifically, how is 1.00 liter of a 0.650 *M* aqueous solution of $CaCl_2$ prepared, and how many grams of $CaCl_2$ are required to do that?

To prepare 1.00 L of a 0.650 *M* solution, 0.650 mol of $CaCl_2$ must be dissolved in 1.00 L of solution. The mass of $CaCl_2$ in grams equivalent to 0.650 mol can be calculated by using the formula mass of $CaCl_2$ as a conversion factor:

$$0.650 \text{ mol CaCl}_2 \left(\frac{111 \text{ g CaCl}_2}{1 \text{ mol CaCl}_2}\right) = 72.2 \text{ g CaCl}_2$$

The solution is prepared by weighing out 72.2 g of $CaCl_2$ and dissolving it in sufficient water to make 1.00 L of solution.

We next calculate the molarity of a $CaCl_2$ solution in which 72.15 g of $CaCl_2$ is dissolved in 200.0 mL of solution. Because molarity is defined as moles per liter, we first calculate the number of moles in 72.15 grams of $CaCl_2$, then convert 200.0 mL into liters, and finally calculate *M*.

Step 1 $72.15 \text{ g CaCl}_2 \left(\frac{1 \text{ mol}}{111.0 \text{ g CaCl}_2}\right) = 0.6500 \text{ mol CaCl}_2$

Step 2 $200.0 \text{ mL} \left(\frac{1 \text{ L}}{1000 \text{ mL}}\right) = 0.2000 \text{ L}$

Step 3 $\frac{0.6500 \text{ mol}}{0.2000 \text{ L}} = \frac{3.250 \text{ mol}}{1.000 \text{ L}} = 3.250M$

Answers to Problems Following In-Chapter Worked Examples

Chapter 1

1.1 $1s^2 2s^2 2p^5$

1.2 (a) K^+; (b) Mg^{2+}; (c) In^{3+}; (d) P^{3-}; (e) O^{2-}; (f) Br^-

1.3 (a) CH_4; (b) PCl_3; (c) CO_2

1.4 (a) 257.4 amu; (b) 231.6 amu; (c) 18.02 amu; (d) 98.09 amu; (e) 88.09 amu

1.5 (a) 142.1 g; (b) 119.0 g; (c) 40.31 g; (d) 86.17 g

1.6

```
          ..
         :Cl:
    ..    |    ..
   :Cl — Si — Cl:
    ..    |    ..
         :Cl:
          ..
```

1.7

```
    ..    ..    ..
   :Cl —  P  — Cl:
    ..    |     ..
         :Cl:
          ..
```

1.8

```
    ..   ..    ..
   :Cl — S  — Cl:
    ..   ..    ..
```

1.9

```
       H   H   H
       |   |   |
   H — C — C — C — H
       |   |   |
       H   H   H
```

1.10 $\ddot{O}=C=\ddot{O}$

1.11

```
       ..  ..
   H — N — N — H
       |   |
       H   H
```

1.12 CCl_4 will have a tetrahedral shape.

1.13 NI_3 will have the shape of a trigonal pyramid.

1.14 OF_2 will have a bent shape.

1.15 Only part *c*: $H-C\equiv N$

1.16 Parts *c* CH_3NH_2 and *e* C_2H_5OH

1.17 No. Liquids having strong attractive forces will not mix with liquids possessing weak attractive forces.

1.18 (a) The secondary forces between CCl_4 molecules are weak London forces, and those between H_2O molecules are strong H-bonds. Solutions between substances possessing such very different secondary forces are not possible.
(b) The secondary forces between C_5H_{12} molecules and between CCl_4 molecules are London forces. Because the secondary forces are the same, solutions of these two substances are possible.

Chapter 2

2.1 (a) $3\ Ca^{2+}(aq) + 6\ Cl^-(aq) + 6\ Na^+(aq) + 2\ PO_4^{3-}(aq) \longrightarrow Ca_3(PO_4)_2(s) + 6\ Cl^-(aq) + 6\ Na^+(aq)$
(b) $3\ Ca^{2+}(aq) + 2\ PO_4^{3-}(aq) \rightarrow Ca_3(PO_4)_2(s)$

2.2 (a) −4; (b) +4

2.3 $C_4H_{10} + 13/2\ O_2 \longrightarrow 4\ CO_2 + 5\ H_2O$
$2\ C_4H_{10} + 13\ O_2 \longrightarrow 8\ CO_2 + 10\ H_2O$

2.4 $C_3H_6O_3 + 3\ O_2 \longrightarrow 3\ CO_2 + 3\ H_2O$

2.5 $\Delta H_{rxn} = +6$ kJ. The reaction is endothermic.

2.6 $K_{eq} = \dfrac{[HI]^2}{[I_2][H_2]}$

2.7 $K_{eq} = 1.1 \times 10^{-1}$

2.8 $K_{eq} = 1 \times 10^2$

2.9 Shift of mass to the left (reactant side)

2.10 0.050 *M*

2.11 0.050 *M*

2.12 pH = 2.52

2.13 pH = 4.89

Chapter 3

3.1 CH_5N and C_2H_5Cl are correct molecular formulas.

3.2 (a) ketone; (b) ester; (c) amide; (d) amine.

3.3

```
CH3
   \
    CH2—CH2             CH3
           \           /
            CH2—CH2
                CH3                 CH2—CH2
                   \               /       \
                    CH2—CH2                 CH3
```

Different conformations

```
    H   H   H   H   H   H
    |   |   |   |   |   |
H — C — C — C — C — C — C — H
    |   |   |   |   |   |
    H   H   H   H   H   H
```

Expanded

$CH_3CH_2CH_2CH_2CH_2CH_3$

Condensed

C—C—C—C—C—C

Skeleton

Line

3.4 $CH_3-CH_2-CH_2-CH_2-CH_2-CH_3$

```
      CH3
      |
CH3—CH—CH2—CH2—CH3

           CH3
           |
CH3—CH2—CH—CH2—CH3

      CH3  CH3                 CH3
      |    |                   |
CH3—CH—CH—CH3          CH3—C—CH2—CH3
                               |
                               CH3
```

3.5

$$\begin{array}{c} \overset{4^\circ\ 1^\circ}{C(CH_3)_3} \\ | \\ \overset{1^\circ}{CH_3}-\overset{2^\circ}{CH_2}-\overset{3^\circ}{CH}-\underset{3^\circ}{CH}-\underset{2^\circ}{CH_2}-\overset{2^\circ}{CH_2}-\overset{2^\circ}{CH_2}-\overset{1^\circ}{CH_3} \\ | \\ {}^{1^\circ}CH_3 \end{array}$$

3.6 (a) 5-Ethyl-2,3,3,8-tetramethylnonane; (b) 4-isopropyl-3-methylheptane

3.7

(a) $CH_3-CH_2-\underset{}{\overset{CH_3}{\overset{|}{CH}}}-\overset{CH_2CH_3}{\overset{|}{CH}}-CH_2-CH_3$

(b) $CH_3-CH_2-\underset{\underset{CH_3}{|}}{CH}-\overset{C(CH_3)_3}{\overset{|}{CH}}-CH_2-CH_2-CH_2-CH_3$

3.8 (a) 1-*t*-Butyl-3-ethyl-5-isopropylcyclohexane; (b) 1-*s*-butyl-1,3-dimethylcyclobutane

Geometrical stereoisomers (Section 3.9) are possible for both compounds.

3.9

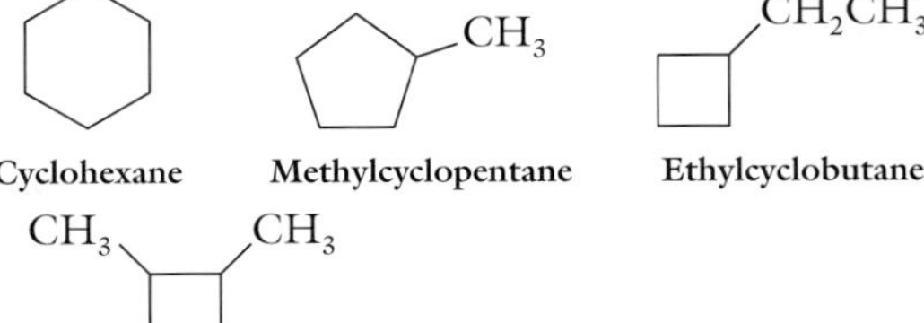

Cyclohexane Methylcyclopentane Ethylcyclobutane

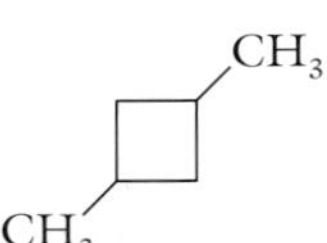

1,2-Dimethylcyclobutane

1,3-Dimethylcyclobutane (ring with CH_3, CH_3)

1,1-Dimethylcyclobutane (ring with CH_3, CH_3)

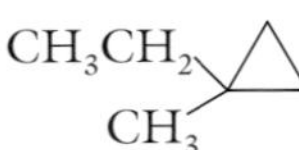

Propylcyclopropane

Isopropylcyclopropane (ring with $CH(CH_3)_2$)

1-Ethyl-1-methylcyclopropane (ring with CH_3CH_2, CH_3)

1-Ethyl-2-methylcyclopropane (ring with CH_3CH_2, CH_3)

1,1,2-Trimethylcyclopropane (ring with CH_3, CH_3, CH_3)

1,2,3-Trimethylcyclopropane (ring with CH_3, CH_3, CH_3)

Geometrical stereoisomers (Section 3.9) are possible for 1,2-dimethylcyclobutane, 1,3-dimethylcyclobutane, 1-ethyl-2-methylcyclopropane, and 1,2,3-trimethylcyclopropane.

(a)

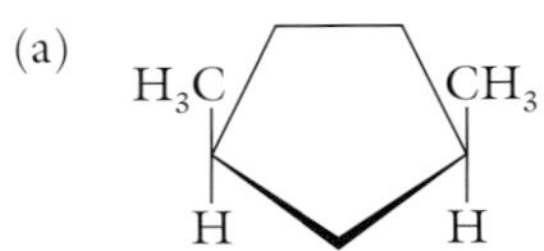

cis-1,3-Dimethylcyclopentane

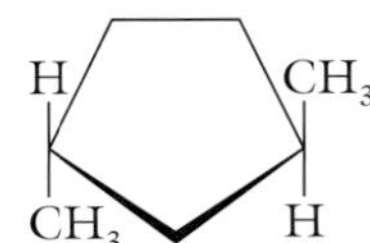

trans-1,3-Dimethylcyclopentane

(b) Cis-trans are not possible.

(c)

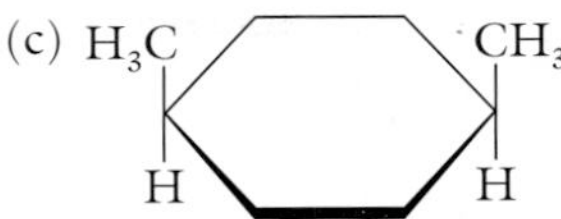

cis-1,4-Dimethylcyclohexane

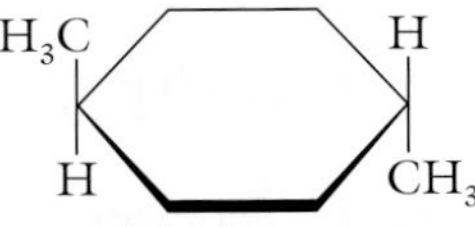

trans-1,4-Dimethylcyclohexane

3.11 (a) Cyclopentane has the higher boiling point because its rigid structure allows tighter packing, which results in greater secondary forces. 2-Methylbutane, being branched and spherical in structure, has much smaller secondary forces.

(b) 3-Methylnonane has the higher boiling point. Both compounds have the same longest chain of nine carbons, but 3-methylnonane has an additional carbon (the 3-methyl branch).

(c) Cycloheptane has the higher boiling point because of its larger molecular mass.

3.12 No. NaCl is an ionic compound. The energy required to separate NaCl into Na^+ and Cl^- cannot be recouped by attraction between the ions and hexane, because hexane is nonpolar.

3.13 (a) $2\ C_5H_{10} + 15\ O_2 \longrightarrow 10\ CO_2 + 10\ H_2O$
(b) $C_9H_{20} + 14\ O_2 \longrightarrow 9\ CO_2 + 10\ H_2O$

3.14

$$CH_3-CH_3 \xrightarrow[\text{UV or heat}]{Cl_2} CH_3-CH_2Cl$$

3.15

$$CH_3-\overset{CH_3}{\overset{|}{CH}}-CH_3 \xrightarrow[\text{UV or heat}]{Br_2}$$

$$CH_3-\overset{CH_3}{\overset{|}{CH}}-CH_2-Br + CH_3-\underset{\underset{Br}{|}}{\overset{CH_3}{\overset{|}{C}}}-CH_3$$

3.16 (a) IUPAC: 1-bromo-2-methylpropane; common: isobutyl bromide. (b) IUPAC: The two possible names are *cis*-1-bromo-2-methylcyclopentane and *cis*-2-bromo-1-methylcyclopentane. Although both names have the same set of numbers (1, 2), the correct IUPAC name is *cis*-1-bromo-2-methylcyclopentane because the numbers are assigned in alphabetical order. There is no common name, because the alkyl group attached to Br is not a simple alkyl group.

Chapter 4

4.1 The unknown compound is an alkane because its molecular formula fits the general formula for an alkane, C_nH_{2n+2}, where $n = 7$.

4.2

$CH_2{=}CH-CH_2-CH_2-CH_2-CH_3$

$CH_3-CH{=}CH-CH_2-CH_2-CH_3$

$CH_3-CH_2-CH{=}CH-CH_2-CH_3$

$CH_2{=}\overset{CH_3}{\overset{|}{C}}-CH_2-CH_2-CH_3$

$CH_3-\overset{CH_3}{\overset{|}{C}}{=}CH-CH_2-CH_3$

$CH_3-\overset{CH_3}{\overset{|}{CH}}-CH{=}CH-CH_3$

$CH_3-\overset{CH_3}{\overset{|}{CH}}-CH_2-CH{=}CH_2$

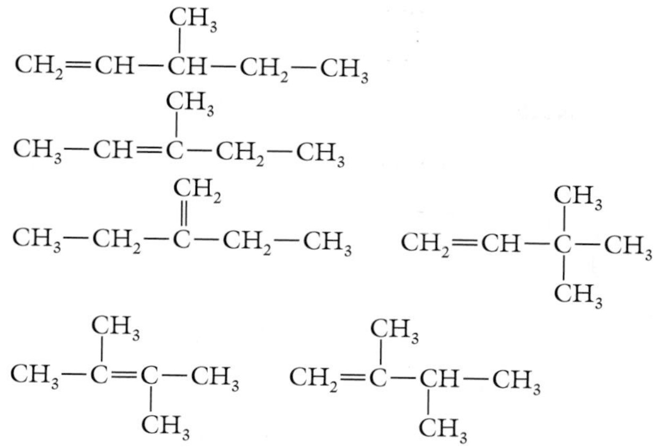

$$CH_2{=}CH{-}CH(CH_3){-}CH_2{-}CH_3$$

$$CH_3{-}CH{=}C(CH_3){-}CH_2{-}CH_3$$

$$CH_3{-}CH_2{-}C(=CH_2){-}CH_2{-}CH_3 \qquad CH_2{=}CH{-}C(CH_3)_2{-}CH_3$$

$$CH_3{-}C(CH_3){=}C(CH_3){-}CH_3 \qquad CH_2{=}C(CH_3){-}CH(CH_3){-}CH_3$$

4.3 (a) 4-Methyl-1-hexene; (b) 1-chloro-4-methyl-2-pentene; (c) 3-*t*-butyl-2-isopropylcyclopentene; (d) 2-methyl-1,4-pentadiene

4.4 (a) No cis-trans isomers

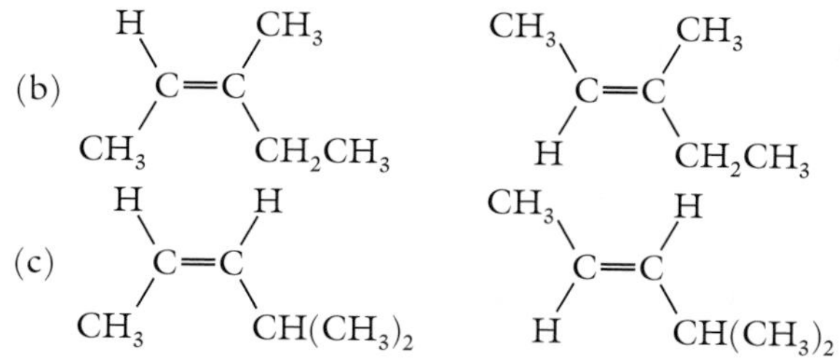

(b) $(H)(CH_3)C{=}C(CH_3)(CH_2CH_3)$ [H and CH_3 on the first carbon; CH_3 and CH_2CH_3 on the second] and $(CH_3)(H)C{=}C(CH_3)(CH_2CH_3)$

(c) $(H)(CH_3)C{=}C(H)(CH(CH_3)_2)$ and $(CH_3)(H)C{=}C(H)(CH(CH_3)_2)$

4.5 (a) $CH_3{-}CH{=}CH{-}CH_3 \xrightarrow[Pt]{H_2} CH_3{-}CH_2{-}CH_2{-}CH_3$

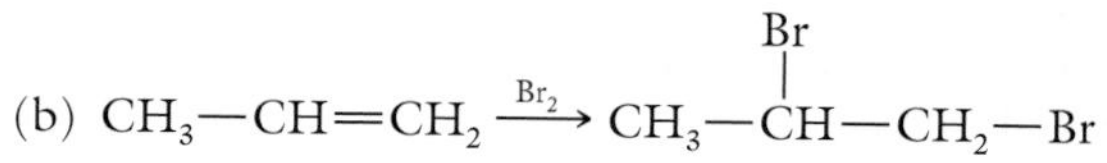

(b) $CH_3{-}CH{=}CH_2 \xrightarrow{Br_2} CH_3{-}CH(Br){-}CH_2{-}Br$

4.6 2-Pentene

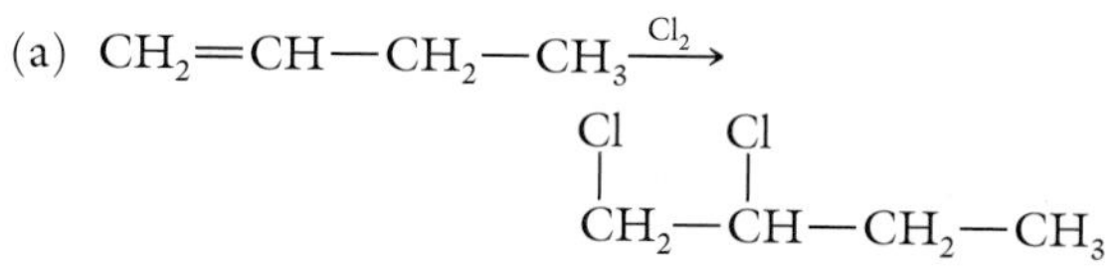

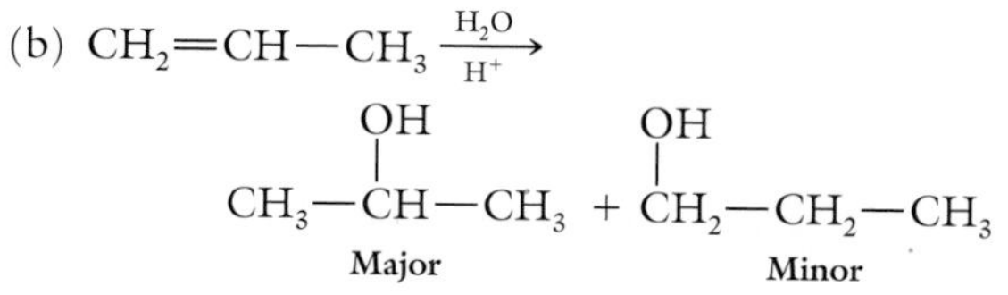

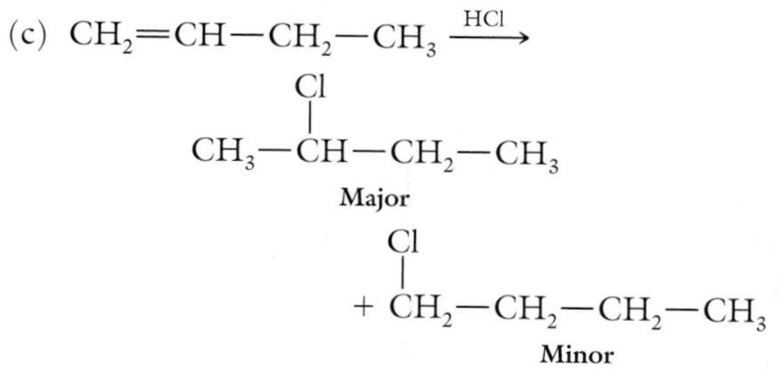

4.7 (a) $CH_2{=}CH{-}CH_2{-}CH_3 \xrightarrow{Cl_2} CH_2(Cl){-}CH(Cl){-}CH_2{-}CH_3$

(b) $CH_2{=}CH{-}CH_3 \xrightarrow[H^+]{H_2O} CH_3{-}CH(OH){-}CH_3$ (**Major**) $+ CH_2(OH){-}CH_2{-}CH_3$ (**Minor**)

(c) $CH_2{=}CH{-}CH_2{-}CH_3 \xrightarrow{HCl} CH_3{-}CH(Cl){-}CH_2{-}CH_3$ (**Major**) $+ CH_2(Cl){-}CH_2{-}CH_2{-}CH_3$ (**Minor**)

4.8 $n\ CH_2{=}CH(CH_3) \xrightarrow{\text{polymerization catalyst}} {-}(CH_2{-}CH(CH_3))_n{-}$

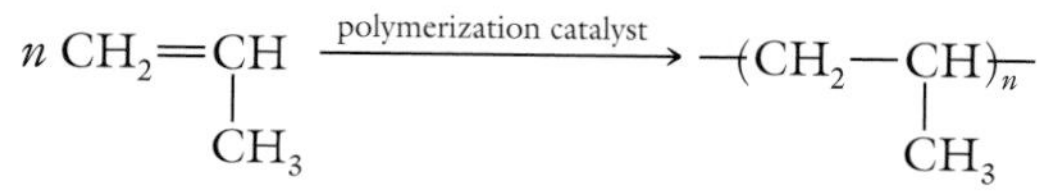

4.9 $C_8H_{16} + 12\ O_2 \longrightarrow 8\ CO_2 + 8\ H_2O$

4.10 $CH_3{-}C{\equiv}C{-}CH_3 \xrightarrow{HCl} CH_3{-}C(Cl){=}CH{-}CH_3 \xrightarrow{HCl} CH_3{-}C(Cl)_2{-}CH_2{-}CH_3$

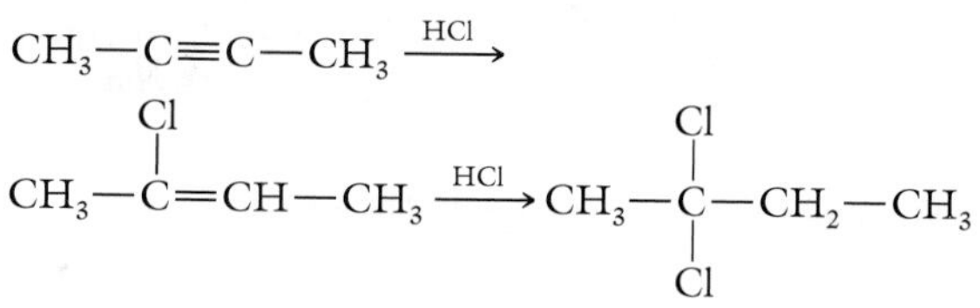

4.11 (a) benzene ring with C(=O)–OH and CH_2CH_3 para (b) benzene ring with Br, CH_3, Br

(c) $CH_2{=}CH{-}CH_2CH_2{-}C_6H_5$

(d) benzene ring with $CH(CH_3)_2$ and HO meta (e) benzene ring with Cl, CH_2CH_3, $CH_2CH_2CH_3$

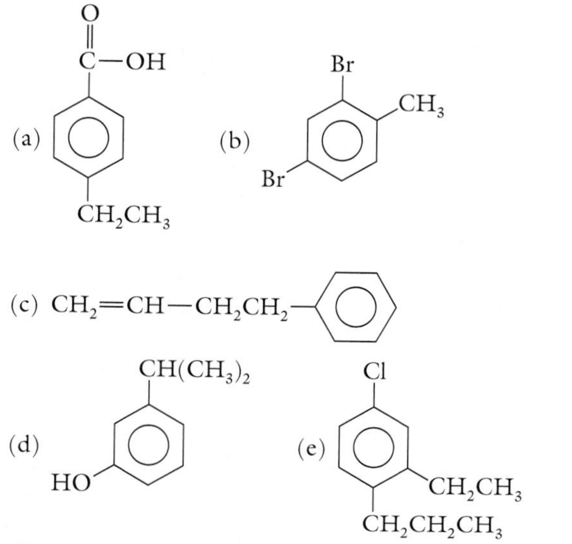

(f) $C_6H_5CH_2Cl$

4.12 (a) No reaction takes place, because alkylation requires a metal halide catalyst.

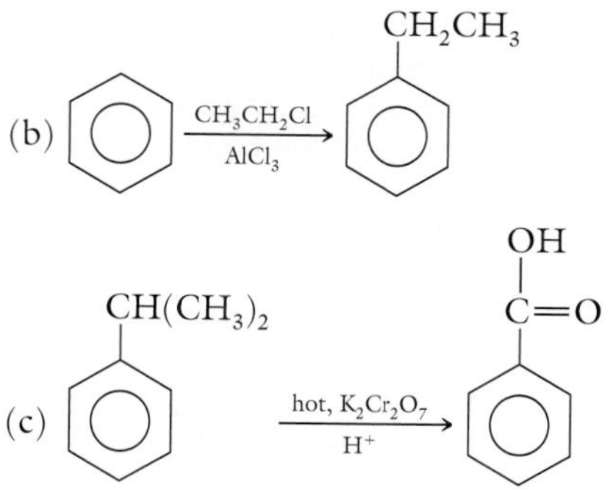

(b) $C_6H_6 \xrightarrow[AlCl_3]{CH_3CH_2Cl} C_6H_5CH_2CH_3$

(c) $C_6H_5CH(CH_3)_2 \xrightarrow[H^+]{\text{hot, } K_2Cr_2O_7} C_6H_5C(=O)OH$

4.13 Benzene does not undergo bromine substitution unless a metal halide is present. Cyclohexane undergoes substitution in the presence of UV radiation (or high temperature):

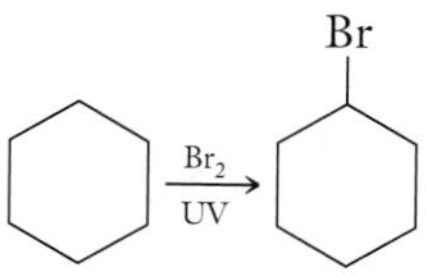

$$\text{cyclohexane} \xrightarrow[UV]{Br_2} \text{bromocyclohexane (Br)}$$

Cyclohexene undergoes addition irrespective of whether there is UV radiation or metal halide:

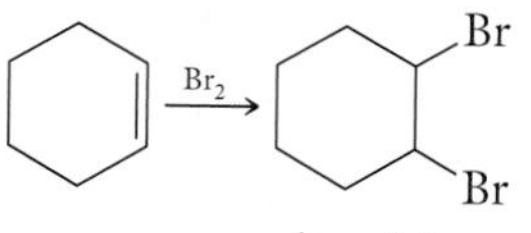

$$\text{cyclohexene} \xrightarrow{Br_2} \text{1,2-dibromocyclohexane (Br, Br)}$$

cis- and *trans-*

Chapter 5

5.1 Structure 1, alcohol; structure 2, ether; structure 3, phenol. Structure 4 is not an alcohol, phenol, or ether; it is an aldehyde.

5.2 Eight isomers:

$CH_3—CH_2—CH_2—CH_2—CH_2—OH$

$CH_3—CH(OH)—CH_2—CH_2—CH_3$

$CH_3—CH_2—CH(OH)—CH_2—CH_3$

$HO—CH_2—CH(CH_3)—CH_2—CH_3$

$CH_3—C(CH_3)(OH)—CH_2—CH_3$ $\quad$ $CH_3—CH(CH_3)—CH(OH)—CH_3$

$CH_3—CH(CH_3)—CH_2—CH_2—OH$ $\quad$ $CH_3—C(CH_3)_2—CH_2—OH$

5.3 Alcohol 1 is $C_5H_{12}O$ with $n = 5$ and $2n + 2 = 12$. Alcohol 2 is C_5H_{10} with $n = 5$ and $2n = 10$.

5.4 (a) Tertiary; (b) primary; (c) secondary; (d) secondary

5.5 (a) 2,5-Dimethyl-3-hexanol; (b) *cis*-2-*t*-butylcyclohexanol; (c) 3-methyl-1,2-butanediol; (d) 4-isopropyl-1-heptanol

5.6 (a) 1,2-Ethanediol has a much higher boiling point (198°C versus 97°C) even though both compounds have nearly the same molecular mass. The presence of two OH groups results in more hydrogen bonding in 1,2-ethanediol. Higher temperatures are needed during boiling to overcome the resulting larger secondary attractive forces in 1,2-ethanediol.

(b) This trend is the same as that observed in all families. Secondary attractive forces and boiling points decrease with branching. Branching decreases the molecular surface area over which molecules attract each other (see Section 12.11).

(c) One compound dissolves in another only when both compounds have similar secondary attractions. Propanol participates in hydrogen bonding and cannot interact with nonpolar compounds. Secondary attractions exist between hexane and butane because both are nonpolar.

5.7

(a) $CH_3CH(OH)CH_3 \xrightarrow[\text{heat}]{H_2SO_4} CH_3CH{=}CH_2 + H_2O$

(b) $CH_3CH_2CH_2CH_2—OH \xrightarrow[\text{heat}]{H_2SO_4} CH_3CH_2CH{=}CH_2 + H_2O$

(c) $CH_3—C(CH_3)_2—OH \xrightarrow[\text{heat}]{H_2SO_4} CH_3—C({=}CH_2)—CH_3 + H_2O$

5.8

(a) $CH_3CH_2—CH(OH)—CH_2CH_3 \xrightarrow{-H_2O} CH_3CH{=}CHCH_2CH_3$
2-Pentene

trans-2-Pentene predominates over *cis*-2-pentene.

(b) 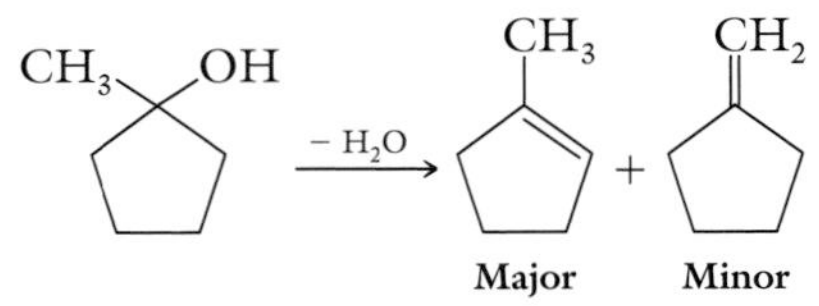 CH_3 OH $\xrightarrow{-H_2O}$ CH_3 + CH_2
Major **Minor**

5.9

(a) $CH_3CH(OH)CH_3 \xrightarrow{Cr_2O_7^{2-} \text{ or } MnO_4^-} CH_3—C({=}O)—CH_3$

(b) 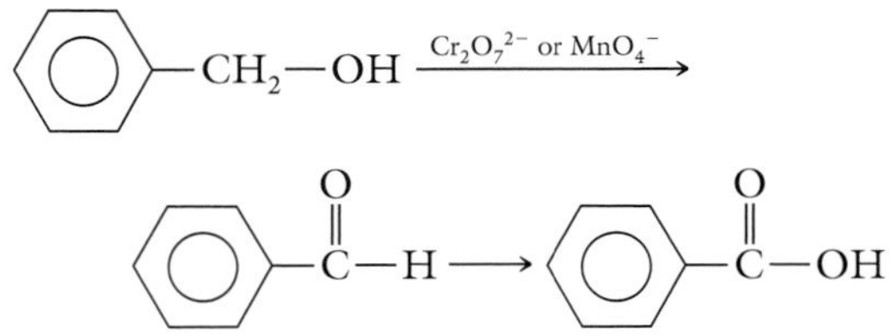 $C_6H_5—CH_2—OH \xrightarrow{Cr_2O_7^{2-} \text{ or } MnO_4^-} C_6H_5—C({=}O)—H \longrightarrow C_6H_5—C({=}O)—OH$

(c) $CH_3—C(CH_3)_2—OH \xrightarrow{Cr_2O_7^{2-} \text{ or } MnO_4^-}$ no reaction
No H

5.10 A positive test with dichromate does not differentiate between 1-butanol (primary alcohol) and 2-butanol (secondary alcohol), because both give positive tests with dichromate (and permanganate).

5.11 (a) 4-Butyl-3-*t*-butylphenol; (b) 3-bromo-2-methylphenol

5.12 Hydrogen bonding is described by representations 2 and 3.

5.13 (a) Butyl *t*-butyl ether; (b) 3-butoxy-3-methyl-1-butene; (c) *s*-butyl isopropyl ether

5.14 The four alcohols of Example 5.2 (butyl alcohol, *s*-butyl alcohol, *t*-butyl alcohol, isobutyl alcohol) are isomeric with the three ethers; all are C_4H_9O.

5.15 (a) *p*-Methylphenol has the higher boiling point because it has an —OH group and can hydrogen bond with itself. Methyl phenyl ether cannot hydrogen bond with itself.

(b) Dipropyl ether is more soluble than 1-hexanol in heptane. Like dissolves in like. The secondary forces in dipropyl ether are the weak London forces, like those in heptane. The secondary forces in 1-hexanol are the strong hydrogen-bonding attractions; there are only very weak attractions between 1-hexanol and the nonpolar heptane.

(c) Diethyl ether and 1-butanol have similar water solubilities. 1-Butanol is slightly more soluble than diethyl ether in water.

5.16

$CH_3CH_2—O—H \xrightarrow[-H_2O]{180°C,\ H_2SO_4} CH_3CH_2—O—CH_2CH_3 + CH_2{=}CH_2$
Minor **Major**

5.17

(a) $CH_3CH_2CH_2SH + NaOH \longrightarrow CH_3CH_2CH_2S^- + Na^+ + H_2O$

(b) $CH_3CH_2{-}S{-}S{-}CH_2CH_3 \xrightarrow{(H)} 2\ CH_3CH_2{-}SH$

(c) $CH_3CH_2CH_2SH + Hg^{2+} \longrightarrow CH_3CH_2CH_2S{-}Hg{-}SCH_2CH_2CH_3 + 2\ H^+$

Chapter 6

6.1 Compound 1, ester; 2, aldehyde; 3, ketone; 4, amide

6.2 There are four aldehydes and three ketones for $C_5H_{10}O$.

$CH_3CH_2CH_2CH_2{-}\overset{O}{\overset{\|}{C}}{-}H$

$CH_3CH_2\overset{CH_3}{\overset{|}{C}H}{-}\overset{O}{\overset{\|}{C}}{-}H$ $\quad CH_3\overset{CH_3}{\overset{|}{C}}HCH_2{-}\overset{O}{\overset{\|}{C}}{-}H$

$(CH_3)_3C{-}\overset{O}{\overset{\|}{C}}{-}H$ $\quad CH_3{-}\overset{O}{\overset{\|}{C}}{-}CH_2CH_2CH_3$

$CH_3CH_2{-}\overset{O}{\overset{\|}{C}}{-}CH_2CH_3$ $\quad CH_3{-}\overset{O}{\overset{\|}{C}}{-}CH(CH_3)_2$

6.3 (a) 2,5-Dimethyl-3-hexanone; (b) 4,4-dimethylpentanal; (c) 3-chlorocyclopentanone; (d) 2-bromo-4-methylbenzaldehyde

6.4 (a) *t*-Butyl cyclohexyl ketone; (b) 3-chlorophenyl propyl ketone

6.5 (a) Hexanal has the higher boiling point because of its higher molecular mass. Secondary attractive forces increase with molecular mass within a family of organic compounds.

(b) 1-Propanal has the higher solubility in water. Increasing molecular mass decreases the fraction of the molecule that is able to hydrogen bond with water.

(c) 2-Propanol has the higher boiling point because it hydrogen bonds with itself. Propanal does not hydrogen bond with itself. The dipole–dipole secondary attractive forces in propanal are weaker than the hydrogen-bonding forces in 2-propanol.

(d) Butanal and 1-butanol have nearly the same solubility in water because both can hydrogen bond with water.

6.6 (a) No oxidation

(b) $CH_3CH_2CH_2CH_2\overset{O}{\overset{\|}{C}}H \xrightarrow{MnO_4^-\text{ or }Cr_2O_7^-} CH_3CH_2CH_2CH_2\overset{O}{\overset{\|}{C}}{-}OH$

6.7 (a) Compounds 2 and 4 are aldehydes and give positive tests with Tollens's reagent. Compounds 1 and 3 are ketones and give negative tests with Tollens's reagent.

(b) Compounds 1, an α-hydroxy ketone, and 4, an α-hydroxy aldehyde, give positive tests with Benedict's reagent. Compound 2, a simple aldehyde, gives a negative test. Compound 3, a simple ketone, gives a negative test.

6.8

(a) 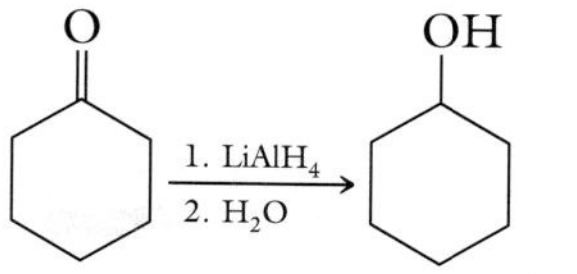

(b) 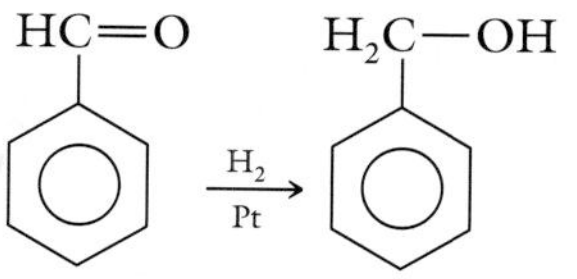

(c) $CH_3\overset{O}{\overset{\|}{C}}CH_2CH_2CH_3 \xrightarrow[2.\ H_2O]{1.\ NADH} CH_3\underset{H}{\underset{|}{\overset{OH}{\overset{|}{C}}}}CH_2CH_2CH_3$

6.9

$CH_3CH_2CH_2\overset{O}{\overset{\|}{C}}H \xrightarrow[H^+]{CH_3OH} CH_3CH_2CH_2\underset{OH}{\underset{|}{\overset{OCH_3}{\overset{|}{C}H}}} \xrightarrow[H^+]{CH_3OH} CH_3CH_2CH_2\underset{OCH_3}{\underset{|}{\overset{OCH_3}{\overset{|}{C}H}}}$

6.10

$CH_3{-}\overset{O}{\overset{\|}{C}}{-}CH_2CH_2CH_2CH_2{-}O{-}H \xrightarrow{H^+}$

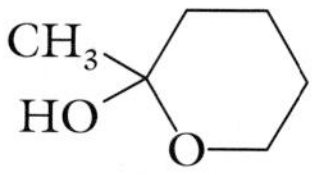

6.11 Compounds 1 and 4, acetal; 2, hemiacetal; 3, something different, a diether.

6.12

(a) $CH_3CH_2O{-}\underset{CH_2C_6H_5}{\underset{|}{C}H}{-}OCH_2CH_3 \xrightarrow{H^+} 2\ CH_3CH_2OH + H\overset{O}{\overset{\|}{C}}{-}CH_2C_6H_5$

(b) 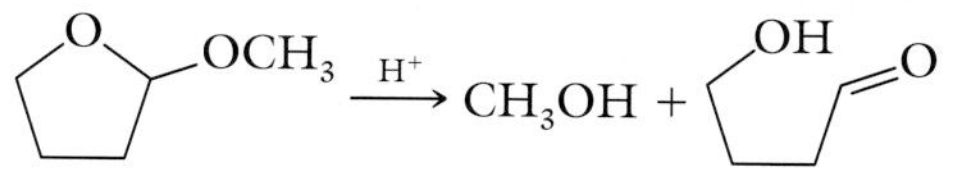

Chapter 7

7.1 Compound 1, ester; 2, aldehyde; 3, amide; 4, acid halide; 5, acid anhydride; 6, carboxylic acid

7.2

(a) 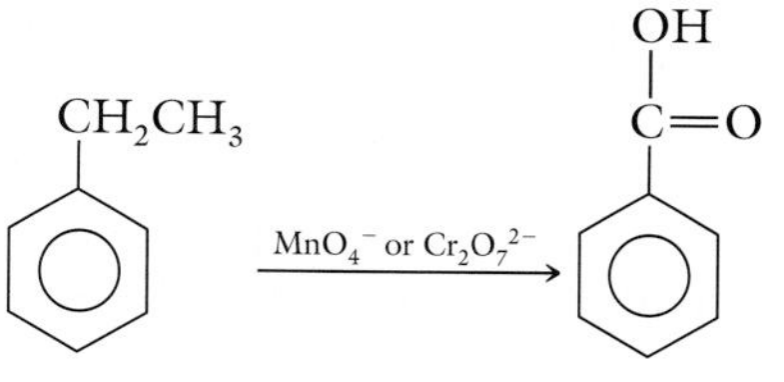

(b) $CH_3CH_2{-}OH \xrightarrow{MnO_4^-\text{ or }Cr_2O_7^{2-}} CH_3\overset{OH}{\overset{|}{C}}{=}O$

(continued on next page)

(c) $CH_3\overset{H}{\overset{|}{C}}{=}OH \xrightarrow{MnO_4^- \text{ or } Cr_2O_7^{2-}} CH_3\overset{OH}{\overset{|}{C}}{=}O$

(d) No reaction takes place, because ketones do not undergo selective oxidation.

7.3 (a) 2, 5-Dimethylheptanoic acid; (b) 4-*t*-butylbenzoic acid or *p*-*t*-butylbenzoic acid; (c) 2-methylpentanedioic acid

7.4

(a) $CH_3CH_2\overset{O}{\overset{\|}{C}}{-}O{-}H + H_2O \rightleftharpoons CH_3CH_2\overset{O}{\overset{\|}{C}}{-}O^- + H_3O^+$

(b) $CH_3CH_2\overset{O}{\overset{\|}{C}}{-}O{-}H + NaOH \longrightarrow CH_3CH_2\overset{O}{\overset{\|}{C}}{-}O^-Na^+ + H_2O$

(c) $CH_3CH_2\overset{O}{\overset{\|}{C}}{-}O{-}H + NaHCO_3 \longrightarrow CH_3CH_2\overset{O}{\overset{\|}{C}}{-}O^-Na^+ + H_2O + CO_2$

(d) no reaction

7.5 (a) Sodium 3-methylbutanoate; (b) aluminum ethanoate (common name: aluminum acetate)

7.6 $CH_3(CH_2)_{14}COOK$ is water soluble, whereas $CH_3(CH_2)_{14}COOH$ and $CH_3(CH_2)_{14}OH$ are water insoluble.

7.7

$CH_3CH_2CH_2\overset{O}{\overset{\|}{C}}{-}OH + HO{-}C_6H_5 \xrightarrow[-\,H_2O]{H^+} CH_3CH_2CH_2\overset{O}{\overset{\|}{C}}{-}O{-}C_6H_5$

7.8 (a) *p*-Ethylphenyl 2-methylbutanoate or 4-ethylphenyl 2-methylbutanoate; (b) pentyl methanoate (common name: pentyl formate); (c) methyl hexanoate

7.9 Polymer is produced only when both reactants are bifunctional. Only the pair in part *c* meets this requirement. The pairs in parts *a* and *b* each have one bifunctional reactant and one monofunctional reactant.

$HO{-}\overset{O}{\overset{\|}{C}}CH_2CH_2CH_2CH_2\overset{O}{\overset{\|}{C}}{-}OH + HO{-}CH_2CH_2{-}OH \xrightarrow[-\,H_2O]{H^+} \left[\overset{O}{\overset{\|}{C}}CH_2CH_2CH_2CH_2\overset{O}{\overset{\|}{C}}{-}O{-}CH_2CH_2{-}O\right]_n$

7.10

(a) $CH_3CH_2CH_2{-}\overset{O}{\overset{\|}{C}}{-}O{-}C_6H_5 \xrightarrow[H_2O]{H^+} CH_3CH_2CH_2{-}\overset{O}{\overset{\|}{C}}{-}OH + HO{-}C_6H_5$

(b) $CH_3CH_2{-}\overset{O}{\overset{\|}{C}}{-}OCH_2CH_2CH_3 \xrightarrow[KOH]{H_2O} CH_3CH_2{-}\overset{O}{\overset{\|}{C}}{-}OK + HOCH_2CH_2CH_3$

7.11

(a) $CH_3CH_2CH_2{-}\overset{O}{\overset{\|}{C}}{-}Br$

(b) $CH_3{-}\overset{O}{\overset{\|}{C}}{-}O{-}\overset{O}{\overset{\|}{C}}{-}CH_3$

7.12

$C_6H_5{-}\overset{O}{\overset{\|}{C}}{-}Cl + CH_3OH \longrightarrow C_6H_5{-}\overset{O}{\overset{\|}{C}}{-}OCH_3 + HCl$

7.13

$HO{-}\underset{OH}{\underset{|}{\overset{O}{\overset{\|}{P}}}}{-}O{-}\underset{OH}{\underset{|}{\overset{O}{\overset{\|}{P}}}}{-}OH \xrightarrow{KOH} KO{-}\underset{OK}{\underset{|}{\overset{O}{\overset{\|}{P}}}}{-}O{-}\underset{OK}{\underset{|}{\overset{O}{\overset{\|}{P}}}}{-}OK$

7.14

$HO{-}\underset{OH}{\underset{|}{\overset{O}{\overset{\|}{P}}}}{-}OH + 2\,CH_3OH \xrightarrow{-\,H_2O} HO{-}\underset{OCH_3}{\underset{|}{\overset{O}{\overset{\|}{P}}}}{-}OCH_3$

Dimethyl phosphate

Chapter 8

8.1 (a) All are amines except compound 5, which is an amide; compound 6 is an amine, not an amide, because the carbonyl carbon is not bonded to nitrogen. (b) Compounds 1, 3, and 7 are primary amines, 4 is a secondary amine, and 2, 6, and 8 are tertiary amines. (c) Only compound 7 is an aromatic amine; the other amines are aliphatic amines.

8.2 $CH_3CH_2CH_2CH_2{-}NH_2$ $\quad CH_3{-}\underset{NH_2}{\underset{|}{CH}}{-}CH_2CH_3$

$(CH_3)_2CH{-}CH_2{-}NH_2$ $\quad CH_3{-}\underset{NH_2}{\underset{|}{\overset{CH_3}{\overset{|}{C}}}}{-}CH_3$

$CH_3CH_2CH_2{-}\underset{H}{\underset{|}{N}}{-}CH_3$ $\quad (CH_3)_2CH{-}\underset{H}{\underset{|}{N}}{-}CH_3$

$CH_3CH_2{-}\underset{H}{\underset{|}{N}}{-}CH_2CH_3$ $\quad CH_3CH_2{-}\underset{CH_3}{\underset{|}{N}}{-}CH_3$

8.3 (a) Common name: *s*-butylethylmethylamine or IUPAC: *N*-ethyl-*N*-methyl-2-butanamine; (b) *CA:* 1,2-butanediamine; (c) *CA: trans*-3-ethyl-*N,N*-dimethylcyclopentanamine; (d) common: *N*-ethyl-3-isopropyl-4-methylaniline; (e) *CA*: 4-ethyl-2,6-dimethyl-2-heptanamine

8.4 (a) $2\,C_6H_5NH_2 + H_2SO_4 \longrightarrow [C_6H_5NH_3^+]_2\,SO_4^{2-}$

(b) $(CH_3CH_2)_2NH + H_2O \rightleftharpoons (CH_3CH_2)_2\overset{+}{N}H_2 + OH^-$

8.5

(a) 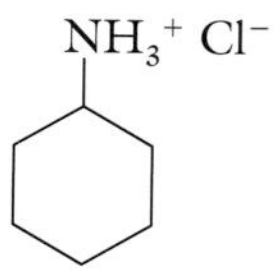 Cyclohexyl ring bearing $NH_3^+\ Cl^-$

(b)
$$CH_3CH_2CH_2CH_2-\overset{+}{N}(CH_3)(H)-CH_3\ HSO_4^-$$

(c) 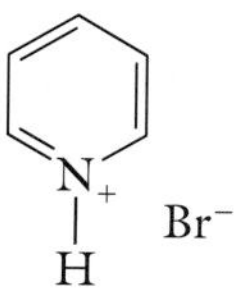 Pyridine ring with $\overset{+}{N}-H$, Br^-

8.6 The order is $CH_3(CH_2)_{14}NH_2 < CH_3(CH_2)_6NH_2 \ll CH_3(CH_2)_{14}NH_3^+Cl^-$. The ionic amine salt is much more soluble than the covalent amines because ions are attracted to water more strongly than are covalent molecules. The lower-molecular-mass amine is more soluble than the higher-molecular-mass amine because the amine group, which hydrogen bonds with water, constitutes a larger part of the molecule.

8.7 Compound 1, tertiary amide; 2, secondary amide; 3 is not an amide—it is an amine and an aldehyde; 4, tertiary amide.

8.8

(a)
$$C_6H_5-\overset{O}{\overset{\|}{C}}-OH + H_2N-CH_3 \longrightarrow C_6H_5-\overset{O}{\overset{\|}{C}}-O^-\ H_3\overset{+}{N}-CH_3$$

(b) The reaction temperature is that given for amide formation, but amide formation does not take place when tertiary amines are used, because there is no hydrogen attached to the nitrogen of the amine. The only possible reaction is acid–base reaction:

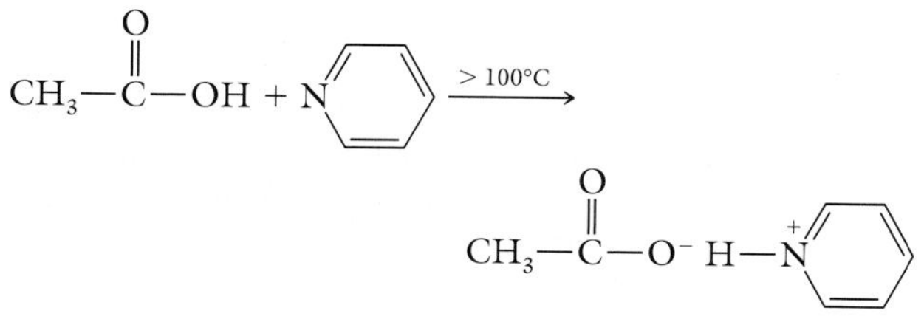

$$CH_3-\overset{O}{\overset{\|}{C}}-OH + N\!\text{(pyridine)} \xrightarrow{>100°C} CH_3-\overset{O}{\overset{\|}{C}}-O^-\ H-\overset{+}{N}\!\text{(pyridine)}$$

(c)
$$C_6H_5-\overset{O}{\overset{\|}{C}}-OH + HN(CH_3)-CH_3 \xrightarrow[-\,H_2O]{>100°C} C_6H_5-\overset{O}{\overset{\|}{C}}-N(CH_3)-CH_3$$

8.9

$$CH_3CH_2-\overset{O}{\overset{\|}{C}}-O-\overset{O}{\overset{\|}{C}}-CH_2CH_3 + HN(CH_3)_2 \longrightarrow CH_3CH_2-\overset{O}{\overset{\|}{C}}-N(CH_3)_2 + CH_3CH_2-\overset{O}{\overset{\|}{C}}-OH$$

8.10 (a) *N*-Butyl-*N*-methylpropanamide or *N*-butyl-*N*-methylpropionamide; (b) *N*-phenylbenzamide; (c) 2-aminopropanal or 2-aminopropionaldehyde; (d) *N*-benzyl-*N*-methyl-2-methylpropanamide or *N*-benzyl-*N*-methylisobutyramide

8.11

(a)
$$CH_3CH_2CH_2-\overset{O}{\overset{\|}{C}}-\overset{H}{\overset{|}{N}}-C_6H_5 \xrightarrow[HCl]{H_2O} CH_3CH_2CH_2-\overset{O}{\overset{\|}{C}}-OH + H_3\overset{+}{N}C_6H_5\ Cl^-$$

(b)
$$CH_3CH_2-\overset{O}{\overset{\|}{C}}-\overset{H}{\overset{|}{N}}CH_2CH_2CH_3 \xrightarrow[NaOH]{H_2O} CH_3CH_2-\overset{O}{\overset{\|}{C}}-O^-\ Na^+ + H_2NCH_2CH_2CH_3$$

Chapter 9

9.1 (a) There is only one 3-hydroxypropanal because none of the carbon atoms have four different substituents.

$$\overset{OH}{\overset{|}{H_2C}}-CH_2-\overset{O}{\overset{\|}{CH}}$$

(b) A pair of enantiomers exists because C2 is a tetrahedral stereocenter.

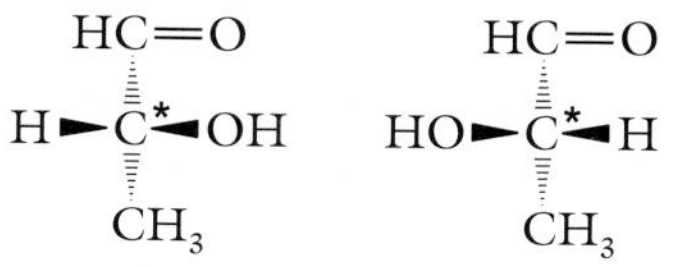

Structure 1: $HC{=}O$ (top), $H\blacktriangleright C^*\blacktriangleleft OH$, CH_3 (bottom)

Structure 2: $HC{=}O$ (top), $HO\blacktriangleright C^*\blacktriangleleft H$, CH_3 (bottom)

9.2 (a) Step 1: Structures 1 and 2 are not superimposable. Step 2: Structures 1 and 2 are not mirror images. Step 3: Rotation of structure 1 by 180° in the plane of the paper yields structure 1′, which is the nonsuperimposable mirror image of 2. Structures 1 and 2 are a pair of enantiomers.

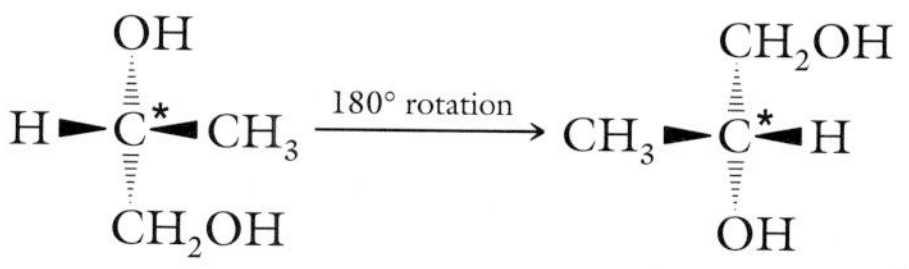

Structure 1: OH (top), $H\blacktriangleright C^*\blacktriangleleft CH_3$, CH_2OH (bottom) $\xrightarrow{180°\ \text{rotation}}$ Structure 1′: CH_2OH (top), $CH_3\blacktriangleright C^*\blacktriangleleft H$, OH (bottom)

(b) Step 1: Structures 3 and 4 are not superimposable. Step 2: Structures 3 and 4 are not mirror images. Step 3: Rotation of structure 3 by 180° in the plane of the paper yields structure 3′, which is the same as structure 4. Structures 3 and 4 are the same.

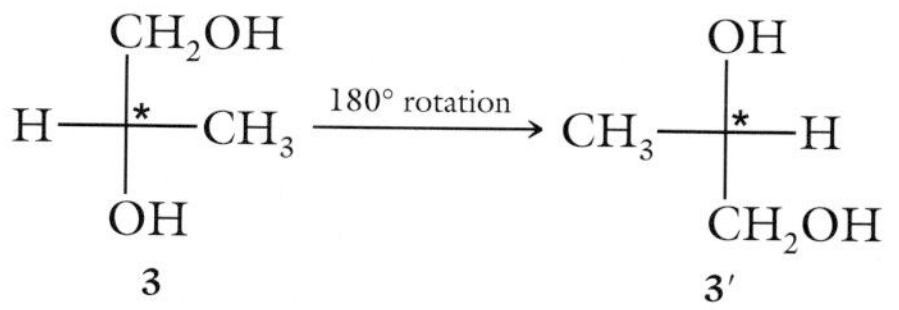

Structure 3: CH_2OH (top), $H-*-CH_3$, OH (bottom) $\xrightarrow{180°\ \text{rotation}}$ Structure 3′: OH (top), CH_3-*-H, CH_2OH (bottom)

9.3 The two carbon substituents are vertical, with the more-substituted substituent (COOH) upward. This enantiomer is the L-enantiomer because the hetero substituent (NH_2) is on the left side of the tetrahedral stereocenter.

9.4 $\alpha = [\alpha]CL = +53° \times 0.030\ g/mL \times 1.5\ dm = +2.4°$

9.5 There are two tetrahedral stereocenters. Only three stereoisomers are possible because the tetrahedral stereocenters possess the same four different substituents:

(*continued on next page*)

COOH / H—C—OH / HO—C—H / COOH (1); COOH / HO—C—H / H—C—OH / COOH (2); COOH / HO—C—H / HO—C—H / COOH (3)

Stereoisomers 1 and 2 are optically active. Stereoisomer 3 is optically inactive because it is a meso compound. Stereoisomers 1 and 2 are enantiomers of each other, and each is a diastereomer of 3. Stereoisomer 3 is diastereomer of both 1 and 2, and vice versa.

9.6 (a) There are no stereoisomers, because the two halves of the ring at C1 are the same.

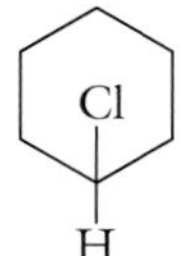

(b) There are four stereoisomers because the two tetrahedral stereocenters do not have the same four different substituents.

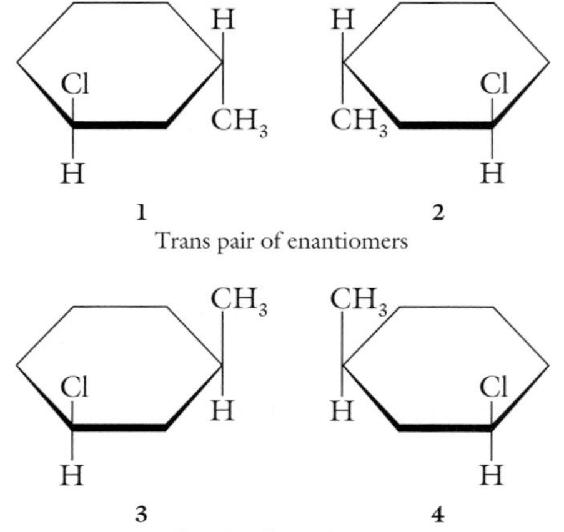

Trans pair of enantiomers (1, 2)

Cis pair of enantiomers (3, 4)

All four compounds are optically active. Each stereoisomer is simultaneously both an enantiomer and a diastereomer. Stereoisomers 1 and 2 are enantiomers of each other, and each is a diastereomer of both 3 and 4. Stereoisomers 3 and 4 are enantiomers of each other, and each is a diastereomer of both 1 and 2.

Chapter 10

10.1 1, aldotriose; 2, ketohexose; 3, aldopentose; 4, aldotetrose

10.2 (a) Diastereomers; they are stereoisomers that are not enantiomers. (b) Not isomers; D-ribose is a pentose and D-sorbose is a hexose. (c) Enantiomers; they are nonsuperimposable mirror-image stereoisomers. (d) Constitutional isomers; D-erythrose is an aldotetrose and L-erythrulose is a ketotetrose. (e) Diastereomers; they are stereoisomers that are not enantiomers.

10.3 Figure 10.2 shows that D-sorbose differs from D-fructose at the configurations at C3 and C4. Draw the cyclic structure of β-D-fructose and then reverse the configurations at C3 and C4.

β-D-Fructose —(Reverse C3 and C4 configurations)→ β-D-Sorbose

10.4

Ketose —(base)→ Aldose —(Cu^{2+})→ (acid)

10.5 The monosaccharides are D-mannose and D-fructose. D-Mannose differs from D-glucose only at the configuration of C2 (Figure 10.1). The linkage between disaccharide residues is an α(1 → 6)-glycosidic linkage from D-mannose to D-fructose. The disaccharide is a reducing sugar and undergoes mutarotation because the hemiacetal linkage of D-fructose is intact.

Chapter 11

11.1 Acids 1 and 3 are found in lipids because both are unbranched and contain an even number of carbons, and the double bond in acid 3 has a cis configuration. Acid 2 is not usually found in lipids, because it is branched.

11.2

CH_2OH—$CHOH$—CH_2OH + HO—$C(=O)(CH_2)_{10}CH_3$ + HO—$C(=O)(CH_2)_7CH{=}CH(CH_2)_5CH_3$ + HO—$C(=O)(CH_2)_6(CH_2CH{=}CH)_2(CH_2)_4CH_3$ $\xrightarrow{-3\,H_2O}$

CH_2—O—$C(=O)(CH_2)_{10}CH_3$
CH—O—$C(=O)(CH_2)_7CH{=}CH(CH_2)_5CH_3$
CH_2—O—$C(=O)(CH_2)_6(CH_2CH{=}CH)_2(CH_2)_4CH_3$

11.3

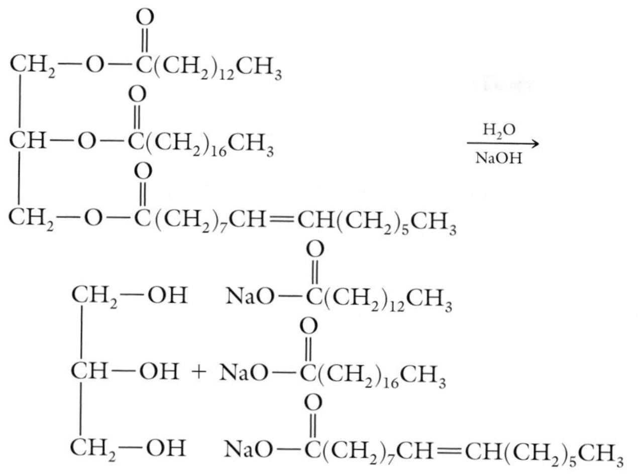

$$\begin{array}{l} CH_2-O-\overset{O}{\overset{\|}{C}}(CH_2)_{12}CH_3 \\ | \\ CH-O-\overset{O}{\overset{\|}{C}}(CH_2)_{16}CH_3 \\ | \\ CH_2-O-\overset{O}{\overset{\|}{C}}(CH_2)_7CH=CH(CH_2)_5CH_3 \end{array} \xrightarrow[NaOH]{H_2O}$$

$$\begin{array}{l} CH_2-OH \\ | \\ CH-OH \\ | \\ CH_2-OH \end{array} + \begin{array}{l} NaO-\overset{O}{\overset{\|}{C}}(CH_2)_{12}CH_3 \\ NaO-\overset{O}{\overset{\|}{C}}(CH_2)_{16}CH_3 \\ NaO-\overset{O}{\overset{\|}{C}}(CH_2)_7CH=CH(CH_2)_5CH_3 \end{array}$$

11.4

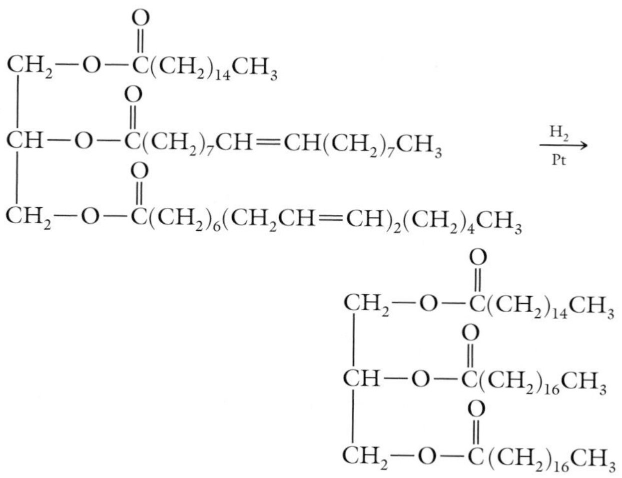

$$\begin{array}{l} CH_2-O-\overset{O}{\overset{\|}{C}}(CH_2)_{14}CH_3 \\ | \\ CH-O-\overset{O}{\overset{\|}{C}}(CH_2)_7CH=CH(CH_2)_7CH_3 \\ | \\ CH_2-O-\overset{O}{\overset{\|}{C}}(CH_2)_6(CH_2CH=CH)_2(CH_2)_4CH_3 \end{array} \xrightarrow[Pt]{H_2}$$

$$\begin{array}{l} CH_2-O-\overset{O}{\overset{\|}{C}}(CH_2)_{14}CH_3 \\ | \\ CH-O-\overset{O}{\overset{\|}{C}}(CH_2)_{16}CH_3 \\ | \\ CH_2-O-\overset{O}{\overset{\|}{C}}(CH_2)_{16}CH_3 \end{array}$$

11.5

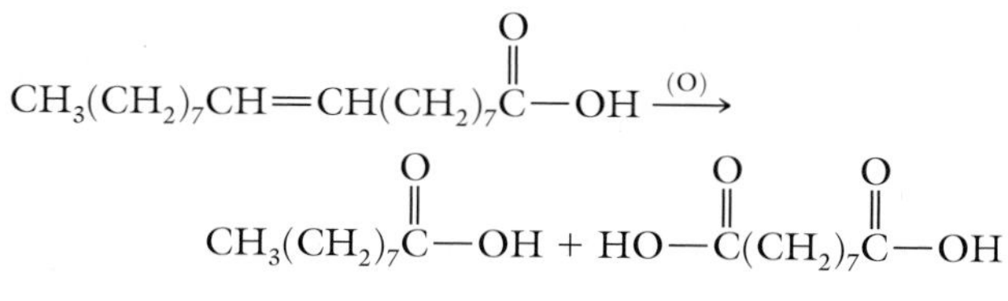

$$CH_3(CH_2)_7CH=CH(CH_2)_7\overset{O}{\overset{\|}{C}}-OH \xrightarrow{(O)}$$

$$CH_3(CH_2)_7\overset{O}{\overset{\|}{C}}-OH + HO-\overset{O}{\overset{\|}{C}}(CH_2)_7\overset{O}{\overset{\|}{C}}-OH$$

11.6

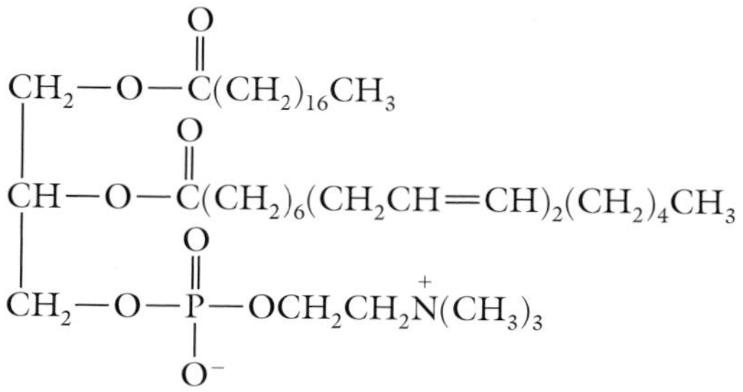

$$\begin{array}{l} CH_2-O-\overset{O}{\overset{\|}{C}}(CH_2)_{16}CH_3 \\ | \\ CH-O-\overset{O}{\overset{\|}{C}}(CH_2)_6(CH_2CH=CH)_2(CH_2)_4CH_3 \\ | \\ CH_2-O-\underset{O^-}{\underset{|}{\overset{O}{\overset{\|}{P}}}}-OCH_2CH_2\overset{+}{N}(CH_3)_3 \end{array}$$

11.7

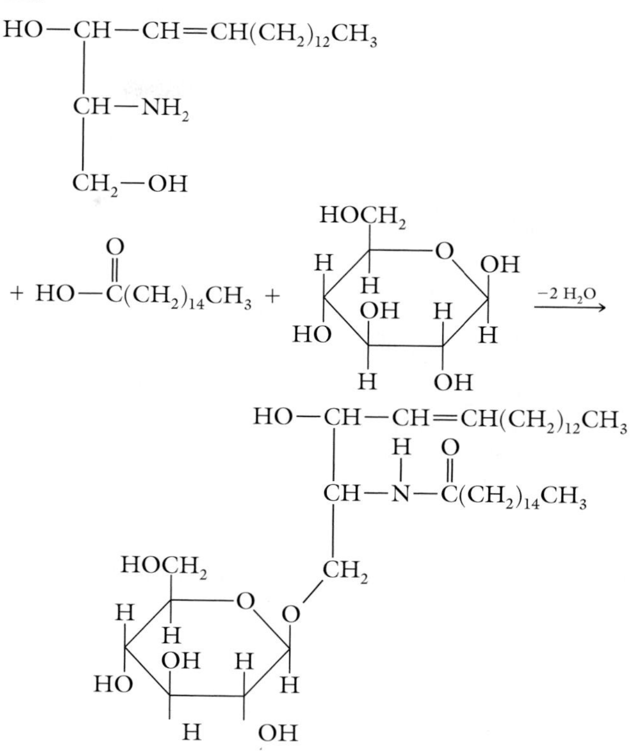

$$\begin{array}{l} HO-CH-CH=CH(CH_2)_{12}CH_3 \\ \quad\ | \\ \quad\ CH-NH_2 \\ \quad\ | \\ \quad\ CH_2-OH \end{array}$$

$$+ HO-\overset{O}{\overset{\|}{C}}(CH_2)_{14}CH_3 + \text{(galactose: HOCH}_2\text{, H, O, OH, H, OH, H, HO, H, H, OH ring)} \xrightarrow{-2\,H_2O}$$

$$\begin{array}{l} HO-CH-CH=CH(CH_2)_{12}CH_3 \\ \quad\ | \\ \quad\ CH-\underset{}{\overset{H}{\overset{|}{N}}}-\overset{O}{\overset{\|}{C}}(CH_2)_{14}CH_3 \\ \quad\ | \\ \quad\ CH_2 \\ \quad\ | \\ \quad\ O-\text{(galactose ring: HOCH}_2\text{, H, O, H, OH, H, HO, H, H, OH)} \end{array}$$

Chapter 12

12.1

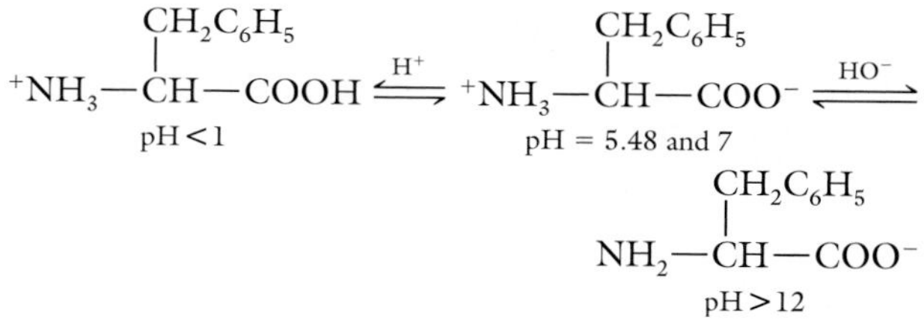

$$\underset{pH<1}{^+NH_3-\underset{}{\overset{CH_2C_6H_5}{\overset{|}{C}H}}-COOH} \underset{}{\overset{H^+}{\rightleftharpoons}} \underset{pH\,=\,5.48\text{ and }7}{^+NH_3-\overset{CH_2C_6H_5}{\overset{|}{C}H}-COO^-} \overset{HO^-}{\rightleftharpoons} \underset{pH>12}{NH_2-\overset{CH_2C_6H_5}{\overset{|}{C}H}-COO^-}$$

12.2

$$\underset{\textbf{Ala}}{H_3\overset{+}{N}-\underset{CH_3}{\underset{|}{C}H}-\overset{O}{\overset{\|}{C}}-O^-} + \underset{\textbf{Lys}}{H_3\overset{+}{N}-\underset{(CH_2)_4\overset{+}{N}H_3}{\underset{|}{C}H}-\overset{O}{\overset{\|}{C}}-O^-}$$

$$+ \underset{\textbf{Phe}}{H_3\overset{+}{N}-\underset{CH_2C_6H_5}{\underset{|}{C}H}-\overset{O}{\overset{\|}{C}}-O^-} \xrightarrow{-2\,H_2O}$$

$$H_3\overset{+}{N}-\underset{CH_3}{\underset{|}{C}H}-\overset{O}{\overset{\|}{C}}-\overset{H}{\overset{|}{N}}-\underset{(CH_2)_4\overset{+}{N}H_3}{\underset{|}{C}H}-\overset{O}{\overset{\|}{C}}-\overset{H}{\overset{|}{N}}-\underset{CH_2C_6H_5}{\underset{|}{C}H}-\overset{O}{\overset{\|}{C}}-O^-$$

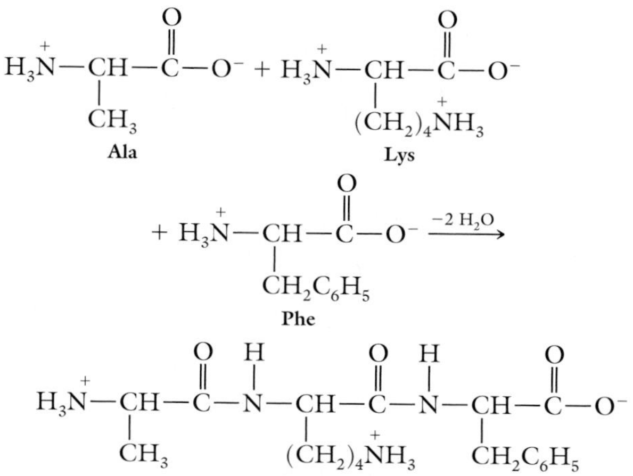

Ala-Lys-Phe
Alanyllysylphenylalanine

12.3 There are 3! = 6 constitutional isomers: Glu-Ile-Lys, Glu-Lys-Ile, Ile-Glu-Lys, Ile-Lys-Glu, Lys-Glu-Ile, Lys-Ile-Glu.

12.4 (a)

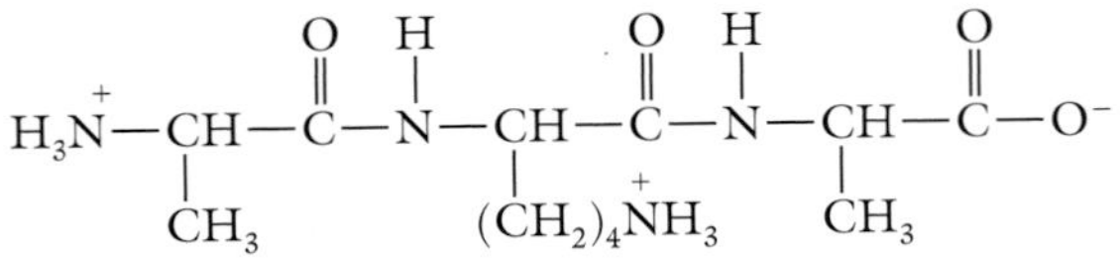

$$\overset{+}{H_3N}-\underset{\underset{CH_3}{|}}{CH}-\overset{O}{\overset{\|}{C}}-\overset{H}{\overset{|}{N}}-\underset{\underset{(CH_2)_4\overset{+}{N}H_3}{|}}{CH}-\overset{O}{\overset{\|}{C}}-\overset{H}{\overset{|}{N}}-\underset{\underset{CH_3}{|}}{CH}-\overset{O}{\overset{\|}{C}}-O^-$$

Ala-Lys-Ala has a charge of 1+ and migrates to the cathode. The pI is on the basic side of that of a peptide containing only neutral residues.

(b)

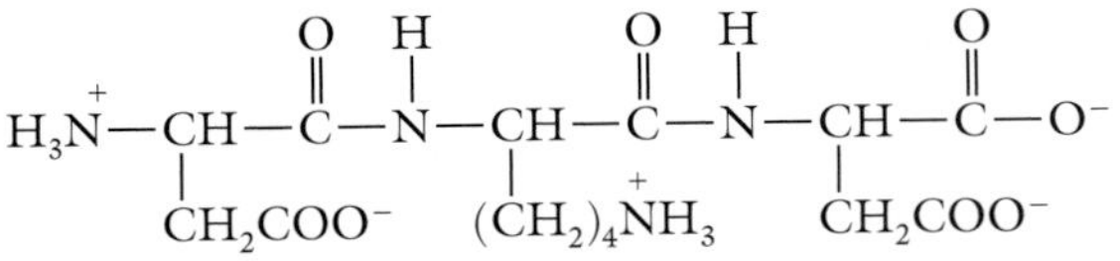

$$\overset{+}{H_3N}-\underset{\underset{CH_2COO^-}{|}}{CH}-\overset{O}{\overset{\|}{C}}-\overset{H}{\overset{|}{N}}-\underset{\underset{(CH_2)_4\overset{+}{N}H_3}{|}}{CH}-\overset{O}{\overset{\|}{C}}-\overset{H}{\overset{|}{N}}-\underset{\underset{CH_2COO^-}{|}}{CH}-\overset{O}{\overset{\|}{C}}-O^-$$

Asp-Lys-Asp has a charge of 1− and migrates to the anode. The pI is on the acid side of that of peptides with only neutral amino acid residues.

Chapter 13

13.1 The name is deoxyguanosine 5′-monophosphate, or dGMP.

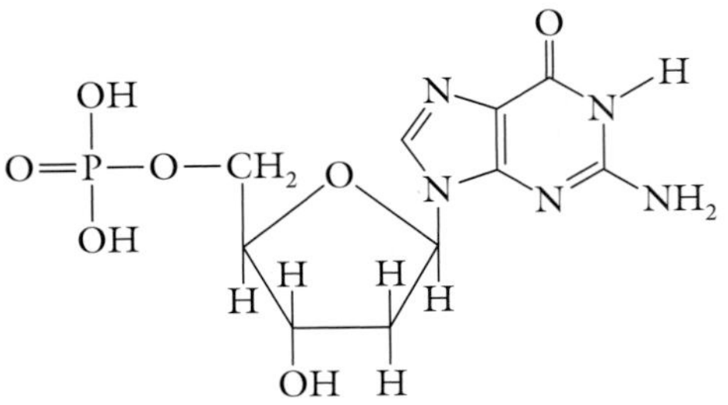

13.2 The components are phosphoric acid, ribose, and uracil. The name is uridine 5′-monophosphate.

13.3

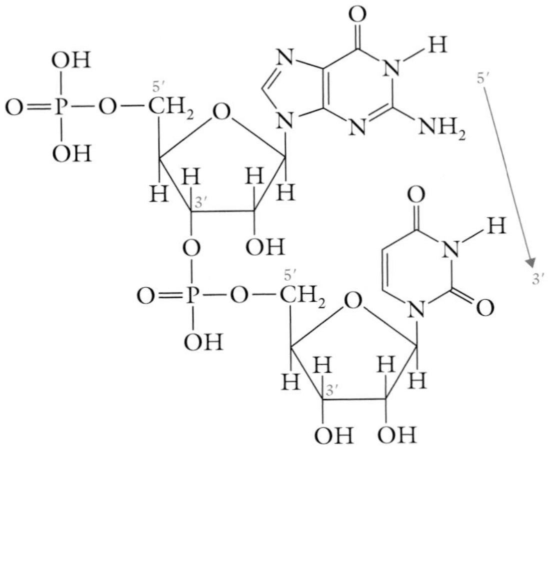

13.4

5′ GGCTAT 3′ **Given strand**

3′ CCGATA 5′ **Complementary strand**

The sequence for the complementary strand is 5′-ATAGCC-3′.

13.5 The complementary parent strand and the new daughter strand are identical and complementary to the ACGTGC sequence. The sequence is 3′-TGCACG-5′ (5′-GCACGT-3′).

13.6 DNA nontemplate strand: GTTGACTTA; RNA: GUUGACUUA

13.7 Anticodon (written 5′ ⟶ 3′): (a) GGA; (b) CUG; (c) CCU; (d) AGC. α-Amino acid: (a) Ser; (b) Gln; (c) Arg; (d) Ala.

13.8

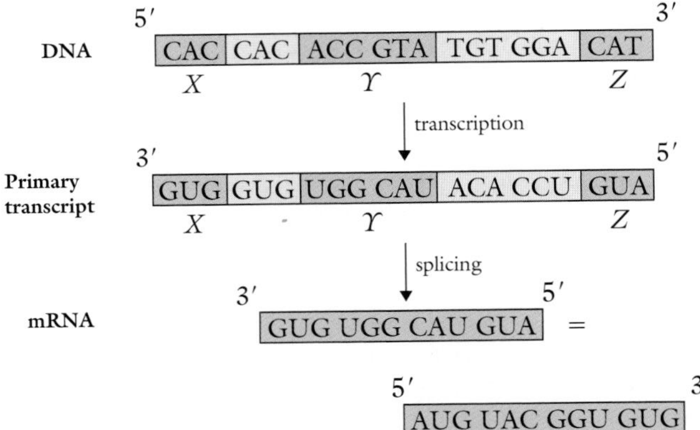

The mRNA written in the 5′ ⟶ 3′ direction is AUG-UAC-GGU-GUG, which codes (Table 21.1) for the polypeptide sequence Met-Tyr-Gly-Val.

13.9 (a) 5′-GAA-3′ in DNA results in 3′-CUU-5′ in mRNA. The codon is 5′-UUC-3′ and codes for Phe. The α-amino acid sequence would be changed: Tyr would be replaced by Phe.

(b) 5′-TTA-3′ in DNA results in 3′-AAU-5′ in mRNA. The codon is 5′-UAA-3′ and codes for Stop. Polypeptide synthesis would be terminated prematurely.

(c) There would be no effects on either the mRNA or the polypeptide. A mutation in an intron is constrained within the intron and has no effect on the base sequence in the exons, because the intron is excised.

Answers to Odd-Numbered Exercises

Chapter 1

1.1 Proton, mass 1.00728 amu, charge, + 1; neutron, mass 1.00867 amu, charge, 0; electron, mass 0.0005486 amu, charge, − 1

1.3 Atomic mass includes the mass of neutrons.

1.5 It would be a cation, charge 3+.

1.7

Element	Group	Period
Li	I	2
Na	I	3
K	I	4
Rb	I	5
Cs	I	6

1.9 Li, Na, K, Rb, and Cs are metals.

1.11 Group I

1.13 Group I. Each possesses one outer electron.

1.15 Group II. Each possesses two outer electrons.

1.17 True

1.19 The formation of either ionic or covalent bonds

1.21 2 (Group I) : 1 (Group VI)

1.23 2 (Group III) : 3 (Group VI)

1.25 Al_2O_3, Al_2S_3, and Al_2Se_3

1.27

```
      :C̈l:
       |
 :C̈l—C—C̈l:
       |
      :C̈l:
```

1.29 H—S̈—H (with lone pairs on S)

1.31 :Ö=C=Ö:

1.33 The four fluorine atoms are situated at the corners of a tetrahedron with the boron atom at its center.

1.35 (a) A tetrahedron, no dipole moment; (b) a tetrahedron, no dipole moment; (c) linear, no dipole moment

1.37 The four C–Cl polar bonds are arranged symmetrically (tetrahedrally) with carbon at the center of the tetrahedron. The symmetrical arrangement of dipoles results in a net dipole moment of zero.

1.39

	London force	Dipole–dipole	Hydrogen bond
CH_4	yes	no	no
$CHCl_3$	yes	yes	no
NH_3	yes	yes	yes

1.41 Acetic acid molecules can form hydrogen bonds among themselves and to water. The attractive forces between acetic acid and water are similar, and therefore acetic acid will dissolve in water.

1.43 They have virtually identical chemical properties.

1.45 Metal

1.47 In chemical reactions, atoms tend to attain the noble-gas outer-shell configuration of eight electrons.

1.49 The chemical properties are identical, but the nucleus of carbon-12 contains six neutrons and that of carbon-13 contains seven neutrons.

1.51

```
     H
     |
 H—C—Ö—H
     |
     H
```

1.53 Elements in the same family have similar chemical properties.

1.55 (a) $C < N < O$; (b) $Sr < Ca < Mg$; (c) $Pb < Ge < C$; (d) $Po < Te < Se$

Chapter 2

2.1 (a) Precipitate forms:

$$Ag^+(aq) + Cl^-(aq) \longrightarrow AgCl(s)$$

(b) Precipitate forms:

$$Mg^{2+}(aq) + 2\ OH^-(aq) \longrightarrow Mg(OH)_2(s)$$

2.3 (a) Precipitate forms:

$$Ba^{2+}(aq) + SO_4^{2-}(aq) \longrightarrow BaSO_4\ (s)$$

(b) Precipitate forms:

$$Al^{3+}(aq) + 3\ OH^-(aq) \longrightarrow Al(OH)_3(s)$$

2.5 Reaction rate a is greater than reaction rate b.

2.7 $E_{back} = +58$ kJ

2.9 No. The reaction takes place in an open system, the CO_2 escapes into the atmosphere, and the system can never come to equilibrium.

2.11 Forward reaction:

$$CaCO_3(s) \longrightarrow CaO(s) + CO_2(g)$$

Back reaction:

$$CaO(s) + CO_2(g) \longrightarrow CaCO_3(s)$$

2.13 $K_{eq} = 0.099$

2.15 Reaction shifts to the right (to product).

2.17 Color shift to blue

2.19 $COCl_2(g) \rightleftharpoons CO(g) + Cl_2(g)$

2.21 $PCl_3(g) + Cl_2(g) \rightleftharpoons PCl_5(g)$

2.23 (a) $K_{eq} = \dfrac{[NO_2][NO_3]}{[N_2O_5]}$

2.25 $K_w = 1.00 \times 10^{-14}$

2.27 A strong base is 100% dissociated in aqueous solution. Examples: NaOH and KOH

2.29 $[H_3O^+] = [Cl^-] = 0.30\ M$

2.31 The pH of pure water at 25°C is 7.00.

2.33 pH = 2.0

2.35 A weak acid is one that is incompletely dissociated in aqueous solution. Examples: acetic and phosphoric acids

2.37

	Acid	Base
(a)	HNO_2	NO_2^-
	H_3O^+	H_2O
(b)	$H_2PO_4^-$	HPO_4^{2-}
	H_3O^+	H_2O

2.39 The conjugate acid prevents shifts to basicity, and the basic form prevents shifts to acidity.

2.41 pH = pK_a = 3.75

2.43 pH = 7.02

2.45 (a) 2.16; (b) 1.72

2.47 Molar ratio of $[HPO_4^{2-}/H_2PO_4^-] = 10/1$.

2.49 (a) pH = 2.47, $[OH^-] = 3.0 \times 10^{-12}$ *M*; (b) pH = 1.60, $[OH^-] = 4.0 \times 10^{-13}$ *M*

2.51 NO is a catalyst. It is a reactant in the first reaction and emerges unchanged as a product in the second reaction.

2.53 If a system in an equilibrium state is disturbed, the system will adjust to neutralize the disturbance and restore the system to equilibrium.

2.55 Ethanol is in excess; therefore it is considered to be the solvent, and water is the solute.

2.57 0.649 *M*

2.59 A conjugate acid–base pair consists of a weak acid and the basic anion resulting from its dissociation.

2.61 When a substance melts, the intermolecular forces are only loosened so that the order is reduced, but the molecules are still largely in contact with one another, and the liquid is about as incompressible as the solid.

2.63 An overall equilibrium constant for a sequence of reactions can be calculated only if a product of the first reaction is a reactant of the next reaction, and so on, for each succeeding reaction.

2.65 The pH at the equivalence point of the titration in Exercise 2.64 could not be 7.00. At the equivalence point, the acid has been completely neutralized and the resulting solution is that of its salt. Because the salt consists of a basic anion, hydrolysis of the anion will cause the solution to be basic. The solution's pH will be greater than 7.00.

Chapter 3

3.1 The number of covalent bonds formed by the atom of an element in forming a compound.

3.3 a

3.5 b

3.7 (a) C—C and C—H single bonds

(b) 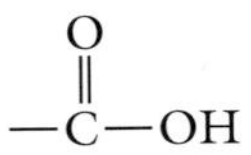 —C(=O)—OH (c) —C(=O)—H

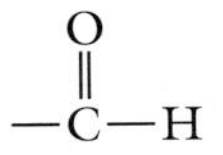

3.9 (a) Alcohol; (b) alkene; (c) ketone; (d) aromatic; (e) ketone; (f) ether; (g) carboxylic acid; (h) amide

3.11 *sp*[3] orbitals are proposed because carbon is tetravalent with four equivalent bonds that have 109.5° bond angles.

3.13

(a) 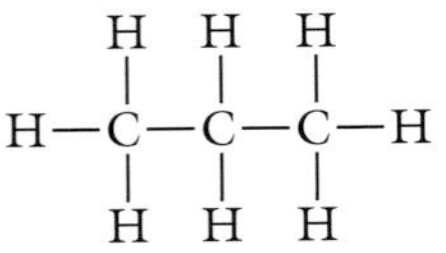 $H_3C—CH_2—CH_3$ (all H shown): H—C(H)(H)—C(H)(H)—C(H)(H)—H

(b) H—C(H)(H)—C(H)(H)—C(H)(H)—C(H)(H)—C(H)(H)—H

(c) H—C(H)(H)—C(H)(H)—C(H)(H)—C(H)(CH₃ as C(H)(H)(H))—C(H)(H)—H

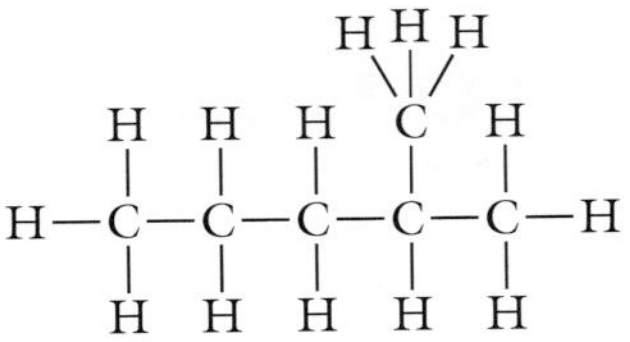

3.15 (a) 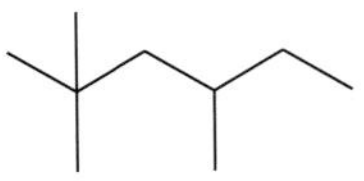 (b) 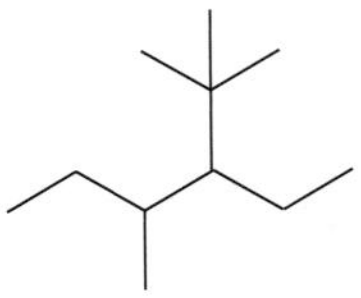 (c) 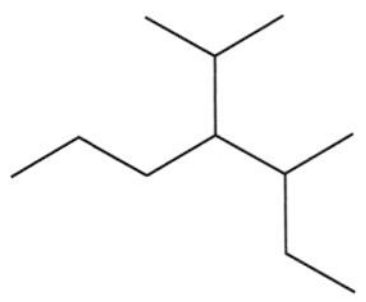 (d)

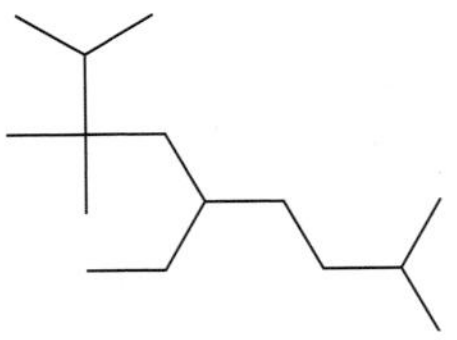

3.17 (a) 1; (b) 2; (c) 2

3.19

$CH_3—CH_2—CH_2—CH_2—CH_2—CH_2—CH_3$

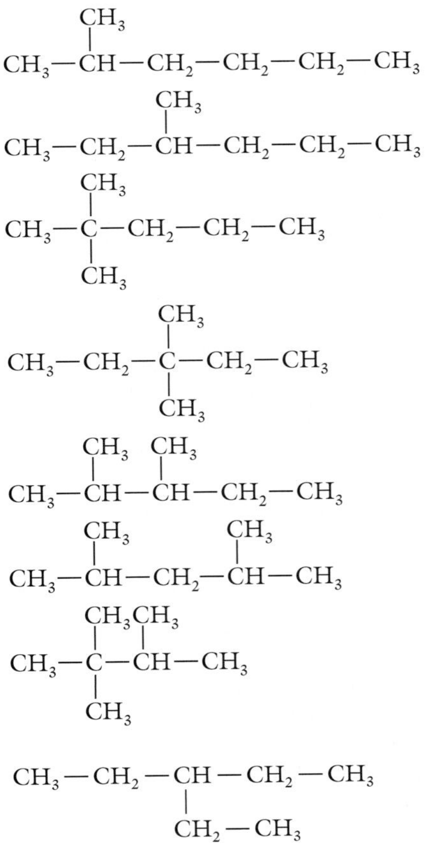

$CH_3—CH(CH_3)—CH_2—CH_2—CH_2—CH_3$

$CH_3—CH_2—CH(CH_3)—CH_2—CH_2—CH_3$

$CH_3—C(CH_3)_2—CH_2—CH_2—CH_3$

$CH_3—CH_2—C(CH_3)_2—CH_2—CH_3$

$CH_3—CH(CH_3)—CH(CH_3)—CH_2—CH_3$

$CH_3—CH(CH_3)—CH_2—CH(CH_3)—CH_3$

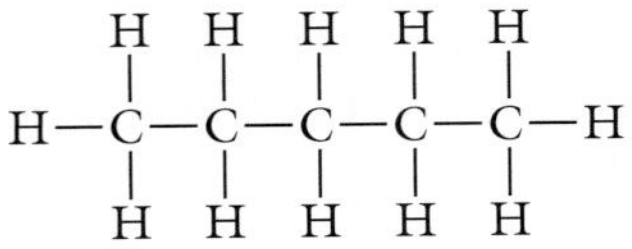

$CH_3—C(CH_3)_2—CH(CH_3)—CH_3$

$CH_3—CH_2—CH(CH_2—CH_3)—CH_2—CH_3$

3.21 (a) 2-methylbutane
(b) 3-methylpentane
(c) 2,2,4-trimethylhexane
(d) 3-isopropyl-4-methylhexane or 3-ethyl-2,4-dimethylhexane

3.23
(a) $CH_3-CH(CH_3)-CH_2-CH_3$
(b) $CH_3-C(CH_3)_2-CH_2-CH(CH_3)-CH_2-CH_3$
(c) $CH_3-CH(CH_3)-C(CH_3)(CH(CH_3)_2)-CH_2-CH_2-CH_3$
(d) $CH_3-C(CH_3)_2-CH_2-C(CH(CH_3)_2)(CH_2CH_3)-CH_2-CH_2-CH_2-CH_3$
(e) $CH_3-CH(CH_3)-CH_2-CH(CH_3CHCH_2CH_3)-CH_2-CH_2-CH_3$

3.25
(a) (b) CH_3, CH_3, $CH(CH_3)_2$
(c) $C(CH_3)_3$, CH_2CH_3

Geometrical stereoisomers (Section 3.8) are possible for b and c.

3.27
(a) 2°, 2°, 2°, 2°, 2° (b) 1° CH_3; 2°, 4°, 2°, 2°; 2° 1° CH_2CH_3
(c) 3° 1° $CH(CH_3)_2$; 3°, 2°, 2°, 2°, 2°, 3°; 1° CH_3

3.29 (a) CH_3 Methylcyclopentane
(b) CH_2CH_3 Ethylcyclobutane

3.31 (a) No cis and trans isomers

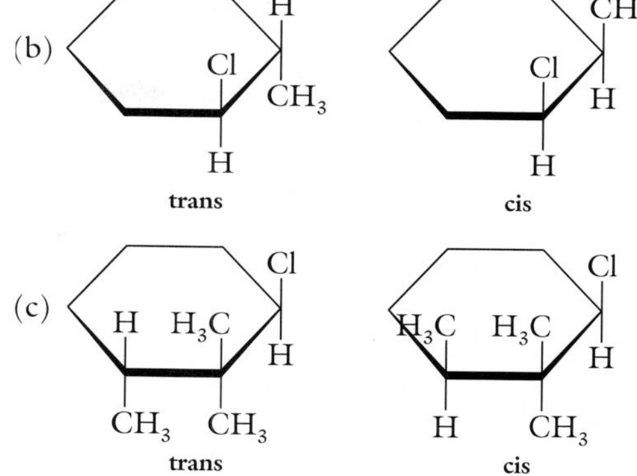

3.33 (a) Hexane has the higher boiling point because it has the higher molecular mass and, therefore, greater secondary forces.
(b) Cyclohexane has the higher boiling point because it has the higher molecular mass and, therefore, greater secondary forces.
(c) Methylcyclohexane has the higher boiling point because it has the higher molecular mass and, therefore, greater secondary forces.
(d) Hexane has the higher boiling point because it is not branched and therefore has greater secondary forces.
(e) Cyclopentane has the higher boiling point because it is cyclic, packs tightly, and therefore has greater secondary forces.

3.35 No reaction for a, b, c, and d
(e) $CH_3-CH_2-CH_2-CH_2-Cl$ $+ CH_3-CH(Cl)-CH_2-CH_3$

3.37 (a, b) The same equation describes both reactions because both compounds have the same molecular formula:

$$C_5H_{12} + 8\ O_2 \longrightarrow 5\ CO_2 + 6\ H_2O$$

(c, d) The same equation describes both reactions because both compounds have the same molecular formula:

$$C_6H_{12} + 9\ O_2 \longrightarrow 6\ CO_2 + 6\ H_2O$$

3.39 Carbon atoms bond to one another in many different patterns.

3.41 $CH_3-CH_2-CH_2-CH_2Cl$
$CH_3-CH(Cl)-CH_2-CH_3$
$CH_3-C(CH_3)_2-Cl$ $\quad$ $CH_3-CH(CH_2Cl)-CH_3$

3.43 (a) 4; (b) 4; (c) 2; (d) 2; (e) 4; (f) 2; (g) 2; (h) 3; (i) 2; (j) 1

3.45 Cyclic. (Chapter 12 will show that another possibility is an acyclic alkene.)

3.47 (a) C_6H_{14}; (b) C_8H_{18}; (c) $C_6H_{11}Cl$; (d) $C_6H_{11}Cl$; (e) C_9H_{20}; (f) C_6H_{12}

3.49 (a) $2\ C_6H_{14} + 19\ O_2 \longrightarrow 12\ CO_2 + 14\ H_2O$

(b)
$$CH_3-\underset{CH_3}{CH}-\underset{CH_3}{CH}-CH_3 \xrightarrow[\text{heat or UV}]{Br_2} CH_3-\underset{CH_3}{CH}-\underset{CH_3}{CH}-CH_2Br + CH_3-\underset{CH_3}{CH}-\overset{CH_3}{\underset{Br}{C}}-CH_3$$

3.51 $CH_3-CH_2-CH_2-CHCl_2$
$+ CH_3-CH_2-CCl_2-CH_3$
$+ CH_2Cl-CHCl-CH_2-CH_3$
$+ CH_2Cl-CH_2-CHCl-CH_3$
$+ CH_2Cl-CHCl-CH_2-CH_2Cl$

3.53 The alkanes in oil possess only the weak London attractive forces, whereas water contains the much stronger hydrogen-bonding forces. Alkane molecules are much more weakly attracted to each other than are water molecules, which results in a lower density for oil.

3.55 Chemical reactions are used to convert alkanes into members of other families.

3.57 The molecules of lipstick and petroleum jelly attract each other because they are both hydrocarbons with the same type of secondary force (London).

3.59 8030 kJ

3.61 $2\ C_6H_{14} + 13\ O_2 \rightarrow 12\ CO + 14\ H_2O$

3.63 The ozone layer is located in the stratosphere, a middle layer of the atmosphere, and protects plant and animal life by absorbing much of the cancer-causing ultraviolet radiation from space. Any decrease in the amount of ozone will let more UV radiation strike Earth's surface and will have detrimental effects on life. Various aerosol-can propellants and refrigeration and air-conditioning coolants escape from Earth and enter the atmosphere. These materials react with ozone and decrease its concentration in the stratosphere.

3.65 There are negligible attractive secondary forces between the highly polar water and the nonpolar alkane coatings. The result is that water does not diffuse through the nonpolar coatings and remains inside the fruit or vegetable.

3.67 Complete combustion yields CO_2 and water, whereas incomplete combustion yields CO and water. The amount of water produced is the same in both reactions. Incomplete combustion uses less O_2. The moles of CO_2 and CO are the same. Incomplete combustion produces less heat because the carbon atoms are not oxidized to their fully oxidized state. Less heat translates into less power in the engine and lower miles per gallon.

3.69 Electrical conduction requires ions. NaCl is an ionic compound and the sodium and chloride ions conduct electricity both when NaCl is in solution or when it is in the molten state. Organic compounds with rare exceptions are covalent, not ionic, and ions are not present either in solution or in the molten state. The conduction of electricity does not occur.

3.71
$$\frac{\text{1-Chloropropane}}{\text{2-Chloropropane}} = \text{reactivity ratio} \times \text{ratio of hydrogens}$$
$$= (1/4) \times (6/2) = 3/4$$
$$\%\ \text{1-Chloropropane} = (3/7) \times 100\% = 42.8\%$$
$$\%\ \text{2-Chloropropane} = (4/7) \times 100\% = 57.2\%$$

3.73 The molecular masses of methane and propane are 16.04 and 44.09, respectively. Thus, the heats of combustion in kilocalories per mole are 13.3 and 12.0, respectively.

3.75
$$\underset{3}{CH_3}-\underset{1}{\overset{O}{\overset{\|}{C}}}-\underset{2}{OCH_3}$$

Chapter 4

4.1 Alkane

4.3 (a) Incorrect; (b) correct; (c) correct

4.5 (a) C_5H_{10}; (b) C_7H_{12}

4.7 Carbons 3, 4, 8, 9

4.9 $CH_2=CH-CH_2-CH_3$ $CH_3-CH=CH-CH_3$ (*cis*- and *trans*-)

$CH_2=\overset{CH_3}{C}-CH_3$ (cyclobutane) (methylcyclopropane, $-CH_3$)

4.11 (a) $CH_2=CH-CH_2-CH_2-CH_3$ and $CH_3-CH=CH-CH_2-CH_3$ (*cis*- and *trans*-)

(b) $CH_3-\overset{CH_3}{C}=CH-CH_3$

(c) (methylcyclobutane, CH_3)

(d) $CH_3-\overset{CH_3}{CH}-CH=CH_2$

4.13 (a) 3; (b) 3; (c) 2

4.15

(a) $CH_2=\overset{CH_3}{C}-CH_2-CH_3$

(b) $CH_3-CH=\overset{CH_2CH_3}{C}-CH_2-CH_3$

(c) $CH_3-\overset{CH_3}{C}=CH-\underset{CH(CH_3)_2}{CH}-CH_2-\overset{CH_3}{CH}-CH_3$

(d) CH_3 CH_3

(e) $CH_2=\overset{CH_3}{C}-\underset{C(CH_3)_3}{CH}-\overset{CH_3}{CH}-CH_3$

(f) $CH_2=CH-CH_2-CH=\overset{CH_3}{C}-CH_3$

4.17 (a) No cis-trans isomers

(b) H, CH_3 / C=C / CH_3, $CH_2CH_2CH_3$

CH_3, CH_3 / C=C / H, $CH_2CH_2CH_3$

(c) H, H / C=C / CH_3, $CH(CH_3)CH_2CH_3$

CH_3, H / C=C / H, $CH(CH_3)CH_2CH_3$

(d) no cis-trans isomers
(e) no cis-trans isomers

(f) H, H, CH_3, CH_3 CH_3, H, H, CH_3

4.19 (a) 3; (b) 1; (c) 2; (d) 4

4.21 (a) CH_3, Br, Br   (b) CH_3, OH

cis and *trans*

(c) $(CH_3)_3CH$ (d) $(CH_3)_3CCl$

(e) Cl, Cl   (f) $CH_3—CH_2—CH(Cl)—CH_3$

(g) $CH_3—CH_2—CH(Br)—CH_2CH_3$ + $CH_3—CH(Br)—CH_2—CH_2CH_3$

(h) $CH_3—CH(Br)—CH(Br)—CH_2CH_3$

(i) $CH_3—CH(Cl)—CH(Cl)—CH_3$

(j) $CH_3—CH_2—CH(OH)—CH_2CH_3$ + $CH_3—CH(OH)—CH_2—CH_2CH_3$

(k) $CH_3—CH_2—CH_2—CH_2CH_3$

4.23 Hexane or methylcyclopentane

4.25 $n\,CH_2{=}CH(CN) \xrightarrow{\text{polymerization catalyst}} {-}\!(CH_2—CH(CN))_n\!{-}$

4.27 $C_6H_{12} + 9\,O_2 \rightarrow 6\,CO_2 + 6\,H_2O$

4.29 Pentane or cyclopentane

4.31 $HC{\equiv}C—CH(CH_3)—CH(CH_3)_2$

4.33 Aromatic compounds have three double bonds alternating with three single bonds in a six-membered-ring structure. Aromatic compounds have high stability, with a very low tendency to undergo addition reactions. Substitution reactions take place without loss of the aromatic structure.

4.35 (a) 1; (b) 2; (c) 1

4.37 (a) 3-Chlorotoluene or *m*-chlorotoluene; (b) 4-bromotoluene or *p*-bromotoluene; (c) 1-benzyl-3-chlorobenzene or *m*-benzylchlorobenzene; (d) 2-bromophenol or *o*-bromophenol; (e) 4-ethylaniline or *p*-ethylaniline; (f) *cis*-1-chloro-2-phenylcyclohexane

4.39 (a) No reaction (b) no reaction

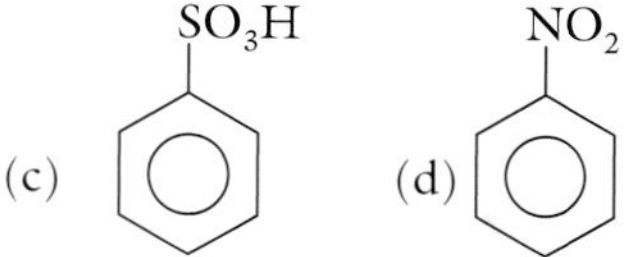

(e) $2\,C_6H_6 + 15\,O_2 \rightarrow 12\,CO_2 + 6\,H_2O$

4.41

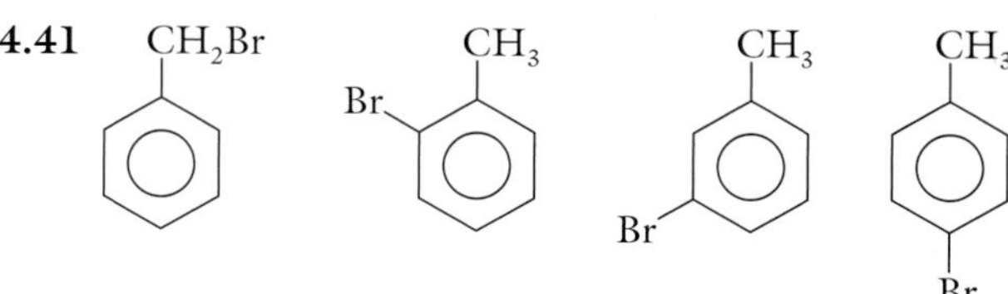

4.43 (a) H, H / C=C / CH_3, $CH_2CH_2CH_3$ H, $CH_2CH_2CH_3$ / C=C / CH_3, H

(b) no cis-trans isomers

(c)

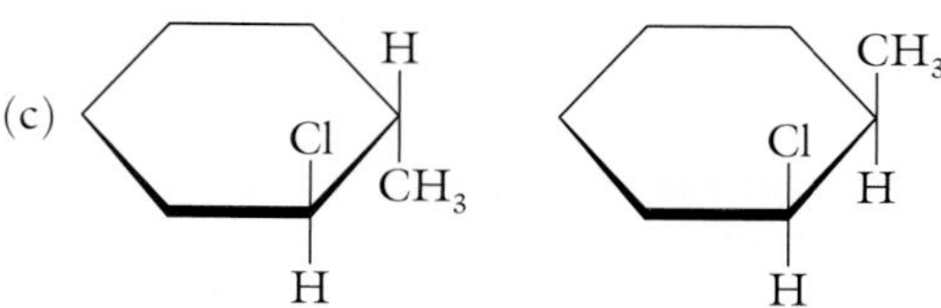

(d) no cis-trans isomers; (e) no cis-trans isomers; (f) no cis-trans isomers

4.45 (a) 2; (b) 2; (c) 4; (d) 2; (e) 2; (f) 4

4.47 All three compounds are nonpolar with very similar boiling points and neglible solubility in water.

4.49 (a) All three compounds burn in air; (b) none of the compounds react; (c) only cyclohexene reacts; (d) only cyclohexene reacts; (e) only cyclohexene reacts.

4.51 $\underset{|}{\overset{\text{Br}}{}}$ $CH_2(Br)—CH(Br)—CH(Br)—CH_2(Br)$

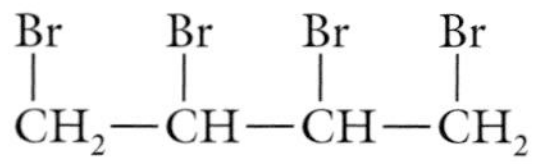

4.53 (a) C_6H_8; (b) C_9H_{12}; (c) C_5H_8; (d) $C_8H_7BrClNO_2$

4.55 Sample A is 1-hexene; sample B is 1-hexyne.

4.57 (a) 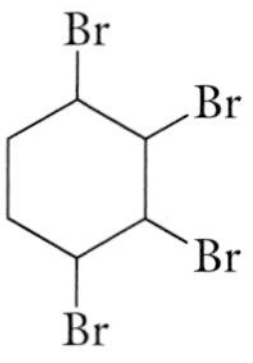 1,2,4,5-tetrabromocyclohexane (b) no reaction

(c) 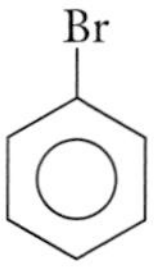 bromobenzene (d) no reaction

(e) no reaction (f) 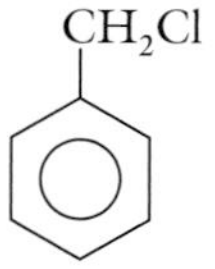 $C_6H_5CH_2Cl$

(g) 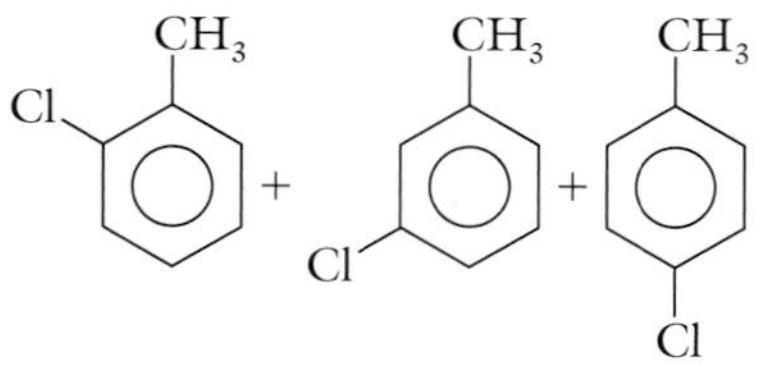 o-chlorotoluene + m-chlorotoluene + p-chlorotoluene

(h) $CO_2 + H_2O$ (i) $CO_2 + H_2O$

(j) no reaction (k) 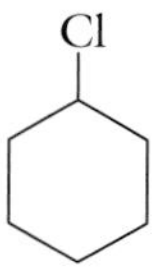 chlorocyclohexane

4.59 1-Hexene

4.61 Mary Jones is correct. The addition of water to propene will proceed according to Markovnikov's rule to produce 2-propanol, not 1-propanol, as the major product.

4.63 Perform the bromine test in a quantitative manner. Weigh out an amount of the shipped compound. Calculate the number of moles, assuming the compound is geraniol. Determine the amount of bromine that reacts with the compound. Calculate the number of moles of bromine used. If the compound is geraniol, the number of moles of bromine will be twice that of the compound. If more than twice the moles of bromine are used, the shipped compound is most likely myrcene. This likelihood can be verified by performing the calculation assuming the compound is myrcene. Myrcene will react with three times the number of moles of bromine.

4.65 Unsaturated fats contain carbon–carbon double bonds; saturated fats contain only single bonds.

4.67 The reaction rate increases with increasing stability of the carbocation formed when H^+ adds to the double bond of each alkene. More-stable carbocations are formed faster than less-stable carbocations. 2-Methylpropene forms a tertiary carbocation, propene forms a secondary carbocation, and ethene forms a primary carbocation. The order of carbocation stability is $3° > 2° > 1°$, which results in the observed order of reactivity of alkenes.

4.69 limonene (CH₃ on ring double bond; $C(CH_3)=CH_2$ substituent) $\xrightarrow{\text{HCl}}$ product with CH_3, Cl on ring carbon and $C(Cl)(CH_3)_2$ substituent

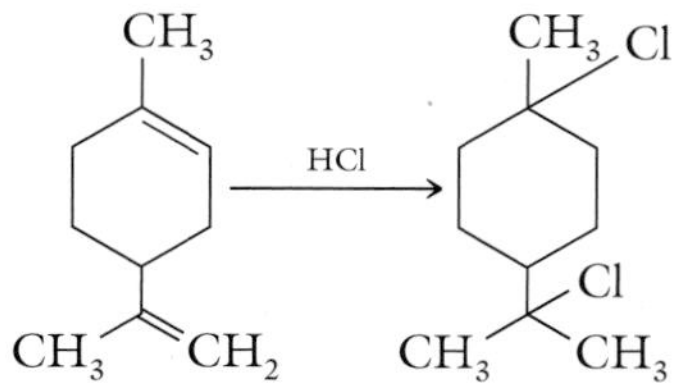

4.71 p-dimethylbenzene (CH_3, CH_3) $\xrightarrow[\text{hot, } H^+]{K_2Cr_2O_7}$ terephthalic acid (COOH, COOH)

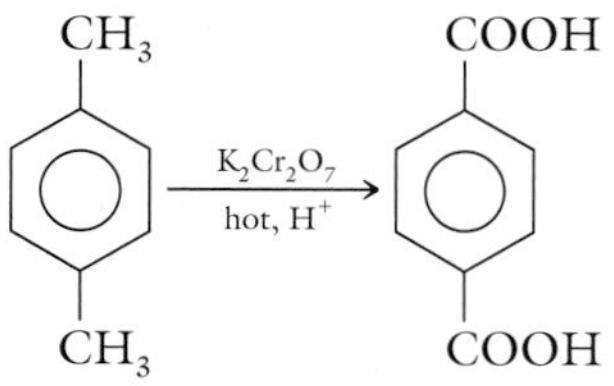

4.73 Limonene contains carbon–carbon double bonds; vanillin does not. Limonene reacts with Br_2 with a color change from red to colorless. There is no reaction with vanillin; the red color of Br_2 is not discharged.

4.75 The unknown is *p*-dimethylbenzene. Each isomer has four positions on the ring available for substitution by Br. Substitution by Br at any one of the four positions in *p*-dimethylbenzene yields the same product, 2-bromo-1,4-dimethylbenzene. Three monobromo products are possible for *m*-dimethylbenzene: 2-bromo-1,3-dimethylbenzene, 1-bromo-2,4-dimethylbenzene, and 5-bromo-1,3-dimethylbenzene. Two monobromo products are possible for *o*-dimethylbenzene: 3-bromo-1,2-dimethylbenzene and 4-bromo-1,2-dimethylbenzene.

4.77 *m*-Methylnitrobenzene is produced by (1) the nitration of benzene and (2) alkylation:

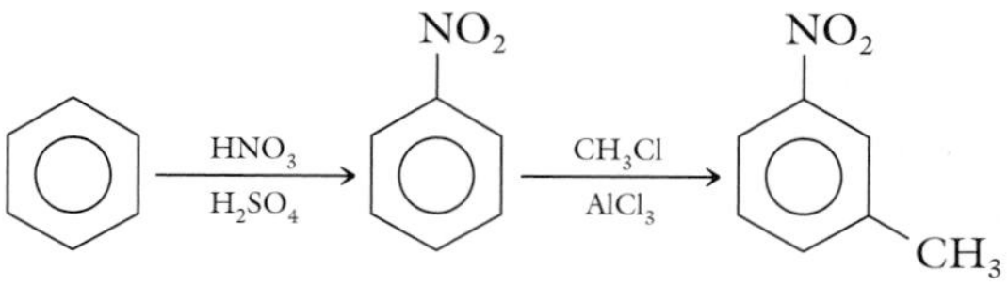

benzene $\xrightarrow[H_2SO_4]{HNO_3}$ nitrobenzene (NO_2) $\xrightarrow[AlCl_3]{CH_3Cl}$ *m*-methylnitrobenzene (NO_2, CH_3)

Alkylation followed by nitration would yield a mixture of the *o*- and *p*-isomers, not the *m*-isomer.

4.79 The C–O bond is too strong to break in the absence of acid. When acid is present, the oxygen is protonated to yield $C—O^+H_2$. The + charge on O weakens the C–O bond and H_2O is expelled.

4.81 The oxidation state of carbon increases as the number of bonds to hydrogen decreases or the number of bonds to oxygen increases or both. Each carbon atom of ethene is bonded to fewer hydrogen atoms than is each carbon atom of ethane.

Chapter 5

5.1 An alcohol has an —OH bonded to a saturated (sp^3) carbon. A phenol has the —OH bonded to a carbon atom of a benzene ring. An ether has an oxygen atom directly bonded to two different carbon atoms.

5.3

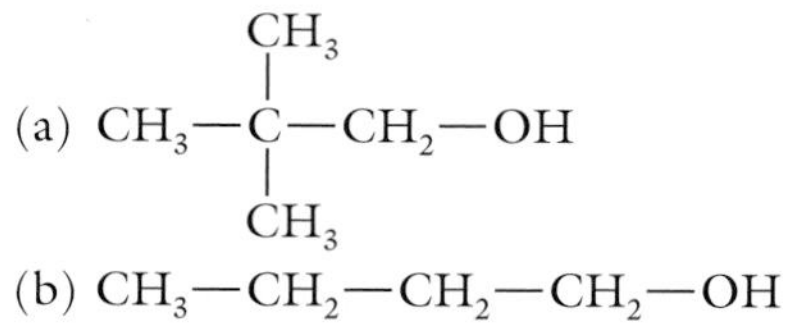

(a) $CH_3—C(CH_3)_2—CH_2—OH$

(b) $CH_3—CH_2—CH_2—CH_2—OH$

5.5 Compound 1, $C_nH_{2n+2}O$; compound 2, $C_nH_{2n}O$

5.7 (a) 2-Methyl-1-butanol, primary; (b) 3-methyl-2-butanol, secondary; (c) 2-methyl-1-butanol, primary; (d) 2-methyl-2-propanol, tertiary

5.9 1,4-Butanediol > 1-pentanol > 2-methyl-2-butanol > hexane. The alcohols have higher boiling points than that of hexane because the secondary forces are stronger—hydrogen bonding compared with London forces. 1-Pentanol's boiling point is higher than that of 2-methyl-2-butanol because the secondary forces are stronger for unbranched than for branched molecules. 1,4-Butanediol has stronger secondary forces compared with the other two alcohols because it has two OH groups compared with one.

5.11 There is no change, because ethanol is not sufficiently acidic to affect blue litmus paper.

5.13 $CH_3CH_2CH_2CH_2{-}OH + HCl \rightleftharpoons CH_3CH_2CH_2CH_2{-}\overset{+}{O}H_2 + Cl^-$

5.15 (a) An acid–base reaction takes place, but dehydration does not, because there is no heat:

$$CH_3CH_2CH_2CH_2{-}OH + H_2SO_4 \rightleftharpoons CH_3CH_2CH_2CH_2{-}\overset{+}{O}H_2 + HSO_4^-$$

(b) no reaction; (c) $CH_2{=}CHCH_2CH_3$

5.17 (a) The primary alcohol is first oxidized to an aldehyde, and then the aldehyde is oxidized to a carboxylic acid:

$$CH_3CH_2CH_2CH_2{-}\overset{\overset{\displaystyle O}{\|}}{C}{-}H \longrightarrow CH_3CH_2CH_2CH_2{-}\overset{\overset{\displaystyle O}{\|}}{C}{-}OH$$

(b) no reaction

5.19 The test does not distinguish between the two possibilities, because both 1-butene and 1-butanol react with permangante solution.

5.21 2-Isopropyl-5-methylphenol

5.23 $p\text{-}CH_3\text{-}C_6H_5{-}O{-}H + H_2O \rightleftharpoons p\text{-}CH_3\text{-}C_6H_5{-}O^- + H_3O^+$

5.25 (a) $C_6H_6O + 7\ O_2 \rightarrow 6\ CO_2 + 3\ H_2O$
(b) $C_6H_5{-}O{-}H + KOH \rightarrow C_6H_5{-}O^- + K^+ + H_2O$

5.27 (a) [benzene ring with OH and, para, OC_2H_5]

(b) $(CH_3)_2CHCH_2{-}O{-}CH_2CH_2CH_3$

(c) $HO{-}CH_2CH_2{-}\underset{}{\overset{\overset{\displaystyle OCH(CH_3)_2}{|}}{C}H}{-}CH_3$

5.29 Ethanol > dimethyl ether > propane

5.31 1-Methylcyclopentene is formed at both temperatures. Although the higher temperature favors ether formation for primary alcohols, ethers are not formed with tertiary alcohols.

5.33 (a) $CH_3{-}\underset{\underset{\displaystyle SH}{|}}{C}H{-}CH_2CH_3$

(b) [cyclopentyl]$-S-S-$[cyclopentyl]

5.35 A, phenol; B, ether; C, alcohol; D, thiol; E, ether

5.37 Compound 1, disulfide; compound 2, alcohol; compound 3, thiol; compound 4, ether

5.39 The formula fits either an alcohol or an ether with either a double bond or a ring.

5.41 (a) $CH_3CH_2CH_2{-}O{-}CH_2CH(CH_3)_2$

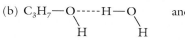

(b) [benzene ring with OH and, meta, $CH_2CH_2CH_3$]

(c) 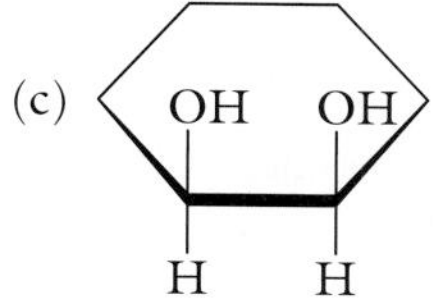

(d) $CH_3{-}\underset{\underset{\displaystyle SH}{|}}{C}H{-}CH_2CH_2CH_3$

(e) $CH_2{=}CH{-}\overset{\overset{\displaystyle OH}{|}}{C}H{-}CH_2{-}CH_3$

5.43 (a) $C_2H_5{-}O(H)\cdots H{-}O{-}C_2H_5$

(b) $C_3H_7{-}O(H)\cdots H{-}O{-}H$ and

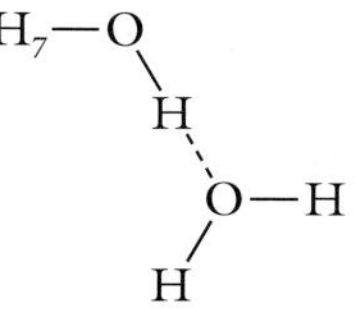

(c) no hydrogen bonding

(d) $CH_3{-}O(CH_3)\cdots H{-}O{-}CH_3$

(e) $C_6H_5{-}O(H)\cdots H{-}O{-}H$ and

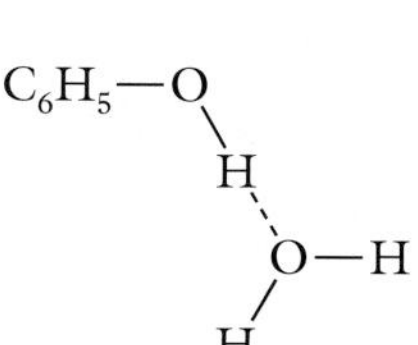

5.45 HCl > phenol > cyclohexanethiol > cyclohexanol

5.47 (a) [benzene ring with ONa]

(b) no reaction

(c) [benzene ring with SNa]

(continued on the next page)

(d) no reaction

(e) cyclohexane ring bearing ONa $+ H_2$

(f) $CH_3CH_2CH_2CH_2—S—Pb—S—CH_2CH_2CH_2CH_3$

(g) $CH_3CH_2CH_2CH_2—S—S—CH_2CH_2CH_2CH_3$

(h) benzene ring—SH

5.49 (a) 1-Propanol; (b) HCl; (c) same; (d) phenol; (e) 1-propanethiol

5.51 Only 1-butanol is eliminated with the permanganate test. None of the other possibilities react with permanganate. The unknown is either *t*-butyl alcohol, ethyl methyl ether, or pentane.

5.53 Phenol. Neither toluene nor cyclohexanol turns blue litmus paper red.

5.55 A,C, and E are close to 109.5°; D = 109.5°; B = 120°

5.57 Hexylresorcinol is a phenol and sufficiently acidic that it turns blue litmus paper red. 3-Hexyl-1,2-cyclohexanediol is an alcohol and is neutral to litmus paper.

5.59 1-Propanol evaporates rapidly from the skin surface, removing heat and creating a cooling effect. 1-Decanol and 2-decanol are much less volatile because of their higher molecular masses, which results in less heat removed from the skin surface.

5.61 Carbocations are intermediates in the reaction. The carbocation formed from 2-propanol is secondary and more stable than the primary carbocation formed from 1-propanol. The more stable carbocation is formed faster, resulting in a faster overall rate for the dehydration reaction.

5.63 Ethylene oxide is relatively unstable due to the distortion of its bond angles from the normal tetrahedral bond angle. Reaction to form an acyclic product with tetrahedral bond angles relieves the instability. The six-membered ring of dioxane possesses tetrahedral bond angles and is as stable as an acyclic compound. The opening of the cyclic structure does not increase stability.

5.65

$$CH_2{=}C(OPO_3H){-}C({=}O){-}OH$$

5.67 2-*t*-Butyl-5-methylcyclohexanol (or 2-*t*-butyl-1-hydroxy-5-methylcyclohexane)

5.69 Electron withdrawal stabilizes the phenoxide anion and enhances the acidity of phenol compounds. Cl is more electronegative than H and its presence on the benzene ring results in greater electron withdrawal in the phenoxide anion formed from *p*-chlorophenol than in phenoxide anion. Thus, acidity is enhanced because *p*-chlorophenoxide anion is more stable than phenoxide anion.

5.71 Tertiary alcohols do not undergo intermolecular dehydration to form ethers.

5.73 $C_2H_6O + 3\ O_2 \longrightarrow 2\ CO_2 + 3\ H_2O$

Chapter 6

6.1 Alcohol or ether

6.3

$$CH_3CH_2CH_2CH_2CH_2{-}C({=}O){-}H$$

$$CH_3CH_2CH_2CH(CH_3){-}C({=}O){-}H \qquad CH_3CH_2CH(CH_3)CH_2{-}C({=}O){-}H$$

$$CH_3CH(CH_3)CH_2CH_2{-}C({=}O){-}H \qquad CH_3CH(CH_3)CH(CH_3){-}C({=}O){-}H$$

$$(CH_3)_3CCH_2{-}C({=}O){-}H \qquad CH_3CH_2C(CH_3)_2{-}C({=}O){-}H$$

6.5 Compound 1, aldehyde; 2, ester; 3, ketone; 4, amide

6.7 a = d = e = sp^3; b = c = sp^2

6.9

(a) $CH_3CH_2CH_2CH(CH_3)CH(CH_3){-}C({=}O){-}H$

(b) $CH_3CH_2{-}C({=}O){-}CH(CH_3)_2$

(c) CH_3CH_2—benzene ring—C(=O)—H, with ring substituent $CH_3CH_2CH(CH_3)$

6.11 (a) 2-Isopropylbutanal; (b) *t*-butyl phenyl ketone; (c) 2-methylcyclopentanone; (d) 3-chloro-4,4-dimethylpentanal

6.13 (a) 3-Pentanone. Both compounds are ketones, but 3-pentanone has a higher molecular mass.

(b) Aldehydes and ketones of the same molecular mass have about the same boiling point.

(c) Aldehydes and ketones of the same molecular mass have about the same water solubility.

(d) Butanal. Aldehydes have much stronger secondary attractive forces, dipole–dipole, compared with the weaker London forces of hydrocarbons.

(e) Butanal. Aldehydes hydrogen bond with water, but there are no attractive forces between pentane and water.

(f) 2-Butanone. A larger fraction of the lower-molecular-mass compound is able to hydrogen bond with water.

(g) 2-Butanol. 2-Butanol has the strongest secondary attractive forces with itself (that is, hydrogen bonding), whereas 2-butanone has the weaker dipole–dipole forces.

(h) The water solubilities are very nearly the same because both ketones and alcohols hydrogen bond with water.

6.15 Compound 1 is not oxidized. Compound 2 is oxidized to CH_3CH_2COOH. Compound 3 is oxidized to the aldehyde, which is further oxidized to the carboxylic acid:

$$CH_3CH_2CH_2CH_2OH \longrightarrow CH_3CH_2CH_2-\overset{\overset{\displaystyle O}{\|}}{C}-H \longrightarrow$$

$$CH_3CH_2CH_2-\overset{\overset{\displaystyle O}{\|}}{C}-OH$$

6.17 (a) Compounds 3 and 4 are both aldehydes and give positive tests with Tollens's reagent. A positive Tollens's test is indicated by the formation of a black precipitate or shiny mirror. Compounds 1 and 2 are not aldehydes and give negative tests.

(b) Compounds 1, an α-hydroxy ketone, and 4, an α-hydroxy aldehyde, give positive tests with Benedict's reagent. A positive test is indicated by a red precipitate. Compound 2, a hydroxy carboxylic acid, gives a negative test. Compound 3, a simple aldehyde, gives a negative test.

6.19 Permanganate does not distinguish between 1-pentanol and pentanal, because both give a positive test.

6.21

(a) $CH_3CH_2\overset{\overset{\displaystyle OH}{|}}{C}HCH_3$

(b) $CH_3CH_2CH_2OH$

(c) no reduction

6.23 $H{:}^- + H_2O \rightarrow H_2 + HO^-$

H^- is a base because it accepts a proton in the reaction.

6.25

$$C_6H_5-\overset{\overset{\displaystyle O}{\|}}{C}-H \xrightarrow{CH_3OH,\ H^+}$$

$$C_6H_5-\underset{\underset{\displaystyle OCH_3}{|}}{\overset{\overset{\displaystyle OH}{|}}{C}}-H \xrightarrow[-H_2O]{CH_3OH,\ H^+} C_6H_5-\underset{\underset{\displaystyle OCH_3}{|}}{\overset{\overset{\displaystyle OCH_3}{|}}{C}}-H$$

6.27

(a) $C_6H_5-\overset{\overset{\displaystyle O}{\|}}{C}-H$ and CH_3CH_2OH

(b) $HO-CH_2CH_2CH_2-\overset{\overset{\displaystyle O}{\|}}{C}-H \longrightarrow$

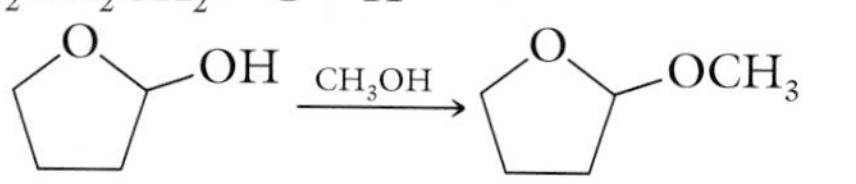

(cyclic) O, OH $\xrightarrow{CH_3OH}$ (cyclic) O, OCH_3

(c) $CH_3CH_2CH_2-\overset{\overset{\displaystyle O}{\|}}{C}-CH_2CH_3 + CH_3CH_2OH$

(d) 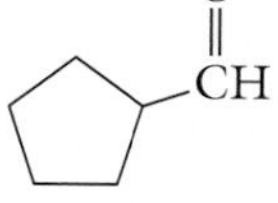 (cyclopentyl)$-\overset{\overset{\displaystyle O}{\|}}{C}H$ $+ CH_3OH$

6.29 (a) The carbon atoms of the C=C double bond as well as the carbon and oxygen atoms of the C=O double bond are sp^2 hybridized; (b) all bond angles are 120°; (c) the C=C is nonpolar, whereas the C=O is polar; (d) both double bonds undergo a variety of addition reactions, although not with exactly all the same reagents.

6.31

(a) $CH_3CH_2CH_2CH_2-\overset{\overset{\displaystyle O}{\|}}{C}-H$ (b) 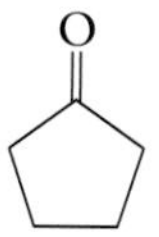 (cyclopentanone, O)

(c) $C_6H_5-\overset{\overset{\displaystyle O}{\|}}{C}-CH_3$ (d) $(CH_3)_3CCH_2-\overset{\overset{\displaystyle O}{\|}}{C}-H$

(e) $CH_3-\overset{\overset{\displaystyle O}{\|}}{C}-C(CH_3)_3$

6.33 (a) 3-Methylcyclopentanone; (b) 2-ethylpentanal; (c) pentanal; (d) 3-chloro-2-pentanone

6.35 (a) 1-Butanol. The alcohol has stronger secondary attractive forces (that is, hydrogen bonding) compared with dipole–dipole forces in the aldehyde.

(b) The water solubilities are very nearly the same because both alcohols and ketones hydrogen bond with water.

(c) Butanal. Aldehydes have stronger secondary attractive forces because the carbonyl group in aldehydes is much more polar than the C–O single bond in ethers.

(d) The water solubilities are very nearly the same because both aldehydes and ethers hydrogen bond with water.

6.37 Diethyl ether ≅ butanal ≅ 1-butanol > pentane. Any compound containing oxygen atoms can hydrogen bond with water, although alcohols do so slightly better than aldehydes, which do so slightly better than ethers.

6.39 Tollens's reagent does not distinguish the two compounds, because both ketones and tertiary alcohols are unreactive with Tollens's reagent.

6.41 Yes

$$HOCH_2-\underset{\underset{\displaystyle OH}{|}}{C}H-\underset{\underset{\displaystyle OH}{|}}{C}H-\underset{\underset{\displaystyle OH}{|}}{C}H-\underset{\underset{\displaystyle OH}{|}}{C}H-CH_2OH$$

6.43

$$CH_3CH_2-\overset{\overset{\displaystyle O}{\|}}{C}-CH_3 \xrightarrow{CH_3OH,\ H^+}$$

$$CH_3CH_2-\underset{\underset{\displaystyle OCH_3}{|}}{\overset{\overset{\displaystyle OH}{|}}{C}}-CH_3 \xrightarrow[-H_2O]{CH_3OH,\ H^+} CH_3CH_2-\underset{\underset{\displaystyle OCH_3}{|}}{\overset{\overset{\displaystyle OCH_3}{|}}{C}}-CH_3$$

Hemiacetal **Acetal**

6.45

(a) 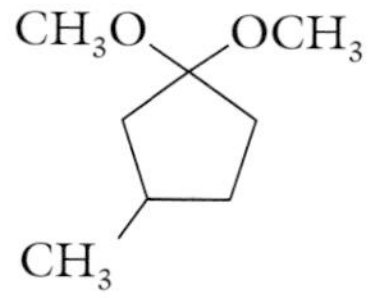 CH_3O OCH_3 (cyclopentane) CH_3

(b) $CH_3-\overset{\overset{\displaystyle OH}{|}}{C}H-C_6H_5$

(c) no reaction (d) $CH_3CH_2CH_2\overset{\overset{\displaystyle O}{\|}}{C}-OH$

(e) no reaction (f) $HOCH_2CH_2C_6H_5$

(g) 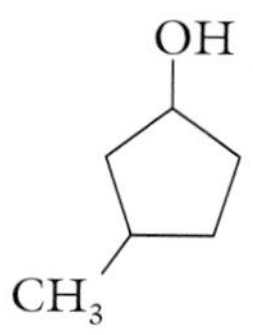 OH (cyclopentane) CH_3 **(cis- and trans-)**

(h) $HO-CH_2CH_2CH_2-\overset{\overset{\displaystyle O}{\|}}{C}-H + CH_3OH$

6.47

$$CH_3-\overset{\overset{\displaystyle O}{\|}}{C}-CH_2CH_3$$

6.49 (a) None (b) $CH_3-\overset{\overset{\displaystyle CH_3}{|}}{CH}-CH{=}O$

(c) cyclopentanone (d) none

6.51 Vanillin turns blue litmus paper red because it is a phenol. Vanillin undergoes oxidation with Tollens's reagent because it is an aldehyde. Menthone is neutral to litmus paper. It gives a negative Tollens's test because it is a ketone, not an aldehyde.

6.53 Hydride reduction reduces only the C=O group; catalytic reduction reduces both the C=O and the C=C groups.

6.55 The cyclic structures are hemiacetals formed from the acyclic structure by reaction between the —OH group at C5 and the carbonyl group (C1). There are two cyclic structures, α-D-glucose and β-D-glucose, formed because there are two positions (up and down) available for the hemiacetal —OH group at C1.

6.57 $C_{21}H_{30}O_2$

6.59 There are two pentanone isomers: 2-pentanone and 3-pentanone. The number in the name is needed to distinguish between the two isomers. There is only one butanone compound because the carbonyl carbon atom for the unbranched C_4 ketone must be the next-to-last carbon atom in the four-carbon chain.

6.61 The unknown is B. A and C would each show eight peaks in the spectrum. Only B shows six peaks.

6.63 The peak for the alcohol O—H (at 3200–3650 cm^{-1}) can differentiate between the two compounds. Testosterone would show that peak, whereas progesterone would not.

6.65 V = butanal, W = 1-butanol, X = 1-butene, Y = 2-butanol, Z = butanone

6.67

$$\underset{2}{HO-CH_2}-\underset{1}{\overset{\overset{\displaystyle O}{\|}}{C}}-\underset{3}{CH_3}$$

6.69 $LiAlH_4$ is not compatible with aqueous systems; it preferentially reacts with water instead of the carbonyl group. NADH does not react with water.

Chapter 7

7.1 Compound 1, something else (ketone and ether); 2, ester; 3, acid halide; 4, something else (ketone and alcohol); 5, carboxylic acid; 6, amide; 7, carboxylic acid and ether; 8, anhydride

7.3

$$CH_3CH_2CH_2CH_2-\overset{\overset{\displaystyle O}{\|}}{C}-OH \qquad CH_3CH_2\overset{\overset{\displaystyle CH_3}{|}}{CH}-\overset{\overset{\displaystyle O}{\|}}{C}-OH$$

$$CH_3\overset{\overset{\displaystyle CH_3}{|}}{C}HCH_2-\overset{\overset{\displaystyle O}{\|}}{C}-OH \qquad (CH_3)_3C-\overset{\overset{\displaystyle O}{\|}}{C}-OH$$

7.5

(a) $C_6H_5CH(CH_3)_2 \xrightarrow{MnO_4^- \text{ or } Cr_2O_7^{2-}} C_6H_5C(OH){=}O + CO_2 + H_2O$

(b) $CH_3CH_2CH_2CH_2CH_2-OH \xrightarrow{MnO_4^- \text{ or } Cr_2O_7^{2-}} CH_3CH_2CH_2CH_2\overset{\overset{\displaystyle OH}{|}}{C}{=}O$

(c) 2-Pentanol undergoes selective oxidation to a ketone (2-pentanone) but the ketone is not oxidized to a carboxylic acid.

7.7 (a) 2-Isopropyl-2,5-dimethylhexanoic acid; (b) 3-chloro-4,4-dimethylpentanoic acid; (c) 3-phenylbutanoic acid; (d) ethylpropanedioic acid

7.9 Hydrogen-bonded dimer of propanoic acid: two $CH_3CH_2C(=O^{\delta-})O-H^{\delta+}$ molecules, each $H^{\delta+}$ hydrogen-bonded to the $O^{\delta-}$ of the other.

7.11 (a) Ethanoic acid. Hydrogen-bond attractions in carboxylic acid dimers are stronger than hydrogen-bond attractions in alcohols. (b) The solubilities are about the same because both have strong hydrogen-bond attractions with water.

7.13 Compound 1 is more acidic than compound 2 because compound 1 is a carboxylic acid, whereas compound 2 is an alcohol and ketone.

7.15

(a) $CH_3\overset{\overset{\displaystyle O}{\|}}{C}-O-H + H_2O \rightleftharpoons CH_3\overset{\overset{\displaystyle O}{\|}}{C}-O^- + H_3O^+$

(b) $CH_3\overset{\overset{\displaystyle O}{\|}}{C}-O-H + NaOH \longrightarrow CH_3\overset{\overset{\displaystyle O}{\|}}{C}-O^-Na^+ + H_2O$

(c) $CH_3\overset{\overset{\displaystyle O}{\|}}{C}-O-H + NaHCO_3 \longrightarrow CH_3\overset{\overset{\displaystyle O}{\|}}{C}-O^-Na^+ + CO_2 + H_2O$

(d) $C_6H_5-O-H + H_2O \rightleftharpoons C_6H_5-O^- + H_3O^+$

(e) $C_6H_5-O-H + NaOH \longrightarrow C_6H_5-O^- + Na^+ + H_2O$

7.17 Propanoic acid predominates in water because the extent of ionization is less than about 1 to 2%. Propanoate ion is the predominant species at pH = 7 because the buffered solution is much more basic compared with the acidity of a carboxylic acid.

7.19 Propanoic acid

7.21

(a) $CH_3CH_2CH_2CH_2CH_2C(=O)-ONa$

(b) $(C_6H_5COO)_2Ca$

7.23 (a) Sodium propanoate has the higher melting point because it is an ionic compound, whereas propanoic acid is a covalent compound. (b) Sodium hexanoate. The attractive forces between the ions of the carboxylate salt and water are stronger than the hydrogen-bond attraction between the carboxylic acid and water. (c) Octanoic acid. The carboxylic acid is a polar covalent compound and interacts more strongly with the nonpolar hexane than does the ionic carboxylate salt.

7.25 Sodium butanoate

7.27 Only compound 3 is a soap. A soap must have both an ionic group and a long hydrocarbon chain.

7.29

(a) $CH_3CH_2CH_2CH_2-C(=O)-OH + HOCH_2CH_3 \xrightarrow[-H_2O]{H^+} CH_3CH_2CH_2CH_2-C(=O)-OCH_2CH_3$

(b) $C_6H_5-C(=O)-OH + HOCH_2CH_2CH_2CH_3 \xrightarrow[-H_2O]{H^+} C_6H_5-C(=O)-OCH_2CH_2CH_2CH_3$

7.31 (a) 4,4-Dimethylpentanoic acid and 2-propanol; (b) *p*-methylbenzoic acid and 2-butanol; (c) butanoic acid and *p*-methylphenol

7.33

(a) $CH_3(CH_2)_2C(=O)-OC(CH_3)_3$

(b) 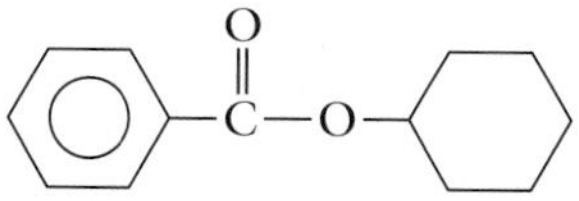 $C_6H_5-C(=O)-O-C_6H_{11}$

7.35 (a) Pentanal's boiling point is higher than that of methyl propanoate because of higher secondary attractive forces. (b) The solubilities in water are similar because both hydrogen bond equally well with water. (c) 1-Pentanol. The hydrogen-bond secondary attractive forces in alcohols are stronger than the dipole–dipole forces in esters. (d) Hexanoic acid is more soluble in water because it hydrogen bonds more extensively with water than does methyl pentanoate. (e) Butanoic acid. The hydrogen-bond secondary attractions in carboxylic acid dimers are much stronger than the dipole–dipole attractions in esters.

7.37 Hexanedioic acid and 1,3-propanediol

7.39

(a) $CH_3CH_2-C(=O)-OCH_2CH_2CH_3 \xrightarrow[H_2O]{H^+} CH_3CH_2-C(=O)-OH + HOCH_2CH_2CH_3$

(b) 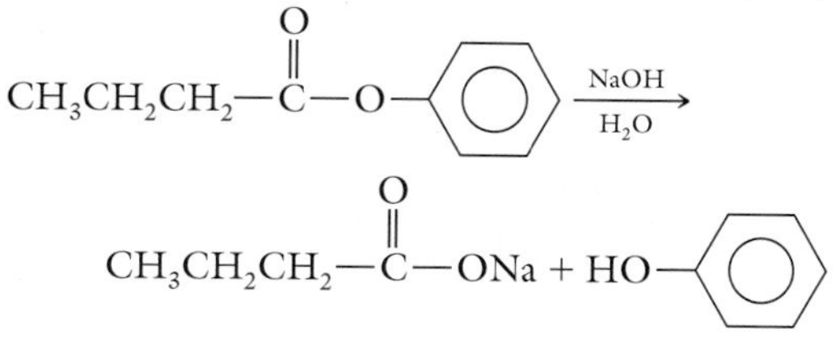 $CH_3CH_2CH_2-C(=O)-O-C_6H_5 \xrightarrow[H_2O]{NaOH} CH_3CH_2CH_2-C(=O)-ONa + HO-C_6H_5$

7.41

(a) $(CH_3)_2CH-C(=O)-Cl$

(b) $CH_3-C(=O)-O-C(=O)-CH_3$

7.43

$C_6H_5-C(=O)-Cl \xrightarrow[-HCl]{H_2O} C_6H_5-C(=O)-OH$

7.45

$HO-P(=O)(OH)-O-P(=O)(OH)-OH + CH_3CH_2CH_2CH_2OH \xrightarrow{-H_2O} HO-P(=O)(OH)-O-P(=O)(OH)-OCH_2CH_2CH_2CH_3$

Butyl diphosphate

7.47

$^-O-P(=O)(O^-)-O-P(=O)(O^-)-O^-$

7.49

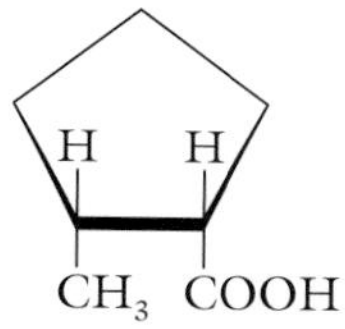

Cyclopentane ring with H and H on adjacent carbons (cis); substituents CH_3 and COOH.

7.51

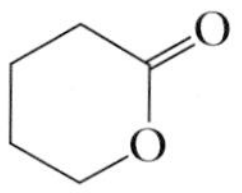

Six-membered cyclic ester (lactone): ring containing O, with C=O adjacent to ring O.

7.53

(a) 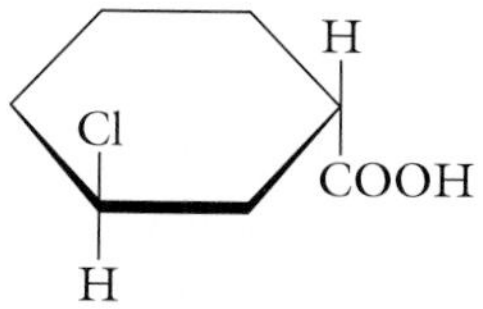 Cyclohexane ring with Cl and H on one carbon, H and COOH on another.

(b) $(CH_3COO)_2Ca$

(c) $C_6H_5-C(=O)-OCH(CH_3)_2$

(d) $CH_3CH_2CH_2CH_2-C(=O)-O-C(=O)-CH_2CH_2CH_2CH_3$

(e) $CH_3CH_2-C(=O)-Cl$

(f) $HO-P(=O)(OH)-O-P(=O)(OH)-OCH_3$

(g) $HO-P(=O)(OCH_3)-OCH_3$

7.55 a = c = sp^3; b = d = sp^2; A = C = 120°; B is close to 109.5°

7.57 Phenol or propanoic acid

7.59

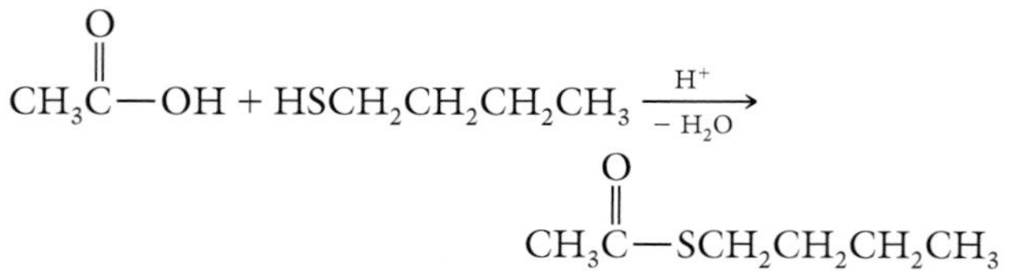

$$CH_3\overset{\overset{O}{\|}}{C}-OH + HSCH_2CH_2CH_2CH_3 \xrightarrow[-\,H_2O]{H^+} CH_3\overset{\overset{O}{\|}}{C}-SCH_2CH_2CH_2CH_3$$

7.61 (a) No reaction

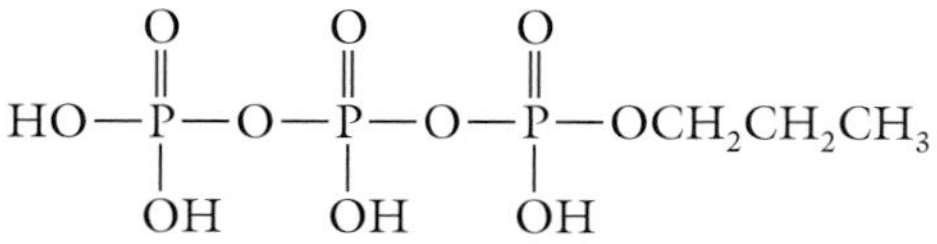

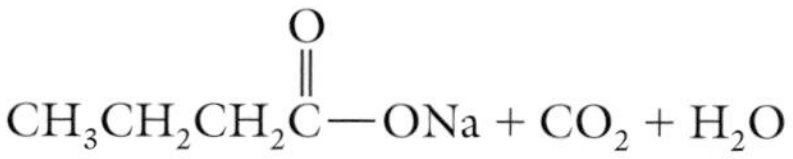

(b) 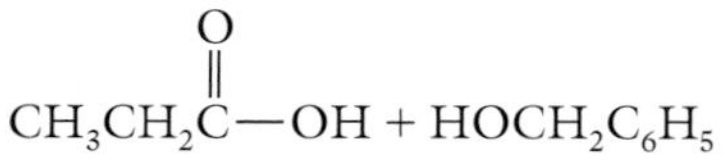 $CH_3CH_2\overset{\overset{O}{\|}}{C}-OH + HOCH_2C_6H_5$

(c) $CH_3CH_2\overset{\overset{O}{\|}}{C}-ONa + HOCH_2C_6H_5$

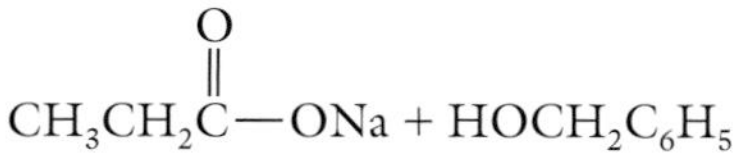

(d) 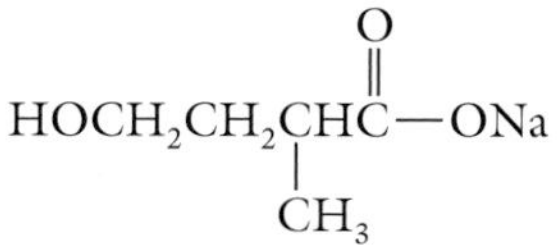 $HOCH_2CH_2\underset{\underset{CH_3}{|}}{C}H\overset{\overset{O}{\|}}{C}-ONa$

(e) $HO-\underset{\underset{OH}{|}}{\overset{\overset{O}{\|}}{P}}-O-\underset{\underset{OH}{|}}{\overset{\overset{O}{\|}}{P}}-O-\underset{\underset{OH}{|}}{\overset{\overset{O}{\|}}{P}}-OCH_2CH_2CH_3$

(f) $CH_3CH_2CH_2\overset{\overset{O}{\|}}{C}-ONa + CO_2 + H_2O$

(g) $(CH_3)_2CHCH_2CH_2\overset{\overset{O}{\|}}{C}-OC_6H_5$

(h) $CH_3CH_2-\overset{\overset{O}{\|}}{C}-OH$

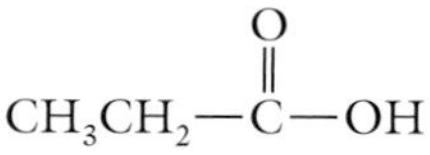

(i) 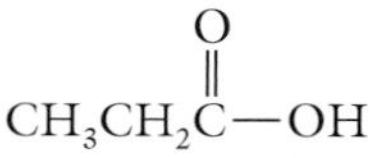 $CH_3CH_2\overset{\overset{O}{\|}}{C}-OH$

(j) 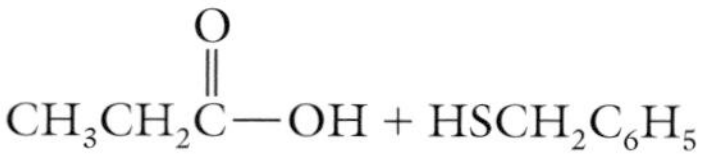 $CH_3CH_2\overset{\overset{O}{\|}}{C}-OH + HSCH_2C_6H_5$

7.63 Add a drop of the unknown to blue litmus paper. The unknown is the carboxylic acid if the litmus paper changes color to red. If there is no color change, the unknown is the ester.

7.65

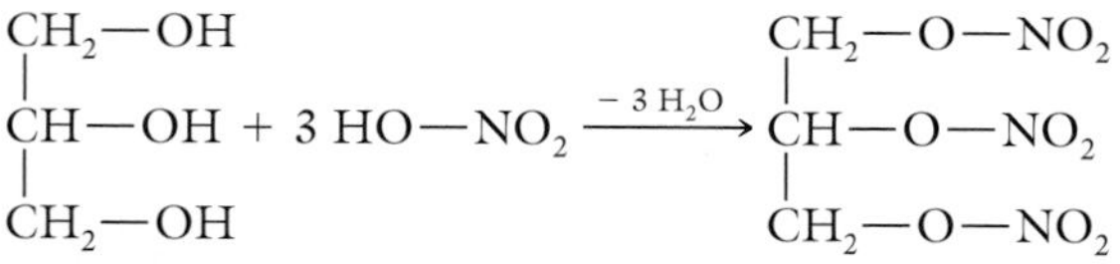

$$\begin{array}{l} CH_2-OH \\ | \\ CH-OH \\ | \\ CH_2-OH \end{array} + 3\,HO-NO_2 \xrightarrow{-\,3\,H_2O} \begin{array}{l} CH_2-O-NO_2 \\ | \\ CH-O-NO_2 \\ | \\ CH_2-O-NO_2 \end{array}$$

7.67 $[H_3O^+] = 4.17 \times 10^{-3}$; pH = 2.38

7.69 Sodium acetate is a salt and 100% ionized in water but acetic acid is only weakly ionized (about 1%). The concentration of ions is much greater for sodium acetate and conductivity depends on the concentration of ions.

7.71

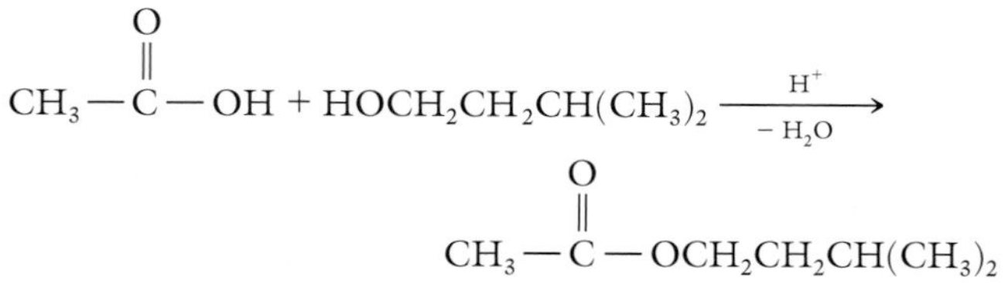

$$CH_3-\overset{\overset{O}{\|}}{C}-OH + HOCH_2CH_2CH(CH_3)_2 \xrightarrow[-\,H_2O]{H^+} CH_3-\overset{\overset{O}{\|}}{C}-OCH_2CH_2CH(CH_3)_2$$

7.73 C > A > B. HCl suppresses the ionization of acetic acid. NaOH converts all of the acetic acid to sodium acetate, which is 100% ionized.

7.75

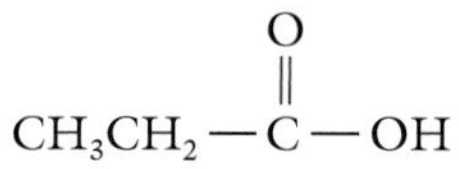

$$CH_3CH_2-\overset{\overset{O}{\|}}{C}-OH$$

7.77

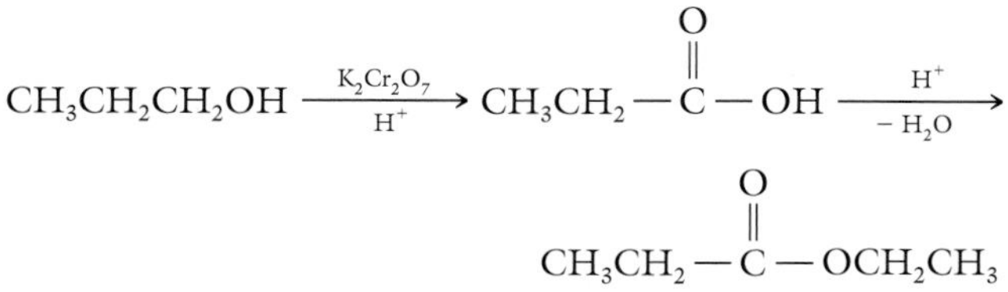

$$CH_3CH_2CH_2OH \xrightarrow[H^+]{K_2Cr_2O_7} CH_3CH_2-\overset{\overset{O}{\|}}{C}-OH \xrightarrow[-\,H_2O]{H^+} CH_3CH_2-\overset{\overset{O}{\|}}{C}-OCH_2CH_3$$

7.79

$$\overset{\overset{O}{\|}}{C}-OCH_2CH_3 \text{ attached to a benzene ring bearing } NH_2 \text{ in the para position}$$

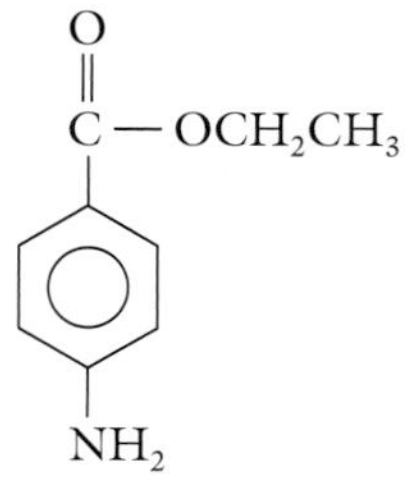

7.81 The trichloroacetate anion is more stable than the acetate anion because of electron withdrawal by the chlorines. Increased stabilization of the anion results in an increased degree of ionization of the carboxylic acid.

7.83

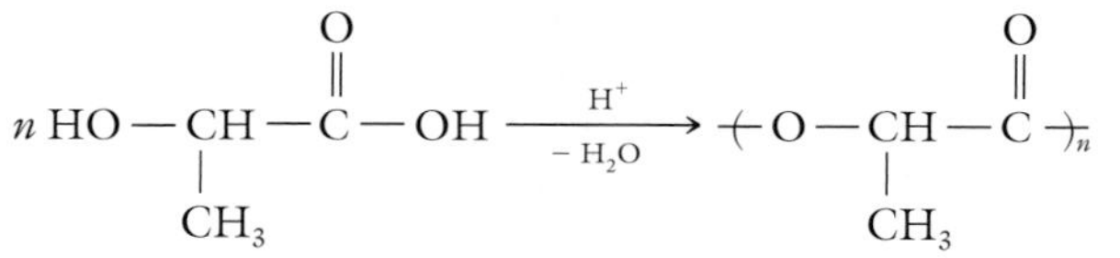

$$n\,HO-\underset{\underset{CH_3}{|}}{CH}-\overset{\overset{O}{\|}}{C}-OH \xrightarrow[-\,H_2O]{H^+} \left(O-\underset{\underset{CH_3}{|}}{CH}-\overset{\overset{O}{\|}}{C}\right)_n$$

7.85 The strong acid functions as a catalyst to increase the rate of hydrolysis. The strong base is a required reactant, not a catalyst. The base is used up during the hydrolysis; the acid is not.

7.87 This analysis is not easy. Both compounds show carbonyl and hydroxyl peaks because both have COOH groups. Aspirin has a second carbonyl group because it is also an ester, but the peak may or may not be separated from the COOH carbonyl peak. If the IR spectrum shows a second carbonyl peak, it identifies aspirin. But the absence of a second carbonyl peak does not rule out aspirin. Although not an easy analysis for the beginning student, an experienced analyst would be able to differentiate the two samples, especially if known samples of aspirin and ibuprofen are available. The analyst would take the IR spectrum of the two known samples and then match them by overlaying them over the spectra of the unknown samples. Almost no two compounds have exactly the same spectrum.

Chapter 8

8.1

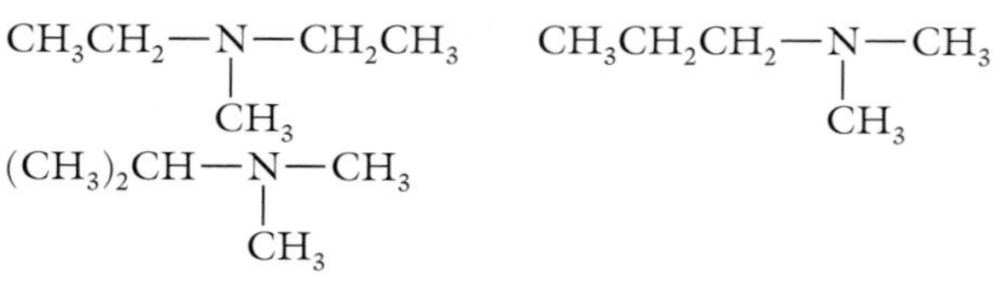

$$CH_3CH_2-\underset{\underset{CH_3}{|}}{N}-CH_2CH_3 \qquad CH_3CH_2CH_2-\underset{\underset{CH_3}{|}}{N}-CH_3$$

$$(CH_3)_2CH-\underset{\underset{CH_3}{|}}{N}-CH_3$$

8.3 1, aliphatic, secondary amine; 2, ketone and aliphatic, primary amine; 3, aromatic amine; both amine groups are tertiary; 4 and 5, amides; 6, neither an amine nor an amide—it is nitrobenzene.

8.5 (a) $CH_3CH_2CH_2CH_2CH_2—N(H)—CH_3$

(b) $(CH_3)_2CH—N(H)—CH_3$

(c) 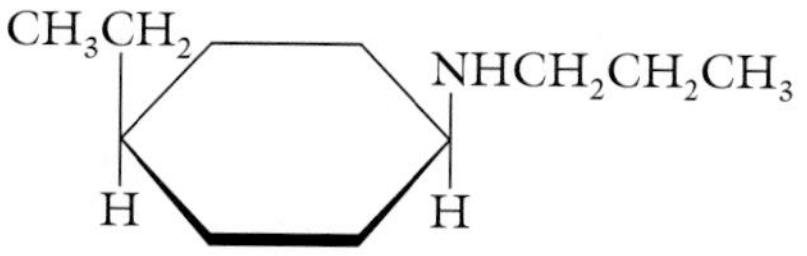 cyclohexane ring with CH_3CH_2 and H on one carbon and $NHCH_2CH_2CH_3$ and H on another

(d) 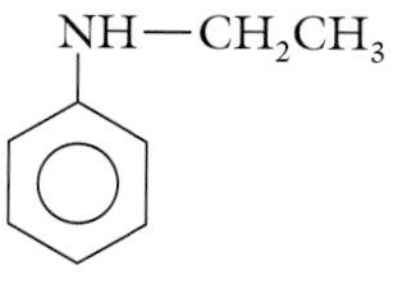 $NH—CH_2CH_3$ attached to a benzene ring

8.7 (a) Ethyldimethylamine;
(b) *trans*-2-methylcyclohexanamine;
(c) *N*-ethyl-4,*N*-dimethyl-2-pentanamine;
(d) *N*-ethyl-*N*-methylaniline

8.9 $C_2H_5—N(H)(H)\cdots\cdots H—N(H)—C_2H_5$

8.11 (a) 1-Butanol. The O–H bond is more polar than the N–H bond, because oxygen is more electronegative than nitrogen, and this polarity results in stronger hydrogen bonding in alcohols than in primary or secondary amines.

(b) The solubilities of butylamine and butanal in water are about the same. The hydrogen bonding of water with the N–H group of butylamine is no stronger than that of water with the carbonyl oxygen atom of butanal.

(c) Butylamine. The hydrogen-bond attractive forces in the primary amine (butylamine) are stronger than the dipole–dipole attractive forces in the tertiary amine (ethyldimethylamine).

(d) Primary and secondary amines have about the same solubility in water. Both have N–H bonds that hydrogen bond with water.

8.13 $CH_3CH_2CH_2NH_2 + HCl \rightarrow CH_3CH_2CH_2NH_3^+\ Cl^-$

8.15 Water = 1-propanol < aniline < propylamine < KOH. Basicity measures the ability of a compound to bond to a proton. Amines are more basic than water and 1-propanol because nitrogen's nonbonded pair of electrons is more available for bonding to a proton, the result of nitrogen being less electronegative than oxygen. Aniline is less basic than propylamine because the benzene ring, through its electron-withdrawing effect, decreases the availability of nitrogen's nonbonded pair of electrons. KOH is the strongest base because the actual base is OH^-. The negative charge on oxygen in this anion greatly increases its ability to accept a proton.

8.17 Propylamine

8.19 $H_2NCH_2CH_2CH_2CH_2CH_2CH_2NH_2 + 2\ HCl \rightarrow$
$Cl^-\ H_3\overset{+}{N}CH_2CH_2CH_2CH_2CH_2CH_2\overset{+}{N}H_3\ Cl^-$

8.21 $(CH_3)_2NH^+\ Cl^-$

(a) 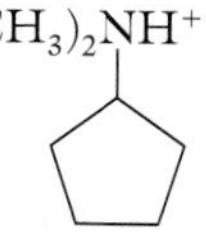 attached to a cyclopentane ring

(b) 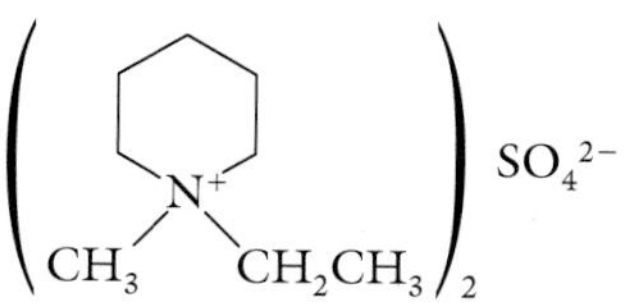 $\left(\text{piperidinium ring with } \overset{+}{N} \text{ bearing } CH_3 \text{ and } CH_2CH_3\right)_2\ SO_4^{2-}$

(c) $CH_3CH_2—\overset{+}{N}(CH_3)(CH_3CH_2CH_2)—H\ Br^-$

8.23 (a) Trimethylammonium chloride has the higher boiling point because the ionic attractive forces of amine salts are stronger than the hydrogen-bond secondary attractive forces of amines.

(b) Hexylammonium chloride has the higher solubility because amine salts interact more strongly with water than does an amine.

8.25 Butylamine

8.27 1, tertiary amide; 2 is not an amide—it is a ketone and amine; 3, primary amide and primary amine; 4, secondary amide

8.29

(a) $C_6H_5—\overset{O}{\overset{\|}{C}}—OH + H_2NCH_2CH_2CH_2CH_3 \xrightarrow{25°C} C_6H_5—\overset{O}{\overset{\|}{C}}—O^-\ H_3\overset{+}{N}CH_2CH_2CH_2CH_3$

(b) $C_6H_5—\overset{O}{\overset{\|}{C}}—OH + H_2NCH_2CH_2CH_2CH_3 \xrightarrow[-H_2O]{>100°C} C_6H_5—\overset{O}{\overset{\|}{C}}—\overset{H}{\overset{|}{N}}CH_2CH_2CH_2CH_3$

(c) $C_6H_5—\overset{O}{\overset{\|}{C}}—Cl + H_2NCH_2CH_3 \xrightarrow{-HCl} C_6H_5—\overset{O}{\overset{\|}{C}}—\overset{H}{\overset{|}{N}}—CH_2CH_3$

(d) $CH_3—\overset{O}{\overset{\|}{C}}—O—\overset{O}{\overset{\|}{C}}—CH_3 + H\overset{CH_3}{\overset{|}{N}}—C_6H_5 \longrightarrow CH_3—\overset{O}{\overset{\|}{C}}—\overset{CH_3}{\overset{|}{N}}—C_6H_5 + CH_3—\overset{O}{\overset{\|}{C}}—OH$

8.31

(a) $(CH_3)_3CCH_2CH_2—\overset{O}{\overset{\|}{C}}—OH + H_2NCH(CH_3)_2$

(b) $2\ CH_3CH_2\overset{O}{\overset{\|}{C}}—OH + H_2NCH_2CH_2CH_2NH_2$

8.33

$HO—\overset{O}{\overset{\|}{C}}—CH_2CH_2CH_2CH_2—\overset{O}{\overset{\|}{C}}—OH + H_2NCH_2CH_2CH_2NH_2$

8.35

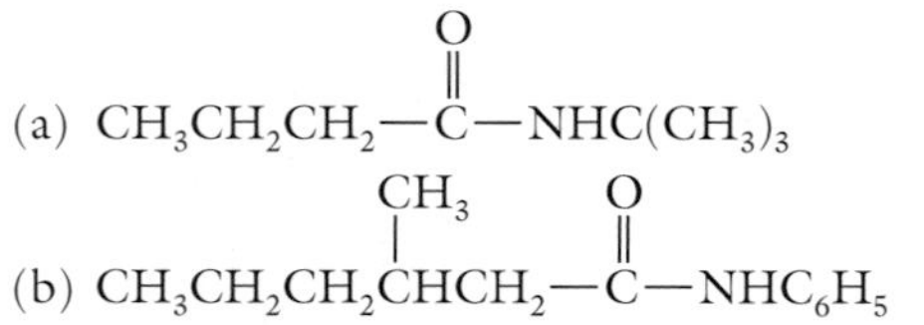

(a) $CH_3CH_2CH_2{-}\overset{O}{\overset{\|}{C}}{-}NHC(CH_3)_3$

(b) $CH_3CH_2CH_2\underset{}{\overset{CH_3}{\overset{|}{C}}}HCH_2{-}\overset{O}{\overset{\|}{C}}{-}NHC_6H_5$

8.37 (a) *N*-Isopropylpentanamide; (b) *N*-phenylpropanamide; (c) *N,N*-diethylbenzamide

8.39 (a) Pentanamide has the higher boiling point because it possesses hydrogen bonding, whereas *N,N*-dimethylpropanamide does not.

(b) The water solubilities are close for the two compounds because both can hydrogen bond with water. Pentanamide may be slightly more soluble because of its N–H bonds, which allow for more hydrogen-bond possibilities with water.

(c) Butanamide has the higher boiling point because of its dipolar ion structure ($^+N{=}C{-}O^-$), in addition to the hydrogen bonding that both compounds possess, which results in the strongest of secondary attractive forces.

(d) The water solubilities are close. Hexanamide may have slightly higher solubility because of the strong interactions of the dipolar structure with water.

(e) Butanamide has the higher boiling point because the secondary forces resulting from its hydrogen bonding and dipolar structure are stronger than the hydrogen-bonding forces in 1-pentanol.

8.41 Water = 1-propanol = ethanamide < propanamine < NaOH. Basicity measures the ability of a compound to bond to a proton. Amines are more basic than water and 1-propanol because nitrogen's nonbonded pair of electrons is more available for bonding to a proton, the result of nitrogen being less electronegative than oxygen. Amides are no more basic than water or alcohols, because the carbonyl group, through its electron-withdrawing effect, decreases the availability of nitrogen's nonbonded pair of electrons. NaOH is the strongest base because the actual base is OH^-. The negative charge on oxygen in this anion greatly increases its ability to accept a proton.

8.43

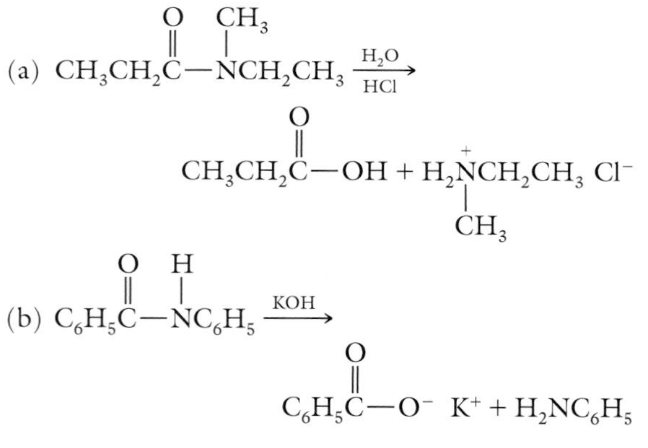

(a) $CH_3CH_2\overset{O}{\overset{\|}{C}}{-}\overset{CH_3}{\overset{|}{N}}CH_2CH_3 \xrightarrow[HCl]{H_2O}$

$CH_3CH_2\overset{O}{\overset{\|}{C}}{-}OH + H_2\overset{+}{\underset{CH_3}{\underset{|}{N}}}CH_2CH_3\ Cl^-$

(b) $C_6H_5\overset{O}{\overset{\|}{C}}{-}\overset{H}{\overset{|}{N}}C_6H_5 \xrightarrow{KOH}$

$C_6H_5\overset{O}{\overset{\|}{C}}{-}O^-\ K^+ + H_2NC_6H_5$

8.45

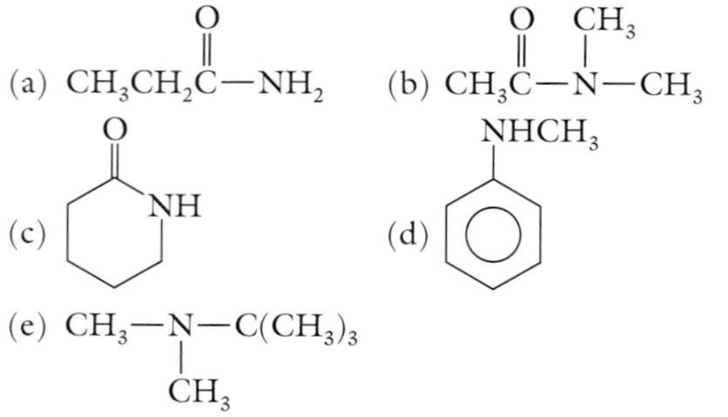

(a) $CH_3CH_2\overset{O}{\overset{\|}{C}}{-}NH_2$ (b) $CH_3\overset{O}{\overset{\|}{C}}{-}\overset{CH_3}{\overset{|}{N}}{-}CH_3$

(c) O NH (d) $NHCH_3$

(e) $CH_3{-}\underset{CH_3}{\underset{|}{N}}{-}C(CH_3)_3$

8.47

(a) $CH_3\overset{O}{\overset{\|}{C}}{-}\overset{CH_3}{\overset{|}{C}}HCH_2CH_2NH_2$

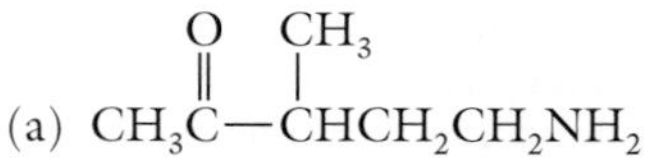

(b) $H_2NCH_2CH_2CH_2CH{=}CH_2$

8.49 (a) a = b = c = sp^3; A = B = 109.5°.
(b) a = b = c = sp^3; A = B = 109.5°.

8.51 (a) Butanamide. Amides possess, in addition to hydrogen bonding, very strong secondary attractive forces because of the dipolar structure of the amide group ($^+N{=}C{=}O^-$).

(b) Butanamide is slightly more water soluble because of the strong interaction of $^+N{=}C{=}O^-$ with water, although both compounds can hydrogen bond with water.

(c) Pentanamide. Amides possess the strongest of all secondary attractive forces because of the dipolar structure of the amide group ($^+N{=}C{=}O^-$), although both compounds possess hydrogen bonding.

(d) Pentanamide is slightly more water soluble because of the strong interaction of $^+N{=}C{=}O^-$ with water, although both compounds can hydrogen bond with water.

(e) Trimethylammonium chloride has the higher boiling point because it is a salt whose ionic attractive forces are stronger than those of a covalent compound.

(f) Sodium butanoate has the higher boiling point because it is a salt whose ionic attractive forces are stronger than those of a covalent compound.

(g) The two compounds have the same solubility because there is one NH_2 group per four carbon atoms in each compound.

8.53 Add a drop of the unknown to red litmus paper. If there is a change from red to blue, the unknown is an amine. If there is no color change, the unknown is an amide.

8.55 Propanol or ethanamide

8.57 Polymer is formed only when both reactants are bifunctional. Only c fulfills this requirement:

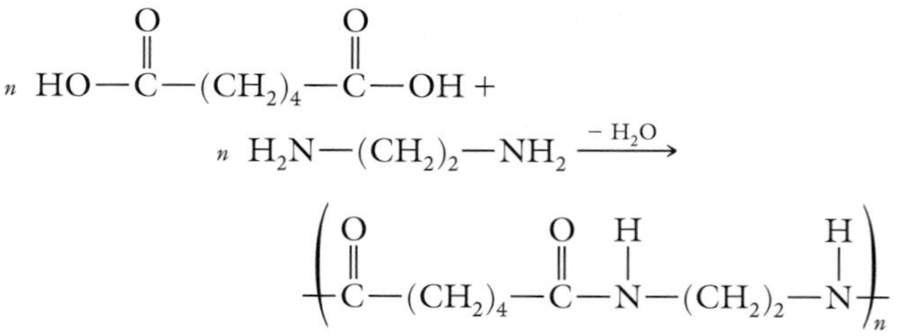

$$n\ HO{-}\overset{O}{\overset{\|}{C}}{-}(CH_2)_4{-}\overset{O}{\overset{\|}{C}}{-}OH\ +$$

$$n\ H_2N{-}(CH_2)_2{-}NH_2 \xrightarrow{-\ H_2O}$$

$$\left(\overset{O}{\overset{\|}{C}}{-}(CH_2)_4{-}\overset{O}{\overset{\|}{C}}{-}\overset{H}{\overset{|}{N}}{-}(CH_2)_2{-}\overset{H}{\overset{|}{N}}\right)_n$$

8.59 (a) 4; (b) 2; (c) 2; (d) 2

8.61 $[H_3O^+] = 4.67 \times 10^{-13}$; pH = 12.33

8.63 Propanoic acid

8.65 Ammonium chloride is not volatile, because it is an ionic compound.

8.67 B > A > C. HCl converts all of the methyl amine into the ammonium salt. NaOH suppresses the reaction of methyl amine with water to form the ammonium salt.

8.69 Physical strength, like boiling and melting points, increases with increasing secondary attractive forces. The secondary attractions between nylon-66 molecules are dipolar ionic forces, which are much greater than the London forces between polyethylene molecules.

8.71 Only compound 2 is a soap. A soap must have both an ionic group and a long hydrocarbon chain (Section 7.7).

8.73 The free amine is a covalent compound with relatively weak secondary attractive forces and volatilizes easily at moderate temperatures. The amine salt does not volatilize at moderate or even high temperatures, because it is an ionic compound with high secondary forces.

8.75 Phenobarbital would show peaks in the region from 1630–1690 cm^{-1} for the amide carbonyl bonds. Methamphetamine does not possess carbonyl bonds and would not show any peaks in that region.

8.77 $(CH_3)_3N + CH_3Cl \longrightarrow (CH_3)_4N^+ \ Cl^-$

8.79 $C_8H_{10}N_4O_2$

8.81 X = ethanal, Y = ethanoic acid, Z = *N*-ethylethanamide

Chapter 9

9.1 $CH_3CH_2CH_2CH_3$; $(CH_3)_3CH$

9.3 $CH_3CH_2OCH_2CH_3$; $CH_3OCH_2CH_2CH_3$; $CH_3OCH(CH_3)_2$

9.5 H, CH_3 / C=C / CH_3, H

9.7 (a) Chiral; (b) chiral; (c) achiral; (d) chiral

9.9 Structures a, b, and d cannot exist as enantiomers; c and e can.

(c) CH_2OH / H►C*◄NH_2 / CH_3 CH_2OH / NH_2►C*◄H / CH_3

(e) C_6H_5 / H►C*◄CH_2NHCH_3 / CH_3 C_6H_5 / CH_3NHCH_2►C*◄H / CH_3

9.11 (a) 3; (b) 1; (c) 1; (d) 3; (e) 1

9.13 (a) D; (b) L; (c) D; (d) L

9.15 Only b and c rotate plane-polarized light. (+)-Phenylalanine rotates in the clockwise direction. The direction for D-glucose is not known from the D configuration; it must be experimentally determined.

9.17 +31.5°

9.19 COOH / NH_2—┼—H / CH_3

9.21 127 g/L

9.23 $[\alpha] = -90°$

9.25 $\alpha = +20.1°$

9.27 (a) One pair of enantiomers, both of which are optically active

$CH(CH_2CH_3)_2$ / CH_3CH_2—*—H / CH_3 $CH(CH_2CH_3)_2$ / H—*—CH_2CH_3 / CH_3

(b) One pair of enantiomers (1 and 2) and one meso compound (3). Structures 1 and 2 are diastereomers of 3, and vice versa. Structures 1 and 2 are optically active; 3 is not.

1: CH_2CH_3 / H—*—CH_3 / CH_3—*—H / CH_2CH_3

2: CH_2CH_3 / CH_3—*—H / H—*—CH_3 / CH_2CH_3

3: CH_2CH_3 / CH_3—*—H / CH_3—*—H / CH_2CH_3

(c) Two pairs of enantiomers (5 and 6, 7 and 8). Enantiomers 5 and 6 are diastereomers of 7 and 8, and vice versa. All stereoisomers are optically active.

5: CH_2CH_3 / H—*—OH / CH_3—*—H / CH_2CH_3

6: CH_2CH_3 / HO—*—H / H—*—CH_3 / CH_2CH_3

7: CH_2CH_3 / HO—*—H / CH_3—*—H / CH_2CH_3

8: CH_2CH_3 / H—*—OH / H—*—CH_3 / CH_2CH_3

9.29 Stereoisomer 3 is the enantiomer of 1; 4 is the enantiomer of 2. Stereoisomer 3 has a specific rotation of −15° and a boiling point of 180–182°C. Stereoisomer 4 has a specific rotation of −26° and a boiling point of 163–165°C.

3: CH_3 / HO—┼—H / CH_3—┼—H / CH_2OH

4: CH_3 / HO—┼—H / H—┼—CH_3 / CH_2OH

9.31 (a) 16 stereoisomers

HC=O / HC*—OH / HC*—OH / HC*—OH / HC*—OH / H_2C—OH

(b) 8 stereoisomers

$H_2N—\overset{*}{C}H(CH_3)—C(=O)—NH—\overset{*}{C}H(CH_2OH)—C(=O)—NH—\overset{*}{C}H(CH_2COOH)—C(=O)—OH$

9.33 (a) One pair of optically active enantiomers.

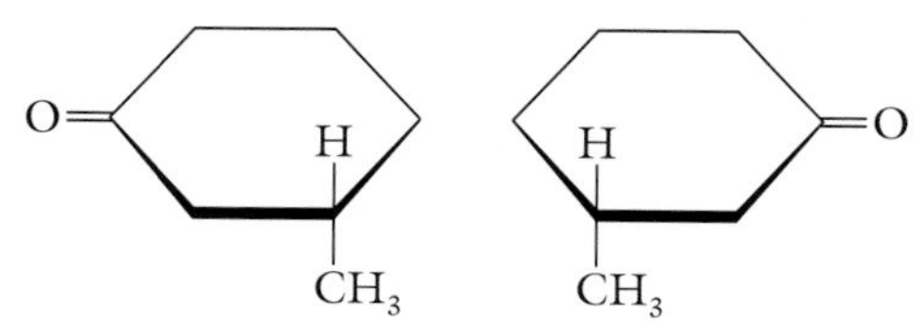

(continued on the next page)

(b) No stereoisomers. (c) No stereoisomers. (d) One pair of trans enantiomers (1 and 2) and one pair of cis enantiomers (3 and 4). All stereoisomers are optically active. Stereoisomers 1 and 2 are each diastereomers of 3 and 4, and vice versa.

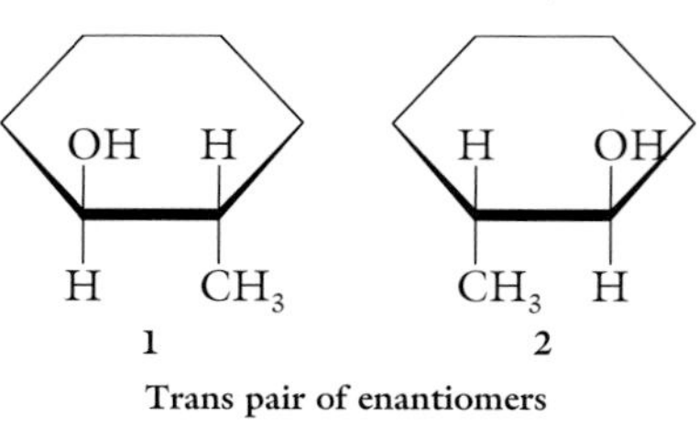

Trans pair of enantiomers

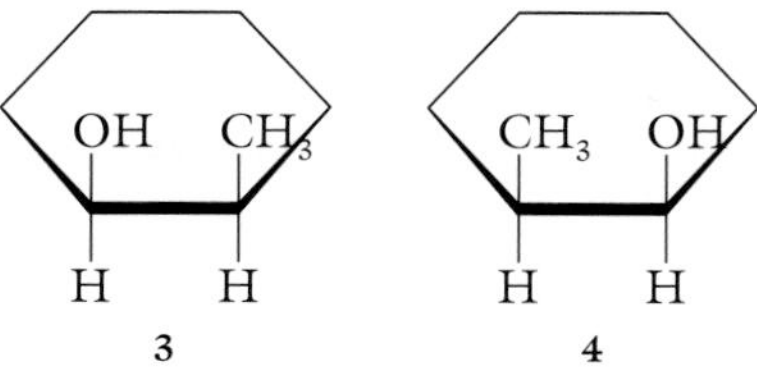

Cis pair of enantiomers

(e) One pair of optically active enantiomers (5 and 6) and one optically inactive meso compound (7). Stereoisomers 5 and 6 are diastereomers of 7, and vice versa.

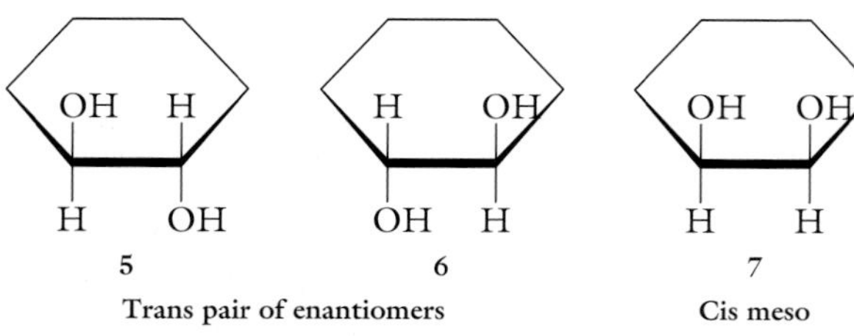

Trans pair of enantiomers Cis meso

9.35 (a) Four tetrahedral stereocenters and 16 stereoisomers

(b) one tetrahedral stereocenter and two stereoisomers

(c) six tetrahedral stereocenters and 64 stereoisomers

9.37 (a) 2; (b) 1; (c) 3; (d) 2; (e) 1; (f) 4; (g) 1; (h) 3; (i) 1; (j) 1; (k) 4; (l) 3; (m) 2; (n) 4; (o) 5

9.39 One pair of cis enantiomers and one pair of trans enantiomers

Cis pair of enantiomers

Trans pair of enantiomers

9.41 The product is the racemic mixture of the two enantiomers of 2-chlorobutane. There is equal probability of formation of the two enantiomers by addition of Cl to C2 from either side of the double bond.

9.43 Chiral recognition is not responsible, because neither isomer is chiral. Biological discrimination based on geometrical isomerism is responsible.

9.45 Square geometry would result in no enantiomers, but there would be three diastereomers:

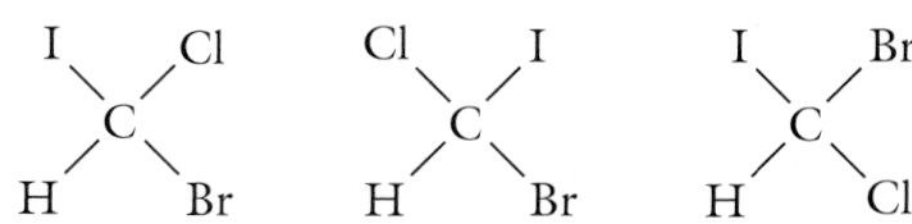

The tetrahedral geometry results in a pair of enantiomers, no diastereomers:

9.47 $\alpha = [\alpha]CL = +223°(0.100)(2.00) = 44.6°$. The delivered product would show an optical rotation of 44.6° if the concentration of (+)-penicillin V were 0.100 g/mL. Because the observed optical rotation is +22.3° instead of 44.6°, the delivered product has only half the specified concentration of (+)-penicillin V. (Another possibility is that the delivered product is contaminated with (−)-penicillin V.)

9.49 The D/L system is not applicable to carvone, because no heteroatom substituent is connected to its tetrahedral stereocenter. (−)-Carvone is (*R*)-carvone.

9.51 (*S*)-Methyldopa

9.53

$$\text{CH}_3\text{C}_6\text{H}_4\text{CH}_3 \xrightarrow[\text{hot, H}^+]{K_2Cr_2O_7} \text{HOOC}\text{C}_6\text{H}_4\text{COOH}$$

Chapter 10

10.1 (a) An aldose is a polyhydroxyaldehyde.

(b) A hexose is a six-carbon polyhydroxyaldehyde or polyhydroxyketone.

(c) A ketopentose is a five-carbon polyhydroxyketone.

(d) An aldotetrose is a four-carbon polyhydroxyaldehyde.

(e) All saccharides in the D family have the D configuration at the tetrahedral stereocenter farthest from the carbonyl carbon—the —OH group is on the right side of this carbon atom in the Fischer projection.

10.3 Eight aldopentoses; four L-aldopentoses

10.5 Aldohexoses have four tetrahedral stereocenters and 16 stereoisomers of which 8 are D-aldohexoses; ketohexoses have three tetrahedral stereocenters and 8 stereoisomers, of which 4 are D-ketohexoses.

10.7 (a) 1; (b) 3; (c) 2; (d) 2; (e) 4

10.9 Haworth projections are used to represent cyclic structures.

10.11 A pyranose ring is a six-membered cyclic hemiacetal ring.

10.13 In α-D-glucose, the —OH group at C1 and the $—CH_2OH$ group at C5 have a trans relation; in β-D-glucose, the two groups have a cis relation.

10.15 [Haworth structure: CH_2OH, H, O, H, H, H, H, HO, OH, OH, OH]

10.17 [Haworth structure: $HOCH_2$, O, OH, H, H, H, CH_2OH, OH, OH]

10.19 An equilibrium is quickly established between the cyclic hemiacetal (α- and β-anomers) and the open-chain aldehyde structures. The aldehyde structure gives the positive Benedict's test. As reaction proceeds, the hemiacetals open up to become the aldehyde, and eventually all of the D-galactose reacts with Benedict's reagent. A positive test is indicated by the disappearance of the blue color of Benedict's reagent and the formation of a red precipitate of Cu_2O.

10.21 (a) 1; (b) 3; (c) 4; (d) 2

10.23 [Haworth structure: CH_2OH, H, O, OH, H, OH, H, HO, H, H, OH] $\xrightarrow[-\ H_2O]{C_2H_5OH,\ H^+}$ [Haworth structure: CH_2OH, H, O, OCH_2CH_3, H, OH, H, HO, H, H, OH]

10.25 [Structure: CH_2OH, HO, OH, H, OH, H, H, =O, OH, H, OH]

10.27 All except b are reducing sugars and undergo mutarotation.

10.29 One aldohexose is linked through the α-anomeric position (C1) of its hemiacetal to the C4 position of the other aldohexose.

10.31 (a) Glucose; (b) glucose, galactose; (c) glucose, fructose; (d) cellobiose cannot be digested by humans

10.33 Figure 18.8 shows the structure of sucrose. Sucrose does not have a hemiacetal group.

10.35 [Haworth structure of gentiobiose with labels: HO, H, O, H, OH, H, HO, H, H, OH, O ← β(1→6)linkage, H, O, H, H, OH, H, HO, OH, H, OH]

Gentiobiose is a reducing sugar and undergoes mutarotation.

10.37 Benedict's test is negative; the unknown is sucrose. Sucrose does not give a positive Benedict's test, but lactose does.

10.39 Both are polymers of D-glucose linked together through α(1→4)-glycosidic linkages, but, in addition, amylopectin branches repeatedly through α(1→6)-glycosidic linkages. Starch, a mixture of amylose and amylopectin, is the storage form of D-glucose in plants. Starch is a nutritional carbohydrate for animals.

10.41 The digestion of amylose, amylopectin, and glycogen by humans yields D-glucose; humans cannot digest cellulose. The digestion of amylose, amylopectin, and cellulose by grazing animals yields D-glucose. Because they are herbivores (plant eaters) and glycogen is present only in animals, grazing animals do not ingest glycogen.

10.43 [Haworth structure: HO, H, O, O, H, OH, H, H, H, $NHCCH_3$, ‖O]

10.45 Plants are the immediate source; photosynthesis is the ultimate source.

10.47 A carbohydrate is (or is derived from) a polyhydroxyaldehyde or polyhydroxyketone.

10.49 Monosaccharides and disaccharides only

10.51 The —OH group on the tetrahedral stereocenter (C5) farthest from the carbonyl group is on the right side in the Fischer projection. The $—CH_2OH$ group at C5 extends upward from the plane of the ring in the Haworth projection.

10.53 (a) Disaccharide; (b) monosaccharide; (c) disaccharide; (d) polysaccharide; (e) polysaccharide; (f) monosaccharide; (g) polysaccharide

10.55 (a) 3; (b) 1; (c) 2; (d) 1; (e) 4; (f) 4; (g) 1

10.57 Storage polysaccharides are the storage forms of molecules used to generate energy or synthesize other needed molecules. Structural polysaccharides are used to construct plant-cell walls and the structures responsible for the macroscopic shape of a plant.

10.59 Grazing animals digest cellulose because microorganisms in their digestive tracts possess cellulase and cellobiase, the enzymes for cleaving the β(1→4)-glycosidic linkages of cellulose. Grazing animals themselves possess amylase and maltase, the enzymes for cleaving the α(1→4)-glycosidic linkages of starch.

10.61 Tetrasaccharide

10.63

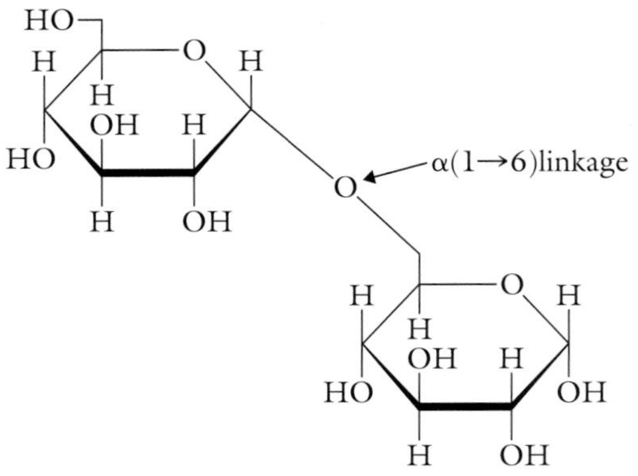

Because it has a hemiacetal group, isomaltose is a reducing sugar and undergoes mutarotation.

10.65

$$6\ CO_2 + 6\ H_2O + \text{sunlight} \xrightarrow{\text{chlorophyll}} (CH_2O)_6 + 6\ O_2$$

10.67 Plants produce oxygen as a by-product of photosynthesis. Plants and animals use oxygen in metabolism to generate energy.

10.69 The glucose residue, $C_6H_{10}O_5$, in amylose and cellulose has a molecular mass of 162 amu. The number of residues is the polysaccharide molecular mass divided by the residue molecular mass. There are 1000 residues for both the amylose and the cellulose of molecular mass 162,000 amu.

10.71 This is an example of chiral recognition whereby a stereoselective enzyme allows the two monosaccharides to link together only in the specified manner—α-D-glucose and β-D-fructose.

10.73 The abdominal distention and cramping, nausea, pain, and diarrhea of lactose intolerance can be minimized by taking lactase with the ingestion of milk and milk products and limiting the amount of such products. There are no long-term irreversible effects of lactose intolerance. Galactosemia has long-term irreversible adverse effects, including mental retardation, impaired liver function, cataracts, and even death, if not detected early in infancy. The only treatment is a lactose-free diet.

10.75 8.0/300, or 27 mg, of saccharin should be equivalent in sweetness to 8.0 g of sucrose because saccharin is 300 times as sweet as sucrose. The manufacturer's directions may have the objective of selling more saccharin or simply ensuring that the customers are satisfied by the sweetness of saccharin when it replaces sucrose.

10.77

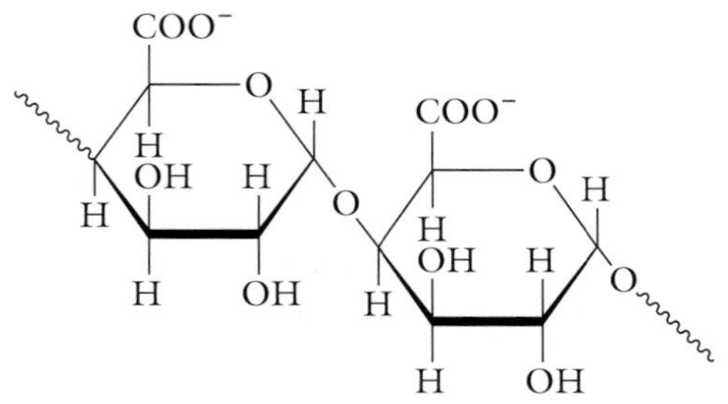

10.79 AB recipients have A and B antigens on their red-blood-cell surfaces and produce no antibodies (neither anti-A nor anti-B). Because the only types of antigens present in any blood type are A and B, no blood type produces an antibody–antigen reaction in AB recipients.

10.81 The daily consumption of sucrose (40 g) contains 160 kcal. Because aspartame is 180 times as sweet as sucrose, only 0.22 g aspartame would be used, which corresponds to 0.88 kcal. The daily savings is 159 kcal.

10.83 Sucrose undergoes hydrolysis to a mixture of fructose and glucose, which have different specific-rotation values from that of sucrose.

10.85

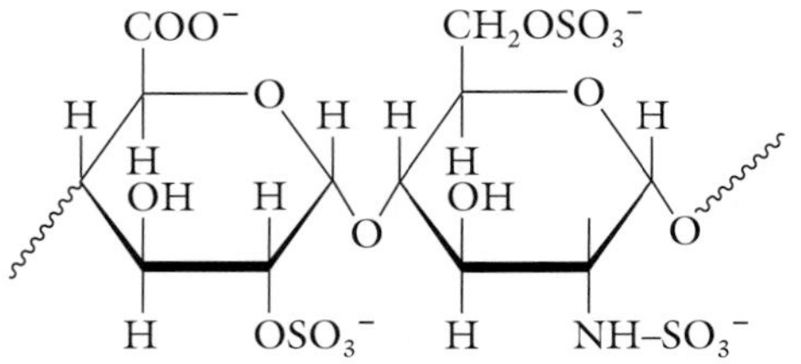

10.87 The α linkages between glucose residues in starch result in helical (coiled) conformations for the starch molecules. There is little intermolecular hydrogen bonding between different starch molecules, which allows much hydrogen bonding between starch and water. The β linkages between glucose residues in cellulose result in extended chain conformations for cellulose molecules. There is extensive hydrogen bonding between different cellulose molecules and very little hydrogen bonding between cellulose and water.

10.89 The equation is the same for both D-glucose and D-fructose:

$$C_6H_{12}O_6 + 6\ O_2 \longrightarrow 6\ CO_2 + 6\ H_2O$$

10.91 One equation is the exact reverse of the other. The conditions required for the two reactions are different. Combustion requires that the reaction be started with a flame or spark. Photosynthesis requires light and chlorophyll.

Chapter 11

11.1 1

11.3 Saturated fatty acids have no C=C double bonds. Unsaturated fatty acids contain cis C=C double bonds; monosaturated contain one double bond, and polyunsaturated contain more than one.

11.5 Compound 2 has the higher melting point. Molecules of compound 2 pack tightly together because of their extended saturated chains, which results in stronger secondary attractive forces. Molecules of compound 1 do not pack tightly and have weaker secondary attractive forces, because the cis double bonds cause the chains to kink.

11.7

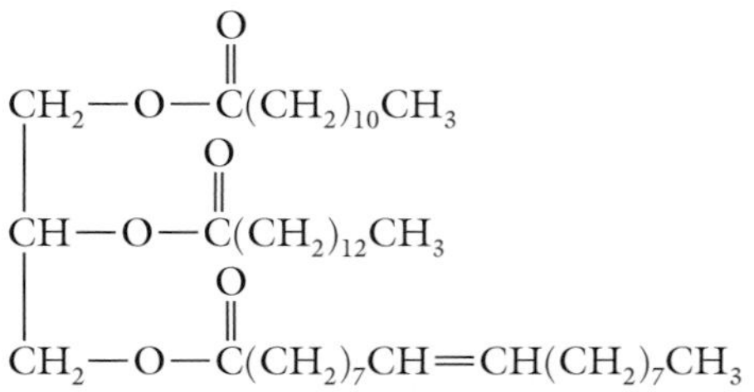

11.9 a > c > b. Melting point decreases with increasing unsaturated fatty acid content.

11.11 (a)

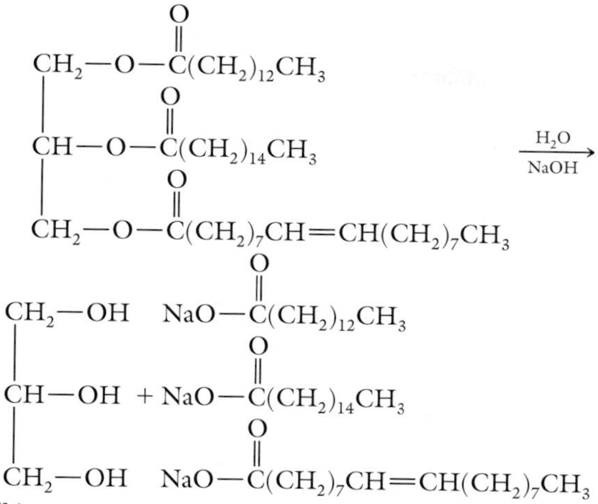

$$\begin{array}{l} CH_2-O-\overset{O}{\overset{\|}{C}}(CH_2)_{12}CH_3 \\ | \\ CH-O-\overset{O}{\overset{\|}{C}}(CH_2)_{14}CH_3 \\ | \\ CH_2-O-\overset{O}{\overset{\|}{C}}(CH_2)_7CH=CH(CH_2)_7CH_3 \end{array} \xrightarrow[NaOH]{H_2O}$$

$$\begin{array}{ll} CH_2-OH & NaO-\overset{O}{\overset{\|}{C}}(CH_2)_{12}CH_3 \\ | & \\ CH-OH & +\,NaO-\overset{O}{\overset{\|}{C}}(CH_2)_{14}CH_3 \\ | & \\ CH_2-OH & NaO-\overset{O}{\overset{\|}{C}}(CH_2)_7CH=CH(CH_2)_7CH_3 \end{array}$$

(b)

$$\begin{array}{l} CH_2-O-\overset{O}{\overset{\|}{C}}(CH_2)_{12}CH_3 \\ | \\ CH-O-\overset{O}{\overset{\|}{C}}(CH_2)_{14}CH_3 \\ | \\ CH_2-O-\overset{O}{\overset{\|}{C}}(CH_2)_7CH=CH(CH_2)_7CH_3 \end{array} \xrightarrow[Pt]{H_2}$$

$$\begin{array}{l} CH_2-O-\overset{O}{\overset{\|}{C}}(CH_2)_{12}CH_3 \\ | \\ CH-O-\overset{O}{\overset{\|}{C}}(CH_2)_{14}CH_3 \\ | \\ CH_2-O-\overset{O}{\overset{\|}{C}}(CH_2)_{16}CH_3 \end{array}$$

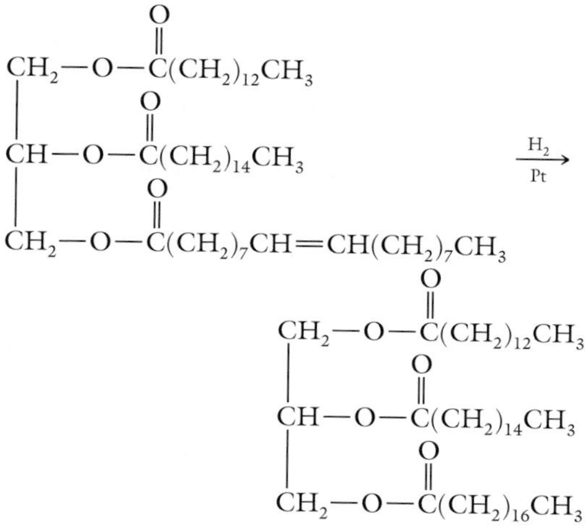

11.13 Butanoic acid is one of the hydrolysis products. Its low molecular mass makes it volatile, and this volatility is responsible for the foul odor.

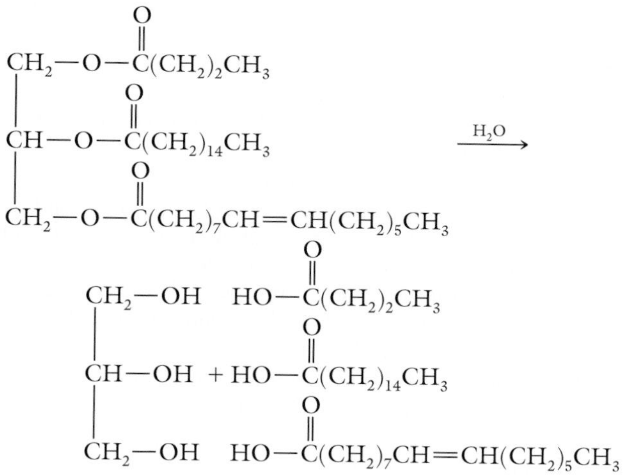

$$\begin{array}{l} CH_2-O-\overset{O}{\overset{\|}{C}}(CH_2)_2CH_3 \\ | \\ CH-O-\overset{O}{\overset{\|}{C}}(CH_2)_{14}CH_3 \\ | \\ CH_2-O-\overset{O}{\overset{\|}{C}}(CH_2)_7CH=CH(CH_2)_5CH_3 \end{array} \xrightarrow{H_2O}$$

$$\begin{array}{ll} CH_2-OH & HO-\overset{O}{\overset{\|}{C}}(CH_2)_2CH_3 \\ | & \\ CH-OH & +\,HO-\overset{O}{\overset{\|}{C}}(CH_2)_{14}CH_3 \\ | & \\ CH_2-OH & HO-\overset{O}{\overset{\|}{C}}(CH_2)_7CH=CH(CH_2)_5CH_3 \end{array}$$

11.15 Reduction of C=C double bonds with H_2 in the presence of a metal catalyst such as Ni or Pt

11.17 2

11.19 Like triacylglycerols, glycerophospholipids have fatty acid ester groups at two of the carbon atoms of glycerol; but, at the third carbon atom, a glycerophospholipid has a phosphodiester group.

11.21

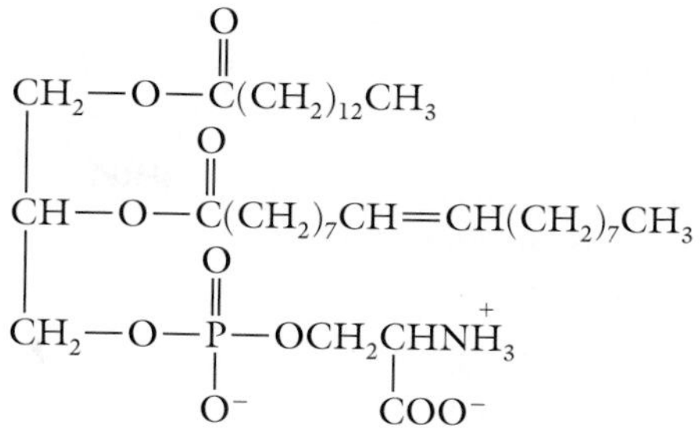

$$\begin{array}{l} CH_2-O-\overset{O}{\overset{\|}{C}}(CH_2)_{12}CH_3 \\ | \\ CH-O-\overset{O}{\overset{\|}{C}}(CH_2)_7CH=CH(CH_2)_7CH_3 \\ | \\ CH_2-O-\overset{O}{\overset{\|}{\underset{O^-}{\underset{|}{P}}}}-OCH_2\underset{COO^-}{\underset{|}{C}}H\overset{+}{N}H_3 \end{array}$$

11.23

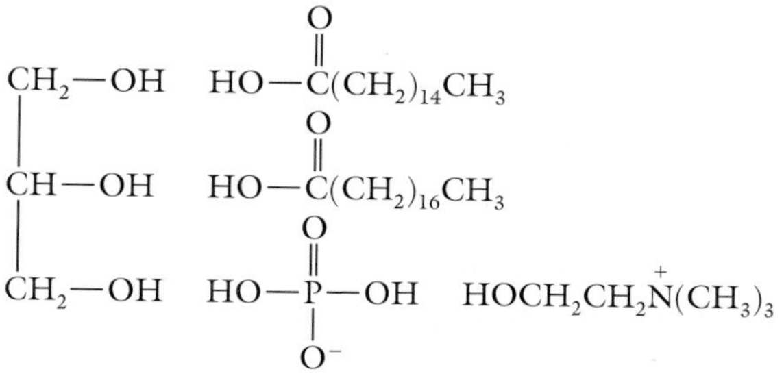

$$\begin{array}{lll} CH_2-OH & HO-\overset{O}{\overset{\|}{C}}(CH_2)_{14}CH_3 & \\ | & & \\ CH-OH & HO-\overset{O}{\overset{\|}{C}}(CH_2)_{16}CH_3 & \\ | & & \\ CH_2-OH & HO-\overset{O}{\overset{\|}{\underset{O^-}{\underset{|}{P}}}}-OH & HOCH_2CH_2\overset{+}{N}(CH_3)_3 \end{array}$$

11.25

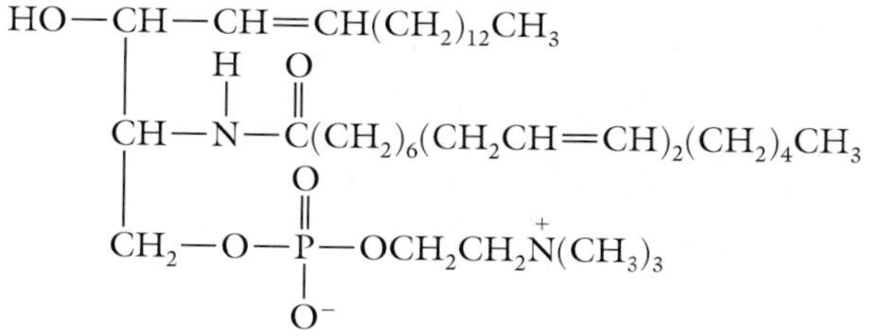

$$\begin{array}{l} HO-CH-CH=CH(CH_2)_{12}CH_3 \\ \quad\;\; | \\ \quad\;\; CH-\overset{H}{\overset{|}{N}}-\overset{O}{\overset{\|}{C}}(CH_2)_6(CH_2CH=CH)_2(CH_2)_4CH_3 \\ \quad\;\; | \\ \quad\;\; CH_2-O-\overset{O}{\overset{\|}{\underset{O^-}{\underset{|}{P}}}}-OCH_2CH_2\overset{+}{N}(CH_3)_3 \end{array}$$

11.27

$$\begin{array}{lll} HO-CH-CH=CH(CH_2)_{12}CH_3 & & \\ \quad\;\; | & & \\ \quad\;\; CH-NH_2 & HO-\overset{O}{\overset{\|}{C}}(CH_2)_7CH=CH(CH_2)_7CH_3 & \\ \quad\;\; | & & \\ \quad\;\; CH_2-OH & HO-\overset{O}{\overset{\|}{\underset{OH}{\underset{|}{P}}}}-OH & HOCH_2CH_2\overset{+}{N}(CH_3)_3 \end{array}$$

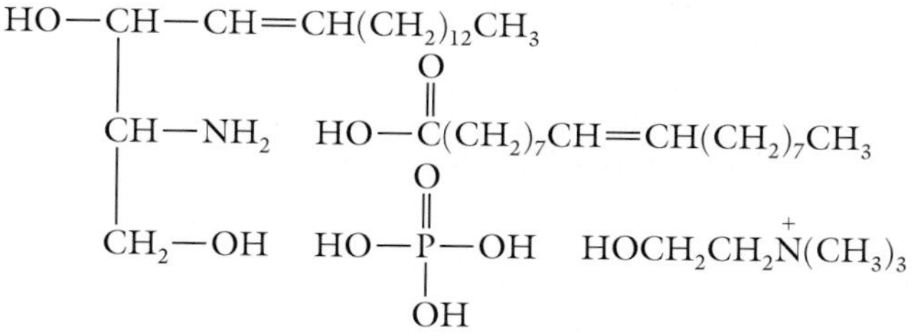

11.29 There is no functional group such as ester or amide whose hydrolysis cleaves the lipid into smaller molecules.

11.31 $2^6 = 64$ stereoisomers

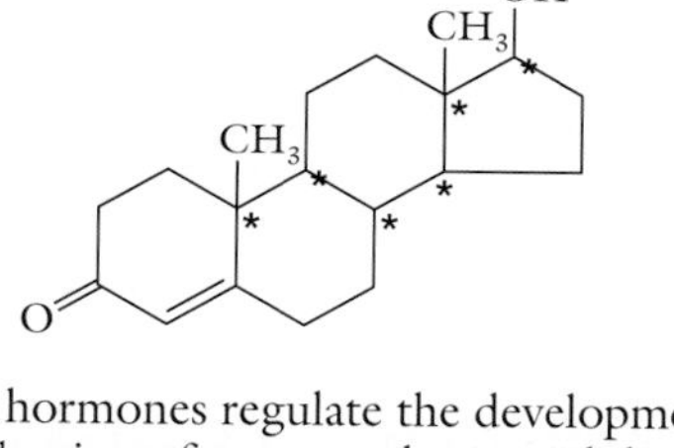

11.33 The sex hormones regulate the development of the sex organs, production of sperm and ova, and development of secondary sex characteristics: lack of facial hair, increased breast size, and high voice in women; facial hair, increased musculature, and deep voice in men.

11.35 Cholesterol

11.37 A carboxylate salt group is attached to the steroid ring structure.

11.39 Hydrolysis of the amide group does not cleave a large part of the molecule; only a small part is cleaved.

11.41 In leukotrienes, the 20-carbon chain of arachidonic acid and its carboxyl group are intact. Prostaglandins are similar to leukotrienes but have a cyclopentane ring formed by bond formation between C8 and C12 of the 20-carbon chain.

11.43 Hormones are substances synthesized in and secreted by endocrine glands and then transported in the bloodstream to target tissues, where they regulate the functions of cells. They are effective at very low concentrations. Eicosanoids are called local hormones because they are not transported in the bloodstream from their sites of synthesis to their sites of action. A local hormone acts in the same tissue in which it is synthesized.

11.45 A vitamin is an organic compound required in trace amounts for normal metabolism, but it is not synthesized by the organism that requires it. Vitamins must be included in the diet.

11.47 Fat-soluble vitamins are fat soluble and water insoluble; water-soluble vitamins are water soluble and fat insoluble.

11.49 See Table 11.4.

11.51 Membranes surround all cells and all organelles.

11.53 Amphipathic lipids (glycerophospholipids, sphingolipids, cholesterol) form the membrane by assembly into the lipid bilayer.

11.55 Unsaturated fatty acid chains are kinked because of the cis double bonds; they do not pack tightly in the lipid bilayer, secondary attractive forces are weaker, and flexibility is increased.

11.57 Hydrophilic molecules and ions are repelled by the hydrophobic regions of the lipid bilayer.

11.59 Energy is required to transport solute against a concentration gradient, from the low-concentration side to the high-concentration side.

11.61 Active transport

11.63 CH_3OH

11.65 (a) No ω number, because there is no C=C; (b) ω-7; (c) ω-6

11.67 Waxes have a variety of protective functions in plants and animals. They protect against parasites and mechanical damage, prevent excessive water loss, and waterproof waterfowl.

11.69 Digestion does not proceed with cleavage of all hydrolyzable groups. A mixture of products is obtained; for example, triacylglycerols yield mostly monoacylglycerol and fatty acids. Hydrolysis with base forms the carboxylate salts of fatty acids instead of the fatty acids.

11.71 a, c, f, g

11.73 a, c, f, g

11.75

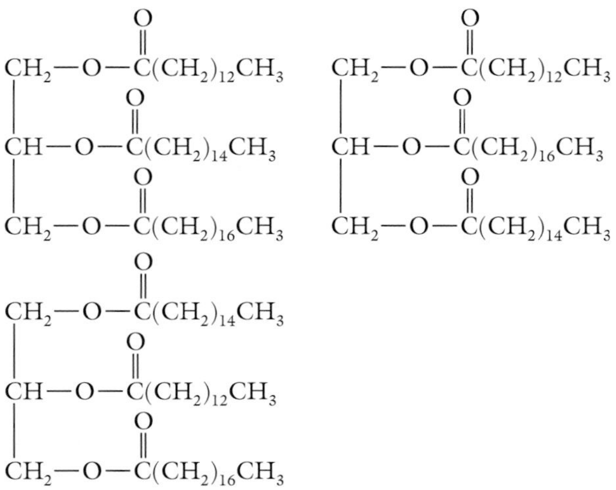

$$\begin{array}{l} CH_2-O-C(=O)(CH_2)_{12}CH_3 \\ CH-O-C(=O)(CH_2)_{14}CH_3 \\ CH_2-O-C(=O)(CH_2)_{16}CH_3 \end{array} \qquad \begin{array}{l} CH_2-O-C(=O)(CH_2)_{12}CH_3 \\ CH-O-C(=O)(CH_2)_{16}CH_3 \\ CH_2-O-C(=O)(CH_2)_{14}CH_3 \end{array}$$

$$\begin{array}{l} CH_2-O-C(=O)(CH_2)_{14}CH_3 \\ CH-O-C(=O)(CH_2)_{12}CH_3 \\ CH_2-O-C(=O)(CH_2)_{16}CH_3 \end{array}$$

11.77 The unsaturated fatty acid component (oleic) of triacylglycerol B yields a carboxylic acid on air oxidation. The carboxylic acid is malodorous because it is sufficiently low in molecular mass to be volatile.

11.79 $2^6 = 64$ stereoisomers

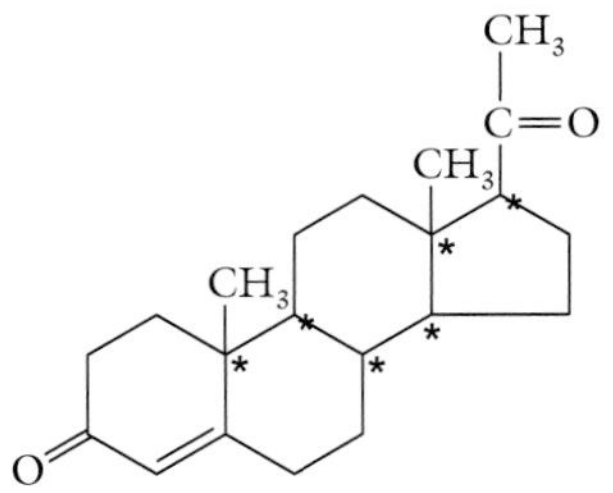

CH_3, C=O, CH_3, CH_3, O (structure with six stereocenters marked *)

11.81 Triacylglycerols are completely hydrophobic; there is no hydrophilic part.

11.83 Active transport

11.85 Melting point is increased with increasing saturated fatty acid composition; it is decreased by a decreasing number of carbons in the fatty acid components. Coconut oil has a high saturated fatty acid composition, but a high percentage of the saturated fatty acids have fewer than 14 carbons. The lower molecular mass lowers the melting point sufficiently that coconut triacylglycerols are liquids at ambient temperatures.

11.87

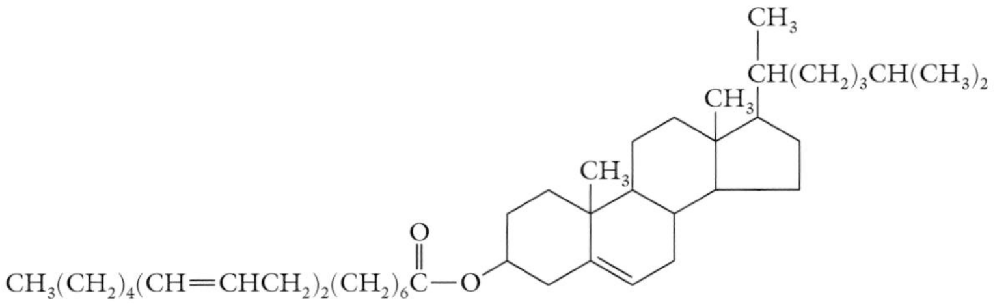

CH_3, $CH(CH_2)_3CH(CH_3)_2$, CH_3, CH_3

$CH_3(CH_2)_4(CH{=}CHCH_2)_2(CH_2)_6C(=O)-O-$

11.89 NaOH hydrolyzes the triacylglycerols (fats) to glycerol and sodium carboxylate salts, which are much more soluble in water than are the triacylglycerols.

11.91 Reverse bilayers are used. Hydrophilic heads are in the middle of the membrane; nonpolar tails are at the two surfaces of the membrane and in contact with the nonpolar heptane.

11.93 Each has a large hydrophobic (hydrocarbon) part and a small hydrophilic (ionic) part.

11.95 Hexane is not present in plants or animals.

11.97 The addition of hydrogen decreases double-bond content but the reaction conditions allow the isomerization of double bonds. The π bond breaks and reforms in the process. Double-bond re-formation yields trans double bonds to a greater extent than it yields cis double bonds because trans double bonds are more stable.

11.99 Testosterone shows a peak for the hydroxyl group in the 3200–3650 cm^{-1} region (Table 14.5). Progesterone has no hydroxyl group and shows no peak in that region.

11.101 Two

11.103 There is a mixture of different triacylglycerols. Some triacylglycerols in the mixture might have myristic, palmitic, and oleic acids, other triacylglycerols might have stearic, oleic, and linolenic acids; other triacylglycerols might have myristic, stearic, and oleic acids, other triacylglycerols might have palmitic, linoleic, and linolenic acids, and so on.

11.105 The catalyzed reaction has a lower E_a. The ΔH values of the two reactions are the same.

Chapter 12

12.1 1, β; 2, α; 3, γ

12.3 (a) Gly; (b) Ser, Thr; (c) Phe, Tyr, Trp; (d) Ile, Thr; (e) Trp, His; (f) Met, Cys; (g) Asp, Glu; (h) Pro

12.5 An acid–base reaction takes place between the carboxyl and amino groups.

12.7 (a)

$$^{+}NH_3-CH(CH_3)-COOH \quad (pH < 1)$$

$$^{+}NH_3-CH(CH_3)-COO^- \quad (pH = pI \text{ and } 7)$$

$$NH_2-CH(CH_3)-COO^- \quad (pH > 12)$$

(b) Solubility is minimum at pI (6.01), increasing at both higher and lower pH values.

(c) No migration at pI and physiological pH; migration to anode at pH > 12, to cathode at pH < 1.

12.9 Serylmethionine

$$H_3\overset{+}{N}-CH(CH_2OH)-C(=O)-NH-CH(CH_2CH_2SCH_3)-C(=O)-O^-$$

12.11 Threonylalanylaspartic acid

$$H_3\overset{+}{N}-CH(HCCH_3OH)-C(=O)-NH-CH(CH_3)-C(=O)-NH-CH(CH_2COO^-)-C(=O)-O^-$$

12.13 Glycylprolylalanine

$$H_3\overset{+}{N}-CH_2-C(=O)-N(\text{pyrrolidine ring})-C(=O)-NH-CH(CH_3)-C(=O)-O^-$$

12.15 Six constitutional isomers: Thr-Cys-Leu, Thr-Leu-Cys, Cys-Leu-Thr, Cys-Thr-Leu, Leu-Cys-Thr, Leu-Thr-Cys.

12.17 L-Lys-L-Ala-L-Glu, D-Lys-D-Ala-D-Glu, L-Lys-L-Ala-D-Glu, D-Lys-D-Ala-L-Glu, L-Lys-D-Ala-D-Glu, D-Lys-L-Ala-L-Glu. The stereoisomer with all L residues would be found in nature.

12.19 The peptide has a charge of 2− and migrates to the anode.

$$H_3\overset{+}{N}-CH(CH_2COO^-)-C(=O)-NH-CH(CH_3)-C(=O)-NH-CH(CH_2COO^-)-C(=O)-O^-$$

12.21 The products are the same—the individual amino acids (Phe, Asp, Lys, and Gly)—in digestion and both acidic and basic hydrolysis. There is a difference between acidic and basic hydrolysis. In acidic hydrolysis, amino groups are charged but carboxyl groups are uncharged. In basic hydrolysis, carboxyl groups are charged but amino groups are uncharged.

12.23 Hydrolysis yields the overall composition of the tripeptide but not the amino acid sequence. There are six possible sequences with the same composition: Ala-Gly-Lys, Ala-Lys-Gly, Gly-Ala-Lys, Gly-Lys-Ala, Lys-Gly-Ala, Lys-Ala-Gly.

12.25 (a)

$$H_3\overset{+}{N}-CH(COO^-)-CH_2-S-S-CH_2-CH(COO^-)-\overset{+}{N}H_3$$

$$2\ H_3\overset{+}{N}-CH(CH_3)-COO^- + H_3\overset{+}{N}-CH(CH_2-S-S-CH_2-CH(\overset{+}{N}H_3)COO^-)-COO^- + 2\ H_3\overset{+}{N}-CH(CH_2C_6H_5)-COO^-$$

(b) no reaction

(c)

$$H_3\overset{+}{N}-CH(CH_3)-CONH-CH(CH_2SH)-CONH-CH(CH_2C_6H_5)-COO^-$$

12.27 Simple proteins contain only peptide molecules. Conjugated proteins contain nonpeptide molecules or ions together with peptide molecules.

12.29 Rotation around a single bond

12.31 a

12.33 (a) None; (b) hydrogen bond; (c) hydrophobic; (d) none; (e) salt bridge; (f) none

12.35 Shielding of nonpolar residues from water; hydrogen bonding between peptide groups; interaction of polar residues with water

12.37 Primary structure

12.39 The functions of structural and contractile proteins require the formation of strong, macroscopic structures. These structures are possible only with fibrous proteins whose shape allows their aggregation by strong secondary attractive forces.

12.41 A double helix has two polypeptide chains coiled around each other; a triple helix has three chains coiled around one another.

12.43 α-Keratins are the structural components of cilia, hair, horn, hoof, skin, and wool.

12.45 The functions of catalytic, transport, regulatory, and protective proteins require that proteins not aggregate together to form macroscopic structures. The globular shape is not conducive to the aggregation of proteins, because there are minimal secondary attractive forces between them. At the same time, globular proteins are solubilized because the hydrophilic nature of their surfaces allows strong hydrogen-bond attractions with water.

12.47 Tertiary structure describes the relation between the different conformational patterns in different local regions of a polypeptide. α-Keratins have the same conformational pattern throughout the polypeptide chain; myoglobin has different conformational patterns in different local regions of the polypeptide.

12.49 The polypeptide is the apoprotein; the heme is the prosthetic group.

12.51 Oxygen is picked up and held at the Fe^{2+} ion of hemoglobin.

12.53 Asp, Glu, His, Lys. The protein is solubilized by attractive interactions of these side groups with the aqueous environment.

12.55 A mutation is an alteration in the DNA structure of a gene that may in turn produce a change in the primary structure of a protein, which in turn may alter its function.

12.57 Side-group polarity is important to the three-dimensional structure and function of hemoglobin. In the detrimental mutation, a polar residue replaces a nonpolar residue. In the mutation with little effect, one nonpolar residue replaces another.

12.59 There will be a tendency for a change in conformation to place that residue on the exterior surface, where attractive interaction with the aqueous environment can take place.

12.61 A native protein has the conformation that exists when the protein functions under physiological conditions.

12.63 Digestion destroys primary structure and produces α-amino acids by the hydrolysis of peptide bonds. Denaturation alters the secondary, tertiary, and quaternary structures but not the primary structure of proteins.

12.65 Ag^+ reacts with sulfhydryl groups of cysteine residues to form metal disulfide bridges that disrupt the protein's native conformation.

12.67 The conformations of polypeptides with lower contents of acidic and basic residues are less dependent on salt-bridge attractions.

12.69

COOH
H_2N—C—H
CH_2CH_2COOH

12.71 (a)

$CH_2CH_2SCH_3$
$^+NH_3$—CH—COOH
pH < 1

$CH_2CH_2SCH_3$
$^+NH_3$—CH—COO^-
pH = pI and 7

$CH_2CH_2SCH_3$
NH_2—CH—COO^-
pH > 12

(b) Solubility is minimum at pI (5.74), increasing at both higher and lower pH values.

(c) No migration at pI and physiological pH; migration to anode at pH > 12, to cathode at pH < 1.

12.73 5! = 120: Val-Phe-Lys-Thr-Pro, Val-Thr-Lys-Phe-Pro

12.75 Four: L-Lys-Gly-L-Ala, D-Lys-Gly-D-Ala, L-Lys-Gly-D-Ala, D-Lys-Gly-L-Ala. L-Lys-Gly-L-Ala would be found in nature.

12.77 Methionylserylisoleucylglycylglutamic acid

$H_3\overset{+}{N}$—CH($CH_2CH_2SCH_3$)—CONH—CH(CH_2OH)—CONH—CH($CHCH_3CH_2CH_3$)—CONH—CH_2—CONH—CH($CH_2CH_2COO^-$)—COO^-

12.79

$H_3\overset{+}{N}$—CH(CH_2COO^-)—C(=O)—N(H)—CH($(CH_2)_4\overset{+}{N}H_3$)—C(=O)—N(H)—CH($(CH_2)_4\overset{+}{N}H_3$)—C(=O)—$O^-$

It has a charge of 1+ and migrates to the cathode. The pI is on the basic side of that of a peptide containing only neutral residues.

12.81 2−

12.83 Gly-Ala-Cys-Gly-Lys-Phe-His-Glu-His-Cys-Met-Gly-Asp

12.85 (a) 4.9; (b) anode

12.87 Pepsin

12.89 Globular proteins. Their function requires water solubility.

12.91 X is not a β-pleated sheet, because it has a low content of Gly, Ala, and Ser. It is not an α-helix, because of the high content of Pro and Glu. X is either a β-turn or a loop.

12.93 Protein conformation and function are generally least affected by mutations in which replacements are by residues of the same size, charge, and polarity: (a) a very small residue is replaced by a very large residue; (b) one basic residue (positive charge at physiological pH) is replaced by another; (c) one neutral polar residue is replaced by another; (d) a basic residue (positive charge at physiological pH) is replaced by an acidic residue (negative charge at physiological pH).

12.95 Residue 1 from lysine; 2 from proline; 3 from glutamic acid.

12.97 (a) Hydrophobic residues (Leu, Phe, Val) are in the interior of the bilayer because it is hydrophobic; (b, c) polar residues (Glu, Lys, Ser) are in the extracellular and intracellular environments because those environments are aqueous.

12.99 The alteration in sickle-cell-anemia hemoglobin must be in regions of the polypeptide chains that greatly affect conformation and function. The difference in residues between beef and human insulins must be in regions of the polypeptide chains that do not significantly alter conformation and function.

12.101 The proteins are much larger molecules.

12.103 The production of animal protein (such as cattle and poultry) is more expensive than the production of vegetable protein (such as rice and beans).

12.105 New hair grows with the person's genetically determined straight shape.

12.107 70 Å

12.109 Insulin would not reach the bloodstream if taken by mouth because it is a polypeptide and would undergo digestion (hydrolysis) in the digestive tract.

12.111 The ratio of acidic-to-basic amino acid residues follows the order pepsin > albumin > lysozyme.

12.113 (a) 4.9; (b) pH > pI; (c) there is no pH at which all acidic and basic groups are uncharged

12.115 Tortoise shells

Chapter 13

13.1 Ribose for RNA; deoxyribose for DNA

13.3

O=P(OH)(OH)—O—CH_2 (deoxyribose ring: O, H, H, H, H, OH, H) attached to N of thymine (O, CH_3, N—H, O)

13.5 Deoxyribose, guanine, phosphoric acid

13.7 Nucleus

13.9 dTMP-dCMP

13.11 DNA does not contain U.

13.13

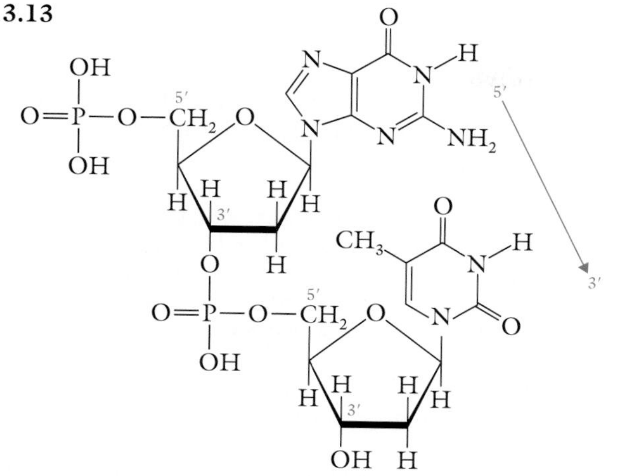

13.15 dAMP, 2 dTMP, dGMP, dCMP

13.17 Histones are proteins that compact the high-molecular-mass DNA double helix into the small volume of the cell nucleus.

13.19 Hydrogen bonding

13.21 d

13.23 30.4 mol % T, 19.6 mol % G

13.25 5′-CGGTTA-3′

13.27 DNA

13.29 A chromosome

13.31 (a) Replication bubbles are unwound parts of the DNA duplex where replication takes place; (b) the leading strand is the DNA strand that is synthesized in a continuous manner and the lagging strand is synthesized in a discontinuous manner; (c) Okazaki fragments are the discontinuous DNA fragments of the lagging strand.

13.33 Replication results in one parent DNA being converted into a pair of daughter DNAs.

13.35 (a) 5′-CGCTAA-3′; (b) 5′-TTAGCG-3′

13.37 RNA

13.39 The template DNA strand is used to produce the primary transcript.

13.41 Various primary transcripts are modified by different chemical reactions to produce rRNA, mRNA, and tRNA.

13.43 DNA: many billions, as high as 100 billion; rRNA: 500,000 to 1 million; mRNA: 100,000 to 1 or 2 million; tRNA: $< 50{,}000$

13.45 15 mol % U, 25 mol % G, 20 mol % C, 40 mol % A

13.47 Polypeptide

13.49 rRNA together with proteins form the ribosomes, in which polypeptide synthesis takes place. mRNA carries the genetic message encoding the polypeptide's amino acid sequence to be synthesized in the ribosome, binds to the ribosome, and acts as the template for polypeptide synthesis. α-Amino acids are supplied for polypeptide synthesis when tRNAs transport them into the ribosome.

13.51 The genetic message carried by mRNA is its sequence of bases, which specifies the sequence of α-amino acids for the polypeptide to be synthesized.

13.53 With very rare exceptions, all species—plant and animal—follow the same genetic code.

13.55 Aminoacyl-tRNA synthetase is the enzyme that catalyzes the joining together of amino acids at aminoacyl and peptidyl sites on mRNA. The reaction is called acyl transfer.

13.57 Three consecutive bases on DNA or RNA

13.59 Primary transcript = 3′-UGU-GUG-GUU-UAC-ACA-CCA-GUA-5′; mRNA = 5′-AUG-CAU-UUG-UGU-3′; polypeptide = Met-His-Leu-Cys

13.61 A mutation is an error in the base sequence of a gene.

13.63 In a substitution mutation, one base substitutes for another in the normal base sequence of DNA. In a frameshift mutation, a base is inserted into the normal base sequence or deleted from it.

13.65 A spontaneous mutation is a mutation resulting from random errors in replication.

13.67 Germ cells are egg and sperm cells; all other cells are somatic cells.

13.69 A cancer is an uncontrolled growth of cells.

13.71 In mutation a, a negatively charged amino acid residue is replaced by an uncharged residue and is thus more harmful; in mutation b, one negatively charged residue is replaced by another. Biological function is most often altered when a residue is replaced by another that is different in size, polarity, or charge.

13.73 (a) Met-Asp-Leu; (b) Met-Glu-Leu; (c) Met-Asp-Leu; (d) Met-Asp-Leu (no effect of mutation in intron)

13.75 An antibiotic is a chemical, usually an organic compound, that kills microorganisms (bacterium, mold, yeast).

13.77 A virus is an infectious, parasitic particle, usually smaller than bacteria, that consists of either DNA or RNA (but not both) encapsulated by a protein coat.

13.79 A vaccine contains a weakened virus or its proteins and is used to prevent viral infections.

13.81 The genome of an organism is altered by transplanting DNA into it from another organism (either the same or a different species).

13.83 Vector DNA is some organism's DNA that is to be altered by the introduction of donor DNA (from some other organism or laboratory synthesis). The altered DNA is the recombinant DNA.

13.85 Nucleotides are building blocks for nucleic acids.

13.87 The 3′-end has an unreacted OH at C3′ of either ribose or deoxyribose; the 5′-end has only one of the oxygen atoms of the phosphate group at C5′ attached to carbon.

13.89 Deoxyribonucleic acids do not contain U.

13.91 dGMP, dTMP, dAMP, and dCMP in the ratio 2:3:1:1.

13.93 DNA is double stranded; RNA is single stranded.

13.95 (a) 31 mol % C, 14 mol % A; (b) 17 mol % C, 38 mol % A

13.97 mRNA = 5′-AUG-CAA-UUU-AGU-3′; polypeptide = Met-Glu-Phe-Ser

13.99 A plant or animal conceived from a germ cell containing recombinant DNA

13.101 To introduce recombinant DNA into an organism to correct a hereditary disease

13.103 G-C base pairs have three hydrogen bonds, whereas A-T base pairs have two. Stability increases as the number of hydrogen bonds increases.

13.105 DNA consists of pairs of DNA molecules held together by the specific base-pairings G-C and A-T. Such base-pairing requires G = C and A = T for the pair (DNA duplex). These ratios are not needed for RNA, because RNA exists as single molecules, not pairs of double-helix molecules.

13.107 The minimum number of nucleotide bases is 438 because three nucleotide bases specify one amino acid residue. There are usually many more nucleotide bases because of the introns that are present. The 438 nucleotide bases are contained only in the exons of the gene.

13.109 The tendency of spontaneous mutations that produce antibiotic-resistant strains of the bacterium increases with increased use of the antibiotic.

13.111 23.7 billion

13.113 A two-base code system based on five different bases would code for a total of 25 different items, sufficient to code for the 20 different amino acids and the initiation and stop signals. Such a genetic code would have very little redundancy and protection against mutations.

13.115 Proto-oncogenes are genes responsible for producing the regulatory proteins responsible for normal cell growth and function. The mutation of proto-oncogenes to oncogenes results in cancer because oncogenes are genes that no longer code for the correct regulatory proteins.

13.117 A reverse transcriptase inhibitor blocks the transcription of the HIV RNA to HIV DNA. A protease inhibitor blocks the cutting up of the proteins produced by the translation of HIV RNA. A fusion inhibitor interferes with the attachment of the virus to the host cell membrane.

Chapter 14

14.1 Biochemistry is the extension of the relation of structure to function to the molecular and subcellular levels of organization.

14.3 In contrast with eukaryotic cells, prokaryotic cells have no subcellular membrane-bounded organelles or structures.

14.5 Oxidative processes take place in the mitochondria of animal cells.

14.7 Protein synthesis takes place at ribosomes.

14.9 DNA in bacterial cells is not bounded by a membrane and is located in a microscopically visible nuclear zone.

14.11 (1) To obtain energy in a chemical form by the degradation of nutrients, (2) to convert nutrient molecules into precursor molecules used to build cell macromolecules, (3) to synthesize cell macromolecules, and (4) to produce or modify the biomolecules necessary for specific functions in specialized cells or both

14.13 Anabolism comprises those processes taking part in the synthesis of biomolecules.

14.15 To reduce nutrient biomolecules to monosaccharides, amino acids, and fatty acids

14.17 Acetyl-*S*-CoA is oxidized to CO_2 and H_2O.

14.19 ATP functions as the carrier of energy to the energy-requiring processes of cells and is the link between catabolism and anabolism.

14.21 Substrate phosphorylation and oxidative phosphorylation

14.23 NAD^+ functions in catabolic reactions, and the reduced form of $NADP^+$ (NADPH) is used in reductive biosynthetic reactions.

14.25 An active site is a region on an enzyme where catalysis takes place. It is a binding site and it has catalytic function.

14.27 Inducible fit describes cases in which a substrate molecule can influence the complementarity of a binding site.

14.29 The vitamins

14.31 The number of moles of substrate that react per unit time per mole of enzyme

14.33 An enzyme whose activity can be modified by combination with specific activators or inhibitors

14.35 High-energy compounds are responsible for driving forward essentially unfavorable chemical reactions.

14.37 Catabolism consists of several distinct stages, the first of which is a collection step in which nutrient biomolecules are degraded to their building blocks, proteins to amino acids, polysaccharides to monosaccharides, and hydrolyzable lipids to fatty acids and glycerol. After that, hexoses, the carbon chains of fatty acids, and most of the amino acids are converted into acetyl-*S*-CoA, which is eventually oxidized to carbon dioxide and water.

14.39 No. The activities of cellular enzymes are regulated so that they can respond appropriately to the immediate metabolic needs of the cell. The activities of allosteric regulatory enzymes are modified by the binding of molecules that are not their substrates; other types of control are by covalent modification or feedback inhibition.

14.41 Yes. A cofactor may be required for achieving an enzyme's correct conformational structural requirements for substrate binding, and a separate coenzyme may be required for the catalytic process.

14.43 The feedback inhibitor is product E. Its concentration is controlled by its acting as an inhibitor of enzyme 2.

14.45 Competitive inhibition is reversed by increasing the concentration of substrate. Noncompetitive inhibition cannot be reversed by increasing substrate concentration.

14.47 No. Example: inhibition by heavy metals such as Ag^+ or Hg^{2+}

14.49 False. Enzymes are catalysts and have no influence on the final changes of the chemical reaction.

Chapter 15

15.1 Glucose is the principal source of energy for metabolism.

15.3 ATP, NADH, lactate (pyruvate under aerobic conditions)

15.5 The enzyme phosphofructokinase is the key regulatory control point in glycolysis.

15.7 In glycolysis, ATP is synthesized in two reactions:

ADP + 1,3-bisphosphoglycerate $\longrightarrow$ ATP + 3-phosphoglycerate

ADP + phosphoenolpyruvate $\longrightarrow$ ATP + pyruvate

15.9 NADH + H^+ + pyruvate $\rightarrow$ NAD^+ + lactate

15.11 Glycolysis would cease if inorganic phosphate were unavailable.

15.13 In the mitochondria

15.15 Yes. It is deactivated by phosphorylation and activated by dephosphorylation.

15.17 Isocitrate $\rightarrow$ α-ketoglutarate + CO_2

α-Ketoglutarate + CoA-SH $\longrightarrow$ succinyl-*S*-CoA + CO_2

15.19 Because the product of the reaction sequence, oxaloacetate, is the first reactant of the same sequence

15.21 Because oxaloacetate is regenerated at the conclusion of each cycle, one molecule of it can effect the oxidation of limitless numbers of acetyl groups.

15.23 The enzyme pyruvate carboxylase is activated by acetyl-*S*-CoA and, in the presence of ATP, catalyzes the formation of oxaloacetate from pyruvate and CO_2.

15.25 Not directly. Oxaloacetate is first synthesized in the mitochondria and reduced to malate, which then enters the cytosol and is reoxidized to oxaloacetate. There, the oxaloacetate reacts with GTP under the influence of the enzyme phosphoenolpyruvate carboxykinase to form phosphoenolpyruvate, CO_2, and GDP.

15.27 Oxaloacetate is first reduced to malate, which can pass through the mitochondrial membrane and enter the cytosol.

15.29 Glycogen is the polymeric storage form of glucose in animal tissue.

15.31 Its hydrolysis has a very large equilibrium constant, is coupled to the overall process, and is the driving force for the fomation of the activated glucose molecule.

15.33 Phosphorolysis results in the formation of glucose-1-PO_4, whereas hydrolysis would yield glucose.

15.35 Glucagon and epinephrine are hormones.

15.37 They both regulate the blood-glucose concentration, but epinephrine also affects blood pressure and heart rate.

15.39 Only liver cells possess the enzyme glucose-6-phosphate phosphatase, which hydrolyzes glucose-6-PO_4 to form free glucose.

15.41 There are three locations along the electron-transport chain where there is sufficient energy for the synthesis of ATP.

15.43 No. The membrane must be intact.

15.45 No. The inner membrane possesses transport proteins specific for only a few substances.

15.47 The maximum total amount of ATP produced under aerobic conditions is 38 mol of ATP per mole of glucose.

15.49 In active muscle cells, the rate of glycolysis is much greater than the rate of the citric acid cycle.

15.51 Active muscle cells produce ATP primarily through glycolysis. For glycolysis to continue at a maximal rate, NAD^+ must be regenerated by the oxidation of NADH and inorganic phosphate must be available for the formation of 1,3-bisphosphoglycerate.

15.53 NADH delivers its reducing power to the electron-transport chain by reducing oxidized cytosolic substances to their reduced counterparts, which can then penetrate the mitochondrial membrane.

15.55 ATP is produced by substrate phosphorylation in glycolysis and by oxidative phosphorylation in the citric acid cycle.

15.57 The rates of both glycogenolysis and glycogenesis will be markedly reduced.

15.59 The conversion of pyruvate into acetyl-*S*-CoA

15.61 If each branch is assumed to contain seven residues and a branch is assumed to form residues, there will be 1000 branches, each with a reducing end—hence, 1000 reducing ends.

Chapter 16

16.1 No. The upper limit for the mass of glycogen stored in the liver and muscles is less than the human daily caloric requirement.

16.3 When the blood-glucose concentration reaches its lowest levels between meals, and glucagon is released

16.5 $\text{R-COO}^- + \text{ATP} + \text{CoA-SH} \rightarrow \text{RCO-}S\text{-CoA} + \text{AMP} + 2\text{P}_i$

16.7 No. There is no specific transport system for fatty acyl-*S*-CoA derivatives in the mitochondrial membrane.

16.9 The first product is:

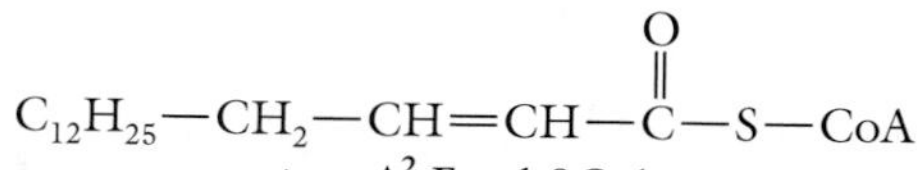

$$C_{12}H_{25}-CH_2-CH{=}CH-\overset{\overset{O}{\|}}{C}-S-CoA$$

trans-Δ^2-Enoyl-*S*-CoA

16.11 The second product is:

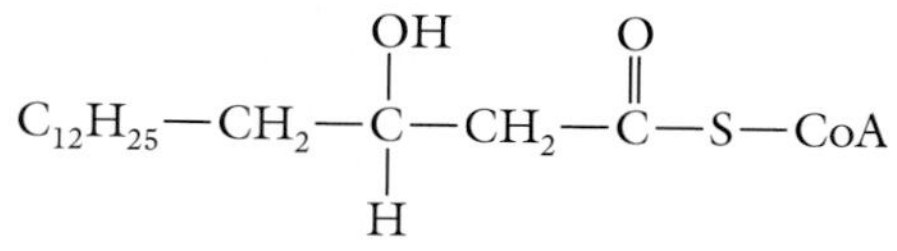

$$C_{12}H_{25}-CH_2-\underset{\underset{H}{|}}{\overset{\overset{OH}{|}}{C}}-CH_2-\overset{\overset{O}{\|}}{C}-S-CoA$$

L-3-Hydroxyacyl-*S*-CoA

16.13 The third product is:

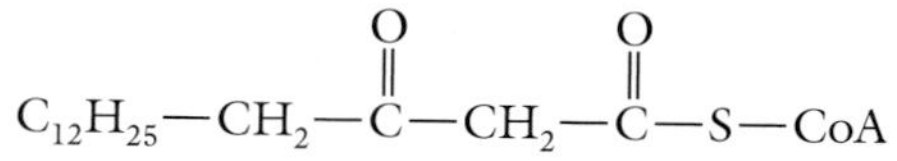

$$C_{12}H_{25}-CH_2-\overset{\overset{O}{\|}}{C}-CH_2-\overset{\overset{O}{\|}}{C}-S-CoA$$

3-Ketoacyl-*S*-CoA

16.15 The reactants in the fourth oxidation step are 3-keto(C_{16})acyl-*S*-CoA and CoA-SH.

16.17 Acetoacetate, D-3-hydroxybutyrate, and acetone

16.19 Ketosis arises during starvation and in diabetes mellitus.

16.21 Fatty acid anabolism takes place in the cytosol, and catabolism takes place in the mitochondria.

16.23 Insulin

16.25 Citrate from the mitochondria

16.27 Acetyl-*S*-CoA is transferred to the α-SH site, and malonyl-*S*-CoA is transferred to the β-SH site.

16.29 CO_2 is lost as HCO_3^-; so the reaction is irreversible and provides the driving force for the reaction. Biotin is the cofactor required for the formation of malonyl-*S*-CoA.

16.31 Before chain lengthening, the α-SH site is occupied by an acetyl group, and a carboxylated acyl derivative is at the β-site.

16.33 The formation of phosphatidate

16.35 The cost of the synthesis of a triacylglycerol is about 15% of the ATP generated in its oxidation.

16.37 In the form of cytidine diphosphoethanolamine

16.39 The hydrolysis of complex cellular glycolipids that normally takes place in the lysosomes does not take place.

16.41 In catabolism, the acyl carrier is CoA-SH, but, in anabolism, the acyl carrier is an —SH protein. Reduction in catabolism employs NADH, but, in anabolism, NADPH is used.

16.43 $8\ \text{Acetyl-}S\text{-CoA} + 7\ \text{ATP}^{4-} + 14\ \text{NADPH} + 7\ \text{H}^+ \rightarrow \text{palmitoyl-}S\text{-CoA} + 14\ \text{NADP}^+ + 7\ \text{CoA-SH} + 7\ \text{ADP}^{3-} + 7\ \text{P}_i^{2-}$

16.45 Yes. The oxidation of the fatty acid chain produces NADH and $FADH_2$, both of which provide reducing power to the electron-transport chain for the synthesis of ATP through oxidative phosphorylation.

16.47 True. The presence of high concentrations of citrate indicates that the cells are in a high-energy state (ATP in high concentration). Equally important is the fact that citrate is a specific allosteric activator of acetyl-*S*-CoA carboxylase, which catalyzes the rate-limiting step in the fatty acid synthase system.

16.49 Fatty acids arise in adipose tissue by the enzymatic hydrolysis of the stored triacylglycerols. The hydrolysis is catalyzed by a lipase that is activated by glucagon. The fatty acids leave adipose cells, become solubilized by being bound to serum albumin, and in that form travel throughout the circulatory system.

16.51 It (1) allosterically activates acetyl-*S*-CoA carboxylase and (2) provides a mechanism for transporting acetyl-*S*-CoA from the mitochondria.

16.53 Sam doesn't realize that excess glucose produces excess acetyl-*S*-CoA which then triggers the synthesis of fatty acids, stored as triglycerides. He must count calories if his aim is to keep weight down.

16.55 The absence of the lysozymal enzyme, *N*-acetylhexosaminidase, causes cellular lysozymes to swell out of control leading to cell death.

16.57 When a person has familial hypercholesterolemia

Chapter 17

17.1 Transamination is a process in which the amino group of an amino acid is interchanged with the carbonyl group of an α-keto acid.

17.3 They are catabolized and used as energy sources.

17.5 The amino groups of ingested amino acids are transferred to α-ketoglutarate by transamination.

17.7 By conversion into urea

17.9 A urotelic animal excretes ammonium ion in the form of urea.

17.11 The excessive lowering of α-ketoglutarate concentrations

17.13 The glucose–alanine cycle

17.15 The reactions constituting the urea cycle begin in the cytosol, continue in the mitochondria, and end in the cytosol.

17.17 Citrulline:

$$H_3\overset{+}{N}-\underset{\displaystyle H}{\overset{\displaystyle COO^-}{C}}-CH_2-CH_2-CH_2-\overset{\displaystyle H}{N}-\overset{\displaystyle O}{\overset{\|}{C}}-NH_2$$

17.19 ATP is not required for the formation of citrulline, the formation of arginine and fumarate from argininosuccinate, and the hydrolysis of arginine to form urea and ornithine.

17.21 Approximately 20% of the energy available in amino acids is used to synthesize urea.

17.23 Their catabolism gives rise to products that can be used to synthesize glucose.

17.25 Their catabolism results in the formation of ketone bodies.

17.27 Acetoacetyl-*S*-CoA and fumarate

17.29 Intermediates of catabolic or anabolic sequences accumulate in cells.

17.31 It leads to severe mental retardation. Treatment requires that phenylalanine be eliminated or severely restricted from the diet of newborns.

17.33 A nonessential amino acid can be synthesized by humans.

17.35 The nonionic water-soluble compound can be excreted without the loss of important anions such as phosphate.

17.37 No. Arginase is an enzyme found only in the livers of terrestrial animals.

17.39 Phenylketonuria, which causes defects in the central nervous system, is the result of a defect in the enzyme that oxidizes phenylalanine to form tyrosine. The hydrolysis of aspartame in the intestine will produce phenylalanine, which phenylketonurics must avoid.

17.41 False. The oxidative deamination of aspartate leads to the formation of pyruvate and ammonium ion. Ammonia cannot exist at physiological pH.

17.43 C could inhibit the A $\longrightarrow$ B step, Y could inhibit the C $\longrightarrow$ D step, and Z could inhibit the C $\longrightarrow$ G step. Another possibility is that Y could inhibit the C $\longrightarrow$ D step, Z could inhibit the C $\longrightarrow$ G step, and the A $\longrightarrow$ B step would be inhibited only if Y and Z are both present.

17.45 False. The end products of leucine catabolism are acetyl-*S*-CoA and acetoacetate. Neither of these compounds is a component of gluconeogenesis.

Chapter 18

18.1 Foods are enzymatically degraded to low-molecular-mass components to prepare them for absorption in the gut.

18.3 Gastrin is a hormone of the stomach stimulated by the entry of protein.

18.5 A zymogen is an inactive enzyme precursor.

18.7 They are transported across intestinal cells into the bloodstream and directly into the portal circulation.

18.9 Gastrin is secreted in the stomach in response to the entry of protein and stimulates the secretion of pepsinogen and HCl.

18.11 If a given protein provides all the required amino acids in the proper proportions and all are released on digestion and absorbed, the protein is said to have a biological value of 100.

18.13 Fatty acids containing more than one unsaturated bond past carbon 9 of a saturated chain, counting from the carboxyl end, cannot be synthesized by humans. Eicosanoids, a family of lipid-soluble organic acids that are regulators of hormones, are synthesized by mammals from arachidonic acid—a polyunsaturated fatty acid that mammals can synthesize with the use of dietary polyunsaturated fatty acids of plant origin as precursors.

18.15 A dietary-deficiency disease is caused by a deficiency in a factor essential to cellular function that can be obtained only through the diet.

18.17 The cells present are erythrocytes, leukocytes, and platelets.

18.19 A lipoprotein consists of a core of hydrophobic lipids surrounded by a shell of amphipathic lipids and proteins.

18.21 The proteins in lipoproteins solubilize lipids, direct specific lipoproteins to particular tissues, and activate enzymes that hydrolyze and unload lipids from lipoproteins.

18.23 The presence of hemoglobin in erythrocytes

18.25 It lowers the affinity of hemoglobin for oxygen.

18.27 Bicarbonate ion is formed in erythrocytes from CO_2 under the influence of the enzyme carbonic anhydrase.

18.29 A high concentration of CO_2 lowers the pH, and the Bohr effect enhances oxyhemoglobin dissociation.

18.31 The blood pH rises to higher than normal levels because CO_2 is being eliminated faster than it is being formed by respiring cells.

18.33 The brain uses mostly glucose and some 3-hydroxybutyrate for energy needs.

18.35 Virtually none

18.37 It must depend on the citric acid cycle for its ATP.

18.39 Glycogen

18.41 Lipases on adipocyte cell surfaces hydrolyze the triacylglycerols of the chylomicrons, allowing the resulting fatty acids to enter the cells.

18.43 Glucagon stimulates lipases within adipocytes.

18.45 It eventually reaches the liver to contribute to gluconeogenesis.

18.47 To produce the ATP necessary to effect the asymmetrical distribution of Na^+ ions around the kidney tubules.

18.49 Excessive amounts of hydrogen ion can be eliminated by increasing the ammonia concentration of the urine, thereby increasing the amount of ammonium ion excreted.

18.51 The liver alone possesses glucokinase, an enzyme able to phosphorylate glucose at the high concentrations present in the blood after a meal.

18.53 It is converted into urea.

18.55 Acetoacetate and 3-hydroxybutyrate

18.57 It synthesizes the bile acids from cholesterol.

18.59 No. For example, the hormone insulin is a protein.

18.61 No. Steroid hormones penetrate cell membranes and are transported to the cell nucleus to modify DNA translation.

18.63 No. The cellular uptake of oxygen by the brain remains constant under all conditions.

18.65 False. Neurotransmitters have a very short life.

18.67 The hypothalamus secretes a specific hypothalamic releasing hormone (HRH) that causes the anterior pituitary gland to secrete thyrotropic hormone.

18.69 A large electrical potential across its cell membranes and the ability to transmit an electrical signal from cell to cell

18.71 ${Na_{in}}^+ < 10$ mM, ${Na_{out}}^+ = 150$ mM; ${K_{in}}^+ = 125$ mM, ${K_{out}}^+ < 5$ mM

18.73 The brain synthesizes hormones and hormone-releasing agents that affect distant organs and tissues.

18.75 The absorptive state is the condition of the body immediately after a meal, when the gastrointestinal tract is full.

18.77 Insulin

18.79 Glucagon

18.81 In the postabsorptive state, blood glucose is supplied by glycogenolysis. When starvation begins, glycogen is unavailable, and gluconeogenesis with the use of glucogenic amino acids from muscle provides the glucose.

18.83 It typically appears early in life and can be controlled by insulin replacement.

18.85 Gluconeogenesis with the use of glucogenic amino acids provides the glucose.

18.87 Because glucose is not available for energy production, fatty acid oxidation becomes the main source of ATP.

18.89 Nitrogen balance is achieved when the intake of protein nitrogen is equal to the loss of nitrogen in the urine and feces.

18.91 This vitamin is not ordinarily essential, because it is synthesized in adequate amounts by intestinal bacterial flora. However, vitamin B_{12} is transported across the intestinal cell membrane as a complex with intrinsic factor; in persons who cannot synthesize this protein, the vitamin cannot enter the bloodstream through intestinal absorption.

18.93 The liver uses excess glucose for the synthesis of fatty acids and cholesterol.

18.95 If it were synthesized as the active proteolytic enzyme chymotrypsin, it would digest the pancreas itself.

18.97 Dietary lipid in the form of chylomicrons is hydrolyzed at the surface of capillary membranes within muscle and adipose tissue. The free fatty acids then penetrate the cells of the tissue to be stored as triacylglycerol or oxidized for energy.

18.99 The protein contained in the wheat is difficult to extract because it is located within an indigestible husk.

18.101 Sensory neurons spontaneously generate action potentials at frequencies that depend on the intensity of an environmental stimulus.

Glossary

accuracy *See* **error.**
acetal An organic compound that contains two —OR or —OAr groups attached to the same carbon atom.
achiral Refers to a molecule or object that is not chiral; that is, it is superimposable on its mirror image.
acid *See* **Brønsted–Lowry acid.**
acid anhydride *See* **carboxylic acid anhydride.**
acid derivative *See* **carboxylic acid derivative.**
acid dissociation constant An equilibrium constant for the dissociation of a weak acid.
action potential After stimulation, an electrical depolarization that moves along a neuron's membrane.
activated complex A transitory molecular structure formed by the collision of two reacting molecules.
activation energy The minimum energy that colliding reactant molecules must possess to reach the activated complex and undergo chemical reaction to form new products.
active site A location on an enzyme's surface where catalysis takes place.
acyclic compound An organic compound that does not contain a cyclic structure.
acyl carrier protein The fatty acid carrier of the fatty acid synthase complex.
acyl group The —CO—R or —CO—Ar group that is found in carboxylic acids and their derivatives.
acyl transfer reaction A reaction that transfers an acyl group from one molecule to another.
addition polymerization The synthesis of a polymer by the self-addition of large numbers of alkene molecules.
addition reaction A reaction in which all the elements of a reactant add to the double or triple bond of a compound.
adipocyte *See* **adipose cell.**
adipose cell The cell in which triacylglycerols are stored.
alcohol An organic compound that contains a hydroxyl group (—OH) attached to an sp^3-hybridized carbon atom.
aldehyde An organic compound that contains a hydrogen and either an alkyl or an aryl group attached to a carbonyl group.
aldose A saccharide that contains an aldehyde group.
aliphatic hydrocarbon A nonaromatic hydrocarbon; that is, either an alkane, alkene, or alkyne.
alkali metal An element in Group I of the periodic table.
alkaline earth metal An element in Group II of the periodic table.
alkane A hydrocarbon that contains only carbon–hydrogen and carbon–carbon single bonds.
alkene An unsaturated hydrocarbon that contains a carbon–carbon double bond.
alkyl group A saturated hydrocarbon group that is attached to another group or to an atom in a molecule.
alkyne An unsaturated hydrocarbon that contains a carbon–carbon triple bond.
α-amino acid An amino acid containing both a carboxyl group and an amino group attached to the same carbon atom.
α-helix A polypeptide or other polymer molecule whose conformation is in the shape of a coiled spring.
α-particle The nucleus of a helium atom emitted by a radioactive substance.
amide An organic compound that contains a nitrogen atom attached to a carbonyl group.
amide bond The bond between the carbonyl carbon atom and the nitrogen atom in an amide group (—CO—N).
amine An organic compound that contains a nitrogen atom that is not attached to a carbonyl group.
amine salt The product of an amine and a strong acid such as HCl.
amino acid residue An α-amino acid that has been incorporated into a peptide.
amino acid sequence The sequence of amino acid residues in a peptide, listed from the N-terminal to the C-terminal residue.
amphipathic molecule A molecule that contains both hydrophilic and hydrophobic groups.
amphoteric compound A compound that is both an acid and a base.
anabolism The reactions of metabolism in which energy is conserved and biomolecules are synthesized.
anhydrous substance A substance that does not contain water.
anion An ion with a negative charge.
anode The positive electrode in a battery or an electrophoretic or other apparatus.
anomers Saccharides that are diastereomers differing only in the configuration at a hemiacetal or acetal carbon.
anticodon A base triplet on tRNA that is complementary to a codon on mRNA.
antioxidant A substance that protects other substances against damage from oxidation.
apoprotein The polypeptide part of a conjugated protein.
aqueous solution A solution with water as the solvent.
aromatic compound An organic compound with high stability due to the presence of six π electrons in a cyclic structure.
aryl group An aromatic group attached through one of its sp^2 carbon atoms to another group or to an atom in a molecule.
atmosphere *See* **standard atmosphere.**
atomic symbol A one- or two-letter symbol for an element or an element's atoms.

basal metabolic rate The measurement of basal metabolic activity.
basal metabolism The minimal metabolic activity of a human at rest whose gastrointestinal tract is empty.
base *See* **Brønsted–Lowry base.**
base (of nucleic acids) A heterocyclic nitrogen base that is a component of a nucleotide.
base dissociation constant An equilibrium constant for the dissociation of a weak base.
base pairing The strong hydrogen-bonding attractive forces between a base on one DNA strand of a DNA double helix and a base on the other DNA strand.
base triplet A set of three consecutive bases in DNA or RNA.
base unit A fundamental unit of measurement for one of the base quantities in the SI system.
β-bend *See* **β-turn**
β-oxidation The first step in fatty acid oxidation.

β-particle An electron emitted from the nucleus of a radioactive substance.
β-turn The conformation of a polypeptide chain in a region where the chain abruptly changes direction.
bifunctional reactant An organic compound with two functional groups per molecule.
binary compound A compound consisting of two elements.
binding site The structural component of an enzyme's catalytic site where substrate is bound by secondary forces.
biochemistry The study of the structures and functions of living organisms at the molecular level.
biological membrane A membrane that surrounds a cell or organelle.
biomolecules The molecules of which living organisms are composed.
boiling point The temperature at which a substance boils when the atmospheric pressure is 760 torr.
bond angle The angle between two bonds that share a common atom.
branched chain Refers to an organic compound in which not all carbon atoms in the molecule are connected one after another in a continuous chain.
Brønsted–Lowry acid Any substance that can donate a proton.
Brønsted–Lowry base Any substance that can accept a proton.
buffer system An aqueous solution containing a Brønsted–Lowry acid with its conjugate base.

calorie The heat absorbed when the temperature of 1.0 g of water rises 1 Celsius degree between 14.5°C and 15.5°C.
carbocation An organic species carrying a positive charge on a carbon atom.
carbohydrate A polyhydroxyaldehyde or a polyhydroxyketone and its derivatives and polymers.
carbonyl group The carbon–oxygen double bond C=O, present in aldehydes and ketones.
carboxyl group The —COOH group present in carboxylic acids.
carboxylic acid An organic compound that contains a carboxyl group.
carboxylic acid anhydride An organic compound that contains the —CO—O—CO— group.
carboxylic acid derivative A compound that can be synthesized from a carboxylic acid or converted into one.
carboxylic acid halide An organic compound that contains a halogen attached to a carbonyl group.
carboxylic ester An organic compound that contains an —OR or —OAr group attached to a carbonyl group.
carboxyl-terminal residue The end of a peptide that contains the carboxyl group.
catabolism Metabolic reactions in which molecules are degraded.
catalysis A process in which the rate of a chemical reaction is increased by the presence of a catalyst.
catalyst A substance that takes part in a chemical reaction and accelerates its rate but emerges unchanged at the reaction's conclusion.
cathode The negative electrode in a battery or an electrophoretic or other apparatus.
cation An ion with a positive charge.
cellular respiration Metabolic reactions in which oxygen is used and carbon dioxide is produced. *See also* **respiration.**
Celsius scale A temperature scale, in degrees, that defines the freezing point of water at 0°C and the boiling point at 100°C.
centimeter A length equal to 1/100 of a meter.
chair conformation The puckered, nonplanar shape of a six-membered ring.
chemical bond An electrical force or attraction strong enough to hold atoms together to form compounds.
chemical change A process through which substances lose their chemical identities and form new substances with new properties.
chemical equation A shorthand representation of a chemical reaction that uses formulas for reactants and products and numbers before components to represent their mole proportions.
chemical equilibrium A state in which the rate at which products form is equal to the rate at which reactants form.
chemical kinetics The study of the rate of a chemical reaction.
chemical property The ability of a pure substance to undergo chemical change.
chemiosmotic mechanism Describes the formation of ATP by a hydronium ion gradient across the inner mitochondrial membrane caused by electron flow through the electron-transport chain.
chiral Refers to a molecule or object that is not superimposable on its mirror image.
chloride shift The movement of chloride ions from plasma into a red blood cell in response to the movement of bicarbonate ions out of the red blood cells.
chromosome A double helix of DNA located in the nucleus of a cell and containing the hereditary information that directs the synthesis of proteins.
chylomicron A small droplet consisting of about 90% triacylglycerol and small amounts of phospholipid, cholesterol, free fatty acids, and protein, formed during the absorption of dietary fat and released into the extracellular space surrounding intestinal cells.
cis isomer The diastereomer of an alkene or cyclic compound in which similar substituents are on the same side of the double bond or ring.
class of organic compounds *See* **family of organic compounds.**
codon A base triplet on mRNA that codes for a specific tRNA carrying a specific α-amino acid.
coenzyme An organic molecule, often derived from a vitamin, that is essential to the functioning of an enzyme.
cofactor Any molecule or ion essential to the functioning of an enzyme.
colloidal particle A particle smaller than 1×10^{-4} cm.
combining power The number of bonds formed by an atom when the atom is present in a covalent compound.
combustion The burning of an element or a compound in air.
competitive inhibitor A compound whose inhibition of enzyme activity can be reversed by increasing the substrate concentration.
complementarity principle A principle that accounts for the selectivity of enzymes. The structure of the substrate must complement the structure of the enzyme's binding site. *See also* **lock-and-key theory.**
complementary Refers to the base pairing of adenine with thymine in DNA, adenine with uracil in RNA, and guanine with cytosine in both DNA and RNA.
complementary strands The two DNA strands in a DNA double helix.
compound A pure substance composed of atoms of two or more elements present in a fixed and definite ratio.

concentration The quantity of a component of a mixture in a unit of mass or a unit of volume of the mixture.
condensation The conversion of the gaseous state into the liquid state.
condensation polymerization The synthesis of a polymer by the reaction between two bifunctional compounds, with the formation of a small-molecule (usually water) by-product.
condensed structural formula An abbreviated structural formula.
condensing enzyme The β-synthase enzyme of the fatty acid synthase complex.
configuration Describes the relative orientations in space of the atoms of a stereoisomer, independent of changes due to rotation about single bonds.
conformation Describes the different orientations of the atoms of a molecule that result only from rotations about its single bonds.
connectivity The order of attachment of the atoms in a molecule.
constitutional isomers Different compounds possessing the same molecular formula but differing in connectivity.
conversion factor *See* **unit-conversion factor.**
cosmic radiation Ionizing radiation emanating from the sun and outer space and consisting mostly of protons.
covalent bond The attractive force holding two atoms together resulting from the sharing of a pair of electrons.
cyclic compound An organic compound that contains a cyclic structure in which the first and last atoms of a chain are connected to each other.

deamination The removal of an amino group from an amino acid.
degree Celsius *See* **Celsius scale.**
degree Fahrenheit *See* **Fahrenheit scale.**
dehydration reaction A reaction that proceeds with the loss of water from within a molecule or between a pair of molecules.
dehydrogenation A reaction that proceeds with the loss of two hydrogen atoms from a molecule.
denaturation A loss of a protein's native conformation brought about by a change in environmental conditions, resulting in loss of physiological function.
density A derived unit defined as mass per unit volume.
deoxyribonucleic acid (DNA) A polynucleotide formed from deoxyribonucleotides.
deoxyribonucleotide A nucleotide that contains deoxyribose, phosphoric acid, and a heterocyclic base bonded together.
derived quantity A unit that is a mathematical relation between two or more base quantities, such as centimeters squared (cm^2) or centimeters per second (cm/s).
derived unit of measurement *See* **derived quantity.**
dextrorotatory compound A compound that rotates plane-polarized light in the clockwise direction.
diastereomers Stereoisomers that are not mirror images of each other.
diatomic molecule A molecule consisting of two atoms, such as O_2.
diffraction A wave property of light that allows it to bend around corners.
diffusion A reduction in a concentration gradient resulting from the random motion of particles.
digestion The hydrolysis of the amide, ester, and other hydrolyzable groups of various foodstuffs in the digestive tract of an organism.
dipolar ion A molecule with an overall zero charge but containing atoms bearing opposite charges.
dipole Any molecule that is electrically neutral overall but electrically asymmetrical—that is, containing separated partial and opposite electrical charges.
dipole–dipole force The secondary attractions between dipoles in different molecules or between dipoles in different portions of the same molecule.
dipole moment A measure of the size of a dipole.
diprotic acid An acid containing two dissociable hydrogen atoms.
disaccharide A saccharide composed of two monosaccharide units bonded together.
dissolution The process of dissolving or of preparing a mixture that will form a solution.
disulfide An organic compound that contains the —S—S— group.
D/L system A system for naming enantiomers.
DNA *See* **deoxyribonucleic acid.**
DNA double helix Two paired DNA molecules wound around each other to form a right-handed double helix.
DNA strand A DNA molecule.
double bond Two bonds between a pair of adjacent atoms, either two carbon atoms or one carbon and one oxygen.
double helix Two polymer molecules wound around each other in a helical manner.
dynamic equilibrium An equilibrium that is the result of two opposing processes both taking place at the same rate.

eicosanoid A nonhydrolyzable lipid derived from arachidonic acid, a polyunsaturated C_{20} fatty acid.
electrode An electrically conductive solid suspended in a conductive medium through which electricity can flow.
electrolyte Any substance that, when dissolved in water, will allow the solution to conduct electricity.
electromagnetic radiation Radiation, such as heat and light, that has its origin in the oscillation of charged particles.
electron configuration The complete description of the organization of the electrons of an atom.
electronegativity The ability of an atom covalently bonded to another atom to draw the bonding electrons toward itself.
electron shell An organization level of atomic electrons defined by a principal quantum number.
electron volt (eV) An energy unit used for radiation (1 eV = 96.5 kJ/mol).
element A substance in which all the atoms have the same atomic number and electron configuration.
elemental symbol A symbolic representation of the name of an element, such as He for helium.
empirical formula *See* **formula, empirical.**
emulsifying agent A substance that stabilizes a suspension of colloidal droplets of one liquid in a continuous phase of another liquid.
emulsion A stable suspension of colloidal droplets of one liquid in a continuous phase of another liquid.
enantiomers Stereoisomers that are nonsuperimposable mirror images of each other.
endothermic reaction A reaction in which heat is absorbed.
end point In a titration, the experimentally determined point at which the unknown acid or base is completely neutralized.
energy The capability to cause a change that can be measured as work.

energy level An atomic energy state defined by a principal quantum number.
enzyme A molecule, usually a protein, that catalyzes a biochemical reaction.
enzyme–substrate complex A temporary combination of enzyme with its substrate before catalysis.
equilibrium *See* **chemical equilibrium.**
equilibrium concentration The concentration of a product or a reactant of a chemical reaction at equilibrium.
equilibrium constant A constant that is calculated from the relation between molar concentrations of products and reactants of a chemical reaction at equilibrium.
equilibrium expression The relation between molar concentrations of products and reactants of a chemical reaction at equilibrium.
equivalence point In a titration, the theoretically expected point at which the unknown acid or base should be completely neutralized.
equivalent mass The equivalent mass of an acid is the formula mass of the acid divided by the number of reacting H^+ ions per mole of acid.
error The difference between the value considered true or correct and the measured value.
erythrocyte A red blood cell.
ester *See* **carboxylic ester.**
esterification The formulation of an ester by the reaction of a carboxylic acid with an alcohol or phenol.
ester linkage The functional group, —CO—O—, present in esters.
ether An organic compound that contains an oxygen atom bonded directly to two different carbon atoms.
eukaryote An organism whose DNA is contained inside the nuclei of the organism's cells.
evaporation The vaporization of a liquid into the atmosphere.
exact number A number that can be considered to have an infinite number of significant figures.
exothermic reaction A chemical reaction that evolves heat.
expanded structural formula A structural formula that shows the bonds of a molecule.
extensive property A property, such as mass or volume, that is directly proportional to the size of the sample.
extracellular fluid The fluid that is outside a cell.

Fahrenheit scale A temperature scale, in degrees, on which the freezing point of water is 32°F and the boiling point is 212°F.
family of organic compounds A large number of different organic compounds that have a characteristic functional group and pattern of behavior in common.
fatty acid A long-chain aliphatic carboxylic acid.
feedback inhibition The inhibition of the first enzyme in a biosynthetic sequence by the last product of the sequence.
fibrous protein A water-insoluble protein whose molecules have an elongated shape with one dimension much longer than the others and with a tendency to aggregate together to form macroscopic structures.
Fischer projection A representation of the bonds at a carbon atom, in which horizontal lines represent bonds extending in front of the plane of the paper and vertical lines represent bonds extending behind the plane of the paper.
flavoprotein A protein combined with a coenzyme containing riboflavin.
fluid-mosaic model The conceptual model that describes the structure and functioning of cell and organelle membranes.
folding The formation of polypeptide conformation by rotations about single bonds.
formula, chemical A representation of how many atoms of each element are in a fundamental unit of a compound.
formula, empirical A representation that gives the smallest whole-number ratio of the atoms of the elements of a compound.
formula, molecular The formula that shows the numbers of the atoms of the elements in a molecule of a compound.
formula, structural The formula that shows how the various atoms in a molecule are bonded together.
formula mass The sum of the atomic masses of the atoms in a formula of a compound in atomic mass units (amu).
formula unit The smallest particle that has the composition of the chemical formula of the compound.
formula weight *See* **formula mass.**
free radical *See* **radical.**
freezing point The temperature at which a substance undergoes the transition from a liquid to a solid.
functional group A specific atom or bond or a specific group of atoms in a specific bonding arrangement.
fused-ring compound An organic compound that contains adjacent rings sharing two or more ring atoms.

γ-ray Radiation of energy higher than that of an X-ray.
gene A part of a chromosome.
genetic code The complete list of mRNA codons and the α-amino acids and other instructions that the codons specify.
genetic disease A disease, caused by a mutation in a germ cell, that is passed from parent to offspring.
genetic message The sequence of codons in mRNA that specifies the sequence of α-amino acids for a polypeptide to be synthesized in a ribosome.
genome The total DNA contained in all the chromosomes of an organism.
geometrical isomers *See* **cis isomer; trans isomer.**
germ cell An egg or sperm cell.
globular protein A water-soluble protein whose molecules have a globelike shape and do not aggregate together to form macroscopic structures.
glycogenesis The process by which glycogen is synthesized from glucose.
glycogenolysis The process by which glucose is obtained by the breakdown of glycogen.
glycoprotein A compound having a saccharide bonded to a protein.
glycoside A saccharide that is an acetal.
gradient The change in value of a physical quantity with distance.
gram A mass equal to 1/1000 of a kilogram.
group, organic *See* **substituent.**
group or family of the periodic table The elements contained in one of the vertical columns of the periodic table.
growth factor A substance, such as a protein or trace element, that regulates cell division.

half-life The time required for a substance to lose one-half of its physical or chemical activity.
Haworth structure (projection) The cyclic structure of a saccharide.
heat A form of energy that moves between two objects in contact that are at different temperatures.
heat of fusion *See* **molar heat of fusion.**
heat of reaction *See* **molar heat of reaction.**

heat of vaporization *See* **molar heat of vaporization.**
hemiacetal An organic compound that contains one each of an —OH group and an —OR or —OAr group attached to the same carbon atom.
heterocyclic compound A cyclic organic compound that contains an oxygen or nitrogen atom in the ring.
heterogeneous mixture A mixture in which there are visual discontinuities in composition.
homeostasis The steady-state physiological condition of the body.
homogeneous mixture A mixture in which there are no visual discontinuities in composition.
hormone A substance synthesized in and secreted by an endocrine gland and then transported in the bloodstream to a target tissue where it regulates a cell function.
hydrate A compound in which water is present in a fixed molar proportion of the other constituents.
hydration A reaction that proceeds with the addition of water to a double or triple bond. In aqueous solutions, the association of water with ions by secondary forces.
hydrocarbon An organic compound that contains only carbon and hydrogen.
hydrogen bond The secondary attractions present in molecules that contain O—H, N—H, or H—F bonds.
hydrolysis reaction The reaction of an organic compound with water resulting in cleavage of the compound into two organic fragments each of which combines with a fragment (H^+ or OH^-) from water.
hydrophilic Refers to a molecule or a part of a molecule that is attracted to water.
hydrophobic Refers to a molecule or a part of a molecule that is repelled by water.
hydroxyl group An —OH group attached to a carbon atom.

inorganic chemistry The study of compounds that contain elements other than carbon.
inorganic compound A compound that contains elements other than carbon.
intermolecular force A secondary force operating between molecules.
intermolecular process A process (physical or chemical) that takes place between molecules.
International System of Units *See* **SI units.**
interstitial fluid Fluid outside cells and not in the blood.
intracellular fluid Fluid that is inside a cell.
intramolecular force A secondary force operating within a molecule.
intramolecular process A process (physical or chemical) that takes place within a molecule.
in vitro Refers to a substance or a process that is outside an organism.
in vivo Refers to a substance or a process that is inside an organism.
ionizing radiation Radiation that enters a medium and creates ions from the molecules therein.
isomerization The conversion of one isomer into another—for example, the conversion of a cis double bond into a trans double bond.
isomers Different compounds that have the same molecular formula.

joule The SI unit of energy (4.184 J = 1 cal).

kelvin The SI unit of temperature; its size is equal to 1/100 of the temperature interval between the freezing point and the boiling point of water.
ketone An organic compound that contains two alkyl or two aryl groups or one alkyl and one aryl group attached to a carbonyl group.
kilocalorie The quantity of heat equal to 1000 calories.
kilogram The SI unit of mass.
kilojoule The quantity of energy equal to 1000 joules.
kinetic energy The energy of a moving body.
kinetics *See* **chemical kinetics.**

leukocyte A white blood cell.
levorotatory compound A compound that rotates plane-polarized light in the counterclockwise direction.
like-dissolves-like rule A solute is soluble in a solvent only if the secondary attractive forces between molecules of solute are similar to the secondary attractive forces between molecules of solvent.
linear chain Refers to an organic compound in which all carbon atoms in the molecule are connected one after another in a continuous chain.
lipid A naturally occurring compound that is soluble in nonpolar or low-polarity solvents.
lipid bilayer Two lipid layers are arranged with the hydrophobic sides in contact with each other and the hydrophilic sides forming the inner and outer surfaces of the membrane, which are in contact with the internal and external aqueous environments.
lipoprotein A protein that has a lipid part either covalently bonded to the protein or held in place by secondary forces.
liter A volume equal to 1000 cm^3, or 1000 mL.
lock-and-key theory A theory to account for the selectivity of enzymes for their substrates. The substrate must fit into a binding site just as a key fits into a lock.
London force The weak secondary attractive forces possessed by all molecules, independent of the presence of permanent dipoles or hydrogen bonds.
longest continuous chain The number of carbon atoms in a molecule that are connected in a successive manner.
loop A region of polypeptide conformation that is less ordered than β-turns and much less ordered than α-helices and β-sheets.

macromolecule *See* **polymer.**
mass A measure of the quantity of matter relative to a reference standard.
mass number The sum of protons and neutrons in an isotope of an element.
matter Anything that has mass and occupies space.
measurement An instrumental determination of a physical quantity.
mechanism of reaction The molecular-level details of how reactants change into products.
melting point The temperature at which a solid is transformed into a liquid.
meso compound A diastereomer that contains tetrahedral stereocenters but is achiral and optically inactive.
metabolism All the chemical reactions that take place in an organism.
metal An element or combination of elements that is shiny, conducts electricity, and, if solid, is malleable.
metalloid An element that has some of the properties of metals and nonmetals.
meter The SI unit of length (1 m = 100 cm).
metric system A decimal system of weights and measures in which base units are converted into smaller or larger multiples by movement of a decimal point; superceded by the International System of Units (SI system).

microgram A mass equal to 1/1000 of a milligram.
microliter A volume equal to 1/1000 of a milliliter.
milliliter A volume equal to 1/1000 of a liter.
millimeter A unit of length equal to 1/1000 of a meter.
mixture Matter consisting of two or more pure substances in varying proportions.
molar concentration A solution's concentration in units of moles of solute per liter of solution; molarity.
molar heat of fusion The amount of heat required to melt 1 mole of a pure solid.
molar heat of reaction The amount of heat generated or absorbed per mole of reactant when a chemical reaction takes place.
molar heat of vaporization The amount of heat required to vaporize 1 mole of a pure liquid.
molarity *See* **molar concentration.**
molar mass The mass of 1 mole of a compound (either molecular or ionic) in grams.
mole An Avogadro number of a substance's formula units.
molecular compound A compound whose atoms are joined by covalent bonds.
molecular formula *See* **formula, molecular.**
molecular mass The mass of one molecule of a molecular compound in atomic mass units (amu).
molecular weight *See* **molecular mass.**
molecule The smallest particle of a molecular compound.
monomer A reactant from which polymers are synthesized.
monoprotic acid An acid containing one dissociable hydrogen atom.
monosaccharide The simplest saccharide; monosaccharides cannot be hydrolyzed into smaller saccharides.
multiple bond A double or triple bond.
mutarotation The change in optical rotation that takes place when a pure α- or β-saccharide with a hemiacetal structure is dissolved in water.
mutation An alteration in the base sequence of a gene that may in turn produce a change in the primary structure of a peptide.

native protein The conformation of a protein under normal physiological conditions.
neuron Any of a number of specialized cells constituting the body's nervous system.
neutralization reaction A reaction between an acid and a base.
noble gas An element in Group VIII of the periodic table.
nomenclature The naming of compounds.
nonbonding electron An electron located in the outer shell of an atom but not participating in bonding to other atoms.
noncompetitive inhibitor A compound whose inhibition of enzyme activity cannot be reversed by increasing the substrate concentration.
nonelectrolyte Any substance that, when dissolved in water, will not allow the solution to conduct electricity.
nonmetal An element that is not shiny, cannot conduct electricity, and is not malleable.
nucleic acid A polynucleotide.
nucleotide A building block (monomer) for nucleic acids.
nucleus of a cell A membrane-enclosed organelle in which the cell's genetic information is stored and nucleic acids are synthesized.

open chain *See* **acyclic compound.**
optical activity The ability of a compound to rotate the plane of plane-polarized light.
optical rotation The rotation of the plane of plane-polarized light.
orbital The region of space in which an electron resides.
orbital hybridization The excitation of an element's ground state to produce a different electronic configuration.
organelle A specialized, self-contained structure surrounded by a membrane and found inside a cell.
organic chemistry The study of compounds that contain carbon.
organic compound A compound that contains carbon.
oxidation of an inorganic compound An increase in the number of oxygen atoms or an increase in the positive charge of the metallic element or both.
oxidation of an organic compound An increase in the number of oxygen atoms or a decrease in the number of hydrogen atoms (or both) bonded to one or more of the carbon atoms of the compound.
oxidative phosphorylation The formation of ATP coupled to the flow of hydronium ions through an ATP synthase in the inner mitochondrial membrane.
oxidizing agent A substance that oxidizes another substance.

peptide A polyamide formed by the bonding together of α-amino acids.
phenol An organic compound that contains a hydroxyl group (—OH) attached to a carbon atom of a benzene ring.
phosphate ester An ester of a phosphoric acid, synthesized in the reaction of a phosphoric acid with an alcohol or phenol.
phospholipid An amphipathic lipid that contains a phosphodiester group, either a glycerophospholipid or sphingophospholipid.
photon A unit of light energy.
physical change A change in a physical property.
physical property Any observable physical characteristic of a substance other than a chemical property.
physical quantity A property that is described by a quantity and a unit.
physical state The state of being either a gas, a liquid, or a solid.
physiological function A function of a living organism or of an individual cell, tissue, or organ of which the organism is composed.
physiological saline solution A solution of sodium chloride with the same osmolarity as that of blood.
pi (π) bond The weaker bond in a double bond.
plasma The liquid part of blood, in which blood cells and other substances are dissolved or suspended.
β-pleated sheet A polypeptide or other polymer molecule whose conformation has side-by-side alignment of adjacent extended chains, in either a parallel or antiparallel arrangement.
polar covalent bond A covalent bond in which the electron pair resides closer to one of the bond partners than the other.
polar molecule A molecule that has a dipole moment.
polyatomic ion An ion, such as OH^- or CO_3^{2-}, made up of two or more atoms.
polyatomic molecule A molecule, such as $C_6H_{12}O_6$ or H_2O, made up of two or more atoms.
polyfunctional molecule A molecule that contains two or more functional groups.
polymer High-molecular-mass molecule produced by bonding together large numbers of smaller molecules.
polymerization The synthesis of polymer from monomer.
polypeptide A peptide that contains large numbers of α-amino acid residues.

polyprotic acid An acid that has two or more dissociable protons.
polysaccharide A polymer that contains large numbers of monosaccharide units.
precipitate A solid that forms in a solution as the result of a chemical reaction.
precipitation The formation of a precipitate.
precision A measure of how close a series of measurements agree with one another.
pressure Force per unit area.
primary carbon A carbon atom directly bonded to one other carbon atom.
primary protein structure The α-amino acid sequence of a polypeptide.
product A substance that is produced by a chemical reaction.
proenzyme A protein that will have enzymatic activity subsequent to some form of activation, such as hydrolysis, of a part of its polypeptide chain.
prokaryote An organism (such as a bacterium) whose DNA is not contained inside a nucleus.
protein An aggregation of two or more identical or different polypeptide molecules (each containing more than 50 α-amino acid residues), often together with ions or molecules other than polypeptides.
protein turnover The net result of the breakdown and the synthesis of proteins.
pure substance *See* **substance, pure.**

quantum A quantity of energy contained by a photon of light.
quaternary carbon A carbon atom directly bonded to four other carbon atoms.
quaternary protein structure The three-dimensional relation among the different polypeptide molecules of a protein.

racemic mixture An equimolar (1 : 1) mixture of the two enantiomers of a pair of enantiomers.
radical A species, such as ·OH, with an unpaired electron.
reactant A substance that undergoes chemical change in a reaction.
reaction, chemical *See* **chemical change.**
reducing agent A substance that reduces another substance.
reduction of an inorganic compound A decrease in the number of oxygen atoms or a decrease in the positive charge of the metallic element or both.
reduction of an organic compound A decrease in the number of oxygen atoms or an increase in the number of hydrogen atoms (or both) bonded to one or more of the carbon atoms of the compound.
renal threshold A blood concentration of a substance above which the substance appears in the urine.
representative element Any element that appears in an A group of the periodic table.
respiration The uptake of oxygen and release of carbon dioxide by either cells or the body.
reversible reaction A chemical reaction that can reach equilibrium by starting with either reactants or products.
ribonucleic acid (RNA) A polynucleotide formed from ribonucleotides.
ribonucleotide A nucleotide that contains ribose, phosphoric acid, and a heterocyclic base bonded together.
ring compound *See* **cyclic compound.**

salt The product (other than water) of the reaction between an acid and a base.
saponification The hydrolysis of an amide or ester with a strong base such as NaOH.
saturated fatty acid A fatty acid with no C=C double bonds.
saturated hydrocarbon A hydrocarbon with only single bonds and no multiple bonds.
scientific notation A method of writing a number as a product of a number between 1 and 10 multiplied by 10^x where x can be either a positive or a negative number.
secondary carbon A carbon atom directly bonded to two other carbon atoms.
secondary force An attractive force between identical molecules (such as H_2O and H_2O) or between different molecules (such as formaldehyde and water) or between different parts of the same molecule (such as a polypeptide).
secondary protein structure The conformation in a local region of a polypeptide molecule.
semipermeable membrane A membrane that will permit passage of particular substances and no others.
sigma (σ) bond The single bond and the stronger bond of double and triple bonds in organic compounds.
significant figure The number of digits in a numerical measurement or calculation that are known with certainty plus one additional digit.
single bond One bond between two atoms.
SI units Units that have the same names as those in the older, metric system but have new reference standards for the base units.
solute The substance present in smaller amount in a solution.
solution A homogeneous mixture of two or more substances that is visually uniform throughout.
solvent The substance present in larger amount in a solution.
somatic cell A cell other than an egg or sperm cell.
specific gravity The ratio of the density of a test liquid to the density of a reference liquid.
specific heat The heat absorbed or lost per Celsius degree change in temperature per gram of substance.
spectroscopy Measurement of the results of the interaction of electromagnetic radiation with atoms and molecules.
sphingolipid An amphipathic hydrolyzable lipid containing sphingosine.
standard, reference A base unit of measurement such as the meter or the kilogram.
standard atmosphere The pressure that supports a column of mercury 760 mm high at zero degrees Celsius.
standard conditions of temperature and pressure (STP) A temperature of zero degrees Celsius and one atmosphere of pressure.
standard solution A solution for which its concentration is known with accuracy.
state of matter A condition in which matter can exist: solid, liquid, or gas.
stereocenter A carbon atom in a molecule at which exchange of two substituents converts one stereoisomer into the other.
stereoisomers Different compounds that have the same connectivity but differ in configuration.
steric hindrance The destabilizing interference that results between two substituents when they are close to another.
stoichiometry The calculation of the quantities of the elements or compounds taking part in a chemical reaction.
STP *See* **standard conditions of temperature and pressure.**
straight chain *See* **linear chain.**
strong acid An acid that is completely dissociated in aqueous solution.

strong base A base that is completely dissociated in aqueous solution.
structural formula *See* **formula, structural.**
structural isomers *See* **constitutional isomers.**
substance, pure An element or a compound; not a mixture.
substituent An atom or group of atoms, such as —Cl or $—CH_3$, in an organic molecule.
substrate The substance that is the object of an enzyme's catalysis.
sugar A monosaccharide or disaccharide.
surface tension A force at the surface of a liquid that reduces the area of the surface.
suspension A mixture in which the solute particles are larger than colloidal in size.

temperature A measure of the hotness or coldness of an object.
tertiary carbon A carbon atom directly bonded to three other carbon atoms.
tertiary protein structure The three-dimensional relation among the different secondary structures in different regions of a polypeptide.
tetrahedral bond angle The 109.5° bond angle at sp^3-hybridized carbon atoms.
tetrahedral stereocenter A carbon atom with four different substituents.
tetrahedron A geometrical shape with four triangular faces of the same size.
theory A fundamental assumption that explains a large number of observations, facts, or hypotheses.
thermal expansion The increase in volume of a substance in response to an increase in its temperature.
thermal property The manner in which a substance responds to changes in its temperature.
trans isomer The diastereomer of an alkene or cyclic compound in which similar substituents are on opposite sides of the double bond or ring.
transition element An element between Group IIA and Group IIIA and in the actinide or lanthanide family.
transition state Another name for the activated complex.
transport (active, facilitated, simple) The passage of a solute across a cell or organelle membrane.
triacylglycerol A triester of glycerol.
triatomic molecule A molecule containing three atoms, such as H_2O or CO_2.
trigonal bond angle The 120° bond angle at the carbon atoms of a double bond.
trigonal stereocenter A carbon atom taking part in a carbon–carbon double bond and having two different substituents.
triprotic acid An acid containing three dissociable hydrogen atoms.
turnover number A quantitative measure of the replacement rate of a cellular component.

unbranched chain *See* **linear chain.**
uncertainty The estimate of the last number in a measurement.
unit-conversion factor A fraction, such as 2.54 cm/in., that allows the conversion of one unit into another unit.
unsaturated fatty acid A fatty acid with one or more C=C double bonds.
unsaturated hydrocarbon A hydrocarbon with a multiple bond.
URL Uniform Resource Location, a Web site address.

vacuum An enclosed space containing no matter.
valence electron An electron in the outermost electron shell of an atom.
valence shell The outermost electron shell of an atom.
valence-shell electron-pair repulsion theory (VSEPR theory) A theory that accounts for the symmetrical distribution of atoms around a central atom in a covalent compound.
vaporization The conversion of a liquid into a gas.
vapor pressure The pressure of a vapor in equilibrium with its liquid.
vitamin An organic compound that is required in trace amounts for normal metabolism but is not synthesized by an organism and must be included in the diet.
volatile liquid A liquid with a high vapor pressure.
volume The capacity of an object to occupy space.

water of hydration Water contained in a pure solid in a specific ratio of moles of water to moles of substance.
weak acid An acid that undergoes incomplete dissociation in aqueous solution; one that has a small dissociation constant.
weak base A base that undergoes incomplete dissociation in aqueous solution; one that has a small dissociation constant.
wedge-bond representation Similar to a Fischer projection except that horizontal bonds are shown as solid wedge bonds and vertical bonds are shown as dashed wedge bonds.
weight The gravitational force on an object relative to some reference standard.

zymogen *See* **proenzyme.**

Index

Note: Page numbers followed by f, t, and b indicate illustrations, tables, and boxed material, respectively. Page numbers preceded by AP refer to appendices.

Families of Organic Compounds

Family	Functional group	Example
alkane	C—C and C—H single bonds	$CH_3—CH_3$ **Ethane**
alkene	$>C{=}C<$	$CH_2{=}CH_2$ **Ethylene**
alkyne	$—C{\equiv}C—$	$CH{\equiv}CH$ **Acetylene**
aromatic	six-membered ring of C atoms with alternating double bonds, each C with one external bond	C_6H_6 (ring of six C, each bonded to H) **Benzene**
alcohol	$—\overset{\vert}{\underset{\vert}{C}}—O—H$	$CH_3CH_2—O—H$ **Ethyl alcohol**
ether	$—\overset{\vert}{\underset{\vert}{C}}—O—\overset{\vert}{\underset{\vert}{C}}—$	$CH_3—O—CH_3$ **Dimethyl ether**
aldehyde	$—\overset{O}{\overset{\Vert}{C}}—H$	$CH_3—\overset{O}{\overset{\Vert}{C}}—H$ **Acetaldehyde**
ketone	$—\overset{\vert}{\underset{\vert}{C}}—\overset{O}{\overset{\Vert}{C}}—\overset{\vert}{\underset{\vert}{C}}—$	$CH_3—\overset{O}{\overset{\Vert}{C}}—CH_3$ **Acetone**
carboxylic acid	$—\overset{O}{\overset{\Vert}{C}}—OH$	$CH_3—\overset{O}{\overset{\Vert}{C}}—OH$ **Acetic acid**
ester	$—\overset{O}{\overset{\Vert}{C}}—O—\overset{\vert}{\underset{\vert}{C}}—$	$CH_3—\overset{O}{\overset{\Vert}{C}}—O—CH_3$ **Methyl acetate**
amine	$—\overset{\vert}{\underset{\vert}{C}}—\overset{H}{\overset{\vert}{N}}—H$	$CH_3—\overset{H}{\overset{\vert}{N}}—H$ **Methyl amine**
amide	$—\overset{O}{\overset{\Vert}{C}}—\overset{H}{\overset{\vert}{N}}—H$	$CH_3—\overset{O}{\overset{\Vert}{C}}—\overset{H}{\overset{\vert}{N}}—H$ **Acetamide**